Hans-Georg Elias
Macromolecules

Macromolecules. Hans-Georg Elias.

ISBN: 978-3-527-31174-3

Related Titles

Krzysztof Matyjaszewski, Yves Gnanou, Ludwik Leibler (eds.)

Macromolecular Engineering

Precise Synthesis, Materials Properties, Applications

4 Volumes

2007

ISBN-13: 978-3-527-31446-1

Maartje F. Kemmere, Thierry Meyer (eds.)

Supercritical Carbon Dioxide in Polymer Reaction Engineering

2005

ISBN-13: 978-3-527-31092-0

Thierry Meyer, Jos Keurentjes (eds.)

Handbook of Polymer Reaction Engineering

2 Volumes

2005

ISBN-13: 978-3-527-31014-2

Marino Xanthos (ed.)

Functional Fillers for Plastics

2005

ISBN-13: 978-3-527-31054-8

Hans-Georg Elias

An Introduction to Plastics

2003

ISBN-13: 978-3-527-29602-6

Edward S. Wilks (ed.)

Industrial Polymers Handbook

Products, Processes, Applications

4 Volumes

2001

ISBN-13: 978-3-527-30260-4

Hans-Georg Elias

An Introduction to Polymer Science

1997

ISBN-13: 978-3-527-28790-1

Hans-Georg Elias

Macromolecules

Volume 3: Physical Structures and Properties

WILEY-VCH

WILEY-VCH Verlag GmbH & Co. KGaA

The Author

Prof. Dr. Hans-Georg Elias
Michigan Molecular Institute
1910 West St. Andrews Road
Midland, Michigan 48640
USA

Volume 1	Chemical Structures and Syntheses (Published in 2005)
Volume 2	Industrial Polymers and Syntheses (Published in 2007)
Volume 3	Physical Structures and Properties
Volume 4	Processing and Application of Polymers

Library of Congress Card No.:
applied for

British Library Cataloguing-in-Publication Data
A catalogue record for this book is available from the British Library.

Bibliographic information published by the Deutsche Nationalbibliothek
The Deutsche Nationalbibliothek lists this publication in the Deutsche Nationalbibliografie; detailed bibliographic data are available in the Internet at http://dnb.d-nb.de.

Printed in the Federal Republic of Germany
Printed on acid-free paper

Printing Strauss GmbH, Mörlenbach
Binding Litges & Dopf Buchbinderei, Heppenheim
Cover Design Gunther Schulz, Fußgönheim

ISBN: 978-3-527-31174-3

Daran erkenn' ich den gelehrten Herrn!
...
Was ihr nicht rechnet, glaubt ihr, sei nicht wahr;
Was ihr nicht wägt, hat für euch kein Gewicht; ...

Johann Wolfgang von Goethe,
Faust II, Act I (Mephistopheles)

Your words reveal to me a man of learning!
...
what you can't calculate, you don't believe is true;
what you can't weigh is of no weight to you ...

Preface

The series "Macromolecules" consists of four volumes. Volume I discusses chemical structures and principles of syntheses of synthetic and some natural macromolecules. Volume II is concerned with raw materials and energy sorces for the polymer industry, monomer syntheses, industrial polymer manufacture, and general properties of individual polymers. The present Volume III treats physical structures and physical properties of both single macromolecules and macromolecular substances, i.e., polymers. The final Volume IV will be concerned with applications of polymers as plastics, fibers, elastomers, coating, thickeners, etc.

The very short first chapter of Volume III introduces a few chemical and technical terms. The subsequent Chapters 2-4 are concerned with structures and properties of *single macromolecules*. Chapter 2 discusses basic *chemical structures* (nomenclature, polymer architectures, chemical configurations) before it embarks on a detailed treatment of *molar mass averages* and *molar mass distributions*. The discussion of *microconformations* in Chapter 3 is followed in Chapter 4 by a treatment of *macroconformations* of single macromolecules (the polymer configurations of physics) which includes remarks on molecular modeling, chain flexibilities, molecule perturbations, and scaling. Because of their importance, *scattering methods* are discussed in a separate Chapter 5.

The next four chapters treat *physical structures in* the *solid state*: in the amorphous state (Chapter 6), in semicrystalline polymers (Chapter 7), in mesophases (Chapter 8), and of polymer molecules in and at interfaces (Chapter 9).

These chapters are followed by a group of three chapters that are concerned with properties of *polymers in solution*: thermodynamics of polymer solutions (Chapter 10); transport of polymer molecules by diffusion, sedimentation, and electrophoresis (Chapter 11); and rheology of dilute polymer solutions (Chapter 12).

The next three chapters are concerned with general properties of the *molten and bulk states* of polymers such as thermal properties including transitions and treansformations (Chapter 13), transport in and through polymers (Chapter 14), and melt viscosity (Chapter 15).

The final group of three chapters comprises *mechanical properties*: elasticity (Chapter 16), viscoelasticity (Chapter 17), and fracture (Chapter 18). Electrical and optical properties will be discussed in Volume IV.

This volume concludes with an Appendix (Chapter 19) that lists SI physical quantities and units, conversion factors for older units, terminologies of concentrations and ratios of physical quantities, and names and chemical constitutions of polymers that are discussed in this volume.

Volume III touches upon many aspects that are of interest to many scientific and technical disciplines ranging from pure and applied chemistry to pure and applied physics, including theoretical mechanics and the processing and use of plastics and other polymeric materials. Each of these fields has its own terminology, its own way of thinking, and its own idiosyncrasies. Chemists, for example, commonly use names of chemical compounds such as "benzene" without specifying whether they mean the molecule or the substance. The word "compound" itself has different meanings for chemists and technologists and words like "monomer," "configuration," "proton," and "density" signify different things for physicists and chemists. I have tried therefore to be consistent by using recommendations of IUPAC and IUPAP, especially when they deviate from common uses of words.

A book that discusses physical properties cannot restrict itself to rigorous physical theories since the necessary simplifications often do not do justice to the many different phenomena that are caused by widely varying chemical structures, and, often treated as nuisance, the sometimes strong effects of type and width of molar mass distributions on physical properties. Therefore, this volume contains not only many derivations of physical equations but also many experimentally observed properties that so far have eluded more rigorous theoretical approaches. Derivations of physical equations are given in detail if they are short and not too complex mathematically. Elaborate mathematical methods are discussed only qualitatively.

I am again indebted to my good friends and former colleagues at Michigan Molecular Institute, Professors Petar R. Dvornic and Steven E. Keinath, who read and checked the final draft of all chapters and made many helpful suggestions.

Midland, Michigan
Fall 2007

Hans-Georg Elias

List of Symbols

Symbols for physical units are strictly those of the International Standardization Organization (ISO). See Chapter 19 (Appendix).

Symbols for physical quantities follow the recommendations of the International Union of Pure and Applied Chemistry (IUPAC) and the International Union of Pure and Applied Physics (IUPAP); exceptions are indicated. In particular, all symbols for physical quantities are slanted, two-letter symbols are used only for dimensionless quantities (for example, Reynolds number), and vectorial quantities are in bold letters. Specific quantities (≡ physical quantity divided by mass) are written in small letters, using the same symbol as for the quantity itself (for example, C_p = isobaric heat capacity, c_p = C_p/m = specific isobaric heat capacity). For "normalized", "reduced", etc., see Appendix.

Indices are slanted if they refer to a quantity that is held constant (for example, C_p = heat capacity at constant pressure). They are written upright if they do not indicate a constant quantity (example: $\overline{M}_\mathrm{n}$ = number-average molar mass).

I.Mills, T.Cvitas, K.Homann, N.Kallay, K.Kuchitsu, Eds., (International Union of Pure and Applied Chemistry, Division of Physical Chemistry), "Quantities, Units and Symbols in Physical Chemistry", Blackwell Scientific Publications, Oxford 1988.

Symbols for Languages

D = German (deutsch),
F = French,
G = (classical) Greek.
L = (classical) Latin.

The Greek letter υ (upsilon) was transliterated as "y" (instead of the customary phonetic "u") in order to make an easier connection to written English (example: πολυς = polys (many)). For the same reason, χ was transliterated as "ch" and not as "kh."

Mathematical Symbols (IUPAC)

=	equal to	>	greater than
≠	not equal to	≥	greater than or equal to
≡	identically equal to	>>	much greater than
≈	approximately equal to	<	less than
~	proportional to (IUPAC: ~ or ∝)	≤	less than or equal to
≙	corresponds to	<<	much less than
→	approaches, tends to	±	plus or minus
Δ	difference	sin	sine of
δ	differential	cos	cosine of
f	function of (IUPAC: *f*)	tan	tangent of
Σ	sum	cot	cotangent of
∫	integral	sinh	hyperbolic sine of
Π	product	arctan	inverse tangent

lg	logarithm to the base 10	(IUPAC: lg or $\log_{10}$)
ln	logarithm to the base e (natural logarithm)	(IUPAC: ln or $\log_e$)

Symbols for Chemical Structures

A: symbol for a monomer or a leaving group (polycondensations)
B: symbol for a monomer or a leaving group (polycondensations)
L: symbol for a leaving molecule, for example, H_2O from the reaction of –COOH + HO–
R: symbol for a monovalent substituent, for example, CH_3– or C_6H_5–
Z: symbol for a divalent unit, for example, –CH_2– or –*p*-C_6H_4–
Y: symbol for a trivalent unit, for example, −C(R)< or −N<
X: symbol for a tetravalent unit, for example, >C< or >Si<
*: symbol for an active site: radical ($^\bullet$), anion ($^\ominus$), cation ($^\oplus$)
*p*Ph *para*-phenylene (in text)
p-C_6H_4 *para*-phenylene (in line formulas)

Averages and Other Markings

– line above letter indicates common average, for example, $\overline{M}_n$ = number-average of molar mass (note: subscript is not italicized since it does *not* represent a physical quantity that is kept constant)

~ tilde indicates a partial quantity, for example, $\tilde{v}_A$ = partial specific volume of component A

[ ] square brackets surrounding the symbol of the substance indicates the amount (of substance) concentration ("mole concentration"), usually in mol/L

⟨ ⟩ angled brackets surrounding a letter indicate spatial averages, for example, $\langle s^2 \rangle$ = mean-square average of radius of gyration (IUPAC)

| | two vertical lines enclosing the symbol for a vectorial quantity indicate the magnitude of that quantity. Example, $|q|$ = magnitude of the scattering vector ***q***

Exponents and Superscripts

Symbols for exponents are slanted if they indicate physical quantities but upright if the symbol indicates a number. Example: exponent α in the intrinsic viscosity = f(molar mass) relationship, $[\eta] = K_v M^{\alpha}$.

°	degree of plane angle [= (π/180) rad]
'	minute of plane angle [= (π/10 800) rad]
"	second of plane angle [= (π/648 000) rad]
o	pure substance
∞	infinite, for example, dilution or molecular weight
m	amount-of-substance related quantity if a subscript is inexpedient. According to IUPAC, ^{m} can be used as either a superscript or a subscript
(q)	qth order of a moment (always in parentheses since it does not represent a power)
‡	activated quantity, for example, $E^‡$ = activation energy
a	general exponent in $P = K_P M^a$ (P = property)
q	general exponent
α	exponent in $[\eta] = K_v M^{\alpha}$
ν	exponent in $\langle s^2 \rangle^{1/2} = K_s M^{\nu}$
ε	exponent in $\eta_0 = K_\eta M^{\varepsilon}$
δ	exponent in $D = K_D M^{\delta}$
ς	exponent in $S = K_S M^{\varsigma}$

Indices and Subscripts

Subscripts are slanted if they refer to physical properties that are held constant. Example: C_p = isobaric heat capacity.

o	standard or original state, for example, T_o = reference temperature
0	state at time zero
1	solvent
2	solute, usually polymer
3	additional component (salt, precipitant, etc.)
∞	final state
A	compound A, for example, M_A = molar mass of substance A
a	group or monomeric unit of A, for example, mass m_a of group a
am	amorphous
B	substance B
B	fracture
b	group, for example, monomeric unit from substance B
bd	bond, especially chain bond
be	effective bond length (= length of monomer unit projected onto the chain direction)
bl	blob
bp	boiling temperature (boiling point)
br	branch, branched
c	chain (L: *catena*), for example, in networks
cl	coil
comb	combination
cr	crystalline
crit	critical
crl	correlation, for example, L_{crl} = correlation length
cryst	crystallization
cu	chain unit, for example, $-CH_2-$ in $\left[\!-O-CH_2-\!\right]_n$
cycl	cyclic
D	related to diffusion
e	entanglement
el	elastic
eff	effective
end	endgroup
eq	equilibrium
exc	excess
F	filler
fl	flexural
G	glass transformation
g	any statistical weight, e.g., n, m, z or x, w, Z
H	hydrodynamically effective property or hydration
h	hydrodynamic average
I	Initiator molecule; I* active initiator species, for example, a radical $I^\bullet$
i	ith component

i	isotactic diad (IUPAC recommends m = *meso*; see Chapter 4)
ii	isotactic triad (IUPAC: mm)
iii	isotactic tetrad (IUPAC: mmm)
inh	inherent (dilute solution viscosity)
is	heterotactic triad (IUPAC: mr)
|is|	sum of heterotactic triads, is = is + si
it	isotactic
j	variable
k	variable
k	chain unit
L	liquid, melt
l	liquid
M	melting
M	monomer molecule
M	matrix (in blends or reinforcd polymers)
Mt	metal
m	monomeric unit in macromolecules
m	molar (also as superscript)
mol	molecule
mon	monomer (if M is confusing)
mu	monomeric unit
n	number average (note: *not* in italics since it does not refer to a physical quantity that is held constant)
P	polymer
p	index for quantity at constant pressure
p	packing
pol	polymer (if P is confusing)
ps	persistence
q	index, defined differently for each section or chapter
q	number of electric charges
R	reactant
r	relative (only in M_r = relative molecular mass = molecular weight)
r	based on end-to-end distance, e.g., α_r = linear expansion coefficient of a coil (with respect to the end-to-end distance)
red	reduced
rel	relative
rep	repeating unit
rlx	relaxation
S	solvating solvent
S	related to sedimentation
s	syndiotactic diad (IUPAC recommends r = *racemo*)
s	related to radius of gyration
seg	segment
si	heterotactic triad (IUPAC recommends rm)
sl	screening length
soln	solution

sph	sphere
ss	syndiotactic triad (IUPAC recommends rr)
T	index for quantity at constant temperature
t	termination
tr	transfer
u	monomeric unit in polymer
u	monomer conversion
V	quantity at constant volume
v	viscosity average (solutions)
w	mass average ("weight average"); note: *not* in italics since it does not refer to a physical quantity that is held constant
x	crosslink(ed)
y	yield (stress-strain)
z	z average
η	viscosity average (melts)

Prefixes of Words (in systematic polymer names in *italics*)

alt	alternating
at	atactic
blend	polymer blend
block	block (large constitutionally uniform segment)
br	branched. IUPAC recommends sh-branch = short chain branch, l-branch = long chain branch, f-branch = branched with a branching point of functionality *f*
cis	cis configuration with respect to C–Cdouble bonds
co	joint (unspecified)
comb	comb
compl	polymer–polymer complex
cyclo	cyclic
ct	cis-tactic
eit	erythrodiisotactic
g	graft
ht	heterotactic
ipn	interpenetrating network
it	isotactic
net	network; μ-net = micro network
per	periodic
r	random (Bernoulli distribution)
sipn	semi-interpenetrating network
sl	screening length
st	syndiotactic
star	star-like. f-star, if the functionality *f* is known; f is then a number
stat	statistical (unspecified distribution)
tit	threodiisotactic
trans	trans configuration with respect to C-C double bonds
tt	trans-tactic

Other Abbreviations

AIBN	*N,N'*-azobisisobutyronitrile
BPO	dibenzoylperoxide
Bu	butyl group (*i*Bu = isobutyl group; *n*Bu = normal butyl group (according to IUPAC, the normal butyl group is not characterized by *n*, which rules out Bu as an unspecified butyl group); *s*Bu = secondary butyl group; *t*Bu = tertiary butyl group)
Bz	benzene or benzyl
C	catalyst; C* = active catalyst or active catalytic center
cell	cellulose residue
Cp	cyclopentadienyl group
DMF	*N,N*-dimethylformamide
DMSO	dimethylsulfoxide
Et	ethyl group
G	gauche conformation
Glc	glucose
GPC	gel permeation chromatography
I	initiator
IR	infrared
L	solvent (liquid)
LC	liquid-crystalline
LS	light scattering
MC	main chain
Me	methyl group
Mt	metal atom
Np	naphthalene
NMR	nuclear magnetic resonance
P	polymer
Ph	phenyl group
Pr	propyl group
SANS	small-angle neutron scattring
SAXS	small-angle X-ray scattering
SC	side chain
SEC	size exclusion chromatography
THF	tetrahydrofuran
UV	ultraviolet

Quantity Symbols (unit symbols: see Chapter 19, Appendix)

Quantity symbols follow in general the recommendations of IUPAC: quantity symbols are always slanted, and vectorial quantities are given in bold letters.

A	absorption ($A = \lg (I_0/I) = \lg (1/\tau_i)$); formerly: extinction
A	area; A_c = cross-sectional area of a chain
A	Helmholtz energy ($A = U - TS$); formerly: free energy
$A^{\ddagger}$	pre-exponential constant (in $k = A^{\ddagger} \exp(-E^{\ddagger}/RT)$)

A_2 second thermodynamic virial coefficient
A_3 third thermodynamic virial coefficient
a thermodynamic activity
a linear absorption coefficient ($a = (1/L) \lg (I_0/I)$)
a_T shift factor in the WLF equation
b bond length; b_{eff} = effective bond length
C number concentration (number of entities per total volume, $C = cN_A/M$);
[C] amount-of-substance concentration of substance C = amount of substance C per total volume = "molar concentration of C"
C transfer constant (always with index, e.g., C_r of a regulator, C_s of a solvent)
C heat capacity (usually in J/K); C_p = isobaric heat capacity (heat capacity at constant pressure p); C_V = isochoric heat capacity (heat capacity at constant volume V); C_m = molar heat capacity (heat capacity per amount-of-substance n)
C electrical capacity
C_N Characteristic ratio in random coil statistics; C_∞ = characteristc ratio at infinitely high molecular weight
c crystallographic bond length = crystallographic length of a repeating unit (usually crystallographic c axis)
c specific heat capacity (usually in J/(g K)); c_p = isobaric specific heat capacity; c_V = isochoric specific heat capacity. Formerly: specific heat
c concentration = mass concentration (= mass-of-substance per total volume) = "weight concentration." IUPAC calls this quantity "mass density" (quantity symbol ρ). The quantity symbol c has, however, traditionally been used for a special case of mass concentration, i.e., mass-of-substance per volume of solution and the quantity symbol ρ for another special case, the mass density ("density") = mass-of-substance per volume of substance.

The mass concentration of a solute 2 is related to its density ρ and volume fraction by $c_2 = \rho_2\phi_2$ if volumes are additive.

$\hat{c}$ velocity of light or sound (depends on chapter)
D diffusion coefficient; D_{rot} = rotatory diffusion coefficient
D tensile compliance
DP often used in literature as symbol for "degree of polymerization". This book uses X instead since slanted (!) two-letter symbols of physical quantities are reserved for dimensionless *transport* quantities (ISO)
d diameter; d_{bl} = diameter of a blob, d_{sph} = diameter of a sphere, etc.
d dimensionality
E energy
E tensile modulus (= modulus of elasticity, Young's modulus); E_f = flexural modulus
$\boldsymbol{E}$ electric field strength (vectorial quantity)
e elementary charge
e cohesion energy density
$\boldsymbol{e}$ component of elongation or shearing (tensor)
$\boldsymbol{F}$ force (vectorial quantity)
f fraction (unspecified); see also x = amount fraction ("mole fraction"), w = mass fraction ("weight fraction"), ϕ = volume fraction

f_o functionality of a molecule
f_{or} Hermans orientation factor
G Gibbs energy ($G = H - TS$); formerly: free enthalpy
G shear modulus, G' = shear storage modulus (real modulus, in-phase modulus, "elastic modulus"), G'' = shear loss modulus (imaginary modulus, 90° out-of-phase modulus, viscous modulus), $G_N{}^o$ = plateau modulus
G statistical weight fraction ($G_i = g_i/\Sigma_i g_i$)
G conductance
g acceleration (due to gravity)
g statistical weight (for example: n, x, w). IUPAC recommends k for this quantity which is problematic because of the many other uses of k. Similarly, K cannot be used for the statistical weight fraction because of the many other meanings of K.
g parameter for the ratio of dimensions of branched macrmolecules to those of unbranched macromolecules of equal molecular weight (branching index); g_h = branching index from hydrodynamic measurements
H height
H enthalpy; ΔH_{mix} = enthalpy of mixing, $\Delta H_{mix,m}$ = molar enthalpy of mixing
h Planck constant ($h = 6.626\ 075\ 5 \cdot 10^{-34}$ J s)
h branching index in hydrodynamics
I electric current
I light intensity
i radiation intensity of a molecule
i variable (ith component, etc.)
J flux (of mass, volume, energy, etc.)
J shear compliance
K general constant; equilibrium constant
K compression modulus
k rate constant (always with index); k_i = rate constant of initiation; k_p = rate constant of propagation, k_t = rate constant of termination, k_{tr} = rate constant of transfer
k_B Boltzmann constant ($k_B = R/N_A = 1.380\ 658 \cdot 10^{-23}$ J K^{-1})
L length (always geometric); L_{chain} = true (historic) contour length of a chain (= number of chain bonds times length of valence bonds); L_{cont} = conventional contour length of a chain (= length of chain in all-trans macroconformation; L_K = length of a Kuhn segment (Kuhnian length); L_{ps} = persistence length; L_{seg} = segment length
L phenomenological coefficient
l length
$\boldsymbol{M}$ moment
M molar mass of a molecule (= physical unit of mass of molecule divided by amount of molecule, for example, g/mol). $\overline{M}_n$ = number-average molar mass; $\overline{M}_w$ = mass-average molar mass; M_{crit} = critical molar mass; $\overline{M}_{R,n}$ = number-average molar mass of reactants (= polymer plus monomer)
M_e entanglement molar mass from Newtonian viscosities ($M_{e,\eta}$) or the plateau modulus ($M_{e,G}$)

M_r relative molar mass = relative molecule mass = molecular weight (physical unit of unity = "dimensionless"); $\overline{M}_{r,n}$ = number-average molecular weight
m mass; m_{mol} = mass of molecule
N number of entities
N_A Avogadro constant ($N_A = 6.022\ 136\ 7 \cdot 10^{23}$ mol^{-1})
n amount of substance (in mol); formerly: mole number
n refractive index in medium; n_1 = refrective index of solvent; n_2 = refractive index of solute
P permeation coefficient ($P = DS$)
P power
P Perrin factor (ellipsoids)
P(q) particle scattering factor
p conditional probability
p pressure
p extent of reaction (fractional conversion); p_A = extent of reaction of A groups
p number of conformational repeating units per completed helical turn
$\boldsymbol{p}$ dipole moment (vectorial quantity)
Q electric charge = quantity of electricity
Q heat
Q parameter in the Q,e copolymerization equation
Q polymolecularity index ("polydispersity index"), for example, $Q = \overline{M}_w/\overline{M}_n$
Q intermediate variable or constant, usually a ratio; varies with section
q intermediate variable or constant, usually a ratio; varies with section
q charge of an ion
q scattering parameter with a magnitude of $|\boldsymbol{q}| = q = (4\ \pi n_1/\lambda_o) \sin (\vartheta/2)$
R molar gas constant (R = 8.314 510 J K^{-1} mol^{-1})
R electrical resistance
R dichroic ratio
R rate of reaction, for example, R_p = rate of propagation
R radius: R_d = Stokes radius (from diffusion coefficient), R_{sph} = radius of equivalent sphere, R_v = Einstein radius (from dilute solution viscosity)
R_θ Rayleigh ratio of scattering intensities
r radius
r spatial end-to-end distance of a chain, usually as $\langle r^2 \rangle^{1/2}$ with various indices; r_{cont} = conventional contour length (= end-to-end distance of a chain in all-trans conformation)
r copolymerization parameter
r_0 initial ratio of amounts of substances in copolymerizations
S entropy; ΔS_{mix} = entropy of mixing, $\Delta S_{mix,m}$ = molar entropy of mixing
S solubility coefficient
S sedimentation coefficient (literature uses mainly s which is the IUPAC symbol for the radius of gyration)
S_{pq} (elastic) compliance tensor (Reuss elasticity constant)
s radius of gyration (IUPAC), shorthand for $\langle s^2 \rangle^{1/2}$ (IUPAC); in the literature often as R_g
s selectivity coefficient (osmotic pressure)

T temperature (always with units). In physical equations always as thermodynamic temperature with unit kelvin; in descriptions, either as thermodynamic temperature (unit: kelvin) or as Celsius temperature (unit: degree Celsius). Mix-ups can be ruled out because the physical unit is always given. IUPAC recommends for the Celsius temperature either t as a quantity symbol (which can be confused with t for time) or θ (which can be confused with Θ for the theta temperature). T_c = ceiling temperature, T_G = glass temperature, T_M = melting temperature

T transparency

t time

t rotational angle around helix axis

U internal energy

U electric potential (voltage drop)

u fractional conversion of monomer molecules (p = fractional conversion of groups; y = yield of substance)

u excluded volume

V volume; V_h = hydrodynamic volume, V_m = molar volume; $\tilde{V}_m$ = partial molar volume

v specific volume; $\tilde{v}$ = partial specific volume

v linear velocity ($v = dL/dt$)

W work

w mass fraction = weight fraction. For example, mass fraction of component 2: $w_2 = m_2/m = x_2/[x_2 + x_1(M_1/M_2)]$

X degree of polymerization of a molecule with respect to monomeric units (not to repeating units! See Y); $\overline{X}_n$ = number-average degree of polymerization of a substance; $\overline{X}_w$ = mass-average degree of polymerization of a substance

x mole fraction (amount-of-substance fraction); x_u = mole fractions of units, x_i = mole fraction of isotactic diads, x_{ii} = mole fraction of isotactic triads, etc.

x_{br} degree of branching

Y refractive index increment (= dn/dc)

Y degree of polymerization with respect to repeating unit

y yield of substance

Z z fraction ($Z_i = z_i/\sum_i z_i$)

z z-statistical weight

z coordination number, number of neighbors

z dissymmetry (light scattering)

z parameter in excluded volume theory

α angle, especially rotational angle of optical activity

α linear thermal expansion coefficient of materials or random coils ($\alpha = (1/L)(dL/dT)$). Note: in literature often as β!

α linear expansion of coils. Indices indicate the type of measurement or property: α_D (from diffusion), α_h (hydrodynamics in general), α_r (end-to-end distance), α_s (radius of gyration), α_v (from viscosity of dilute solutions)

α degree of crystallinity (with index for method: X = X-ray, d = density, etc.)

α electrical polarizability of a molecule

$[\alpha]$ "specific" optical rotation

β	angle
β	compressibility coefficient
β	cubic thermal expansion coefficient [$\beta = (1/V)(\mathrm{d}V/\mathrm{d}T)$]; in literature often as α
β	integral of excluded volume
Γ_H	parameter of preferential solvation (preferential hydration)
γ	angle
γ	surface tension, interfacial energy
γ	crosslinking index
$\dot{\gamma}$	shear rate (velocity gradient)
δ	loss angle
δ	solubility parameter
δ	chemical shift
ε	linear extension [$\varepsilon = (L - L_0)/L_0$]; nominal strain (Cauchy strain); ε' = true strain (Hencky strain)
ε	energy per molecule, cohesive energy
ε	expectation
ε_r	relative permittivity (formerly: dielectric constant)
ζ	ratio R_h/R_s of hydrodynamic radius and radius of gyration, for example, ζ_{sph} of spheres, ζ_{cl} of coils
η	dynamic viscosity, e.g., η_0 = viscosity at rest (Newtonian viscosity),
	η_1 = viscosity of solvent, η_e = extensional viscosity
	$\eta_r = \eta/\eta_1$ = relative viscosity,
	$\eta_i = (\eta - \eta_1)/\eta_1$ = relative visc. increment (= specific viscosity η_{sp}) *
	$\eta_{inh} = (\ln \eta_r)/c$ = inherent viscosity (= logarithmic visc. number)
	$\eta_{red} = (\eta - \eta_1)/(\eta_1 c)$ = reduced viscosity (= viscosity number η_{sp}/c)
	$[\eta] = \lim \eta_{red,c \to 0}$ = limiting visc. number (= intrinsic viscosity)
	* IUPAC recommends "relative viscosity increment" and the symbol η_i. However, the symbol η_i is easily confused with the symbol η_i for the viscosity of the substance *i*.
Θ	characteristic temperature, especially theta temperature
θ	torsional angle (conformational angle in macromolecular science)
ϑ	angle, especially scattering angle or torsional angle (organic chemistry)
κ	isothermal (cubic) compressibility
κ	enthalpic interaction parameter in solution theory
Λ	aspect ratio = axial ratio of rods (length/diameter) or rotational ellipsoids (main axis/secondary axis)
λ	wavelength in medium, $\lambda = \lambda_o/n$ (λ_o = wavelength of incident light)
λ	thermal conductivity
λ	strain ratio = draw ratio ($\lambda = L/L_0$)
μ	chemical potential
μ	moment of a distribution
μ	dipole moment
μ	Poisson ratio
ν	moment of a distribution, related to a reference value
ν	kinetic chain length
ν	frequency

ν effective amount concentration of network chains
v velocity
Ξ zip length
ξ friction coefficient. IUPAC recommends f which conflicts with the same IUPAC symbol for a function
Π osmotic pressure
π mathematical constant *pi*
ρ density (= mass/volume of the same matter), for example, mass of substance A per volume of substance A. ρ is also used by IUPAC for other densities, for example, for the number density (= number of entities per volume of matter)
ρ electric volume resistivity (= volume resistance)
σ nominal mechanical stress; σ_{11} = normal stress, σ_{21} = shear stress, σ' = true stress
σ (number) standard deviation
σ hindrance parameter (steric factor)
σ cooperativity
σ electrical conductivity
ς degree of coupling of chains in Schulz-Zimm distributions; $\varsigma = \overline{M}_n/(\overline{M}_w - \overline{M}_n)$
τ bond angle, valence angle
τ relaxation time
τ shear stress (= σ_{21})
τ_i light transmission; τ_{it} = internal transmission; τ_{et} = external transmission
Φ Flory parameter; Φ_Θ = Flory constant (theta state)
$[\Phi]$ "molar" optical rotation
ϕ volume fraction; ϕ_f = free volume fraction
ϕ angle
$\varphi(r)$ potential between two segments that are separated by a distance r
χ Flory-Huggins interaction parameter
Ψ Simha factor for ellipsoids
ψ entropic interaction factor in the theory of polymer solutions
Ω angle
Ω thermodynamic probability
Ω skewness of a distribution
ω angular velocity, angular frequency

Table of Contents

1 Introduction

Macromolecular substances, commonly called **polymers**, are composed of high-molecular weight chemical compounds that exist in a great variety of chemical and physical structures. These structures lead to many chemical and physical properties that cannot be obtained from matter composed of chemical elements or low-molecular weight chemical compounds. It is these properties that make macromolecules indispensable as information carriers, structural elements, transport molecules, etc. in living matter and as materials for mechanical, electrical, optical, etc. applications.

Low-molecular weight natural and synthetic chemical compounds are used in nature and industry mainly because of their *chemical* properties, for example, as chemical reactants (e.g., chlorine), energy carriers (e.g., hydrocarbons in combustion engines, sugars in living beings), or active substances (hormones, pharmaceuticals, herbicides, etc.). Chemical properties also dominate the action of many natural high-molecular weight substances (**biopolymers**) whether as chemical matrices (e.g., nucleic acids), catalysts (enzymes), or reserve foods (glycogen in animals, poly(alkylene acid)s in bacteria, amylose in plants).

Other biopolymers are important because of their *physical* properties, for example, as structural materials such as collagen in animals, cellulose and lignins in plants, and fibers for human use (silk, wool, cotton). Most synthetic polymers are used for their physical properties, especially for their mechanical ones; examples are plastics, fibers, and elastomers. Far less important in terms of tonnage is the use of polymers in solution; examples are thickeners, paints, and additives for engine oils. An even smaller consumption by volume (but not by value) is for electrical and optical applications, such as insulators, electrical resists, optical lenses, optical light-emitting devices, and the like (Volume IV).

Polymers are synthesized from **monomers** (chemical substances consisting of monomer molecules) by **polymerization**, i.e., the conversion of monomers to polymers. Participants in these reactions are monomer molecules M; polymer molecules P_i with various degrees of polymerization, $i \geq 2$; and reactant molecules (monomer or polymer) R_m or R_n with degrees of polymerization, $m, n \geq 1$; L = leaving molecules, for example, water in the polycondensation of hexamethylene diamine and adipic acid to polyamide 6.6 Schematically, using IUPAC nomenclature (see Volume I, p. 157 for details and Volume II for industrial syntheses):

chain polymerization	$P_i + M \rightarrow P_{i+1}$	; often called "addition polymerization" or "chain-growth polymerization". Example: polymerization of styrene by anions, cations, or free radicals
polyaddition	$R_m + R_n \rightarrow R_{m+n}$	; no traditional English name. Example: synthesis of polyurethanes
polycondensation	$R_m + R_n \rightarrow R_{m+n} + L$	; often as "step-growth polymerization". Example: synthesis of nylon 6.6
polyelimination	$P_i + M \rightarrow P_{i+1} + L$	; no traditional English name. Example: polymerization of *N*-carboxy anhydrides of α-amino acids

For mechanical applications, polymers are usually subdivided into plastics, elastomers, and fibers. The distinguishing critera are the physical transition temperatures of the polymer, either the melting temperature, T_M, of crystalline polymers or the glass temperture, T_G, of amorphous ones, i.e., the temperatures at which a polymer transforms from either a crystalline solid to a melt (T_M) or from a frozen-in amorphous state to a truly amorphous liquid (T_G).

A **plastic** is a synthetic polymer or a chemical derivative of a natural polymer that is used as a working material below its melting temperature (if crystalline) or its glass temperature (if amorphous). Chemically uncrosslinked plastics are known as **thermoplastics**. They can be heated without decomposition above these characteristic temperatures; in principle, the cycle plastic → melt → plastic can be repeated many times without loss of properties. Examples of major thermoplastics are poly(ethylene), poly(propylene), poly(vinyl chloride), and poly(styrene).

A **thermosetting** polymer is a low-molecular weight polymer that crosslinks extensively on thermal processing and becomes a **thermoset**. It cannot be reprocessed to its original state. Examples are phenolic resins and unsaturated polyesters.

In both thermoplastics and thermosets, the use temperature is always *lower* than the melting temperature (if crystalline) or the glass temperature (if amorphous). An example is high-density poly(ethylene), a semicrystalline polymer with a glass temperature of −123°C, a continuous service temperature of 70-80°C (long time) or 90-120°C (short time), and a melting temperature of ca. 135°C.

Elastomers are *chemically* weakly crosslinked polymers that result from the chemical crosslinking of **rubbers** which are high-molecular weight, low T_G precursors (for the terminology, see Volume IV). The use temperature of an elastomer is always *higher* than the glass temperature but far lower than its chemical decomposition temperature. An example of an elastomer is crosslinked natural rubber (a poly(isoprene)), the so-called vulcanized rubber, which consists of chemically interconnected poly(isoprene) chains.

Thermoplastic elastomers are *physically* weakly crosslinked polymers. An example is poly(styrene)–*block*–poly(isoprene)–*block*–poly(styrene) consisting of two "hard" blocks of large poly(styrene) segments that frame a large "soft" poly(isoprene) segment. "Hard" and "soft" refer here to the glass temperature T_G relative to the use temperature T_{use} and not to surface hardnesses. The poly(styrene) block is "hard" because its $T_G > T_{use}$; the poly(isoprene) block is "soft" because its $T_G < T_{use}$.

Fibers are usually "one-dimensional" thermoplastics; fibers from thermosets are rare. **Elastomeric fibers** are obtained from low T_G precursors that are physically crosslinked.

The most important property of a macromolecule is its molecular weight, which may range up to quadrillions for soluble polymers (Table 1-1) and to "infinity" for crosslinked polymers. Molecular weights of soluble synthetic macromolecules are usually between several thousands and several millions.

Many natural macromolecular substances are molecularly uniform, i.e., they consist of macromolecules of the same molecular weight. In contrast, all synthetic macromolecular compounds are mixtures of macromolecules of various molecular weights. They have a molecular weight distribution and their molecular weights are therefore always averages (Section 2.3).

The macromolecules themselves may be "linear" like strings composed of repeating units or cyclic like rings and also branched (star-like, comb-like, dendritic, random) or

Table 1-1 Molecular weights M_r, diameter d of chains, conventional contour lengths r_{cont} (= maximum physical lengths of macromolecular chains), and actual shapes of some macromolecules.
na = not applicable.

Macromolecule	M_r	r_{cont}/nm	d/nm	Chemical structure
Deoxyribonucleic acids				
Lung fish	69 000 000 000 000	34 700 000 000	2.0	linear chains (double helix)
Human	2 000 000 000 000	1 000 000 000	2.0	linear chains (double helix)
Yeast	9 000 000 000	4 600 000	2.0	linear chains (double helix)
Bacterium subtilis	2 000 000 000	1 000 000	2.0	linear chains (double helix)
Polyoma SV 40 virus	3 000 000	1 000	2.0	cyclic molecule
Wheat				
Starch	64 000 000	na		mixture (amylose/amylopectin)
Amylose	2 100 000		0.74	linear chain
Amylopectin	122 000 000			branched chain
Poly(ethylene)				
Ultra-high molecular weight	3 000 000	272 000	0.49	linear chain
Conventional low density	100 000	9 100	0.49	branched chain
Poly(ethylene terephthalate)				
Textile fiber	< 20 000		1.08	linear chain
Industrial fiber	< 36 000		1.08	linear chain
Bottle grade	< 50 000		1.08	linear chain

crosslinked with respect to their chemical constitution (Section 2.1). Monomeric units may also be connected to each other in various *chemical* configurations (Section 2.2).

Constitutions and chemical configurations, in combination with so-called conformations (Chapter 3), lead to various shapes of molecules in solution (Chapter 4) and the solid state (Chapters 6-9) which are often explored by scattering methods (Chapter 5).

The physical properties of macromolecular substances are not only controlled by their chemical structure (constitution, i.e., type of monomeric units (often called "monomers" by physicists), molecular weight, molecular-weight distribution, chemical configuration) but also by their physical structure (micro- and macroconformation, often called (physical) configurations by physicists), and the mobility of polymer segments and molecules. Polymers may be amorphous (Chapter 6), crystalline (Chapter 7), or mesomorphic (Chapter 8). Their physical structures differ in bulk (Chapters 6-8), in solution (Chapter 10), and in surfaces (Chapter 9).

The interplay between chemical and physical structures leads to many different phenomena such as various transport processes in solution (Chapter 11) and in the solid state (Chapter 14) as well as vastly different viscosities in solution (Chapter 12), in melts (Chapter 15), and in polymeric solids (Chapter 17). Mechanically, polymers may behave elastically (Chapter 16), show viscoelasticity (Chapter 17), and may fracture in a brittle or ductile manner (Chapter 18). They also have distinct electrical and optical properties which are discussed in Volume IV of this series together with the application of polymers as plastics, fibers, elastomers, coatings, and the like.

The interplay between chemical and physical structures on one hand and properties on the other is usually investigated by two different approaches. Chemists synthesize polymers with different chemical structures which are then characterized by various physical testing procedures. Such technological tests are usually experimentally simple but physically complex. Chemical structures and physical test results are then correlated, usually by empirical or semiempirical rules. Subsequently, the resulting data are used to predict physical properties under complex practical conditions.

Physicists usually single out the simplest physical structure and derive from this structure predictions for physical properties in various physical states, often with intellectually demanding theories that require fairly abstract mathematical procedures. This approach is mathematically elegant but neglects by necessity the specific effects that are caused by various chemical structures.

The problem can be simplified as follows. Physicists try to find the *function* that describes the dependence of the property P on the structure S, for example, P = AS, P = A + BS, P = A exp(– BS), etc. Structural parameters are then reduced to such a minimum that they still allow one to describe the desired essential characteristics of the targeted property. An example is the volume of macromolecules as a function of molecular weights.

In contrast, chemists and technologists are generally interested in the *parameters* A and B that provide the material with its distinct properties. An example is the effect of polymer constitution on the volume of different types of macromolecules that have the same molecular weight. The observed properties are then discussed in terms of physical theories and models albeit often without questioning whether the simplifying theoretical assumptions still apply.

This book discusses both types of approaches albeit to various extents, depending on the topic. It describes various physical theories and models. However, rigorous and/or lengthy mathematical treatments are sometimes replaced by less rigorous approaches. On the other hand, material-specific aspects are treated in greater detail as is customary in textbooks of polymer physics.

Literature to Chapter 1

HISTORY

W.H.Stockmayer, B.H.Zimm, When Polymer Science Looked Easy, Ann.Rev.Phys.Chem. **35** (1984) 1

H.Morawetz, Polymers: The Origins and Growth of a Science, Wiley-Interscience, New York 1985 **35** (1984) 1

Y.Furukawa, Inventing Polymer Science. Staudinger, Carothers, and the Emergence of Macromolecular Chemistry, University of Pennsylvania Press, Philadelphia (PA) 1998

H.-G.Elias, Macromolecules, Volume I: Chemical Structures and Syntheses, Section 1.3: Development of the Macromolecular Hypothesis, Wiley-VCH, Weinheim 2005

HANDBOOKS

Houben-Weyl, Methoden der organischen Chemie (E.Müller, Ed.), 4th ed., Vol. XIV (2 parts), Thieme, Stuttgart 1961-63; 5th ed., Vol. **E 20**, Makromolekulare Stoffe (3 parts), Thieme, Stuttgart 1987

G.Allen, J.C.Bevington, Eds., Comprehensive Polymer Science, Pergamon, Oxford, 7 vols. (1989), First Supplement (1992), Second Supplement (1996)

J.E.Mark, Ed., Physical Properties of Polymers Handbook, AIP Press, Williston (VT), 1996

J.C.Salamone, Ed., Polymeric Materials Encyclopedia, CRC Press, Boca Raton (FL) 1996, 12 vols.; short version: Concise Polymeric Materials Encyclopedia, CRC Press, Boca Raton (FL) 1999

J.E.Mark, Ed., Polymer Data Handbook, Oxford Univ. Press, New York 1999

E.S.Wilks, Ed., Industrial Polymers Handbook (relevant chapters of Ullmann's Encyclopedia of Industrial Chemistry, 5th ed.), Wiley-VCH, Weinheim 2001 (5 vols.)
H.F.Mark, Encyclopedia of Polymer Science and Engineering, Wiley, Hoboken (NJ), 3rd ed. (2003-2007), 12 volumes
J.L.Atwood, J.W.Steed, Eds., Encyclopedia of Supramolecular Chemistry, Dekker, New York 2004

DATA COLLECTIONS
O.Griffin Lewis, Physical Constants of Linear Homopolymers, Springer, Berlin 1968
W.J.Roff, J.R.Scott, J.Pacitti, Handbook of Common Polymers, Butterworth, London 1971
J.E.Mark, Ed., Physical Properties of Polymers Handbook, AIP Press, Woodbury (NY) 1996
J.Brandrup, E.H.Immergut, E.A.Grulke, Eds., Polymer Handbook, Wiley, New York, 4th ed. 2003 (2 vols.)
N.A.Waterman, M.F.Ashby, Eds., The Materials Selector, Chapman & Hall, Boca Raton (FL) 1999
D.J.David, A.Misra, Relating Materials Properties to Structure. Handbook and Software for Polymer Calculations and Materials Properties, Technomic Publ., Lancaster (PA) 2000
B.Ellis, Polymers. A Property Database, Chapman & Hall/CRC, London 2000 (CD-ROM)
–, The Wiley Database of Polymer Properties, Wiley, Hoboken (NJ) 2002 (selected data from J.Brandrup, E.H.Immergut, E.A.Grulke, Eds., Polymer Handbook (see above)
–, CAMPUS®, diskettes with properties of plastics that are produced by ca. 30 companies belonging to the CAMPUS system
–, PLASPEC, Fachinformationszentrum Chemie, Berlin (data bank, ca. 7000 plastics)
–, Plastics Databases, ASM International, Materials Park (OH)

BIBLIOGRAPHIES
O.A.Battista, The Polymer Index, McGraw Hill, New York 1976
J.T.Lee, Literature of Polymers, Encyclopedia of Polymer Science and Engineering, 2nd ed., **9** (1987) 62
R.T.Adkins, Ed., Information Sources in Polymers and Plastics, K.G.Saur, New York 1989

TEXTBOOKS OF POLYMER PHYSICS
U.Eisele, Introduction to Polymer Physics, Springer, Berlin 1990
R.W.Cahn, P.Haasen, E.J.Kramer, Eds., Materials Science and Technology, Vol. 12, E.L.Thomas, Ed., Structure and Properties of Polymers, VCH, Weinheim 1993
U.W.Gedde, Polymer Physics, Chapman and Hall, London 1995
M.Doi, Introduction to Polymer Physics, Oxford Univ. Press, New York 1996
G.R.Strobl, The Physics of Polymers, Springer, Berlin, 2nd ed. 1997
A.Grosberg, Ed., Theoretical and Mathematical Models in Polymer Research, Academic Press, San Diego 1998
M.Rubinstein, R.Colby, Polymer Physics, Oxford Univ.Press, New York 2003
J.E.Mark, Ed., Physical Properties of Polymers, Cambridge Univ. Press, Cambridge, 3rd ed. 2004
J.E.Mark, K.Ngai, W.W.Graessley, L.Mandelkern, E.T.Samulski, J.L.Koenig, G.D.Wignall, Physical Properties of Polymers, Cambridge Univ.Press, Cambridge, 3rd ed. 2004
T.Kawakatsu, Statistical Physics of Polymers. An Introduction, Springer, Berlin 2005
L.H.Sperling, Introduction to Physical Polymer Science, Wiley, Hoboken (NJ), 4th ed. 2006

2 Chemical Structure

2.1 Constitution

2.1.1 Definitions

Macromolecules are large molecules (G: *makros* = large; L: *molecula* = small mass, diminutive of *moles* = mass) consisting of hundreds to millions of atoms. They are produced from **monomers** by **polymerizations** (p. 1) which connect many monomer molecules by chemical bonds (G: *polys* = many; *meros* = part). The chemical composition of the resulting **monomer(ic) units** (= **mers**) is either the same as that of constituent monomer molecules (examples I-IV, VI, XI of Table 2-1) or not (examples V, VII-X). Common names of macromolecules often consist of names of monomer molecules prefixed by "poly". Examples are the polymers from monomers IV and V that have the same unit $-O-CH_2-CH_2-$ but different names. The systematic name of this polymer is poly(oxyethylene); however, systematic names are often more complex (see Table 2-2).

Table 2-1 Examples of monomer molecules, monomeric units, and repeating units refer to macromolecular substances: PE = poly(ethylene), PM = poly(methylene), PP = poly(propylene), PEOX = poly(ethylene oxide), PEG = poly(ethylene glycol). PA 6 = polyamide 6 = poly(ε-caprolactam), PA 6.6 = polyamide 6.6 = nylon 6.6 = poly(hexamethylene adipamide), S = polymeric sulfur.

	Monomer molecule(s)	Monomeric unit(s)	Repeating unit	Symbol
I	$CH_2{=}CH_2$ →	$-CH_2-CH_2-$	$-CH_2-$	PE
II	CH_2N_2 →	$-CH_2-$	$-CH_2-$	PM
III	$CH_2{=}CH(CH_3)$ →	$-CH_2-CH(CH_3)-$	$-CH_2-CH(CH_3)-$	PP
IV	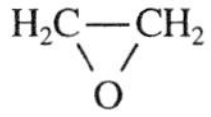 →	$-O-CH_2-CH_2-$	$-O-CH_2-CH_2-$	PEOX,
V	$HOCH_2CH_2OH$ →			PEG
VI	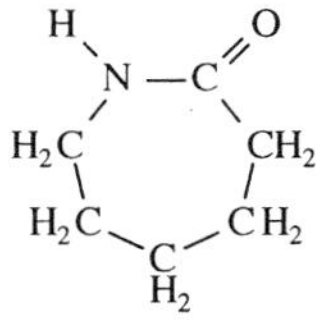 →	$-NH(CH_2)_5C(=O)-$	$-NH(CH_2)_5C(=O)-$	PA 6
VII	$H_2N(CH_2)_5COOH$ →			
VIII	$H_2N(CH_2)_6NH_2$ } →	{ $-NH(CH_2)_6NH-$	$-NH(CH_2)_6NHC(=O)(CH_2)_4C(=O)-$	PA 6.6
IX	$HOOC(CH_2)_4COOH$ }	$-CO(CH_2)_4CO-$ }		
X	$H_2N(CH_2)_6NHCO(CH_2)_4COOH$ →	$-NH(CH_2)_6NHC(=O)(CH_2)_4C(=O)-$	$-NH(CH_2)_6NHC(=O)(CH_2)_4C(=O)-$	PA 6.6
XI	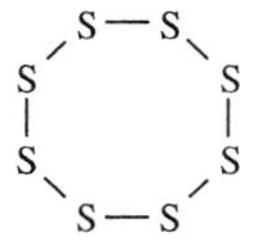 →	$-(S)_8-$	$-S-$	S

Macromolecules with the same type of monomeric or repeating units may carry various **endgroups**. In polyamide 6.6, $H{+}NH(CH_2)_6NH{-}CO(CH_2)_4CO{+}_nOH$, from the stoichiometric polycondensation of hexamethylenediamine (VIII) and adipic acid (IX), endgroups are one amino group, H_2N–, and one carboxylic group, –COOH, per molecule. Macromolecules from free-radical polymerization usually carry initiator fragments, groups from chain transfer reactions, and the like. Since the proportion of endgroups is small and their type often unknown, endgroups are usually ignored.

The number of *repeating units* per macromolecule is defined as the **degree of polymerization**, X, of the macromolecule. Repeating units are often identical with monomeric units (examples II-VII, X), but repeating units may also be greater (VIII-IX) or smaller (I, XI). Multiplication of the degree of polymerization by the molar mass of the repeating unit delivers the **molar mass of the macromolecule**, M. Physical methods usually deliver molar masses (physical unit: mass per amount-of-substance) whereas chemists usually report *relative molar masses*, called **molecular weights** (see Section 2.3).

Macromolecules are also called **polymer molecules**. A collection of macromolecules of the same type constitutes a **macromolecular substance** (L: *substantia*, from *sub* = under, *stare* = to stand), commonly known as a **polymer** (G: *polys* = many, *meros* = part). The term "polymer" is often used as shorthand for "polymer molecule" and/or in a narrow sense for a molecule or substance composed of many *equal* or similar mers.

In polymer physics, monomeric units are often simply called "monomers." This use of the word "monomer" is not only imprecise (a monomer is a substance and not a unit of a molecule) but also misleading since a polymer may indeed contain monomer, i.e., as residue from polymerization.

A monomeric unit may consist of one **chain unit** or more than one chain unit. For example, the hexamethylenediamine unit $-NH(CH_2)_6NH-$ (VIII in Table 2-1) contains 8 chain units and correspondingly 8 **chain atoms** (6 carbon atoms and 2 nitrogen atoms). In the terminology of organic chemistry, neither the carbon atoms nor the nitrogen atoms of this monomeric unit are *substituted* because hydrogen is not defined as a substituent. Oxygen groups O= of carbonyl groups >C=O are also not considered substituents but methyl groups $-CH_3$ of propylene units $-CH_2-CH(CH_3)-$ are.

2.1.2 Simple Polymer Chains

Monomeric units may be strung together linearly to **linear chains** or non-linearly to branched and crosslinked polymers (Section 2.1.3). In **homochains**, all chain atoms are identical; examples are poly(ethylene) and poly(sulfur) (Table 2-1). **Heterochains** contain two or more types of chain atoms; examples are poly(oxyethylene), ${+}OCH_2CH_2{+}_n$, and poly(hexamethylene adipamide), ${+}NH(CH_2)_6NH{-}CO(CH_2)_4CO{+}_n$.

Polymers are called **uniform** if their molecules are identical with respect to constitution and degree of polymerization; examples are enzymes. Practically all synthetic polymers are **non-uniform**, however, since their molecules differ in the number, constitution, and/or arrangement of monomeric units. This book calls polymers **constitutionally non-uniform** if the same monomer leads to different types of monomeric units (**constitutional isomerism**) and/or different arrangements of monomeric units (**regioisomerism**). An example of the former is the polymerization of isoprene, $CH_2{=}C(CH_3){-}CH{=}CH_2$, to 1,4- or 1,2-isoprene units, $-CH_2-C(CH_3){=}CH-CH_2-$ and $-CH_2-C(CH_3)(CH{=}CH_2)-$, respectively. An example of regioisomerism is the arrangement of $-CH_2-CHR-$ units in head-to-tail ($-CH_2-CHR-CH_2-CHR-$), head-to-head ($-CHR-CH_2-CH_2-CHR-$), or tail-to-tail position ($-CH_2-CHR-CHR-CH_2-$). In **molecularly non-uniform** polymers, macromolecules differ in their degrees of polymerization, i.e., these polymers have distributions of their molecular weights (see Section 2.3).

Molecularly non-uniform polymers are often called "polydisperse" in the literature. However, "disperse" is used in natural science exclusively for the distribution of matter in another matter, i.e., for *multiphase* systems (G: *dispergare* = to distribute). "Monodisperse" is an oxymoron since "disperse" relates to "more than one" and what is already multiple ("disperse") cannot be "mono."

In literature, constitutional formulas of polymers are always idealized and so are the names of polymers. In general, **generic names** are used, i.e., names of monomers (not names of monomeric *units*) prefixed with "poly." In this book, names of monomers in generic polymer names are always written in parentheses. In the literature, this is usually done only for complex monomer names. **Systematic names** are based on polymer constitution; they are mainly used for archival purposes as are also **CAS Registry Numbers**. Examples of polymer names are shown in Table 2-2 (see also Volume I, p. 30).

Table 2-2 Monomers, monomeric units, and generic and systematic names of polymers.

Monomer	Constitution of		Polymer name based on	
name	monomer	monomeric unit	monomer	constitution
Ethene	$CH_2=CH_2$	$-CH_2-CH_2-$	poly(ethene)	poly(ethylene)
-	-	$-CH_2-CH_2-CH_2-$	-	poly(propane-1,3-diyl)
Propene	$CH_2=CH(CH_3)$	$-CH_2-CH(CH_3)-$	poly(propene)	poly(propylene)
1-Butene	$CH_2=CH(C_2H_5)$	$-CH_2-CH(C_2H_5)-$	poly(1-butene)	poly(butene-1,4-diyl)
Styrene	$CH_2=CH(C_6H_5)$	$-CH_2-CH(C_6H_5)-$	poly(styrene)	poly(1-phenylethylene)
Butadiene	$CH_2=CH-CH=CH_2$	$-CH_2-CH=CH-CH_2-$	1,4-poly(butadiene)	poly(1-butene-1,4-diyl)
Acetylene	$CH\equiv CH$	$-CH=CH-$	poly(acetylene)	poly(ethene-1,2-diyl)
Formaldehyde	HCHO	$-O-CH_2-$	poly(formaldehyde)	poly(oxymethylene)

2.1.3 Polymer Architecture

Linear Polymers

For historical reasons, straight-chain macromolecules such as poly(ethylene) or poly(oxymethylene) (Table 2-1) are called **linear chains** since it was assumed by Hermann Staudinger, the father of macromolecular chemistry, that such molecules are completely stretched in both crystals and solutions. "Linear" now denotes the constitutionally one-dimensional assembly of chain units, i.e., the absence of branches (see below). **Cyclic macromolecules** ("**ring polymers**") are not considered linear molecules.

Homopolymers are derived from one species of monomer, **copolymers** from $i \geq 2$ species, for example, $i = 2$ (**bipolymers**), $i = 3$ (**terpolymers**), $i = 4$ (**quaterpolymers**), and $i = 5$ (**quinterpolymers**). "Copolymer" is often used as a synonym for bipolymer.

The sequence of various monomeric units in copolymers depends on the types of monomers and polymerizations. Bipolymers have two borderline cases, **diblock polymers** and **alternating bipolymers,** which frame **gradient bipolymers** (= **tapered bipolymers**), **segmented bipolymers**, **periodic bipolymers**, and **statistical bipolymers** (Table 2-3). In the latter type, **random bipolymers** with Bernoulli statistics (= Markov zeroth

Table 2-3 Types of bipolymers with monomeric units a and b.

Name	Constitution (schematic)
Alternating bipolymer	...abababababababababababababababababababababab...
Periodic bipolymer (example 1)	...abbabbabbabbabbabbabbabbabbabbabbabbabbabb...
Periodic bipolymer (example 2)	...aabbaabbaabbaabbaabbaabbaabbaabbaabbaabbaabb...
Random bipolymer (Bernoulli statistics)	...abaabbababbaaaaaabaabaabbbbaaaabaabbaaaabaabbb...
Gradient bipolymer	$a_n$baaaaaaaabbaaaaabaaabababbabbbaabbbbaabbbbbbb_m
Segmented bipolymer (large n and m)	$a_nb_ma_nb_ma_nb_m$
Diblock polymer	a..................................ab.................................b
Graft bipolymer	....a–a–a.......a–a–a...a–a–a..............a–a–a.......a–a–a... b_n b_m b_l b_k b_h

order statistics) of distribution of monomeric units can be distinguished from statistical bipolymers with first, second, ... order Markov statistics. Random copolymers are always molecularly non-uniform because compositions and degrees of polymerization of molecules as well as sequential arrangements of monomeric units vary from molecule to molecule. All these bipolymers and higher copolymers are linear polymers whereas **graft bipolymers** are polymers with branches that have been grafted onto or from linear polymers.

Block copolymers [a_nb_m, $a_nb_mc_p$, etc.], alternating bipolymers [$(ab)_n$], and periodic copolymers [$(abb)_n$, $(aabb)_n$, $(abc)_n$, etc.] can be obtained constitutionally uniform by suitable polymerization conditions though they may not necessarily be molecularly uniform. Note that "**block *copolymer***" refers to a polymer that is obtained by a copolymerization of a mixture of different types of monomers (n A + m B $\rightarrow$ a_nb_m) whereas "**block polymers**" are obtained by coupling of preformed polymer "blocks" ($a_n + b_m \rightarrow a_nb_m$) or by sequential polymerization (a_n + m B $\rightarrow$ a_nb_m).

Branched Polymers

The term "branched molecule" has different meanings in organic and macromolecular chemistry. In organic chemistry, a molecule is called branched if its substituents consist of a side chain with the same chain atoms as the main chain. An example is 3-ethylpentane, $CH_3CH_2CH(C_2H_5)CH_2CH_3$, where the ethyl side chain has the same chain atoms as the pentane main chain. In macromolecular chemistry, however, a macromolecule with the monomeric unit $-CH_2-CH(C_2H_5)-$ is considered to be linear (i.e., not branched) because the side chain $-C_2H_5$ is already present in the monomer, $CH_2{=}CH(C_2H_5)$. For this reason, bipolymers of ethene, $CH_2{=}CH_2$, with 1-olefins, $CH_2{=}CH(C_nH_{2n+1})$, are called *linear* low-density poly(ethylene)s (n = 4, 6, or 8).

Branched polymers in the meaning of polymer chemistry result only from syntheses, either targeted or at random. **Star molecules** (Fig. 2-1) are synthesized by either growth from a core with $f \geq 3$ functional sites or by coupling of three or more chains to a core molecule. The core may by a small molecule such as ammonia, NH_3, or a large entity such as a latex particle. Star molecules may thus have between three and hundreds of arms with tens to thousands of monomeric units each.

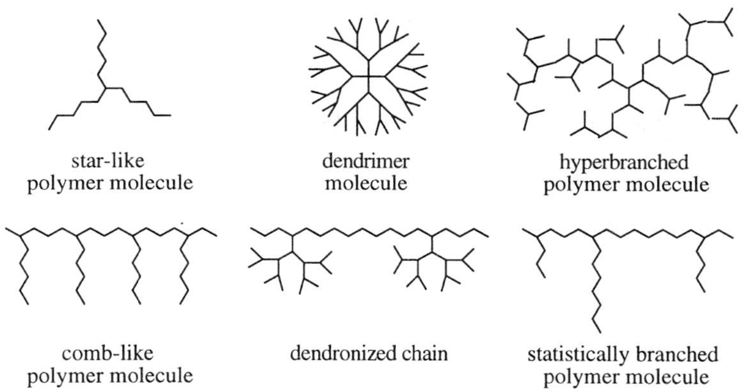

Fig. 2-1 Schematic representation of some types of branched polymer molecules.

Dendrimer molecules are star molecules with regular "branches-upon-branches" (G: *dendron* = tree). The "layers" of branch units are called "generations." For example, the dendrimer molecule of Fig. 2-1 has a tetrafunctional core and three generations of trifunctional branching units which leads to a total of 4 + 8 + 16 = 28 branching points and 32 endgroups. Controlled syntheses have led to practically molecularly uniform dendrimers with up to nine generations and molecular weights of almost a million. The shapes of such dendrimer molecules need not be sphero-symmetrical.

Random reactions of functional groups of monomer molecules such as AB_2, AB_3, etc., with their own type or with bifunctional molecules AB lead to randomly branched polymer molecules if A can react only with B but A cannot react with A or B with B. The resulting polymers are called **hyperbranched.**

Comb molecules with side chains of equal length at regular distances along the backbone chain are obtained by polymerization of so-called **macromonomers** which are monomers with very long substituents, for example, $CH_2{=}C(CH_3)[COO(CH_2)_iCH_3]$ with i = 20. Comb molecules with irregularly distributed side chains of unequal lengths are obtained by grafting monomers onto or from preformed backbone chains.

Statistically branched polymer molecules with long side chains result from intermolecular chain transfer reactions during polymerization whereas intramolecular chain transfer reactions lead to short side chains (see Volume I). Examples are

$$—CH_2—\underset{CH_2CH_3}{CH}—(CH_2)_x—\underset{(CH_2)_4CH_3}{CH}—(CH_2)_y—$$

short branches
(x, y >> 1)

$$—CH_2—\underset{(CH_2)_zH}{CH}—(CH_2)_x—\underset{(CH_2)_qH}{CH}—(CH_2)_y—$$

long branches
(x, y, z, q >> 1)

Macromolecules with Higher Constitutional Dimensions

Linear macromolecules can be viewed as *constitutionally* one-dimensional entities since extended chains have lengths that are much larger than their diameters. Correspondingly, constitutionally two- and three-dimensional macromolecular compounds exist as well as intermediary forms. The resulting **polymer architecture** ranges from

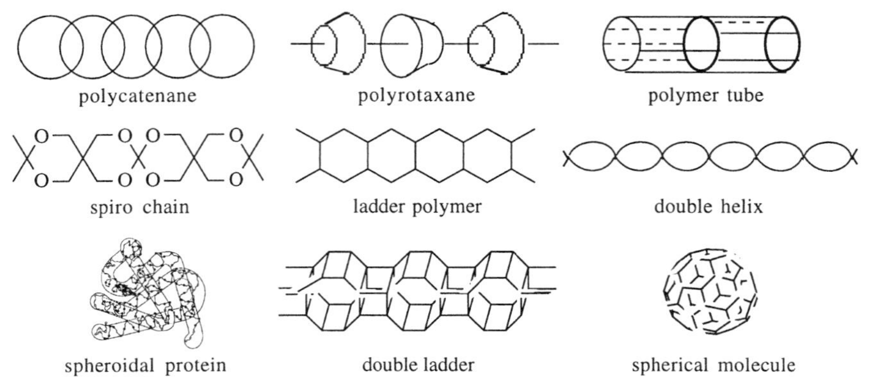

Fig. 2-2 Some types of macromolecules of higher constitutional dimensionality (not to scale).

combinations of non-bonded rings to bonded ladder, tube, and spherical structures (Fig. 2-2). However, the geometric shapes of these assemblies are determined not only by their constitution but also by their chemical configurations (Section 2.2), microconformations (Chapter 3), and macroconformations (Chapter 4).

Polycatenanes are interlocked rings that are not connected by chemical bonds (L: *catena* = chain). **Polyrotaxanes** resemble pearl necklaces in which rotatable hollow cylinders are stringed on polymer chains that are sealed at both ends by a bulky chemical groups (not shown in Fig. 2-2) (L: *rota* = wheel). **Polymer tubes** are hollow cylinders with a wall of chemically bound atoms; examples are carbon nanotubes.

Spiro chains consist of two single chains that share common chain atoms at regular intervals (G: *speira* = coil) whereas **ladder polymers** can be viewed either as linearly fused rings or as two parallel chains that are interconnected in regular intervals by chemical bonds. In **double ladder chains**, two ladder polymers are bound to each other in regular intervals leading to tube-like structures; the silicate apophyllite is an example.

Phyllo polymers, usually called **layer** or **parquet polymers**, form two-dimensional lattices (G: *phyllon* = leaf) (Fig. 2-3); an example is graphene. **Sheet polymers** consist of several such layers (stacked graphene molecules of graphite, mica). **Tecto polymers (lattice polymers)** have regular three-dimensional lattices consisting of chemically bound atoms (G: *tekton* = carpenter, L: *tectum* = roof). Irregular interconnections of polymer chains lead to **crosslinked polymers** which may be either **thermosets** with short chain sections between crosslinks and $T_G > T$ or **elastomers** with long sections between crosslinks and $T_G < T$ with where T_G = glass temperature and T = application temperature.

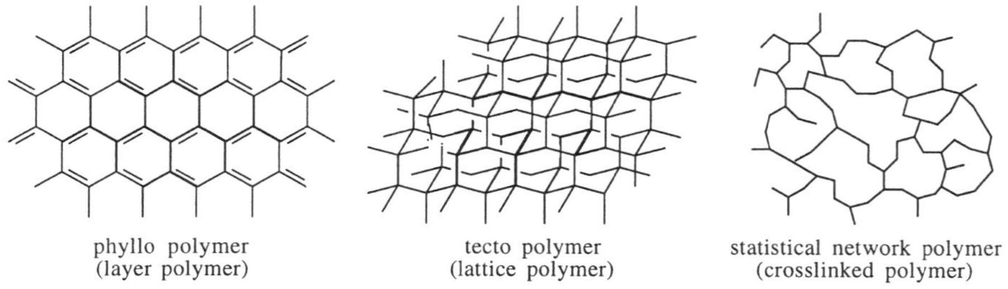

Fig. 2-3 Some two- and three-dimensional polymers (schematic).

2.2 Chemical Configuration

2.2.1 Basic Terms

Isomers are molecules of identical chemical composition (type and number of atoms) that differ in the arrangement of atoms (G: *isos* = equal, *meros* = part). **Constitutional isomers** vary in the succession of their atoms; examples are butane, $CH_3CH_2CH_2CH_3$, and isobutane, $CH_3CH(CH_3)_2$. **Stereoisomers** have the same succession of atoms but different spatial arrangements; examples are D-alanine and L-alanine with the same constitution, H_2N–*$CH(CH_3)$–COOH, but with a different spatial arrangement of their four substituents NH_2, H, CH_3, and COOH, around the central atom *C. An isomer is either a constitutional isomer *or* a stereoisomer; it can never be both.

Stereoisomers are subdivided according to their symmetry properties (enantiomers versus diastereomers (Fig. 2-4)) or their energy barriers (configurational versus conformational isomers (Fig. 2-5)). Two stereoisomers are either enantiomeric *or* diastereomeric to each either; they can never be both.

Enantiomers relate to each other like mirror and mirror image. They are like a left hand and a right hand and are therefore always *chiral* (G: *cheir* = hand). **Diastereoisomerism** is found in isomers that possess at least two stereogenic centers of which some are enantiomeric and the others configurationally identical (Fig. 2-4). Diastereomeric *molecules* do not have mirror isomerism. They may be chiral like threose and erythrose or achiral like *cis*- and *trans*-1,2-dibromoethene (see Volume I, Fig. 4-1).

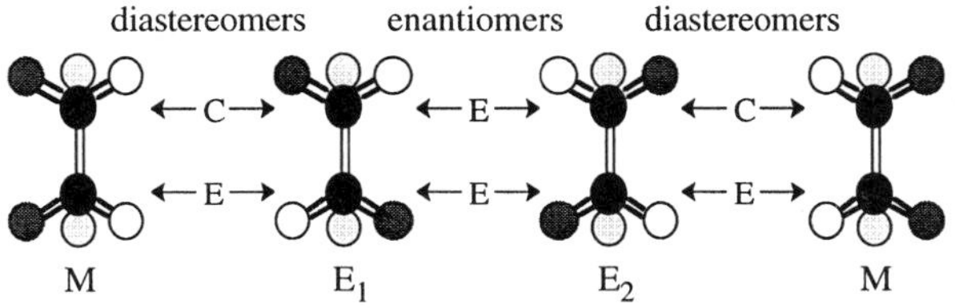

Fig. 2-4 Enantiomers and diastereomers of a molecule of the type RR'R"C–CRR'R". Units –CRR'R" are either enantiomeric to each other (E) or configurationally identical (C). E_1, E_2 = enantiomeric molecules, M = meso compound.

Chemistry divides stereoisomers into configurational and conformational isomers according to the height of their energy barrier. A (*chemical*) **configuration** (L: *com* = together, *figura* = shape) is the spatial arrangement of atoms or atomic groups about a central atom or a multiple bond that can be transformed only into another configuration by overcoming a large energy barrier. Examples of such *kinetically stable* **configurational isomers** are isotactic and syndiotactic poly(propylene)s and the cis- and trans-isomers of 1,4-poly(butadiene) (Fig. 2-5), the latter being **torsional** or **geometric isomers**.

Fig. 2-5 Examples of configurational and geometric isomers. Left: isotactic and syndiotactic poly(propylene); right: *cis*- and *trans*-1,4-poly(butadiene). Bonds are in (——), above (wedge), or below (hashed) the paper plane. For types of representation of molecules see Fig. 2-6.

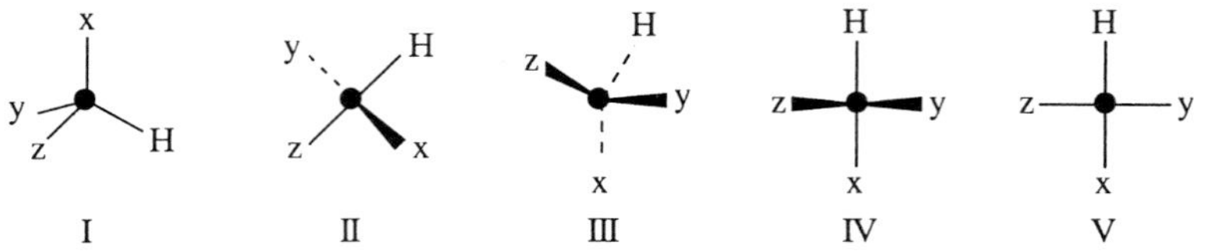

Fig. 2-6 Two-dimensional representations of the three-dimensional molecule CzyxH (with C as ●). I: Perspective. II-IV: Wedges indicate bonds above the paper plane; solid lines, in the paper plane; and broken lines, below the paper plane. The two-dimensional Fischer projection, V, corresponds to the "spatial" representation III or the pseudo-spatial representation IV.

Conformational isomers are distinguished from configurational ones by a low energy barrier for the transition from one conformation into another (Chapter 3). Thus, conformational isomers are interconverted rapidly into each other. Since both conformational and configurational isomers are characterized by different spatial positions of atoms or groups of atoms, polymer physics does not distinguish between these two types of spatial arrangements and calls their statistics "configurational statistics" as was common up to the 1940s.

2.2.2 Chemical Configurational Statistics

Chemical configurational statistics describes the statistical sequence of configurations around central atoms or multiple bonds. In organic chemistry, one is interested in absolute configurations; in macromolecular chemistry, in relative ones.

Absolute configurations around stereogenic centers (formerly: asymmetric central atoms) are described by two conventions. In the **D,L-system**, the optically right-rotating (+)-glyceraldehyde is given the D-configuration. The **R,S-system** does not need a reference substance. Substituents ("ligands") are rather assigned seniorities that depend on the position of the atoms of the substituents in the Periodic Table. Iodine has the highest seniority; a lone electron pair, the lowest (Volume I, Section 4.1.6).

"Absolute configurations" of organic chemistry refer to configurations about each center of stereoisomerism relative to the ligand with the lowest seniority. In contrast, "relative configurations" of macromolecular chemistry are determined by moving along the chain from one end to the other.

Examples are the chemical configurations of the two chain units $-CH(CH_3)-$ and $-CH(C_2H_5)-$ of the constitutional repeating unit, $-CH(CH_3)-CH(C_2H_5)-$, of poly(2-pentene). Here, the two chain atoms 1 and 2 can be (a) both defined, (b) one defined and the other undefined, or (c) both undefined. Correspondingly, stereorepeating units are distinguished from tactic repeating units as shown in Fig. 2-7.

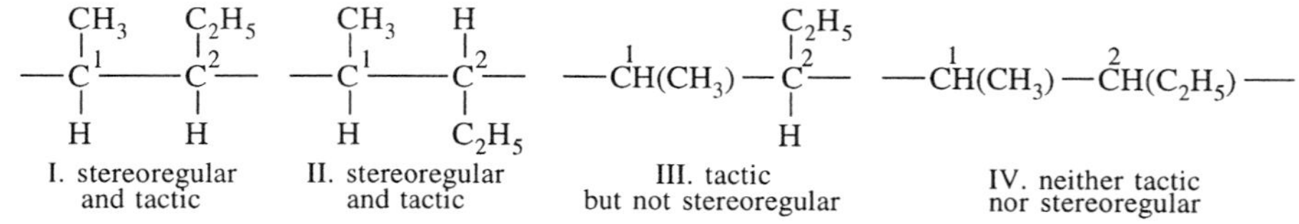

Fig. 2-7 Definitions of "stereoregular" and "tactic" monomeric units (example: poly(2-pentene)).

Stereorepeating units are configurational repeating units in which *all* stereogenic centers are defined. Correspondingly, a **stereoregular polymer** consists of molecules composed of only one type of stereoregular repeating units that are all interconnected in the same way (Formulas I and II in Fig. 2-7).

In a **tactic repeating unit**, at least one stereogenic center is defined whereas the other centers are not (Formula III in Fig. 2-7). The molecules of a **tactic polymer** thus always contain the same types of tactic repeating units which are coupled in the same manner. Stereoregular polymers are always tactic but tactic polymers are not necessarily stereoregular since not all centers of stereoisomerism need to be defined.

An example is poly(propylene, $\text{+}CH_2\text{—}CH(CH_3)\text{+}_n$, in which the simplest stereorepeating units are identical to the simplest corresponding tactic repeating units. Each constitutional diad of this polymer has two types of tactic repeating units, isotactic and syndiotactic. The smallest **isotactic repeating unit** is the isotactic diad consisting of two monomeric units that are constitutionally and configurationally identical (G: *isos* = equal). **Isotactic molecules** thus contain only molecules with such isotactic units.

The smallest **syndiotactic repeating unit** is the syndiotactic diad composed of two enantiomeric configurational monomer units (G: *syn* = together, *dios* = two). **Syndiotactic polymer molecules** contain only syndiotactic diads.

Tactic poly(propylene)s contain one center of stereoisomerism per monomeric unit; they are **monotactic**. Stereoregular poly(2-pentene)s (Fig. 2-7) have two defined centers of stereoisomerism per monomeric unit; they are **ditactic**.

Each monomeric unit belongs to two tactic diads, 3 tactic triads, 4 tactic tetrads, etc. The tacticity is characterized by the symbols "i" and "s" . There are two types of diads (i, s), four types of triads (ii, is, si, ss), eight types of tetrads (iii, iis, isi, sii, ssi, sis, iss, sss), etc. The triads "is" and "si" are called **heterotactic** (Fig. 2-8).

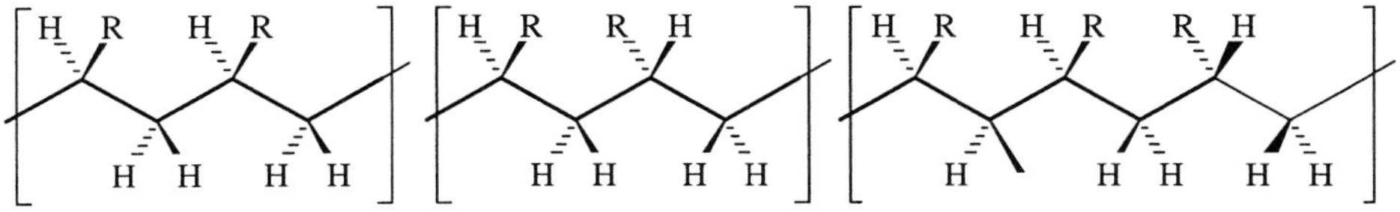

Fig. 2-8 Isotactic diad "i", syndiotactic diad "s", and heterotactic triad "is" of a polymer with the monomeric unit $\text{–}CHR\text{–}CH_2\text{–}$. The other heterotactic triad ("si") is not shown.

Note that in isotactic polymer molecules, $\text{+}CH_2\text{—}CHR\text{+}_n$, composed of constitutional repeating units with *two* chain atoms, all substituents R are "on the same side" of the molecule (in this figure, above the paper plane) whereas in syndiotactic molecules, they alternate in space (one R above the paper plane, the next R below the paper plane, and so forth). This is true for all types of polymer molecules if the constitutional repeating unit consists of even numbers of chain atoms. However, spatial positions of R alternate in stereoprojections of isotactic polymer molecules if constitutional repeating units consist of odd numbers of chain atoms; they are still on the same side in Fischer projections.

For historic reasons, literature uses mainly the symbols "m" (from "meso") instead of "i", and "r" (from "racemic") instead of "s". The symbol "m" is justified since an infinitely long, all isotactic polymer molecule (negligible endgroups) is indeed a meso compound because one-half of the molecule has the absolute [R] configuration and the other half the [S] configuration. However, a polymer composed of completely syndiotactic molecules is not "racemic". All molecules are rather composed of alternating [R] and [S] configurations; they are not mixtures of molecules with opposite absolute configurations. Since "r" is so ingrained as a symbol, it has been proposed to call syndiotactic molecules "racemo" instead of "racemic" but this does not remove the wrong implication (see Volume I).

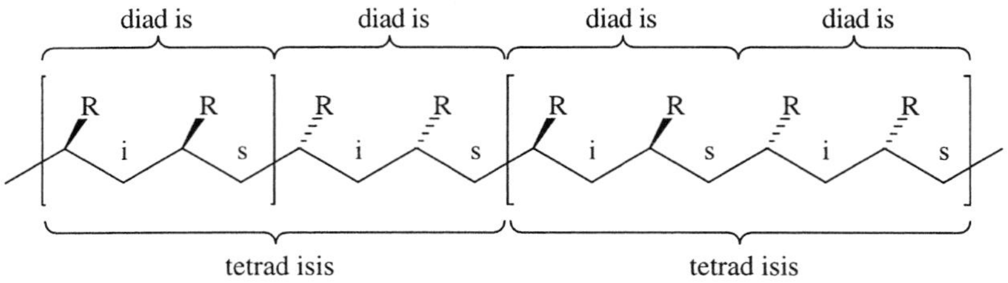

Fig. 2-9 Section of a heterotactic polymer molecule, $\text{-[-CHR—CH}_2\text{-]-}_n$, composed of four heterotactic triads, –is–. The configurational repeating unit is not –is– but rather –isis–.

Ideally configured polymer molecules (100 % isotactic or 100 % syndiotactic) have only one type of configurational diad ("i" or "s"); this diad is also the configurational repeating unit. However, in truly heterotactic polymers (all "is" or "si"), repeating units are *not* identical with heterotactic triads (Fig. 2-9). The configurational repeating unit consists of four configurational diads, -[-isis-]-_n, rather than two. Alternatively, the configurational repeating unit may be depicted as $\text{-[-isi-}\textit{alt}\text{-sis-]-}_n$.

A completely isotactic polymer consists only of isotactic diads "i", triads "ii", tetrads "iii", etc. The *amount fractions* ("mole fractions") of all types of isotactic J-ads are therefore $x_i = x_{ii} = x_{iii} = ... \equiv 1$; for syndiotactic J-ads, one has similarly, $x_s = x_{ss} = x_{sss} = ... \equiv 1$. Heterotactic polymers consist of 100 % heterotactic units "is"; hence $x_i = x_s = 1/2$, $x_{ii} = x_{ss} = 0$, and $x_{is} = x_{si} = 1/2$. The *sum of all amount fractions* ("mole fractions") of configurational units of a given J-ad (J = di, tri, tetra, etc.) equals unity. Thus, $x_i + x_s \equiv 1$ for all diads, $x_{ii} + x_{is} + x_{si} + x_{ss} \equiv 1$ for all triads, etc.

General relationships exist between different types of J-ads (J = di, tri, tetra, etc.) since every triad, tetrad, pentad, etc., is composed of several diads, etc. Configurational diads and triads are related by $x_i = x_{ii} + (1/2)(x_{is} + x_{si})$ and $x_s = x_{ss} + (1/2)(x_{si} + x_{is})$; for relationships between other types of J-ads see Volume I. All of these relationships apply only to constitutionally uniform chains, which may be either linear ones of infinite length or cyclic ones of any length.

So-called stereoregular polymers are usually not 100 % sterically pure. They are characterized by the percentages of various types of J-ads and average sequence lengths of isotactic and syndiotatic homosequences. An isotactic sequence ...i_i... begins with a heterotactic triad ...si... and ends with a hetereotactic triad ...is... The number-average degree of polymerization, X, of isotactic sequences is thus given by $\overline{X}_{I,n} = 2\, x_i/(x_{is} + x_{si})$ and the number-average degree of polymerization of all tactic sequences by $\overline{X}_{\text{seq},n} = (1/2)[\overline{X}_{I,n} + \overline{X}_{S,n}] = 1/(x_{is} + x_{si})$.

2.3 Molar Masses and Degrees of Polymerization

2.3.1 Overview

Atomic masses, m_a, are very small experimental quantities. An example is the atomic mass of the carbon isotope ^{12}C, $m_a(^{12}\text{C}) = 1.992\,648\,2 \cdot 10^{-26}$ kg. This carbon nuclide serves as the standard and has been assigned the atomic mass number, $A = 12$. The

atomic mass constant, m_u, is thus $m_u = m_a(^{12}C)/A \approx 1.660\,540\,2{\cdot}10^{-27}$ kg. Its value is not exact because it depends on the Avogadro constant, N_A, an experimental quantity.

By definition (ISO), the atomic mass constant equals the **unified atomic mass unit,** $m_u \equiv$ u, sometimes symbolized by "amu". The unified atomic mass unit is also called the **dalton** (with abbreviation Da) but this name is being discouraged because it has not been approved by the international *Conférence Générale des Poids et Mesures* and it is thus neither an SI base unit nor an SI derived unit (see Appendix). To make matters worse, the "Dalton" (with capital D) has been used as the unit of molecular weight and/or molar mass, neither of which has the unit of a mass (see below).

The **molecular mass** of a molecule, m_{mol}, equals the sum of the atomic masses of its atoms. These small quantities are never determined experimentally in polymer science. For example, a poly(methylene) molecule with a degree of polymerization of $X = 10^5$, $^1H(^{12}C^1H_2)_{100\,000}{}^1H$, consists of 100 000 ^{12}C-atoms and 200 002 ^{1}H-atoms. This molecule has a molecular mass of $m_{mol} = 2.327\,358\,3{\cdot}10^{-21}$ kg since the atomic masses are $m_a(^{12}C) = 1.992\,648\,2{\cdot}10^{-26}$ kg and $m_a(^1H) = 1.673\,533\,9{\cdot}10^{-27}$ kg, respectively.

Instead, constitutional sizes of macromolecular *substances* are characterized by molar masses, molecular weights, or degree of polymerizations. The **molar mass of a substance** is the product of the molecular mass, m_{mol}, and the Avogadro constant, N_A, which indicates the number of chemical entities per mole of substance. Thus, the poly(methylene) mentioned above has a molar mass of $M = m_{mol}N_A$ = (2.327 358 3·10^{-21} kg) × (6.022 136 7·10^{23} mol^{-1}) ≈ 1 401 567 g/mol. The molar mass of a substance can thus be calculated from the ratio of the mass of the substance to the amount-of-substance, $M = m/n$. Its value is usually reported in g/mol (and not in kg/mol) because molar mass and molecular weight then become numerically identical.

The molar mass is a physical quantity that is obtained by almost all physical "molecular weight" methods such as membrane osmometry (Section 10.3.2), analytical ultracentrifugation (Section 11.3), static light scattering (Section 5.2), etc. An exception is mass spectroscopy (Volume I, Section 3.4.1) where one measures the ratio of the molecular mass and electric charge of ionized molecules, m_{mol}/Q. m_{mol} is reported as the ratio of molecular mass (in kg) and atomic mass constant (in kg), $m^* = m_{mol}/m_u$, and Q as the ratio of electric charge (in C) and elementary charge e (in C) so that $z = Q/e$ = number of electric charges per ion. The ratio, m^*/z, is dimensionless; its numerical value is identical with the dimensionless molecular weight. Instead, it is often reported in "dalton" (no plural!) which has the physical unit of a mass and is much smaller (see above)..

It seems that people assume the "dalton" to be a modern physical unit for the molecular weight. The erroneous use of "daltons" in mass spectroscopy has spread to other molecular weight and molar mass methods. However, light scattering (and many other physical methods) delivers molar masses (in mass per amount-of-substance such as g/mol) but neither molecular weights (dimensionless) nor daltons (physical unit of a mass).

The **molecular weight** of a molecule is defined as $M_r \equiv m_{mol}/m_u$, i.e., the ratio of the mass of a molecule of a substance, m_{mol}, to the atomic mass constant. It is called a **weight** since "weight" is the product of mass and acceleration in free fall. "Weight" can be used here because all molecules of a substance experience the same acceleration. However, "weight" must be used only in a consistent manner. One can thus speak of a *weight* average of molecular *weight* but not of a *weight* average of molar *mass*.

The **degree of polymerization** of a macromolecule, X, is a quantity from polymerization theory. It cannot be determined by direct experiment but can be calculated from the molar mass, M, if the molar mass, M_{rep}, of the constitutional repeating unit, the number, N_{end}, of endgroups per molecule, and the molar mass(es), M_{end}, of the endgroup(s) are known. If only one type of endgroup is present, one obtains for the molar mass M of the polymeric *substance*

(2-1) $$M = XM_{\text{rep}} + N_{\text{end}}M_{\text{end}}$$

Only a few natural polymeric substances are constitutionally and molecularly uniform, i.e., they consist of macromolecules of exactly the same constitution and thus also of the same molar mass. In contrast, most biopolymers and all synthetic polymers are rather molecularly non-uniform even if they are constitutionally uniform, i.e., their molecules differ in the number of repeating units although the constitution of repeating units and endgroups as well as the number of endgroups per molecule do not vary with molecular size. These polymeric substances thus have molar mass distributions (Section 2.4) that can be determined experimentally by many methods.

However, instead of molar mass distributions, one often uses only averages of molar masses or moments of distributions. These parameters differ in the statistical weights that are assigned to the molecules of the substance; for example, number- or mass-statistical weights (Section 2.3.2). Each distribution can be described by an infinite number of averages or moments but only a few of them are experimentally accessible or theoretically important.

2.3.2 Statistical Weights

Molecularly non-uniform polymers contain polymer molecules of various molar masses. The proportion of molecules of the same molar mass can be "weighed" in various ways according to the prevalence of their numbers, masses, lengths, volumes, etc. The most important statistical weights for the purpose of this section are the number-, mass- (weight-), z-, and (z+1)-statistical weights because they are interconnected via molar masses. Length-, area-, and volume-related statistical weights are discussed in Volume IV.

Chemical reactions are usually related to either the number, N_1, of the reacting molecules of type 1 or the amount of substance, n_1 (in moles). Both quantities are connected by the Avogadro constant, $N_A \approx 6.023 \cdot 10^{23}$ mol^{-1}, via $n_1 \equiv N_1/N_A$. Note that the Avogadro constant is *not* "the Avogadro *number*" since it has the unit of an inverse amount-of-substance.

Alternatively, amount-of-substance concentrations (also called amount concentrations, "molar concentrations", and "mole concentrations"), $[1] = n_1/V$, are used to indicate the amount-of substance, n_1, per volume V of solution *after* the mixing of 1 and solvent (see the definitions of the various types of concentrations in the Appendix, Chapter 19).

The molar mass of a substance, M_1, composed only of molecules of type 1, is defined as the product of the mass, $m_{\text{mol},1}$, of a molecule of type 1 multiplied by the Avogadro constant, $M_1 \equiv m_{\text{mol},1}N_A$. Since the mass of all molecules 1 is given by $m_1 = N_1 m_{\text{mol},1}$, it follows that the molar mass of a *molecularly homogeneous* substance 1, M_1, is given by the ratio of the mass, m_1, and amount of all molecules of type 1, $n_1 \equiv N_1/N_A$:

(2-2) $$M_1 = m_{\text{mol},1}N_A = (m_1/N_1)N_A = m_1/n_1$$

The molar mass of a *molecularly non-uniform* polymer, i.e., a polymer with a distribution of molecules with various degrees i of polymerization, is given by the ratio of the sum of all masses of molecules, m_i, of types i = 1, 2, 3, ..., divided by the sum of all amounts, n. Since this molar mass averages amounts (which are directly proportional to numbers), the resulting molar mass is the **number-average molar mass,** $\overline{M}_n$:

(2-3) $$\overline{M}_n \equiv m/n = \Sigma_i m_i/\Sigma_i n_i = \Sigma_i n_i M_i/\Sigma_i n_i = \Sigma_i N_i M_i/\Sigma_i N_i$$

Experimentally, one often determines masses of molecules, $m_i = (m_{mol})_i N_i$, and not their amounts, n_i. The **mass-average molar mass,** $\overline{M}_w$, is then defined in analogy to the expression for the number-average molar mass, $\Sigma_i n_i M_i/\Sigma_i n_i$, as

(2-4) $$\overline{M}_w \equiv \Sigma_i m_i M_i/\Sigma_i m_i$$

In general, averages of molar masses can be written in terms of **statistical weights** g (= n, w, etc.) or **statistical weight fractions,** $G_i = g_i / \sum_i g_i$,

(2-5) $$\overline{M}_g \equiv \sum_i g_i M_i / \sum_i g_i = \sum_i G_i M_i$$

Many other average molar masses can be defined, some of which can be measured directly whereas others play a role in the dependence of certain physical properties on molar masses. The most important of the corresponding statistical weights g are

(2-6)	{n–1}-statistical weight:	$\{n-1\}_i$	$\equiv n_i M_i^{-1}$	$\equiv m_i M_i^{-2}$	$\equiv \{m-2\}_i$
(2-7)	number-statistical weight:	n_i	$\equiv n_i$	$\equiv m_i M_i^{-1}$	$\equiv \{m-1\}_i$
(2-8)	mass-statistical weight:	m_i	$\equiv n_i M_i$	$\equiv m_i$	$\equiv \{m\}_i$
(2-9)	z-statistical weight:	z_i	$\equiv n_i M_i^2$	$\equiv m_i M_i$	$\equiv \{m+1\}_i$
(2-10)	{z+1}-statistical weight:	$\{z+1\}_i$	$\equiv n_i M_i^3$	$\equiv m_i M_i^2$	$\equiv \{m+2\}_i$

The terms in braces do not represent mathematical operations in which a number is subtracted from or added to a physical quantity. They are rather symbols for physical quantities. Only the symbols to the left of the first identity sign are commonly used.

The z-statistical weight got its name because it was first determined by centrifuge methods (G: zentrifuge). $\{z+1\}$ and $\{n-1\}$ were introduced as the next higher or lower statistical weights to z and n, respectively; they are not characterized by special symbols. The series of statistical weights can be expanded to higher or lower statistical weights; experimental methods for these statistical weights are unknown, however.

Expressions (2-6)–(2-10) imply molecularly uniform species i. For *molecularly non-uniform* species i (example: fractions i with distributions of molar masses), numerical calculations show that molar masses M_i in Eqs.(2-6)–(2-10) have to be replaced by molar mass averages that correspond to the multiplying statistical weight. For example, $m_i = n_i M_i$, (Eq.(2-8)), has to be replaced by $m_i = n_i \overline{M}_{n,i}$ (Eq.(2-8a)) and $z_i = m_i M_i$ (Eq.(2-9)) by $z_i = m_i \overline{M}_{w,i} = n_i \overline{M}_{n,i} \overline{M}_{w,i}$ (Eq.2-9a)). This rule applies to each statistical weight and not only to molar masses, molecular weights, and degrees of polymerization but also to other properties such as lengths, areas, and volumes. Neglect of these substitutions can lead to grave numerical errors even for narrow distributions.

(2-6a)	$\{n-1\}_i$	$\equiv (n_i)^{-1}(\overline{M}_{n-1,i})^{-1}$	$\equiv m_i(\overline{M}_{n-1,i})^{-1}(\overline{M}_{n,i})^{-1}$	
(2-7a)	n_i	$\equiv n_i$	$\equiv m_i(\overline{M}_{n,i})^{-1}$	$\equiv z_i(\overline{M}_{n,i})^{-1}(\overline{M}_{w,i})^{-1}$
(2-8a)	m_i	$\equiv n_i \overline{M}_{n,i}$	$\equiv m_i$	$\equiv z_i(\overline{M}_{w,i})^{-1}$
(2-9a)	z_i	$\equiv n_i \overline{M}_{n,i}\overline{M}_{w,i}$	$\equiv m_i \overline{M}_{w,i}$	$\equiv z_i$
(2-10a)	$\{z+1\}_i$	$\equiv n_i \overline{M}_{n,i}\overline{M}_{w,i}\overline{M}_{z,i}$	$\equiv m_i \overline{M}_{n,i}\overline{M}_{w,i}$	$\equiv z_i \overline{M}_{z,i}$

Instead of statistical weights (amounts, masses, etc.,) one can also use the corresponding **statistical weight fractions** such as amount fractions ("mole fractions") x_i, mass fractions ("weight fractions") w_i, z-fractions Z_i, etc. With the definitions of these fractions and Eqs.(2-7a)–(2-9a), one obtains for the three major types of statistical-weight fractions:

(2-11) $x_i = n_i/\Sigma_i n_i = n_i/n$

(2-12) $w_i = m_i/\Sigma_i m_i = m_i/m = n_i \overline{M}_{n,i}/(n\overline{M}_n) = x_i \overline{M}_{n,i}/\overline{M}_n$

(2-13) $Z_i = z_i/\Sigma_i z_i = z_i/z = m_i \overline{M}_{w,i}/(m\overline{M}_w) = x_i \overline{M}_{n,i}\overline{M}_{w,i}/(\overline{M}_n\overline{M}_w)$

$\overline{M}_{n,i}$, $\overline{M}_{w,i}$, and $\overline{M}_{z,i}$ are the number-, weight-, and z-averages of molar masses of fractions i and $\overline{M}_n$, $\overline{M}_w$, and $\overline{M}_z$ the corresponding molar-mass averages of the total polymer. For molecularly uniform fractions i, molar masses reduce to $\overline{M}_{n,i} = \overline{M}_{w,i} = \overline{M}_{z,i} \equiv M_i$ and Eqs.(2-11)–(2-13) become

(2-11a) $x_i = n_i/\Sigma_i n_i = n_i/n$

(2-12a) $w_i = m_i/\Sigma_i m_i = m_i/m = n_iM_i/(n\overline{M}_n) = x_iM_i/\overline{M}_n$

(2-13a) $Z_i = z_i/\Sigma_i z_i = m_iM_i/(m\overline{M}_w) = w_iM_i/\overline{M}_w = x_iM_i^2/(\overline{M}_n\overline{M}_w)$

Comparison of Eqs.(2-11)-(2-13) and Eqs.(2-7)-(2-9) shows that interconversions of statistical weights or statistical weight fractions of polymer fractions always require knowledge of the corresponding averages of molar masses of the whole polymer. This can also be seen from dimensional analysis since $w_i = x_iM_i/\overline{M}_n$ and *not* $w_i = x_iM_i$, and $Z_i = x_iM_i^2/(\overline{M}_n\overline{M}_w)$ and *not* $Z_i = x_iM_i^2$. These common errors can be traced to the use of molecular weights instead of molar masses. As noted above and in Chapters 5 and 10-12, physical methods never deliver molecular weights but always molar masses (exception: mass spectroscopy).

2.3.3 Simple Averages of Molar Masses

The molar mass M of a *molecularly uniform* polymer is given by the degree of polymerization, the molar mass of the repeating unit, and the sum of the molar masses of the endgroups, Eq.(2-1). The following discussion assumes that contributions of endgroups to the molar mass of the polymer can be neglected (high molar masses). This assumption does not apply to oligomers and highly branched polymers and the equations to follow have to be amended for the presence of endgroups.

It is furthermore assumed that polymers are constitutionally uniform with chemically identical repeating units. However, all synthetic polymers are molecularly non-uniform ("polydisperse"), i.e., they consist of molecules with different degrees of polymerization. Experimentally, such polymers may consist of fractions with average degrees of polymerizations, $\overline{X}_{g,i}$.

The **number-average molar mass,** $\overline{M}_n$, of a polymer composed of molecularly non-uniform species i with number-average molar masses, $\overline{M}_{n,i}$, is defined as

$$(2\text{-}14) \qquad \overline{M}_n \equiv \sum_i x_i \overline{M}_{n,i} = \frac{\sum_i n_i \overline{M}_{n,i}}{\sum_i n_i} = \frac{\sum_i m_i}{\sum_i (m_i / \overline{M}_{n,i})} = \frac{1}{\sum_i (w_i / \overline{M}_{n,i})}$$

where $x_i = N_i/\sum_i N_i = n_i/\sum_i n_i$ = amount fraction ("mole fraction") of species i, $n_i \equiv N_i/N_A$ = amount (in "moles") of species i, $m_i = n_i \overline{M}_{n,i}$ = mass of fraction i (Eq.(2-8a)), and $w_i = m_i/\sum_i m_i$ = mass fraction ("weight fraction") of species i. Furthermore: $\sum_i N_i \equiv N$, $\sum_i n_i \equiv n$, $\sum_i x_i \equiv 1$, $\sum_i m_i \equiv m$, and $\sum_i w_i \equiv 1$.

According to the definition of a mass-average and Eqs. (2-12a) and (2-8a), the **mass-average** and **z-average of molar mass** of a substance composed of molecularly non-uniform species i are

$$(2\text{-}15) \qquad \overline{M}_w \equiv \sum_i w_i \overline{M}_{w,i} = \frac{\sum_i m_i \overline{M}_{w,i}}{\sum_i m_i} = \frac{\sum_i n_i \overline{M}_{n,i} \overline{M}_{w,i}}{n \overline{M}_n} = \frac{\sum_i x_i \overline{M}_{n,i} \overline{M}_{w,i}}{\overline{M}_n}$$

$$(2\text{-}16) \qquad \overline{M}_z \equiv \sum_i Z_i \overline{M}_z = \frac{\sum_i z_i \overline{M}_{z,i}}{\sum_i z_i} = \frac{\sum_i m_i \overline{M}_{w,i} \overline{M}_{z,i}}{m \overline{M}_w} = \frac{\sum_i x_i \overline{M}_{n,i} \overline{M}_{w,i} \overline{M}_{z,i}}{\overline{M}_w}$$

For polymers with molecularly uniform species i, $M_i = \overline{M}_{n,i} = \overline{M}_{w,i} = \overline{M}_{z,i}$, and Eqs.(2-13)-(2-15) reduce to

$$(2\text{-}14a) \qquad \overline{M}_n = \sum_i x_i M_i = \frac{\sum_i w_i}{\sum_i (w_i / M_i)} = \frac{1}{\sum_i (w_i / M_i)}$$

$$(2\text{-}15a) \qquad \overline{M}_w = \sum_i w_i M_i = \frac{\sum_i x_i M_i^2}{\sum_i x_i M_i} = \frac{\sum_i x_i M_i^2}{\overline{M}_n}$$

$$(2\text{-}16a) \qquad \overline{M}_z = \sum_i Z_i M_i = \frac{\sum_i w_i M_i^2}{\sum_i w_i M_i} = \frac{\sum_i w_i M_i^2}{\overline{M}_w} = \frac{\sum_i x_i M_i^3}{\overline{M}_n \overline{M}_w}$$

In calculations of molar mass averages of polymers composed of molecularly non-uniform fractions, Eqs.(2-14a)-(2-16a) are often used instead of the correct Eqs.(2-14)-(2-16). This procedure may lead to considerable errors. Assuming $M_i = \overline{M}_{n,i} = \overline{M}_{w,i} = \overline{M}_{z,i}$ will lead to a correct $\overline{M}_n$ but a far too low $\overline{M}_w$ and $\overline{M}_z$. Calculations with $M_i = \overline{M}_{w,i}$ will find a correct $\overline{M}_w$ but a too high $\overline{M}_n$ and a too low $\overline{M}_z$ (see Volume I, p. 87).

As shown by dimensional analysis of Eq.(2-15a), a grave error is committed if relative molecular masses (= molecular weights) are used instead of molar masses and the weight-average molecular weight is expressed as $\overline{M}_{R,w} = \sum_i x_i M_{R,i}^2$.

2.3.4 Hydrodynamic Averages of Molar Masses

Hydrodynamic measurements deliver hydrodynamic quantities H (diffusion coefficients, intrinsic viscosities, etc.) that depend on a power h of the molar mass:

$$(2\text{-}17) \qquad H = K_h M^h$$

K_h and h are usually empirical constants that are independent of molar masses but depend on constitution and chemical configuration of the polymer as well as on the solvent and temperature.

Simple Hydrodynamic Averages

The hydrodynamic property of a molecularly inhomogeneous polymer is a hydrodynamic average, $\overline{H}_\mathrm{g}$, that is controlled by the statistical weight g (= n, m, z, etc.) or the statistical weight fraction $G = x$, w, Z, etc., respectively.

$$(2\text{-}18) \qquad \overline{H}_\mathrm{g} = \frac{\sum_i g_i H_i}{\sum_i g_i} = \frac{\sum_i g_i K_\mathrm{h} M_i^\mathrm{h}}{\sum_i g_i} = K_\mathrm{h} \sum_i G_i M_i^\mathrm{h} \quad ; \quad G_i \equiv g_i / \sum_i g_i$$

The g-average $\overline{H}_\mathrm{g}$ of the hydrodynamic property depends on the g-hydrodynamic average of the molar mass according to $\overline{H}_\mathrm{g} = K_\mathrm{h} \overline{M}_\mathrm{h,g}^\mathrm{h}$, similarly to Eq.(2-17). Solving for $\overline{M}_\mathrm{h,g}$ and introducing the resultat in Eq.(2-18) shows that the g-hydrodynamic average of molar mass is an **exponent average**:

$$(2\text{-}19) \qquad \overline{M}_\mathrm{h,g} = \left(\frac{\overline{H}_\mathrm{g}}{K_\mathrm{h}}\right)^{1/h} = \left(\frac{\sum_i g_i M_i^\mathrm{h}}{\sum_i g_i}\right)^{1/\mathrm{h}} = \left(\sum_i G_i M_i^\mathrm{h}\right)^{1/\mathrm{h}}$$

The hydrodynamic average $\overline{M}_\mathrm{h,g}$ of the molar mass is thus the h-root of the h-th moment of the g-distribution of molar masses. It is usually obtained from calibrations of the hydrodynamic property with polymers of known molar mass. For h = 1, it reduces to a simple one-moment average; for example, a mass average if $G_i = w_i$. An example is the so-called viscosity-average of molar mass (Section 12.3.1).

Complex Hydrodynamic Averages

Molar masses can sometimes be obtained from combinations of two hydrodynamic properties. An example is the Svedberg-Gleichung, $M_\mathrm{SD} = K_\mathrm{SD} S D^{-1}$ (p. 387), which allows the calculation of a molar mass, M_SD, from sedimentation and diffusion coefficients, S and D, without any calibrations. Molar masses can also be calculated from intrinsic viscosities, $[\eta]$, (p. 405) and either sedimentation or diffusion coefficients, $M_\mathrm{Sv} = K_\mathrm{Sv} S^{3/2} [\eta]^{1/2}$ (p. 388) and $M_\mathrm{Dv} = K_\mathrm{Dv} D^{-3} [\eta]^{-1}$. In these equations, K_SD, K_Sv, and K_Dv are system-specific constants that do not need calibrations.

Because S, D, and $[\eta]$ depend on a power of the molar mass,

$$S = K_\mathrm{S} M^{\varsigma} \qquad D = K_\mathrm{D} M^{\delta} \qquad [\eta] = K_\mathrm{v} M^{\alpha}$$

hydrodynamic molar masses M_SD, M_Sv, and M_Dv can be written as

$$(2\text{-}20) \qquad M_\mathrm{SD} = K_\mathrm{SD} S D^{-1} = K_\mathrm{SD} K_\mathrm{S} K_\mathrm{D}^{-1} M^{\varsigma} M^{-\delta}$$

$$(2\text{-}21) \qquad M_\mathrm{Sv} = K_\mathrm{Sv} S^{3/2} [\eta]^{1/2} = K_\mathrm{Sv} K_\mathrm{S}^{3/2} K_\mathrm{v}^{1/2} M^{(3/2)\varsigma} M^{(1/2)\alpha}$$

$$(2\text{-}22) \qquad M_\mathrm{Dv} = K_\mathrm{Dv} D^{-3} [\eta]^{-1} = K_\mathrm{Dv} K_\mathrm{D}^{-3} K_\mathrm{v}^{-1} M^{-3\delta} M^{-\alpha}$$

The left and right sides of Eq.(2-20)-(2-22) must have the same physical units. Since exponents α, ς, and δ as well as constants K_{SD}, K_{Sv}, and K_{Dv} are independent of molar masses, exponents are interrelated according to the **exponent rule** which is *independent of any theory* about shapes and interactions of molecules:

$$(2\text{-}23)\qquad 1 = \varsigma - \delta = (3/2)\,\varsigma + (1/2)\,\alpha = -3\,\delta - \alpha$$

$$(2\text{-}24)\qquad \alpha = 2 - 3\,\varsigma = -(1 + 3\,\delta)$$

For dimensional reasons, the products of constants on the right side of Eqs.(2-20)-(2-22), $K_{SD}K_SK_D^{-1}$, $K_{Sv}K_S^{3/2}K_v^{1/2}$, and $K_{Dv}K_D^{-3}K_v^{-1}$, must equal unity. This can be seen easily for Eq.(2-20): from $M/(M^{\varsigma-\delta}) \equiv 1$, it follows $K_{SD}K_S/K_D = 1$.

The molar masses calculated from the expressions on the left sides of Eqs.(2-20)–(2-22) are *absolute values* since none of the various constants K_{SD}, K_S, K_D, K_{Sv}, K_v, and K_{Dv} requires any assumption about shapes of molecules, frictional coefficients, interactions between macromolecules and solvents, and the like. For example, both S and D and the constant $K_{SD} = RT/(1 - \tilde{v}_2\rho_1)$ are direct experimental quantities (R = molar gas constant, T = thermodynamic temperature, $\tilde{v}_2$ = partial specific volume of the polymer, ρ_1 = density of solvent). The constants $K_{Sv} = [(6^2/20^{1/2})\,\pi N_A][\eta_1/(1 - \tilde{v}_2\rho_1]^{3/2}$ and $K_{Dv} = [20/(6^4\,\pi^2 N_A{}^2)][RT/\eta_1]^3$ also do not require any assumption (η_1 = dynamic viscosity of solvent).

Although these molar masses are "absolute" (no calibration needed), they still vary with the solvent for molecularly non-uniform polymers, which seems to be a contradiction. The reason is the use of various statistical weights g for sedimentation coefficients and g' for diffusion coefficients (g, g' = n, w, z, etc.) whereas intrinsic viscosities, $[\eta]$, are always weight averages (see Volume I, p. 98):

$$\overline{S}_g = \textstyle\sum_i g_iS_i \,/\, \sum_i g_i \qquad \overline{D}_{g'} = \sum_i g'_i\, D_i \,/\, \sum_i g_i' \qquad [\eta] = \sum_i w_i[\eta]_i \,/\, \sum_i w_i$$

Introduction of these equations into Eqs.(2-20)-(2-22) and consideration of Eq.(2-24) delivers expressions for hydrodynamic molar masses which show clearly the effects of statistical weights, g and g', and the thermodynamic quality of solvents (via α) on the resulting averages and numerical values of molar masses:

$$(2\text{-}20a)\qquad \overline{M}_{\overline{S}_g\overline{D}_{g'}} = K_{SD}\frac{\overline{S}_g}{\overline{D}_{g'}} = \frac{K_{SD}K_S}{K_D}\left(\frac{\sum_i g_iM_i^{\varsigma}}{\sum_i g_i}\right)\left(\frac{\sum_i g_i'}{\sum_i g_i' M_i^{\delta}}\right) = \left(\frac{\sum_i g_i'}{\sum_i g_i}\right)\left(\frac{\sum_i g_iM_i^{(2-\alpha)/3}}{\sum_i g_i' M_i^{-(1+\alpha)/3}}\right)$$

$$(2\text{-}21a)\qquad \overline{M}_{\overline{S}_g\text{v}} = K_{Sv}\overline{S}_g^{3/2}[\eta]^{1/2} = K_{Sv}K_S^{3/2}K_v^{1/2}\left(\frac{\sum_i g_iM_i^{(2-\alpha)/3}}{\sum_i g_i}\right)(\textstyle\sum_i w_iM_i^{\alpha})^{1/2}$$

$$(2\text{-}22a)\qquad \overline{M}_{\overline{D}_{g'},\text{v}} = K_{Dv}\frac{1}{(\overline{D}_{g'})^3[\eta]} = \frac{K_{Dv}}{K_D^3K_v}\left(\frac{(\sum_i g_i')^3\sum_i w_i}{(\sum_i g_i' M_i^{-(1+\alpha)/3})^3(\sum_i w_iM_i^{\alpha})}\right)$$

For example, the combination of $\overline{S}_n$ and $\overline{D}_w$ always delivers $\overline{M}_n$, regardless of the value of the exponent α whereas the combination of $\overline{S}_n$ and $\overline{D}_n$ leads to $\overline{M}_{n-1}$ for $\alpha = 2$. Some examples are shown in Table 11-3, p. 387.

2.3.5 Averages of Degrees of Polymerization

Average degrees of polymerization are defined in analogy to average degrees of molar masses (see Eqs.(2-14)-(2-16)),

$$(2\text{-}25)\qquad \overline{X}_n \equiv \frac{\sum_i n_i \overline{X}_{n,i}}{\sum_i n_i} = \frac{\sum_i (m_i / \overline{M}_{n,i})\overline{X}_{n,i}}{\sum_i (m_i / \overline{M}_{n,i})} = \frac{1}{\sum_i (w_i / \overline{X}_{n,i})}$$

$$(2\text{-}26)\qquad \overline{X}_w \equiv \frac{\sum_i m_i \overline{X}_{w,i}}{\sum_i m_i} = \frac{\sum_i n_i \overline{M}_{n,i} \overline{X}_{w,i}}{\sum_i n_i \overline{M}_{n,i}} = \frac{\sum_i x_i \overline{M}_{n,i} \overline{X}_{w,i}}{\overline{M}_n}$$

$$(2\text{-}27)\qquad \overline{X}_z \equiv \frac{\sum_i z_i \overline{X}_{z,i}}{\sum_i z_i} = \frac{\sum_i m_i \overline{M}_{w,i} \overline{X}_{z,i}}{\sum_i m_i \overline{M}_{w,i}} = \frac{\sum_i x_i \overline{M}_{n,i} \overline{M}_{w,i} \overline{X}_{z,i}}{\sum_i x_i \overline{M}_{n,i} \overline{M}_{w,i}}$$

where one notices that molar masses are introduced if a type of statistical weight is replaced by another one (see Eqs.(2-6a) through (2-13a)). Since degrees of polymerization refer exclusively to monomeric or repeating units whereas molar masses always include endgroups, different expressions are obtained for molecularly non-uniform and uniform species i with or without endgroups (Table 2-4).

Table 2-4 Average degrees of polymerization.

Definition of average	Non-negligible endgroups, non-uniform species i	No endgroups, non-uniform species i	No endgroups, uniform species i
$\overline{X}_n \equiv \sum_i x_i \overline{X}_{n,i}$	$\sum_i x_i \overline{X}_{n,i}$	$\sum_i x_i X_i$	$\sum_i x_i X_i$
$\overline{X}_w \equiv \sum_i w_i \overline{X}_{w,i}$	$\dfrac{\sum_i x_i \overline{M}_{n,i} \overline{X}_{w,i}}{\overline{M}_n}$	$\dfrac{\sum_i x_i \overline{X}_{n,i} \overline{X}_{w,i}}{\overline{X}_n}$	$\dfrac{\sum_i x_i \overline{X}_i^2}{\overline{X}_n}$
$\overline{X}_z \equiv \sum_i Z_i \overline{X}_{z,i}$	$\dfrac{\sum_i x_i \overline{M}_{n,i} \overline{M}_{w,i} \overline{X}_{z,i}}{\overline{M}_n \overline{M}_w}$	$\dfrac{\sum_i x_i \overline{X}_{n,i} \overline{X}_{w,i} \overline{X}_{z,i}}{\overline{X}_n \overline{X}_w}$	$\dfrac{\sum_i x_i \overline{X}_i^3}{\overline{X}_n \overline{X}_w}$

2.3.6 Averages of Other Properties

Expressions for molar masses and degrees of polymerization, Sections 2.3.2 through 2.3.5, cannot be adopted for other properties P by simply replacing M or X by P. For example, the weight-average $\overline{P}_w$ of a property P of a substance composed of molecularly uniform species is given by

$$(2\text{-}28)\qquad \overline{P}_w \equiv \frac{\sum_i m_i P_i}{\sum_i m_i} = \frac{\sum_i n_i M_i P_i}{\sum_i n_i M_i} \quad \text{and not by} \quad \overline{P}_w \equiv \frac{\sum_i n_i P_i^2}{\sum_i n_i P_i}$$

For number-, surface-, and mass-averages of particles (spheres, discs, circular cylinders), see Volume IV, Section 3.3.2.

2.3.7 Moments of Distributions

Some experimental methods do not deliver simple arithmetic averages of properties but more complicated ones. Examples are averages of molar masses that are calculated from sedimentation and diffusion coefficients (Section 2.3.4). These averages are best written as combinations of moments of distribution functions (see also Section 11.3.4).

In mechanics, the **first moment** of a force, $\nu^{(1)} \equiv \boldsymbol{g}(P - P_o)$, is defined as a vector product of the force (for example, $\boldsymbol{g}$) and distance of the axis to the origin of the force (for example, $P - P_o$). The **second moment**, $\nu^{(2)}$, is then the product of the force and the square of the distance, etc.

Moments cannot only be defined for interrelationships between forces and distances but in general for distributions of any other property P. The q-th moment, $\nu_g^{(q)}$, of the g-distribution of P with respect to a reference value P_o is therefore

(2-29) $$\nu_g^{(q)}(P) = \Sigma_i\, g_i(P_i - P_o)^q/(\Sigma_i\, g_i) = \Sigma_i\, G_i(P_i - P_o)^q$$

The **order of the moment**, q, can adopt any value, positive or negative, whole numbers or fractions, rational or irrational values, etc. A moment of a property distribution may thus have a different physical unit than a property or its average.

Statistical weight fractions, G, may be amount (x), weight (w), Z fractions, etc., whereas properties P may include molar masses, M; degrees of polymerization, X; diffusion coefficients, D; etc. In principle, reference values are arbitrary. However, there are no negative molar masses or degrees of polymerization, etc., so that the reference value is taken as zero. The resulting moments are symbolized by μ instead of ν:

(2-30) $$\mu_g^{(q)}(P) = (\Sigma_i\, g_i P_i^q)/(\Sigma_i\, g_i) = \Sigma_i\, G_i P_i^q$$

For example, the number-average degree of polymerization, $\overline{X}_n$, can be written as the first moment of the amount distribution (number distribution) of degrees of polymerization, X, with respect to the origin, assuming molecularly uniform species i:

(2-31) $$\overline{X}_n = (\Sigma_i\, n_i X_i)/(\Sigma_i\, n_i) = \Sigma_i\, x_i X_i = \mu_n^{(1)}(X)$$

The second moment of the amount distribution of degrees of polymerization is then

(2-32) $$\mu_n^{(2)}(X) = (\Sigma_i\, n_i X_i^2)/(\Sigma_i\, n_i) = \Sigma_i\, x_i X_i^2$$

The weight-average degree of polymerization is defined as

(2-33) $$\overline{X}_w = \frac{\Sigma_i\, m_i X_i}{\Sigma_i\, m_i} = \frac{\Sigma_i\, n_i M_i X_i}{\Sigma_i\, n_i M_i}$$

so that for molecularly uniform species i, where $M_i = M_u X_i$, Eq.(2-33) may be written as the ratio of the second to the first moment of the amount distribution of X:

(2-34) $$\overline{X}_w = \frac{\Sigma_i\, m_i X_i}{\Sigma_i\, m_i} = \frac{\Sigma_i\, n_i M_i X_i}{\Sigma_i\, n_i M_i} = \frac{\Sigma_i\, n_i X_i^2}{\Sigma_i\, n_i X_i} = \frac{\mu_n^{(2)}(X)}{\mu_n^{(1)}(X)}$$

2.3.8 Molecular Non-Uniformity

Inspection of Eqs.(2-14)-(2-16) for molar masses and Eqs.(2-25)-(2-27) for degrees of polymerization shows that numerical values of these quantities increase with higher statistical weights, Eqs.(2-6)-(2-10), and statistical weight fractions, Eqs.(2-11)-(2-13), respectively. One can therefore always write the inequalities

$$(2\text{-}35) \qquad \overline{M}_{\{z+1\}} \geq \overline{M}_z \geq \overline{M}_w \geq \overline{M}_n \geq \overline{M}_{\{n-1\}}$$

$$(2\text{-}36) \qquad \overline{X}_{\{z+1\}} \geq \overline{X}_z \geq \overline{X}_w \geq \overline{X}_n \geq \overline{X}_{\{n-1\}}$$

The width of a distribution of molar masses (or degrees of polymerization) can therefore always be characterized by the ratio of two subsequent averages such as $\overline{M}_w/\overline{M}_n \geq 1$ or $\overline{X}_w/\overline{X}_n \geq 1$ which are usually called **polymolecularity indices,** $Q_{w,n}$. Because end-groups are included in M but not in X, $Q_{w,n}(X)$ is always larger than $Q_{w,n}(M)$ except for cyclic polymers where it is identical. The polymolecularity index, $Q_{w,n}$, is an *absolute* measure of the width of Schulz-Flory distributions (Section 2.4.5).

In the literature, the polymolecularity index has been called "polydispersity (index)" (with abbreviation PDI, PD, or PI). However, "disperse" has a different scientific meaning (see p. 9). "Polymolecular" is literally wrong since every polymeric substance consists of many ("poly") molecules.

The polymolecularity index is not very sensitive to the width of narrow distributions. A better measure of the width of distributions is the **number-standard deviation**

$$(2\text{-}37) \qquad \sigma_{n,M} \equiv \overline{M}_n[(\overline{M}_w/\overline{M}_n) - 1)^{1/2} \approx (\overline{X}_w\,\overline{X}_n - \overline{X}_n{}^2)^{1/2} \qquad \textbf{(Herdan relation)}$$

The number-standard deviation is an *absolute* measure of the width of Gaussian distributions but only a relative one for that of other distributions (Section 2.4.2) since

$$(2\text{-}38) \qquad Q_{w,n} \equiv \overline{M}_w/\overline{M}_n \approx \overline{X}_w/\overline{X}_n = 1 + (\sigma_{n,X}/\overline{X}_n)^2$$

2.3.9 Constitutional Non-Uniformity

Molecules of bipolymers may differ from each other by both relative compositions and absolute numbers, N_a and N_b, of monomeric units. Each bipolymer molecule with molar mass M consists of an a-fraction with *total* molar mass M_a and a b-fraction with *total* molar mass M_b. These molar masses are the molar masses of a- and b-blocks in diblock polymers and the cumulative molar masses of a- and b-monomeric units in statistical bipolymers.

Bipolymers consist of N_i bipolymer molecules with molar mass M_i. The number-average molar masses of the a- and b-segments (= blocks) are therefore

$$(2\text{-}39) \qquad \overline{M}_{n,a} = \sum_i N_i M_{a,i} / \sum_i N_i \quad ; \quad \overline{M}_{n,b} = \sum_i N_i M_{b,i} / \sum_i N_i$$

where the number-average molar mass of the bipolymer, $\overline{M}_n$, equals $\overline{M}_{n,a} + \overline{M}_{n,b}$:

$$(2\text{-}40) \qquad \overline{M}_n = \frac{\sum_i N_i M_i}{\sum_i N_i} = \frac{\sum_i N_i M_{a,i} + \sum_i N_i M_{b,i}}{\sum_i N_i} = \overline{M}_{n,a} + \overline{M}_{n,b}$$

Sequence Lengths of Segments

The **number-average degree of polymerization of a-sequences** is defined as

(2-41) $$\overline{X}_{n,a-block} \equiv \frac{\sum_i N_{a-block,i} X_{a-block,i}}{\sum_i N_{a-block,i}}$$

This quantity can be calculated from experimental properties as follows. The nominator of Eq.(2-41) equals the total number, $N_a = \Sigma_i N_{a,i}$, of all a-units. The denominator can be obtained by remembering that each a-homosequence begins with a ~b–a– bond and ends with an –a–b~ bond. The number of a-sequences (= a-blocks) must therefore equal one-half of the number $N_{a/b+b/a} \equiv N_{|a/b|}$ of all (a/b + b/a) bonds:

(2-42) $$\Sigma_i N_{a\text{-}block,i} = (1/2)\, N_{|a/b|}$$

Eq.(2-42) is exact for cyclic polymer molecules and a good approximation for linear polymer molecules with high degrees of polymerization.

The number $N_{|a/b|}$ can be expressed by the corresponding mole fraction, $x_{|a/b|}$, which is defined as the ratio of the total number of a–b bonds, $N_{|a/b|}$, to the total number of all chain bonds, $N_{bond} = N_{a/a} + N_{|a/b|} + N_{b/b}$. The latter number is the product of the number of copolymer molecules, N_{cop}, and the number of chain bonds, $N_{bd} = \overline{X}_n - 1$, which is one less than the number-average degree of polymerization of a linear molecule.

(2-43) $$x_{|a/b|} \equiv \frac{N_{|a/b|}}{N_{a/a} + N_{|a/b|} + N_{b/b}} = \frac{N_{|a/b|}}{(\overline{X}_n - 1)\, N_{cop,n}}$$

The number of all a-units is given by $N_a = m_a N_A / M_a$ and the number of all bipolymer molecules by $N_{cop} = (m_a + m_b) N_A / \overline{M}_n$. Number-average degrees of polymerization and molar mass are interrelated by

(2-44) $$\overline{X}_n = (N_a + N_b)/N_{cop} = \overline{M}_n [(w_a/M_a) + (w_b/M_b)]$$

where M_a and M_b are the molar masses of units a and b. Introduction of all of these terms in Eq.(2-41) delivers

(2-45) $$\frac{1}{\overline{X}_{a-block,n}} = \frac{x_{|a/b|}}{2} \left\{ 1 - \frac{M_a}{w_a \overline{M}_n} + \frac{w_b M_a}{w_a M_b} \right\}$$

The number-average degree of polymerization of a-blocks, i.e., the average "length" of sequences, can thus be calculated from the weight fractions, w_a and w_b, and the molar masses, M_a and M_b, of the units a and b; and the number-average molar mass, $\overline{M}_n$, of the bipolymer if the mole fraction $x_{|a/b|}$ of the two cross-bonds a–b and b–a is known. The calculation thus requires an analytical method for the determination of the proportion of cross-bonds, e.g., by nuclear magnetic resonance spectroscopy.

Instead of number-average degrees of polymerization of a- and b-blocks, $\overline{X}_{a-block,n}$ and $\overline{X}_{b-block,n}$, a **run number** is often used. The run number, $\overline{R}_n$, indicates the total number of blocks per 100 monomeric units. It can be calculated from copolymerization parameters and the ratio of instantaneous monomer concentrations, [B]/[A], in copolymerizations (see Volume I, p. 399).

(2-46) $$\overline{R}_{\text{n}} = \frac{100}{[\overline{X}_{\text{a-block,n}} + \overline{X}_{\text{b-block,n}}]/2}$$

Mass-average Molar Mass of Segments

The mass-average molar mass of a bipolymer is not the sum of the mass averages of its a- and b-units. The mass-average of a mixture of a-homopolymers and b-homopolymers is given by

(2-47) $$\overline{M}_{\text{w}} = \frac{\sum_i m_i M_i}{\sum_i m_i} = \frac{\sum_i n_i M_i^2}{\sum_i n_i M_i} = \frac{\sum_i N_i M_{\text{a},i}^2 + \sum_i N_i M_{\text{b},i}^2}{\sum_i N_i M_{\text{a},i} + \sum_i N_i M_{\text{b},i}}$$

whereas the mass-average molar mass of a-segments in bipolymers is defined as

(2-48) $$\overline{M}_{\text{a,w}} = \frac{\sum_i m_{\text{a},i} M_{\text{a},i}}{\sum_i m_{\text{a},i}} = \frac{\sum_i N_{\text{a},i} M_{\text{a},i}^2}{\sum_i N_{\text{a},i} M_{\text{a},i}}$$

Since $\sum_i N_i M_{\text{a},i}^2 \neq \sum_i N_{\text{a},i} M_{\text{a},i}^2$, a cross term must be introduced that considers a/b- and b/a-bonds:

(2-49) $$\overline{M}_{\text{w}} = w_{\text{a}} \overline{M}_{\text{a,w}} + w_{\text{b}} \overline{M}_{\text{b,w}} + 2\, \overline{M}_{\text{la/bl,w}} \quad ; \quad \overline{M}_{\text{la/bl,w}} = \frac{\sum_i N_i M_{\text{a},i} M_{\text{b},i}}{\sum_i N_i (M_{\text{a},i} + M_{\text{b},i})}$$

Unfortunately, Eq.(2-49) is useless since the cross term cannot be determined directly. However, a lengthy calculation shows that the following inequality applies:

(2-50) $$w_{\text{a}} \overline{M}_{\text{a,w}} + w_{\text{b}} \overline{M}_{\text{b,w}} \leq \overline{M}_{\text{w}} \leq \frac{w_{\text{a}} \overline{M}_{\text{a,w}} + w_{\text{b}} \overline{M}_{\text{b,w}}}{1 - 2 w_{\text{a}} w_{\text{b}}}$$

$\overline{M}_{\text{la/bl,w}}$ equals zero for a mixture of two homopolymers. The mass-average molar mass of the mixture is the sum of the weighted mass-average molar masses of the two homopolymers (left equality of Eq.(2-50)). However, the mass-average molar mass of a bipolymer is given by the right equality of Eq.(2-50) if there is no compositional variation from chain to chain.

2.3.10 Determination of Molar Masses

Experimental methods for the determination of molar masses and molecular weights deliver various averages, usually for restricted molar mass ranges (Table 2-5). They can be classified as absolute, equivalent, and relative methods.

Absolute methods allow one to determine molar masses or reduced molecular masses from measured quantities without any assumption about the chemical or physical struc-

Table 2-5 Types of methods, averages, and approximate ranges of experimental methods for the determination of molecular sizes (molar masses, molecular weights).

A = Absolute method, R = relative method (requires calibration), E = equivalent method (constitution must be known), n = number average, v = viscosity average, w = mass (weight) average, z = z-average. § With certain assumptions, * upper limit ca. 10^6 g/mol.

Method	Type	Average	Range in g/mol	Section
Light scattering, static	A	w	> 100	5.2
Viscometry of dilute solutions	R §	v	> 200	12
X-ray or neutron small angle scattering	A	w	> 500	5.4, 5.5
Combined sedimentation and diffusion	A	various	> 1 000	11.2.2
Size exclusion chromatography	R	n, w, z	> 1 000	14.4.3
Viscometry of melts	R	w	> 1 000	15
Field flow fractionation	R	n, w, z	> 1 000	11.4
Membrane osmometry *	A	n	> 5 000	10.3.2
Ebullioscopy, cryoscopy	A	n	< 20 000	10.3.3
Endgroup determination (titration)	E	n	< 40 000	volume I
Vapor phase osmometry	A	n	< 50 000	10.3.4
Mass spectroskopy	A	n, w, z	< 100 000	volume I
Sedimentation equilibrium	A	w, z §	< 1 000 000	11.3.5
Light scattering, dynamic	R	z §	< 10 000 000	11.1.2

ture of the polymer. The resulting numerical values are independent of experimental conditions for methods delivering simple averages of molar masses (e.g., membrane osmometry, static light scattering) but dependent on solvent quality if complex averages are obtained (Section 2.3.4); examples are molar masses from measurements of sedimenation and diffusion coefficients (Section 11.3.4). Some methods employ calibrations (vapor phase osmometry, static light scattering) but this is for convenience and would be unnecessary if all relevant instrumental data were known.

Equivalent methods require the knowledge of the chemical structure of the polymer. The only known method determines endgroups. Upper limits of M are ca. 8000 by ^{13}C NMR, 40 000 by titration, 100 000 by microanalysis of iodine contents, 200 000 for radioactive endgroups, and 1 000 000 for intensely colored or fluorescent endgroups.

Relative methods are affected by both the chemical and physical structure of the polymer as well as its interaction with solvents; these methods always need calibrations with polymers of the same constitution and known molecular weight and molecular weight distribution. Examples are viscometry and size exclusion chromatography.

The choice of a method depends primarily on the desired information and secondarily on the working range, available amount of substance, required time of analysis, and necessary preparation of samples. Absolute and relative methods usually require measurements at various polymer concentrations. For each concentration, an **apparent molar mass** or **apparent molecular weight**, $\overline{M}_{g,app}$, is calculated, using expressions that apply strictly to infinite dilution.

Apparent molar masses (weights) are then extrapolated to zero polymer concentration, usually by plotting $1/\overline{M}_{g,app}$ as a function of the mass concentration, c in g/mL. Examples can be found in Sections 5.3. (static light scattering) and 10.4.1 (membrane osmometry). Special types of extrapolation and evaluation are required for solutions of polyelectrolytes (Section 10.6.3) and self-associating neutral polymers (Section 10.5).

2.4 Distribution Functions

2.4.1 Types

Distribution functions of properties of polymers are normalized mathematical functions that describe the variation of property values with molar mass or molecular weight. Such functions may be discontinuous or continuous and they may be differential or integral (Fig. 2-10). In either of these cases, statistical weights must be defined.

Continuity. All distributions of molar masses and degrees of polymerization are physically *discrete* (*discontinuous*) since degrees of polymerization can only be whole, positive numbers. The distributions are step distributions (Fig. 2-10, left) and either differential (Fig. 2-10, upper left) or integral (Fig. 2-10, lower left).

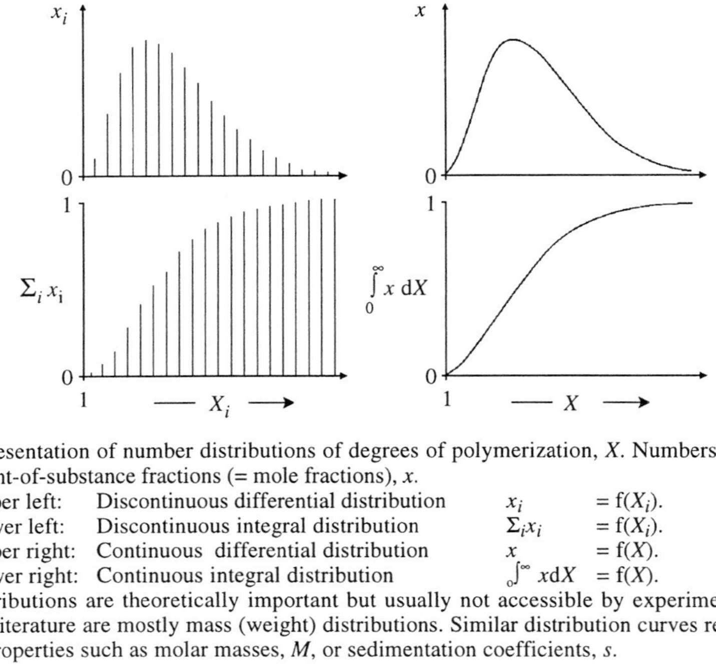

Fig. 2-10 Representation of number distributions of degrees of polymerization, X. Numbers are expressed as amount-of-substance fractions (= mole fractions), x.

Upper left:	Discontinuous differential distribution	x_i	$= f(X_i)$.
Lower left:	Discontinuous integral distribution	$\Sigma_i x_i$	$= f(X_i)$.
Upper right:	Continuous differential distribution	x	$= f(X)$.
Lower right:	Continuous integral distribution	$\int_0^\infty x\mathrm{d}X$	$= f(X)$.

Number distributions are theoretically important but usually not accessible by experiment. The distributions of literature are mostly mass (weight) distributions. Similar distribution curves result for other physical properties such as molar masses, M, or sedimentation coefficients, s.

In theoretical work, discontinuous distributions are usually replaced by *continuous* distributions (right side of Fig. 2-10) which can also be either differential or integral. This substitution is mathematically elegant and allows many mathematical manipulations but neglects the discrete nature of molar mass distributions. The caveat is that the actual physical interval between entities, the molecular weight of a repeating unit (typically in the tens or hundreds), is negligible compared to *high* average molecular weights that may range into the millions.

Summation. *Differential* distributions describe the population of a species with a certain property, for example, the mole fraction, x_i, of molecules with a degree of polymerization, $X_i = 1050$, in a polymer with $1 \leq X_i \leq 5000$ (Fig. 2-10, upper left and right). The property may either be a single value, such as $X_i = 1050$, or a range of properties, such as $1000 \leq X_i \leq 1100$. Discontinuous differential distributions (Fig. 2-10, upper left) are sometimes called *frequency distributions* but this term may also apply to other weighing factors such as mass fractions. *Discontinuous integral* (= *cumulative*) distributions sum up populations to a certain property value (Fig. 2-10, lower left).

Fractionation. The fractionation of a non-uniform polymer with respect to constitution or molar mass delivers a discontinuous differential distribution. Conversion to a continuous distribution must consider that each fraction is not uniform but also has a distribution of, e.g., molar masses. Hence, the degree of polymerization of such a fraction is an average.

In first approximation, one half of this fraction has a composition below the average and the other half, above. For calculation of the integral composition, not the full weight fraction w_1 of fraction 1 has to be used but only $w_1/2$. The integral composition of fraction 2 is then $w_1 + (w_2/2)$, etc. (Table 2-6).

A plot of $\Sigma_i w_i$ as function of X_i delivers the discontinuous integral distribution, which, after connection of the data points, converts to the continuous integral distribution. Graphic differentiation of this curve leads to the continuous differential distribution. The latter distribution cannot be obtained directly from the discontinuous differential distribution because of the non-uniformity of fractions.

Distribution functions. Experiments usually deliver mass distributions of molar masses (or weight distributions of molecular weights) whereas reaction mechanisms lead to number distributions. The former can be transformed into the latter by the equations of Sections 2.3.3-2.3.5.

Polymer science employs different types of distribution functions which mostly carry the name of their discoverers. The next sections describe some mathematical consequences of these functions. Their use in polymerization equilibria and reaction mechanisms is discussed in Volume I of this series.

Table 2-6 Example of the conversion of a discontinuous differential distribution of degrees of polymerization to a discontinuous integral one.

Fraction i	Degree of polymerization X_i	Mass fraction w_i	Cumulative mass fraction $\Sigma_i w_i$
1	15	0.0532	0.0266
2	31	0.0740	0.0902
3	50	0.0622	0.1583
4	76	0.0864	0.2326
etc.	etc.	etc.	etc.

2.4.2 Gaussian Distribution

The Gaussian distribution is the best known distribution function. It describes the **error law** for random errors in independent experiments. Because of its widespread application, it is also called **normal distribution**. The shape of the distribution resembles that of a bell, hence the name **bell curve** for differential Gaussian distributions in which the plotted statistical weight corresponds to the statistical weight of the function (Fig. 2-11).

Gaussian distributions allow positive and negative values of properties. In a strict sense, they are not applicable to degrees of polymerization, X, since X can never be negative. However, Gaussian distributions can be used for X (or M) without significant error if the position of the maximum in a differential distribution as well as the width of the distribution leads to diminishingly small contributions of negative values of properties.

For mole fractions, x, as a function of the degree of polymerization, X, the Gaussian differential number-distribution, $x(X)$, is given by

$$\text{(2-51)} \qquad x = \frac{1}{(2\pi)^{1/2}\sigma_n} \exp\left[\frac{-(X - X_{median})^2}{2\sigma_n^2}\right] = \frac{1}{(2\pi)^{1/2}\sigma_n} \exp\left[\frac{-(X - \overline{X}_n)^2}{2\sigma_n^2}\right]$$

This distribution function ranges from positive to negative infinity of X. At the median, 50 % of the population have X values higher than X_{median} and 50 %, lower. Since the distribution of mole fractions is symmetric about the median for Gaussian number-distributions (Fig. 2-11), the median is also the number-average of X, i.e., $\overline{X}_n$.

The number-standard deviation, σ_n, describes the width of the distribution. It can be calculated from number and weight averages as follows. The property X_i deviates from the number-average, $\overline{X}_n$, by an average "error" of s_n. For n_i errors with values of X_i each, the square of the average error of a single value is given by

$$\text{(2-52)} \qquad s_n^2 = \frac{\sum_i n_i (X_i - \overline{X}_n)^2}{\sum_i n_i} \equiv \sigma_n^2$$

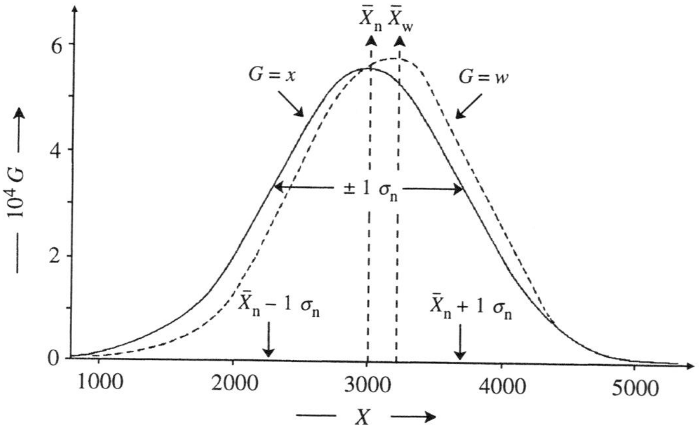

Fig. 2-11 Gaussian number distributions of degrees of polymerization, $g = f(X)$, of a polymer with $\overline{X}_n = 3000$ and $\overline{X}_w = 3170$, i.e., a number-standard deviation of $\sigma_n = 714$ (Eq.(2-53)).

Solid curve: continuous differential number distribution, $G = x = f(X)$. The curve is symmetric about its median, X_{median}, (left dotted vertical line); median and number-average degree of polymerization, $\overline{X}_n$, are identical.

Broken curve: the same Gaussian number distribution plotted as continuous differential weight distribution, $G = w = f(X)$. The curve is no longer symmetric about its median (right dotted line). Conversely, a Gaussian weight distribution would be symmetric about $\overline{X}_w$ if plotted as $w = f(X)$ but not if plotted as $x = f(X)$.

Solving Eq.(2-52) delivers $\sigma_n^2 \Sigma_i n_i = \Sigma_i n_i X_i^2 - 2\overline{X}_n \Sigma_i n_i X_i + \overline{X}_n^2 \Sigma_i n_i X_i$. Dividing by $\Sigma_i n_i X_i$, introducing $\overline{X}_n \equiv \Sigma_i n_i X_i / \Sigma_i n_i$, and $\overline{X}_w \equiv \Sigma_i n_i X_i^2 / \Sigma_i n_i X_i$, and solving for σ_n delivers the **number-standard deviation**:

$$\sigma_n = (\overline{X}_w \overline{X}_n - \overline{X}_n^2)^{1/2} = \overline{X}_n [(\overline{X}_w / \overline{X}_n) - 1]^{1/2} \tag{2-53}$$

The number-standard deviation is an absolute measure of the width of a Gaussian number distribution since a value of $\overline{X}_n \pm 1\ \sigma_n$ always corresponds to an amount fraction (mole fraction) of 68.26 %, a value of $\overline{X}_n \pm 2\ \sigma_n$ to a fraction of 95.44 %, and a value of $\overline{X}_n \pm 3\ \sigma_n$ to a fraction of 99.73 %. The example of Fig. 2-11 has a number-average degree of polymerization of $\overline{X}_n$ = 3000 and a number-standard deviation of σ_n = 714: 68.26 % of all molecules are thus in the range $2286 \leq \overline{X}_n \leq 3714$.

Differential Gaussian number-distributions turn asymmetric if weight fractions, $w = xX/\overline{X}_n$, replace mole fractions x as statistical weights (see Fig. 2-11). Eq.(2-51) becomes

$$w = \frac{X}{(2\pi)^{1/2}\sigma_n \overline{X}_n} \exp\left[\frac{-(X - \overline{X}_n)^2}{2\sigma_n^2}\right] \tag{2-54}$$

This equation is a Gaussian number-distribution that is expressed by a distribution of weight fractions; it is *not* the Gaussian weight-distribution! A true **Gaussian weight-distribution** of weight fractions would rather be based on the weight-average of the degree of polymerization, $\overline{X}_w$, and the standard deviation, σ_w:

$$w = \frac{X}{(2\pi)^{1/2}\sigma_w} \exp\left[\frac{-(X - \overline{X}_w)^2}{2\sigma_w^2}\right] \tag{2-55}$$

$$\sigma_w = (\overline{X}_z \overline{X}_w - \overline{X}_w^2)^{1/2} = \overline{X}_w [(\overline{X}_w / \overline{X}_n) - 1]^{1/2} \tag{2-56}$$

With the weight fraction as statistical weight, this Gaussian weight distribution is symmetric about its median which is the weight average degree of polymerization, $\overline{X}_w$.

Other distributions are always asymmetric about the median. Their mass fractions w_+ for the range $\overline{X}_w + \sigma_w$ thus differ from the mass fractions for the range $\overline{X}_w - \sigma_w$ (Table 2-7). In contrast to Gaussian distributions, w_+ and w_- vary with the ratio, $\overline{X}_w/\overline{X}_n$. The same phenomenon is found for logarithmic normal and Schulz-Zimm number distributions if the mole fraction x is the chosen parameter.

Table 2-7 Variation of weight fractions w_- of polymers in the range $\overline{X}_w - \sigma_w$ and weight fractions w_+ in the range $\overline{X}_w + \sigma_w$ with the ratio $\overline{X}_w/\overline{X}_n$ for various types of mass distributions: Gaussian (this section), logarithmic normal (Section 2.4.3), and Schulz-Zimm (Section 2.4.5) [1].

$\overline{X}_w/\overline{X}_n$	Gaussian		Logarithmic normal		Schulz-Zimm	
	w_-	w_+	w_-	w_+	w_-	w_+
1.1	0.3413	0.3413	0.427	0.290	0.386	0.304
1.5	0.3413	0.3413	0.571	0.251	0.441	0.274
2.0	0.3413	0.3413	0.662	0.232	0.477	0.260
5.0	0.3413	0.3413	$w_- + w_+ > 1$!		0.553	0.242

2.4.3 Logarithmic Normal Distribution

Differential logarithmic normal distributions (**LN distributions**) of mole fractions have the same mathematical shape as the corresponding Gaussian (= normal) distributions, except that the variable X is replaced by the natural logarithm of X:

$$(2\text{-}57) \qquad x = \frac{1}{(2\pi)^{1/2}\,\sigma_n^*} \exp\left[\frac{-(\ln X - \ln X_{med})^2}{2(\sigma_n^*)^2}\right]$$

The function $x = f(\ln X)$ corresponds to the error law of geometric averages. The resulting distribution curve is symmetric about the median, $\ln X_{med}$, but X_{med} is no longer identical to $\overline{X}_n$ (see below). In LN distributions, the important variable is the *ratio* of degrees of polymerization, since $(\ln X - \ln X_{med}) = \ln (X/X_{med})$ whereas in Gaussian distributions it is the difference, $(X - X_{med})$ (compare Eq.(2-51)).

Differential logarithmic normal distributions can be generalized, for example, for the weight distribution of degrees of polymerization,

$$(2\text{-}58) \qquad w = \frac{1}{(2\pi)^{1/2}\,\sigma_w^*}\left(\frac{X^A}{BX_{med}^{A+1}}\right) \exp\left[\frac{-(\ln X - \ln X_{med})^2}{2(\sigma_w^*)^2}\right]$$

where $B = \exp\,[(1/2)(\sigma_w{}^*)^2(A + 1)^2]$. In polymer science, two types of LN distributions are commonly used: the **Wesslau distribution** with $A = 1$ and $B = 1$ and the **Lansing-Kraemer distribution** with $A = 0$ and $B = \exp\,[(1/2)(\sigma_w{}^*)^2]$.

Figure 2-12, top, shows a continuous logarithmic-normal weight distribution of mole fractions according to Eq.(2-57). The resulting distribution curve is skewed because its abscissa is given in degrees of polymerization, X. Use of the natural logarithm of X instead of X would produce a bell curve.

The overall shape of the curve does not change appreciably if weight fractions instead of mole fractions are plotted against the degree of polymerization (Fig. 2-12, bottom). For both $w = f(X)$ and $x = f(X)$, the maximum of the curve is neither identical with $\overline{X}_n$ nor with $\overline{X}_w$.

The value of the median, X_{med}, is obtainable from g-average degrees of polymerization, $\overline{X}_g$, and the weight-standard deviation, $\sigma_w{}^*$, of the logarithmic normal distribution,

$$(2\text{-}59) \qquad \overline{X}_g = X_{med} \exp\,[\{2\,A + C\}(\sigma_w{}^*)^2/2]$$

where $C = 1$ for g = n, $C = 3$ for g = w, and $C = 5$ for g = z. For viscosity-average degrees of polymerization, $\overline{X}_v$, $(2\,A + C)$ in Eq.(2-59) is replaced by $\{2(A + \alpha) + 1\}$ where α is the exponent in the relationship, $[\eta] = K_v M^\alpha$ (Section 12.3.1). The weight-standard deviation is related to ratios of average degrees of polymerization by

$$(2\text{-}60) \qquad \exp\,(\sigma_w{}^*)^2 = \overline{X}_w/\overline{X}_n = \overline{X}_z/\overline{X}_w$$

For logarithmic normal weight distributions, $\exp\,(\sigma_w{}^*)^2$ is proportional to the ratio of two subsequent simple average degrees of polymerization.

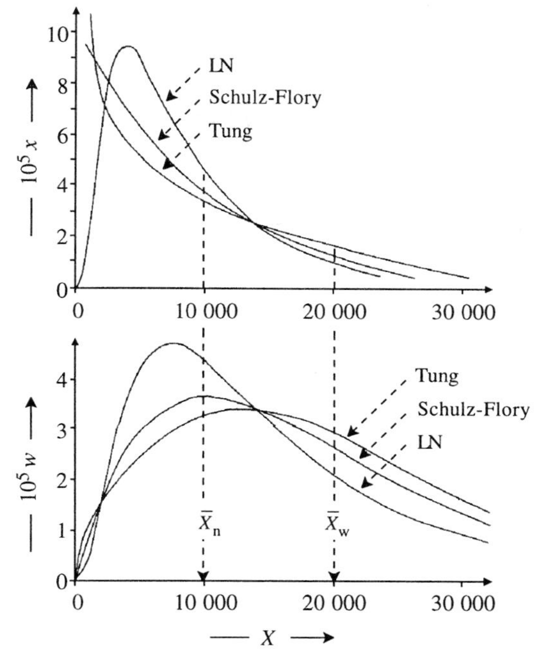

Fig. 2-12 Continuous differential number distributions of degrees of polymerization, X, of a polymer with $\overline{X}_w$ = 20 000 and $\overline{X}_n$ = 10 000, shown as distributions of mole fractions (top) and corresponding weight fractions (bottom), assuming distributions of the logarithmic normal, Schulz-Flory (Section 2.4.5), or Tung type (Table 2-8). The corresponding Gaussian distribution (Section 2.4.2) would be a bell curve for numbers but a skewed curve for weights (see Fig. 2-11). A Poisson distribution (Section 2.4.4) with the same number-average degree of polymerization of $\overline{X}_n$ = 10 000 would be so narrow that $\overline{X}_w \approx$ 10 001 would fall on the same dotted line as $\overline{X}_n$ = 10 000.

2.4.4 Poisson Distribution

Poisson distributions of degrees of polymerization result if a constant number of polymer chains begin to grow simultaneously and monomer molecules add to these chains at random and independently of preceding additions. Such distributions are approached in ideal living polymerizations (for the derivation, see Volume I).

The amount and weight fractions of Poisson number distributions are interrelated by

(2-61) $$x = \frac{(\overline{X}_n - 1)^{X-1} \exp(1 - \overline{X}_n)}{(X-1)!} \quad ; \quad v \equiv \overline{X}_n - 1 \quad ; \quad (X-1)! = \Gamma(X)$$

(2-62) $$w = \frac{X(\overline{X}_n - 1)^{X-1} \exp(1 - \overline{X}_n)}{(X-1)!\,\overline{X}_n} = \frac{X v^{X-1} \exp(-v)}{(X-1)!\,(v+1)}$$

whereas number- and weight-average degrees of polymerization are given by

(2-63) $$\overline{X}_w / \overline{X}_n = 1 + (1/\overline{X}_n) - (1/\overline{X}_n)^2$$

The polymolecularity index, $\overline{X}_w/\overline{X}_n$, depends only on the number-average degree of polymerization (Eq.(2-63)) and not on any other average (compare with Eqs.(2-60) and (2-56)). It approaches unity with increasing number-average degree of polymerization. Poisson distributions are therefore very narrow distributions.

2.4.5 Schulz-Zimm and Schulz-Flory Distributions

Schulz-Zimm distributions (= gamma distribution of statisticians) of degrees of polymerization arise if a time-independent number of active chains randomly adds monomer molecules until individual chains are deactivated (Volume I). Contrary to Poisson distributions, initially present active chains do not need to be preserved for the duration of the process. Active chains are also not required to be started at the same time.

The width of such distributions is controlled by the degree of coupling, ς, which indicates the number of independently grown chains that are coupled to a dead chain. The degree of coupling equals 2 if two chains are coupled to one chain.

Differential **Schulz-Zimm distributions (SZ distribution)** of mole fractions, x, and weight fractions, w, are given by (for the derivation, see Volume I)

(2-64) $$x = \frac{(\varsigma/\overline{X}_n)^{\varsigma+1} X^{\varsigma-1} \overline{X}_n \exp(-\varsigma X/\overline{X}_n)}{\Gamma(\varsigma+1)} \quad ; \quad w = \frac{(\varsigma/\overline{X}_n)^{\varsigma} X^{\varsigma} \exp(-\varsigma X/\overline{X}_n)}{\Gamma(\varsigma+1)}$$

where $\Gamma(\varsigma + 1)$ is the gamma function of $(\varsigma + 1)$. It follows that ratios of subsequent arithmetic averages decrease with increasing degrees of coupling:

(2-65) $$\overline{X}_n/\varsigma = \overline{X}_w/(\varsigma + 1) = \overline{X}_z/(\varsigma + 2)$$

Schulz-Zimm distributions can be distinguished from logarithmic normal distributions by the ratio of averages of degrees of polymerization. According to Eq.(2-65), one always has $(\overline{X}_n + \overline{X}_z)/\overline{X}_w = 2$ for Schulz-Zimm distributions whereas such an expression is obtained for logarithmic normal distrutions only if $\overline{X}_z : \overline{X}_w : \overline{X}_n = 3:2:1$.

Schulz-Zimm distributions reduce to the **Schulz-Flory distribution (SF distribution**; in the United States known as **Flory distribution**) for $\varsigma = 1$ and high degrees of polymerization. For $x = f(X)$, the differential number SF distribution leads to an exponentially decreasing curve (Fig. 2-12, top) whereas the corresponding logarithmic normal distribution passes through a maximum. Both SF and LN distributions show maxima for $w = f(X)$ (Fig. 2-12, bottom). The position of the maximum corresponds to the number-average degree of polymerization for Schulz-Zimm distributions but not for any other type of Schulz-Zimm distribution such as the Schulz-Flory distribution.

SF distributions of X are obtained from many polymerization processes that involve *high-molecular weight* polymers, for example, simple bifunctional equilibrium polycondensations, free-radical polymerizations with termination by disproportionation of active chains, and random chain scissions. Since these processes are "most probable", the Schulz-Zimm distribution is also called the **most probable distribution**. Because it shows up normally for many processes, it is known in some countries as the **normal distribution** (of polymers) which is not to be confused with the Gaussian (normal) distribution.

2.4.6 Generalized Exponential Distributions

*G*eneralized *ex*ponential distributions (**GEX distributions**) are exponential distributions that contain the variable X with a power ε. These **stretched exponentials** deliver very versatile distribution functions.

Polymer science uses many empirical GEX distributions for both molar masses and physical properties such as fracture strength. Distributions of degrees of polymerization can often by described by the **Kubin distribution**,

$$w = \frac{\gamma \beta^{(\varepsilon+1)/\gamma} X^{\varepsilon} \exp(-\beta X^{\gamma})}{\Gamma[(\varepsilon+1)/\gamma]} \qquad (2\text{-}66)$$

in which β determines the position, ε the width, and γ the skewness of the distribution. The various average degrees of polymerization are calculated from

$$\overline{X}_g = \frac{\beta^{-1/\gamma} \Gamma[(\varepsilon+1+a)/\gamma]}{\Gamma[(\varepsilon+a)/\gamma]} \qquad (2\text{-}67)$$

where a = 0 for g = n, a = 1 for g = w, and a = 2 for g = z. The Kubin distribution contains a number of other distributions as special cases (Table 2-8).

There are many other exponential distributions that are not special cases of the Kubin distribution. Well known examples are the following empirical distributions with A, b, C, D, e, and X_0 as adoptable constants:

(2-68) $w = \exp(-X/X_0)^{b}$ **Rammler-Bennett distribution**

(2-69) $w = 1 - \exp[(X_0 - X)^{b}/A]$ **Weibull distribution**

(2-70) $w = AX^{b} \exp[C - D(\ln X)^{e}]$ **Miltz-Rom distribution**

Table 2-8 Special cases of the Kubin distribution.

Name	γ	ε	β	Note
Pearson distribution	−1			
Logarithmic normal distribution	0	∞		limiting values
χ^2 distribution	1		1/2	
Schulz-Flory distribution	1	1	$1/\overline{X}_n$	
Schulz-Zimm distribution	1	ς	$\varsigma/\overline{X}_n$	
Tung distribution	γ	$\gamma - 1$		

2.4.7 Determination of Molar Mass Distributions

Distributions of degrees of polymerization can be obtained by various experimental methods (Table 2-9). Most of these methods are relative procedures, i.e., they require calibrations with polymer standards. Often used are size exclusion chromatography for analytical purposes (Volume I) and fractionation by stepwise addition of a non-solvent for the preparation of fractions (Section 10.2.3).

Some of the methods listed in Table 2-9 deliver differential distributions whereas others furnish integral ones. Some methods lead to discontinuous distributions but others to continuous ones. Variables are not necessarily the desired mass fractions and degrees of polymerization so that conversion procedures have to be employed.

Mass spectroscopy, for example, delivers discontinuous differential mass distributions of ratios of molecular masses and electric charges that have to be converted to molecular weights, M_r (p. 17). The discontinuous distributions can be approximated by continuous ones if the molecular weight of the monomeric unit is small compared to the molecular weight of the species, $M_{r,mu} < M_r$.

Size-exclusion chromatography furnishes continuous differential curves of a concentration-dependent parameter (refractive index, infrared absorption, etc.) as a function of the elution volume per time, V_e. This elution curve is then converted to a distribution of molar masses M with a calibration function, $V_e = f(M)$ (Section 14.4.3 and Volume I, Section 3.4.2).

Alternatively, one can measure "continuously" the average molar mass of molecules in a defined elution volume, for example, directly by light scattering or indirectly by solution viscometry. Since these measurements relate to finite volumes and not to infinitesimally small ones, the resulting experimental curve appears as a continuous differential one although it is a discontinuous differential with respect to molar masses or properties that are proportional to molar masses. Each of the measured volume elements contains a distribution of molar masses which has to be considered for the conversion of the differential-discontinuous curve into a differential continuous one.

Table 2-9 Methods for the determination of distributions of molar mass or molecular weights.
A = Analytical, D = desorption from a matrix, S = in solution, SP = in solution with a precipitant, ST = in solution with temperature variation, M = in melt, P = preparative.

Method	Type	State	Section
Mass spectroscopy (MALDI)	A	D	Volume I
Size exclusion chromatography (SEC, GPC)	A, P	S	14.4.3
Dynamic light scattering	A	S	11.1.2
Fractionation	P	SP	10.2.3
Cloud point titration	A	SP	10.2.3
Cloud temperature analysis	A	ST	10.2.3
Sedimentation (velocity or equilibrium)	A	S	11.3
Field flow fractionation (FFF)	A	S	11.4
Flow birefringence	A	S	11.2.2
Storage modulus and other viscoelastic properties	A	M	17.5.3

Historical Notes to Chapter 2

For the discovery of the macromolecular character of polymers, see Volume I, Section 1.3; for the discovery of tacticity, see Volume I, Section 4.1.1, and Volume II, Section 6.2.6.

G.Herdan, Nature **163** (1949) 139
Herdan relation

Literature to Chapter 2

2.0.0 NOMENCLATURE
IUPAC, Commission on Macromolecular Nomenclature, Compendium of Macromolecular Nomenclature, Blackwell, London 1991 ("Purple Book")
IUPAC, Commission of Macromolecular Nomenclature, Glossary of Basic Terms in Polymer Science (IUPAC Recommendations 1996), Pure Appl. Chem. **68**/12 (1996) 2287

2.0.1 GENERAL ANALYTICAL METHODS
L.S.Bark, N.S.Allen, Eds., Analysis of Polymer Systems, Appl.Sci.Publ., Barking, Essex 1982
J.Mitchell, Jr., Ed., Applied Polymer Analysis and Characterization, Hanser, Munich 1987
T.R.Crompton, Analysis of Polymers, Pergamon, Oxford 1989
J.R.White, D.Campbell, Polymer Characterization. Physical Techniques, Chapman & Hall, New York 1989
G.Allen, J.C.Bevington, Eds., Comprehensive Polymer Science; C.Booth, C.Price, Eds., Vol. 1, Polymer Characterization, Pergamon Press, Oxford 1989
J.I.Kroschwitz, Ed., Polymers: Characterization and Analysis, Wiley, New York 1990 (reprints of entries in Encyclopedia of Polymer Science and Engineering)
H.G.Barth, M.I.James, Eds., Polymer Characterisation, Blackie Academic, Glasgow 1993
J.L.Koenig, Spectroscopy of Polymers, Elsevier, Amsterdam, 2nd ed. 2000

2.1 CONSTITUTION and 2.2 CHEMICAL CONFIGURATION, for older literature see Volume I
G.R.Newkome, C.N.Moorefield, F.Vögtle, Dendritic Molecules. Concepts, Syntheses, Perspectives, VCH, Weinheim 1996
P.R.Dvornic, D.A.Tomalia, Recent Advances in Dendritic Polymers, Curr.Opin.Colloid Interface Sci. **1** (1996) 221
L.H.Sperling, S.C.Kim, Eds., IPNs Around the World, Wiley, New York 1997
R.F.T.Stepto, Ed., Polymer Networks - Principles of Their Formation, Structure, and Properties, Blackie Academic, Glasgow 1997
J.A.Semlyen, Ed., Large Ring Molecules, Wiley, New York 1997
O.A.Matthews, A.N.Shipway, J.F.Stoddart, Dendrimers–Branching out from Curiosities into New Technologies, Progr.Polym.Sci. **23** (1998) 1
M.K.Mishra, S.Kobayashi, Star and Hyperbranched Polymers, Dekker, New York 1999
A.D.Schlüter, J.P.Rabe, Dendronized Polymers, Angew.Chem. **112** (2000) 860; Angew.Chem.Int. Ed.Engl. **39** (2000) 864
A.Ciferri, Ed., Supramolecular Polymers, Dekker, New York 2000
G.R.Newkome, C.N.Moorefield, F.Vögtle, Dendrimers and Dendrons. Concepts, Syntheses, Applications, Wiley-VCH, Weinheim 2001
R.D.Archer, Inorganic and Organometallic Polymers, Wiley, New York 2001
J.M.J.Fréchet, D.A.Tomalia, Eds., Dendrimers and Other Dendritic Polymers, Wiley, New York 2002
N.Hadjichristidis, S.Pispas, G.A.Floudas, Block Copolymers, Wiley, New York 2003
J.L.Atwood, J.W.Steed, Eds., Encyclopedia of Supramolecular Chemistry, Dekker, New York 2004
R.C.Advincula, W.J.Brittain, K.C.Kaster, J.Rühe, Polymer Brushes, Wiley-VCH, Weinheim 2005

2.3.a AVERAGES AND MOMENTS

H.-G.Elias, R.Bareiss, J.G.Watterson, Mittelwerte des Molekulargewichtes und anderer Eigenschaften, Adv.Polym.Sci.-Fortschr.Hochpolym.Forschg. **11** (1973) 111 (= averages of molecular weights and other properties)

H.-G.Elias, Polymolecularity and Polydispersity in Molecular Weight Determinations, Pure Appl. Chem. **43**/1-2 (1975) 115

R.E.Bareiss, Polymolecularity Correction Factors, in J.Brandrup, E.H.Immergut, E.A.Grulke, Eds., Polymer Handbook, Wiley, New York, 4th ed. 1999

2.3.b DETERMINATION OF MOLAR MASSES
(general literature; for special methods, see relevant sections)

R.U.Bonner, M.Dimbat, F.H.Stross, Number Average Molecular Weights, Interscience, New York 1958

D.V.Quayle, Molecular Weight Determination of Polymers by Electron Microscopy, Brit.Polym.J. **1** (1969) 15

S.R.Rafikov, S.Pavlova, I.I.Tverdokhlebova, Determination of Molecular Weights and Polydispersity of High Polymers, Acad.Science USSR, Moscow 1963; Israel Program of Scientific Translation, Jerusalem 1964

P.E.Slade, Jr., Polymer Molecular Weights, Dekker, New York 1975 (2 Vols.)

N.C.Billingham, Molar Mass Measurements in Polymer Science, Halsted Press, New York 1977

A.R.Cooper, Ed., Determination of Molecular Weight, Wiley 1990

S.R.Holding, E.Meehan, Molecular Weight Characterization of Synthetic Polymers (RAPRA Review Report), Plastics Design Library, Norwich (NY) 1997

2.4 DISTRIBUTION FUNCTIONS

J.Aitchison, J.A.C.Brown, The Lognormal Distribution, Cambridge University Press, Cambridge 1969

L.H.Peebles, Molecular Weight Distributions in Polymers, Interscience, New York 1971

Reference to Chapter 2

[1] J.G.Watterson, H.-G.Elias, J.Macromol.Sci.-Chem. **A 5** (1971) 459

3 Microconformations

3.1 Introduction

Polymer molecules of defined constitution and configuration are characterized by defined bond lengths b between two chain atoms and defined bond angles τ between three chain atoms. The spatial positions of chain atoms are fixed in polymer crystals but not in dilute polymer solutions where microbrownian motion allows them to move from one position to another.

An example is the chain atom 3 in Fig. 3-1 whose position is controlled by the constant bond lengths $b_{2\text{-}3}$ and $b_{3\text{-}4}$ and the constant bond angle $\tau_{2\text{-}3\text{-}4}$ but which still can rotate on a circle around the extension of the bond axis $b_{1\text{-}2}$. The chain atom 3 can adopt principally an infinite number of positions on this circular path. In reality, some positions are preferred energetically because of repulsion or attraction by neighboring atomic groups and/or electron pairs (see below).

These preferential positions are called **conformations** (small molecules) or **microconformations** (macromolecules) (L: *forma* = shape; *com* = with, together (in Latin, *com* becomes *con* before all consonants except *b*, *gn*, *h*, *l*, *m*, *p*, and *r*)). The type and sequence of microconformations determines the macroconformation (Chapter 4).

All molecules of an ideal crystalline polymer have the same macroconformation composed of repetition of sequences of microconformations which in turn are controlled by intra- and intermolecular interactions between atomic groups. Because of the packing of molecules in crystals, a macroconformation is only converted into another one if sufficient energy is available for the change of one crystal modification into another.

In dilute polymer solutions, microconformations change rapidly (see below). A polymer chain thus contains more or less irregular sequences of microconformations and the overall shape of a polymer chain is no longer regular. However, the constant bond angles between chain atoms as well as interactions between substituents lead to short preferential directions of chain segments: the chain has a **persistence** (Section 4.3.5).

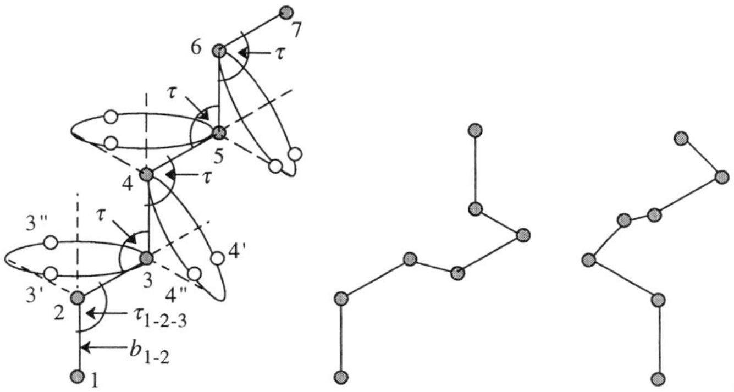

Fig. 3-1 Left: Possible rotations of chain atoms of a poly(methylene) chain, $\text{+}CH_2\text{+}_n$, around the imaginary extension of the previous chain bond. Chain atoms are shown in the often energetically favored trans position ●; possible gauche positions are indicated by ○. For the definition of trans and gauche, see below. For simplicity, hydrogen atoms are not shown.

Center and right: projection of the same chain atoms in the plane of the paper for two chains with random succession of trans and gauche positions of chain atoms.

3.2 Local Conformations

3.2.1 Definitions

Simple molecules of the type A–B–D are described completely by their geometry: the two bond lengths b_{A-B} and b_{B-D} and the bond angle τ_{A-B-D}. An example is the methane molecule, H–CH_2–H, with b_{C-H} = 0.1094 nm and τ_{H-C-H} = 109°18'.

For a complete geometrical description of molecules of the type A–B–D–E, such as the ethane molecule, H–CH_2–CH_2–H, one needs to know not only the three bond lengths, b_{A-B}, b_{B-D}, and b_{D-E}, and the two bond angles, τ_{A-B-D} and τ_{B-D-E}, but also the **torsional angle** θ **(conformational angle, rotational angle, dihedral angle)**.

The torsional angle of A and E about the bond B–D is the angle between the projections of bonds A–B and D–E on a plane that is normal (i.e., at an angle of 90°) to the bond B–D (Fig. 3-2). The torsional angle is thus the angle between the planes A–B–D and B–D–E.

The torsional angle determines the relative spatial position of "bound" substituents (ligands, atomic groups, atoms) to "unbound" ones. For example, the 3 H atoms of the C^1 of ethane, H_3C^1–C^2H_3, are bound to C^1 but the adjacent H atoms at C^2 are "unbound" (to C^1!) and vice versa.

In polymer science, torsional angles are measured from –180° to +180° and not from 0° to 360° as in organic chemistry. Torsional angles are positive if the rotation of the front atom A around the central bond B–D of ABDE needs the smaller of the two rotational angles (in this case, a turn to the right) to coincide with the back atom E (see G^+ in Fig. 3-2).

Some of the infinite number of torsional angles correspond to energy minima or maxima. Ethane with 3 bound H atoms has 3 equal energy minima at 0°, +120°, and –120° and 3 equal energy minima at +60°, –60°, and 180° (Fig. 3-3). In macromolecular science, the conformation at 0° is called **trans** and that at 180°, **cis**; in organic chemistry, it is just the opposite. The organic chemistry convention is impractical for polymer purposes because "trans" is often the energetically most stable conformation and cis conformations are difficult to picture for chains where ○ corresponds to chain atoms ●~~.

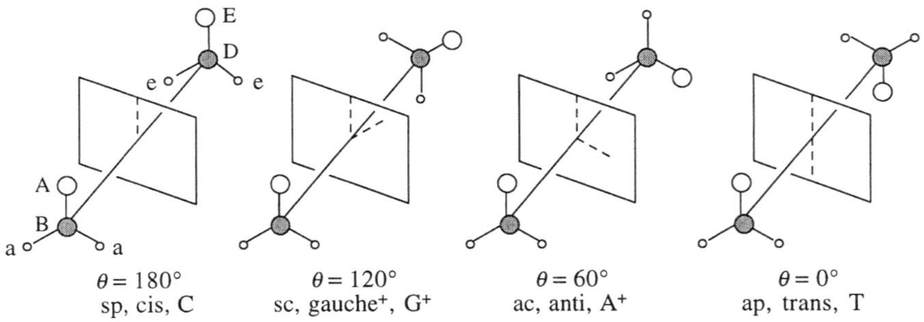

Fig. 3-2 Definition of torsional angles θ (macromolecular convention) as the projection - - - - of bonds A–B and D–E (●—○) on a plane that is normal to the bond B–D (●—●). Positions with negative torsional angles (A^- at θ = –60° and G^- at θ = –120°) are not shown.

Ethane H_3C–CH_3:	a = a = e = e = H	A = E = H	B = D = C (carbon atom)
Butane H_3C–CH_2–CH_2–CH_3:	a = a = e = e = H	A = E = CH_3	B = D = C (carbon atom)

ac = anticlinal, ap =antiperiplanar, sc = synclinal, sp =synperiplanar.

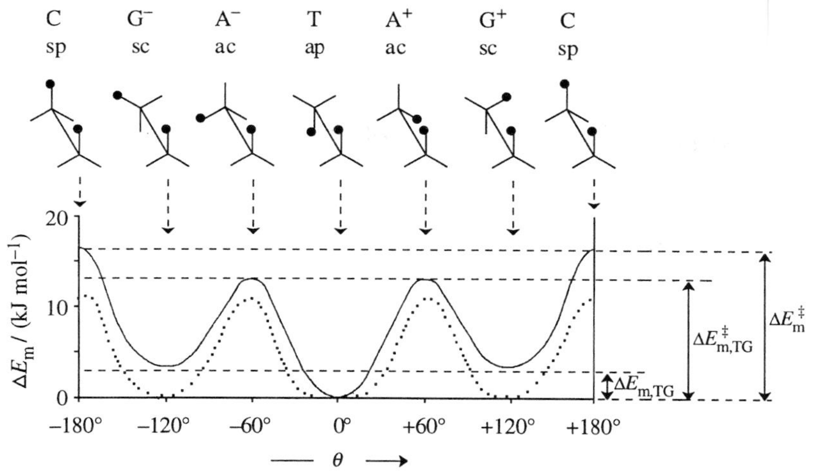

Fig. 3-3 Ideal microconformations (top) and rotational barriers (bottom) about the C-C bond of ethane, CH_3–CH_3, (dotted curve) and the central C-C bond of butane, CH_3CH_2–CH_2CH_3, (solid curve) as a function of the torsional angle θ (macromolecular convention for C, G, A, and T). In ethane, • is an H- atom, in butane a CH_3 group. The energies to the right refer to butane.

In butane, the two energy minima at +120° and −120° are not as deep as the minimum at 0° and the two energy maxima at +60° and −60° are not as high as the maximum at ±180°. The minima at ±120°, G^+ and G^-, are called **gauche** (F: *gauche* = left, clumsy (older French: askew)) and the maxima at ±60°, A^+ and A^-, **anti**. In organic chemistry, gauche is called "synclinal" (sc) and anti, "anticlinal" (ac).

Rotational potentials of tetravalent atoms such as carbon are not always at the ideal positions of 0°, ±60°, ±120°, and ±180°. However, non-ideal conformations are referred to the same as ideal ones if they deviate from ideal positions by no more than 30°. A conformation at +10° is therefore also a trans conformation (ideally, at 0°).

Small molecules in energetically distinguished spatial positions can be considered defined chemical species since they appear as such when studied by fast methods. Examples are the cis and trans conformers of ethane and the cis, anti, gauche, and trans conformers of butane. These species are called **conformers, conformational isomers**, or **conformational stereoisomers** or, if open-chained, also **rotational isomers** or **rotamers**. In polymer science, the same terms refer to the *local* conformations in a chain.

3.2.2 Rotational Potentials

Energy maxima and minima in potential curves are the resultant of attractive and repulsive forces between ligands of chain atoms (Volume I, p. 141). For the three maxima of *ethane* at C, A^+, and A^-, this is often interpreted as steric repulsion although the bound H atoms are just within the van der Waals distance of unbound ones (Section 3.3.3). Other calculations indicate that the three energy minima at T, G^+, and G^- are not caused by repulsion at C, A^+, and A^- but are rather due to attractions between bonding and antibonding orbitals. A molar activation energy of $\Delta E^{\ddagger}_{m}$ = 11.3 kJ/mol is needed to overcome the **rotational barriers** at C, A^+, and A^- in ethane.

Like ethane, *butane*, C^1H_3-C^2H_2-C^3H_2-C^4H_3, also has a threefold **rotational potential**. However, only two of the three ligands (C^1H_3 H, and H) of chain atom C^2 are identical (similarly for C^3), which results in two types of energy maxima instead of one for ethane. At ±180°, the two methyl groups A and E in Fig. 3-2 are nearest to each other which leads to the highest energy maximum. At ±60°, methyl groups face hydrogen atoms (A faces one of the two e's in Fig. 3-2) which produces lower energy maxima at the two anti positions (Fig. 3-3). These energy maxima are separated by energy minima at 0° (trans) and ±120° (gauche). The difference between these two types of energy minima is called molar **conformational energy** or **potential energy**; for butane, it is $\Delta E_{m,TG}$ = 2.9 kJ/mol. The molar activation energies for overcoming rotational barriers are $\Delta E^{\ddagger}_{m,TG}$ = 13 kJ/mol (trans → gauche) and $\Delta E^{\ddagger}_{m,TG}$= 15.9 kJ/mol (trans →cis).

Trans and gauche rotamers have different vibrational spectra which can be measured by Raman scattering. The potential energy $\Delta E_{m,TG}$ is obtained from the temperature dependence of the intensities of G and T bands, $I_G/I_T \sim \exp[-\Delta E_{m,TG}/(RT)]$ and the fraction ϕ_G of molecules in the two gauche states from $\phi_G = [2\, I_G/I_T]/[1 + 2\,(I_G/I_T)]$.

For *pentane*, C^1H_3–C^2H_2–C^3H_2–C^4H_2–C^5H_3, one has to consider **conformational diads**, i.e., the two consecutive chain conformations at C^2–C^3 and C^3–C^4, respectively (Fig. 3-4). Each of these two diads can exist in one trans and two gauche conformations. Consequently, there are four types of conformational diads with different rotational barriers: TT < TG^+ = TG^- = G^-T = G^+T < G^+G^+ = G^-G^- < G^+G^- = G^-G^+. For calculations of conformations of carbon chains, one usually assumes that diads G^+G^- and G^-G^+ are not present (see also Table 3-1).

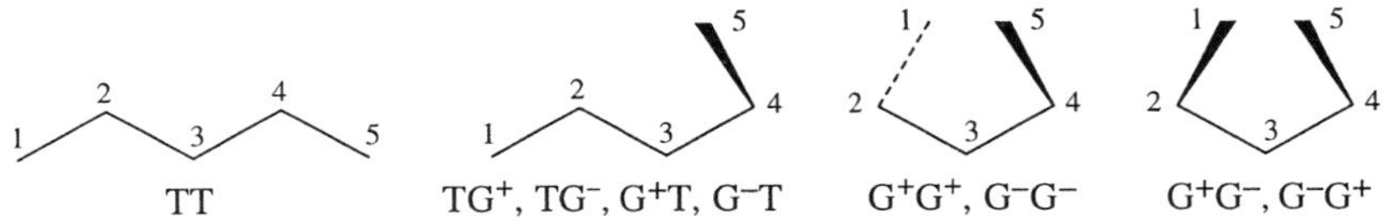

Fig. 3-4 Conformational diads of pentane C^1H_3-C^2H_2-C^3H_2-C^4H_2-C^5H_3.

Since rotational barriers between different conformational states are fairly small, conformers of alkanes in the fluid phase are interconverted relatively fast (see Volume I, Table 4-1). The necessary energy is provided by the collision of two molecules. However, only a small fraction of thermal energy is transferred on collison to the other molecule, on average at T = 298.15 K, ca. $RT/2 \approx$ 1.24 kJ per mole of degree of freedom.

Because of the Maxwell-Boltzmann energy distribution, just a small fraction of collisions can thus transfer sufficient energy to overcome the potential barrier. Instead, most collisions lead to vibrations of no more than ± 20° about the potential minimum. As a consequence, the majority of molecules resides at room temperature at a minimum of potential energy. For this reason, small molecules can be treated as if they exist in discrete conformational states, i.e., as conformers.

3.2.3 Effect of Constitution

Effects of constitution on rotational barriers have already been discussed in Volume I so that only the most important effects are listed here, such as types and lengths of bonds and size and number of ligands.

In most cases, adjacent ligands repel each other. The spatial requirements are controlled by the van der Waals radii of ligands and not by their atomic radii.

- *Single versus multiple bonds.* Rotational barriers are greater for stereoisomers than for conformational isomers. For example, the molar rotational barrier, $\Delta E^{\ddagger}_{m}$, at the C–C bond of ethane, $H_3C–CH_3$, is much smaller than that of the C=C double bond of ethene, $H_2C=CH_2$: 11.3 kJ/mol *versus* 272 kJ/mol.
- *Ring molecules.* Six-membered rings require far higher activation energies for ring inversion than four-membered ones. Examples are cyclobutane, C_4H_8, and cyclohexane, C_6H_{12}: 6.2 kJ/mol *versus* 43.2 kJ/mol, as well as oxetane, C_3H_6O, and tetrahydropyran, $C_5H_{10}O$: 0.4 kJ/mol *versus* 39.8 kJ/mol.
- *Bond lengths.* The longer the "chain" bond, the smaller is the rotational barrier. An example is the series CH_3CH_3 .... CH_3SiH_3 .... SiH_3SiH_3 where bond lengths increase from 0.154 nm (C–C) to 0.193 nm (C–Si) and 0.234 nm (Si–Si) whereas molar rotational barriers, $\Delta E^{\ddagger}_{m}/(\text{kJ mol}^{-1})$, decrease from 11.3 to 7.1 to 4.2.
- *Size of ligands.* The larger the van der Waals volumes of groups, the greater is the rotational barrier. Examples are C_6H_5OH *versus* CH_3OH with 13.0 kJ/mol *versus* 1.6 kJ/mol, and the series $CH_3C(CH_3)_3$, $CH_3CH(CH_3)_2$, $CH_3CH_2CH_3$, and CH_3CH_3 with $\Delta E^{\ddagger}_{m}/(\text{kJ mol}^{-1})$ of 19.7, 16.3, 14.2, and 11.3, respectively.
- *Number of ligands.* Rotational barriers decrease with decreasing number of ligands. An example is $H_3C–CH_3$ *versus* $H_3C–OH$ where $\Delta E^{\ddagger}_{m}/(\text{kJ mol}^{-1})$ is 11.3 *versus* 1.6, respectively.

In *polar environments*, chains with electronegative substituents or free electron pairs as ligands try to adopt the greatest number of gauche interactions between these ligands. Because of this **gauche effect**, many polymer chains do not crystallize in ...TTTT... sequences as poly(ethylene) does but rather in macroconformations of the type $(TG^+)_n$, $(TTG^-)_n$, etc. In most cases, such ordered macroconformations do not survive dissolution (but see Fig. 4-2 in Chapter 4).

The gauche effect causes poly(oxymethylene) (POM) to crystallize in all-gauche macroconformations, $(G^+)_n$ and $(G^-)_n$, and poly(oxyethylene) (PEOX) and poly(glycine) (PG) in both $(TTG^+)_n$ and $(TTG^-)_n$. Because of the steric hindrance by the methyl group, isotactic poly(propylene oxide) (PPOX) crystallizes in the $(T)_n$ macroconformation despite the presence of unpaired electron pairs.

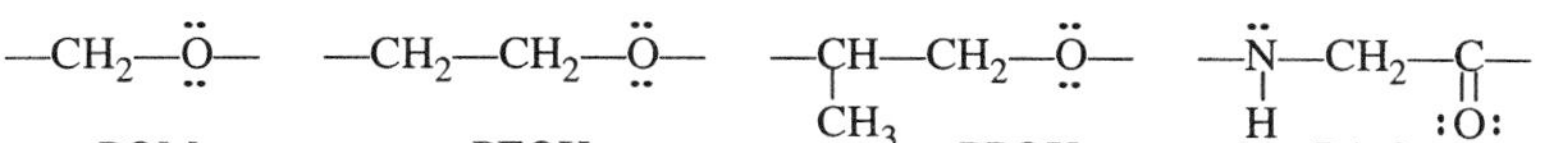

POM PEOX PPOX PA 2

The gauche effect does not arise from internal effects of microconformations per se but stems from the interaction of the solute with its polar or polarizable environment. An example is the linear variation of the difference of the molar conformational energies between trans and gauche conformations, $\Delta E_{m,TG}$, of 1,2-dichloroethane, $ClCH_2–CH_2Cl$, with the inverse relative permittivity (= reciprocal dielectric constant) of the environment ($C_2H_4Cl_2$ in the gaeous state, as neat liquid, or in solution) (Fig. 3-5). This energy difference is very negative in the gaseous state; negative for $C_2H_4Cl_2$ in apolar solvents such as *n*-hexane, carbon tetrachloride, or carbon disulfide; zero (by definition) in the liquid state of 1,2-dichloroethane; and positive for $C_4H_2Cl_2$ in methanol solution.

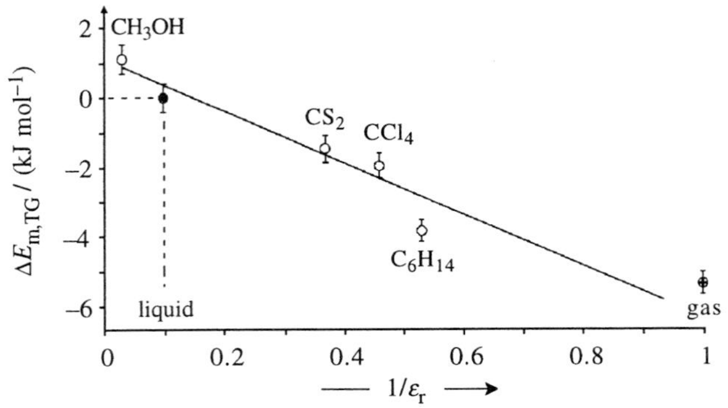

Fig. 3-5 Energy difference of trans and gauche conformations, $\Delta E_{m,TG}$, of 1,2-dichloroethane as a function of the inverse relative permittivity ε_r of the medium. Data for the gaseous state (⊕), the neat liquid (●), and in various solutions (O). The energy difference in the neat liquid was set to zero.

The increase of gauche conformations of polar solutes with increasing relative permittivities is shown for *meso*-2,4-hydroxypentane in Table 3-1. In all solvents, the percentage of G^+G^- and G^-G^+ diads is zero.

The population of microconformations with higher energy increases with increasing temperature since rotational barriers are more easy to overcome. For example, the energy difference of gauche and trans conformations of a partially deuterated isotactic poly(methyl methacrylate) is $\Delta E_{m,TG}$ = 1.46 kJ/mol at −30°C. At this temperature, about twice as many trans conformers as gauche conformers are present whereas at infinitely high temperature, the trans/gauche ratio is unity (Fig. 3-6),

3.3 Sequences of Microconformations

3.3.1 Introduction

Stereoregular polymer molecules consist of sterically equal repeating units. In crystalline polymers, this regularity leads to an equivalent regular sequence of microconformations if thermally caused rotations of units are either completely supressed by tight packing of chains or, if present, such rotations are completely synchronous.

Table 3-1 Solvent dependence of the percentage of conformational diads of *meso*-2,4-hydroxypentane, CH_3–CHOH–CH_2–CHOH–CH_3, at 40°C.

Solvent Name	ε_r	Percentage of conformational diads TT	TG^+, G^-T	TG^-, G^+T	G^+G^+, G^-G^-	G^+G^-, G^-G^+
Carbon tetrachloride	2.2	70	10	10	10	0
Dichloromethane	8.9	90	10	0	0	0
Pyridine	12.4	45	48	7	0	0
Dimethyl sulfoxide	46.7	30	60	10	0	0
Deuterium oxide	78.4	5	70	25	0	0

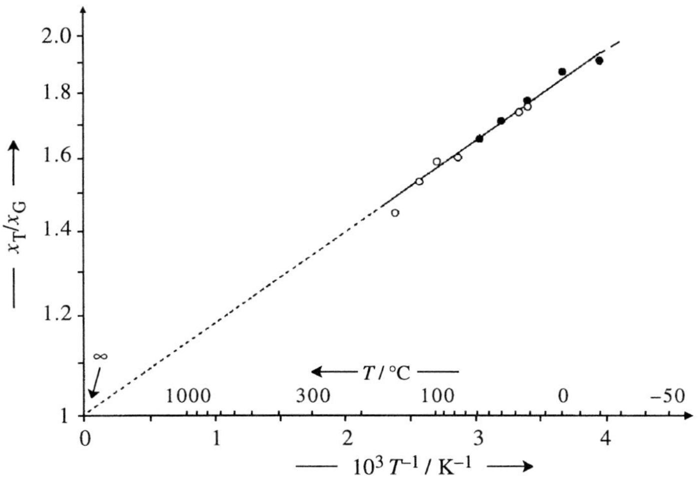

Fig. 3-6 Temperature dependence of the ratio of mole fractions of trans and gauche conformations, x_T/x_G, of an isotactic poly(methyl methacrylate) with $x_i = 0.93$ in deuterochloroform (●) and *o*-dichlorobenzene (○). Data obtained from geminal NMR coupling constants [1]. $x_T + x_G \equiv 1$.

In dilute polymer solutions, regular sequences of microconformations usually do not survive so that the macroconformation of a chain consists of random arrangements of all types of microconformations (Chapter 4). However, there are two exceptions. First, regular sequences are maintained if strong intramolecular attractive bonds exist between adjacent or neighboring monomeric units, such as hydrogen bonds or π-π interactions. Second, an ordering of solvent molecules may impose order on dissolved polymer chains, e.g., by intermolecular lateral self-association (see Sections 4.1.2 and 10.5). In the vast majority of cases, these regular sequences of microconformations comprise only a few monomeric units and not longer sequences or the whole macromolecule.

Regular sequences of microconformations lead to two types of regular macroconformations: zigzag chains and helices. Zigzag chains comprise the macroconformations $(T)_n$ and $(TTGG)_n$ (with $G = G^+$ or $G = G^-$). An example of the former is poly(ethylene); examples of the latter are syndiotactic poly(propylene) and modification II of poly(glycine), $\text{+}NH\text{–}CH_2\text{–}CO\text{+}_n$. Zigzag chains are found only in the crystalline state; they will be discussed in Chapter 7. Helical macromolecules can exist both in the crystalline state and in dilute solution.

3.3.2 Helices

A helix (G: *elix* = spiral) is a spiral which may be conical as the shell of the snail *Helix pomatia* or a wood screw or cylindrical as a metal screw. Of course, helical macroconformations of macromolecules are always cylindrical.

Helices exist in various types, depending on the kind of substituents and their interactions (Fig. 3-7). They are usually characterized by the number of constitutional repeating units, N_{rep}, contained in N_s turns where N_{rep} and N_s are integral numbers. An example is the 3_1 helix of isotactic poly(propylene) with 3 monomeric units per 1 turn.

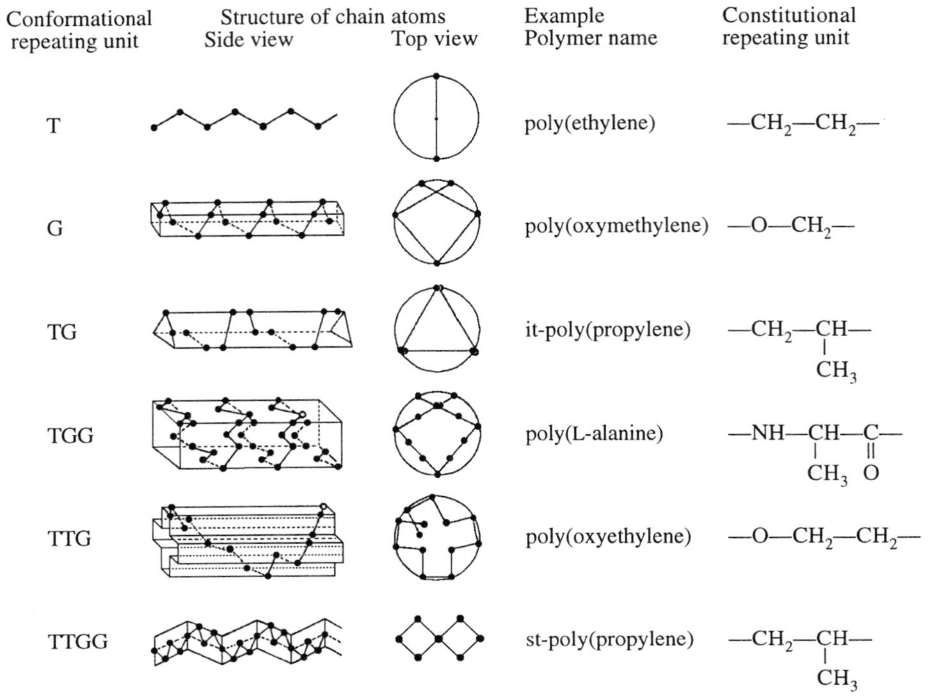

Conformational repeating unit
Structure of chain atoms
Side view
Top view
Example Polymer name
Constitutional repeating unit
T
poly(ethylene)
—CH2—CH2—
G
poly(oxymethylene)
—O—CH2—
TG
it-poly(propylene)
—CH2—CH—
CH3
TGG
poly(L-alanine)
—NH—CH—C—
CH3 O
TTG
poly(oxyethylene)
—O—CH2—CH2—
TTGG
st-poly(propylene)
—CH2—CH—
CH3

Fig. 3-7 Schematic representation of various types of macroconformations of polymer molecules [2]. Gauche conformations of these polymer *molecules* always have the same sign, for example TG^+ or TG^-. Polymeric *substances* may be 1:1 mixtures of chains with macroconformations of opposite signs for G, for example, the 1:1 mixture of $(TG^+)_n$ and $(TG^-)_n$ of it-poly(propylene).

IUPAC recommends characterizing spatial repeating units by $AN_c{*}N_{rep}/N_s$ where A = type of repetition along the long axis (A = t for translation, A = s for screw repetition), N_c = number of chain atoms per constitutional or configurational repeating unit, respectively, N_s = number of turns until the relative starting position is obtained, and N_{rep} = number of conformational repeating units per N_s. N_s and N_{rep} are always whole numbers; * and / are separators.

Poly(ethylene), $\text{+}CH_2\text{–}CH_2\text{+}_n$, crystallizes in the all-trans conformation. It has $N_c = 2$ carbon atoms per $N_{rep} = 1$ conformational repeating unit; after $N_s = 1$ turn, the relative starting position is reached. Poly(ethylene) thus exists in the macroconformation t2*1/1; it is a "1_1-helix." As poly(methylene), it has the symbol t1*2/1 ("2_1-helix").

Isotactic poly(propylene) molecules, $\text{+}CH_2\text{-}CH(CH_3)\text{+}_n$, with the macroconformations $(TG^+)_n$ or $(TG^-)_n$ form 3_1 helices (Figs. 3-7 and 3-8) with $N_c = 2$ chain atoms per configurational repeating unit and 3 constitutional repeating units per conformational repeating unit; the symbol is therefore s2*3/1. Correspondingly, the helical syndiotactic poly(propylene) (Fig. 3-7) has the symbol s4*2/1 since it has $N_c = 4$ chain atoms per configurational repeating unit.

The height, $h = c/N_{rep}$, of a conformational repeating unit, projected onto the helix axis, is calculated from the length of the identity period and the number of conformational repeating units, N_{rep}. An example is it-poly(propylene) with h = 0.65 nm/3 = 0.217 nm. The rotational angle α around the helix axis per conformational repeating unit is thus $\alpha = 2\,\pi\,N_s/N_{rep}$; for it-poly(propylene), $\alpha = 2\,\pi/3$.

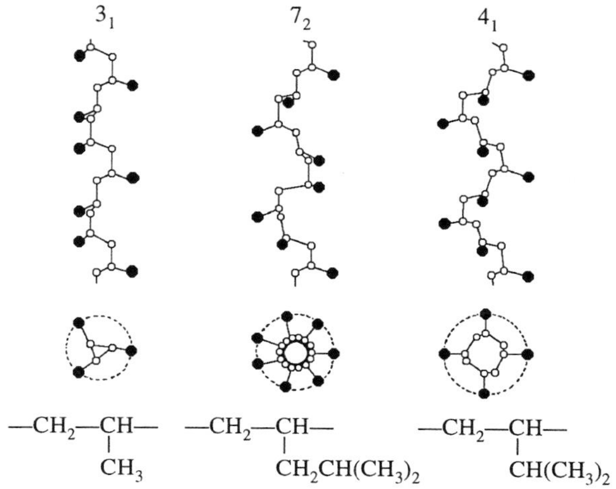

Fig. 3-8 Helical structures of three it-poly(1-olefin)s with the same conformational repeating unit TG$^+$. From left to right: it-poly(propylene), it-poly(4-methyl-1-pentene), and it-poly(3-methyl-1-butene). Symbols: O chain atom, ● substituent; hydrogen atoms at chain atoms are not shown.

Spatial repeating units and conformational repeating units are not necessarily identical. Polymer chains with the same conformational repeating unit can form very different types of helices. For example, it-poly(propylene), it-poly(4-methyl-1-pentene), and it-poly(3-methyl-1-butene) all have the same conformational repeating unit TG but crystallize as 3_1-, 7_2-, and 4_1-helices, repectively (Fig. 3-8).

By definition, helices are right-handed if, on looking along the cylinder axis, the screw turns away clockwise from the observer and left-handed if it turns away counterclockwise. Chains with TG$^+$ sequences have the opposite handedness to that of chains with TG$^-$ sequences, etc. (Fig. 3-9). Since TG$^+$ and TG$^-$ are energetically equal, both are present in crystalline polymers in the same proportion.

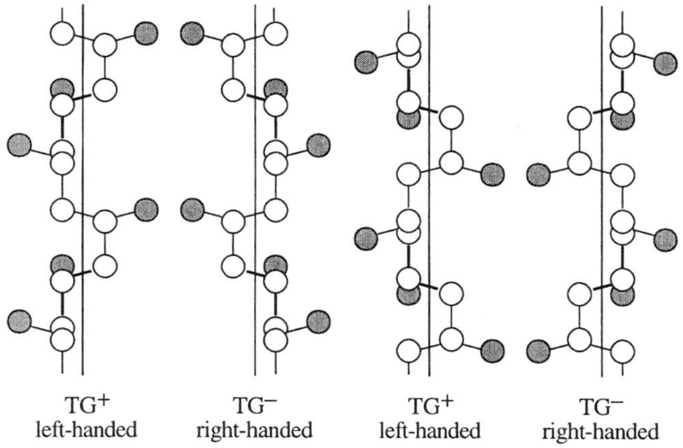

Fig. 3-9 Helical sequences of isotactic poly(propylene). O Chain atoms, ● methyl groups; H atoms on chain units CH$_2$ and CH are not shown. Thick black lines indicate bonds above the paper plane; thin black lines, bonds below the paper plane. See also Section 7.1.8.

Double helices consist of two chains that are wound around each other. An example is the double helix of deoxyribonucleic acids (Volume I) which exists in both the crystalline and the dissolved state. Double helices are also present in the crystalline states of the polysaccharides amylose and carrageen (Volume II) and probably also in the crystalline states of it-poly(methyl methacrylate) and poly(*p*-hydroxy benzoic acid).

Triple helices are formed by the protein collagen and the polysaccharides poly(β-1,3-D-xylan) of certain green algae and poly(β-1,3-D-glucan)s of bacteria, fungi, and algae (Volume II). Triple helices also exist in some synthetic ribonucleic acids, such as poly(rU)-*block*-poly(rA)-*block*-poly(rU).

There are also cases of so-called **superhelices** (= **supersecondary structures**) where a helix is wound around another helix or double helix.

3.3.3 Constitution and Configuration Dependence

Macromolecules in crystalline states try to pack themselves as tightly as the van der Waals volumes of their atoms would allow. The resulting macroconformations usually do not survive dissolution but there are exceptions (see Section 3.4.2 and Fig. 4-1). The following remarks thus refer only to macroconformations in the crystalline state.

Poly(ethylene), $\text{+}CH_2\text{–}CH_2\text{+}_n$, has a bond length of L_{C-C} = 0.154 nm and a valence angle $\angle_{C\text{-}C\text{-}C}$ = 111.5° of carbon atoms from which one can calculate for all-trans conformations that the distance between unbound hydrogen atoms is a maximum 0.254 nm. This distance corresponds approximately to the sum of van der Waals radii of hydrogen atoms which is ca. 0.264 nm. The methylene groups of poly(ethylene) can thus accommodate themselves in the all-trans macroconformation, the zigzag chain.

In the all-trans macroconformation of poly(tetrafluoroethylen), $\text{+}CF_2\text{–}CF_2\text{+}_n$ the distance between two unbound fluorine atoms would be smaller than the sum of their van der Waals radii of 0.31 nm. Poly(tetrafluoroethylene) chains thus cannot exist in zigzag chains with all-trans macroconformations. The fluorine atoms rather twist somewhat which increases the torsional angle to 16° from 0° of an ideal all-trans conformation: at temperatures below 19°C, poly(tetrafluoroethylene) forms a 13_6 helix (Fig. 3-10).

Fig. 3-10 Conformational repeating units of poly(ethylene) (PE; 2_1 helix) and poly(tetrafluoroethylene) (PTFE; 13_6 helix at T < 19°C). Conformational angles are θ = 0° (PE) and θ = 16° (PTFE).

The substituents R of isotactic polyvinyl compounds, $\text{+}CH_2\text{–}CHR\text{+}_n$, are even larger than fluorine atoms which forces the chains to adopt the $(TG)_n$ macroconformation instead of $(T)_n$. The steric requirements also enlarge the bond angles $\angle_{C\text{-}C\text{-}C}$ between chain atoms from 1115° of poly(ethylene) (R = H) to 114° for it-poly(propylene) (R = CH_3) and 116° for it-poly(styrene) (R = C_6H_5).

Torsional angles of *isotactic* poly(propylene) (PP), poly(1-butene) (PB), and poly(5-methyl-1-heptene) (PMH) alternate between 0° (T) and 120° (G); all these polymer chains crystallize as 3_1 helices. The shorter the distance between chain atoms and bulky substituents, the more diagonally expanded is the chain (see Fig. 3-8). Examples are the 3_1 helices of it-PP, it-PB, and it-PMH versus the 7_2 helix of it-poly(4-methyl-1-pentene) (P4MP) and the 4_1 helix of it-poly(3-methyl-1-butene) (PMB) (see Fig. 3-11).

These polymer molecules have chains composed of prochiral repeating units (Section 4.1.4 in Volume I), which are enantiomeric to each other. In a polymeric substance, positive and negative gauche microconformations must therefore be present in the same proportion. Since subsequent gauche microconformations cannot reverse their signs for steric reasons, chains must be composed of either left-handed, $(TG^+)_n$, or right-handed, $(TG^-)_n$, macroconformations which are present therefore in a 50:50 proportion.

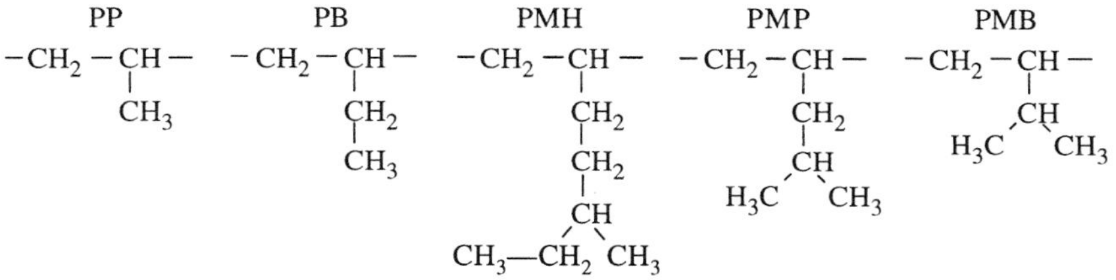

	PP	PB	PMH	PMP	PMB
	3_1 helix	3_1 helix	3_1 helix	7_2 helix	4_1 helix
$\theta =$	±0°, +120°	±0°, +120°	±0°, +120°	−13°, +110°	−24°, + 96°

Fig. 3-11 Helix structures and conformational angles of some isotactic poly(1-olefin)s.

However, isotactic polymers with very large substituents in prochiral repeating units may also form helices with a single handedness. Examples are

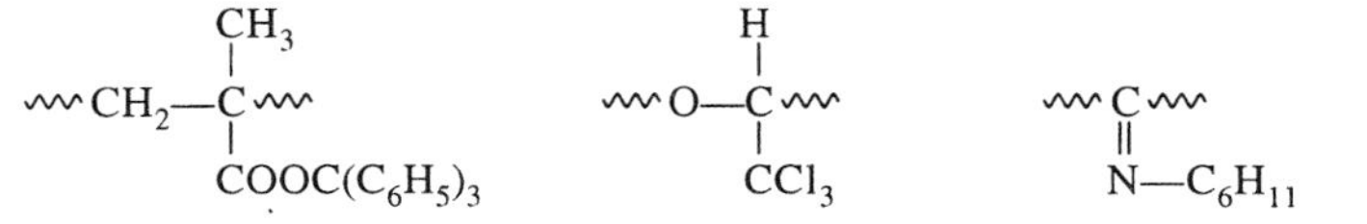

poly(trityl methacrylate) poly(chloral) poly(*N*-hexylisocyanide)

Substituents R of *syndiotactic* vinyl polymers, $\text{+CH}_2\text{–CHR+}_n$, in all-trans conformation are much farther away from each other than the same substituents in isotactic polymers (Volume I, Fig. 4-10). All-trans conformations with torsional angles of 0°/0° are therefore often the most energy-poor macroconformation of syndiotactic polymers. Examples are st-poly(vinyl chloride) (R: Cl), st-poly(acrylonitrile) (R: CN), and st-1,2-poly(butadiene) (R: $CH{=}CH_2$). In other syndiotactic vinyl polymers, a conformational sequence 0°,0°,–120°,–120° is more advantageous. For example, st-poly(propylene) usually adopts a $(TTGG)_n$ macroconformation but may also crystallize in $(T)_n$ which does not differ much in energy from $(TTGG)_n$.

Poly(vinyl alcohol), $\text{+CH}_2\text{–CH(OH)+}_n$, carries a hydroxyl group at every second chain atom. Adjacent OH groups can form intramolecular hydrogen bridges which causes crystalline isotactic poly(vinyl alcohol) to exist in the all-trans macroconformation and not as a helix like other isotactic polymers, $\text{+CH}_2\text{–CHR+}_n$. Syndiotactic poly(vinyl alcohol), on the other hand, forms a helix and not a zigzag chain.

The smaller the number of nonbonded ligands, the lower is the rotational barrier; an example is $H_3C–CH_3$ ($\Delta E^{\ddagger}_{m}$ = 11.3 kJ/mol) versus $–CH_2–COCH_2–$ ($\Delta E^{\ddagger}_{m}$ = 3.4 kJ/mol) versus $H_3C–OH$ ($\Delta E^{\ddagger}_{m}$ =1.6 kJ/mol) (see Volume I, Table 5-2). Molecules containing oxygen chain atoms are thus more flexible than those with just methylene groups.

However, the C–O bond is shorter than the C–C bond (0.144 nm versus 0.154 nm). Substituents R in polymers +O–CHR+_n are thus nearer to each other than they are in polymers $\text{+CH}_2\text{–CHR+}_n$, which forces helices to widen. As a result, isotactic poly(acetaldehyde), $\text{+O–CH(CH}_3\text{)+}_n$, crystallizes as a 4_1 helix whereas its counterpart it-poly(propylene), $\text{+CH}_2\text{–CH(CH}_3\text{)+}_n$, forms a 3_1 helix.

Effects of bond orientation are especially strong in unsubstituted heterochains. Poly(oxymethylene), $\text{+O–CH}_2\text{+}_n$, crystallizes in the gauche conformation, $(G)_n$, but poly(oxyethylene), $\text{+O–CH}_2\text{–CH}_2\text{+}_n$, as $(TTG)_n$; both $(G)_n$ and $(TTG)_n$ are helices. Poly(glycine) II, $\text{+NH–CO–CH}_2\text{+}_n$, forms a 7_2 helix like poly(oxyethylene) but its helix is deformed because of hydrogen bonds between amide groups. In it-poly(propylene oxide), $\text{+O–CH}_2\text{–CH(CH}_3\text{)+}_n$, methyl groups repel each other which decreases bond orientations: this polymer crystallizes in an all-trans conformation.

Strong electrostatic effects may also lead to anti (A) and cis (C) conformations in addition to T, G^+, and G^-. For example, $(A^-TA^+T)_n$ is the macroconformation with the lowest energy for *trans*-1,4-poly(butadiene), $\text{+CH}_2\text{–CH=CH–CH}_2\text{+}_n$. Chains of poly(dimethylsiloxane), $\text{+O–Si(CH}_3\text{)}_2\text{+}_n$, crystallize in the macroconformation $(CT)_n$; the energy of the macroconformation $(T)_n$ is higher by 13.3 kJ/mol.

Polymer molecules composed of a single type of *chiral configurational repeating unit* can form both left-handed and right-handed helices. These two types of helices are diastereomers, however, and thus not energetically equivalent. As a result, one handedness is always preferred by helices with chiral repeating units. For example, poly([*S*]-1-olefin)s and poly(D-saccharide)s always form left-handed helices and their antipodes always right-handed ones. All natural deoxyribonucleic acids, except the Z modification, and most poly(L-α-amino acid)s exist in right-handed helices.

Macromolecules from monomeric antipodes deliver helical polymers with opposite handednesses. If L-monomeric units lead to right-handed helices, D-monomeric units will form left-handed ones and *vice versa*. These antipodes are energetically equivalent and therefore have the same spectral properties. For example, amide-I bands of *crystalline* poly(L-α-amino acid)s and poly(D-α-amino acid)s are always at the same wavelength as are the corresponding amide-II bands (Table 3-2). Copolymers from D- and L-monomeric units have other conformations and thus different amide bands.

Table 3-2 Amide bands and handedness of helices of L-, D-, and D,L-poly(α–amino acid)s [3].

Poly(α-amino acid)	Handedness		Amid I band in σ/cm^{-1}			Amid II band in σ/cm^{-1}		
	L	D	L	D	D,L	L	D	D,L
Poly(methionine)	right	left	1650	1650	1657	1544	1544	1548
Poly(γ–benzylglutamate)	right	left	1650	1650	1662	1546	1546	1555
Poly(tyrosine)	right	left	1659	1659	1663	1546	1546	1550
Poly(β–methylaspartate)	left	right	1666	1666	-	1550	1550	-
Poly(β–ethylaspartate)	right	left	1659	1659	-	1548	1548	-
Poly(β–benzylaspartate)	left	right	1668	1668	-	1550	1550	-

3.4 Optical Activity

3.4.1 Fundamentals

In crystalline polymers, handednesses of chiral structures (and thus also of helices) can be determined by X-ray structural analysis. In solutions, a distinction between enantiomers is possible only if the probe interacts differently with either of the enantiomers. The probe must therefore also have a handedness. If the probe is another chemical compound, then sample and probe will form diastereomers that are energetically different and can thus be separated.

The probe may also be circularly polarized light that will interact differently with various electron configurations of otherwise chemically identical molecules. Molecules with only one type of stereogenic center will turn the plane of polarized light; they are **optically active**. In polymer molecules, the effect of chiral monomeric units is further reinforced if the units reside in helical structures which are chiral themselves.

Circular Dichroism

Linearly polarized light can be visualized as an overlay of two circularly polarized waves of opposite rotations but equal amplitudes. Before entering an optically active substance, electrical vectors of these two circularly polarized waves will form circles on the x-y plane; these circles have equal diameters because the two amplitudes are of equal magnitude. However, left and right circularly polarized light is absorbed differently on passing through an optically active substance because of the asymmetric electron configuration of the molecules of the latter. The radii of the two circles, and thus the two amplitudes of left and right circularly polarized light, are no longer the same.

Combination of the different amplitudes of the two circularly polarized light waves produces elliptically polarized light. The inverse tangent of the ratio of small and large axes of this ellipse is defined as **ellipticity** Θ, a quantity that depends on the absorptions A_L and A_D of left and right circularly polarized light.

The **molar ellipticity**, $[\Theta]$, relates the ellipticity to the length L of the path through the medium (e.g., the length of a cuvette) as well as to the molar concentration, $[M_u]$, of monomeric units (e.g., in mole per liter):

$$(3\text{-}1) \qquad [\Theta] = \frac{\Theta}{L[\mathrm{M_u}]} = \frac{2.303\,(A_L - A_R)\cdot 180°}{4\,\pi\,L[\mathrm{M_u}]}$$

The wavelength dependence of ellipticity is called **circular dichroism** (**CD**). CD spectra are measured over the same range of wavelengths as absorption spectra. Both spectra have the same shape since ellipticities are controlled by the difference of absorptions (Eq.(3-1)). However, molar ellipticities may be positive (**positive Cotton effect**) or negative (**negative Cotton effect**) (Fig. 3-12).

The different absorption of the two circularly polarized components of light also leads to different refractive indices, n_L and n_D. One component of the light thus moves faster through the medium than the other. The resulting phase shift is proportional to the difference, $n_L - n_D$, the so-called **circular birefringence**.

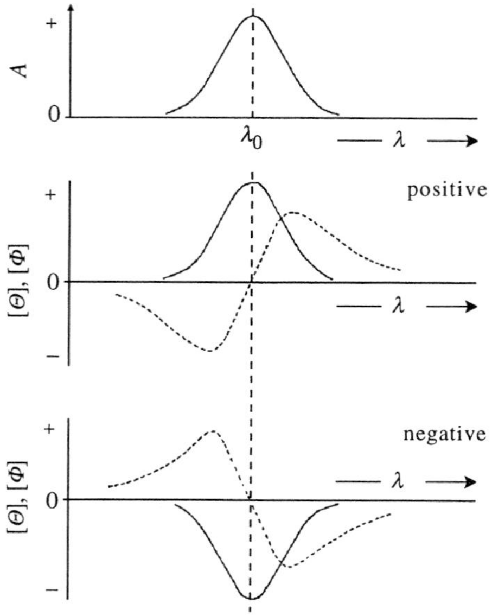

Fig. 3-12 Wavelength dependence of absorption A, ellipticity [*Θ*], and molar optical rotation [*Φ*] of substances with positive or negative Cotton effects. Top: absorption; center and bottom: — circular dichroism (ellipticity), - - - optical rotatory dispersion (ORD).

Optical Activity

The phase shift on combining the two circularly polarized components causes the main axis of the ellipse to become no longer parallel to the direction of polarization of the entering polarized light. If the substance does not absorb light, then the axial ratio of the ellipse is so small that the exciting light can be considered **linearly polarized**. One can then say that the *plane* of polarized light has rotated.

The observed **optical rotation**, α, for matter in a cuvette of length L at light of wavelength, λ_0, in vacuum, is directly proportional to the difference of refractive indices:

(3-2) $$\alpha = 180° \, (n_L - n_R)(L/\lambda_0)$$

Experimental results are often related to the mass concentration c and expressed as **specific optical rotation** (or **specific activity**), using various physical units,

(3-3) $$[\alpha] = \alpha/(Lc)$$

as **molar optical rotation** (for example, in degree cm^2 mol^{-1}),

(3-4) $$[\Phi] = \alpha/(L[\mathrm{M_u}]) = \alpha M_u/(Lc)$$

or, related to the refractive index, n, and the three spatial directions, as **effective molar optical rotation** (for example, in degree cm^2 mol^{-1}),

(3-5) $$[\Phi]_{\mathrm{eff}} = 3\,[\Phi]/(n^2 + 2)$$

where M_u = molar mass of monomeric unit, $[M_u]$ = molar concentration of monomeric units, and c = mass concentration of polymer. Note that literature uses various physical units and numeric constants for L, c, $[M_u]$, and M_u.

Optical Rotatory Dispersion

The wavelength dependence of optical rotation is called **optical rotatory dispersion (ORD)**. Since optical rotations are zero at the maximum of absorbance (Fig. 3-12), measurements of ORD have been mostly replaced by those of circular dichroism.

However, optical rotations are also observed outside of absorption bands. The dependence of specific rotation, $[\alpha]$, on wavelength, λ, can often be described by the empirical **Drude equation** in which k_i = **rotation constants** and λ_i = **dispersion constants**:

(3-6) $$[\alpha] = \sum_i \frac{k_i}{\lambda^2 - \lambda_i^2} \quad ; \quad i = 1, 2 \ldots$$

In many cases, experimental results can be described by a one-term Drude equation (i = 1). The constants k_1 and λ_1 are then obtained from plots of either $1/[\alpha] = f(\lambda^2)$ or $[\alpha]\lambda^2 = f([\alpha])$.

The one-term Drude equation describes fairly well the wavelength dependence of the specific rotation of dissolved coil molecules. For helical molecules, the two-term **Moffitt-Yang equation** is a better choice. In this equation, the first term on the right side indicates the configurational contribution of stereogenic centers and the second term the conformational contribution of the helix;

(3-7) $$[\Phi]_{eff} = \frac{3\,M_u}{n^2 + 2}[\alpha] = a_o\left(\frac{\lambda_o^2}{\lambda^2 - \lambda_0^2}\right) + b_o\left(\frac{\lambda_o^2}{\lambda^2 - \lambda_0^2}\right)^2$$

3.4.2 Structural Dependence

All optically active systems are chiral but the converse is not necessarily true. Optical activity depends not only on the chemical structure (constitution and configuration) but also on the physical structure (microconformation and macroconformation) which in turn is a function of solvent structure, polymer concentration, and temperature.

Observed effects are often difficult to reconcile with a structure because these effects always rely on differences. Dilute solutions of malic acid are left-rotating but concentrated solutions are right-rotating because of the increasing self-association of malic acid. The type of rotation may also depend on the temperature: in $CHCl_3$, the helix of poly(α-propyl-L-aspartate), $+\!NH–CH(CH_2COOC_3H_7)–CO\!+_n$, is right-handed at 30°C but left-handed at 60°C.

Poly(α-amino acid)s and Proteins

The optical activity of a stereogenic center is determined mainly by its nearest neighbors. Endgroups of polymers therefore affect the optical activity of non-self-associating,

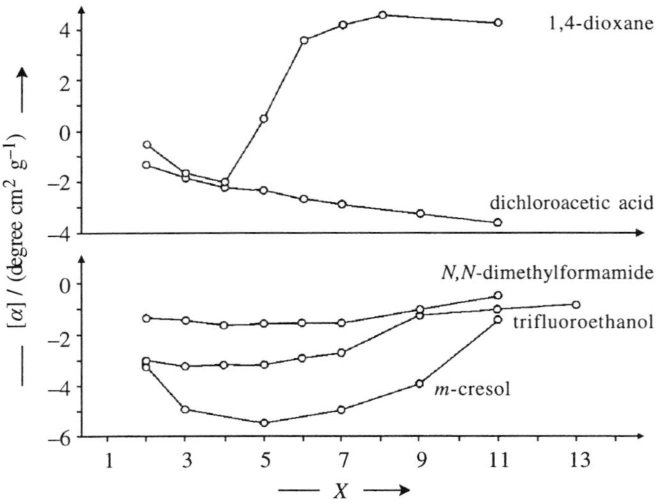

Fig. 3-13 Specific rotation [α] (D line, 25°C) at constant concentration c as a function of the degree of polymerization, X, of oligo(γ-methyl-L-glutamate)s in various solvents [4, 5].

Concentrations in m-cresol: 0.01 g/mL ($2 \leq X \leq 5$) or c = 0.0050 g/mL ($X > 5$); in dichloroacetic acid: 0.02 g/mL; in 1,4-dioxane: 0.02 g/mL; in N,N-dimethylformamide: 0.01 g/mL; and in trifluoroethanol: (0.001-0.01) mol/L.

coil-forming polymer molecules only at low degrees of polymerization. An example is oligo(γ-methyl-L-glutamate), $H[NH–CH(CH_2CH_2COOCH_3)–CO]_nOH$, in dichloroacetic acid where the specific rotation decreases with increasing degree of polymerization, X, and seems to approach a constant value at high degrees of polymerization (Fig. 3-13).

For solutions in 1,4-dioxane, specific rotations decrease from dimer to tetramer but start to increase dramatically at X = 5. The reason is the onset of formation of α-helices with 3.6 peptide units per turn. Such a helix must be stabilized by at least 2 hydrogen bonds which requires 7 peptide units, hence the observed leveling off at $X \approx 7$.

With N,N-dimethylformamide and trifluoroethanol as solvents, weak minima are observed for [α] = f(X) whereas m-cresol leads to a strong minimum. In each case, specific rotations remain negative up to X = 11-13, indicating an absence of helices.

At fairly low degrees of polymerization, practically all contributions to specific activities stem from the configuration of stereogenic centers as expressed by the a_o value of the Moffitt-Yang equation. At higher degrees of polymerization, conformational contributions by helices become important as shown by the decrease in the Moffitt-Yang b_o values with higher degrees of polymerization which seem to approach a constant value b_∞ at high degrees of polymerization (Fig. 3-14).

In general, both b_o and b_∞ are somewhat dependent on the solvent (via refractive index, see Eq.(3-7)); they also vary slightly with temperature. For poly(α-amino acid)s, $\text{+}NH–CHR–CO\text{+}_n$, a b_o = –6500 degree cm^2 mol^{-1} was found to be independent of the chemical structure of α-amino acid units, the type of solvent, and the temperature. From this datum, it was concluded that the protein serum albumin must be 46 % helical since its b_o value is –2900 degree cm^2 mol^{-1}. In general, helix contents of proteins by circular dichroism or optical rotatory dispersion of solutions agree fairly well with those from X-ray measurements in the crystalline state (Table 3-3).

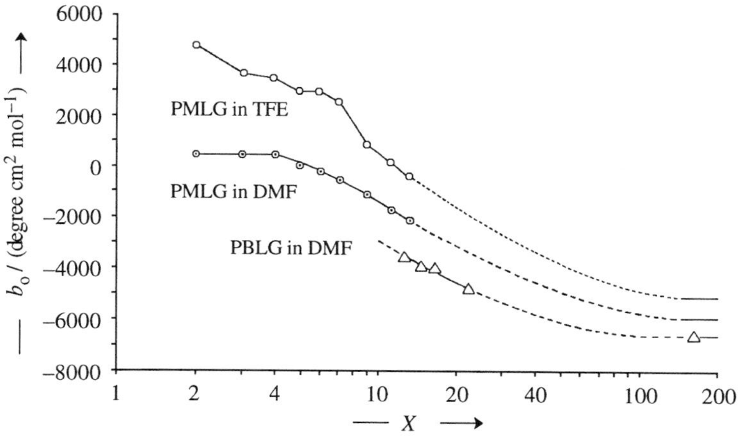

Fig. 3-14 Dependence of the conformational constant b_o on the degree of polymerization, X, of poly-(γ-methyl-L-glutamate)s (PMLG) in 2,2,2-trifluoroethanol (TFE) [4] and of poly(γ-benzyl-L-glutamate) (PBLG) in N,N-dimethylformamide (DMF) at 25°C [6]. b_o values in 2,2,2-trifluoroethanol were shifted upwards by 3000 degree cm^2 mol^{-1}. ------ Data for PMLG of unknown, but high, degree of polymerization; - - - assumed dependence.

Estimates of helix contents allow one to study the effect of solvents on macroconformations of dissolved proteins by comparing them with the percentage of helix structures from X-ray studies of crystalline proteins (Table 3-3). Both methods assume a "3-state" model, i.e., regions of helical, pleated sheet, and coil structures that are distinct and sharply separated from each other.

The assumption of distinct macroconformational "states" is supported by observations of helix-coil transitions (Section 3.5.21). However, calculations of helix contents from b_o data are not without problems since (a) short helical sections do not fully contribute to b_o, (b) L-α-amino acid units may also reside in left-handed helices, which changes the sign of b_o, and (c) mixtures of right-handed and left-handed helices may exist.

Poly(1-olefin)s

Like proteins, molar optical rotations of dissolved optically active poly(1-olefin)s are little effected by the type of solvent. In contrast to poly(α-amino acid)s and proteins, their effective molar optical rotation can be described by the 1-term Drude equation (Table 3-4).

Table 3-3 Percentage of macroconformations in proteins according to various methods.

Macroconformation	Percentage of macroconformations in dilute salt solution				
	Carboxypeptidase		α–Chymotrypsin		
	X-ray	CD	X-ray	CD	ORD
Helix	23	26	8	20	15
Pleated sheet	18	18	22	20	85
Random coil	59	56	70	60	

Table 3-4 Constants λ_1 and k_1 of the one-term Drude equation, Eq.(3-6), for solutions of various poly(1-olefin)s, $+CH_2-CHR+_n$, and their hydrogenated monomers as model compounds. Polymerization of monomers by anions (A), cations (C), free radicals (R), or Ziegler catalysts (Z).

Type	Monomer Name	Substituent R	λ_1/nm Model	λ_1/nm Polymer	k_1/(degree cm^2 mol^{-1}) Model	k_1/(degree cm^2 mol^{-1}) Polymer
Z	[*S*]-3-Methyl-1-pentene	$CH(CH_3)(C_2H_5)$	176	167	–113	1143
Z	[*S*]-4-Methyl-1-hexene	$CH_2CH(CH_3)(C_2H_5)$	170	165	3078	104
K	[1*R*,3*R*,4*S*]-1-Methyl-4-isopropylcyclohex-3-yl-vinyl ether	$OC_6H_3(CH_3)(CH(CH_3)_2)$	155	165	–1144	–2169
A	[(–)-*N*-Propyl-*N*-α-phenyl-ethyl]acrylamide	$CON(C_3H_7)(CH(CH_3)(C_6H_5))$	155	165	–1144	–2169
R	{[*S*]-2-Methylbutyl}-methacrylate	$COOCH_2CH(CH_3)(C_2H_5)$	191	188	59	53

The optical rotations of poly(1-olefin)s furthermore decrease with increasing temperature which is interpreted as "melting" of relatively long left-handed helical sections. According to the same model calculations, lengths of relatively short right-handed helical sections should not change appreciably with temperature.

Both observations indicate that molecules of these poly(1-olefin)s are not present as 100 % helix structures but as random coils composed of helical and non-helical segments (see also Fig. 4-1, III). The average number of monomeric units in helical conformations can be estimated from the conformational energy. This energy, ΔE, is given by one-half of the Gibbs energy of the "reaction" LL + DD $\rightleftarrows$ LD + DL between a left-handed conformational diad LL and a right-handed conformational diad DD where E_{LL}, E_{DD}, E_{LD}, and E_{DL} are the corresponding conformational energies:

$$(3\text{-}8) \qquad \Delta E = -(1/2)(E_{LL} + E_{DD}) + (1/2)(E_{LD} + E_{DL})$$

Chains with achiral configurational monomeric units have equal energies, $E_{LD} = E_{DL}$, whereas $E_{DL} \neq E_{LD}$ for those with chiral main chain or side chain units. The average number of monomeric units in helical sequences is

$$(3\text{-}9) \qquad N_{hc} = \frac{1 + \exp(-\Delta E/RT)}{\exp(-\Delta E/RT)}$$

According to these calculations, tactic poly(1-olefin)s exist in hydrocarbon solutions as partial helices (see also Fig. 4-1, III), which is supported by IR spectroscopy (see Fig. 4-2). *it*-poly([*S*]-4-methyl-1-hexene), $+CH_2-CH(CH_2CH(CH_3)(CH_2CH_3)+_n$, contains on average 31 monomeric units in left-handed helices whereas right-handed helices are on average only 2.2 units long. In *it*-poly(4-methyl-1-pentene) with the achiral monomeric unit $-CH_2-CH(CH_2CH(CH_3))-$, on the other hand, both left-handed and right-handed helices are of equal length, ca. 12 monomeric units. The sequence length does not vary with the polymer concentration in solutions of non-self-associating polymers because of the dominance of intramolecular polymer effects and polymer-solvent interactions.

Copolymers

Copolymers composed of alternating monomeric units with opposite chirality are optically inactive because the chirality of units is compensated intramolecularly. However, such copolymers may form helices with a single handedness in certain helicogenic solvents. An example is poly(L-*alt*-D-leucine) which exists as an optically active π-helix in its solution in benzene.

The molar optical rotations of poly([*S*],[*R*])-1-olefin)s are usually hyperbolic functions of the optical purity of their monomeric units (Fig. 3-15). They are thus larger than indicated by the additivity rule. It is not clear whether this effect is caused by long tactic blocks of monomeric units, by mixtures of [*S*]- and [*R*]-polymers, or by helical chiral units or sequences that force non-chiral units to adopt a helical structure.

Polymers from achiral monomeric units are never optically active even if their chains adopt a helical macroconformation which makes them chiral. Since these chains are enantiomers, left- and right-handed helices exist in equal proportions and the optical activity of the *polymer* is zero. Sometimes, such racemates can be separated chromatographically on chiral columns.

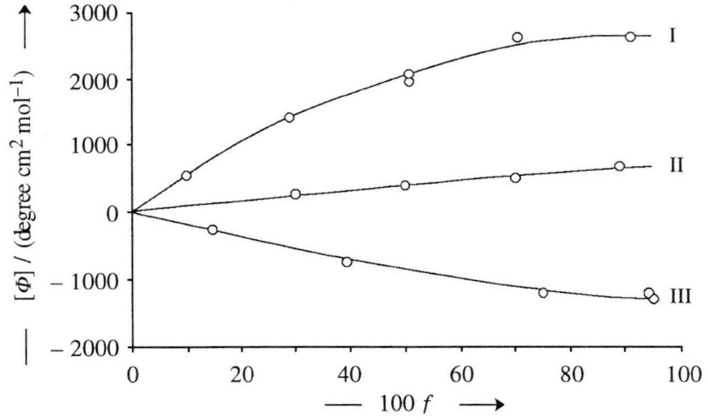

Fig. 3-15 Molar optical rotations of poly(1-olefin)s, $-[CH_2-CHR^*]_n-$, with chiral substituents as a function of the optical purity *f* of their monomeric units [7]. I and II: [*S*]; III: [*R*], $R^* = CH_2CH(CH_3)(C_2H_5)$ (I), $(CH_2)_2CH(CH_3)(C_2H_5)$ (II), or $CH(CH_3)(CH_2)_2CH(CH_3)_2$ (III).

3.5 Transformation of Macroconformations

3.5.1 Phenomena

Microconformations, and thus also macroconformations, are affected by the environment such as temperature, pressure, solute-solvent interactions, and even the physical structure of the solvent itself. In random distributions of microconformations, environmental changes cause only small changes in macroconformations and therefore also of molecule dimensions, optical activities, etc. However, very large changes can occur in macromolecules with very long regular sequences of microconformations (Fig. 3-16); these changes are very pronounced at high degrees of polymerization (Section 3.5.2).

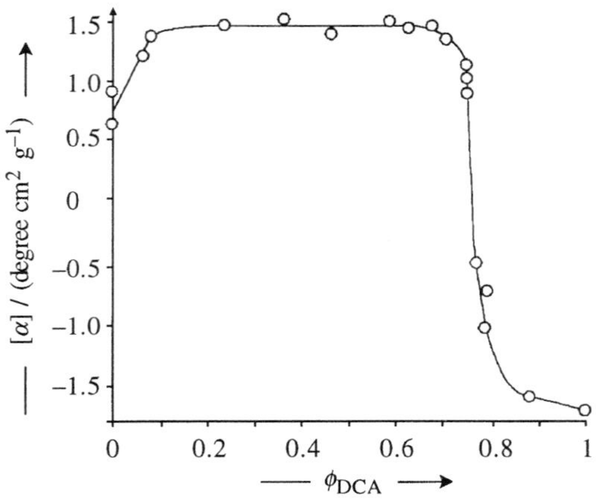

Fig. 3-16 Specific rotation of a poly(γ-benzyl-L-glutamate) ($\overline{M}_w$ = 350 000 g/mol) as a function of the volume fraction ϕ_{DCA} of dichloroacetic acid in mixtures of ethylene dichloride and dichloroacetic acid at 20°C [8]. By permission of Elsevier Science, Oxford, England.

For example, increasing concentrations of dichloroacetic acid (DCA) in mixtures of ethylene dichloride + dichloroacetic acid as a solvent cause the specific rotation of poly-(γ-benzyl-L-glutamate) (PBLG), $+\!\!\!\!-$NH-CH($CH_2CH_2COOCH_2C_6H_5$)-CO$-\!\!\!\!+_n$, to increase and then become constant before it drastically drops to negative values at DCA contents of ca. 75 %. Since ethylene dichloride is a helicogenic solvent for PBLG, the small initial increase of [α] is ascribed to a change of helix structure (expansion?) and the strong later decrease to a helix-coil transition.

3.5.2 Thermodynamics

Optical activities vary slightly with temperature for low-molecular weight polymers but sharply for high-molecular weight ones (Fig. 3-17). These changes must be cooperative effects since each microconformation in a conformational homosequence is affected by its two neighbors. The conformational change itself may be trans/gauche as in poly(1-olefin)s or polypeptides (Fig. 3-16) or cis/trans as in poly(proline).

An all-A macroconformation can be transformed to an all-B macroconformation in a series of consecutive equilibrium reactions, for example, for a sequence of four consecutive intramolecular chain units:

(3-10)

$$...AAAA... \overset{K_i}{\rightleftharpoons} ...BAAA... \overset{K}{\rightleftharpoons} ...BBAA... \overset{K}{\rightleftharpoons} ...BBBA... \overset{K}{\rightleftharpoons} ...BBBB...$$

In the simplest case, an initiation (formation of a nucleus) with an equilibrium constant K_i is followed by three steps with the same equilibrium constant, K. The ratio, $\sigma = K_i/K$, measures the **cooperativity** of the conformational transformation. At $\sigma < 1$ ($K > K_i$), segments prefer the conformation of their neighbors: diads AA and BB are more probable than diads AB and BA (**positive cooperativity**). No cooperativity is present at $\sigma = 1$, i.e., $K = K_i$. **Anticooperativities** (= **negative cooperativities**), $\sigma > 1$, are not known.

Thermodynamically controlled nucleations are microscopically reversible. Since $K_i = \sigma K$ for ...AAA... ⇄ ...ABA..., the equilibrium constant for ..BBB... ⇄ ...BAB... must be $K_i' = \sigma K^{-1}$. In both cases, the microconformation to be transformed has two identical neighbors. However, terminal microconformations have only one neighbor, and σ values at chain ends, σ_{end}, must therefore by different from those within the chain, σ_{int}. Co-operativities σ may also depend on the type of microconformation, i.e., A or B. In many cases, $\sigma_{end} \approx \sigma_{int}$ is found, however.

Equilibrium constants K of growth are usually assumed to be identical but different from that of initiation:

$$(3\text{-}11) \qquad K_i \equiv \frac{[\mathrm{BAAA}]}{[\mathrm{AAAA}]} \neq \frac{[\mathrm{BBAA}]}{[\mathrm{BAAA}]} = \frac{[\mathrm{BBBA}]}{[\mathrm{BBAA}]} = \frac{[\mathrm{BBBB}]}{[\mathrm{BBBA}]} \equiv K$$

The equilibrium concentration of ...BBBB... is therefore

$$(3\text{-}12) \qquad [\mathrm{BBBB}] = K_i K^3 [\mathrm{AAAA}] = \sigma K^4 [\mathrm{AAAA}]$$

At $\sigma K^4 = 1$, ...AAAA... and ...BBBB... are present in equal concentrations. If, in addition, $K >> 1$, then $1/\sigma^{1/4}$ must be much greater than 1 and $\sigma << 1$. The intermediates ...BAAA..., ...BBAA..., and ...BBBA... are therefore present in much smaller concentrations than the homosequences ...AAAA...and ...BBBB... A $\sigma = 10^{-4}$ would lead to [BBBB] = 10 [BBBA] = 100 [BBAA] = 1000 [BAAA]. Hence, a helix is converted almost completely or not at all.

For thermodynamically controlled helix transformations, the decisive quantity is the product σK^N. Depending on the number N of microconformations, various expressions are obtained for the fraction f_B of microconformations B.

For very short chains, chain lengths and sequence lengths of homoconformations are identical. In an all-or-none process, the conversion of *one* microconformation is given by $\sigma^{1/N} << 1$ (see above). If all microconformations are converted with equal probability, then the conversion *per chain* is given by $N\sigma^{1/N} << 1$ and the fraction f_B of the newly formed B-type microconformations is given by

$$(3\text{-}13) \qquad f_B = \frac{\sigma K^N}{1 + \sigma K^N}$$

For very long chains with helical and non-helical sequences, sequence lengths of homosequences become independent of the degree of polymerization. The degree of conversion is then

$$(3\text{-}14) \qquad f_B = \frac{1}{2} + \frac{K-1}{2[(K-1) + 4\sigma K]^{1/2}}$$

Thus, the equilibrium constant at the center of conversion ($f_B = 1/2$) is independent of σ and always $K = 1$. The conversion is sharper, the smaller the value of σ.

Complicated equations apply to $f_B = f(\sigma, K)$ if the degree of polymerization is neither very high nor very low. In this case, the sharpness of the transition increases with increasing degree of polymerization (Fig. 3-17).

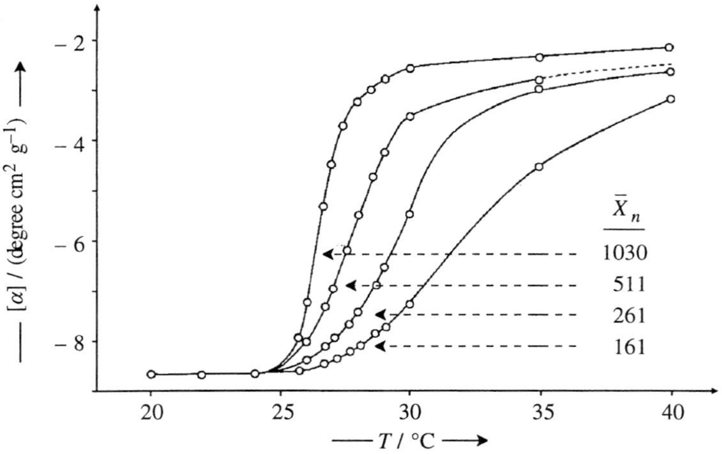

Fig. 3-17 Temperature dependence of the specific optical rotation of poly(ε-carbobenzyloxy-L-lysine)s with different number-average degrees of polymerization in *m*-cresol at a wavelength of 436 nm [9].

σ thus measures the effects of the ends of helical sequences since the monomeric units here are in a different environment from that of those in the center. For proteins and poly(α-amino acid)s, σ is very small, ca. $3 \cdot 10^{-3}$ to $1 \cdot 10^{-5}$ (Table 3-5). Since ends of helices are not preferred, two separate helical sequences try to unite. In a chain section ~h_4–n_i–h_3~, the two helical segments h_4 and h_3 thus try to form one helical segment h_7, ~h_7–n_i~, etc.

Equilibrium constants indicate whether a monomeric unit prefers to be in a helix ($K > 1$) or not ($K < 1$). In proteins, α-amino acid units from proline, serine, glycine, and asparagine are helix breakers whereas units from lysine, tyrosine, aspartic acid, threonine, arginine, cysteine, and phenylalanine are indifferent. All other α-amino acid units prefer to be in helical conformations (see Volume I).

3.5.3 Kinetics

The kinetics of conformational transformations has been relatively little explored. However, studies of the helix-coil transition of poly(α-amino acid)s and polynucleotides have revealed high rate constants of 10^6 s^{-1} to 10^7 s^{-1}. These high values are caused most likely by the high cooperativity of such chains.

Table 3-5 Thermodynamic parameters K and σ of helix/coil transformations of monomeric units of poly(α-amino acid)s and poly(nucleotide)s.

Monomer	T/°C	K	$10^5\,\sigma$	Monomer	T/°C	K	$10^5\,\sigma$
Glycine	60	0.63	1	L-Phenylalanine		1.00	180
L-Serine	60	0.74	8	L-Leucine	60	1.09	330
L-Alanine	0	0.96					
L-Alanine	60	1.01	80	Adenine/thymine (1:1)		0.5	10
L-Alanine	80	0.99		Guanine/cytosine (1:1)		2.0	10

Much smaller values of 10^{-6} s^{-1} to 1 s^{-1} were observed for the denaturation of proteins which includes helix-coil transitions. Hence, these low values must result from conformational transitions in non-helical regions of protein molecules.

Obviously, a rotation around a single chain bond in non-helical regions requires a movement of a large part of the rest of the molecule (Fig. 3-18, I) which is very difficult in viscous media. A coupled rotation around two bonds, similar to the movement of a skipping rope (Fig. 3-18, II), would be expected to have a higher activation energy than the same movement in a similar low-molecular weight compound that is not burdened by a long tail.

Fig. 3-18 Rotational transitions at one bond (I) and at two bonds (II).

The problem was studied with diacetylpiperazine (III) as a model for piperazine polymers (IV). These compounds were chosen because their peptide bonds –N–CO– are partial double bonds. The relatively slow rotation around the peptide bonds can be studied by proton magnetic resonance spectroscopy because different absorption bands are observed for neighboring groups in the cis or trans position to the peptide bonds. The temperature dependence of band intensities delivers the Gibbs activation energy, $\Delta G^{\ddagger}$, of the process.

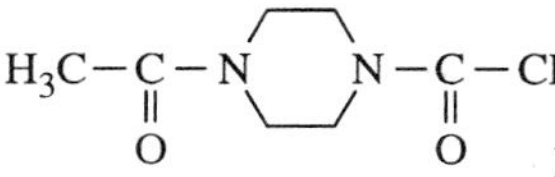
III

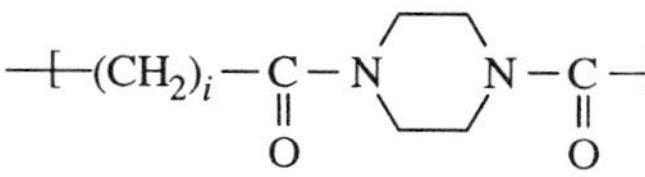
IV

Experimentally, the same activation energy of $\Delta G^{\ddagger}$ = 76 kJ/mol was found for both the model compound III and polymers IV with i = 2, 4, or 8 methylene groups. Two explanations of this finding have been advanced: either a counterrotation of neighboring groups relieves the stress caused by the rotation or the activation energy is stored in the polymer molecule so that it can be used for a rotation around another bond.

Historical Notes to Chapter 3

For the history of polymer stereochemistry and the discovery of helical structures, see Volume I, Chapter 5, Historical Notes.

An excellent review of the discovery of optical activity and the development of theories and terminologies can be found in the paper by J.K.O'Loane, Optical Activity in Small Molecules, Non-enantiomorphous Crystals, and Nematic Liquid Crystals, Chem.Rev. **30** (1980) 41-61.

Literature to Chapter 3

3.0 NOMENCLATURE

IUPAC Commission on Nomenclature of Organic Chemistry, 1974 Recommendations, Section E, Fundamental Stereochemistry, Pure Appl. Chem. **45** (1976) 11

International Union of Pure and Applied Chemistry, Macromolecular Division, Commission on Macromolecular Nomenclature, Compendium of Macromolecular Nomenclature, Blackwell Scientific, Oxford 1991

3.2 LOCAL CONFORMATIONS (incl. GENERAL STEREOCHEMISTRY)

S.Mizushima, Structure of Molecules and Internal Rotation, Academic Press, New York 1956

E.L.Eliel, Stereochemistry of Carbon Compounds, McGraw-Hill, New York 1962

M.Hanack, Conformation Theory, Academic Press, New York 1965

W.Orville-Thomas, Ed., Internal Rotation in Molecules, Wiley, New York 1974

E.L.Eliel, N.L.Allinger, S.J.Angyal, G.A.Morrison, Conformational Analysis, Am.Chem.Soc., Washington (D.C.) 1981

3.3 SEQUENCES OF MICROCONFORMATIONS

F.A.Bovey, Chain Structure and Conformation of Macromolecules, Academic Press, New York 1982

3.4 OPTICAL ACTIVITY

C.Djerassi, Optical Rotatory Dispersion, McGraw-Hill, New York 1960

B.Jirgensons, Optical Rotatory Dispersion of Proteins and Other Macromolecules, Springer, Berlin 1969

P.Pino, F.Ciardelli, M.Zandomeneghi, Optical Activity in Stereoregular Synthetic Polymers, Ann.Rev.Phys.Chem. **21** (1970) 561

P.Crabbé, ORD and CD in Chemistry and Biochemistry, Academic Press, New York 1972

3.5 TRANSFORMATION OF MACROCONFORMATIONS

D.Poland, H.A.Scheraga, Theory of Helix-Coil Transitions in Biopolymers–Statistical Mechanical Theory of Order-Disorder Transitions in Biological Macromolecules, Academic Press, New York 1970

C.Sadron, Ed., Dynamic Aspects of Conformation Changes in Biological Macromolecules, Reidel, Dordrecht, Netherlands 1973

R.Cerf, Cooperative Conformational Kinetics of Synthetic and Biological Chain Molecules, Adv.Chem.Phys. **33** (1975) 73

A.Teramoto, H.Fujita, Conformation-Dependent Properties of Synthetic Polypeptides in the Helix-Coil Transition Region, Adv.Polym.Sci. **18** (1975) 65

A.Teramoto, H.Fujita, Statistical Thermodynamic Analysis of Helix-Coil Transitions in Polypeptides, J.Macromol.Sci.-Rev.Macromol.Chem. **C 15** (1976) 165

References to Chapter 3

[1] K.Matsuzaki, F.Kawazu, T.Kanai, Makromol.Chem. **183** (1982) 185, Fig. 5

[2] S.-I-Mizushima, T.Shimanouchi, J.Am.Chem.Soc. **86** (1964) 3521, Figs. 2 and 4

[3] H.Yamamoto, K.Inouye, T.Hayakawa, Polymer **18** (1977) 1288, data of Table 1

[4] M.Goodman, I.Listowsky, Y.Masuda, F.Boardman, Biopolymers **1** (1963) 33, Table I

[5] M.Goodman, E.E.Schmidt, D.A.Yphantis, J.Am.Chem.Soc. **84** (1962) 1288, Table I

[6] P.Rohrer, H.-G.Elias, Makromol.Chem. **151** (1972) 281, Tables 1 and 3

[7] P.Pino, F.Ciardelli, G.Montagnoli, O.Pieroni, J.Polym.Sci. **B** (Polym.Lett.) **5** (1967) 307, data of Table 1

[8] J.T.Yang, Tetrahedron **13** (1961) 143, Table 5

[9] M.Matsuoka, T.Norisuye, A.Teramoto, H.Fujita, Biopolymers **12** (1973) 1515, data of Fig. 7

4 Macroconformations

4.1 Overview

4.1.1 Introduction

The shape of an isolated macromolecule is determined by its **molecular conformation** (= **macroconformation**) which in turn depends on the type, proportion, and sequence of microconformations within the molecule. Microconformations are controlled by the chemical constitution and configuration (Chapter 3) and also by the environment since shapes of single macromolecules can be studied only in dilute solution. Even for the simplest chemical structures, it is not possible to determine experimentally sequences of microconformations.

Proportions of the various types of microconformations are obtained by spectroscopy, radii of gyration by scattering methods (Chapter 5), and volumes of molecules by hydrodynamic methods (Chapters 11 and 12); sequence lengths of microconformations are only accessible for optically active macromolecules (p. 58). All methods deliver time-averaged shapes or volumes of molecules in substances, not individual molecules. An additional complication is the non-uniformity of polymers with respect to constitution, configuration, and/or molecular weight.

Two large groups of macromolecules can be distinguished with respect to their molecular shapes and sizes. One group comprises more or less compact macromolecules such as spheroidal proteins or rodlike nucleic acids (Fig. 2-2). Dimensions of these molecules are controlled by their constitution, molecular weight, and shape (Section 4.2). Their behavior in solution follows directly from their shapes and sizes.

The situation is quite different for linear chains, star molecules, dendrimers, hyperbranched polymers, etc. (Fig. 2-1). Here, a great many different theoretical approaches exist for the simplest structure, the linear chain, (Sections 4.3-4.5) and far fewer ones for other types of molecules (Sections 4.6-4.7). There are also few generally accepted theoretical approaches to the dependence of molecular shapes on chemical structure.

Even for simple linear chains, complex mathematical expressions are obtained if *chemical* approaches are used that are based on chemical structure (number and volume of chain atoms, lengths of chain bonds, valence angles, rotational potentials, size and interactions of substituents) (Section 4.3.3). *Physical* approaches omit practically all chemical details and concentrate on a few essential structural parameters such as the degree of polymerization. In the simplest models, chains are assumed to be infinitely thin but not infinitely flexible. These chains cannot bend back 180° but retain a certain stiffness. i.e., a *persistence* (p. 96).

This type of approach to a many-facetted system is called **coarse-graining**. A finely grained model would show many details, a coarse-grained one, only a few. For example, a map of a country on the scale 1: 20 000 000 would show major rivers and cities whereas one on the scale 1: 100 000 would also indicate creeks and villages. The former map is course-grained and the latter, fine-grained. Similarly, a *granular material* has no continuous distribution of matter. In this sense, molecules are definitely granular. In coarse-graining of polymer molecules, one would, for example, replace space-filling models by just the chain atoms and their chain bonds while neglecting substituents, etc.

4.1.2 Macroconformations in Solution

Coarse-graining is a powerful approach but may lead to inaccurate conclusions about molecular structures if its limitations are neglected. For example, dissolved polymer molecules cannot always be pictured as more or less flexible strings of constant thickness since some chain sections may have helical structures but others may not (Fig. 4-1).

In the solid state, polymers may be (Section. 3.3):

- crystalline with macromolecules in intramolecularly stabilized helices;
- crystalline with macromolecules in intermolecularly stabilized helices;
- crystalline with macromolecules in folded or unfolded zigzag chains; or
- amorphous with macromolecules in the shape of random coils.

On dissolution, helix structures may be completely or partially preserved or not at all, depending on the interaction of macromolecules with the solvent and on the solvent structure. In the first case, solutions contain relatively stiff macromolecules that are rodlike at low molecular weights but wormlike at higher ones (Fig. 4-1, I and II). In the two other cases, random coils are obtained, either with some preserved helical sections (Fig. 4-1, III) or none at all (Fig. 4-1, IV). Crystalline polymers with zigzag chains as well as amorphous polymers always lead to random coils if lateral self-associations are absent.

Complete Helices in Solution

Internal stabilization of double helices and intramolecular stabilization of single helices require polar macromolecules. These helices survive in solution if interactions between polymer and solvent are relatively weak. An example is the double helix of deoxyribonucleic acids (DNA) that is held together by internal hydrogen bridges between complementary bases as well as by π-π interactions of aromatic groups (Volume I, p. 518 ff.). These helices remain intact in dilute salt solutions (Fig. 4-1, I). Single helices of poly(γ-benzyl-L-glutamate) are stabilized by intramolecular hydrogen bonds; they survive in 1,4-dioxane (Fig. 4-1, II) but not in dichloroacetic acid (Fig. 4-1, IV).

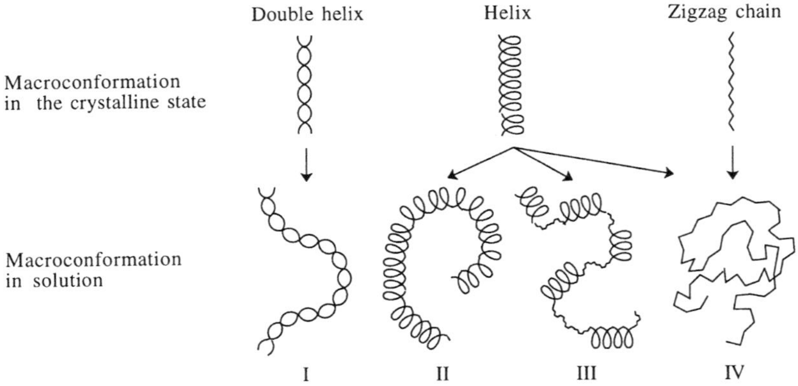

Fig. 4-1 Shapes of linear polymer molecules in crystalline and dissolved states. Examples:
I = Deoxyribonucleic acids in dilute salt solutions at 25°C;
II = Poly(γ-benzyl-L-glutamate) in 1,4-dioxane or *N,N*-dimethylformamide at 25°C;
III = Poly(oxyethylene) in water at 25°C, at-poly(methyl methacrylate) in acetonitrile at 44°C;
IV = Poly(ethylene) in xylene at 160°C, poly(γ-benzyl-L-glutamate) in dichloroacetic acid at 25°C.

The stabilizing effect of packing molecules in crystals is no longer present in dilute solution. The amplitudes of thermal vibrations around conformational resting positions increase and helices become somewhat more flexible and start bending. These helices and also stiff non-helical chains behave like worms (Section 4.3.9) and become coil-like at high molecular weights (Figs. 4-1, IV; 4-9; and 4-26).

Helical Sections in Solution

Coarse graining ignores the possible presence of helical sections in random-coil forming macromolecules (Fig. 4-1, III) although there is evidence for such macroconformations.

Observations on solutions of optically active chiral poly(1-olefin)s indicate the presence of helical sections. In the crystalline state, such helix structures are stabilized by packing effects. These polymers dissolve only in apolar solvents with similar structures as the monomeric units of the polymers. The acting forces between two polymer groups are thus very similar to those between either a polymer group and solvent molecules or between solvent molecules themselves. Dissolutions are neither driven by solvations nor by gauche effects (Volume I, p. 144). Conformational changes must therefore be mainly due to entropy effects and not enthalpic ones.

Syndiotactic poly(propylene) (st-PP) crystallizes in 50 % $[TTG^+G^+]_n$ and 50 % $[TTG^-G^-]_n$ macroconformations. The crystalline IR band from these macroconformations disappears on melting and is only very weak in CCl_4 solution (Fig. 4-2). However, it is very strong in benzene solution which indicates the presence of helical structures.

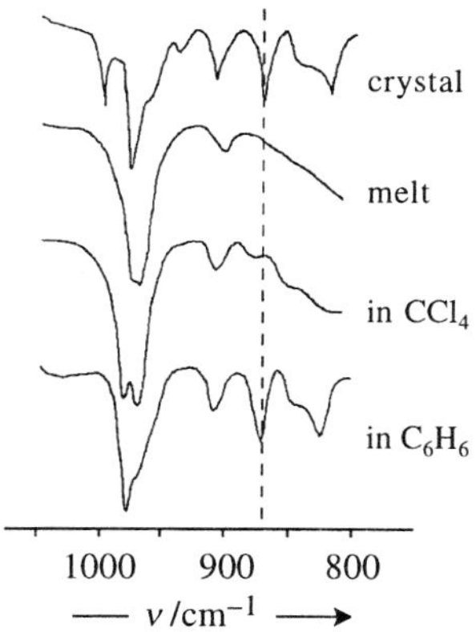

Fig. 4-2 Parts of the infrared spectra of a syndiotactic poly(propylene). The intensity of the "crystalline band" (- - -) at ν = 868 cm^{-1} measures the total concentration of left-handed and right-handed helices, $[TTGG]_n$ [1].

Spectra from top to bottom:
- in the crystalline state (67 % crystalline) at 37°C [1];
- in the melt at 170°C [2];
- in 4 wt% CCl_4 solution at 25°C [2];
- in 4 wt% benzene solution [2].

CCl_4 (ε_r = 2.24) and C_6H_6 (ε_r = 2.28) are both apolar solvents but the molecules of the former are disordered in the liquid state whereas the molecules of the latter pack like rolls of coins. It seems that this ordering promotes the formation of helical sequences in st-PP that are stabilized by lateral association. The self-association of st-poly(propylene) was proven by the typical dependence of $1/M_{app} = f(c)$ (Section 10.5).

The intensity of the "crystalline" band of st-poly(propylene) in benzene decreases with increasing temperature until all helical segments are dissolved at 57°C (Fig. 4-3).

Short helical segments also seem to be present in aqueous solutions of poly(oxyethylene); this polymer crystallizes as a 7_2 helix (Section 7.1.6). Such short helical segments

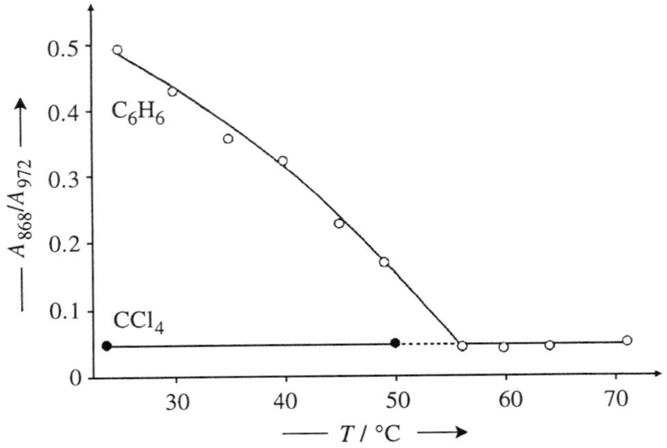

Fig. 4-3 Temperature dependence of the ratio of absorptions A at 868 cm^{-1} ("crystalline" band) and 972 cm^{-1} (reference band) of a 0.004 g/mL solution of a syndiotactic poly(propylene) in benzene (O) and in carbon tetrachloride (●) [3].

also exist in acetonitrile solutions of at-poly(methyl methacrylate); this polymer is amorphous in the solid state. The overall shape of all of these dissolved macromolecules is that of a random coil since a few flexible sections are sufficient to convert a "stiff" helix into a coil-like structure, albeit not one with a constant diameter of its sections (Fig. 4-4).

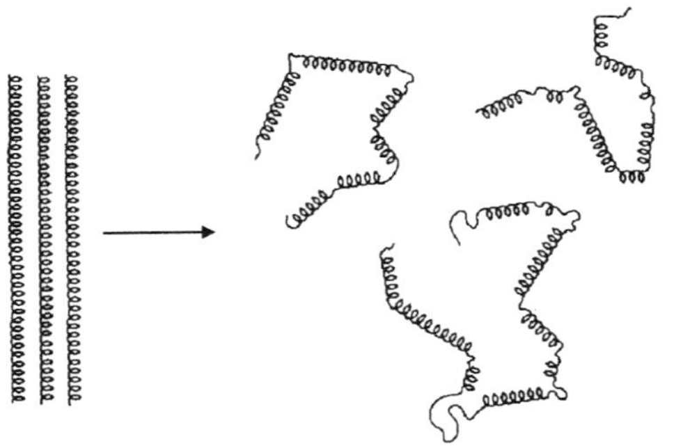

Fig. 4-4 A few "wrong" microconformations are sufficient to convert rod-like helices into coil-like structures on melting or dissolution.

4.2 Compact Molecules

4.2.1 Introduction

Isolated macromolecules and their associations may exist not only as loose coils but also as compact physical structures that resemble Euclidean bodies such as spheres, ellipsoids or rods. Such bodies are isotropic if they are homogeneously filled with matter and "hard" if they resist deformation. Isotropic and "hard" Euclidean bodies are defined as *compact*; such bodies have defined external dimensions and homogeneous interiors.

Spherical, ellipsoidal, and rod-like macromolecules are mostly not compact. The distributions of their segments range from coil-like density distributions to tight packing with or without internal order. Their surfaces may also range from "smooth" to rough. These bodies can be characterized by their external dimensions, for example, by electron micrographs of molecules that are deposited on a surface. However, sample preparation for electron microscopy may deform the shape of such macromolecules.

The dimensions of these molecules can be determined without any assumptions directly by scattering methods (Chapter 5) and indirectly by hydrodynamic methods (Chapters 11 and 12), using models. Scattering methods deliver the average radius of gyration of macromolecules or, to be more exact, the **z-average of the square of the radius of gyration**, $\langle s^2\rangle_z$ (Section 5.3.2) of the molecules of the polymeric substance,

$$(4\text{-}1) \qquad \langle s^2\rangle_z \equiv \frac{\sum_i z_i R_i^2}{\sum_i z_i} = \sum_i Z_i R_i^2$$

where R_i is the distance of the unit i (atom, group, etc.) with mass m_i from the center of gravity of the molecules (spheroids, ellipsoids, rods, random coils, etc.) and z and Z, respectively, the statistical weight (p. 19 ff.). An example is a two-dimensional linear chain composed of 16 chain units (Fig. 4-5).

By definition, averages of spatial quantities are symbolized by angled brackets, $\langle\ \rangle$, and not by a line over the symbol of the physical quantity. Radii of gyration are never measured as simple values of s or $\langle s\rangle$ but always as averages over their squares, $\langle s^2\rangle$. For convenience, $\langle s^2\rangle^{1/2}$ is often symbolized by s, however.

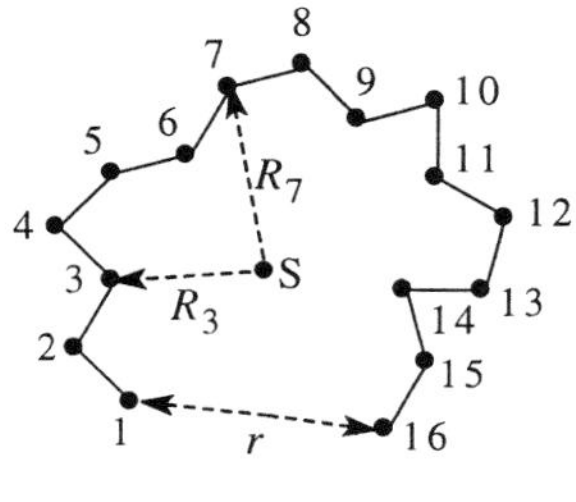

Fig. 4-5 A two-dimensional chain consisting of 16 consecutive chain units which are at distances R_i from the center of gravity, S. The chain has an end-to-end distance r between the end units 1 and 16.

4.2.2 Spheroids

Spheres

All points on the surface of a sphere are at the same distance to the center of the sphere but the distribution of polymeric material within the sphere may be homogeneous or inhomogeneous. One thus distinguishes between **compact spheres** (example: latex particles), **hollow spheres** (example: the protein apoferritin), and **solvated spheres** with or without a homogeneous internal distribution (examples: some enzymes and also some dendrimer molecules).

Compact and hollow spheres are characterized by their radius, R_{sph}, their surface area, $A_{sph} = 4\,\pi\,R_{sph}^2$, and their volume, $V_{sph} = (4/3)\,\pi\,R_{sph}^3$.

Since $V_{sph} = m_{sph}/\rho_{sph} = m_{sph}v_{sph}$, the radius of a compact sphere can be calculated from the mass, m_{sph}, and the density ρ_{sph} or specific volume v_{sph} of its matter. For spherical macromolecules in solution, the radius is available from the molar mass, $M = m_{mol}N_A$, of the macromolecule and its partial specific volume, $\tilde{v}_2 \equiv (\delta V/\delta m_2)_{m_1,T,p}$, where m_1 and m_2 are the masses of solvent (1) and polymer (2), respectively; m_{mol} = the mass of one macromolecule; V = volume of solution, m_1 = mass of solvent, T = temperature, p = pressure, and N_A = Avogadro constant:

$$R_{sph} = \left(\frac{3\,V_{sph}}{4\pi}\right)^{1/3} = \left(\frac{3\,v_{sph}m_{sph}}{4\pi}\right)^{1/3} = \left(\frac{3\,\tilde{v}_2 M}{4\pi N_A}\right)^{1/3} \tag{4-2}$$

Geometric radii, R_{sph}, and radii of gyration, s, are identical for hollow spheres with infinitely thin shells since the total mass is at a distance R_{sph}, from the center of the sphere. In compact spheres, however, the number of mass elements at a distance between r and $(r + dr)$ from the center is proportional to the surface area, A_{sph}. The square of the radius of gyration of compact spheres, s^2,

$$s^2 = \frac{\int_{r=0}^{r=R_{sph}} 4\pi r^4 dr}{\int_{r=0}^{r=R_{sph}} 4\pi r^2 dr} = \frac{3}{5} R_{sph}^2 \tag{4-3}$$

is therefore always smaller by a factor of $(3/5)^{1/2}$ than the square of the geometric radius, R_{sph}^2. Table 4-1 contains the relationships between the mean-square radii of gyration and other Euclidean bodies.

The molar mass dependence of the radius of gyration of compact spheres is obtained from Eq.(4-3), the volume $V = 4\pi R^3/3$, and the general relationship, $V = m/\rho = M/(\rho N_A)$:

$$s = \left(\frac{3}{5}\right)^{1/2}\left(\frac{3}{4\pi\rho N_A}\right)^{1/3} M^{1/3} = K_{sph}M^{1/3} \tag{4-4}$$

Table 4-1 Mean-square radii of gyration, s^2, and excluded volumes, u, of various Euclidian bodies.

Particles	s^2	u
Spheres with a total volume V		
Hollow with radii R_e (external) and R_i (internal)	$(3/5)(R_e^5-R_i^5)/(R_e^3-R_i^3)$	$8\ V$
Hollow with $R_e = R$ and $R_i \rightarrow 0$	R^2	$8\ V$
Compact with radius R	$(3/5)\ R^2$	$8\ V$
Compact ellipsoids with half axes a, b, and c; length L; radius R; and thickness d		
Ellipsoids with half-axes a, b, and c	$(1/5)(a^2 + b^2 + c^2)$	
Prolate rotational ellipsoids with $R << L$	$(1/5)[(L/2)^2 + 2\ R^2]$	$(3/8)\ \pi\ (L/R)V$
Oblate rotational ellipsoids with $R >> d$	$(1/5)[R^2 + 2\ (d/2)^2]$	$(3/2)\ \pi\ (R/d)V$
Compact rods with length L, radius R, half axes a and b, and circumference C		
Circular rods (with orientational angle γ)	$(L^2/12) + (R^2/2)$	$8V[1+(L/C)\sin\gamma]$
Ellipsoidal rods with half axes a and b	$(L^2/12) + (a^2 + b^2)/4$	
Compact discs with half axes a and b and a negligible thickness d		
Circular discs with radius R	$R^2/2$	$\pi\ (R/d)V$
Ellipsoidal discs	$(a^2 + b^2)/4$	

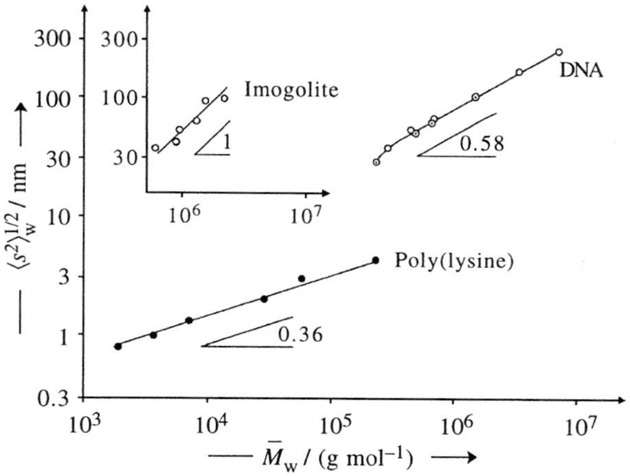

Fig. 4-6 Square root of the mass-average mean-square radius of gyration as a function of the mass-average molar mass of spherical poly(α,ε-lysine)s in DMF at room temperature [4], rod-like imogolites in dilute salt solution at 30°C [5], and coil-like double helices of deoxyribonucleic acids (DNA) in buffer solutions at 20°C [6,7].

The radius of gyration of compact spheres thus increases with the cubic root of the molar mass. A slightly higher value of 0.36 was observed for hyperbranched poly(α,ε-L-lysine)s in *N,N*-dimethylformamide (Fig. 4-6), probably because of a slight swelling of the sphere-like molecules.

A hard sphere (or any other hard Euclidian body) occupies space that cannot be taken up by another sphere (or body). Centers of two equal-sized spheres with radii R and volumes $V = 4\,\pi\,R^3/3$ can approach each other only to a distance of $d = 2\,R$ (Fig. 4-7). A sphere thus has an *external volume* u that is excluded for all other spheres. This excluded volume can be easily calculated as $u = 4\,\pi\,d^3/3 = 32\,\pi\,R^3/3 = 8\,V$. Excluded volumes of other types of compact particles are listed in Table 4-1.

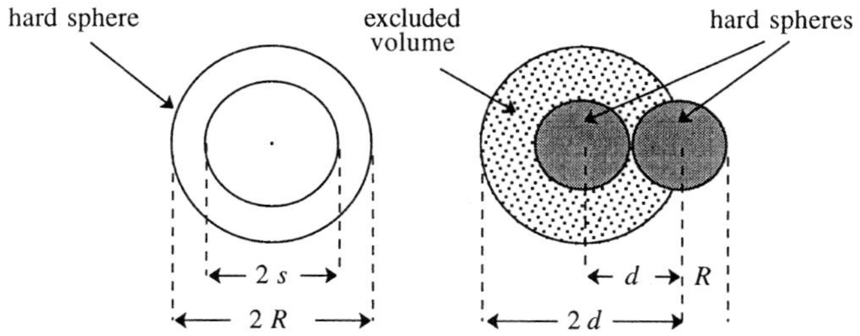

Fig. 4-7 Hard spheres with radii R; diameters $d = 2\,R$; volumes $V = (4\,\pi/3)\,R^3$; radii of gyration, $s = (3/5)^{1/2}\,R$; and external excludud volumes, $u = 8\,V$.

Ellipsoids

Ellipsoids are spheroids with three half-axes a, b, and c that occupy volumes of $V_{ell} = (4\,\pi\,abc)/3$. An ellipsoid with the same volume as a sphere of radius R_{sph} has an equivalent radius of $(R_{sph})_{eq} = (abc)^{1/3}$.

Ellipsoids of revolution have two equal half-axes that are smaller than the third in prolate ellipsoids ($a > b = c$) and greater than the third in oblate ones ($a < b = c$). A prolate ellipsoid thus has a length of $L = 2\ a$ and two equal, smaller radii of $R = b = c$ whereas an oblate ellipsoid has a thickness of $d = 2$ and two equal, larger radii of $R = b = c$.

These ellipsoids are usually characterized by their axial ratios, $\Lambda = a/c$. At $a/c = 1$, both prolate and oblate ellipsoids convert to spheres. Prolate ellipsoids are transformed to rods for $a/c \to \infty$ whereas oblate ones become ellipsoidal platelets for $a/c \to 0$.

Table 4-1 lists expressions for the radii of gyration of hard (compact or hollow) ellipsoids whereas Table 4-2 includes expressions for the dependence of the mean-square radius of gyration on molar mass.

Table 4-2 Exponents ν and ε of molar masses in expressions for the dependence of the mean-square radii of gyration on molar mass, $\langle s^2 \rangle = K_s^2 M^{2\nu} = K_s^2 M^{1+\varepsilon}$. For molecularly non-uniform polymers, z-average mean-square radii of gyration, $\langle s^2 \rangle_z$, deliver various g-average molar masses (last column).
1) ν and ε depend on the thickness of the shell; 2) ν and ε vary with axial ratios; 3) ν and ε vary with the thermodynamic goodness of the solvent.

Particle	ν	ε	$\overline{M}_g$
Compact, isotropic sphere	1/3	−1/3	$(\sum_i Z_i M_i^{2/3})^{3/2}$
Hollow sphere, infinitely thin shell	1/2	0	
Hollow sphere, thick shell 1)	$1/3 \leq \nu \leq 1/2$	$-1/3 \leq \varepsilon \leq 0$	
Ellipsoid of revolution, compact 2)	$1/3 \leq \nu \leq 1/2$	$-1/3 \leq \varepsilon \leq 0$	
Cylinder, compact, elliptical 2)	$1/2 \leq \nu \leq 1$	$0 \leq \varepsilon \leq 1$	
Cylinder, compact, circular, thin	1	1	$(\overline{M}_z \overline{M}_{z+1})^{1/2}$
Disc, compact, elliptical, thin	1/2	0	$\overline{M}_z$
Coil, flexible, unperturbed	1/2	0	$\overline{M}_z$
Coil, flexible, perturbed 3)	$1/2 \leq \nu \leq 0.588$	$0 \leq \varepsilon \leq 0.176$	

4.2.3 Rods

A circular rod (cylinder) of length L and radius R has a volume of $V = \pi R^2 L$ whereas a square rod of length L and length a of sides has a volume of $V = a^2 L$.

In each case, the center of mass is in the middle of the rod. For infinitely thin rods, the number of mass elements at a distance x to $(x + \mathrm{d}x)$ to the center is proportional to $\mathrm{d}x$ where $0 \leq x \leq L/2$.

For cylindrical rods with radius R, one also has to consider the radial distance y which may take values of $0 \leq y \leq R$. The mean-square radius of gyration of cylindrical rods is therefore

(4-5)
$$s^2 = \int_0^{L/2} \int_0^R \frac{2\,\pi y \mathrm{d}y[x^2 + y^2]\mathrm{d}x}{\pi R^2 L/2} = \frac{L^2}{12} + \frac{R^2}{2}$$

Radii of gyration can thus be obtained from the lengths and radii of rigid, compact rods which are available from electron micrographs. They can also be calculated from radii R, molar masses $M = \rho V N_A$, and densities ρ with errors of less than ±1 % if the rods are cylindrical with ratios of $L/R > 25$:

(4-6) $$\langle s^2 \rangle^{1/2} = \left(\frac{L^2}{12}\right)^{1/2} = \frac{1}{12^{1/2}}\left(\frac{V}{\pi R^2}\right) = \frac{M}{12^{1/2}\pi R^2 \rho N_A}$$

Molar mass dependencies of radii of gyration can be written as exponential expressions for all types of molecule shapes (spheres, rods, coils, etc.) (see Fig. 4-6):

(4-7) $$\langle s^2 \rangle^{1/2} \equiv s = K_s M^{\nu} = K_s M^{(1+\varepsilon)/2} \quad ; \quad \nu \equiv (1 + \varepsilon)/2$$

The magnitude of exponents ν and ε is controlled by the shape of the macromolecules (Table 4-2); for derivations, see the following sections. For polymers with molar mass distributions, corrections are necessary that depend on the averages of $\langle s^2 \rangle$ and M as well as the type and width of the molar mass distribution (see also Section 4.5.5).

For thin rigid rods, the exponents are $\nu = \varepsilon = 1$; an example is the behavior of the aluminum silicate imogolite, $(OH)_3Al_2O_3SiOH$, in the molecular weight range shown in Fig. 4-6. At high molecular weights, double helices of deoxyribonucleic acids appear as perturbed random coils with $\varepsilon = 0.58$ (Figs. 4-6 and 4-13) although shorter sections of the molecule behave as fairly rigid rods.

External excluded volumes u vary with the shape of molecules (Table 4-1). For stiff rods, they depend on the orientation angle γ of the long axes which is zero for a parallel packing. The excluded volume of stiff rods of equal lengths L and equal circumferences C is $u = 8\ V[1 + (L/C) \sin \gamma]$ (Table 4-1) where $C = 2\pi R$ for circular rods with radius R and $C = 4\ a$ for square rods with side length a.

4.3 Molecular and Atomistic Modeling

4.3.1 Introduction

Most synthetic macromolecules are not Euclidean bodies but have string-like structures that can be modeled by various methods. Some of these models will be discussed in greater detail, especially the molecular ones, whereas others are far too complex to be treated in this volume. The latter methods will be summarized below in order to provide some understanding of their basic assumptions, approaches, and results.

Physical structures of molecules and their assemblies as well as their static and dynamic properties can be simulated mathematically by molecular or atomistic modeling. In low-molecular weight chemistry, such methods are important for the calculation of spectra, three-dimensional molecular structures, and chemical reactivities, especially for developing pharmaceutically effective compounds. In polymer science, the same methods are used and in addition also some that are based on the chain characteristics of linear polymers. In general, mathematical modeling of polymers is more complex than that of low-molecular weight molecules because of the number of participating atoms, the spatial size of the molecules, and the resulting temporal effects. For example, a generation 10 polyamidoamine PAMAM dendrimer of molecular weight 934 720 consists of 147 396 atoms. The situation becomes more complex if systems of many molecule are studied, such as melts or solids.

For example, properties of low-molecular weight liquids can be simulated by using volume elements that contain 10^3-10^6 atoms. Such small volume elements simulate the properties of liquids because the sum of the interactions between atoms in the interior of the volume element dwarfs the contributions by the atoms on the surface. For atomic radii of ca. 0.1 nm, such a cubic volume element comprises 1-1000 nm^3 and has side lengths of 1-10 nm. The characteristic lengths of atoms or molecules (atomic radii or bond lengths) are small compared to the dimensions of the volume element.

High-molecular weight compounds have not only larger characteristic lengths but also many different ones. Though bond lengths are the same as in low-molecular weight compounds (ca. 0.15 nm), persistence lengths are much greater (ca. 1-200 nm) (Table 4-7) and radii of gyration even larger (10-1000 nm). Hence, in order to simulate a polymer melt, much larger volume elements with many more atoms must be chosen than for low-molecular weight liquids. Consequently, the required atomistic calculations surpass the capacities of today's supercomputers.

Polymers also present another problem. In mono-atomic liquids such as helium, fluctuations subside in ca. 10^{-15} seconds (ca. 1 femtosecond) if the liquid is at a temperature far away from a phase transition. But vibrations of bond lengths require ca. 10^{-13} seconds and changes of microconformations ca. 10^{-11} seconds. Melts of molecules composed of 500 chain atoms have relaxation times of 10^{-5} seconds which is 10^{10} greater than the fluctuation time of mono-atomic liquids.

For this reason, simulations of polymer melts require very long computation times, even with supercomputers. The calculation of a single macroconformation of a single poly(methylene) molecule consisting of just 1000 $-CH_2-$ units required 24 hours on a Silicon Graphics 260 GTX supercomputer. Simulations of polymer melts are therefore restricted to short chains (which prohibits studies of entanglements) or to coarse-graining, using, for example, pearl necklace models (see below).

4.3.2 Methods

Group Increment Methods

Group increment methods do not belong to atomistic or molecular simulations but are empirical and semiempirical methods for the prediction of intrinsic, processing, and product (article) properties of polymers. They are mentioned here for completeness but cannot be discussed in detail because they are very elaborate.

Group increment methods have been long known in low-molecular weight chemistry. The oldest and simplest example is the calculation of the molar mass of a molecule from the atomic masses of its constituent atoms (Dalton 1801). Another old method is the calculation of the molar dielectric polarization according to Mosotti (1850) and Clausius (1879). The most important group increment methods for polymers were developed by van Krevelen, Bicerano, and Askadskii.

The **van Krevelen method** is a group-additive method in which properties such as the refractive index, melting temperature, zero-shear viscosity, etc., are calculated from increments that are specific for each type of constitutive groups of a polymer. The group contributions are empirical values that were obtained from the known property values of many polymers. The contributions are added with due consideration of the degree of polymerization, if relevant. Examples of constitutive groups are chlorine (–Cl) and methyl ($-CH_3$) substituents, methylene groups ($-CH_2-$), amide groups (–NH–CO–), 1,4-phenylene groups ($-C_6H_4-$), ester groups (–O–C(=O)–), etc. The method is easy to use but is restricted to known structural elements and cannot be used for copolymers.

The **Bicerano method** employs contributions by atoms and bonds, called connectivity indices, instead of those by atomic groups. Most properties are calculated from two atomic connectivity indices (zeroth-order connectivities) and two bond connectivities (first-order connectivities). The method allows the prediction of extensive and intensive properties of homopolymers and copolymers composed of the elements C, H, N, O, F, Cl, Br, Si, and S.

The **Askadskii method** treats polymer repeating units as sets of anharmonic oscillators. Oscillations result from thermal motions of atoms of these units; they are affected by the valence of atoms, hydrogen bonds, dipole-dipole interactions, and dispersion forces. The critical temperature of these oscillations is assumed to be the glass temperature of the polymer or the temperature of the onset of chemical decomposition. The method allows one to estimate properties of homopolymers and random and alternating copolymers composed of the same elements as those used in the Bicerano method as well as those that also contain B, P, As, Sn, Pb, and I.

All three methods are very helpful to polymer practioners because they allow one to estimate many physical properties of yet to be synthesized polymers. The caveat is that polymer properties depend not only on the chemical structure of macromolecules but also on the physical properties which are imposed on polymeric materials by processing (Volume IV).

Lattice Methods

Lattice methods simulate real polymer chains by comparing them with abstract models such as the random walk of a drunkard which corresponds to the two-dimensional structure of an infinitely thin chain. These methods are physically unrealistic (Table 4-3) but are constructs that nevertheless model certain features correctly such as the spatial dimensions of coil-like polymer chains in the so-called unperturbed state.

An example is the **exact enumeration** of all self-avoiding walks on a lattice which corresponds to chains with excluded volumes. Because of the rapidly increasing number of possibilities (Section 4.4.4), exact enumerations are presently known only up to N = 39 steps. The properties (for example, macroconformations and properties caused by them) are then extrapolated to $N \rightarrow \infty$ by suitable mathematical methods.

Table 4-3 Comparison of the properties of a poly(ethylene) chain with a self-avoiding random walk (SAW) on a two-dimensional lattice.

Property	Poly(ethylene) chain	Self-avoiding walk
Dimension	continuous space	discrete lattice
Entities	CH_2 groups	lattice points
Angle to next entity	109.47° (if ideal C-C-C)	0° and 90° (see Fig. 4-19)
Potential barrier	depends on conformation	independent of conformation
Long-range interaction	attraction and repulsion	repulsion

Lattice methods are useful not only for equilibrium properties but also for dynamic processes such as self-diffusion. These processes are often modeled by **Monte Carlo processes** which are enumerations from the "top" and not from the bottom like exact enumerations. The number N of steps is often chosen from the range $10^2 \leq N \leq 10^5$. For a certain N, a number of macroconformations and properties depending on them is generated at random. The values for various N are then extrapolated to $N \to \infty$. The randomness of this procedure causes "experimental" errors but the extrapolation to ∞ from $10^2 \leq N \leq 10^5$ is more reliable than the one from $15 \leq N \leq 35$ by exact enumeration.

Static Monte Carlo processes start with a desired distribution of probabilities and generate a number of statistically independent samples. Dynamic Monte Carlo methods employ stochastic processes (usually Markov trials) to obtain a number of correlated samples. The equilibrium distribution of these samples is then the desired probability distribution.

For example, Monte Carlo processes allow one to simulate molecular jump processes such as self-diffusion in melts or solidification of melts to amorphous glasses. All movements of one entity (a bead, a chain unit, a monomeric unit, etc.) and some movements of two entities are shown in Fig. 4-8.

Lattice-free Methods

Polymer chains can be simulated by subdividing them into segments that are jointed freely or with some restrictions with respect to their length, thickness, interactions, etc. (Section 4.4.4). For example, infinitely thin segments selected by Monte Carlo processes can be made to rotate about imaginary axes such as the segment i around the axis ---- between segments i+1 and i–1 to a new position i* (Fig. 4-9, I).

In the **pearl necklace model**, polymer chains are approximated by a series of beads with diameter d that are joined by weightless bonds of length b (Fig. 4-9, II). Beads may be chain units such as the methylene units of poly(methylene) chains, $\text{+CH}_2\text{+}_n$, or effective units such as the ethylene units of $\text{+CH}_2\text{–CH}_2\text{+}_n$. The beads have a volume of $\pi d^3/6$ and an excluded volume (p. 71) that is simulated by the ratio d/b.

Some dynamic properties of polymers can be simulated by the very simplifying model of an **elastic dumbbell** (Fig. 4-9, III). The mass of the chain is united here in two beads that are connected by a weightless spring with a variable spring constant.

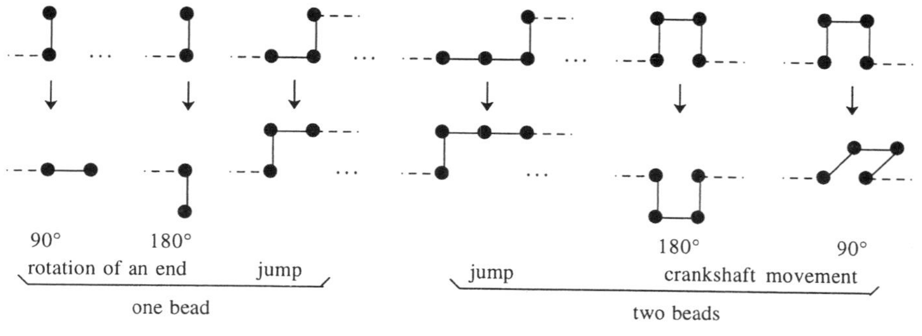

Fig. 4-8 All possible movements of one bead and some local joint movements of two beads on a hypercubical lattice. ··· Starting plane.

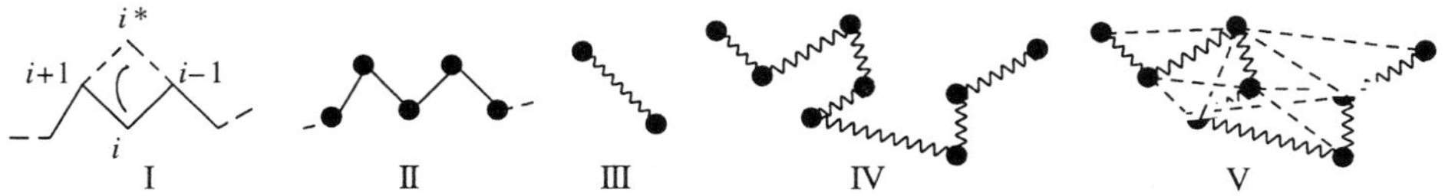

Fig. 4-9 Lattice-free models for the simulation of dynamic properties of polymer chains. (I) Freely jointed chain with rotation about bonds; (II) pearl-necklace model; (III) elastic dumbbell; (IV + V)) bead-spring models without (IV) and with some (V) hydrodynamic interactions (- - -).

More realistic is the **bead-spring model** which simulates polymer molecules by many beads that are interconnected by springs (Fig. 4-9; IV shows just one bond and V shows seven bonds of a long chain). In this model, there are either no hydrodynamic interactions between beads (**Rouse model**) or such interactions are considered (**Kirkwood-Riseman model**). These models are used for Monte Carlo calculations as well as for simulations by Brownian and molecular dynamics.

In **Brownian dynamics**, the temporal change of the spatial distribution of chain segments is simulated by the random flight of the Brownian movement of a particle. The variation of probability densities with time is calculated from the sum of all potential energies which are composed of a Lennard-Jones potential for the beads and a harmonic potential for the spring. The proportionality constant is the diffusion coefficient which is controlled by the thermal movement of the beads.

Molecular mechanics employs a force field (Section 4.3.3) to calculate the effect of mechanical stresses on bond lengths, bond angles, and torsional angles. Because of the many possible macroconformations of molecules, the resulting potential functions are very complex. For this reason, and because no potential barriers are surpassed, usually only local energy minima are obtained and not global ones. The resulting potential energy of the molecule or the molecular system is then introduced into the equations of classical mechanics (deformation as a function of stress, etc.).

With suitable mathematical processes, **energy minimizations** change systematically the variables of the force field (bond angles, etc.) to the macroconformations with the lowest potential energies. Since the computing effort increases with the square of the number of participating atoms, force fields are usually simplified, the reach of forces is curtailed, etc. The structures resulting from such energy minimizations correspond to macroconformations at a temperature of 0 K.

Molecular dynamics (**MD**) does not treat molecules as rigid entities. The method is based on the classical Newtonian equations for the movement of a system consisting of N entities (atoms, segments, etc.), starting from a set of initial conditions. The entities have an initial momentum and are affected by forces from the other entities. The motions of the entities with respect to location and time are described by differential equations that are solved numerically to yield finite difference equations.

The system is provided with kinetic energy which allows jumps over potential barriers. The new state is then calculated as a function of the old one, usually for time intervals of femtoseconds (1 fs = 10^{-15} s). This interval is chosen because the highest vibration frequency is ca. 10^{14} s^{-1} (C–H stretch). Such time intervals require computing times of hours to days. The dynamics of changes in conformations and properties is then followed for picoseconds (10^{-12} s) to nanoseconds (10^{-9} s) which corresponds to computing times of weeks to even months.

4.3.3 Atomistic Force Fields

For many modelings, potential energies have to be known. These energies are calculated from force fields (DREIDING, AMBER, CHARMM, etc.; see end of chapter) which may be based on simple valence-dependent energies E_{val} and energies E_{non} between unbound atoms or may also consider coupled valence-dependent energies $E_{val,c}$.

The various force fields also differ in the mathematical functions for force fields as well as in the corresponding parameters. Some approaches are restricted to single molecules ("gaseous state") whereas others can analyze systems of molecules (crystals, amorphous states). Some available computer programs also allow one to choose between various force fields.

The simple valence-dependent energy E_{val} is composed of four types of energy, with each type representing the sum of the contributions of all atoms (see below). These four types are the energy E_{bd} for the stretching of a valence bond (2 participating atoms), E_τ for the widening of a bond angle (3 participating atoms), E_θ for the deformation of a torsional angle (4 participating atoms), and E_ω for an inversion (4 participating atoms) (Fig. 4-10).

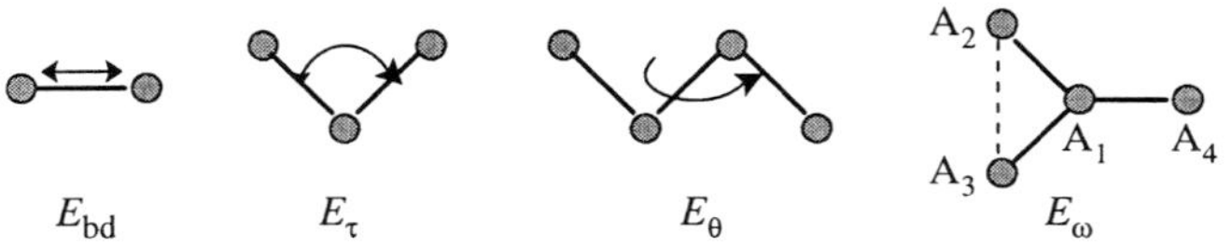

Fig. 4-10 Simple valence-dependent energies for the stretching of valence bonds (E_{bd}), widening of bond angles (E_τ), deformation of torsional angles at valence bonds (E_θ), and inversions (E_ω). The latter type of energy is required to keep the bonds A_1-A_2, A_1-A_3, and A_1-A_4 in one plane, i.e., A_1 can be positioned above or below the plane A_2-A_3-A_4. - - - Virtual bond in improper torsions (see text).

Energies E_{non} between unbound atoms may be caused by van der Waals interactions (E_{vdW}), electrostatic effects (E_q), and/or hydrogen bonds (E_H). They may be intra- or intermolecular and act between two or more unbound atoms (Fig. 4-11).

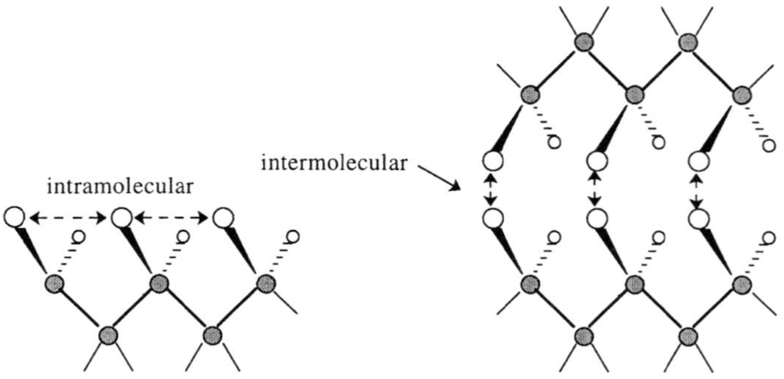

Fig. 4-11 Schematic representation of intra- and intermolecular interactions between unbound atoms, symbolized by O. Interactions may be of the van der Waals type, electrostatic, or hydrogen bonds.

Coupled valence-dependent energies $E_{val,c}$ arise from the combination of two or more simple valence-dependent energies (Fig. 4-12).

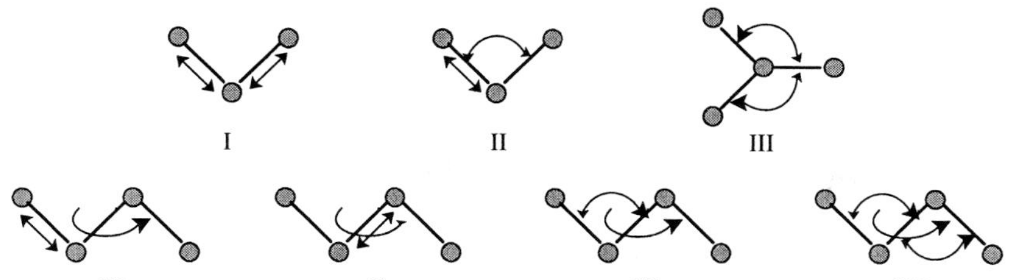

Fig. 4-12 Coupled valence-dependent energies: two bond extensions (I), bond extension plus deformation of valence angle (II), deformation of two valence angles (III), coupling of bond extension and torsion at an adjacent (IV) or the same bond (V), coupling of torsion and deformation of bond angle at one bond (VI) or at two bonds (VII).

The total potential energy is the sum of all of these energies:

(4-8) $$\Delta E = E_{val} + E_{non} + E_{val,c} = E_{bd} + E_{\tau} + E_{\theta} + E_{\omega} + E_{vdW} + E_{q} + E_{H} + E_{val,c}$$

The calculation of these energies depends on the chosen force field. It is shown below for DREIDING because this force field treats atoms of the same chemical element as identical whether or not they are in chains, rings, or substituents as other force fields do (AMBER, CHARMM, CFF91, etc.). Atoms of the same element may have different electron configurations, however, and therefore also different covalent radii, van der Waals radii, and ionic radii. The electron configuration of carbon, for example, may be sp^3 (tetrahedral), sp^3 in resonance systems, sp^2 (trigonal), or sp^1 (linear).

The energy E_{bd} for the *stretching of valence bonds* from length b_o to a length b is modeled by DREIDING and AMBER as the sum of the energies of all bonds. The bonds are treated as simple harmonic oscillators:

(4-9) $$E_{bd} = \Sigma_b\,(1/2)\,K_b(b - b_o)^2 \quad ; \quad K_b = \text{force constant}$$

Other force fields add higher terms, for example, $K_{b'}(b - b_o)^3 + K_{b''}(b - b_0)^4$ by CFF91.

Since the length b may take any value from b_o to infinity, all of these approaches lead to infinitely high energies E_b. Real bonds are cleaved if they are stretched to a certain length, however, which is considered by introducing a cut-off value. In DREIDING, Eq.(4-9) may also be replaced by a Morse function, $E_{bd} = E_b\{\exp[-(b_{anh} - b_0)] - 1\}^2$, which automatically introduces a separation energy, E_b, and includes an anharmonic term b_{anh} near the equilibrium value b_o.

The energy E_{τ} for *enlarging bond angles* to θ from θ_o is treated by AMBER in analogy to Eq.(4-9), using the valence angle τ instead of the bond length. CFF91 uses the same expression but adds higher terms. These forms do not deliver a value of zero for $\tau \to 180°$, however. DREIDING thus employs a harmonic cosine expression:

(4-10) $$E_{\tau} = \Sigma_{\tau}\,(1/2)\,K_{\tau}(\cos\tau - \cos\tau_o)^2 \quad ; \quad K_{\tau} = \text{force constant}$$

The energy E_{θ} for the *deformation of torsional angles* from θ_o to θ is calculated via

(4-11) $$E_{\theta} = \Sigma_{\theta}\,(1/2)\,\Delta E^{\ddagger}\{1 - \cos[N_{sym}(\theta - \theta_0)]^2\} \quad ; \quad \Delta E^{\ddagger} = \text{rotational barrier}$$

where N_{sym} = symmetry of the force field (2, 3, or 6). CFF91 uses instead the sum of three terms, $\Delta E_i^{\ddagger}\{1 - \cos(i\theta - \theta_{0,i})\}$, with i = 1, 2, and 3.

DREIDING describes *inversions* by

(4-12) $E_\omega = \Sigma_\omega (1/2) K_\omega(\omega - \omega_0)^2$; K_ω = force constant

where ω = angle between the bond A_1-A_2 and the plane A_1-A_3-A_4. In bioorganics, inversions are treated as improper torsions, however.

In CHARMM, ω is the improper torsion angle of A_1-A_2 with respect to A_3-A_4 (see Fig. 4-10) where the angle between the planes A_1-A_2-A_3 and A_2-A_3-A_4 is the dihedral angle of the virtual bond A_2-A_3. AMBER replaces $(\omega - \omega_0)^2$ by $\{1 - \cos[i(\vartheta - \vartheta_0)]\}$ where ϑ = angle between planes A_1-A_2-A_4 and A_1-A_3-A_4 and i = 2 (planar) and i = 3 (tetrahedral), respectively.

Van der Waals interactions between unbound atoms (p. 42) that are separated by a distance L are described usually by a **Mie (Lennard-Jones) 12-6 potential**:

(4-13) $E_{vdW} = \Sigma_{vdW} (AL^{-12}) - (BL^{-6})$; A, B = constants

A Lennard-Jones 9-6 potential would give more accurate results but the 12-6 potential is attractive since it leads to higher computational speeds if the L^{-12} term is calculated as $L^{-6}*L^{-6}$. Alternatively, an exponent (–6) potential may be used: $E_{vdW} = \Sigma_{vdW} A \exp(-CL) - BL^{-6}$.

Energies of *electrostatic interactions* between atoms with electric charges q_A and q_B at a distance $L_{A\text{-}B}$ are calculated via

(4-14) $E_q = \Sigma_q K_q(q_A q_B/(\varepsilon_r L_{A\text{-}B})$

where ε_r = relative permittivity of medium (ε_r = 1 for single molecules ("vacuum"!)) and K_q = factor for the interconversion of physical units since the distance L is given in SI units whereas charges q are usually introduced with units of the electromagnetic theory (electrostatic units, esu) with Fr as abbreviation for the unit Franklin of electrostatic charge, electromagnetic units (emu), Gaussian (Gau; esu for electrostatics, emu for electrodynamics), and atomic units (au)).

The *energy of a hydrogen bond* of length L and at an angle θ_{DHA} between donor D, hydrogen atom H, and acceptor A is calculated from

(4-15) $E_H = \Sigma_H K_H[(C/L^{12}) - (C'/L^{10})] \cos^4 \theta_{DH}$; C, C' = constants

In contrast to other force fields, DREIDING does not use coupled valence-dependent energies. For example, the coupling between bond extension and deformation of valence angle (Fig. 4-12, II) can be calculated from $E_{b\theta} = \Sigma_{b\tau} K_{b\tau}(b - b_0)(\tau - \tau_0)$. The larger the number of coupling terms, the greater the number of adaptable parameters and the computing time.

The various force fields use molecular data that differ from each other and from the "average" value of literature (Table 4-4). There does not seem to be a comprehensive comparison of polymer properties that have been calculated by various force fields.

Table 4-4 Covalent radii R_{co}, van der Waals radii R_{vdW} (= 1/2 of the length of van der Waals bonds), and ionic radii R_{ion} as "average values" [8] and in the DREIDING force field [9].
1 Å = 0.1 nm.

Atom	R_{co}/Å Avg.	R_{co}/Å DREIDING	R_{vdW}/Å Avg.	R_{vdW}/Å DREIDING	R_{ion}/Å Average	
H	0.37	0.330	1.32	1.598	1.54 (H^-)	
C (sp^3 tetrahedral)	0.77	0.770	1.67	1.949		
C (sp^3 (if resonance))		0.700				
C (sp^2 trigonal)		0.670				
N (sp^3 tetrahedral)	0.75	0.702		1.831		0.13 (N^{3+})
O (sp^3 tetrahedral w/o resonance)	0.73	0.660	1.50	1.702	1.26 (O^{2-})	
F	0.71	0.611	1.55	1.736	1.19 (F^{1-})	
Si (sp^3)	1.18	0.937	2.10	2.135	1.70 (Si^{4-})	0.42 (Si^{4+})
S (sp^3)	1.02	1.040	1.80	2.015	1.70 (S^{2-})	

4.4 Unperturbed Coils of Linear Chains

The simplest chemical structure of a macromolecule is that of a long chain of identically interconnected chain units. Examples are poly(methylene), $\text{+CH}_2\text{+}_n$, with chain units $-CH_2-$ and polymeric sulfur, +S+_n, with chain units –S–. The more or less free rotation around chain bonds allows chain units of isolated macromolecules to distribute themselves in space as much as neighboring chain units will allow. Such chains adopt the macroconformation of a three-dimensional random coil. An example is the **random coil** that is formed by the double helices of very high-molecular weight deoxyribonucleic acids (Fig. 4-13).

It is unclear why polymer science calls these entities "random coils" or just "coils" since they do not resemble the "coils" of common language (see next page) but rather very loose, non-spherical balls of sewing threads or tangles of intertwined wires.

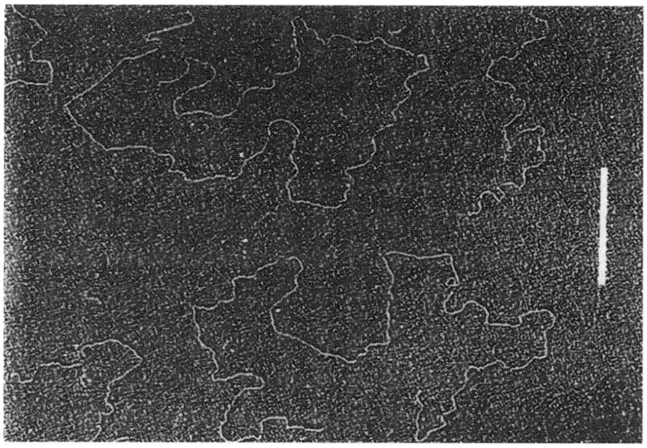

Fig. 4-13 Electron micrograph of chains of deoxyribonucleic acids that have been spread on a planar surface [10]. The structures correspond to a projection of three-dimensional random coils on a plane.
With kind permission by Academic Press, London.

What we now call "coil" was first introduced into the scientific literature in 1934 by Werner Kuhn as "Knäuel" which indicates in German an intertwined mass such as a ball of wool or a tangle of wires. It is unclear how the German "Knäuel" became the English "coil" since "coils" in common language are a series of loops or connected spirals such as three-dimensional curves about a common pole. It is in the latter sense that "coil" is used in the biosciences, i.e., as a synonym of "helix." The "coiled coil" of biosciences is a helix that is encased by another helix and the "supercoil" is a helix (or double helix) that is in the macroconformation of the "random coil" of polymer science.

4.4.1 Contour Lengths

In the simplest case, a linear chain consists of N_c identical chain units that are connected by $N_{bd} = N_c - 1$ chain bonds with equal bond lengths b. The largest *mathematically* possible chain length is thus $L_{chain} = N_{bd}b$ (Fig. 4-14) which is independent of the bond angles between successive chain atoms. This chain length is called **contour length** in the older literature because it truly traces the contour of a chain that is stretched so much that the valence angles between chain atoms become 180°. In this book, this length is referred to as **historical contour length** in order to avoid confusion with what is presently called "contour length."

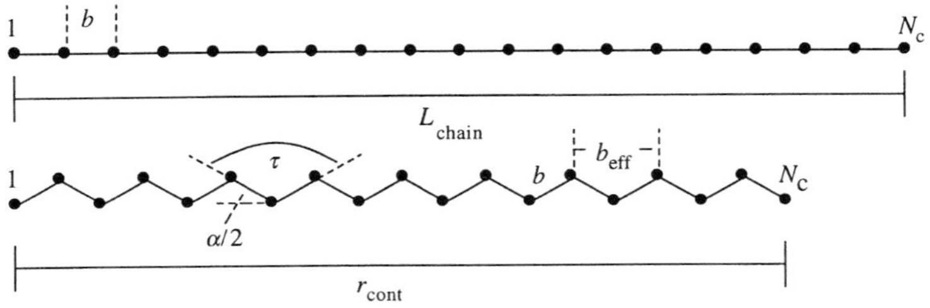

Fig. 4-14 Historical contour length L_{chain} (top) and conventional contour length r_{cont} (bottom) of a chain in the all-trans conformation. This chain consists of N_c = 19 chain atoms, N_{bd} = 18 chain bonds of length b, and N_{eff} = 9 effective chain bonds with effective bond lengths b_{eff}. Effective bond lengths of vinyl polymers, $\text{-[}CH_2\text{–}CHR\text{]}_n\text{-}$, correspond to crystallographic bond lengths c (see Chapter 7). $\alpha = 180° - \tau$ is the supplement of the bond angle (valence angle), τ.

The present-day contour length, r_{cont}, is the maximally possible *physical* length of a linear chain. It is usually shorter than the historical contour length since valence angles τ between three chain atoms are smaller than 180° in most cases. This length is called "contour length" by IUPAC although it does not measure the contour of the chain.

The IUPAC contour length is a **conventional contour length** that equals a maximum physical length if all chain atoms are in the trans conformation (see Chapter 3). However, this theoretical maximum contour length is not necessarily the true maximum physical length since all-gauche macroconformations may be energetically favored over all-trans ones (Section 3.3.3). The conventional contour length of an all-gauche chain would be smaller than that of an all-trans one (p. 44).

The conventional contour length r_{cont} of an all-trans chain with effective bond lengths b_{eff}, is the maximum **end-to-end distance** of a physically completely stretched chain:

(4-16) $$r_{cont} = N_{bd}b \sin(\tau/2) = N_{eff}b_{eff}$$

Effective bond lengths are projections of chain bonds on the end-to-end vector of a physically fully stretched chain. In vinyl polymers, the effective bond length is usually obtained from the crystallographic length b_{cr} of a monomeric unit; it comprises two chain bonds (see Fig. 4-14). The crystallographic length of a vinyl monomeric unit is $b_{eff} = b_{cr} = 0.254_6$ since b = 0.154 nm and τ = 111.5°.

Physically completely stretched chains require periodic sequences of microconformations. They are very rare in crystalline polymers and do not exist at all in solution since conformational energies do not differ appreciably among the various types of microconformations and also not from the thermal energy. As a result, sequences of microconformations are aperiodic and constantly changing so that the resulting random coil structures are averages over space and time, i.e., snapshots (see Fig. 4-13).

4.4.2 End-to-end Distance and Radius of Gyration

Random coils of linear macromolecules are characterized by their theoretical end-to-end distances, r, and experimental radii of gyration, s. The end-to-end distance is the spatial distance between the two ends of a linear chain (Figs. 4-5 and 4-15); it has no physical meaning for cyclic or branched molecules. This quantity can be only determined experimentally for chains with specially marked endgroups (e.g., fluorescent groups, etc.). It is usually a theoretical quantity that is calculated with various models as the square root of the spatial average mean-square end-to-end distance, $\langle r^2 \rangle^{1/2}$.

The square root of the mean-square radius of gyration, $\langle s^2 \rangle^{1/2}$, is always experimentally accessible. This quantity is a spatial average of unit positions in rigid particles and a *space-and-time* average of those in random coils with their rapidly changing macroconformations. In molecularly non-uniform polymers, it is also an average of the various molar masses. For example, scattering methods deliver the z-average, $\langle s^2 \rangle_z$ (Chapter 5).

The shapes of the DNA chains in Fig. 4-13 are snapshots since any distinct macroconformation of many possible ones exists only for a very short time. For example, a poly(methylene) chain, $\text{+CH}_2\text{+}_i$, with a degree of polymerization of i = 20 002 and three types of microconformations (T, G^+, G^-) can exist in $3^{i-2} = 3^{20\,000} = 10^{9452}$ different short-lived macroconformations.

The time for the interconversion of microconformations can be calculated from

(4-17) $\quad k = (k_B T/h) \exp(-\Delta E^{\ddagger}/RT)$ **Eyring equation**

where $k_B \approx 1.381 \cdot 10^{-23}$ J K^{-1} (Boltzmann constant) and $h \approx 6.626 \cdot 10^{-34}$ J s (Planck constant). At T = 298.15 K (= 25°C) and a rotational barrier of $\Delta E^{\ddagger} = \Delta E_{TG}{}^{\ddagger}$ = 10 kJ/mol (see Fig. 3-4), 5 % of trans microconformations are converted to gauche and *vice versa* within $t_{5\%} = 0.05/k \approx 1.4 \cdot 10^{-16}$ s.

None of the many possible momentary macroconformations of a macromolecule has a simple shape: isolated chain molecules are neither rods nor rotational ellipsoids or spheres (Fig. 4-15). These momentary shapes cannot be determined experimentally but can be calculated theoretically as follows.

The center of gravity of the molecule is put onto the center of a Cartesian coordinate system. The molecule is then oriented in such a way that the main axes of inertia are

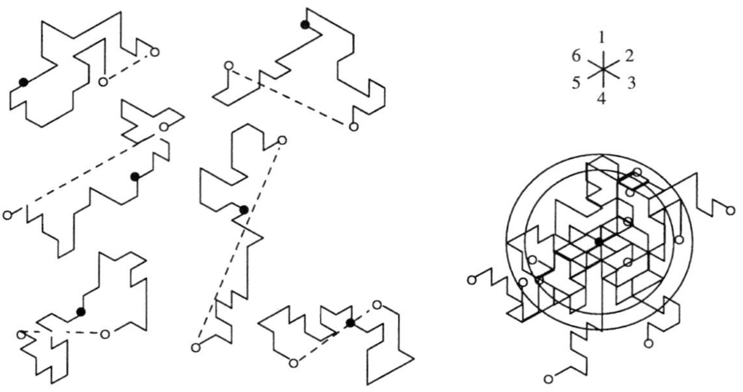

Fig. 4-15 Left: Six 2-dimensional momentary macroconformations of a chain with $N_c = 31$ chain atoms and $N_{bd} = 30$ chain bonds. Successive bond directions were determined by throwing a dice (see upper right), neglecting overlapping directions (for example, a "4" that followed a "1"). Central chain atoms 16 are marked by ● and end-to-end distances r by O- - - -O.

Right: Superposition of these 6 chains on their central chain atoms 16. Circles correspond to time-averaged diameters of one chain, indicating $2\,\langle s^2\rangle_o^{1/2}$ (inner circle) and $2\,\langle r^2\rangle_o^{1/2}$ (outer circle) where s = radius of gyration and r = end-to-end distance.

With kind permission by Springer-Verlag, Heidelberg [11].

identical to coordinate axes. The vector radius, $\boldsymbol{R}_i$, of each mass point (for details, see Fig. 4-17) can be expressed by three orthogonal components $(\boldsymbol{R}_i)_1$, $(\boldsymbol{R}_i)_2$, and $(\boldsymbol{R}_i)_3$:

$$(4\text{-}18) \qquad (\boldsymbol{R}_i^2)_1 + (\boldsymbol{R}_i^2)_2 + (\boldsymbol{R}_i^2)_3 = R_i^2$$

A similar approach can be used for the radius of gyration:

$$(4\text{-}19) \qquad s^2 = (s_1^2 + s_2^2 + s_3^2)/3$$

The three components are identical for equivalent, isotropic spheres. However, random coils of linear macromolecules have three unequal components according to theoretical calculations. Monte Carlo calculations for non-overlapping steps (p. 76) indicate ratios of the main components of 11.8:2.7:1 for low degrees of polymerization and 12.07:2.72:1 for high ones. The instantaneous shape of a coil molecule is thus neither that of a sphere (ratio 1:1:1) nor that of a rotational ellipsoid where two axes are equal and different from the third one (Section 4.2.3). It rather resembles a kidney.

Cyclic and branched polymer molecules have more symmetric shapes. For example, the components of the average radii of gyrations of cyclic poly(dimethylsiloxane)s of infinitely high molecular weight are only 5.9:2.6:1. The shapes of medium-sized dendrimer molecules are ellipsoidal but become more like spheres at very high molecular weights (Section 4.7.4).

The *time-averaged* three-dimensional shape of a coil from a molecularly uniform linear macromolecule resembles that of a sphere (Fig. 4-15, right). Random coils can therefore be treated as **equivalent spheres**. However, considerable proportions of chain atoms reside outside the shells (Fig. 4-15, right) that are defined by the average radius of gyration, $s_o \equiv \langle s^2\rangle_o^{1/2}$, and the average end-to-end distances, $r_o \equiv \langle r^2\rangle^{1/2}$.

4.4.3 Types of Coils

Depending on the range of interactions between chain segments, two types of random coils can be distinguished: unperturbed and perturbed ones. In the simplest case, only **short-range interactions** exist between molecularly adjacent chain units and/or segments (Fig. 4-16). A **segment** is defined as a short section of a chain; its length is usually left open.

Long-range interactions are defined as interactions between spatially (but not molecularly) adjacent segments that are separated from other chain segments of the *same* chain by many chain units. In very dilute solutions, both long-term and short-term interactions are intramolecular. In concentrated solutions, coils start to overlap (Section 6.3) and long-range interactions are both intramolecular within a polymer molecule and intermolecular between segments of different polymer molecules. In both cases, long-range interactions are intersegmental. "Long-range" and "short-range" thus do not refer to spatial ranges of forces *per se* but to the number of segments *within* a chain over which these forces are acting.

Long-range interactions may be repulsive because of the space required by the atoms of chain segments, interactions between two dipoles of the same sign, etc. In long-range interactions, the space of one chain segment is then excluded for all other segments of the same chain: the coil has an *internal* **excluded volume**. In concentrated solutions, coil molecules also have an *external* excluded volume just like hard bodies (Fig. 4-7).

Internal excluded volumes expand the coil which therefore becomes **perturbed**. If repulsive and attractive forces of long-range interactions balance each other, excluded volumes disappear and the coil is then **unperturbed**.

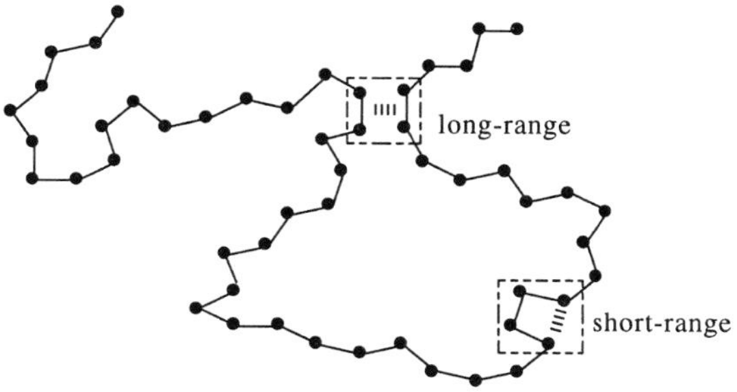

Fig. 4-16 Long-range and short-range (local) intramolecular interactions IIII between chain segments.

4.4.4 Molecular Models

The macroconformation of a linear chain can be described by various atomistic and non-atomistic models. Atomistic models take atoms as the smallest entities whereas non-atomistic models also consider larger entities such as monomeric units or Kuhnian segments (Section 4.4.5). Chemical atomistic models differ in the number of interacting chain atoms and the type of the factors that control the macroconformation of the chain (Table 4-5). Quantum chemical models also consider electrons. All chemical models are unspecific with respect to interaction forces (van der Waals, dipole-dipole, etc.).

Table 4-5 Chemical models. Covalent bonds — between participating chain atoms ●.

Name of model →	Freely jointed chain	Rotating chain		Rotational isomeric state
		freely rotating	restricted rotation	

Participating chain atoms	2	3	4	5
Control by	bond length	valence angle	conformer	conformer diad
Bond angle τ	any	defined	defined	defined
Torsional angle θ	any	any	defined average	several defined
End-to-end distance r	r_{oo}	r_{of}	r_{or}	r_o

Random-flight Chain

The simplest random-flight chain consists of many infinitely thin segments that follow each other randomly in space. Segments are undefined but must by larger than persistence lengths (Sections 3.1 and 4.4.9). The random-flight chain is thus not a molecular model.

The random-flight chain resembles the random movements that a particle or molecule experiences by Brownian movements in three-dimensional space; a two-dimensional analog is the random walk of a drunkard on a plane (Fig. 4-17, left). These movements do not leave behind traces; the analog distribution of chain segments (Fig. 4-17, right) thus assumes infinitely thin chain segments.

Such random flights can take place in any spatial dimension d, not just in three-dimensional space ($d = 3$) or on a plane ($d = 2$). However, calculations need to specify the lattice on which the movements take place (square, cubic, hexagonal, etc.).

A walker may walk on a line ($d = 1$) to the left or to the right with equal probability (Fig. 4-18). On a two-dimensional square lattice ($d = 2$), a walker has four choices (North, East, South, West) whereas on a cubic lattice ($d = 3$), six possibilities exist (North, East, South, West, up, down). Even more possibilities are present for so-called hypercubic lattices with $d \geq 4$. In each of these cases, the probability of one step is $1/(2\ d)$. After N_{seg} steps, the walker has the choice of $N_{mc} = (2\ d)^{N_{seg}}$ different but equally probable walking patterns, i.e., macroconformations.

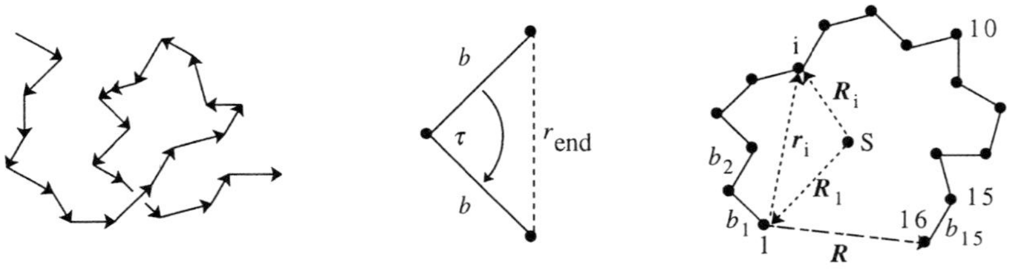

Fig. 4-17 Left: planar random flight with steps of different lengths. Center: "chain" consisting of three chain atoms with bond lengths b, a bond angle τ, and an "end-to-end distance" r_{end}. Right: planar chain with 15 segments, vectors $\boldsymbol{R}_i$ from the center of gravity, S, to chain atoms i, vectors $\boldsymbol{r}_i$ between the first chain atom and chain atom i, and vector $\boldsymbol{R}$ between chain ends.

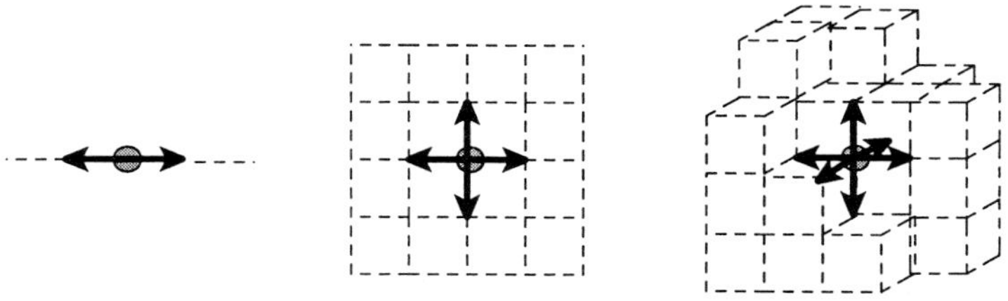

$d = 1$ 2 possibilities ($F = 2$) $d = 2$ 4 possibilities ($F = 4$) $d = 3$ 6 possibilities ($F = 6$)

Fig. 4-18 Steps on a line, a square two-dimensional lattice, and a cubic lattice. F = lattice factor.

Paths of two-dimensional walks, three-dimensional Brownian movements, etc., may cross (Fig. 4-17, left) which corresponds to infinitely thin polymer chains (**phantom chains**). Since real polymer chains have finite thicknesses, they cannot cross each other at the same *point* but may do so above or below. The mathematical analog to a real polymer chain is therefore the **self-avoiding walk (SAW)**, not the random walk.

Starting at the origin of an SAW, $C = 4$ walks are possible for $N = 1$ step on a two-dimensional square lattice (Fig. 4-19, left). Since the second step of $N = 2$ total steps cannot be reversed in SAWs, a first step d4→d3 can be followed only by d3→d2, d3→c3, or d3→e3, but not by d3→d4. Hence, the lattice factor reduces to $F = 3$ from $F = 4$ and the number of walking patterns is lowered to $C = 4{\cdot}3 = 12$ from $C = 4{\cdot}4 = 16$.

For the same reason, only $C = 4{\cdot}3{\cdot}3 = 36$ arrangements exist for $N = 3$ instead of $4{\cdot}4{\cdot}4 = 64$. Starting with $N = 4$ steps, one has to avoid the formation of squares of steps, etc. The counting of steps becomes very cumbersome for $N \geq 4$ which is the reason why the present world record for exact numbers of self-avoiding steps, N_{mc}, on two-dimensional square lattices is only $N = 51$ (see Table 4-6 for some numbers). Numbers N_{mc} of self-avoiding steps are drastically lower than those of phantom random walks. The same is true for different types of lattices (hexagonal, etc.) and dimensionalities other than 2.

The number of self-avoiding steps (macroconformations) is written as

$$(4\text{-}20) \qquad N_{mc}(\text{SAW}) = \mu^{N_{seg}} \quad ; \quad \mu = \text{connective constant}$$

in analogy to the expression for phantom chains, $N_{mc} = (2\,d)^{N_{seg}}$.

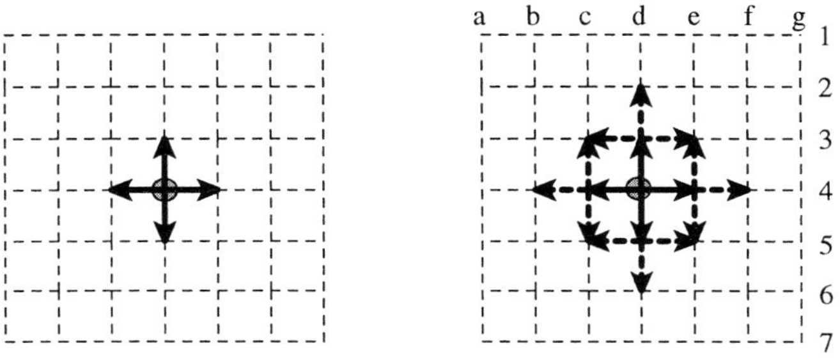

Fig. 4-19 Self-avoiding steps on a two-dimensional square lattice. Four equally possible steps are possible from the origin d4 to d3, e4, d5, and c4 but only three equally possible steps are allowed from each of these four positions since the reverse steps d3→d4, e4→d4, d5→d4, and c4→d4 correspond to crossings of the original paths.

Table 4-6 Number of macroconformations (= walks) on a two-dimensional square lattice for phantom chains (N_{mc}) and SAW chains ($N_{mc,SAW}$) as a function of the number N_{seg} of segments (steps). The numbers for phantom chains and SAWs with N_{seg} = 39 are known exactly but are rounded for easy comparison (for example, SAW with N_{seg} = 39: $N_{mc,SAW}$ = 113 101 676 587 853 932.

N_{seg}	Number of macroconformations Phantom chains	Self-avoiding walks
1	4	4
2	16	12
3	64	36
4	256	100
10	1 048 576	44 100
39	$\approx 3.022 \cdot 10^{23}$	$\approx 1.131 \cdot 10^{17}$

The average number of the next possible steps before a potential crosspoint is measured by a parameter μ which is called the **connective constant**. For two-dimensional lattices, the connective constant has values of 2.638 159 ± 0.000 002 (square lattice) and 1.847 759 ± 0.000 006 (hexagonal lattice), both of which are much lower than 2 d = 4 of phantom chains. The connective constant of a hypercubic lattice (d = 4) is 4.683 93 ± 0.000 06 and therefore also considerably lower than the corresponding value of 2 d = 8 for phantom chains in four-dimensional space.

Instead of counting exactly the number of steps, one can also use Monte Carlo statistics (p. 76) to compute a certain fraction of self-avoiding walks for a certain number N_{seg} and then average over all trials. The result is a modified Eq.(4-20),

$$(4\text{-}20a) \qquad N_{MC}(\text{SAW}) = A\mu^{N_{seg}} N_{seg}^{\gamma}$$

where μ = 1, A = constant, and γ = critical exponent that depends only on the dimensionality and not on the lattice type. These Monte Carlo values are $\gamma_1 = 1$ (d = 1), γ_2 = 43/32 = 1.343 75 ≈ 4/3 (d = 2), γ_3 = 1.1619 ≈ 7/6 (d = 3), and $\gamma_4 = 1$ ($d \geq 4$). No mathematical theory exists for these exponents. The γ_3 value agrees well with the exponent 2ν = 1.176 of the molar-mass dependence of the mean-square radius of gyration of random coils in good solvents (Section 4.5.4), i.e., the **Flory constant** ν = 0.588.

These enormous numbers of possible macroconformations have to be taken into account for the calculation of properties of polymer melts. Computer simulations of melt properties are therefore extremely time consuming; they may take days on very fast computers. At present, such calculations for solid polymers are therefore restricted to crystalline polymers where local structures are important and not the global ones. One can thus use shorter chains or smaller lattices.

Freely Jointed Chain

A freely jointed chain is the simplest *molecular* model of a coiled molecule. The macromolecule is modeled as a hypothetical chain consisting of N infinitely thin segments with equal lengths b that may occupy any allowed spatial position with equal probability. The segment is not specified. It may be the length of the bond between two chain atoms or may consist of several monomeric units.

Freely jointed chains of N bonds of length b are controlled by bond lengths; bond angles and torsional angles are arbitrary (Table 4-5). The simplest structure is a "chain" consisting of just two bonds that form a bond angle τ (Fig. 4-17, center). The "end-to-end distance" of such a "chain" is given by the cosine law:

$$(4\text{-}21) \qquad r_{\text{end}}^2 = 2\,b^2 - 2\,b^2 \cos \tau$$

Eq.(4-21) applies also to the corresponding spatial averages $\langle r^2 \rangle_{\text{oo}}$ and $\langle \cos \tau \rangle$ which leads to $\langle r^2 \rangle_{\text{oo}} = 2\,b^2 - 2\,b^2 \langle \cos \tau \rangle$. In chains with N chain bonds instead of two, N replaces 2. Since all angles τ are equally probable, "spatial" averages $\langle \cos \tau \rangle$ become zero and Eq. (4-21) converts to the expression for the mean-square radius of gyration:

$$(4\text{-}22) \qquad \langle r^2 \rangle_{\text{end}} = Nb^2$$

Eq.(4-22) indeed describes the correct dependence of the mean-square radius of gyration of freely jointed chains on the number N and length b of chain bonds. However, the derivation is physically incorrect because the chain is not on a plane but resides in three-dimensional space. Instead of bond lengths b, one has to consider bond vectors $\boldsymbol{R}_i$. The end-to-end vector $\boldsymbol{R}$ of the chain is given by the sum of the various bond vectors $\boldsymbol{R}_i$ and the square of the end-to-end vector accordingly by the sum of the products:

$$(4\text{-}23) \qquad \begin{aligned} \boldsymbol{R}^2 = \; & \boldsymbol{R}_1\boldsymbol{R}_1 + \boldsymbol{R}_1\boldsymbol{R}_2 + \boldsymbol{R}_1\boldsymbol{R}_3 + \boldsymbol{R}_1\boldsymbol{R}_4 + \boldsymbol{R}_1\boldsymbol{R}_5 + \ldots \\ & \boldsymbol{R}_2\boldsymbol{R}_1 + \boldsymbol{R}_2\boldsymbol{R}_2 + \boldsymbol{R}_2\boldsymbol{R}_3 + \boldsymbol{R}_2\boldsymbol{R}_4 + \boldsymbol{R}_2\boldsymbol{R}_5 + \ldots \\ & \boldsymbol{R}_3\boldsymbol{R}_1 + \boldsymbol{R}_3\boldsymbol{R}_2 + \boldsymbol{R}_3\boldsymbol{R}_3 + \boldsymbol{R}_3\boldsymbol{R}_4 + \boldsymbol{R}_3\boldsymbol{R}_5 + \ldots \\ & \boldsymbol{R}_4\boldsymbol{R}_1 + \boldsymbol{R}_4\boldsymbol{R}_2 + \boldsymbol{R}_4\boldsymbol{R}_3 + \boldsymbol{R}_4\boldsymbol{R}_4 + \boldsymbol{R}_4\boldsymbol{R}_5 + \ldots \end{aligned}$$

Products $\boldsymbol{R}_i\boldsymbol{R}_j$ with the same indices ($i,j = i,i$; $i,j = j,j$) equal the square of the bond length, b^2. In freely jointed chains, all vector products $\boldsymbol{R}_i\boldsymbol{R}_j$ are the same if their indices $i \neq j$ differ by the same value. On averaging all products $\boldsymbol{R}_i\boldsymbol{R}_j$ ($i \neq j$), the square of the end-to-end vector, $\boldsymbol{R}^2$, becomes the mean-square end-to-end distance, $\langle r^2 \rangle$, because the end-to-end distance is defined as the length of the end-to-end vector R. Eq.(4-23) becomes

$$(4\text{-}24) \qquad \langle r^2 \rangle \text{end} = Nb^2 + 2\,(N-1)\langle \boldsymbol{R}_1\boldsymbol{R}_2 \rangle + 2\,(N-2)\langle \boldsymbol{R}_1\boldsymbol{R}_3 \rangle + \ldots + 2\,\langle \boldsymbol{R}_1\boldsymbol{R}_N \rangle$$

In freely jointed chains, two randomly chosen bond vectors $\boldsymbol{R}_i$ and $\boldsymbol{R}_j$ ($i \neq j$) can frame any angle α_{ij}. All scalar products such as $\langle \boldsymbol{R}_i\boldsymbol{R}_j \rangle = b_i b_j \langle \cos \alpha_{ij} \rangle$, etc., become zero and Eq.(4-24) reduces to

$$(4\text{-}25) \qquad \langle r^2 \rangle_{\text{end}} = Nb^2 \equiv \langle r^2 \rangle_{\text{oo}}$$

where the index oo indicates the end-to-end distance of a freely jointed chain.

Freely Rotating Chains

Freely jointed chains with arbitrary bond angles are not very realistic since real bond angles can only adopt values that are dictated by the chemical constitution. Because of constant bond angles, τ, and complementary angles, $\alpha = 180° - \tau$, products $\langle \boldsymbol{R}_i\boldsymbol{R}_j \rangle$ in Eq.(4-24) no longer equal zero as in freely jointed chains. The product $\langle \boldsymbol{R}_i\boldsymbol{R}_j \rangle$ rather becomes $b^2 \cos \alpha$ for two successive vectors $\boldsymbol{R}_i\boldsymbol{R}_{i+1}$.

Real chains also have torsional angles (Fig. 3-1). For three successive vectors, the torsional angle is the angle between the projections of vectors $\boldsymbol{R}_i$ und $\boldsymbol{R}_{i+2}$ on the plane that is normal to the vector $\boldsymbol{R}_{i+1}$ (Chapter 3.1). The product $\boldsymbol{R}_i\boldsymbol{R}_{i+2}$ is thus the projection of bond 3 on bond 1 (Fig. 4-20) which can be subdivided into two components: one component, $R_{\parallel} \cos \alpha$, parallel to vector 2 and another component, $R_{\perp} \cos \alpha$, normal to this vector. The *spatial* direction of the latter component is given by the torsional angle (conformational angle) θ (see Fig. 3-2).

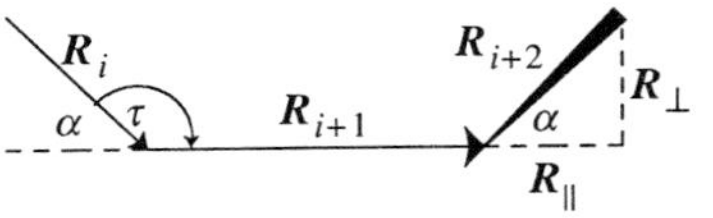

Fig. 4-20 Separation of the bond vecto $\boldsymbol{R}_{i+2}$ into 2 components that are parallel and normal, respectively, to the preceding vector $\boldsymbol{R}_{i+1}$. The direction of the component $R_{\perp} \cos \alpha$ is controlled by the torsional angle θ.

A special case is the **freely rotating chain** where chain bonds can rotate freely around the projections of the preceding bonds. In freely rotating chains, the normal component of the bond vector $\boldsymbol{R}_{i+2}$ becomes on average, $R_{\perp} \cos \alpha = 0$. The same value is obtained for chains with three *energetically equal* microconformations that are symmetrical to each other such as T (0°), G^+ (120°), and G^- (–120°). In this case, the freely rotating bond i+2 extends *on average* only to the bond i+1 and one has to consider only the parallel component $R_{\parallel} \cos \alpha = b \cos \alpha$. This component can then be projected back to the bond i which leads again to $b \cos \alpha$. The average of the product $\boldsymbol{R}_1\boldsymbol{R}_3$ thus adopts a value of $\langle \boldsymbol{R}_1\boldsymbol{R}_3 \rangle = (b \cos \alpha)(b \cos \alpha) = b^2 \cos^2 \alpha$.

The same approach is used for all other products $\boldsymbol{R}_i\boldsymbol{R}_j$. The average of $\boldsymbol{R}_1\boldsymbol{R}_4$ becomes $\langle \boldsymbol{R}_1\boldsymbol{R}_4 \rangle = b^2 \cos^3 \alpha$, etc., and Eq.(4-24) converts for freely rotating chains to

$$(4\text{-}26) \qquad \begin{aligned} \langle r^2 \rangle_{of} = {} & Nb^2 + 2\,(N-1)b^2 \cos \alpha + 2\,(N-2)b^2 \cos^2 \alpha \\ & + 2\,(N-3)b^2 \cos^3 \alpha + ... + 2\,(N-i)b^2 \cos^i \alpha \\ & + ... + 2\,b^2 \cos^{N-1} \alpha \end{aligned}$$

For infinitely long freely rotating chains with the bond angle, $\tau = 180 - \alpha$, the solution of this series delivers (see Appendix A-4.1, Eq.(A 4-7))

$$(4\text{-}27) \qquad \langle r^2 \rangle_{of} = Nb^2 \left(\frac{1-\cos \tau}{1+\cos \tau} \right) = N(b_{eff})^2 \qquad ; \quad \text{for } N \to \infty$$

The constant term $(1 - \cos \tau)/(1 + \cos \tau)$ can be united with the square of the bond length, b^2, to the square of an **effective bond length**, $(b_{eff})^2$.

The mean-square end-to-end distance of a freely rotating chain is larger than that of a freely jointed chain if the bond angle τ is greater than 90°. An example is a freely rotating carbon chain with $\tau = 109.5°$ where the mean-square end-to-end distance, $\langle r^2 \rangle_{of} \approx 2\,Nb^2$ is about twice as large as that of a freely jointed chain with $\langle r^2 \rangle_{oo} = Nb^2$.

For chains with finite degrees of polymerization, a considerably more complicated expression is obtained, for example with $\alpha = 180° - \tau$:

$$(4\text{-}28) \qquad \langle r^2 \rangle_{\text{of}} = Nb^2 \left[\frac{1+\cos\alpha}{1-\cos\alpha} - \frac{2}{N}\left(\frac{1-(\cos\alpha)^N}{(1-\cos\alpha)^2} \right)\cos\alpha - \frac{2}{N}(\cos\alpha)^N \right]$$

Chains with Restricted Rotation

Rotations about chain bonds are rarely free but are usually restricted because chain atoms reside on average in discrete microconformations, for example, T, G^+, and G^- (Chapter 3). In the simplest case of a chain with restricted rotation, all potential barriers of a bond are independent of each other. The smallest unit of such a chain consists therefore of 3 chain bonds and 4 chain atoms (Table 4-5). The mean-square end-to-end distance of such chains is obtained by matrix calculations similar to that for freely jointed and freely rotating chains, writing each chain bond as a vector. For very long chains with symmetric rotational potentials, the result is

$$(4\text{-}29) \qquad \langle r^2 \rangle = Nb^2 + 2\,(N-1)\,\langle \boldsymbol{b}_1\boldsymbol{b}_2 \rangle + 2\,(N-2)\,\langle \boldsymbol{b}_1\boldsymbol{b}_3 \rangle + 2\,(N-3)\,\langle \boldsymbol{b}_1\boldsymbol{b}_4 \rangle + \ldots$$

Eq.(4-29) resembles the general Eq.(4-24) for freely rotating chains. However, one has to consider here not only two successive chain bonds but also distinct conformations so that spatial averages $\langle \boldsymbol{b}_i\boldsymbol{b}_j \rangle$ have to be replaced by the product of three vectorial quantities, $\boldsymbol{b}^{\mathrm{T}}\langle \boldsymbol{t}^{j-i} \rangle$ where T indicates the transposition of the matrix.

The result for infinitely long chains with restricted rotation (index or) shows that an additional term with the constant average conformational angle θ has to be considered:

$$(4\text{-}30) \qquad \langle r^2 \rangle_{\text{or}} = Nb^2 \left(\frac{1-\cos\tau}{1+\cos\tau} \right)\left(\frac{1+\langle\cos\theta\rangle}{1-\langle\cos\theta\rangle} \right) = N(b'_{\text{eff}})^2$$

Similarly to Eq.(4-27), angular terms can be united with b^2 to the square of an effective bond length, $(b'_{\text{eff}})^2$. However, conformational angles θ have no physical meaning if they are calculated from experimental quantities $\langle r^2 \rangle_{\text{or}}$, N, b, and τ because the microconformations are neither equal nor independent of each other.

Eqs.(4-29) and (4-30) were obtained with stochastic methods. They diverge for the limiting conditions $\tau = 180°$ and $\theta = 0°$ and can therefore not be used to calculate conventional contour lengths where $\theta = 0°$ for all-trans conformations.

Rotational Isomeric State

The rotational isomeric state model (**RIS model**) considers that successive microconformations are not independent of each other but rather affected by the type of the preceding microconformation. An example is the pentane effect where a G^+ microconformation cannot be followed by a G^- microconformation and vice versa (Fig. 3-4).

In the simplest RIS model, coil dimensions are controlled by *pairs* of microconformations (Table 4-5) which requires 4 chain bonds and 5 chain atoms, respectively. Eq.(4-30) thus has to be expanded *formally* by a new parameter, Q_{pair}, which describes the effects that are caused by conformational diads in addition to the effects by single microconformations *via* $\langle\cos\theta\rangle$. However, $\langle\cos\theta\rangle$ in Eq.(4-30) is an adaptable quantity and not an independently measurable one. It is therefore expedient to combine the two conformational terms, Q_{pair} and $(1 + \langle\cos\theta\rangle)/(1 - \langle\cos\theta\rangle)$, to a new quantity, the **steric factor** σ (Section 4.4.5) which describes all those deviations from freely rotating chains that are caused by conformational statistics. This quantity is introduced as a square so that it can be compared to the effective bond length b_{eff} in Eq.(4-27).

The calculation for infinitely long chains delivers for the mean-square end-to-end distance of such a chain the expression

$$(4\text{-}31) \qquad \langle r^2\rangle_o = Nb^2\left(\frac{1-\cos\tau}{1+\cos\tau}\right)\left(\frac{1+\cos\theta}{1-\cos\theta}\right)Q_{pair} = Nb^2\left(\frac{1-\cos\tau}{1+\cos\tau}\right)\sigma^2 = \langle r^2\rangle_{of}\,\sigma^2$$

which describes the end-to-end distance of an **unperturbed chain**, i.e., a chain that is not perturbed by the presence of excluded volumes.

The calculation of Q_{pair} by the RIS model corresponds *mathematically* to that of a so-called **Ising chain**, i.e., the one-dimensional Ising model (**Lenz-Ising model**). An Ising chain consists of an arrangement of entities that interact which each other; in polymers, these entities are the chain bonds. Each entity can be present in various states; i.e., chain bonds in various microconformations. The total energy of an Ising chain is the sum of all interaction energies between neighboring entities.

In polymers, the total energy is the sum of all conformational energies. It equals the product of the number of all bonds minus the two bonds at the chain ends, $N - 2$, and the average energy of a bond provided that the conformational energy of a bond is independent of that of other bonds and all of these energies are equal. The total number of possible conformations of such a chain is $(N_{cf})^{N-2}$ if each bond can exist in N_{cf} types of microconformations (for example, $N_{cf} = 3$ for T, G^+, and G^-).

Since two adjacent bonds may influence each other, the energy $E(\alpha) = E(180°-\theta)$ of a bond must be replaced by pairwise energies where α' and α are the conformations at the bonds $i - 1$ and i, respectively:

$$(4\text{-}32) \qquad E(\alpha) = \sum_{i=2}^{N-1} E_i(\alpha_{i-1},\alpha_i) = \sum_{i=2}^{N-1} E_{\alpha'\alpha,i}$$

The energies of the various pairs of conformations (here: TT, TG^+, TG^-) are obtained experimentally from contour diagrams of conformational energies (Fig. 4-21) or are calculated from force fields by molecular mechanics (Section 4.3.3). These energies provide the statistical weights $u_{\alpha'\alpha,i}$ for the presence of the different bonds.

High-energy conformations are more probable at higher temperatures than at low temperatures (Fig. 3-6). In the simplest case, the temperature dependence of statistical weights, $u_{\alpha'\alpha,i}$, is described by the product of a Boltzmann term, $\exp(-E_{\alpha'\alpha,i}/RT)$, and a pre-exponential factor, A:

$$(4\text{-}33) \qquad u_{\alpha'\alpha,i} = A\exp(-E_{\alpha'\alpha,i}/RT)$$

The pre-exponential factor A indicates the shape of the energy well. For poly(ethylene), this well is symmetric and the pre-exponential factor is $A = 1$ for the three microconformations T (0°), G^+ (+120°), and G^- (–120°). For many other polymers, the well is not symmetric which leads to $A \neq 1$.

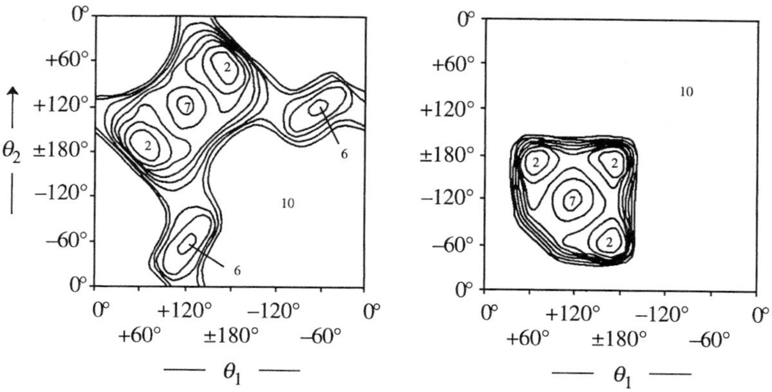

Fig. 4-21 Contour diagrams of conformational energies as a function of torsional angles θ_1 and θ_2 of two successive chain bonds for isotactic (left) and syndiotactic (right) poly(propylene) [12]. Numbers indicate energies in kcal/mol (1 kcal = 4.184 kJ).

it-Poly(propylene) has two energy minima at $\theta_1 = 60°$ and $\theta_2 = 180°$ (TG^+) and $\theta_1 = 180°$ and $\theta_2 = 60°$ (G^-T) that belong to left-handed and right-handed helices, respectively. st-Poly(propylene) shows three energy minima: one at $\theta_1 = 180°$ and $\theta_2 = 180°$ (TT) and the other two for TG^+ and TG^-.

The statistical weights are then written as a matrix, $\mathbf{U}_i$, for the various pairs of bonds. Poly(ethylene) with three microconformations T, G^+, and G^- has 3^2 of such pairs. Such normalized matrices are shown in Fig. 4-22 for the energies and the statistical weights of the pairs of microconformations at its melting temperature of 140°C.

Energies $E_{\alpha',\alpha,i}$/(kJ mol^{-1})

0° (T), +120° (G^+), −120° (G^-)

$$\mathbf{U}_i = \begin{bmatrix} 0 & 2.09 & 2.09 \\ 0 & 2.09 & 10.46 \\ 0 & 10.46 & 2.09 \end{bmatrix} \begin{matrix} 0° \ (T) \\ +120° \ (G^+) \\ -120° \ (G^-) \end{matrix}$$

Statistical weights $u_{\alpha,\alpha,i}$

0° (T), +120° (G^+), −120° (G^-)

$$\mathbf{U}_i = \begin{bmatrix} 1 & 0.54 & 0.54 \\ 1 & 0.54 & 0.05 \\ 1 & 0.05 & 0.54 \end{bmatrix} \begin{matrix} 0° \ (T) \\ +120° \ (G^+) \\ -120° \ (G^-) \end{matrix}$$

Fig. 4-22 Energies and statistical weights of conformational diads of poly(ethylene) at 140°C. Horizontal: bond i; vertical, bond $i-1$.

Poly(ethylene), $\text{+}CH_2\text{–}CH_2\text{+}_n$, contains only one type of conformational sequence, $–CH_2–CH_2–CH_2–$. For most other polymers, two or more types need to be considered. For example, poly(dimethylsiloxane), $\text{+}Si(CH_3)_2\text{–}O\text{+}_n$, consists of two types of chain units, $–O–Si(CH_3)_2–O–$ and $–Si(CH_3)_2–O–Si(CH_3)_2–$.

An even greater number of conformational sequences is present in poly(vinylidene fluoride), $\text{+}CH_2\text{–}CF_2\text{+}_n$, because this polymer contains not only two different types of chain units but also shows regioisomerism, i.e., the joining of monomeric units head-to-tail, head-to-head, and tail-to-tail (see Volume I, p. 45). These constitutional features lead to six different types of conformational sequences: $–CH_2CF_2CH_2–$, $–CF_2CH_2CF_2–$, $–CH_2CH_2CF_2–$, $–CF_2CH_2CH_2–$, $–CF_2CF_2CH_2–$, and $–CH_2CF_2CF_2–$. Because vectors are involved, statistical weights are different even if conformational diads are chemically identical; an example is $–CH_2CH_2CF_2–$ and $–CF_2CH_2CH_2–$.

The statistical weight f of a chain conformation is simply the product of the statistical weights of the various types. At its melting temperature of 140°C, poly(ethylene) consists of 60 % trans conformers T and 40 % gauche conformers $G^+ + G^-$. However, there are only 29 % conformational diads TT but 56 % trans-gauche diads (TG^+, TG^-, G^-T, and G^+T) and 15 % gauche-gauche diads. The pentane effect (Fig. 3-4) reduces the proportion of hetero-gauche diads G^+G^- and G^-G^+ to just 3 %:

$$f_T = 0.60 \qquad f_{G^+} = f_{G^-} = 0.20$$

$$f_{TT} = 0.29 \qquad f_{TG^+} = f_{TG^-} = 0.14 \qquad f_{G^+G^+} = f_{G^-G^-} = 0.06$$
$$f_{G^+T} = f_{G^-T} = 0.14 \qquad f_{G^+G^-} = f_{G^-G^+} = 0.015$$

The summation of all possible chain conformations delivers the so-called **configurational partition function**, Z_{conf}:

(4-34) $$Z_{conf} = \sum_{(\alpha)} \Omega_{(\alpha)} = \sum_{(\alpha)} \prod_{i=2}^{N-1} u_{\alpha'\alpha,i}$$

This distribution function is the normalization factor for the expression for the mean-square end-to-end distance,

(4-35) $$\langle r^2 \rangle_o = Z^{-1}\mathbf{G}_1\mathbf{G}_2 \ldots \mathbf{G}_N$$

where each matrix, $\mathbf{G}_i$, of the matrix product, $\mathbf{G}_1\mathbf{G}_2 \ldots \mathbf{G}_N$, consists of terms, $u_{\alpha'\alpha}\mathbf{F}(\alpha)$, that contain themselves matrices $\mathbf{F}_i$, for example,

(4-36) $$\mathbf{G}_i \equiv \begin{bmatrix} u_{\alpha',\alpha}\mathbf{F}(\alpha) & u_{\alpha',\beta}\mathbf{F}(\beta) & \cdots \\ u_{\beta',\alpha}\mathbf{F}(\alpha) & u_{\beta',\beta}\mathbf{F}(\beta) & \cdots \\ \vdots & \vdots & \ddots \end{bmatrix}_i$$

The first matrices, $\mathbf{G}_1$ and $\mathbf{F}_1$, and the last matrices, $\mathbf{G}_N$ and $\mathbf{F}_N$, have special forms (see literature 4.3b in "Literature to Chapter 4".

End-to-end distances calculated by the RIS method frequently do not agree with experimental ones. For example, end-to-end distances of poly(methylene) can only be reproduced if the bond angle C–C–C is chosen as 112° although the bond angle in crystalline poly(ethylene) is 111.5°C and the exact dihedral angle of a C–C–C section is 109°28'. Also, conformational angles have to be set as $\theta_{G+} = +127.5°$ and $\theta_{G-} = -127.5°$ and not as ideal values of $\theta_{G+} = +120°$ and $\theta_{G-} = -120°$.

These deviations may be caused by various effects. For example, energy differences between microconformations must be known to ±0.4 kJ/mol or better. Although energy diagrams are usually given for the vacuum and should thus reflect true unperturbed dimensions, experiments also have to deal with solvent effects (in solution) and packing of chains (in crystals and melts) (for the difference between unperturbed dimensions and dimensions in theta states, see Section 10.4.2). Furthermore, energy diagrams are valid only at 0 K while experimental data refer to higher temperatures.

4.4.5 Flexibility of Chains

Flexibilities of chains are controlled by transitions between microconformations *i* and *j*, for example, T ⇄ G^+. A conformer is **statically flexible** if the conformation energy (= potential energy) per amount of conformer is smaller than the molar thermal energy, i.e., $\Delta E_{m,ij} < RT$. At 25°C, this amounts to $\Delta E_{m,ij}$ = 2.49 kJ mol^{-1}. A statically flexible conformer has many accessible minima of conformational energy.

Conformers are **dynamically flexible** if the rotational barrier is not substantially higher than the molar thermal energy, $\Delta E_{m,ij}^{\ddagger} \leq RT$. Because microconformations are interconverted fast and at random, an *individual* dynamically flexible macromolecule is continuously changing its shape and overall dimensions although at the same time these parameters stay spatially and temporally constant for the *collection* of such macromolecules, the macromolecular substance.

Flexibilities (or stiffnesses) of chains are characterized by many parameters: steric factors, characteristic ratios, Kuhnian lengths, and persistence lengths.

Steric Factor

The steric factor σ is defined as the ratio of the mean-square end-to-end distance, $\langle r^2\rangle_o^{1/2}$, of an unperturbed chain, Eq.(4-31), to that of the mean-square end-to-end distance, $\langle r^2\rangle_{of}^{1/2}$, of a freely rotating chain, Eq.(4-27). For infinitely high molecular weights, this leads to

$$(4\text{-}37) \qquad \sigma^2 = \frac{\langle r^2\rangle_o}{\langle r^2\rangle_{of}} = \frac{Nb^2(1-\cos\tau)(1+\cos\tau)^{-1}(1+\cos\theta)(1-\cos\theta)^{-1}}{Nb^2(1-\cos\tau)(1+\cos\tau)^{-1}} = \frac{(1+\cos\theta)}{(1-\cos\theta)}$$

Steric factors are independent of the degree of polymerization. They describe the steric hindrance to rotation and are therefore also called **hindrance parameters**. Since they are controlled only by the average conformational angle, θ, they are also known as **conformational factors**. For example, hypothetical carbon chains consisting of only one type of microconformation would have the conformational factors of

$\sigma = 0$	if all microconformations are cis	(C)	(τ = 180°);
$\sigma = 0.58$	if all microconformations are gauche	(G^+ or G^-)	(τ = 120°);
$\sigma = 1.73$	if all microconformations are anti	(A^+ or A^-)	(τ = 60°);
$\sigma = \infty$	if all microconformations are trans	(T)	(τ = 0°).

Hindrance parameters of flexible carbon chains vary between 1.7 and 3.3 (Table 4-7).

For polymer systems with approximately equal polymer-solvent interactions, steric factors increase with the increasing size of substituents as shown by the series of poly(alkyl methacrylate)s, $\text{+}CH_2C(CH_3)(COOR)\text{+}_n$, with various substituents R where σ = 1.9 for R = CH_3 but σ = 3.3 for R = $C_{22}H_{45}$ (Table 4-7). In the latter case, this would correspond to a lowering of the average conformational angle to 33.7° from 55.5° or an increase in the proportion of trans microconformations and a corresponding decrease of the gauche ones.

Since hindrance parameters reflect the average conformational angle, they can only be compared for chains with the same chain atoms. Hence, one cannot conclude that cellulose chains (σ = 2.0) are about as flexible as carbon chains with σ = 2.0.

Table 4-7 Hindrance parameters σ, characteristic ratios C_∞, and persistence lengths L_{ps}. Calculations with a C–C–C bond angle of τ = 112°. Θ = theta temperature (unperturbed dimensions).

Polymer	Solvent	T/°C		σ	C_∞	L_{ps}/nm
Poly(butadiene), 1,4-*cis*	Decalin®	55		1.63	4.9	
Poly(butadiene), 1,4-*trans*	various	50		1.23	5.8	
Poly(ethylene)	1-chloronaphthalene	140		1.77	6.87	0.61
Poly(isobutylene)	benzene	24		1.8	6.5	0.59
Poly(styrene), at-	butyl formate	–9	Θ	2.17	9.4	
	cyclohexane	34	Θ	2.18	9.5	0.90
	methyl cyclohexane	68	Θ	2.17	9.4	0.84
Poly(methyl methacrylate), at-	1-butyl chloride	40.8	Θ	1.87	8.40	0.72
	butanone	25		1.89	7.9	0.69
Poly(butyl methacrylate), at-	butanone	25		1.9	7.9	0.69
Poly(decyl methacrylate), at-	butanone	25		2.4	12.7	1.05
Poly(docosyl methacrylate), at-	butanone	25		3.3	23.9	1.91
Cellulose	Cadoxen®	25		2.0		
	N,*N*-dimethyl acetamide + 9 % LiCl	30		6.71	92	11
Amylose	nitromethane	22.5		2.75		
Poly(hexylisocyanate)	hexane	25				42
Poly(1,4-benzamide)	*N*,*N*-dimethyl acetamide + 3 % LiCl	30				50
Deoxyribonucleic acid	0.2 mol/L NaCl in water	20				63
Schizophyllan	water	25				200
Poly(acrylic acid), at-	1,4-dioxane	30		1.83	6.7	0.65
Poly(sodium acrylate), at-	1.5 mol/L NaBr in water	15		2.38	11.3	1.04

Characteristic Ratio

Hindrance parameters are not very good measures of flexibilities of polymer chains because they require the calculation of end-to-end distances, $\langle r^2\rangle_{\text{of}}{}^{1/2}$, of freely rotating chains. One thus needs to know the number N of chain bonds, the lengths of chain bonds b (usually from crystallographic data), and the bond angles τ.

Bond lengths can be assumed to be constant and independent of the thermodynamic state because bond energies are very high (40-400 kJ/mol). However, bond angles of chains in crystalline states and solutions are not necessarily identical. According to spectroscopy and determination of heats of combustion of cyclic molecules, a deformation of the C–C–C bond angle by 5.6° requires only 2 kJ/mol and by 10°, only 7 kJ/mol. These values have the same order of magnitude as conformational energies (Section 3.2); they are *not* independent of the state of the polymer chain.

The same is true for bond angles between chain atoms. It is thus convenient to define a new parameter, the **characteristic ratio** C_N, as the product of the hindrance parameter σ and the bond angles term (1 – cos τ)/(1 + cos τ), using Eqs.(4-23) and (4-17):

(4-38) $$C_N \equiv \sigma^2 \left(\frac{1-\cos\tau}{1+\cos\tau} \right) = \frac{\langle r^2\rangle_\text{o}}{Nb^2} = \frac{\langle r^2\rangle_\text{o}}{\langle r^2\rangle_\text{oo}}$$

In general, C_N increases with increasing number N of chain bonds and then becomes independent of N (Fig. 4-22). For certain helical conformations, e.g., that of st-poly-(methyl methacrylate), calculated values of C_N may pass through a maximum.

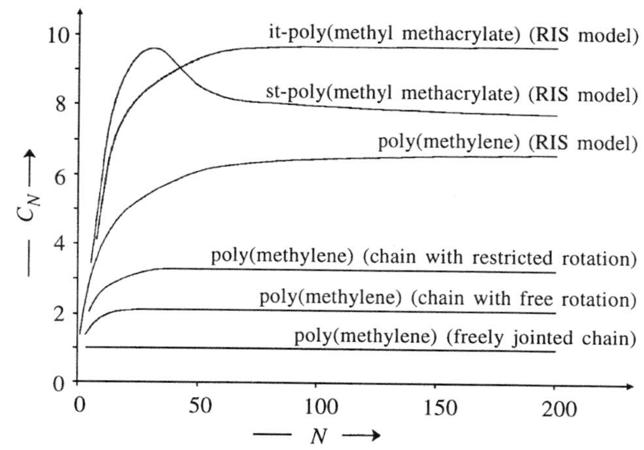

Fig. 4-22 Calculated characteristic ratio C_N as a function of the number N of chain bonds for four molecular models of poly(methylene) chains (freely jointed chain, chain with free rotation, chain with restricted rotation, RIS model) [13] and for isotactic and syndiotactic poly(methyl methacrylate) [14]. For poly(methylene), angles of chain bonds, C-C-C, were assumed as 112° and conformational energies, $(E_G - E_T)/(\text{kJ mol}^{-1})$, as 2.09 (restricted rotation) and 8.30 (RIS), respectively.

Kuhnian Length

End-to-end distances of unperturbed coils increase with larger bond lengths b, bond angles τ, and hindrance parameters σ (Eq.(4-31)). These parameters act as if longer segments of length L_K are present in smaller numbers N_K. The unperturbed coil (and other coils (p. 125)) may thus be replaced by a so-called **Kuhnian coil** (**Kuhn's equivalent coil, ersatz coil**) where the mean-square end-to-end distance is written as the product of the number of Kuhnian segments, N_K, and the square of **Kuhnian lengths**, L_K,

(4-39) $\quad \langle r^2 \rangle_o = N_K L_K^2$

in analogy to the expression, $\langle r^2 \rangle_{oo} = Nb^2$, for the freely jointed chain (Eq.(4-22)).

The product, $N_K L_K$, must equal the conventional contour length, $r_{cont} = N_{eff} b_{eff}$ (see Fig. 4-14). Kuhnian lengths L_K of unperturbed coils can therefore be calculated from $L_K = \langle r^2 \rangle_o / r_{cont}$; they are directly related to persistence lengths (next section). The greater the Kuhnian length, the smaller the thermodynamic flexibility of the chain, and the greater its stiffness. Stiff chains no longer form random coils at small and medium degrees of polymerization; they are wormlike (Section 4.4.9). At very high degrees of polymerization, they approach the macroconformation of a random coil.

Assume a poly(ethylene) molecule with $N = 2000$ C–C bonds of length $b = 0.154$ nm. As a freely jointed chain, the random coil of this molecule would have a mean-square end-to-end distance of $\langle r^2 \rangle_{oo} = Nb^2 = 47.432$ nm^2. The mean-square end-to-end distance of the corresponding unperturbed random coil is calculated from $\langle r^2 \rangle_o = C_N \langle r^2 \rangle_{oo}$ (Eq.(4-38)) as $\langle r^2 \rangle_o = 6.87 \times 47.432$ nm$^2 \approx 325.858$ nm^2.

Such a molecule has $N_{eff} = 1000$ effective bonds of effective bond length $b_{eff} = 0.2546$ nm (see Fig. 4-14); its conventional contour length is therefore $r_{cont} = N_{eff} B_{eff} = 1000 \times 0.2546 = 245.6$ nm. The Kuhnian length L_K is calculated from the conventional contour length, $r_{cont} = N_K L_K$, and the mean-square end-to-end distance of the unperturbed coil, $\langle r^2 \rangle_o \approx 325.858$ nm^2, with the help of Eq.(4-39) as $L_K = \langle r^2 \rangle_o / N_K L_K \approx 325.858$ nm^2/245.6 nm ≈ 1.327 nm. The molecule thus contains $N_K = N_K L_K / L_K = 245.6$ nm/(1.327 nm) ≈ 185 Kuhnian segments.

Persistence Length

The constancy of bond angles in freely rotating chains (p. 89) and in chains with restricted rotation (p. 91) gives chains a certain persistence (Section 3.1) against flexing (L: *persistere* = *per* (intensive), *sistere* = to stand still) which, at high persistence, leads to wormlike chains (Setion 4.4.9).

Persistence increases with the size of substituents, which leads to greater hindrance parameters and characteristic ratios (Table 4-7). They are especially large if bond angles approach 180° since a chain with $\tau = 180°$ is a stiff rod.

The reach of the persistence can be characterized by a **persistence length** that must be a function of the characteristic ratio, C_N, which is also a measure of the stiffness of chains. The characteristic ratio of unperturbed coils is proportional to the mean-square end-to-end distance, $\langle r^2 \rangle_o$, which depends on the number N of chain bonds of length b (Eq.(4-38)). Since unperturbed coils are larger than coils of freely jointed chains (Eq.(4-25)), one can write

(4-40) $$\langle r^2 \rangle_o > \langle r^2 \rangle_{oo} = Nb^2$$

The persistence length, L_{ps}, is defined as the average sum of all projections of bonds $j \geq i$ on the bond i in an infinitely long chain. Since L_{ps} is greater than the bond length b, the expansion of an unperturbed coil relative to that of a freely jointed chain can be expressed by

(4-41) $$\langle r^2 \rangle_o = Nb(2\,L_{ps} - b) = 2\,NbL_{ps} - Nb^2$$

for the limiting case of infinitely long chains ($N \rightarrow \infty$). Eq.(4-41) reduces to that of the freely jointed chain if the persistence length equals the bond length , i.e., $\langle r^2 \rangle_{oo} = Nb^2$ if $L_{ps} = b$. The characteristic ratio, Eq.(4-38), in the limit of $C_N \rightarrow C_\infty$, is therefore

(4-42) $$C_\infty \equiv \left(\frac{\langle r^2 \rangle_o}{Nb^2} \right)_\infty = \frac{2L_{ps}}{b} - 1$$

Persistence lengths can thus be calculated from characteristic ratios, C_∞, (Table 4-7) and bond lengths, b. Typical bond lengths are b = 0.154 nm for carbon chains and b = 0.425 nm for chains with 1,4-glucose units.

Carbon chains have small persistence lengths of $0.6 \leq L_{ps}/\text{nm} \leq 1.9$ which correspond to only a few monomeric units. Larger persistence lengths are observed for wormlike chains of stiff molecules such as the double helices of deoxyribonucleic acids, the helix of the polysaccharide schizophyllan, and tobacco mosaic virus (see p. 107).

4.4.6 Radius of Gyration

End-to-end distances are theoretically important. For linear chains, they can be calculated by various methods (Section 4.4.4). However, they have no physical meaning for cyclic polymer molecules, which have no chain ends at all, and also not for branched macromolecules which have more than two chain ends. Experimentally, end-to-end distances can be determined only in very special cases.

A direct experimental quantity is the **radius of gyration**, s, which is mostly observed as the spatial mean square, $\langle s^2 \rangle$, of the distribution of the squares of all radii, s_i, usually as the second moment of the mass distribution (Eq.(4-1) and Eq.(A 4-14)).

For infinitely long phantom chains (freely jointed (index oo), freely rotating (of), with restricted rotation (or), various RIS chains (o)), the same relationship between mean-square end-to-end distances and mean-square radii of gyration has been calculated by lengthy vector analysis:

(4-43) $$\langle r^2 \rangle_y = 6 \langle s^2 \rangle_y \qquad y = \text{oo, of, or, o}$$

For all of these chains, the mean-square end-to-end distance is proportional to the number of chain bonds, N, and the square of an effective bond length. Introducing the second equality of Eq.(4-31) into Eq.(4-43), replacing the number of chain bonds, N, by the number of chain atoms, $N_a = N + 1 \approx N$, and expressing $N_a = M/M_u$ by the molar mass, M, of the polymer molecule and the molar mass, M_u, of the chain unit, leads to

(4-44) $$\langle s^2 \rangle_o^{1/2} = \left[\left(\frac{b^2}{6\, M_a} \right) \left(\frac{1-\cos\tau}{1+\cos\tau} \right) \sigma^2 \right]^{1/2} M^{1/2} = K_{s,o} M^{1/2} = K_{s,o} M^{\nu}$$

The constant $K_{s,o}$ is independent of the molar mass but specific for the polymer-solvent system because it depends on the parameters b, M_u, τ, and σ. The ratio, $\langle s^2 \rangle_o/M$, is therefore a constant for a given polymer-solvent-temperature system. It is experimentally accessible by scattering experiments on melts (Chapter 6) and solutions (Chapter 5). Measurements in solution require the use of so-called **theta solvents** which are thermodynamically bad solvents. At a certain temperature, the **theta temperature** Θ (Section 10.4.2), all interactions between polymer segments are compensated by those between polymer segments and solvent molecules in such a way that the polymer chain behaves as a **phantom chain**, i.e., as an unperturbed chain without excluded volume. In thermodynamically good solvents, on the other hand, polymer chains have excluded volumes which expand the chains and lead to perturbed coils (Section 4.5).

The dependence of the square root of the unperturbed radius of gyration on molar mass, $s_o \sim M^{\nu}$ with $\nu = 1/2$, is often found for very large ranges of molar masses. An example is atactic poly(styrene) in the theta solvent cyclohexane at 34.5°C where Eq.(4-44) is obeyed by molar masses between 5000 and 4 000 000 g/mol (Fig. 4-25). In the thermodynamically good solvent toluene at 15°C, coils are expanded and the molar mass exponent ν increases to ca. 0.59 from 0.50 (see Section 4.5).

At low molar masses, radii of gyration are substantially smaller than demanded by Eq.(4-44) (Fig. 4-23). Because such oligomer chains are short, they cannot adopt ideal coil statistics but rather behave as wormlike chains with $\nu > 1/2$ (Section 4.4.9). They also do not produce excluded volumes, which causes coil dimensions to become independent of polymer-solvent interactions.

At the same molar mass, completely stretched cyclic macromolecules are only half as long as completely stretched linear ones. On average, this should also apply to all other macroconformations so that $\langle s^2 \rangle_{o,ring} = K_{s,o}(M/2)^{1/2}$ if $\langle s^2 \rangle_{o,lin} = K_{s,o}M^{1/2}$. Thus, in the unperturbed state, the radius of gyration of cyclic molecules should be smaller by the factor $(1/2)^{1/2} = 0.707$ than that of linear molecules of the same molar mass (Fig. 4-23).

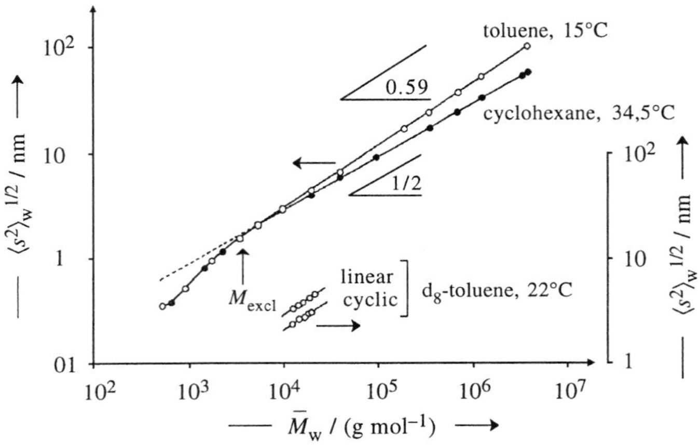

Fig. 4-23 Mass-average radius of gyration as a function of the mass-average molar mass of practically molecularly uniform atactic poly(styrene)s in toluene and cyclohexane [15, 16]. Note that the radii of large cyclic molecules are only one-half of that of linear ones. ↑ Onset of excluded volume effect.

Ring formation restricts the population and sequence of microconformations. Therefore, the theta temperatures of cyclic macromolecules are lower than those of linear ones with the same constitution. An example is high-molar mass poly(styrene) in cyclohexane with Θ = 28.5°C (cyclic) *versus* Θ = 34.5°C (linear).

Physical structures of isolated unperturbed coils are essentially dominated by short-range forces which in turn depend on local chain structures and thus on tacticities. Consequently, isotactic and syndiotactic polymers of the same constitution have different theta temperatures in the same solvent. Correspondingly, unperturbed radii of gyration may differ by as much as 20 %. In good solvents, on the other hand, long-range forces dominate and tacticities affect neither the theta temperature nor the ratio, $\langle s^2 \rangle / M$.

4.4.7 Coil Density

The radius of gyration, $\langle s^2 \rangle_o^{1/2}$, of an unperturbed coil is much smaller than the conventional contour length but considerably greater than the radius of gyration or the geometric radius of a compact sphere (Table 4-8). Coils must be therefore very low-density, loose entities but with what distribution of segments?

The key to understanding coil densities is the observation that arrangements of chain segments in unperturbed coils resemble the random three-dimensional Brownian movement of entities (Section 4.4.4). Distributions of the squares of the lengths of steps of such paths are well known: they correspond to Gaussian distributions (Fig. 2-11). Chain segments are thus not distributed homogeneously in the coil: segment densities are rather high in the center of the coil but far lower at the periphery (Figs. 4-15 and 4-24).

Peter J. Debye approximated the distribution of the local number concentrations, $C = N_{seg}/V$, of segments by a **Gaussian function**, $C = A \exp(-B^2R^2)$, where R = distance of segments from the center of gravity and A, B = model constants. He also assumed that such coils are spherically symmetric with respect to time. The volume of a very thin spherical layer with thickness dR is $4\,\pi R^2 \mathrm{d}R$; it contains $\mathrm{d}N = C(4\,\pi R^2 \mathrm{d}R)$ segments.

Table 4-8 Historic contour length L, end-to-end distance r, radius of gyration s, and spherical radius R of a poly(methylene) molecule, $H(CH_2)_{20000}H$. Bond length $b_{C\text{-}C}$ = 0.154 nm, effective (= crytallographic) bond length b_{eff} = 0.254 nm, bond angle τ =112°, $\langle\cos\theta\rangle$ = 0.5 (corresponds to θ = 60°), partial specific volume $\tilde{v}$ = 1 mL/g. a) With Eq.(4-43); b) RIS model.

Dimension		Equation or table	Symbol	Dimension in nm calculation	experiment
Contour length, historical	L	$L_{cont} = Nb$	L_{cont}	3080	-
Contour length, conventional all-trans	r	Eq.(4-16)	r_{cont}	2553	-
Freely jointed chain	r	Eq.(4-25)	$\langle r^2\rangle_{oo}^{1/2}$	21.8	-
	s	Eq.(4-25) a)	$\langle s^2\rangle_{oo}^{1/2}$	8.9	-
Freely rotating chain	s	Eq.(4-27) a)	$\langle s^2\rangle_{of}^{1/2}$	13.2	-
Chain with restricted rotation	s	Eq.(4-30) a)	$\langle s^2\rangle_{or}^{1/2}$	17.7	-
Unperturbed chain b)	s	Eq.(4-31) a)	$\langle s^2\rangle_{o}^{1/2}$	21.6	23.3
Compact sphere	s	Table 4-1	s_{sphere}	3.74	-
	R	Eq.(4-2)	R_{sphere}	4.82	-

The number N_{seg} of *all* segments in a sphere is obtained from the integration over all such layers from $R = 0$ to $R = \infty$:

(4-45) $$N_{seg} = \int_0^\infty dN = \int_0^\infty 4\pi R^2 C dR = 4\pi R^2 \int_0^\infty R^2 \exp(-B^2 R^2) dR = \pi^{3/2} A / B^3$$

A section of a layer with the area R^2 contains $dN = 4\pi R^2 C dR$ segments. The average of all such squares is the mean-square radius of gyration. Integration of Eq.(4-45) and subsequent introduction of $C = A \exp(-B^2R^2)$ delivers

(4-46) $$\langle s^2\rangle_o = \int_0^\infty R^2 dN / \int_0^\infty dN = \int_0^\infty 4\pi R^4 C dR / \int_0^\infty 4\pi R^2 C dR = 3\pi^{3/2} A / (2 N_{seg} B^5)$$

The combined Eqs.(4-45) and (4-46) allow one to obtain the model constants A and B:

$$A = N_{seg}[3/(2\pi\langle s^2\rangle_o)]^{3/2} \quad \text{and} \quad B^2 = 3/(2\langle s^2\rangle_o)$$

The number of segments, N_{seg}, equals the degree of polymerization, X, if the segments are the monomeric units. The number concentration C of monomeric units thus varies with their distance R to the center of gravity as

(4-47) $$C = A \exp(-B^2/R^2) = X[3/(2\pi\langle s^2\rangle_o)]^{3/2} \exp[-3R^2/(2\langle s^2\rangle_o)]$$

The concentration C is highest at $R = 0$, i.e., at the center of gravity (Fig. 4-24). It decreases with increasing distance from the center (Figs. 4-24 and 4-15). The higher the molecular weight, the smaller the number-concentration of monomeric units at the center of gravity (insert of Fig. 4-24).

At the center of gravity, monomeric units are present in higher concentrations, C_o, in unperturbed coils than in perturbed coils, i.e., in polymers in good solvents. However, it is just the reverse at large distances from the center of gravity (Fig. 4-24). Hence, good solvents decrease the segment density in the exterior of the coil and increase it at the periphery, leading to an increase of the coil volume.

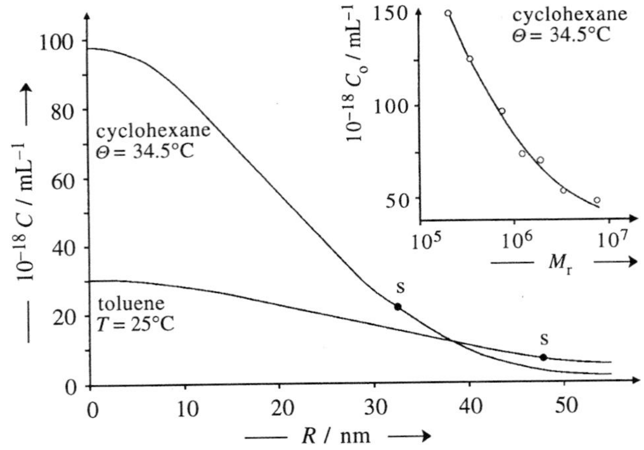

Fig. 4-24 Dependence of the number concentration C of monomeric units of a single molecule of poly(α-methyl styrene), $-[CH_2-C(CH_3)(C_6H_5)]_n-$, on the distance R from the center of gravity. Calculations for a molecule with a molecular weight of M_r = 1 190 000 in the theta solvent cyclohexane at the theta temperature Θ = 34.5°C and in the good solvent toluene at T = 25°C. s = position of the radius of gyration: $\langle s^2 \rangle_o^{1/2}$ = 32.4 nm in cyclohexane, $\langle s^2 \rangle^{1/2}$ = 48.0 nm in toluene.

Insert: number concentration C_o of monomeric units at the center of gravity as a function of the logarithm of the molecular weight for poly(α-methyl styrene)s in cyclohexane at Θ = 34.5°C.

At the center of gravity, the radius R is zero and the concentration of monomeric units is $C_0 = [3/(2\,\pi)]^{3/2} X \langle s^2 \rangle_o^{-3/2}$ (Eq.(4-47)). The mean-square radius of gyration of unperturbed chains is proportional to the molecular weight M, $\langle s^2 \rangle_o \sim M$ (Eq.(4-44)). Since $M \sim X$, the concentration $C_0 = const\ M^{-1/2}$ of monomeric units at the center of gravity thus decreases with the square root of the molecular weight (insert in Fig. 4-24)

Concentrations of monomeric units in unperturbed coils are very low. For example, a monomeric unit of poly(α-methyl styrene) occupies a molar van der Waals volume of $V_{u,m}$ = 74 mL/mol which corresponds to a volume of $V_{u,m}/N_A$ = 1.23·10^{-22} mL. At the center of gravity of the unperturbed coil of a macromolecule with the molar mass M = 1 190 000 g/mol, the volume fraction of monomeric units is therefore ϕ_{seg} = (1.23·10^{-22} mL)×(97.7·10^{18} mL^{-1}) = 0.012. In solution, only 1.2 % of the coil volume at the center of gravity is occupied by monomeric units whereas 98.8 % are solvent molecules. In melts (Chapter 6), these 98.8 % are segments of other polymer molecules.

The Gaussian function is asymptotically exact for the limiting case of very long freely jointed chains. The number concentration C of monomeric units can never become zero because Gaussian curves extend to infinity (Section 2.4.2). However, real coil molecules do have an outer "border" where C = 0 since distances R from the center of gravity cannot exceed historic contour lengths, $L_{chain} = Nb$.

However, there is a mathematical distribution function that fulfills the physical requirement of finite extensions of coil molecules. This function was first derived by W.Kuhn and F.Grün and then modified by P.J.Flory. It considers a chain with N bonds that is projected on an axis. If $\boldsymbol{b}_{xj}$ is the projection of a single bond and N_j the number of bonds in an interval $\boldsymbol{b}_{xj} + \delta\boldsymbol{b}_{xj}$, then the total projection x must be $\Sigma_j N_j \boldsymbol{b}_{xj}$. For $\Sigma_j N_j = N$, the calculation delivers

(4-48) $$\Sigma_j N_j \boldsymbol{b}_{xj} = Nb\ \pounds(\beta)$$

where $\pounds(\beta) = \coth \beta - 1/\beta$ is the **Langevin function** of β and β a Lagrangian multiplier.

Setting $\beta = \pounds^*(\Sigma_j N_j b_{xj} / Nb) = \pounds^*(r/N_{seg}L_{seg})$, Eq.(4-48) is written frequently as the inverse Langevin function $\pounds^*$ with the fractional extension $r/N_{seg}L_{seg}$ where N_{seg} = number of segments of length L_{seg} of a volume-less freely rotating chain with an end-to-end distance $r = |\mathbf{r}|$ that is the absolute value of the end-to-end vector of the chain.

According to Flory, the "correct $\pounds^*$ distribution" of vectors between chain ends is

(4-49) $$p(r) = (A\beta / rL_{seg}^2)(\beta^{-1}\sinh\beta)^{N_{seg}} \exp\left(-\beta r / L_{seg}\right)$$

where A = normalization constant. The "incorrect $\pounds^*$ distribution" of Kuhn and Grün has a factor A' / L_{seg}^3 instead of $A\beta / rL_{seg}^2$. Both distributions deliver $p(r) = 0$ for $r = N_{seg}L_{seg}$ and both reduce to the Gaussian distribution for $r/N_{seg}L_{seg} < 1/3$. The two distributions do not differ much if N_{seg} is very large. In most cases, one can therefore use the Gaussian distribution. An exception is the case of highly extended elastomers where relatively short segments between crosslinking sites are strongly stretched which causes $r/N_{seg}L_{seg}$ to approach unity.

4.4.8 Distribution of End-to-End Distances

Because Brownian movements cause populations and sequences of microconformations to change rapidly, molecularly uniform polymers consisting of coil molecules possess at any moment a distribution of end-to-end distances. According to Eq.(A 4-12) in the Appendix to this chapter, one obtains for the distribution function $p(r_x)$ of the x-component of the end-to-end distance r in the one-dimensional case:

(4-50) $$p(r_x) = \left(\frac{1}{2\pi\langle r_x^2\rangle}\right)^{1/2} \exp\left(-\frac{r_x^2}{2\langle r_x^2\rangle}\right)$$

This one-dimensional distribution function can be generalized for the three-dimensional case of an unperturbed chain. The **radial distribution function** of end-to-end distances describes the probability $p(R)$ of finding a chain end in a spherical layer at the distance R from the center of gravity:

(4-51) $$p(R) = 4\pi R^2\left(\frac{3}{2\pi\langle r^2\rangle_o}\right)^{3/2} \exp\left(-\frac{3r^2}{2\langle r^2\rangle_o}\right)$$

This radial distribution function has a maximum at a distance R that corresponds to the most probable end-to-end distance. The value of R_{max} is obtained by differentiating Eq.(4-51) and setting $dp(R)/dR = 0$,

(4-52) $$\frac{dp(R)}{dR} = 8\pi R\left(\frac{3}{2\pi\langle r^2\rangle_o}\right)^{3/2}\left(1 - \frac{3R^2}{2\langle r^2\rangle_o}\right) \exp\left(-\frac{3R^2}{2\langle r^2\rangle_o}\right) = 0$$

to give $R_{max} = (2\langle r^2\rangle_o/3)^{1/2}$.

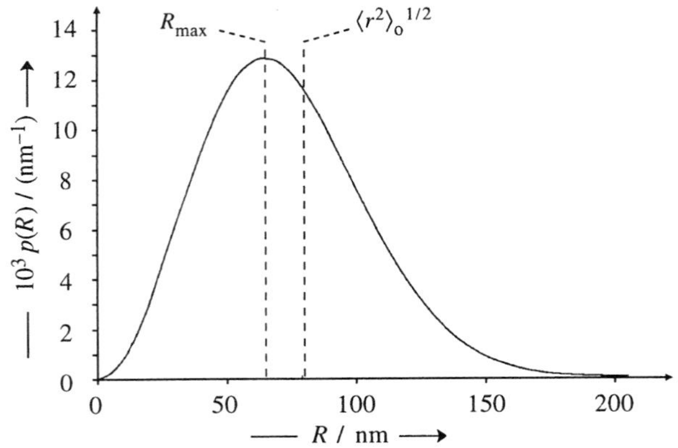

Fig. 4-25 Probability $p(R)$ of finding a chain end at the distance R from the center of gravity. Calculation for a freely jointed chain of poly(α-methyl styrene) with M_r = 1 190 000 in the theta solvent cyclohexane Θ = 34.5°C. $\langle s^2 \rangle_o^{1/2}$ = 32.4 nm, $\langle r^2 \rangle_o^{1/2} = 6^{1/2} \langle s^2 \rangle_o^{1/2}$ = 79.36 nm.

The position R_{max} of the maximum of the radial distribution function (= the most *probable* end-to-end distance) does not coincide with the position of the *average* end-to-end distance, $\langle r^2 \rangle_o^{1/2}$ (Fig. 4-25). Furthermore, the probability of finding *both* chain ends at the center of the coil is zero. However, the probability of finding *a segment* is highest at this position (Fig. 4-24).

These equations apply to freely jointed chains composed of N segments of length b. For unperturbed chains, Nb^2 has to be replaced by $N_{b,u}b_{eff}^2XC_N$ where $N_{b,u}$ = number of bonds per monomeric unit, b_{eff} = effective length of such a bond, X = degree of polymerization, and C_N = characteristic ratio.

In the three-dimensional case, the probability density $W(r)$ of finding both chain ends in the same volume element is given by

(4-53) $$W(r) = \left(\frac{3}{2\pi N_{b,u} b_{eff}^2 X C_N} \right)^{3/2}$$

4.4.9 Wormlike Chains

Locally, polymer chains are not completely flexible. Even a monomeric unit of an infinitely thin freely rotating chain can occupy only those positions that are allowed by the rigid bond angle of the preceding unit (see Fig. 3-1). Additional restrictions are imposed by large substituents, helical chain structures, and the like.

Such **semiflexible** to **rigid chains** resist coiling. They rather curl like a worm or a not too long garden hose and can be visualized as elastic wires with flexural energies.

Wormlike chains (**Kratky-Porod chains**) are described by their **persistence lengths,** $L_{ps} \equiv b/(1 + \cos \tau)$, where b = bond length and τ = bond angle. The mean-square end-to-end distance is then calculated for a freely rotating chain with vanishingly small bond lengths ($b \to 0$), bond angles near 180° ($\tau \to \pi$), and the number of bonds approaching infinity ($N \to \infty$).

In the exact equation for a freely rotating chain (see Appendix to Chapter 4),

(A 4-7) $$\langle r^2\rangle_{of} = Nb^2\left[\frac{1-\cos\tau}{1+\cos\tau}+\frac{2\cos\tau}{N}\left(\frac{1-(-\cos\tau)^N}{(1+\cos\tau)^2}\right)\right]$$

the historical contour length L_{chain} is expressed by the number N_{ps} of segments with a persistence length L_{ps}, i.e., $L_{chain} = Nb = N_{ps}L_{ps}$, and the bond length b by $b = L_{ps}(1 + \cos\tau)$ which leads after rearranging to

(4-54) $$\langle r^2\rangle_{of} = L_{chain}L_{ps}(1-\cos\tau) + 2\,L_{ps}{}^2(\cos\tau)[1-\{1-(L_{chain}/NL_{ps})\}^N]$$

The term in braces is rearranged to

(4-55) $$\left\{1-\frac{L_{chain}}{NL_{ps}}\right\}^N = \left[\left\{1-\frac{L_{chain}}{NL_{ps}}\right\}^{NL_{ps}/L_{chain}}\right]^{L_{chain}/L_{ps}} \quad ; \ \lim_{x\to\infty}[1-\{1/x\}]^x = e^{-1}$$

The right side of Eq.(4-55) is an exponential, $[1-\{1/x\}]^x$, with $x = NL_{ps}/L_{chain}$ and a limit of $1/e$ for $x \to \infty$. In the limit of $\tau \to 180°$, one obtains $\cos\tau = -1$ and $(1-\cos\tau) \to +2$ and the freely rotating chain converts to the wormlike chain with $\langle r^2\rangle_{worm}$ replacing $\langle r^2\rangle_{of}$:

(4-56) $$\langle r^2\rangle_{worm} = 2\,L_{ps}L_{chain} - 2\,L_{ps}{}^2\{1-\exp(-L_{chain}/L_{ps})\}$$

(4-56a) $$\langle r^2\rangle_{worm} = 2\,N_{ps}L_{ps}{}^2[1-(1/N_{ps}) + (1/N_{ps})\exp(-N_{ps})]$$

These equations show that the Kratky-Porod wormlike chain is described by just two parameters, either L_{chain} and L_{ps} (Eq.(4-56)) or L_{ps} and N_{ps} (Eq.(4-56a)). Contrary to all other molecular models, wormlike chains do not require the knowledge of the number of segments for a full description (Eq.(4-56)).

Kratky-Porod chains have two limiting cases: random coils with flexible chains and stiff chains:

Random coils with flexible chains. For these chains, historic contour lengths, $L_{chain} = Nb$, are much larger than persistence lengths. The condition $Nb/L_{ps} = N_{ps} >> 1$ leads to $\exp(-N_{ps}) \to 0$ and Eq.(4-56a) simplifies to $\langle r^2\rangle_{worm} = 2\,N_{ps}L_{ps}{}^2$. Since historic contour lengths can be written in terms of persistence lengths or Kuhnian lengths, $L_{chain} = N_{ps}L_{ps} = N_K L_K$, and also $\langle r^2\rangle_o = N_K L_K{}^2$ (Eq.(4-39)), one sees that Kuhnian lengths are twice as large as persistence lengths, $L_K = 2\,L_{ps}$.

Stiff chains. The persistence length of such chains is much larger than the historic contour length, $L_{chain} = Nb$. Developing the exponential $\exp(-N_{ps})$ of Eq.(4-56a) into a series, $\exp(-N_{ps}) = 1 - N_{ps} + (N_{ps}{}^2/2) - \ldots$, terminating the series after the third term, and inserting the result in Eq.(4-56a) delivers $\langle r^2\rangle_o{}^{1/2} = N_{ps}L_{ps}$. The end-to-end distance of an infinitely stiff chain thus equals the length of a rigid rod.

The mean-square radius of gyration is obtained by a similar procedure:

(4-57) $$\langle s^2\rangle_{worm} = L_{ps}{}^2[(y/3) - 1 + (2/y) - (2\,y^{-2})\{1-\exp(-y)\}] \quad ; \quad y = L_{chain}/L_{ps}$$

In the limit $L_{chain}/L_{ps} = y \to 0$, one obtains the radius of gyration of an infinitely thin rod. Expanding $\exp(-y) = 1 - y + (y^2/2!) - (y^3/3!) + (y^4/4!) - \ldots$ in a series, inserting the series in Eq.(4-57), and using $y = L_{chain}/L_{ps}$ results in the expression for the mean-square radius of gyration of an infinitely thin rod (cf. Eq.(4-5)):

$$(4\text{-}58) \qquad \langle s^2 \rangle_{of} \approx L_{ps}^2 y^2/12 = (L_{chain})^2/12$$

A universal function is obtained from Eq.(4-57) with $L_{chain} = N_K L_K$ and $L_K = 2\,L_{ps}$:

$$(4\text{-}59) \qquad \langle s^2 \rangle_{worm}/(4\,L_{ps}^2) = (N_K/6) - (1/4) + [1/(4\,N_K)] - [1/(8\,N_K^2)][1 - \exp(-2\,N_K)]$$

Plotting $\langle s^2 \rangle_{worm}/(4\,L_{ps}^2)$ as a function of the number N_K of Kuhnian segments delivers a curve that is universal for all wormlike chains (Fig. 4-26). At low molecular weights (small N_K), one obtains the limiting line for rigid rods (slope of 2) and, at higher molecular weights (large N_K), the limiting line for unperturbed coils (slope of 1). At $1 \leq N_K \leq 10$, rods convert to coils and the exponent ν in $\langle s^2 \rangle_o = K_s M^\nu$ changes from ν = 1 (rigid rods) to ν = 1/2 (unperturbed coils). Note that these chains have finite thicknesses whereas the Kratky-Porod model applies to infinitely thin chains. This causes an error which can be neglected if the persistence length is much larger than the chain diameter.

Hence, very few Kuhnian segments are necessary to convert a rod to a coil. The required molecular weights are also small as shown by the numerical values of reduced relative molar masses, $M_{r,L} = M_r/L_{chain}$, i.e., the relative molar mass M_r per historic contour length, L_{chain} (Table 4-9). The only true macromolecular rod known is the tobacco mosaic virus, a helical deoxyribonucleic acid that is studded with protein molecules. All other "rodlike" molecules are stiff but not real rods.

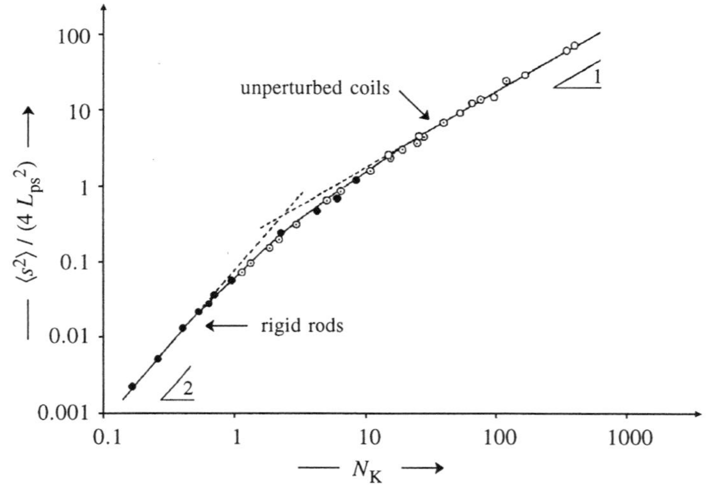

Fig. 4-26 Reduced mean-square radius of gyration as a function of the number N_K of Kuhnian segments of wormlike chains. L_{ps} = persistence length. Plot according to Eq.(4-58) [17].
O Poly(1-phenyl-1-propyne) in cyclohexane at Θ = 36°C [18];
ʘ Poly(hexylisocyanate) in hexane at 25°C [19];
● Schizophyllan (triple helix of a β-1,3-D-glucan) in 0.01 mol/L NaOH at 25°C [20].

Table 4-9 Diameters $d_{sp} = 4\ v(M/L_{chain})/(\pi N_A)^{1/2}$ from specific volumes v and d_η from intrinsic viscosities (Chapter 12), persistence lengths L_{ps}, and reduced molar masses M/L_{chain} of rigid macromolecules. Imogolite is a tubelike aluminum silicate that contains 10 Gibbsite units $SiO_2 \cdot Al_2O_3 \cdot 2\ H_2O$ per constitutional repeating unit (M_u = 3962 g/mol). Schizophyllan (Volume II, p. 411) and xanthan (Volume I, p. 551; Volume II, p. 407) are polysaccharides. Most data from a compilation of [21].

Polymer	Solvent	$\frac{T}{°C}$	$\frac{d_{sp}}{nm}$	$\frac{d_\eta}{nm}$	$\frac{L_{ps}}{nm}$	$\frac{M/L_{chain}}{g/(mol\ nm)}$
Tobacco mosaic virus	aqueous buffer solution	25	18		(∞)	133 000
Imogolite	1.2 % acetic acid in water	30	2		170	6170
Schizophyllan	water	25	1.68	2.6	200	2150
	0.01 mol/L NaOH in H_2O	25	1.7		150	2170
Poly(γ-benzyl-L-glutamate)	*N,N*-dimethylformamide	25	1.56	1.7	150	1450
Xanthan	0.1 mol/L NaCl in H_2O	25	2.2		120	1940
Deoxyribonucleic acid	0.2 mol/L NaCl in H_2O	20	2.5		61	1950
Poly(*p*-benzamide)	96 % H_2SO_4				50	198
Poly(hexylisocyanate)	toluene	25	1.24	1.6	37	740
	dichloromethane	20			21	740
Poly(*p*-phenylene terephthalamide)	98 % sulfuric acid		0.51	0.6	18	198
Hydroxypropylcellulose	dimethylacetamide				6.5	720
Acetoxypropyl cellulose	dibutyl phthalate	36	1.19		5.9	821
Cellulose triacetate	trifluoroacetic acid	25	0.95		5.3	560
Poly(isobutylene)	benzene	24		0.73	0.59	241

Kratky-Porod chains are volumeless wormlike chains that have to be replaced by, for example, helical wormlike chains if chain diameters and persistence lengths are comparable. An example is poly(isobutylene) (Table 4-9). Such chains of finite thickness produce external volumes and thus perturbed chains.

4.5 Perturbed Coils of Linear Chains

4.5.1 Excluded Volumes

Real chains are not infinitely thin phantom chains, which leads to three problems. First, all scattering methods do not deliver the mean-square radius of gyration, $\langle s^2 \rangle$, *per se* but an apparent mean-square radius of gyration, $\langle s^2 \rangle_{app} = \langle s^2 \rangle + s_a^2$, that contains a contribution by the cross-sectional area of the chain, s_a^2. The term s_a^2 can be neglected for high chain volumes, i.e., for high degrees of polymerization of thin linear chains. It becomes significant for linear oligomers and for bulky cross-sections and probably also for some dendrimers.

Second, a non-negligible cross-sectional area prevents a segment of a high-molecular weight chain to occupy a space that is already occupied by another segment of the same chain: the chain is perturbed. Because of the persistence of chains, this perturbation is not caused by short-range interactions (adjacent segments) but by long-range interactions (Fig. 4-16). These "long-range" forces also act in single chains so that their excluded volume is an *intramolecularly* **excluded volume**.

Third, the volume requirement of a segment is also controlled by interactions between segments. Repulsive forces cause excluded volumes to increase; attractive forces cause them to decrease. These forces furthermore compete with those between segments and solvent molecules so that excluded volumes depend on the thermodynamic goodness of solvents for the polymer.

4.5.2 Expansion Factors

Excluded volumes expand polymer coils, which can be described by expansion factors $\alpha \geq 1$ for the radius of gyration, s, or the end-to-end distance, r:

$$(4\text{-}60) \qquad \langle s^2 \rangle = \alpha_s{}^2 \langle s^2 \rangle_o \qquad ; \qquad \langle r^2 \rangle = \alpha_r{}^2 \langle r^2 \rangle_o$$

The larger the perturbation by the expansion, the larger are α_s and α_r, and the "better" is the solvent. Unperturbed coils are characterized by $\alpha_s{}^2 = \alpha_r{}^2 = 1$.

Eqs.(4-60) define expansion factors for the linear expansion of radii and end-to-end distances. In reality, these factors are averages over all spatial dimensions since polymer coils are not spherical, Eq.(4-19), and expand differently in different directions. The distribution of polymer segments thus deviates from that of a Gaussian distribution which is the reason why expansion factors of radii of gyration, α_s, differ from those of end-to-end distances, α_r, or from those of hydrodynamic measurements, α_h.

In order to calculate the excluded volume of a segment, u_{seg}, one segment is placed at the center of the coordinate system and another one on an infinitely thin shell with a surface of $O = 4\,\pi R^2$ at a distance R from that center. The probability of finding two segments at the same location must decrease exponentially with their distance from the center of the molecule. The proportionality factor is the potential function $\varphi(R)$ that is normalized for the same energy, $k_B T$, where k_B = Boltzmann constant. The potential function depends only on the distance (= radius) R from the center but not on the direction.

$$(4\text{-}61) \qquad u_{seg} = 4\,\pi \int_0^{\infty} [1 - \exp(-\varphi(R))/(k_B T)]\, R^2 \mathrm{d}R$$

u_{seg} is the so-called **binary cluster integral** which describes the volume of one segment that is excluded for all other segments. It is a polymer-specific quantity that is independent of the molecular weight but has a physical meaning only if the potential function is smaller than the mean-square radius of gyration.

In order to obtain a universal function, one defines an **excluded volume parameter** z:

$$(4\text{-}62) \qquad z \equiv (4\,\pi)^{-3/2} [M/\langle s^2 \rangle_o]^{3/2} (u_{seg}/M_{seg}{}^2) M^{1/2}$$

$M/\langle s^2 \rangle_o$ is independent of the molecular weight for polymers with the same chemical structure since $(\langle s^2 \rangle_o/M)^{1/2} = K_{s,o}$ (Eq.(4-44)). The excluded volume parameter is therefore proportional to the square root of the molecular weight.

The parameter z cannot be measured directly since segments, excluded volumes u_{seg}, and molecular weights M_{seg} are not clearly defined. Therefore, neither Eq.(4-62) nor theoretically calculated functions $\alpha_s = \mathrm{f}(z)$ can be tested by independent experiments.

So far, the search for a theoretical, universal, closed function $\alpha_s = f(z)$ for the whole range of α_s values has been unsuccessful. However, limiting values for very small and very large z are known. For very small z one can make three assumptions: (1) the probability for the distribution of distances between segments follows a Gaussian distribution, similar to that of segments in undisturbed coils; (2) the potential for the interaction of segments is additive; and (3) the pair potential follows the expression

$$\exp(-\varphi(r)/k_BT) = 1 - u_{seg}\delta(\boldsymbol{r}) \approx \exp[-u_{seg}\delta(\boldsymbol{r})] \tag{4-63}$$

where $\boldsymbol{r}$ is the vectorial distance between two segments, and $\delta(\boldsymbol{r})$ is the three-dimensional Dirac delta function. According to this theory, expansion factors α_s and α_r can be expressed by a power series of z with exactly known coefficients:

$$\alpha_s^2 = 1 + (134/105)\,z + [(536/105) - (1247/1296)\,\pi]z^2 + \ldots \tag{4-64}$$

$$\alpha_r^2 = 1 + (140/105)\,z + [(32/105) - (97/1296)\,\pi]z^2 + 6.459\,z^3 + \ldots \tag{4-65}$$

The coefficients indicate the probability of segment-segment contacts: the first coefficient of Eq.(4-64), $134/105 \approx 1.276$, that of a segment with another one; the second coefficient, $(536/105) - (1247/1296)\,\pi \approx 2.082$, that of a segment with two other ones; etc.

The series, Eqs.(4-64) and (4-65), converge very slowly. For example, the first three terms of Eq.(4-64) deliver a *negative* value of $\alpha_s^2 \approx -27.21$ for $z = 4$. Hence, these series can therefore only be used for small values of z, i.e., $z \leq 0.10$ for α_s and $z \leq 0.15$ for α_r (Fig. 4-27). In this range, α_s can be represented by $\alpha_s^3 = 1 + 2\,z$ with less than 3 % deviation from the exact value. Most experimental values are at much larger z values, however, usually in the range $1 \leq z \leq 5$.

Many semi-empirical equations have been proposed for this range, for example, the Flory equation, $\alpha_s^5 - \alpha_s^3 = K_F z$, and the Tanaka equation, $\alpha_s^5 = 1 + 1.90\,z$. A good fit is obtained by computer calculations that furnished $K_\alpha^5 = 2.90\,z$ for $z \to \infty$.

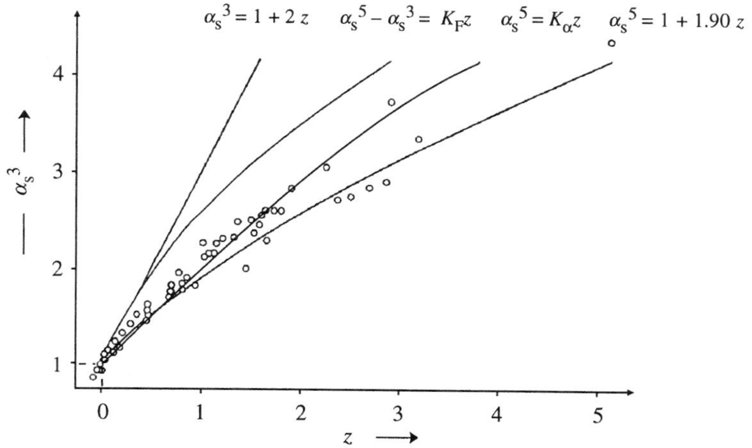

Fig. 4-27 Cubic expansion factor of the radius of gyration of poly(styrene) in various solvents as a function of the z parameter. O Experiments, — theoretical equations listed at the top of this figure.
From left to right: exact solution for very small values of z; Flory equation with $K_F = 2.276$; computer calculation with $K_\alpha = 2.90$; and half-empirical Tanaka equation.

4.5.3 Helical Wormlike Chains

The theories and calculations presented in the preceding section neglect the fine structure of polymer chains, for example, helical chains. These aspects are considered in the **Yamakawa-Stockmayer-Shimada theory** (**YSS theory**) of helical wormlike chains (**HW theory**). The HW theory describes not only the behavior of helical wormlike chains in solution but also the dynamics of polymer molecules with randomly distributed dipole moments, i.e., the dielectric relaxation of chains. The derivations are fairly complex (see Literature to Chapter 4) so that just the results will be discussed.

The HW theory is based on four parameters: the differential geometric curvature κ_0 and the torsion τ_0 of the characteristic helix at its energy minimum; a stiffnes parameter Λ^{-1}; and a shift factor, $M_L = M/L_{chain} = M/(Nb)$ which is the molar mass per historic contour length, the product of the number N and length b of chain bonds. The first two parameters can be obtained by independent measurements such as X-ray data of crystalline polymers (Chapter 7). Since only the last two parameters are adjustable, this is a quasi two-parameter theory (QTP theory).

This theory replaces the series $\alpha_s{}^2 = f(z)$, Eq.(4-64), for the dependence of the square of the expansion factor of the radius of gyration, $\alpha_s{}^2$, Eq.(4-65), on the excluded volume parameter, $\tilde{z}$, Eq.(4-62) by the **Domb-Barrett equation**

(4-66) $$\alpha_s{}^2 = f(\tilde{z}) = [1 + 10\tilde{z} + (70\,\pi/9 + 10/3)\tilde{z}^2 + 8\,\pi^{3/2}\tilde{z}^3]^{2/15} \times [0.933 + 0.067\exp(-0.85\tilde{z} - 1.39\tilde{z}^2)]$$

which writes $\alpha_s{}^2$ as a function of an modified excluded volume parameter, $\tilde{z}$:

(4-67) $$\tilde{z} = \left(\frac{3}{2\,\pi}\right)^{3/2}\left(\frac{\Lambda M_L}{C_\infty}\right)^{3/2}\left(\frac{u_{seg}}{M_{seg}^2}\right)M^{1/2}$$

In the limit $\Lambda L_{chain} \rightarrow \infty$, i.e., for a long flexible chain, the characteristic length becomes $C_\infty = 6\,\langle s^2\rangle_o \Lambda/L_{chain}$. With $L_{chain} = M/M_L$, Eq.(4-67) reduces to Eq.(4-62), i.e., the expression for the excluded volume parameter z,.

The theory describes the dependence of the expansion factor, α_s, on the modified excluded volume parameter, $\tilde{z}$, by a universal curve for both flexible coils of non-helical polymers such as atactic poly(styrene) and poly(methyl methacrylate) and wormlike helical chains such as poly(isobutylene) (Fig. 4-28).

4.5.4 Molar Mass Dependence

Instead of trying to theoretically calculate the molar mass dependence of the radii of gyration s from expansion factors α_s or excluded volume parameters z or $\tilde{z}$, one can also approach the problem semi-empirically. Combining the expression for the molar mass dependence of the unperturbed mean-square radius of gyration, $\langle s^2\rangle_o = K_{s,o}^2 M$, Eq.(4-44), with the definition of the expansion factor, $\langle s^2\rangle = \alpha_s^2\langle s^2\rangle_o$, Eq.(4-60), yields

(4-68) $$\langle s^2\rangle = \alpha_s^2 K_{s,o}^2 M$$

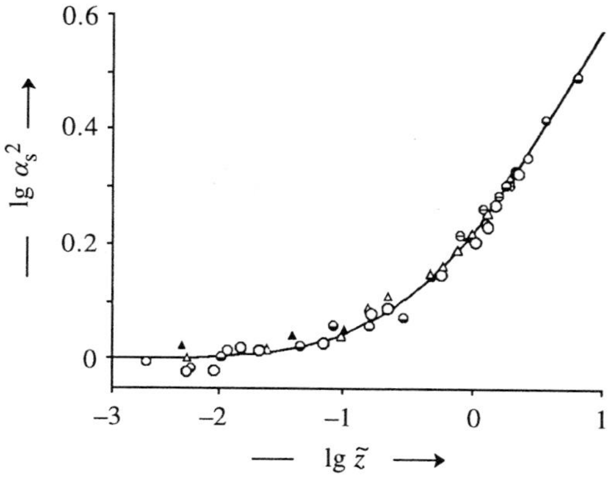

Fig. 4-28 Logarithm of the square of the expansion factor of the radius of gyration, lg α_s^2, as a function of the logarithm of the modified excluded volume parameter, $\tilde{z}$ [22].

Experimental data: (O) poly(isobutylene) in heptane at 25°C; (●) at-poly(styrene) in toluene at 15°C; atactic (Δ) and isotactic (▲) poly(methyl methacrylate) in acetone at 25°C; and (⊖) poly(dimethylsiloxane) in toluene at 25°C. Solid line: YSS theory.

By permission of the American Chemical Society, Washington (DC).

Since the expansion factor is a function of the excluded volume factor, $\alpha_s = f(z)$, and the excluded volume factor varies with the molar mass, $z = f(M)$, one must also have $\alpha_s = f(M)$. By definition, the expansion factor of perturbed coils must be greater than unity. It must also depend on some power of the molar mass which can be written as $\alpha_s^2 = K_\alpha^2 M^{2\nu-1}$ with ν as an empirical constant. Inserting this expression for α_s in Eq.(4-68) delivers

$$(4\text{-}69) \qquad \langle s^2 \rangle^{1/2} = K_\alpha K_{s,o} M^\nu = K_s M^\nu \quad ; \quad \nu \geq 1/2$$

where the exponent ν must be greater than 1/2 since $\langle s^2 \rangle_o^{1/2} = K_{s,o} M^{1/2}$.

The magnitude of the exponent ν can be estimated by assuming that the same mean force field acts on each segment. This theory is known as **mean-field theory** which is not fully justified since the averaging over all energetic contributions makes no explicit assumption for the average force field.

The theory assumes two opposing effects. *Repulsion forces* between segments expand the coils, which leads to a smaller number of possible microconformations. These forces are opposed by elastic forces that try to increase the number of microconformations. The Gibbs energy, $\Delta G = \Delta G_{rep} + \Delta G_{el}$, is thus the sum of two factors.

The contribution by the **repulsion**, ΔG_{rep}, consists of four parts:

- The repulsion increases with increasing average concentration C_K of Kuhnian segments in the coil, $C_K = N_K/V_{coil}$. The coil is pictured as an **equivalent sphere** with a volume of $V_{coil} = (4\,\pi\, s^3)/3$ where the effective radius is the radius of gyration, $s \equiv \langle s^2 \rangle^{3/2}$. Note that the actual radius of a *compact* sphere is larger than its radius of gyration (Table 4-1).
- The repulsion increases with the number N_K of Kuhnian segments and also
- with increasing thermal energy $k_B T$.

There must be a net contribution by the polymer-solvent interaction because it is larger than the contributions by polymer-polymer and solvent-solvent interactions. Dimensional analysis shows that this net contribution must have the physical unit of an effective volume, $V_{eff} = K_{eff}L_K^3$, with L_K = length of a Kuhnian segment because otherwise the physical unit of the repulsion term, ΔG_{rep}, would not match the physical unit of the elastic term.

The total contribution by repulsion is given by the product of these four partial contributions:

$$(4\text{-}70) \qquad \Delta G_{rep} = C_K N_K k_B T V_{eff} = (3\ K_{eff}/4\ \pi) k_B T N_K^2 L_K^3 s^{-3}.$$

The **elastic retraction force**, ΔG_{el}, is assumed to be equal the Gibbs energy of rubber elasticity (see Section 16.4.3), which is strictly valid only for a three-dimensional network consisting of many network chains but is applied here to a polymer coil consisting of $N_c = 1$ network chain of functionality $f = 2$ (two endgroups) which expands in the three spatial directions with the same expansion factor, $\alpha_s \equiv \lambda_x = \lambda_y = \lambda_z$. According to Eq.(16-44), the elastic retraction force for a large expansion, $\alpha_s >> 1$, of a single molecule ($N_c = 1$) is given by $\Delta G_{el} = (k_B T/2)(3\ \alpha_s^2 - 3)$. Introducing $\alpha_s^2 = s^2/s_o^2$ and $s_o^2 = r_o^2/6 = N_K L_K^2/6$ leads to $\Delta G_{el} = (k_B T/2)(18\ N_K^{-1} L_K^{-2} s^2 - 3) = 9\ k_B T N_K^{-1} L_K^{-2} s^2$.

The total Gibbs energy is therefore

$$(4\text{-}71) \qquad \Delta G = \Delta G_{rep} + \Delta G_{el} = (3\ K_{eff}/4\ \pi)\ k_B T\ N_K^2 L_K^3 s^{-3} + 9\ k_B T N_K^{-1} L_K^{-2} s^2$$

The energy minimum is obtained from the first derivative of ΔG with respect to the radius of gyration:

$$(4\text{-}72) \qquad \partial \Delta G/\partial s = -\ 3\ (3\ K_{eff}/4\ \pi)\ k_B T\ N_K^2 L_K^3 s^{-4} + 18\ k_B T\ N_K^{-1} L_K^{-2} s = 0$$

Solving for s, introducing $N_K = M/M_K$, and combining all constants to a single constant K_s delivers

$$(4\text{-}73) \qquad s \equiv \langle s^2 \rangle^{1/2} = \{6^{-1/5}[3\ K_{eff}/(4\ \pi)]^{1/5} L_K M_K^{-3/5}\} M^{3/5} = K_s M^{3/5} = K_s M^{\nu}$$

The mean-field theory thus predicts that the radius of gyration of perturbed coils increases with the 0.6th power of the molar mass. This exponent $\nu = 3/5$ is often called the **Flory exponent**. It characterizes a fractal dimension (Section 4.8.2).

Approximately the same value of $\gamma_3/2 = 1.167/2 \approx 0.584$ is obtained from self-avoiding walks in three spatial dimensions (p. 88). Improved mathematical models deliver slightly different numerical values for ν, for example, $\nu = 0.588$ by renormalization which is a mathematical procedure for the calculation of properties by doubling, quadrupling, octupling, etc., of essential parameters. For polymer coils, such a parameter is the segment length whose doubling, quadrupling, etc., changes the calculated excluded volume until a limiting value is reached after some steps. According to renormalization theory, exponents ν can adopt only two values, $\nu = 0.500$ for unperturbed coils and $\nu = 0.588$ for perturbed ones, both in the limit of infinite molar masses.

Experimentally, a value of $\nu \approx 0.590$ is obtained for, e.g., atactic poly(styrene) with $M > 10\ 000$ g/mol in the good solvent toluene at 15°C (Fig. 4-23) which agrees excellently with the predicted value of $\nu = 0.588$. However, many polymers in good solvents often show $0.500 < \nu < 0.588$, usually because of narrow ranges of medium molar masses.

Table 4-10 Exponents ν of the dependence of the radius of gyration on molar mass, M, Eq.(4-73). CD = square of the coefficient of determination; x_s = mole fraction of syndiotactic diads. * Ditto for poly(D,L-β-methyl-β-propiolactone). Θ = theta temperature.

Polymer	$10^{-3}M/(\text{g mol}^{-1})$	x_s	Solvent		T/°C	ν	CD	Ref.
Poly(methyl methacrylate), at	5.5 - 2830	0.79	acetonitrile	Θ	44	0.501	1.000	[23]
Poly(styrene), at	5.4 - 3900	0.59	cyclohexane	Θ	34.5	0.501	1.000	[15]
Poly(α-methyl styrene), at	342 - 7500	0.40	cyclohexane	Θ	34.5	0.499	0.999	[24]
	768 - 7500	0.40	*trans*-Decalin®	Θ	9.5	0.492	0.998	[24]
	204 - 7500	0.40	toluene		25	0.577	0.997	[24]
Poly(styrene), at	5.4 - 3900	0.59	toluene		15	0.590	1.000	[15]
Poly(D-β-hydroxybutyrate) *	86.5 - 9100	it+at	trifluoroethanol		25	0.603	0.999	[25]
Deoxyribonucleic acid	200 - 6000	-	water (buffer)		20	0.58		[6,7]
Poly(ethylene), linear	19 - 771	-	1,2,4-$C_6H_3Cl_3$		135	0.590		[26]

For large ranges of sufficiently high molar masses, exponents ν of polymers in good solvents always seem to approach ν ≈ 0.59 ± 0.01 in good agreement with theoretical predictions (Table 4-10). This exponent applies for flexible, semiflexible, and rigid chains, provided that $M \to \infty$.

Since long-range forces dominate in good solvents, coil expansions are practically not influenced by local effects such as various microconformations by different tacticities. For example, the same exponent ν is found in good solvents for poly(D-β-hydroxybutyrate) and poly(D,L-β-methyl-β-propiolactone), both of which have the same constitution $\text{+O–CH(CH}_3\text{)–CH}_2\text{–CO+}_n$) although the former was isotactic and the latter consisted of stereoblocks and atactic segments (Table 4-10). This behavior contrasts with that in the unperturbed state where short-range forces dominate.

The expansion of the coil is caused by the increase of the excluded volume which is a function of the molar mass and can be described by the z-parameter, Eq.(4-62). At low molecular weights, only relatively few segments are present which cannot generate excluded volumes. Such a low-molecular weight coil thus behaves as an unperturbed one despite being in a thermodynamically good solvent.

In principle, one should be able to obtain unperturbed volumes from perturbed ones by suitable extrapolations. At $z < 0.1$, one can approximate the dependence of the expansion factors on z by $\alpha_s^3 = 1 + 2z$ (p. 109). Expression of α_s by Eq.(4-60) and z by Eq.(4-62) delivers the **Baumann-Stockmayer-Fixman equation** which is an offshoot of the Stockmayer-Fixman theory of dilute polymer solutions:

$$(4\text{-}74)\qquad \left(\frac{\langle s^2\rangle}{M}\right)^{3/2} = K_{s,o}^3 + 2\left(\frac{1}{4\pi}\right)^{3/2}\left(\frac{u_{seg}}{M_{seg}^2}\right)M^{1/2} \quad ; \quad K_{s,o}^3 = \left(\frac{\langle s^2\rangle_o}{M}\right)^{3/2}$$

A plot of $\left(\langle s^2\rangle/M\right)^{3/2} = f(M^{1/2})$ for good solvents should deliver $\left(\langle s^2\rangle_o/M\right)^{3/2}$ for the unperturbed state. This function works in a limited range of molar masses (Fig. 4-29) albeit with great uncertainty because the basic assumptions are too simple, especially $\alpha_s^3 = 1 + 2z$ (see p. 109).

The same can be said about the **Baumann-Kurata-Stockmayer equation**, (Eq.(4-75)), which is based on the Kurata-Stockmayer theory:

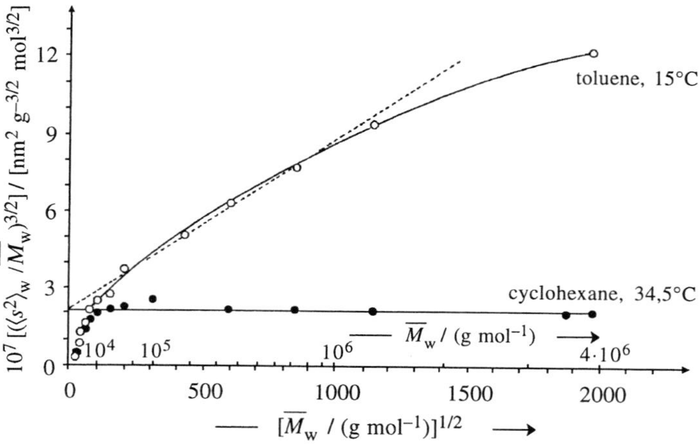

Fig. 4-29 Plot according to Eq.(4-74) for linear atactic poly(styrene)s in the good solvent toluene and the theta solvent cyclohexane (data of Fig. 4-23).

$$(4\text{-}75)\qquad \frac{\langle s^2\rangle}{M} = K_{s,o}^2 + B\frac{M}{\langle s^2\rangle^{1/2}}K_{s,o}^2 \qquad ; \quad K_{s,o}^2 = \frac{\langle s^2\rangle_o}{M}$$

Eqs.(4-74) and (4-75) as well as Eq.(4-69) need to be corrected for the molecular non-uniformity (polymolecularity, "polydispersity"), i.e., the distribution of molar masses of polymer molecules.

4.5.5 Non-Uniformity Corrections

Theories are usually derived for molecularly homogeneous polymers ($\overline{X}_w/\overline{X}_n \equiv 1$) whereas experiments on synthetic polymers always deal with molecularly non-uniform macromolecular substances ($\overline{M}_w/\overline{M}_n > 1$). Since mean-square radii of gyrations are usually obtained as z-averages but molar masses as either number or mass averages, constants K_s' and K_s'' in the functions $\langle s^2\rangle_z = K_s'\,\overline{M}_w^{2\nu}$ and $\langle s^2\rangle_z = K_s''\,\overline{M}_n^{2\nu}$ have to be corrected for non-uniformity (= polymolecularity). This is rarely done, however, because the effect is underestimated or experimental data are missing.

The question is: which molar mass average, M_{corr}, corresponds to the z-average mean-square radius of gyration or what is the polymolecularity correction factor if, for example, mass-average molar masses are used in $\langle s^2\rangle_z = f(M)$?

For convenience, the exponent ν in Eq.(4-69), $\langle s^2\rangle^{1/2} = K_s M^{\nu}$, is replaced by $(1+\varepsilon)/2$ which delivers $\langle s^2\rangle_z = K_s^2\,\overline{M}_{corr}^{1+\varepsilon}$. Rearrangement and use of the definition of the z-average, $\langle s^2\rangle_z \equiv \sum_i w_i M_i \langle s^2\rangle_i/(\sum_i w_i M_i)$, leads to

$$(4\text{-}76)\qquad \overline{M}_{corr} = \left(\frac{\langle s^2\rangle_z}{K_s^2}\right)^{1/(1+\varepsilon)} = \left(\frac{\sum_i w_i M_i \langle s^2\rangle_i}{K_s^2 \sum_i w_i M_i}\right)^{1/(1+\varepsilon)} = \left(\frac{\sum_i w_i M_i^{2+\varepsilon}}{\sum_i w_i M_i}\right)^{1/(1+\varepsilon)}$$

Eq.(4-76) shows that for $\varepsilon = 0$, i.e., for unperturbed coils, the average of the molar mass that corresponds to $\langle s^2\rangle_z$ is the z-average molar mass and certainly not the mass average. For all other molecule shapes, i.e., $\varepsilon \neq 0$, the corresponding molecular weight is a complex average that depends on the (2+ε)th moment of the molar mass distribution.

The expression for the corresponding mean-square average radius of gyration is obtained in a similar way as that of Eq.(4-76):

$$(4\text{-}77) \qquad \langle s^2 \rangle_{\text{corr}} = \left(\frac{\sum_i w_i \langle s^2 \rangle_i^{1/(1+\varepsilon)}}{\sum_i w_i} \right)^{1+\varepsilon}$$

In the unperturbed state ($\varepsilon = 0$), $\langle s^2 \rangle_{\text{corr}} = \langle s^2 \rangle_{\text{w}}$ has to be used if the molar mass is a mass-average. For $\varepsilon \neq 0$, the average of $\langle s^2 \rangle_{\text{corr}}$ is not so simple. In order to correctly compare experimental radii of gyration, $\langle s^2 \rangle_{\text{z}}$, with mass-average molar masses, $\overline{M}_{\text{w}}$, one has to introduce a correction factor $q_{\text{z,w}}$:

$$(4\text{-}78) \qquad \langle s^2 \rangle_{\text{z}} = K_{\text{s}}^2 \overline{M}_{\text{w}}^{1+\varepsilon} q_{\text{z,w}}$$

Correction factors are obtained from the expressions for molar mass distributions (see Literature to Chapter 4). For mass distributions of molar masses of the Schulz-Zimm type (SZ) with degrees of coupling, $\varsigma = \overline{M}_{\text{n}}/(\overline{M}_{\text{w}} - \overline{M}_{\text{n}})$ (Eq.(2-65)), and for logarithmic normal distributions (LN), correction factors are

$$(4\text{-}79) \qquad q_{\text{z,w}} = \frac{\Gamma(\varsigma + \varepsilon + 3)}{(\varsigma + 1)^{\varepsilon + 2}\Gamma(\varsigma + 1)} \quad \text{(SZ)} \quad ; \quad q_{\text{z,w}} = \left(\frac{\overline{M}_{\text{w}}}{\overline{M}_{\text{n}}} \right)^{(\varepsilon^2 + 3\varepsilon + 2)/2} \quad \text{(LN)}$$

Correction factors $q_{\text{z,w}}$ are quite large, even for polymers with relatively narrow molar mass distributions (Table 4-11). For example, for an SZ distribution with $\overline{M}_{\text{w}}/\overline{M}_{\text{n}} = 1.1$, the correction factor is already ca. 12 % for perturbed coils ($\alpha = 0.764$, $\nu = 0.588$).

Table 4-11 Correction factors $q_{\text{z,w}}$ for the dependence of the z-average mean-square radius of gyration on the mass-average molar mass for Schulz-Zimm distributions (SZ) with a coupling factor of $\varsigma = 2$ and for logarithmic normal distributions (LN) with various uniformities, $\overline{M}_{\text{w}}/\overline{M}_{\text{n}}$.

Correction factors are calculated for hard spheres ($\alpha = 0$), unperturbed random coils ($\alpha = 1/2$), perturbed random coils ($\alpha = 0.764$), freely draining coils ($\alpha = 1$), and rigid rods ($\alpha = 2$). α is the exponent of the intrinsic viscosity-molar mass relationship, Eq.(10-12), which is related to the exponents ν (Eq.(4-69)) and ε by $\alpha = 3\nu - 1 = (1/2) + (3/2)\varepsilon$.

$\overline{M}_{\text{w}}/\overline{M}_{\text{n}}$		Molecular non-uniformity factors $q_{\text{z,w}}$ for									
		SZ distributions with					LN distributions with				
	$\alpha \rightarrow$	0	0.500	0.764	1.000	2.000	0	0.500	0.764	1.000	2.000
	$\nu \rightarrow$	0.333	0.500	0.588	0.667	1.000	0.333	0.500	0.588	0.667	1.000
	$\varepsilon \rightarrow$	−0.333	0	0.176	0.333	1.000	−0.333	0	0.176	0.333	1.000
1.1		1.050	1.091	1.117	1.144	1.289	1.054	1.110	1.130	1.160	1.331
1.2		1.091	1.167	1.216	1.266	1.556	1.107	1.200	1.263	1.328	1.728
1.4		1.155	1.286	1.373	1.464	2.020	1.206	1.400	1.538	1.688	2.744
1.7		1.218	1.412	1.423	1.680	2.574	1.343	1.700	1.972	2.283	4.913
2.0		1.264	1.500	1.664	1.838	3.000	1.470	2.000	2.428	2.940	8.00
5.0		1.410	1.800	2.082	2.393	4.680	2.445	5.000	7.84	12.23	125

4.5.6 Temperature Dependence of Radii of Gyration

Unperturbed random coils in endothermic solutions convert to perturbed coils with increase in temperature. Several factors act in unison. Contacts between polymer segments and solvent molecules increase: segments stiffen and the coils expand. In helical wormlike chains, helical segments "melt"; the change G →T leads to more extended segments. On the other hand, potential barriers are more easy to overcome and the relative proportion of rotamers with higher energy increases, for example, gauche.

Because of the interplay between enthalpic and entropic factors, radii of gyration and end-to-end distances enlarge with increasing temperature first strongly and then less strongly. Since the number of possible microconformations is greater, the change should be more pronounced for polymers with high molar masses. All of these qualitative predictions have been confirmed experimentally (Fig. 4-30). Below the theta temperature, expansion factors of high-molar mass polymers first drop drastically and then seem to approach a limiting value.

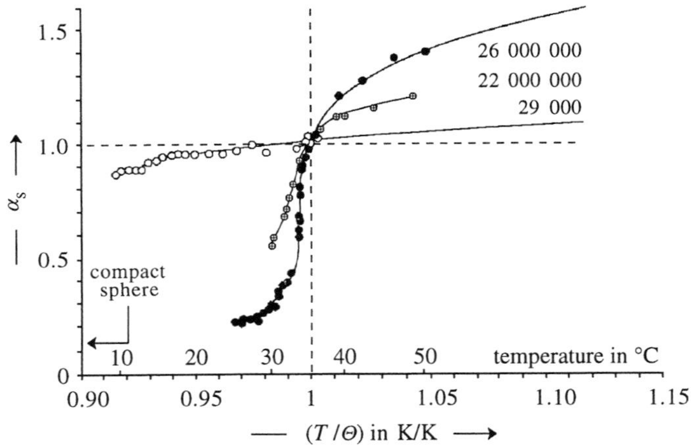

Fig. 4-30 Expansion factors α_s as a function of the relative temperature, T/Θ, for cyclohexane solutions of poly(styrene)s with mass-average molar masses, $\overline{M}_w/(\text{g mol}^{-1})$, of $26 \cdot 10^6$ [27], $22 \cdot 10^6$ [28], and $29 \cdot 10^3$ [29]. The arrow indicates the value for a compact sphere. Measurements require extremely low concentrations in order to prevent aggregation of polymer molecules at $T < \Theta = 308.6$ K.

Both mean-field theories (Section 4.5.4) and scaling theories (Section 4.8) predict for temperatures near the theta temperature Θ that the radius of gyration is proportional to a power of a reduced temperature, $(T - \Theta)/\Theta$:

(4-80) $\langle s^2 \rangle^{1/2} \sim M^{3/5} [(T - \Theta)/\Theta]^{1/5}$ (for $T > \Theta$)

(4-81) $\langle s^2 \rangle_o^{1/2} \sim M^{1/2}$ (for $T = \Theta$)

(4-82) $\langle s^2 \rangle^{1/2} \sim M^{1/3} [(T - \Theta)/\Theta]^{1/3}$ (for $T < \Theta$)

Eqs.(4-80) and (4-81) have been confirmed experimentally. Eq.(4-82) assumes that polymer coils collapse to compact globules at temperatures that are sufficiently below the theta temperature of the polymer/solvent system.

For the highest molar mass of $\overline{M}_w = 26 \cdot 10^6$ g/mol, the lowest measured radius of gyration was $\langle s^2 \rangle_z^{1/2} \approx 40$ nm at 25°C (Fig. 4-33). Since $\langle s^2 \rangle_z = (3/5)\, R_{z,sph}^2$ for compact spheres (Table 4-2), this value of $\langle s^2 \rangle_z^{1/2}$ would correspond to a radius of $R_{z,sph} \approx 31$ nm for a compact sphere. However, the partial specific volume of $\tilde{v}_2 = 0.952$ mL/g leads with $R_{sph} = [(3\ \tilde{v}_2 M)/(4\ \pi\ N_A)]^{1/3}$ to a far lower value of $R_{sph} = 21.4$ nm. It is not clear whether this discrepancy is caused solely by residual solvent in the "sphere" (lowest experimental temperature still too high) or whether there is also an influence of molecular non-uniformity (neglect of non-uniformity corrections).

4.6 Cyclic Macromolecules

Cyclic macromolecules have smaller radii of gyration than linear ones of the same constitution, chemical configuration, and molar mass since a completely stretched ring can only be half as long as the historic contour length of a linear molecule. For unperturbed dimensions, one can thus write $\langle s^2 \rangle_{o,lin}^{1/2} = K_{s,o} M^{1/2}$ for linear molecules (Eq.(4-44)) and $\langle s^2 \rangle_{o,ring}^{1/2} = K_{s,o}(M/2)^{1/2}$ for cyclic ones. The constant $K_{s,o}$ is the same for linear and cyclic macromolecules of the same molar mass since it contains only the parameters b, M_u, τ, and σ (see Eq.(4-44)) which are the same for sufficiently large rings and linear chains as well as the same interaction parameter K_{eff} for polymer-solvent, polymer-polymer, and solvent-solvent interactions. In the unperturbed state, a relative contraction of $q_{ring} = \langle s^2 \rangle_{o,ring}^{1/2} / \langle s^2 \rangle_{o,lin}^{1/2} = (1/2)^{1/2} \approx 0.707$ can thus be expected and is indeed found (Fig. 4-23).

The situation is more complicated for cyclic macromolecules in good solvents. The main problem is the scarcity of experimental data for well-characterized high-molecular weight cyclic macromolecules.

4.7 Branched Macromolecules

4.7.1 Introduction

Branched macromolecules have smaller radii of gyration than linear ones with the same constitution, chemical configuration, and molar mass, as can be easily visualized. The shrinking of spatial volumes is characterized by a **shrinking factor (branching parameter)**, $g_s \equiv \langle s^2 \rangle_{br} / \langle s^2 \rangle_{lin}$, as the ratio of the mean-square radii of gyration of branched and linear macromolecules for the same molar mass and solution conditions (different theta temperatures!). Note: the smaller the g_s, the *greater* is the branching!

Branching parameters are controlled by three factors: (1) molecular architecture, (2) number of branching points per molecule, and (3) intramolecular interactions.

Molecular architectures are affected by six parameters: (I) functionality of branching points, (II) number of branching points per molecule, (III) symmetry of the distribution of branching points with respect to the core, (IV) presence of branches-upon-branches, (V) distance between branching points, and (VI) lengths of subunits, i.e., segments between branching points or from a branching point to an endgroup. Literature often speaks loosely about statistically or randomly branched molecules without more detailed information. In many cases, this information is either not known or only inferred from the chemical synthesis of the polymer.

Theories of branching parameters rely on topological models, not on molecular features. In the simplest models, one assumes that (1) the polymer is molecularly uniform (all molecules have the same molar mass), (2) intramolecular excluded volumes are zero, (3) intramolecular interactions are absent, and (4) branching points are vanishingly small or have the same constitution as the segments between branching points. Hence, branching indices should be controlled only by the molecular architecture.

Radii of gyration are calculated from random-flight statistics. Because of the lengthy calculations, only the results are given here. For the literature, see Historical Notes to this chapter as well as the books and reviews in Literature to Chapter 4.

4.7.2 Star Molecules

The branching parameter g of star molecules with a degree of polymerization of $X = N_{\text{arm}}X_{\text{arm}}$, consisting of N_{arm} arms with an (average) degree of polymerization, X_{arm}, i.e., negligible cores, is determined only by the number f of arms per star molecule if the stars are in dilute solution in theta solvents but is independent of molecular weight:

(4-83a) $g_s \equiv \langle s^2\rangle_{\text{br}}/\langle s^2\rangle_{\text{lin}} = (3f-2)/f^2$; arms of equal length

(4-83b) $g_s \equiv \langle s^2\rangle_{\text{br}}/\langle s^2\rangle_{\text{lin}} = 6f/[(f+1)(f+2)]$; Gaussian statistics of arm lengths

The branching parameter should thus *decrease* if the number of equal-sized arms *increases* in molecules of constant molecular weight (Fig. 4-31) (the definition of g_s is counterintuitive). Molecules therefore become more compact with increasing branching. At high numbers of arms, branching parameters should approach $g_s \to 3/f$ for arms of equal length but $g_s \to 6/f$ for those with a Gaussian distribution of arm lengths.

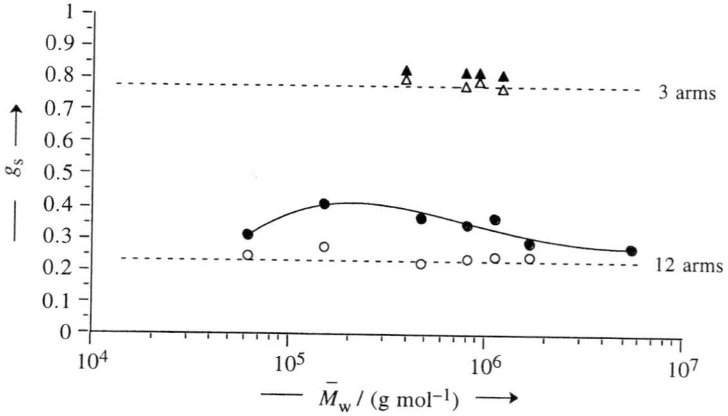

Fig. 4-31 Branching parameters of 3-arm and 12-arm atactic poly(styrene)s in cyclohexane (▲, ●; theta solvent) or toluene (Δ, O; good solvent), both at 35°C. Data of [30]. Dotted lines: prediction by Eq.(4-83a): $g_s \approx 0.778$ (3 arms) and $g_s \approx 0.236$ (12 arms). For scattering theory, see p. 156.

3-arm and 12-arm poly(styrene)s in the theta solvent cyclohexane have higher branching parameters than predicted theoretically for random-flight statistics (Fig. 4-31). The deviation from theory cannot be due to the fact that the actual theta temperature for star-like polymers differs somewhat from that of linear ones with the same monomeric units (Θ = 35°C) since effects on coil dimensions become marked only at very high molar masses (see Fig. 4-30). With increasing molar mass (i.e., length of arms), data for 12-arm poly(styrene)s in cyclohexane pass through a maximum and approach the theoretical value at high molar masses because only long arms can exist in the unperturbed state (at $\overline{M}_w = 5.5 \cdot 10^6$ g/mol, arms are ca. 4400 styrene units long).

In the good solvent toluene, branching parameters of 3-arm and 12-arm poly(styrene)s follow the theory for unperturbed chains although the chains of the arms are supposedly perturbed. It is open to question whether these observations can be generalized for all star molecules, regardless of their chemical constitution. For example, experimental g_s values of 3-arm poly(ethylene)s with equally long arms in the good solvent 1,2,4-trichlorobenzene are somewhat lower than theoretical ones (see Fig. 4-32).

Branching parameters for other distributions of chain lengths are more complicated. However, the g factor of 3-arm molecules, (A_2ZA'), with 2 equal arms and one unequal is given only by the ratio of molar masses of these two types of arms:

(4-83c) $$g_s = 1 - [6\,(M_A/M_{A'})]/[2 + (M_A/M_{A'})]^3$$

For this kind of 3-arm molecules, experimental data on poly(ethylene)s in good solvents scatter about the theoretical line for random-flight statistics (see Fig. 4-33).

4.7.3 Comb Molecules

Theory assumes regular comb molecules with a backbone of degree of polymerization, X_{bb}, to which N_{arm} equally sized arms (teeth) with degrees of polymerization, X_{arm}, are connected via branching points with functionalities f (usually, $f = 3$). Backbone and arms shall consist of the same type of monomeric units and branching points in the chain shall be identical with regular chain units.

According to this theory, branching parameters depend only on the number of arms per molecule, N_{arm}, and the mass fraction of the backbone, $w_{bb} \approx X_{bb}/(X_{bb} + N_{arm}X_{arm})$:

(4-84) $$g_{s,comb} = \langle s^2 \rangle_{comb} / \langle s^2 \rangle_{lin} = w_{bb}^3 + \left(\frac{2N_{arm}+1}{N_{arm}+1} \right) w_{bb}^2 (1-w_{bb}) + \left(\frac{N_{arm}+2}{N_{arm}} \right) w_{bb}(1-w_{bb})^2 + \left(\frac{3N_{arm}-2}{N_{arm}^2} \right) (1-w_{bb})^3$$

The branching parameter $g_{s,comb}$ of combs with arms of equal length varies with the ratio of degrees of polymerization of arms and backbone, $X_{arm}/X_{backbone}$ (Fig. 4-32). Obviously, comb molecules with just one arm are star molecules with three arms. If arms of 3-arm molecules are equally long, then the backbone of the corresponding comb molecule is twice as large as its arm. For such a comb with $N_{br} = 1$ and $q = X_{arm}/X_{bb}$ = 1/2, Eq.(4-84) predicts $g_{s,comb} \approx 0.778$, the same value as Eq.(4-83a) (Fig. 4-32).

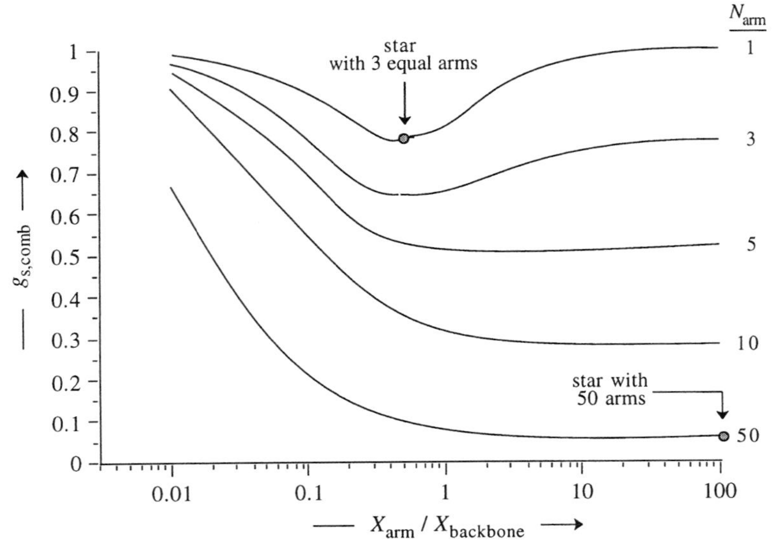

Fig. 4-32 Comb molecules: calculated effect of the number N_{arm} of equally long arms, each with a degree of polymerization of X, on the variation of the branching parameter g_s with the ratio of degrees of polymerization of arms and backbone, $X_{arm}/X_{backbone}$. Calculations with Eq.(4-84).

A comb molecule with $N_{arm} = 1$ becomes a linear molecule for both $q \rightarrow 0$ (diminishing length of arm) and $q \rightarrow \infty$ (diminishing length of backbone); i.e., $g_s \rightarrow 1$ (Fig. 4-32). On the other hand, a comb molecule becomes an N-star if the arms are considerably longer than the backbone. This limit is practically fulfilled for a comb with $N_{arm} = 50$ and $X_{arm}/X_{bb} = 100$ (Fig. 4-32) where Eq.(4-84) delivers $g_2 \approx 0.0594$ and the exact parameter for a 50-arm star is $g_s = 0.0592$ (Eq.(4-83a)).

Experimental $g_{s,comb}$ values of comb poly(ethylene)s are considerably higher than predicted ones (Fig. 4-33). This observation is not surprising because arm lengths of four of the five comb polymers in Fig. 4-36 were only 5-7 % of that of the backbone. The deviation from theory is less pronounced if the relative arm length increases to 24 % of the backbone.

Fig. 4-33 also contains some data for so-called H-branched poly(ethylene) molecules of the type $(A_i)_f[C_j](A_i)_f$ where C_j is a connector with j monomeric units and ends of functionality f and A_i the end units with degree of polymerization, i. In terms of monomeric units, investigated molecules varied between $(6)_2 27(6)_2$ and $(10)_5 92(10)_5$. Despite the relatively short lengths of end units and connector units (which should result in strong deviations from random-flight statistics) good agreement was found between experimental and predicted values (Fig. 4-33). Theoretical g_s values for these molecules were calculated from

$$(4\text{-}85) \qquad g_{s,H} = x_c^3 + 3x_c^2(1-x_c) + \frac{3}{2}\left(\frac{f+1}{f}\right)x_c(1-x_c)^2 + \left(\frac{3f-1}{2f^2}\right)(1-x_c)^3$$

where f = functionality of the ends of the connector and $x_c = X_{conn}/(2\,fX_A)$ with degrees of polymerization of the connector, X_{conn}, and the ends, X_A.

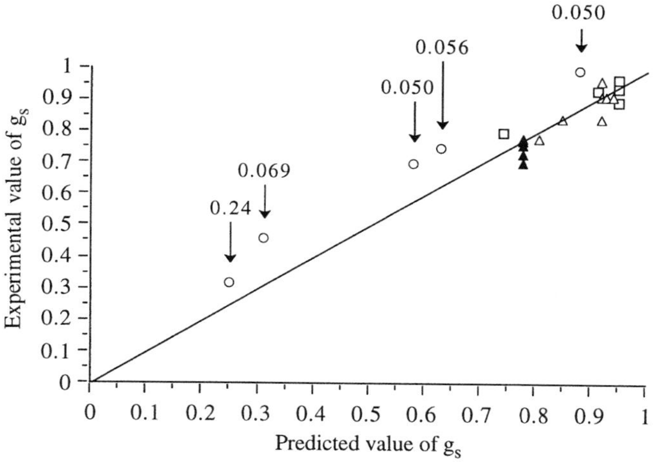

Fig. 4-33 Experimental and predicted values of the branching parameter g_s for differently branched poly(ethylene)s in the good solvent 1,2,4-trichlorobenzene at 135°C. Data of [26]. (▲) Star molecules with a small core Z and 3 equal arms (ZA_3); (△) star molecules with a small core Z, 2 equal arms A, and 1 different arm A' (A_2ZA'); (□) H-type molecules with a long linear connector L that carries two arms A at each end (A_2LA_2); (○) comb molecules with a long backbone L and various side chains A (LA_i). Numbers indicate X_{arm}/X_{bb} for comb poly(ethylene)s.

4.7.4 Dendrimers

Dendrimers are regular branch-upon-branch polymers (p. 11) that consist of a core, branching units, and endgroups. Examples are polyamidoamines (PAMAM) and polypropyleneimines (PPI) with linear-symmetric, elongated cores and symmetric layers of branching units as well as poly(α,ε-D-lysine)s (PLYS) with asymmetric cores and branching units (Table 4-12).

No accepted theory seems to exist which describes the dependence of the radius of gyration on the symmetry of cores, functionalities of core and branching units, number of generations, lengths of branching units, and the like.

Table 4-12 Examples of dendrimers.

Type	Core	Branching units	Endgroups
PAMAM	$>N-(CH_2)_2-N<$	$-(CH_2)_2-C(=O)-NH-(CH_2)_2-N<$	$-H$, $-COOCH_3$
PPI	$>N-(CH_2)_4-N<$	$-(CH_2)_3-N<$	$-CH_2CN$, $-H$
PLYS	$-C(=O)-CH(-(CH_2)_4-NH-)(-NH-)$	$-C(=O)-CH(-(CH_2)_4-NH-)(-NH-)$	$-H$, $-OH$

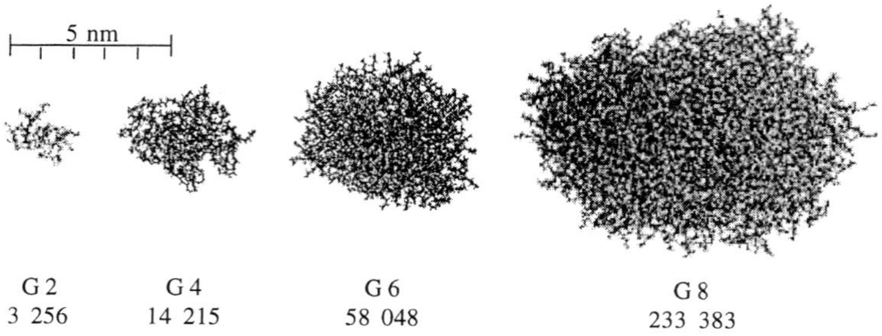

Fig. 4-34 Instantaneous snapshots of generations (G) 2, 4, 6, and 8 of PAMAM dendrimers after long molecular dynamics simulations at 300 K (see p. 77), employing integration steps of 1 fs and using 200 ps for equilibration and 200 ps for data collection [31a]. Numbers are molecular weights. All figures are to the same scale.
With kind permission by American Chemical Society, Washington (DC).

Two-dimensional drawings of dendrimers with symmetric cores suggest spherical shapes of molecules (Fig. 2-1). But cores are usually not sphero-symmetric (Table 4-12). Even a trivalent nitrogen atom as the core is not symmetric since it has four "arms": three valence bonds to other atoms plus a free electron pair. Structures of such cores must therefore cause dendrimers to be aspherical, at least for the early generations.

According to computer simulations, *instantaneous* shapes of PAMAM dendrimers of generations 3-11 are indeed somewhat ellipsoidal (Fig. 4-34). The ellipticity as a measure of the deviation from the spherical shape can be expressed as eccentricity or as asphericity (Fig. 4-35). Eccentricities are aspect ratios of minor to major axes (I_z/I_x and I_z/I_y); asphericities are calculated from $\delta = 1 - 3\,[(I_xI_y + I_xI_z + I_yI_z)/(I_x + I_y + I_z)^2]$.

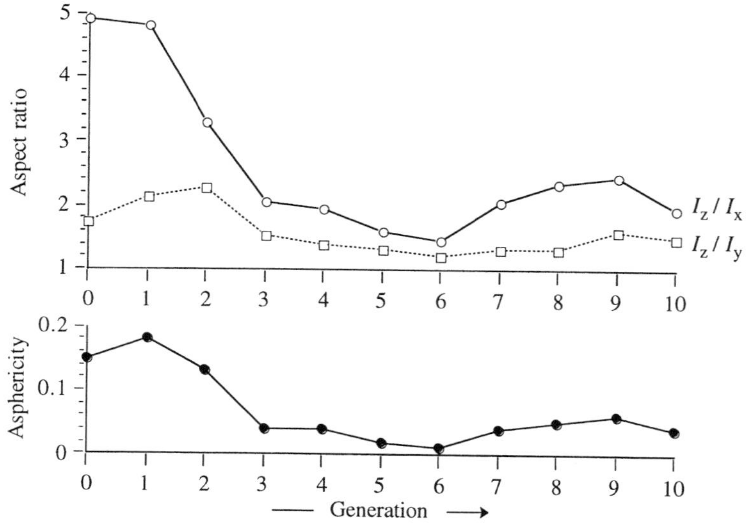

Fig. 4-35 Time-averaged ellipticities of PAMAM dendrimers expressed as aspect ratios (top) and asphericities (bottom), both by molecular dynamics calculations [31b]. Snapshots were taken at 0.5 ps intervals and equilibrated for 200 ps (generation 3-8) or 50 ps (generation 9-10). Data were collected for 200 ps (generations 3-8) and 50-100 ps (generations 9-10), respectively.
With kind permission by American Chemical Society, Washington (DC).

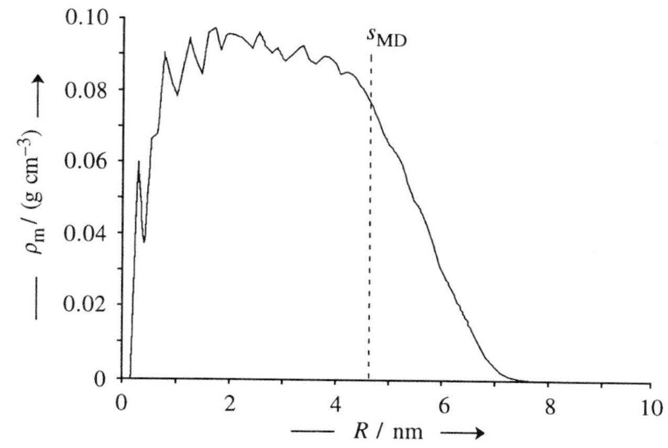

Fig. 4-36 Variation of density ρ_u of monomeric units of the G-9 PAMAM dendrimer with the radius R from the center of mass at $R = 0$ according to molecular dynamics calculations [31c]. Data correspond to the molecule in vacuum. - - - Radius of gyration.

Eccentricities (aspect ratios) and asphericities of low-generation PAMAM dendrimers are high (Fig. 4-35). The two aspect ratios also differ considerably but become more similar for higher generations. Both aspect ratios and asphericities decrease from G-1/G-2 to G-6, pass through a yet unexplained minimum, and then increase again to a maximum at G 9 before they decrease again to G 10. Hence, high-generation dendrimers become more sphero-symmetric. However, many other calculations come to other conclusions.

Molecular dynamics also allows one to calculate the variation of densities ρ_u with the distance R from the center of mass. For G-1 to G-3 of PAMAM, this density increases sharply to a maximum and then decreases; for G-1, the maximum is at $R \approx 0.15$ nm. Densities of G-4 to G-10 increase with increasing R, pass through an extended plateau, and then decrease (see Fig. 4-36 for G-9). The onset of the decrease corresponds approximately to the radius of gyration from molecular dynamics (see also Fig. 4-37).

Because of the ellipsoidal shape and the somewhat "puffy" internal structure of dendrimer molecules, radii of gyration of G-1 to G-9 from molecular dynamics simulations are higher than those calculated for compact spheres from bulk densities (Fig. 4-37). For G-10, both densities become almost identical. although aspect ratios are not unity and asphericities are not zero (Fig. 4-35).

Radii of gyration, $s_{MD} \equiv \langle s^2 \rangle^{1/2}$ by molecular dynamics (MD) are lower than experimental $\langle s^2 \rangle_z^{1/2}$ by both small-angle X-ray scattering (SAXS) and small-angle neutron scattering (SANS) (Fig. 4-37). These differences are systematic: with one exception, s_{SANS} is always greater than s_{SAXS}. For G-3 to G-9, both s_{SANS} and s_{SAXS} are higher than s_{MD} from molecular dynamics but for G-10, s_{SAXS} is identical with s_{MD}.

It seems that the differences between s_{SAXS}, s_{SANS}, and s_{MD} are not so much caused by experimental errors, differences in evaluating data, and theoretical assumptions but are inherent in the samples, systems, and methods themselves. Molecular dynamics data (p. 77) refer to structures in vacuum which corresponds to a poor solvent whereas all experimental radii of gyration were obtained in the good solvents CH_3OH and CD_3OD, which may swell molecules differently. For example, the G-8 PAMAM dendrimer had an $s_{SANS} = 4.38$ nm in D_2O but an $s_{SANS} = 3.95$ nm in $D(CD_2)_4OD$.

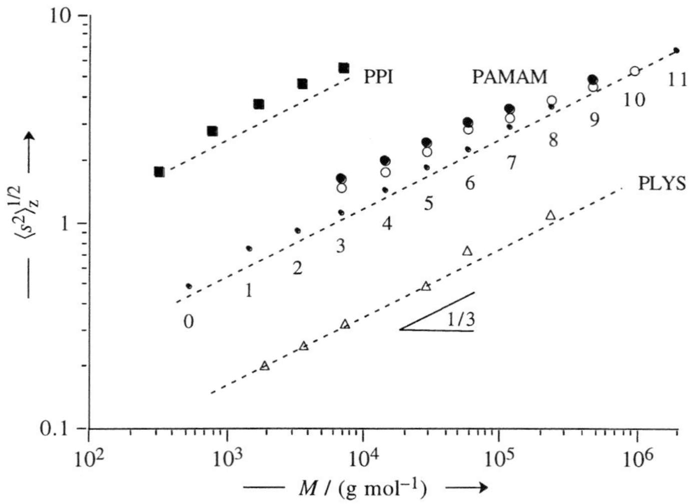

Fig. 4-37 Molar mass dependence of radii of gyration of dendrimers by Guinier analysis of X-ray small-angle scattering (SAXS; ○, △), neutron small-angle scattering experiments (SANS, ●,■), and molecular dynamics simulations (•). For clarity, radii of gyration of PLYS were multiplied by 1/4 and those of PPI by 4. Numbers refer to PAMAM generations (G 11 is hypothetical).

Broken lines indicate trend lines for compact spheres (slope: 1/3); the PAMAM line is exact for the known bulk density whereas those for PPI and PLYS indicate only slopes.

○,● PAMAM: polyamidoamines in CH_3OH (SAXS) [32] and in CD_3OD (SANS) [33]; 25°C;
• PAMAM: computer simulation (time-averaged data) [31c];
△ PLYS: poly(α,ε-D-lysine)s in *N,N*-dimethylformamide (SAXS) [4];
■ PPI: poly(propyleneimine)s with NH_2 endgroups in D_2O (SANS) [34].

Higher generation dendrimers may also not have ideal compositions since addition of a new generation may not be complete in chemical synthesis. For example, the molecular non-uniformity of a G-10 PAMAM was found to be $\overline{M}_w/\overline{M}_n$ = 1.03.

Experimental methods also probe distributions of different structural elements, i.e., electrons by SAXS, atomic nuclei by SANS, and mass elements by light scattering (Section 5.1). The effect is probably small but may still be significant.

Little is known about the size and internal structure of other types of dendrimers (Figs. 4-34-4-36). Poly(propyleneimine) dendrimers with extended cores have been investigated only up to molecular weights of ca. 10 000. Poly(α,ε-D-lysine)s practically behave like compact spheres for the whole molecular weight range of 2000-200 000.

4.8 Scaling

4.8.1 Local and Global Views

Classical molecular theories of shapes and dimensions of molecules consisting of "infinitely" thin chains rely on the similarity of the statistical distribution of segments and the statistics of thermal random flights of particles in three-dimensional space (Sections

4.3-4.5). The number of steps corresponds to the number of segments and the degree of polymerization, respectively, and the step length to the length of segments or chain bonds. The desired dependence of the end-to-end vector of the random flight and the end-to-end distance of chains in unperturbed coils on one hand and the number of steps and the degree of polymerization on the other is controlled by step or bond lengths, i.e., *local quantities*.

The mathematical model for *unperturbed coils* is the **Brownian chain** which is modeled as a Gaussian distribution of an infinite number of segments, N_{seg}, with vanishing segment lengths ($L_{seg} \rightarrow 0$) (Section 4.4.7). Such a chain has a mean-square end-to-end distance of $\langle r^2 \rangle_o = N_{seg} L_{seg}^2$. The segment length remains molecularly undefined; it may be the bond length b of a freely jointed chain (Eq.(4-22)), an effective bond length b_{eff} (Eq.(4-27)), etc. Since the historical contour length L_{chain} is the product of the number and length of segments, i.e., $L_{chain} = N_{seg} L_{seg} = \langle r^2 \rangle_o / L_{seg}$, and $L_{seg} \rightarrow 0$ for a Brownian chain, it follows $L_{chain} \rightarrow \infty$.

A Brownian chain can represent a real unperturbed chain only in a limited range of segment lengths L, i.e., $L_{pers} < L < \langle r^2 \rangle_o^{1/2} \equiv r_o$. The lower limit is given by the persistence length L_{pers} as a measure of the local stiffness whereas the upper limit is controlled by the end-to-end distance r_o. The lower limit usually corresponds to the monomeric unit, except for very long side chains (Table 4-13).

Table 4-13 Characteristic ratios C_∞ (Eq.(4-38)), persistence lengths L_{ps} (Eq.(4-42)), molecular weights $M_{r,pers}$, and degrees of polymerization X_{pers}, calculated from L_{pers}. C_∞ and L_{ps} refer to infinite degrees of polymerization; actual values of L_{ps} are therefore lower (see Table 4-9).

PE = poly(ethylene), PHDMA = poly(hexadecyl methacrylate), PMMA = at-poly(methyl methacrylate), PS = at-poly(styrene).

Polymer	Solvent	Θ / °C	$10^4 r_o / M^{1/2}$ / nm	C_∞	L_{ps} / nm	$M_{r,pers}$	X_{pers}
PE	1-octanol	180	1095	7.1	0.624	32.5	≈ 1
	p-xylene	100	950	5.3	0.485	26.1	≈ 1
PS	cyclohexane	34	660	9.5	0.809	150	≈ 1.5
PMMA	3-octanone	72	560	6.65	0.589	111	≈ 1
	2-methyl-4-pentanone	–42	500	5.25	0.481	92.5	≈ 1
PHDMA	heptane	21	620	25.1	2.010	1050	≈ 3

The mathematical model for perturbed chains is the **Kuhnian chain** (p. 97) which models expanded chains as N_K Kuhnian segments of length L_K. The Kuhnian chain can model both unperturbed (p. 97) and perturbed coils since the only model condition is that the product $N_K L_K$ equals the conventional contour length, $N_K L_K = r_{cont} = N_{eff} b_{eff}$. Hence, the Kuhnian length of a perturbed chain, $L_K = \langle r^2 \rangle / r_{cont}$, can be obtained from the end-to-end distance $\langle r^2 \rangle$. In turn, the end-to-end distance is related to the radius of gyration by $\langle s^2 \rangle = \alpha_s^2 \langle s^2 \rangle_o$ (Eq.(4-60)) and the radius of gyration to the molecular weight by $\langle s^2 \rangle = K_s^2 M^{2\nu}$ (Eq.(4-69)). The change of the exponent from $2\nu = 1$ for unperturbed coils to $2\nu = 6/5$ for perturbed ones indicates the presence of long-range interactions *within* the macromolecule.

All models and theories discussed so far are based on local quantities such as segments. Alternatively, one can also model objects such as coils *globally*, for example, with respect to the similarities of their shapes. One is looking then for **self-similarities** that manifest themselves in power expressions of properties. These expressions allow one to scale up or scale down properties of objects. Hence, the process is called **scaling**.

4.8.2 Fractals

The characteristic of an Euclidean body is independent of its size; Euclidian bodies of the same shape are therefore self-similar. For example, an isotropic sphere with density ρ and radius R has a mass of $m = \rho V = (4\,\pi\rho/3)\,R^3$. Doubling the radius leads to an eight-fold increase of the mass to $m = (4\,\pi\rho/3)(2\,R)^3 = 8\,(4\,\pi\rho/3)\,R^3$. For both spheres, the mass is still proportional to the 3rd power of the radius. The exponent 3 in $m \sim R^3$ thus represents the scaling **geometric dimension,** $d = 3$, of the sphere with respect to mass. A *smooth* surface of a sphere scales correspondingly with the square of its radius, $A \sim R^2$; the scaling geometric dimension with respect to the surface is $d = 2$.

The mass m (and therefore also the molar mass M) of Euclidean bodies is thus proportional to a power d of a characteristic length L (= radius R, etc.), i.e., $m \sim L^d$ and $M \sim L^d$, respectively. The spatial dimension of such bodies is a whole number. It is $d = 3$ for isotropic spheres, $d = 2$ for plane squares, and $d = 1$ for straight lines.

Dimensionalities are whole numbers for regular objects but fractions for irregular ones which are therefore called (geometric) **fractals** (L: *fractus*, from *frangere* = to break). Macroscopic examples are mountain ranges, snow flakes, and meandering rivers.

Polymers are fractal objects; examples are perturbed coils, dendrimer molecules, and crosslinked polymers. For example, the radius of gyration is proportional to the νth power of the molar mass, $s \sim M^\nu$ (Eq. (4-60)). The scaling of this fractal object leads to $M \sim s^{1/\nu} = s^{\bar{d}_m}$ where $1/\nu \equiv \bar{d}_m$ in analogy to the dimension d of geometric objects, $M \sim L^d$. The quantity $\bar{d}_m$ is called the **fractal dimension** or **Hausdorff dimension**.

Fractal objects are called **mass fractals** if they follow $m \sim L^{\bar{d}_m}$ (or $M \sim L^{\bar{d}_m}$). They possess inhomogeneous density distributions but have smooth surfaces. **Surface fractals** $A \sim L^{\bar{d}_a}$ are characterized by homogeneous densities and rough surfaces. There are also **time fractals**, for example, in diffusion processes where subsequent steps depend on the previous ones. This leads to time-dependent diffusion coefficients which are important quantities for photoconductivities.

Whether a fractal object is a mass fractal or a surface fractal can be determined from the dependence of the scattering intensity I on the scattering angle ϑ (see Chapter 5). The scattering intensity, $I \sim (Q_\vartheta)^P$, is proportional to a power P of the scattering factor, $Q_\vartheta = q\,\sin(\vartheta/2) = (4\,\pi/\lambda)\,\sin(\vartheta/2)$ (Eq.(5-33)). The exponent P is known as the **Porod slope**, $P = -2d + \bar{d}_a$, where d is the geometric dimension and $\bar{d}_a$ the surface fractal.

Examples are **sol-gel processes** in which inorganic monomers are converted to chemically crosslinked gels (Volume I, p. 570), such as silicon derivatives to glasses (amorphous), ceramic masses (polycrystalline), or glass ceramics (crystallites in an amorphous matrix). The resulting gels are either mass fractals (called "polymers") or surface fractals (usually called "colloids"). "Polymers" are obtained from the acid-catalyzed two-step polycondensation of $Si(OR)_4$ with a little water whereas "colloids" result from base-catalyzed polycondensations with a large excess of H_2O. After surpassing the gel point, all further crosslinking reactions are affected by the structure of the entities that were formed before the onset of gelation. The structure of these fractals controls the properties of the solids that are obtained after drying and firing.

A "colloid" has a uniform density; its mass fractal therefore equals the Euclidean dimension: $\bar{d}_m = d = 3$. Its surface fractals $\bar{d}_a$ vary between 2 (smooth surface) and 3 (very rough surface). The Porod slope is therefore $-4 \leq P \leq -3$.

The mass fractal $\bar{d}_m = 1/\nu$ of a "polymer" is determined by the structure of the particles, i.e., the exponent ν in the dependence of the radius of gyration on the molar mass, $s = K_s M^\nu$. On precipitation from solution, a linear polymer is in a bad solvent and must therefore have at least $\nu = 1/2$. However, the true value of ν must be smaller for two reasons: (a) on precipitation, polymer molecules with $M < \infty$ are already below their theta temperature and (b) molecules at the gel point are not linear but branched with a branching index $g_s < 1$ (see Section 4.7.1); this leads to $\nu < 1/2$.

On passing through the gel point, "polymers" will therefore collapse to a tighter structure and their shape approaches that of a sphere. The mass fractal of "polymers" must therefore be $2 \leq \bar{d}_m \leq 3$ since $\nu = 1/3$ for spheres (Eq.(4-4)). Because of the smooth surface of "polymers", their surface fractal is given by the Euclidean dimension $d = 3$ ($\bar{d}_a = d = 3$). The Porod slope of "polymers" is therefore $-3 \leq P \leq -1$.

4.8.3 Self-similarity

Euclidean bodies and fractals are self-similar; on enlarging or reducing, an object and a section of it have the same characteristics. For example, a sufficiently large section of an infinitely large polymer coil has the same characteristics as the whole coil (Fig. 4-38). However, real polymer chains may be self-similar for certain length scales only.

An unperturbed chain can be modeled as Kuhnian chain consisting of N_K Kuhnian segments of length N_K and a mean-square end-to-end distance of $\langle r^2 \rangle_o = N_K L_K^2 = const.$, i.e., as a Brownian chain. For vanishingly small segment lengths, $L_K \rightarrow 0$, the conventional contour length of such a chain approaches infinity:

$$\text{(4-86)} \qquad \lim_{L_K \to 0} r_{cont} = N_K L_K = \langle r^2 \rangle_o / L_K \rightarrow \infty$$

The fractal properties of a Brownian chain therefore range from vanishingly small segment lengths to infinitely large end-to-end distances. A Brownian chain is thus self-similar for all length scales.

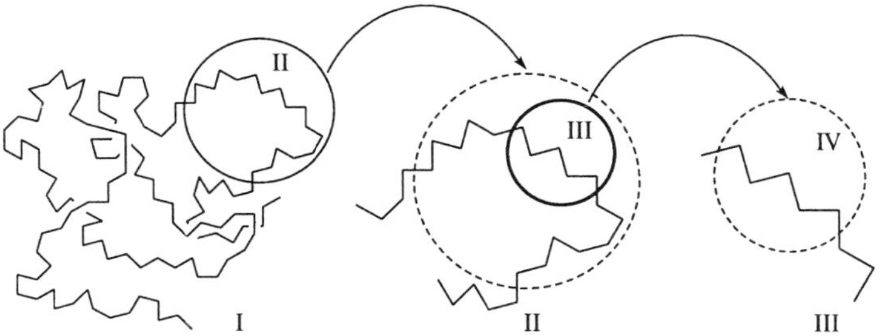

Fig. 4-38 Self-similarity of random coils (schematic). *On average*, section II of a coil I is self-similar to the total coil I with respect to coil characteristics. The self-similarity disappears if section III of the coil II approaches the persistence length $L_{ps} = L_K/2$ (IV) (see pp. 98 and 105).

The situation is different for perturbed chains where the radius of gyration depends on the molar mass according to $s \equiv \langle s^2 \rangle^{1/2} = K_s M^{\nu} \approx K_s M^{3/5}$ (Eq.(4-73)). This equation was derived by mean-field theory, assuming that the volume of a perturbed coil can be approximated by that of an equivalent sphere whose radius equals the radius of gyration of the coil (p. 111). Since $s^{1/\nu} \sim M$ and s is a characteristic length, the exponent $1/\nu = \bar{d}_m$ must be a mass fractal (p. 126).

A Brownian chain with $\nu = 1/2$ has a Hausdorff dimension of $\bar{d}_m = 2$ and thus the characteristic of an area and not that of a rectifiable curve. For a perturbed chain with $\nu = 3/5$, the fractal dimension is ca. 1.67. However, the mean-field theory assumes an equivalent sphere as a model for coils but isotropic spheres and coils are certainly not self-similar. How then are Hausdorff dimensions $\bar{d}_m$ related to geometric dimensions?

According to the mean-field theory, the reduced Gibbs energy, $\Delta G/k_B T$, is the sum of two terms (Eq.(4-71)). The elastic term is modeled as a spring, $\Delta G_{el}/k_B T = (1/s_o^2)(s^2/N)$, where $1/s_o^2$ = spring constant. The term $\Delta G_{rep}/k_B T$ describing the repulsive interaction is obtained as follows. For a d-dimensional space, the number concentration of N monomeric units per chain is N/s^d. The number of the pair-pair interactions is therefore proportional to $(N/s^d)^2$. Multiplication by the excluded volume u delivers the total interaction energy, $u(N/s^d)^2$. In d-dimensional space s^d, the total repulsive interaction energy is therefore $s^d u(N/s^d)^2 = uN^2/s^d$.

The reduced Gibbs energy becomes

$$\Delta G/k_B T = \Delta G_{el}/k_B T + \Delta G_{rep}/k_B = s^2/Ns_o^2 + uN^2/s^d \quad (4\text{-}87)$$

This expression is minimized by differentiating ΔG with respect to s and setting the resulting sum equal to zero. The result $(2/Ns_o^2)s = duN^2 s^{-d-1}$ is solved for N to give $N = (dus_o^2)^{-1/3} s^{(d+2)/3}$. Since the number N of monomeric units per chain is directly proportional to the molar mass M and $M \sim s^{\bar{d}_m}$, the Hausdorff dimension is $\bar{d}_m = (d + 2)/3$.

The fractal dimension of perturbed chains is 1 for the 1-dimensional case, 4/3 for the 2-dimensional one and 5/3 for the 3-dimensional one. In the 4-dimensional case, the value $\bar{d}_m = 2 = 1/\nu$ is recovered for unperturbed coils. The Hausdorff dimension is therefore an upper critical dimension above which excluded volumes become irrelevant.

Note that Flory's mean-field theory neglects correlations between forces, which results in a too high interaction energy. On the other hand, elastic energies are also too high because they are related to unperturbed coils instead of perturbed ones. To a first approximation, both effects cancel each other and the mean-field theory delivers $\nu = 3/5 = 0.600$ instead of the correct $\nu = 0.588$.

To summarize: dimensions of mass fractals are 1 for straight lines, 1-2 for curved lines, 2 for planar surfaces, 2-3 for rough surfaces, and 3 for hard spheres or cubes. Unperturbed coils of linear chains have a fractal dimension of 2 and perturbed coils of linear chains one of ca. $1/0.588 \approx 1.70$. For randomly branched polymers, a value of 2.0 is reported for good solvents and a value of 2.5 for bad ones, the same as for gels.

Dendrimer and enzyme molecules seem to be surface fractals. To a first approximation, the distribution of masses in their interiors seems to be isotropic whereas their surfaces appear as "rough" because of the many different groups in the "surface." For several enzymes, value of $\bar{d}_a \approx 2.4$ have been reported. Polyamidoamine (PAMAM) dendrimers with branching units $-CH_2CH_2CONHCH_2CH_2N<$ seem to have a fractal dimension of $\bar{d}_a \approx 2.42$ but polyether dendrimers with the branching unit $-CH_2OCH_2C\leftarrow$ are packed more tightly with $\bar{d}_a \approx 1.96$.

A-4 Appendix to Chapter 4

A-4.1 Freely Rotating Chains

The calculation of the mean-square end-to-end distances of freely rotating chains starts with Eq.(4-26), where N = number of chain units, b = bond length, and $\tau = 180 - \alpha$ = valence angle, assuming saturated linear homochains,

(4-26) $$\langle r^2\rangle_{of} = Nb^2 + 2\,(N-1)b^2 \cos\alpha + 2\,(N-2)b^2 \cos^2\alpha + 2\,(N-3)b^2 \cos^3\alpha + ... + 2\,(N-i)b^2 \cos^i\alpha + ... + 2\,b^2 \cos^{N-1}\alpha$$

Eq.(4-26) can be simplified by setting $\cos\alpha \equiv \mathrm{x}$:

(4-26a) $$\langle r^2\rangle_{of} = Nb^2 + 2\,(N-1)b^2\mathrm{x} + 2\,(N-2)b^2\mathrm{x}^2 + 2\,(N-3)b^2\mathrm{x}^3 + ... + 2\,(N-i)b^2\mathrm{x}^i + ... + 2\,b^2\mathrm{x}^{N-1}$$

The second and the following terms of this series are united in a sum that can be split into two individual sums:

(A 4-1) $$\langle r^2\rangle_{of} = Nb^2 + 2\,b^2 \sum_{i=1}^{N-1}(N-i)\mathrm{x}^i = Nb^2 + 2\,b^2 N \sum_{i=1}^{N} \mathrm{x}^i - 2\,b^2 \sum_{i=1}^{N} i\mathrm{x}^i$$

Because $1 \le i \le N$, the second to last sum of Eq.(A 4-1) is a finite series which can be split into two new sums:

(A 4-2) $$\sum_{i=1}^{N} \mathrm{x}^i = \mathrm{x}\sum_{i=0}^{N-1} \mathrm{x}^i = \mathrm{x}\sum_{i=0}^{\infty} \mathrm{x}^i - \mathrm{x}\sum_{i=N}^{\infty} \mathrm{x}^i = \mathrm{x}\left(\sum_{i=0}^{\infty} \mathrm{x}^i - \mathrm{x}^N \sum_{i=0}^{\infty} \mathrm{x}^i\right)$$

Because $\mathrm{x} \equiv (\cos\alpha) < 1$, both sums can be transformed to the same infinite geometric series, $1 + \mathrm{x} + \mathrm{x}^2 + ... = 1/(1-\mathrm{x})$.

Similarly, the last sum of Eq.(A 4-1) can be written as

(A 4-3) $$\sum_{i=1}^{N} i\mathrm{x}^i = \mathrm{x}\sum_{i=1}^{N-1} i\mathrm{x}^{i-1} = \mathrm{x}\left[\sum_{i=1}^{\infty} i\mathrm{x}^{i-1} - \mathrm{x}^N \sum_{i=1}^{\infty} (N+i)\mathrm{x}^{i-1}\right]$$

which leads to the infinite series, $1 + 2\mathrm{x} + 3\mathrm{x}^2 + ... = 1/(1-\mathrm{x})^2$.

The two sums of Eq.(A 4-1) can therefore be substituted with

(A 4-4) $$\sum_{i=1}^{N} \mathrm{x}^i = \mathrm{x}\left(\frac{1-\mathrm{x}^N}{1-\mathrm{x}}\right) \quad \text{and} \quad \sum_{i=1}^{N} i\mathrm{x}^i = \mathrm{x}\left(\frac{1-\mathrm{x}^N}{(1-\mathrm{x})^2} - \frac{N\mathrm{x}^N}{1-\mathrm{x}}\right)$$

and Eq.(A 4-1) becomes

(A 4-5) $$\langle r^2\rangle_{of} = Nb^2 + 2\,Nb^2\mathrm{x}\left(\frac{1-\mathrm{x}^N}{1-\mathrm{x}}\right) - 2\,b^2\mathrm{x}\left(\frac{1-\mathrm{x}^N}{(1-\mathrm{x})^2}\right) + 2\,b^2\mathrm{x}\left(\frac{N\mathrm{x}^N}{1-\mathrm{x}}\right)$$

Since valence angles α are greater than 90°, complementary angles will be smaller than 90°; hence, cos $\alpha < 1$. For very large N, $x^N = (\cos\alpha)^N$ will approach zero. The first bracketed term in Eq.(A 4-5) becomes $1/(1-x)$, the second one, $1/(1-x)^2$, and the third one, 0. Eq.(A 4-5) changes to

$$\text{(A 4-6)} \qquad \langle r^2 \rangle_{\text{of}} = Nb^2\left[1+\frac{2x}{1-x}-\frac{2x}{N(1-x)^2}\right] = Nb^2\left[\frac{1+x}{1-x}-\frac{2x}{N(1-x)^2}\right]$$

Finally, with $x \equiv \cos\alpha = \cos(180° - \tau)$ and $1 + \cos\alpha = 1 - \cos\tau$, it converts to

$$\text{(A 4-7)} \qquad \langle r^2 \rangle_{\text{of}} = Nb^2\left[\frac{1-\cos\tau}{1+\cos\tau}+\frac{2\cos\tau}{N}\left(\frac{1-(-\cos\tau)^N}{(1+\cos\tau)^2}\right)\right] \approx Nb^2\left(\frac{1-\cos\tau}{1+\cos\tau}\right)$$

where the rightmost expression is the approximation for $N \to \infty$.

More complicated expressions are obtained for unsaturated chains, heterochains, chains with cyclic structures in the main chain, etc.

A-4.2 Distribution of End-to-End Distances

The distribution of end-to-end distances of linear phantom chains of equal lengths is calculated in analogy to the Maxwellian distribution of velocities of ideal gas molecules.

Let $p(r_x)$ be the distribution function of component x of the end-to-end distance and $p(r_y)$ and $p(r_z)$ be those of the y and z components, respectively. Because of the isotropy of space, $p(r_x) = p(-r_x)$ and $p(r_x) = f(r_x^2)$, and similarly for $p(r_y)$ and $p(r_z)$.

The three spatial distribution functions are coupled for small numbers N of chain bonds since $r_x^2 + r_y^2 + r_z^2 = b^2$ for a "polymer" consisting of $N = 1$ monomeric unit with a bond length b between the two "chain" atoms. The larger N is, the less interdependent are the three distribution functions $p(r_x)$, $p(r_y)$, and $p(r_z)$. They can be viewed as being independent of each other if $N \to \infty$ and $r^2 \ll r_{\text{cont}}^2$, the square of the conventional contour length. The total probability $F(r^2)$ of finding a unit in space is then simply the product of the probabilities for the three spatial directions, $p(r_x)\cdot p(r_y)\cdot p(r_z) = f(r_x^2)\cdot f(r_y^2)\cdot f(r_z^2)$.

This probability cannot depend on the spatial direction but must be a function of the squares of the end-to-end distances, $r^2 = r_x^2 + r_y^2 + r_z^2$. Therefore,

$$\text{(A 4-8)} \qquad f(r_x^2)\cdot f(r_y^2)\cdot f(r_z^2) = F(r^2) = f(r_x^2 + r_y^2 + r_z^2)$$

Condition (A 4-8) can be satisfied by only one mathematical function; for example, for the x direction (and similarly for the y and z directions):

$$\text{(A 4-9)} \qquad p(r_x) = f(r_x^2) = \exp(-d r_x^2)$$

The minus sign is caused by the condition $p(r_x) \to 0$ if $r_x \to \infty$.

The constant d is obtained as follows: The distribution function needs to be normalized

(A 4-10) $$p(r_x)\mathrm{d}r_x = \int_{-\infty}^{+\infty} \exp(-\mathrm{d}r_x^2)(\pi/d)^{1/2} = 1$$

The second moment of this distribution function must also be the mean-square of the components of r, i.e., $\langle r_x^2\rangle = \langle r^2\rangle/3 = Nb^2/3$:

(A 4-11) $$\langle r_x^2\rangle = \int_{-\infty}^{+\infty} r_x^2 p(r_x)\mathrm{d}r_x = \int_{-\infty}^{+\infty} r_x^2 \exp(-\mathrm{d}r_x^2) = \pi^{1/2}(2\,d^{3/2})^{-1} = Nb^2/3$$

Division of Eq.(A 4-10) by Eq.(A 4-11) and introduction of the result in Eq.(A 4-9) leads to the Gaussian distribution of end-to-end distance in the one-dimensional case:

(A 4-12) $$p(r_x) = \left(\frac{3}{4\,\pi\,Nb^2}\right)^{1/2} \exp\left(-\frac{3\,r_x^2}{2\,Nb^2}\right) = \left(\frac{1}{2\,\pi\langle r_x^2\rangle}\right)^{1/2} \exp\left(-\frac{r_x^2}{2\langle r_x^2\rangle}\right)$$

A-4.3 End-to-End Distance and Radius of Gyration

Freely jointed chains, freely rotating chains, chains with restricted rotation and other unperturbed linear chains are all phantom chains since they behave as if they are infinitely thin. All types of linear phantom chains have the same relationship between their end-to-end distances r and radii of gyration s.

The calculation of $s = \mathrm{f}(r)$ assumes a chain composed of N chain units with identical masses m_i where i indicates the running number of the chain unit. The mass of the chain unit is concentrated in a point. Each mass point is connected to two other mass points by massless bonds of length b. The mass points and thus the chain segments between any two mass points are distributed in three-dimensional space, forming a random coil (for a two-dimensional arrangement, see Fig. 4-17). The center of gravity of the coil is defined as the first moment of the radius of gyration; at this point, the vector from the center of gravity to the zeroth mass point is, of course, $\boldsymbol{R}_0 = 0$ which leads to $\Sigma_i\, m_i\boldsymbol{R}_i = 0$.

$\boldsymbol{R}_1$ is the vector from the center of gravity, S, to the first mass point and $\boldsymbol{R}_i$ the vector from S to the ith mass point. The vector between the first mass point 1 and the ith mass point is $\boldsymbol{r}_i$ (see p. 86, Fig. 4-17, right) so that $\boldsymbol{R}_i = \boldsymbol{R}_1 + \boldsymbol{r}_i$ with $1 \leq i \leq N$. The sum of all vectors is therefore $\Sigma_i\, \boldsymbol{R}_i = N\boldsymbol{R}_1 + \Sigma_i\, \boldsymbol{r}_i$. This sum must be zero since all mass points are identical. Therefore,

(A 4-13) $$\boldsymbol{R}_1 = -(1/N_c)\,\Sigma_i\, \boldsymbol{r}_i$$

The mean-square radius of gyration is defined as the second moment of the mass distribution of all radii R_i, i.e., $\langle s^2\rangle_{oo} = \langle\Sigma_i\, m_iR_i^2\rangle/\Sigma_i\, m_i$ for a freely jointed chain. Since each mass point has the same mass; the total number N of the mass points is therefore given by $N = (\Sigma_i\, m_i)/m_i$ and the mass of one mass point by $m_i = \Sigma_i\, m_i/N = m/N$.

Averaging may be done first for all sums and then for all products, or first for all products and then for all sums, resulting in

(A 4-14) $$\langle s^2\rangle_{oo} = \frac{\langle \sum_i m_i R_i^2\rangle}{\sum_i m_i} = \frac{\sum_i m_i \langle R_i^2\rangle}{\sum_i m_i} = \frac{\sum_i \langle R_i^2\rangle}{N}$$

Introduction of $\boldsymbol{R}_i = \boldsymbol{R}_1 + \boldsymbol{r}_i$ delivers

(A 4-15) $$\langle s^2\rangle_{oo} = (1/N)\ \Sigma_i\ (\boldsymbol{R}_1 + \boldsymbol{r}_i)(\boldsymbol{R}_1 + \boldsymbol{r}_i) = \boldsymbol{R}_1{}^2 + (2/N_c)\ \boldsymbol{R}_1\ \Sigma_i\ \boldsymbol{r}_i + (1/N)\ \Sigma_i\ r_i{}^2$$

$\boldsymbol{R}_1$ is expressed by Eq.(A 4-13) which is written for pairs of i and j where *i and* j have the same meaning and are only introduced for mathematical convenience:

(A 4-16) $$\langle s^2\rangle_{oo} = (1/N^2)(\Sigma_i\Sigma_j\ \boldsymbol{r}_i\boldsymbol{r}_j) - (2/N^2)\ \Sigma_i\Sigma_j\ \boldsymbol{r}_i\boldsymbol{r}_j + (1/N)\ \Sigma_i\ r_i{}^2$$

(A 4-16a) $$\langle s^2\rangle_{oo} = (1/N)\ \Sigma_i\ r_i{}^2 - (1/N^2)\ \Sigma_i\Sigma_j\ \boldsymbol{r}_i\boldsymbol{r}_j$$

The product $\boldsymbol{r}_i\boldsymbol{r}_j$ is the product of the lengths of the two vectors $\boldsymbol{r}_i$ and $\boldsymbol{r}_j$ and the cosine of the angle ω between them. It is solved by remembering that the mass points 1, i, and j form a triangle to which the cosine rule applies, i.e., $a^2 = b^2 + c^2 - 2bc \cos \omega$, or, in the present notation, $r_{ij}{}^2 = r_i{}^2 + r_j{}^2 - 2\ \boldsymbol{r}_i\boldsymbol{r}_j$, since $i = j$ and therefore $\omega = 0$. After solving for $\boldsymbol{r}_i\boldsymbol{r}_j$, Eq.(4-16a) becomes

(A 4-16b) $$\langle s^2\rangle_{oo} = (1/N)\ \Sigma_i\ r_i{}^2 - (1/(2\ N^2))\ \Sigma_i\Sigma_j\ (r_i{}^2 + r_j{}^2 - r_{ij}{}^2)$$

which with $\Sigma_i\Sigma_j\ r_i{}^2 = \Sigma_i\Sigma_j\ r_j{}^2 = N\ \Sigma_i\ r_i{}^2$ converts to

(A 4-17) $$\langle s^2\rangle_{oo} = [1/(2\ N^2)]\ \Sigma_i\Sigma_j\ \langle r_{ij}{}^2\rangle$$

In this equation, $r_{ij}{}^2$ is written as the spatial average $\langle r_{ij}{}^2\rangle$ of the end-to-end distance of a chain of $|j - i|$ elements of length b, i.e., $\langle r_{ij}{}^2\rangle = |j - i|\, b^2$, using vertical bars to indicate absolute numerical values without regard to sign. According to Eq.(4-25), b^2 of freely jointed chains can be expressed by $b^2 = \langle r^2\rangle_{oo}/N$ and Eq.(A 4-17) becomes

(A 4-18) $$\langle s^2\rangle_{oo} = [1/(2\ N^2)]\ \Sigma_i\Sigma_j\ |j - i|\ \langle r^2\rangle_{oo}/N$$

The product of the sums of the absolute difference $|j - i|$ can be solved separately for each sum. Summation over all values of j delivers Eq.(A 4-19) and the summation over all values of i^2 leads to Eq.(A 4-19a),

(A 4-19) $$\begin{aligned}\sum_{j-1}^{j=N} |j - i| &= \sum_{j-1}^{i} |i - j| + \sum_{j=i+1}^{N} |j - i| \\ &= i^2 - (1/2)i(i+1) + (1/2)(N-i)(N+i+1) - i(N-i) \\ &= i^2 - iN + (1/2)N^2 + (1/2)N - i\end{aligned}$$

(A 4-19a) $$\Sigma_i\ i^2 = 1^2 + 2^2 + \dots N^2 = N(N + 1)(2\ N + 1)/6.$$

which results for $N >> 1$ in

(A 4-20) $$\sum_{i=1}^{i=N} \sum_{j=1}^{j=N} |j - i| = (N^3 - N)/3 \approx N^3/3$$

Introduction of Eq.(A 4-20) into Eq.(A 4-18) delivers the desired relationship between $\langle s^2 \rangle_{oo}$ and $\langle r^2 \rangle_{oo}$:

(A 4-21) $$\langle s^2 \rangle_{oo} = \frac{1}{2\,N^2} \cdot \frac{N^3}{3} \cdot \frac{\langle r^2 \rangle_{oo}}{N} = \frac{\langle r^2 \rangle_{oo}}{6}$$

Historical Notes to Chapter 4

Random flight statistics:
K.Pearson, Nature **77** (1905) 294 (two-dimensional random walk)
A.A.Markoff, Wahrscheinlichkeitsrechnung, Teubner, Leipzig 1912 (a book on probability theory that contains the most general form of random flight statistics)
Lord Rayleigh, Phil.Mag. **37** (1919) (three-dimensional random flight statistics for very small or very large numbers of steps)

Application of random-flight statistics to distributions of end-to-end distances of coils:
W.Kuhn, Kolloid-Z. **68** (1934) 2
E.Guth, H.F.Mark, Monatsh.Chem. **65** (1934) 93

Freely jointed chain:
W.Kuhn, Kolloid-Z. **68** (1934) 2 (chain molecules form bean-like coils. The mean-square end-to-end distance is proportional to a power of the molecular weight, $\langle r^2 \rangle = K_r M^{\nu}$. ν is calculated as 1/2 for freely jointed chains and estimated as 0.61 for chains with excluded volumes (present value: $0.588 \approx 0.6$)).

Freely rotating chain:
H.Eyring, Phys.Rev. **39** (1932) 746 (derivation of Eqs.(4-26) and (4-27))
E.Guth, H.Mark, Monatsh.Chem. **65** (1934) 93; Z.Elektrochem. **43** (1937) 683

Chain with restricted rotation:
C.Sadron, J.chim.phys. **43** (1946) 145; J.Polym.Sci. **3** (1948) 812
H.Benoit, J.chim.phys. **44** (1947) 18; J.Polym.Sci. **3** (1948) 376
H.Kuhn, J.Chem.Phys. **15** (1947) 843
W.J.Taylor, J.Chem.Phys. **15** (1947) 412; **16** (1948) 257

Wormlike chain:
O.Kratky, G.Porod, Rec.Trav.chim.Pays-Bas **68** (1949) 1106
G.Porod, Monatsh.Chem. **80** (1949) 251

Helical wormlike chain:
H.Yamakawa, W.H.Stockmayer, J.Chem.Phys. **57** (1972) 2843
H.Yamakawa, J.Shimada, J.Chem.Phys. **83** (1985) 2607
J.Shimada, H.Yamakawa, J.Chem.Phys. **85** (1986) 591

Radius of gyration as a function of the end-to-end distance:
P.Debye, J.Chem.Phys. **14** (1946) 636
B.H.Zimm, W.H.Stockmayer, J.Chem.Phys. **17** (1949) 1301

Branching parameters:
Star molecules:
B.H.Zimm, W.H.Stockmayer, J.Chem.Phys. **17** (1949) 1301
B.H.Zimm, R.W.Kilb, J.Polym.Sci. **37** (1959) 19
Star molecules with unequal arms, combs:
G.C.Berry, T.A.Orofino, J.Chem.Phys. **46** (1964) 1614
H-type molecules:
W.W.Graessley, S.T.Milner, quoted by [26]

Elastic dumbbell:
W.Kuhn, Kolloid-Z. **68** (1934) 2

Pearl necklace model:
P.E.Rouse, J.Chem.Phys. **21** (1953) 1272
Generalization by
K.S.Schweizer, J.Chem.Phys. **91** (1989) 5802, 5822; J.Non-Cryst.Solids **131** (1991) 643

Pearl necklace model with hydrodynamic interactions:
J.G.Kirkwood, J.Riseman, J.Chem.Phys. **16** (1948) 565
B.H.Zimm, J.Chem.Phys. **24** (1956) 269

Bibliography of Force Fields

AMBER (**A**ssisted **M**odel **B**uilding and **E**nergy **R**efinement): P.Weiner, P.A.Kollman, J.Comput. Chem. **2** (1981) 287; S.J.Weiner, P.A.Kollman, D.A.Case, U.C.Singh, C.Ghio, G.Algona, S.Profeta Jr., P.Weiner, J.Am.Chem.Soc. **106** (1984) 765; S.J.Weiner, P.A.Kollman, D.T.Nguyen, D.A.Case, J.Comput.Chem. **7** (1986) 230 (nucleic acids and proteins)

CHARMM (**C**hemistry of **Har**vard **M**acromolecular **M**echanics): B.R.Brooks, R.E.Bruccoleri, B.D.Olafson, D.J.States, S.Swaminathan, M.Karplus, J.Comput.Chem. **4** (1983) 187 (nucleic acids and proteins)

CFF91 (**C**onsistent Valence **F**orce **F**ield 19**91**): "Discover" program of Molecular Simulations, Inc. (formerly: Biosym, Inc.): (a) J.A.Mapic,, U.Dinur, A.T.Hagler, Proc.Natl.Acad.Sci.USA **85** (1988) 5350; (b) J.A.Mapic, M.-J.Hwang, T.P.Stockfisch, U.Dinur, M.Waldman, C.S.Ewig, A.T.Hagler, J.Comput.Chem. **15** (1994) 162; (c) M.-J.Hwang, T.P.Stockfisch, A.T.Hagler, J.Am.Chem.Soc. **116** (1994) 2515

CVFF (**C**onsistent **V**alence **F**orce **F**ield): T.Halicioglu, M.Pound, Phys.Sta.Sol. [A] **30** (1975) 619; A.T.Hagler, S.Lifson, P.Dauber, J.Am.Chem.Soc. **101** (1979) 5122

DISCOVER®: Computer program with AMBER, CVFF, CFF91, ESFF (Molecular Simulations, Inc., San Diego) (includes Biosym, Inc., since Fall 1996)

DREIDING (in honor of A.S.Dreiding, University of Zurich): S.L.Mayo, B.D.Olafson, W.A.Goddard III, J.Phys.Chem. **94** (1990) 8897 (organic molecules and inorganic molecules of groups 1-3)
ECEPP: M.J.Sippl, G.Nemethy, H.A.Scheraga, J.Phys.Chem. **88** (1984) 6231 (biological macromolecules); ECEPP-05: Y.A.Arnautova, A.Jagielska, H.A.Scheraga, J.Phys.Chem. **B 110** (2006) 5025

ESFF (**E**xtensible **S**ystematic **F**orce **F**ield): S.Shi, L.Yan, Y.Yang, J.Fisher-Shaulsky, T.Thacher, J.Comput.Chem. **24** (2003) 1059

MM2/MMP2 (**M**olecular **M**echanics): N.L.Allinger, J.Am.Chem.Soc. **99** (1977) 8127; U.Burkert, N.L.Allinger, Molecular Mechanics (ACS Monograph 177), Am.Chem.Soc., Washington, DC 1982; N.L.Allinger, J.Comput.Chem. **8** (1987) 581 (inorganic and organic molecules)

OPLS (**O**ptimized **P**otentials for **L**iquid **S**imulations): W.L.Jorgensen, J.Tirado-Rives, J.Am.Chem.Soc. **110** (1988) 1657 (proteins and nucleic acids).

PCFF (extension of CFF for polymers): see CFF91(c) and H.Sun, S.J.Mumby, J.R.Maple, A.T.Hagler, J.Am.Chem.Soc. **116** (1994) 2978; H.Sun, J.Comput.Chem. **15** (1994) 752; H.Sun, Macromolecules **28** (1995) 701

Literature to Chapter 4

4.1 OVERVIEW: Polymers in Solution

V.N.Tsvetkov, V.Ye.Eskin, S.Ya.Frenkel, Structure of Macromolecules in Solution, Butterworths, London 1970

H.Yamakawa, Modern Theory of Polymer Solutions, Harper and Rowe, New York 1971

H.Morawetz, Macromolecules in Solution, Interscience, New York, 2nd ed. 1975

W.C.Forsman, Ed., Polymers in Solution. Theoretical Considerations and Newer Methods of Characterization, Plenum, New York 1983

H.Fujita, Polymer Solutions, Elsevier, Amsterdam 1990

W.W.Graessley, Polymeric Liquids and Networks. Volume I, Structure and Properties, Garland Science, New York (2004)

T.A.Witten (with P.A.Pincus), Structured Fluids: Polymers, Colloids, Surfactants, Oxford University Press, New York 2004

4.3a MOLECULAR AND ATOMISTIC MODELING (general)

D.M.Hirst, A Computational Approach to Chemistry, Blackwell, Oxford 1990

A.Aharony, D.Stauffer, Introduction to Percolation Theory, Taylor and Francis, Philadelphia, 2nd ed. 1993

J.M.Haile, Molecular Dynamics Simulation, Wiley, New York 1994

W.T.Coffey, Yu.P.Kalmykov, J.T.Waldron, Eds., The Langevin Equation. With Applications in Physics, Chemistry and Electrical Engineering, World Sci.Publ., River Edge (NY) 1996

D.Frenkel, B.Smit, Understanding Molecular Simulation: From Algorithms to Applications, Academic Press, San Diego 1996

M.F.Schlecht, Molecular Modeling on the PC, Wiley, New York 1998

K.Machida, Principles of Molecular Mechanics, Wiley, New York 1999

F.Jensen, Introduction to Computational Chemistry, Wiley, Chichester 1999

A.R.Leach, Molecular Modelling. Principles and Applications, Prentice-Hall, Englewood Cliffs (NJ), 2nd ed. 2001

K.Binder, D.W.Heermann, Monte Carlo Simulation in Statistical Physics. An Introduction, Springer, Berlin, 4th ed. 2002

C.J.Cramer, Essentials of Computational Chemistry. Theory and Models, Wiley, Chichester 2002

E.Lewars, Computational Chemistry. Introduction to the Theory and Applications of Molecular and Quantum Mechanics, Kluwer, Dordrecht 2003

A.Hinchliffe, Molecular Modelling for Beginners, Wiley, Chichester 2003

D.P.Landau, K.Binder, A Guide to Monte Carlo Simulations in Statistical Physics, Cambridge University Press, New York, 2nd ed. 2005

4.3b MOLECULAR AND ATOMISTIC MODELING (polymers)

K.F.Freed, Renormalization Group Theory of Macromolecules, Wiley, New York 1987

R.J.Roe, Ed., Computer Simulation of Polymers, Prentice Hall, Englewood Cliffs (NJ) 1991

J.Bicerano, Ed., Computational Modeling of Polymers, Dekker, New York 1992

E.A.Colbourn, Ed., Computer Simulation of Polymers, Longman, Harlow (Essex) 1994

B.R.Gelin, Molecular Modeling of Polymer Structures and Properties, Hanser, Munich 1994

C.Monnerie, U.W.Suter, Atomistic Modeling of Physical Properties, Adv.Polym.Sci. **116** (1994)

W.L.Mattice, U.W.Suter, Conformational Theory of Large Molecules. The Rotational Isomeric State Model in Macromolecular Systems, Wiley, New York 1994

K.Binder, Ed., Monte Carlo and Molecular Dynamics Simulations in Polymer Science, Oxford Univ. Press, New York 1995

M.Rehahn, W.L.Mattice, U.W.Suter, Rotational Isomeric State Models in Macromolecular Systems, Adv.Polym.Sci. **131/132** (1997)
A.Grosberg, Theoretical and Mathematical Models in Polymer Research, Academic Press, San Diego (CA) 1998
H.-D.Höltje, W.Sippl, D.Rognan, G.Folkers, Molecular Modeling. Basic Principles and Applications, Wiley-VCH, Weinheim, 2nd ed. 2003 (mainly peptides, no synthetic polymers)
M.Kotelyanskii, D.N.Theodorou, Simulation Methods for Polymers, Dekker, New York 2004
V.Galiatsatos, Ed., Molecular Simulation Methods for Predicting Polymer Properties, Wiley, New York 2005

4.4a and 4.5 LINEAR CHAINS: UNPERTURBED AND PERTURBED CHAINS

J.G.Kirkwood, Macromolecules (Collected Works of JGK), Gordon and Breach, New York 1967
P.J.Flory, Statistical Mechanics of Chain Molecules, Interscience, New York 1969; reprint: Hanser, Munich 1989
A.J.Hopfinger, Conformational Properties of Macromolecules, Academic Press, New York 1973
C.Williams, F.Brochard, H.L.Frisch, Polymer Collapse, Ann.Rev.Phys.Chem. **32** (1981) 433
N.Madras, G.Slade, The Self-Avoiding Walk, Birkhäuser, Basel 1993; see also G.Slade, Random Walks, American Scientist **84** (1996) 146
C.Vanderzande, Lattice Models of Polymers, Cambridge University Press, New York 1997
L.Schäfer, Excluded Volume Effects in Polymer Solutions, Springer, Berlin 1999
E.J.Janse Van Rensburg, The Statistical Mechanics of Interacting Walks, Polygons, Animals and Vesicles, Oxford University Press, Oxford 2000

4.4b WORMLIKE AND STIFF CHAINS

V.N.Tsvetkov, Rigid Chain Polymers. Hydrodynamic and Optical Properties in Solution, Plenum, New York 1989
G.L.Brelsford, W.R.Krigbaum, Experimental Evaluation of the Persistence Length for Mesogenic Polymers; in A.Ciferri, Ed., Liquid Crystallinity in Polymers, VCH, Weinheim 1991, p. 61
F.E.Arnold, Jr., F.E.Arnold, Rigid-Rod Polymers and Molecular Composites, Adv.Polym.Sci. **117** (1994) 257
H.Yamakawa, Helical Wormlike Chains in Polymer Solutions, Springer, Berlin 1998

4.6 CYCLIC MACROMOLECULES

J.A.Semlyen, Ed., Cyclic Polymers, Elsevier Sci.Publ., New York 1986

4.7 BRANCHED MACROMOLECULES

W.W.Graessley, Entangled Linear, Branched and Network Polymer Systems–Molecular Theories, Adv.Polym.Sci. **47** (1982) 67
W.L.Mattice, Masses, Sizes, and Shapes of Macromolecules from Multifunctional Monomers, in G.R.Newkome, C.N.Moorefield, F.Vögtle, Eds., Dendritic Molecules. Concepts, Synthesis, Perspectives, VCH Verlag, Weinheim 1996, Chapter I

4.8 SCALING

B.B.Mandelbrot, Fractals: Form, Chance and Dimensions, Freeman, New York 1977; The Fractal Geometry of Nature, Freeman, New York 1982
M.K.Kosmas, K.F.Freed, On Scaling Theories of Polymer Solutions, J.Chem.Phys. **69** (1978) 3647
P.G. de Gennes, Scaling Concepts in Polymer Physics, Cornell Univ. Press, Ithaca, New York 1979
L.Pietronero, E.Tosatti, Fractals in Physics, North Holland, Amsterdam 1986
K.R.Freed, Renormalization Group Theory of Macromolecules, Wiley, New York 1987
J.des Cloizeaux, G.Jannink, Les Polymères en Solution: Leur Modélisation et Leur Structure, Les Éditions de Physique, Les Ulis Cedex (France) 1987; Polymers in Solution. Their Modelling and Structure, Clarendon Press, Oxford 1990
H.H.Kaye, A Random Walk Through Fractal Dimensions, VCH, Weinheim 1989
D.Avnir, The Fractal Approach to Heterogeneous Chemistry: Surfaces, Colloids, Polymers, Wiley, New York 1989
M.Takayasu, Fractals in the Physical World, Manchester Univ. Press, Manchester 1989
D.W.Schaefer, Polymers, Fractals, and Ceramic Materials, Science **243** (1989) 1023
A.Blumen, H.Schnörer, Fractals and Related Hierarchical Models in Polymer Science, Angew.Chem.Int.Ed.Engl. **29** (1990) 113; –, Angew.Chem. **102** (1990) 158

R.J.Creswick, H.A.Farach, C.P.Poole, Jr., Introduction to Renormalization Group Methods in Physics, Wiley, New York 1991

F.C.Moon, Chaotic and Fractal Dynamics. An Introduction for Applied Scientists and Engineers, Wiley, New York 1992

References to Chapter 4

[1] J.Boor, Jr., E.A.Youngman, J.Polym.Sci. **A 4** (1966) 1861, Fig. 6

[2] B.H.Stofer, PhD Thesis 4577, ETH Zürich 1970, Fig. 17

[3] B.H.Stofer, H.-G.Elias, Makromol.Chem. **157** (1972) 245, Fig. 6

[4] S.M.Aharoni, N.S.Murthy, Polym.Commun. **24** (1983) 132, Table 1

[5] N.Donkai, H.Inagaki, K.Kanjiwara, M.Urukawa, M.Schmidt, Makromol.Chem. **186** (1985) 2623, Table 1

[6] P.Doty, B.Bruce McGill, S.A.Rice, Proc.Natl.Acad.Sci. **44** (1958) 432 (reported by B.Jirgensons, Natural Organic Macromolecules, Pergamon Press, Oxford 1962, p. 309)

[7] R.Pecora, Science **251** (1992) 893, Table 1

[8] O.A.Neumüller, Römpps Chemie-Lexikon, Franckh'sche Verlagshandlung, Stuttgart, 8th ed. 1979

[9] S.L.Mayo, B.D.Olafson, W.A.Goddard III, J.Phys.Chem. **94** (1990) 8897, Tables I, III, VII

[10] D.Lang, H.Bujard, B.Wolff, D.Russell, J.Mol.Biol. **23** (1967) 163, Plate II

[11] H.-G.Elias, Grosse Moleküle, Springer-Verlag, Berlin 1985; –, Mega Molecules, Springer-Verlag, Berlin 1987; –, Megamolekulnyi, Leningrad "Khimiya", Leningradskoe Otdelenje 1990, all Figure 14

[12] G.Natta, P.Corradini, I.W.Bassi, Gazz.Chim.Ital. **89** (1959) 784

[13] R.L.Jernigan, P.J.Flory, reported by P.J.Flory, Statistical Mechanics of Chain Molecules, Interscience, New York, 1969, p. 147, Fig. 9; P.J.Flory, Science **188** (1975) 1268, Fig. 9

[14] D.H.Yoon, P.J.Flory, Polymer **16** (1975) 645

[15] F.Abe, Y.Einaga, T.Yoshizaki, H.Yamakawa, Macromolecules **26** (1993) 1884, Tables I, II, V

[16] M.Ragnetti, D.Geiser, H.Höcker, R.C.Oberthür, Makromol.Chem. **186** (1985) 1701, Table 1

[17] T.Norisuye, Prog.Polym.Sci. **18** (1993) 543; data of [15]-[16] in Fig. 6

[18] T.Hirao, A.Teramoto, T.Sato, T.Norisuye, T.Masuda, T.Higashimura, Polym.J. **23** (1991) 925, Table 1

[19] H.Murakami, T.Norisuye, H.Fujita, Macromolecules **13** (1980) 345, Table 1

[20] T.Kashiwagi, T.Norisuye, H.Fujita, Macromolecules **14** (1981) 1220, Table 1

[21] T.Sato, A.Teramoto, Adv.Polym.Sci. **126** (1996) 85

[22] M.Yamada, M.Osa, T.Yoshizaki, H.Yamakawa, Macromolecules **30** (1997) 7166, Fig. 2

[23] Y.Fujii, Y.Tamai, T.Konishi, H.Yamakawa, Macromolecules **24** (1991) 1608, Tables I, II, IV

[24] I.Noda, K.Mizutani, T.Kato, T.Fujimoto, M.Nagasawa, Macromolecules **3** (1970) 787, Table I

[25] Y.Miyaki, Y.Einaga, T.Hirosuye, H.Fujita, Macromolecules **10** (1977) 1356, Tables I, II

[26] N.Hadjichristidis, M.Xenidou, H.Iatrou, M.Pitsikalis, Y.Poulos, A.Avgeropoulos, S.Sioula, S.Paraskeva, G.Velis, D.J.Lohse, D.N.Schulz, L.J.Fetters, P.J.Wright, R.A.Mendelson, C.A.Garcia-Franco, T.Sun, C.J.Ruff, Macromolecules **33** (2000) 2424, Tables 1-8

[27] S.-T.Sun, I.Nishio, G.Swislow, T.Tanaka, J.Chem.Phys. **73** (1980) 5973, Fig. 2

[28] D.Nerger, M.Eisele, K.Kajiwara, Polym.Bull. **10** (1983) 182, Fig. 1

[29] M.Nierlich, J.P.Cotton, B.Farnoux, J.Chem.Phys. **69** (1978) 1379, Table I

[30] N.Khasat, R.W.Pennisi, H.Hadjichristidis, L.J.Fetters, Macromolecules **21** (1988) 1100, Tables I and II

[31] P.K.Maiti, T.Çağın, G.Wang, W.A.Goddard, III, Macromolecules **37** (2004) 6236; (a) Fig. 5; (b) Table 3 and Figs. 4a and 4b, (c) taken from Fig. 6, (d) Table 2

[32] T.J.Prosa, B.J.Bauer, E.Amis, Macromolecules **34** (2001) 4897

[33] E.J.Amis et al., 29th ACS Central Regional Meeting, Midland (MI), 1997-05-28/30; quoted by P.R.Dvornic, S,Uppuluri, in J.M.J.Fréchet, D.A.Tomalia, Eds., Dendrimers and Other Dendritic Polymers, Wiley, New York 2002

[34] R.Scherrenberg, B.Coussens, P. van Vliet, G.Eduouard, J.Brackman, E. de Brabander, K.Mortensen, Macromolecules **31** (1998) 456, Table I

5 Scattering Methods

5.1 Overview

Dust particles glitter in sunlight because light is scattered by the particles. The intensity of this **Tyndall effect** is determined by particle size and shape. Similarly, light is also scattered by molecules. The incident electromagnetic waves shift atomic nuclei and their electrons in opposite directions. The resulting dipoles follow the oscillating electric field with the same frequency and emit electromagnetic radiation. The scattering intensity increases with the number of dipoles per molecule, which allows one to determine molecular weights. Scattering is also controlled by the distribution of dipoles in the scattering molecule which allows one to determine its spatial size and shape.

The theory of scattering of electromagnetic waves applies to all wavelengths (Fig. 5-1), and thus to all energy quanta including visible light, X-ray photons, neutrons, and electrons. Like all elementary particles, neutrons and electrons have dual particle-wave character. The masses and speeds of particles are interconnected by the Planck equation, $E = h\nu = hc/\lambda$, and the Einstein equation, $E = mc^2$, where E = energy (in J = m^2 kg s^{-2}), $h = 6.626\ 075\ 5 \cdot 10^{-34}$ J s = Planck constant, ν = frequency (in s^{-1}), c = speed of light in the medium (in m s^{-1}), λ = wavelength in the medium (in m), and m = mass (in kg).

For polymers, four ranges are especially important: scattering of visible light (wavelengths between ca. 300 nm and 700 nm), X-ray scattering (0.02-2 nm), neutron scattering (slow neutrons: 0.1-1 nm), and electron scattering (ca. 0.01 nm). In solution, **static light scattering (LS)** measures the difference of polarizabilities of macromolecules and solvents, **small-angle X-ray scattering (SAXS)** the difference of electron densities of molecules, and **small-angle neutron scattering (SANS)** the difference of coherent neutron scattering lengths of monomeric units and solvent molecules. Static light scattering photometers are relatively inexpensive and can be found in many laboratories. Equipment for SAXS is much more expensive and that for SANS is so extraordinarily costly that very few institutions can afford it.

Depending on the dimensions d of the particle relative to the wavelength λ_o of incident electromagnetic radiation and the difference between the refractive indices of the particle (n_p) and medium (n_o), three ranges can be distinguished:

- **Rayleigh scattering**: $d < \lambda_o/20$ and $n_p \approx n_o$;
- **Debye scattering**: $d \approx \lambda_o$ and $n_p \approx n_o$;
- **Mie scattering**: $d > \lambda_o$ and $n_p > n_o$ or $n_p < n_o$.

Macromolecular solutions and melts scatter in the Rayleigh and Debye ranges, which is the subject of this chapter. In these experiments, incident and scattered radiation have the same frequencies (i.e., energy): the scattering is "elastic."

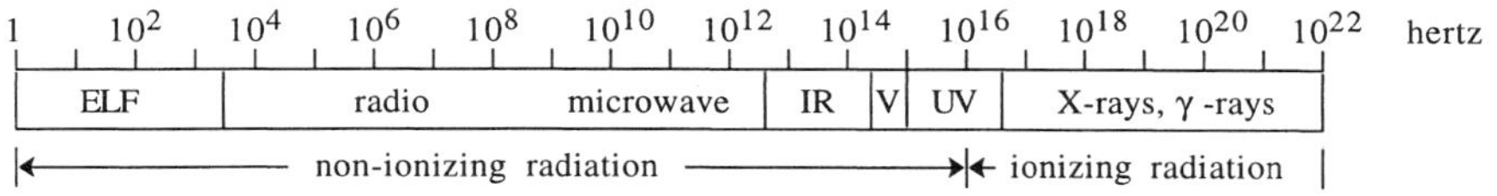

Fig. 5-1 Frequency (in hertz) of electromagnetic radiation. ELF = Extra-low frequency (overhead power lines), IR = infrared, V = visible light, UV = ultraviolet, γ = gamma radiation.

5.2 Static Light Scattering: Rayleigh Range

5.2.1 Fundamentals

According to **Beer's Law**, the intensity $I_o = P_o/A$ (= energy flux per area) of incident primary light is decreased by the intensity I_s of the scattered light on transmission through a scattering medium,

$$(5\text{-}1) \qquad I = I_o - I_s = I_o \exp(-\tau L)$$

where L = path in the medium and τ = absorption coefficient of scatttered light. The total intensity does not change, $I_o = I + I_s$. Light scattering is therefore a conservative absorption and not a consumptive one like the light absorption by colored solutions.

Scattering intensities I_s of pure liquids and dilute polymer solutions are ca. 1/10 000 to 1/50 000 of the primary intensity I_o. Because of the small difference $I_o - I_s$, scattering intensities are measured directly and not as the difference $I_o - I$ of the intensities of entering and exiting light as it is done in turbidometry (nephelometry).

Light is scattered because Brownian movements cause gases, clear solvents, and dilute solutions to be nonhomogeneous on a molecular scale. In polymer solutions, scattering intensities arise from three contributions, $I_s = I_c + I_d + I_a$. The largest effect stems from fluctuations of local polymer concentrations, I_c. A much smaller effect arises from density fluctuations of the solvent itself, I_d, and another small effect, I_a, from fluctuations of anisotropies of solute molecules. The solvent effect is eliminated by subtracting I_d from I_s, i.e., $I_s - I_d = I_c'$. The resulting apparent intensitiy I_c' is then corrected for the "depolarization" caused by fluctuations of anisotropy to give I_c (see Eq.(5-9)).

The theoretically simplest case is the scattering behavior of dilute gases. The "particles" of the gas (atoms, molecules) are much smaller than the wavelength λ_o of the incident light. Since the number concentration N/V of gas particles is small, each particle will scatter independently of the others. There are neither interactions between particles nor interactions of particles with their environment since the "solvent" is the vacuum.

In a scattering experiment, a monochromatic primary light beam of intensity I_o is scattered in a small volume V of the medium, which results in a scattering intensity I_ϑ at a distance L from the primary light source that is observed at an angle ϑ from the incident beam. The observed experimental quantity is the **Rayleigh ratio** R_ϑ which is not a ratio *per se* but the product of the normalized scattering intensity and an inverse length:

$$(5\text{-}2) \qquad R_\vartheta \equiv \left(\frac{I_\vartheta}{I_o}\right)\left(\frac{L^2}{V}\right)$$

Experimentally, Rayleigh ratios depend on four characteristics of the investigated system: polarization of molecules (I), wavelength (II) and polarization (III) of incident light, and concentration of molecules (IV). Their effect on the Rayleigh ratio can be evaluated as follows (for the exact derivation, see Literature to Chapter 5, 5.1 General).

(I) In *dilute gases*, scattering is caused only by the polarizability α of gas molecules. In these molecules, the alternating electrical field of the incident light generates dipoles that pulsate in phase the stronger the greater the polarizability. Since the polarizability is proportional to the electric field strength $\boldsymbol{E}$ and the intensity is proportional to the square of the electric field strength, $I_\vartheta \sim E^2$, according to the **Poynting theorem**, it follows that scattering intensities are proportional to the square of the polarizability, $I_\vartheta \sim \alpha^2$.

(II) The electrical field strength is also proportional to the inverse square of the speed of light, $\boldsymbol{E} \sim 1/\hat{c}^2$, which in turn is related to the wavelength λ via the angular frequency, $\omega = 2\,\pi\,\hat{c}/\lambda$. Since $I_\vartheta \sim E^2 \sim 1/\hat{c}^4$, scattering intensities are therefore inversely proportional to the fourth power of the wavelength, $I_\vartheta \sim 1/\hat{c}^4 \sim 16\,\pi^4/\lambda^4$.

(III) The observable Rayleigh ratio, R_ϑ', furthermore depends on the polarization of the incident light. Unpolarized (natural) light is a superposition of linear horizontally and linear vertically polarized light, each with the same intensity.

Vertically polarized light traveling in the x-direction will induce a dipole that pulsates in the particle in the z-direction. The field strength of scattered light is largest vertically to the dipole axis but zero in the direction of this axis (Fig. 5-2). The field strength must therefore be proportional to sin ϑ_v where ϑ_v is the angle between the dipole axis and the direction of observation. The intensity of vertically polarized scattered light is therefore independent of the observation angle in the xy-plane (Fig. 5-2, bottom left).

The dipole pulsates in the y-direction for horizontally polarized light. Again, the highest intensity is observed vertical to the dipole axis while it is zero in the y-direction. The field strength is proportional to sin ϑ_h where ϑ_h = angle between the dipole axis and the direction of observation.

One half of the incident light is vertically polarized and the other half horizontally. Since scattering intensities are proportional to the square of field strengths and field

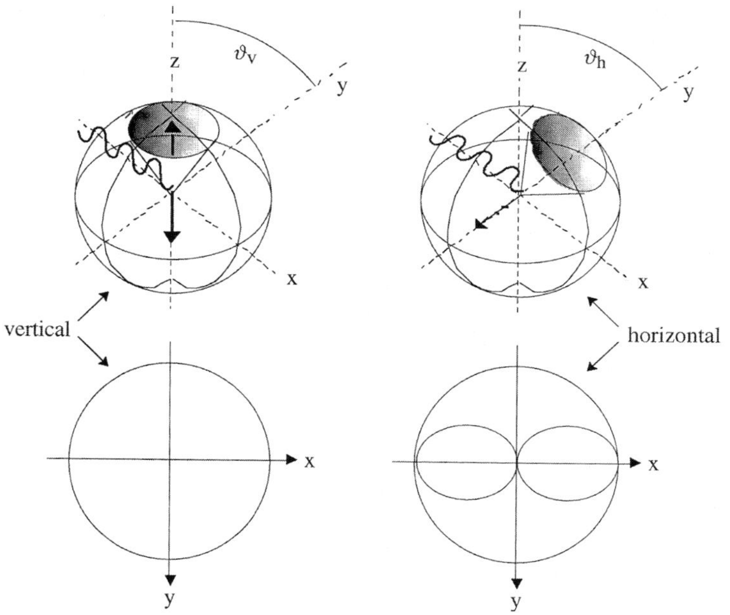

Fig. 5-2 Scattering diagrams of small isotropic particles for vertically (left) and horizontally (right) polarized incident light. Top: polarized light in x-direction (wavy lines) creates pulsating dipoles (thick arrows) in the z-direction (vertically polarized light) and y-direction (horizontally polarized light). Bottom: polar diagrams of scattering intensities in the xy plane.

strengths proportional to the sine of the angles, the scattering intensity must be proportional to $(\sin^2\vartheta_v + \sin^2\vartheta_h)/2$, i.e., $I_\vartheta \sim (\sin^2\vartheta_v + \sin^2\vartheta_h)/2 = (1 + \cos^2\vartheta)/2$. Because the Rayleigh ratio becomes zero at $\vartheta = 90°$ for horizontally polarized light, scattering experiments use only either unpolarized or vertically polarized light.

(IV) The Rayleigh ratio is also higher, the greater the number N of scatterers per volume V, i.e., the number concentration (= number density) of the solute, $C = N/V$.

For unpolarized light, the intensity of the scattered light thus depends on (1) the intensity of the incident radiation, I_o, (2) the square of the distance from the primary light source, L^2, and (3) the product of factors I-IV mentioned above (for the exact derivation, see textbooks of physics):

(5-3) $$I_\vartheta = \frac{16\,\pi^4\alpha^2 I_o V}{L^2\lambda^4}\left(\frac{\sin^2\vartheta_h + \sin^2\vartheta_v}{2}\right)C = \frac{16\,\pi^4\alpha^2 I_o V}{L^2\lambda^4}\left(\frac{1+\cos^2\vartheta}{2}\right)C$$

(5-4) $$R_\vartheta \equiv \left(\frac{I_\vartheta}{I_o}\right)\frac{L^2}{V} = \frac{16\,\pi^4\alpha^2}{\lambda^4}\left(\frac{1+\cos^2\vartheta}{2}\right)C$$

For gases, Rayleigh ratios R_ϑ can be obtained directly from the intensity of the scattered light. For solutions, the Rayleigh ratio is given by the difference of scattering intensities of solution and solvent. For dilute solutions, the wavelength of the scattered light in solution, $\lambda = \lambda_o/n_1$, can be approximated by the wavelength of the incident light, λ_o, with n_1 = refractive index of the solvent.

Except for the polarizability of particles, α, all other parameters of Eq.(5-4), I_ϑ,, I_o, L, λ, ϑ, and C, can be measured directly. Polarizabilities can be calculated from relative permittivities (formerly: dielectricity constants) according to **Clausius-Mosotti**,

(5-5) $$\alpha = \frac{3}{4\,\pi\,C}\left(\frac{\varepsilon_r - \varepsilon_{r,o}}{\varepsilon_r + 2\,\varepsilon_{r,o}}\right)$$

where $\varepsilon_r = \varepsilon/\varepsilon_o$ = relative permittivity of substance, ε = permittivity of the substance and ε_o = permittivity of the vacuum.

In *dilute gases*, the relative permittivity of the substance, ε_r, is that of molecules in vacuum; the relative permittivity of the vacuum itself is $\varepsilon_{r,o} = 1$. In *dilute solution*, ε_r is the relative permittivity of the solution and $\varepsilon_{r,0} = \varepsilon_{r,1}$ the relative permittivity of the solvent (index 1). In dilute solutions, $\varepsilon_r \approx \varepsilon_{r,1}$, and Eq.(5-5) converts to

(5-6) $$\alpha \approx \frac{3}{4\,\pi\,C}\left(\frac{\varepsilon_r - \varepsilon_{r,1}}{3\,\varepsilon_{r,1}}\right)$$

According to **Maxwell**, relative permittivities can be replaced by the squares of the corresponding refractive indices, i.e., $\varepsilon_{r,1} \approx n_1^2$ and $\varepsilon_r - \varepsilon_{r,1} \approx n^2 - n_1^2$. Refractive indices of dilute solutions vary with the solute mass concentration c, $n = n_1 + (\mathrm{d}n/\mathrm{d}c)c$, where $\mathrm{d}n/\mathrm{d}c$ = **refractive index increment**. Squaring delivers $n^2 = n_1^2 + 2\,n_1(\mathrm{d}n/\mathrm{d}c)c + (\mathrm{d}n/\mathrm{d}c)^2c^2$. The last term of this equation can be neglected because $\mathrm{d}n/\mathrm{d}c$ is in the range [(–0.2)–(+0.2)]/(mL/g) and c is usually smaller than 0.1 g/mL.

Polarizabilities can thus be calculated from the experimentally observable quantities C, n_1, and dn/dc:

$$(5\text{-}7) \qquad \alpha \approx \frac{3}{4\pi C}\left(\frac{n^2 - n_1^2}{3 n_1^2}\right) \approx \frac{1}{4\pi C}\left(\frac{2 n_1 (\mathrm{d}n/\mathrm{d}c)c}{n_1^2}\right)$$

Introducing Eq.(5-7) into Eq.(5-4), using $\lambda = \lambda_o/n_1$, and replacing the number concentration of the solute, $C = N/V = cN_A/M$, by the mass concentration, c, delivers

$$(5\text{-}8) \qquad R_\vartheta = \left(\frac{4\pi^2 n_1^2 (\mathrm{d}n/\mathrm{d}c)^2}{N_A \lambda_o^4}\right)\left(\frac{1+\cos^2\vartheta}{2}\right) cM = K_{LS,\vartheta} cM$$

The **optical constant** (= **contrast factor**), $K_{LS,\vartheta}$, contains the Avogadro constant, N_A, and the experimental quantities n_1, dn/dc, and λ_o. For high molecular weights, refractive index increments dn/dc are independent of degrees of polymerization because contributions by endgroups can be neglected. In this case, optical constants are system constants that depend only on polymer constitution, solvent, and temperature.

For depolarized scattered light, directly measured Rayleigh ratios R_ϑ,, Eqs.(5-2), (5-4), or (5-8), must be multiplied by the **Cabannes factor**

$$(5\text{-}9) \qquad f_C = \frac{6 + 6\Delta_U}{6 - 7\Delta_U} \quad ; \quad \Delta_U = \frac{I_{h,soln} - I_{h,solv}}{I_{v,soln} - I_{v,solv}}$$

This factor is calculated from the horizontal (index h) and vertical (index v) scattering intensities of the solution (soln) and solvent (solv), respectively.

The molar mass M of the solute can thus be calculated from experimental values of R_ϑ, K_ϑ, and c (Eq.(5-8) without assuming a model. However, this equation was derived for gases at very low pressure, i.e., infinite dilution. For polymers at concentration c, M is an **apparent molar mass**, M_{app}, that needs to be extrapolated to zero polymer concentration in order to obtain the true molar mass, M (Section 5.2.3).

For particles with dimensions larger than ca. 0.05 λ_o, apparent molar masses depend also on the scattering angle and need to be extrapolated to zero scattering angle, $\vartheta \to 0$ (Section 5.3). The angular dependence allows one to calculate the radius of gyration.

Scattering intensities of molecules are additive. According to Eq.(5-8), a polymer at c $\to 0$ and $\vartheta \to 0$ (i.e., $(1 + \cos^2\vartheta)/2 \to 1$ since $\cos\vartheta \to 1$), consisting of i molecules of molar masses M_i present in concentrations c_i, will thus have a Rayleigh ratio of

$$(5\text{-}10) \qquad R_0 = \textstyle\sum_i R_{0,i} = \sum_i K_{LS,0} c_i M_i = K_{LS,0} \sum_i c_i M_i = K_{LS,0} c \overline{M}_w$$

since $\overline{M}_w \equiv \sum_i c_i M_i / \sum_i c_i$ and $\sum_i c_i \equiv c$. Extrapolation of $K_{LS}c/R_\vartheta$ to $c \to 0$, $\vartheta \to 0$ delivers the inverse mass-average molar mass, $\overline{M}_w$:

$$(5\text{-}11) \qquad \frac{K_{LS,0}c}{R_0} = \frac{1}{\overline{M}_w} \quad \text{(for } c \to 0,\ \vartheta \to 0\text{);} \qquad K_{LS,0} = \frac{4\pi^2 n_1^2 (\mathrm{d}n/\mathrm{d}c)^2}{N_A \lambda_o^4}$$

Conventional light scattering instruments employ incoherent primary light with wavelengths that are selected by color filters, usually 435.8 nm or 546.1 nm. Because of the incoherence, scattering angles are accessible only in the range 37.5-142.5°.

Modern light-scattering photometers employ lasers as light source, for example, the He-Ne laser with $\lambda_o = 632.8$ nm. Since laser beams are coherent, R_ϑ can be measured at such small scattering angles that $R_\vartheta \approx R_0$, usually at 7° (Fig. 5-2). However, low-angle laser light scattering (LALLS) at $\vartheta \to 0$ does not deliver any information about dimensions of scattering particles. Conversely, such measurements are fairly insensitive to the presence of dust particles, a major experimental problem in conventional static light-scattering experiments with natural light sources. Dust particles are much larger than polymer molecules. They contribute considerably to scattering intensities at large scattering angles (see Section 5.3), even at very low dust concentrations.

5.2.2 Copolymers

Refractive index increments of polymer molecules, dn/dc, are independent of molar masses if the contribution of endgroups is negligible. They vary from molecule to molecule in random copolymers, graft copolymers, and other constitutionally non-uniform polymers consisting of "a" and "b" monomeric units. As a variable, they cannot be part of the optical constant K_{LS}, and Eq.(5-10) must therefore be written as

$$(5\text{-}12) \qquad R_0 = \sum_i R_{0,i} = K'_{LS,0} \sum_i (\mathrm{d}n/\mathrm{d}c)_i^2 c_i M_i \quad ; \quad K'_{LS,0} = \frac{4\pi^2 n_1^2}{N_A \lambda_o^4}$$

Refractive index increments, $(\mathrm{d}n/\mathrm{d}c)_i$, and concentrations c_i of components i cannot be measured independently. Only the average refractive index increment, $\overline{(\mathrm{d}n/\mathrm{d}c)}_{cp}$, and the concentration c of the copolymer itself is experimentally accessible. Eq.(5-12) thus becomes

$$(5\text{-}13) \qquad R_0 = K' \overline{(\mathrm{d}n/\mathrm{d}c)}_{cp}{}^2 c\, \overline{M}_{w,cp,app}$$

with $\overline{M}_{w,cp,app}$ = apparent mass-average molar mass that is calculated from the experimental values of R_0, K', c, and $\overline{(\mathrm{d}n/\mathrm{d}c)}_{cp}$ (for $c \to 0$!). Equating Eqs.(5-12) and (5-13), and introducing $w_i \equiv c_i/c$, $(\mathrm{d}n/\mathrm{d}c)_i \equiv Y_i$, and $\overline{(\mathrm{d}n/\mathrm{d}c)}_{cp} \equiv Y_{cp}$ delivers the apparent mass-average molar mass of the copolymer:

$$(5\text{-}14) \qquad \overline{M}_{w,cp,app} = \frac{\sum_i (\mathrm{d}n/\mathrm{d}c)_i^2 w_i M_i}{\overline{(\mathrm{d}n/\mathrm{d}c)}_{cp}^2} = \frac{\sum_i Y_i^2 w_i M_i}{Y_{cp}^2}$$

The deviation of the refractive index increment $(\mathrm{d}n/\mathrm{d}c)_i \equiv Y_i$ of the ith type of molecule from the average refractive index increment $\overline{(\mathrm{d}n/\mathrm{d}c)}_{cp} \equiv Y_{cp}$ of the copolymer depends on the difference $\Delta w_{a,i} \equiv w_{a,cp} - w_{a,i}$ between the average mass fraction $w_{a,cp}$ of a-units in the copolymer and the mass fraction $w_{a,i}$ of a-units in the copolymer molecule i and on the difference $(\mathrm{d}n/\mathrm{d}c)_a - (\mathrm{d}n/\mathrm{d}c)_b \equiv \Delta Y_{a,b}$ of the refractive index increments of the two units "a" and "b" of the copolymer:

(5-15) $(dn/dc)_i - \overline{(dn/dc)}_{cp} = [(dn/dc)_a - (dn/dc)_b][w_{a,cp} - w_{a,i}]$

(5-16) $Y_i - Y_{cp} = \Delta Y_{a,b} \times \Delta w_{a,i}$

Combination of Eqs.(5-14) and (5-15) results in

$$\overline{M}_{w,cp,app} = \sum_i w_i M_i + 2\frac{\Delta Y_{a,b}}{Y_{cp}} \sum_i w_i M_i \Delta w_{a,i} + \left(\frac{\Delta Y_{a,i}}{Y_{cp}}\right)^2 \sum_i w_i M_i \Delta w_{a,i}^2 \tag{5-17}$$

The first sum of this equation is the mass-average molar mass of the copolymer, $\overline{M}_w$. The second sum is the first moment of the z-distribution of the property Δw_a and the third sum the corresponding second moment as one can see by setting $w_i = m_i / \sum_i m_i$, multiplicating by $\sum_i m_i M_i / \sum_i m_i M_i$, separating the terms, and remembering that $z_i = m_i M_i$ (see Eq.(2.9)):

$$\sum_i w_i M_i \Delta w_{a,i} = \left(\frac{\sum_i m_i M_i \Delta w_{a,i}}{\sum_i m_i M_i}\right)\left(\frac{\sum_i m_i M_i}{\sum_i m_i}\right) = v_z^{(1)} \overline{M}_w \tag{5-18}$$

$$\sum_i w_i M_i \Delta w_{a,i}^2 = \left(\frac{\sum_i m_i M_i \Delta w_{a,i}^2}{\sum_i m_i M_i}\right)\left(\frac{\sum_i m_i M_i}{\sum_i m_i}\right) = v_z^{(2)} \overline{M}_w \tag{5-19}$$

With $\Delta Y \equiv (Y_a - Y_b)/Y_{cp} = \Delta Y_{a,b}/Y_{cp}$, Eq.(5-17) can be written as

$$\overline{M}_{w,cp,app} = \overline{M}_w [1 + 2\, v_z^{(1)} \Delta Y + v_z^{(2)} \Delta Y^2] \tag{5-20}$$

The mass-average molar mass of a copolymer can thus be obtained from light scattering experiments in solvents with strongly different refractive indices n_1 (and therefore different values of dn/dc), which deliver different values of $Y_{cp} = (dn/dc)_{cp}$ and thus different values of $\overline{M}_{w,cp,app}$ (Fig. 5-3). The true mass-average molar mass, $\overline{M}_{w,cp}$, of the copolymer is the value of $\overline{M}_{w,cp,app}$ at $\Delta Y = (Y_a - Y_b)/Y_{cp} = 0$.

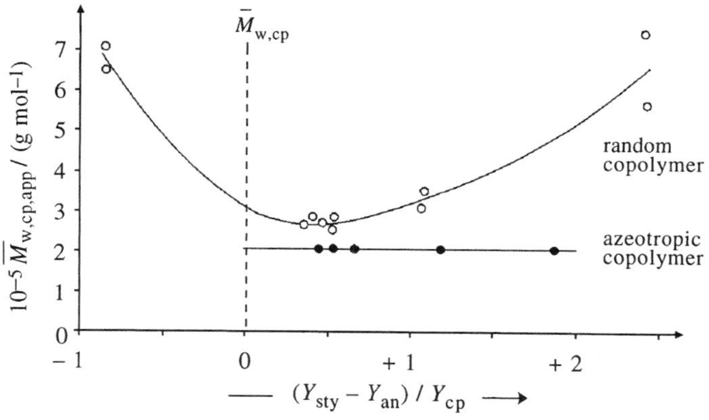

Fig. 5-3 Dependence of the apparent mass-average molar masses (at $c \to 0$, $\vartheta \to 0$) of a random and an azeotropic styrene/acrylonitrile copolymer on the refractive index parameter, $(Y_{sty} - Y_{an})/Y_{cp}$ [1].

Constitutionally uniform copolymers, such as those obtained by azeotropic copolymerization, have $\Delta w_{a,i} = 0$; the first and second moments of the compositional distribution are therefore also zero (Eq.(5-20)). Independently of the refractive index of the solvent, the same mass-average molar mass is obtained.

5.2.3 Concentration Dependence

Previous sections were concerned with randomly distributed small, isotropic molecules that move independently of each other. In very dilute solutions, temporal and local fluctuations of solute concentrations and solvent densities are independent of each other to a first approximation. As a result, the scattering intensity of the solute is simply the difference of the scattering intensities of the solution and solvent.

At higher concentrations, momentary (α) and average ($\langle\alpha\rangle$) polarizabilities are no longer identical. The former differs from the latter by $\Delta\alpha$, and the square of probabilities, Eq.(5-7), becomes $\alpha^2 = (\langle\alpha\rangle + \Delta\alpha)^2 = \langle\alpha\rangle^2 + 2\,\langle\alpha\rangle\Delta\alpha + (\Delta\alpha)^2$. All volume elements have the same average polarizabilitiy, $\langle\alpha\rangle$, which therefore adds nothing to the contribution by fluctuation. The average fluctuation, $\Delta\alpha$, about the average polarizability is also zero. Only $(\Delta\alpha)^2$ contributes somewhat to the fluctuation.

The theoretical calculation of $(\Delta\alpha)^2$ – or to be more precise, that of the spatial average $(\Delta\alpha)^2$ – is very elaborate (see Literature to Chapter 5). The essential result is an expansion of Eq.(5-10) by the spatial average of the square of concentration fluctuation, $\langle(\Delta c)^2\rangle$. Since these fluctuations are proportional to the concentration, Eq.(5-10) must be normalized in order to obtain the correct physical units. Furthermore, $(1 + \cos^2 \vartheta)/2 = 1$ at $\vartheta = 0$. Since $R_0 = K_{LS,0} c \overline{M}_w$ at $c \rightarrow 0$, Eq.(5-11) converts to

$$(5\text{-}21) \qquad R_0 = K_{LS,0}[\langle(\Delta c)^2\rangle/c^2]c\,\overline{M}_w$$

Thermodynamics relates fluctuations of concentrations to changes in the concentration dependence of Gibbs energies, G, and that of the chemical potential of the solvent, μ_1, respectively (see Chapter 10),

$$(5\text{-}22) \qquad \langle(\Delta c)^2\rangle = k_B T/(\partial G^2/\partial c^2)_{p,T} = k_B T\,\tilde{V}_{1,m} c(N/V)/(-\,\partial\mu_1/\partial c)$$

where $\tilde{V}_{1,m}$ = partial molar volume of the solvent 1 and $k_B = R/N_A$ = Boltzmann constant.

The concentration dependence of the chemical potential of *non-self-associating* polymers can be expressed by a series (Section 10.4)

$$(5\text{-}23) \qquad -\,\partial\mu_1/\partial c = RT\tilde{V}_{1,m}(\overline{M}_w^{-1} + 2\,A_2 c + 3\,A_3 c^2 + \ldots)$$

in which the coefficients A_2, A_3, ... are the second, third, ... **virial coefficients**. The first virial coefficient is identical to the inverse mass-average molar mass. The second virial coefficient, A_2, is a measure of the thermodynamic interactions between two molecules and the third virial coefficient, A_3, that between three molecules. For molecularly non-uniform polymers, second and third virial coefficients from light scattering differ from the ones obtained from colligative methods such as osmometry (Section 10.4). For example, the second virial coefficient constitutes an average

(5-24) $$A_{2,\text{LS}} = \sum_i \sum_j w_i M_i w_j M_j A_{ij} / (\sum_i w_i M_i)^2$$

Introduction of Eqs.(5-22) and (5-23) into Eq.(5-21) delivers the concentration dependence of Kc/R_0 at the scattering angle $\vartheta = 0$, here indexed for light scattering:

(5-25) $$K_{\text{LS},0}c/R_0 = (1/\overline{M}_\text{w}) + 2\,A_{2,\text{LS}}c + 3\,A_{3,\text{LS}}c^2 + \ldots \;; \qquad (\text{at } \vartheta \to 0)$$

5.2.4 Mixed Solvents

Light scattering experiments are conducted occasionally in mixed solvents, either because one-component solvents do not dissolve the polymer (Sections 10.1.3 and 10.1.4) or because one wants to study preferential interactions of the polymer 2 with one of the components 1 or 3 of the mixed solvent. Such a mixed solvent may consist of two solvents, a solvent and a non-solvent, or even two non-solvents that form a mixed solvent (Section 10.1.3).

Mixed solvents cause both a thermodynamic and an optical problem. With respect to thermodynamics, fluctuations of local polymer concentrations (Section 5.2.3) are now accompanied by fluctuations of local solvent compositions. In addition, polymer molecules interact often preferentially with one of the solvents 1 or 3. In the literature, this **preferential solvation** is often called preferential *adsorption* although it does not involve a phase boundary as true adsorptions do.

Because of the coupled fluctuations of three components, the refractive index increment of the solution is no longer independent of the polymer concentration as in a two-component solution of polymer 2 in solvent 1. However, contributions of the two components 1 and 3 of the mixed solvent can be separated from that of polymer 2 if one works at constant chemical potential of the mixed solvent, $\Delta\mu = const.$

This can be achieved by dialyzing the solution of polymer 2 in 1+3 against the mixed solvent 1+3. In equilibrium, all three chemical potentials are the same, $\Delta\mu_1 = \Delta\mu_2 = \Delta\mu_3$, and the refractive index of 1+3 is now n_μ whereas the refractive index increment of the solution becomes $(\mathrm{d}n/\mathrm{d}c)_\mu$. The polymer concentration is now the polymer concentration in the dialysis equilibrium, c_μ, and not the original concentration c in the mixed solvent. Eq.(5-25) converts to

(5-26) $$\frac{K^*(\mathrm{d}n/\mathrm{d}c)_\mu^2 c_\mu}{R_0} = \frac{1}{\overline{M}_\text{w}} + 2\,A_{2,\text{LS}}c_\mu + \ldots \quad ; \quad K^* = \left(\frac{4\,\pi^2 n^2}{N_\text{A}\lambda_0^4}\right)\left(\frac{1+\cos^2\vartheta}{2}\right)$$

For small molecules, Rayleigh ratios R_0 are independent of the observation angle ϑ and the extrapolation of the left side of Eq.(5-26) to $c_\mu \to 0$ delivers the true inverse mass-average molar mass. However, the $A_{2,\text{LS}}$ of Eq.(5-26) is no longer identical with the $A_{2,\text{LS}}$ of Eq.(5-25).

The situation is different if one does not work at constant chemical potential (i.e., using different compositions of 1+3 at different polymer concentrations) but keeps instead the volume fraction $\phi_1 = 1 - \phi_3$ of component 1 constant. At infinite dilution, the intercept of $Kc/R_0 = \mathrm{f}(c)$ is then no longer the inverse mass-average molar mass, $1/\overline{M}_w$, but an inverse apparent molar mass, $1/M_{app} \equiv 1/(Y^2 \overline{M}_w)$. Similarly, the initial slope of the function $K^*(\mathrm{d}n/\mathrm{d}c)^2_\phi c/R_0 = \mathrm{f}(c)$ contains no longer the true 2nd virial coefficient, $A_{2,LS}$, but an apparent 2nd virial coefficient, $A_{2,LS}/Y^2$:

$$\text{(5-27)} \qquad \frac{K^*(\mathrm{d}n/\mathrm{d}c)^2_\phi c}{R_0} = \frac{1}{Y^2}\left(\frac{1}{\overline{M}_w} + 2\,A_{2,LS}c + \ldots\right)$$

The deviation factor Y^2 is a function of the change of the refractive index n of the solution with the volume fraction ϕ_1 of component 1 of the mixed solvent, $\mathrm{d}n/\mathrm{d}\phi_1$, the change of n with the polymer concentration c at constant volume fraction ϕ_1 of the mixed solvent, $(\mathrm{d}n/\mathrm{d}c)_\phi$, and the coefficient $\aleph$ of the preferential solvation:

$$\text{(5-28)} \qquad Y = 1 + \aleph\frac{\mathrm{d}n/\mathrm{d}\phi_1}{(\mathrm{d}n/\mathrm{d}c)_{\phi_1}} = 1 + \aleph\Re \quad ; \quad \Re \equiv (\mathrm{d}n/\mathrm{d}\phi_1)/(\mathrm{d}n/\mathrm{d}c)_{\phi_1}$$

Refractive indices do not vary with the composition of the mixed solvent if solvent components are isorefractive, $n_1 = n_3$. Hence, $\mathrm{d}n/\mathrm{d}\phi_1 = 0$ in Eq.(5-28) and therefore also $Y = 1$, i.e., experiments with isorefractive solvent components deliver true mass-average molar masses and true second virial coefficients.

For non-isorefractive components of mixed solvents, coefficients $\aleph = \mathrm{f}(\phi_1)$ can be obtained if $\overline{M}_w$ is known from measurements in a single solvent and $\mathrm{d}n/\mathrm{d}\phi_1$ and $(\mathrm{d}n/\mathrm{d}c)_\phi$ are determined by refractometry. Alternatively, Eqs.(5-26) and (5-27) may be combined and $\aleph$ calculated from $(\mathrm{d}n/\mathrm{d}c)_\mu = \mathrm{f}(\mathrm{d}n/\mathrm{d}\phi_1)$ for $c_\mu \approx c$:

$$\text{(5-29)} \qquad (\mathrm{d}n/\mathrm{d}c)_\mu = (\mathrm{d}n/\mathrm{d}c)_\phi + \aleph(\mathrm{d}n/\mathrm{d}\phi_1)$$

According to Eq.(5-29), coefficients $\aleph$ of preferential solvation describe the difference between the volume $V_{1,u}$ of component 1 near the polymer chain and the average volume V_1 of component 1 per mass m_u of monomeric units. They may be positive or negative (Table 5-1); positive values indicate that molecules of type 1 reside preferentially near the polymer chain.

Volumes V_1 can be expressed by masses m_1 of components 1 $V_1 = m_1/\rho_1$ if one assumes that densities ρ_1 of component 1 are the same near the polymer chain and far from it. Masses m_1 are related to molar masses M_1 and numbers of molecule, N_1, by $m_1 = N_1M_1/N_A$. Because of the change of density with composition, molar volumes $V_{1,m} = M_{1i}/\rho_1$ have to be replaced by partial molar volumes, $\tilde{V}_{1,m}$. The coefficient $\aleph$ of preferential solvation can be expressed therefore by

$$\text{(5-30)} \qquad \aleph = \frac{V_{1,u} - V_1}{m_u} = \frac{m_{1,u} - m_1}{\rho_1 M_u n_u} = \frac{(N_{1,u} - N_1)M_1}{\rho_1 M_u N_u} = \frac{(N_{1,u} - N_1)}{N_u}\cdot\frac{\tilde{V}_{1,m}}{M_u} = \mathfrak{I}\frac{\tilde{V}_{1,m}}{M_u}$$

where $\mathfrak{I} = (N_{1,u} - N_1)/N_u$ describes the preferential solvation as excess, $N_{1,u} - N_1$, of the number of solvent molecules of type 1 per number N_u of monomeric units relative to the average in the system.

Table 5-1 Apparent molar masses $Y^2 \overline{M}_w$, refractivity factor $\mathfrak{R}$, coefficient $\aleph$ of preferential solvation, and parameter $\mathfrak{I}$ of preferential solvation according to light-scattering measurements on poly(styrene) (PS) with $\overline{M}_w$ = 413 000 g/mol [2, 3] and poly(2-hydroxyethyl methacrylate) (PHEMA) with $\overline{M}_w$ = 225 000 g/mol [3, 4] in mixed solvents 1 + 3 with volume fractions $\phi_1 = 1 - \phi_3$.

$V_{1,m}$/(mL mol^{-1}) = 88.90 (benzene) and 74.79 (1-propanol), respectively; M_u/(g mol^{-1}) = 104.15 (PS) and 130.14 (PHEMA).

Polymer	Solvent 1	Solvent 3	ϕ_1	$\frac{Y^2 \overline{M}_w}{\text{g mol}^{-1}}$	$\frac{\mathfrak{R}}{\text{g mL}^{-1}}$	$\frac{\aleph}{\text{mL g}^{-1}}$	$\mathfrak{I}$
PS	benzene	cyclohexane	1.00	413 000	1	0	0
			0.75	446 000	0.680	0.058	0.07
			0.50	478 000	0.552	0.137	0.16
			0.35	477 000	0.498	0.150	0.18
			0.25	449 000	0.439	0.097	0.11
PHEMA	1-propanol	water	1.00	225 000	1	0	0
			0.95	210 000	0.358	–0.095	–0.17
			0.80	170 000	0.303	–0.432	–0.75
			0.60	185 000	0.385	–0.242	–0.42
			0.40	260 000	0.439	0.171	0.30
			0.20	280 000	0.480	0.240	0.42

The preferential solvation of poly(styrene), $\mathfrak{I}$, by benzene in mixtures of the good solvent benzene and the very bad solvent cyclohexane ($T < \Theta$) passes with increasing cyclohexane content through a maximum at $\phi_3 = 0.5$ (Fig. 5-4). Additional measurements showed that this function does not vary with the molar mass.

The situation is quite different for poly(2-hydroxyethyl methacrylate) in 1-propanol (1) if the precipitant water (3) is added. With increasing ϕ_3, values of $\mathfrak{I}$ first become negative and pass through a minimum before increasing again. It is unclear whether $\mathfrak{I} \to 1$ (as shown in Fig. 5-4), $\mathfrak{I} = 1/2$, or even $\mathfrak{I} \to 0$ at $\phi_3 \to 1$. The negative value of $\mathfrak{I}$ at $\phi_3 = 0.2$ indicates that at this solvent concentration water and not 1-propanol is preferentially solvating the polymer, i.e., the hydroxyl groups of the side chain.

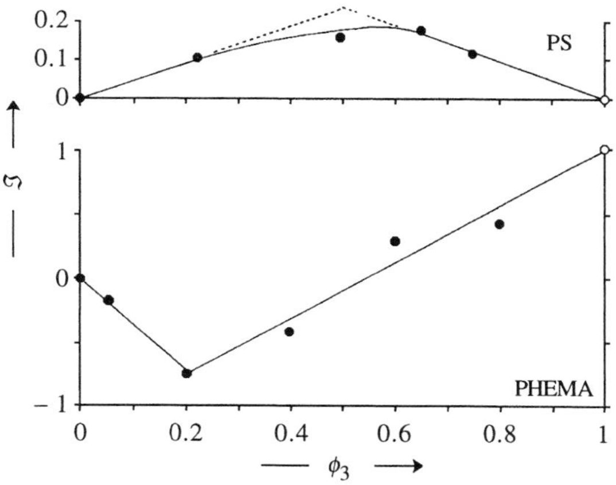

Fig. 5-4 Preferential solvation $\mathfrak{I}$ of a poly(styrene) (PS) and a poly(2-hydroxyethyl methacrylate) (PHEMA) as a function of the volume fraction of component 3 of a mixed solvent (data of Table 5-1). ● Experimental, ○ extrapolated.

5.3 Static Light Scattering: Debye Range

5.3.1 Introduction

The previous section was concerned with molecules whose spatial dimensions are small compared to the wavelength λ_o of incident light. Such molecules behave as if they have only one scattering center. At molecule dimensions greater than ca. 0.05 λ_o, however, molecules must be represented by several scattering centers (Fig. 5-5, left). Light scattered by these centers at the same scattering angle $\vartheta > 0°$ arrives at the observer considerably out of phase. The phase shift Δ depends on the cosine of the scattering angle:

$$\Delta = \overrightarrow{DB} = \overrightarrow{AB} - \overrightarrow{AD} = \overrightarrow{AB}(1 - \cos \vartheta) \tag{5-31}$$

Light waves emitted by the various scattering centers interfere because they originate from the same light source. The interference is measured by the phase shift Δ which is controlled by the scattering angle ϑ. The destructive interference is zero at $\vartheta = 0°$ but increases with increasing scattering angle (Fig. 5-5, right). For the same molecular shape, it increases with the molar mass. It also varies with the shape of the molecule.

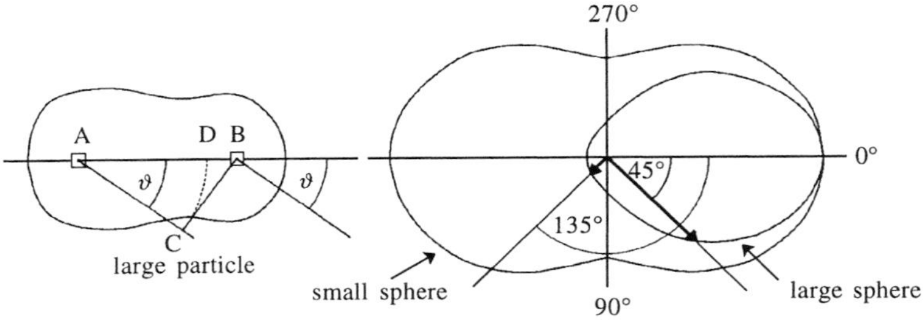

Fig. 5-5 Left: phase shift caused by scattering at two scattering centers A and B of a large molecule. Right: scattering diagrams of spheres by unpolarized incident light of wavelength λ_o, assuming a diameter of less than 0.05 λ_o for the small sphere and a diameter of $\lambda_o/2$ for the large one.

The simplest way to gather information about spatial dimensions of larger polymer molecules is thus to measure the ratio z of scattering intensities at two different angles, usually at 45° and 135°, i.e., the **dissymmetry** R_{45}/R_{135}. Dissymmetries are measures of characteristic lengths L of molecules such as the diameter $L = d$ of spheres, the length L of rigid rods, or the end-to-end distance $L = \langle r^2 \rangle^{1/2}$ of random coils (Fig. 5-6). In order to calculate particle *dimensions* from dissymmetries, particle *shapes* must be known from other sources of information.

5.3.2 Scattering Functions

No assumptions about particle shapes are needed if the scattering is measured for a large range of scattering angles. The angular dependence of the normalized scattering intensity from a single molecule (infinite dilution) is described by a **particle scattering factor**, $P(q)$.

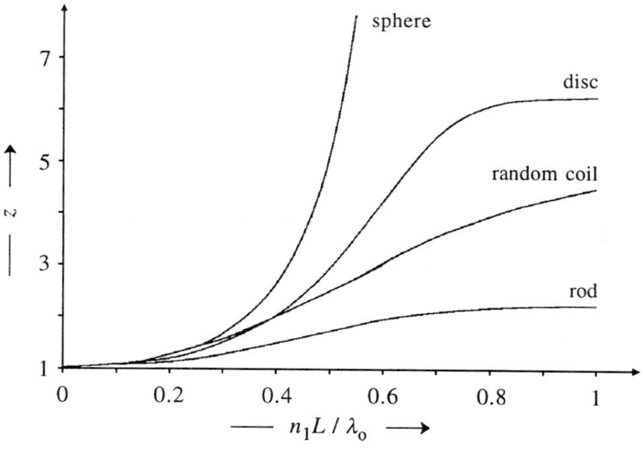

Fig. 5-6 Dissymmetry $z = R_{45}/R_{135}$ as a function of the dimensionless length parameter, $n_1L/\lambda_o = L/\lambda$, of various uniform molecules. L = major dimension (= length L of rods, diameter d of spheres and discs, end-to-end distance of unperturbed random coils, $\langle r^2\rangle_o^{1/2}$). n = refractive index of medium, n_1 = ditto of solvent, λ_o = wavelength of incident light, $\lambda = \lambda_o/n \approx \lambda_o/n_1$ = wavelength in the medium.

(5-32) $$P(q) = \frac{\text{experimental scattering intensity at angle } \vartheta}{\text{scattering intensity at angle } \vartheta \text{ without interference}} = \frac{R_q}{R_0}$$

The particle scattering factor is not a function of the scattering angle itself (and therefore not written as $P(\vartheta)$) but as a function of the scattering vector $\boldsymbol{q}$ (and therefore written as $P(q)$). Vector analysis has shown that the magnitude of $\boldsymbol{q}$ is

(5-33) $$|q| = (4\pi n_1/\lambda_o) \sin(\vartheta/2) = Q_\vartheta \quad ; \quad \lambda = \lambda_o/n_1$$

With a particle scattering factor $P(q)$ for the molecule and an analogous scattering factor $P'(q)$ for the polymer-solvent interaction, Eq.(5-25) for $\vartheta \rightarrow 0$ converts to

(5-34) $$\frac{K_{LS}c}{R_q} = \frac{1}{\overline{M}_w P(q)} + 2\frac{A_{2,LS}}{P'(q)}c + \ldots \quad ; \quad K_{LS}c/R_q \equiv 1/\overline{M}_{w,app}$$

The inverse apparent mass-average molar masses, $1/\overline{M}_{w,app}$, must be extrapolated to zero polymer concentration *and* zero scattering angle in order to obtain the true mass-average molar mass, $\overline{M}_w$. For the extrapolation, the dependence of the particle scattering factor $P(q)$ on the scattering vector has to be known.

The calculation of the **particle scattering factor**, $P(q)$, assumes a large particle containing N scattering centers in a uniform electric field. The distance between scattering centers i and j is r_{ij}. In each scattering center, the incident electromagnetic radiation creates a vibrating dipole that emits scattered light. The resulting electric fields at points i and j are characterized by wave vectors $\boldsymbol{k}$ that form a vector $\boldsymbol{q} = \boldsymbol{k}_i - \boldsymbol{k}_j$ whose magnitude is the variable $|q|$ of Eq.(5-33).

Summation of scattering intensities of all N scattering centers delivers the particle scattering factor of a spatially fixed, rigid particle. In isotropic solutions, however, such a particle may adopt all possible positions in space. The average particle scattering factor of such randomly oriented rigid particles is given by the **Debye equation**, Eq.(5-35), which was derived in 1915 for X-ray scattering but applies also to light scattering, regardless of the particle shape:

$$(5\text{-}35) \qquad P(q) = \sum_{i=1}^{i=N} \sum_{j=1}^{j=N} \frac{\sin qr_{ij}}{qr_{ij}}$$

Macromolecules with flexible chains continuously change the distances r_{ij} between their scattering centers so that an additional temporal averaging is required. In this case, $P(q)$ is the average particle scattering factor for all macroconformations.

The double sum in Eq.(5-35) can be solved for $qr_{ij} << 1$, i.e., for small distances r_{ij} or small angular variables q, respectively, and thus for large wavelengths λ_o or small scattering angles ϑ. In this case, the sine can be developed in a MacLaurin series:

$$(5\text{-}36) \qquad \sin qr_{ij} = qr_{ij} - \frac{(qr_{ij})^3}{3!} + \frac{(qr_{ij})^5}{5!} - \ldots$$

and the particle scattering factor becomes

$$(5\text{-}37) \qquad P(q) = \frac{1}{N^2} \sum_{i=1}^{i=N} \sum_{j=1}^{j=N} \left(1 - \frac{(qr_{ij})^2}{3!} + \frac{(qr_{ij})^4}{5!} - \ldots\right)$$

For small values of qr_{ij}, the first two terms of the double sum suffice. The square of the distance, r_{ij}^2, is replaced by the mean-square distance, $\langle r_{ij}^2 \rangle$, which is related to the mean-square radius of gyration by $\langle s^2 \rangle = [1/(2\,N^2)]\,\Sigma_i \Sigma_j \langle r_{ij}^2 \rangle$. For *small* angles, Eq.(5-37) converts to $P(q) = 1 - [q^2 \langle s^2 \rangle /3]$ which, because $1 - x \approx (1 + x)^{-1}$, changes to the **Zimm equation**

$$(5\text{-}38) \qquad P(q) = 1 - [q^2 \langle s^2 \rangle /3] \rightarrow \frac{1}{P(q)} = 1 + \frac{q^2 \langle s^2 \rangle}{3} - \ldots \; ; \; |q| = (4\,\pi\,n_1/\lambda_o) \sin(\vartheta/2)$$

The Zimm equation allows one to calculate the mean-square radius of gyration from the *initial* slope of $1/P(q) = f(\sin^2(\vartheta/2))$, regardless of the shape of particles. Because it is restricted to $\langle r_{ij}^2 \rangle$, it does not deliver information about the shape of particles; this information can be gained from the higher members of the series, $\langle r_{ij}^4 \rangle$, $\langle r_{ij}^6 \rangle$, etc. The simplifying assumptions also restrict Eq.(5-38) to $\langle s^2 \rangle^{1/2}/\lambda \leq 0.05$, i.e., $\langle s^2 \rangle^{1/2} \leq \lambda/20$. For necessary corrections, see p. 169 (for data with/without corrections, see p. 162 and 169).

For molecularly non-uniform polymers, mean-square radii of gyration from light scattering experiments are averages. The type of average is obtained as follows. According to Eq.(5-38) left, $\langle s^2 \rangle$ is proportional to $1 - P(q)$. In turn, the particle scattering function equals $R_\vartheta/(K_\vartheta c \overline{M}_w)$ at infinite dilution (Eq.(5-34)). Summation over all contributions of components i of a non-uniform polymer and introduction of $c_i = m_i/V$ and $z_i = m_i/\overline{M}_{w,i}$ (Eq.(2-9a)) shows that $\langle s^2 \rangle$ is obtained as the z-average:

$$(5\text{-}39) \qquad P(q) = \frac{R_q}{K_{LS} c \overline{M}_w} = \frac{\sum_i K_{LS} c_i \overline{M}_{w,i} R_{q,i}}{\sum_i K_{LS} c_i M_{w,i}} = \frac{\sum_i z_i R_{q,i}}{\sum_i z_i} \equiv \overline{P}_z(q)$$

5.3.3 Zimm Plot

A relatively safe procedure for the extraction of radii of gyration from light-scattering data is to insert Eq.(5-33) into Eq.(5-38) and combine the result with Eq.(5-34):

$$(5\text{-}40)\quad \frac{K_{LS}c}{R_q} = \frac{1}{\overline{M}_w P(q)} + 2\frac{A_{2,LS}}{P'(q)}c + \ldots \cong \frac{1}{\overline{M}_w}\left(1 + \frac{16\pi^2 n_1^2}{3\lambda_0^2}\langle s^2\rangle \sin^2\frac{\vartheta}{2}\right) + 2\frac{A_{2,LS}}{P'(q)}c + \ldots$$

Eq.(5-40) allows simultaneous extrapolations of scattering data to $\vartheta \to 0$ and $c \to 0$ if $K_{LS}c/R_q$ is plotted as a function of $[\sin^2(\vartheta/2) + kc]$ (Fig. 5-7). The arbitrary constant k serves to disentangle data and prepare a clean gridlike diagram (Fig. 5-7). The grid may be curved for block polymers (Fig. 5-8), self-association, polyelectrolytes, etc.

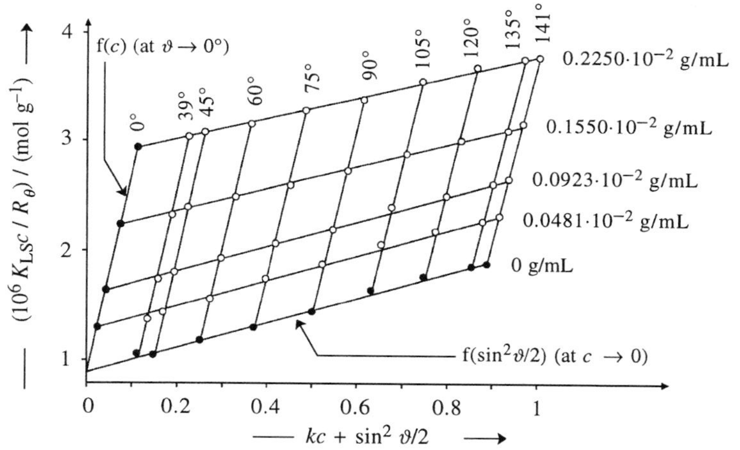

Fig. 5-7 Zimm plot for a poly(vinyl acetate) in butanone at 25°C, using k = 50 mL/g.

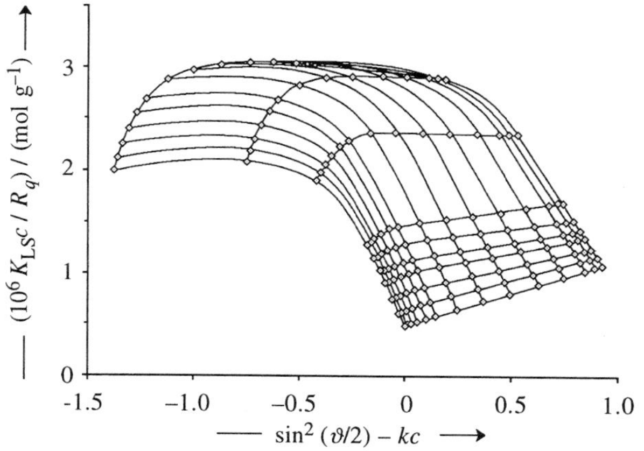

Fig. 5-8 Zimm plot for a triblock poly(methyl methacrylate)-*block*-poly(styrene)-*block*-poly(methyl methacrylate) in toluene at 30°C [5]. $x_{styrene}$ = 0.41; $\overline{M}_w$ = 1.47·10^6 g/mol, k = 5 mL/g.
By permission of American Chemical Society, Washingtom (DC).

5.3.4 Effect of Particle Shape

Information about particle shapes is obtained from the particle scattering function, $P(q) = f(\vartheta)$. Since derivations of these functions for the various particle shapes are lengthy, only basic assumptions and final results are shown here.

Spheres

The scattering function $P(q) = f(q) = f(4\pi n_1/\lambda_o) \sin(\vartheta/2))$ of an isotropic compact sphere with radius R was derived by Rayleigh:

$$(5\text{-}41) \qquad P(q) = \left[\frac{3}{q^3 R^3}(\sin qR - qR\cos qR)\right]^2$$

For large values of qR, the inverse particle scattering factor has an asymptote of

$$(5\text{-}42) \qquad \lim_{qR\to\infty} \frac{1}{P(q)} = \frac{2q^4R^4}{9} = \frac{512\pi^4 n_1^4 R^2}{9\lambda_o^4}\sin^4\frac{\vartheta}{2}$$

For small values of qR, Eq.(5-41) can be developed in a series:

$$(5\text{-}43) \qquad \frac{1}{P(q)} = 1 + \frac{1}{5}(qR)^2 + \frac{4}{175}(qR)^4 + \frac{47}{23625}(qR)^6 + \ldots$$

Comparison of coefficients of Eqs.(5-43) and (5-38, right) delivers $\langle s^2\rangle = (3/5)\,R^2$.

Rods

Debye and Neugebauer were the first to calculate the particle scattering factor for thin rigid rods of large length $L = Nb$ and radius $R \ll L$. The rod is modeled as an assembly of N+1 scattering centers that are linearly arranged and connected by bonds of length b. Along the rod, scattering centers i and j are separated from each other by r_{ij}. There are $2(N+1)$ distances with $r_{ij} = 0$; $2N$ distances with $r_{ij} = b$; $2(N-1)$distances with $r_{ij} = 2b$; and $2(N+1-k)$ distances with $r_{ij} = kL$. The Debye Eq.(5-35) thus becomes

$$(5\text{-}44) \qquad P(q) = \frac{1}{(N+1)^2}\sum_{k=0}^{N} 2(N+1-k)\frac{\sin kLq}{kLq}$$

Because $N \gg 1$, N+1 can be replaced by N and the sum by the integral:

$$(5\text{-}45) \qquad P(q) = \frac{2}{N}\int_0^N \frac{\sin kLq}{kLq} - \frac{2}{LqN^2}\int_0^N \sin(kLq)\mathrm{d}k$$

The first integral cannot be solved analytically but is found in standard mathematical tables. The second integral is obtained from $kLq \equiv$ y and using $1 - \cos y = 2\sin^2(y/2)$:

$$(5\text{-}46) \qquad P(q) = \frac{2}{qL}\int_0^{qL} \frac{\sin y}{y}\mathrm{d}y - \left[\frac{\sin(qL/2)}{qL/2}\right]^2$$

For small values of qL, Eq.(5-46) may be developed in a series:

(5-47) $$P(q) = 1 - \frac{(qL)^2}{36} + \frac{(qL)^4}{1800} - \frac{(qL)^6}{141\,120} + \ldots$$

Comparison of coefficients of Eqs.(5-47) and (5-38, left side) delivers $\langle s^2 \rangle = L^2/12$ which is identical with Eq.(4-5) for infinitely thin rods.

For large values of qL, the asymptote is

(5-48) $$\lim_{qL\to\infty} P(q) = \frac{\pi}{qL} - \frac{2}{(qL)^2}$$

No closed expression exists for the particle scattering factor of molecularly uniform rods. However, such an expresion was found for thin rods of constant diameter and a Schulz-Flory distribution of lengths (and thus of molar masses):

(5-49) $$P(q) = \frac{2}{q\sum_i w_i L_i} \arctan\left(\frac{q}{2}\sum_i w_i L_i\right)$$

For large values of qL, the function $Kc/R_\vartheta = f(q)$ has an asymptote with a slope of $\pi L_i/M_i$ that is proportional to the ratio of length and molar mass.

Random Coils

The particle scattering factor of random coils with a Gaussian distribution of segments and a radius of gyration s was first derived by both Peter Debye and Werner Kuhn:

(5-50) $$P(q) = \frac{2}{q^4 \langle s^2 \rangle^2}\left[\{q^2\langle s^2 \rangle - 1\} + \exp(-q^2\langle s^2 \rangle)\right]$$

Introduction of $P(q) = R_q/K_{LS}cM$ (Eq.(5-34)) and $M = \overline{M}_w$ leads to

(5-51) $$\frac{q^4 R_q}{K_{LS}c} = \frac{2M}{\langle s^2 \rangle^2}\left[q^2\langle s^2 \rangle - 1 + \exp(-q^2\langle s^2 \rangle)\right]$$

This function has three ranges which can be seen clearly for very high molar masses: the **Guinier range** for $qs < 1$, an intermediate range at $qs \approx 1$, and an asymptotic range for $qs >> 1$ (Fig. 5-9).

For small molar masses, only the Guinier range is observed. For small values of $q^2\langle s^2 \rangle$, $\exp(-q^2\langle s^2 \rangle) = e^x$ can be developed in a series

$$e^x = 1 + x/(1!) + x^2/(2!) + x^3/(3!) + x^4/(4!) \ldots$$

and Eq.(5-50) becomes

(5-52) $$P(q) = \frac{2}{q^4 \langle s^2 \rangle^2}\left[\{q^2\langle s^2 \rangle - 1\} + 1 - q^2\langle s^2 \rangle + \frac{(q^2\langle s^2 \rangle)^2}{2} - \frac{(q^2\langle s^2 \rangle)^3}{6} + \ldots\right]$$

(5-53) $$P(q) = 1 - \frac{q^2\langle s^2 \rangle}{3} + \frac{(q^2\langle s^2 \rangle)^2}{12} - \ldots \quad \text{or} \quad \frac{1}{P(q)} = 1 + \frac{q^2\langle s^2 \rangle}{3} - \frac{(q^2\langle s^2 \rangle)^2}{36} + \ldots$$

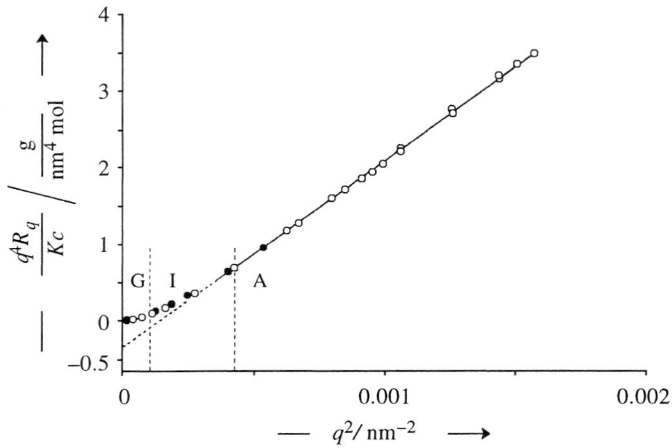

Fig. 5-9 Plot according to Eq.(5-51) for an atactic poly(styrene) with $\overline{M}_w$ = 8 780 000 g/mol in cyclohexane at Θ = 34.5°C [6]. Measurements at 436 nm (O) and 546 nm (●), respectively. G = Guinier range, I = intermediate range, A = asymptotic range.

The *Guinier range* (G) delivers the mass-average molar mass and the z-average mean-square radius of gyration (Section 5.3.2). The *asymptotic range* (A) is mostly inaccessible. According to Henri Benoit, an obtainable asymptotic range will deliver both the z-average and the number-average molar mass:

$$(5\text{-}54) \qquad \lim_{q^2 \to \infty} \frac{q^4 R_q}{K_{LS}c} = -\frac{2\overline{M}_z^2}{\langle s^2 \rangle_z^2 \overline{M}_n} + \frac{2\overline{M}_z}{\langle s^2 \rangle_z} q^2$$

In contrast to the Guinier range and the asymptotic range, the *intermediate range* is affected by the type of molar mass distribution. For Schulz-Zimm distributions, the average particle scattering factor was calculated as

$$(5\text{-}55) \qquad P(q) = \frac{2}{\varsigma(\varsigma+1)}\left(\frac{\varsigma+2}{q^2 s^2}\right)^2 \left[\left(\frac{q^2 s^2}{\varsigma+2}+1\right)^{-\varsigma} + \frac{\varsigma}{\varsigma+2} q^2 s^2 - 1\right]$$

where $\varsigma = \overline{X}_n/(\overline{X}_w - \overline{X}_n)$ = degree of coupling of chains during polymerization (Eq.(2-65). In the special case of a Schulz-Flory distribution, $\overline{X}_w/\overline{X}_n = 2$, $\varsigma = 1$, and Eq.(5-55) is reduced to $1/P(q) = 1 + (q^2s^2/3)$ (Eq.(5-38), right side).

The degree of coupling becomes smaller than unity for $\overline{M}_w/\overline{M}_n > 2$. As a result, $1/P(q)$ is no longer linearly dependent on q^2s^2; the function $1/P(q) = f(q^2s^2)$ rather bends upwards. For $\overline{M}_w/\overline{M}_n < 2$, it bends downward. These effects become noticeable for $q > 1$ as can be seen from a plot of $q^{1/2}P(q)$ as a function of $q^{1/2}$ (Fig. 5-10).

The mean-square radius of gyration from light scattering, small angle X-ray, and small angle neutron scattering is in reality an apparent value, $\langle s^2 \rangle_{app} = \langle s^2 \rangle + s_A^2$, since it also contains the contribution s_a^2 of the chain cross-sectional area. For example, the experimental value $\langle s^2 \rangle_{app}^{1/2}$ = 0.475 nm for a poly(styrene) with $\overline{M}_w$ = 578 g/mol was reduced to $\langle s^2 \rangle^{1/2}$ = 0.339 nm after subtracting s_A = 0.136 nm.

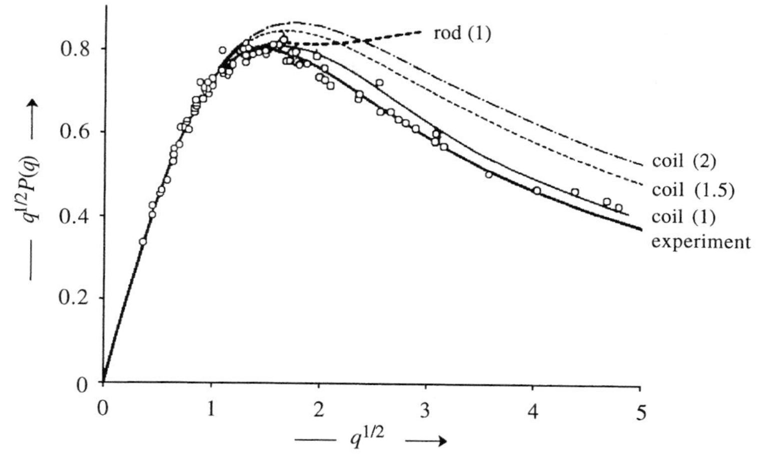

Fig. 5-10 Effect of molecular non-uniformity and particle shape on the angular dependence of the particle scattering factor, here as $q^{1/2}P(q) = f(q^{1/2})$ [7]. (O) Experimental data for poly(D-β-hydroxybutyrate)s of various molar masses in trifluoroethanol at 25°C. Coil: Theoretical curves for coils with molecular non-uniformities of $\overline{M}_w / \overline{M}_n$ = 1, 1.5, and 2. Rod (- - -): Theoretical curve for uniform rods. By permission of American Chemical Society, Washington (DC).

In general, the same evaluations for $P(q)$ are used for unperturbed and perturbed coils so that the mean-square radius of gyration of the latter is just increased by the square of the linear expansion coefficient, $\langle s^2 \rangle = \alpha_s^2 \langle s^2 \rangle_o$. Note, however, that Eq.(5-50) was derived for Gaussian coils;corrections for the non-Gaussian character of perturbed coils are rarely applied, if ever.

Worm-like Chains

The particle scattering factor of worm-like and stiff chains, respectively, with historic contour lengths L_{chain} and persistence lengths L_{ps} (Section 4.4.9) was calculated for phantom chains as

(5-56)

$$P(q) = \frac{2}{u^2}[u - 1 + \exp(-u)] + \frac{8}{15}\frac{L_{ps}}{L_{chain}}[1 - \exp(-u)] + \frac{14}{15}\frac{L_{ps}}{L_{chain}}\left(\frac{1}{u}\right)[1 - 2\exp(-u)]$$

where u is a modified square of the scattering parameter $\boldsymbol{q}$:

(5-57) $$u = \frac{L_{chain}L_{ps}}{3}q^2 = \frac{L_{chain}L_{ps}}{3}\left(\frac{4\pi}{\lambda}\sin\frac{\vartheta}{2}\right)^2 \quad ; \quad \lambda = \lambda_o/n_1$$

The contour length $L_{chain} = N_K L_K$ can be expressed by the number N_K and length L_K of Kuhnian segments (p. 97). Persistence lengths are half as large as Kuhnian segments, $L_{ps} = L_K/2$ (p. 98). According to Eq.(4-59), mean-square radii of gyration, $\langle s^2 \rangle$, of worm-like chains can be expressed by persistence lengths and numbers of Kuhnian segments, which allows one to calculate $P(q)$ and qs and *vice versa*.

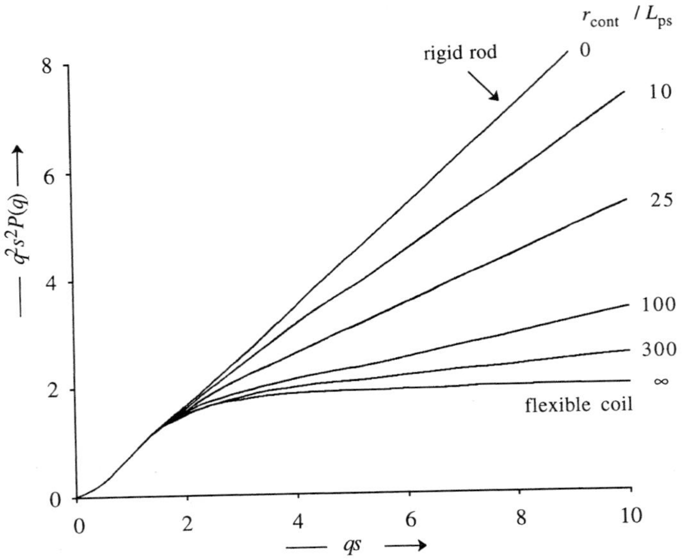

Fig. 5-11 Dependence of $q^2s^2P(q)$ on qs for linear chains with various ratios r_{cont}/L_{ps} of conventional contour lengths r_{cont} and persistence lengths L_{ps} [8a]. Data are based on the Kratky-Porod worm-like chain which is a phantom chain; s corresponds therefore to $\langle s^2\rangle_o^{1/2}$. $r_{cont}/L_{ps} = 0$ indicates a rigid rod and $r_{cont}/L_{ps} = \infty$ a flexible coil. A value of $qs = 10$ corresponds to $\langle s^2\rangle_o^{1/2} = 0.75\ \lambda_o$ at $n_1 = 1.5$.

The scattering behavior of worm-like chains can be demonstrated best by a **Kratky diagram** in which $q^2s^2P(q)$ is plotted as a function of qs (Fig. 5-11). As indicated qualitatively in Section 4.4.9, scattering parameters of worm-like chains reside between those of flexible coils and rigid rods. Worm-like chains become rigid rods for large persistence lengths compared to conventional contour lengths ($r_{cont}/L_{ps} \rightarrow 0$) and flexible coils for small ones ($r_{cont}/L_{ps} \rightarrow \infty$).

Values of $q^2s^2P(q)$ at $qs < 2$ do not allow one to distinguish between rods and coils (Fig. 5-11) (for example, $qs = 2$ corresponds to $s \approx 95$ nm if $\vartheta = 90°$, $\lambda_o = 632.8$ nm, and $n_1 = 1.5$). For very large ratios of r_{cont}/L_{ps} (i.e., flexible coils), $q^2s^2P(q)$ approaches a value of 2 (Eq.(5-50)). For very small ratios of r_{cont}/L_{ps} (i.e., rigid rods), the function $q^2s^2P(q) = \mathrm{f}(qs)$ becomes a straight line, $q^2s^2P(q) = [\pi/(2\cdot 3^{1/2})]qs$, as can be seen from Eq.(5-48) and $s^2 = L^2/12$ (Eq.(4-5)).

Star Molecules

Particle scattering factors have been calculated for star-like polymer molecules with f arms of equal length and an infinitely small core. Each arm is supposed to be long enough to form an unperturbed coil with a Gaussian distribution of segments in space, i.e., excluded volumes are supposed to be absent. According to this theory, the particle scattering factor, $P(q)$, depends on both f and qs according to

$$(5\text{-}58) \qquad P(q) = \frac{2}{q^2s^2}\left[q^2s^2 - f + f\exp\left(-\frac{q^2s^2}{f}\right) + \frac{f(f-1)}{2}\left\{1 - \exp\left(-\frac{q^2s^2}{f}\right)\right\}^2\right]$$

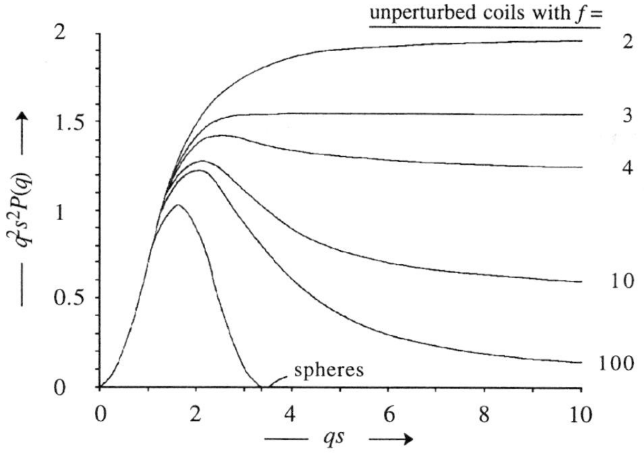

Fig. 5-12 Kratky diagrams for molecularly uniform linear molecules with $f = 2$ and star molecules with $f = 3$, 4, 10, or 100 arms, all as phantom chains. Compact sphere for comparison: the first maximum is followed by a weak minimum, then another maximum (not shown), etc.; see also Section 5.4.2). For branching parameters calculated from scattering data, see p. 118.

In contrast to linear macromolecules, values of $q^2s^2P(q) = f(qs)$ of star molecules always pass through maxima with increasing qs. The maximum is very weak for $f = 3$ (not visible in Fig. 5-12) but becomes more pronounced with increasing numbers of arms (Fig. 5-12). For $f \rightarrow \infty$, the function $q^2s^2P(q) = f(qs)$ approaches that of a sphere, showing a series of maxima. Such maxima are present for all star molecules but are either very weak ($f = 3$) or can be seen only at $qs > 10$ for $3 < f < 100$ (not shown). It follows that such maxima are not produced by branching *per se* but rather by the tendency of arms to form overlapping coils which increases with the number and length of arms.

Comb Molecules

Comb molecules consist of a main chain with a degree of polymerization, X, to which $2 \leq f \leq X$ side chains with degrees of polymerization N_i are attached in regular or irregular intervals ($f = 1$ corresponds to a 3-star molecule). Chain units of side chains may or may not have the same constitution as the chain units of the main chain. The total degree of polymerization of the comb molecule is $X_{tot} = X + fN_i$ and the mole fraction of monomer units in the main chain, $x = X/X_{tot}$.

Scattering functions have been derived for three major types of comb molecules. Type I is a comb molecule with side chains of equal length and constant number of chain units between branching sites. Type II maintains equal lengths of side chains but allows one to vary the number of chain units between branching sites. Type III is the most general (no restrictions with respect to length and distribution of side chains). Type I combs are obtained from living polymerizations of macromonomers whereas Type III combs result from free-radical grafting of monomers onto polymer chains.

Scattering functions of all three types are fairly complicated to derive and not easy to evaluate experimentally. For example, the derivation of the scattering function of Type I comb molecules assumes phantom chains with very long backbones and side chains. The spatial distribution of segments is supposed to be Gaussian. The resulting particle scattering factor depends on the number f of side chains per molecule, the mole fraction $x = X/X_{tot}$ of monomeric units in the main chain, and a scattering parameter $Q = q^2s^2$ where $q = (4\,\pi n_1/\lambda_o)\sin(\vartheta/2)$ and s = radius of gyration:

$$(5\text{-}59) \qquad P(q) = \frac{2}{Q^2}\left\{Q - 1 + \exp(-xQ) + AY + BY^2\right\} \quad ; \quad Y = 1 - \exp\left(\frac{-Q(1-x)}{f}\right)$$

$$A = \left[f + \frac{2[1 - \exp\{-Qxf/(f+1)\}]}{1 - \exp[-Qx/(f+1)]}\right]$$

$$B = \left[\frac{[(f-1)\exp(Qx/(f+1)) - 1] - [1 - \exp(-Qx(f-1)/(f+1))]}{[1 - \exp(-Qx/(f+1))]^2}\right]$$

Systematic experimental scattering data of comb molecules do not seem to exist but there are hydrodynamic data that can be correlated with scattering data since the intrinsic viscosity, $[\eta]$, is related to the radius of gyration by $[\eta] = (\Phi/M)s^3$ where Φ = Flory parameter (Section 12.3.8).

5.4 Small-Angle X-Ray Scattering

5.4.1 Fundamentals

Static light scattering allows one to determine radii of gyration, $\langle s^2\rangle^{1/2}$, in the range $0.05 \leq \langle s^2\rangle_z^{1/2}/\lambda \leq 0.5$, i.e., with blue light (λ_o = 436 nm) in the range $22 \leq \langle s^2\rangle_z^{1/2}/\text{nm} \leq 220$ and with He-Ne laser light (λ_o = 632.8 nm) in the range $32 \leq \langle s^2\rangle_z^{1/2}/\text{nm} \leq 320$. These ranges cover the radii of gyration of many macromolecules. However, one usually works in the Guinier range which delivers the radius of gyration but does not inform about the shape of molecules.

X-rays have much shorter wavelengths than light waves, which allows one to determine radii of gyration down to ca. 1 nm. The wavelengths of X-rays are much smaller than the radii of gyration of polymer molecules, whereas in light scattering it is just the opposite. For example, the wavelength of Cu-K_α radiation is λ_o = 0.154 nm which is more than 3000 times smaller than that of blue light of λ_o = 436 nm. As a consequence, X-rays probe much finer details of polymer structures than light waves.

The theory of scattering of electromagnetic waves applies to all wavelengths; hence, the dependence of particle scattering factors on wavelength, molar mass, and radius of gyration is maintained. Only the optical constant K in Eqs.(5-11) and (5-34) is changed because small angle X-ray scattering (**SAXS**) is controlled by electron densities and not by polarizabilities (and therefore refractive indices) as light scattering (LS) is:

$$(5\text{-}60) \qquad K_{\text{SAXS}} = \frac{(\Delta\rho_e/m_e)^2}{(\hat{c}^4/e^4)N_A} \quad ; \quad K_{\text{LS}} = \frac{4\pi(\mathrm{d}n/\mathrm{d}c)^2}{\lambda^4 N_A}$$

where $\Delta\rho_e = (N_e/M) - (\tilde{v}_2 C_{e,1})$ = excess electron density of the polymer; N_e = number of electrons per polymer molecule; M = molar mass of polymer; $\tilde{v}_2$ = partial specific volume of polymer; $C_{e,1}$ = number-concentration (= number density) of electrons in the solvent; m_e = 9.109 534 (47)·10^{-28} g = mass of electron; $\hat{c} = \hat{c}_o/n$ = speed of light in medium; $\hat{c}_o$ = speed of light in vacuum; n = refractive index of medium; N_A = Avogadro constant; and $e \approx 1.6022 \cdot 10^{-19}$ C = elementary charge (= charge of electron).

In order to achieve the same effect, the parameter $q = (1/3)[(4\ \pi n_1)/\lambda_o]^2\ [\sin^2(\vartheta/2)]$ must have the same value in LS and SAXS. For example, what can be observed by light scattering at λ_o = 436 nm and ϑ = 90° in a solvent of refractive index n_1 = 1.45 (i.e., λ = 436 nm/1.45 ≈ 300 nm) needs to be measured by SAXS with λ = 0.1 nm at a very small angle of ϑ = 0.027°! Clearly, the precise measure of such small angles and experimental problems such as focussing makes SAXS a very demanding technique.

At the very small required angles, the dependence of the particle scattering factor $P(q)$ on wavelength and scattering angle can be approximated by the **Guinier equation:**

$$(5\text{-}61) \qquad P(q) \cong \exp\left(-\frac{q^2 s^2}{3}\right) \quad ; \quad q^2 = [(4\ \pi/\lambda)\sin(\vartheta/2)]^2$$

Regardless of particle shape, a plot of ln $P(q)$ = f(q^2) should thus deliver a straight line with a slope of $-s^2/3$ if one indeed works in the Guinier range.

However, the Guinier range is not always accessible. For example, branched polymer molecules have relatively high particle densities at small spatial dimensions which makes it impossible to observe the linear dependence of ln $P(q)$ on q^2 at sufficiently small values of q as required by the Guinier function (Fig. 5-13, left). One then has to use other evaluation procedures such as the method of Zimm where $1/P(q)$ is plotted against q^2 (Fig. 5-13, right).

At about the same molar mass, scattering data of the dendrimer with its high segment density are well represented by a Guinier function but not by the Zimm method. Conversely, data of the much more open-structured hyperbranched polyol are well represented by a Zimm plot but not by the Guinier representation.

Guinier plots, ln $P(\vartheta)$ = f(q^2), should deliver straight lines for $q^2 \to 0$ if polymers are uniform with respect to molar mass, shape, and electron distributions, all attributes of higher generation PAMAM dendrimers. Scattering curves of such polymers are consistent with those calculated from the scattering of isotropic spheres (solid line in Fig. 5-13, left) and those from molecular dynamics calculations (see p. 77 and p. 122 ff.).

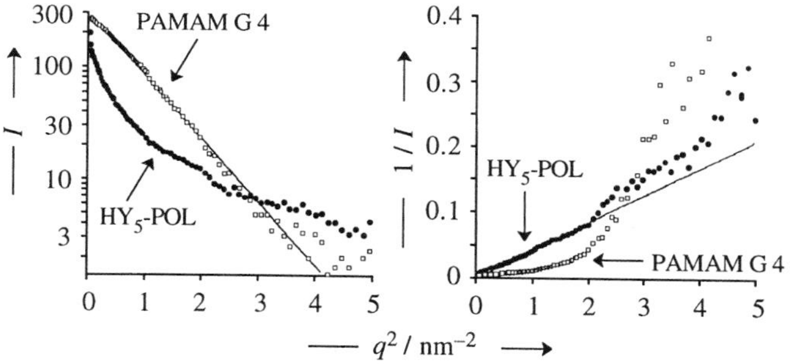

Fig. 5-13 Guinier plot (left) and Zimm plot (right) of X-ray intensities I (left, log scale) and their inverse values (right) as function of the square q^2 of the scattering vector $\mathbf{q}$ for 1 % CH_3OH solutions of a G 4 polyamidoamine dendrimer (□ PAMAM) and a "generation 5" of a hyperbranched polyol (● HY_5-POL) [9a]. Intensities are represented in arbitrary units, hence I and not $P(q)$.

Table 5-2 Effect of evaluation method on calculated radii of gyration, *s*/nm, from small-angle neutron scattering (N), small-angle X-ray scattering (X), or static light scattering (L).

Measurements on star poly(butadiene)s (S_N-PB) with N = 32 or 128 arms [10], polyamidoamine dendrimers (PAMAM) with 4 or 10 generations [9], and a hyperbranched polyol (HY_5-POL) [9].

Evaluations assuming isotropic spheres or applying Guinier, Kratky, or Zimm plots with (U) or without (-) Ullman corrections (see p. 169).

Polymer	Molar mass in g/mol	Method	Radius of gyration, $\langle s^2\rangle^{1/2}$, in nanometers N Isotropic sphere	N, X Guinier	N Kratky	N, X Zimm (-)	N Zimm (U)	L Zimm (-)
S_{128}-PB	715 000	SANS	-	8.6	10.9	11.1	10.5	-
S_{32}-PB	256 000	SANS, L	-	8.7	9.8	11.4	10.1	10.0
PAMAM G-10	934 000	SAXS	5.74	6.00	-	9.49	-	-
PAMAM G-4	7 230	SAXS	2.25	2.41	-	4.11	-	-
HY_5-POL	14 600	SAXS	2.36	2.53	-	4.36	-	-

Information about fine structures of polymer molecules can thus be obtained from very precise small-angle X-ray scattering data at very low scattering angles. Use of larger ranges of q does not necessarily improve the interpretation of scattering data because of experimental uncertainties, molecular non-uniformity of polymers, presence of excluded volumes, and/or non-ideal fine structures (Section 5.4.2 ff.). Furthermore, most evaluation methods use mathematical and/or physical approximations. The Zimm equation, Eq.(5-38), for example, replaces $1-(q^2\langle s^2\rangle/3)$ by $1/[1+(q^2\langle s^2\rangle/3)]$ which is valid only within 1 % or less if $q^2\langle s^2\rangle < 0.3$. Furthermore, the conversion from theoretical end-to-end distances to experimental radii of gyration assumes unperturbed flexible coils. As a result, evaluations of scattering data by Zimm plots inflate data if $q^2\langle s^2\rangle$ is greater than ca. 1. Better results are obtained by using Ullman corrections or a Berry plot (see p. 169). Table 5-2 compares radii of gyration that were obtained from different evaluations of scattering data.

In general, it can be concluded that SAXS has no advantage over static light scattering with respect to molar mass determinations. But is is an indispensable method for the evaluation of the fine structure of macromolecules.

5.4.2 Scattering Functions

SAXS uses very small wavelengths λ_0 which produce very large magnitudes of scattering vectors, $\boldsymbol{q} = (4\,\pi n_1/\lambda_0)\sin(\vartheta/2)$. In the limiting case, $\boldsymbol{q} \rightarrow \infty$, the Debye equation, $P(q) = (2/q^4s^4)[\exp(-q^2s^2) - (1-q^2s^2)]$ (Eq.(5-50)), asymptotically approaches a limiting value of $\lim_{q\rightarrow\infty} P(q) = 2/[q^2s^2\{1-(1/q^2s^2)\}]$ (note that these equations were derived for unperturbed coils!). In the corresponding Kratky plot, $q^2s^2P(q) = f(qs)$, one obtains an asymptote (Fig. 5-10, lowest curve) and the limiting value of $q^2P(q)$ becomes

$$\text{(5-62)} \qquad \lim_{q\rightarrow\infty} q^2P(q) = \frac{2}{s^2[1-(1/q^2s^2)]} \approx \frac{2}{s^2}$$

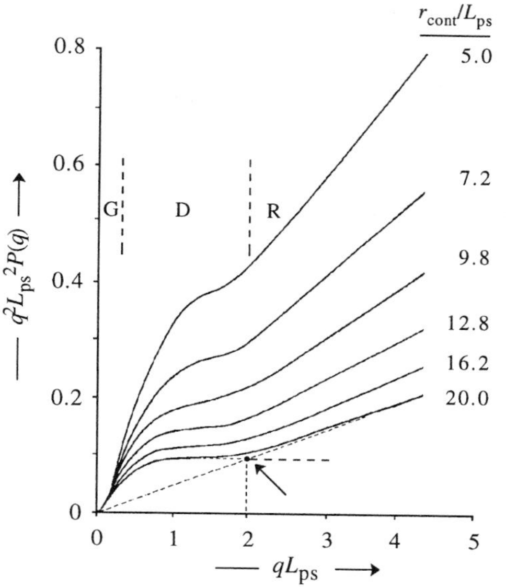

Fig. 5-14 Kratky diagram for wormlike chains with varying ratios, r_{cont}/L_{ps}, of conventional contour lengths, r_{cont}, to persistence lengths, L_{ps} [8b]. The plot uses $q^2L_{ps}{}^2P(q) = f(qL_{ps})$ instead of the common $q^2s^2P(q) = f(qs)$. Calculations with Eq.(4-28) for freely rotating chains with valence angles of $\tau = 180° - \alpha$ and cos $\alpha = 0.9$. $\mathbf{q}$ = wave vector and q = magnitude of this vector.

G = Guinier range, D = Debye range with intermediate and asymptotic range, R = rodlike segments. The arrow at $qL_{ps} = 2$ indicates the transition from the asymptote of the Debye equation for unperturbed random coils to rodlike segments of wormlike molecules.

For *molecularly uniform* random coils, this value of s^2 should be identical with those obtained from Guinier or Zimm plots (for molecularly *non-uniform* polymers, see Eq.(5-54)). Model calculations show for molecularly non-uniform wormlike chains without excluded volumes that this asymptote is approached if the ratio of the conventional contour length to persistence length is ca. 20 (Fig. 5-14).

Eq.(5-62) and Fig. 5-14 are based on the Debye equation, Eq.(5-50), which applies to the total structure of unperturbed random coils. However, the smaller the wavelength, the more the scattering vector probes local structures, i.e., segments and persistence lengths in random coils. The latter type of scattering is not covered by the Debye equation.

Rods of sufficient Kuhnian length L_K can be described by $q^2P(q) = (\pi q/L_K) - (2/L_K{}^2) \approx (\pi/L_K)q$, Eq.(5-48), whereas coils have an asymptote of $\lim_{q\to\infty} q^2P(q) = 2/s^2$ (Eq.(5-62)), if the Debye equation is plotted according to Kratky as $q^2P(q) = f(q)$ (Fig. 5-10). The straight line for rods and the asymptote for coils intersect at $q = q^*$ with $(\pi/L_{ps})q^* = 2/s^2$. For long rods ($L_K >> R$) the radius of gyration is proportional to the Kuhnian length, $s^2 \approx L_K{}^2/12$ (Eq.(4-5)). The Kuhnian length thus becomes $L_K = 24/(\pi q^*)$ and the persistence length, $L_{ps} = L_K/2$, converts to $L_{ps} = 12/(\pi q^*)$.

These equations should be used with caution. On one hand, characteristic lengths L_K and L_{ps} have to be large enough to fulfill the conditions for the scattering function of rods. On the other hand, Kuhnian lengths are only twice as large as persistence lengths (p. 105) if the polymer molecules form random coils; for rods, persistence lengths approach the lengths of rods.

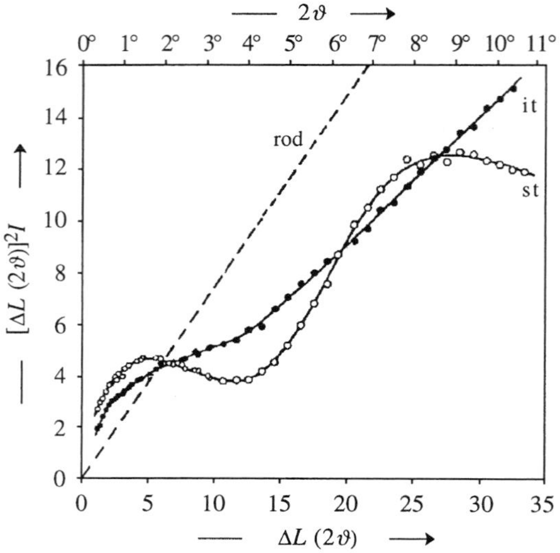

Fig. 5-15 Kratky diagram of the scattering behavior of an isotactic and a syndiotactic poly(methyl methacrylate) in the good solvent benzene [11]. ΔL = 179 mm = distance from specimen to the observational plane, 2ϑ = scattering angle, I = intensity. - - - see also lowest curve in Fig. 5-14.

The simplifying assumptions of scattering theory are often not followed by real polymer chains. Examples are the Kratky diagrams of an isotactic and a syndiotactic poly(methyl methacrylate) (Fig. 5-15): the pattern of the former could not be assigned to a particular particle fine structure whereas that of the latter is indicative of a helical structure. Patterns with maxima and minima are characteristic for particles with densely clustered scattering elements such as spheres, highly branched polymer molecules, helices, etc. Because of such patterns at $qs > 1$, radii of gyration are underestimated by $P(q) = \exp(-q^2s^2/3)$ (Guinier), and overestimated by $P(q) = 1 - (q^2s^2/3)$ (Zimm) (Table 5-2).

Branched Polymers

Experimental scattering functions are relatively easy to represent by theoretical models if the distribution of scattering elements is homogeneous (for example, in compact, isotropic spheres) or follows simple mathematical equations (for example, unperturbed random coils with a Gaussian distribution of segments). Scattering data are much more difficult to interpret if these conditions do not apply and/or polymers are not molecularly uniform. In these cases, different evaluation methods deliver different radii of gyration from data at $q \to 0$ (Table 5-2). Most evaluation methods are approximations for small values of q, which is sufficient for the determination of mass-average molar masses and z-average mean-square radii of gyration but not for evaluation of fine structures: at $q/\text{nm}^{-1} \approx 10^{-3}$-$10^{-2}$, only total structures are registered; at $q/\text{nm}^{-1} \approx 10^{-1}$, segments; and at $q/\text{nm}^{-1} \approx 1$, substituents, including branch structures in, for example, dendrimers.

The interpretation of scattering data is more difficult for polymers with unknown distributions of scattering elements. Examples are dendrimers with their branch-upon-branch structures (pp. 11, 121; Volume I: pp. 52, 619). Here, the presence of maxima in the scattering curve is indicative of the presence of sphere-like structures (Fig. 5-16).

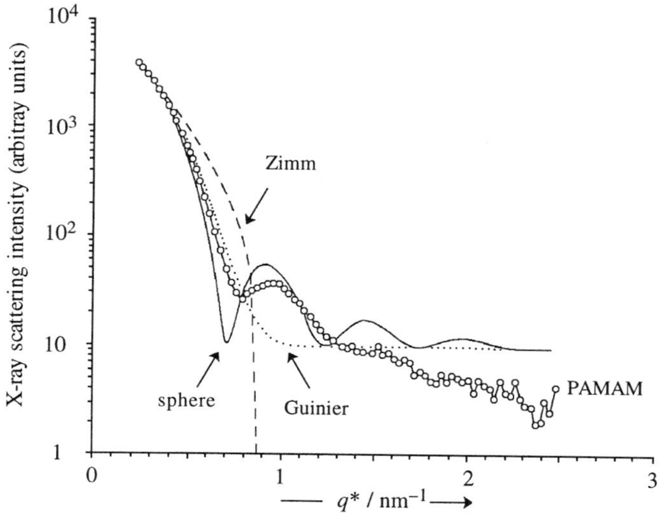

Fig. 5-16 X-ray scattering intensities of a 1 wt% methanolic solution of a 9th generation polyamido-amine dendrimer [9b]. ○ Experimental; lines: data at $q < 0.4/\text{nm}^{-1}$ fitted to spheres (s = 4.92 nm), Guinier equation (s = 5.13 nm), and Zimm (s = 7.49 nm). In this graph, the scattering parameter is defined as $q^* = (4\ \pi/\lambda) \sin \vartheta$ and not as $q = (4\ \pi/\lambda) \sin \vartheta/2$.

More illuminating is a plot of $q^4 I = f(q)$ (Fig. 5-17) where I = scattering intensity. The plot shows four maxima with decreasing heights which rules out a Gaussian distribution of scattering elements (only 1 maximum, see Fig. 5-16) and a Gaussian distribution with a hole in the center (2 maxima). Four maxima of *equal* height are expected for isotropic spheres but the theoretical maxima are also much too high and they do decrease with increasing q^*. The latter effect can be simulated by modeling the dendrimer molecule as a sphere with radius R and an electron density ρ_e that is constant in the interior between $R = 0$ and $0.75\ R$ but decreases linearly with R to $\rho_e = 0$ from $0.75\ R$ to R.

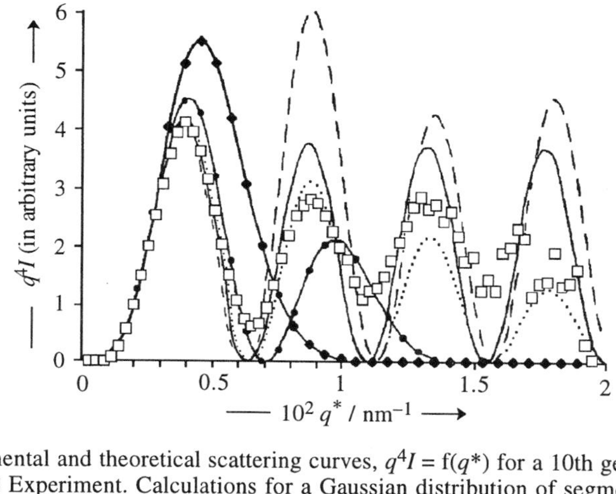

Fig. 5-17 Experimental and theoretical scattering curves, $q^4 I = f(q^*)$ for a 10th generation PAMAM dendrimer [9c]. □ Experiment. Calculations for a Gaussian distribution of segments, either regular (—◆—) or in the thick shell of a hollow sphere (—●—); a hollow sphere (- - -), an isotropic sphere (——), and a "rough" sphere with a gradient of a density distribution (·····).

5.5 Small-Angle Neutron Scattering

5.5.1 Fundamentals

Small-angle neutron scattering (**SANS**) uses elastic scattering of incident neutrons at neutrons of the target molecule to gain information about the physical structure of the target. Incident neutrons are produced by nuclear fission or spallation. Fission of uranium-235 is most commonly used. Per fission, it produces 2-3 neutrons from which one is needed to sustain the chain reaction and only 1-2 are available for SANS.

Much more powerful is spallation. In this method, particle accelerators and synchrotrons produce intense, high-energy beams of protons that bombard a heavy nucleus such as tantalum. The high kinetic energy of protons allows them to overcome the so-called "Coulomb wall" of long-range electrostatic and short-range nuclear forces of the target nucleus which is shattered (Middle English: *spall* = to crumble or chip). The resulting beam of ejected thermal or "cold" (slower) neutrons is then directed at the sample where it is scattered at the neutrons of the atoms of the sample molecules. Neutrons penetrate matter deeply because they are electrically neutral: in condensed matter, their free path length is in the millimeter range.

Small-angle neutron scattering measures the scattering of slow neutrons (diameter: 0.2-2 nm) at atomic nuclei (diameter: 10^{-5}-10^{-6} nm). In contrast to small-angle X-ray scattering but similarly to light scattering, wavelengths of probes are thus much larger than scattering elements. However, wavelengths of neutrons are comparable to the lengths of chemical bonds, which leads to interference of scattered waves from different atoms of the same molecule. Scattering intensities thus vary with the scattering angle, which allows one to determine the radius of gyration of molecules as small as methane.

Scattering elements are practically only atomic nuclei which act as point scatterers because their diameters are only ca. one millionth of that of neutrons. The scattered wave is characterized by the coherent scattering length b which is the square root of the probability that an entering neutron is scattered within an angle of 1 steradian per flux of neutrons. The neutron flux is defined as the number of neutrons that pass through a plane in unit time.

The derivation of the equations of SANS from first principles is complex and cannot be presented in this book. However, the resulting equations can be written (with minor corrections) with the same symbols as those for light scattering since the theory of scattering of electromagnetic waves applies to all wavelengths. Hence, the dependence of particle scattering factors on wavelength, molar mass, and radius of gyration is maintained (see Eq.(5-34)):

$$\text{(5-63)} \qquad \frac{K_{\text{SANS}}c}{R_q} = \frac{1}{\overline{M}_{\text{w}}P(q)} + 2\frac{A_{2,\text{LS}}}{P'(q)}c + \ldots$$

where $R_q \equiv (I_\vartheta / I_\text{o})(L^2/V)$ is the Rayleigh ratio (Eq.(5-2)), $|q| = (4\,\pi\, n_1/\lambda_\text{o}) \sin(\vartheta/2)$ the magnitude of the scattering vector (Eq.(5-33)), and $P(q)$ the one-particle scattering function of the polymer. Only the optical constant (contrast factor) K in Eqs.(5-11) and (5-34) is changed because small angle neutron scattering is controlled by the difference of coherent neutron scattering lengths of monomeric units and solvent molecules.

The contrast factor K_{SANS} is defined as

(5-64) $$K_{SANS} \equiv N_A \left[\frac{\sum_i b_{u,i}}{M_u} - \left(\frac{\tilde{V}_{u,m}}{\tilde{V}_{1,m}} \right) \frac{\sum_i b_{1,i}}{M_1} \right]^2$$

where N_A = Avogadro constant; $b_{u,i}$ and $b_{1,i}$, respectively, are bound atom coherent scattering lengths of all nuclei i of monomeric units (u) and solvent molecules (1); M_u and M_1 are the molar masses of monomeric units and solvent molecules; and $\tilde{V}_{u,m}$ and $\tilde{V}_{1,m}$ are the partial molar volumes of u and 1 in a volume V per amount-of-substance n.

Bound atom coherent scattering lengths are also called neutron scattering lengths. They are calculated from coherent scattering cross sections, $\sigma_C = 4\,\pi b^2$, which are in turn obtained from the difference $\sigma_C = \sigma_T - \sigma_I$ of total scattering cross sections, $\sigma_T = 4\,\pi |b|^2$, and incoherent scattering cross sections, σ_I. Incoherent scattering cross sections arise from two types of incoherent scattering: the spin incoherence of nuclei possessing a spin and the isotopic incoherence caused by the various isotopes of an element.

Table 5-3 shows the values of b of the most common isotopes of organic molecules that have been calculated from $\sigma_C = \sigma_T - \sigma_I = 4\,\pi b^2$. With the exception of the hydrogen isotope ^{1}H, all isotopes in organic molecules have positive coherent neutron scattering lengths. Neutron scattering lengths of isotopes of the same element vary widely (see ^{1}H and ^{2}D), which causes large differences in b of an element and its major isotope.

These differences in atomic scattering cross sections give rise to large differences in the so-called scattering length densities ρ of protonated and deuterated monomeric units (Table 5-4) which have the unit of an inverse area

(5-65) $$\rho \equiv N_A b_u / \tilde{v}_{u,m} M_u = A_u^{-1}$$

Scattering length densities are expressed in lengths per volume and are thus densities in the terminology of physics but not in that of chemistry (see Note on p. 168). They have the physical unit of an inverse area, i.e., they are inverse scattering cross sections, A_u^{-1}. The latter are often given in barns (1 barn = 10^{-24} cm^2) since they are so "large."

Table 5-3 Total scattering cross sections (σ_T), incoherent scattering cross sections (σ_I), coherent scattering cross sections (σ_C), and neutron scattering lengths b of isotopes (with atomic mass number of nuclide, A) in organic-chemical compounds [12a, 13]. P = isotopic abundance of nuclide in element, N = number of types of other nuclides in element, b^* = b of element with natural abundance.

Element	A	$\frac{P}{\%}$	N	$\frac{10^{24}\,\sigma_T}{\text{cm}^2}$	$\frac{10^{24}\,\sigma_I}{\text{cm}^2}$	$\frac{10^{24}\,\sigma_C}{\text{cm}^2}$	$\frac{10^{12}\,b}{\text{cm}}$	$\frac{10^{12}\,b^*}{\text{cm}}$
H	1	99.985		81.67	79.91	1.76	−0.374	-
	2	0.015		7.64	2.04	5.6	0.667	-
C	12	98.90	1	5.555	0	5.555	0.665	0.6646
N	14	99.634	1	11.52	0.49	11.03	0.937	0.9362
O	16	99.762	2	4.235	0	4.235	0.580	0.5803
F	19	100	0	4.018	0	4.018	0.565	0.5650
Si	28	92.23	2	2.178	0.01	2.168	0.411	0.4153
Cl	35	75.77	1	21.63	5.2	16.43	1.166	0.9577

Table 5-4 Scattering length densities $\rho = A_u^{-1}$ of protonated monomeric units of polymers and their fully deuterated counterparts, all probably at 25°C [12b, 13].

Polymer or solvent	Constitution of protonated monomeric unit	$10^{10}\, A_u^{-1}/\text{cm}^{-2}$ protonated	deuterated	$10^{10}\,(A_{u,d}^{-1} - A_{u,h}^{-1})$ in cm^{-2}
Poly(ethylene)	$-CH_2-CH_2-$	−0.316	8.24	8.56
Poly(butadiene), 1,4-	$-CH_2-CH{=}CH-CH_2-$	0.647	6.823	6.18
Poly(methyl methacrylate)	$-CH_2-C(CH_3)(COOCH_5)-$	1.069	7.03	5.96
Poly(oxyethylene)	$-O-CH_2-CH_2-$	0.64	6.46	5.82
Poly(styrene)	$-CH_2-CH(C_6H_5)-$	1.413	6.50	5.09
Poly(dimethylsiloxane)	$-O-Si(CH_3)_2-$	0.06	4.66	4.60

Chemistry and physics often use different terminologies. In physics, "density" denotes a quantity of something per another quantity, often with the symbol ρ. Examples are the mass density ($\rho = m/V$) in kg m^{-3}, the surface density ($\rho_A = m/A$) in kg m^{-2}, the number density ($\rho = N/V$) in m^{-3}, the charge density ($\rho = Q/A$) in C m^{-3}, and the scattering length density ρ (see above) in m/m^3 = m^{-2}.

Chemistry usually reserves "density" and its symbol ρ for mass densities in which both the mass m and the volume V refer to the *same* substance; an example is the density of a liquid 1 as the ratio of its mass per volume ($\rho_1 = m_1/V_1$). Mass densities in which the mass of a substance is divided by the volume of another matter are called "mass concentrations" in chemistry (usually just as "concentrations" or wrongly as "weight concentrations"); an example is the mass concentration of a solid 2 per volume V of a solution of 2 in a solvent 1 ($\rho_2 = m_1/V$) (see Chapter 19, Appendix).

5.5.2 Experimental Results

Because of the drastically different scattering lengths of ^{1}H and ^{2}H (= deuterium (D)), SANS experiments are performed with deuterated polymers in protonated solvents or vice versa. Such selective deuterations allows one to "see" the contributions of individual groups to scattering, for example, that of backbones or substituents (Fig. 5-18).

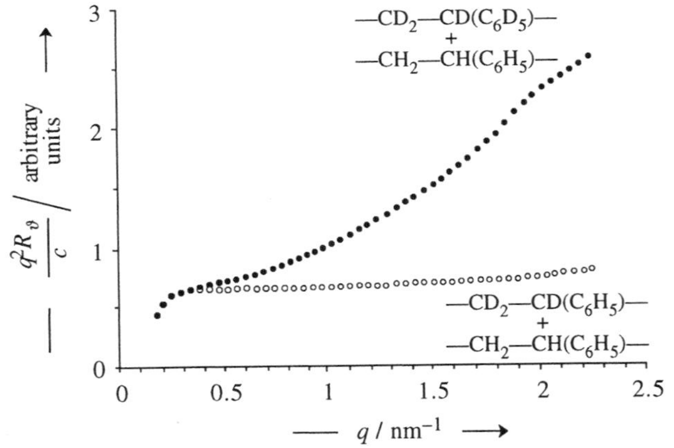

Fig. 5-18 Scattering parameter q^2R_ϑ/c as a function of the magnitude q of the scattering vector for the mixture of a conventional poly(styrene) with either a completely deuterated poly(styrene) (top) or a partially deuterated poly(styrene) (bottom) [14].

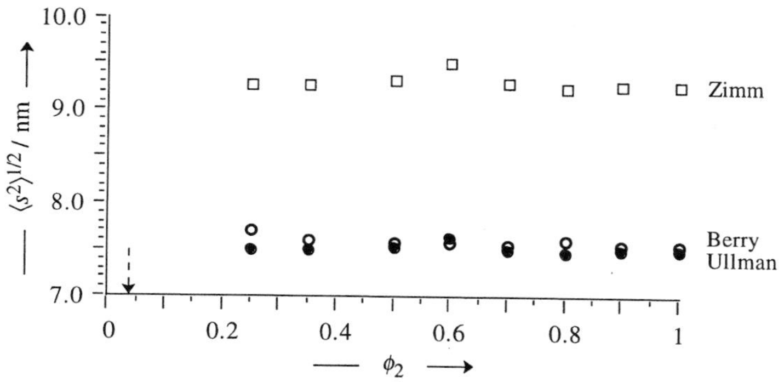

Fig. 5-19 Radius of gyration (no average given) of a hydrogenated poly(butadiene) (= "poly(ethylene)") with ca. 2 ethyl groups per 100 skeletal carbon atoms and $\overline{M}_w$ = 28 000 g/mol as a function of its volume fraction ϕ_2 in the good solvent deutero-nonadecane at 150°C [15]. Evaluation of data according to Zimm, Berry, and Ullman, respectively. ←-- Overlap concentration (see p. 184).

Much more common in polymer science is the use of SANS to determine dimensions of polymers in melts (as mixture of protonated and deuterated polymers; Chapter 6) or solutions (deuterated polymer in protonated solvent or vice versa) (Fig. 5-19).

The radius of gyration of the polymer of Fig. 5-19 is independent of its volume fraction ϕ_2 in the range $0.25 \leq \phi_2 \leq 1$. The data also show that the values of $\langle s^2 \rangle$ depend on the evaluation method: radii of gyration from Zimm plots are always much greater than those from Berry plots or those corrected according to Ullman (see also p. 160) because one is already outside of the range of the Zimm approximation (see p. 150). Ullman corrections and Berry plots gave identical results.

In Berry plots, $(Kc/R_q)^{1/2}$ instead of Kc/R_q is plotted against $kc + \sin^2 \vartheta/2$ in Zimm-like diagrams. They are used if third virial coefficients become noticeable. Third virial coefficients, A_3, measure 3-body interactions which depend on 2-body interactions and thus on second virial coefficients, A_2. Statistical thermodynamics predicts $A_3M = \kappa(A_2M)^2$ with $\kappa = 5/8$ for compact spheres and $\kappa = 1/3$ for perturbed coils. With $\kappa = 1/3$, $\overline{M}_w Kc/R_0 = 1 + 2 A_2 \overline{M}_w c + 3 A_3 \overline{M}_w c^2 + ...$ (Eq.(5-25) for LS, X, SANS) becomes $(Kc/R_0)^{1/2} = (1/\overline{M}_w)^{1/2}(1 + A_2 \overline{M}_w c)$. Ullman corrections are obtained from published tables (see Historical Notes).

Instead of polymer-solvent mixtures, one can also use a deuterated polymer in a protonated one or vice versa. If both polymers have the same constitution, molar mass, and molar mass distribution, this would amount to a tracer experiment (Chapter 6).

SANS has also been used to study the structure of solution-grown, melt-grown, and melt-quenched polymer crystals, the solid-state structure of block polymers and polymer blends, and the deformation and relaxation of linear polymers, block polymers, and polymer networks.

In all of these experiments, molecular characteristics of deuterated polymers (constitution, chemical configuration, molar mass, molar mass distribution) must exactly match that of their protonated counterparts. Even then, the method may not give clear-cut results because deuterated and protonated polymers do have different properties, for example, different theta temperatures (p. 333), because they are different chemical entities. The influence of this chemical difference can be eliminated by using various concentrations of the tracer and extrapolating the results to zero tracer concentration. For cost reasons, this is usually never done, however.

Historical Notes to Chapter 5

J.W.S.Rayleigh, Philos.Mag. XLI (1871) 107; reprint in: Scientific Papers by Lord Rayleigh (John William Strutt), Volume 1, pp. 87-99 (Paper 8), Dover, New York 1964
Derivation of the scattering equation for small particles (gases).

G.Mie, Ann.Phys. **25**/4 (1908) 377-445
Derivation of the scattering equation for very large particles.

P.Debye, Ann.Phys. **46** (1915) 809; Phys.Z. **31** (1930) 348
Derivation of Eq.(5-35).

A.Guinier, C.R.Acad.Sci. [Paris] **204** (1937) 1115; Thèses, Série A, Paris, Nr. 1854 (1939)
Derivation of Eq.(5-61).

B.H.Zimm, J.Chem.Phys. **16** (1948) 1093, 1099
Zimm plot

G.C.Berry, J.Chem.Phys. **44** (1966) 4550
Modification of Zimm plot (Berry plot)

R.Ullman, J.Polym.Sci.-Polym.Lett.Ed. **21** (1983) 521
Correction tables for Zimm plot if data are outside Zimm's approximation (originally for SANS)
R.Ullman, J.Polym.Sci.-Polym.Phys.Ed. **23** (1985) 1477
Corrections for various types of molar mass distributions

Literature to Chapter 5

5.1 GENERAL
J.S.Higgins, R.S.Stein, Recent Developments in Polymer Applications of Small Angle Neutron, X-Ray, and Light Scattering, J.Appl.Cryst. **11** (1978) 346
L.A.Feigin, D.I.Svergun, in G.W.Taylor, Ed., Structure Analysis by Small-Angle X-Ray and Neutron Scattering, Plenum, New York 1987
P.Lindner, Th.Zemb, Eds., Neutron, X-Ray and Light Scattering. Methods Applied to Soft Condensed Matter, Elsevier Science, Amsterdam 1991 (proceedings of a workshop)
O.Glatter, Modern Methods of Data Analysis in Small Angle Scattering and Light Scattering, Kluwer Academic, Dordrecht 1995
W.Brown, K.Mortensen, Eds., Scattering in Polymeric and Colloidal Systems, Gordon and Breach, Newark (NJ) 2000
R.-J.Roe, Methods of X-ray and Neutron Scattering in Polymer Science, Oxford University Press, Oxford 2000

5.2 and 5.3 STATIC LIGHT SCATTERING
H.C. van de Hulst, Light Scattering by Small Particles, Wiley, New York 1957
M.Kerker, The Scattering of Light and Other Electromagnetic Radiation, Academic Press, New York 1969
M.B.Huglin, Ed., Light Scattering from Polymer Solutions, Academic Press, London 1972
B.Chu, Laser Light Scattering, Academic Press, New York 1974
W.Burchard, Static and Dynamic Light Scattering from Branched Polymers and Biopolymers, Adv.Polym.Sci. **48** (1983) 1
P.Kratochvil, Classical Light Scattering from Polymer Solutions, Elsevier, Amsterdam 1987
W.Brown, Ed., Light Scattering, Clarendon Press, Oxford 1996
R.Borsali, Scattering Properties of Multicomponent Polymer Solutions: Polyelectrolytes, Homopolymer Mixtures and Diblock Copolymer, Macromol.Chem.Phys. **197** (1996) 3947

5.4 SMALL-ANGLE X-RAY SCATTERING
A.Guinier, G.Fournet, Small Angle Scattering of X-Rays, Wiley, New York 1955
H.Brumberger, Ed., Small Angle X-Ray Scattering, Gordon and Breach, New York 1967
O.Glatter, O.Kratky, Eds., Small Angle X-Ray Scattering, Academic Press, Oxford 1982 and 1996
R.-J.Roe, Methods of X-ray and Neutron Scattering in Polymer Science, Oxford University Press, Oxford 1999

5.5 SMALL-ANGLE NEUTRON SCATTERING
W.Marshall, S.W.Lovesey, Theory of Thermal Neutron Scattering, Clarendon Press, Oxford 1971
B.J.M.Willis, Chemical Applications of Neutron Scattering, Oxford Univ. Press, Oxford 1973
A.Maconnachie, R.W.Richards, Neutron Scattering and Amorphous Polymers, Polymer **19** (1978) 739
R.Ullman, Small Angle Neutron Scattering of Polymers, Ann.Rev.Mater.Sci. **10** (1980) 261
G.Kostorz, Neutron Scattering, in G.Kostorz, Ed., Treatise on Materials Science and Technology, Volume 15, Academic Press, New York 1982
S.W.Lovesey, The Theory of Neutron Scattering from Condensed Matter, Oxford Univ. Press, Oxford 1984 (2 volumes)
M.Bee, Quasielastic Neutron Scattering, Hilger, Bristol, UK 1988
J.S.Higgins, H.C.Benoit, Polymers and Neutron Scattering, Clarendon Press, Oxford 1994 (Oxford Series on Neutron Scattering in Condensed Matter, Volume 8)

References to Chapter 5

[1] H.Benoit, Ber.Bunsenges.Phys.Chem. **70** (1966) 286, Fig. 4
[2] C.Strazielle, H.Benoit, J.chim.phys. **58** (1961) 678; $Y^2 \overline{M}_w$ and $\mathfrak{R}$ taken from [3]
[3] C.Strazielle, in M.B.Huglin, Ed., Light Scattering from Polymer Solutions, Academic Press, London 1972, p. 654
[4] Z.Tuzar, P.Kratochvil, Coll.Czech.Chem.Commun. **32** (1967) 3358; values of $Y^2 \overline{M}_w$ and $\mathfrak{R}$ taken from [3]
[5] T.Tanaka, T.Kotaka, K.Ban, M.Hattori, H.Inagaki, Macromolecules **10** (1977) 960, Fig. 4
[6] Y.Miyaki, Y.Einaga, H.Fujita, Macromolecules **11** (1978) 1180, data of Fig. 6
[7] S.Akita, Y.Einaga, Y.Miyaki, H.Fujita, Macromolecules **9** (1976) 774, Fig. 7
[8] A.Peterlin, J.Polym.Sci. **47** (1960) 403, (a) Fig. 1, (b) Fig. 5
[9] T.J.Prosa, B.J.Bauer, E.J.Amis, D.A.Tomalia, R.Scherrenberg, J.Polym.Sci. **B** (Polym.Phys.) **35** (1997) 2913, (a) Fig. 3, (b) Fig. 11
[10] L.Willner, O.Jucknischke, D.Richter, J.Roovers, L.-L.Zhou, P.M.Toporowski, L.J.Fetters, J.S.Huang, M.Y.Lin, N.Hadjichristidis, Macromolecules **27** (1994) 3821
[11] R.Kirste, W.Wunderlich, Makromol.Chem. **73** (1964) 240, Fig. 1; W.Wunderlich, R.G.Kirste, Ber.Bunsenges.Phys.Chem. **68** (1964) 646, Fig. 3
[12] R.W.Richards, Scattering Properties: Neutrons, in G.Allen, J.C.Bevington, Eds., Comprehensive Polymer Science; Vol. 1, Volume Eds. C.Booth, C.Price, Polymer Characterization, p. 133, (a) data of Table 1 (amended), (b) data of Table 2
[13] S.M.King, www.isis.rl.ac.uk/largescale/loq/documents/sans.htm (December 1995, fixed September 2003), accessed 13 December 2006
[14] M.Rawiso, R.Duplessix, C.Picot, Macromolecules **20** (1987) 630, data of Fig. 8a
[15] S.Westermann, L.Willner, D.Richter, L.J.Fetters, Macromol.Chem.Phys. **201** (2000) 500, data of Table 1

6 Disordered Condensed Systems

6.1 Amorphous Polymers

6.1.1 Structure

Crystalline polymers convert to polymer melts at temperatures above their melting temperatures T_M (Section 13.3). These melts often have very large viscosities that are decades greater than those of melts of low-molecular weight compounds (Chapter 15). Phenomenologically, melts of high-molecular weight polymers at temperatures $T > T_M$ resemble soft rubbers morethan liquids.

Tough, rubber-like materials also result if non-crystalline, glass-like polymers are heated to temperatures above their so-called glass (transition) temperatures, T_G (Section 13.5). These materials are sometimes also called "melts" although they do not result from the melting of crystalline materials (a thermodynamic process) but from the softening of non-crystalline matter (a kinetic process).

Cooling of melts of non-crystallizing polymers to temperatures $T < T_G$ and sometimes also a *very rapid* cooling of melts of crystallizing polymers below T_M results in matter without discernible long-range order, so-called **amorphous polymers** (G: *a-* = without; *morphe* = shape, form). These materials are considered "solids" because they do not change shape under their own weight. Examples are containers made from crystal-clear amorphous atactic poly(styrene), called crystal poly(styrene) (Volume II, p. 271).

The question then is: what is the physical structure of melts and amorphous polymers and especially, what is the physical structure of individual macromolecules in melts and amorphous polymers? In both melts and amorphous polymers, long-range order is absent as revealed by X-ray diffraction. Since individual chain molecules adopt their unperturbed dimensions in theta solvents, the question is what happens if one goes from very dilute theta solutions ("infinite" dilution) to melts ("infinite" concentration)? Do coils retain their spatial identity and pack like soft individual spheres or do they overlap to form a kind of "felt"?

This question can be answered by studying the change of radii of gyration from infinite dilution to "infinite" concentration, i.e., the melt (Section 6.2). In the melt itself, only short-range interactions exist between chain segments, and one would expect therefore that radii of gyration of polymer coils in melts are identical with those of unperturbed coils in solution. In turn, radii of gyration should not change on transition from the melt to the amorphous state if long-range order is absent in both of them. The alternative, preservation of the same long-range order in both the melt and the amorphous state can be ruled out if the coil has the same radius of gyration in the melt and in the unperturbed state in solution since the latter is definitely characterized by the absence of long-range order (p. 85).

Small angle neutron scattering showed that the radius of gyration of an atactic poly(styrene) did not change if its melt was cooled rapidly below the glass temperature (Fig. 6-1, Table 6-1): hence, the melt and the glass must have the same physical structure. Most interestingly, crystalline it-poly(propylene) and its melt also exhibited the same radius of gyration and so did semi-crystalline poly(ethylene) and its melt (Fig. 7-17; see p. 208 for a detailed discussion of this phenomenon).

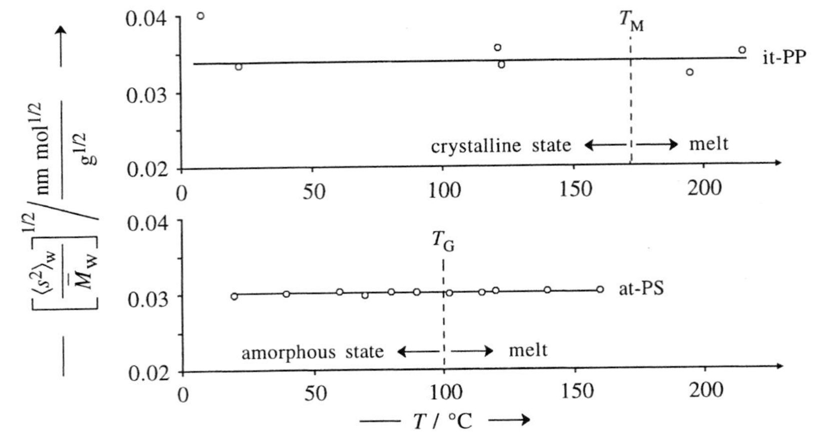

Fig. 6-1 Temperature dependence of reduced mass-average radius of gyration of (top) an isotactic poly(propylene) (T_G = –7°C) in the crystalline and the molten state (averaged data of four polymers with different mass-average molar masses in the range 145 000 ≤ $\overline{M}_w$/(g mol^{-1}) ≤ 1 500 000) [1a] and bottom) an atactic poly(styrene) with $\overline{M}_w$ = 98 000 g mol^{-1} in the amorphous state and in the melt [2].

Many reduced radii of gyration, $(\langle s^2\rangle_w/\overline{M}_w)^{1/2}$, by small angle neutron scattering are nearly identical in the melt and in theta solutions (Table 6-1). In melts, polymer molecules must therefore adopt their unperturbed dimensions; long-range order is certainly absent. However, there are indications that amorphous polymers and amorphous regions of semicrystalline polymers have some local order. Most remarkable is that radii of gyration vary only a little for theta, molten, and crystalline states (see p. 208).

Table 6-1 Reduced radii of gyration of polymers in theta solvents at $c \to 0$ (some from light scattering data and others from intrinsic viscosities) and in the molten, glassy, and crystalline states from small-angle neutron scattering data using small concentrations of deuterated polymers in the corresponding protonated polymers of approximately the same molar mass [3,4]. RIS: calculated with the rotational isomeric state model; nc = non-crystallizing. *) Corrected data of [4].
1) From the hydrogenation of a 1,4-poly(butadiene) with 7 % 1,2-vinyl groups. 2) From the hydrogenation of a 1,4-poly(isoprene) with 7 % 3,4 additions. 3) So-called atactic poly(propylene).

Polymer	$[\langle s^2\rangle_w/\overline{M}_w]^{1/2}$/(nm mol$^{1/2}$ g$^{-1/2}$)				d ln $[\langle s^2\rangle_o/\text{nm}^2]$/d$[T/\text{K}]$		
	Θ state	melt	glass	crystal	Θ state	RIS	melt
Poly(ethylene)	0.0473	0.0482		0.045	−12	−11	
Poly(ethylene-*co*-(1-butene)) 1)	0.0452	0.0473					−12
Poly(ethylene-*alt*-propylene) 2)	0.0381	0.0397			−10	−11	−12
Poly(propylene), it	0.0340	0.0340		0.036	−42	−25 ± 15	0
Poly(methyl ethylene), at 3)	0.0333	0.0336		nc	−20 ± 10	−9 ± 9	−2*
Poly(ethyl ethylene), at	0.0281	0.0271			−14 ± 10	0 ± 1	+3*
Poly(isobutylene)	0.030	0.0305	0.031	nc			
Poly(styrene), at	0.0275	0.0280	0.0278	nc	−12		0
Poly(methyl methacrylate), at	0.030		0.031	nc			
, st	0.028		0.030		+16	+20	
, it	0.024		0.029				
Poly(oxyethylene)		0.042		0.052			

In X-ray diagrams of semi-crystalline polymers (Section 7.1.4), for example, halos are observed that result from scattering of amorphous regions and are therefore indicative of fluctuating electron densities. The dependencies of peak positions on the sizes of density fluctuations are not known, but to a first approximation these fluctuations may be correlated with Bragg spacings d in the Bragg equation, Eq.(7-1), $N\lambda_o = 2\,d \sin\theta$, where λ_o = wavelength of X-ray irradiation, θ = scattering angle, and N = order number of reflections. The values of d so calculated from θ, λ_o, and N are smaller than the actual ones but are of the correct order of magnitude.

A plot of Bragg spacings d of different polymers against the cross-sectional areas of their chains, A_c, reveals that the data seem to belong to three different groups I-III (Fig. 6-2). About one-half of the polymers tested produced only a single halo (I, II, or III) whereas the other half showed two halos (either I+II or II+III). Group I halos correspond to van der Waals distances which vary only slightly with cross-sectional areas (Fig. 6-2).

However, many polymers do not have Group I halos. Bragg spacings of Group II vary strongly with cross-sectional areas as is evident for poly(1-olefin)s. This group also includes many unsubstituted polymers, which rules out an effect of regular packing of pendant groups which is often invoked as the source of such amorphous halos. Group III comprises polyamides, poly(vinylidene chloride), etc. (see caption to Fig. 6-2).

Judging from the variety of polymer structures in Groups II and III, there are also no apparent effects of microconformations or electron-rich groups. It seems that there is some kind of packing regularity in the amorphous regions of semi-crystalline polymers which is usually not considered in discussions of macroconformations of solid polymers.

Far-reaching orientations seem to be absent in amorphous polymers and amorphous regions of semi-crystalline polymers. For example, a sphagetti model of amorphous polymer structures predicts densities of amorphous polymers to be no more than 65 % of their crystalline counterparts. However, values of 85-95 % were observed experimentally (Table 6-2) which agrees with computer simulations according to which up to 88 % of lattice sites of primitive cubic lattices can be occupied by monomeric units without the formation of ideal or disturbed bundles of chain units.

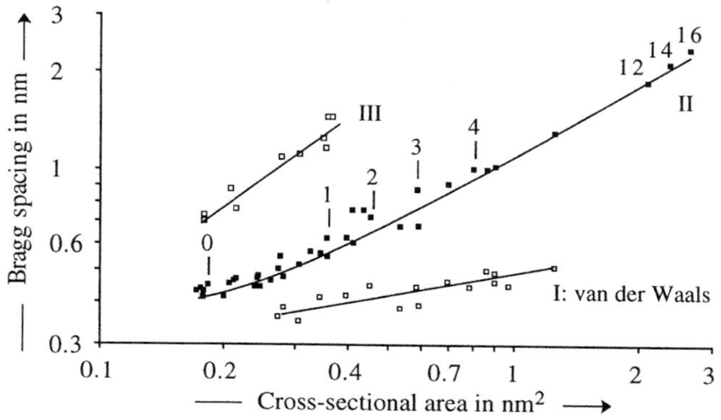

Fig. 6-2 Bragg spacings d of amorphous halos of partly crystalline polymers as a function of cross-sectional area of chains [5]. Numbers refer to the number N of carbon atoms in linear side chains of partly isotactic poly(1-olefin)s, $\text{+}CH_2\text{–}CH[(CH_2)_NH]\text{+}_n$. Other data points belong to a variety of very different polymers such as styrene polymers, polyamides, polyvinyls, halogen-containing polyvinyls, polyethers, cellulose, poly(dimethylsiloxane), etc.

Electron micrographs of amorphous polymers sometimes show spherical entities. These "nodules" have diameters of 2-4 nm (poly(styrene)) or ca. 8 nm (poly(ethylene terephthalate)). In most cases, they seem to be artefacts caused by the fracturing of samples, mistakes in the preparation of samples for electron microscopy, insufficient focussing of the electron microscope, and the like. Nodules are not observed if polymers are rapidly quenched to –165°C.

6.1.2 Density

Absence of long-range order does not exclude short-range order as indicated by Fig. 6-2. A certain parallelization of chain segments is indeed fairly probable since it has been detected in liquid alkanes by X-rays.

However, chains pack less efficiently if long-range order is absent. In general, densities of amorphous polymers, ρ_{am}, are therefore smaller than those of their crystalline counterparts, ρ_{cr} (Table 6-2). it-Poly(4-methyl-1-pentene) is an exception. Ratios ρ_{am}/ρ_{cr} of polymer densities in amorphous and crystalline states do not differ markedly from those of low-molecular weight liquids. This indicates that short-range orders of high-molecular weight compounds do not differ much from low-molecular weight ones.

Amorphous polymers have an additional temperature-dependent fluctuation of anisotropies. The extent of this local segmental orientation is similar to that of simple organic liquids but much smaller than that of isotropic phases of nematic liquid crystals (Chapter 8). Such orientations extend to no more than ca. 0.5 nm according to NMR data of atactic poly(styrene).

Cooling the melt to temperatures below the glass temperature T_G causes viscosities to increase dramatically from ca. (10^2-10^6) Pa s at $T > T_G$ to ca. 10^{12} Pa s at $T \approx T_G$. Polymer segments become far less mobile and can no longer pack efficiently, mainly because chains have finite thicknesses and certain persistences.

Table 6-2 Densities ρ_{cr} (crystalline) and ρ_{am} (amorphous or liquid) of polymers and solvents in the crystalline state at the temperature T_{cr} and in the amorphous state (polymers) or in the liquid state (solvents) at temperature T_{am} [6]. * After conversion to the respective melting temperatures T_M with the help of cubic thermal expansion coefficients.

Polymer	Crystal system	$\frac{T_{am}}{°C}$	$\frac{T_{cr}}{°C}$	$\frac{T_M}{°C}$	$\frac{\rho_{am}}{g\ cm^{-3}}$	$\frac{\rho_{cr}}{g\ cm^{-3}}$	$\frac{\rho_{am}}{\rho_{cr}}$
Poly(ethylene) (I)	orthorhombic			144	0.887	0.999	0.888
Poly(isobutylene)	orthorhombic	25	25	44	0.915	0.937	0.977
Poly(isoprene), *cis*-1,4 (β)	orthorhombic			64	0.906	1.000	0.906
Poly(styrene), it-	trigonal			240	1.04	1.12	0.929
Poly(oxyethylene)	monoclinic	25	25	66	1.124	1.235	0.910
Poly(ethylene terephthalate)	triclinic	18	18	264	1.337	1.498	0.893
Poly(bisphenol A carbonate)	monoclinic			230	1.196	1.315	0.910
Poly(dimethylsiloxane)	monoclinic	–90	–90	-	0.98	1.07	0.916
Poly(4-methyl-1-pentene), it	tetragonal			238	0.838	0.828	1.012
Benzene	orthorhombic	20	–3	5.5	0.879	1.022	0.88 *
Cyclohexane	cubic	20	–8	6.5	0.779	0.837	0.95 *
Carbon tetrachloride	hexagonal	20	–44	–23	1.594	1.788	0.95 *

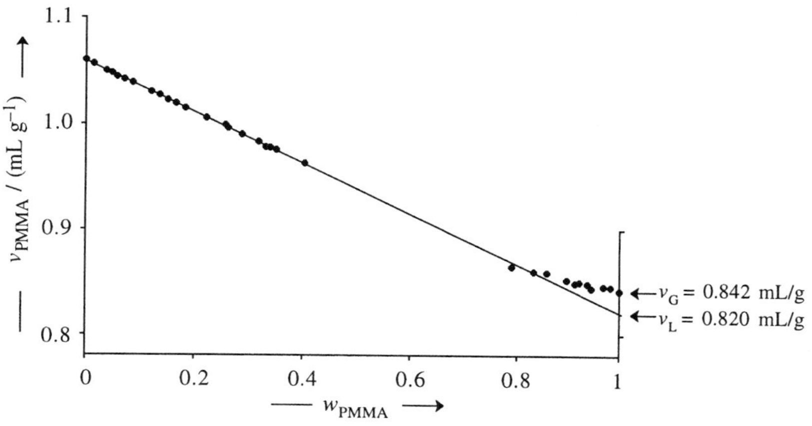

Fig. 6-3 Specific volume of an at-poly(methyl methacrylate), v_{PMMA}, in its solutions in methyl methacrylate at 25°C as a function of the mass fraction of the polymer, w_{PMMA} [7].

The specific volume of amorphous polymers will therefore be greater than the specific volume of the corresponding liquid polymer at the same temperature (Fig. 6-3). The amorphous polymer contains "empty spaces" which are regions of ca. atomic diameter. Hence, the polymer has a so-called **free volume**.

The volume fraction, ϕ_{WLF}, of this free volume in the solid polymer is calculated from the specific volumes of the glass, v_G, and the melt, v_L, at $w_{polymer} = 1$:

(6-1) $$\phi_{WLF} = (v_G - v_L)/v_G$$

The free-volume fraction of solid PMMA assumes a value of $\phi_{WLF} \approx 0.0261$ (Fig. 6-3). Similar values of $\phi_{WLF} \approx 0.026$ are found for other amorphous polymers (Table 6-3). The same free-volume fraction is a parameter in the so-called WLF equation for dynamic glass temperatures, hence the index WLF (Section 13.5.5).

Table 6-3 Volume fractions of various free volumes of amorphous polymers at the glass temperature, calculated with crystalline densities at 0°C, not 0 K.

Polymer	T_G/°C	ϕ_{empty}	ϕ_{exp}	ϕ_{WLF}	ϕ_{fluc}
Poly(styrene), at	100	0.375	0.127	0.025	0.0035
Poly(vinyl acetate), at	27	0.348	0.14	0.028	0.0023
Poly(methyl methacrylate), at	105	0.335	0.13	0.026	0.0015
Poly(butyl methacrylate), at	20	0.335	0.13	0.026	0.0010
Poly(isobutylene)	–73	0.320	0.125	0.026	0.0017

Besides ϕ_{WLF}, several other "free volumes" are defined and discussed. The so-called empty volume, $\phi_{empty} = (v_G - v_{vdW})/v_G$, relates to the specific volume v_G at a temperature T and the specific volume v_{vdW} that is calculated from van der Waals radii.

This "empty volume" is only partially available for thermal movements of segments since monomeric units cannot occupy all empty spaces. The volume fraction available for thermal expansion, $\phi_{exp} = (v_{G,0} - v_{cr,0})/v_{G,0}$, is calculated from the specific volumes of glasslike and crystalline polymers, $v_{G,0}$ and $v_{cr,0}$, both at 0 K.

Measurements of sound velocities deliver the fraction ϕ_{fluc} of a fluctuation volume that describes the movement of the center of mass of a molecule which is caused by the thermal movement of segments.

6.2 Polymer Melts

6.2.1 Microconformations

Microconformations of *small molecules* are controlled by intramolecular forces, exclusively in the gaseous state and almost exclusively in the crystalline state. Melting of a crystal changes the packing of small molecules but does not change their intramolecular interactions much: *n*-butane has practically the same conformational energy ΔE_{TG} in the gaseous and the liquid state, 3.35 versus 3.22 kJ/(mol bond).

Polymer chains with periodic sequences of microconformations such as $[T]_n$, $[TG^+]_n$, $[TTG^+]_n$, $[G^+]_n$, etc., are present in crystalline states as either helices or zigzag chains, both with rod-like shapes (see Chapter 7) which are stabilized by tight packing. Packing of zigzag chains may also involve intermolecular hydrogen bridges such as in polyamides or cellulose (Chapter 7).

Increasing temperature causes crystalline molecules or their segments to vibrate (sometimes also to rotate) about their positions at rest. Finally, the crystal starts to melt (Chapter 13). In the melt, linear chains of flexible macromolecules are no longer forced to assume long regular periodic sequences of microconformations, whether as whole molecules or in sections thereof, because thermal energies are large enough to overcome rotational barriers. The chains adopt the macroconformation of random coils. Melting produces a less favorable enthalpic state which is however overcompensated by a gain in the entropy parameter $T\Delta S$. In the melt, each macromolecule forms a macroconformation consisting mainly of randomly distributed trans and gauche microconformations. Since trans converts to gauche and *vice versa* in ca. 10^{-13} s (Volume I, p. 142), macroconformations change rapidly. Hence, the observed macroconformations of molecules of the molten substance are spatial and temporal averages.

The melt does not contain large regular sequences of microconformations. For example, crystalline syndiotactic poly(propylene) consists of helical $[TTGG]_n$ sequences. In the melt, these helices are no longer present as indicated by the disappearance of the "crystalline" infrared band on melting (Fig. 4-2).

X-ray diffraction of melts of crystalline or amorphous polymers also does not indicate any long-range order. Data from small-angle neutron scattering experiments ruled out long sequences of trans-conformations in poly(ethylene); they also indicated only weak orientation correlations of ca. 0.4 nm. These correlations are not much larger than the crystallographic bond length of 0.254 nm that exists in crystalline linear poly(ethylene)s. The crystallographic bond length of poly(ethylene) is the distance between carbon atoms C^1 and C^3 in chains ~C^1–C^2–C^3~ with bond angles of 111.5° for –C^1–C^2–C^3– and C–C bond lengths of 0.154 nm (p. 82).

6.2.2 Macroconformations

Melts of linear polymers do not consist of neatly packed spheroidal molecules but rather of interpenetrating coils as the following reasoning shows. The distribution of local number concentrations of segments, $C = N_{seg}/V$, in a single random coil molecule can be approximated by a Gaussian function (Section 4.4.7). The segment density decreases with increasing chain length, including that at the center of mass. However, macroscopic densities of polymer melts do not vary with polymer molecular weight, except for the effect of endgroups at low molecular weights. It follows that random coils must overlap in melts and that this overlap increases with the size of the molecule.

The extent of overlapping can be estimated by comparing the mass density of a single polymer molecule and that of the corresponding polymeric substance, i.e., the assembly of many polymer molecules. The mass density of *one* molecule, $\rho_{mol} = m_{mol}/V_{mol}$, is given by the ratio of its molecular mass, $m_{mol} = M/N_A = N_{seg}M_{seg}/N_A$, and molecular volume V_{mol}, where M = molar mass, N_{seg} = number of segments per molecule, M_{seg} = molar mass of a segment, and N_A = Avogadro constant. The molecule volume can be modeled as that of an equivalent sphere with a radius given by the radius of gyration, i.e., $V_{mol} = (4/3)\,\pi\,\langle s^2\rangle_o^{3/2}$. For unperturbed dimensions, the mean-square radius of gyration can be expressed by the mean-square end-to-end distance, $\langle r^2\rangle_o = 6\,\langle s^2\rangle_o$, (Eq.(4-43)), which is given by the number N_{seg} and effective length b_{seg} of segments, $\langle r^2\rangle_o = N_{seg}b_{seg}^2$ (Eq.(4-22)). The mass density of one molecule in a melt is therefore

$$\text{(6-2)} \qquad \rho_{mol} = \frac{m_{mol}}{V_{mol}} = \frac{M/N_A}{(4/3)\,\pi\langle s^2\rangle_o^{3/2}} = \frac{3\cdot 6^{3/2}\,M_{seg}}{4\,\pi N_A N_{seg}^{1/2} b_{eff}^3}$$

An example is a poly(styrene) molecule, $\text{+}CH_2\text{–}CH(C_6H_5)\text{+}_N$, consisting of $N_u = 10^4$ monomeric units $\text{–}CH_2\text{–}CH(C_6H_5)\text{–}$ with a molar mass of M_u = 104.15 g/mol and a crystallographic length of b_{eff} = 0.254 nm. Such a molecule has a mass density of ρ_{mol} = 0.370 g/cm^3 which is much lower than the mass density of the molten polymer, ρ = 0.962 g/cm^3 at 217°C, and that of the amorphous polymer, ρ = 1.049 g/mL at 25°C. The "missing" mass per molecular volume of a condensed polymer must therefore consist of the mass of segments of other polymer molecules that reside within the volume of the polymer molecule under consideration.

Because segments of other chain molecules with the same chemical structure are present in the volume V_B occupied by molecule B, V_B must be smaller than the volume V_G that the molecule would occupy without such restriction such as in a thermodynamically good solvent. The chain B thus wants to expand but is hampered in this effort by neighboring chains that also want to expand. Since all chains have the same constitution and configuration, all intermolecular attraction and repulsion forces have the same magnitude. Expansion and compression balance each other and each molecule adopts its unperturbed dimension which is the same in the melt and in the theta state for many polymers (Table 6-1).

Indeed, many polymers have the same radius of gyration in melts (directly measured by SANS) and in theta solutions (calculated from intrinsic viscosities, $[\eta]_\Theta$, see below) (Table 6-1). However, different signs of the temperature dependence of radii of gyration were sometimes found for both methods, which will be discussed in Section 6.2.4.

6.2.3 Entanglement of Polymer Chains

The overlap of loose polymer coils in melts causes polymer chains to entangle with other polymer chains. No method is known that allows one to detect the presence of entanglements in melts at rest, let alone to measure entanglement concentrations. The presence of entanglements is rather deduced from rheological properties such as diffusion in melts (Section 14.2.2), melt viscosities (Section 15.3.1), shear storage moduli of melts (Section 17.5.4), and relaxation in neutron spin echo experiments where entanglements impede the free flow or relaxation of individual chains. Such measurements deliver the **entanglement molar mass** M_e, i.e., the molar mass between two entanglement contacts. M_e must be an average but the average is not defined.

The structure of these physical crosslinks is not specified but there are at least three types (Fig. 6-4). In cohesion contacts, segments of different chains are bound to each other by physical bonds (e.g., dipole-dipole interactions) between adjacent chain units which leads to either nematic structures (Section 8.2.2) or fringed micelles (Section 7.2.1). Most entanglements are probably purely topological like hooks or loops; these entanglements do not require physical bonds between different chains.

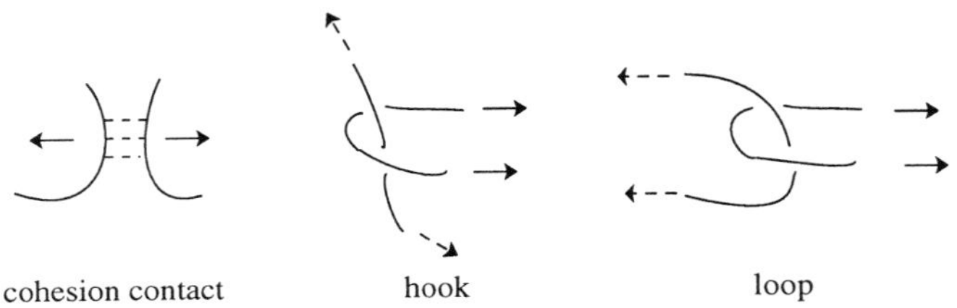

Fig. 6-4 Physical crosslinks that become effective in ⟶ direction but not in --→ direction. Cohesion contacts are probably responsible for the low "entanglement" molar mass LCPs (see below).

Entanglements are formed by sufficiently long chains, i.e., above an entanglement molar mass, M_e. Typical values of $M_e/(\text{g mol}^{-1})$ are 220 for the main chain liquid crystalline polymer (LCP), $\text{+O–C}_6\text{H}_3(\text{CH}_3)\text{–N=N}(\rightarrow\text{O})\text{–C}_6\text{H}_3(\text{CH}_3)\text{–O–CO–}(\text{CH}_2)_i\text{–CO+}_n$; 860 for linear poly(ethylene); 3700 for 1,4-*cis*-poly(isoprene); 13 500 for atactic poly(styrene), and 178 000 for so-called polymeric sulfur. Values of M_e can also be calculated from experimental data (ρ, M, $\langle s^2\rangle_o$) since a simple universal relationship

$$(6\text{-}3) \qquad M_e/\rho = K_p L_p^{\,3}$$

exists between M_e, polymer density ρ, and packing length L_p (Fig. 6-5). The packing length, $L_p \equiv V_{oc}/\langle r^2\rangle_o$, is defined as the ratio of occupied volume, $V_{oc} \equiv M/(\rho N_A)$, and unperturbed mean-square end-to-end distance, $\langle r^2\rangle_o = 6\,\langle s^2\rangle_o$, where $\langle s^2\rangle_o$ = mean-square radius of gyration (either experimental from SANS or calculated as $\langle s^2\rangle_\Theta$ from experimental $[\eta]_{Q)}$), and M = molar mass of polymer.

The proportionality constant K_p in Eq.(6-3) is independent of the molar mass and of the polymer constitution for polymers as diverse as poly(ethylene)s, various poly(1-olefin)s and poly(diene)s, poly(tetrafluoroethylene), different poly(acrylate)s, and poly(methacrylates), and even polymeric sulfur. It is also independent of the polymer tacticity and applies to the extraordinarily large temperature range of 25-380°C.

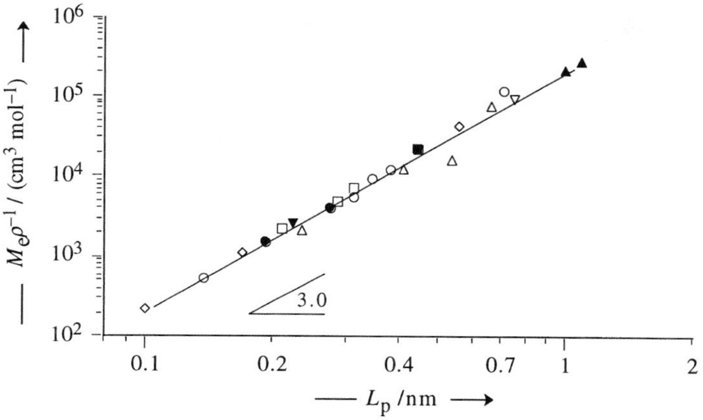

Fig. 6-5 Reduced entanglement molar mass, M_e/ρ, of various polymers as a function of their packing length L_p [8]. Measurements at 25°C (O), 32-38°C (●), 100°C (Δ), 120°C (▲), 140-160°C (◇), 190°C (□), 250°C (■), 300°C (∇). M_e was obtained from the plateau modulus, Eq. (17-35).

6.2.4 Temperature Dependence of Radii of Gyration

Above degrees of polymerization of ca. 100, reduced unperturbed radii of gyration, $(\langle s^2\rangle_o/M)^{1/2}$, become independent of molar masses (Fig. 4-32), provided that both $\langle s^2\rangle_o$ and M have the same average (most commonly both as mass ("weight") average). Since scattering methods deliver mean-square radii of gyration as z-average and molar masses as mass average, non-uniformity corrections are necessary. These "polymolecularity" correction factors are substantial even for relatively narrow molar mass distributions: for unperturbed coils with Schulz-Zimm molar mass distributions, they amount to 9.1 % at $\overline{M}_w/\overline{M}_n$ = 1.1 and 16.7 % at $\overline{M}_w/\overline{M}_n$ = 1.2 (Table 4-11).

In melts, several types of temperature dependences of radii of gyration, $\langle s^2\rangle_{o,w}^{1/2}$, or reduced radii of gyration, $[\langle s^2\rangle_{o,w}/\overline{M}_w]^{1/2}$, have been observed by SANS; examples are shown in Fig. 6-6. These quantities were found to be independent of temperature for atactic poly(styrene) and isotactic poly(propylene) (Fig. 6-1). But they increased with temperature for aPEE and decreased for aPME (both Fig. 6-6) as well as for PEP and PEB. The latter two are hydrogenation products of a poly(isoprene) (hence, a "poly(ethylene-*alt*-propylene)", PEP) and a poly(butadiene) with 1,4- and 1,2-units (hence, a "poly(ethylene-*co*-1-butene)", PEB). The variation of the sign of d ln $\langle s^2\rangle_{o,w}^{1/2}/dT$ with constitution has not been explained in the literature.

There are also differences in the temperature dependence of radii of gyration that have been obtained directly from small-angle neutron scattering using deuterated polymers of the same constitution as tracers and those that were calculated from intrinsic viscosities $[\eta]_\Theta$ of the same polymers in various solvents at theta conditions (see Section 12.3.2). For a poly(ethyl ethylene) (aPEE), the s_o in melts increased with temperature (SANS data) while s_o from theta solutions decreased (viscosity data) (Fig. 6-7).

It is generally assumed that melt data from SANS are correct (because of the chemical identity of tracer and substrate) and that intrinsic viscosity data are somehow influenced by so-called specific solvent effects. But this is not so: a protonated substance and its

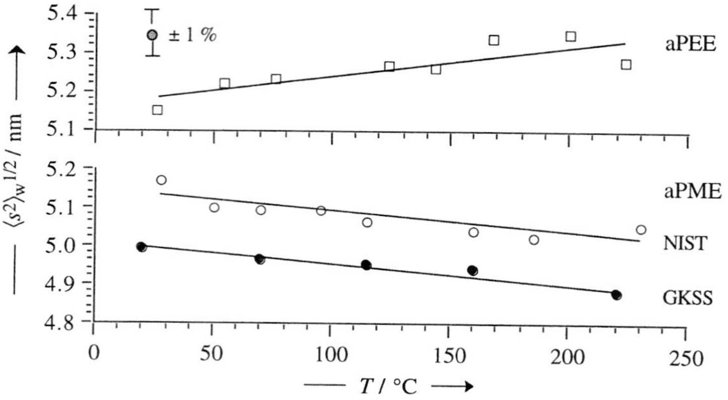

Fig. 6-6 Temperature dependence of weight-average radii of gyration of melts of a narrow distribution, atactic poly(1-methylethylene) (aPME, $\overline{M}_w$ = 23 300 g/mol) and a narrow distribution atactic poly(1-ethylethylene) (aPEE, $\overline{M}_w$ = 36 000 g/mol), from SANS with deuterated tracers [4b]. aPME measured at NIST (USA) and GKSS (Germany). The difference between the NIST and GKSS data is probably due to differences in calibration.

aPME = $\text{+}CH_2\text{–}CH(CH_3)\text{–}CH_2\text{–}CH(CH_3)\text{+}_n$ = hydrogenation product of a poly(2-methyl-1,3-pentadiene) (PMPD), $\text{+}CH_2\text{–}C(CH_3){=}CH_2\text{–}CH(CH_3)\text{+}_n$ with more than 99.8 % 1,4-content, obtained by anionic polymerization of 2-methyl-1,3-pentadiene, $CH_2{=}C(CH_3)\text{–}CH{=}CH(CH_3)$. T_G = –5°C. In the primary literature, aPME is called "atactic poly(propylene)," aPP, although it is not a propylene-based polymer and the tacticity of aPME itself was not determined (i.e., after hydrogenation of PMPD).

aPEE with the constitution $\text{+}CH_2\text{–}CH(C_2H_5)\text{+}_n$ is the hydrogenation product of a 1,2-poly(butadiene) with >98 % vinyl content, $\text{+}CH_2\text{–}CH(CH{=}CH_2)\text{+}_n$, which was obtained by anionic polymerization of 1,3-butadiene, $CH_2{=}CH\text{–}CH{=}CH_2$. T_G = –27°C. In the primary literature, aPEE is sometimes called "atactic poly(butene)" although it is not a polymerized butene.

The problem is that polymer names do not distinguish clearly between source-based and structure-based polymer names.

deuterated counterpart are *not* chemically identical and they *do* have different properties. For example, theta temperatures are quite different for atactic poly(styrene) in C_6H_{12} (Θ = 34.5C) and in C_6D_{12} (Θ = 40°C). Similarly, mixtures of a protonated polymer and its deuterated counterpart such as $\text{+}CH_2\text{–}CH(CH_3)\text{+}_n$ and $\text{+}CD_2\text{–}CD(CD_3)\text{+}_n$ must be treated as a solution of the former in the latter or vice versa.

Solutions, whether polymers in solvents or deuterated polymers in their protonated counterparts, may be exothermic or endothermic (Section 10.2.3) and show correspondingly upper or lower critical mixing temperatures (and thus 2 theta temperatures). Theta temperatures are always near demixing temperatures and one expects the temperature coefficients of $\langle s^2\rangle_{0,w}^{1/2}$ for polymers in dilute solutions (at $c \to 0$) to be positive for endothermic systems but negative for exothermic ones (see Section 6.3.2). Athermic polymer solutions with $d \ln \langle s^2\rangle_{0,w}^{1/2}/dT = 0$ are rare so that this case has not been found.

The situation is different for radii of gyration of molten protonated/deuterated polymers. These experiments are always performed at fairly high tracer concentrations with polymers with relatively low molar masses of 20 000-50 000 g/mol, and at temperatures that are presumbly far away from demixing temperatures. The sign of $d \ln \langle s^2\rangle_{0,w}^{1/2}/dT$ can therefore be positive, negative, or zero and these signs do not necessarily have to be identical with those from intrinsic viscosity data.

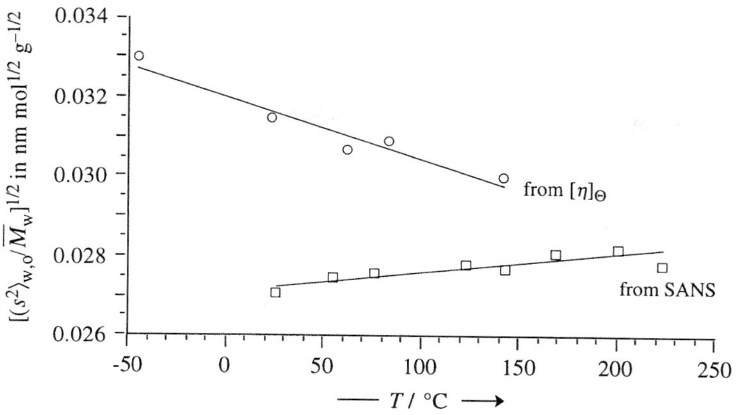

Fig. 6-7 Temperature dependence of reduced radii of gyration of an aPEE from SANS in melts (data of Fig. 6-6) and from intrinsic viscosities in various theta solvents [9].

6.3 Semi-Concentrated Solutions

6.3.1 Concentration Ranges

In dilute dispersions, particles are isolated from each other. Their volume fraction ϕ can be increased until a critical packing density ϕ_{max} is reached. For hard spheres, this density is 0.754 for hexagonal dense packing and 0.637 for statistically dense packing.

Random coils in solution reach their "maximum packing density" if their "surfaces" start to contact each other. Since coils are very loose entities without defined surfaces, they will start to interpenetrate each other to some extent. The critical "packing" concentration is rather a **(coil) overlap concentration** c^*, also called **cross-over concentration** because the solution crosses over from a dilute solution to a semi-dilute one.

At $c^* > c$, the same volume must take up more polymer segments than at $c^* < c$. Since coil molecules in theta solvents already possess their unperturbed dimensions and cannot be compressed further, at least not in the limit of $M \to \infty$, theta dimensions remain constant with further increase of concentration,

In thermodynamically good solvents, coil dimensions are much larger than in theta solvents for the same constitution and molecular weight. Thus, with increasing polymer concentration, these coils will be compressed more and more until they reach their unperturbed dimensions at the **critical entanglement concentration** c^{**}.

Hence, one can thus distinguish five different concentration regions:

$0 = c_2$	infinitely dilute solutions
$0 < c_2 \leq c_2^*$	dilute solutions
$c_2^* \leq c_2 \leq c_2^{**}$	semi-dilute solutions
$c_2^{**} < c_2 = \rho_2\phi_2$	concentrated solutions
$c_2 = \rho_2$	polymer melts ($\phi_2 \equiv 1$)

where $c_2 \equiv m_2/V = m_2/(V_1 + V_2)$ is the mass concentration of the polymer, $\rho_2 = m_2/V_2$ the bulk density of the polymer, and $\phi_2 = V_2/(V_1 + V_2)$ the volume fraction of the polymer in solution, assuming additivity of volumes of polymer 2 and solvent 1. The transitions at c_2* and c_2** are not sharp since coils do not possess defined surfaces.

Perturbed coils are assumed to be equivalent spheres with a controlling radius of gyration, s. The overlap concentration of such coils is $c_s^* = m_{mol}/V_{mol} = (3\,M)/(4\,\pi\,s^3 N_A)$ where $m_{mol} = M/N_A$ is the mass of the molecule and V_{mol} its volume. Introduction of $s = K_s M^{\nu}$, Eq.(4-73), delivers the overlap concentration c_s* at which the density of the chain in the coil equals the overall chain density in solution:

$$c_s^* = \frac{3M}{4\pi N_A s^3} = \frac{3}{4\pi K_s^3 N_A} M^{1-3\nu} = K_s^* M^{1-3\nu} \tag{6-4}$$

A poly(styrene) of molar mass $M = 1 \cdot 10^6$ g/mol has an overlap concentration of $c_s^* \approx 3.6 \cdot 10^{-3}$ g/mL in the good solvent CS_2 ($\nu = 3/5$; $K_s = 1.2 \cdot 10^{-9}$ cm (mol/g)$^{3/5}$ at 25°C) and $c_s^* \approx 14.7 \cdot 10^{-3}$ g/mL in the theta solvent cyclohexane ($\nu = 1/2$; $K_{s,o} = 3 \cdot 10^{-9}$ cm (mol/g)$^{1/2}$ at 34.5°C). All radii of gyration shown in Fig. 5-19 for a hydrogenated poly(butadiene) in $C_{19}D_{40}$ at 150°C were measured at volume fractions that are much higher than the overlap volume fraction of $\phi_2^* \approx 0.034$.

Overlap concentrations can also be estimated for other coil dimensions such as hydrodynamic radii from dilute solution viscosity (Section 12.3.2). In theta solvents at infinite dilution, these measurements deliver intrinsic viscosities, $[\eta]_\Theta$, which are related to so-called Einstein radii by $R_{v,\Theta} = [(3\,M[\eta]_\Theta)/(10\,\pi N_A)]^{1/3}$. Experimental data showed that this radius is smaller than the Stokes radius $R_{D,\Theta}$ from diffusion by a factor of $R_{D,\Theta}/R_{v,\Theta} = 1.07$. The Stokes radius, in turn, is smaller than the radius of gyration by a factor of $s_o/R_{D,\Theta} = 1.28$ (see Table 12-6). The overlap concentration in theta solvents, c_s*, is therefore practically identical with the inverse intrinsic viscosity, $1/[\eta]_\Theta$, which is a measure of the volume that is occupied by the unit mass of the molecule:

$$c_s^* = \frac{3M}{4\pi N_A s_o^3} = \frac{10}{4(1.28 \cdot 1.07)^3 [\eta]_\Theta} \approx \frac{0.973}{[\eta]_\Theta} \approx \frac{1}{[\eta]_\Theta} = c_v^* \tag{6-5}$$

For a poly(styrene) with $M = 1 \cdot 10^6$ g/mol in cyclohexane at 34°C, $[\eta]_\Theta = 84.6$ mL/g which leads to $c_s^* = 0.0118$ g/mL which agrees fairly well with $c_s^* \approx 0.0147$ g/mL from the radius of gyration.

Overlap concentrations can also be related to number concentrations, C_s, instead of mass concentrations. This quantity is defined as that concentration in which just one molecule with the degree of polymerization, X, resides in a cube with a side length equalling the radius of gyration. Since $s = K_{s,X} X^{\nu}$ (Eq.(4-73) with $M = XM_u$ and $K_{s,X} = K_s M_u^{\nu}$) and, in good solvents, $\nu \approx 3/5$ (Eq.(4-73)) (exact: $\nu = 0.588$), one obtains

$$C_s^* = \frac{X}{s^3} = \frac{X}{K_{s,X}^3 X^{3\nu}} = \frac{1}{K_{s,X}^3 X^{3\nu-1}} \approx K_{s,X}^{-3} X^{-4/5} \tag{6-6}$$

6.3.2 Temperature Dependence of Coil Dimensions

Temperature changes lead to changes in the population of the various microconformations and thus to changes in coil dimensions and consequently to a variation of $\langle s^2 \rangle_{o,w}^{1/2}$ and $\langle s^2 \rangle_{o,w}^{1/2}/\overline{M}_w$ with temperature at $c_2 \to 0$ (Table 6-1 and Section 6.2.4). At higher concentrations, this effect is negligible compared to the changes of segment-segment, segment-solvent, and solvent-solvent interactions with temperature.

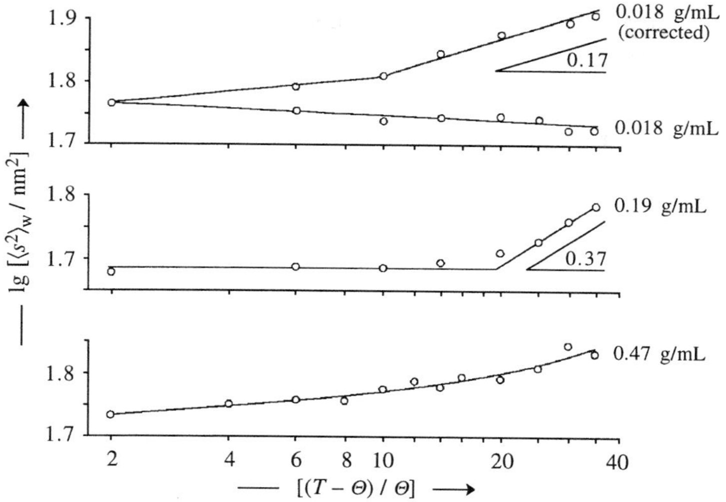

Fig. 6-8 Logarithm of mass-average mean-square radius of gyration, $\langle s^2 \rangle_w$, as a function of the logarithm of reduced distance to the theta temperature, $(T - \Theta)/\Theta$, for three concentrations of a poly(styrene) in cyclohexane [10a]. $\overline{M}_w$ = 75 700 g/mol, $\overline{M}_n$ = 49 000 g/mol. Dilute solution: 0.018 g/mL d_8-PS; semi-concentrated solution: 0.025 g/mL d_8-PS in a solution of total 0.19 g/mL of d_8-PS + h_8-PS; concentrated solution: 0.020 g/mL d_8-PS in a solution of total 0.47 g/mL d_8-PS + h_8-PS.

Unperturbed dimensions become perturbed and thus larger at $T > \Theta$ for endothermic solutions but at $T < \Theta$ for exothermic ones. At constant concentration, radii of gyration should thus increase with temperature for endothermic solutions (Fig. 6-8, bottom and center) but decrease for exothermic ones.

However, a *de*crease is sometimes observed for *low* concentrations of the same endothermic system (Fig. 6-8, top). This effect arises because data were not corrected for the influence of the second virial coefficient, A_2, a thermodynamic quantity that describes binary interactions (p. 146 and Section 10.4). Corrected mean-square radii of gyration are obtained from

$$(6\text{-}7) \qquad \langle s^2 \rangle_{c \to 0} = \langle s^2 \rangle_c (1 + 2\, A_2\, \overline{M}_w c)$$

and these corrected radii do indeed increase with lg $[(T - \Theta)/\Theta]$ (Fig. 6-8, top). Similar increases of lg $\langle s^2 \rangle_w$ with lg $[(T - \Theta)/\Theta]$ are found for higher polymer concentrations albeit with different powers than predicted by theory (see next section).

6.3.3 Blobs

The entanglement molar masses of polymers mentioned on p. 180 indicate a considerable number of chain units N_{cu} (= skeletal atoms plus their substituents) between two entanglements: 61 for poly(ethylene), 216 for *cis*-1,4-poly(isoprene), and 260 for atactic poly(styrene). The very high entanglement molar mass of polymeric sulfur is a fictional quantity because so-called polymeric sulfur consists of only 37 % polymers at 300°C, and 54 % S_8 rings and 9 % other small cyclics (Volume II, Fig. 12-13).

The sections between two entanglements are therefore large enough to maintain coil-like characteristics. They are coils-in-coils (**blobs**) that fulfill the requirements of self-similarity (Section 4.8.2). A polymer chain consisting of X_u monomeric units of molar mass M_u can thus be viewed as a necklace consisting of N_{bl} blobs of molar mass M_{bl} and degree of polymerization, $X_{bl} = M_{bl}/M_u$ where $M = X_u M_u = N_{bl} M_{bl}$. The diameter d_{bl} of a blob is the distance L_c between two entanglements (Fig. 6-9).

It is convenient to express overlap concentrations as volume fractions ϕ^* and not as mass concentrations c^*. Because $c_2^* = m_2^*/V = V_2^* \rho_2/V = \phi_2^* \rho_2$ and $M = M_u X$, Eq.(6-4) can also be written as $\phi_2^* = [(3\ M(4\ \pi\ N_A \rho_2)^{-1}][X s^{-3}] = K_c' X s^{-3}$. This equation for the whole coil must also be applicable to a blob albeit with another proportionality constant, so that $\phi_2^* = K_c X_{bl}(s_{bl})^{-3}$.

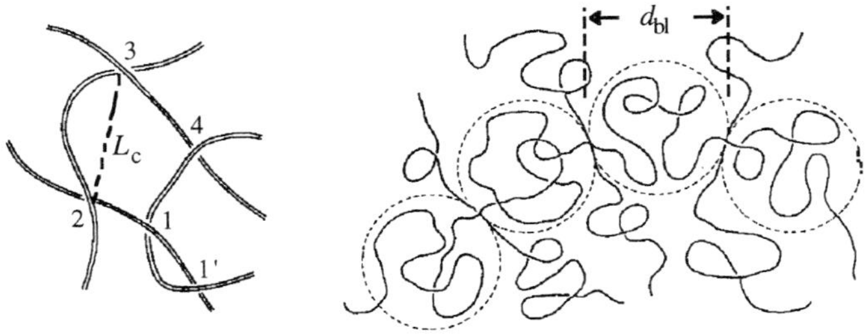

Fig. 6-9 Left: entanglement of large chain molecules (schematic). Contact points 1 and 1' indicate the entanglement of segments of two chains. Contact points 2, 3, and 4 are likewise part of entanglements 2-2', 3-3', and 4-4' (not shown). Right: molecule sections between entanglements behave as quasi-independent blobs (broken circles).

Blob diameters d_{bl} must be directly proportional to radii of gyration s_{bl} of blobs near critical volume fractions ϕ_2^* of polymers for overlap, i.e., $d_{bl} = Q_{bl}(s_{bl})$ where Q_{bl} is a proportionality constant (see below). Entanglement densities are low near ϕ_2^* so that blobs are relatively large. In *good* solvents, excluded volumes exist for monomeric units *within* blobs to the same extent as for the whole polymer so that $s_{bl} = K_{bl}' M_{bl}^{3/5}$ because $s = K_s M_2^{3/5}$. It is convenient to use degrees of polymerization, $X = M_2/M_u$ and $X_{bl} = M_{bl}/M_u$, instead of M_2 so that $s_{bl} = K_{bl} X_{bl}^{3/5}$ (with $K_{bl} = K_{bl}' M_u^{3/5}$).

Blobs in good solvents are increasingly compressed with increasing total volume fraction of polymer, $\phi_2 > \phi_2^*$. The proportionality factor Q_{bl} in $d_{bl} = Q_{bl}(s_{bl})$ can therefore *not* by independent of the volume fraction of the polymer. It can also not be directly proportional to ϕ_2 since the compression is more and more difficult with increasing ϕ_2. The factor is thus assumed to be $Q_{bl} = (\phi_2/\phi_2^*)^y$ where y is a negative number. The diameter of the blob is therefore $d_{bl} = (\phi_2/\phi_2^*)^y (s_{bl})$.

The exponent y is obtained as follows. Introduction of $\phi_2^* = K_c X_{bl}/(s_{bl})^3$ and $s = K_{bl} X_{bl}^{3/5}$ in $d_{bl} = (\phi_2/\phi_2^*)^y (s_{bl})$ delivers $d_{bl} = [K_c^{-y} K_{bl}^{3y+1}](\phi_2)^y X_{bl}^{(3/5)+(4y/5)}$. By definition, blob diameters are dependent on degrees of polymerization, $d_{bl} \sim X_{bl}^0$. The exponent of X_{bl} thus becomes $(3/5) + (4y/5) = 0$ which delivers $y = -3/4$.

d_{bl} is a length and ϕ_2 and X_{bl} are dimensionless. The factor $[K_c^{-y} K_{bl}^{3y+1}]$ must therefore be a length; it will be called b. Increasing volume fractions ϕ_2 causes a rapid decrease of blob diameters since $d_{bl} = b\phi_2^{-3/4}$.

Because of excluded volumes in blobs, diameters of blobs must depend on degrees of polymerization similar to that of coils themselves, i.e., $d_{bl} = bX_{bl}^{3/5}$ and, because $d_{bl} = b\phi_2^{-3/4}$, also $X_{bl} = \phi_2^{-5/4}$. At $\phi_2 = 1$ one obtains $X_{bl} = 1$ and $d_{bl} = b$. In melts, the degree of polymerization of blobs themselves is therefore unity, i.e., the diameter of blobs corresponds to an (effective) monomer length b.

A coil consists of only a few blobs that do not produce additional excluded volumes. The necklace of blobs thus behaves like an unperturbed Kuhnian chain with $\langle s^2\rangle_o = N_K L_K^2$ (p. 97) in which the number N_K of Kuhnian segments is represented by the number N_{bl} of blobs and the length of Kuhnian segments by the diameter d_{bl},. The mean-square radius of gyration, $\langle s^2\rangle_o$, refers to the necklace of blobs and must be identical with the mean-square radius of gyration $\langle s^2\rangle$ of the chain itself.

The blob diameter d_{bl} is thus a **screening length** L_{sl} that screens the excluded volume *within* the blob. A correlation thus exists at $L_{sl} < d_{bl}$ for the natural tendency to exclude other segments from the blob but not for $L_{sl} > d_{bl}$. The screening length is therefore also called the **correlation length**.

Introduction of $N_{bl} = X/X_{bl}$, $X_{bl} = \phi_2^{-5/4}$, and $d_{bl} = b\phi_2^{-3/4}$ into the mean-square radius of gyration, $\langle s^2\rangle = N_{bl}d_{bl}^2$, delivers

(6-8) $$\langle s^2\rangle = b^2 X \phi_2^{-1/4} \quad \text{or} \quad \langle s^2\rangle/M = (b^2/M_u)\phi_2^{-1/4}$$

As predicted, logarithms of reduced mean-square radii of gyration decrease with increasing logarithm of volume fractions of polymers until they reach the unperturbed reduced mean-square radius of gyration at $\phi_2 = 1$ (Fig. 6-10). However, the experimental slope is only –0.146 instead of the theoretical one of –0.25.

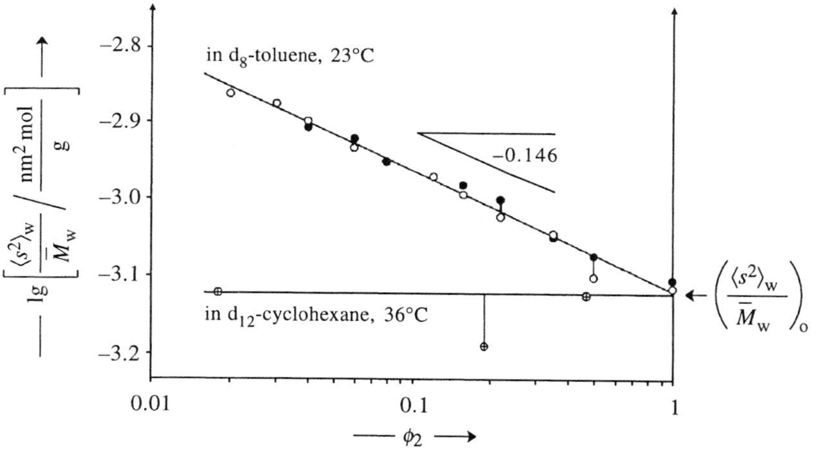

Fig. 6-10 Double-logarithmic plot of $\langle s^2\rangle_w/\overline{M}_w = f(\phi_2)$ for mixtures of poly(styrene) and deutero-poly(styrene) with $\overline{M}_w \approx 114\ 000$ g/mol each in the good solvent d_8-toluene (○,●: two series measured in a time interval of several months) [11] and a poly(styrene with $\overline{M}_w \approx 75\ 500$ g/mol in the theta solvent d_{12}-cyclohexane (⊕) [10b], both by SANS.

Scaling theory also predicts an increase of $\langle s^2\rangle$ with the one-fourth power of the temperature increment, $(T-\Theta)/\Theta$:

(6-9) $$\langle s^2\rangle \sim \frac{X}{c^{1/4}}\left(\frac{T-\Theta}{\Theta}\right)^{1/4}$$

However, such a temperature dependence may cover only a small temperature range since a polymer in a solvent usually has two theta temperatures. An example is at-poly-(styrene) in *t*-butyl acetate with Θ_{UCST} = 10°C (endothermic) and Θ_{LCST} = 109.3°C (exothermic) (see also Fig. 10-21). The polymer has the same radius of gyration at both Θ_{UCST} and Θ_{LCST}. The radius of gyration increases if $T > \Theta_{UCST}$ and also at $T < \Theta_{LCST}$. At a certain temperature T_{ath}, $\langle s^2 \rangle$ must therefore pass through a maximum. At this temperature, dilute polymer solutions behave as athermic ones.

At higher polymer concentrations, coil segments are distributed more uniformly and the assumptions of scaling theory are no longer applicable. Coils finally adopt their unperturbed dimensions at volume fractions $\phi_2 > \phi_2^{**} = \phi_2^{*}[\langle s^2 \rangle / \langle s^2 \rangle_\Theta]$ where ϕ_2^{**} is given by the product of overlap volume fraction, ϕ_2^{**}, and the ratio of mean-square radii of gyration in good and theta solvents.

6.3.4 Radii of Gyration of Star Molecules

In moderately concentrated solutions, long arms of star molecules can be modeled as blobs composed of linear chains. A star molecule is thus envisioned as a core that is surrounded by blobs. Cores are negligible at $\phi_2 >> \phi_2^*$ and one would therefore expect the same functionality as in Eq.(6-8), i.e., $s \sim X_{arm}^{1/2} \phi_2^{-1/8}$ for good solvents (ν = 3/5).

At low polymer concentrations, radii of gyration of a star polymer with 18 poly(isoprene) arms were found by SANS to be practically independent of the polymer concentration (Fig. 6-11). Above a critical concentration of $\phi_2 \approx 0.09$, s decreased with ϕ_2 with approximately the predicted exponent of –0.125 in lg s = f(lg ϕ_2). The corresponding 64-arm molecule also showed concentration-independent values of s at low concentrations but a slight increase of s with ϕ_2 at higher ones. The latter effect is probably caused by a crowding of arms near the core which forces the arms to become more stretched and the star molecule to assume a more brush-like structure (Section 9.5.2).

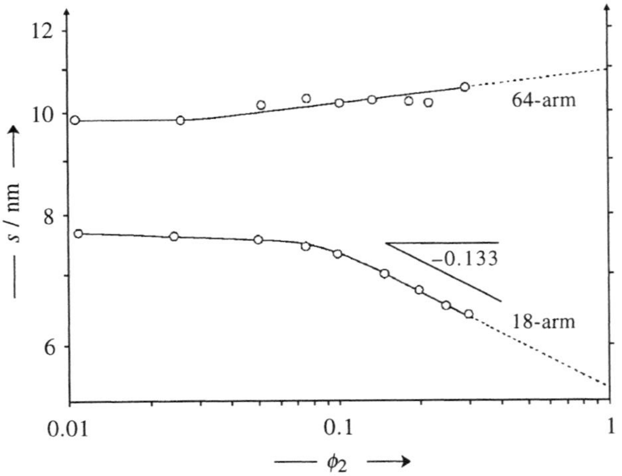

Fig. 6-11 Radius of gyration s as a function of the volume fraction ϕ_2 of an 18-arm poly(isoprene) star with a carbosilane core and a 64-arm poly(butadiene) star with a carbosilane dendrimer core, both at 25°C in cyclohexane as solvent [12].

Historical Notes to Chapter 6

UNPERTURBED DIMENSIONS

P.J.Flory, Principles of Polymer Chemistry, Cornell University Press, Ithaca, NY, 1953, p. 602
Predicts that coil-forming linear polymer molecules adopt their unperturbed dimensions in the amorphous state.

J.S.King, W.Boyer, G.D.Wignall, R.Ullman, Macromolecules **18** (1985) 709
First experimental proof of unperturbed dimensions in the amorphous state.

ENTANGLEMENTS

W.F.Busse, J.Phys.Chem. **36** (1932) 2862
Assumes the presence of chain entanglements in the amorphous state of polymers by either van der Waals forces (i.e., cohesion contacts) or mechanical hooks and loops (Fig. 6-4).

R.Signer, H.Gross, Helv.Chim.Acta **17** (1934) 59; s.a. T.Svedberg, K.O.Pedersen, The Ultracentrifuge, Pt. IV, Chapter B, Clarendon Press, Oxford 1940
First experimental proof of physical networks in concentrated solutions of linear polymer chains (sedimentation rate is independent of molar mass at these concentrations).

L.R.G.Treloar, Trans.Faraday Soc. **36** (1940) 538
Explains stress-strain behavior of rubbers and elastomers as an effect of polymer entanglements according to Busse (at medium strains) and crystallization (at high strains).

F.H.Müller, Kolloid-Z. **95** (1941) 138, 306
Deformation of amorphous polymers is caused by site-hopping and deformation.

M.S.Green, A.V.Tobolsky, J.Chem.Phys. **14** (1946) 80
Chain molecules in melts form a network with temporary crosslinks.

Literature to Chapter 6

R.N.Haward, Occupied Volume of Liquids and Polymers, J.Macromol.Sci.-Macromol.Revs. **C 4** (1970) 191
G.S.Y.Yeh, Morphology of Amorphous Polymers, Crit.Revs.Macromol.Sci. **1** (1972) 173
R.N.Haward, The Physics of the Glassy State, Interscience, New York 1973
R.E.Robertson, Molecular Organization of Amorphous Polymers, Ann.Rev.Mater.Sci. **5** (1975) 73
R.F.Boyer, Structure of the Amorphous State in Polymers, in M.Goldstein, R.Simha, Eds., Structure of Amorphous Solids, Ann.N.Y.Acad.Sci. **279** (1976) (p. 279)
G.Allen, S.E.B.Petrie, Eds., Physical Structure of the Amorphous State, Dekker, New York 1977
R.A.Komoroski, Ed., High Resolution NMR Spectroscopy of Synthetic Polymers in Bulk, VCH, Weinheim 1986
M.Doi, S.F.Edwards, The Theory of Polymer Dynamics, Chapter 5, Overlap Concentration, Oxford University Press, London 1986
S.E.Keinath, R.L.Miller, J.K.Rieke, Eds., Order in the Amorphous "State" of Polymers, Plenum, New York 1987
R.G.Larson, Constitutive Equations for Polymer Melts and Solutions, Butterworth, Boston 1988
M.Kröger, Rheologie und Struktur von Polymerschmelzen, Wissenschaft und Technik Verlag, Berlin 1995 (rheology and structure of polymer melts)
G.V.Koslov, G.E.Zaikov, Structure of the Polymer Amorphous State, VSP, Utrecht 2004
K.Binder, W.Kob, Glassy Materials and Disordered Solids, World Scientific, Hackensack (NJ) 2005

References to Chapter 6

[1] (a) D.G.H.Ballard, P.Cheshire, G.W.Longman, J.Schelten, Polymer **19** (1978) 379 (taken from Fig. 7); (b) D.G.H.Ballard, J.Schelten, Dev.Polym.Charact. **2** (1980) 31, Fig. 5

[2] According to [1a] and [3], the poly(styrene) in Fig. 6-1 is *crystalline* but the reference [D.G.H.Ballard, G.D.Wignall, J.Schelten, Eur.Polym.J. **9** (1973) 965], quoted by [1a], speaks of *amorphous* poly(styrene). Presumbly, the poly(styrene) of references [1a] and [3] was a so-called crystal poly(styrene) which is an amorphous, atactic poly(styrene) with high gloss.

[3] Amended data of Table 4 of G.D.Wignall, Encycl.Polym.Sci.Eng. **10** (1987) 112, see also [2]

[4] A.Zirkel, V.Urban, D.Richter, L.J.Fetters, J.S.Huang, R.Kampmann, N.Hadjichristidis, Macromolecules **25** (1992) 6148, (a) Table IV, (b) Tables I and II

[5] R.L.Miller, R.F.Boyer, J.Polym.Sci.-Polym.Phys.Ed. **22** (1984) 2043, modified Fig. 1

[6] R.E.Richardson, Ann.Rev.Mater.Sci. **5** (1975) 73, Table 1 (expanded)

[7] D.Panke, W.Wunderlich, Makromol.Chem. **167** (1973) 351, Fig. 1

[8] L.J.Fetters, D.J.Lohse, W.W.Graessley, J.Polym.Sci.-Polym.Phys. **B 37** (1999) 1023, selected data of Tables I-IV

[9] G.Moraglio, G.Gianotti, F.Danusso, Eur.Polym.J. **3** (1967) 251;

[10] R.W.Richards, A.Maconnachie, G.Allen, Polymer **19** (1978) 266, (a) Fig. 4 ($c \to 0$) and (b) Table 3 (dilute, semi-concentrated, and concentrated solutions)

[11] J.S.King, W.Boyer, G.D.Wignall, R.Ullman, Macromolecules **18** (1985) 709, Tables II and III

[12] L.Willner, O.Jucknischke, D.Richter, J.Roovers, L.-L.Zhou, P.M.Toporowski, L.J.Fetters, J.S.Huang, M.Y.Lin, N.Hadjichristidis, Macromolecules **27** (1994) 3821, Table 4

7 Crystalline States

7.1 Crystal Structures

7.1.1 Definitions of a Crystal

Ancient philosophers assumed rock crystals, the transparent crystals of quartz (SiO_2), to be a hard, dry form of ice and called it therefore "crystal" (L: *crystallum*; G: *crystallos* = ice). In the mid-1800s, the term "crystal" was then applied to all materials with external planar surfaces that are arranged at constant angles to each other.

At the end of the 1800s, "crystal" was redefined as a homogeneous, anisotropic, solid material. A crystal is "homogeneous" because its physical properties do not change on translation in the directions of crystal axes. It is also "anisotropic" because its physical properties vary on rotation of the crystal (G: *anisos* = unequal; G: *tropikos*, from G: *tropē* = a turn).

"Crystal" was again redefined at the beginning of the 20th century, now as a solid body with long-range three-dimensional order in a three-dimensional lattice. A lattice consists of smaller units, the **unit cells**, whose three-dimensional repetition (**translation**) produces the crystal. Unit cells are parallelepipeds composed of **lattice points (lattice sites**) which are geometric points that represent an atom or a group of atoms. According to this definition, crystals differ from amorphous substances by symmetry and the periodicity which is caused by the regular arrangement of lattice points. On irradiation of crystals with X-rays, such regularities lead to diffractograms with sharp reflections.

In the 1970s, Roger Penrose discovered two-dimensional tile-like arrangements of matter in which so-called **prototiles** are arranged non-periodically, i.e., without translational symmetry. This was followed by discoveries of three-dimensional **quasicrystals** in metallic alloys (1982), dendrimer-based liquid crystals, and certain polymer blends (2007). These quasicrystals often have five-, ten-, or twelve-fold rotational symmetries.

Unlike classic crystals, quasicrystals do not consist of repeating unit cells. But they do give X-ray diffractograms that are as sharp as those of classic crystals. Quasicrystals must therefore have long-range order despite absence of short-range order from repeating unit cells. In 1992, the *International Union of Crystallography* thus defined a crystal as a material with discrete diffraction patterns that consist of individual reflections. Therefore, a crystal need not have periodicity or symmetry.

7.1.2 Lattice Structures

The building blocks of crystals may be atoms, ions, atomic groups, segments, or even molecules. In ideal crystals, all building blocks are identical and arranged in a time-independent, perfect, three-dimensional lattice. Atoms and small ions occupy lattice sites directly: they form atomic **point lattices**. For larger entities, the lattice point is identified as the center of gravity of the building block. Such entities also create point lattices, for example, the ethylene groups $-CH_2-CH_2-$ of poly(ethylene) (see Fig. 7-5). Lattice points may also be occupied by molecules themselves, for example, spheroidal protein molecules in so-called **molecule lattices**. **Super lattices** are composed of densely packed large spheres such as latex particles.

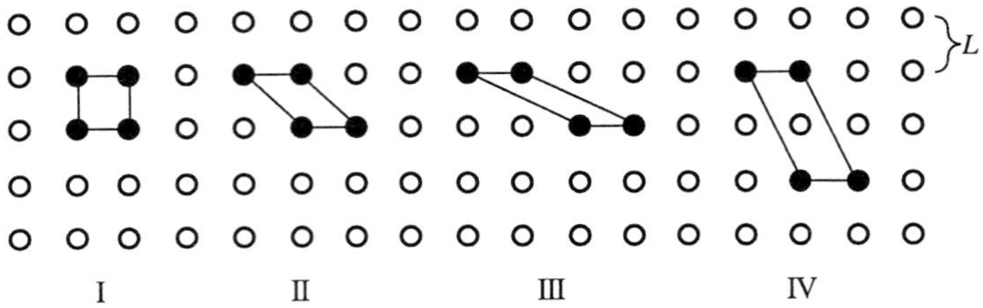

Fig. 7-1 Primitive unit cells I - III and a centered unit cell IV in a two-dimensional lattice plane. L = distance between corner points on a square lattice.

A two-dimensional point lattice is called a **lattice plane** (**net plane**). The smallest periodic unit of a lattice plane is the two-dimensional unit cell, a congruent polygon with at least three non-colinear corner points. **Primitive** (**simple**) **unit cells** contain only corner points as lattice points and **centered unit cells** both corner and center points (Fig. 7-1).

Classic crystals contain parallel lattice planes. In this three-dimensional case, unit cells consist of the simplest **parallelepipeds** that are given by lattice points (G: *para* = beside, for; *allos* = another; *epipedos* = plane (from *epi* = upon, over, after; *pedon* = earth, soil). Like their two-dimensional counterparts, three-dimensional unit cells are either primitive or centered. In the latter case, center points may be either at the intersection of space diagonals (i.e., **body-centered**) or at the intersection of planar diagonals and here either **face-centered** (at the centers of all cell faces) or **base-centered** (at the centers of two opposite faces).

Parallelepipeds are characterized by three **lattice constants** a, b, and c which describe the lengths of elementary vectors $\boldsymbol{a}$, $\boldsymbol{b}$, and $\boldsymbol{c}$. Two vectors each enclose an **(axial) angle**: γ (between $\boldsymbol{a}$ and $\boldsymbol{b}$), β (between $\boldsymbol{a}$ and $\boldsymbol{c}$), and α (between $\boldsymbol{b}$ and $\boldsymbol{c}$). The three lattice constants and three angles lead to seven possible crystal systems and 14 possible Bravais lattices (Table 7-1). Quasicrystals do not belong to any of these systems.

For chain molecules, the c-axis normally gives the direction of the chain axis which is characterized by short chemical bonds between chain atoms; a- and b-directions are generated by longer physical bonds between chains or their segments. Since chemical bond lengths are much shorter than physical ones, cubic lattices are not formed by chain molecules, many of which crystallize in either the orthorhombic or monoclinic system.

Table 7-1 Definitions of crystallographic properties. P = primitive (simple), F = face-centered, B = base-centered, I = body-centered (symbol "I" from the German "innenzentriert"). For the explanation of symmetry symbols, see Volume I, p. 112. *) Also: isometric. **) Also: rhombohedral; sometimes as a subdivision of the hexagonal system.

Optical system	Crystal system	Axes	Angles	Symmetry	Bravais lattice P	F	B	I
Isotropic	cubic *)	$a = b = c$	$90° = \alpha = \beta = \gamma$	m3m (O_h)	+	+		+
Uniaxial	hexagonal	$a = b \neq c$	$90° = \alpha = \beta$ $\gamma = 120°$	6/mmm (O_{6h})	+			
	tetragonal	$a = b \neq c$	$90° = \alpha = \beta = \gamma$	4/mmm (D_{4h})	+			+
Biaxial	orthorhombic	$a \neq b \neq c \neq a$	$90° = \alpha = \beta = \gamma$	mmm (D_{2h})	+	+	+	+
	monoclinic	$a \neq b \neq c \neq a$	$90° = \alpha = \gamma$ $\beta \neq 90°$	2/m (C_{2h})	+		+	
	triclinic	$a \neq b \neq c \neq a$	$\alpha \neq \beta \neq \gamma$	$\bar{1}$ (C_i)	+			
	trigonal **)	$a = b = c$	$90° \neq \alpha = \beta = \gamma$	$\bar{3}$m (D_{4d})	+			

Table 7-2 Radii of atoms for different attractive forces [1]. Data in nanometers. See also Table 4-4.

Element	Radius of atoms in nanometer						
	Covalent simple	Covalent double	Covalent triple	Ionic positive	Ionic negative	Metallic	van der Waals
1 H	0.030			0.00	0.208		0.117
3 Li				0.060		0.1549	
6 C	0.0772	0.0667	0.0630	0.015	0.260		0.170
7 N	0.074	0.062	0.055	0.011	0.174		0.157
8 O	0.074	0.062	0.055	0.009	0.140		0.150
9 F	0.072	0.060		0.007	0.136		0.147
14 Si	0.117	0.107	0.100	0.041	0.271	0.1357	0.210
15 P	0.110	0.100	0.093	0.034	0.059	0.128	0.180
16 S	0.104	0.094	0.087	0.029	0.053	0.127	0.185
17 Cl	0.099	0.089		0.026	0.181		0.178

Volumes V of unit cells are calculated from vectors $\boldsymbol{a}$, $\boldsymbol{b}$, und $\boldsymbol{c}$, and thus from lattice constants a, b, and c, and the angles α, β, and γ.

Cubic	$V = \boldsymbol{a}^3$
Tetragonal	$V = \boldsymbol{a}^2\boldsymbol{b}$
Orthorhombic	$V = \boldsymbol{abc}$
Monoclinic	$V = \boldsymbol{abc} \sin \beta$
Hexagonal	$V = \boldsymbol{abc} \sin 60°$
Trigonal	$V = \boldsymbol{a}^3[1 + 3 \cos^2 \alpha + 2 \cos^3 \alpha]$
Triclinic	$V = \boldsymbol{abc}\,[1 + 2 \cos \alpha \cdot \cos \beta \cdot \cos \gamma - \cos^2 \alpha - \cos^2 \beta - \cos^2 \gamma]^{1/2}$

Crystal structures are controlled by the constitution and configuration of polymer molecules and especially by the packing of molecules in the crystal. These quantities are controlled by the space requirements of atoms (Table 7-2) and thus by forces between atoms as well as bond and conformational angles.

7.1.3 Symmetry Properties

Like all other objects, symmetry properties of crystals are described by three simple and two complex symmetry elements and the corresponding five symmetry operations (Table 7-3 and Volume I, p. 112) which leads to 32 crystal classes and 230 space groups. For polymers, **identity** has to be added to the three symmetry elements and **translation** to the corresponding three symmetry operations.

Simple symmetry operations correspond to "dimensionalities," i.e., "0-dimensional" (**point** or **center**), "1-dimensional" (**line** or **axis**), "2-dimensional" (**area** or **plane**), and "3-dimensional" (**identity**). The corresponding symmetry operations are **inversion** (0), **rotation** (1), **reflection** (2), and **translation** (3) along the chain axes. Complex symmetry operations combine two simple symmetry operations (Table 7-3). For example, screw axes result from combining translation and rotation and glide mirror axes from combining translation with reflection at a plane.

Table 7-3 Symmetry elements and symmetry operations.

Symmetry element	Symbol	Symmetry operation
Simple		
Inversion center	I	inversion at a point
Proper axis of rotation	C	rotation about an axis
Mirror plane of symmetry	σ	reflection at a plane that cuts through the object
Identity	-	translation as repetition of identity
Complex		
Rotation-inversion axis	-	rotation-inversion (combination of inversion and rotation)
Mirror axis	S	rotation-reflection (combination of rotation and reflection)
Screw axis	-	translational rotation (combination of translation and rotation)
Glide mirror axis	-	translational reflection (combination of translation and reflection)

7.1.4 Crystal Structure by X-Ray Diffraction

X-ray diffractometry is the most important method for the determination of crystal structures, which include the molecular structure (position of individual atoms and chain conformation) and the spatial arrangement and packing of chains or bundles thereof. Since so-called crystalline polymers are rarely single crystals and never onehundred percent crystalline, X-ray diffraction is also used in polymer science to determine the degree of crystallinity and the way crystallites are packed and oriented.

X-rays are produced by bombarding a metal anode, for example, copper or molybdenum, with high-energy electrons. Most of the kinetic energy of the electrons is converted to heat, but of it is consumed for the generation of X-rays. The fast incident electrons knock off an inner orbital electron from an atom of the target metal. The knocked-off electron is replaced by an outer electron which loses potential energy (ΔE) in the process, i.e., it emits a photon with a wavelength $\lambda = ch/\Delta E$ where c = speed of light and h = Planck constant. For example, the wavelength of the Cu-K_α radiation is 0.15405 nm (= 1.5405 Å).

Photons are elastically diffracted at matter, i.e., without transfer of energy, for example, at lattices if the distance between lattice planes (net planes) is comparable to the wavelength. Each irradiated atom acts like a transmitter. Since waves from various net planes differ in phase, they can interfere constructively (if in phase) or destructively (if not in phase) which gives rise to X-ray reflections.

Two net planes G_1 and G_2 may be at the distance d (Fig. 7-2). The incident wave L hits atom A at an angle θ and the parallel wave L_1 the atom A_2 in the same net plane. The phase shift is therefore $PA_2 + A_2Q = 2\,d \sin \theta$.

Waves interfere constructively if they arrive simultaneously at the plane. This requires the phase shift to equal the wavelength λ_o of the incident radiation and a multiple of the order number N of reflections, respectively. N adopts only whole numbers corresponding to interferences of first, second, ... order. This is described by **Bragg's law** (W.H.Bragg (father), W.L.Bragg (son); joint Nobel prize 1915):

(7-1) $$N\lambda_o = 2\,d \sin \theta$$

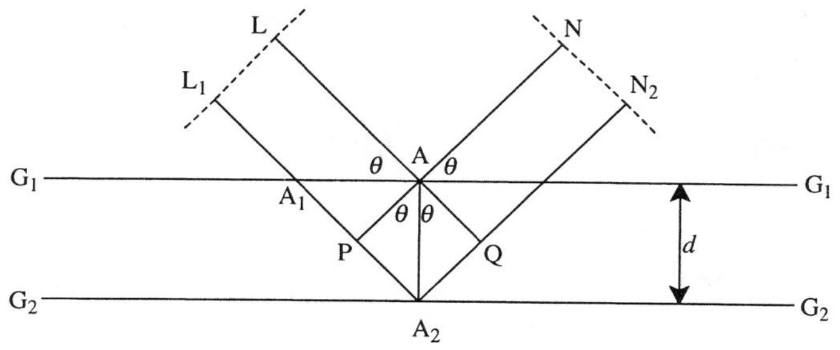

Fig. 7-2 Illustration of Bragg's law (see text).

The reflections with the strongest intensity usually gets the order number $N = 1$ since it originates from the most abundant lattice elements. Diffraction at a single lattice element, for example, a unit cell, is very weak indeed but the structure of this element repeats periodically many times in three dimensions in a crystal. The amplitudes of the diffracted X-rays are thus proportional to the number of lattice elements in the crystal.

The major part of X-rays travels straight through the crystal. A smaller part is diffracted and seen as black reflections on photographic films. The intensity of these reflections is now measured photoelectrically.

In order to determine the crystal structure of *single crystals*, Max von Laue let "white" X-rays (i.e., mixture of various wavelengths) fall vertically on an NaCl single crystal. A photographic plate behind the crystal produced a diffraction pattern consisting of black spots whose position and intensity reveals the crystal structure. The **von Laue method** can be used if one of the crystal axes of a single crystal is perpendicular to the incident beam, which is the case for NaCl.

In most single crystals, crystal axes are *a priori* unknown and certainly not perpendicular to the incident beam. Today, single crystals are investigated mostly by the **Bragg rotating crystal method**. Here, 0.1-1 mm large single crystals are placed on a goniometer plate which is rotated until a strong reflection is observed at a certain **Bragg angle**. Further rotation produces other reflections at other Bragg angles.

Because the placement on the plate fixes one crystal axis, orientations of the other lattice planes are also fixed. The resulting reflections appear therefore as spots on photographic films and not as circles or arcs as in fiber diagrams (see below).

The rotating crystal method is used to determine the structure of single crystals of globular enzymes that are doped with heavy metal ions as contrast enhancers. Truly three-dimensional single crystals of synthetic polymers can be obtained in principle by polymerization of monomer single crystals. The only known examples are polymers of diacetylene. All other polymer "single crystals" are thin platelets (Section 7.2.2).

Most crystalline materials are *polycrystalline*, however. They consist of assemblies of crystallites that are more or less arranged at random so that all their net planes are also distributed at random. These materials are examined by the **Debye-Scherrer powder method** which originally employed crystal powders. For polymers, test bars are used.

In the powder method, a monochromatic primary beam finds sufficient numbers of lattice planes that satisfy the Bragg condition because the lattice planes of the various crystallites are arranged completely at random. The many small crystallites with their many different orientations produce a system of coaxial radiation cones with the com-

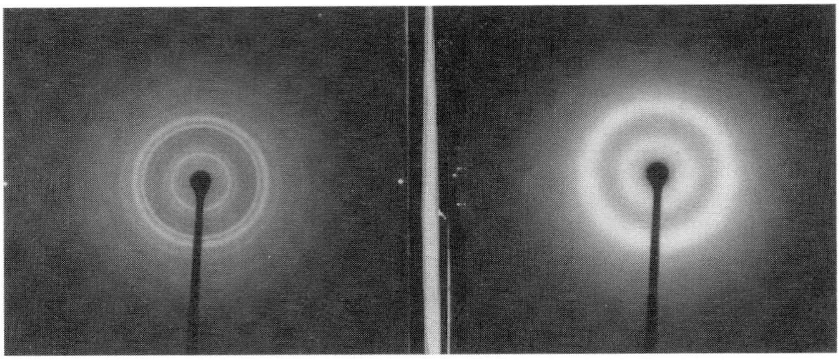

Fig. 7-3 X-ray diagrams of an unstretched, semicrystalline, isotactic poly(styrene) (left) and an amorphous atactic poly(styrene) (right).

mon tip in the center of the probe. A vertical cut through the cone delivers a series of concentric circles (planar photographic films) or sectors (concentric films).

Debye-Scherrer X-ray diagrams of semicrystalline polymers show relatively strong rings that are produced by crystalline reflections (Fig. 7-3, left). In addition, weaker rings and a strong background can also be seen. Such weak rings and strong backgrounds are also observed for amorphous polymers (Fig. 7-3, right). The weak rings are called **halos**; they are caused by short-range ordering phenomena (see Fig. 6-2).

Polymers always show a relatively strong background scattering which is caused mainly by the scattering of air, to some extent by thermal movements in crystallites, and also by **Compton scattering**, an inelastic X-ray scattering caused by free or weakly bound electrons that is independent of the physical state of the specimen.

In stretched *fibers* or *films*, molecule axes are oriented to a large degree. X-rays entering the specimen perpendicular to the stretch direction thus produce relatively sharp reflections which are arranged in layers (Fig. 7-4). The diffraction patterns resemble those obtained from rotating single crystals but, in contrast to single crystals, fibers need not be rotated since the molecule axes are already in the c direction. For historic reasons, such patterns are called **fiber diagrams**. They are also observed for stretched films.

Reflections on the zeroth layer plane, called **equatorial reflections**, correspond to net planes that are parallel to molecule axes. **Meridional reflections** result from net planes that are perpendicular to molecule axes on the plane that halves the equator.

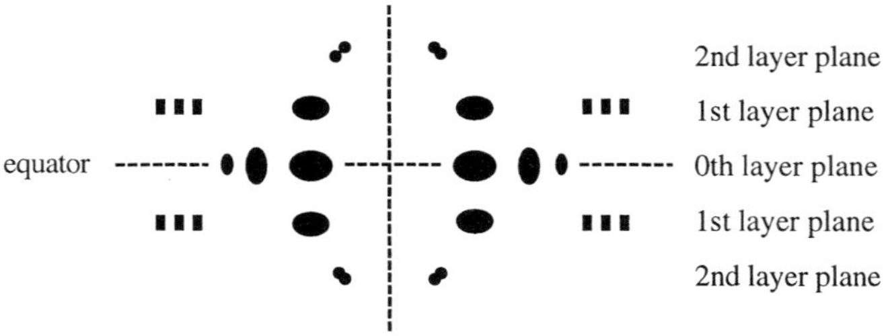

Fig. 7-4 Schematic representation of a fiber diagram of a polymer that crystallizes as a 3_1 helix showing three layers because the fourth, seventh, ... chain unit is in the same relative position as the first one, giving rise to three layer lines on each side of the equator. Such diagrams are found for stretched films of isotactic poly(propylene)s (see [2]). Spots degenerate to arcs if crystallites are insufficiently oriented. Unoriented crystallites generate circles.

7.1.5 Lattice Constants

Crystal structures of polymers are controlled by the macroconformation and packing of polymer molecules. According to the **equivalence principle**, monomeric units of linear chains assume geometrically equivalent positions in crystals. The chains themselves obey the principle of the smallest intramolecular change of conformation and adopt the macroconformation with the lowest energy that is compatible with the equivalence principle. Effects of chain packing are usually secondary.

In the thermodynamically preferred crystal modification I of poly(ethylene), $+CH_2-CH_2+_n$, chains are present in the macroconformation of all-trans (zigzag) chains. In the chain direction, each third, fifth, ... methylene group occupies the same relative position as the first one (Fig. 7-5). The C–C bond length of 0.154 nm and the C–C–C bond angle of 111.5° (larger than the tetrahedral angle of 109°28'!), lead to a lattice constant of c = 0.2546 nm (= crystallographic bond length = distance between C^1 and C^3 in $-C^1-C^2-C^3-$). In a series of subsequent unit cells, chains are parallel to each other. A short periodicity thus exists in the a-direction for the first, third, fifth, ... chain (a = 0.742 nm) and in b-direction for the first, second, third, ... chain (b = 0.495 nm). Since all lattice constants are different but all lattice angles are 90°, modification I of poly(ethylene) belongs to the orthorhombic crystal system (Table 7-1). Poly(ethylene) also has other crystal modifications (Table 7-4).

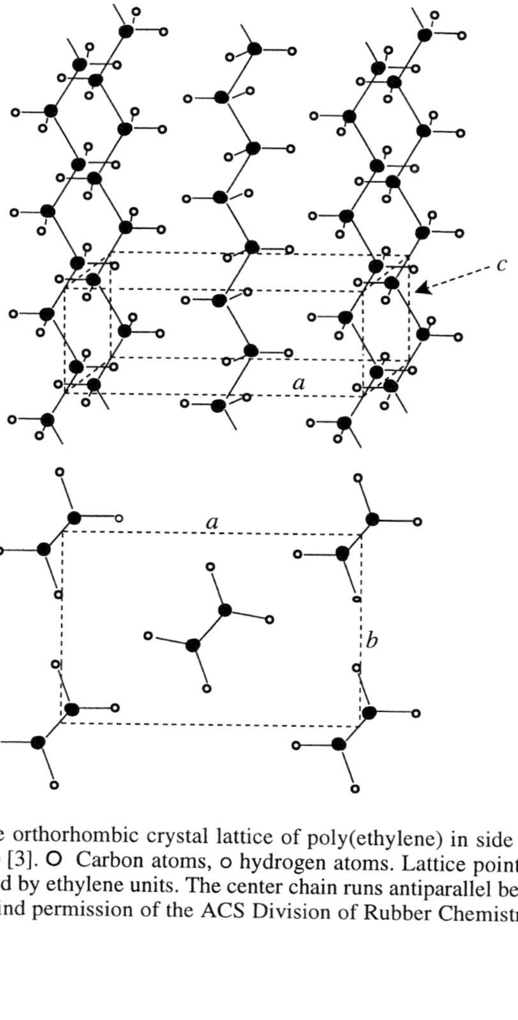

Fig. 7-5 Section of the orthorhombic crystal lattice of poly(ethylene) in side view (top) and cross-sectional view (bottom) [3]. O Carbon atoms, o hydrogen atoms. Lattice points at corners of square prism (- - -) are occupied by ethylene units. The center chain runs antiparallel because of chain folding (Section 7.2.2). With kind permission of the ACS Division of Rubber Chemistry, Akron (OH).

Table 7-4 Crystal structures and modifications (*α*, *β*, *γ*, I, II, III). N_u = number of monomeric units per unit cell; *a*, *b*, *c* = lattice constants; *α*, *β*, *γ* = unit cell angles; A_c = cross-sectional area of cell.
PA 6 = poly(*ε*-caprolactam), PA 66 = poly(hexamethylene adipamide), PB = poly(1-butene), PE = poly(ethylene), PEOX = poly(oxyethylene), PG = poly(glycine), PIB = poly(isobutylene), POM = poly(oxymethylene), PP = poly(propylene), PPOX = poly(propylene oxide), PS = poly(styrene), PTFE = poly(tetrafluoroethylene), PVC = poly(vinyl chloride).
it = isotactic, st = syndiotactic. * Fiber axis (other than *c*).

Polymers		N_u	$\frac{a}{nm}$	$\frac{b}{nm}$	$\frac{c}{nm}$	$\frac{\alpha}{°}$	$\frac{\beta}{°}$	$\frac{\gamma}{°}$	$\frac{A_c}{nm^2}$	Helix	System
PE	I	2	0.742	0.495	0.254	90	90	90	0.183	1_1	orthorhombic
	II	2	0.809	0.253*	0.479	90	90	107.9	0.186	1_1	monoclinic
PVC, st		4	1.040	0.530	0.510	90	90	90	0.272	2_1	orthorhombic
PTFE	I	15	0.566	0.566	1.950	90	90	119.3	0.276	15_7	hexagonal
	II	13	0.952	0.559	1.706	88	90	92	0.296	13_6	triclinic
PP, st	I	8	1.450	1.12	0.740	90	90	90		2_1	orthorhombic
it	I (α)	12	0.665	2.09	0.650	90	99.5	90	0.378	3_1	monoclinic
it	II (β)	18	1.908	1.101	0.649	90	90	90	0.351	3_1	orthorhombic
it	III (γ)	12	0.650	2.140	0.650	89	100	99	0.342	3_1	triclinic
PIB		16	0.688	1.191	1.860	90	90	90	0.434	8_5	orthorhombic
PB, it	I	18	1.770	1.770	0.651	90	90	90	0.452	3_1	trigonal
	II	44	1.485	1.485	2.060	90	90	90	0.554	11_3	tetragonal
	III	8	1.238	0.892	0.745	90	90	90	0.554	4_1	orthorhombic
PS, it		18	2.19	2.19	0.665	90	90	120	0.700	3_1	trigonal
POM	I	9	0.446	0.446	1.730	90	90	90	0.172	9_5	trigonal
PEOX	I	28	0.803	1.304	1.948	90	125.4	90	0.216	7_2	monoclinic
PPOX, it		4	1.052	0.468	0.710	90	90	90	0.245	2_1	orthorhombic
PG	I	2	0.477	0.477	0.70	90	90	66	0.185	2_1	monoclinic
	II	3	0.48	0.48	0.93	90	90	120	0.200	3_1	hexagonal
PA 6	α	8	0.960	1.718*	0.805	90	68.6	90	0.179	2_1	monoclinic
	β	1	0.48	0.48	0.86	90	90	120	0.202	1_1	hexagonal
PA 6.6	I	1	0.49	0.54	1.73	48	77	63	0.176	1_1	triclinic

Lattice constants *a* and *b* depend little on temperature and lattice constants *c* not at all because lengths of covalent chain bonds and valence angles between those bonds are practically temperature invariant. Interchain forces do change somewhat with temperature and so do lattice constants *a* and *b*. For example, *b* of poly(ethylene) increases by ca. 7 % between −196°C and +138°C.

Lattice constants often allow one to directly deduce the macroconformation of polymer chains in crystals. For example, the lattice constant *c* = 0.51 nm of syndiotactic poly(vinyl chloride), $\text{+}CH_2\text{–}CHCl\text{+}_n$, is double the value of *c* = 0.254 nm of poly(ethylene), $\text{+}CH_2\text{–}CH_2\text{+}_n$ (Table 7-4). Because of syndiotacticity, only every first, third, fifth, ... CHCl group is in the same relative configuration and so is every first, third, fifth, ... CH_2 group (i.e., every first, fifth, ninth ... chain atom). An st-poly(vinyl chloride) in the all-trans conformation therefore must have twice the lattice constant of poly(ethylene). The *c* = 1.860 nm of poly(isobutylene), on the other hand, is not a simple multiple of the *c* = 0.254 nm of poly(ethylene). Hence, poly(isobutylene) does not crystallize in an all-trans conformation; it rather forms an 8_5 helix.

Microconformations are controlled by the space requirements of substituents *near* the polymer chain. However, the packing of atomic groups *within* the unit cell is controlled by the total size of the substituent. This packing is described by the cross-sectional area,

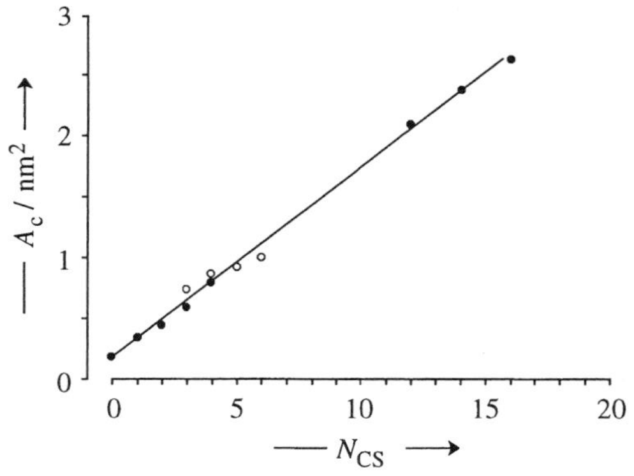

Fig. 7-6 Cross-sectional areas A_c of isotactic poly(1-olefin)s, $\text{+CH}_2\text{–CHR+}_n$ as a function of the number N_{CS} of carbon atoms in side-chains with linear (○) or branched (●) substituents R.

$A_c = V/(N_c c)$, where V = volume of unit cell, N_c = number of chains per unit cell, and c = lattice constant in the chain direction.

For example, the cross-sectional area of it-poly(olefin)s (and thus a and b) increases linearly with the number of *all* carbon atoms in side chains whether they are branched or not (Fig. 7-6). Simple zigzag chains have similar cross-sectional areas: poly(ethylene) 0.183 nm^2, polyamide 6 0.179 nm^2, and 1,4-*trans*-poly(butadiene) 0.179 nm^2. Cross-sectional areas of cis polymers are greater than those of trans polymers, for example, 0.207 nm^2 (1,4-*cis*-poly(butadiene)) and 0.280 nm^2 (1,4-*cis*-poly(isoprene)).

The packing of chains in crystals affects the crystalline density of polymers and thus melting temperatures, melting enthalpies, and melting entropies. For example, the three crystal modifications of poly(1-butene) have densities ρ/(g cm^{-3}) of 0.95 (I), 0.91 (II), and 0.90 (III) and melting temperatures T_M/°C of 142 (I), 130 (II), and 108 (III). Poly(oxymethylene) with ρ/(g cm^{-3}) ≈ 1.51 has a melting temperature of 184°C but poly(oxyethylene) with ρ/(g cm^{-3}) ≈ 1.23 only one of 69°C (Chapter 13).

7.1.6 Lattice Structures

Lattice constants are controlled by the macroconformation (and thus by constitution and configuration) which, in turn, also determines the packing of molecules.

Zigzag chains (all-trans macroconformations) are usually only adopted by unsubstituted, apolar chains such as poly(ethylene) and some syndiotactic vinyl polymers with not so large, polar substituents such as $\text{+CH}_2\text{–CHR+}_n$ with R = Cl, CN, and CH=CH$_2$.

Heteroatoms have unbound electron pairs. In chains, heteroatoms thus decrease interactions between electron clouds of bonds between chain atoms which leads in polar environments to the **gauche effect** (Volume I, p. 144). Examples of polymer chains with gauche effects are

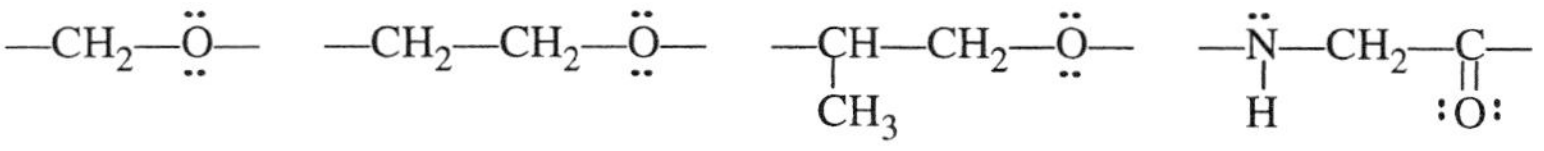

POM PEOX PPOX PG

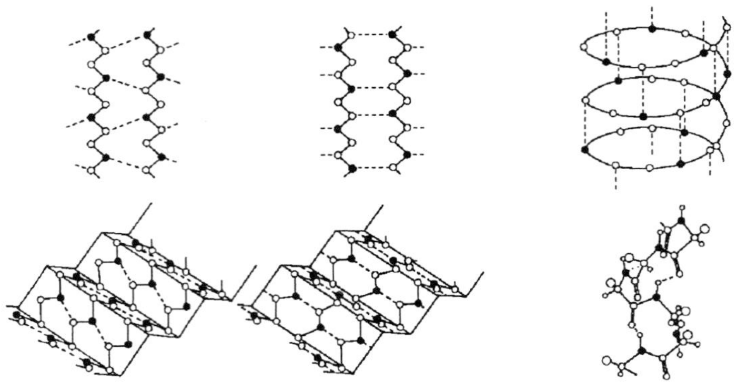

Fig. 7-7 Secondary structures of poly(α-amino acid)s, polypeptides, and proteins. O Methyl groups in helices, o carbon atoms in helices and pleated sheets, o= carbonyl oxygen, o hydrogen atoms in helices, ● nitrogen atoms, – – – hydrogen bridges. Hydrogen atoms in sheets are not shown.

Because of the gauche effect, modification I of poly(oxymethylene) (POM) crystallizes in all-gauche macroconformation $(G)_n$ and modification I of poly(oxyethylene) (PEOX) in macroconformation $(TTG)_n$. These macroconformations lead to a 9_5 helix of POM and a 7_2 helix of PEOX (see Fig. 13-14). However, repulsion between neighboring methyl groups suppresses the gauche effect in isotactic poly(propylene oxide) (PPOX). The bond orientation decreases and PPOX crystallizes in the T_n macroconformation.

Modification I (β structure) of poly(glycine) (PG), $\text{+NH–CH}_2\text{–CO+}_n$, forms zigzag chains but these chains do not reside in a plane. Because of intermolecular hydrogen bridges, they rather form pleated sheets (Fig. 7-7). Such pleated sheets are also present in polyamides 6 and 6.6 (Fig. 7-8). However, modification II of poly(glycine) forms 3_1 helices which are deformed because of intramolecular hydrogen bridges.

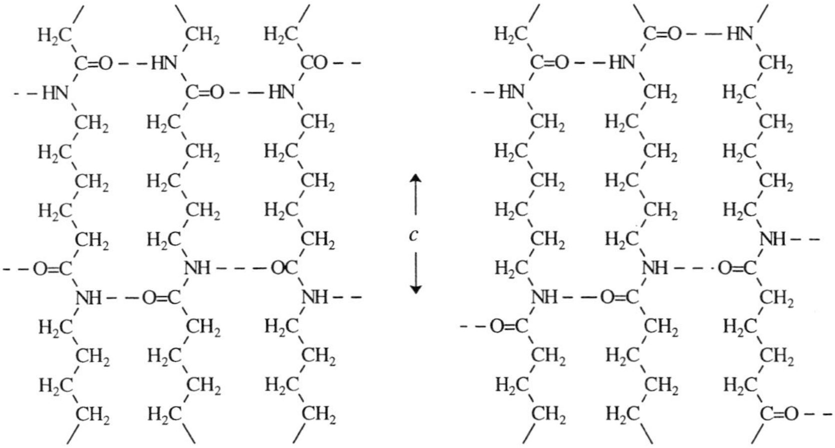

Fig. 7-8 Pleated sheet structures of polyamide 6 (left) and polyamide 6.6 (right). Isoclinal structures have the same distance between hydrogen bonds in the c direction, anticlinal structures do not.

7.1.7 Polymorphism

In crystallography, polymorphism is defined as the ability of identical molecules to assume various crystal modifications that are distinguished by different lattice constants and/or angles and thus different unit cells.

Polymorphism is generated by either different macroconformations of molecules or by different packing of molecules with the same macroconformation. Both differences are caused by slight changes in crystallization conditions such as temperature, pressure, shearing, nucleating agents, etc.

Polymorphism is relatively frequent for chain molecules where it always occurs in the presence of isoenergetic states. For example, the thermodynamically stable crystal modification I of poly(ethylene) is orthorhombic but stretching leads to the monoclinic modification II (Table 7-4).

Isotactic poly(propylene) forms three different crystal modifications (Table 7-5). Most prominent is the monoclinic modification I with equal proportions of left-handed and right-handed 3_1 helices (see also p. 48). Depending on the literature, modification II is either hexagonal, trigonal, or orthorhombic. It contains helices with only one single screw sense, either left-handed or right-handed. In both modifications, chains thus have the same macroconformation, and polymorphism is caused by differences in the packing of chains. Modification III is triclinic; it also consists of 3_1 helices.

The three modifications of isotactic poly(1-butene), on the other hand, contain different helix types. Hence, this polymorphism is caused by different macroconformations (Table 7-5) which leads to larger differences in melting temperatures. However, melting enthalpies of the various modifications do not differ much, which points to larger differences in melting entropies.

Table 7-5 Crystal modifications, crystal structures, helix types, densities ρ, melting temperatures T_M, and melting enthalpies ΔH_M of some polymers according to measurements by many authors. The widely varying melting temperatures are caused by differences in tacticities and/or molecular weights.

Modification		Crystal structure	Helix type	$\frac{\rho}{\text{g cm}^{-3}}$	$\frac{T_M}{°\text{C}}$	$\frac{\Delta H_M}{\text{kJ mol}^{-1}}$
it-Poly(propylene)						
I	(α)	monoclinic	3_1	0.931 - 0.949	165 - 221	5.8 - 11.0
II	(β)	hexagonal	3_1	0.923 - 0.940	147 - 183	4.2
		trigonal	3_1	0.940 - 0.950	170 - 200	4.0 - 8.2
		orthorhombic	3_1	0.922	192	
III	(γ)	triclinic	3_1	0.946		
it-Poly(1-butene)						
I		trigonal	3_1	0.950	126 - 139	6.9 - 13.9
		hexagonal	3_1	0.96	132	6.1
II		tetragonal	11_3	0.886 - 0.902	120 - 130	4.1 - 8.3
III		orthorhombic	4_1	0.897 - 0.907	106 - 110	6.5
st-Poly(1-butene)						
I		orthorhombic	2_1	0.937 - 0.946	300 - 306	17.3
II		monoclinic	5_3	0.964		

7.1.8 Isomorphism

Isomorphism is observed if two different types of monomeric units can replace each other in a crystal lattice. Two chains with the same chirality and conformation are always isomorphous, for example, two helices of the same polymer with the same sequence of conformations such as ...$TG^{+}TG^{+}TG^{+}$...

Two isomorphous chains are furthermore **isoclinal** (G: *klinein* = to lean) if bond vectors have the same positive or negative orientation in each chain (same orientation of substituents) (Fig. 7-9). In two **anticlinal** chains, bond vectors have the opposite orientation (opposite orientation of substituents) (Fig. 7-9, see also Fig. 7-8). Isoclinal and anticlinal pairs of chains are always isomorphous.

Two chains with equal macroconformations but opposite chiralities are enantiomorphous (Fig. 7-9). An example is isotactic polymer chains with macroconformations ...$TG^{-}TG^{-}TG^{-}$... (right-handed helix) and $G^{+}TG^{+}TG^{+}T$... (left-handed helix) Enantiomorphous chains may be either isoclinal or anticlinal.

Isomorphism is possible in bipolymers if the corresponding homopolymers possess analogous crystal modifications, similar lattice constants, and the same helix type. For example, both the γ modification of it-poly(propylene) and the modification I of it-poly(1-butene) have the same helix type (3_1), the same value of c (0.650 versus 0.651 nm), and the same lattice angle α (Table 7-5). 1-Butene units may thus replace propylene units in crystal lattices.

Isomorphism is especially found in many helix-forming macromolecules since helix structures lead to "channels" in crystal lattices in which different substituents fit right in. The same phenomenon is also called **allomerism**. It is for this reason that one company sells its crystalline copolymers from two or more olefinic monomers as "allomers" .

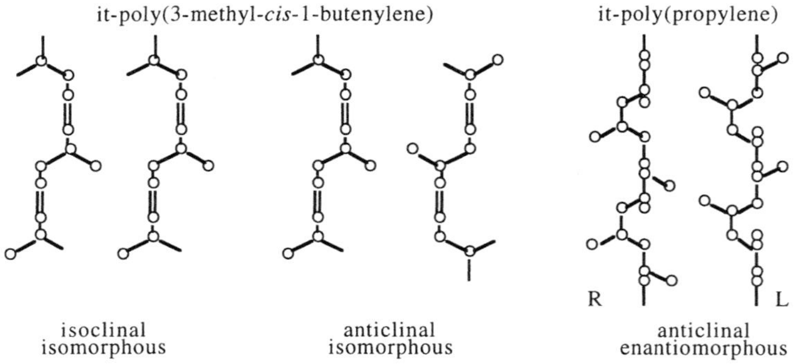

Fig. 7-9 Relative arrangements of chains in crystal lattices (see text).

7.1.9 Unit Cells

Unit cells are relatively small (Table 7-4). Reflections from unit cells are therefore observed at relatively large angles according to Bragg's law. Because of the small dimensions of unit cells, these reflections are therefore called **short periodicities**. Crystalline long-chain macromolecules also show in addition **long periodicities** at smaller angles.

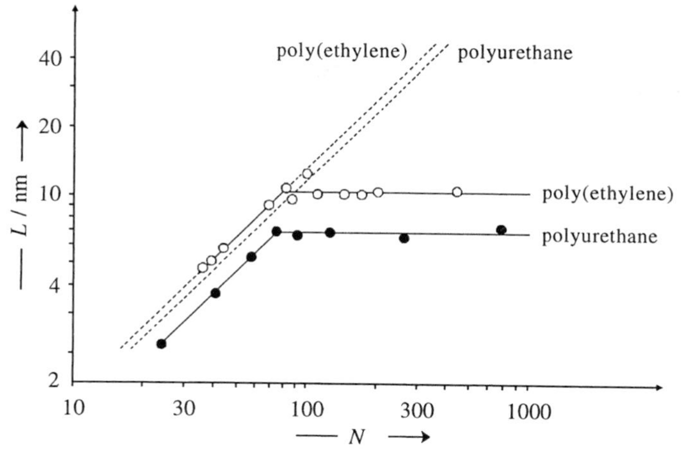

Fig. 7-10 Double-logarithmic plot of long periodicities L versus the number N of chain units of polyurethanes, $HOCH_2CH_2OCH_2CH_2O[CONH(CH_2)_6NHCOOCH_2CH_2OCH_2CH_2O]_nH$ and alkanes [4]. - - - - Calculated for all-trans conformations vertical to the base plane.

For alkanes, $H(CH_2)_NH$, with $N \leq 70$ chain units, long periodicities in the c direction (chain axis) equal the conventional contour lengths L of molecules as calculated from bond lengths and bond angles, assuming trans conformations (Fig. 7-10). Long periodicities here comprise the complete molecule.

Above $N \approx 80$, long periodicities become constant and independent of chain length and degrees of polymerization. Since conventional contour lengths continue to increase, chains must fold back (Section 7.2.2).

The same phenomenon is observed for polyurethanes. At small N, the long periodicity is smaller than the conventional contour length because the molecule axis is not vertical to the base plane but at a certain angle.

7.2 Structure of Crystallites

7.2.1 Fringed Micelles

In the early days of macromolecular science, X-ray diffractograms of gelatin showed both crystalline reflections and amorphous halos, which was interpreted as the coexistence of perfect crystalline and completely amorphous regions (**2-phase model**). Both the position of reflections by small-angle X-ray scattering and line-broadening of reflections by X-ray diffraction pointed to crystallite sizes of 10-80 nm.

These crystallite sizes were smaller than conventional contour lengths that were calculated from molar masses. For poly(oxymethylene)s, it was also observed that with increasing molar mass short periodicities from unit cells remained but long periodicities disappeared. This effect was assumed to be caused by the absence of higher order, i.e., absence of very regular lattices. The alternative, regular lattices with lattice defects, was ruled out because these polymers looked neither macrocrystalline nor microcrystalline.

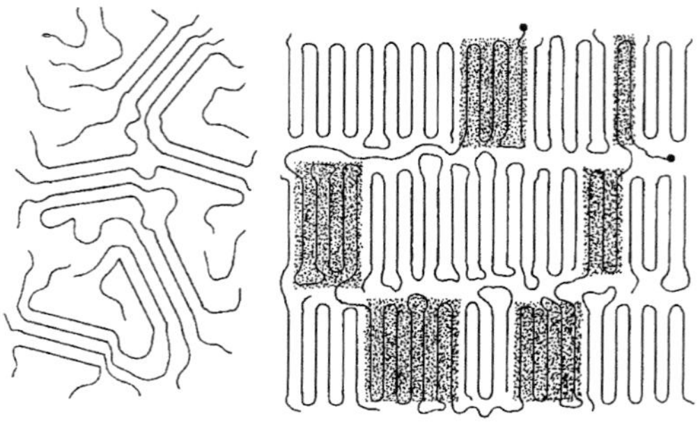

Fig. 7-11 Structure of semicrystalline polymers in which a polymer chain runs through several crystalline regions. Left: historic representation of fringed micelles. Right: folded micelle consisting of regions of folded polymer segments that are interconnected by short "amorphous" segments. A single polymer chain may consist of several "blobs" (grey) (see discussion on p. 209).

These observations led to the so-called **fringed micelle model** in which a polymer chain resides in several crystalline regions that are connected and separated by amorphous ones (Fig. 7-11, left). The model was able to explain not only X-ray diffractograms but also (a) macroscopic densities of polymers that are smaller than those calculated from unit cells because of the existence of amorphous regions, (b) arcs in diffractograms as a consequence of orientation of crystallites, (c) broad melting ranges of polymers as a result of different crystallite sizes, (d) optical birefringence of stretched polymers as an indication of oriented molecule chains in amorphous regions, and (e) heterogeneity of **semicrystalline polymers** in chemical reactions and physical processes as a consequence of their better accessibility in the amorphous "phase."

7.2.2 Polymer Single Crystals

Electron diffractograms of crystallized gutta-percha indicated as early as 1938 that polymer chains do not run through several fringed micelles (Fig. 7-11, left) but are rather backfolded (Fig. 7-11, right). This phenomenon was rediscovered 19 years later and got much more attention after three researchers (see p. 235) found independently of each other that very thin rhombic platelets crystallized from very dilute solutions of the newly available linear poly(ethylene) (Fig. 7-12). The platelets were often several micrometers long but only 7-20 nm thick. Later, similar platelets were also found to form from dilute solutions of many other polymers such as stereoregular poly(1-olefin)s, polyamide 6, poly(oxymethylene), cellulose derivatives, and amylose.

Since the thickness of the platelets is smaller than the conventional contour length of large polymer molecules, chains must refold in the crystal as shown schematically in Fig. 7-11. Electron diffraction of the platelets showed sharp point-like reflections which indicated **polymer single crystals**. The following decades saw a "lively discussion" (to put it mildly) about the nature of the folds (sharp or not, immediate reentry or not, etc.) and internal platelet structure. Similar structures were also obtained from crystallization of melts albeit not as thin platelets but as stacked **lamellae**. In analogy to "fringed micelles", these chain structures were therefore also called **fold(ed) micelles**.

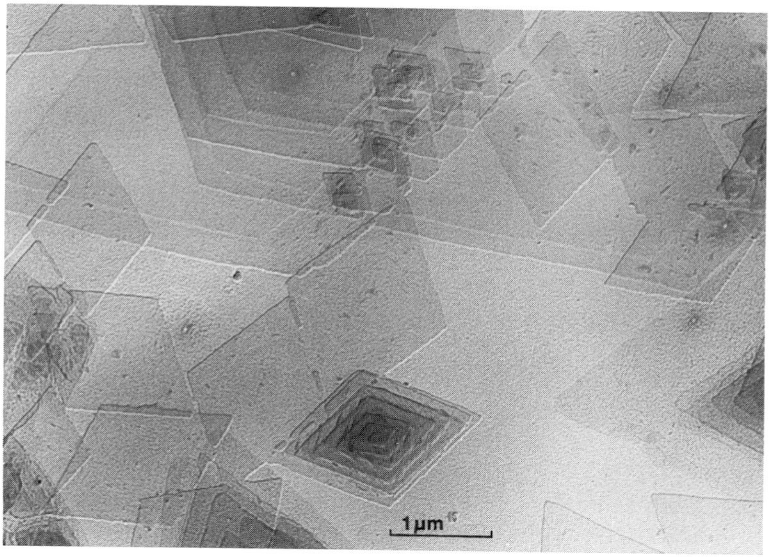

Fig. 7-12 Electron micrograph of poly(ethylene) single crystals [5]. The platelet below center shows a screw dislocation.

Details about the physical structure of single crystals and lamellae emerged slowly, partly, because some physical methods were either not available or not sophisticated enough early on and partly, because the investigated polymers were not comparable, for example, with respect to molar mass distributions (narrow versus broad), which led to different models that all claimed to be general. What follows is therefore not a historical narrative but a discussion of the features that emerged.

In the ideal case, single crystals and lamellae consist both of layers of parallel chain segments (**stems**) that are normal (or nearly so) to the base plane and and a top surface layer consisting of more or less sharp chain **folds** (Fig. 7-11). Electron microscopy showed the height of monolayers of large molecules to be much smaller than their chain length. Since chain axes of poly(ethylene) molecules are not parallel to platelet surfaces but rather parallel to side faces of platelets, chains must fold. This view of the orientation of stems is supported by electron micrographs of fissures in single crystals (Fig. 7-13).

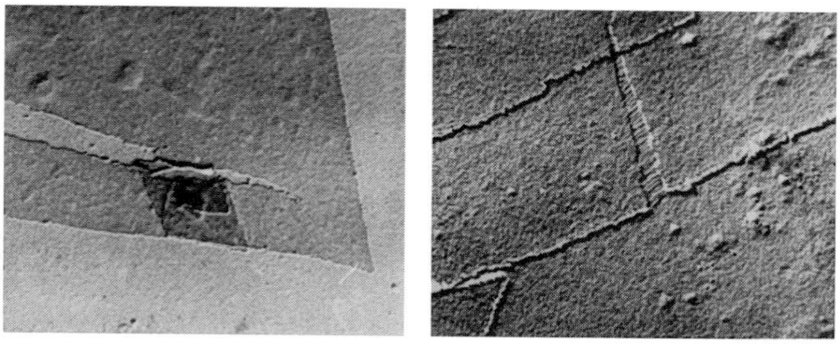

Fig. 7-13 Fissures in poly(ethylene) single crystals indicate direction of chain segments [6].
Left: A split practically parallel to the edge of the single crystal stops at the diagonal of the platelet because the vertical chain planes change direction. Right: A split practically perpendicular to the side of the platelet pulls out fibrils consisting of bundles of polymer segments.

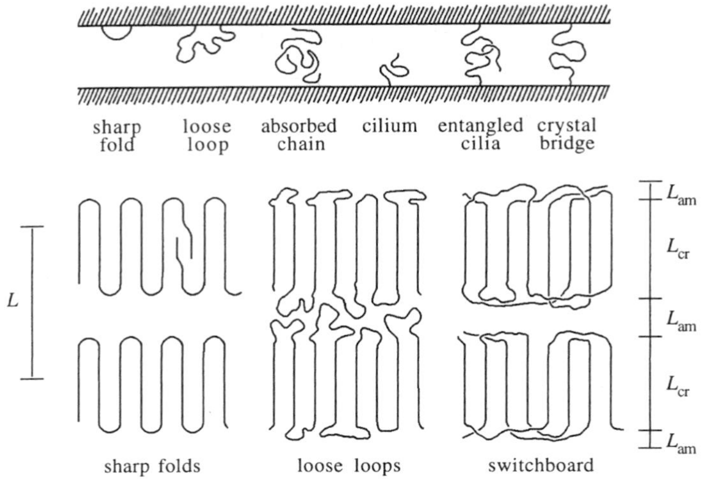

Fig. 7-14 Schematic representation of features of single crystals (monolayers) and stacks of lamellae (multilayers). Top: chain segments and chains between two sharp lamella surfaces. Bottom: Models for monolayers and multilayers shown as two lamellae with parallel stems with "amorphous" surface layers (if monolayer of stems) or interlayers (if multilayers of stacked lamellae).

Bottom left: sharp folds, adjacent reentries of stems, dislocation of chain ends.
Bottom center: loose chain folds with adjacent reentry of stems.
Bottom right: switchboard with remote reentry of stems.

L = Length of crystallographic long period. L_{am} = thickness of non-crystalline ("amorphous") top layers (monolayer of stems) or interlayers (stacks on monolayers), L_{cr} = height of lamella.

7.2.3 Structure of Folds

Single crystals consist of monolayers of vertically (or nearly so) arranged stems that contain some endgroups and dislocations but are nearly 100 % X-ray crystalline (Fig. 7-14). The monolayers are topped on both sides by non-crystalline surface layers. Multilayers consist of stacked monolayers (**lamellae**) that are separated by non-crystalline interlayers. Surface and interlayers may consist of sharp chain folds, loose loops connecting two stems, absorbed polymer chains that are not part of the lamella, and single or entangled **cilia** (chain ends). Also, two lamellae may be connected by crystal bridges.

Surface layers of monolamellae are different from interlayers of multilamellae. In solution-grown *single crystals* of the straight-chain alkane $H(CH_2)_{198}H$ (M = 2779.3 g/mol), chain folds are completely regular and sharp, which corresponds to a sequence ~$T_nGGTGGT_n$~ of trans and gauche microconformations with 6 methylene groups per fold. Such a loop is also seen in a lamella of a high-molar mass poly(ethylene) chain (Fig. 7-15). The loop has a conformational energy of 16.5 kJ/mol which agrees nicely with the conformational energy that is calculated for 6-7 microconformations with conformational energies of $\Delta E_{T\text{-}G} \approx 2.5$ kJ/(mol bond) each (p. 43 ff.). The height of the loop is therefore less than 1 nm.

The loops of poly(hexamethylene adipamide), $\text{+}NH(CH_2)_6NHCO(CH_2)_4CO\text{+}_n$, also consist of methylene groups. The loops of polyamide 4, $\text{+}NH(CH_2)_3CO\text{+}_n$, and polyamide 4.6, $\text{+}NH(CH_2)_4NHCO(CH_2)_4CO\text{+}_n$, contain –NH–CO– groups, however.

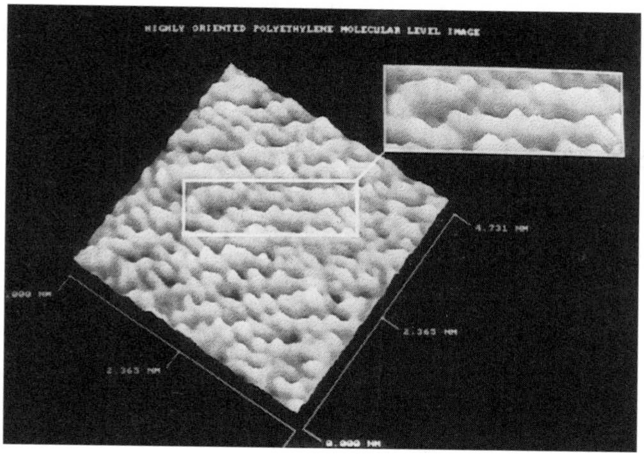

Fig. 7-15 Atomic force microscopy shows a sharp, hairpin-like loop in a folded chain of a highly oriented poly(ethylene) [7]. The loop contains 6 methylene groups that show up as "pearls."
By permission of the American Chemical Society, Washington (DC).

In chain-folded lamellae, polymer chains leave the lamella and reenter it somewhere so that the immediate neighbors of a stem are antiparallel. For the reentrance, several models have been proposed (Fig. 7-14): sharp folds with adjacent reentry, loose loops with adjacent reentry, and the switchboard model with remote reentry. In general, low-molar mass polymers with narrow molar mass distributions crystallize with sharp folds from dilute solution whereas high-molar mass polymers with broad molar mass distributions tend to form remote entries, i.e., switchboards. Even melt-crystallized low-molar mass poly(ethylene) is 94 % crystalline (Fig. 7-16). Poly(ethylene)s with higher molar masses show not only a strong increase of disorder in interlayers between lamellae but also increasing contents of amorphous, rubber-like regions.

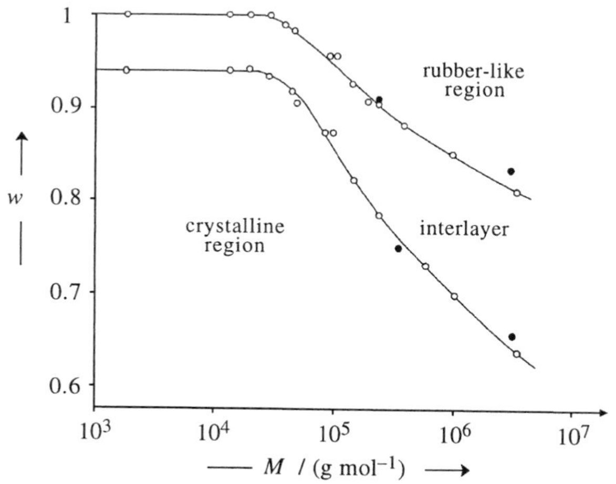

Fig. 7-16 Cumulative mass fractions of crystalline regions (stems), interlayers, and amorphous (rubber-like) regions by broad-line nuclear magnetic resonance of ^{1}H (o) and ^{13}C (●) as a function of the molar mass of melt-crystallized poly(ethylene)s [8].

7.2.4 Structure of Lamellae

Model calculations using random-flight statistics have also been used to gain insight into the problem of reentry of chain segments in lamellae and sharp folds *versus* loose loops. For the switchboard model, these calculations delivered much higher densities of "amorphous" surface layers than those of lamellae. This anomaly can be avoided if one assumes that at least 2/3 of the chains fold back sharply. The model predicts that mean-square radii of gyration of crystalline polymers, $\langle s^2\rangle_{cr}$, can be higher than unperturbed mean-square radii of gyration in melts, $\langle s^2\rangle_o$, if the fraction f_{cr} of the crystalline regions exceeds ca. 57 %:

(7-2) $$\langle s^2\rangle_{cr} = (1/3)(2 f_{am} + f_{am}^{-1})\langle s^2\rangle_o \quad ; \quad f_{cr} + f_{am} \equiv 1$$

In the range $0.5 < f_{am} < 1$, mean-square radii of gyration of semi-crystalline polymers are predicted to be identical with those in melts within ± 5 %.

Indeed, $\langle s^2\rangle_{cr} > \langle s^2\rangle_o$ has been found experimentally for rapidly crystallizing low-molar mass poly(ethylene)s (M < 25 000 g/mol) if melts were quenched (Fig. 7-17). For the much more slowly crystallizing higher molecular weights, quenching led to the same unperturbed radii of gyration as in melts. Radii of gyration were much higher than the unperturbed ones if the poly(ethylene)s were crystallized under pressure. These polymers had large fold lengths that were independent of molar mass albeit smaller than their conventional contour lengths.

These experimental observations are explained as follows. Quenching of poly(ethylene) melts does not suppress crystallization completely. Instead, small lamellae are formed randomly but backfolding is difficult for high-molar mass molecules because the melt is very viscous and solidifies rapidly. Individual molecules thus run through several lamellae, each of which consists of bundles of stems. The resulting structure resembles that of fringed micelles in an amorphous matrix (Fig. 7-11 left) except that the "micelles" are now randomly oriented "folded micelles" (Fig. 7-11, right).

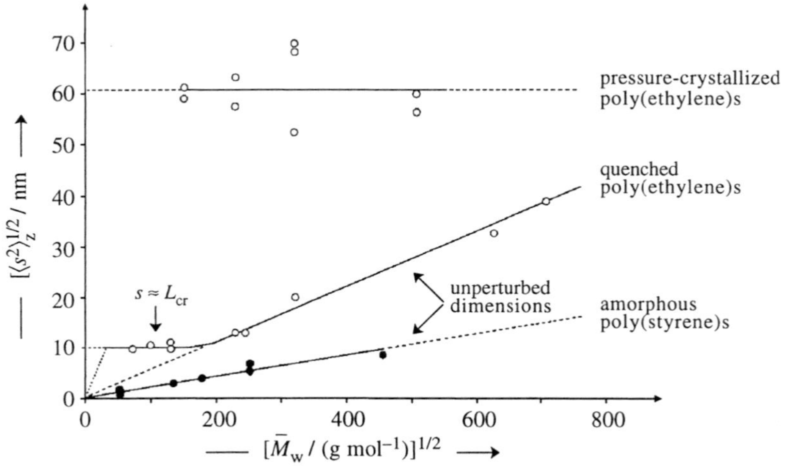

Fig. 7-17 Radii of gyration of pressure-crystallized or quenched poly(ethylene)s as a function of the square-root of molar mass. Atactic poly(styrene)s for comparison. Measurements by SANS [9].

These folded micelles can thus be modeled as Kuhnian ersatz coils (p. 97) in which stiff blobs (here the lamella sections) are interconnected by flexible segments (Fig. 7-11). Similarly to a random coil consisting of equal flexible segments, the ersatz coil adopts unperturbed dimensions, which causes the radius of gyration to be a function of the square root of the molar mass.

Low-molar mass polymers have lower melt viscosities, which leads to greater mobilities of chain sections and thus to greater chances for backfolding. The length of the stems and thus the thickness of the lamellae increases with undercooling (see Fig. 7-23). Quenching thus leads to fairly thick lamellae and, at low molar masses, to relatively few folds per molecule. Heights of lamellae produced from low-molar mass polymers are therefore much greater than the radii of gyration of Kuhnian ersatz coils, and the molecule behaves as a rod with a mean-square radius of gyration of $\langle s^2 \rangle \approx L_{cr}{}^2/12$, Eq.(4-5), where L_{cr} = lamella height (see Fig. 7-14). Since lamella heights are independent of molar masses, radii of gyration will be constant and equal to lamella heights (horizontal section with $s \approx L_{cr}$ for quenched poly(ethylene) in Fig. 7-17.

It is therefore highly probable that one molecule becomes part of several lamellae if the crystallization proceeds in melts or concentrated solutions. Such **interlamellar connections** or **crystal bridges** were first shown to exist in the material from the joint crystallization of the mixture of a poly(ethylene) and paraffin (Fig. 7-18). The proportion of crystal bridges increases with increasing molar mass since it is then much more probable that chain sections are incorporated in different lamellae before the chain has a chance to fold. Melt-crystallized polymers thus have relatively high amorphous contents, especially at high molar masses (Fig. 7-16).

Crystal bridges and parts of amorphous surface layers of polymer crystallites can be removed by grinding. Mechanical forces break chemical bonds and the resulting powders are therefore of fairly low molar mass. However, they are highly crystalline and are therefore called **microcrystalline polymers**. So-called microcrystalline celluloses are not obtained by mechanical grinding but by hydrolytic degradation (Volume II, p. 389).

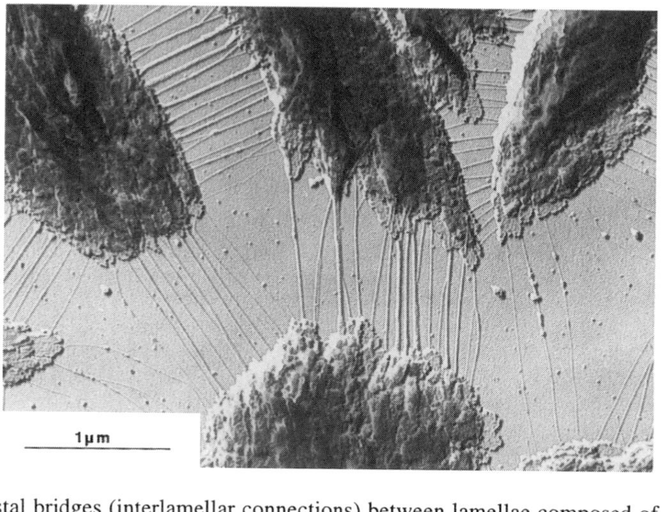

Fig. 7-18 Crystal bridges (interlamellar connections) between lamellae composed of poly(ethylene) molecules [10]. The starting material was a crystallized mixture of poly(ethylene) and paraffin from which the paraffin was extracted.

7.3 Crystallization

The tendency of polymers to crystallize in various morphologies is controlled by the macroconformation of polymer molecules or their assemblies. Polymeric spheres such as enzymes or latices form superlattices. Rigid molecules unite to rods that pack laterally. Flexible molecules chain fold to lamellae which then assemble to various superstructures, depending of crystallization conditions (Section 7.4).

Crystallization requires the presence of homogeneous or heterogeneous crystallization nuclei that act as seeds. Crystallization rates are determined by both the rates of formation of nuclei and crystallite growth which in turn depend on the degree of supercooling. The temperature for the best crystal growth is always higher than the temperature for the fastest formation of heterogeneous nuclei. Of course, crystallization rates also depend on the viscosity of the system: crystallizations from dilute solutions or melts differ considerably. Polymer constitutions, configurations, and conformations influence crystallization rates markedly, by factors up to ca. 500 000. Last but not least: rates are also affected by molar masses and the types and widths of molar mass distributions.

7.3.1 Formation of Nuclei

Homogeneous Nucleation

Crystallizations are initiated by homogeneous or heterogeneous crystallization nuclei. In **homogeneous (thermal) nucleation**, thermal movements cause molecules and molecule segments to cluster *spontaneously* to loose, unstable pre-nuclei that add further molecules and molecule segments, respectively, and become stable crystal **nuclei**.

The homogeneous formation of nuclei is *primary* (three-dimensional) since surfaces are enlarged in all three spatial directions (Fig. 7-19), whereas heterogeneous nucleation may be secondary or tertiary. Homogeneous nucleation is also *sporadic* since nuclei are formed in timely succession.

Primary nuclei become stable after they reach a certain size. Their forerunners, the loose aggregates of molecules and molecule segments, are constantly formed and destroyed. Gibbs energies ΔG_i of formation of aggregatess consisting of i lattice units are given by the difference between Gibbs surface energies, ΔG_S, and Gibbs crystallization energies, ΔG_cryst:

(7-3) $$\Delta G_i = \Delta G_\text{S} - \Delta G_\text{cryst}$$

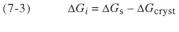

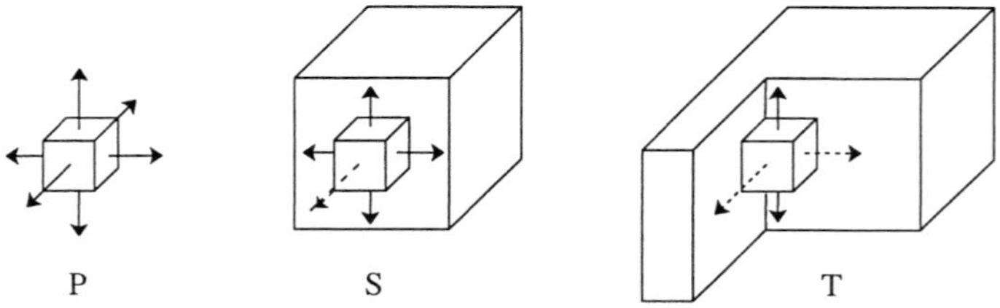

Fig. 7-19 Primary (P), secondary (S), and tertiary (T) formation of nuclei. Surfaces of nuclei are enlarged in → directions but not in - -› directions.

For a spherical pre-nucleus of radius R with a Gibbs surface energy of $\Delta G_{s,A}$ per area and a Gibbs crystallization energy of $\Delta G_{cryst,V}$ per volume, the Gibbs energy is

$$(7\text{-}4) \qquad \Delta G_{i,R} = 4\,\pi R^2 \Delta G_{s,A} - (4\,\pi/3)\,R^3 \Delta G_{cryst,V}$$

and thus with $R^3 \sim N_{seg}$ for any shape consisting of N_{seg} segments of molecules:

$$(7\text{-}5) \qquad \Delta G_{i,seg} = K' N_{seg}^{2/3} \Delta G_{s,A} - K'' N_{seg} \Delta G_{cryst,V}$$

Surface energies and crystallization energies have opposite signs so that $\Delta G_{i,seg}$ becomes negative only after a critical size $(R^3)_{crit}$ or $N_{seg,crit}$ is surpassed (Fig. 7-20). Above these critical sizes, unstable pre-nuclei become stable nuclei that grow further, for example, to lamellae (p. 205) or spherulites (p. 219 ff.).

Homogeneous nucleation is very rare. For polymers, it has been observed only for strongly supercooled melts of poly(pivalolactone), $\text{+O–CH}_2\text{–C(CH}_3)_2\text{–CO+}_n$, and poly(chlorotrifluoroethylene), $\text{+CClF–CF}_2\text{+}_n$. However, homogeneous nucleation may happen during polymerization, albeit only by oligomer molecules since monomer molecules are too small to form pre-nuclei. Critical sizes of nuclei are 2-10 nm.

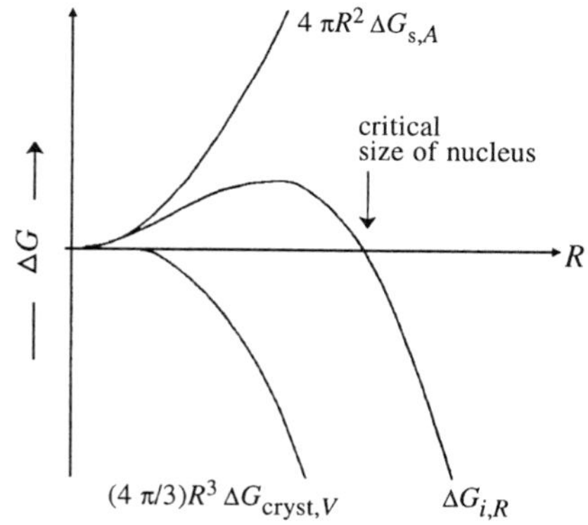

Fig. 7-20 Gibbs surface energy $\Delta G_{\sigma,A}$, Gibbs crystallization energy $\Delta G_{cryst,V}$, and Gibbs energy $\Delta G_{i,R}$ for the formation of spherical nuclei with radii R.

Heterogeneous Nucleation

Heterogeneous nucleation is caused by external surfaces such as dust, container surfaces, and added nucleation agents. These nuclei are already present at the start of nucleation. Heterogeneous nucleation is therefore **simultaneous** and **athermal** and either secondary or tertiary (Fig. 7-19). Concentrations of heterogeneous nuclei vary between $N = 1$ nucleus/cm^3 and $N = 10^{12}$ nuclei/cm^3.

Heterogeneous nuclei may also be present in the crystallizing polymer itself. On melting of semicrystalline polymers with broad melting ranges, not all crystallites may be dissolved at temperatures above the conventional melting temperature. Some of the higher melting crystallites survive and become athermal crystallization nuclei that simultaneously initiate crystallization upon subsequent cooling of the melt (Fig. 7-21).

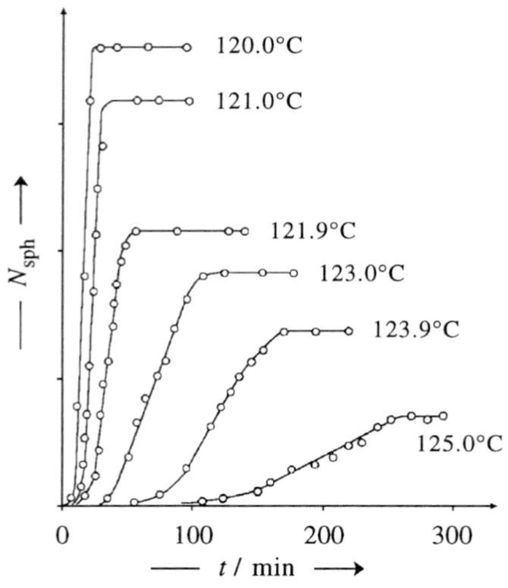

Fig. 7-21 Time dependence of the number N_{sph} (in arbitray units) of generated spherulites during the crystallization of molten poly(decamethylene terephthalate) [11]. At low temperatures, nuclei are formed simultaneously (spontaneously), at higher temperatures, successively (sporadic). By permission of Elsevier Science, Oxford.

Left-over nuclei are responsible for the so-called **memory of melts** which is the phenomenon that spherulites (= spherical superstructures, Section 7.4) reappear at the same spot after the specimen has been molten and recrystallized. They appear at the same spot because nuclei remain intact and cannot diffuse away because of high melt viscosity.

Nucleation Agents

Nuclei are often present in very small concentrations which leads to slow crystallization. Final physical properties are obtained only after long times which, for example, necessitates long cycle times in injection molding of thermoplastics. Small concentrations of nuclei also lead to large spherulites that in turn worsen mechanical properties.

Crystallizations are therefore often controlled by adding nucleation agents. In principle, *internal* homogeneous nucleations can be obtained by adding nuclei of the same polymer. In the droplet method, a crystalline polymer is ground to very fine particles. Addition of high concentrations of these seed particles overrides the effects of external nuclei and guarantees simultaneous nucleations. For industrial purposes, this method is impractical because thermoplastic materials are tough and not brittle which makes grinding to very fine particles very energy consuming.

Another method for internal homogeneous nucleations employs the tendency of chain molecules to chainfold. Bundles of stiff macromolecules are good nucleation agents but their chains do not fold well. Incorporation of flexible segments in such stiff macromolecules produces chain-folded lamellae in which the flexible segments reside preferentially in folds, which eases the formation of nuclei.

Industry uses *external* nucleation agents. Examples for poly(olefin)s are alkali, alkaline-earth, aluminum, and titanium salts of organic carboxylic, sulfonic, and phosphoric acids as well as planar aromatic ring systems such as flavanthrone and copper phthalocyanines. Nucleating agents for polyamides are quartz, graphite, carbon black, titanium dioxide, and alkali halides whereas for aromatic polyesters, carbon black and sulfates of bivalent metals are important. Even polymers act as nucleating agents for other polymers: poly(ethylene)s or polyamide 6 nucleate the crystallization of it-poly(propylene).

Nucleating agents must be insoluble in the polymer; their particle sizes should be as small as possible (usually 1-10 μm) and their melting temperatures should exceed that of the polymer. All nucleation agents increase the crystallization rate and reduce the size of spherulites. Some also generate other crystal modifications such as the β form of it-poly-(propylene) instead of the common α form. A special class of nucleating agents are the so-called **clarifiers** that improve the transparency of polymers, which is highly desirable for blow-molded bottles and extrusion-molded films. Examples of clarifiers for the commercially most important polymer, isotactic poly(propylene), are dibenzylidene sorbitol and 1,3,5-benztrisamides with various substituents (see also Volume IV).

The action of nucleating agents depends not only on their wettability by polymer melts but probably also on their surface structure. At least some effective nucleating agents have grooves on their particle surfaces that force adsorbed polymer segments to adopt extended structures which, in turn, are a prerequisite for chain folding.

Crystallization from Solution

In secondary nucleation, segments of flexible polymer chains add to surfaces whereas in tertiary nucleation they become incorporated in corners and grooves (Fig. 7-19). Secondary nucleation as well as supercooling the melt below the melting temperature controls chain folding and thus the variable lamella height L_c (Fig. 7-22). The cross-sectional area L_dL_b of stems does not change.

The deposition of a chain segment of length L_c on the surface of the particle of the nucleation agent increases the surface by 2 L_cL_d from the two side areas of the stem and by 2 L_dL_b from the top and bottom areas L_dL_b. The Gibbs energy is increased by the Gibbs surface energies σ_s for each area L_cL_d and $\sigma_f = L_f\Delta H_f$ for each area L_dL_b where ΔH_f = surface enthalpy per volume. This energy gain by new surfaces is counteracted by a loss of Gibbs crystallization energy ΔG_{cryst} per segment volume $L_cL_bL_d$, Eq.(7-6).

(7-6) $$\Delta G_i = 2\,L_bL_d\sigma_f + 2\,L_cL_d\sigma_s - L_cL_bL_d\Delta G_{cryst}$$

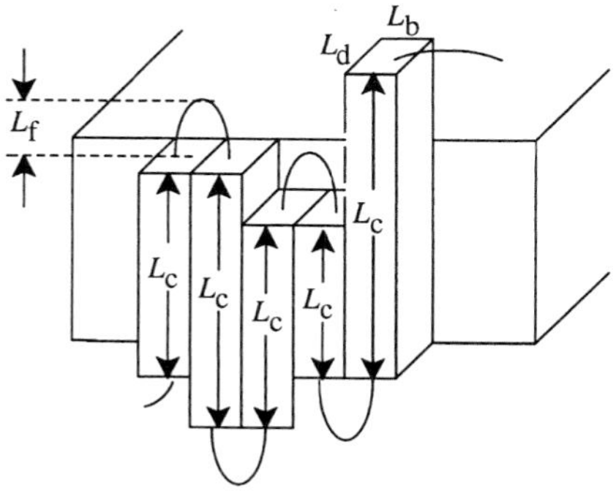

Fig. 7-22 Secondary nucleation by deposition of chain segments with various segment lengths L_c and equal stem diameters L_dL_b. L_f = height of fold.

Differentiating Eq.(7-6) and setting the result to equal zero delivers the critical (minimum) lamella height, $L_{c,0} = 2\ \Delta\sigma_e/\Delta G_{cryst}$, at which the Gibbs crystallization energy balances the formation of a top area, i.e., the addition of the first segment. However, since the change of Gibbs energy is zero for such an addition, the resulting nucleus of this size can never be stable. In order to grow from the first segment to a stable crystallite, Gibbs energies need to be somewhat negative and stem lengths a little greater than $L_{c,0}$. However, this additional length ΔL is small and can be neglected in the following.

The Gibbs crystallization energy per volume of a straight chain segment is $\Delta G_{cryst} = \Delta H_{M,u} - T_{cryst}\Delta S_{M,u}$. Since such a crystal has a melting temperature of $T_{M,o} = \Delta H_{M,u}/\Delta S_{M,u}$, the critical lamella height is

$$(7\text{-}7) \qquad L_{c,o} = L_{c,0} = \frac{2\,\sigma_e}{\Delta G_{cryst}} = \frac{2\,\sigma_e T_{M,u}}{\Delta H_{M,u}(T_{M,o} - T_{cryst})} \qquad \text{(neglecting } \Delta L\text{)}$$

The critical lamella height $L_{e,0}$ is therefore inversely proportional to the supercooling $T_{M,o} - T_{cryst}$ which has been confirmed experimentally, for example, for poly(ethylene) (PE), poly(oxymethylene) (POM), and poly(4-methyl-1-pentene (P4MP1) (Fig. 7-23).

The lowest value of $1/(T_{m,o} - T_{cryst})$ is $1/T_{m,o}$ since T_{cryst} cannot be smaller than 0 K. At $1/T_{m,o}$, values of $L_{c,o}$/nm are 3.8 (PE), 4.0 (POM), and 7.6 (P4MP1). The number of repeating units per length of stem is calculated from $N_{rep} = L_{c,o}/c$ where c = crystallographic bond length (= crystallographic length of repeating unit) (c/nm = 0.254 (PE), 1.73 (POM), and 1.38 (P4MP1)). At $T_{m,o}$, the number of repeating units per height of the lamella is therefore 15 (PE), 21 (POM), and 38.5 (P4MP1).

The dependence of lamella heights on supercooling, $T_{m,o} - T_{crys}$, seems to be specific for polymers with dominating repulsion between chains (PE, P4MP1, POM). However, lamella heights of polyamides are independent of supercooling, obviously because of the existence of hydrogen bonds between chain segments (see Fig. 7-8). Polyamides 3, 6.6, 6.10, 6.12, and 12 have 16 hydrogen bonds per lamella height but polyamides 10.10 and 12.12 only 12. For nylon 6.6, this corresponds to 3.5 repeating units per lamella height but for PA 10.10 and 12.12 to 3 repeating units. As mentioned before, loops also have different compositions: only CH_2 in PA 6.6 but –NH–CO– in PA 4.6.

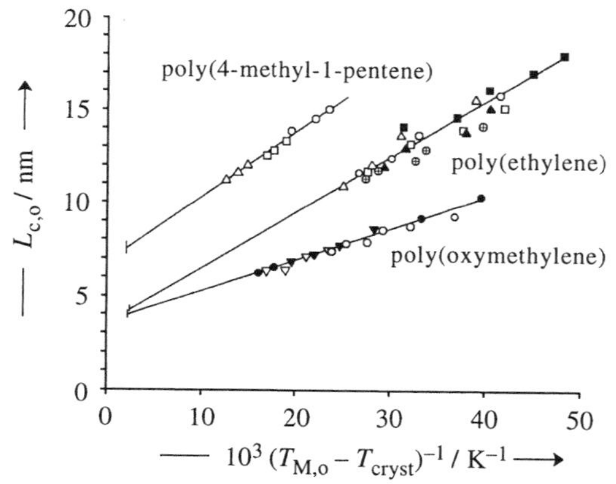

Fig. 7-23 Lamella heights $L_{c,o}$ as a function of the inverse supercooling temperature according to experiments in various solvents [12]. Vertical slashes I indicate values of $L_{c,o}$ at $1/T_{m,0}$ if $T_{cryst} = 0$ K. With kind permission of IUPAC, Research Triangle Park (NC).

Crystallization from Melts

Lamella heights of solution-crystallized polymers are relatively small, usually only 5-20 nm (Figs. 7-10 and 7-23). Melt-crystallized polymers have much larger lamella heights, up to several hundred nanometers, depending on the cooling rate. Obviously, lamella heights are diffusion controlled.

The high concentration of chain segments in melts impedes diffusion, especially of polymers with molar masses above the entanglement molar mass (Chapter 14). Hence, chain segments have problems finding their regular spots in lamellae to form regular folds, loops are more irregular and of different heights, and surfaces become rough. The roughness of lamella surfaces is revealed by the dependence of surface energy on the degree of polymerization (Fig. 7-24).

Surface energies of solution-crystallized alkanes and poly(ethylene)s increase with increasing degree of polymerization, X, and finally become independent of ,X which indicates a more or less smooth surface. In contrast, surface energies of melt-crystallized poly(ethylene)s continue to increase with increasing X. The two curves start to drift away from each other at $2\ N = X \approx 80$, i.e., at double the value at which crystallographic long periods become independent of the degree of polymerization, X (Fig. 7-10).

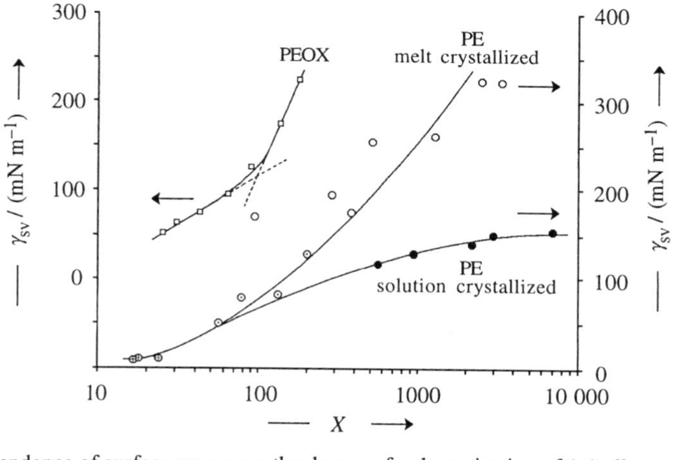

Fig. 7-24 Dependence of surface energy on the degree of polymerization of (⊕) alkanes (both solution and melt crystallization), poly(ethylene)s (O,⊙ melt crystallization; ● solution crystallization), and poly(oxyethylene)s (□ melt crystallization) [13].

7.3.2 Crystallization Rates

Nuclei can add chain segments only in a certain temperature interval: at $T > T_M$, crystallites melt and at $T < T_G$, diffusion of chain segments is no longer possible. With increasing temperature, crystallization rates thus pass through a maximum at $T_{cryst,max}$ = (0.80-0.87) $T_{M,o}$ where $T_{m,o}$ = melting temperature of perfect crystals (Fig. 7-25).

Crystallization processes can be subdivided into a primary and a secondary time interval. The primary interval ranges from the start of crytallization until the specimen is totally filled with crystalline material, for example, lamellae or spherulites. However, the specimen is not 100 % crystalline at the end of this period. It still contains non-crys-

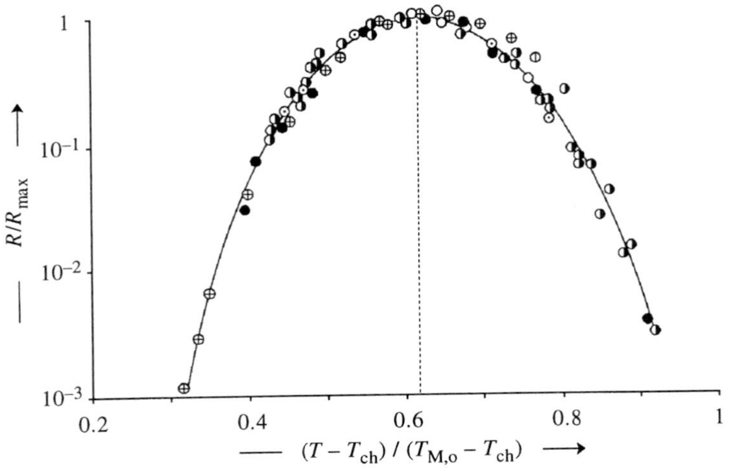

Fig. 7-25 Reduced linear growth rate, R/R_{max}, of spherulites of various polymers as a function of a reduced temperature [14]. R_{max} = maximum growth rate; T_{ch} = characteristic temperature (ca. 50 K below the glass temperature, T_G, where all segmental movements stop); $T_{M,o}$ = thermodynamic melting temperature. By kind permission of Elsevier Science, Oxford.

talline material that may continue to crystallize in the second time interval in which freshly formed lamellae thicken, lattices become more perfect, etc.

The time dependence of *primary* (simultaneous) *crystallization* is usually described by the **Avrami equation** which was originally derived for the crystallization of metals. The equation assumes that N nuclei in a total volume V_0 become N crystalline entities, each with a volume V_i. At any time, there is a probability $p_i = 1 - (V_i/V)$ that a crystallizable segment (atom, molecule, etc.; here: segment) does *not* reside in any one of the entities i. The total probability that this crystallizable segment is *not* in any entity at all equals the volume fraction $\phi_F = V_F/V_0$ of all non-crystallized segments in the partially crystallized melt, i.e., the product of all probabilities p_i, $p = \prod_i p_i = \prod_i [1 - (V_i/V)]$ (with $1 \leq i \leq N$), so that $\ln p = \sum_i \ln [1 - (V_i/V)]$.

The natural logarithm can be developed in a series, $\ln (1 - y) = - y - y^2/2 - \ldots$, since the volume V_i is much smaller than the total volume, $(V_i/V = y << 1)$. Higher terms of the series are very small; they can be neglected so that $\ln p = - \sum_i (V_i/V) = - (1/V) \sum_i V_i$ and $p = \exp [- (1/V) \sum_i V_i]$. Introduction of the number-average, $\overline{V}_n = (\sum_i V_i)/N$, of the volumes of entities and their number concentration, $C = N/V$, delivers $p = \exp [- C \overline{V}_n]$.

The volume of the total specimen is V_0 at the beginning of the experiment, V during the crystallization, and V_∞ at the end of primary crystallization. The density of the melt before crystallization, $\rho_0 = m_0/V_0$, and the density of the non-crystallized (liquid) fraction are assumed to be identical. The non-crystallized part therefore has a volume fraction of $\phi_L = V_L/V_0 = m_L/m_0$ and a mass fraction of $m_L/m_0 = \exp [- C \overline{V}_n]$.

The volume $V = V_L + V_S$ of the specimen at the time t of crystallization is the sum of the volume $V_L = m_L/\rho_L$ of the remaining melt (= liquid L) and the volume $V_S = m_S/\rho_S$ of the crystallized fraction (= solid S). The mass of solids is $m_S = 0$ at $t = 0$, $m_S = m_0 - m_L$ at time t, and $m_S = m_0$ at the end of the primary crystallization ($t \to \infty$). Since $\rho_L = m_0/V_0$ and $\rho_S = m_0/V_\infty$, the volume at time t is $V = V_\infty + (m_L/m_0)(V_0 - V_\infty)$ which can be written as $(V - V_\infty)/(V_0 - V_\infty) = \exp [- C \overline{V}_n]$.

The number average $\overline{V}_n$ of volumes of entities is obtained as follows. The concentration of nuclei is independent of time in *simultaneous nucleations*, $C = C_o$. At a time t, all nuclei have the same volume but these volumes increase with time. For example, the length L of rods increases with t but their cross-sectional area A stays the same. The height H of circular discs remains constant but their radii R increase. Spheres have a growing radius R, etc. Simple rate equations result, if L and R increase linearly with time:

Rods:	$\overline{V}_n = AL$;	with $L = k_1t$	$\rightarrow$ $\overline{V}_n = Ak_1t$	$= K_1t$
Discs:	$\overline{V}_n = \pi HR^2$;	with $R = k_2t$	$\rightarrow$ $\overline{V}_n = \pi Hk_2{}^2t^2$	$= K_2t^2$
Spheres:	$\overline{V}_n = (4\,\pi/3)R^3$;	with $R = k_3t$	$\rightarrow$ $\overline{V}_n = (4\,\pi/3)k_3{}^3t^3$	$= K_3t^3$

Introduction of these expressions for $\overline{V}_n$ in $(V - V_\infty)/(V_0 - V_\infty) = \exp\,[-\,C_o\,\overline{V}_n]$ leads to the Avrami equation (with **Avrami constants *K*** and **Avrami exponents** z):

(7-8) $$(V - V_\infty)/(V_0 - V_\infty) = \exp\,[-\,C_oK_jt^z] = \exp\,[-\,Kt^z]$$

This stretched exponential can be linearized by taking the double logarithm:

(7-9) $$\ln\,\{-\,\ln\,[(V - V_\infty)/(V_0 - V_\infty)]\} = \ln K + z \ln t$$

A typical Avrami plot is shown in Fig. 7-26. The intersection of the two linear sections in the right diagram indicates the transition from primary to secondary crystallization. The Avrami exponents z of the polymer in Fig. 7-26 decrease with increasing temperature from z = 2.20 at –3°C to z = 0.66 at +15°C (theory for primary: z = 1-3).

In *sporadic nucleations*, concentrations C of nuclei increase with time, for example, according to $C = k_ct$. The new nuclei may evolve either outside or inside the already existing crystallites. The latter possibility appears to be a double occupation but it is not since it does not change the volume fraction that is available for crystallization.

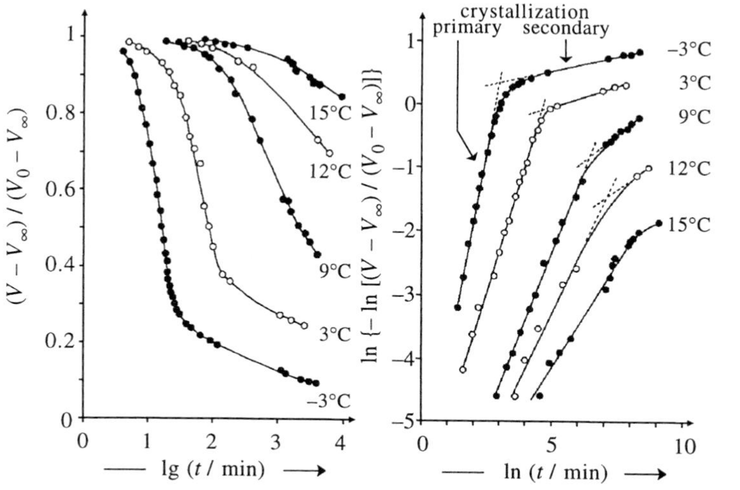

Fig. 7-26 Crystallization kinetics of a poly(butadiene) with 80 % 1,4-trans units. Left: time dependence of relative contractions [15]. Right: Avrami plot according to Eq.(7-8). Estimated Avrami coefficients are 2.20 at –3°C and 0.66 at 15°C.

Table 7-6 Avrami exponents z and Avrami constants K for sporadic (homogeneous) and simultaneous (heterogeneous) nucleations without (-) or with diffusion control. A = cross-sectional area, H = height, C = concentration, k = rate constant.

Nucleus	Spatial dimension	z sporadic -	z sporadic diff.-contr.	z simultaneous -	K sporadic -	K simultaneous -
Rod	1	2	3/2	1	$(1/2)\ ACk_1$	AC_ok_1
Disc	2	3	4/2	2	$(1/3)\ \pi HCk_2^2$	$\pi HC_ok_2^2$
Sphere	3	4	5/2	3	$(1/3)\ \pi Ck_3^3$	$(4/3)\ \pi C_ok_3^3$
Sheaf-like		≥ 6		5		

In the sporadic formation of spherical nuclei, each nucleus has the same chance of formation. For the time interval $(t - \tau)$ to $(t - \tau + d\tau)$, it is $d\tau/t$. $\overline{V}_n = (4/3)\ \pi\ (k_3t)^3$ for simultaneous nucleation thus has to be replaced by Eq.(7-10) for a sporadic one:

$$\text{(7-10)} \qquad \overline{V}_n = (4/3)\ \pi\ k_3^3 \int (t-\tau)^3 (d\tau/t) = (1/3)\ \pi (k_3 t)^3$$

For spheres, expressions for $\overline{V}_n$ are very similar except for the numerical factor that is 4/3 for instantaneous and 1/3 for sporadic nucleations. However, concentrations of nuclei are different. In instantaneous nucleations, concentrations are constant ($C = C_o$) but in sporadic nucleations, they increase with time, for example, linearly ($C = k_ct$). The term $\exp(-C\overline{V}_n)$ becomes $\exp(-k_ctK_3't^3)$ and the Avrami exponent z increases from 3 to 4.

Theoretical Avrami coefficients for simultaneous or sporadic crystallizations to simple nuclei (rods, discs, spheres) are positive whole numbers (Table 7-6) but experimental values are often fractions (Fig. 7-26). There are several theoretical or experimental reasons for these discrepancies.

A common experimental error is to include measurements beyond the application range of theory which is valid only between zero crystallization and the point at which growing crystallites start to touch each other. Theoretical values for simultaneous nucleations are also lower limits; upper limits are ≤ 2 (rods), ≤ 3 (discs), and ≤ 4 (spheres) (Table 7-6). Crystallites may also have shapes other than rods, discs, or spheres; for example, they may be sheaf-like. Different experimental methods may also cause different Avrami coefficients z: dilatometry measures the growth of spherulites whereas calorimetry also covers also the growth of lamellae in spherulites.

Crystallization rates vary strongly with chemical structures of polymers. Symmetrical polymers such as poly(ethylene) usually crystallize fast whereas polymers with bulky substituents or chain units have low crystallization rates. For example, at a crystallization temperature of $(T_M - 30)$/K, linear poly(ethylene) has a linear crystallization rate of ca. 5000 μm/min whereas that of conventional poly(vinyl chloride) is only ca. 0.01 μm/min.

Since crystallization rates depend on both nucleation and crystallite growth, crystallization rates are usually higher for simultaneous (athermic) nucleations than for sporadic (thermal) ones. For example, on cooling melts very fast below the melt temperature, poly(ethylene terephthalate)s can be prepared as practically amorphous materials. The fast crystallizing poly(ethylene)s, on the other hand, have never been obtained as amorphous substances even if the melts are chilled by liquid nitrogen.

7.4 Morphology

Depending on crystallization conditions, primary lamellae often combine to form larger entities that may range from spherical structures to sheaf-like ones. These "superstructures" have a more or less ordered interior and range in size from micrometers to millimeters. Whether or not such structures are formed depends on the polymer concentration, the molar mass and molar mass distributions, the presence and type of nucleating agents, the degree and velocity of supercooling, and the presence of flow gradients.

7.4.1 Spherulites

Crystallization of melts can lead to polycrystalline sphere-like entities that are called spherulites (G: *sphaira* = sphere, *lithos* = stone). The term "spherulite" is also applied to the round "two-dimensional" entities in crystallized thin films because they are like cross-sections of three-dimensional spherulites (Fig. 7-27).

Spherulites with diameters between five micrometers and several millimeters can be viewed by light microscopy and those with diameters of less then 5 μm by electron microscopy or small-angle light scattering. In polarized light, spherulites appear as "Maltese crosses" (Fig. 7-27, left) which are characteristic fof interference effects. Maltese crosses emanate because velocities of light differ in various parts of spherulites which behave as crystals with radial optical symmetry; such entities have four positions for extinction. Spherulites are indeed radial-symmetric as shown for "two-dimensional" spherulitic structures by phase contrast microscopy (Fig. 7-27, right) and for three-dimensional ones by microtome cuts through the center of spherulites.

Differences in light velocities result from differences in refractive indices. Positive spherulites are defined as those with the highest refractive index in the radial direction whereas negative spherulites have the highest refractive index in the tangential direction.

The optical behavior delivers information about the microstructure of spherulites. Examples are oriented poly(ethylene) fibers in which the light velocity is smaller in the fiber direction than in the other two. Light parallel to the fiber direction here leads to a larger refractive index. Molecule axes must therefore be predominantly parallel to the fiber axis. Since poly(ethylene) forms negative spherulites, its molecule axes must be vertical to the spherulite radii (Fig. 7-28).

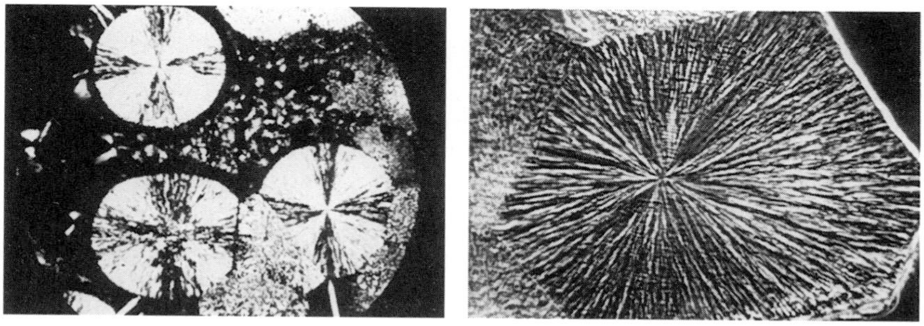

Fig. 7-27 Spherulites of isotactic poly(propylene) as viewed by a polarized light microscope (left) and phase contrast microscope (right) [16].

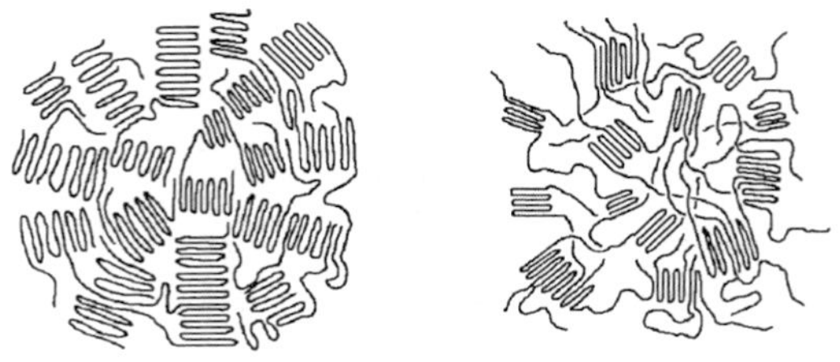

Fig. 7-28 Schematic representation of spherulite structures. If the main axis of polarizability is in the direction of the chain axis, then the left spherulite has a negative birefringence and the right spherulite a positive one. The left spherulite is positive if the polarizability is vertical to the main axis.

In poly(vinylidene chloride) (PVDC), it is just the opposite: the refractive index is smaller in the chain direction than perpendicular to it. Since PVDC forms positive spherulites, its molecule axes must also be tangential to the spherulite radii. Such behavior is often found for polymers with strongly polarizable groups such as polyesters and polyamides. The same polymer may also form both positive and negative spherulites, even simultaneously. An example is polyamide 6.6 where negative spherulites have higher melting temperatures than positive ones.

Spherulites lead to opaque films and sheets if their diameters are greater than ca. one-half of the wavelength of incident light *and* the material contains inhomogeneities with respect to density and/or refractive index. For example, spherulitic poly(ethylene) is opaque but spherulitic poly(4-methyl-1-pentene) is glass clear even if it contains as many spherulites of the same size as poly(ethylene).

Spherulites are generated if gross crystallization rates are the same in all three spatial directions but different crystallization rates prevail in various directions in the spherulite interior. The interior structure of spherulites is therefore never perfect and crystallizations may proceed within the spherulites themselves even if the whole volume of the specimen is completely filled with spherulites. This secondary crystallization (Fig. 7-26, right) manifests itself by an increase of X-ray crystallinity with time (Section 7.5.5).

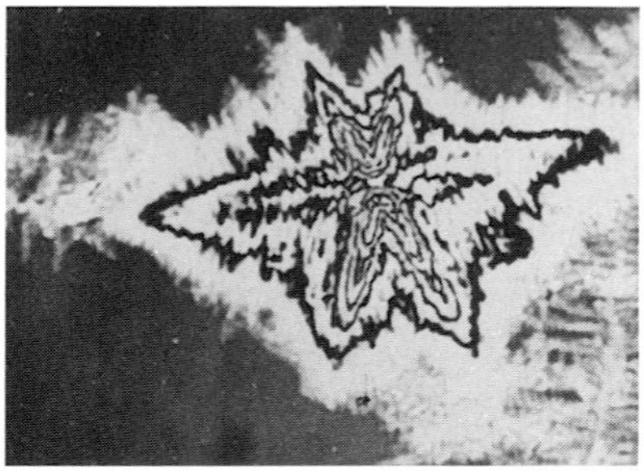

Fig. 7-29 Dendrite obtained by crystallization of poly(ethylene) from a dilute xylene solution at 70°C [17].

7.4.2 Other Superstructures

Light or electron microscopy sometimes detects snowflake-like structures (Fig. 7-29). The interior of these **dendrites** consists of crystalline and amorphous regions that were caused by different crystallization rates. For example, amorphous parts of poly(olefin)

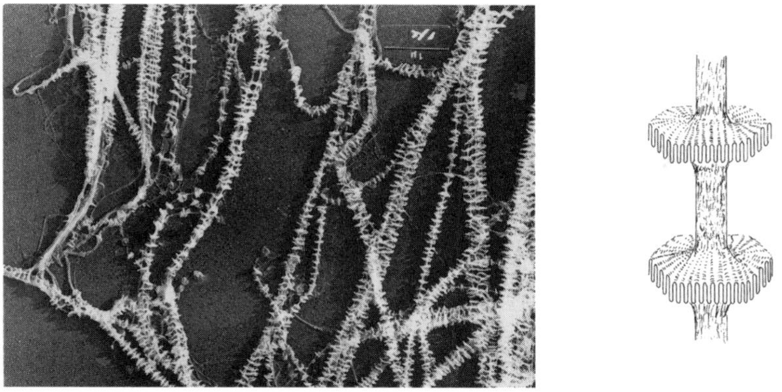

Fig. 7-30 Left: electron micrograph of a shish kebab from the crystallization of a 5 % solution of a linear poly(ethylene) ($\overline{M}_w$ = 153 000 g/mol, $\overline{M}_n$= 12 000 g/mol) in xylene at 102°C [18a]. Right: chains in a shish kebab (schematic) [18b]. By kind permission of Steinkopff Verlag, Darmstadt.

dendrites can be oxidized and removed by nitric acid. The remaining crystalline parts have lamella structures with equal lamella heights.

Different crystallization rates in different spatial directions lead to structures that resemble **shish kebabs** (Turkish: *sis* = skewer, *kebap* = roast meat). These structures consist of secondary crystallites that grew vertically oriented on fiber-like primary crystallites (Fig. 7-30). The formation of these structures is a special case of **epitaxial growth**, the oriented growth of a crystalline substance on another one.

Shish kebabs result if polymers crystallize from strongly stirred dilute polymer solutions. In these solutions, polymer chains align themselves in the flow gradient of the extensional flow, associate laterally, and form fibrils. The fibrils assemble to bundle-like nuclei in which chains are aligned parallel to the fiber axes according to X-ray and electron diffraction as well as optical birefringence.

The flow gradient is much smaller between the nuclei so that the crystallization from the remaining solution is not affected by the flow gradient. The polymer chains rather crystallize by chain folding in lamellae that are vertical to the fibrils (Fig. 7-30).

7.4.3 Factors Controlling Crystallization

The same polymer can crystallize in very different morphologies as shown in Fig. 7-31 for a polyamide 6:

Rapid quenching of dilute solutions results in clusters of spheres that are practically amorphous (Fig. 7-31, top left).

Crystallization from *very dilute solutions* leads to lamellae. Polymer segments prefer to attach themselves "one-dimensionally" to the side faces of lamellae but may also do so to the surface of lamellae, leading to steps or spirals (Fig. 7-31, bottom left).

Lamella heights do not vary with the molar mass but are usually affected by the degree of supercooling. Lamella heights depend on the preferred macroconformation in solution. Amylose tricarbanilate, for example, crystallizes from dilute solutions in 1,4-dioxane/ethanol in folded chains but from pyridine/ethanol in folded helices.

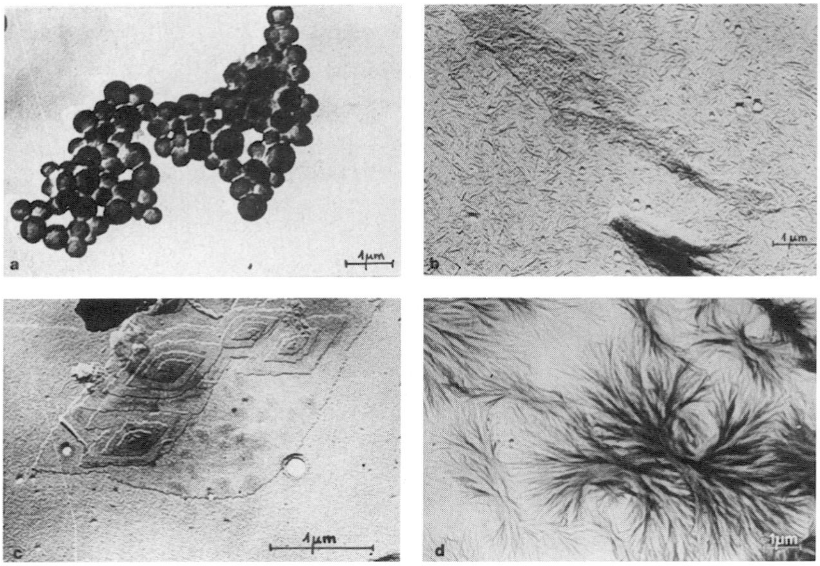

Fig. 7-31 Electron micrographs of supermolecular structures of a polyamide 6 [19].

Top left:	260°C solution in glycerol quenched by pouring into glycerol at 20°C.
Bottom left:	260°C solution in glycerol cooled fast at 40 K/min.
Top right:	260°C solution in glycerol cooled slowly at 1-2 K/min.
Bottom right:	slow evaporation of a formic acid solution at room temperature.

Polymer molecules may self-associate in *semidilute solutions*. The resulting supermolecular units may crystallize as bundles of crystallites and, at even higher concentrations, as fibrils (Fig. 7-31, top right), networks of crystallites, or even dendrites (Fig. 7-31, bottom right).

Lamellae are the first entities that are formed from *melts at rest*. They then cluster to bundles and finally to spherulites whose interiors consist of fibrils (Fig. 7-32). These structures arise because spherulite formation is a kind of fractionating crystallization. High-molecular weight, linear polymer molecules crystallize first while strongly branched and low-molecular weight fractions need stronger supercooling and are therefore excluded from the growth zone. These fractions assemble in an intermediate zone where they suppress the crystallization which in turn results in a preferential growth in the growth zone to a fibrillar structure.

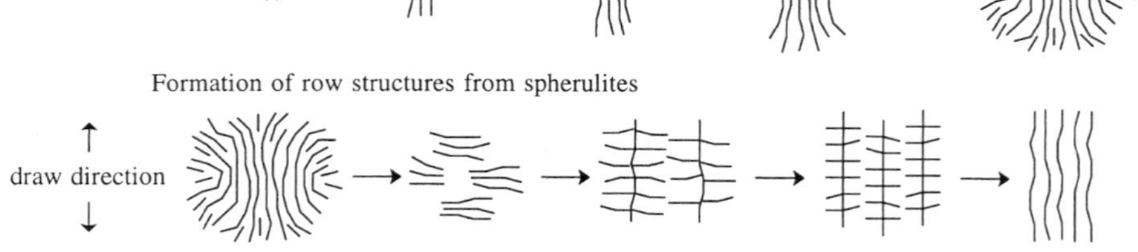

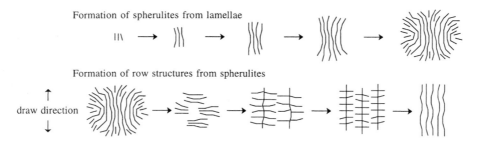

Fig. 7-32 Formation of spherulites and row structures. Double arrow: shear or strain direction.

Crystallization of supercooled melts with a strong temperature gradient leads to dendrites. The liquid phase between dendrites often solidifies to a microcrystalline structure.

Nucleation may also start at surfaces of seeds or of fibers that are embedded in melts (see Volume IV). The subsequent **trans crystallization** is strongly affected by diffusion processes and leads to *epitaxial growth*.

Special morphologies are obtained from crystallizations in the presence of *flow processes*. In nature, such processes are responsible for the formation of cellulose and silk fibers, blood coagulation, and mechanical denaturation of proteins. In industry, such crystallizations are exploited for flash spinning, formation of high-modulus fibers, and production of synthetic papers (Volume IV).

The shearing of crystallizing polymer melts causes spherulites to deform (Fig. 7-32, bottom). Epitaxial growth on lamellae then produces shish kebab structures (Fig. 7-30) and finally chain orientation in the draw direction. At high shear stresses, crystallization seems to be induced by a continuous series of nucleations along the shear gradient which leads to row structures of oriented lamellae and chains.

Like spherulites, **row structures** consist of folded chains in lamellae. Unlike spherulites, lamellae are arranged perpendicular and not radial to the flow or strain direction.

Shish kebab-like fiber structures are also obtained by strong turbulent shearing of dilute solutions of crystallizing polymers. For example, such processes may lead to growth rates of 160 cm/min and to poly(ethylene) fibers with lengths of ca. 2000 m and strengths of ca. 40 % of theory (Chapter 18).

Crystallization of melts *under pressure* increases lamella heights considerably (Fig. 7-33) while reducing the thickness of the surface layer. In these **extended chain crystals**, polymer chains are *not* completely stretched, i.e., the lamella height is smaller than the conventional contour length. However, their radius of gyration is independent of their molar mass (Fig. 7-17) if $\overline{M}_w > 30\ 000$ g/mol.

Fig. 7-33 Extended chain crystals of a poly(ethylene) of $\overline{M}_n = 78\ 300$ g/mol after crystallization at 225°C and 480 MPa [20]. The specimen had an X-ray crystallinity of 99 %.

7.5 Crystallinity

7.5.1 Ideal Crystals

Macromolecules rarely form large classical crystals, i.e., those with macroscopic dimensions in which planar surfaces are arranged at constant angles. Fairly large crystals can be obtained from spheroidal proteins but these crystals contain **super lattices** with up to 95 % water in which lattice points are occupied by protein *molecules* and not by monomeric units. The channels and nooks between protein molecules are often so large that heavy metal ions can diffuse into and reside in the protein molecules without disturbing the three-dimensional protein structure. For example, the unit cell of myoglobin (156 amino acid units) contains two molecules, each of which accommodates a heavy metal ion. This effect is utilized in X-ray structure determination since the loading by heavy metal ions allows one to determine the electron density distribution and thus the three-dimensional structure of protein molecules in such crystals.

Super lattices are also formed by densely packed spherical latex particles. However, lattices with large spherical domains of polymer blocks in an amorphous matrix are not considered super lattices but mesophases (Section 8.5).

Large crystals of chain molecules are very rare. One of the few examples is centimeter sized crystals of poly(oxy-2,6-diphenyl-1,4-phenylene) that contained up to 35 % of the solvent tetrachloroethane from which they were crystallized.

7.5.2 Crystallizability and Crystallinity

In ideal crystals, all chain units are present in crystallographically equivalent positions in an ideal lattice. In real crystals, kinetics prevents some units from occupying their ideal positions. The crystal lattice then either contains defects and/or the more or less ideal lattice structures extend only to relatively short distances, sometimes to less than 2 nanometers. Matter between crystalline regions is more or less disordered.

It is therefore imported to distinguish between crystallinity and crystallizability. **Crystallizability** is a thermodynamic quantity that depends only on temperature and pressure. It indicates the maximum degree of physical order a polymer can ever attain.

Crystallinity is always smaller than crystallizability. It is controlled by crystallization kinetics and thus by nucleation, rate and degree of supercooling, viscosity, entanglement, oriental flow, and the like. Different experimental methods detect different types and degrees of order and disorder and deliver therefore different percentages of crystallinity (Section 7.5.5 ff.). Polymers can never achieve perfect (100 %) crystallinity and are therefore *always* **semicrystalline**.

Semicrystalline polymers are never in thermodynamic equilibrium. According to the **phase rule**, $P + F = C + 2$, a single component ($C = 1$) with two degrees of freedom ($F = 2$; temperature and pressure) can exist in thermodynamic equilibrium in only one phase ($P = 1$). It follows for semicrystalline polymers that crystalline and non-crystalline regions must be connected, i.e., that a single chain molecule must be present in both types of regions which are consequently *not* "phases" in the thermodynamic sense. In contrast to true 2-phase systems, these "phases" of semicrystalline polymers cannot be separated by physical methods.

7.5.3 Lattice Defects

Crystalline lattices of polymers may contain the following point defects:

- **End groups** usually do not fit into lattices created by monomeric units.
- **Kinks** are conformational errors that arise if chain segments are shifted parallel to the long axis of chain segments. The shift is smaller than the distance between two chains. A typical kink has the conformation ...TTTTG$^+$TG$^-$TTTT...
- **Jogs** are strong shifts, i.e., conformational errors where the parallel shift of chain segments is larger than the distance between chain segments. An example is the jog ...TTTTG$^+$TTTTG$^-$TTTT... Jogs and kinks shorten zigzag chains and twist helices. Their travel through lattices requires large spatial chain movements.
- **Reneker defects** consist of both conformational mistakes and changes in bond angles (see Fig. 7-34). Like kinks and jogs, Reneker defects also shorten chains. In contrast to kinks and jogs, Reneker defects can move in a chain without changing the relative position of the chain in the crystal lattice.

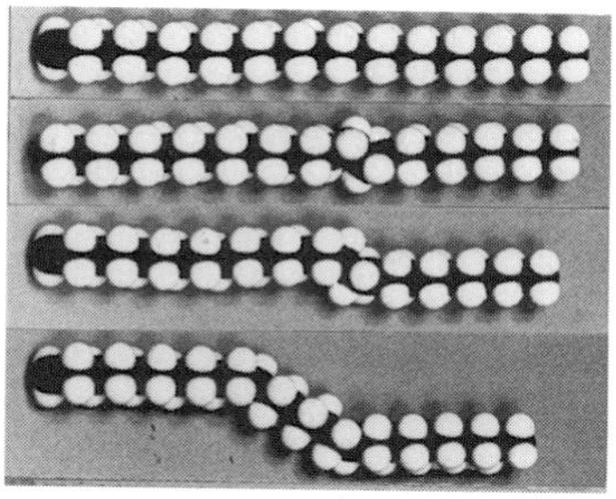

Fig. 7-34 Lattice defects in poly(ethylene). From top to bottom: all-trans conformation (for comparison), Reneker defect, kink, and jog.

7.5.4 One-Phase and Two-Phase Models

Depending on crystallization conditions, crystalline regions in polymers can form more or less ordered superstructures. The order in these structures is usually described by the **degree of crystallinity**, using either a 1-phase or a 2-phase model. In the **1-phase model**, it is assumed that deviations from 100 % crystallinity are caused just by lattice defects. The **2-phase model** postulates the coexistence of totally crystalline and totally amorphous regions and the complete absence of regions that are neither 100 % crystalline nor 100 % amorphous. In both cases, "phase" is not used in the thermodynamic meaning of the word (p. 224).

Degrees of order can be described by either model. For example, the lamellae of the polyamide 6 of Fig. 7-31, bottom left, have less than 50 % crystallinity according to the 2-phase interpretation of X-ray measurements although they look like polymer single crystals which they are according to electron diffraction. The same density can be interpreted as being caused by a relatively high proportion of amorphous regions (2-phase model) or as a small fraction of point defects (1-phase model) (Table 7-7).

Table 7-7 Comparison of crystallinities of poly(ethylene)s by the 2-phase or 1-phase model.

Morphology	Density in g/cm³	Specific volume in cm³/g	Crystallinity in percent 2-phase model	Crystallinity in percent 1-phase model
100 % crystalline	1.000	1.000	100	0
	0.981	1.020	89	1.9
	0.971	1.030	83	2.9
100 % amorphous	0.852	1.174	0	-

Polymer science uses mainly the 2-phase model and expresses degrees of crystallinity either as weight fraction, w_{cr}, or volume fraction, ϕ_{cr}. Both quantities are interconnected by the density of the specimen, ρ, and the density of the 100 % crystalline polymer which is usually obtained from X-ray measurments (see Section 7.5.6):

$$(7\text{-}11) \qquad w_{cr} = (\rho_{cr}/\rho)\phi_{cr}$$

The magnitudes of w_{cr} and ϕ_{cr} vary with the type of measurement (see Section 7.5.5 ff.). Degrees of crystallinity are therefore not defined by the degrees of order such as number, weight, surface, volume, etc., averages but according to the experimental method. One thus distinguishes X-ray crystallinities from density crystallinities, infrared crystallinities, and the like. The degrees of crystallinity by the various methods sometimes agree (example: poly(ethylene)) but more often they do not (Table 7-8).

Table 7-8 Degrees of crystallinity in weight percent according to the 2-phase model.

Method	Celluloses Cotton	Mercerized cotton	Rayon	Wood pulp	Poly(ethylene terephthalate) not oriented	oriented
Physcial methods						
Infrared spectroscopy	62	-	42	-	61	59
X-Ray diffraction	73	51	38	60	29	2
Density	64	36	25	50	20	20
Chemical methods						
Acid hydrolysis	90	80	45	86	-	-
Formylation	79	65	61	69	-	-
Deuteration	58	41	32	45	-	-

7.5.5 X-Ray Diffraction

Crystallinities by X-ray diffraction are calculated from the intensities of reflections and halos as a function of the Bragg angle 2 θ (Fig. 7-35) and interpreted by the 2-phase model. After separation of the background intensity, the intensity I_{am} of the amorphous halo is obtained by drawing a line from the lowest angles (because there are usually no crystalline reflections) to the highest one by connecting the minima of intensities (see Fig. 7-35). The degree of crystallization, $w_{cr,X}$, is then calculated from

(7-12) $w_{cr,X} = I_{cr}/(I_{cr} + K_X I_{am})$

where I_{cr} is the intensity of crystalline reflections. The proportionality factor K_X depends on both the observation angle and the specific function; it can be determined by comparing the specimen with either completely amorphous or "completely crystalline" polymers of the same chemical structure. Amorphous polymers are sometimes obtained by rapid quenching of their melts. Amorphous cellulose is produced by grinding cellulose in ball mills. Occasionally, melts may also serve as amorphous standards.

Discrete crystalline reflections result from sufficiently high concentrations of three-dimensionally ordered regions which are at least 2-3 nm long in each of the three spatial directions. The intensity of the reflections is a measure of the X-ray crystallinity. The width of the reflections depends on both the size of crystallites and local lattice variations. For very small crystallites, selective diffraction converts to scattering. Hence, very small crystallites do not contribute to measured X-ray crystallinities which is the reason why X-ray crystallinities are sometimes smaller than infrared crystallinities (Table 7-8).

In principle, the sizes of crystallites can be calculated from the widths of crystalline reflections. However, such calculations may be questionable since positions of monomeric units may shift during crystallization because chain segments may deposit on nuclei or lamellae before these segments reach their ideal lattice positions for kinetic reasons. This effect leads to local variations of lattice constants and hence to broadening of reflections, an effect similar to that caused by thermal movements.

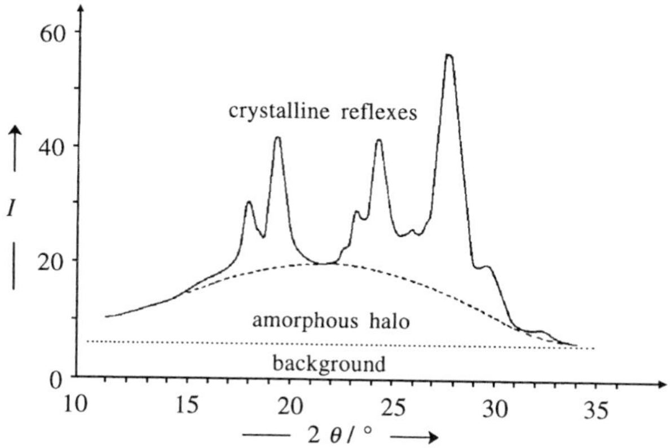

Fig. 7-35 X-ray intensity I as a function of the Bragg angle 2 θ for a poly(ethylene terephthalate) showing crystalline reflections, the amorphous halo, and background scattering [21]. The halo line is drawn through those minima between reflections that are separated by more than 3° (no crystallinity).

7.5.6 Density

The density crystallinity of polymers is calculated from three densities, ρ (specimen), ρ_{am} (100 % amorphous), and ρ_{cr} (100 % crystalline):

(7-13) $$w_{cr,d} = \frac{\rho_{cr}}{\rho}\left(\frac{\rho - \rho_{am}}{\rho_{cr} - \rho}\right)$$

The density of the specimen is determined directly, for example, in a density gradient of a salt solution or two miscible organic solvents. These liquids must wet the specimen but should neither swell nor dissolve it. The density gradient is produced mechanically in such as way that the density of the liquid varies linearly, convex, concave, etc. with the height of the column. The specimen floats at a position that corresponds to its density.

The density of the hypothetical 100 % crystalline polymer is calculated from the number N_i of atoms of atomic mass A_i that reside in the unit cell of volume V:

$$\rho_{cr} = (\Sigma_i \; N_i A_i)/(N_A V) \tag{7-14}$$

Amorphous polymers are produced by grinding solids, quenching melts, etc. (see p. 227). Alternatively, one can try to extrapolate the specific volumes of the melt, $v_{am} = 1/\rho_{am}$, from high temperatures through the melting temperature to the temperature at which the density of the specimen was determined.

Densities of crystalline polymers, ρ_{cr}, are usually greater than those of their amorphous counterparts, ρ_{am}, because segments are more tightly packed in the former than in the latter. The ratio ρ_{am}/ρ_{cr} is often 0.85-0.95 and seems to depend on the cross-sectional area of the chain according to $\rho_{am}/\rho_{cr} = KA_c^{0.18}$.

The largest differences between amorphous and crystalline specimens are observed for linear polymers without substituents such as poly(ethylene) and polyamide 6.6 if they crystallize in the all-trans conformation. Smaller density differences are shown by helix-forming macromolecules because of the much looser packing of chains in the crystalline state. The helical poly(4-methyl-1-pentene) even has a smaller crystalline density than that in the amorphous state (0.812 g/cm^3 versus 0.838 g/cm^3).

7.5.7 Calorimetry

Enthalpies vary with crystallinities. The calorimetric degree of crystallinity can thus be calculated from the specific melt enthalpies Δh of the specimen and Δh_{cr} of its 100 % crystalline counterpart,

$$w_{cr,h} = \Delta h/\Delta h_{cr} \tag{7-15}$$

where Δh is determined by differential scanning calorimetry (Section 13.1.4). Since totally crystalline specimens are never available, specific melt enthalpies Δh_{cr} are obtained from low-molar mass model compounds, melting point depression on addition of diluents, or extrapolation to 100 % crystallinity of specific melt enthalpies Δh_{cr} of specimens of different degrees of crystallinity.

7.5.8 Infrared Spectroscopy

Infrared spectra of semicrystalline organic polymers often show bands between 650 and 1500 cm^{-1} that are absent in the melt (Fig. 4-2) and in amorphous polymers. These vibrational bands are caused by deformations of certain microconformations in ordered regions. Hence, infrared spectroscopy does not deliver direct information about crystallinity but about the presence and proportion of certain local conformations.

The degree of crystallinity by infrared, $w_{cr,ir}$, measures the proportion of these microconformations in regular sequences, assuming again a 2-phase morphology with clear and distinct boundaries between crystalline and amorphous regions. The weight fraction of crystalline regions is calculated from the intensities of incident light, I_o, and transmitted light, I, the density ρ and thickness L of the specimen, and the absorption coefficient α_{cr} of the crystalline part:

$$(7\text{-}16) \qquad w_{cr,ir} = (\alpha_{cr}\rho L)^{-1} \log (I_o/I)$$

It is good practice to calculate the degree of crystallinity of a specimen from more than one crystalline band because a macromolecular substance may crystallize in various crystal modifications (Section 7.1.5) and two or more of these modifications may be present in the same specimen. Since an IR band is specific for only one of these modifications, the degree of crystallinity may be underestimated if it is calculated from the absorption of just one band.

7.5.9 Indirect Methods

Industry often uses indirect methods to determine degrees of crystallinity because such methods are simple and fast and/or do not require hard-to-obtain data on 100 % crystalline and 100 % amorphous specimens. These methods rely on the fact that physical and chemical processes proceed differently in crystalline and in non-crystalline regions. They really measure accessibilities and not crystallinities per se.

Commonly used physical processes are the absorption of water vapor by hydrophilic polymers or the diffusion of dyes into polymers. Chemical processes comprise hydrolysis, formylation, deuteration (deuterium exchange), and periodate oxidation. Such indirect methods are commonly used for cellulosic materials (see p. 226, Table 7-8).

The main problem with these methods is that the probe often changes the physical structure of the specimen that it is investigating. For example, entry of water vapor or liquid water into a solid hydrophilic specimen may lead to swelling which improves the accessibility to certain regions. The degree of crystallinity so obtained is no longer that of the orginal specimen but that of the probed one.

7.6 Orientation

Stretching of fibers, films, and sheets may cause crystallites and/or chain segments in amorphous regions to orient themselves in the strain direction. The degree of orientation can be measured by X-ray wide-angle diffraction, infrared dichroism, optical birefringence, polarized fluorescence, and sound velocity.

Such methods are often time-consuming and costly. Common industrial practice is therefore to take the **draw ratio**, L/L_o, as a measure of the orientation of chain segments, molecules, and crystallites. Clearly, this is not a good measure of orientation since drawing may lead only to viscous flow and not to molecular-level type orientation at all. In addition, draw ratios are not only affected by drawing conditions but also by the history of the specimen.

7.6.1 X-Ray Interference

With increasing draw ratios L/L_o, circular reflections in wide-angle X-ray diagrams transform first to arcs and then to point reflections that are normal to the draw direction (Fig. 7-36). The inverse length of the arcs is a measure of the orientation of crystallites (more precisely: the lattice planes). Arcs at various positions in the X-ray diagram correspond to various lattice planes. An orientation factor f can thus be defined for each of the three space coordinates. This factor is related to the angle of orientation of crystallites, β, by the **Hermans orientation function**:

(7-17) $$f_{or} = (1/2)\,(3\,\langle\cos^2\beta\rangle - 1)$$

where β is the angle between the draw direction and the optical main axis of segments.

The orientation angle is 1 for complete orientation in the chain direction ($\beta = 0$), $-1/2$ for complete orientation perpendicular to the chain direction ($\beta = 90°$), and 0 for a random orientation. The sum of the orientation factors in the three spatial directions, x, y, and z, becomes zero if the optical axes of the crystallites are vertical to each other ($f_x + f_y + f_z = 0$). Uniaxially oriented polymers are characterized by a single value of f.

7.6.2 Optical Birefringence

Materials possess three refractive indices n_x, n_y, and n_z along the three main axes x, y, and z. In isotropic materials, these refractive indices are equal ($n_x = n_y = n_z$) whereas at least one of them differs in anisotropic materials (for example, $n_x = n_y \neq n_z$). The difference between two unequal refractive indices is called **optical birefringence** Δn; for example $\Delta n_{yz} = n_y - n_z$. Optical birefringence is observed only in transparent or translucent materials, either in solids or in flowing solutions (Section 11.2).

Differences in refractive indices are caused by differences in polarizabilities. For example, an alkane possesses a greater polarizability along the chain than perpendicular to it because electron mobilities are greater in the chain direction.

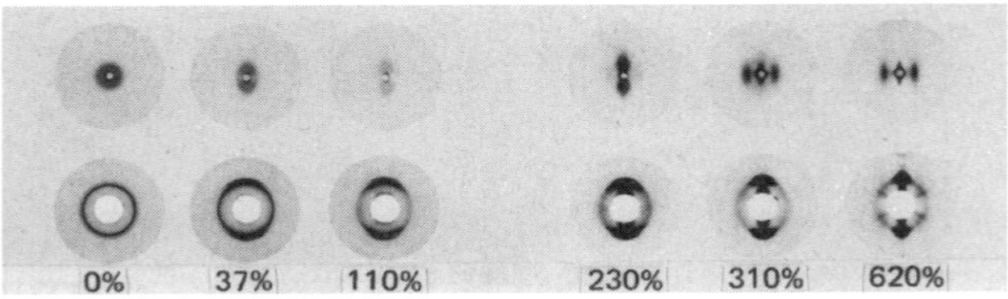

Fig. 7-36 Small angle X-ray (top) and wide angle X-ray interferences (bottom) of a drawn poly(ethylene) [22]. Draw direction is from left to right. Numbers indicate the draw ratio, L/L_o, in percent.

Birefringence develops in polymeric solids if chains are oriented or are under stress. The total birefringence of a solid is composed of birefringences Δn_i of various "phases" i, the so-called form birefringence, Δn_f, and the stress birefringence, Δn_{st}:

(7-18) $$\Delta n = (\Sigma_i\ \phi_i \Delta n_i) + \Delta n_f + \Delta n_{st}$$

Each phase contributes to the total birefringence according to its volume fraction ϕ_i and birefringence Δn_i. Such phases may be the amorphous and crystalline phases of semicrystalline polymers, the domains in block polymers (Section 8.5), filler particles and the surrounding polymer, or strongly plasticized regions in less plasticized ones.

The total birefringence increases linearly with the orientation factor f from measurements of the propagation of sound waves (Section 7.6.3). At the same degree of orientation, polar polymers show stronger birefringence than apolar ones (Fig. 7-37).

Form birefringences Δn_f is caused by distortions of the electrical field at the interface of two phases, if the size of at least one of the phases is comparable to the wavelength of light. Form birefringence is important for semicrystalline polymers, block polymers, and polymer molecules in a flow field.

Stress causes polymers to become birefringent even if there is only one phase. The magnitude of *stress birefringence* Δn_{st} depends on the applied mechanical stress and the anisostropy of monomeric units. It is especially pronounced in poly(styrene) with its strongly anisotropic phenyl side groups. The effect is visible as reddish-bluish fringes in stressed articles from clear atactic poly(styrene)s, even in unpolarized light.

Stress birefringence is usually investigated by polarized light. The observed interference colors are strongest at an angle of 45° to the oscillating light waves. The colors of the fringes follow each other in the **fringe order** δ/λ which is the ratio of **retardation**, $\delta = (n_{\parallel} - n_{\perp})d$, and wavelength λ where $n_{\parallel}$ and $n_{\perp}$ are refractive indices parallel and perpendicular to the strain direction and d = thickness of the specimen.

Stress birefringence is especially important for the design of parts from plastics, since these break at the position of highest stress. It is often characterized by the **stress-optical coefficient**, $C = \Delta n/\Delta\sigma$, which is the ratio of the two principal refractive indices to the difference of principal stresses. Similarly, a **strain-optical coefficient** is used, $k = \Delta n/\Delta\varepsilon$, where $\Delta\varepsilon$ is the difference of elongations ε.

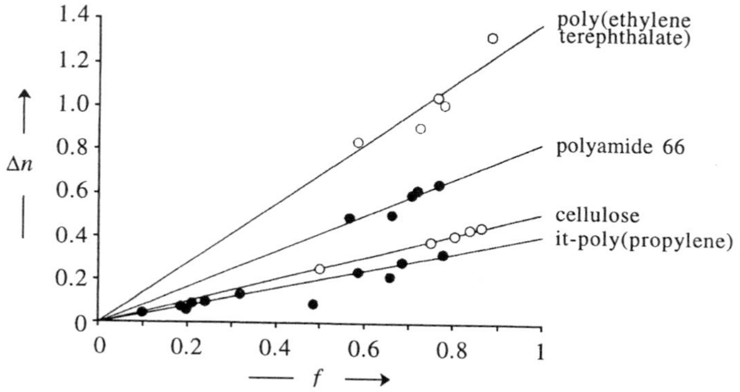

Fig. 7-37 Optical birefringence Δn as a function of the orientation factor f from the measurements of ultrasound propagation [23]. By kind permission of the Textile Research Institute, Princeton (NJ).

7.6.3 Propagation of Ultrasound

The propagation of ultrasound in polymers depends on intramolecular bond lengths between chain atoms and on intermolecular distances between chains. It can be used to measure the orientation of chains, which is defined as the orientation angle β between the strain axis and the fiber axis, respectively, and the average direction of chains.

Experimentally, one calculates the spatial mean-square sound speed, $\langle c^2 \rangle$, from the orientation angle β and the velocities of sounds parallel ($c_{\parallel}$) and normal (perpendicular) ($c_{\perp}$) to the fiber direction:

$$(7\text{-}19) \qquad \frac{1}{\langle c^2 \rangle} = \frac{\langle \cos^2 \beta \rangle}{c_{\parallel}^2} + \frac{1 - \langle \cos^2 \beta \rangle}{c_{\perp}^2}$$

Typical values for polymers are $c_{\parallel} \approx 1.5$ km/s and $c_{\perp} \approx (7\text{-}10)$ km/s.

Sound velocities c_u in unoriented specimens can be measured directly but the measurement of c, $c_{\parallel}$, and $c_{\perp}$ in oriented specimens requires considerable effort. However, the orientation factor f can be obtained by the following considerations.

Unoriented specimens have Hermans orientation factors of $f_{or} = 0$ and thus $\langle \cos^2 \beta \rangle = 1/3$ (Eq.(7-17)); the sound velocity here will be c_u. A specimen with orientations of chain segments $\parallel$ and $\perp$, that behaves as if it has no segmental orientation at all, will therefore also have $\langle \cos^2 \beta \rangle = 1/3$ and $\langle c^2 \rangle = \langle c_u^2 \rangle$ so that Eq.(7-19) becomes

$$(7\text{-}20) \qquad c_{\perp}^2 = \frac{2 c_u^2 c_{\parallel}^2}{3 c_{\parallel}^2 - c_u^2} \approx \frac{2 c_u^2}{3}$$

since $3\ c_{\parallel}^2$ is much larger than c_u^2. Such specimens are characterized by $3\ c_{\parallel}^2 >> c_u^2 = c^2$. The first term of the right side of Eq.(7-19) can be neglected, which leads to $1/c^2 = (1 - \langle \cos^2 \beta \rangle / c_{\perp}^2$. Combination of this expression with $c_{\perp}^2 = 2 c_u^2/3$, Eq.(7-20), delivers an expression for the orientation angle, $\langle \cos^2 \beta \rangle = 1 - [(2\ c_u^2)/(3\ c^2)]$, and, with Eq.(7-17), the equation

$$(7\text{-}21) \qquad f_{or} = (1/2)\ [3\ \langle \cos^2 \beta \rangle - 1] = 1 - (c_u^2/c^2)$$

Orientation factors f_{or} can thus be calculated from the measured sound velocity c of the oriented specimen and the known sound velocity c_u of the unoriented polymer, which allows the on-line determination of orientation during film and fiber production.

Orientation factors f_{or} from measurements of ultrasound velocities is linearly related to optical birefringence (Fig. 7-37). If optical orientation factors equal acoustic ones, the amorphous phase is either absent or has the same orientation as the crystalline one.

7.6.4 Infrared Dichroism

Light is absorbed if the vibration direction of the electric vector of light equals the vibration direction of the light-absorbing chemical group. The intensity A of an absorption band of an oriented polymer thus depends on the direction of the electric vector of the incident light beam relative to its direction of orientation. The magnitude of absorption depends on the polarization of the incident light. The **dichroic ratio** R is thus calculated from the absorptions A or intensities I of transmitted light parallel (index $\parallel$) and normal (index $\perp$) to the orientation or fiber direction:

(7-22) $$R = A_{\parallel}/A_{\perp} = \ln (I_0/I_{\parallel}) / \ln (I_0/I_{\perp})$$

Dichroic ratios can be obtained if a dipole of a uniaxially drawn polymer oscillates normal to the chain axis. Examples are hydrogen bonds between amide groups of polyamides (see Fig. 7-8).

The orientation factor f is calculated from the dichroic ratios of the specimen, R, and that of a polymer with complete orientation of the groups, R_{∞}:

(7-23) $$f = \left(\frac{R-1}{R+2}\right)\left(\frac{R_{\infty}+2}{R_{\infty}-1}\right)$$

The method can be applied to both "amorphous" and "crystalline" IR bands. However, each band must always be checked to determine whether observed changes do indeed come from orientations or are caused by conformational changes without changes of chain directions.

7.6.5 Polarized Fluorescence

Orientation of polymer chains with incorporated pendent fluorescent groups causes these groups to change their positions, which alters the intensity of fluorescence. Since most polymers do not contain such groups, one mixes them with ca. 10^{-4} wt% of a low-molecular weight organic fluorescent dye. In general, these dyes are unable to enter the crystalline regions of polymers so that the method is only sensitive to orientations within amorphous regions.

Incident polarized parallel light produces polarized fluorescent light whose intensity depends on the angle β between the molecule axes and the strain direction. It is assumed that the axes of dye molecules and chain segments have the same direction and that addition of the dye does not change the morphology of the polymer. During the lifetime of the excited state, chromophores of the fluorescent dye should not rotate, which is probably the case because of the high viscosity of the specimen.

The observed intensity depends on the fourth power of the cosine of the angle β between the draw direction and molecule axis, $I_{\parallel} = const\ \langle\cos^4 \beta\rangle$, if the polarizer and analyzer are parallel to the draw direction. The intensity of the fluorescent light in the direction of polarization normal to the draw direction is $I_{\perp} = const\ [\langle\cos^2 \beta\rangle - \langle\cos^4 \beta\rangle]$ for uniaxial drawing. The orientation factor f is then calculated from the ratio $R_F = I_{\parallel}/I_{\perp}$, similarly to the calculation of the orientation factor from the dichroic ratio, Eqs.(7-23) and (7-22).

Historical Notes to Chapter 7

HELICES

E.Sauter, Z.physik.Chem. **B 18** (1932) 417

Based on X-ray data from single crystals of poly(oxymethylene), "helicogyres" are proposed as macroconformations of chains. The helicogyres are depicted as two-dimensional tubs (D: Wanne). According to E.Sauter, Z.physik.Chem. **B 21** (1933) 186, this polymer forms a "strongly wound meander chain" (D: "stark gewundene Mäanderkette") with screw axis (D: "Schraubenachse"). A space-filling model (Fig. 2 of this paper) shows the helical structure but the term "helix" is not used.

H.Lohmann, in H.Staudinger, Die hochmolekularen organischen Verbindungen. Kautschuk und Cellulose, Springer, Berlin 1932, p. 287-332

Suggests a "meander structure" for poly(oxyethylene), based on X-ray data (from G = *maiandros*, a strongly winding river in Phrygia, an ancient country in west-central Asia Minor. The river is now called "Menderes" in Turkish). According to Sauter (loc.cit. 1933), the meander structure is not two-mensionally plane but "strongly wound" with a "screw axis" (i.e., a helix!).

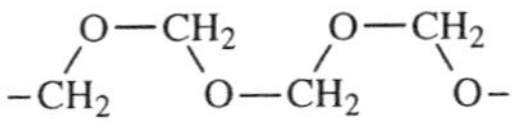

bathtub representation of poly(oxymethylene)

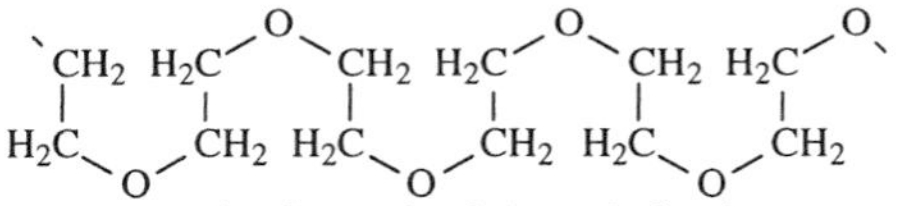

meander form of poly(oxyethylene)

C.D.Darlington, Proc.Royal Soc. London [B] **118** (1935) 33-96

Suggests that the "coiling and uncoiling" of chromosomes is caused by "invisible molecular spirals." This choice of words is probably the reason why the "helix" of polymer scientists is called a "coil" by molecular biologists. The latter meaning of "coil" corresponds to the common use of this word as an individual spiral or ring or "a series of connected spirals or concentric rings formed by gathering or winding" (American Heritage Dictionary).

C.S.Hanes, New Phytologist **36** (1937) 101

Suggests that amylose crystallizes in spirals; experimentally confirmed by R.E.Rundle, R.R.Baldwin, J.Am.Chem.Soc. **65** (1943) 554.

C.W.Bunn, Proc.R.Soc. **180** (1942) 67

Postulates spiral forms for main chains of vinyl polymers with sterically equivalent monomeric units. Experimental confirmation for poly(1-olefin)s by G.Natta, Acc.Naz.Lincei.Mem. **4** (sez. 2) (1955) 61; -, J.Polym.Sci. **16** (1955) 143; -, Makromol.Chem. **16** (1955) 213; -, Accad.Naz.Lincei., Mem. **4** (sez. 2) (1955) 73; G.Natta, P.Pino, P.Corradini, F.Danusso, E.Mantica, G.Mazzanti, G.Moraglio, J.Am.Chem.Soc. **77** (1955) 1708; G.Natta, P.Corradini, J.Polym.Sci. **20** (1956) 251.

M.L.Huggins, Chem.Revs. **32** (1943) 195

Ordered regions of proteins are assumed to contain peptide sections with screw axes (i.e., helical structures). Experimental proof of α-helix by
L.Pauling, R.B.Corey, H.R.Branson, Proc.Natl.Acad.Sci. (USA) **37** (1951) 205.

J.D.Watson, F.H.C.Crick, Nature [London] **171** (1953) 737, 964

Proposed the double helix for deoxyribonucleic acids based on experimental results of various authors (R.Franklin, M.H.F.Wilkins, etc.). F.H.C.Crick, J.D.Watson, M.H.F.Wilkins, Nobel prize 1962

G.N.Ramachandran, G.Kartha, Nature **176** (1955) 593

Experimental proof of the triple helix structure of collagen.

FRINGED MICELLES

O.Gerngross, K.Herrmann, W.Abitz, Z.physik.Chem. **B 10** (1930) 371

Observed crystalline reflections and amorphous halos of gelatin which were explained by the model of fringed micelles.

CHAIN FOLDING

K.H.Storks, J.Am.Chem.Soc. **60** (1938) 1753

Electron diffraction of very thin, stretched films of aliphatic polyesters and gutta-percha (*trans*-1,4-poly(isoprene)) indicates chain folding of polymers (fold length: 20 nm).

Single crystals were rediscovered 15-19 years later by three groups of researchers:

W.Schlesinger, H.M.Leeper, J.Polym.Sci. **11** (1953) 203 (single crystals of α-gutta-percha)
R.Jaccodine, Nature **176** (1955) 305 (spiral growth of low-molecular weight poly(ethylene))
P.H.Till, J.Polym.Sci. **24** (1957) 301 (single crystals of high-molecular weight poly(ethylene)

Chain folding was rediscovered 19 years later:

E.W.Fischer, Z.Naturforschg. **12a** (1957) 753 (spherulites are lamellar and not fibrillar)
A.Keller, Philos.Mag. **2** (1957) 1171 (picture of single crystal)

ORIENTATION FUNCTION

The Hermans orientation function is also known as the averaged second Legendre polynomial. In Russian scientific literature, it is called Tsvetkov-Hermans orientation function.

P.H.Hermans, P.Platzek, Kolloid-Z. **88** (1939) 68
V.N.Tsvetkov, Acta physicochim. USSR **16** (1942) 132
J.J.Hermans, P.H.Hermans, D.Vermaas, A.Weidinger, Rec.Trav.Chim.Pays-Bas **65** (1946) 427
P.H.Hermans, Contributions to the Physics of Cellulosic Fibres, Elsevier, Amsterdam 1946, p. 133

Literature to Chapter 7

7.0 GENERAL

IUPAC Macromolecular Division, Definitions of Terms Relating to Crystalline Polymers, Pure Appl.Chem. **61** (1989) 769
B.Wunderlich, Macromolecular Physics, Academic Press, New York, 3 vols., 1973 ff.
H.Tadokoro, Structure of Crystalline Polymers, Wiley, New York 1979
F.A.Bovey, Chain Structure and Conformation of Macromolecules, Academic Press, New York 1982
I.H.Hall, Ed., Structure of Crystalline Polymers, Elsevier Appl.Sci.Publ., London 1984
M.Senechal, Quasicrystals and Geometry, Cambridge Univ.Press, Cambridge, UK 1995
C.Janot, Quasicrystals. A Primer, Oxford Univ.Press, New York, 2nd ed. 1996
A.M.Kosevich, Ed., The Crystal Lattice. Phonons, Solitons, Dislocations, Wiley-VCH, Weinheim 1998
A.McPherson, Introduction to Macromolecular Crystallography, Wiley, Hoboken (NJ) 2002 (proteins only)
R.A.Pethrick, C.Viney, Eds., Techniques for Polymer Organization and Morphology Characterization, Wiley, Hoboken (NJ) 2003

7.1.0.a X-RAY DIFFRACTION

A.Guinier, X-Ray Diffraction in Crystals, Imperfect Crystals and Amorphous Bodies, W.H.Freeman, San Francisco 1963
B.K.Vainsthein, Diffraction of X-Rays by Chain Molecules, Elsevier, Amsterdam 1966
L.E.Alexander, X-Ray Diffraction Methods in Polymer Science, Wiley, New York 1969
K.Kakudo, N.Kasai, X-Ray Diffraction by Polymers, Elsevier, Amsterdam 1972
G.H.W.Milburn, X-Ray Crystallography. An Introduction to the Theory and Practice of Single-Crystal Structure Analysis, Butterworth, London 1972
F.J.Baltá-Calleja, C.G.Vonk, X-Ray Scattering of Synthetic Polymers, Elsevier, Amsterdam 1989
G.H.Stout, L.H.Jensen, X-Ray Structure Determination. A Practical Guide, Wiley, New York, 2nd ed. 1989
M.F.C.Ladd, R.A.Palmer, Structure Determination by X-ray Crystallography, Plenum, New York, 3rd ed. 1993

7.1.0.b INFRARED SPECTROSCOPY
S.Krimm, Infrared Spectra of High Polymers, Fortschr.Hochpolym.Forschg. **2** (1960) 51

7.1.0.c ELECTRON DIFFRACTION
E.W.Fischer, Electron Diffraction, in B.Ke, Ed., Newer Methods of Polymer Characterization, Interscience, New York 1964

7.1.0.d ELECTRON MICROSCOPY
H.J.Purz, E.Schulz, Die Elektronenmikroskopie in der Polymerforschung, Acta Polym. **30** (1979) 377

7.1.0.e MICROSCOPY
S.Y.Hobbs, Polymer Microscopy, J.Macromol.Sci.-Revs.Macromol.Chem. **C 19** (1980) 221
D.A.Hemsley, The Light Microscopy of Synthetic Polymers, Oxford Univ.Press, New York 1985
L.C.Sawyer, D.T.Grubb, Polymer Microscopy, Chapman and Hall, New York 1987
A.E.Woodward, Atlas of Polymer Morphology, Hanser, Munich 1989

7.1.7 POLYMORPHISM, 7.1.8 ISOMORPHISM, 7.1.9 UNIT CELLS
F.Danusso, Macromolecular Polymorphism and Stereoregular Synthetic Polymers, Polymer [London] **8** (1967) 281
G.Allegra, I.W.Bassi, Isomorphism in Synthetic Macromolecular Systems, Adv.Polym.Sci. **6** (1969) 549
R.Hosemann, The Paracrystalline State of Synthetic Polymers, Crit.Revs.Macromol.Sci. **1** (1972) 351
A.I.Kitaigorodsky, Molecular Crystals and Molecules, Academic Press, New York 1973
J.E.Spruiell, E.S.Clark, X-Ray Diffraction. Unit Cell and Crystallinity, Methods Exp.Phys. **16 B** (1980) 1
B.Wunderlich, M.Möller, J.Grebowicz, H.Baur, Conformational Motion and Disorder in Low and High Molecular Mass Crystals, Adv.Polym.Sci. **87** (1988) 1

7.2 CRYSTAL STRUCTURES
P.H.Geil, Polymer Single Crystals, Wiley, New York 1963
D.A.Blackadder, Ten Years of Polymer Single Crystals, J.Macromol.Sci. (Revs.) **C 1** (1967) 297
J.Willems, Oriented Overgrowth (Epitaxy) of Macromolecular Organic Compounds, Experientia **23** (1967) 409
R.A.Fava, Polyethylene Crystals, J.Polym.Sci. **D 5** (1971) 1
R.H.Marchessault, B.Fisa, H.D.Chanzy, Nascent Morphology of Polyolefins, Crit.Revs.Macromol. Sci. **1** (1972) 315
A.Keller, Morphology of Lamellar Polymer Crystals, in C.E.H.Bawn, Ed., Macromolecular Science (= Physical Chemistry Series One, Vol. 8 (1972), MTP International Review of Science)
R.J.Samuels, Structured Polymer Properties, Wiley, New York 1974
J.-I.Wang, I.R.Harrison, X-Ray Diffraction. Crystallite Size and Lamellar Thickness by X-Ray Methods, Methods Exp.Phys. **16 B** (1980) 128
D.C.Bassett, Principles of Polymer Morphology, Cambridge Univ.Press, Cambridge 1981
A.E.Woodward, Atlas of Polymer Morphology, Hanser, Munich 1988
P.Coradini, G.Guerra, Polymorphism in Polymers, Adv.Polym.Sci. **100** (1992) 183
A.E.Woodward, Understanding Polymer Morphology, Hanser, Munich 1995

7.3 CRYSTALLIZATION
L.Mandelkern, Crystallization of Polymers, McGraw-Hill, New York 1964
A.Sharples, Introduction to Polymer Crystallisation, Arnold, London 1966
B.Wunderlich, Macromolecular Physics, Vol. **2**, Crystal Nucleation, Growth, Annealing, Academic Press, New York 1976
K.A.Mauritz, E.Baer, A.J.Hopfinger, The Epitaxial Crystallization of Macromolecules, J.Polym.Sci.-Macromol.Rev. **13** (1978) 1
D.W.van Krevelen, Crystallinity of Polymers and the Means to Influence the Crystallization Process, Chimia **32** (1978) 279
R.L.Miller, Ed., Flow Induced Crystallization of Polymers(= Midland Macromolecular Monographs **6**), Gordon and Breach, New York 1979

G.S.Ross, L.J.Frolen, Nucleation and Crystallization (of Polymers), Methods Exp.Phys. **16 B** (1980) 339
A.J.Hugh, Mechanisms of Flow Induced Crystallization, Polym.Eng.Sci. **22** (1982) 75
J.C.Wittman, B.Lotz, Epitaxial Crystallization of Polymers on Organic and Polymer Substrates, Progr.Polym.Sci. **15** (1990) 909
K.Armitstead, G.Goldbeck-Wood, Polymer Crystallization Theories, Adv.Polym.Sci. **100** (1992) 219
M.Dosière, Ed., Crystallization of Polymers, NATO ASI Series C. Mathematical and Physical Sciences, Kluwer Academic, Boston 1993
Y.Long, R.A.Shanks, Z.H.Stachurski, Kinetics of Polymer Crystallization, Progr.Polym.Sci. **20** (1995) 651
J.Schultz, Polymer Crystallization, ACS Washington (DC) 2001
D.A.Ivanov, S.Magonov, Polymer Crystallization: Observations, Concepts and Interpretations, Springer, Berlin 2003
L.Mandelkern, Crystallization of Polymers, 2nd ed., Cambridge Univ.Press, Cambridge, UK; Vol. I, Equilibrium Concepts (2003); Vol. II, Kinetics and Mechanism (2005); Vol. 3, Structure, Morphology, and Properties (in preparation)

7.6 ORIENTATION

G.L.Wilkes, The Measurement of Molecular Orientation in Polymeric Solids, Adv.Polym.Sci. **8** (1971) 91
C.R.Desper, Technique for Measuring Orientation in Polymers, Crit.Revs.Macromol.Sci. **1** (1973) 501
I.M.Ward, Ed., Structure and Properties of Oriented Polymers, Halsted Press, New York 1975
H.Kawai, S.Nomura, Characterization and Assessment of Polymer Orientation, Dev.Polym.Charact. **4** (1983) 211
S.Fakirov, Ed., Oriented Polymer Materials, Hüthig und Wepf, Heidelberg 1996

References to Chapter 7

[1] B.Wunderlich, Macromolecular Physics, Academic Press, New York, Vol. I (1973), selected data of Table II.3
[2] R.J.Samuels, Structured Polymer Properties, Wiley, New York 1974, Figs. 2-4
[3] G.Natta, P.Corradini, Rubber Chem.Technol. **33** (1960) 703, Fig. 11
[4] W.Kern, J.Davidovits, K.J.Rauterkus, G.F.Schmidt, Makromol.Chem. **43** (1961) 106, Fig. 3
[5] P.H.Lindenmeyer, private communication
[6] P.H.Lindenmeyer, J.Polym.Sci. **C 1** (1963) 5, Figs. 6 and 7
[7] F.Lin, D.J.Meier, Langmuir **10** (1994) 1660, Fig. 7
[8] R.Kitamaru, F.Horii, K.Murayama, Macromolecules **19** (1986) 637, Fig. 10
[9] D.G.H.Ballard, G.W.Longman, T.L.Crowley, A.Cunningham, J.Schelten, Polymer **20** (1979) 399, Figures 4, 5, and 7
[10] F.J.Padden, Private communication, cf. H.D.Keith, F.J.Padden, Jr., R.G.Vadimsky, J.Appl.Phys. **42** (1971) 4585
[11] N.Sharples, Polymer **3** (1962) 250, data of Fig. 1
[12] A.Nakajima, F.Hameda, Pure Appl.Chem. **31**/(1-2) 1972, 1, Fig. 6; *cf.* B.Sedláček, Ed., IUPAC, Macromol.Microsymp. VIII and IX, Butterworths, London 1972, p. 1
[13] After a compilation of J.H.Magill, Macromol.Chem.Phys. **199** (1998) 2365, Table 1; Alkane data: D.Turnbull, R.L.Cormia, J.Chem.Phys. **34** (1961) 830
[14] A.Gandica, J.H.Magill, Polymer **13** (1972) 595, Fig. 2
[15] L.Mandelkern, F.A.Quinn, as reported by L.Mandelkern in R.H.Doremus, B.W.Roberts, D.Turnbull, Eds., Growth and Perfection of Crystals, Wiley, New York 1958, p. 467, data of Fig. 12
[16] R.J.Samuels, Structured Polymer Properties, Wiley, New York 1974, Figs. 2-31 and 3-22
[17] B.Wunderlich, private communication

[18] A.J.Pennings, J.M.M.A. van der Mark, A.M.Kiel, Kolloid-Z. **237** (1970) 336, (a) Fig. 2, (b) Fig. 14 b
[19] C.Ruscher, E.Schulz, private communication
[20] B.Wunderlich, B.Prime, private communication
[21] A.Jeziorny, S.Kepka, J.Polym.Sci. **B 10** (1972) 257, Fig. 3
[22] H.Hendus, private communication
[23] H.M.Morgan, Textile Res.J. **32** (1962) 866, Fig. 1

8 Mesophases

8.1 Introduction

8.1.1 Types of Mesophases

Mesomorphic substances have submicroscopic physical structures between those of crystals with long-range three-dimensional order and liquids or amorphous substances without any such order (G: *mesos* = middle, *morphe* = shape). Three classes of mesophases can be distinguished: liquid crystals, plastic crystals, and condis crystals.

In **liquid crystals (LCs)**, mesomorphic low-molecular weight molecules or polymer segments are ordered like their counterparts in true crystals but flow like regular liquids. LCs were the first examples of mesomorphic substances that were discovered which is the reason why "liquid crystal" and "mesophase" are often used interchangeably.

Liquid-crystalline behavior is shown by anisotropic low-molecular weight substances and by macromolecules with anisotropic segments (**liquid crystalline polymers, LCPs**). These anisotropic units are called **mesogens** (G: *genes* = to cause) (Section 8.2.1). LCs and LCPs form domains in which mesogens are oriented with respect to their molecule or segment long axes, respectively. The domains themselves are either only partially ordered relative to each other or not at all.

Thermotropic mesophases (Section 8.4) are formed by heating crystalline LCs or LCPs or cooling their melts. Certain thermotropic mesophases also result from heating or cooling other thermotropic mesophases (see below). **Lyotropic mesophases** (Section 8.3) are generated by some mesogenic compounds above their critical concentrations (G: *tropos* = a turn, change from *trepein* = to turn; *thermos* = hot; *lysis* = a loosening).

Plastic crystals are so to speak the reverse of liquid crystals: they show order with respect to spatial positions of domains but disorder of molecules (LCs) and segments (LCPs). The only known examples are spheroidal low-molecular weight compounds in densely packed cubic lattices. In contrast to liquid crystals, plastic crystals are always isotropic. Because plastic crystals have slip planes but no strong attractive forces between their molecules, they are easily deformable, sometimes even under their own weight; hence the name "plastic crystals."

Condis crystal is a rarely used short name for *con*formationally *dis*ordered *crystal*. These crystals contain several conformational isomers side-by-side while retaining the order of ideal crystals with respect to position and orientation of monomeric units.

Condis crystals of macromolecules with relatively rigid backbones or sidechains may form liquid crystals at higher temperatures or in suitable solvents. Examples are the hexagonal high-pressure phase of poly(ethylene) extended chain crystals and the modification II of 1,4-*trans*-poly(butadiene).

8.1.2 Other Mesophases

Liquid crystalline structures result from the parallel asssembly of relatively rigid molecules or polymer segments. The formation of LCs and LCPs is therefore a special kind of ordered intermolecular or intersegmental self-association.

Self-association may also be disordered with respect to segment long axes (Section 10.5). Low-molecular weight examples are the **micellizations** of tensides in water. **Tensides** are surface-active amphoteric chemical substances with hydrophilic "heads" and hydrophobic tails. In dilute solution, they form spherical **micelles** that contain ca. 10-100 molecules. In a spherical micelle, a hydrophobic interior consisting of randomly arranged tails is surrounded by a hydrophilic "surface" layer of molecule heads (see Figs. 5-9 and 5-12 in Volume II). At higher concentrations, rod-like and lamellar micelles are formed that at even higher concentrations develop liquid-crystalline structures.

Such micellizations occur not only with hydrophilic/hydrophobic segments but also in diblock polymers A_n-B_m in which A_n blocks consisting of n monomeric units of type A are incompatible with B_m blocks from m monomeric units B (Section 8.5). Depending on the ratio n/m, spherical, cylindrical, or lamellar **domains** (and intermediate types) are generated. Micellization of triblock molecules $A_nB_mA_n$ leads to physical crosslinks between domains. The substance cannot separate into two distinct phases, though, because the blocks are chemically coupled.

Such domains are also formed by **ionomers** which are copolymers of hydrophobic monomers with small proportions of ionic comonomers (Section 8.6). In the solid state, the ionic groups of ionomers associate in pairs, multiplets, ion droplets, clusters, etc. These ionic domains are much smaller than those formed by solid block polymers.

The combination of rigid mesogenic structures with amphiphilic ones leads to chemical structures that can be called "**amphotropic**" (Section 8.3.5).

Liquid crystals, micelles, microphases, and ionic domains are border type physical structures of matter in which one molecular characteristic dominates over all others. Combination of different chemical structures in varying proportions can therefore lead to physical structures which combine various features.

8.2 Mesomorphic States

In liquid crystals, mesogens form domains of micrometer size. Within each domain, molecule axes have a preferential direction. However, these preferential arrangements vary from domain to domain (Fig. 8-1). Liquid crystals and liquid-crystalline polymers are therefore categorized either according to the type of mesogen (Section 8.2.1) or according to the arrangement of mesogens in domains and the resulting appearance of mesophases (Section 8.2.2). These nomenclatures evolved historically and are therefore not very systematic.

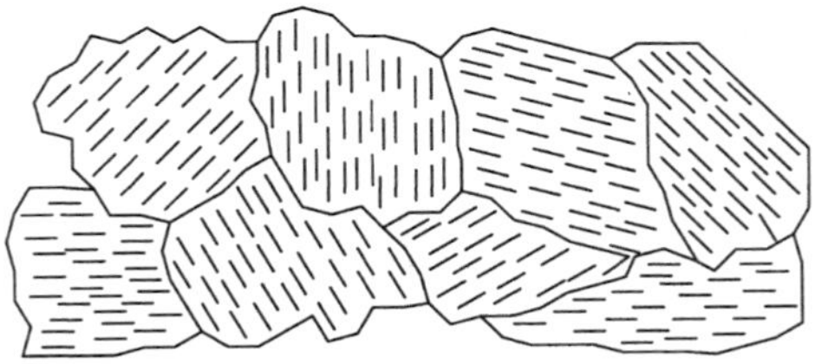

Fig. 8-1 Domains of mesogens, here as nematic arrangements of calamatic liquid crystals.

8.2.1 Mesogens

Mesogens are relatively small (nanometer size), rigid, anisotropic molecules or molecule segments that associate along their long axes to liquid-crystalline structures. These structures may be rod-like (**calamitic**; G: *kalamos* = reed), disc-like (**discotic**; L: *discus*; G: *diskos* = round, from *dikein* = to throw), board-shaped (**sanidic**; the mineral sanidine (moonstone) belongs to the orthoclase group) forms slab-like crystals), or **cross-shaped**. In low-molecular weight molecules, the whole molecule is often the mesogen. Examples are the rod-like pentacene and the disc-like cholesterol.

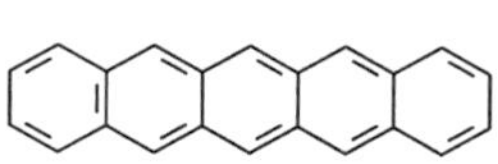

pentacene

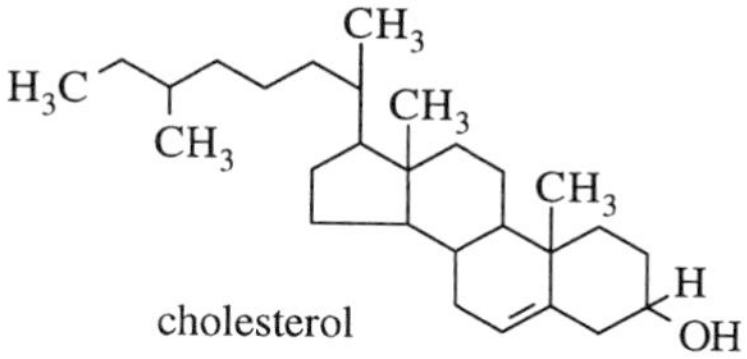

cholesterol

In most LCs and LCPs, however, mesogens are sections of molecules, not the molecules themselves. An LC molecule always contains at least one mesogen, an LCP molecule always several mesogens. These mesogens are either truly rigid or have such a constitution that a rotation around a "chain" bond of the mesogen either leaves the main axis undisturbed as in ester or amide bonds (which have partial double bond character) or restores the main axis by a crankshaft motion (bottom of this page). Mesogenic units are therefore semiflexible. In contrast, the main axis of flexible segments is not restored on rotation; hence, such segments do not form mesogens.

The linearity of the main axis of mesogens is preserved if the mesogens contain 1,4-connected (para connected) phenylene groups (example I) or carry in the para position (partially) conjugated groups such as –COO–, –CONH–, –N=N–, etc. (examples II and III). In III, even a crankshaft motion around the $-C_6H_4-COO-$ bond (dotted arrow) would not change the position of the main axis very much.

Mesogens are also present in poly(*p*-phenylene ethylene), $\text{+}C_6H_4-CH_2CH_2\text{+}_n$, since in melts of this compound methylene groups prefer to be in the trans conformation and not gauche so that a crankshaft motion restores the main axis of the chain. Such motions are not possible in poly(*p*-phenylenetrimethylene), $\text{+}C_6H_4-CH_2CH_2CH_2\text{+}_n$, and similar compounds with uneven numbers of methylene groups per repeating unit. Flexible segments are also generated by angled groups such as ether bonds (V) and ortho and meta substituted phenylene groups (IV and VI; see also next page).

semiflexible units

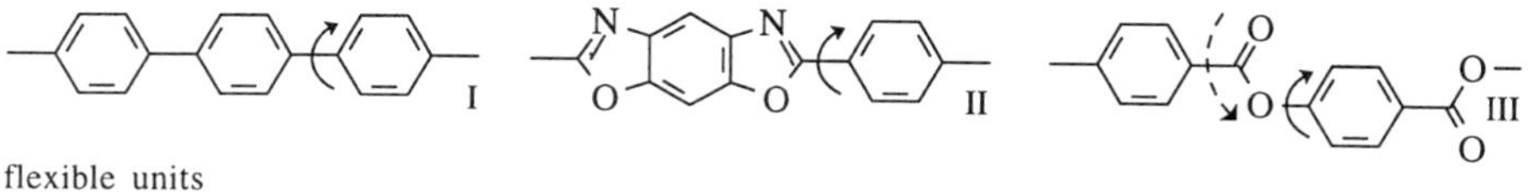

flexible units

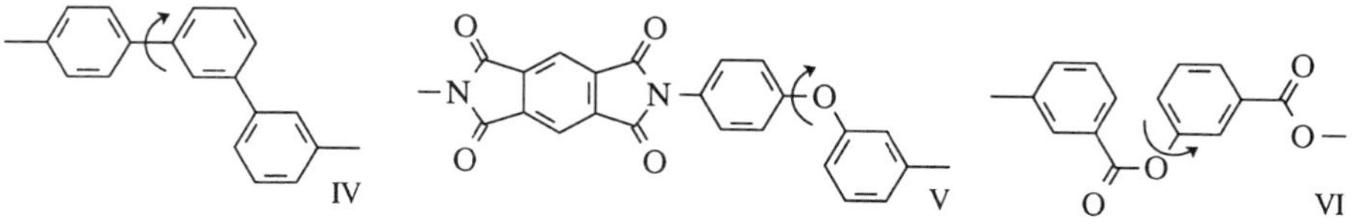

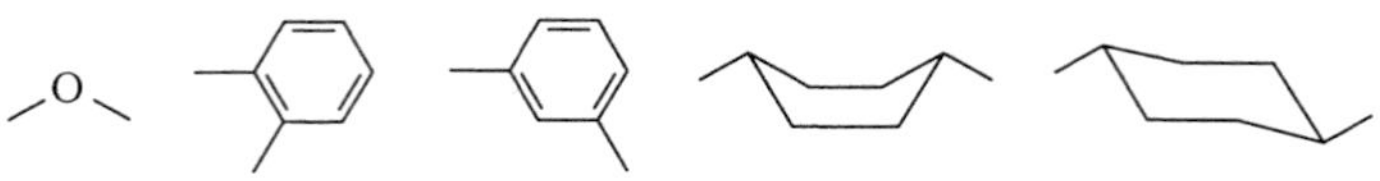

ether group — 1,2- (ortho) 1,3- (meta) phenylene group — cis trans 1,4-cyclohexylene group

1,4-Phenylene groups can also be replaced by 1,4-cyclohexylene groups. In an isotropic phase, cyclohexylene groups of 1,4-cyclohexane dicarboxylic acid are 34 % in cis and 66 % in trans positions. For packing reasons, however, these groups are 100 % in the trans position in crystalline and liquid-crystalline states.

A mesogen must therefore be sufficiently rigid over a certain minimum length L in order to become a mesogen. The decisive quantity is not a critical length L *per se* but the ratio of L to diameter d. For example, theoretical calculations showed this critical ratio to be $L/d \approx 4$ for calamitic mesogens (see also Fig. 8-6).

Geometric axial ratios of $L/d \geq 4$ of mesogens are sufficient to stabilize mesophases by more or less the parallel arrangement of main mesogen axes. The resulting degree of orientation of these axes in mesogenic domains is controlled by both the length of mesogens and the lengths of LC or LCP molecules.

At $L/d < 4$, geometric stabilization by intermolecular repulsion and space requirements of calamitic mesogens is insufficient to stabilize liquid-crystalline domains, and additional weak orientation-dependent attractive forces are needed. Examples are intersegmental hydrogen bonds.

Axial ratios of helical macromolecules in bulk and in helicogenic solvents are practically always greater than 4. Examples of lyotropic liquid crystals are deoxyribonucleic acids in water and poly(α-amino acid)s and poly(alkylisocyanate)s in certain organic solvents. In all these cases, $L_{mesogen} < L_{molecule}$ (see p. 107).

8.2.2 Order in Domains

Alignment of mesogens in domains leads to different refractive indices parallel and perpendicular (normal) to the polarization direction of the incident light: domains become birefringent. Optical properties of mesophases can thus be used to identify the physical structure of LCs and LCPs. However, anisotropic, birefringent, liquid phases are not necessarily liquid-crystalline materials since such properties can also be caused by higher melting crystallites in partially molten polymers or by gelation with crystallite formation in concentrated solutions.

Calamitic LCs and LCPs are subdivided into smectic, nematic, and cholesteric mesophases. Under the polarization microscope, **smectic mesophases** show fanlike entities, and low-molecular weight smectic mesophases have a soaplike feel (G: *smegma* = soap). The soaplike character is a consequence of the two-dimensional, layer-like arrangement of mesogens (Fig. 8-2).

At present, 12 different types of smectic mesophases are known. Four of them are orthogonal phases with average directions of long mesogen axes normal to two-dimensional layers (S_A, $S_{B,hex}$, $S_{B,cryst}$, S_E). The mesogen axes of smectic type S_C and seven other smectic types of mesogens are inclined with respect to the layer planes. Structures are either pseudohexagonal (S_F, S_G, S_I, S_J) or orthorhombic (S_D, S_H, S_K).

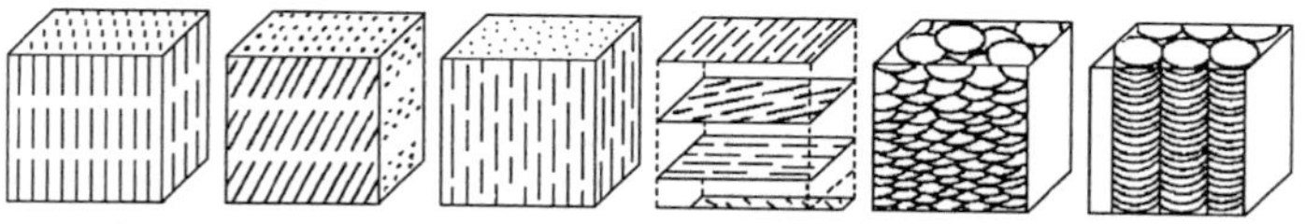

Fig. 8-2 Schematic representation of mesophases of low-molecular weight molecules.

Smectic mesophases S_A and S_C consist of layers of parallel rodlike mesogens that are normal to the layer planes (S_A) or at some angle (S_C, etc.). The polarization microscope shows fan-like (S_A) or schlieren (S_C) textures (D: plural of Schliere = streak).

Smectic mesophases $S_{B,hex}$ (= H_B) and S_F have perfect hexagonal arrangements of mesogen long axes. The rods are normal to the layer plane in $S_{B,hex}$ and inclined at an angle in S_F.

Nematic liquid crystals form thread-like structures (G: *nema* = thread) since they are only *one-dimensionally* ordered and not two-dimensionally ordered like smectic ones. The long axes of rod-like mesogens are still parallel in each domain but the preferential orientations of the domains are distributed at random (Fig. 8-1).

In LCPs, mesogens may be either in main chains or in side chains (Fig. 8-3). In both main-chain and side-chain LCPs, cylindrical mesogens can rotate about their long axes; these LCPs are optically uniaxial systems. Lath-like mesogens in sidechains may be oriented relative to each other; such systems are optically biaxial. Nematic mesophases do not form layers; they are rather swarmlike.

Discotic mesophases are either nematic discotic or columnar discotic (Fig. 8-2); the latter are also called "canonic phases" (G: *kanon* = rod, rule). In columnar ones, mesogens are packed like rolls of coins whereas arrangements of mesogens in nematic ones resemble heaps of coins.

Cholesteric liquid crystals were first detected in cholesterol, hence the name. Here, chiral mesogens are arranged helically in nematic domains. The pitch of the helix is usually in the micrometer range. As a result, optical light causes Bragg scattering. Many cholesteric LCs thus show pretty iridescent colors.

Liquid-crystal polymers are furthermore subdivided into **main-chain LCPs (MCLCPs)** and **side-chain LCPs (SCLCPs)**. Both MCLCPs and SCLCPs may be smectic, nematic, or cholesteric and contain calamitic, discotic, or lath-like mesogens. SCLCPs may also be interconnected to networks.

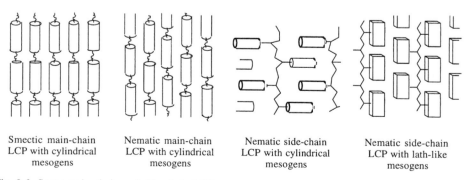

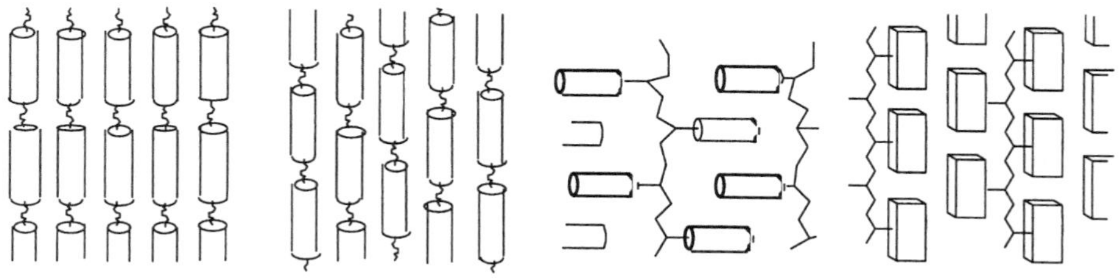

Fig. 8-3 Some main-chain and side-chain LCPs.

8.3 Lyotropic Liquid-Crystalline Polymers

8.3.1 Survey

Lyotropic LCPs may contain mesogens in either the main chains or the side chains. In the latter case, mesogens are preferably bound to the main chains by flexible spacers such as long alkylene or ester groups.

Most lyotropic LCPs are synthesized by man but some are found in nature such as tobacco mosaic virus (TMV) or deoxyribonucleic acids (DNA). Dilute aqueous salt solutions of TMV ($c > 2$ %) phase-separate spontaneously into two solutions: an upper dilute isotropic one and a lower concentrated birefringent one. At concentrations of more than ca. 6 %, aqueous NaCl solutions of double helices of DNA exhibit the properties of cholesteric mesophases. These properties are caused by mesogens and not by whole molecules *per se* since DNA forms random coils at high molecular weights (Fig. 4-6).

Poly(α-amino acid)s, $\text{+NH–CH(CH}_2\text{CH}_2\text{COOR)–CO+}_n$, such as poly($\gamma$-benzyl-L-glutamate) (PBLG) (R = $CH_2C_6H_5$) and poly(γ-methyl-L-glutamate) (PMLG) (R = CH_3), also produce lyotropic mesophases in helicogenic solvents, for example, PMLG in 12:5 mixtures of methylene chloride and ethyl acetate at $c > 15$ %. These fairly concentrated lyotropic solutions have lower viscosities than less concentrated isotropic ones since oriented mesogens easily slide past each other on shearing. PMLG can be dry spun (i.e., into air; see Volume IV) from lyotropic solutions to silk-like fibers since the liquid-crystalline structure is frozen-in on removal of solvent. The high orientation of mesogen main axes provides the fibers with good tensile strengths and low extensions. A pilot production of PBLG in the 1970s was terminated a few years later because the polymer was not cost competitive.

More successful are some other fibers that are spun from the nematic liquid-crystalline state: Kevlar®, X-500®, Technora®, PBO, and PBTZ.

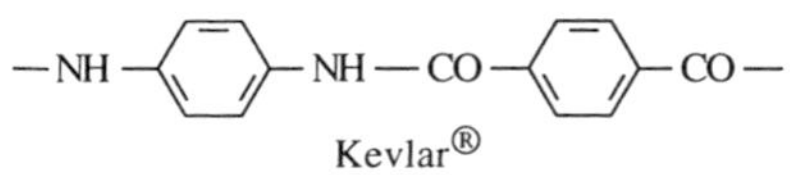

Kevlar®

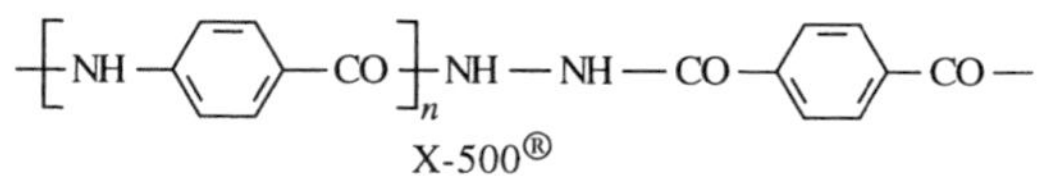

X-500®

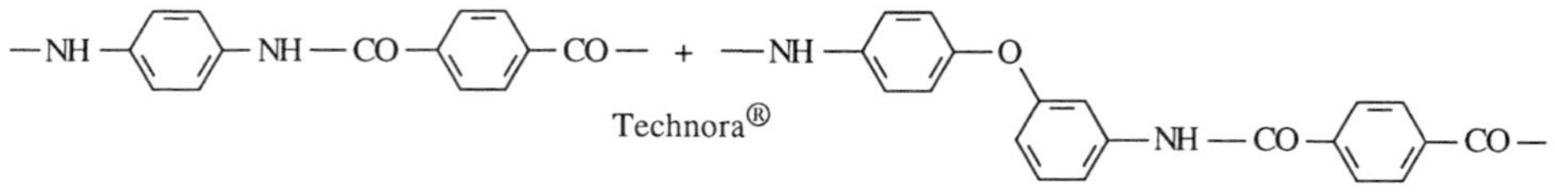

Technora®

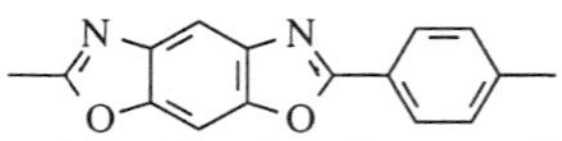

poly(p-phenylene benzoxazole)
PBO

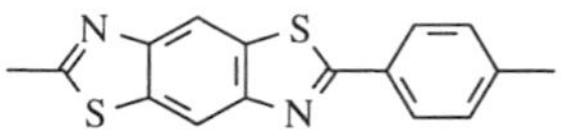

Poly(p-phenylene benzthiazole)
PBTZ

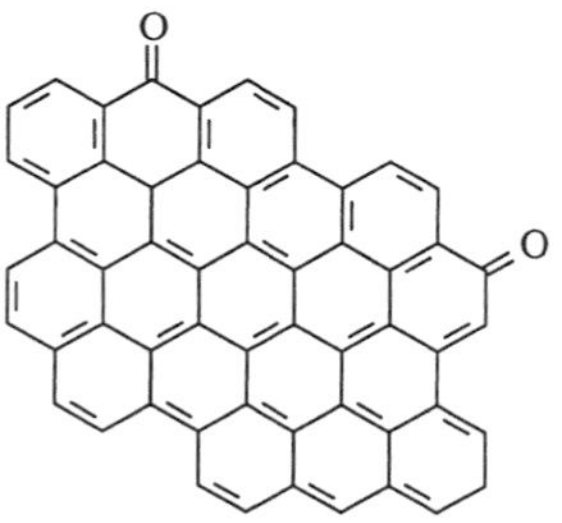

a polynuclear hydrocarbon found in tar

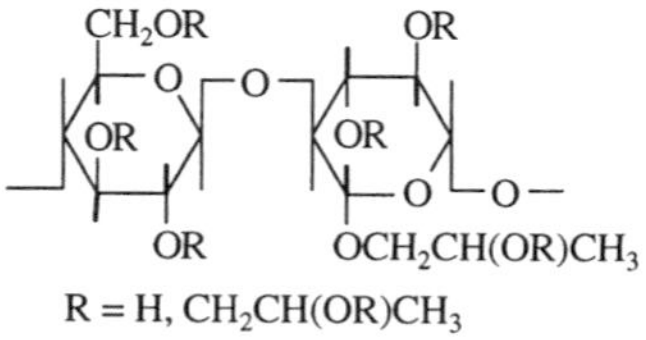

R = H, $CH_2CH(OR)CH_3$

hydroxypropylcellulose

Poly(*p*-phenylene terephthalamide) (PPTA; Kevlar®, Arenka®, Vniivlon®) dissolves only in solvents that break the strong hydrogen bridges between repeating units. The industrial solvent of choice was first concentrated sulfuric acid; a newer solvent is chlorinated *N*-methylpyrrolidone. Kevlar® is wet-spun into a precipitating bath from the nematic state in 20 wt% polymer solution in H_2SO_4. The resulting fibers have high strengths even without stretching (see p. 255). Kevlar® fibers are offered as tire cords (Kevlar 29 with $E \approx 60$ GPa) and high-modulus fibers (Kevlar 49 with $E \approx 140$ GPa).

The 1:1 copolyamide from terephthaloyl chloride and a 50:50 mixture of *p*-phenylene diamine and 3,4'-diaminodiphenyl ether (Technora® HM-50) has similar properties to those of Kevlar® but is spun to fibers from 6 wt% isotropic polymer solutions in *N*-methylpyrrolidone + $CaCl_2$. The fibers are then drawn at 460-500°C which produces the nematic structure on cooling. Similar high tensile fibers are obtained from the polyamidehydrazide X-500®.

Fibers with even higher tensile moduli of up to 320 GPa are obtained from poly(*p*-phenylenebenzoxazole) (PBO) and poly(*p*-phenylenebenzthiazole) (PBTZ) that both exist in a "cis" form (as shown for PBO) and a "trans" form" (as shown for PBTZ). These polymers are produced by polycondensation in phosphoric acid and are spun directly from these solutions.

While most of the fibers are spun from nematic liquid-crystalline polymer solutions, films are cast from isotropic ones. After removal of the solvent, nematic structures are produced by heating the films at temperatures between the glass temperature and the higher isotropization temperature (nematic LCP ⇄ isotropic melt).

Nematic-cholesteric lyotropic mesophases are fairly rare. An example is hydroxypropyl cellulose (HPC) which dissolves in water to dilute solutions at temperatures below 41°C; phase separation starts at $T > 41$°C (Fig. 8-4). At room temperature and polymer weight fractions greater than 41 %, mesophases appear. These mesophases have brilliant iridescent colors and must therefore be cholesteric. The required chirality is produced by the asymmetric carbon atoms *C of oxypropylene units $-O^*CH(CH_3)-CH_2-$, the substituents of cellulose chains.

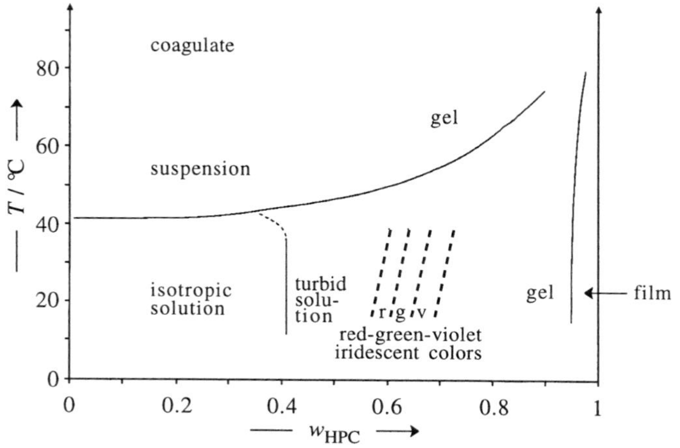

Fig. 8-4 Phase diagram of aqueous solutions of hydroxypropyl cellulose. By kind permission of American Chemical Society, Washington, (DC) [1].

LCPs with cholesteric mesogens can also be obtained by bipolymerization of LC polymer forming monomers with small proportions of chiral comonomers. An example is the co-condensation of small proportions of valine, $H_2NCH(CH(CH_3)_2)COOH$, with *p*-aminobenzoic acid, $H_2N(p\text{-}C_6H_4)COOH$. Cholesteric mesophases may also be produced by adding optically active low-molecular weight compounds to lyotropic nematic mesophases of polymers. An example is the addition of (+)-1-methylcyclohexanone to solutions of poly(*p*-benzamide).

Smectic-lyotropic LC polymers are apparently unknown; they should have very high viscosities. Synthetic polymers with discotic mesogens are laboratory curiosities. Such polymers do exist in nature, though, for example, as polynuclear hydrocarbons in tar and asphalt. Discotic mesogens are also present in coke.

8.3.2 Phase Separation

Stiff rods in very dilute solutions are totally disordered; such solutions are isotropic. At higher concentrations, space requirements cause rods to start to align themselves along their long axes. Solutions become anisotropic if the average distance between two rods is smaller than the length of the rods. There is a critical concentration at which isotropic solutions convert to anisotropic ones and two phases begin to form (Fig. 8-4) This concentration can be calculated from second virial coefficients or with the help of lattice theories of solutions.

Onsager Theory

According to Lars Onsager, excluded volumes of dissolved polymer molecules can be calculated from second virial coefficients A_2 in dilute solution, for example from the concentration dependence of reduced osmotic pressures, $\Pi/(RTc_2) = (1/\overline{M}_n) + A_2c + ...$ (see Section 10.4.3 for coils), since A_2 is a measure of two-body interactions.

Calamatic liquid crystalline polymers contain rod-like segments with excluded volumes of $u = 8\ V[1 + (L/U)\sin\gamma]$ (Table 4.1) where γ = average angle of orientation. Circular cylinders have circumferences $U = 2\,\pi R$, volumes $V = \pi R^2L$, diameters $d = 2\,R$, axial ratios $\Lambda = L/d$, and thus excluded volumes of $u = 2\,\pi\,(L^3/\Lambda^2) + 2\,(L^3/\Lambda)\sin\gamma$.

In ideal nematic mesophases, all rods are exactly parallel with respect to their long axes; the angle of orientation is zero. The excluded volume in nematic mesophases is therefore much smaller than that in isotropic solutions. Nematic mesophases have therefore much higher concentrations than isotropic solutions with which they are in equilibrium. However, the viscosity of nematic solutions is much smaller than that of isotropic solutions since parallel rods slide past each other easily.

The transition from isotropic solutions with randomly oriented rods to nematic mesophases with domains of parallel rods occurs at a critical volume fraction ϕ_{crit}. This transition is of the first thermodynamic order according to the Onsager theory. The volume fractions at the nematic-isotropic transition were calculated to be $\phi_{nem} \cong 4.5/\Lambda$ in the nematic phase and $\phi_{iso} \cong 3.3/\Lambda$ in the isotropic one.

This description of the nematic/isotropic phase separation is qualitatively correct but applies only to infinitely long rods in infinitely small concentrations since it is based on the 2nd virial coefficient and neglects the effects of the 3rd (and higher) virial coefficients at higher concentrations, i.e., interactions of three and more bodies. The theory thus considers mutual orientations of *two* segments (and thus A_2) but not the conjoint orientation of *many* segments. Indeed, critical concentrations for nematic-isotropic transitions are higher than predicted by the Onsager theory (see Fig. 8-6).

Flory Theory

The lattice theory of Paul Flory is more successful. In this theory, a rod-like molecule is placed on a two-dimensional lattice at an angle γ to the preferential direction of molecules in a nematic domain. The molecule is thus "disoriented." Since the long axis of the molecule is not parallel to the axes of the rectangular lattice, it is replaced by a series of stepwise segments. The long axis of the segments coincides with the preferential direction; laterally, a segment is as wide as a lattice site (Fig. 8-5). The size of a lattice site is identical with the size of a solvent molecule. The number of segments per molecule thus equals the axial ratio, $\Lambda = L/d$. The disorientation factor for a rod in a two-dimensional lattice is $Y = \Lambda \sin \gamma$ and in the general case

(8-1) $$Y = B + A(\Lambda - C) \sin \gamma$$

The constant A equals unity for square rods and $4/\pi$ for cylindrical ones. The factor B is zero for a perfect orientation of rods and unity for a random one. The parameter C is either 0 or 1, depending on the particular model.

The simple lattice theory assumes that nematic phases are generated solely by the arrangements and space requirements of rod-like segments, i.e., entropic effects. An additional effect of mixing enthalpy is represented by a van Laar term, similarly to the approach by the lattice theory of random coils (Section 10.2.2).

The derivation of the expression for the Gibbs mixing energy per mole lattice site is too lengthy for the scope of this book. However, the result may be compared to that of the lattice theory for random coils (see Chapter 10). In both cases it is assumed that each polymer segment and each solvent molecule occupies one lattice site. For random coils, the "segment" is usually a monomeric unit, whereas for rods it is one of the y segments of the macromolecule where $y = \Lambda \sin \gamma$ with $\Lambda = L/d$ = axial ratio and γ = angle (see Fig. 8-5). For rods (index 2) in solvents (index 1), simple lattice theory calculates the Gibbs mixing energy per amount (mole) of lattice site as

(8-2) $$\Delta G_{\text{mix,m}}/RT = \chi\phi_1\phi_2 + \phi_1 \ln \phi_1 + (1/\Lambda)\phi_2 \ln \phi_2 - [\phi_1 + \phi_2(y/\Lambda)] \ln [1 - \{1 - (y/\Lambda)\}\phi_2] - (1/\Lambda)\phi_2[1 - y + \ln (\Lambda y^2)]$$

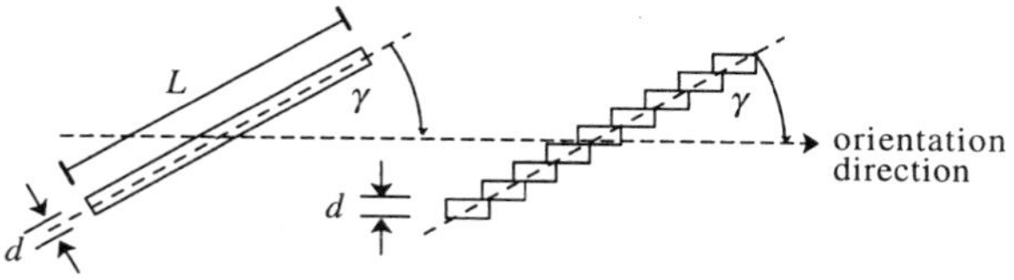

Fig. 8-5 Approximation of a rod with length L and diameter d by a series of staggered segments. The rod resides at an angle γ to the orientation direction.

which can be compared with Eq.(10-24) for random coils with degrees of X_2:

(8-3) $$\Delta G_{\text{mix,m}}/RT = \chi\phi_1\phi_2 + \phi_1 \ln \phi_1 + (1/X_2)\phi_2 \ln \phi_2 \quad ; \quad \phi_1 = 1 - \phi_2$$

The first summand on the right side of Eqs.(8-2) and (8-3) is the enthalpic van Laar term with the Flory-Huggins polymer/solvent interaction parameter, χ (see p. 309). All other summands are entropic terms. Both equations describe the Gibbs mixing energy as a function of the volume fraction of the solvent, ϕ_1, and the molecule size, i.e., the degree of polymerization X_2 of random coils (Eq.(8-3)) and the axial ratio $\Lambda \equiv L/d$ of rods (Eq.(8-2)). The equation for rods also contain a "disorientation factor", $y = \Lambda \sin \gamma$.

The angle γ becomes 90° if the disorientation factor y equals the axial ratio Λ. In this case, segments are not only all oriented but also identical with monomeric units: $\Lambda = L/d$ equals the degree of polymerization, X_2. Eqs.(8-2) and (8-3) become identical except for the additional disorientation term in the case of rods:

(8-4) $$\Delta G_{\text{mix,m}}/RT = \chi\phi_1\phi_2 + \phi_1 \ln \phi_1 + (1/\Lambda)\phi_2 \ln \phi_2 - (1/\Lambda)\phi_2[1 - \Lambda + \ln (\Lambda^3)]$$

The disorientation factor can be assumed to minimize the Gibbs mixing energy. Differentiation of Eq.(8-2) with respect to y, setting the result to equal zero, and solving the resulting expression for ϕ_2 delivers

(8-5) $$\phi_2 = [\Lambda/(\Lambda - y)][1 - \exp (- 2/y)]$$

This equation has two solutions for high concentrations (large ϕ_2) of long rods (large Λ): an upper one corresponding to a maximum of $\Delta G_{\text{mix,m}}$ and a lower one with a minimum value of y^* that corresponds to a minimum concentration ϕ_2^*. If in the latter case the first derivative is set to equal zero, $\mathrm{d}\phi_2/\mathrm{d}y = 0$, and the result solved for Λ, one obtains

(8-6) $$\Lambda = y^* + (1/2)\, y^{*2}[\exp (2/y^*) - 1]$$

Introduction of Eq.(8-6) into Eq.(8-5) for $y = y^*$ and $\phi_2 = \phi_2^*$ delivers

(8-7) $$\phi_2^* = 1 - [1 - (2/y^*)] \exp (- 2/y^*)$$

Developing exp $(- 2/y^*)$ in a series, replacing y^* by $y^* = (1/2)(\Lambda - 1) - (1/3y^* + 1/6y^{*2} + ...)$, and neglecting terms with $1/y^*$ leads to

(8-8a) $$\phi_2^* \cong 1 - [1 - 4/(\Lambda - 1)]^2 + ...$$

Numerical calculations show that this series can be approximated by the closed expression

(8-8b) $$\phi_2^* \cong 8\,(1 - 2\,\Lambda^{-1})/\Lambda$$

For an axial ratio $\Lambda = 20$, Eq.(8-8b) predicts $\phi_2^* = 0.360$ whereas the simple lattice theory, Eq.(8-8a), delivers $\phi_2^* = 0.377$ and an improved theory a value of $\phi_2^* = 0.364$.

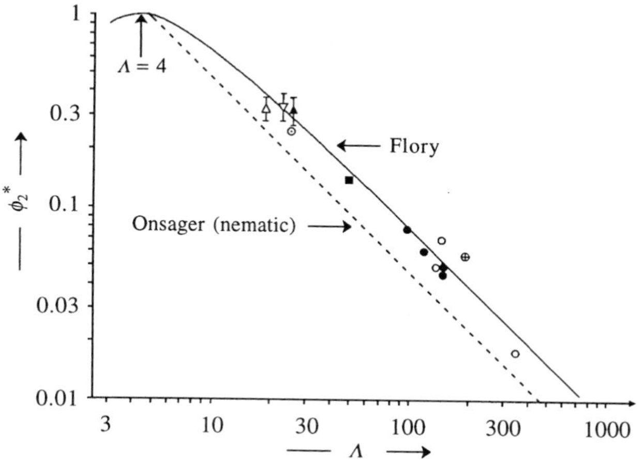

Fig. 8-6 Critical volume fractions ϕ_2* for the phase separation isotropic ⇄ nematic as a function of the axial ratio Λ of mesogens (= true axial ratio L/d of rods or Kuhnian axial ratios L_K/d of worm-like (semi-flexible) molecules). $L_K = 2\ L_{ps}$; for persistence lengths L_{ps}, see Table 4-9.
● Poly(γ-benzyl-L-glutamate)s with different molecular weights in *m*-cresol, O ditto in mixtures of *N,N'*-dimethylformamide and methanol; ⊕ poly(γ-phenyltrimethylene-L-glutamate); ʘ deoxyribonucleic acid (M_r = 110 000) in aqueous NaCl solution; ■ poly(*p*-benzamide) in H_2SO_4; ▲ cellulose acetate; Δ,▼ various cellulose ethers. —— Prediction by Eq.(8-8b); - - - Onsager theory.

Eq.(8-8) is well obeyed by rigid rods and semi-flexible molecules if the latter are characterized by a Kuhnian axial ratio $\Lambda_K = L_K/d$ (Fig. 8-6). These axial ratios can be calculated from molecular data as follows :

Circular rods with molar masses $M = XM_u$, diameters d, and lengths $L = XL_u$ have axial ratios of $\Lambda = L/d = XL_u/d = (ML_u)/(dM_u)$ where X = degree of polymerization, M = molar mass, M_u = molar mass of repeating unit, and L_u = length of repeating unit (as projection on the molecule axis). For molecularly non-uniform polymers, mass-averages are to be used for molar masses and degrees of polymerization.

The diameter d of molecules with circular cross-sections is calculated from the volume $V_u = \pi R^2 L_u = \pi (d^2/4)L_u$, molar masses $M_u = m_u N_A$ of repeating units, and densities $\rho = m_u/V_u$ of substances:

(8-9) $$d = (2/\pi^{1/2})[M_u/(\rho N_A L_u)]^{1/2}$$

whereas the axial ratio is calculated as

(8-10) $$\Lambda = XL_u/d = (\pi^{1/2}/2)(\rho N_A)^{1/2}(L_u/M_u)^{3/2}M$$

The form factor $(\pi^{1/2}/2)$ of cylindrical rods becomes unity for square rods with cross-sections a^2.

For *semi-flexible chains*, Kuhnian lengths L_K have to be used instead of rod lengths XL_u. Apparently, it does not matter whether the chains are worm-like or consist of rigid segments with flexible hinges or spacers (Fig. 8-7). One has only to assume that the flexibility of hinges and spacers is independent of the orientation of rigid segments in the anisotropic phase.

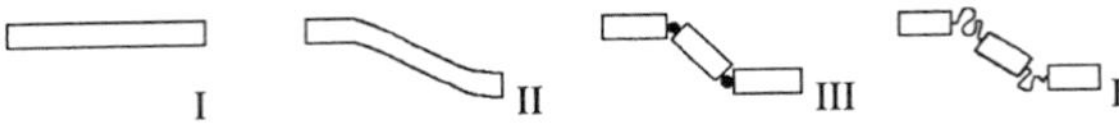

Fig. 8-7 Models for rigid and semi-flexible molecules. From left to right: rigid rod, worm-like chain, rod-like segments with flexible hinges, and rod-like segments with spacers.

In unperturbed Kuhnian chains, the mean-square end-to-end distance of an unperturbed chain, $\langle r^2\rangle_o = Nb^2C_N$, Eq.(4-38), consisting of N monomeric units with bond lengths b and characteristic ratios C_N, is replaced by an ersatz chain with N_K Kuhnian segments of Kuhnian length L_K, $\langle r^2\rangle_o = N_K L_K{}^2$, Eq.(4-39). The axial ratio of a Kuhnian segment is

(8-11) $$\Lambda_K = L_K/d$$

and therefore independent of the molar mass of the polymer provided that chains are long enough to be modeled as Kuhnian chains. Comparison of axial ratios $\Lambda = XL_u/d$, Eq.(8-10), and $\Lambda_K = L_K/d$, Eq.(8-11), with $XL_u \equiv r_{cont} \equiv N_K L_K$ shows furthermore that the axial ratio of a Kuhnian chain is always smaller than that of a completely stiff chain, $\Lambda_K = (1/N_K)\,\Lambda$, by a factor of $1/N_K$. Since the axial ratio controls the volume fraction at which phase separation sets in, Eq.(8-8b), it follows that semi-flexible LC chains phase separate at higher concentrations than rigid ones.

8.3.3 Chemical Potentials

Phase separation of polymer solutions occurs at a critical volume fraction $\phi_{2,crit}$ of the polymer which is the volume fraction at which the second derivative of the chemical potential of the solvent 1 with respect to the volume fraction of the polymer 2 becomes zero, $\partial^2\Delta\mu_1/\partial\phi_2{}^2 = 0$ (Section 10.2.3). The chemical potential $\Delta\mu_i$ of a component i is the partial molar Gibbs energy of this component, i.e., the first derivative of the Gibbs energy with respect to the amount-of-substance n_i, $\mu_i \equiv \tilde{G}_{i,m} \equiv (\partial G_i / \partial n_i)_{T,p,n_{j\neq i}}$. It is usually given as the difference to the pure state of this component, i.e., as $\Delta\mu_i \equiv \mu_i - \mu_{i,0} \equiv \Delta\tilde{G}_{i,m} \equiv (\partial\Delta G_i / \partial n_i)_{T,p,n_{j\neq i}}$.

For the calculation of chemical potentials of rod-like molecules in solution. Eqs.(8-2) and (8-5) are written for amounts $n_i = N_i/N_A$ (i = 1, 2) instead of volume fractions ϕ_i using $\phi_1 = N_1/N_g$ and $\phi_2 = N_2\Lambda/N_g$. The modified Eq.(8-5) (for the anisotropic phase) is then combined with the similarly modified Eq.(8-2). The resulting expression for ΔG_{mix} is differentiated with respect to n_1 to give the chemical potential $\Delta\mu_{1,n}$ for the solvent in the anisotropic (nematic) phase. Similarly obtained are the chemical potentials of the solvent in the isotropic phase ($\Delta\mu_{1,i}$) and that of the polymer in the nematic phase ($\Delta\mu_{2,n}$) and in the isotropic phase ($\Delta\mu_{2,i}$), both relative to the completely ordered phase:

(8-12) $$\Delta\mu_{1,n}/RT = \chi\phi_2{}^2 + \ln(1-\phi_2) + [(y-1)/\Lambda]\phi_2 + 2/y$$

(8-13) $$\Delta\mu_{1,i}/RT = \chi\phi_2{}^2 + \ln(1-\phi_2) + [1-\Lambda^{-1}]\phi_2$$

(8-14) $$\Delta\mu_{2,n}/RT = \chi\Lambda(1-\phi_2)^2 + \ln(\phi_2/\Lambda) + (y-1)\phi_2 - \ln y^2 + 2$$

(8-15) $$\Delta\mu_{2,i}/RT = \chi\Lambda(1-\phi_2)^2 + \ln(\phi_2/\Lambda) + (\Lambda-1)\phi_2 - \ln\Lambda^2$$

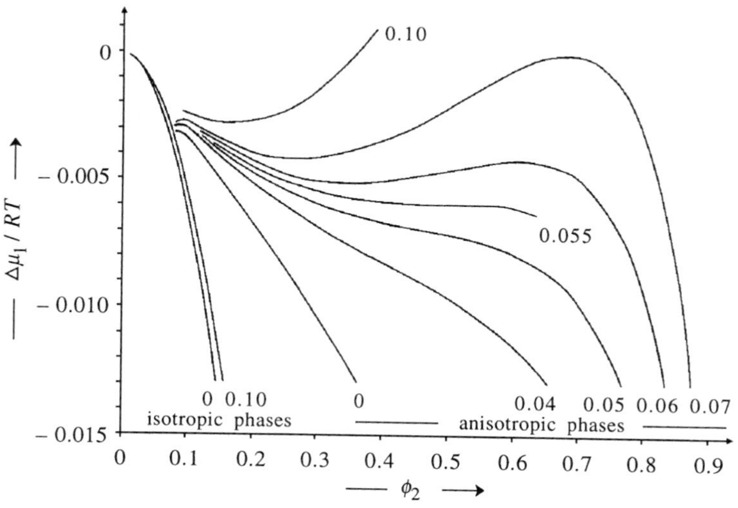

Fig. 8-8 Reduced chemical potential of the solvent as a function of the volume fraction of rod-like polymer molecules with axial ratios L/d = 100 at different interaction parameters $0 \leq \chi \leq 0.10$. Calculations with Eqs.(8-12) and (8-13) [2].

The chemical potential of the solvent is independent of the shape of the solute. Eq.(8-13) for isotropic solutions of rods is therefore identical with Eq.(10-25) for solutions of random coils. Chemical potentials of polymers are different for solutions of rods and coils, respectively (cf. Eqs.(8-15) and (10-26)).

The chemical potential of the solvent in the *isotropic* phase decreases with increasing volume fraction of rods first slowly, than faster (Fig. 8-8). The presence of enthalpic interactions, as expressed by $\chi > 0$, increases chemical potentials only slightly compared to the athermic case ($\chi = 0$).

Anisotropic phases are always more concentrated than isotropic ones as seen by their volume fractions at the same chemical potential. As a consequence, molecularly non-uniform polymers are fractionated: molecules with higher molar masses (longer rods) move almost exclusively to the anisotropic phase. It is for this reason that one cannot produce "molecular composites" of reinforcing rod-like molecules in coil-like polymers such as poly(p-phenylenebenzthiazole) in polyamide 6 by solvent evaporation from and coagulation of their dilute isotropic solutions. On concentrating such solutions, critical phase separation temperatures are surpassed and rods and coils move to separate phases.

In contrast to isotropic phases, anisotropic ones are strongly affected by interaction parameters χ (Fig. 8-8). A small value of χ is sufficient to move $\Delta\mu_{1,n}/RT$ to considerably higher values of ϕ_2 (cf., for example, $\Delta\mu_{1,n}/RT = -0.01$ at $\phi_2 = 0$ and $\phi_2 = 0.04$). For the example of Fig. 8-8 ($\Lambda = 100$), a practically constant value of $\Delta\mu_{1,n}/RT = -0.0059$ is observed for $\chi = 0.055$ in the range $0.41 < \phi_2 < 0.55$. At even higher values of χ, maxima and minima are seen in $\Delta\mu_1/RT = f(\phi_2)$. At constant $\Delta\mu_1/RT$, minima followed by maxima indicate the coexistence of two anisotropic phases.

This theoretical prediction has been verified experimentally (Fig. 8-9). At low concentrations of poly(γ-benzyl-L-glutamate) in helicogenic solvents, only one isotropic phase I was observed. At higher concentrations, a heterogeneous phase I + LC appeared, consisting of isotropic and liquid-crystalline domains. The transition I $\rightarrow$ I + LC moves

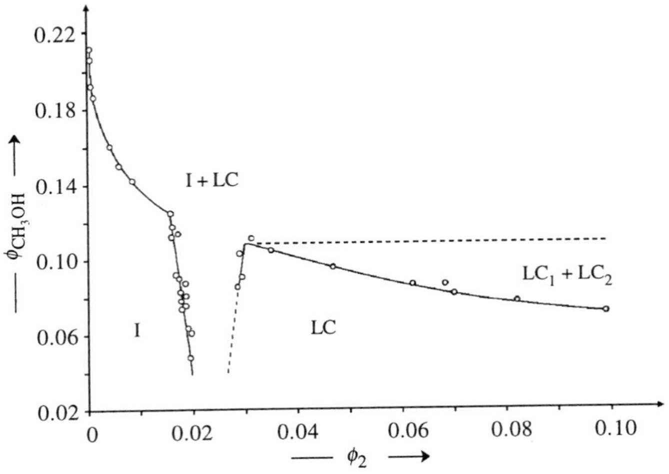

Fig. 8-9 Volume fractions ϕ_{CH_3OH} of the precipitant methanol that are required to induce phase separations in *N,N*-dimethylformamide solutions of poly(γ-benzyl-L-glutamate) (Λ = 350) [3].
I = Isotropic solution, LC, LC_1, LC_2 = liquid-crystalline phases. The volume fraction of precipitant required for the onset of phase separation and the inverse precipitation temperature, respectively, are proportional to the interaction parameter χ.

rapidly to greater volume fractions ϕ_2 of the polymer with decreasing concentration of precipitant CH_3OH (decreasing χ). At $\phi_{CH_3OH} < 0.12$ and $0.017 < \phi_2 < 0.03$, the broad (I + IC) range degenerates to a narrow "chimney" that separates the isotropic phase I from the liquid-crystalline phase LC. The theory also predicts and experiment confirms a heterogeneous phase $LC_1 + LC_2$ composed of two liquid crystalline structures that exist above a certain value of ϕ_2 (see Fig. 8-9).

Theory predicts the "chimney" to be parallel to the χ axis, i.e., parallel to ϕ_{CH_3OH} (Fig. 8-9) or the inverse precipitation temperature. Experimentally, either tapered or curved chimneys are often found. This difference between theory and reality is caused by several assumptions that are included in the theory, such as identical interaction parameters in both isotropic and anisotropic phases which is generally not true. Furthermore, persistence lengths $L_{ps} = L_K/2$ are not independent of temperatures T or interaction parameters χ but rather decrease with increasing T or decreasing χ which causes the chimney to bend at higher concentrations.

According to theory, heterogeneous phases $LC_1 + LC_2$ always appear if $\Lambda > 50$ or $\Lambda_K > 50$. Such phases should also be formed for certain ranges of interaction parameters, apparently for $0 \leq \chi \leq 0.12$.

Solvents for lyotropic LCs are usually low-molecular weight compounds with low melting temperatures and without liquid-crystalline properties. Phase diagrams for the usual temperature ranges thus show only the transition isotropic ⇄ liquid-crystalline (usually: nematic) and, in special cases, transitions between two different liquid-crystalline phases such as smectic ⇄ nematic. At lower temperatures, crystallizing components lead to crystalline phases. At low temperatures, high LCP concentrations, and fast cooling, glassy phases are sometimes observed.

More complicated phase diagrams are obtained from solutions of LCPs in LCs (Fig. 8-10, m = 3). At high temperatures, only an isotropic phase i exists in the whole concen-

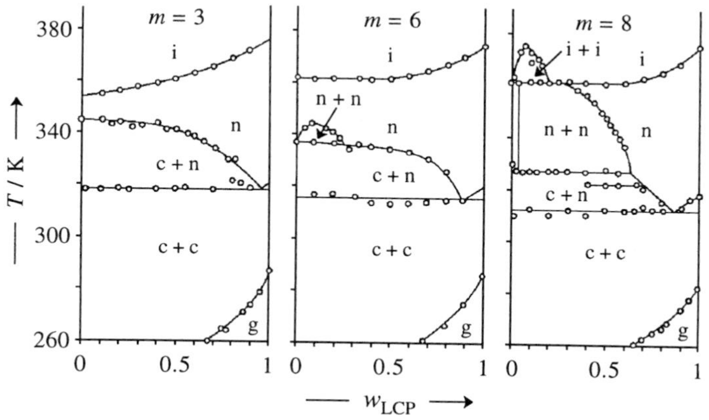

Fig. 8-10 Phase diagrams of mixtures of a liquid-crystalline side-chain polymer LCP and a low-molecular weight LC of similar constitution [4].
LCP: $(CH_3)_3Si[OSiRCH_3]_{72}OSi(CH_3)_3$ with R = $(CH_2)_4O(p\text{-}C_6H_4)COO(p\text{-}C_6H_4)OCH_3$
LC: $C_2H_5(CH_2)_4O(p\text{-}C_6H_5)COO(p\text{-}C_6H_4)(CH_2)_mH$ with m = 3, 6, or 8.
i = Isotropic phase, i+i = two isotropic phases, n = nematic phase, n+n = two nematic phases, c+n = a crystalline and a nematic phase, c+c = two crystalline phases, g = glass.

tration range $0 < w_2 < 1$ of the polymer 2. At lower temperatures, the LCP forms a nematic lyotropic phase n, again for the whole concentration range. The greater the weight fraction w_2 of the polymer, the broader is the temperature range of this phase.

At still lower temperatures, an inhomogeneous phase c + n appears that is composed of crystalline low-molecular weight LC and nematic LCP. At even lower temperatures, a heterogeneous phase c + c of crystalline LC and crystalline LCP exists. At very low temperatures and high LCP concentrations, a glassy state is found.

The phase diagram gets more complicated for $m = 6$. At low LCP concentrations, an additional heterophase n + n with two types of nematic domains (LC + LCP) appears in a very small temperature interval. The temperature-concentration range of this double-nematic phase gets larger if the number of methylene groups in the low-molecular weight LC is increased to 8 (Fig. 8-10, right). In addition, a new heterogeneous phase appears at low w_{LCP} and high T that consists of two types of isotropic regions.

8.3.4 Orientation of Mesogens

The average orientation of mesogens in domains of mesophases of both main-chain and side-chain liquid crystalline polymers is described by the Hermans orientation factor $f_{or} = (1/2)\,[3\,\langle\cos^2\beta\rangle - 1]$ in complete analogy to the orientation of crystallites in semicrystalline polymers. The angular term indicates the spatially averaged deviation of the mesogen long axes from the symmetry axis of the distribution of orientations of these axes in domains. The orientation factor is unity if mesogen long axes are parallel to the symmetry axis ($\beta = 0°$) and $-1/2$ if all axes are normal to the symmetry axis ($\beta = 90°$).

Polymer molecules are defined as rods if the conventional contour length is smaller than the Kuhnian length, $L_{cont} < L_K$. Rods with short contour lengths cannot form many contact points for attraction or repulsion, however, so that mutual orientations of rods

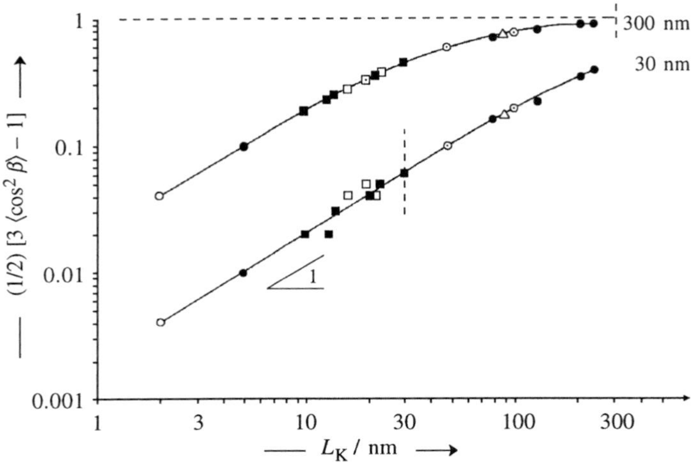

Fig. 8-11 Hermans orientation factors $f_{or} = (1/2)\ [3\ \langle \cos^2 \beta \rangle - 1]$ as a function of the Kuhnian length L_K for polymer molecules with conventional contour lengths of $L_{cont} = 30$ nm and $L_{cont} = 300$ nm, respectively [5]. Double logarithmic plot for clarity. At low L_K
● Polyamide; ○ phenylsiloxane ladder polymers; ⊙ poly(alkyl isocyanate)s; ⊕ deoxyribonucleic acids; △ cellulose polymers; ▲ poly(styrene), poly(methyl methacrylate), or poly(ethylene).

along their long axes are small (Fig. 8-11, lower curve). The ability to form such contact points increases with increasing contour length so that $f_{or} = AL_K$ at $L_{cont} = const$ where A = constant. Polymer chains with the same Kuhnian length thus have an A times greater orientation factor if their conventional contour length is A times longer, provided Kuhnian lengths are relatively small. At greater Kuhnian lengths, internal flexibilities of rod-like mesogens become noticeable. f_{or} is no longer proportional to L_K and approaches $f_{or} = 1$ at large Kuhnian lengths. These relationships apply to both rigid rods such as deoxyribonucleic acids and poly(alkyl isocyanate)s, semi-flexible polymers such as aliphatic polyamides, and flexible molecules such as poly(ethylene).

The orientation of rigid segments in nematic domains is utilized to prepare polymer fibers and films with exceptionally high elastic moduli E and fracture strengths s_B. Moduli of commodity plastics are usually 0.2-3.4 GPa whereas those in the longitudinal direction of fibers from LCPs can reach 320 GPa (Table 8-1).

The main contribution to the high elastic moduli and fracture strengths in the fiber direction comes certainly from the highly oriented mesogens. For Kevlar®, hydrogen bonds (see below) are usually held responsible for the excellent mechanical properties, but no such bonds exist in PBTZ with even better values of E and σ_B (Table 8-1). Kevlar® fibers are also said to have a core-shell structure.

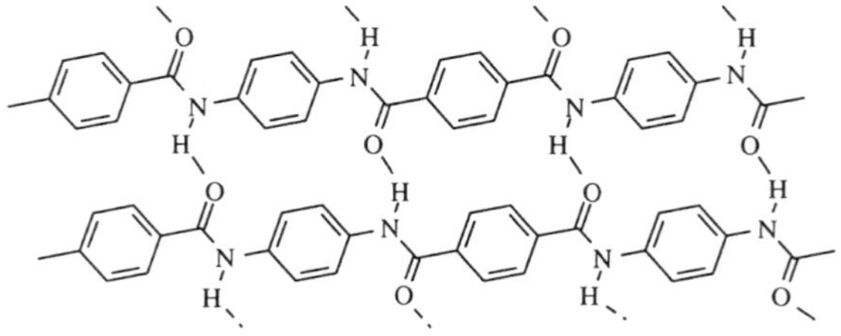

Table 8-1 Tensile moduli E and fracture strengths σ_B of thermotropic (TT) and lyotropic (LT) liquid crystalline polymers. Data for longitudinal (L) and transverse (T) directions in fibers and films as well as for the isotropic state (I).

X7G and Vectra® 950 are aromatic copolyesters with monomeric units from *p*-hydroxybenzoic acid (I), terephthalic acid (II), ethylene glycol (III), and/or 6-hydroxy-2-naphthoic acid (IV). The aramid Kevlar® is poly(*p*-phenylene terephthalamide) from terephthalic acid and 1,4-phenylene diamine. PBTZ is a composite of 30 wt% poly(2,6-benzobisthiazolediyl-1,4-phenylene) in poly(2,5-benzimidazole); HT = heat-treated. Data for low-density poly(ethylene) (LDPE), high-density poly(ethylene) (HDPE) and poly(hexamethylene adipamide) (PA 6.6, nylon 6.6) are shown for comparison.

Polymer		E/GPa			σ_B/MPa		
		L	T	I	L	T	I
-	LDPE			0.15			23
-	HDPE			1.0			30
-	PA 6.6	13		2.5	1000		74
TT	X7G	54	1.4	2.2	151	10	63
TT	Vectra®	11	2.6	5.0	144	54	97
LT	Kevlar 49®, fiber	138	7		2800		
	, film, biaxial	8.3	0.6				
LT	PBTZ, filament, HT	320	17	62	4200	680	700
	, film, uniaxial	270			2000		
	, film, biaxial	34			550		

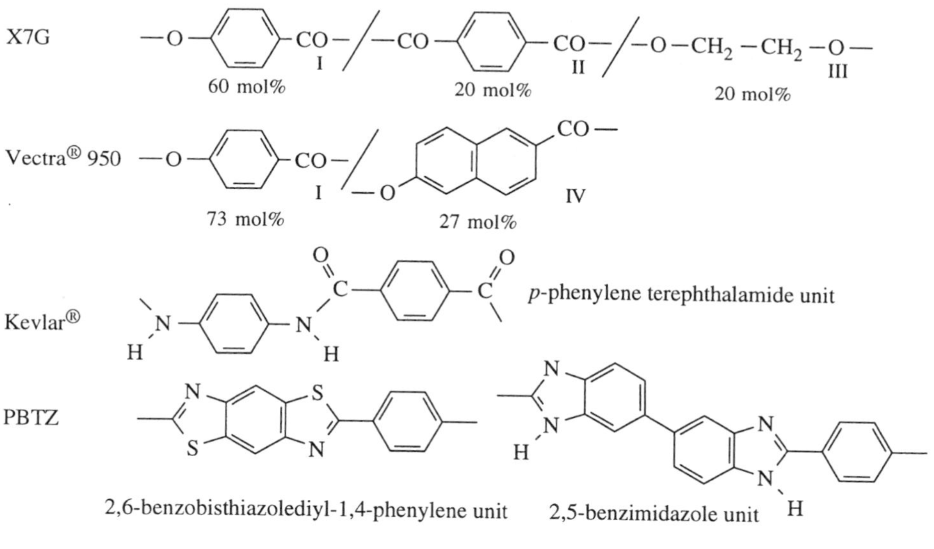

8.3.5 Amphotropic Liquid Crystals

Lyotropic LCs and LCPs usually contain mesogenic segments with weak inter-mesogenic and/or solvent interactions. Stronger interactions are expected for amphotropic LCPs that contain rigid mesogens and hydrophobic and hydrophilic groups (G: *amphoteros* = either of two). Examples are amphotropic side-chain LCPs with hydrophilic groups in the main chain or at the ends of side chains (Fig. 8-12).

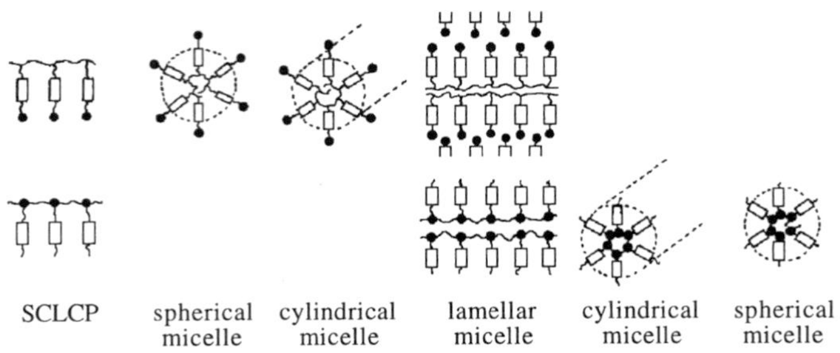

Fig. 8-12 Micellar structures of amphotropic side-chain liquid-crystalline polymers (SCLCP) with hydrophobic mesogenic segments (▭) in side-chains and hydrophilic headgroups (●) at the end of side chains (top) [6a]. The proportion of mesogenic segments increases from left to tght.

Such amphotropic liquid crystals form micelles in dilute aqueous solutions but organnize themselves in liquid-crystalline structures at higher concentrations (Fig. 8-13). The shape of these LC structures depends on the relative volume requirements of mesogens, similar to the behavior of block polymers (Section 8.5).

Mesogens of monomeric (LC) and polymeric (LCP) chemical compounds possess the same hydrophilic segment $-COO(CH_2CH_2O)_8CH_3$ and similar hydrophobic segments, i.e., $CH_2{=}CH(CH_2)_8-$ in the monomer and $-CH_2-CH_2(CH_2)_8-$ in the polymer, in the latter on a siloxane chain, $\text{+}O{-}Si(R)(CH_3)\text{+}_n$, where R is the hydrophobic segment. Between ca. 0°C and ca. 50°C, both LC and LCP form dilute isotropic aqueous solutions that phase separate into two isotropic solutions at $T > 50$°C. Below 0°C (Fig. 8-13), the solution phase separates to ice and liquid solute (M or P) and at lower T to ice and a crystalline solute. At higher solute concentrations and 0-50°C, hexagonal phases appear with a larger range for P than for M because of larger cooperative effects.

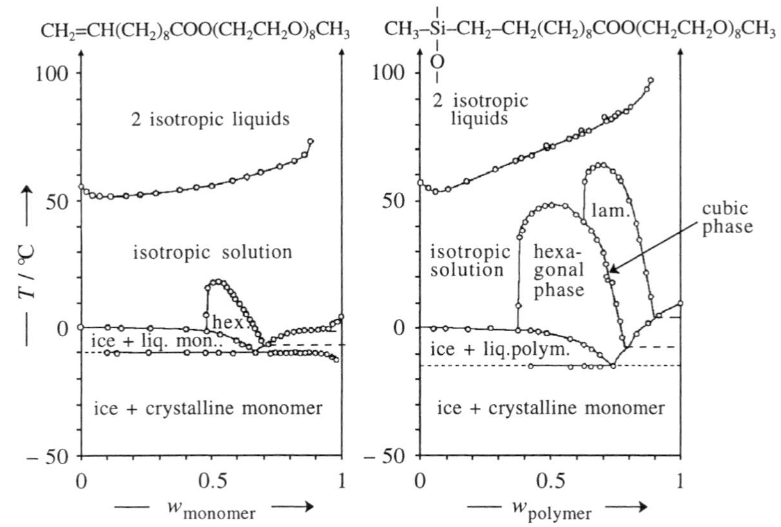

Fig. 8-13 Phase diagrams of aqueous solutions of a liquid-crystalline monomer (LC, left) and a liquid-crystalline polymer (LCP, right) with similar chemical structures (see formulae on top of the graph) [6b]. hex = Hexagonal phase, lam = lamellar phase. A cubic phase exists in a very narrow region between the hexagonal and lamellar LCP phase.

8.4 Thermotropic Liquid-Crystalline Polymers

Thermotropic liquid crystals can only exist if their decomposition temperatures T_{dec} are higher than the temperatures T_{cm} for the transition crystal ⇄ mesophase. The temperature interval $T_{dec} - T_{cm}$ is smaller the more rigid the molecules and the higher the molecular weights. Pentacene only sublimates and hexa(*p*-phenylene) decomposes without melting. Rigid rods do not form thermotropic LCPs at all; examples are deoxyribonucleic acids, poly(alkyl isocyanates), and poly(α-amino acid)s.

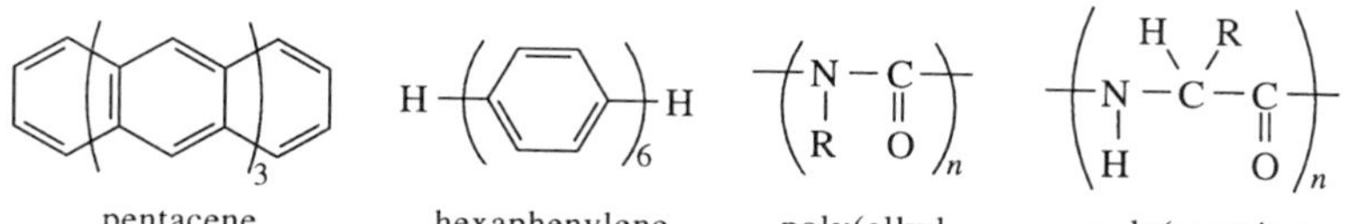

8.4.1 Structural Requirements

Mesophases should be stabilized by repulsion if the axial ratios Λ of rigid mesogens are greater than ca. 4 (Fig. 8-6). But truly stiff mesogens of thermotropic molecules become so stable that such chemical compounds decompose before they melt (see above). Thermotropic polymers can therefore be expected if the axial ratios are smaller than ca. 4 and interactions between mesogens are stabilized by additional weak attractive forces between mesogens. However, such forces should not lead to crystallization.

The rigidity of mesogens can be lessened by incorporating chemical groups that disturb or prevent formation of crystallites: (I) bulky substituents, (II) non-linear chain units, or (III) flexible chain units (Fig. 8-14), for example:

Nematic states of *main-chain liquid-crystalline polymers* (MCLCPs) are fairly easy to obtain since they require only a parallelization of rodlike mesogens. However, the resulting one-dimensional order is relatively easily disturbed by broad molecular weight distributions which leads to broad thermal transitions crystalline ⇄ nematic liquid crystalline.

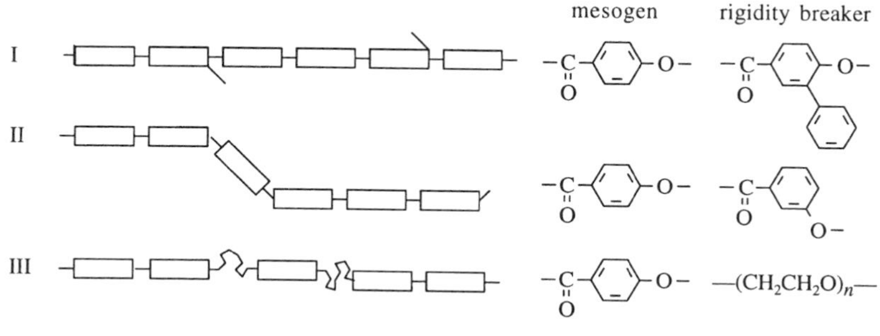

Fig. 8-14 Breaking of rigid mesogens by incorporating rigidity breakers.

Thermotropic nematic LCPs are known industrially as **self-reinforcing thermoplastics** since rod-like segments in nematic domains act similarly to fibers (glass fibers, etc.) in

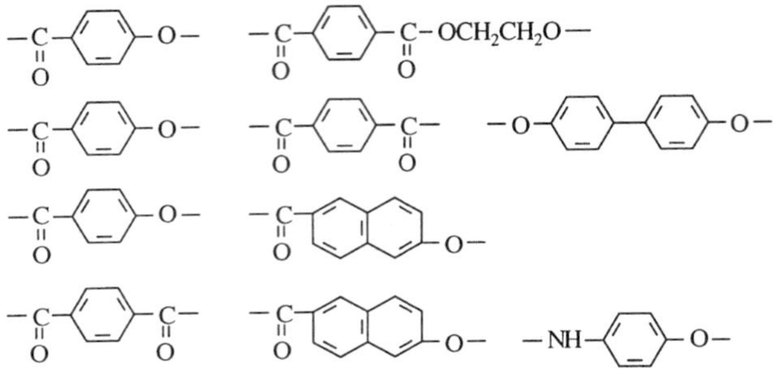

thermoplastics (Volume IV). The industrial polymers Xydar® and Vectra® use type II non-linear chain units such as –O–(1,4-C_6H_4)–(1,4-C_6H_4)–O– or –CO–(2,6-$C_{10}H_6$)–O– to break up the rigidity of mesogens (see also p. 255). The first self-reinforcing thermoplastic, X7G, with type III rigidity breakers, is no longer produced since similar properties can be obtained by far less expensive glass-fiber reinforced saturated polyesters.

Smectic liquid-crystalline states can be obtained from rigid LC main-chain polymers only if rodlike mesogens are present in periodic sequences. Since such states are characterized by layer structures, this is most easily achieved from equal-sized mesogens that are separated from each other along the chain by *long* flexible spacers.

Side-chain liquid-crystalline polymers (SCLCPs) result if mesogenic side groups are connected via flexible spacers (such as oligomethylene segments) to non-mesogenic main chains. In some cases, this can be achieved by direct polymerization of monomers with mesogenic side-groups, but in general such polymerizations do not succeed because of the strong steric hindrance by side chains. One then has to first polymerize a main-chain backbone containing reactive groups to which mesogenic side chains are coupled later. Some thermotropic SCLCPs are compared to MCLCPs in Fig. 8-15.

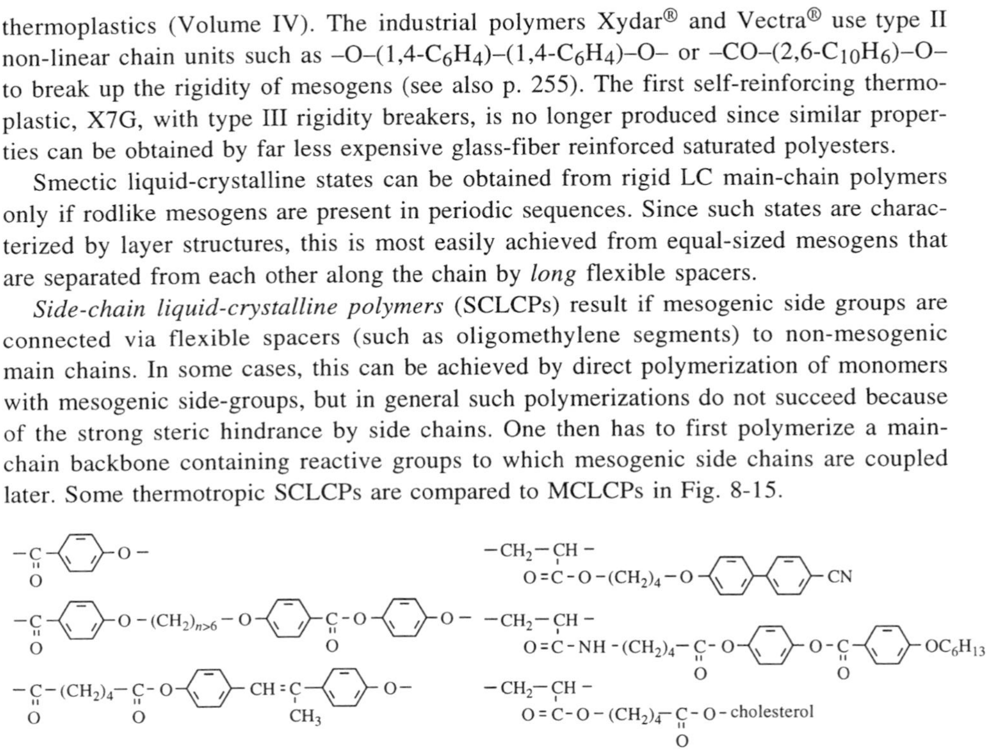

Fig. 8-15 Some thermotropic LCPs. Left: main-chain LCPs; right: side-chain LCPs. From top to bottom: nematic, smectic, cholesteric.

8.4.2 Physical Structure of SCLCPs

In side-chain LCPs, the physical structures of the main chains differ considerably from that of side-chain mesogens. An example is a methacrylic ester polymer with substituents R' = $(CH_2)_6O(p\text{-}C_6H_4)COO(p\text{-}C_6H_4)O(CH_2)_6H$ in $\text{+}CH_2\text{–}C(CH_3)(COOR')\text{+}_n$. In the isotropic state at $T > 127°C$, these polymers exist as unperturbed coils with a reduced radius of gyration of $(\langle s^2\rangle/\overline{M}_w)_o^{1/2} = 0.0171$ nm according to small-angle neutron scattering. The reduced radius of gyration remains practically the same at $T = 80°C$

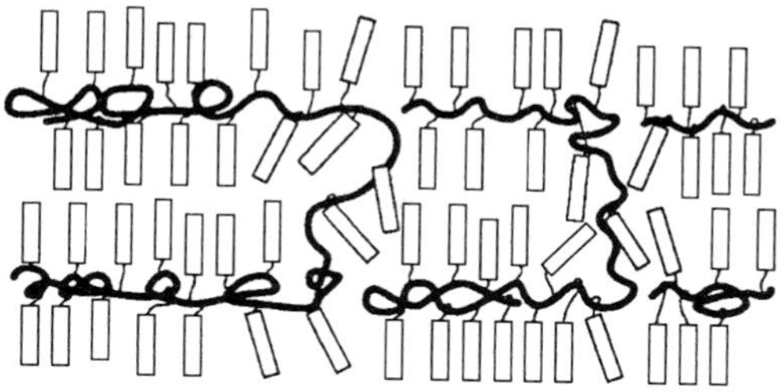

Fig. 8-16 Random coil structure of the main chain in the smectic mesophase of an SCLCP [7]. By kind permission of Elsevier Science, Oxford.

where the side groups are now in the nematic state. However, these reduced radii of gyration in the two neat states (isotropic and nematic) are much smaller than the reduced radius of gyration of the same polymer in theta solutions: 0.0171 nm versus 0.0216 nm.

The parallelization of side-chain mesogens thus produces much more compact coil structures. An even greater parallelization is found for the smectic state where the axial ratio $L_{\|}/d_{\perp}$ of polymer molecules drops to ca. 0.2 from ca. 0.8 in the nematic state. Fig. 8-16 illustrates such a random coil structure of main chains of an SCLCP with side chains in the smectic state.

The Hermans orientation factor of mesogens is usually $0.85 < f < 0.95$ in the smectic state and $0.45 < f < 0.65$ in the nematic state. The orientation factor decreases with increasing approach to the clearing temperature T_{NI} (nematic ⇄ isotropic) (Fig. 8-17).

The Hermans orientation factors f of mesogens of the polymers of Fig. 8-17 are practically independent of the length of the flexible spacers, i.e., the number of CH_2 groups. f values of polymers are considerably and systematically lower than those of low-molar mass compounds. This is probably not caused by the polymeric structure *per se* but by the fact that only every fourth siloxane unit of the polymer carries a mesogen which lowers the packing density of mesogens in LCP relative to that in LC.

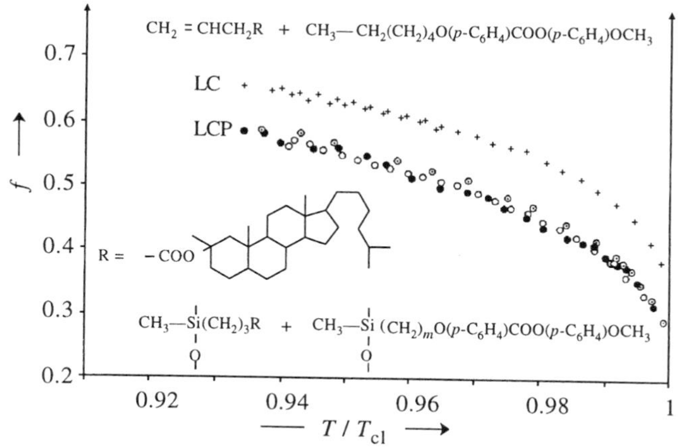

Fig. 8-17 Hermans orientation factor f as a function of the reduced inverse isotropization temperature (clearing temperature) nematic ⇄ isotropic, T/T_{cl} [8].

Bottom: polysiloxane copolymers with $m = 3$ (●), 4 (○), or 6 (⊙) CH_2 groups.

Top: mixtures of similar low-molecular weight chemical compounds (+).

8.4.3 Properties of Mesophases

In mesophases, mesogens of LCs are highly mobile. Because of this, application of electric or magnetic fields causes mesogens of all domains to align themselves within milliseconds or microseconds in the field direction. As a consequence, thin layers become clear and transparent. The resulting electro-optical effect is utilized in digital watches and other instruments.

At the same mesogen length, LCPs have considerably greater chain lengths than LCs. Hence, alignments by electric or magnetic fields require seconds rather than microseconds (Fig. 8-18). The required time increases with the lengths of both mesogens and polymer molecules, which makes LCPs unsuitable for electro-optical indicators. Side-chain LCPs do serve for thermo-optical storage, however: a laser beam increases the local temperature which causes a phase transformation. The changed local order is "permanently" frozen-in on cooling below the glass temperature. Such storage components can replace silver halide films in holography. Resolutions are ca. 0.3 μm.

The high orientation times of LCPs can be utilized in the manufacture of molded parts. High shear gradients cause mesogens in nematic phases to orient themselves in the shear direction, which leads to low apparent viscosities. Injection molding and fiber spinning from nematic states thus saves energy.

Rapid cooling of nematic states of LCPs to temperatures $T < T_G$ freezes in nematic states because of the high orientation times of mesogens. The resulting materials are liquid-crystalline glasses with improved mechanical properties in the orientation direction (see Table 8-1, p. 255) which serve as self-reinforcing thermoplastics or high-modulus fibers. For practical applications, orientation of mesogens must not be very strong if only weak intermolecular attraction forces exist between mesogens since materials with such highly oriented mesogens would defibrillate or delaminate easily.

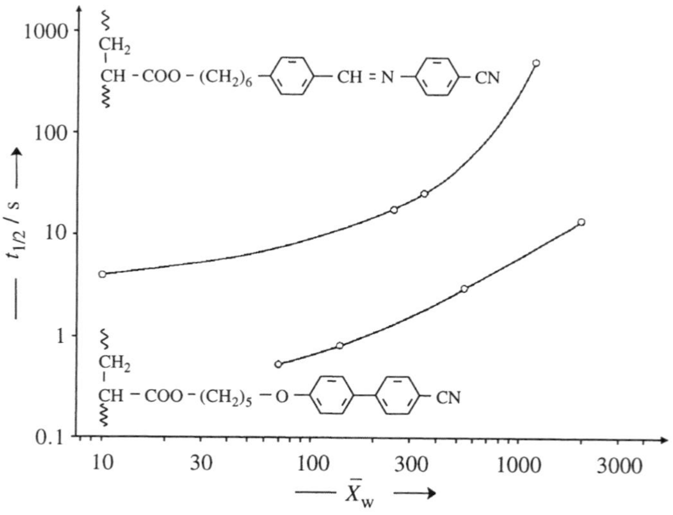

Fig. 8-18 Half-lives of orientation of side-chain liquid-crystalline polymers as a function of the mass-average degree of polymerization after applying an electric potential of $U = 200$ V at temperature differences of $T_{cl} - T = 10$ K ($T_{cl} = 154$°C) (top) and 45 K ($T_{cl} = 120$°C) (bottom) [9].

8.5 Block Polymers

8.5.1 Survey

Block polymers consist of polymer molecules with two or more, different, linearly connected large "blocks" of monomeric units. Each block is constitutionally and/or configurationally uniform. The blocks are produced by coupling preformed blocks or, in block *co*polymers, by sequential polymerization processes (p. 10).

In most cases, blocks are thermodynamically immiscible and try to demix in the melt. Chemically different blocks also interact differently with the same solvent. If one of the blocks is insoluble in the solvent and the other one so very soluble that the block polymer remains dissolved, then the block polymer molecules will form micelles. Industrially, two large groups of applications of block polymers can be distinguished: polymeric amphiphiles (I) and thermoplastic elastomers (II).

I. **Polymeric amphiphiles** have lyotropic domains (Section 8.5.3). They are diblock or multiblock polymers in which the various blocks have different affinities to the substance(s) to which they are applied. **Polymeric surfactants** are polymers consisting of hydrophilic and hydrophobic blocks. Examples are multiblock copolymers consisting of water-soluble ethylene oxide blocks, $-(OCH_2CH_2)_m-$, and water-insoluble propylene oxide blocks, $-(OCH_2CH(CH_3))_n-$. These multiblock polymers associate in water; their solutions have high viscosities.

Another subgroup of polymeric amphiphiles are **compatibilizers** for blends of two immiscible polymers, 1 + 2. Compatibilizers are diblock polymers A_nB_m in which blocks A_n are compatible with polymer 1 and blocks B_m are compatible with polymer 2. The compatibilizers reside at the interfaces between the domains of 1 and 2.

II. **Thermoplastic elastomers** have thermotropic domains (Section 8.5.2). They are multiblock polymers with at least three blocks per molecule. An example is the triblock polymer $S_mBu_nS_m$ with a "soft" poly(butadiene) center block ($T_G < T$) and two "hard" poly(styrene) end blocks ($T_G > T$) where "hard" and "soft" do not refer to the resistence against deformation but to the relative range of the glass temperature T_G. At ratios of ca. $2\ m < n$, styrene blocks form "hard" domains in the "soft" poly(butadiene) matrix, i.e., they behave like physical crosslinks in an elastomer. At temperatures $T_{G,S} > T$, poly(styrene) domains soften and the triblock polymer can be processed like a thermoplastic, hence the name.

This class of materials also comprises multiblock polymers with "hard" poly(ethylene terephthalate) segments, $-[OCH_2CH_2OOC(p\text{-}C_6H_4)CO]_m-$, and "soft" poly(tetrahydrofuran) segments, $-[OCH_2CH_2CH_2CH_2]_n-$, as well as some polyurethanes with "hard" aromatic group-containing segments and "soft" polyether segments.

8.5.2 Thermotropic Domains

Polymers A_n and B_m are generally not miscible (Section 10.2.3). Corresponding blocks A_n and B_m in block polymers $A_n–B_m$, $A_n–B_m–A_n$, etc., thus try to demix. However, because blocks are coupled, complete demixing is impossible and the blocks rather form domains (called "**microphases**") of one type of block in a matrix of the other type (Fig. 8-19). The resulting morphology depends on the ratio A_n/B_m.

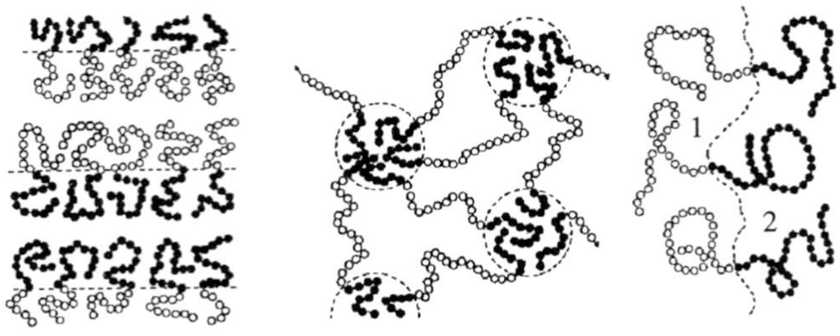

Fig. 8-19 Some morphologies of block polymers.
Left: Lamellae in a diblock polymer with equal volumes of blocks A_n and B_m.
Center: Spheres of small end blocks of a triblock polymer in the continuous microphase of the large center blocks.
Right: Diblock polymers at an interface between polymers 1 and 2, acting as a compatibilizer.

The simplest case is a diblock polymer A_nB_m, with two non-crystallizing blocks A_n and B_m that do not form mesogens. Triblock polymers $A_{n/2}B_mA_{n/2}$ can be treated as diblock polymers since a block $A_{n/2}$ corresponds to one-half of an A block in A_nB_m.

Each block $B_{\rm m}$ and A_n (or $A_{n/2}$) tries to form random coils. Because of the incompatibility of $B_{\rm m}$ and A_n, blocks also try to segregate. If both types of blocks have similar volume requirements, all A_n, blocks will cluster in one layer and all $B_{\rm m}$ blocks in another. Because of the incompatibility of the two types of blocks, each A layer faces another A layer and each B layer another B layer. The diblock polymer A_nB_m thus organizes itself in a layered structure consisting of alternating A and B lamellae in which each layer is just two coil diameters thick (Fig. 8-19, left).

The blocks can no longer pack in lamellae if the two types of blocks have vastly different volume requirements. Coil-forming smaller blocks cannot be packed in lamellae without violating the requirement for densest packing and they can also not deviate from the overall macroconformation of a random coil because both deviations would be energetically unfavorable. The smaller blocks thus remain as random coils but the coils cluster together in spheres in a continuous matrix of the random coils of the larger blocks (Fig. 8-19, center).

The morphology of block polymers is thus controlled by the volume requirements of blocks and the demand for densest packing of the entangled coil molecules. This leads to the limiting cases of alternating lamellae (same volume requirements of both types of blocks) and spheres in a continuous matrix (volume requirement of one type of block much smaller than that of the other type). If molecules of the minority component need space that is neither very large nor very small compared to that of the molecules of the majority component, they will form cylinders as a kind of one-dimensionally shrunk lamellae or one-dimensionally stretched spheres. Such lamellae, cylinders, and spheres have indeed been observed by electron microscopy (see drawing in Fig. 8-21, p. 264).

The stability ranges of spherical, cylindrical, and lamellar structures can be calculated theoretically with the help of random-flight statistics of coils. The calculations require the assumption of a constant density of domains. This is practically always the case since coils in the amorphous state have bulk moduli (Section 16.2.2) of more than 10^9 Pa. A density change of 10 % would require a pressure of more than 10^8 Pa ($\approx$ 1000 atm)!

In its simplest form, the theory predicts that the change of Gibbs energy of a block polymer, $\Delta G_{\rm bl}$, relative to that of a random bipolymer, $\Delta G_{\rm cp}$, i.e., $\Delta G^* = \Delta G_{\rm bl}/\Delta G_{\rm cp}$, only depends on the volume fraction ϕ of one of the two monomeric units A or B; the unperturbed radius of gyration, $\langle s^2\rangle_o^{1/2}$, of the coils of the corresponding block A_n or B_m; and the thickness d of the interlayer between the domains formed by the assembled A_n and B_m blocks:

(8-16)
$$\left.\begin{array}{ll}\text{lamellae:} & \Delta G^*_{\text{lam}} = [(1-\phi)^2/\phi]^{1/3}Q \\ \text{cylinders:} & \Delta G^*_{\text{cyl}} = 3^{1/3}(\phi - 1 - \ln\phi)^{1/3}Q \\ \text{spheres:} & \Delta G^*_{\text{sph}} = 5.4\,(\phi + 2 - 3\,\phi^{1/3})^{1/3}Q\end{array}\right\} Q = \left(\frac{d\phi^2}{2\cdot 3^2\,(1-\phi)\langle s^2\rangle_o^2}\right)^{1/3}$$

At volume fractions of $\phi_A < 0.21$ for diblock polymers, spherical domains from A blocks in a continuous matrix of B blocks should have the smallest Gibbs energy, whereas spherical B domains in an A matrix should prevail at $\phi_A > 0.79$ (Fig. 8-20). These spherical domains are *physical crosslinking* regions that bestow elastomeric properties on triblock polymers $A_{n/2}$–B_m–$A_{n/2}$ if A blocks have glass temperatures of $T_{G,A} > T$ and B blocks $T_{G,B} < T$ where T = temperature of application. At $T > T_{G,A}$, domains of A segments soften and the polymer can be processed like a thermoplastic.

Alternating lamellae are predicted to be stable in the range $0.35 < \phi_A < 0.65$, whereas cylindrical A domains should be present in the range $0.2 < \phi_A < 0.35$ and cylindrical B domains in the range $0.65 < \phi_A < 0.8$.

Since the theory applies to thermodynamic equilibria of molecularly uniform polymers with equal space requirements for the two blocks A_n and B_m and sharp boundaries between the microphases, it is not surprising that experimentally found stability regions are sometimes broader and sometimes smaller than predicted. For example, broad molecular weight distributions lead to wider stability ranges of spherical domains whereas the stability range of lamellar domains becomes more narrow (see Volume IV, Section 7.5.2). In addition, new types of microphases may appear that are not predicted by the simple theory (Fig. 8-21).

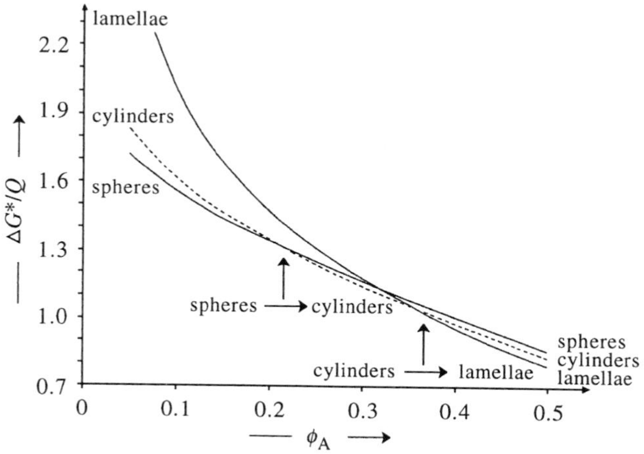

Fig. 8-20 Reduced Gibbs energy $\Delta G^*/Q$ as a function of the volume fraction ϕ_A of random-coil-forming A_n blocks in diblock polymers A_nB_m with equal space requirements for A_n and B_m blocks. Calculations with Eqs.(8-16) [10].

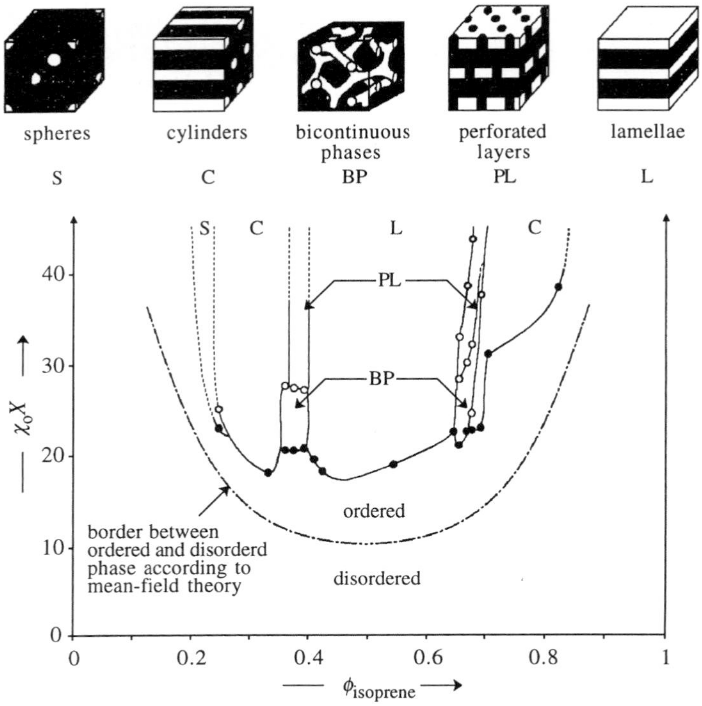

Fig. 8-21 Product $\chi_0 X$ of the total degree of polymerization, X, of a poly(isoprene)-*block*-poly(styrene) and Flory-Huggins interaction parameter χ_0 (see Section 10.2.2) as a function of the volume fraction of isoprene units [11]. According to mean-field theory, $\chi_0 X$ is a measure of phase stability.
With permission by American Chemical Society, Washington (DC).

In *thermodynamic equilibrium*, domains of diblock and triblock polymers with uniform block lengths should be distributed uniformly in space, i.e., they should form **superlattices**. Three types of superlattices can be distinguished:

- *Lamellae* form one-dimensional lattices with lattice points arranged linearly at equal distances. Small-angle X-ray patterns show diffraction lines with inverse Bragg distances in the ratio 1:2:3.
- *Cylinders* produce two-dimensional (hexagonal) lattices. Cylinders are packed laterally and parallel to the surface. Inverse Bragg distances are in the ratio $1 : \sqrt{3} : \sqrt{4} : \sqrt{7}$.
- *Spheres* lead to three-dimensional body-centered or face-centered lattices with inverse Bragg distances of $1 : \sqrt{2} : \sqrt{3} : \sqrt{4}$. Primitive cubic lattices are not found.

At constant chemical composition, the sizes of lamellar or cylindrical domains may vary with the thickness of the lamellae or the length of the cylinders, respectively. Such length changes are possible without segment diffusion through incompatible matrices. Cylindrical and lamellar phases may thus indeed be in thermodynamic equilibria.

The situation is different for spherical domains in which blocks are present as random coils whose dimensions vary little with temperature. Domain sizes can thus vary only by changing the number of blocks per domain, for example, if microphases are formed by lowering the temperature. Such a change is only possible if the sphere-forming blocks diffuse through the incompatible matrix of the other type of block. Since that is unlikely, the system will remain in the state in which it was at the onset of microphase separation.

Sinple thermodynamic theory also neglects other thermodynamic contributions. For example, bonds connecting the two types of blocks must be in the interfaces between the microphases, which leads to a contribution to the positional entropy. There must also be a contribution by a Gibbs surface energy which increases with the Flory-Huggins interaction parameter χ (see Fig. 8-21 for a plot of $\chi_0 X = f(\phi_i)$).

All of these thermodynamic and kinetic factors lead to more types of domain structures and more complicated phase diagrams than envisioned by simple theory. Sometimes, bicontinuous phases are found in a very narrow range between the ranges for spheres and cylinders; an example is the bicontinuous phase, BP, with space group $Ia\overline{3}d$ in the ranges $0.36 \leq \phi_{\text{isoprene}} \leq 0.39$ and $0.65 \leq f_{\text{isoprene}} \leq 0.68$ (Fig. 8-21). This diblock polymer also has an additional domain structure PL with perforated layers but not the sometimes observed "bicontinuous diamond structure." The last two domain types are possibly metastable.

Experimentally observed thicknesses d_L of lamallae correspond to theoretically predicted ones (Fig. 8-22).

Interlayers between microphases are $d_I \approx$ 2nm according to X-ray and neutron small-angle scattering experiments (Fig. 8-22). Nuclear magnetic resonance relaxation spectra confirmed these thicknesses and showed that interfaces are not sharp but somewhat fuzzy. However, experimental thicknesses are between those that are predicted by the various theories (Fig. 8-22).

Much more complicated structures are observed for triblock polymers $A_mB_nC_p$ with three different types of blocks, A_m, B_n, and C_p. At a constant volume ratio, $V_A/V_C = 1$, increasing proportions of B blocks lead first to spherical domains in a continuous matrix, then to cylindrical domains of the so-called "ball-in-the-box" type, and finally to well-ordered lamellae with spherical domains.

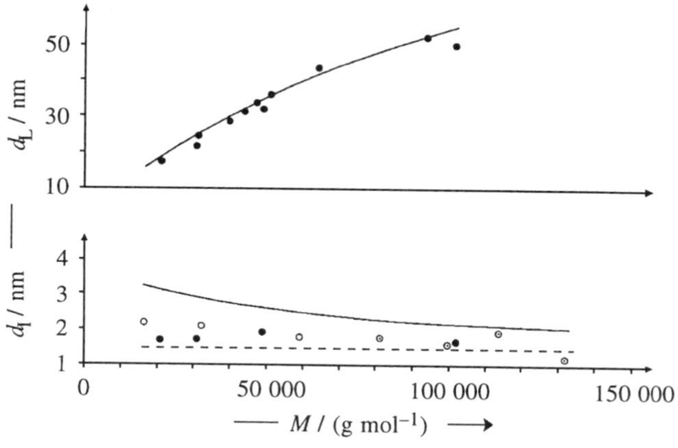

Fig. 8-22 Molecular weight dependence of thicknesses d_L of lamellae and thicknesses d_I of interfaces between block types for lamellar (●), cylindrical (⊙), and spherical (○) structures of poly(isoprene)-*block*-poly(styrene). Date from small-angle X-ray [12,13] and small-angle neutron [14] measurements. Predictions by theories: ------ [15], - - - - [16].

8.5.3 Lyotropic Structures

A solvent for a diblock polymer may be a good solvent for both types of blocks ("neutral" or "indifferent" solvent) or a good solvent for one type of block and a bad solvent or non-solvent for the other type of block ("selective" solvent).

In the ideal case, an indifferent solvent is equally thermodynamically good for both types of blocks; it will then expand equally the random coils of both types of blocks. With increasing relative proportion of the minority component, one can thus expect for moderately good solvents first spherical, then cylindrical, and finally lamellar domains, similarly to the situation in the thermotropic case.

The situation is different for selective solvents. Here, the sequence of appearance of the various types of domain structures depends on the ratio of block sizes. According to theoretical calculations for a ratio of $M_A/M_B = 4$ of molecular weights of A and B blocks, spherical domains of B in a continuous matrix of A are formed in the absence of solvents. At a polymer concentration of $\phi_P \geq 70$ % in a selective solvent for the minority B blocks, lamellar structures are more stable then spherical domains (Fig. 8-23, right) because random coils of B blocks are more swollen then those of A blocks and the space requirements of B and A random coils are the same, despite the large difference in molar masses.

At $\phi_P = 50$ % in a selective solvent for B, cylindrical domains are the most stable. Concentrating 50 % solutions to 70 % solutions converts cylindrical domains to lamellar ones without problems since B blocks do not need to diffuse through a matrix of A blocks. However, a further concentration from 70 % to 100 % does not necessarily produce spherical domains of B in the matrix of A.

For 50 % solutions, the sequence C > S > L for polymers with $M_A/M_B = 4$ is replaced by S > L > C for polymers with $M_A/M_B = 2$. At 70 % solutions, lamellae dominate for both $M_A/M_B = 2$ and $M_A/M_B = 4$. The situation is quite different if selective solvents are used for the majority A component (left sides of graphs).

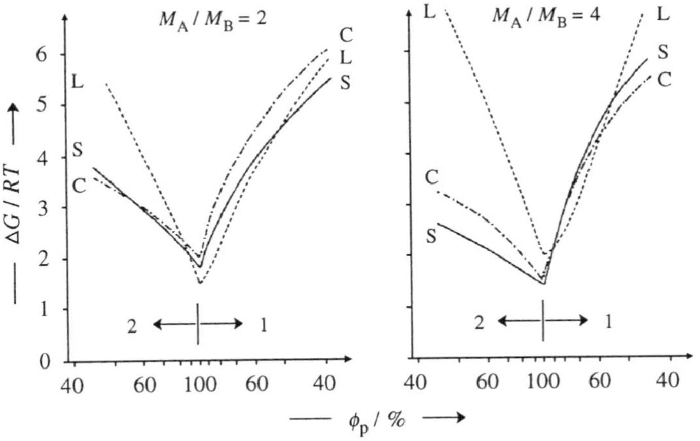

Fig. 8-23 Reduced Gibbs energy, $\Delta G/RT$, of the formation of domains [spherical (S, —), cylindrical (C, - · -), lamellar (L, - - -)] as a function of the volume fraction ϕ_p of diblock polymers with relative block lengths of $M_A/M_B = 2$ or 4 in selective solvents 1 for A blocks and 2 for B blocks, respectively [10]. B is the minority component for the right sides of graphs and A for the left sides.

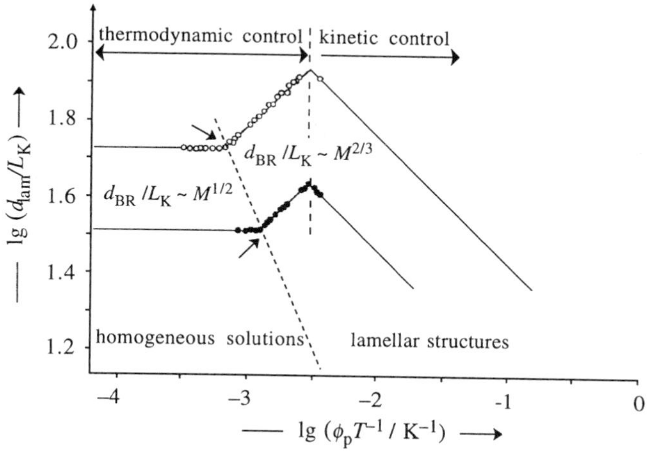

Fig. 8-24 Logarithm of reduced Bragg distance of lamellae, d_{lam}/L_K, as a function of the logarithm of the ratio ϕ_p/T for two styrene-isoprene diblock polymers (S/I) with molar masses of 94 000 g/mol (O) and 31 000 g/mol (●), respectively, in the indifferent solvent toluene [17].

d_{BR}= Bragg distance, L_K = Kuhnian length of chain segments, ϕ_p = volume fraction of polymer, T = thermodynamic temperature. Arrows indicate critical values for the formation of lamellae.

Increasing proportions of one of the blocks of diblock polymers thus do not necessarily lead to sequences spheres → cylinders → lamellae in lyotropic block polymer systems. Instead, equilibrium morphologies rather depend not only on the ratio of volumes of the two types of blocks, V_A/V_B, but also on the selectivity and volume fraction of the solvent (Fig. 8-23), the temperature, and the molar mass of the block polymer.

An example is the variation of reduced Bragg distances, d_{BR}/L_K, with the reduced volume fraction, ϕ_P/T, of a diblock polymer in an indifferent solvent where L_K = Kuhnian length of polymer segments and T = thermodynamic temperature (Fig. 8-24). At very low values of ϕ_P/T, the polymer is in a homogeneous solution: d_{BR}/L_K is independent of ϕ_P/T but increases with the square root of molar mass. Above a certain critical value of ϕ_P/T, lg (d_{BR}/L_K) increases linearly with lg (ϕ_P/T); this critical value decreases with increasing molar mass. For example, the polymer with M = 31 000 g/mol has a critical value of lg ($\phi_P\ T^{-1}/K^{-1}$) = –2.9 which corresponds at 300 K to ϕ_P = 0.378. Above this critical value, lamellae form and the reduced Bragg distance is now proportional to $M^{2/3}$. At a still higher value of lg ($\phi_P\ T^{-1}/K^{-1}$) = –2.52, values of d_{BR}/L_K start to decrease with ϕ_P/T. This second critical volume fraction is independent of the molar mass; it corresponds to ϕ_P = 0.905 at 300 K and ϕ_P = 1 at 331 K. Above these values of ϕ_P, physical structures are controlled kinetically and not thermodynamically.

8.6 Ionomers

Ionomers are copolymers with high proportions of hydrophobic monomeric units and small proportions of comonomers that contain ionic groups in their main or side chains. Four types of ionomers are industrially important: Surlyn®, EAA copolymer®, Nafion®, and Thionic®. The first three of these copolymers are produced by bi-

$—CH_2—CH_2—$ + < 10 mol-% $—CH_2—C(CH_3)(COOH)—$ Surlyn®

$—CH_2—CH_2—$ + < (3.5-20) mol-% $—CH_2—CH(COOH)—$ EAA Copolymer®

$—CF_2—CF_2—$ + few mol-% $—CF_2—CF(O[CF(CF_3)CF_2O)]_n(CF_2)_2SO_2H)—$ Nafion®

$—CH_2—CH_2—$ / $—CH_2—CH(CH_3)—$ + 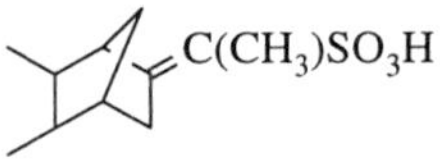 Thionic®

polymerization of the corresponding monomers whereas Thionic® results from the after-sulfonation (< 5 mol%) of the primary terpolymer. EAA copolymer® and Nafion® are used as acids whereas Surlyn® and Thionic® are traded as the sodium or zinc salts, respectively.

The introduction of ionic groups leads to ion association in solid polymers and subsequently to various types of ionic structures (Fig. 8-25). An **ion pair** consists of an anion (here as part of a polyanion) and its low-molecular weight counterion. An **ion triplet** may consist of a cation between two polymer-bound anions; it is the simplest type of an **ion multiplet**. Ion pairs and ion multiplets can associate to **ion droplets** that in turn form greater **ion clusters** and **ion domains**.

Ion clusters and ion droplets contain anions from various polymer chains, i.e., they act as reversible physical crosslinks. Ionomers are therefore "reversible thermosets" that behave like thermosets at temperatures below the glass temperature of their hydrophobic chain segments and like processable thermoplastics at temperatures above the glass temperature where the ionic bonds dissociate.

The action of ionic crosslinking sites is decided not by the stoichiometry of ion pairs but by the coordination number of ions, which is 6 for both $Na^{\oplus}$ and $^{\oplus}Zn^{\oplus}$. Sodium ions $Na^{\oplus}$ with 1 positive charge are therefore as good "crosslinkers" as zinc ions $^{\oplus}Zn^{\oplus}$ with 2 positive charges. Free acids such as free sulfonic acid groups in Thionic® therefore do not associate and do not behave as "reversible thermosets."

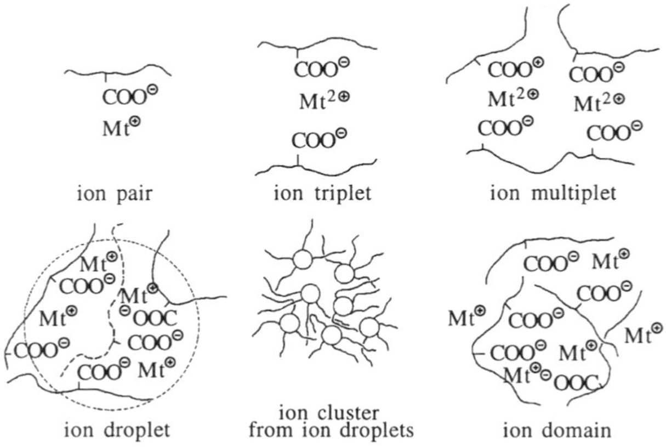

Fig. 8-25 Models for ion associations and domains in ionomers.

Historical Notes to Chapter 8

DISCOVERY OF LIQUID-CRYSTALLINE COMPOUNDS
F.Reinitzer, Monatsh.Chem. **9** (1888) 421 (two-step melting of cholesterol benzoate)
O.Lehmann, Flüssige Kristalle, Engelmann, Leipzig 1904 (introduction of the term "liquid crystals" (D: flüssige Kristalle)
M.G.Friedel, Ann.phys. (Paris) **18** (1922) 273 (introduction of the term "mesomorphic")

DISCOVERY OF LIQUID-CRYSTALLINE POLYMERS
C.Robinson, Trans.Faraday Soc. **52** (1956) 571 (Poly(γ-benzyl-L-glutamate): first lyotropic LCP)
A.Roviello, S.Sirigu, J.Polym.Sci.Lett. **13** (1975) 455 (first thermotropic LCP)
W.J.Jackson, Jr., H.F.Kuhfuss, J.Polym.Sci.-Polym.Chem. **14** (1976) 2043 (first industrial LCP)

PHASE SEPARATION OF LC POLYMERS
L.Onsager, Ann.N.Y.Acad.Sci. **51** (1949) 627
Entropic stabilization of liquid crystals by parallelization of rods; phase separation calculated from 2nd virial coefficients and excluded volumes of rigid rods.

W.Maier, A.Saupe, Z.Naturforschg. **14a** (1959) 882, **15a** (1960) 287
Enthalpic stabilization of liquid crystals by attractive forces.

P.J.Flory, Proc.R.Soc. [London] **A 234** (1956) 73
Entropic stabilization by parallelization of rods; calculations with lattice theory of rigid rods.

MORPHOLOGY OF BLOCK COPOLYMERS
D.J.Meier, J.Polym.Sci. **C 26** (1969) 81
Pioneering theory of formation and morphology of microdomains in solid block copolymers.

Literature to Chapter 8

8.1-8.4 LIQUID CRYSTALS (general or low-molecular weight)
IUPAC, Definitions of Basic Terms Relating to Low-Molar Mass and Polymer Liquid Crystals, Pure Appl.Chem. **73** (2001) 845
G.W.Gray, P.A.Winsor, Liquid Crystals and Plastic Crystals. Physico-Chemical Properties and Methods of Investigation, Ellis-Horwood, Chichester 1974 (2 volumes)
P.G.de Gennes, The Physics of Liquid Crystals, Clarendon Press, Oxford 1974
H.Kelker, R.Hatz, Handbook of Liquid Crystals, Verlag Chemie, Weinheim 1980
G.W.Gray, J.W.Goodby, Smectic Liquid Crystals, Hill, London 1984
G.W.Gray, Ed., Thermotropic Liquid Crystals, Wiley, New York 1987
P.J.Collings, Liquid Crystals: Nature's Delicate Phase of Matter, Princeton Univ.Press, Princeton (NJ) 1990
B.Bahadur, Liquid Crystals: Applications and Uses, World Scientific, Singapore 1990-1992 (3 vols.)
P.G. de Gennes, J.Prost, The Physics of Liquid Crystals, Clarendon Press, Oxford, UK 1993
D.Demus, J.W.Goodby, G.W.Gray, H.W.Spiess, V.Vill, Eds., Handbook of Liquid Crystals, Wiley-VCH, Weinheim 1998 (4 vols., Vol. 3: High Molecular Weight Liquid Crystals)
G.W.Gray, V.Vill, H.W.Spiess, D.Demus, J.W.Goodby, Eds., Physical Properties of Liquid Crystals, Wiley-VCH, Weinheim 1999
S.Singh, Liquid Crystals: Fundamentals, World Scientific, River Edge (NJ) 2002
T.J.Sluckin, D.A.Dunmur, M.Stegemeyer, Eds., Crystals that Flow–Classic Papers from the History of Liquid Crystals, Taylor and Francis, London 2004

8.1-8.4 LIQUID CRYSTALS (polymers in general)
N.A.Platé, V.P.Shibaev, Comb-Shaped Polymers and Liquid Crystals, Khimia, Moscow 1980 (in Russian); Plenum, New York 1987 (in English)

Yu.B.Amerik, B.A.Krentsel, Chemistry of Liquid Crystals and Mesomorphic Polymer Systems, Khimia, Moscow 1981 (in Russian).
A.Ciferri, W.R.Krigbaum, R.B.Meyer, Eds., Polymer Liquid Crystals, Academic Press, New York 1982
P.J.Flory, Molecular Theory of Liquid Crystals, Adv.Polym.Sci. **59** (1984) 1
N.March, M.Tosi, Ed., Polymers, Liquid Crystals, and Low-Dimensional Solids, Plenum, New York 1984
H.Finkelmann, G.Rehage, Liquid Crystal Side Chain Polymers, Adv.Polym.Sci. **60/61** (1984) 99
M.G.Dobb, J.E.McIntyre, Properties and Applications of Liquid-Crystalline Main-Chain Polymers, Adv.Polym.Sci. **60/61** (1984) 61
L.L.Chapoy, Ed., Recent Advances in Liquid Crystalline Polymers, Elsevier Appl.Sci.Publ., London 1985
A.Blumstein, Ed., Polymeric Liquid Crystals, Plenum, New York 1985
C.B.McArdle, Side Chain Liquid Crystal Polymers, Blackie, Glasgow, UK, 1989
W.W.Adams, R.K.Eby, D.E.McLemore, Eds., The Materials Science and Engineering of Rigid-Rod Polymers, Materials Research Society, Pittsburgh 1989
A.Ciferri, Ed., Liquid Crystallinity in Polymers, VCH, New York 1991
A.M.White, A.H.Windle, Liquid Crystalline Polymers, Cambridge Univ. Press, Cambridge 1992
A.A.Collyer, Liquid Crystal Polymers: From Structure to Application, Elsevier Appl. Science, London 1992
N.A.Platé, Ed., Liquid-Crystal Polymers, Plenum, New York 1993
P.G. de Gennes, J.Prost, The Physics of Liquid Crystals, Oxford Univ. Press, New York 1995
A.M.Donald, A.H.Windle, S.Hanna, Liquid Crystalline Polymers, Cambridge Univ. Press, Cambridge, UK, 2nd ed. 2006

8.3 LYOTROPIC LIQUID-CRYSTALLINE POLYMERS

V.N.Tsvetkov, Rigid-Chain Polymers. Hydrodynamic and Optical Properties in Solution, Plenum, New York 1989
M.G.Northolt, D.J.Sikkema, Lyotropic Main Chain Liquid Crystal Polymers, Adv.Polym.Sci. **98** (1990) 115
T.Sato, A.Teramoto, Concentrated Solutions of Liquid-Crystalline Polymers, Adv.Polym.Sci. **126** (1996) 85

8.4 THERMOTROPIC LIQUID-CRYSTALLINE POLYMERS

B.Wunderlich, J.Grebowicz, Thermotropic Mesophases and Mesophase Transitions of Linear, Flexible Macromolecules, Adv.Polym.Sci. **60/61** (1984) 1
A.E.Zachariades, R.S.Porter, Eds., Structure and Properties of Oriented Thermotropic Liquid Crystalline Polymers in the Solid State, Dekker, New York 1988
J.Economy, K.Goranov, Thermotropic Liquid Crystalline Polymers for High Performance Applications, Adv.Polym.Sci. **117** (1994) 221
T.S.Chung, Thermotropic Liquid Crystal Polymers. Thin Film Polymerization, Characterization, Blends, and Applications, Technomics, Lancaster (PA) 2001
M.Warner, E.M.Terentjev, Liquid Crystal Elastomers, Oxford Univ.Press, New York 2003

8.5 BLOCK POLYMERS

A.Nohay, J.E.McGrath, Block Copolymers: Overview and Critical Survey, Academic Press, New York 1976
B.R.M.Gallot, Preparation and Study of Block Copolymers with Ordered Structures, Adv.Polym.Sci. **29** (1978) 87
I.Goodman, Ed., Developments in Block Copolymers, Vol. **1** (1982) ff., Appl.Sci.Publ., Barking, Essex
D.J.Meier, Ed., Block Copolymers. Science and Technology, Harwood Academic Publ., New York 1983
M.J.Folkes, Ed., Processing, Structure and Properties of Block Copolymers, Elsevier, New York 1985
G.Riess, G.Hurtrez, P.Bahadur, Block Copolymers, Encycl.Polym.Sci.Eng., Wiley, New York, 2nd ed., **2** (1985) 324
N.R.Legge, G.Holden, H.E.Schroeder, Eds., Thermoplastic Elastomers, Hanser, Munich 1987; 2nd ed.: G.Holden, N.R.Legge, R.P.Quirk, H.E.Schroeder, Eds., Hanser-Gardner, Munich 1996

S.Datta, D.J.Lohse, Eds., Polymeric Compatibilizers: Uses and Benefits in Polymer Blends, Hanser-Gardner, Cincinnati (OH) 1996

P.Alexandridis, B.Lindman, Eds., Amphiphilic Block Copolymers: Self-Assembly and Applications, Elsevier, Amsterdam 1997

I.W.Hamley, The Physics of Block Copolymers, Oxford Univ. Press, Oxford 1999

N.Hadjichristidis, S.Pispas, G.Floudas, Block Copolymers, Wiley, Hoboken (NJ) 2002

I.W.Hamley, Ed., Developments in Block Copolymer Science and Technology, Wiley, Hoboken (NJ) 2004

8.6 IONOMERS

M.Pineri, A.Eisenberg, Eds., Structure and Properties of Ionomers, NATO ASI Series, Reidel, Dordrecht 1987

S.Schlick, Ed., Ionomers. Characterization, Theory, and Applications, CRC Press, Boca Raton (FL) 1996

M.R.Tant, K.A.Mauritz, G.L.Wilkes, Ionomers: Synthesis, Structure, Properties and Applications, Chapman and Hall, London 1997

A.Eisenberg, J.-S.Kim, Introduction to Ionomers, Wiley, New York 1998

References to Chapter 8

[1] R.W.Werboyj, D.G.Gray, Macromolecules **13** (1980) 69, Fig. 2

[2] P.J.Flory, Proc.R.Soc. [London] **A 234** (1956) 73, Fig. 4 (modified)

[3] A.Nakayama, T.Hayashi, M.Ohmori, Biopolymers **6** (1968) 973, data of Fig. 6

[4] H.Benthack-Thoms, H.Finkelmann, Makromol.Chem. **186** (1985) 1895, Figs. 1, 3-5

[5] V.N.Tsvetkov, E.I.Rjumtsev, I.N.Shtennikova, in A.Blumstein, Ed., Liquid Crystalline Order in Polymers, Academic Press, New York 1978, Table 1

[6] B.Lühmann, H.Finkelmann, G.Rehage, Angew.Makromol.Chem. **123/124** (1984) 217, (a) Fig. 1 (modified), (b) Figs. 2 and 3

[7] V.Tsukruk, J.H.Wendorff, Trends Polym.Sci. **3** (1995) 82, Fig. 2

[8] H.Finkelmann, G.Rehage, Adv.Polym.Sci. **60/61** (1984) 99, data of Fig. 15

[9] N.A.Platé, V.P.Shibaev, Comb-Shaped Polymers and Liquid Crystals, Plenum, New York 1987, data of page 372

[10] D.J.Meier, Michigan Molecular Institute, private communication; see also private communication to J.M.G.Cowie, in I.Goodman, Ed., Dev. Block Copolym. **1** (1982), Figs. 6 and 7

[11] A.K.Khandpur, S.Förster, F.S.Bates, I.W.Hamley, A.J.Ryan, W.Bras, K.Almdal, K.Mortensen, Macromolecules **28** (1995) 8796, Fig. 13; for theoretical calculations, see also M.W.Matsen, F.S.Bates, Macromolecules **29** (1996) 1091, Fig. 4

[12] R.Mayer, Polymer **15** (1974) 137

[13] T.Hashimoto, M.Shibagamu, H.Kawai, Macromolecules **13** (1980) 1237

[14] R.W.Richards, J.L.Thomason, Polymer **24** (1983) 1089

[15] D.J.Meier, J.Polym.Sci. **C 26** (1969) 81

[16] E.Helfand, Macromolecules **8** (1975) 552; E.Helfand, Z.R.Wasserman, Macromolecules **9** (1976) 879; **11** (1978) 960

[17] T.Hashimoto, M.Shibayama, H.Kawai, Macromolecules **16** (1983) 1093, data of Fig. 14

9 Polymers in and at Interfaces

9.1 Surface of Polymers

9.1.1 Fundamentals

Interfaces are faces between two solids, two non-miscible liquids, or a solid and a non-dissolving liquid. The interface between a solid or a liquid and the surrounding gas is commonly called the **surface** of the solid or the liquid.

Like surfaces of other materials (metals, glasses, etc.), surfaces of polymers are usually not "clean" but contain in general adsorbed, absorbed, and/or adherent substances such as oxygen, carbon dioxide, water, fats, etc. The removal of these impurities and the preparation of absolutely clean surfaces is an art. However, absolutely clean surfaces are required if one wants to study the effects of chemical and physical structure of substrates on interfacial and surface tensions, adhesion, and other surface properties.

Real surfaces (and also interfaces) are also not "smooth" but "rough." **Roughness *r*** is defined as the ratio of the true (submicroscopic) surface to the geometric surface, $r \geq 1$. Freshly cleaved mica has a roughness of $r \approx 1$ while polished metal surfaces have values of r = 1.5-2.

The chemical composition of surfaces and interfaces is not necessarily identical with that of the interiors of materials. For polymers, it is never identical (Section 9.1.2).

The compositions and topographies of surfaces can now be determined by many physical methods that are based on very different physical phenomena and procedures (Table 9-1). Compositional profiles can be obtained by combining the results of methods with various penetration depths.

Table 9-1 Some methods for the determination of chemical compositions and topographies of polymer surfaces. a) ESCA = *e*lectron *s*pectroscopy for *c*hemical *a*nalysis.

Method Abbreviation	Name	Penetration in nm	Information
AFM	atomic force microscopy	0.1	topography
STM	scanning tunneling microscopy	0.1	topography
EL	ellipsometry	0.1	thickness, composition
XR; NR	X-ray (or neutron) reflectometry/scattering	0.2	roughness, composition
SIMS	secondary ion mass spectroscopy	< 1	composition
ST	surface tension	≤ 1	hydrophobicity
XPS, ESCA[a)]	X-ray photoelectron spectroscopy	< 5	vibration spectra
SEM	scanning electron microscopy	> 10	composition
ATR-IR	attenuated total reflection IR analysis	1000	composition

9.1.2 Surface Composition

Investigations by various methods have shown that (a) bulk and surface compositions usually differ and (b) surface and interface compositions may vary with time.

Table 9-2 Groups and monomer units that reside preferentially at the polymer-air surface.

Polymer(s)	Surface enriched by
Poly(2-vinylpyridine)	$-CH_2-$, $>CH-$
Aromatic polyimides	$>N-CO-$
Poly(styrene)-*block*-poly(oxyethylene)	$-CH_2-CH(C_6H_5)-$
Poly(vinyl chloride) + poly(α-methyl styrene-*co*-acrylonitrile)	$-CH_2-CHCl-$
Poly(dimethylsiloxane) + poly(bisphenol A-sulfone)	$-O-Si(CH_3)_2-$
Poly(dimethylsiloxane) + poly(bisphenol A-carbonate)	$-O-Si(CH_3)_2-$

In *equilibrium*, polymer segments with the lowest Gibbs surface energies tend to reside in surfaces against air although these segments are not necessarily the most polar or the most hydrophilic ones (Table 9-2). The situation is different for *interfaces* since interfacial energies vary with the composition of the partner (air, water, metal surfaces, other polymers). In addition, there are kinetic effects (thermal history, crystallization kinetics, diffusion, etc.).

The compositions of polymer *melts* vary with time because of the different diffusion rates of components. An example is the mixture of protonated and deuterated poly-(styrene) of approximately the same molecular weight where the volume fraction of the latter in the surface almost doubled after 45 days (Fig. 9-1). The highest volume fraction of d_8-PS is found at a distance of ca. 16 nm from the melt/air surface, i.e., at ca. one-half of the radius of gyration of this polymer, $\langle s^2 \rangle_o^{1/2} = 28$ nm. Note that deuterated polymers are often used as tracers in their non-deuterated counterparts in SANS experiments because it is assumed that the "marginal" chemical difference does not affect the results.

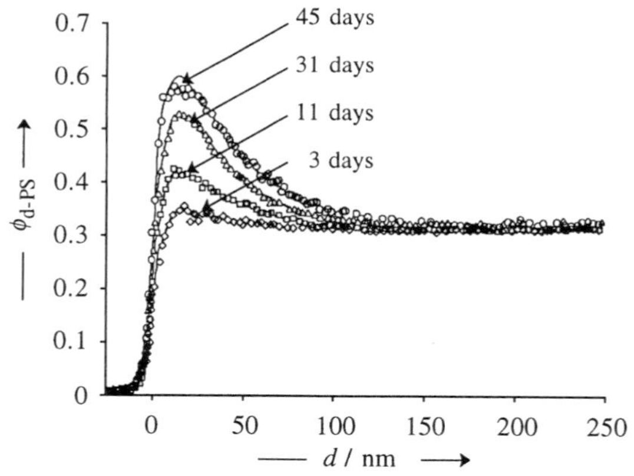

Fig. 9-1 Increase of the volume fraction $\phi_{d\text{-PS}}$ of deuterated poly(styrene) at the surface of the melt of a mixture of deuterated and non-deuterated poly(styrene) ($\phi_{d\text{-PS}} = 0.33$) after several days at 184°C (T_G = 100°C) [1]. With permission of the American Chemical Society, Washington, DC.

The surface composition and structure of polymer solids is usually established on solidification. However, certain groups are always enriched in the surface in both copolymers and mixtures of homopolymers. For example, dimethylsiloxane groups (DMS) reside preferentially in the polymer/air surface in mixtures of poly(dimethylsiloxane)

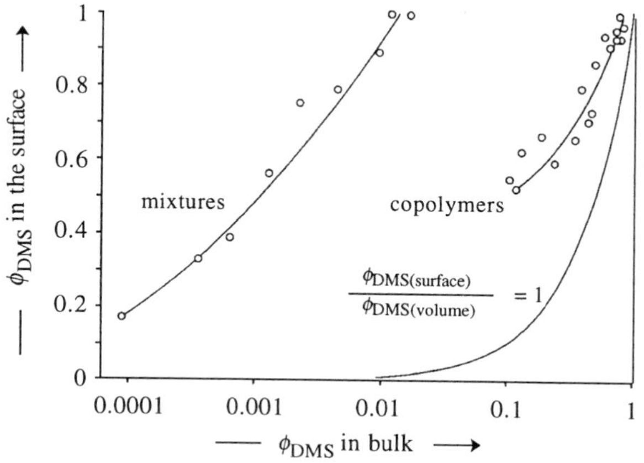

Fig. 9-2 Volume fraction of dimethylsiloxane (DMS) groups of copolymers containing DMS and bisphenol A sulfone groups or mixtures of the corresponding homopolymers in the surface as a function of the corresponding volume fraction in the bulk [2]. Data by XPS (penetration depths < 5 nm).

(PDMS) and poly(bisphenol A sulfone) as well as in copolymers with DMS and bisphenol A sulfone units (Fig. 9-2). In such mixtures, a volume fraction of $\phi_{PDMS} = 1$ % in the bulk suffices to saturate the surface completely with DMS groups ($\phi_{DMS} = 1$).

9.2 Interfacial Tension

The reversible formation or enlargement of an interface (or surface) requires work. For isothermal processes, the ratio work/area, γ, is identical with the Helmholtz energy ΔA per area; it is measured in J/m^2 = N/m and given in mN/m for liquids and mJ/m^2 for solids. As a force per length, it is an **interfacial tension** for interfaces and a **surface tension** for surfaces. The latter was formerly given in dyn/cm (1 dyn/cm ≡ 1 mN/m).

Interfacial tension exists because polymer segments at an interface can form fewer physical bonds with each other than they can in the bulk where they are surrounded on all sides by identical segments. The energy of these "missing" bonds is the **interfacial energy** (interfaces) and **surface energy** (surfaces), respectively (see below).

The term "surface tension" originated with the observation that a hanging drop stretches before it drops (L: *tendere* = to stretch out). "Surface tension" is now used when liquids are involved, be it for the surface tension of liquids themselves or for the determination of energies per surface of solids as in the Zisman method (Section 9.2.6). Since no stretching is involved in the measurement of energy per surface area of solids, the term "surface free energy" is often felt to be more appropriate for solids than "surface tension." However, "free energy" is an outdated name for "Helmholtz energy" so that the "free surface energy" is really a "Helmholtz energy per surface area." Note that IUPAC and IUPAP use mostly "surface tension" regardless of the state of matter.

9.2.1 Determination of Interfacial Tension

Surface tensions of low-molecular weight liquids and many dilute polymer solutions can be measured by various methods. Only a few of these methods are suitable for concentrated polymer solutions or polymer melts because of (a) high viscosities *per se*, (b) the dependence of these viscosities on shear rates, and (c) often also a variation of viscosities with time (Section 12.1.2). Not all dynamic methods are usable since measured surface tensions depend on the velocity of measurement. All static methods can be used such as the hanging drop method and the Wilhelmy plate method.

In the *Wilhelmy plate method*, a polymer plate or a polymer-coated plate is partially dipped into a wettable liquid. If the lower edge of the plate is just touching the surface of the liquid, then the force on the plate (in newton) just equals $\gamma_{lv}L_{per}$ where γ_{lv} = surface tension (in N/m) of the liquid (l) in equilibrium with its saturated vapor (v) and L_{per} = perimeter of the plate (in m). The method is used for the determination of the surface tension between a solid polymer and a liquid.

The shape of a *hanging drop* is controlled by surface tension and gravity. The drop is photographed and its diameters measured at different distances. In hydrodynamic equilibrium, calculated shape factors for all distances must all agree.

In the *contacting drop* method, a drop of the liquid is placed on a plane solid surface where it does not spread completely but forms a contact angle ϑ between liquid and solid (Fig. 9-3). The shape of the spread drop is controlled by interfacial tensions γ_{ij} between solid (s), liquid (l), and vapor (v) where (i, j = s, l, v; i ≠ j). On shifting the drop by an area ΔA, the Gibbs surface energy G_A is changed (see Fig. 9-3) by

$$\text{(9-1)} \qquad \Delta G_A = \Delta A(\gamma_{sl} - \gamma_{sv}) + \Delta A\,\gamma_{lv} \cos(\vartheta - \Delta\vartheta)$$

In equilibrium, $\lim_{\Delta A \to 0} \mathrm{d}\Delta G_a/\mathrm{d}\Delta A = 0$ and Eq.(9-1) becomes the **Young equation**,

$$\text{(9-2)} \qquad \gamma_{sv} - \gamma_{sl} = \gamma_{lv} \cos \vartheta$$

since $\Delta\vartheta/\Delta A$ behaves as a second order differential and becomes zero in the limiting case $\Delta A \to 0$. The **contact angle** ϑ between the solid surface and the drop is a measure of the spreading of the drop: completely at $\vartheta = 0°$ and not at all at $\vartheta = 180°$. The cosine of the contact angle, cos ϑ, measures the **wetting** of the surface.

The observed contact angle ϑ is not the true contact angle between *substance* s and vapor v but the contact angle of the *material* that depends on the roughness of the surface. Because of the roughness of the surfaces of amorphous and semi-crystalline polymers, contact areas are greater than geometric areas. This tendency to enlarge the surface is effected by both cohesion and adhesion.

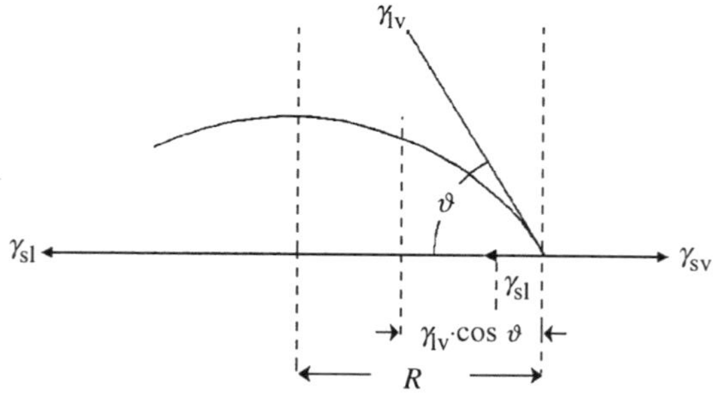

Fig. 9-3 Contact angle ϑ, section R of radius, and interfacial tensions γ_{sl}, γ_{sv}, and γ_{lv} (see text).

Cohesion dominates for poorly spreading liquids ($\vartheta > 90°$). The enlargement of the surface, as caused by its roughness, is balanced here by an increase of the contact angle, $\vartheta_{rough} > \vartheta$.

Adhesion dominates for well-spreading liquids ($\vartheta < 90°$). The liquid spreads over a greater rough area than a smooth one: the experimental contact angle ϑ_{rough} is smaller than the true one, thus $\vartheta_{rough} < \vartheta$ and also $r = (\cos \vartheta_{rough}/\cos \vartheta) > 1$.

Rough surfaces are not only present on solid polymers but also on polymer melts. Here, Brownian movements cause some segments of the molecule to stick out of the "smooth" melt surface. According to neutron and X-ray diffraction, the resulting differences between "peaks" and "valleys" are ca. 0.5-1 nm.

9.2.2 Time Effects

Freshly formed surface layers are not in equilibrium so that one can distinguish between dynamic (non-equilibrium) and static (equilibrium) interfacial tensions.

The time dependence of interfacial and surface tensions depends on the temperature as well as the constitution of monomeric units and endgroups, molar mass distribution, the type and physical structure of the substrate, and, for solutions, also on the viscosity η_1 of the solvent. These quantities as well as the polymer concentration control the surface tension γ_v and the viscosity η of the solution as well as the "radius" R of the deposited drop (see Fig. 9-3) and the radius R_0 of the drop before wetting. The time dependence before wetting can be described by the change of R/R_0 with reduced time, $(\gamma_v/\eta_1 R_0)t$, a dimensionless quantity (Fig. 9-4).

For example, the wetting of glass or PTFE by PIB in decalin increases initially with time but is independent of the substrate. Finally, R/R_0 and thus also the contact angle ϑ of PTFE (but not that of glass) becomes independent of time with ϑ reaching 98°, i.e., PIB solutions do not wet poly(tetrafluoroethylene) very well.

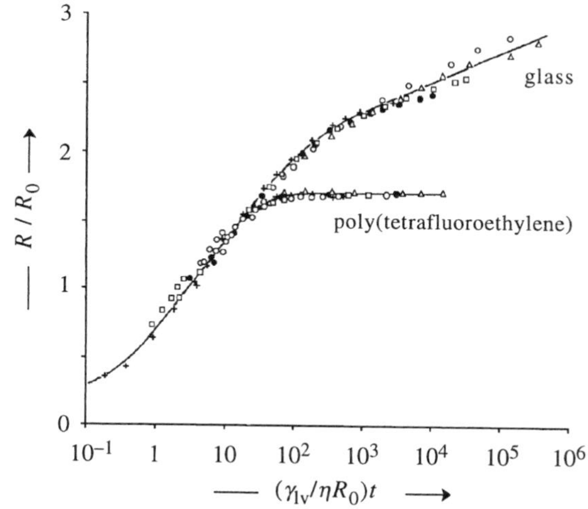

Fig. 9-4 Relative size of droplets, R/R_0, on glass and on poly(tetrafluoroethylene) (PTFE) as a function of reduced time, $(\gamma_v/\eta R_0)t$ [3]. Data for decalin™ solutions of a poly(isobutylene) (PIB) at various concentrations (○,●,△,□,+) at 23°C. By permission of Steinkopff-Verlag, Darmstadt.

The situation is quite different for glass as the substrate. Initially, glass behaves like PTFE, but at longer times R/R_0 continues to increase linearly with the logarithm of the time parameter (Fig. 9-4) whereas cos ϑ asymptotically approaches unity, i.e., ϑ becomes 0° (not shown). At long times, glass is completely wetted by poly(isobutylene) solutions in decalin.

The time dependence of wetting is affected by the type of endgroups (Fig. 9-5), obviously because different endgroups congregate differently in the uppermost surface layer. The contact angle of water droplets on poly(styrene) surfaces at 40°C does not change with time if endgroups are hydrogen atoms H. If endgroups are SiF_3, however, the contact angle is much greater initially than that for H as endgroups. With time, the same values are approached: hydrophobic SiF_3 groups are slowly buried in the interior of poly(styrene). Hydrophilic endgroups COOH, on the other hand, give initially the same contact angle as H endgroups. The contact angle then decreases and the wettability correspondingly increases because more and more COOH groups move to the surface which slowly becomes more hydrophilic.

These observations show that surfaces of "glass-like" polymers are not rigid but may change their compositions with time, depending on the wetting agent. In the case of poly(styrene) with COOH endgroups, water acts as a plasticizer and the surface becomes more rubbery as shown by scanning electron microscopy.

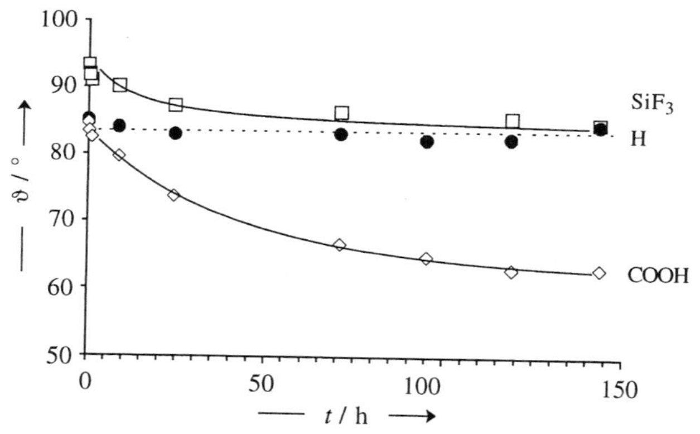

Fig. 9-5 Contact angles of water droplets on poly(styrene) surfaces (PS) as a function of the time the PS surfaces were in contact with water vapor at 40°C before the droplet was placed on the surface [4]. Poly(styrene)s with endgroups H (●), SiF_3 (□), or COOH (◇). By permission of the Materials Research Society, Warrendale (PA).

9.2.3 Surface Tension of Polymer Melts

Surface tensions γ_{lv} of liquid polymers depend on both their chemical structure (monomeric units, endgroups, molar masses, configuration) as well as temperature. Experiments showed that surface tensions of low-molar mass polymers depend on the 2/3rd power of the inverse molar mass (Fig. 9-6):

(9-3) $$\gamma_{lv} = \gamma_{lv,\infty} - K_{end}\overline{M}_n^{-2/3}$$ **(LeGrand-Gaines equation)**

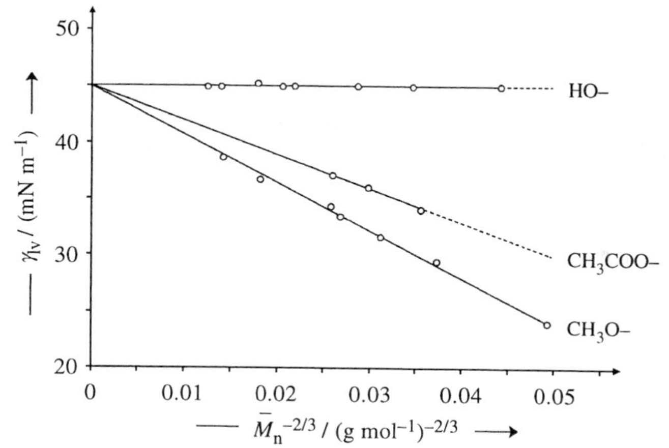

Fig. 9-6 Surface tension γ_{lv} of low-molar mass liquid poly(oxyethylene)s, $R(OCH_2CH_2)_nOR$, with endgroups RO- = HO-, CH_3COO-, or CH_3O- at 24°C as a function of $1/\overline{M}_n^{2/3}$ [5]. A value of $\overline{M}_n^{-2/3}$ = 0.01 $(g\ mol^{-1})^{-2/3}$ corresponds to only $\overline{M}_n$ = 1000 g mol^{-1}!

The slope of $\gamma_{lv} = f(\overline{M}_n^{-2/3})$ for liquid poly(oxyethylene)s in air is more negative, the more apolar the endgroups. The intercept $\gamma_{lv,\infty}$ is a polymer-specific constant that is independent of molar mass and type of endgroups.

Eq.(9-3) is often used to obtain surface tensions for infinite molar masses of polymers although theory predicts that $\gamma_{lv} = f(\overline{M}_n^{-2/3})$ applies only to low molar masses whereas high molar masses should obey $\gamma_{lv} = f(\overline{M}_n^{-1})$. However, the difference between the two types of extrapolated surface tensions is less than 1 % if $\overline{M}_n$ > 10 000 g/mol. In the following, γ_{lv} refers only to high molar masses.

Surface tensions $\gamma_{lv}/(mN\ m^{-1}\ K^{-1})$ at 20°C range from 18.4 for poly(oxyhexafluoropropylene) to 49.3 for poly(propylene isophthalate), (see also Table 9-3). It does not seem that there is an obvious relationship between the surface tension at infinite molar mass and molecular parameters such as polarity or packing density.

Surface tensions decrease linearly with increasing temperature. Typical values of $d\gamma_{lv}/dT$ range from –0.048 mN/(m K) for poly(dimethylsiloxane) to –0.076 mN/(m K) for poly(ethylene terephthalate) and –0.095 mN/(m K) for poly(oxyethylene).

Table 9-3 Surface tensions γ_{lv} of liquid polymers at 20°C and 150°C and interfacial tensions of liquid polymer 1 versus liquid polymer 2 at 150°C. * 180°C.

Polymer 1	$\gamma_{lv,1}/(mN\ m^{-1})$		$\gamma_{ll}/(mN\ m^{-1})$ of polymer 1 against the following polymer 2							
Name	20°C	150°C	PDMS	it-PP	PBMA	PE	PS	PMMA	PVAC	PEOX
PDMS	21.3	13.6	0	3.0	3.8	5.4	6.0		7.4	9.8
PTFE	25.6	9.4*								
itPP		22.1	3.0	0		1.1	5.1			
PVAC	36.5	27.9	7.4		2.8	11.0	3.7		0	
PE	35.7	28.1	5.4	1.1	5.2	0	5.7	9.5	11	9.5
PS	40.7	30.8	6.0	5.1		5.7	0	1.6	3.7	
PMMA	41.1	31.2			1.8	9.5	1.6	0		
PEOX	45	33.0	9.8			9.5				0

9.2.4 Interfacial Tension between Polymer Melts

Interfacial tensions between two polymeric liquids are always smaller than the surface tensions of liquid polymers against air (Table 9-3). Again, no clear picture arises for the effect of molecular properties on interfacial tension, probably because little is known about the average group composition of the liquid layer of polymer 1 that is in immediate contact with the corresponding liquid layer of polymer 2.

For example, the interfacial tension of apolar poly(ethylene), $\text{+}CH_2\text{–}CH_2\text{+}_n$, versus apolar it-poly(propylene), $\text{+}CH_2\text{–}CH(CH_3)\text{+}_n$, is only 1.1 mN/m whereas that of apolar poly(ethylene) versus polar poly(oxyethylene), $\text{+}O\text{–}CH_2\text{–}CH_2\text{+}_n$, is 9.5 mN/m. However, the interfacial tension between the two polar polymers poly(oxyethylene) and poly(dimethylsiloxane), $\text{+}O\text{–}Si(CH_3)_2\text{+}_n$, is also high, 9.8 mN/m.

9.2.5 Surface Tension of Polymer Solutions

Like surface tensions of low-molecular weight surfactants (Volume II, Fig. 5-8), surface tensions of solutions of low-molecular weight amphiphilic polymers decrease with increasing solute concentration and become constant above a critical micelle concentration CMC. An example is poly(oxyethylene)-*block*-poly(oxypropylene)-*block*-poly(oxyethylene) with $\overline{M}_w$ = 8800 g/mol at 60°C (Fig. 9-7). At concentrations above the CMC, surfaces are completely covered with molecules of the amphiphilic solute, whose hydrophilic ends are in the water and whose hydrophobic ends are in the air.

At lower temperatures and higher molar mass, this polymer show a stepwise variation of $\gamma_{lv} = f(\lg w_2)$. At both lower tempratures and higher molar mass, segmental mobility is decreased. The effect is therefore probably due to a kinetically controlled reorganization of molecules in the liquid surface where the hydrophilic blocks must be in water and the hydrophobic groups in air.

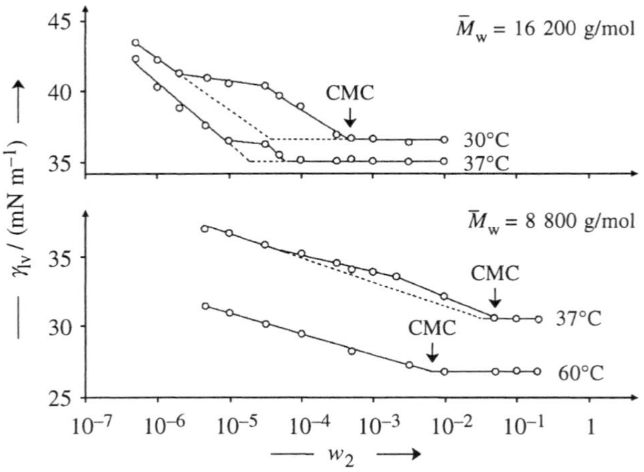

Fig. 9-7 Dependence of surface tensions γ_{lv} of aqueous solutions of poly(oxyethylene)-*block*-poly(oxypropylene)-*block*-poly(oxyethylene)s, $\text{+}OCH_2CH_2\text{+}_m\text{+}OCH_2CH(CH_3)\text{+}_n\text{+}OCH_2CH_2\text{+}_m$, (m/n = 2.65) on the mass fraction w_2 of polymers [6].

9.2.6 Critical Interfacial Tension Solid-Liquid

Interfacial tension γ is a force per length, hence an energy per area (N m^{-1} = J m^{-2}). The **interfacial energy** γ_{sv} between a solid (s) and its vapor (v) can thus be subdivided into the **surface free energy** γ_s^o of the solid (i.e., the energy per area) and the **spreading pressure** (= pressure times length) Π_{sp} of the saturated solvent vapor on the solid polymer surface in equilibrium:

(9-4) $$\gamma_{sv} = \gamma_s^o + \Pi_{eq}$$

Spreading pressures Π_{eq} become zero in vacuum and also at diminishing contact angles. At finite contact angles, spreading pressures may be considerable, for example, Π_{eq} = 14 mN/m for water on poly(ethylene).

The surface free energy γ_s^o of the solid is an important material constant. Since it is not directly accessible, many methods have been proposed for its determination. The method of choice is most often the **Zisman method** which determines a **critical surface tension** γ_{crit} of wetting. In this method, the cosines of contact angles of a series of liquids on a polymer plate are plotted against the surface tension γ_{lv}^o of that liquid according to

(9-5) $$\cos \vartheta = [1 - K\gamma_{crit}] + K\gamma_{lv}^o \quad ; \quad K = \text{constant}$$

and extrapolated to $\cos \vartheta = 1$, i.e., $\vartheta = 0$ (complete spreading) where $\gamma_{lv}^o = \gamma_{crit}$. The contacting liquids have to be of a similar type (all non-polar, all polar, or all hydrogen-bonding), otherwise, data points would be scattered considerably (Fig. 9-8).

A somewhat clearer picture emerges if one sets $K \equiv -K'/\gamma_{lv}^o$ and plots γ_{lv}^o as a function of $\gamma_{lc}^{\emptyset}(1 - \cos \vartheta)$ according to

(9-5a) $$\gamma_{lc}^o = \gamma_{crit} + K'\gamma_{lc}^{\emptyset}(1 - \cos \vartheta)$$

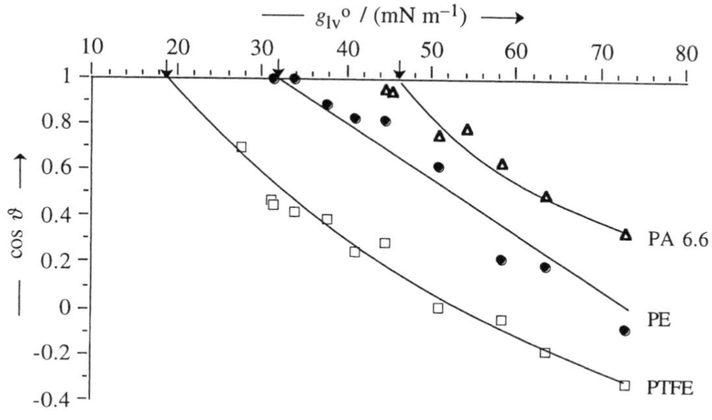

Fig. 9-8 Zisman plot according to Eq.(9-5) of poly(hexamethylene adipamide) (PA 6.6), low-density poly(ethylene) (PE), and poly(tetrafluoroethylene) (PTFE) [7]. Liquids were apolar, polar, or hydrogen-bonding. Arrows indicate literature data for PA 6.6, LDPE, and PTFE. Lines are "best fit."

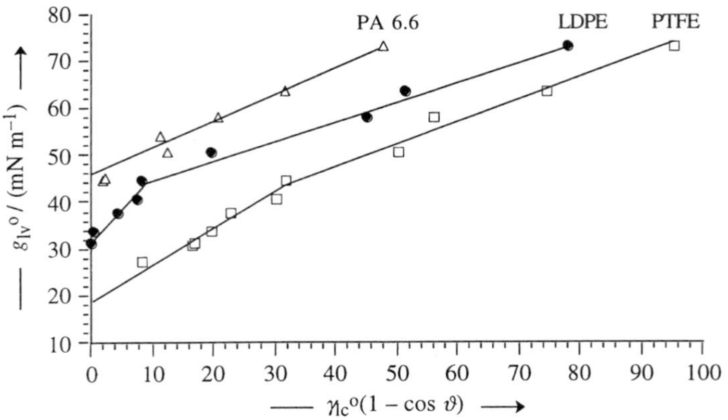

Fig. 9-9 Determination of critical surface tensions using Eq.(9-5a). Data of Fig. 9-8. The intercepts for PE and PTFE agree with the data reported in literature, i.e., γ_{crit} = 46 mN m^{-1} for PA 6.6, γ_{crit} = 31 mN m^{-1} for LDPE, and γ_{crit} = 18.5 mN m^{-1} for PTFE.

A plot according to Eq.(9-5a) delivers two straight lines for each polymer because $\gamma_{lc}^{\emptyset}(1 - \cos \vartheta)$ as a variable instead of $\cos \vartheta$ separates the $\gamma_{lc}^{\emptyset}$ data of the less polar solvents from those of the more polar ones (Fig. 9-9). Eq.(9-5a) has the additional advantage in that it can be compared directly with the Young equation.

Writing Eq.(9-2) as $\gamma_{sv} - \gamma_{sl} = \gamma_{lv}^{o} \cos \vartheta = -\gamma_{lv}[-1 + (1 - \cos \vartheta)]$ and solving for γ_{lv}^{o} delivers

$$(9\text{-}5b) \qquad \gamma_{lv}^{o} = (\gamma_{sv} - \gamma_{sl}) + \gamma_{lv}^{o}(1 - \cos \vartheta)$$

so that γ_{crit} of Eq.(9-5a) is now identified by Eq.(9-2) as $\gamma_{sv} - \gamma_{sl}$.

Table 9-4 compares experimental surface tensions γ_{lv} of liquid polymers with experimental critical surface tensions γ_{crit} of their solids by the Zisman method and theoretical estimates of the surface Helmholtz energy γ_{s}^{o} (= surface free energy). In general, these three values are not very different. Therefore, γ_{sl} must be very small if $\gamma_{crit} = \gamma_{sv} - \gamma_{sl}$ and if γ_{sv} can be identified as γ_{s}^{o}.

Critical surface tensions of all known solid organic polymers are lower than the surface tension of water at 20°C, γ_{lc} = 72 mN/m (Table 9-4). Hence, all polymer surfaces are not wetted by water very well.

Critical surface tensions of fluorinated polymers are especially low (Table 9-4), and here particularly those of poly(tetrafluoroethylene) (Teflon). These polymers are not only not wetted by water but also not by oils and fats such as the glycerol esters of fatty acids (in butter, olive oil, etc.) which have surface tensions of γ_{lc} = 20-30 mN/m. Teflon coated pans and pots thus reduce or even prevent sticking. Poly(1,4-phenylene sulfide) acts similarly but its critical surface tension does not seem to have been reported in the scientific literature.

Some inorganic materials such as quartz and titanium dioxide have high critical surface tensions. This presents a problem if they are used as fillers or reinforcing fibers in organic polymers. The problem can be ovecome to some extent by the use of so-called coupling agents (Volume IV, Section 3.3.3).

Table 9-4 Liquid surface tensions, γ_{lc}, critical surface tensions of polymers, γ_{crit}, and estimated surface free energies, $\gamma_s{}^o$, of polymers at 20°C. Literature gives γ_s in mJ m^{-2} ≡ mN m^{-1}. * 25°C.

Polymer		γ_{lc} / mN m^{-1}	γ_{crit} / mN m^{-1}	$\gamma_S{}^o$ / mN m^{-1}
Poly(heptadecafluorodecyloxymethylstyrene)	-		6	
Poly(hexafluoropropylene)	-		16.2	
Poly(dimethylsiloxane)	PDMS	21.3	24	24
Poly(tetrafluoroethylene)	PTFE	25.6	18.5	19.1
Poly(vinylidene fluoride)	PVDF		25	30.3
Poly(vinyl fluoride)	PVF		28	36.7
Poly(propylene), isotactic	PP	29.4	31	30.2
Poly(propylene oxide)	PPOX	30.7	32	
Poly(chlorotrifluoroethylene)	PCTFE	30.9	31	
Poly(ethylene), low density	LDPE		31.5	
Poly(ethylene), high density	HDPE	35.7	33.1	
Poly(methyl methacrylate)	PMMA		39	40.2
Poly(vinyl chloride)	PVC		39	41.5
Poly(methyl methacrylate), atactic	PMMA	41.1	39	
Poly(styrene), atactic	PS	40.7	43	42.0
Poly(ethylene terephthalate)	PET	44.5	43	41.3
Poly(ethylene oxide)	PEOX	45*	43	
Cellulose (from cotton)	-		44	
Wool	-		45	
Poly(hexamethylene adipamide)	PA 6.6	46.4	46	43.2
Sodium silicate	-		47	
Urea-formaldehyde resin	UF		61	
Quartz (silicon dioxide)	-		78	
Anatase (titanium dioxide)	-		91	
Sulfur	-		128	

9.3 Thin Layers of Polymers

Insoluble chemical compounds spread on liquid surfaces (**hypophases**) which can be investigated by using a Pockels-Langmuir trough (Volume I, p. 386). Three sides of the surface liquid area in this trough are given by the walls of the trough whereas the fourth side is controlled by a barrier that floats on the surface of the liquid, usually water. On applying a certain amount of spreadable material on the surface area of the water between the barrier and the three walls, the barrier is pushed back by the surface pressure of the spreading material. These measurements are not easy to do because the liquid surface has to be painstakingly clean and the small amounts of material produce only very small pressures.

Given enough available surface area, insoluble liquid material such as oils or monomers will spread on the water surface until a monomolecular layer is formed; the surface pressure will then be small. Reducing the surface area by pushing back on the barrier will increase the surface pressure, first slightly and then dramatically (Fig. 9-10) because molecules in the spread layer become more densily packed. Finally, the whole surface layer collapses.

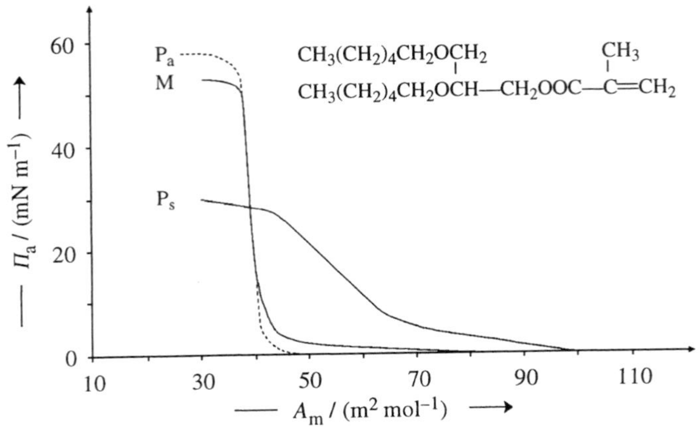

Fig. 9-10 Surface pressure Π_a on water as a function of the molar surface A_m (area per amount of monomer molecules or monomeric units) of a monomeric lipid M (formula in insert) or the corresponding polymeric lipid by polymerization in the surface layer (P_a) or separately in solution (P_s) [8].

Spread monomer molecules M_a can be polymerized to the polymer P_a. In the range of strong increase of surface pressure Π_a with decreasing molar surface, A_m, Π_a varies with A_m for both M_a and P_a alike (Fig. 9-10). However, polymer layers P_a collapse at somewhat lower surfaces A_m and higher surface pressures Π_a than monomer layers M_a since monomeric units are less mobile than monomer molecules.

The situation is different for polymers P_s that were obtained by polymerization in solution. Spreading these polymers on the water surface generates a surface pressure at a larger surface. The compression to smaller surfaces A_m is also far less abrupt and the collapse occurs at larger areas A_m and smaller surface pressures Π_a. This happens because increasing surface pressures cause polymer coils to interpenetrate each other more strongly. However, the surface structure of these polymers P_s is less compact than that of polymers P_a from the direct polymerization in the surface layer which leads to less resistance against collapse.

Surface pressures Π_a (= energy/area = force/length) correspond to two-dimensional osmotic pressures. At very small surface concentrations $c_a = m_a/A$ (mass of spread molecules per area A), surface pressures are proportional to the numbers of molecules. At higher polymer concentrations, effects of two-body interactions become important, i.e., second virial coefficients A_2, followed by effects of 3-body interactions (third virial coefficients), etc. In analogy to the three-dimensional case (Section 10.4.4), one can thus write for the concentration dependence of reduced surface pressures

(9-6) $$\Pi_a/(RTc_a) = (1/\overline{M}_n) + A_{2,a}c_a + ...$$

or, with $c_a/\overline{M}_n = [P_a]$ and $A_{2,p} = A_{2,a}\overline{M}_n$,

(9-6a) $$\Pi_a/(RT[P_a]) = 1 + A_{2,p}\overline{M}_n c_a + ...$$

The second virial coefficient $A_{2,a}$ here refers to the mass concentration of polymer molecules in the surface. It is not identical with the second virial coefficient A_2 from the

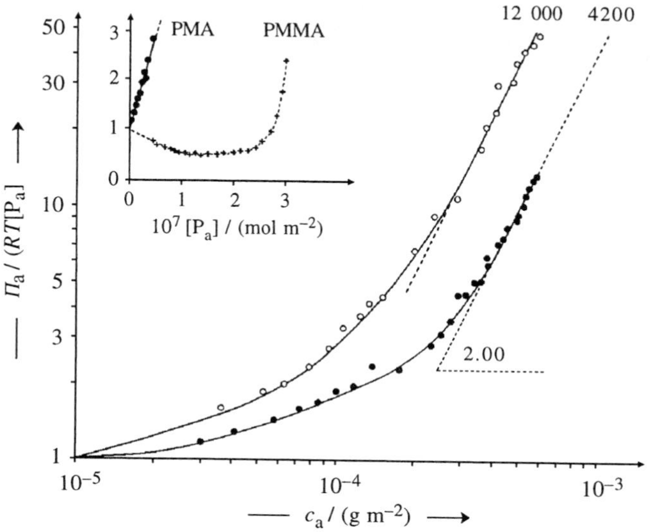

Fig. 9-11 Logarithm of the reduced surface pressure, $\Pi_a / (RT[\mathrm{P_a}])$, as a function of the logarithm of the surface concentration c_a of two poly(methyl acrylate)s (PMA) with molar masses of 4200 g/mol (●) and 12 000 g/mol (○) on water against air at 15°C [9].

Insert: $\Pi_a / (RT[\mathrm{P_a}])$ as a function of the molar surface concentration $[\mathrm{P_a}]$ for the PMA with M = 4200 g/mol (●) and a poly(methyl methacrylate) (PMMA, +) with M = 3300 g/mol at low concentrations on water against air at 15°C [9].

analogous equation for membrane osmometry. The 2nd virial coefficient $A_{2,p}$ has the unit of a specific area, i.e., area/mass.

An example is poly(methyl acrylate), PMA, on water in air where the reduced surface pressure, $\Pi_a/(RT[\mathrm{P_a}])$ increases in direct proportion to the molar surface concentration of polymer molecules, $[\mathrm{P_a}]$ (insert in Fig.9-11), indicating a positive second virial coefficient $A_{2,p}$. For poly(methyl methacrylate) (PMMA), the initial slope is negative which probably does not signal a negative second virial coefficient but a self-association of PMMA molecules at low polymer surface concentrations (see Section 10.5).

At higher concentrations, lg $[\Pi_a/(RT[\mathrm{P_a}])]$ increases linearly with lg c_a (Fig. 9-11):

(9-7) $$\lg [\Pi_a M/(RTc_a)] = \lg K_\omega + \lg c_a^{\,\omega} \qquad \text{(at higher concentrations } c_a)$$

Higher molar masses lead to lower critical concentrations but the exponent ω is independent of the molar mass.

Scaling predicts $\omega = 1/(2\,y - 1)$ where y is the exponent in the dependence of the two-dimensional radius of gyration, s_a, on the degree of polymerization X, $s_a = KX^y b$, with b = length of monomeric unit and K = constant. Curves for all molar masses fall on the same line if lg $[\Pi_a M/(RTc_a)] = f[\lg (c_a/c_a^*)]$ where c_a^* is a critical surface concentration (see also Fig. 10-23 for the three-dimensional case).

The parameter $y = (1 + \omega^{-1})/2$ has been calculated theoretically by Monte Carlo statistics, self-avoiding walks on two-dimensional lattices, mean-field theory, etc. For spreading on "good" solvents, all theories agree that y has a value of 3/4, either exactly or with good approximation. The exponent ω in Eq.(9-7) should thus equal 2 which is indeed observed for poly(methyl methacrylate) on water (Fig. 9-11). Similar values have also been found for the hydrophilic (but water insoluble) polymers poly(tetrahydrofuran), $\left[\!-O(CH_2)_4-\!\right]_n$, and poly(vinyl acetate), $\left[\!-CH_2-CH(OOCCH_3)-\!\right]_n$.

Theoretical predictions for ω for spreading on "bad" solvents vary between 2/3 (mean-field theory with only ternary interactions) and 1/2 (ideal random-flight statistics on a plane). It seems that calculations for self-avoiding walks on a honeycomb lattice give the most reliable result of $y = 4/7 \approx 0.571$. Spreading of poly(methyl methacrylate) on water against air delivered $y \approx 0.53$, i.e., $\omega \approx 16.7$.

9.4 Adsorption on Surfaces

9.4.1 Fundamentals

Dissolved macromolecular compounds adsorb on solid surfaces quite differently than low-molecular weight compounds. Small molecules can be approximated as more or less spherical entities that form a single contact with the surface. The number of contacts per area thus determines the percentage of the covered area. In order to determine this area, it suffices to measure adsorption isotherms and their temperature dependence.

Rodlike macromolecules with only one adsorbable endgroup can also form only one contact with the surface. The axis of the molecule is then either normal or somewhat inclined to the adsorption plane. Rodlke molecules with many adsorbable groups along the chain can form many contact points and the main axis of the rod will then lie parallel to the surface.

Chain molecules behave quite differently (Fig. 9-12). Many potentially adsorbable segments in this case neither guarantee an especially strong adsorption *per molecule* nor any adsorption at all. Instead, adsorption is controlled by many different factors.

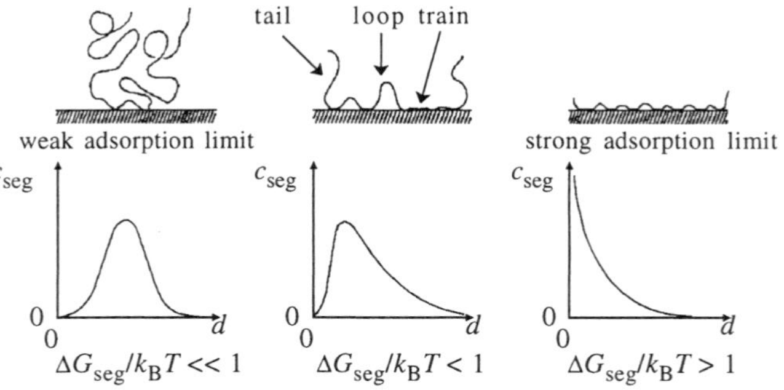

Fig. 9-12 Top: weak and strong adsorption limits of coil-like macromolecules. Adsorbed chain segments can form tails, loops, or trains.

Bottom: segment concentration c_{seg} of adsorbed coils as function of the distance d to the surface for different reduced adsorption energies per segment, $\Delta G_{seg}/k_BT$.

Adsorptions are commonly characterized by **exchange energies** or **binding energies** per segment, $\Delta G_{seg} = \Delta H_{seg} - T\Delta\Delta S_{seg}$. The adsorption enthalpy ΔH_{seg} describes the energy gain and the term $-T\Delta\Delta S_{seg} = -T(\Delta S_{conf} - \Delta S_{solv})_{seg}$ the net loss of entropy. The entropy change is caused by two factors: the strong entropy loss caused by the decrease of the number of possible macroconformations of the molecule and the small entropy gain when many adsorbed solvent molecules are replaced by one polymer segment.

The exchange energy is small for most adsorptions of polymer chains, i.e., the coil forms only a few contact points per chain with the surface. In this **weak adsorption limit**, adsorbed chains form tails, loops, or trains (Fig. 9-12). Such small exchange energies *per segment* lead to large binding energies *per molecule*. Weak adsorption limits may thus lead to large adsorbed amounts.

The other limiting case is the **strong adsorption limit** in which many contact points are formed per molecule. Here, loops are small and tails short. The resulting adsorption layers are not very thick (Fig. 9-12).

9.4.2 Methods

Adsorptions can be studied by many different methods that supplement each other. The adsorbed mass is most accurately determined by scintillography of radioactively marked polymers. Nuclear magnetic resonance and electron spin resonance spectroscopy deliver not only adsorbed amounts but also information about the conformation of adsorbed segments.

Layer thicknesses and polymer concentrations in the adsorbed layers are obtainable by ellipsometry, the change of elliptically polarized light after reflection at the adsorbed surface layer. Layer thicknesses can also be determined by measuring the forces between two surfaces that are both covered with the adsorbed polymer (see Section 9.5.1). Hydrodynamic methods such as measurements of viscosities are much more indirect and often difficult to interpret.

9.4.3 Time Dependence

Adsorption of polymer molecules proceeds more slowly than that of small molecules because (a) diffusion coefficients are smaller and (b) immediately formed macroconformations of adsorbed molecules are not in equlibrium and rearrange themselves later. The "adsorption equilibrium" of polymers on smooth surfaces (films) is obtained in minutes to hours but may even take days to reach at rough surfaces (powders). The adsorbed mass of polymer per area and the thicknesses of adsorbed layers continue to increase with time until equilibrium values are reached at long times (Fig. 9-13). The polymer concentration in the adsorbed layer continues to decrease with time before it becomes constant.

The processes are interpreted as follows. In very dilute solutions, macromolecules approach the solid surface as coils that are "three-dimensionally" adsorbed as such at the surface (Fig. 9-12, left) where they liberate a few solvent molecules. With increasing time, more and more molecules are adsorbed until the adsorption layer, as measured by the adsorbed mass of polymer per area, m/A, finally becomes constant at very long times (Fig. 9-13, but see also Fig. 9-14). If it is energetically favorable, more and more segments try to contact the solid surface and the polymer molecule spreads on the surface until it is finally "two-dimensionally" adsorbed in the strong adsorption limit.

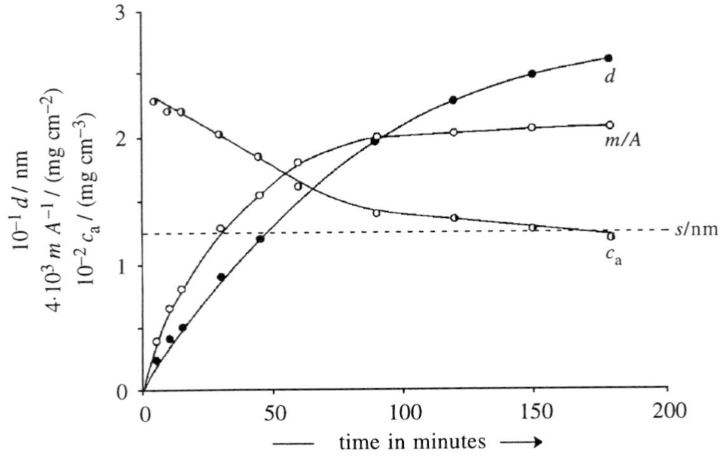

Fig. 9-13 Time dependence of the thickness d of adsorbed polymer layer, adsorbed mass m per area A, and concentration c_a in the adsorbed layer. Data for the adsorption of a poly(styrene) ($\overline{M}_w$ = 176 000 g/mol) on a chromium surface from 5 mg/mL solutions in cyclohexane [10]. The unperturbed radius of gyration for such a macromolecule is $\langle s^2 \rangle_o^{1/2}$ = 12.2 nm (dotted line) and the critical overlap concentration for a solution in cyclohexane is c_s^* = 35 mg/mL (Eq.(6-3)).

The scenario is different for the weak adsorption limit. Five minutes after the start of the adsorption, the thickness of the adsorbed layer and the adsorbed mass per surface area are both small but the concentration c_a of the polymer in the surface layer is high (Fig. 9-13), ca. 230 mg/cm^3, which is higher than the critical entanglement concentration for this polymer (c^* = 35 mg/mL). Since all adsorbed entangled polymer molecules compete for the surface, the thickness of the layer increases with time. At long times, the adsorbed mass per area becomes approximately constant but, since the thickness d continues to increase, the concentration in the adsorbed layers continues to decrease.

9.4.4 Adsorption Equilibria

The complex interplay between the polymer concentration in solution and in the adsorbed layer, degree of polymerization, adsorption energy per segment, and Flory-Huggins interaction parameters for interactions between polymer and solvent has been illustrated by calculations with a quasi-lattice theory for the strong adsorption limit (adsorption energy: $\Delta G_{seg} = 1\,k_BT$) (Fig. 9-14).

For dilute solutions ($10^{-9} < \phi_p < 10^{-3}$), logarithms of fractions f_a of equilibrium surface coverage by low-molecular weight polymers (X = 1 and X = 10) increase linearly with increasing logarithms of volume fractions ϕ_p of polymers in solution until a critical volume fraction ϕ_{crit} is reached (Fig. 9-14). The slope of lg f_a = f(lg ϕ_p) decreases with increasing X. At $\phi_p = 10^{-7}$, surface coverage increases strongly with increasing degree of polymerization from $f_a = 10^{-6}$ (X = 1) to $f_a \approx 8 \cdot 10^{-5}$ (X = 10), and $f_a \approx 0.7$ (X = 50).

Above a critical volume fraction ($\phi_{p,crit} \approx 10^{-3}$ for X = 1), surface coverages vary less strongly with f_a. For adsorptions from the melt (ϕ_p = 1), the surface is completely covered (f_a = 1) with molecules that can have only one contact with the surface (i.e.,

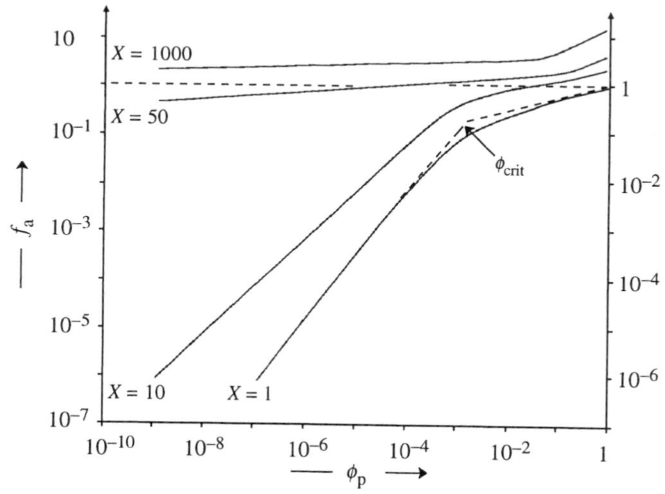

Fig. 9-14 Calculated fractional surface coverage f_a of polymers with degrees of polymerization $X = V_{polymer}/V_{solvent}$ (expressed as a ratio of volumes) from polymer solutions with volume fractions ϕ_p of polymers [11], assuming a Gibbs energy of $\Delta G_{seg} = 1\ k_BT$ and a polymer-solvent interaction parameter of $\chi_o = 1/2$ (see Chapter 10). The critical volume fraction ϕ_{crit} is indicated for $X = 1$ only.

those with $X = 1$). For higher degrees of polymerization, fractions f_a exceed unity, i.e., surfaces are covered by more than one molecular layer of polymer molecules.

Complete coverage ($f_a = 1$) is obtained at $\phi_p = 1$ for $X = 1$ but at $\phi_p \approx 2 \cdot 10^{-2}$ for $X = 10$ and ca. 10^{-4} for $X = 50$. For a degree of polymerization of $X = 1000$, complete coverage ($f_a = 1$) is obtained at extremely low polymer concentrations, $\phi_p << 10^{-10}$ (not shown). The coverage of chromium by the poly(styrene) with $X \approx 1700$ of Fig. 9-13 is $m/A \approx 5$ mg/m^2 which corresponds to 2-5 equivalent monolayers of polymer molecules.

For high degrees of polymerization, fractional surface coverages f_a vary only a little with polymer volume fractions ϕ_p as shown by the data for $X = 1000$. Such small variances of f_a with ϕ_p appear as a plateau in experimental investigations that usually cover only a small range of polymer concentrations.

Adsorption isotherms, $f_a = f(\phi_p)$, are sharp for molecularly uniform polymers (Fig. 9-14) but rounded for molecularly non-uniform ones because of overlap of various values of X (not shown). Desorption isotherms are always sharp because polymers desorb only if their solution concentrations are far below their detection limits.

Adsorption of polymers from thermodynamically poor solvents is characterized by relatively small interactions between polymer segments and fairly strong polymer/adsorbent interactions. Adsorptions thus decrease with increasing goodness of the solvent.

For example, poly(styrene) adsorbs well on chromium from its solutions in cyclohexane (a thermodynamically bad solvent) but not at all from its solutions in the good solvent 1,4-dioxane. The more polar the polymer and the adsorbent, the more contacts are formed between the polymer molecules and the surface and the more flat and compact will be the polymer surface layer. For example, the adsorbed layer of poly(oxyethylene), $\text{+}OCH_2CH_2\text{+}_n$, on chromium is only ca. 2 nm thick but the PEOX molecules are so tightly packed that the refractive index of the adsorbed layer is identical with that of the crystalline polymer.

9.5 Polymer Brushes

Tethered chains refers to many polymer chains whose ends are attached via one chemical or physical bond each to a common segment, molecule, or other substrate (Fig. 9-15). The term "**brush**" refers to high concentrations of tethered chains; an example is rod-like molecules that are physically or chemically bound to a solid surface (Fig. 9-15, top left).

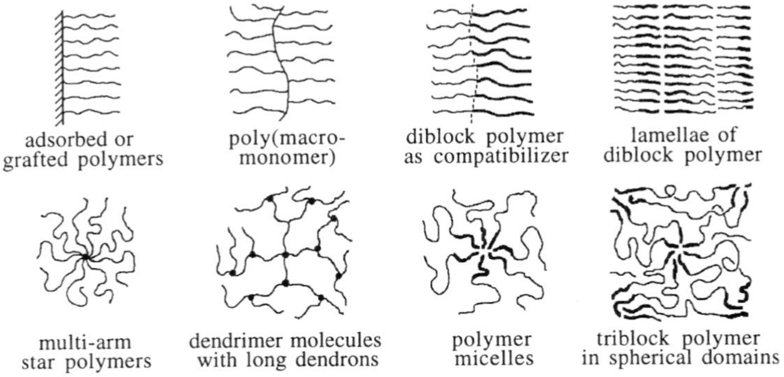

Fig. 9-15 Schematic representation of some tethered chains. Diblock polymers may act as compatibilizers between two solid surfaces (as shown) or emulsifiers between two liquid surfaces in microemulsions of vesicles (not shown). For the domain formation of diblock and triblock polymers, see Section 8.5. "Long dendron" of dendrimers refers to the distance between branching points.

9.5.1 Repulsive Forces

In *equilibrium*, tethered chain molecules of brushes do not form random coils but rather stretch away from their anchor points in order to prevent or diminish electrostatic interactions between their monomeric units. This stretching does not require the application of mechanical stresses or electric fields.

Repulsive forces between tethered chains have been shown to exist between poly(styrene) blocks of a poly(2-vinylpyridine)$_{60}$-*block*-poly(styrene)$_{60}$ that were adsorbed on mica platelets. The polar 2-vinylpyridine units cause the poly(2-vinylpyridine) blocks to adsorb strongly on the polar mica surface (strong adsorption limit, Fig. 9-12, top right). Apolar poly(styrene) blocks are not adsorbed. They extend from the flatly adsorbed poly(2-vinylpyridine) blocks (strong adsorption limit, see Fig. 9-12) into the solution in a brush-like manner.

Two mica platelets are then pressed together with the adsorbed layers facing each other and the force is measured between opposing layers of poly(styrene) blocks. For poly(styrene) blocks in toluene, repulsive forces between adsorbed layers are still considerable at a distance of 30 nm between mica plates which is the lower limit of the linear variation of $F/L = \mathrm{f}(d)$ (Fig. 9-16). This distance is considerably larger than twice the radius of gyration of free poly(styrene) molecules with $X = 60$ in toluene ($2\ s \approx 4.6$ nm) and practically identical with the sum of conventional contour lengths of 2 fully stretched poly(styrene) molecules ($2\ r_{\mathrm{cont}} \approx 31$ nm). Poly(styrene) blocks must therefore be considerably elongated.

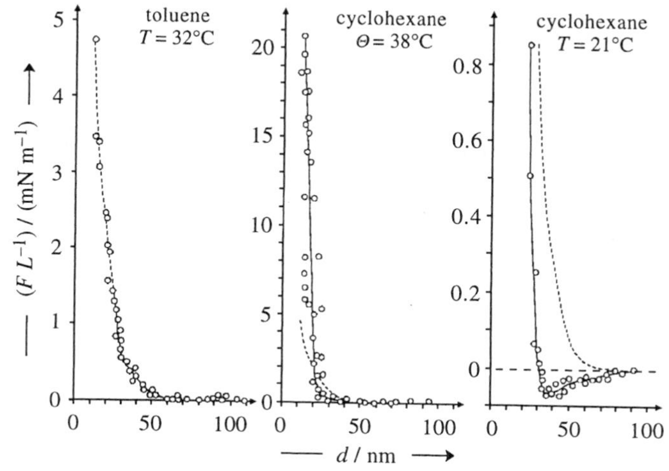

Fig. 9-16 Force per length F/L (= energy per area) between poly(styrene) blocks of poly(2-vinyl pyridine)$_{60}$-*block*-poly(styrene)$_{60}$ adsorbed on mica as a function of the distance d between two parallel mica platelets [12]. For poly(styrene), toluene at 32°C is a good solvent and cyclohexane at 38°C almost a theta solvent (Θ = 34°C) or a bad solvent (21°C). A free poly(styrene) molecule with X = 60 in toluene at 25°C would have a radius of gyration of ca. 2.1 nm and a conventional contour length of 15.3 nm (completely stretched chain in all-trans conformation).

9.5.2 Blob Theory

The stretching of tethered chains has been modeled by assuming that tethered chains are neither fully stretched to their conventional contour length r_{cont} (p. 82) nor are present as conventional random coils. The tethered chain is rather depicted as consisting of a series of stacked blobs (Fig. 9-17) which, for non-tethered chains, are minicoils within the true polymer coils (p. 185).

Chains are assumed to consist of X interconnected "cubic" monomeric units of (effective!) length b; the conventional contour length of such a chain is $r_{\text{cont}} = Xb$. The tethered chains consist of stacked blobs; the geometric length of these stacks is L (Fig. 9-17). The lowest blobs of each stack are attached to the planar surface at equal distances a from each other. A layer volume a^2L thus contains Xb^3 monomeric units which are evenly distributed throughout the layer. The volume fraction of monomeric units in the layer is the same in each volume element. It is therefore given by Eq.(9-8):

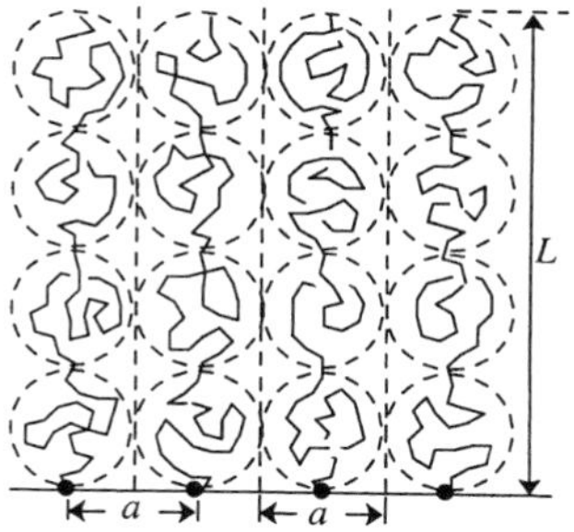

Fig. 9-17 Tethered chains are modeled as series of blobs with diameters a. The chains are adsorbed or grafted on a planar surface at distances a; they stretch to a length L [13]. By permission of Springer Verlag, Berlin.

(9-8) $\phi_u = Xb^3/a^2L$

The total Gibbs energy of tethered chains is the sum of contributions by intersegmental contacts and by an elastic part, $\Delta G = \Delta G_{int} + \Delta G_{el}$. Densely absorbed (or grafted) tethered chains are strongly coiled in melts or theta solvents. The coiling increases intersegmental contacts and the interaction energy ΔG_{int}. In good solvents, chains try to expand. However, tethering to the substrate surface and lateral competition by other chains allow such expansions only perpendicular to the substrate surface, which increases the thickness of the adsorbed layer. This expansion lowers the segment density which reduces ΔG_{int} but increases ΔG_{el}.

Contributions of ΔG_{int} and ΔG_{el} to $\Delta G = \Delta G_{int} + \Delta G_{el}$ can be calculated in two different ways. In method I, one assumes $\Delta G_{int} \sim k_BTX\phi_u k_{excl}$ where k_B = Boltzmann constant, k_BT = thermal energy, X = degree of polymerization, ϕ_u = volume fraction of monomeric units in the layer, and $k_{excl} = 1 - 2\chi$, a parameter for the effect of excluded volume in good solvents which is given by the Flory-Huggins interaction parameter χ (Section 10.2.2). This parameter for the interaction between polymer and solvent often assumes values of exactly or nearly 1/2 (Fig. 10-6). χ is therefore assumed to be 0.5, hence $k_{excl} = 0$ and $\Delta G_{int}/k_BT \sim X\phi_2$.

The elastic contribution ΔG_{el} is assumed to be that of an entropic "spring", hence, $(k_BT/\langle r^2\rangle_o)L^2$ with the spring constant $k_BT/\langle r^2\rangle_o$. A quantity of work k_BT is thus required to double the end-to-end-distance $\langle r^2\rangle_o^{1/2} = (Xb^2)^{1/2}$. With these expressions and Eq.(9-8), method I calculates the reduced Gibbs energy as

(9-9) $(\Delta G/k_BT)_I \sim X\phi_2 + L^2/\langle r^2\rangle_o = (a^2L/b^3)\phi_u^2 + L^2/(Xb^2)$

where ~ indicates that proportionality constants of the magnitude "1" are neglected.

Example: Monomeric units are certainly not cubes but the volume a^3 of a cube approximates the volume well. A circular cylinder with diameter a and height a has a volume of $\pi(a/2)^2a = \pi a^3/4$. Its volume is thus ca. 75.85 % of that of a cube with the same side length, i.e., ≈ 100 %.

Approach II also assumes $1 - 2\chi \approx 1$. It models the stretched chain with degree of polymerization X as a series of linearly connected blobs. The interaction energy per chain, $\Delta G_{int} = N_{bl}k_BT$, is assumed to be the product of thermal energy k_BT and the number N_{bl} of blobs per chain. The number of blobs is $N_{bl} = X/X_{bl}$, i.e., the ratio of degrees of polymerization of chains, X, and of bonds, $X_{bl} = \phi_u^{-5/4}$ (p. 186, last line). Degree of polymerization X and volume fraction ϕ_u of units are furthermore connected by $\phi_u = Xb^3/a^2L$, Eq,(9-8). Hence, the reduced interaction energy is

(9-10) $\Delta G_{int}/k_BT \approx N_{bl} \approx X/X_{bl} \approx X\phi_u^{5/4} \approx (a^2L/b^3)\phi_u^{9/4}$

Excluded volumes exist *within* blobs in *good solvents* although the total tethered chain has a Gaussian character since it resides in a solvent-swollen brush, i.e., in a moderately concentrated solution. The elastic contribution $\Delta G_{el}/k_BT$ is therefore not $L^2/\langle r^2\rangle_o$ as in Eq.(9-9) but $\Delta G_{el}/k_BT = L^2/\langle r^2\rangle$.

Using Eq.(6-8) (with $\phi_2 = \phi_u$), $\langle s^2\rangle = b^2X\phi_u^{-1/4}$, the elastic contribution becomes $\Delta G_{el}/k_BT = (L^2/Xb^2)\phi^{1/4}$ and the reduced Gibbs energy is

(9-11) $(\Delta G/k_BT)_{II} = (a^2L/b^3)\phi_u^{9/4} + (L^2/Xb^2)\phi_u^{1/4}$ (good solvents)

Because the same correlation effects exist for segment-segment interactions and chain elasticities, each term of the right side of Eq.(9-11) is multiplied by $\phi_u^{1/4}$ as compared to Eq.(9-9).

Eqs.(9-9) and (9-11) deliver the same expression for the thickness L of the adsorbed layer if Gibbs energies are minimized by introducing $\phi_2 = Xb^3/a^2L$ (Eq.(9-8)), differentiating ΔG with respect to L, and setting $\Delta G/\mathrm{d}L = 0$ which results in

(9-12) $$L = Xb(b/a)^{2/3}$$

The stretched chain becomes fully extended only if the distance a between contact points shrinks to the effective monomer length (crystallographic length) b, i.e., if the chains are packed in a kind of crystallographic array. In general, $a > b$, and the length L is smaller than Xb. However, L is still larger than the radius of gyration, $s \sim X^{3/5}$, of an isolated coil in a good solvent since L is proportional to X and not to $X^{3/5}$.

In *theta solvents*, binary contacts between monomeric units disappear but 3-body interactions do not (Sections 8.3.2 and 10.4.2). Instead of Eq.(9-9) for good solvents, one obtains

(9-13) $$\Delta G/k_BT \sim (a^2L/b^3)\phi_u^3 + L^2/(Xb^2)$$

and, after minimizing,

(9-14) $$L \approx Xb(b/a)$$

Since $b/a < 1$, tethered chains in theta solvents cannot extend as much as such chains in good solvents (Eq.(9-11)) but their extended lengths are still greater than their unperturbed dimensions.

The situation becomes more complicated if contact points are on curved surfaces instead of planar ones. Examples are multi-armed star molecules and polymer micelles (Fig. 9-15). The many arms of multi-arm stars can no longer be modeled as square cylinders but must be approximated by circular cones whose diameters enlarge from the contact points to the free ends of the tethered chains.

Historical Notes to Chapter 9

A.Pockels, Surface tension, Nature **43** (1891) 437.
For a review of Pockels' work, see Wo.Ostwald, Kolloid-Z. **48** (1932) 1.
Development of Pockels' trough, now commonly called Langmuir trough.

Literature to Chapter 9

9.1 INTERFACES and SURFACES (general overviews)
W.J.Feast, H.S.Munro, Polymer Surfaces and Interfaces, Wiley, New York 1987
J.N.Israelachvili, Intermolecular and Surface Forces, Academic Press, London, 2nd ed. 1992
I.C.Sanchez, Ed., Physics of Polymer Surfaces and Interfaces, Butterworths-Heineman, Stoneham (MA) 1992
C.-M.Chan, Polymer Surface Modification and Characterization, Hanser, Munich 1993
E.Eisenriegler, Polymers Near Surfaces, World Scientific, Singapore 1993
F.Garbassi, M.Morra, E.Occhiello, Polymer Surfaces. From Physics to Technology, Wiley, New York 1994
R.P.Wool, Polymer Interfaces, Structure and Strength, Hanser, Munich 1995
G.J.Fleer, M.A.Cohen Stuart, J.M.H.M.Scheutjens, T.Cosgrove, B.Vincent, Polymers at Interfaces, Chapman and Hall, London 1995
J.J.Pesek, M.Matyska, R.Abuelafiya, Eds., Chemically Modified Surfaces: Recent Developments, Royal Soc.Chem., London 1996
K.L.Mittal, K.-W.Lee, Polymer Surfaces and Interfaces, VSP, Zeist, Netherlands, 1997
A.W.Adamson, A.P.Gast, Physical Chemistry of Surfaces, Wiley, New York, 6th ed. 1997
R.A.L.Jones, R.W.Richards, Polymers at Surfaces and Interfaces, Cambridge Univ. Press, Cambridge (UK) 1999
K.Esumi, Ed., Polymer Interfaces and Emulsions, Dekker, New York 1999
P.Chen, Ed., Molecular Interfacial Phenomena of Polymers and Biopolymers, CRC Press, Boca Raton (FL) 2006
C.Miller, P.Neogi, Interfacial Phenomena, Equilibrium and Dynamic Effects, CRC Press, Boca Raton (FL), 2nd ed. 2007

9.1.2 SURFACE COMPOSITION
D.T.Clark, ESCA Applied to Polymers, Adv.Polym.Sci. **24** (1977) 125
H.J.Purz, E.Schulz, Electron Microscopy in Polymer Science. Review, Acta Polym. **30** (1979) 377
G.Beamson, D.Briggs, High Resolution XPS of Organic Polymers, Wiley, New York 1992
D.A.Bonnell, Ph.N.Ross, Eds., Scanning Tunneling Microscopy and Spectroscopy, VCH, Weinheim 1993
L.Sabbatini, P.G.Zambonin, Eds., Surface Characterization of Advanced Polymers, VCH, Weinheim 1993
D.Sarid, Scanning Force Microscopy, Oxford Univ. Press, New York 1994 (revised edition)
M.W.Urban, Vibrational Spectroscopy of Molecules and Macromolecules on Surfaces, Wiley, New York 1994
H.Zhuang, J.A.Gardella, Jr., Spectroscopic Characterization of Polymer Surfaces, MRS Bulletin **21**/1 (1996) 43 (MRS = Materials Research Society (US))
T.P.Russell, Characterizing Polymer Surfaces and Interfaces, MRS Bulletin **21**/1 (1996) 49
J.C.Vickerman, Ed., Surface Analysis. The Principal Techniques, Wiley, New York 1997
D.Brune, R.Hellborg, H.J.Whitlow, O.Hunderi, EdS., Surface Characterization, Wiley, New York 1997
D.Briggs, Surface Analysis of Polymers by XPS and Static SIMS, Cambridge Univ. Press, Cambridge, UK 1998

9.2 INTERFACIAL TENSION
G.L.Gaines, Jr., Surface and Interfacial Tension of Polymer Liquids–A Review, Polym.Eng.Sci. **12** (1972) 1
S.Wu, Interfacial and Surface Tensions of Polymers, J.Macromol.Sci.-Revs.Macromol.Chem. **C 10** (1974) 1
W.A.Zisman, Relation of the Equilibrium Contact Angle to Liquid and Solid Constitution, in F.F.Fowkes, Ed., Contact Angle, Wettability, and Adhesion, Adv.Chem.Ser. **43** (1964) 1
W.Bascom, The Wettability of Polymer Surfaces and the Spreading of Polymer Liquids, Adv.Polym.Sci. **85** (1988) 89
J.C.Berg, Ed., Wettability, Dekker, New York 1993
R.P.Wool. Polymer Interfaces: Structure and Strength, Hanser, Munich 1995
G.T.Dee, B.B.Sauer, Recent Advances in the Study of Surface Tension of Polymer Liquids, Trends Polym.Res. **5** (1997) 230

F.Garbassi, M.Morra, E.Occhiello, Polymer Surfaces: From Physics to Technology, Wiley, Chichester, 2nd ed. 1998

9.3 THIN LAYERS OF POLYMERS

D.J.Crisp, Surface Films of Polymers, in J.F.Danielli, K.G.A.Pankhurst, A.C.Riddifort, Eds., Surface Phenomena in Chemistry and Biology, Pergamon, New York 1958

M.C.Petty, Langmuir-Blodgett Films. An Introduction, Cambridge Univ. Press, Cambridge, UK, 1996

9.4 ADSORPTION ON SURFACES

Yu.S.Lipatov, L.M.Sergeeva, Adsorption of Polymers (in Russian), Naukova Dumka, Kiew 1972; English edition, Wiley, New York 1974

S.G.Ash, Polymer Adsorption at the Solid/Liquid Interface, in D.H.Everett, Ed., Colloid Science **1**, Chemical Society, London 1973

E.Killmann, J.Eisenlauer, M.Korn, The Adsorption of Macromolecules on Solid/Liquid Interfaces, J.Polym.Sci.-Polym.Symp. **61** (1978) 413

P.Stroeve, E.Frances, Molecular Engineering of Ultrathin Polymeric Films, Elsevier Appl.Sci., London 1987

9.5 BRUSHES

S.T.Milner, Polymer Brushes, Science **251** (1991) 905

A.Halperin, M.Tirrell, T.P.Lodge, Tethered Chains in Polymer Microstructures, Adv.Polym.Sci. **100** (1992) 31

References to Chapter 9

[1] X.Zhao, W.Zhao, J.Sokolov, M.H.Rafailovich, S.A.Schwarz, B.J.Wilkens, R.A.L.Jones, E.J.Kramer, Macromolecules **24** (1991) 5991, Fig. 6

[2] M.Gorelova, V.Levin, A.Pertsin, Macromol.Symp. **44** (1991) 317, data of Fig. 1

[3] A.Zosel, Colloid Polym.Sci. **271** (1993) 680, Fig. 9

[4] J.T.Koberstein, MRS Bulletin **21** (1996) 16, data of Fig. 3

[5] G.L.Gaines, Jr., Polym.Eng.Sci. **12** (1972) 1, data of Fig. 7

[6] A.V.Kabanov, I.R.Nazarova, I.V.Astafieva, E.V.Batrakova, V.Yu.Alakhov, A.A.Yaroslavov, V.A.Kabanov, Macromolecules **28** (1995) 2303, Figs. 1b and 1c

[7] R.C.Bowers, W.A.Zisman, in E.Baer, Ed., Engineering Design for Plastics, Reinhold, New York 1964, p. 689, Tables 3-5

[8] A.Laschewsky, H.Ringsdorf, J.Schneider, Angew.Makromol.Chem. **145/146** (1986) 1, p. 8

[9] R.Vilanova, D.Poupinet, F.Rondelez, Macromolecules **21** (1988) 2880, Figs. 3, 5, and 6

[10] E.Killmann, J.Eisenlauer, M.Korn, J.Polym.Sci.-Polym.Symp. **61** (1977) 413, data of Figs. 1a-1c

[11] D.J.Meier, private communication. Calculations with the theory of J.M.H.M.Scheutjes and G.J.Fleer, J.Phys.Chem. **83** (1979) 1619; **84** (1980) 178; Adv. Colloid Interface Sci. **16** (1982) 361. G.J.Fleer, J.Lyklema, Chapter 4 in G.D.Parfitt, C.H.Rochester, Eds., Adsorption from Solution at the Solid/Liquid Interface, Academic Press, London 1983

[12] G.Hadziioannou, S.Patel, S.Granick, M.Tirrell, J.Am.Chem.Soc. **108** (1986) 2869, Figs. 2, 5, and 6

[13] A.Halperin, M.Tirrell, T.P.Lodge, Adv.Polym.Sci. **100** (1992) 31, Fig. 2

10 Thermodynamics of Polymer Solutions

10.1 Chemical Thermodynamics

10.1.1 Introduction

The sizes and shapes of dissolved small molecules do not differ very much from those of the surrounding solvent molecules and are also not strongly affected by solute-solvent interactions which are rather local and not far-reaching. The structure of solutions also does not vary much with increasing solute concentration, except for solutions of amphiphilic compounds.

Dissolved polymers, however, behave differently than their low-molar mass counterparts. Rod-like macromolecules, for example, can form domains in which their long axes are parallel to each other (Chapter 8). Coil-like macromolecules have kidney-like shapes (Section 4.4.2) with rather low segment concentrations (Section 4.4.8) and can be modeled as "isotropic", solvent-filled spheres. The large volumes of such coils cause flexible polymer chains to overlap at relatively low concentrations (Section 6.3) which leads to physical crosslinking by chain entanglement at higher concentrations. Additional effects are caused by so-called specific solute-solvent interactions.

This chapter discusses the thermodynamic classification of solutions (Section 10.1.2), the solubility of polymers and its prediction (Section 10.1.3 ff.), statistical thermodynamics and phase separation of polymer solutions (Section 10.2), osmotic pressure (Sections 10.3) and virial coefficients (Section 10.4), self-association of polymers (Section 10.5), polyelectrolytes (Section 10.6), and polymer gels (Section 10.7).

10.1.2 Thermodynamic Classification of Solutions

The solubility of a substance in a solvent is controlled by solute-solvent interactions relative to solute-solute and solvent-solvent ones (enthalpic effect) and the number and arrangement of solute molecules and their segments, respectively, relative to those of solvent molecules (entropic effect). The resulting changes of enthalpy and entropy are described by the **second law of thermodynamics**,

(10-1) $$\Delta G = \Delta H - \Delta(TS) = \Delta U + \Delta(pV) - \Delta(TS) = \Delta A + \Delta(pV)$$

where G = Gibbs energy, H = enthalpy, U = internal energy, A = Helmholtz energy (formerly: free energy), S = entropy, p = pressure, V = volume, and T = thermodynamic temperature. Δ indicates the change of these quantities on mixing solute and solvent. Isobaric processes of condensed systems often show $\Delta G \approx \Delta A$ since volume changes can be frequently neglected (but not always!).

Solute and solvent are present in solution in different amounts-of-substance (in moles). The change of Gibbs energy with the amount of component i is the **partial molar Gibbs energy** $\tilde{G}_{i,\mathrm{m}}$ or **chemical potential** μ_i of this component:

(10-2) $$\tilde{G}_{i,\mathrm{m}} \equiv (\partial G / \partial n_i)_{T,p,n_{j\neq i}} \equiv \mu_i$$

The differential of the molar Gibbs energy of component i is given by

$$(10\text{-}3)\qquad dG_{i,\mathrm{m}} = (\partial\tilde{G}_{i,\mathrm{m}}/\partial p)\mathrm{d}p + (\partial\tilde{G}_{i,\mathrm{m}}/\partial T)\mathrm{d}T + (\partial\tilde{G}_{i,\mathrm{m}}/\partial x_i)\mathrm{d}x_i$$
$$= \tilde{V}_{i,\mathrm{m}}\mathrm{d}p - \tilde{S}_{i,\mathrm{m}}\mathrm{d}T + RT\,\mathrm{d}\ln a_i$$

(see textbooks of chemical thermodynamics) where $\tilde{V}_{i,\mathrm{m}}$ = partial molar volume, $\tilde{S}_{i,\mathrm{m}}$ = partial molar entropy, and a_i = activity. The complete differential of the molar Gibbs energy G_m is

$$(10\text{-}4)\qquad \mathrm{d}G_\mathrm{m} = \Sigma_i\,\tilde{G}_{i,\mathrm{m}}\mathrm{d}n_i + \Sigma_i\,n_i\mathrm{d}\,\tilde{G}_{i,\mathrm{m}}$$

The requirement that the left side and the first term of the right side of Eq.(10-4) must be identical according to Eq.(10-2) leads to the **Gibbs-Duhem equation**;

$$(10\text{-}5)\qquad \Sigma_i\,n_i\mathrm{d}\,\tilde{G}_{i,\mathrm{m}} = 0 = \Sigma_i\,n_i\mathrm{d}\mu_i$$

For isothermal-isobaric processes ($\mathrm{d}p = 0$, $\mathrm{d}T = 0$), Eq.(10-3) delivers after integration and conversion of activities to chemical potentials:

$$(10\text{-}6)\qquad \mu_i = \mu_{i,\mathrm{o}} + RT\ln a_i = \mu_{i,\mathrm{o}} + RT\ln x_i\gamma_i$$

The integration constant $\mu_{i,\mathrm{o}}$ is the chemical potential of the pure chemical substance i that is often written as the product of the mole fraction x_i and activity coefficient γ_i. The contribution to the chemical potential by the mole fraction is called the **ideal chemical potential** and that by the activity coefficient the **excess chemical potential**:

$$(10\text{-}7)\qquad \Delta\mu_i = \mu_i - \mu_{i,\mathrm{o}} = RT\ln x_i + RT\ln\gamma_i = \Delta\mu_i^{\mathrm{id}} + \Delta\mu_i^{\mathrm{exc}}$$

In a similar manner, Gibbs energies and entropies can be split into ideal and excess contributions, for example, for the mixing of two components:

$$(10\text{-}1\mathrm{a})\qquad \Delta G_\mathrm{mix} = \Delta H_\mathrm{mix} - T\Delta S_\mathrm{mix} = \Delta H_\mathrm{mix} - T\Delta S_\mathrm{mix,id} - T\Delta S_\mathrm{mix,exc}$$

According to the various contributions of enthalpic and entropic terms, solutions can be classified as ideal, athermal, regular, irregular, or pseudo-ideal (Table 10-1). The first four of these classes of solutions are common to low- and high-molecular weight solutes. The fifth class is specific for polymer coils in solution because it allows experimental access to unperturbed radii of gyration (Section 4.4).

An ideal solution has neither a mixing enthalpy nor an excess mixing entropy but only (a very small) ideal mixing entropy. Pseudo-ideal (**theta**) solutions are dilute solutions that behave like an ideal solution at a characteristic temperature, the **theta temperature** Θ, but not at *all* temperatures. For polymers in dilute solution at $T = \Theta$, $-T\Delta S_\mathrm{mix,id}$ is very small so that $\Delta G_\mathrm{mix} \approx 0$. The magnitude of the theta temperature is controlled by a polymer/solvent interaction parameter χ and its variation with temperature. Only one theta temperature is observed if χ varies monotonous with T but two theta temperatures exist if $\chi = \mathrm{f}(T)$ has an extremal value (pp. 323, 333, 337).

Table 10-1 Thermodynamic types of solutions. ΔG_{mix} of all types of solutions contains a contribution by a positive mixing entropy, $\Delta S_{mix,id}$, i.e., a negative $T\Delta S_{mix,id}$.

Type	ΔG_{mix}	ΔH_{mix}	$\Delta S_{mix,exc}$	Miscibility
ideal	$-T\Delta S_{mix,id}$	0	0	at all T
athermal	negative	0	positive	at all T
regular	negative	negative	0	at all T
irregular A	negative	negative	positive	at all T
B	negative at $T > T_{crit}$ positive at $T < T_{crit}$	positive	positive	at $T \geq T_{crit}$
C	positive	positive	negative	at no T
D	negative at $T < T_{crit}$ positive at $T > T_{crit}$	negative	negative	at $T \leq T_{crit}$
pseudo-ideal (theta)	$-T\Delta S_{mix,id} \approx 0$	$\Theta\Delta S_{mix,exc}$	$\Delta H_{mix}/\Theta$	at Θ

10.1.3 Solubility Parameters

The solubility of a polymer in a solvent or its miscibility with another polymer cannot be calculated from molecular data. However, the concept of the **solubility parameter** often allows a good guess at which polymer dissolves in which solvent.

At a temperature T, a solvent exists as a liquid and does not become a gas because of cohesion forces between its molecules (hydrogen bridges, dipole-dipole interactions, dispersion forces). The cohesion energy is ε_{11} per binary contact and $z\varepsilon_{11}/2$ per single solvent molecule 1, since each solvent molecule is surrounded by z other solvent molecules. The **cohesive energy density** Υ_1 is defined as the cohesive energy per volume,

(10-8) $$\Upsilon_1 = z(\varepsilon_{11}/2)/V_{1,m} = N_A z\varepsilon_{11}/(2\ V_{1,m}) = \delta_1^2$$

where $V_{1,m} = N_A V_{1,mol}$ = molar volume of the solvent and $V_{1,mol}$ = volume of one solvent molecule. The square root of the cohesive energy density is defined as the **solubility parameter** δ_1 of the solvent.

Solubility Parameters of Solvents

To separate a pair of solvent molecules, a separation energy is necessary which is $E_{1,m} = V_{1,m}\delta_1^2 = N_A z\varepsilon_{11}/2$ *per mole of molecules*, i.e., per one-half mole of binary contacts. This separation energy corresponds to the molar evaporation energy. The solubility parameter δ_1 of a solvent can therefore be obtained from the difference of the experimental molar evaporation enthalpy and the work that has to be done against the external pressure of the surrounding air.

Solubility parameters δ_1 are traditionally reported in (cal cm^3)$^{1/2}$ (Hildebrand unit, a non-SI unit). According to Table 10-2, values of δ_1/(cal cm^3)$^{1/2}$ vary between 7.48 (heptane) and 23.43 (water) which correspond to values of δ_1 = 15.30 (J cm^{-3})$^{1/2}$ for heptane and δ_1 = 47.94 (J cm^{-3})$^{1/2}$ (water).

Table 10-2 Solubility parameters δ_1 (solvent), δ_d (contribution by dispersion forces), δ_p (contribution by polar forces), and δ_h (contribution by hydrogen bonds) of solvents at 300 K = 26.85°C in Hildebrand units. 1 Hildebrand = 1 $(cal/cm^3)^{1/2}$ = 2.046 $(J/cm^3)^{1/2}$.

Solvent	δ_1	δ_d	δ_p	δ_h
Apolar solvents				
Heptane	7.48	7.48	0	0
Cyclohexane	8.21	8.21	0	0.10
Carbon tetrachloride	8.65	8.65	0	0
Lewis acids				
Chloroform	9.33	8.75	1.65	2.8
Dichloromethane	9.73	8.72	3.1	3.0
t-Butanol	10.8	7.7	2.8	7.1
m-Cresol	11.52	9.14	2.35	6.6
Methanol	14.60	7.42	6.1	11.0
Glycerol	17.65	8.5	5.9	14.3
Water	23.43	7.6	8.0	20.9
Lewis Bases				
Diethyl ether	7.72	7.09	1.42	2.49
Ethyl acetate	8.90	7.72	2.59	3.52
Benzene	9.19	8.99	0	0.98
Methyl ethyl ketone	9.29	7.82	4.40	2.49
Tetrahydrofuran	9.49	8.22	3.05	3.5
Acetone	9.82	7.58	5.08	3.42
1,4-Dioxane	10.02	9.29	0.88	3.62
Pyridine	10.61	9.29	4.30	2.88
Dimethylacetamide	11.12	8.2	5.6	5.0
Hexamethylphosphortriamide	11.35	9.0	4.2	5.5
Acetonitrile	11.95	7.50	8.8	3.0
N,N-Dimethylformamide	12.14	8.5	6.7	5.5
Dimethylsulfoxide	13.04	9.0	8.0	5.0
Formamide	17.9	8.4	12.8	9.3

Solubility parameters are often written as so-called "three-dimensional" solubility parameters, i.e., δ_1 is thought to be composed of contributions by dispersion forces (δ_d), dipole forces (δ_p), and hydrogen bonds (δ_h): $\delta_1^2 = \delta_d^2 + \delta_p^2 + \delta_h^2$. Contributions by dipole forces are practically independent of the type of solvent (Table 10-2). However, the splitting of polar forces in δ_p and δ_h is questionable since both quantities arise from interactions between Lewis acids and Lewis bases.

Solubility Parameters of Polymers

Because of their large cohesive energies per molecule, macromolecules cannot be vaporized without decomposition. Their solubility parameters are therefore often taken as equal to those of similar low-molecular weight model compounds. A better approach, however, is to use a series of homologous model compounds and extrapolate their solubilities to infinitely small molar volumes of their endgroups, $V_{m,end}$, or to infinitely high molar volumes of monomeric units, $V_{m,u}$ (Fig. 10-1).

Alternatively, solubility parameters can be obtained from intrinsic viscosities $[\eta]$ of linear polymers in dilute solution or from volume fractions ϕ_2 of crosslinked polymers

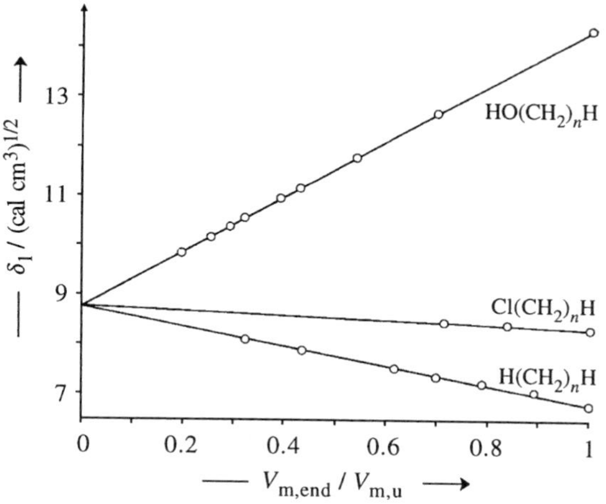

Fig. 10-1 Solubility parameters of alkanes, 1-chloroalkanes, and 1-hydroxyalkane (linear aliphatic alcohols) as a function of the ratio of molar volumes of endgroups and monomeric units [1].

in swelling solvents (Fig. 10-2). The larger the polymer-solvent interaction, the more expanded are polymer coils and the greater are intrinsic viscosities of dissolved polymers and the degree of swelling of crosslinked ones. The solubility parameter that delivers the maximum intrinsic viscosity and the minimum volume fraction of the polymer in the gel is therefore the solubility parameter of the polymer (Fig. 10-2).

Intrinsic viscosities and degrees of swelling deliver quite reasonable solubility parameters for apolar polymers such as natural rubber which consists predominantly of 1,4-*cis*-isoprene units, $-CH_2-C(CH_3)=CH-CH_2-$ (Volume II, p. 246). For moderately polar polymers, and even more so for strongly polar ones, different classes of solvents deliver different solubility parameters. An example is crosslinked poly(styrene-*co*-divinylbenzene) with $\delta_1/(\text{cal cm}^{-3})^{1/2} = 9.5$ (in aromatics), 8.4 (in esters), and 7.3 (in ethers).

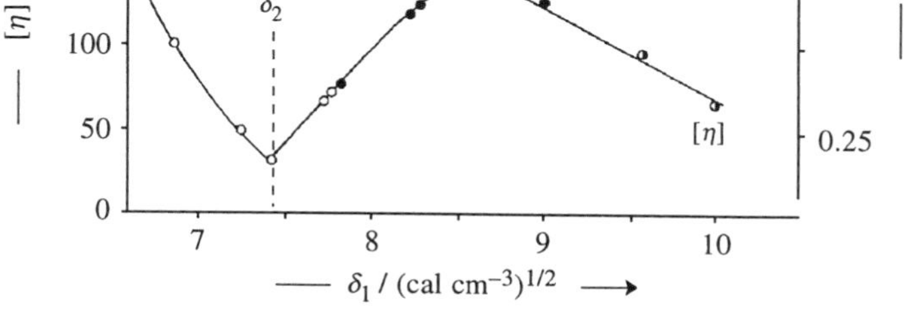

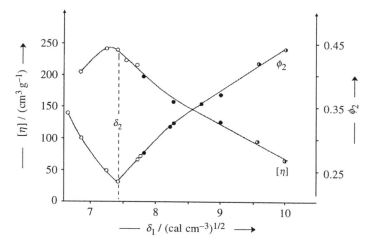

Fig. 10-2 Intrinsic viscosities $[\eta]$ of a dissolved non-crosslinked natural rubber and volume fractions ϕ_2 of a swollen crosslinked natural rubber as a function of the solubility parameter δ_1 of aliphatic hydrocarbons (O), esters (●), and ketones (half-filled circles) [2].

Enthalpy of Mixing

The mixing of a polymer 2 and a solvent 1 can be considered a quasi-chemical reaction between a pair 1-1 of solvent molecules with degrees of polymerization $X_1 = 1$ each and a pair 2-2 of monomeric units of the polymer which leads to two pairs of 1-2 "product" at 100 % "reaction". The **interchange energy** per pair is therefore

$$(10\text{-}9) \qquad \Delta\varepsilon = \varepsilon_{12} - (1/2)(\varepsilon_{11} + \varepsilon_{22})$$

According to quantum mechanical calculations for *dispersion forces*, interaction energies ε_{12} between a solvent molecule and a monomeric unit equal the geometric mean of the two homo interaction energies, $\varepsilon_{12} = (\varepsilon_{11}\varepsilon_{22})^{1/2}$. According to Eq.(10-8), the solvent has an interaction energy of $\varepsilon_{11} = [2\ V_{1,\text{m}}/(N_\text{A}z)]\delta_1{}^2 = K_1\delta_1{}^2$. In analogy, one can write $\varepsilon_{22} = K_2\delta_2{}^2$ for the polymer. If solvent and polymer have the same molar volumes, $V_{1,\text{m}} = V_{2,\text{m}}$, and if solvent molecules and monomeric units have the same number z of neighbors, then $K_1 = K_2 = K$. Introduction of all these parameters into Eq.(10-9) leads to

$$(10\text{-}10) \qquad \Delta\varepsilon/K = -\,(\delta_1 - \delta_2)^2/2$$

For equally strong 1-1 and 2-2 interactions, the difference $\delta_1 - \delta_2$ of solubility parameters becomes zero. In this case, mixing enthalpies are also zero because they are proportional to interchange energies, $\Delta H_\text{mix} \sim -\,\Delta\varepsilon$. Gibbs energies of mixing remain negative, however, since mixing increases the entropy of the system.

The less similar the interactions 1-1 and 2-2, the more positive becomes the mixing enthalpy until it can no longer be compensated by $-T\Delta S_\text{mix}$ with a positive mixing entropy ΔS_mix. Hence, above a certain magnitude of $|\delta_1 - \delta_2|$, polymer 2 and solvent 1 can no longer be mixed. This range $|\delta_1 - \delta_2|$ is fairly small for apolar polymers in apolar solvents but usually larger for polar polymers in polar solvents (Table 10-3).

Table 10-3 Experimental solubility ranges.. 1 $(\text{cal/cm}^3)^{1/2}$ = 2.046 $(\text{J/cm}^3)^{1/2}$ = 1 H = 1 Hildebrand.

Polymer	Polymer $\delta_2/(\text{cal cm}^{-3})^{1/2}$	Solvent $\delta_1/(\text{cal cm}^{-3})^{1/2}$ apolar	polar
Poly(dimethylsiloxane)	7.5	8.3 ± 0.8	8.9 ± 0.7
Poly(isobutylene)	8.0	7.9 ± 0.6	9.0 ± 0.9
Poly(styrene), at-	9.1	9.3 ± 1.3	9.0 ± 0.9
Poly(methyl methacrylate), at-	9.1	10.8 ± 1.2	10.9 ± 2.4
Poly(vinyl acetate), at-	9.4	10.8 ± 1.9	11.6 ± 3.1
Cellulose trinitrate	10.8	11.9 ± 0.8	11.2 ± 3.4
Poly(acrylonitrile), at-	12.5	—	13.1 ± 1.4

Experimental data differ from theoretical predictions because theory makes some assumptions that are not necessarily met: (1) the condition $\Delta G_\text{mix} \leq 0$ is necessary but not sufficient; (2) $\Delta H_\text{mix} < 0$ is never true; (3) only dispersion forces are considered but not polar interactions; and (4) the theory applies to the *mixing* of a liquid polymer and a solvent but not to the *dissolution* of a solid polymer for which a melting enthalpy has to be considered for crystalline polymers and a freezing-in enthalpy for amorphous ones. For example, highly crystalline linear poly(ethylene) (PE; δ_2 = 8.0 H), $+CH_2-CH_2+_n$, dissolves in decane, $H(CH_2-CH_2)_5H$ (δ_1 = 7.8 H), only near the melting temperature of PE.

Crystallinity is also responsible for the phenomenon that a polymer may first dissolve but later precipitate at the same temperature. In such a case, the original polymer dissolves easily because of its low crystallinity but, in the resulting high dilution, equilibrium is easily established and the polymer crystallizes out with a higher crystallinity than the original one. A famous case is the so-called **retrogradation** of starch solutions (Volume II, p. 383) which is caused by the crystallization (due to helix formation) of amylose in the amylose/amylopectin mixture of starch.

Solubilities are better predicted by the so-called three-dimensional solubility parameters. Since dispersion parameters δ_d vary little with the type of solvent (Table 10-1), one constructs solubility diagrams by plotting δ_h against δ_p for many solvents. In this diagram, one then notes the corresponding values of δ_d and marks all points which are solvents for the polymer and constructs contour lines for all dissolving solvents. In general, solubilities increase with increasing δ_d at constant δ_p and δ_h. A liquid with a δ_d value at the contour line will thus be most likely a solvent for the polymer.

Some polymers dissolve in mixtures of two non-solvents, especially in a mixture of a non-solvent I with $\delta_I < \delta_2$ and a non-solvent II with $\delta_{II} < \delta_2$ (Table 10-4). On the other hand, a mixture of two solvents may be non-dissolving. An example is poly(acrylonitrile) (δ_2 = 12.8 H) that dissolves in both *N,N'*-dimethylformamide (δ_1 = 12.1 H) and malonic dinitrile (malonodinitrile; δ_1 = 15.1 H) but not in the mixture of these two liquids. Another example is cellulose acetate (δ_2 = 9.56 H) which is soluble in aniline (δ_1 = 11.0 H) and glacial acetic acid (δ_1 = 10.4 H) but not in the mixture of the two.

10.1.4 Molecular Considerations

The behavior of polymers in mixed non-solvents shows that global approaches like the concept of solubility parameters or statistical thermodynamics (Section 10.2.) are not refined enough for subtle differences in solvent-solvent, solute-solvent, and solute-solute interactions. These global approaches do not address self-associations of polymer and/or solvent molecules, solvation of polymer molecules in binary systems, and preferential solvations of polymer molecules by one liquid in mixtures of two non-solvents.

Table 10-4 Solubility of polymers with δ_2 in mixtures of non-solvents with δ_I and δ_{II}. δ values are given in traditional Hildebrand units; 1 Hildebrand = 1 H = 1 $(cal/cm^3)^{1/2}$ = 2.046 $(J/cm^3)^{1/2}$.

Polymer	δ_2/H	Soluble in mixtures of I and II			
		non-solvent I	δ_I/H	non-solvent II	δ_{II}/H
Poly(chloroprene)	8.2	diethyl ether	7.4	ethyl acetate	9.1
Poly(styrene), at-	9.3	cyclohexane	8.2	acetone	9.8
Poly(vinyl chloride), at-	9.5	acetone	9.8	carbon disulfide	10.0
Cellulose nitrate	10.6	diethyl ether	7.4	ethanol	12.7
Poly(methyl methacrylate), at-	11.1	methanol	14.5	water	23.4
Poly(acrylonitrile), at-	12.8	nitromethane	12.6	water	23.4

Self-association of Solvent Molecules

Some curious solubility effects are caused by self-association of solvent molecules. An example is the behavior of high-molecular weight atactic poly(styrene) (δ_2 = 9.1 H) which forms dilute solutions in butanone (δ_1 = 9.3 H) or *N,N'*-dimethylformamide (δ_1 = 12.1 H) but not in acetone (δ_1 = 9.8 H). It is known that dipole-dipole interactions between keto groups cause acetone molecules to dimerize in the gaseous state. Some of the acetone molecules will therefore also form dimers or even other associates in the liquid state. In such dimers, keto groups are screened by methyl groups and can no longer solvate the phenyl groups of poly(styrene).

Addition of cyclohexane (δ_1 = 8.2 H; a non-solvent for high-molecular weight poly(styrene) at 25°) to dilute acetone solutions of poly(styrene) dilutes the acetone and reduces the probability of dimerization of keto groups. Free keto groups can thus solvate phenyl groups, and poly(styrene) dissolves in such butanone-cyclohexane mixtures.

In highly concentrated solutions of poly(styrene) in acetone, poly(styrene) itself is the diluent that decreases the tendency of keto groups to self-associate. One can therefore prepare 40 % acetone solutions of poly(styrene) but not 1 % ones. Butanone, on the other hand, is "internally diluted" by the additional methylene group. It associates far less (if at all) and is therefore a solvent for poly(styrene) for all concentration regions.

Solvation

"Solvation" is a somewhat nebulous term that includes all solute-solvent interactions from changes in the physical structure of the solvent near the solute molecules to the so-called polymer-solvent compounds.

Polymer-solvent compounds are known as **polymer-solvent complexes, intercalates**, or **crystal solvates**. For example, crystalline poly(oxyethylene) (PEOX) forms such compounds with *p*-dihalogen benzenes (but not with ortho and meta isomers) which fit into the lateral "chimney" that is formed by the 7_2 helices of PEOX molecules (Fig. 13-14). This chimney is pretty spacious because the helix has an identity period of 1.93 nm and a diameter of 0.38 nm.

Although the global conformation of PEOX molecules in water is that of a random coil, some of its monomeric units reside in helical sections (cf. Fig. 4-1, III). According to Raman and infrared measurments as well as calculations by molecular dynamics, these helical sections have the structure of an 11_2 helix. This wider helix contains ca. 2.9 water molecules per oxyethylene unit which probably stabilizes the helix.

Water molecules are firmly bound to the helix; i.e., the physical structure of this water layer (solvate) differs from that of free water. Such different structures have also been shown by NMR for water that is bound in spheroidal enzyme molecules. This water is so firmly bound that one can calculate the water bound by hydration just by measuring the hydrodynamic volume of enzyme molecules in water (see Section 12.3.4).

Solvations can also be determined by measuring sound velocities. Sound velocities decrease linearly with increasing polymer concentration which is mainly caused by polymer-solvent interactions. According to these investigations, the solvate hull of poly(oxyethylene) contains per monomeric unit 0.74 water molecules, 0.20 toluene molecules, or 0.12 cyclohexane molecules (these are *not bound* molecules). For poly(butadiene), the numbers are 0.10 (ethyl benzene), 0.21 (cyclohexane), and 0.51 (hexane).

Preferential Solvation

In mixed solvents, one type of molecule will reside more preferentially near polymer segments than the other type. This preferential solvation can be described by a parameter Γ that indicates the excess volume of the preferentially bound solvent per mass of polymer. This parameter is accessible from light scattering of dialyzed solutions at constant chemical potential (Section 5.2.4) or from direct evaluations as follows.

Preferential solvation causes the refractive index n near a polymer chain to deviate from the average refractive index of the solution. The conventionally obtained molar mass is then an apparent one (M_{app}), even at infinite dilution where there is no influence of second virial coefficient. The true molar mass M can be calculated from the refractive index increment dn/dc of the polymer at constant solvent composition and the change dn/dϕ_1 of n of the mixed solvent with the volume fraction ϕ_1 of component 1:

(10-11) $$M_{app} = M[1+\Gamma\{(\mathrm{d}n/\mathrm{d}\phi_1)/(\mathrm{d}n/\mathrm{d}c)\}$$

Compositions of mixed solvents near polymer chains must depend on the density of the distribution of chain segments. This assumption and modeling perturbed polymer coils as equivalent spheres leads to a semi-empirical expression for $\Gamma = \mathrm{f}(M)$ where Γ_∞ = preferential solvation at infinitely high molar mass M, X = degree of polymerization, b = bond length, τ = bond angle, σ = hindrance parameter, M_u = molar mass of monomeric unit, α_s = expansion factor of radius of gyration, s, and A = system-specific constant.

(10-12) $$\Gamma = \Gamma_\infty + KM^{-1/2} \quad \text{or} \quad \Gamma M^{1/2} = K + \Gamma_\infty M^{1/2}$$

(10-13) $$K = A\left(\frac{3}{4\pi}\right)^{3/2} \frac{X}{\langle s^2\rangle^{3/2}} = \frac{AM_\mathrm{u}^{1/2}}{\alpha_\mathrm{s}^3}\left(\frac{3}{4\pi b^2(1-\cos\tau)(1+\cos\tau)^{-1}\sigma^2}\right)^{3/2}$$

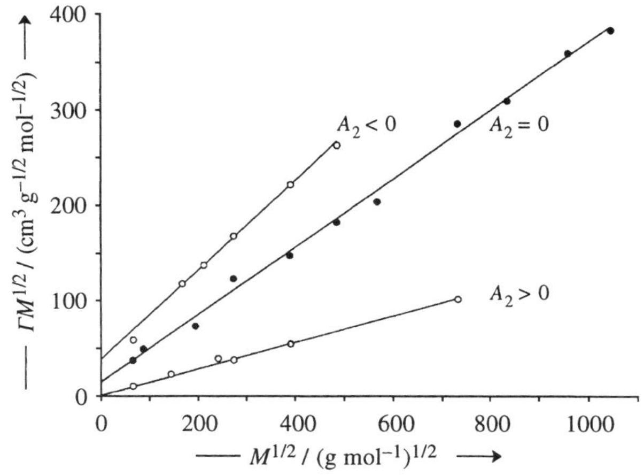

Fig. 10-3 Plot of $\Gamma M^{1/2} = \mathrm{f}(M^{1/2})$ (Eq.(10-12), right) for the preferential solvation of poly(styrene) in mixtures of the good solvents CCl_4 or C_6H_6 with the precipitant CH_3OH [3].

$A_2 < 0$: carbon tetrachloride + methanol (79 : 21 = V:V)
$A_2 = 0$: benzene + methanol (77.8 : 22.2 = V:V)
$A_2 > 0$: benzene + methanol (90 : 10 = V:V)

The constant K is inversely proportional to the expansion factor α_s^3. Thermodynamic goodness, as measured by the second virial coefficient A_2 (see Section 10.4), thus leads to smaller K values, i.e., smaller intercepts in plots of $\Gamma M^{1/2} = f(M^{1/2})$ (Fig.10-3).

10.1.5 Rate of Dissolution

The dissolution of solid (semi-crystalline or amorphous) polymers is preceded by an "induction period" before the concentration of the polymer in the liquid phase increases linearly with time (Fig. 10-4). This induction period is caused by the following effects.

The solid polymer is at a temperature below the melting temperature (if semicrystalline) or below the glass temperature (if amorphous). The dissolving solvent must thus penetrate the densely packed polymer molecules whose segments are surrounded by segments of other polymer molecules which the solvent needs to replace. Solvent molecules thus enter intersegmental spaces but these spaces are smaller than the diameters of solvent molecules. As a result of the solvent penetration, these spaces become larger and the upper layer of the polymer swells even if the liquid is a theta solvent in which polymer molecules have the same dimensions as in the amorphous state (Chapter 6).

A polymer molecule of the upper layer of a solid polymer can only enter the solvent or solution when all segments of adjacent polymer molecules are replaced by solvent molecules. Hence, the dissolution of solid polymers is preceded by a quasi induction period (Fig. 10-4) in which a swollen surface layer is formed.

This layer acts as a barrier for other solvent molecules that want to enter the remaining "dry" polymer. Strong stirring transfers swollen polymer coils from the solid to the solution which results in a steady state with a constant dissolution rate of the polymer and a linear penetration rate dL/dt of the solvent into the polymer, respectively. dL/dt is proportional to the mutual diffusion coefficient D of polymer/solvent and inversely proportional to the thickness d of the swollen surface layer, Eq.(10-14).

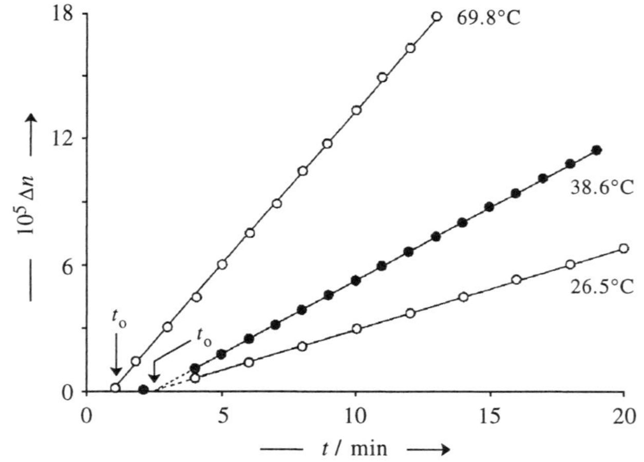

Fig. 10-4 Increase of refractive index of solutions as a measure of polymer concentration during the dissolution of a fractionated poly(styrene) in stirred toluene at various temperatures [4]. Induction times t_o decrease with increasing temperature.

(10-14) $\mathrm{d}L/\mathrm{d}t = const\ (D/d)$

The induction time is controlled by the diffusion law, $t_o = d^2/(6\ D)$ (Section 14.3.2).

An example is the dissolution of an amorphous poly(styrene) which, at 25°C, had a linear dissolution rate of $\mathrm{d}L/\mathrm{d}t = 6.4 \cdot 10^{-5}$ cm/s, an induction time of 7 minutes, and a measured thickness of the swollen layer of d = 1.6 mm. This thickness is far larger than the diameter of unperturbed or perturbed poly(styrene) molecules, possibly because the swollen layer consists of entangled macromolcules and the induction time also relates to the time that is needed to disentangle the molecules.

The described process is idealized since polymers may have clefts either from preparing test specimens or from **stress cracking** during swelling (see Section 18.3.5). Matter around clefts is under stress so that penetration of matter near clefts by solvent molecules releases strain energy which leads to cracks in the specimen if the solvent is simply swelling and to separation of whole chunks of polymer if the solvent is dissolving.

10.2 Statistical Thermodynamics

10.2.1 Introduction

Thermodynamics is concerned with relations between heat and other forms of energy in macroscopic systems, as related to masses or volumes. Chemical thermodynamics applies the laws of thermodynamics to molecules and especially to reactions between molecules. The resulting relationships are independent of the shape of molecules which are anyway not important for small molecules because, for example, mixing entropies of two small low-molecular weight compounds can be treated in the same way as the mixing of red and white spheres, regardless of their size.

However, mixing of macromolecules to polymer blends or macromolecules and solvents to polymer solutions does depend on the shape and size of these molecules. A simple example is the variation of the two critical overlap concentrations c^* and c^{**} (p. 182 ff.) with the shape of macromolecules (Table 10-5).

Table 10-5 Overlap concentrations c^* (dilute ⇄ semi-concentrated) and c^{**} (semi-concentrated ⇄ concentrated regimes) for different types of molecules. d = diameter, L = length, L_{pers} = persistence - length, M = molar mass, s = radius of gyration, $[\eta]$ = intrinsic viscosity (see also Section 6.3).

Shape of molecules	c^*	c^{**}
Random coils	$\dfrac{3\,M}{4\,\pi N_A s^3} \approx \dfrac{1}{[\eta]}$	$\dfrac{0.477}{[\eta]}$
Worm-like chains	$\dfrac{2^{3/2}\,M}{N_A(L_{pers}L)^{3/2}}$	$\dfrac{0.243\,M}{N_A(L_{pers}\,/\,L)^{1/4}\,d(L_{pers}L)}$
Rigid rods	$\dfrac{2^{3/2}\,M}{N_A L^3}$	$\dfrac{0.243\,M}{N_A d L^2}$

10.2.2 Lattice Theory

The miscibility of two types of molecules (such as a polymer and a solvent) is controlled by the Gibbs energy of mixing, $\Delta G_{mix} = \Delta H_{mix} - T\Delta S_{mix}$. The thermodynamic parameters ΔH_{mix} and ΔS_{mix} can be calculated by statistical thermodynamics, for example, for rods (Section 8.3.2) and for random coils (this section).

Lattice theory treats solutions as three-dimensional lattices composed of $N_g = N_1X_1 + N_2X_2$ lattice sites (Fig. 10-5). Each lattice site is occupied by either a solvent molecule or a polymer segment, for example, a monomeric unit. The solution contains N_2 polymer molecules of degree of polymerization $X_2 > 1$ and N_1 solvent molecules of size X_1, usually assumed as unity. The theory can also be applied to polymer blends that consist of two polymers with degrees of polymerization of $X_1 > 1$ and $X_2 > 1$, respectively.

The number N_{12} of contacts between polymer and solvent molecules is calculated by the **Flory-Huggins theory** from the number N_g of all lattice sites, the number z of nearest neighbors of a unit, and the probability that neighboring lattice sites are occupied by either solvent molecules 1 or polymer units 2. These probabilities are identical with volume fractions ϕ_1 of solvent molecules and ϕ_2 of monomeric units.

The same force field is assumed to act at each polymer segment and at each solvent molecule. The Flory-Huggins theory is therefore called a **mean-field theory** although it does not calculate a force field. Fig. 10-5 shows that the assumption of a mean force field can be true only for moderately to highly concentrated polymer solutions but not for dilute ones in which polymer molecules form islands in a sea of solvent molecules. The theory thus describes the behavior of polymer-polymer mixtures (polymer blends) better than that of dilute polymer solutions.

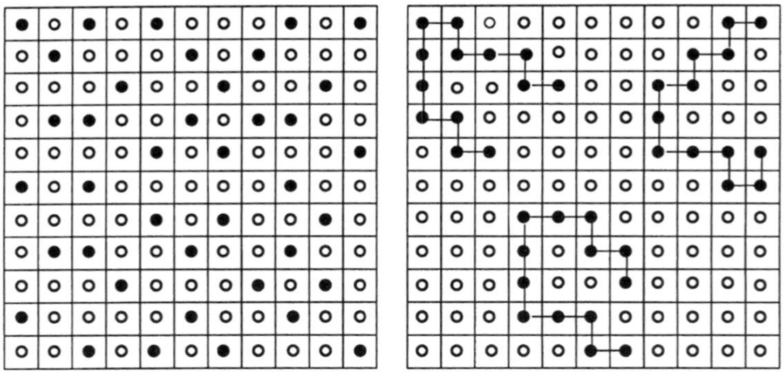

Fig. 10-5 Schematic representation of (left) dissolved low molecular weight solute molecules (●) and (right) monomeric units of a polymer 2 with $X = 13$ (●) in a low-molecular weight solvent 1 (○), both with a volume fraction of $\phi_2 \approx 0.322$.

Enthalpy of Mixing

The mixing of solvent molecules and monomeric units is modeled as a quasi-chemical reaction with an exchange energy of $\Delta\varepsilon = \varepsilon_{12} - (1/2)(\varepsilon_{11} + \varepsilon_{22})$, similar to the calculation of solubility parameters, Eq.(10-9). The "reaction" is assumed to proceed without volume change.

The exchange energy is then given by the product of the exchange energy, $\Delta\varepsilon_{12}$, and the number N_{12} of pairs 1-2 which is the product of the total number N_g of lattice sites, the number z of nearest neighbors of a 1-2 pair, and the probability that the next lattice site is occupied by a solvent molecule. These probabilities are expressed by the corresponding volume fractions ϕ_1 and ϕ_2, and the enthalpy of mixing is therefore

$$(10\text{-}15) \qquad \Delta H_{mix} = N_{12}\Delta\varepsilon = N_g z \phi_1 \phi_2 \Delta\varepsilon$$

The **Flory-Huggins interaction parameter** χ is defined as the product of the exchange energy and the number of neighbors divided by the thermal energy, $k_B T$:

$$(10\text{-}16) \qquad \chi \equiv z\Delta\varepsilon/(k_B T) = \Delta H_{mix}/[N_g \phi_1 \phi_2 k_B T]$$

In the original Flory definition, the right side of Eq.(10-16) was multiplied by the degree of polymerization of solvent molecules, X_1, in order to allow a description of polymer(1)-polymer(2) blends.

For purely enthalpic systems, the **enthalpy of mixing**, ΔH_{mix}, is given by an equation of the **van Laar type**. Division of ΔH_{mix} by the amount concentration $n_g = N_g/N_A$ delivers with $R = k_B N_A$ the molar enthalpy of mixing, $\Delta H_{mix,m}$:

$$(10\text{-}17) \qquad \Delta H_{mix} = N_g \chi k_B T \phi_1 \phi_2 \quad ; \quad \Delta H_{mix,m} \equiv \Delta H_{mix}/n_g = \chi RT \phi_1 \phi_2$$

Interaction Parameter

The dimensionless Flory-Huggins interaction parameter χ describes the **thermodynamic goodness** of the solvent for the polymer. For soluble polymers, it can be obtained by various methods: at small polymer volume fractions ϕ_2 from the concentration dependence of apparent molar masses from light scattering or sedimentation equilibrium experiments; in the range $0 \leq \phi_2 \leq 0.3$ from osmotic pressure measurements or critical solution temperatures; in the range $0.3 \leq \phi_2 \leq 0.9$ from vapor sorption; and in the limiting case $\phi_2 \to 1$ by gas-liquid chromatography. For crosslinked polymers, χ is obtained from swelling equilibria.

The theory postulates an independence of χ on ϕ_2 but this is found only for apolar or weakly polar polymers in apolar or weakly polar solvents such as *cis*-1,4-poly(isoprene) in benzene (Fig. 10-6). The reason for experimentally observed concentration dependences of χ is the simplifying theoretical assumption that solution volumes do not change on mixing 1-1 with 2-2 to give 1-2. Because $\chi = f(1/T)$ according to Eq.(10-16) and $\Delta G/T = \Delta H/T - \Delta S$ according to the second law of thermodynamics, interaction parameters χ are postulated to be purely enthalpic quantities. However, a "reaction" 1-1 + 2-2 $\rightleftarrows$ 2 (1-2) also changes rotations and vibrations of 1 and 2 in 1-2 "complexes" so that the mixing enthalpy ΔH_{mix} is really an internal energy of mixing, ΔU_{mix}. The Flory-Huggins interaction parameter should therefore also contain an entropic parameter in addition to the enthalpic one. Indeed, χ does change with temperature according to $\chi \sim a + b/T$ which indicates the presence of entropic contributions.

The experimentally found dependence of χ on ϕ_2 is described empirically as $\chi = \chi_o + K\phi_2 + K'\phi_2^2$ where K and K' are adoptable, positive or negative constants for the poly-

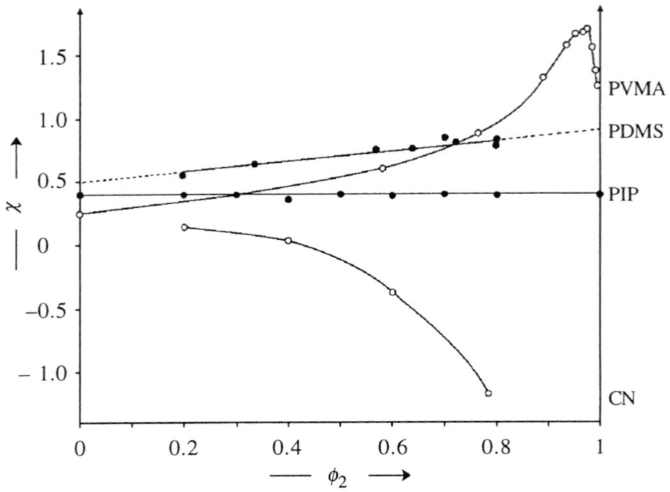

Fig. 10-6 Flory-Huggins interaction parameter χ as a function of the volume fraction ϕ_2 of poly(vinyl methoxyacetal) in water at 25°C (PVMA); poly(dimethylsiloxane) in benzene at 20°C (PDMS); *cis*-1,4-poly(isoprene) in benzene at 20°C (PIP); cellulose nitrate (with a degree of substitution of 2.6; see Volume I, p. 402) in acetone at 20°C (CN).

mer-solvent-temperature system at $T = const$. This function can therefore describe both positive and negative values of χ and also maxima and minima in $\chi = f(\phi_2)$. For example, the maximum in $\chi = f(\phi_2)$ for poly(vinyl methoxyacetal) is caused by hydrophobic effects whereas the negative χ values for cellulose nitrate in acetone stem from lyotropic liquid crystals.

Parameters χ_o are mostly positive because interactions 1-1, 2-2, and 1-2 are dominated by dispersion forces (see Table 10-1) which in turn depend on the product of polarizabilities. χ is controlled by the thermodynamic goodness of the solvent for the polymer. For infinitely high molecular weights, a critical value of $\chi = 1/2$ is predicted for systems near phase separation by the Flory theory for the non-combinatorial entropy (see Eq.(10-33)). Values of $\chi = 1/2$ are indeed observed for infinitely small concentrations of polymers in thermodynamically bad solvents, for example, poly(dimethylsiloxane) in benzene at 20°C (Fig. 10-6) and at-poly(styrene) in cyclohexane at 34°C.

Entropy of Mixing

The entropy of mixing, $\Delta S_{mix} = S_{comb}(N_1,N_2) - S_{comb,o}$, is the sum of two terms: the combinatorial entropy of mixing and the entropy of disorientation.

Combinatorial entropies arise because solvent molecules and polymer segments can be arranged in various ways relative to each other (see Fig. 10-5). Since this entropy describes various *physical* configurations of a polymer chain (i.e., macroconformations in the lingo of polymer chemists), it is also called **configurational entropy**.

In ideal solutions, no energy is gained or lost if a solvent molecule is replaced by a polymer segment or *vice versa*, i.e., the enthalpy of mixing is zero. Since all interactions 1-1, 1-2, and 2-2 are of equal magnitude, all environment dependent entropic contributions do not add anything to the change of entropy on mixing: translational and internal rotational and vibrational entropies do not change. However, solution components may be arranged in many different ways giving rise to a combinatorial entropy.

The **combinatorial entropy** $S_{comb}(N_1, N_2) = k_B \ln \Omega$ of polymer solutions with $X_1 = 1$ and $X_2 >> 1$ is calculated by the Flory-Huggins theory from the thermodynamic probability Ω that v_i different arrangements exist for the ith chain. The first monomeric unit of a chain can be on any lattice site but the second unit can only occupy sites adjacent to the first unit, the third unit $z - 1$ sites adjacent to the second unit, etc.

The first chain to be placed on the lattice has $v_1 = N_g z(z-1)^{X_2-2}$ choices if one assumes that the third unit and all other ones can indeed choose between any of the $z - 1$ lattice sites. In reality, fewer than $z - 1$ sites are available since a site may already be occupied by a previously placed unit.

After filling the lattice with $i - 1$ polymer chains, $N_f = N_g - (i-1)X_2$ lattice sites remain unoccupied and the probability of finding a free site is approximately N_f/N_g. The ith polymer chain can thus be arranged in $v_i = N_f(N_f/N_g)^{X_2-1} z(z-1)^{X_2-2}$ different ways.

The thermodynamic probability Ω_i is proportional to $\Pi_i v_i$, which is the product of all v_i values for all N_2 chains of equal length. This product has to be divided by $N_2!$ since many chains can be in the same macroconformation but one can consider only one molecule per type of macroconformation.

Each chain can furthermore be placed on the lattice head first or tail first. Depending on the symmetry number σ, one counts therefore σ^{N_2} too many combinations where $\sigma = 1$ for distinguishable chains and $\sigma = 2$ for non-distinguishable ones. It follows

(10-18) $\Omega = (\sigma^{N_2}/N_2!)\ \Pi_i v_i, \qquad (\text{for } 1 \leq i \leq N_2)$

Introducing $v_i = N_f(N_f/N_g)^{X_2-1} z(z-1)^{X_2-2}$ (see above), $N_f = N_g - (i-1)X_2$, and Stirling's approximation $x! \approx (2\pi x)^{1/2} x^x \exp(-x)$ for $x >> 1$ (i.e., $x! \approx (x/e)^x$), one obtains

(10-19) $$\Omega = \frac{N_g!}{N_1!(N_2X_2)!}\left(\frac{N_2X_2}{N_g}\right)^{N_2(X_2-1)}\left[\frac{X_2 z(z-1)^{X_2-2}}{\sigma \exp(X_2-1)}\right]^{N_2}$$

(10-20) $$\Omega = \frac{N_g!}{N_1!(N_2X_2)!}\left(\frac{N_2X_2}{N_g}\right)^{N_2(X_2-1)}[\Omega_{rel}]^{N_2}$$

Introduction of Stirling's approximation and the definition of volume fractions, $\phi_1 \equiv N_1X_1/N_g$ and $\phi_2 \equiv N_2X_2/N_g$, of lattice components results in

(10-21) $S_{comb}(N_1,N_2) = k_B \ln \Omega = -k_B(N_1 \ln \phi_1 + N_2 \ln \phi_2) + k_B N_2 \ln \Omega_{rel}$

where the first term is the **entropy of mixing** of amorphous polymer/solvent,

(10-22) $\Delta S_{mix} = -k_B(N_1 \ln \phi_1 + N_2 \ln \phi_2)$

whereas the second term, $k_B N_2 \ln \Omega_{rel} = S_{comb,or} = S_{comb}(N_1,0) + S_{comb}(0,N_2)$, is the **disorientation entropy** which describes the entropy of coil molecules relative to the entropy of these molecules in a perfect crystal.

For blends of two amorphous polymers, $S_{comb,or}$ is a complex quantity since both types of polymer molecules can adopt many different macroconformations. For polymer solutions, one only has to consider the macroconformations of the N_2 polymer molecules; solvent molecules simply fill the remaining empty lattice sites. Since $N_1 = 1$, the disorientation entropy becomes $S_{comb,or} = S_{comb}(0,N_2) = k_B N_2 \ln \Omega_{rel}$.

The **entropy of mixing of amorphous polymers per amount of lattice site** is given by Eq.(10-22) and $N_1 = \phi_1 N_g / X_1$, $N_2 = \phi_2 N_g / X_2$, $N_g = n_g N_A$, and $N_A k_B = R$:

$$(10\text{-}23) \qquad \Delta S_{mix,m} = \Delta S_{mix}/n_g = -R(X_1^{-1}\phi_1 \ln \phi_1 + X_2^{-1}\phi_2 \ln \phi_2)$$

Gibbs Energy of Mixing

The molar Gibbs energy of mixing is obtained from the second law of thermodynamics, $\Delta G_{mix,m} = \Delta H_{mix,m} - T\Delta S_{mix,m}$, and Eqs.(10-17) and (10-23):

$$(10\text{-}24) \qquad \Delta G_{mix,m} = \Delta H_{mix,m} - T\Delta S_{mix,m} = RT[\phi_1\phi_2\chi + X_1^{-1}\phi_1 \ln \phi_1 + X_2^{-1}\phi_2 \ln \phi_2]$$

For equal-sized components, $X_1 = X_2$, such as two solvents, the reduced molar Gibbs energies of mixing are symmetric about $\phi_2 = 1/2$ with respect to volume fractions ϕ_2 as can be seen for $X_2 = 1$ and $\chi = 0.5$ and $\chi = 1.8$, respectively (Fig. 10-7). The function $\Delta G_{mix,m}/RT = f(\phi_2)$ becomes asymmetric for $X_2 \neq X_1$. Hence, solutions of polymers in low-molecular weight solvents are thermodynamically different from mixtures of two solvents because of the vast difference in possible macroconformations.

Chemical Potentials

The first derivative of the Gibbs energy of mixing with respect to the amount n_1 of solvent is defined as the **chemical potential** of the solvent, $\Delta\mu_1$, and the corresponding derivative with respect to the amount n_2 of the polymer as the chemical potential $\Delta\mu_2$ of the polymer. Before differentiation, volume fractions are converted to mole fractions of the same type of component, i.e., $\phi_1 = n_1 N_A X_1 / N_g$, $\phi_1 = 1 - \phi_2$, and $\phi_2 = n_2 N_A X_2 / N_g$.

The chemical potential per mole of repeating unit of polymer, $\Delta\mu_u$, is $\Delta\mu_2$ divided by $(V_{1,m}/V_{u,m})X_2$. For $X_1 = 1$, the reduced chemical potentials of the solvent (index 1), polymer (index 2), and polymer repeating units (index u) are therefore

$$(10\text{-}25) \qquad \Delta\mu_1 / RT = [\partial \Delta G_{mix} / \partial n_1] / RT = \chi\phi_2^2 + \ln(1-\phi_2) + (1 - X_2^{-1})\phi_2$$

$$(10\text{-}26) \qquad \Delta\mu_2 / RT = [\partial \Delta G_{mix} / \partial n_2] / RT = \chi X_2 (1-\phi_2)^2 + (X_2 - 1)(1-\phi_2) + \ln \phi_2$$

$$(10\text{-}27) \qquad \Delta\mu_u / RT = (V_{u,m} / V_{1,m})[\chi(1-\phi_2)^2 + (X_2 - 1)(1-\phi_2)X_2^{-1} + X_2^{-1} \ln \phi_2]$$

Chemical potentials can therefore be obtained from the concentration dependence of the molar Gibbs energy of mixing. According to Eqs.(10-4) and (10-5), the complete differential is $d\Delta G_{mix} = \Delta\mu_1 dn_1 + \Delta\mu_2 dn_2$ which can be integrated to give

$$(10\text{-}28) \qquad \Delta G_{mix} = n_1\Delta\mu_1 + n_2\Delta\mu_2 = (\phi_1 N_g \Delta\mu_1 + \phi_2 N_g X_2^{-1}\Delta\mu_2)/N_A$$

$$(10\text{-}29) \qquad \Delta G_{mix}(N_A/N_g) = \Delta\mu_1 - (\Delta\mu_1 - X_2^{-1}\Delta\mu_2)\phi_2$$

using the definition of volume fractions, ϕ_i, and amounts of substances, n_i. The function $\Delta G_{mix}(N_A/N_g) = f(\phi_2)$ delivers the chemical potential of the solvent, $\Delta\mu_1$ for $\phi_2 \rightarrow 0$ and the chemical potential of the polymer, $\Delta\mu_2$, for $\phi_2 \rightarrow 1$.

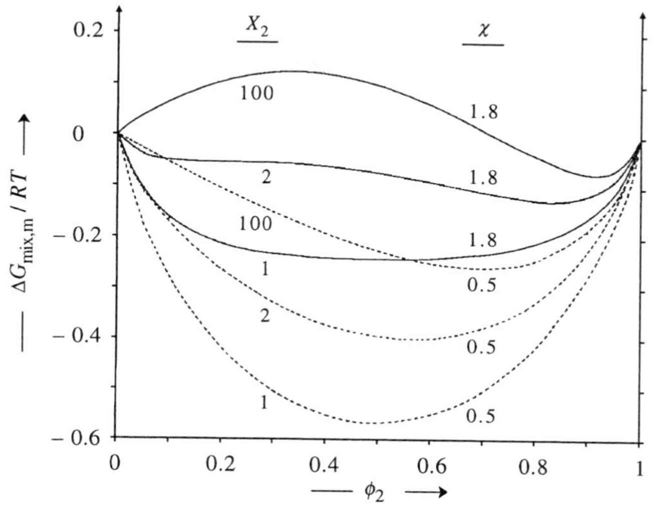

Fig. 10-7 Reduced molar Gibbs energy of mixing, $\Delta G_{mix,m}/RT$, as a function of volume fractions ϕ_2 of polymers of different degrees of polymerization X_2 and different interaction parameters χ with solvents of degree of polymerization $X_1 = 1$. Calculated from Eq.(10-25).

Summary

The Flory-Huggins theory delivers an excellent description of the behavior of *solvents* in polymer-solvent systems as shown for the sorption of solvents by polymers at temperatures $T > T_G$ (Fig. 10-8). The reduced chemical potential of the solvent, $\Delta\mu_1/RT$, was obtained here from partial pressures of the solvent above polymer solutions or above the pure liquid, its molar volume, and the compressibility of the gas phase. Interaction parameters χ were independent of concentration and temperature, as predicted.

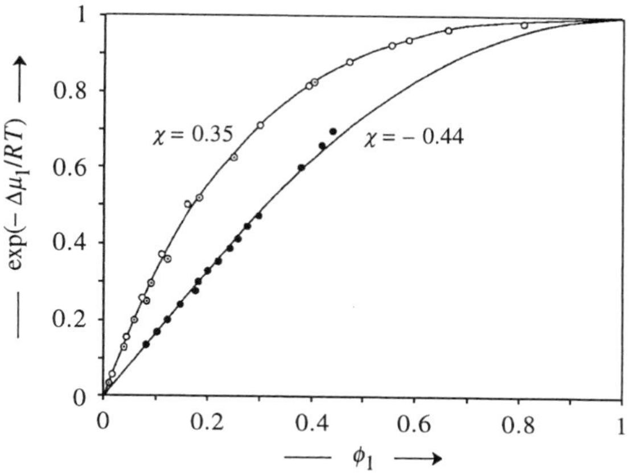

Fig. 10-8 Sorption parameter $\Delta\mu_1/RT$ as a function of the volume fraction $\phi_1 = p_1/p_{1,o}$ of the solvent. p_1 = partial pressure of solvent in the polymer-solvent system, $p_{1,o}$ = partial pressure of the solvent above the liquid solvent. Solid curves were calculated with interaction parameters $\chi = 0.35$ for poly(styrene) (T_G = 100°C) in ethyl benzene at 130°C and 178°C (upper curve) and $\chi = -0.44$ for poly(vinyl acetate) (T_G = 30°C) in chloroform at 45°C [5].

Quite different is the behavior of *polymers*. The simple Flory-Huggins theory discussed above describes the behavior of solutions of coil molecules *qualitatively* correctly (see below) but, contrary to theory, interaction parameters are generally found to be concentration dependent (Fig. 10-6) because specific polymer-solvent interactions (solvations) and polymer self-associations are neglected. The theory generally fails for dilute solutions because polymer segments are not homogeneously distributed in space (see Fig. 10-5), i.e., the number of free lattice sites is not given by $N_f = N_g - (i - 1)X_2$.

The theory also does not consider that the flexibility of thin polymer chains allows intramolecular polymer contacts which reduce the number of intermolecular ones. Because of these contacts, surface fractions would be better measures than volume fractions but surface fractions cannot be measured. Also neglected is the fluctuation of local segment concentrations.

The theory also assumes additivity of volumes, which neglects volume changes on mixing and the role of the so-called free volumes (p. 177), also known as packing effects. In contrast, **equation-of-state theories** consider that both components and their mixtures can be compressed. The thermodynamic functions are then written in terms of normalized, dimensionless volumes, $\tilde{V}_m = V_m / V_m^*$, where V_m^* = characteristic parameter for each polymer. These theories are complex and have not found wide applications.

10.2.3 Phase Separation

Solutions of Amorphous Polymers

In the examples of Fig. 10-7, reduced molar Gibbs energies of mixing, $\Delta G_{mix,m}/RT$ remain negative for the whole concentration range if the interaction parameter is $\chi = 1/2$. However, the curves are only symmetric about this value if $X_2 = 1$. The larger the degree of polymerization, the more the minimum shifts to higher polymer concentrations ϕ_2.

At the higher interaction parameter of $\chi = 1.8$, two flat minima and a flat maximum appear already at the low degree of polymerization of $X_2 = 2$. At the degree of polymerization of $X_2 = 100$, these minima correspond to negative values of $\Delta G_{mix,m}/RT$ in the concentration ranges $0.728 \leq \phi_2 \leq 1$ and $0 \leq \phi_2 \leq 10^{-35}$ (not visible in Fig. 10-7).

In a multiphase system, the chemical potential of a component must be the same in each phase. Hence, a binary system with two components 1 and 2 in phases ' and " must obey $\mu_1' = \mu_1''$ and $\mu_2' = \mu_2''$ and therefore also the condition $\Delta\mu_1' = \mu_1' - \mu_{1,0} = \mu_1'' - \mu_{1,0} = \Delta\mu_1''$ for phase ' and correspondingly for phase ". However, chemical potentials are identical only if two points of the function $\Delta G_{mix,m} = f(\phi_2)$ have a common tangent (cf. Eq.(10-24)).

The bottom part of Fig. 10-9 shows calculated curves $\Delta G_{mix,m} = f(\phi_2)$ for four temperatures. Of these, curves for temperatures 260 K, 300 K, and 350 K have two minima each. In a diagram $T = f(\phi_2)$, temperatures at the six contact points tangents/curve form a curve that is called a **binodal** (L: *bi* = two, *nodus* = knob, knot) (Fig. 10-9, top). The binodal separates the stable one-phase region above from the non-stable two-phase region below.

Binodals of polymers with broad molecular weight distributions are difficult to calculate because chemical potentials depend strongly on molecular weights and their distributions. More easy to calculate is the **spinodal** which is defined by the two inflection

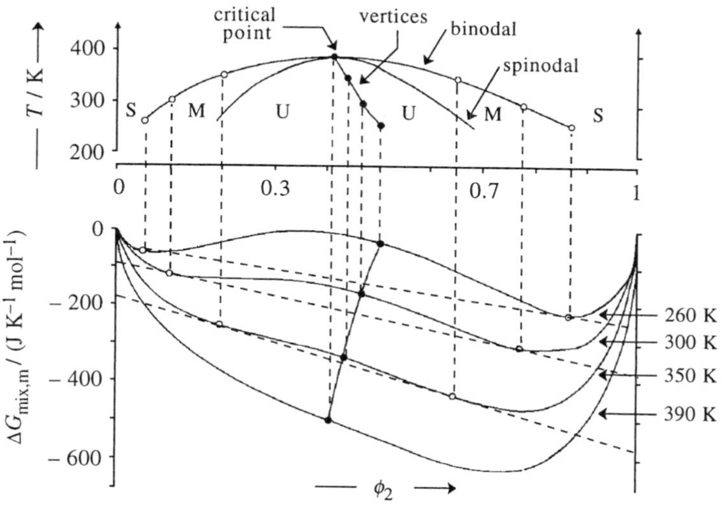

Fig. 10-9 Top: binodal, spinodal, and stable (S), metastable (M), and unstable (U) regions of a polymer/solvent system. Bottom: molar Gibbs energy of mixing as a function of the volume fraction of the solute ($X = 2$) in a solvent ($X_1 = 1$) at four temperatures. Calculations with Eq.(10-24) and $\chi = 0.3 + [(450\text{ K})/T]$. o Contact points of tangents, ● points calculated with Eq.(10-31).

points of the function $\Delta G_{\text{mix,m}} = f(\phi_2)$, i.e., by $\partial^2(\Delta G_{\text{mix,m}})/\partial\phi_2^2 = \partial\Delta\mu_1/\partial\phi_2 = 0$. Differentiation of Eq.(10-25) delivers

$$(10\text{-}30) \qquad \partial\Delta\mu_1/\partial\phi_2 = RT[2\,\chi\phi_2 - (1-\phi_2)^{-1} + (1 - X_2^{-1})] = 0 \qquad \text{(spinodal)}$$

The spinodal subdivides each of the two non-stable regions left and right between the vertices and below the binodal into a metastable region M and an unstable region U. The demixing of unstable regions is controlled by thermodynamics; an unstable region demixes spontaneously into two continuous phases that form an interpenetrating network. The phase separation of a metastable region is kinetically controlled; after a nucleation, the minority phase is dispersed in the majority phase.

The highest temperature on the binodal and spinodal, respectively, is the **critical point** of the system ($\phi_{2,\text{crit}} \approx 0.414$ in Fig. 10-9) at which maximum, minimum, and inflection point become identical. The critical point is given by setting the second derivative of the function $\Delta\mu_1 = f(\phi_2)$ equal to zero, i.e., from Eq.(10-30)

$$(10\text{-}31) \qquad \partial^2\Delta\mu_1/\partial\phi_2^2 = RT[2\,\chi - (1-\phi_2)^{-2}] = 0$$

Solving Eqs.(10-30) and (10-31) for χ and equating the results shows that the critical volume fraction of the polymer decreases with increasing degree of polymerization:

$$(10\text{-}32) \qquad \phi_{2,\text{crit}} = 1/(1 + X_2^{1/2})$$

Eqs.(10-32) and (10-31) deliver the critical interaction parameter:

$$(10\text{-}33) \qquad \chi_o = [(1 + X_2^{1/2})^2]/[2\,X_2] \approx (1/2) + (1/X_2)^{1/2} \qquad (\text{if } \chi \neq f(\phi_2))$$

Critical interaction parameters depend only on the degree of polymerization if interaction parameters are independent of volume fractions ϕ_2. They approach $\chi_o = 1/2$ for $X_2 \to \infty$. Concentration-dependent interaction parameters lead to $\lim \chi_{X_2 \to \infty} \neq 1/2$.

Quasi-binary Systems

All of the systems discussed above are truly binary: both components are molecularly uniform. However, all synthetic polymers are molecularly non-uniform, i.e., they have molar mass distributions; their polymer-solvent systems are always *quasi*binary.

Phase separations of quasibinary systems differ from those of binary systems as can be seen from **cloud curves** that indicate the "precipitation" (to be exact: onset of formation of two phases) that occurs if the temperature is changed or a precipitant is added. They are a special case of phase separation since they indicate the phase equilibrium in which the concentration of the precipitated phase approaches zero.

In binary systems, cloud curves and **coexistence curves** are identical so that the maximum of the cloud curve, T_{max} at $w_{2,max}$ (Fig. 10-10), coincides with the maximum of the coexistence curve T_{crit} at $w_{2,crit}$, i.e., that temperature at which the two coexisting phases contain equal mass fractions of polymer. For quasibinary systems such as the one in Fig. 10-10, this is no longer true since cloud point and coexistence curves refer to different fractions of the polymer.

For quasibinary systems with Schulz-Flory distributions of molar masses, Eq.(10-32) for the critical volume fraction and Eq.(10-33) for the critical interaction parameter of binary systems have to be replaced by Eqs.(32a) and (10-33a):

(10-32a) $\phi_{2,crit} = 1/[1 + (\overline{X}_w / \overline{X}_z{}^{1/2})]$

(10-33a) $\chi_{crit} = (1/2)[1 + \overline{X}_z{}^{1/2}\,\overline{X}_w{}^{-1}][1 + \overline{X}_z{}^{-1/2}]$

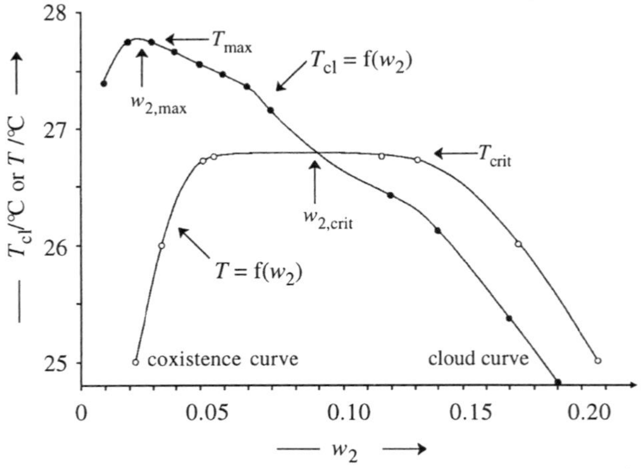

Fig. 10-10 Cloud temperatures T_{cl} of a poly(styrene) with $\overline{M}_n = 210\ 000$ g/mol and $\overline{M}_z : \overline{M}_w : \overline{M}_n = 2.4{:}1.65{:}1$ as a function of the mass fraction w_2 of the polymer in cyclohexane at 28°C (●) and the coexistence curve of a 6 % solution of the same polymer, i.e., the mass fraction w_2 of the polymer in the two coexisting phases at various temperatures T (O) [6]. Both curves are spinodals.

The differences $\phi_{2,\max} - \phi_{2,\text{crit}}$ and $w_{2,\max} - w_{2,\text{crit}}$ or $T_{\max} - T_{\text{crit}}$, respectively, may serve as indicators of the width of molar mass distributions of polymers with Schulz-Flory distributions. Other equations apply for different types of distributions.

Fractionation

The dependence of critical volumes and binodals, respectively, on the degree of polymerization is utilized for the **precipitation fractionation** of amorphous polymers with respect to molar masses. Lowering the temperature of an endothermic quasibinary solution causes polymer molecules with the highest molar mass to separate from the rest of the solution. The "precipitation" of this polymer fraction is a *phase separation* into a highly concentrated **gel phase** (the "precipitate") and a highly diluted **sol phase**; it is also called **coacervation** (L: *co* = together, *acervare* = to heap).

The efficiency of fractionations by lowering the temperature of a solution can be estimated by replacing $(\phi_2/X_2) \ln \phi_2$ in Eq.(10-24) with $\Sigma_i (\phi_i/X_i) \ln \phi_i$. Differentiation of the modified equation delivers the chemical potential of the ith polymer component, setting $\phi_2 = \Sigma_i \phi_i$:

$$(10\text{-}34) \qquad \Delta\mu_i/RT = \chi X_i(1-\phi_2)^2 - (X_i - 1) + X_i[1 - (1/\overline{X}_n)]\phi_2 + \ln \phi_i$$

The volume ratio of the ith component in the two phases ' and " (with $\mu_i' = \mu_i''$) is given by a very simple relationship:

$$(10\text{-}35) \qquad \phi_i''/\phi_i' = \exp(qX_i)$$

The degree of polymerization, X_i, is a complicated average that depends on the type of distribution. The parameter q can be calculated theoretically from the interaction parameter χ and the degrees of polymerization $\overline{X}_n$ in phases ' and ". Because of the many approximations required, q is usually regarded as an adjustable parameter.

The parameter q usually has small values. The poly(styrene) of Fig. 10-10 has a logarithmic normal distribution (LND) and a number-average degree of polymerization of $\overline{X}_n = \overline{M}_n/M_u = 210\,000/108 = 1944$. With $\overline{X}_w^2/\overline{X}_z = \overline{X}_n = X_2 = X_i$ for LNDs (Eq.(2-60)), a value of $\phi_{2,\text{crit}} \approx 0.022$ is obtained from Eq.(10-32). For a phase ratio of ϕ_2''/ϕ_2' = 5.36 at 26°C, Eq.(10-35) delivers $q = 7.5 \cdot 10^{-4}$.

The efficiency of a precipitation fractionation is estimated as follows. In equilibrium, both phases have volumes of V'' and V', respectively. According to Eq.(10-35), the volume fraction of molecules with a degree of polymerization X_i in phase " is

$$(10\text{-}36) \qquad \phi_i'' = \frac{V''\phi_i''}{V''\phi_i'' + V'\phi_i'} = \frac{(V''/V')\exp(qX_i)}{1 + (V''/V')\exp(qX_i)}$$

A fractionation is thus the more effective, the smaller the fraction of component i in the precipitated (concentrated) phase; this is achievable only if phase ratios V''/V' are very small. One thus lowers the temperature of a *dilute* polymer solution in a thermodynamically bad solvent until the solution becomes turbid and a small amount of a gel phase settles at the bottom. The fraction is separated and the temperature is lowered again to yield the next fraction, etc.

Any polymer can thus be divided in fractions with different molar masses from which the molar mass distribution of the original polymer can be calculated (see Table 2-6). These fractions do not necessarily have narrower molar mass distributions than the starting polymer; the distribution may even be much broader.

Precipitation temperatures of quasibinary solutions very often lie in experimentally unfavorable temperature ranges. Fractionations are therefore usually not performed in polymer-solvent sytems *per se* but by adding a weak non-solvent to a 1 % polymer solution in a bad solvent. Good fractionations are obtained if the initial precipitate is dissolved by heating the system and then letting cool again. Additional fractions are obtained by successive addition of precipitant and separation of fractions.

Fractionations do not give satisfactory results if a weak precipitant is combined with a good solvent or a strong precipitant with either a good or (worse) a bad solvent.

Polymers can also be fractionated by dissolution. In **dissolution fractionations**, polymers are adsorbed from solution onto a thin carrier (metal foil, quartz sand, etc.) and dried. The thin surface film is then eluted (L: eluere = to wash out) at constant temperature by solvent/non-solvent mixtures with increasing solvent content. The first fractions are therefore of low molar mass, just opposite to that in precipitation fractionation.

An elegant variant of this procedure is known as the **Baker-Williams method** which uses a column with a heated jacket in which a temperature gradient is maintained. The efficiency of separation here is increased by the simultaneous action of concentration and temperature gradients.

Cloud Point Titrations

A dilute polymer solution becomes cloudy after a volume fraction ϕ_3 of a precipitant is added. Experimentally, it was found that this volume fraction is a linear function of the logarithm of the initial volume fraction ϕ_2 in the concentration range $10^{-5} < \phi_2 < 10^{-2}$ (Fig. 10-11). Subsequent theoretical calculations showed that this behavior corresponds to a dependence $\chi = f(\ln \phi_2)$ of the interaction parameter.

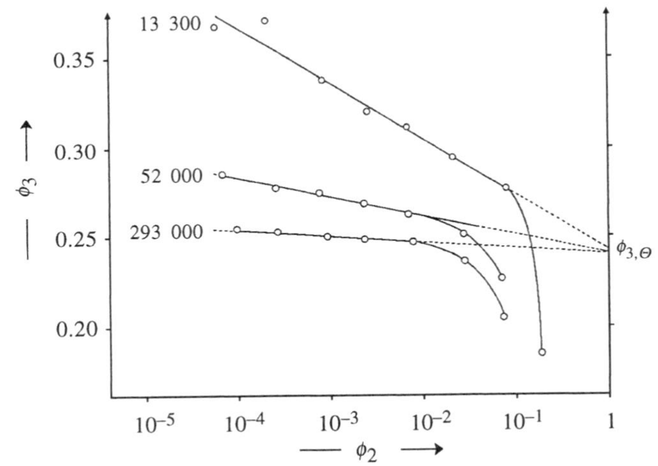

Fig. 10-11 Cloud point titration of benzene solutions of poly(styrene)s with various mass-average molar masses with methanol as a precipitant at 25°C [7]. $\phi_{3,\Theta}$ = theta composition.

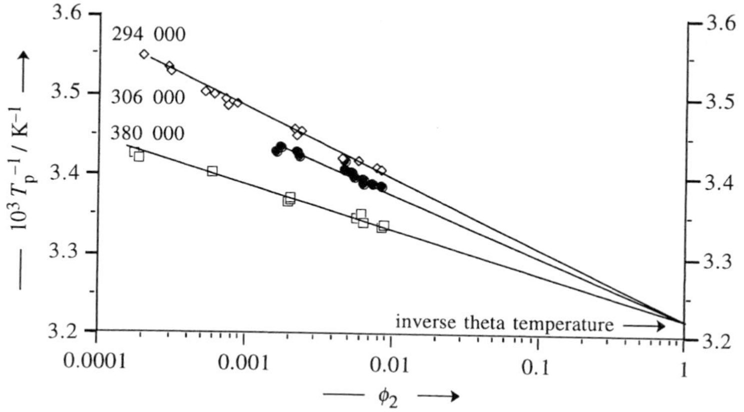

Fig. 10-12 Dependence of inverse cloud point temperatures, T_p, on the logarithm of the volume fraction ϕ_p of poly(α-methyl styrene)s in cyclohexane [8].

Extrapolation of ϕ_3 to lg $\phi_2 \rightarrow 0$ yields the theta composition $\phi_{3,\Theta} = 1 - \phi_{2,\Theta}$. Similarly, theta temperatures of polymer can be determined by extrapolating inverse temperatures of the beginning of phase separations as a function of the logarithm of the polymer concentration to lg $\phi_2 \rightarrow 0$, using data in the range $-4 <$ lg $\phi_2 < -2$ (Fig 10-12).

Precipitation Fractionation of Polymers

Conventional chain copolymerizations and copolycondensations deliver polymers that are non-uniform with respect to both chemical composition and molar mass. Consecutive fractions from precipitation fractionations thus may not show a systematic variation of composition or molar mass with the number of fractions. Table 10-6 shows such fractions arranged according to increasing mole fractions of vinyl chloride units. For composition distributions, effective fractions w_i* rather than the weight fractions w_2 have to be taken because all fractions have molar mass and composition distributions.

Table 10-6 Fractionation of a 45:55 (mol/mol) vinyl acetate-vinyl chloride copolymer, arranged according to increasing mole fraction x_{VC} of vinyl chloride units in successive fractions 1, 2, 3, ... with masses m_i, mass fractions w_i, and effective integral fractions, $\Sigma_i w_i$* [9].

Fraction	x_{VC}	m_i/mg	$10^2 w_i$	$10^2 \Sigma_i w_i$*	Fraction	x_{VC}	m_i/mg	$10^2 w_i$	$10^2 \Sigma_i w_i$*
2	0.363	41.0	5.32	2.660	13	0.595	38.0	4.93	55.613
1	0.364	56.0	7.27	8.955	7	0.625	72.5	9.41	62.874
5	0.412	78.5	10.19	17.683	9	0.636	51.0	6.62	70.798
3	0.414	43.5	5.65	25.600	8	0.638	63.5	8.24	78.228
4	0.510	61.5	7.98	32.414	10	0.642	32.0	4.15	84.425
6	0.577	64.5	8.37	40.591	14	0.665	56.0	7.27	90.136
15	0.587	26.5	3.44	46.495	12	0.676	48.0	6.23	96.885
11	0.595	38.0	4.93	50.681					
					Total	-	770.5	100.00	

The lowest fraction of this sequence (no. 2) has a weight fraction w_2 and an effective weight fraction $w_2{}^* = w_2/2$. The sums of the next effective weight fractions are then $w_2 + (1/2)\ w_1$, $w_2 + w_1 + w_5/2$, and so on. Distributions of molar masses are obtained in the same manner after arranging fractions according to their molar mass.

Suitable solvent/non-solvent pairs allow one to fractionate copolymers with respect to either molar mass or chemical composition. Unsuitable solvent/non-solvent pairs may mimic constitutionally uniform polymers.

Polymer Mixtures in Solution

The solubility of a mixture of two polymers 2 and 3 in a common solvent 1 is controlled by three interaction parameters, χ_{21}, χ_{31}, and χ_{23}. Theoretical calculations of spinodals showed for high polymer concentrations that miscibilities depend mainly on the polymer-polymer interaction parameter χ_{23}. At low polymer concentrations, the differences $\chi_{23} - \chi_{21}$ and $\chi_{23} - \chi_{31}$ are important.

In dilute solution, at-poly(styrene) ($\delta_2/(\text{cal/cm}^3)^{1/2} = 9.1$) mixes with at-poly(vinyl methyl ether) (9.6) in toluene (8.9), benzene (9.2), and perchloroethylene (9.3) but not in chloroform (9.3) or methylene chloride (9.73). At high concentrations, insolubility of polymer mixtures in one solvent usually also means insolubility in all others.

Conversely, one can speculate that a highly negative interaction parameter χ_{21} for a highly concentrated polymer solution will indicate that polymer 2 will mix with other polymers 3 in the same solvent 1. For example, cellulose nitrate has a strong negative interaction parameter χ_{21} in acetone (Fig. 10-6); it is indeed miscible with many other polymers in concentrated solutions.

Polymer Blends

The lattice theory of polymer-solvent systems can also be applied to mixtures of amorphous polymers (**polymer blends**). In such blends, polymer 1 with $X_1 >> 1$ reduces the number of arrangements of units of polymer 2 with $X_2 >> 1$. In contrast to polymer-solvent mixtures, molar entropies of mixing can therefore never become positive in blends. The entropy term $-T\Delta S_{\text{mix,m}} = RT[X_1{}^{-1}\phi_1 \ln \phi_1 + X_2{}^{-1}\phi_2 \ln \phi_2]$ is only slightly negative and can never compensate the enthalpy term $\Delta H_{\text{mix,m}} = RT[\chi\phi_1\phi_2]$ if the interaction parameter is positive. In most common cases, Gibbs energies of mixing become positive and the system demixes unless this is delayed or prevented by kinetics.

There are only a few polymer-polymer blends with negative interaction parameters. Such systems have strong attractive forces between their components (see Volume IV).

In the literature, "not miscible" usually does not mean "non-miscible in the whole concentration range" but only "non-miscible in the practically important concentration range" (see also Fig. 10-9 for miscibilities at very low and very high concentrations of one component). Also, the thermodynamic term "miscibility" should not be confused with the phenomenological term "compatibility." Because of high viscosities, a blend composed of two non-miscible polymers may demix only slowly (sometimes *very* slowly) and the system may *appear* to be miscible, hence "compatible."

The classification compatible/incompatible may also depend on the observation method. Incompatible polymer blends are often opaque because they contain sufficiently large domains with sizable differences in refractive indices. Optically clear blends, on the other hand, may consist of very small domains (i.e., they are indeed incompatible) but these domains may show up only under the electron microscope. Incompatible blends with sufficiently large domains may show two glass temperatures (Section 13.5) that do not change with composition.

Incompatible blends are dispersions of one polymer in the other. Since all dispersions are thermodynamically unstable, incompatible blends are stabilized by the addition of diblock polymers that act as **compatibilizers**. Such diblock polymers will reside with one type of their blocks in one phase and with the other type of blocks in the other phase (p. 262). The blocks need not be chemically identical with their host phases; it suffices if they are merely compatible. An incompatible polymer blend, poly(A)-*blend*-poly(B), can be made compatible by adding poly(A)-*block*-poly(C) as a compatibilizer if the poly(C) block is compatible with block poly(B) or can form mixed crystals with it.

Constitutional identity of polymer chains does not guarantee compatibility. For example, poly(methyl methacrylate) chains grafted onto glass spheres are incompatible with poly(methyl methacrylate) proper because its tethered chains are stretched and densely packed (Section 9.5) and cannot accommodate interpenetrating chains. Compatibility is possible for sufficiently low grafting densities and small graft lengths.

Critical Miscibility Temperatures

Solutions and blends of non-crystallizable polymers exhibit different phase diagrams, depending on the temperature dependence of the interaction parameters χ (Fig. 10-13). Systems with positive K_T in $\chi = \chi_\infty + (K_T/T)$ are endothermal. In these systems, interaction parameters decrease with *in*creasing temperature (diagram I in Fig. 10-13) and the system demixes with *de*creasing temperature. Two liquid phases form below the binodal and one liquid phase above. The maximum of the binodal, U, is the **upper critical solution temperature** (UCST).

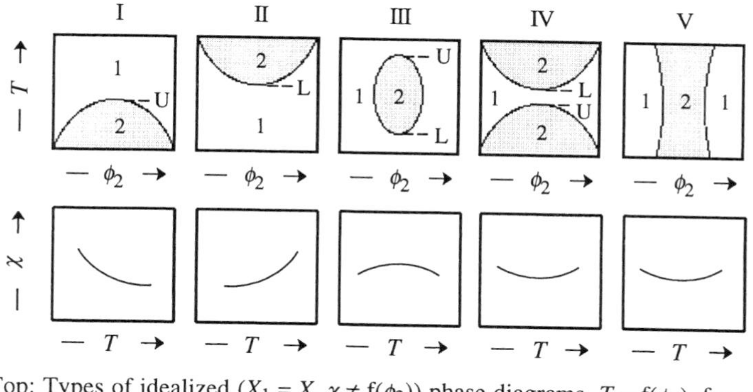

Fig. 10-13 Top: Types of idealized ($X_1 = X$, $\chi \neq f(\phi_2)$) phase diagrams, $T = f(\phi_2)$, for polymer solutions or polymer blends with one-phase (1) and two-phase (2) regions and upper (U) and lower (L) critical solution temperatures. Type III is known as a closed miscibility gap and type V as an hour-glass diagram. Real systems are not symmetrical because of $X_1 \neq X_2$ and $\chi = f(\phi_2)$.

Bottom: temperature dependence of interaction parameters. Examples:

	Polymer solutions	*Polymer blends*
I	Poly(styrene) + cyclohexane	Poly(butadiene) (deuterated + protonated)
II	Poly(ethylene) + hexane (at 5 bar)	Poly(styrene) + poly(vinyl methyl ether)
III	Poly(oxyethylene) + water	Poly(methyl methacrylate) + polycarbonate A
IV	Poly(styrene) (low molar mass) + acetone	
V	Poly(styrene) (high molar mass) + acetone	

Exothermal systems with positive K_T mix at lower temperatures but demix with *increasing* temperature (II in Fig. 10-13). The minimum of the binodal, L, is the **lower critical solution temperature** (**LCST**). LCSTs correspond to entropically induced phase separations and UCSTs to enthalpically induced ones.

Designations LCST and UCST do *not* refer to the absolute position of the critical temperature. A lower critical solution temperature LCST can therefore be higher than an upper critical solution temperature UCST (Fig. 10-13, IV).

With increasing temperature, interaction parameters χ decrease for UCSTs (Fig. 10-13, I) but increase for LCSTs (Fig. 10-13, II). A minimum in $\chi = f(T)$ leads to LCST > UCST (Fig. 10-13, IV) and a maximum to a closed miscibility loop with LCST < UCST (Fig. 10-13, III).

The type of phase diagram depends on the solvent and the range of molar masses. at-Poly(styrene) in cyclohexane has a Type IV phase diagram with LCST > UCST (Fig. 10-14, left) where the gap between the LCST and UCST becomes smaller with increasing molecular weight of the polymer. Conversely, at-poly(styrene) in acetone shows a decreasing gap with decreasing molecular weight (Fig. 10-14, center) that leads to the hour glass diagram of Fig. 10-13, V. Poly(oxyethylene) in water, on the other hand, has a closed miscibility loop (Fig.10-13, III and Fig. 10-14, right) but an hour glass diagram in *t*-butyl acetate (Fig. 10-13, V).

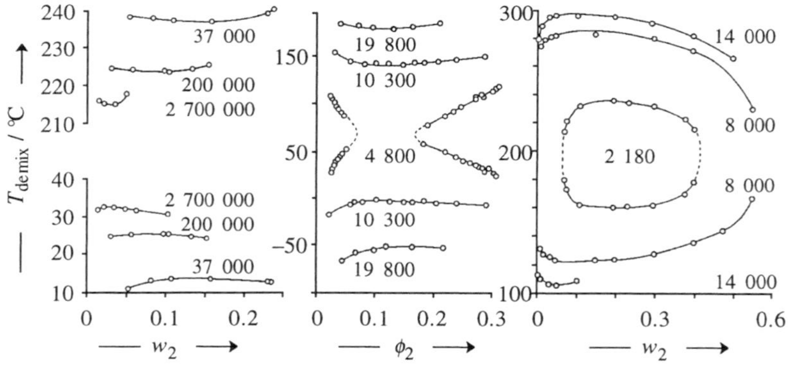

Fig. 10-14 Demixing temperatures T_{demix} as a function of mass fractions w_2 or volume fractions ϕ_2 of polymers. Left: poly(styrene)s in cyclohexane [10]; center: poly(styrene)s in acetone [11]; right: poly(oxyethylene)s in water [12, 13].

Most systems amorphous polymer + solvent and amorphous polymer 1 + amorphous polymer 2 have only UCSTs (Fig. 10-13, I), probably because the accessible temperature range is too small. Poly(styrene) in cyclohexane is one of the few systems with both a UCST and an LCST at normal pressure (Fig. 10-14).

Under pressure, some systems with UCSTs show LCSTs above the normal boiling temperature of the solvent. Such systems contract when the dense polymer and the highly expanded solvent are mixed: the entropy of mixing becomes negative. The system becomes soluble between UCST and LCST (Fig. 10-13, IV) and $\chi = f(T)$ passes through a minimum.

Acetone solutions of poly(styrene) show an hour glass behavior (Fig. 10-13, V) in which the gap between UCST and LCST narrows with *decreasing* molar mass (Fig. 10-

14, center). Such a behavior is commonly blamed on "specific" polymer-solvent interactions albeit without any explanation. However, acetone is a self-associated solvent (p. 304) that does not form dilute solutions of high-molar mass poly(styrene).

The case UCST > LCST is relatively common for water-soluble polymers such as poly(vinyl alcohol), poly(vinyl methyl ketone), methyl cellulose, and poly(L-proline). Heating moderately concentrated aqueous solutions of poly(oxyethylene) causes helical segments of this polymer to desolvate, an effect that increases with increasing temperature. Polymer molecules within the miscibility gap thus do not have the same macroconformation as they do outside the gap. Another example is poly(*N*-isopropyl acrylamide) in water where polymer molecules form long hydrated helical segments below the LCST but exist as hydrophobic coils above the LCST.

The temperature at the critical point of the phase diagram is identical with the **theta temperature** of a polymer with infinite molecular weight as can be seen from the dependence of inverse cloud temperatures on polymer concentrations (Fig. 10-14). Poly(styrene) in cyclohexane therefore has two theta temperatures, one at Θ = 34.5°C (UCST) and one at 213°C (LCST) (see also critical points for M_r = 2 700 000 in Fig. 10-14, left).

Table 10-7 shows double theta temperatures for other polymer-solvent systems.

Table 10-7 Polymer-solvent systems for which two theta temperatures are known [14].

Polymer and solvent	Θ/°C		Polymer and solvent	Θ/°C	
Poly(styrene), at-, in			1,4-poly(butadiene), 93 % cis, in		
n-propyl acetate	−80	178	ethyl propyl ketone	−22	237
i-amyl acetate	−49	220	diethyl ketone	14	208
t-butyl acetate	−35	109	propylene oxide	35	141
i-propyl acetate	−27	107			
butyl formate	−9	36	poly(α-methyl styrene), at, in		
cyclohexane	34	213	butyl chloride	−10	139
methyl acetate	43	114			
methyl cyclopentane	75	144	poly(methacrylic acid), at, in		
			methanol	26	151

Solutions of Crystalline Polymers

Phase separations of solutions of amorphous polymers deliver two liquid phases, the high concentration **gel phase** with higher molecular weights and the low concentration **sol phase** with lower molecular weights. These phase separations allow the fractionation of polymers with respect to molecular weights.

Phase separations of solutions of semicrystalline polymers P also yield two phases: a sol phase L consisting of a *solution* of dissolved polymer molecules P' in the solvent S and a two-phase mixture consisting of a *dispersion* of crystalline polymer P" in a polymer solution L' from polymer molecules P'" in S. This polymer P" cannot be fractionated with respect to molar mass because melting enthalpies of crystalline polymers are independent of the molecule size above a fairly low molar mass. Since melting enthalpies depend on chemical structure, fractionation is possible according to constitution (for example, branching) and chemical configuration (tacticity).

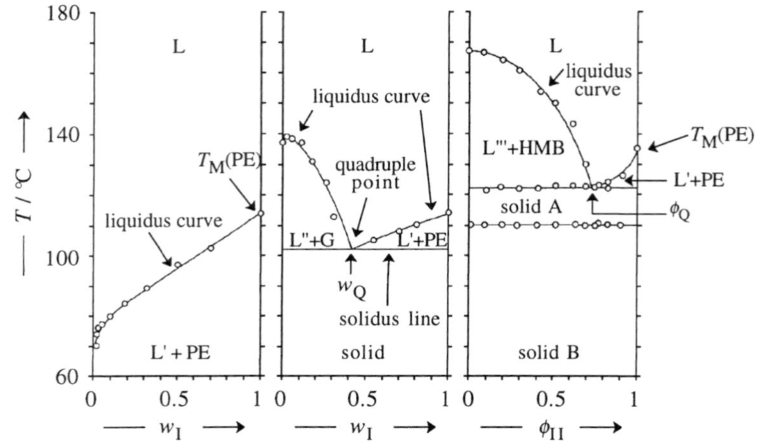

Fig. 10-15 Phase diagram of two poly(ethylene)s I (left and center) and II (right) in various solvents. L, L', L", L''' = solutions of I or II in the solvent; G = gel; w_Q = eutectic point. The horizontal line in the center graph is the solidus line that separates L" + G and L' + PE, respectively, from the solid.

Left: in good solvent xylene (mixture of ortho, meta, and para compounds with T_M: –47°C (meta), –25°C (ortho), 13°C (para) [15].

Center: in bad solvent amyl acetate (T_M = –70.8°C) [15].

Right: in crystallizing solvent hexamethylbenzene (T_M = 165°C) [16].

Lowering the temperature of a 120°C solution of a poly(ethylene) in the *good* solvent xylene causes a separation of the solution into two phases: a dilute solution L of the polymer and a heterogeneous 2-phase region which consists of a dispersion of a crystalline poly(ethylene) fraction PE in a sol phase that is a dilute solution L' of the polymer in the solvent (Fig. 10-15, left). The lower the original weight fraction of the polymer, the lower is the temperature for phase separation.

The same polymer in the *bad* solvent amyl acetate also undergoes a phase separation along the **liquidus curve** (Fig. 10-15, center) but into two different regions left and right of the so-called **eutectic point** at the mass fraction w_Q. At $w_I > w_Q$, a 2-phase region L' + PE appears again. At $w_I < w_Q$, another 2-phase region appears that consists of a dilute polymer solution L" and a highly concentrated polymer gel G. The eutectic point is a quadruple point that separates the phases L, L" + G, and L' + PE above the **solidus line** from a 2-phase solid below that line.

Still another behavior is observed for an ultrahigh molecular weight poly(ethylene) in the bad solvent hexamethylbenzene (HMB) that crystallizes in the same temperature range as the polymer. The phase separation leads here to two regions: at high initial polymer concentrations $\phi_{II} > \phi_Q$ to a 2-phase system L' + PE composed of a crystalline poly(ethylene) PE that is dispersed in a polymer solution L' and at concentrations $\phi_{II} < \phi_Q$ to a 2-phase system composed of crystalline HMB in a polymer solution L'''. Below the solidus line at ca. 122°C, an unspecified "solid A" exists (should be HMB crystals + eutectic left of w_Q and eutectic and crystalline PE right of w_Q). Below 110°C, a phase transformation occurs from "solid A" to "solid B."

The mixture at point w_Q is called **eutectic** because it has a lower melting temperature than its two components (G: *eu* = good, *tekein* = to melt). The eutectic behaves like a one-phase chemical compound but is a mixture of compounds.

The eutectic point of a system with two components (C = 2) at constant temperature and composition is a **quadruple point** according to the **Gibbs phase rule**, P + F = K + 2, since four phases (P = 4) are in equilibrium: solution, crystallized components I and II, and, at low pressure, the saturated vapor above the solution. Hence, the degree of freedom is zero (F = 0) and the eutectic point is invariant. However, the system PE + HMB (Fig. 10-15, right) is not a binary system, only a quasibinary one, since even completely linear poly(ethylene) is a mixture of poly(ethylene) molecules with very different degrees of polymerization. In addition, the poly(ethylene)s of Fig. 10-15 are branched, PE II a little, PE I much more (compare melting temperatures T_M(PE)).

Variations of melting temperatures caused by differences in the degrees of branching of the molecules are utilized to determine the extent of short-chain branching of linear low-density poly(ethylene) (LLDPE) by **temperature rising elution fractionation (TREF)**. On cooling an LLDPE solution in the presence of glass spheres, the least branched LLDPE molecules are deposited first on the spheres because their crystallization delivers the most perfect crystallites. Later deposits consist of less perfect crystals from more highly branched molecules. A solvent stream then first extracts the highly branched molecules and later the less branched ones (Fig. 10-16). The degree of branching is determined by infrared spectroscopy.

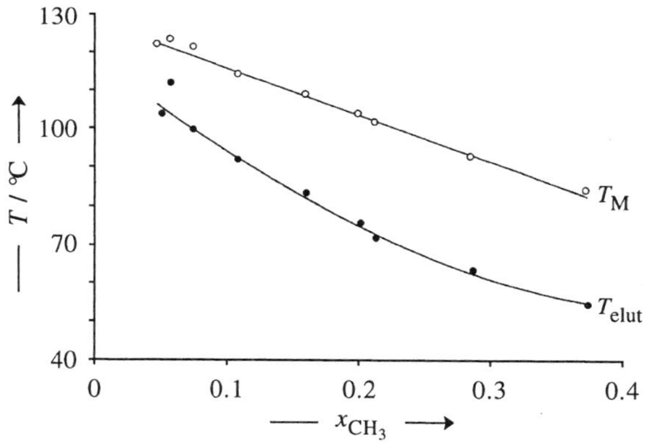

Fig. 10-16 Melting and elution temperatures (by 1,2,4-trichlorobenzene) of fractions of an LLDPE (from ethene and 1-butene) as a function of the mole fraction of carbon atoms in methyl groups [17].

10.3 Osmotic Pressure

10.3.1 Fundamentals

According to chemical thermodynamics, the differential of Gibbs energy is given by the change of partial molar Gibbs energy, $\tilde{G}_{i,m}$, with pressure p, temperature T, and amount-of-substance x_i:

$$(10\text{-}37a)\quad \mathrm{d}\tilde{G}_{i,m} = (\partial \tilde{G}_{i,m}/\partial p)\mathrm{d}p + (\partial \tilde{G}_{i,m}/\partial T)\mathrm{d}T + (\partial \tilde{G}_{i,m}/\partial x_i)\mathrm{d}x_i$$

$$(10\text{-}37b)\quad \mathrm{d}\tilde{G}_{i,m} = \tilde{V}_{i,m}\mathrm{d}p - \tilde{S}_{i,m}\mathrm{d}T + RT\,\mathrm{d}\ln a_i$$

In membrane osmometry, a solution is separated from the pure solvent by a membrane. The membrane is semipermeable, i.e., permeable for the solvent but not for the solute. Since the activity a_1 of the solvent in the solution differs from the activity of the pure solvent $a_{1,o}$, a pressure difference is established between the two liquids. For an isothermal process ($dT = 0$) in equilibrium ($d\tilde{G}_{i,m} = 0$), this pressure difference equals the osmotic pressure difference, $d\Pi$. On integration, Eq.(10-37b) converts to

$$(10\text{-}37c) \qquad \tilde{V}_{1,m}\Pi = -RT \ln a_1 = -\Delta\mu_1.$$

Activities a_1 of *ideal* solutions are identical with mole fractions $x_1 = 1 - x_2$ for the whole concentration range. For *dilute* ideal solutions ($n_1 > n_2$; $m_1 > m_2$; $V_1 > V_2$), solvent activities a_1 can be expressed by solute mole fractions, $\ln a_1 \approx \ln(1 - x_2) \approx -x_2$, which leads to $\tilde{V}_{1,m}\Pi = RTx_2$. The mole fraction x_2 of the solute can be replaced by the mass concentration c_2 since $c_2 = m_2/(V_1 + V_2) \approx m_2/V_1 = n_2M_2/V_1 = n_2M_2/(n_1\tilde{V}_{1,m}) \approx x_2M_2/\tilde{V}_{1,m}$ The resulting equation describes **van't Hoff's law** according to which the reduced osmotic pressure, Π/c_2, *at infinite dilution* ($c_2 \to 0$) is inversely proportional to the molar mass M_2 of the solute:

$$(10\text{-}38) \qquad \lim_{c_2 \to 0} \frac{\Pi}{c_2} = \frac{RT}{M_2}$$

The osmotic pressure of such a solution equals the sum of osmotic pressures of all components, $\Pi = \Sigma_i \Pi_i$. Eq.(10-38) applies to each component i of a molecularly non-uniform polymer, i.e., $\Pi = \Sigma_i \Pi_i = \Sigma_i RT(c_i/M_i)$. Comparison of the latter equation with Eq.(10-38) delivers $M_2 = c_i/[\Sigma_i (c_i/M_i)]$ and, after introducing $c_2 = \Sigma_i c_i$ and $c_i = n_iM_i/V$, also $M_2 = \Sigma_i n_iM_i/(\Sigma_i n_i) \equiv \overline{M}_n$. For non-uniform polymers, M_2 is therefore the *number-average molar mass* $\overline{M}_n$ of the solute.

The law of van't Hoff applies only to infinitely dilute solutions. Molar masses calculated with van't Hoff's law from osmotic pressures at finite concentrations are therefore only *apparent* number-average molar masses, $\overline{M}_{n,app}$, that need to be extrapolated to infinite dilution. These extrapolations are different for non-associating polymers (Section 10.4) and associating ones (Section 10.5).

10.3.2 Membrane Osmometry

Semipermeable Membranes

Membrane osmometry is the most important absolute method for the direct determination of number-average molar masses of polymers. The method measures the equilibrium pressure difference between a solution and its solvent that are separated by a semipermeable membrane which lets only solvent molecules pass through. Such semipermeable membranes for organic solvents are most often films of regenerated cellulose that are known by different names (Cellophane® 600, Gel cellophane®, Ultracella filter®, etc.). Membranes for aqueous solutions consist of cellulose acetate or cellulose nitrate. Aggressive solvents require membranes of porous glass.

Simple membrane osmometers work statically. At the beginning of the experiment, solution and solvent are not in equilibrium. Solvent thus flows from the solvent chamber through the membrane to the solution chamber (or vice versa, depending on fill heights) until equilibrium is established, which may take days, depending on the solvent volume that needs to be shifted. The final pressure difference is the osmotic pressure Π.

Automatic membrane osmometers reduce the time required for the establishment of equilibrium by an engineering trick. An increase of pressure difference by transport of solvent from the solvent chamber to the solution chamber is here immediately compensated by an automatic adjustment of liquid volumes, using a servo mechanism. In such dynamic experiments, equilibrium is obtained in 10-30 minutes.

According to Eq.(10-38), osmotic pressures are smaller, the higher the molar mass. They become less accurate at high molar masses, which imposes an upper limit of $\overline{M}_n = 1 \cdot 10^6$ to $2 \cdot 10^6$ g/mol. The lower limit is given by the absence of semipermeability of membranes for low molar masses; it is often in the range 3000-5000 g/mol.

Non-semipermeable Membranes

The success of membrane osmometry depends on the semipermeability of the membrane which in turn is a function of the molar mass and molar mass distribution as well as the macroconformation of the solute. Random coil molecules have only small chain diameters, which allows them to pass through the labyrinth of interconnected pores if the molar masses of macromolecules are sufficiently low. In osmotic equilibria at such leaky membranes, permeable components of the polymer are distributed on both sides of the membrane according to their activities in static measurements. The resulting **Donnan equilibrium** is no longer the true equilibrium for a non-permeating solute and the resulting osmotic pressure is neither the true osmotic pressure of the polymer nor the osmotic pressure of the non-permeating fraction.

Partial or complete permeation of solutes through the membrane is often discernible if in static measurements "from below" (initial $\Delta p < \Pi$) measured pressure differences increase, pass through a maximum, and then decrease. This effect is caused by the opposing effects of solvent transport into the solution chamber and solute transport from the solution chamber to the solvent chamber.

Since practically no solute could permeate through a leaky membrane in very short experimental times (such as in dynamic osmometry), it is often assumed that the resulting osmotic pressure must represent the true osmotic equilibrium pressure for a non-leaky membrane. This assumption is erroneous, however, because experiments with leaky membranes are governed by the laws of irreversible thermodynamics and not by equilibrium thermodynamics.

Irreversible thermodynamics considers the total volume flow, $J_V = L_p \Delta p + L_{pD} \Pi$, from the solvent chamber to the solution chamber. The flow is caused by a hydrostatic pressure difference Δp between solvent and solution and an osmotic pressure Π. Since dynamic osmometry determines pressure differences at a volume flow $J_V = 0$, one obtains

$$(10\text{-}39) \qquad \Delta p \text{ (at } J_V = 0) = -(L_{pD}/L_p)\Pi = s\Pi \quad ; \quad s \equiv -L_{pD}/L_p$$

The proportionality constants L_p and L_{pD} are the **phenomenological coefficients** or **Onsager coefficients**. The negative ratio of phenomenological coefficients is called the **Staverman coefficient** s, **reflection coefficient**, or **selectivity coefficient**.

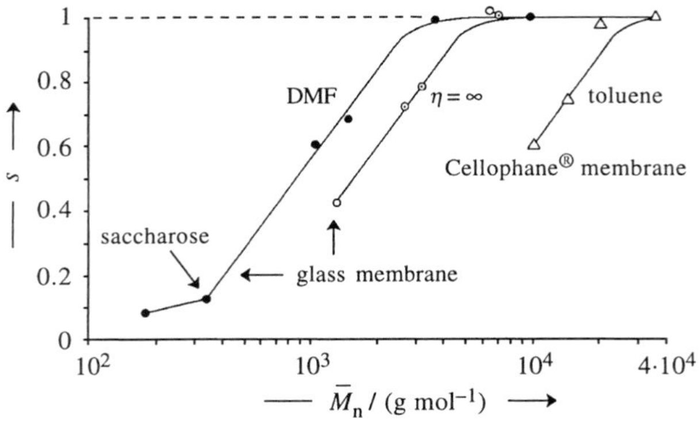

Fig. 10-17 Dependence of Staverman coefficients s on molar masses. (●) Experiments with levulose, saccharose, and poly(ethylene glycol)s (PEG) in *N*,*N*'-dimethylformamide at 25°C with glass membranes [18] and (Δ) poly(styrene)s in toluene at 30°C with cellophane membranes [19]. Curve for $\eta = \infty$: experimental data for PEGs (⊙) and poly(α-methyl styrene)s (O) in various solvents after extrapolation to infinite solvent viscosity [20].

At the end of the osmotic experiment, the solution chamber still had the original saccharose concentration and no saccharose was detected in the solvent chamber.

For truly semipermeable membranes, one has $-L_p = +L_{pD}$ and thus $s = 1$ whereas for leaky membranes $|L_p| > L_{pD}$, and therefore $s < 1$. Staverman coefficients are zero for totally permeable membranes. Hence, a zero volume flow in dynamic osmotic measurements with leaky membranes never delivers the theoretical osmotic pressure but only a fraction s of Π, even at $t \to 0$ (Fig. 10-17).

Staverman coefficients have not been calculated theoretically. They depend on molar mass, molar mass distribution, type of solvent (viscosity), temperature, and type of membrane but not on solute constitution (Fig. 10-17). For a membrane-solvent-temperature system, Staverman coefficients equal unity above a critical number-average molar mass but decrease linearly with decreasing logarithm of molar mass of the solute (Fig. 10-17). In this range, inverse Staverman coefficients are proportional to inverse solvent viscosities. Dynamic membrane-osmotic experiments are thus preferably performed in *viscous* solvents, which is contrary to the expectation that low-viscosity solvents allow a faster approach to equilibrium without noticeable solute permeation.

Number-average molar masses can be obtained by membrane osmometry with leaky membranes if solutions are first completely dialyzed with the same membrane that is later used in membrane osmometry of the non-permeating, high-molar mass fraction of the polymer (gives $\overline{M}_n$(H)). The number-average molar mass $\overline{M}_n$(L) of the low-molar mass fraction is determined by vapor pressure osmometry (Section 10.3.4). The number-average molar mass $\overline{M}_n$ of the specimen is then calculated from $\overline{M}_n$(H) and $\overline{M}_n$(L).

10.3.3 Ebullioscopy and Cryoscopy

Dissolved chemical compounds increase the boiling temperature of the solvent (as measured by ebullioscopy (L: *ebullire* = to boil over; G: *skopein* = to watch)) and decrease its freezing temperature (as measured by cryoscopy (G: *kryos* = cold)). Like membrane osmometry, ebullioscopy and cryoscopy are **colligative methods** that are based on solution thermodynamics (L: *colligare*; *com* = together, *ligare* = to tie). They thus lead to similar dependences of boiling temperature increases and freezing temperature decreases on the molar mass of solutes as osmotic pressures. For this reason, ebullioscopic and cryscopic measurements are called *osmotic measurements* in biomedicine although no osmotic pressure is measured.

Ebullioscopy relates changes of boiling temperatures to polymer concentrations and molar masses. In thermodynamic equilibrium, the change of Gibbs energy is zero: $\mathrm{d}\Delta G = \Delta V \mathrm{d}p - \Delta S \mathrm{d}T = 0$. At the boiling temperature of the solvent. T_{bp}, the entropy change is $\Delta S = \Delta H_{T,p}/T_{bp}$ for an isothermal-isobaric process. It follows that $\Delta H_{bp} = T_{bp}\Delta V(\mathrm{d}p/\mathrm{d}T)$.

At T_{bp}, the volume of a mass of gas, V_{gas}, is much greater than the volume of the liquid of the same mass, $\Delta V = V_{gas} - V_{liq} \approx V_{gas}$. To a first approximation, the solvent follows the ideal gas law, $pV_{gas} = RT_{bp,o}$, and the molar enthalpy is therefore $\Delta H_{bp,1,m} = T_{bp,1}(\mathrm{d}p/\mathrm{d}T)(RT_{bp,o}/p)$. Applying **Raoult's law**, using $\Delta p/p = x_2 = n_2/(n_1 + n_2) \approx n_2/n_1 = m_2M_1/m_1M_2 = m_2M_1/\rho_1V_1M_2 \equiv c_2M_1/\rho_1M_2$, and replacing differentials by differences, one obtains for infinite dilution, $c_2 \to 0$ (see also Eq.(10-38)):

$$(10\text{-}40) \qquad \left(\frac{\Delta H_{bp,1,m}\rho_1}{M_1T_{bp,1}}\right)\frac{\Delta T_{bp}}{c_2} = K_{bp}\frac{\Delta T_{bp}}{c_2} = \frac{RT_{bp}}{M_2} \quad ; \quad K_{bp} = \frac{\Delta H_{bp,1,m}\rho_1}{M_1T_{bp,1}}$$

Large increases of boiling temperatures require small ebullioscopic constants, K_{bp}. Solvents must therefore have high boiling temperatures $T_{bp,1}$, large molar masses M_1, low liquid state densities ρ_1, and/or small molar vaporization enthalpies, $\Delta H_{bp,1,m}$.

The reduced boiling point increase, $\Delta T_{bp}/c_2$, is inversely proportional to the molar mass of the solute, M_2 (Eq.(10-40)) which is a number average. Like Eq.(10-38), Eq.(10-40) applies to infinitely small concentrations of solutes (extrapolation required).

An equation similar to Eq.(10-40) applies to **cryoscopy** where the ebullioscopic constant K_{bp} is replaced with the cryoscopic constant, $K_M = \Delta H_{M,1,m}\rho_1/(M_1T_{M,1})$, and the boiling temperature $T_{bp,1}$ with the melting temperature $T_{M,1}$. This constant contains the molar melting enthalpy $\Delta H_{M,1,m}$ of the solvent:

$$(10\text{-}40a) \qquad \lim_{c_2 \to 0} K_M\frac{\Delta T_M}{c_2} = \frac{RT_M}{M_2} \quad ; \quad K_M = \frac{\Delta H_{M,1,m}\rho_1}{M_1T_{M,1}}$$

Effects measured by ebullioscopy and cryoscopy are much smaller than those measured by membrane osmometry. At a concentration of 0.01 g/mL polymer with $M = 10^5$ g/mol in a thermodynamically ideal solution, one would measure at 100°C an osmotic pressure that is equivalent to a water column of 3.2 cm height whereas the boiling temperature would be increased by only ca. 10^{-5} K. Upper limits for both ebullioscopy and cryoscopy are therefore $M \approx 20\,000$ g/mol.

The upper limit can sometimes be boosted to ca. 1 000 000 g/mol in cryoscopy if low molar mass liquid crystalline substances (LCs) are used as solvents since LCs have much lower melting enthalpies.

Ebullioscopy and cryoscopy are fairly time-consuming methods that are prone to errors such as foam formation, delays of boiling, supercooling, etc. Hence, the molecular weight method of choice for lower molar masses is vapor pressure osmometry.

10.3.4 Vapor Phase Osmometry

Vapor phase osmometry (= thermoelectric or vaporometric measurements) is based on the following principle. In a thought experiment, a drop of a solution of a non-volatile solute in a volatile solvent is placed on a temperature-measuring device such as a thermistor. The space surrounding the thermistor is saturated with solvent vapor. At the beginning of the experiment, drop and vapor have the same temperature but solvent vapor will condense on the solution drop since the vapor pressure of the solution is smaller than the vapor pressure in the surrounding space. The condensation releases heat of condensation and the temperature of the drop increases until the temperature difference ΔT_{th} between the solution drop and solvent vapor compensates the difference of vapor pressures so that the chemical potential of the solvent is the same in both phases. Hence, similar to ebullioscopy, one can write $K\Delta T_{th} = RTc_2/M_2$ with $K = \Delta H_{j,m}\rho/(TM_1)$ where M_2 is the molar mass of the solute and M_1 that of the solvent.

The method would have a solid thermodynamic foundation if it were possible to thermally separate solution drop and solvent vapor. Since they *are* in thermal contact, temperature differences between drop and vapor try to even out by convection, radiation, and conduction. These actions cause more solvent vapor to condense on the solution drop until a steady state is reached with a temperature difference ΔT. Hence, for $c_2 \rightarrow 0$, equation $K\Delta T_{th} = RTc_2/M_2$ with $K = \Delta H_{j,m}\rho/(TM_1)$ has to be replaced by

(10-41) $$(K/k_E)\Delta T = RTc_2/M_2 \quad ; \quad \Delta T = k_E\,\Delta T_{th} \quad ; \quad K = \rho\Delta H_{j,m}/(TM_1)$$

Since k_E is difficult to calculate theoretically, it is usually obtained by calibration with substances of known molar mass. However, such calibrations often lead to varying values for molar masses M_2 of unknown substances because drops of various solutions may have different sizes (surface tension), and heat may be lost because of unfavorable chamber geometry, the fastening of thermistors, etc.

At finite concentrations c_2, Eq.(10-41) delivers apparent molar masses that have to be extrapolated to zero concentration (Section 10.4). Vapor phase osmometry is a fairly fast method. Commercial instruments allow one to determine number-average molar masses up to ca. 50 000 g/mol.

10.4 Virial Coefficients

10.4.1 Fundamentals

The law of van't Hoff (p. 326) applies only to infinite dilutions. However, even ideal solutions have a (very small, see below) concentration dependence of reduced osmotic pressures, Π/c_2, since ideal entropies of mixing are not zero. Π/c_2 is constant only for theta solutions at *low* polymer concentrations (see below).

An expression for the concentration dependence of reduced osmotic pressures of non-ionic solutions of non-self-associating polymers is obtained as follows. According to statistical thermodynamics, the natural logarithm of solvent activity, a_1, can always be developed in a series with whole positive exponents of mole fractions of solutes, x_2:

(10-42) $- \ln a_1 = x_2 + x_2^2 + x_2^3 + ...$

Introduction of $\tilde{V}_{1,m}\Pi = - RT \ln a_1$ (Eq.(10-37c)) and replacement of the mole fraction by $x_2 = \tilde{V}_{1,m}c_2 / M_2$ leads to

(10-43) $$\frac{\Pi}{c_2} = RT\left[\frac{1}{M_2} + \frac{\tilde{V}_{1,m}}{M_2^2}c_2 + \frac{\tilde{V}_{1,m}^2}{M_2^3}c_2^2 + \ ...\right] = RT[A_1 + A_2c_2 + A_3c_2^2 + ...]$$

where A_1, A_2, A_3, ... are the first, second, third, ... **virial coefficients**. Sometimes, RT is included in virial coefficients. Eq.(10-43) then becomes $\Pi/c_2 = RT/M_2 + A_2'c_2 + A_3'c_2^2 +$ ... with $A_2' = RTA_2$ and $A_3' = RTA_3$.

The word "virial" has its origin in the so-called virial theorem which was widely used in the 19th century. The theorem equated the average of $mv^2/2$ with the average of $(Xx + Yy + Zz)/2$ where m = mass of particles, v = velocity, x,y,z = coordinates of particles, and X,Y,Z = components of forces that act on the particles. The average of $(Xx + Yy + Zz)/2$ was called "virial" (L: *vis* = force). The virial can be developed in a series whose coefficients are the virial coefficients.

The function $\Pi/c_2 = f(c_2)$ delivers the first virial coefficient $A_1 = 1/M_2$ as the intersection of the ordinate at $c_2 = 0$ (Fig. 10-18) where M_2 is the number-average molar mass of molecularly non-uniform polymers. Molar masses M_2 calculated from $\Pi/c_2 = RT/M_2$ for finite concentrations c_2 are *apparent* number-average molar masses that contain contributions by the second, third, ... virial coefficients (see Eq.(10-43)).

The initial slope of $\Pi/c_2 = f(c_2)$ delivers the second virial coefficient A_2 albeit only for non-associating, non-ionic systems (see Sections 10.5.2 and 10.5.3). The third virial coefficient can be obtained by curve-fitting or by plotting $[(\Pi/c_2)_i - (\Pi/c_2)_j]/(c_{2i} - c_{2,j})$ as function of $(c_{2i} + c_{2,j})$ using pairs of $\Pi/c_2 = RT/M_2$ (Eq.(10-38)) for concentrations i and j and subsequent subtraction.

A_2 and A_3 (and A_2' and A_3') are complex averages. The average of the second virial coefficient is $(A_2)_{OP} = \Sigma_i\Sigma_j\, w_iw_jA_{ij}$ for colligative methods (osmotic pressure OP, ebullioscopy, cryoscopy) but $(A_2)_{LS} = \Sigma_i\Sigma_j\, w_iM_iw_jM_jA_{ij},/(\Sigma_i\, w_iM_i^2)$ for light scattering.

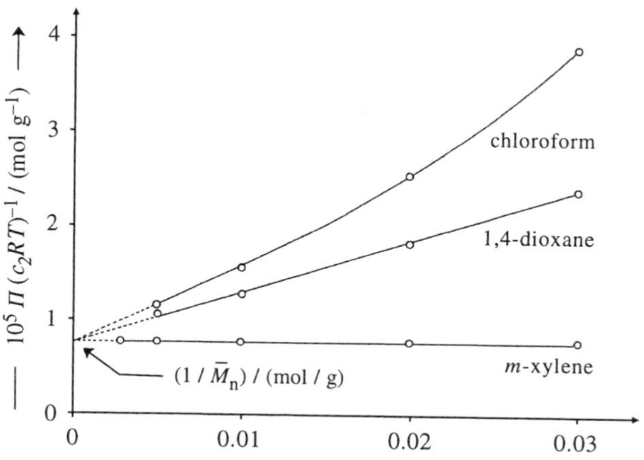

Fig. 10-18 Concentration dependence of reduced osmotic pressure of a poly(methyl methacrylate) at 20°C [21]. At this temperature, *m*-xylene is a theta solvent (defined by $A_2 = 0$). 1,4-dioxane solutions show only a second virial coefficient A_2 whereas chloroform solutions deliver both A_2 and A_3.

10.4.2 Lattice Theory

Second virial coefficients are controlled by interactions between two bodies, third virial coefficients by three bodies, etc. Virial coefficients therefore depend both on the size and shape of the solute and on its interactions with the solvent.

For two equal-sized, solid *hard spheres* with volumes V_{mol} per molecule in thermodynamically ideal solutions, second virial coefficients are obtained from excluded volumes $u = 8\ V_{mol}$ (p. 71) as $A_2 = 4\ N_A V_{mol}/M_2^2$. The second virial coefficient of *rigid rods* that are arranged at orientation angles γ is $A_2 = 2\ L^2 d \sin \gamma$ (see p. 73) whereas the third virial coefficient of these rods is proportional to $L^3 d^2[\ln (L/d)]^3$.

Simple lattice theory calculates second and third virial coefficients of *random coils* as follows. The chemical potential of the solvent per mol of solvent molecules is $\Delta\mu_1 = -\tilde{V}_{1,m}\Pi$ (Eq.(10-37c)) and also $\Delta\mu_1 = RT[\chi\phi_2^2 + \ln(1-\phi_2) + (1-X_2^{-1})\phi_2]$ (Eq.(10-25)). Equating both expressions and developing the logarithmic term in a series, $\ln(1-\phi_2) = -\phi_2 - \phi_2^2 - \phi_2^3 - ...$, delivers $-\tilde{V}_{1,m}\Pi = -RT[X_2^{-1}\phi_2 + \{(1/2) - \chi\}\phi_2^2 + (1/3)\phi_2^3 + ...]$. Further introduction of $X_2 = M_2/M_u$ and $\phi_2 = V_2/(V_1 + V_2) \approx V_2/V_1 = v_2 c_2$ results in

$$\frac{\Pi}{c_2} = RT\left(\frac{1}{M_2}\cdot\frac{M_u v_2}{\tilde{V}_{1,m}} + \frac{[(1/2)-\chi]v_2^2}{\tilde{V}_{1,m}}c_2 + \frac{v_2^3}{3\,\tilde{V}_{1,m}}c_2^2 + \ldots\right) \quad (10\text{-}44)$$

Because of $v_2 = V_2/m_2 = V_u/m_u = v_u$ and $M_u = m_u/n_u$, the product $M_u v_2$ equals the molar volume of polymer segments u, $V_{u,m} = V_u/n_u$. At small concentrations, partial molar volumes become identical with molar volumes, $\tilde{V}_{1,m} \approx V_{1,m}$. A lattice site can be occupied by either a solvent molecule or a polymer segment, which leads to $V_{u,m} = V_{1,m}$ and therefore also to $M_u v_2/\tilde{V}_{1,m} = 1$.

The theory delivers for the first virial coefficient the same expression as phenomenological thermodynamics, $A_1 = 1/M_2$, but the second virial coefficient is given by $A_2 = [(1/2) - \chi]v_2^2/\tilde{V}_{1,m}$ and the third virial coefficient by $A_3 = v_2^3/(3\ \tilde{V}_{1,m})$.

The second virial coefficient becomes zero at a Flory-Huggins interaction parameter of $\chi = 1/2$ and the polymer is in a **theta state** in a **theta solvent** at a **theta temperature** Θ, also called the **Flory temperature**. This theta temperature of solutions corresponds to the Boyle temperature of gases. Note that $A_2 = 0$ does not imply $A_3 = 0$. A zero initial slope of $\Pi/c_2 = f(c_2)$ does not necessarily imply a negative A_2 (Section 10.5).

At the theta temperature, infinitely thin chains adopt their unperturbed dimensions in a locally and globally homogeneous continuum of solvent molecules. However, these conditions do not always apply. In such cases, theta dimensions deviate from unperturbed dimensions because theta dimensions reflect global interactions whereas unperturbed dimensions arise from the absence of local long-range interactions.

The ability of a solvent to cause a chain molecule to adopt its unperturbed dimensions is therefore controlled by the constitution and chemical configuration of polymer molecules as well as the temperature because these factors affect long-range intramolecular interactions between polymer segments as well as interactions between polymer segments and solvent molecules. Mixed solvents often do not represent homogeneous continua because of preferential solvations (p. 303 ff.) whereas self-association of linear macromolecules is caused either endgroup interactions or by long-range intermolecular interactions of segments of different molecules (Section 10.5).

Table 10-8 Theta temperatures Θ (at $A_2 = 0$ or from phase equilibria) and temperatures T_u at which unperturbed dimensions are obtained according to the molar mass dependence of radii of gyration s, diffusion coefficients D, and intrinsic viscosities $[\eta]$ [14]. x_s = mole fraction of syndiotactic diads.

Polymer	Solvent	Θ/°C	T_u/°C from $\langle s^2 \rangle$	T_u/°C from D	T_u/°C from $[\eta]$
Poly(styrene), linear	cyclohexane, d_{12}	40			
Poly(styrene), linear, d_8	cyclohexane, d_{12}	36			
Poly(styrene), linear	cyclohexane	34.5	34.5	34.5	34.5
Poly(styrene), linear, d_8	cyclohexane	30			
Poly(styrene), cyclic	cyclohexane	28.5			40.0
Poly(octadecyl methacrylate)	butyl acetate	10.5	≈ 40	25	13
Poly(α-methyl styrene), $x_s = 0.95$	cyclohexane	32.3			
Poly(α-methyl styrene), $x_s = 0.83$	cyclohexane	34.3			
Poly(α-methyl styrene), $x_s = 0.65$	cyclohexane	36.9			

At least 1200 theta systems are known but there is very little systematic work with respect to the effects of chemical structure of polymers and solvents. For the "standard system" linear poly(styrene)-cyclohexane, theta temperatures and temperatures T_u for unperturbed dimensions agree within ± 0.5°C (Table 10-8). Theta temperatures vary if the polymer and/or the solvent is deuterated. There is also a small, systematic variation of theta temperatures with tacticity, which may explain the scatter of some literature data.

A special problem arises for polymer molecules with long side chains that are constitutionally very different because global macroconformations of coils may not be self-similar with local ones (Table 10-8). Differences between Θ and T_u can also be expected for random coil-forming macromolecules with local helical sections (see pp. 67, 304).

Little systematic work has been done to elucidate the role of the solvent. Theta temperatures of poly(ethylene) increase with increasing chain length of alkanes as solvent and so do those of poly(dimethylsiloxane) (Table 10-9). Theta temperatures of poly(ethylene) become smaller for higher alcohols while those of poly(cyclohexyl methacrylate) pass through a minimum. The dependences are controlled by upper and lower critical solution temperatures and by solubility parameters (Fig. 10-19),

Table 10-9 Theta temperatures (in °C) of poly(ethylene) (PE), poly(styrene) (PS), poly(cyclohexyl methacrylate) (PCMA), and poly(dimethylsiloxane) (PDMS) as a function of the number N of carbon atoms of solvent molecules.
A = alkanes $H(CH_2)_iH$ ($N = i$) ; AL = alcohols $H(CH_2)_iOH$ ($N = i$)
CA = cycloalkanes c-$(CH_2)_i$ ($N = i$) ; AA = alkyl acetates $HCH_2C(O)O(CH_2)_iH$ ($N = i + 3$)

N	PE A	PE AL	PS CA	PS CA	PS AA	PS AA	PCMA AL	PDMS A
5	85		20	154	43	114	23	
6	133		34	213	–44	139		–173
7	174		17		–80	178	9	–173
8	210		12					–143
9		180	16				18	–113
10			16				20	
11		153					23	

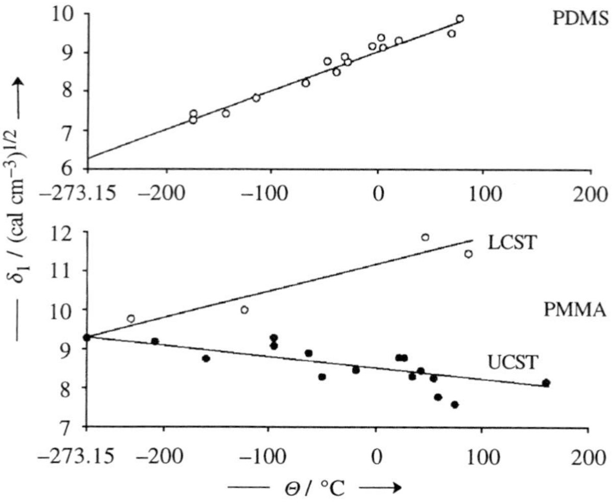

Fig. 10-19 Solubility parameters of solvents, δ_1, as a function of theta temperatures Θ of poly(dimethylsiloxane) (PDMS) and atactic poly(methyl methacrylate) (PMMA), the latter for upper (●) and lower (O) critical mixing temperatures. Extrapolation of solubility parameters of solvents to Θ = –273.15 °C delivers the solubility parameter of the polymer.

Theta temperatures are also affected by polymer architecture. Cyclic poly(styrene) has a lower theta temperatures than its linear analog (Table 10-8). Theta temperatures (UCST) of star molecules are the lower, the shorter the arms and the higher the number of arms per molecule (Fig. 10-20), which is caused by an increasing segment density near the core of the molecule.

It is unclear how and to what extent theta temperatures are affected by tacticities. The theta temperature of poly(α-methyl styrene)s in cyclohexane decreased somewhat with increasing mole fraction of syndiotactic diads but no variation was found for isotactic and atactic poly(1-pentene)s in a series of solvents.

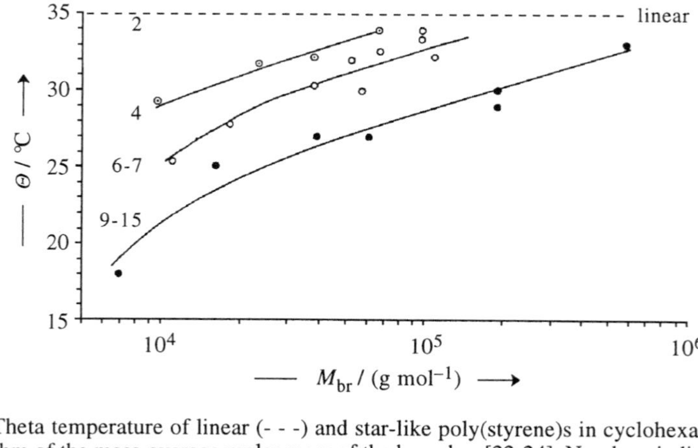

Fig. 10-20 Theta temperature of linear (- - -) and star-like poly(styrene)s in cyclohexane as function of the logarithm of the mass-average molar mass of the branches [22-24]. Numbers indicate number of arms per molecule. Lines are empirical.

10.4.3 Effect of Excluded Volume

Thermodynamically good solvents lead to positive second virial coefficients. The expansion of coils in such solvents (Section 4.5) is controlled by the space requirement of segments, the interactions between segments and segments, and the interactions between segments and solvent molecules. The thermodynamic probability Ω of the possible arrangements of segments in a lattice thus includes the enthalpic contribution and one needs only to calculate the entropic part of the Gibbs energy of mixing.

The first molecule to be placed can use the whole volume V of the lattice. However, because of the excluded volume u, only a volume $(V - u)$ is available for the second molecule, a volume $(V - 2\,u)$ for the third one, and a volume $(V - iu)$ for the ith one. The product of all these arrangements is proportional the the probability Ω:

(10-45) $$\Delta G_{\text{mix}} = -T\Delta S_{\text{mix}} = -k_{\text{B}}T \ln \Omega = -k_{\text{B}}T \ln \left[const \prod_{i=0}^{N_2-1}(V - iu)\right]$$

Solving the logarithm leads to a sum instead of a product:

(10-46) $$\Delta G_{\text{mix}} = -k_{\text{B}}T\left[N_2 \ln V + \sum_{i=0}^{N_2-1} \ln(1 - (iu/V))\right] + const'$$

Since $y \equiv iu/V << 1$ for dilute solutions, the logarithm is developed in a series $(1-y) = -y - ...$ and $\ln(1-(iu/V))$ becomes $-iu/V$. Since $u/V = constant$, the summation is only over all i, delivering approximately $N_2^2/2$.

It follows that $\Delta G_{\text{mix}} = -k_{\text{B}}T[N_2 \ln V - (N_2^2/2)(u/V)] + const'$. Differentiation with respect to volume yields $\partial\Delta G_{\text{mix}}/\partial V = -k_{\text{B}}TN_2/V - k_{\text{B}}TN_2^2u/(2\,V^2)$. Insertion of $N_2/V = c_2N_{\text{A}}/M_2$ and $R = k_{\text{B}}N_{\text{A}}$ leads to $\partial\Delta G_{\text{mix}}/\partial V = -RTc_2/M_2 - RTc_2^2N_{\text{A}}u/(2\,M_2^2)$.

Because of $\tilde{V}_{1,\text{m}} = \partial V/\partial n_1$, one also has $\partial\Delta G_{\text{mix}}/\partial V = (1/\tilde{V}_{1,\text{m}})(\partial\Delta G_{\text{mix}}/\partial n_1) = -\Pi$ and therefore also $\Pi/c_2 = RT\{(1/M_2) + [(N_{\text{A}}u)/(2\,M_2^2)]c_2\}$. Comparison of this equation with Eq.(10-43) shows that the second virial coefficient A_2 is proportional to $u = 32\,\pi\,R_{\text{th}}^3/3$ (p. 71) and therefore directly proportional to the thermodynamic radius R_{th}:

(10-47) $$A_2 = (N_{\text{A}}u)/(2\,M_2^2) = 16\,\pi\,N_{\text{A}}R_{\text{th}}/(3\,M_2^2)$$

10.4.4 Effect of Molar Mass

Partial molar volumes $\tilde{V}_{1,\text{m}}$ of solvents, specific volumes v_2, and Flory-Huggins interaction parameters χ (Eq.(10-16)) of high-molar mass polymers ($M >$ ca. 10^4 g/mol) are independent of molar masses. Hence, according to Eqs.(10-43) and (10-44), second and third virial coefficients should not depend on molar mass either.

However, molar mass dependencies of A_2, A_3, and the so-called reduced third virial coefficient, $g = A_3/(A_2^2M_2)$, are found experimentally (Fig. 10-21). All of these dependences can be described by power laws, $g = (K_3/K_2^2)M_2^{\text{z}-(1+2\text{d})}$, $A_2 = K_2M_2^{\text{z}}$, and $A_3 = K_3M_2^{\text{d}}$, where K_2, K_3, z, and d are constants for a polymer-solvent-temperature system.

The exponent z can be expressed by molecular parameters. Eq.(10-43) is rewritten as

(10-48) $$\Pi/c_2 = (RT/M_2)[1 + A_2M_2c_2 + A_3M_2c_2^2 + ...]$$

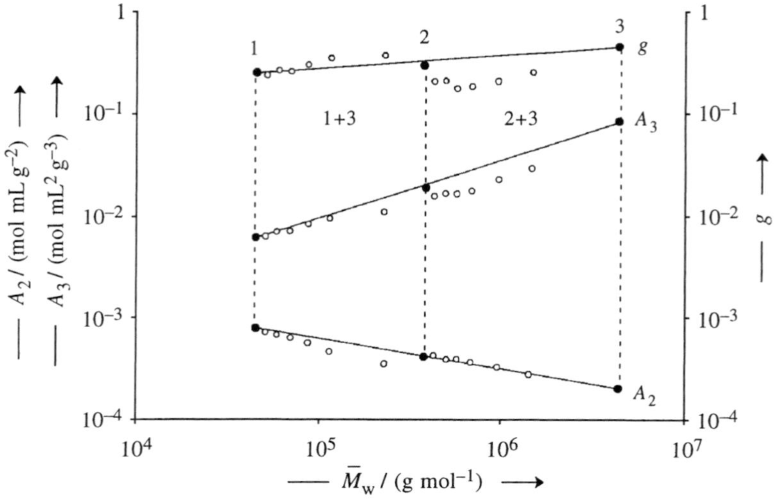

Fig. 10-21 Second virial coefficients A_2, third virial coefficients A_3, and reduced third virial coefficients g as functions of mass-average molar masses $\overline{M}_w$ of three narrow distribution poly(styrene)s (●) 1 ($\overline{M}_w$/(g mol^{-1}) = 44 700) , 2 (381 000), and 3 (4 340 000) and their mixtures 1 + 3 and 2 + 3 in various proportions [25]. Measurements in benzene at 25°C. Proportionalities are $A_2 \sim \overline{M}_w^{-0.28}$, $A_3 \sim \overline{M}_w^{0.58}$, and $g \sim \overline{M}_w^{0.14}$.

Simple lattice theory does not consider overlapping of polymer coils, an effect that sets in at relatively low concentrations c_2* (p. 183 ff.) where it affects second virial coefficients. A_2 is thus expressed by a_2/c_2* where the constant a_2 is an effective virial coefficient, $a_2 = (M_2c_2^*)A_2$, as shown by the comparison of Eqs.(10-49) and (10-48):

(10-49) $\Pi/c_2 = (RT/M_2)[1 + (a_2/c_2^*)c_2 + ...]$

The critical overlap concentration is given by $c_s^* = K_s^*M^{1-3\nu}$ (Eq.(6-4)). The second virial coefficient A_2 should thus be proportional to a power z = (3ν – 2) of the molar mass. The exponent z = –0.28 in Fig. 10-21 delivers a ν = 0.573 which agrees relatively well with the theoretical value of ν = 0.588 (p. 112) and the Monte Carlo value of ν = 0.584 (z = –0,248).

A_2 values of mixtures 1+3 and A_3 values of mixtures 2+3 deviate from the values for "nearly uniform" polymers since $A_2 = K_2M_2^z$ and the expression for A_3 was not corrected for molar mass distribution. Such corrections amount to 4.7-6.7 % for polymers with Schulz-Flory distributions and $1.1 \leq \overline{M}_w/\overline{M}_n \leq 2.0$. Depending on the mixing ratio, values of A_2 may decrease with increasing $\overline{M}_w/\overline{M}_n$, pass through a maximum, etc.

10.4.5 Effect of Temperature

According to lattice theory, second virial coefficients A_2 and interaction parameters χ are connected by $A_2 = [(1/2) - \chi]v_2^2/\tilde{V}_{1,m}$, Eq. (10-44), where (1/2) is the entropic term (see Eq.(10-25)) and χ the enthalpic one. However, interaction parameters χ do contain entropic contributions (p. 310) so that χ should be written as a sum of enthalpic (H) and entropic (S) contributions, $\chi = \chi_H + \chi_S$. The two entropic contributions can be united to a new entropic contribution, $\psi = (1/2) - \chi_S$.

If the remaining enthalpic contribution χ_H represents an enthalpy (i.e., an energy), then ψ must also have the physical unit of energy. If it is an energy, then it cannot be an entropy since only ψ/T has the physical unit of entropy. Gibbs energies become zero at $T = \Theta$, so that $\Delta G = \Delta H - \Theta\Delta S = 0$ and therefore also $\chi_H - \Theta(\psi/T) = 0$. It follows that $\chi_H = \Theta(\psi/T)$ and therefore also

(10-50) $$\chi = (1/2) - \psi + (\psi\Theta/T)$$

The temperature dependence of interaction parameters should therefore be caused only by the entropy parameter ψ. At the theta temperature, $\Theta/T = 1$ and χ should become unity (see Fig. 10-6, however).

According to Eq.(10-50), theta temperatures are obtainable from the temperature dependence of interaction parameters χ but this requires many experiments. It is much easier is to determine theta temperatures from cloud curves. For *binary* systems, the critical demixing temperature T_{crit} is identical with the maximum of the cloud curve for systems with UCST (Fig. 10-10) and with the minimum for those with LCST. Using $\chi = \chi_{crit}$, $T = T_{crit}$, $\overline{X}_w = \overline{X}_z$, Eq.(10-33a), and Eq.(10-50) leads to an equation for the dependence of the demixing temperature on the degree of polymerization,

(10-51) $$\frac{1}{T_{crit}} = \frac{1}{\Theta} + \frac{1}{\Theta\psi}\left(\frac{1}{2X_2} + \frac{1}{X_2^{1/2}}\right) \approx \frac{1}{\Theta} + \frac{1}{\Theta\psi}\cdot\frac{1}{X_2^{1/2}}$$

where the equation to the right holds approximately for high degrees of polymerization. The method can also be used for quasibinary systems although the maximum of the cloud curve is not identical with the critical demixing temperature.

From Eq.(10-50) and $A_2 = [(1/2) - \chi]v_2^2/\tilde{V}_{1,m}$ from Eq.(10-44), one obtains for the temperature dependence of the second virial coefficient:

(10-52) $$A_2 = \left(\frac{\psi v_2^2}{\tilde{V}_{1,m}}\right)\left(\frac{T - \Theta}{\Theta}\right)$$

At $T > \Theta$, second virial coefficients should therefore increase slightly concave with temperature. However, each polymer-solvent system has two theta states and therefore also two theta temperatures, which means that $A_2 = f(T)$ must pass through a maximum (Fig. 10-22). The maximum of this curve is at that temperature at which the solution behaves athermally. For high molar masses, the curve is symmetric about the A_2 axis.

10.4.6 Osmotic Pressure of Semi-concentrated Solutions

In thermodynamically good solvents, osmotic pressures increase strongly with concentration (Fig. 10-18). At moderately high polymer concentrations, the first term in brackets of Eq.(10-44) is small compared to the second term while the third term may still be negligible. Eq.(10-44) thus converts to

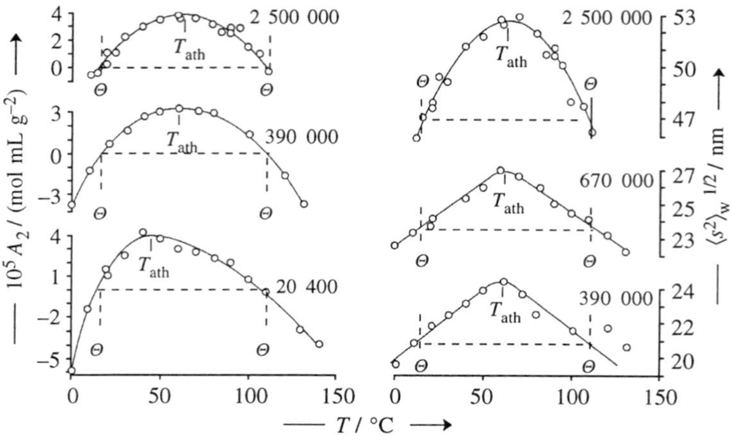

Fig. 10-22 Temperature dependence of second virial coefficients (left) and radii of gyration (at $c_2 \rightarrow 0$) (right) of *t*-butyl acetate solutions of poly(styrene)s with various molar masses [26].

(10-53) $\quad \Pi/(c_2RT) = \{[(1/2) - \chi]v_2^2/\tilde{V}_{1,m}\}c_2$

Hence, according to lattice theory, reduced osmotic pressures Π/c_2 should be (a) independent of the molar mass and (b) directly proportional to the polymer concentration. Experiments confirm the first prediction but not the second one as can be seen from the following considerations.

At polymer concentrations $c_2^* > c_2$, polymer coils are overlapping (Section 6.3). The **critical osmotic overlap concentration**, $c_2^* = c_{osm}^*$, is easily obtained as $c_{osm}^* = 1/(A_2M_2)$ from dimensional analysis. Usually, viscometric overlap concentrations, $c_v^* \approx 1/[\eta]_\Theta$, are used instead (Eq.(6-5)).

In thermodynamically good solvents, overlapping compresses polymer coils. Excluded volumes, on the other hand, expand coils since no other polymer segment can be placed on a lattice site that is already occupied by a segment. The function $\Pi M_2/(cRT) = f(c/c_{osm}^*)$ can then be obtained as follows.

Reduced osmotic pressures can be approximated at higher concentrations by $\Pi M_2/(c_2RT) = 1 + A_2M_2c_2 \approx A_2M_2c_2$. According to Eq.(10-47), the second virial coefficient is proportional to the excluded volume from which it follows

(10-54) $\quad \Pi M_2/(c_2RT) = (N_Au/2)(c_2/M_2)$

Just below the critical overlap concentration, coils do not overlap and can be treated as equivalent coils with volumes of $V = u/8$ (Section 4.2.2). The radius of the equivalent sphere is assumed to be the radius of gyration of the coil, $s \equiv \langle s^2 \rangle^{1/2}$. The excluded volume is therefore $u = 8\ V = (32\ \pi/3)\ s^3$.

The radius of gyration is proportional to the ν-th power of the molar mass, $s = K_vM_2^\nu$ (Eq.(4-69)). Insertion of the expressions for s and u into Eq.(10-54) delivers for the reduced osmotic pressure $\Pi M_2/(c_2RT) = (16\ \pi/3)\ N_AK_s^3M_2^{3\nu-1}c_2$. Mean-field theory (p. 111) predicts the exponent ν to be 3/5 for good solvents. The critical osmotic overlap concentration then becomes

$$(10\text{-}55) \qquad c_{\text{osm}}^{*} = \left(\frac{3\,M_2 \Pi}{16\,\pi c_2 RTN_{\text{A}} K_{\text{s}}^{3}} \right) \cdot \frac{1}{M_2^{3\nu-1}} = K^{*} M_2^{-4/5}$$

The dimensionless parameter $(M_2\Pi)/(c_2RT)$ can be written in terms of the extent of overlap, c_2/c_2^*, while recognizing that this parameter must have the same numerical value just above and just below the critical concentration. According to Eq.(10-55), c_2/c_2^* must therefore be scaled by $1/(3\nu - 1)$:

$$(10\text{-}56) \qquad (M_2\Pi)/(c_2RT) = \text{f}(c_2/c_{\text{osm}}^*) = const\,(c_2/c_{\text{osm}}^*)^{1/(3\nu-1)} = const\,(c_2/c_{\text{osm}}^*)^{\text{m}}$$

In the unperturbed state, $\nu = 1/2$, and the scaling exponent adopts a value of m = 2 which is indeed found experimentally (Fig. 10-23). For good solvents, lattice theory predicts m = 1.25 ($\nu = 3/5$) and renormalization theory m = 1.309 ($\nu = 0.588$) whereas the experiment delivered m = 1.325. Scaling says nothing about the value of the *const*.

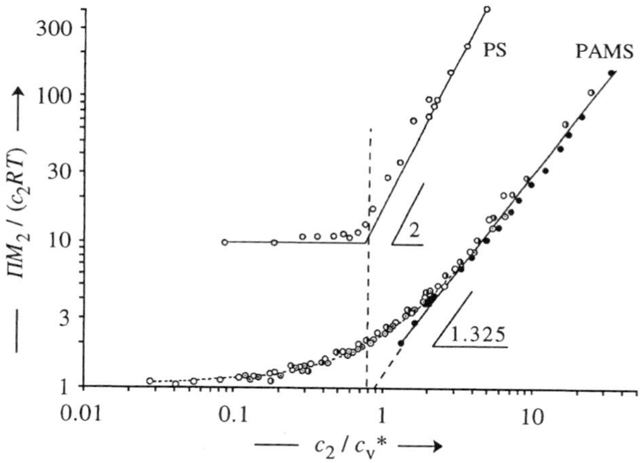

Fig. 10-23 Dependence of reduced osmotic pressures, $\Pi M_2/(v_2RT)$, on reduced concentrations c_2/c_v^* (from viscosity data) for poly(styrene)s (PS) in the theta solvent cyclohexane at 34°C [27] (data shifted upwards by a factor of 10) and for poly(α-methyl styrene)s (PAMS) in the good solvent toluene at 25°C [28]. Osmotic overlap concentrations are $c_v^* = 0.8\,c_{\text{osm}}^*$.

10.5 Association and Self-association

10.5.1 Fundamentals

Some macromolecules assemble in solution to larger entities that can be thought of as "physical molecules." Their formation will be called **multimerization** in order to distinguish it from the formation of macromolecules themselves by **polymerization**, a term that is also used in the biosciences for multimerization. The reversible multimerization to soluble physical molecules will be named **association** and the irreversible one, **aggregation**. The literature often uses "association" and "aggregation" as synonyms.

The association of several like molecules is also known as **self-association** and that of constitutionally or configurationally different ones as **complex formation**. **Stereocomplexes** are associates or aggregates of sterically different but constitutionally identical polymers. In many cases, formation of stereocomplexes is incomplete and results in a solvent-swollen physical network. These networks will be called **gels** like their chemically crosslinked and solvent-swollen chemical counterparts (Section 10.7).

A chemical polymer molecule will be called a **unimer (molecule)** if it can associate with other chemical polymer molecules to a **multimer**.

Self-associations may be molecule or segment related. In **molecule-related associations**, numbers of associating groups per molecule are independent of the degree of polymerization of the macromolecule. An example is the association of hydroxyl endgroups of poly(ethylene glycol)s, $HO(CH_2CH_2O)_{X-1}CH_2CH_2OH$, in apolar solvents. Each molecule has just two endgroups regardless of the degree of polymerization, X. For these associations, equilibrium constants are based on amounts of substances (moles).

Segment-related associations are controlled by the number of associating segments per molecule. An example is the number of longer syndiotactic sequences in an atactic poly(propylene), which increases with the molecular weight at constant average constitution and chemical configuration. Equilibrium constants are here related to mass concentrations and not to molar concentrations.

Both types of association can be **open**, i.e., without upper limit for the number of unimers that are assembled in a multimer (Section 10.5.2) or **closed** if the multimerization is of the type $N\,M_I \rightleftarrows M_N$ (Section 10.5.3).

10.5.2 Open Self-association

Open self-associations are associations of unimer molecules with molar masses M_I to physical dimers, trimers, ..., N-mers with molar masses M_{II}, M_{III}, ... M_N, ... Unimers are in consecutive equilibria with their multimers: $2\,M_I \rightleftarrows M_{II}$, $M_{II} + M_I \rightleftarrows M_{III}$, $M_{III} + M_I \rightleftarrows M_{IV} \rightleftarrows M_{II} + M_{II}$, etc. With increasing polymer concentration c, proportions of higher multimers increase and inverse normalized apparent number-average molar masses of multimers, $\overline{M}_{n,I}/\overline{M}_{n,app}$, decrease if virial coefficients are absent or relatively small. The contributions of second virial coefficients to inverse apparent number-average molar masses become greater, the higher the molar mass of the unimer.

Examples are poly(ethylene glycol)s in benzene (Fig. 10-24) where the association of the endgroups via hydrogen bonds is affected only by the nearest chain units:

$$2 \sim\!(CH_2{-}CH_2{-}O)_n{-}CH_2{-}CH_2{-}O\overset{H}{\underset{H}{\diamond}}O{-}CH_2{-}CH_2{-}(O{-}CH_2{-}CH_2)_m\!\sim$$

The observed concentration dependence of $\overline{M}_{n,I}/\overline{M}_{n,app}$ cannot be interpreted as a consequence of a decrease of equilibrium constants with increasing molar mass as can be seen from the following derivation.

The total amount concentration of all species is $[M] = [M_I] + [M_{II}] + [M_{III}] + ...$ In equilibrium, amount concentrations $[M_i]$ of species i are controlled by molecule-related equilibrium constants (index n) of the open association (index o), for example, for the trimerization, ${}^nK_{III,o} = [M_{III}]/([M_{II}][M_I]) = [M_{III}]/({}^nK_{II,o}[M_I]^3)$.

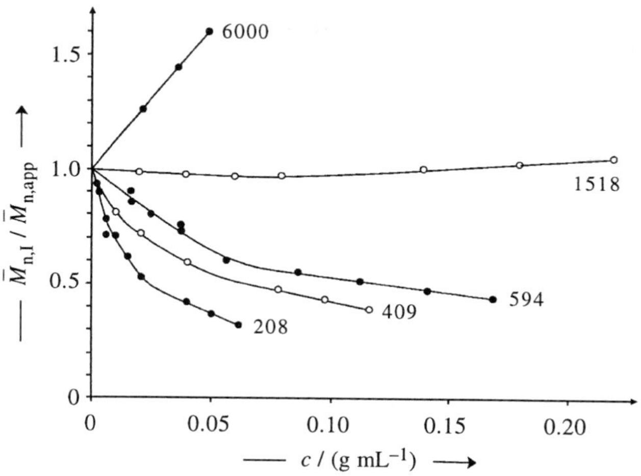

Fig. 10-24 Concentration dependence of inverse normalized apparent number-average degrees of polymerization of poly(ethylene glycol)s $HO(CH_2CH_2O)_nH$ in benzene at 25°C [29]. Numbers indicate number-average molar masses of polymers. Initial slopes of curves for low molar masses do *not* indicate second virial coefficients but are caused by self-association (see below).

Equilibrium constants are often independent of the degree of association, N: ${}^nK_o = {}^nK_{II,o} = {}^nK_{III,o} = ... = {}^nK_{N,o} = ...$ For these consecutive equilibria, the total amount concentration [M] of unimers + multimers is thus

$$(10\text{-}57) \qquad [M] = [M_I]\{1 + ({}^nK_o[M_I]) + ({}^nK_o[M_I])^2 + ...\} = [M_I](1 - {}^nK_o[M_I])^{-1}$$

since ${}^nK_o[M_I] = [M_{II}]/[M_I] < 1$ (see Volume I, p. 218).

If the system is in a quasi-theta state ($A_2 = 0$), then the apparent number average molar mass, as calculated from the van't Hoff equation, is $RTc_2/\Pi = (\overline{M}_n)_{app,\Theta}$, at a total solute concentration c_2. Since the molar concentration of all species is $[M] \equiv c_2/(\overline{M}_n)_{app,\Theta}$. Eq.(10-57) can be rewritten as

$$(10\text{-}58) \qquad [M_I]^{-1} = {}^nK_o + (\overline{M}_n)_{app,\Theta}c^{-1}$$

The weight concentration $c_2 = c$ of all species is obtained from $[M_i] = c_i/\overline{M}_{n,i}$ as

$$(10\text{-}59) \qquad c = c_I + c_{II} + c_{III} + ... = [M_I]\,\overline{M}_{n,I}(1 + 2\,{}^nK_o[M_I] + 3\,({}^nK_o[M_I])^2 + ...)$$

$$(10\text{-}60) \qquad c = [M_I]\,\overline{M}_{n,I}/(1 - {}^nK_o[M_I])^2$$

Because this series is of the type $1 + 2x + 3\,x^2 + ...$, it can be converted to $1/(1 - x)^2$ since $x = {}^nK_o[M_I] < 1$ (see also Volume I, pp. 218).

From Eqs.(10-58) and (10-60), one thus obtains for the concentration dependence of apparent *number-average* molar masses in the absence of virial coefficients

$$(10\text{-}61) \qquad \overline{M}_{n,app,\Theta}/\overline{M}_{n,I} = 1 + {}^nK_o[c/\overline{M}_{n,app,\Theta}]$$

Fig. 10-25 shows a corresponding plot for the data of Fig. 10-24.

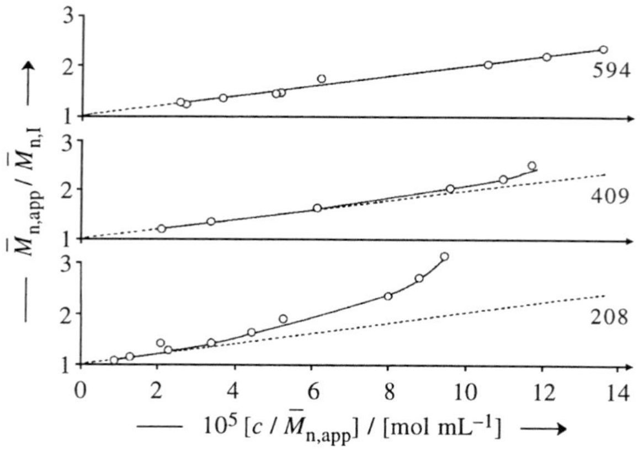

Fig. 10-25 Plot of the data of Fig. 10-24 according to Eq.(10-61), assuming that equilibrium constants are independent of molar masses. Experimentally, $\overline{M}_{n,app}$ is obtained instead of $\overline{M}_{n,app,\Theta}$. Broken lines correspond to $(A_2)_{av} = 0$.

As required, a plot of $\overline{M}_{n,app,\Theta}/\overline{M}_{n,I} = f(c/\overline{M}_{n,app,\Theta})$ delivers unity as the intercept (Fig. 10-24). Initial slopes are parallel and thus independent of unimer molar masses, i.e., equilibrium constants (${}^{n}K_o \approx 2$ mL/mol) are independent of unimer molar masses. The negative initial slopes of the plots in Fig. 10-24 are therefore *not* indicative of negative second virial coefficients. Instead, such negative values of A_2 rather show up as upswings in plots of $\overline{M}_{n,app,\Theta}/\overline{M}_{n,I} = f(c/\overline{M}_{n,app,\Theta})$ (Fig. 10-25) whereas they are camouflaged by the stronger effect of association in Fig. 10-24. The second virial coefficient of the polymer with $\overline{M}_n = 594$ g/mol is zero; polymers with higher molar masses have positive second virial coefficients.

For the apparent *mass-average* molar mass, a similar equation applies:

$$(10\text{-}62) \qquad \overline{M}_{w,app,\Theta} = \overline{M}_w + 2\,{}^{n}K_o \overline{M}_{n,I}[c/\overline{M}_{n,app,\Theta}]$$

However, mass-average molar masses can only be obtained from apparent mass-average molar masses if apparent number-averages are also known.

The situation is different for *segment-related* open associations where equilibrium constants depend on *mass* concentrations. For apparent *mass-average* molar masses, one obtains for both molecularly uniform and molecularly non-uniform unimers

$$(10\text{-}63) \qquad \overline{M}_{w,app,\Theta} = \overline{M}_{w,I} + {}^{w}K_{N,o}\overline{M}_{w,I}c \quad ; \quad {}^{w}K_{N,o} \equiv c_N/(c_{N-1}c_I)$$

For apparent *number-average molar masses*, a simple equation is obtained only for molecularly uniform unimers but not for molecularly non-uniform ones:

$$(10\text{-}64) \qquad \overline{M}_{n,app,\Theta} = \overline{M}_{n,I}\,{}^{w}K_o c[\ln(1 + {}^{w}K_o c)]^{-1}$$

Equilibrium constants ${}^{w}K_o$ can be converted to ${}^{n}K_o$ via ${}^{w}K_o = N\,{}^{n}K_o/[(N-1)\overline{M}_{n,I}]$ but this is not very useful since association numbers N are concentration dependent in open associations, in contrast to segment-related closed associations (Section 10.5.3).

All equations shown above apply to open associations in theta systems; they consider only the relative change of concentration of independent entities with respect to mass concentrations as caused by open self-association. All other interactions are characterized by a positive or negative second virial coefficient A_2. For number- and mass-average methods, such as osmotic pressure (OP) and light scattering (LS), one can write

(10-65) $(\overline{M}_{n,app})^{-1} = (\overline{M}_{n,app,\Theta})^{-1} + (A_{2,OP})c$

(10-66) $(\overline{M}_{w,app})^{-1} = (\overline{M}_{w,app,\Theta})^{-1} + 2\,(A_{2,LS})c$

10.5.3 Closed Self-association

Closed associations involve only two types of species, unimers with the molar mass M_I (and the corresponding molar mass averages) and N-mers with the molar mass M_N (and the corresponding molar mass averages). Like open associations, they may be molecule- or segment-based. Depending on the type of closed association, and sometimes also on the type of distribution function, various relationships between the various averages exist according to statistical calculations.

For *molecule-based* closed associations, one obtains

(10-67) $\overline{M}_{n,N} = N\,\overline{M}_{n,I}$ all types of distributions

(10-68) $\overline{M}_{w,N} = \overline{M}_{w,I} + (N-1)\,\overline{M}_{n,I}$ all types of distributions

and for *segment-based* closed associations:

(10-69) $\overline{M}_{n,N} = \overline{M}_{n,I} + (N-1)\,\overline{M}_{w,I}$ only Schulz-Flory type associations

(10-70) $\overline{M}_{w,N} = N\,\overline{M}_{w,I}$ all types of distributions

Fig. 10-26 shows $\overline{M}_{w,N}/\overline{M}_{n,N} = f(\overline{M}_{w,I}/\overline{M}_{n,I})$ for various values of N.

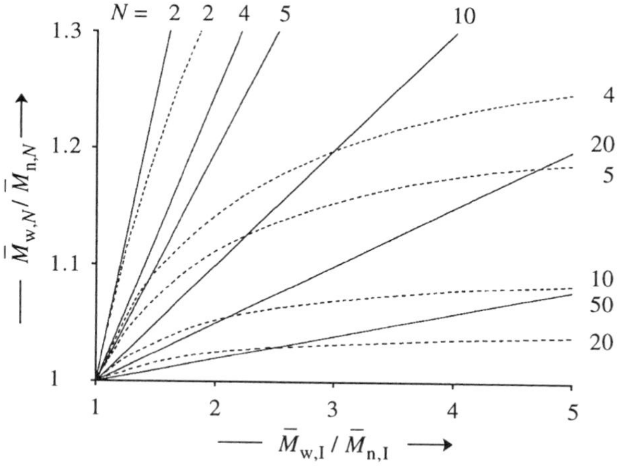

Fig. 10-26 Association index $\overline{M}_{w,N}/\overline{M}_{n,N}$ of N-mers as a function of the non-uniformity (polymolecularity) index of unimers at various degrees of association, N, for molecule-based self-associations of unimers with any type of molar mass distribution function (——) and for segment-based self-associations (- - -) with Schulz-Flory types of molar mass distributions [30].

For both molecule- and segment-based closed associations, association-based non-uniformity indices, $U_{w,n,ass} \equiv \overline{M}_{w,N}/\overline{M}_{n,N}$, are always smaller than the molecule-based non-uniformity indices, $U_{w,n,mol} \equiv \overline{M}_{w,I}/\overline{M}_{n,I}$, since the variation of molar masses now relates to the entities themselves and no longer to the system. For a selfassociating polymer with $N = 10$ and $U_{w,n,mol} = 2$ of the unimer, values of $U_{w,n,ass}$ drop to 1.1 for molecule-based and to 1.05 for segment-based closed associations (Fig. 10-26). Entity distributions of closed associations are thus very narrow even for unimers with broad molar mass distributions.

Concentration dependences of apparent molar masses of closed self-associations differ characteristically from those of open self-associations. Since equilibrium constants of molecule-based associations are defined as ${}^{n}K_{c} \equiv [M_N]/[M_I]^N$ and those of segment-based associations as ${}^{w}K_{c} \equiv [c_N]/[c_I]^N$, closed associations have all the characteristics of a cooperative process. Below a **critical micelle concentration (CMC)**, concentrations of N-mers are very small so that multimerization *practically* (but not theoretically!) sets in only above CMC (Fig. 10-27). This transition is gradual and therefore difficult to determine; it is also different for number-average and mass-average molar mass methods.

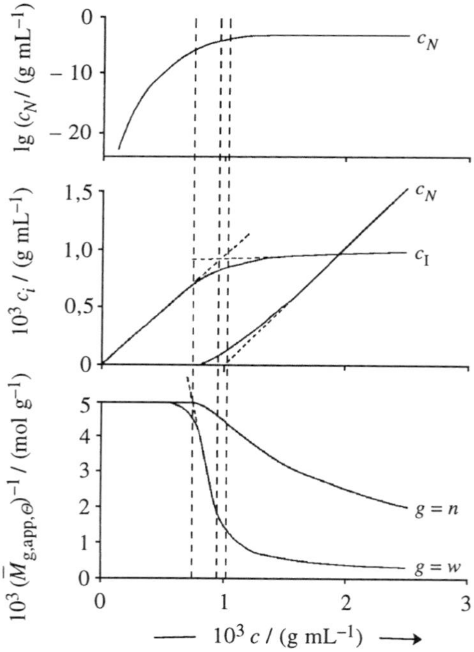

Fig. 10-27 Calculated dependence of (top) logarithms of multimer concentrations, (center) concentrations of multimers and unimers, and (bottom) inverse apparent number- and mass-average molar masses, respectively, on the total polymer concentration for a molecule-based closed association in the theta state [31]. Calculations for $\overline{M}_{w,I} = \overline{M}_{n,I} = 200$ g/mol, $N = 21$ and ${}^{n}K_{c} = 10^{45}$ (L mol^{-1})$^{N-1}$. Different tangents, intersections, etc., result in different CMCs (broken lines).

In the theta state, inverse apparent molar masses do not vary with concentration c at small c. Above the CMC, they decrease sharply and then less so until they asymptotically approach the value of M_N. In contrast to open associations, this behavior of apparent

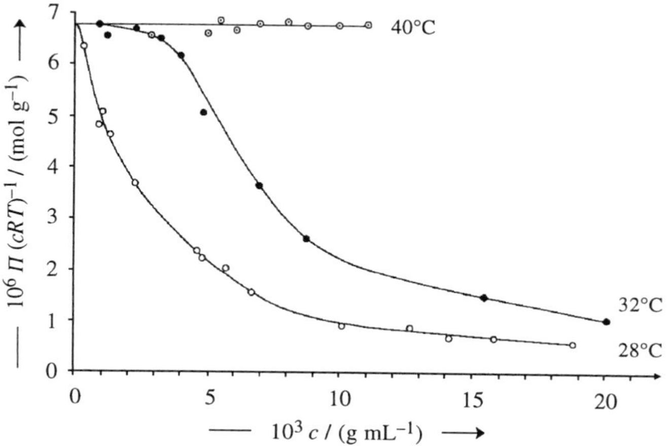

Fig. 10-28 Concentration dependence of reduced osmotic pressures, $\Pi/(cRT) \equiv 1/\overline{M}_{n,app}$, of a poly(styrene)-*block*-poly(2-vinyl pyridine) with $\overline{M}_{n,I}$ = 150 375 g/mol and w_{sty} = 0.32 in toluene [32]. The system behaves as a theta system at 40°C and as a closed self-association at 32°C since poly(2-vinyl pyridine) blocks are insoluble in toluene at the latter temperature. For 28°C, one cannot decide without additional data whether the system behaves as a closed or as an open self-association.

molar masses is not caused by the change of N and c_N with concentration c but just by the change of the ratio of numbers of N-mers and unimers. The ratio $\overline{M}_{w,app} / \overline{M}_{n,app}$ is therefore *not* a measure of the association index of the N-mers, $\overline{M}_{w,N} / \overline{M}_{n,N}$.

Very small critical micelle concentrations can often not be detected by conventional plots of inverse apparent molar masses as a function of total concentration (see data for 28°C in Fig. 10-28). In such cases, plots of, for example, $1/\overline{M}_{w,app} = f(c)$ of closed associations do not differ qualitatively from those of open associations (compare with Fig. 10-24). The two types of self-association can be distinguished, however, if both apparent number- and mass-average molar masses are known for the same total concentration c. For this case, calculations have shown for closed molecule-based associations that $\overline{M}_{w,app,\Theta}$ is a linear function of the inverse apparent number-average molar mass, $\overline{M}_{n,app,\Theta}$:

$$(10\text{-}71) \qquad \overline{M}_{w,app,\Theta} = \overline{M}_{w,I} + N\overline{M}_{n,I} - N(\overline{M}_{n,I})^2[\overline{M}_{n,app,\Theta}]^{-1}$$

Eq.(10-71) has been confirmed experimentally for low-molar mass amphiphilic substances in water. Such closed associations can also be found for block polymers because a good solvent for one type of the blocks is usually a bad solvent for the other type (Fig. 10-28). Little is known about the thermodynamics of self-association of block polymers. For example, the self-association of poly(styrene)-*block*-poly(isoprene) in organic solvents is controlled enthalpically which contrasts with the one entropically controlled one found for low-molar mass amphiphiles. It is unclear whether this can be generalized.

In principle, diblock polymers can form either intramolecular or intermolecular self-associations. In *intramolecular associates* (= **unimolecular micelles**), a shell of the soluble block surrounds the core of the insoluble one (Fig. 10-29, left). In *intermolecular associates*, shells and cores are both comprised of blocks from several molecules. Intermolecular associates may be spherical or wormlike (Fig. 10-29, right).

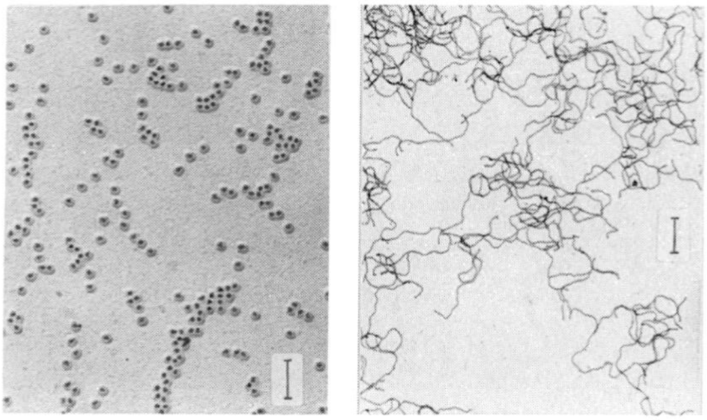

Fig. 10-29 Associates of a poly(styrene)-*block*-poly(isoprene). Left: collapsed intramolecular micelles from dimethylacetamide, scale: 200 nm (black dot: OsO_4 complex of isoprene units) [33a]. Right: wormlike intermolecular micelles; scale: 1000 nm [33b]. With kind permission by Royal Society of Chemistry, London [a] and Elsevier Science, Oxford [b].

In such systems, light scattering sometimes indicates the presence of larger particles that do not add much to osmotic pressures because of their small concentration. Such associations can be described by two consecutive associations, one between unimers and *N*-mers and the other between N-mers and higher molar mass P-mers. There are indications that the "small" *N*-mers are spherical and the larger P-mers rod- or wormlike. Such elongated micelles may associate laterally (segment-related) or via their ends (molecule-related). In the first case, the association would be of the closed type, in the latter case, P would vary with the solute concentration. Such multimers would have broad distributions of particle masses which would make interpretations of indirect methods very difficult; examples are diffusion and viscosity measurements.

Closed associations do not always lead to compact "micelle-type" particles. For example, the self-association of poly(γ-benzyl-L-glutamate)s is of the closed type in several helicogenic solvents. However, the Gibbs energy of self-association from the temperature dependence of equilibrium constants depends inversely on the number-average molar mass which is an indication of an association via endgroups which has been verified by IR and NMR spectroscopy. Radii of gyration, on the other hand, correspond to those of random coils. It follows that this association of helical (rodlike) molecules probably proceeds via endgroups to cyclic associates.

10.5.4 Complexation of Polymers with Small Molecules

Binding of small molecules B to polymer molecules poly(U) with N binding sites per molecule leads to multiple equilibria. Examples are the binding of dyestuffs to synthetic polymer molecules or the binding of α-amino acids to enzyme molecules (Fig. 10-30). The resulting UB complexes can contain 1, 2, 3 ... N B-molecules per polymer molecule; each U-B bond is in an equilibrium with an equilibrium constant $K = [UB]/([U][B])$. The concentration of UB complexes, [UB], is proportional to the probability p_U that one group U is complexed and inversely proportional to $(1 - p_U)$ that it is not complexed:

$$[UB]/[U] = K[B] = p_U/(1 - p_U) \quad (10\text{-}72)$$

In the simplest case, all N potential binding sites are equally active and independent of each other. The number N_U of complexed sites is then $N_U = Np_U$ which, when introduced in Eq.(10-72), leads to three equations that are named after their originators:

(10-73)	$1/N_U = (1/N) + (NK)^{-1}(1/[B])$	Klotz equation
(10-74)	$N_U/[B] = KN - KN_U$	Scatchard equation
(10-75)	$[B]/N_U = (KN)^{-1} + [B]/N$	Langmuir equation

Plotting of variables N_U and [B] according to any of these equations delivers straight lines which give the unknown values of K and N if the binding sites are identical and independent of each other.

The number N of binding sites is relatively easy to obtain if saturation occurs at sufficiently high concentrations of B (Fig. 10-30). However, this is rarely the case because equilibrium constants are usually small. It is then difficult to decide from plots of the hyperbolic Eq.(10-72) or its three linearizations, Eqs.(10-723)-(10-75), whether complexations are really independent of each other with the same equilibrium constant K. This condition can be tested by calculating equilibrium constants K_1, K_2, K_3, ... for the binding of the first, second, third, ... molecule by using nonlinear curve fitting. A non-identity of the various equilibrium constants signals a non-equivalency of binding sites.

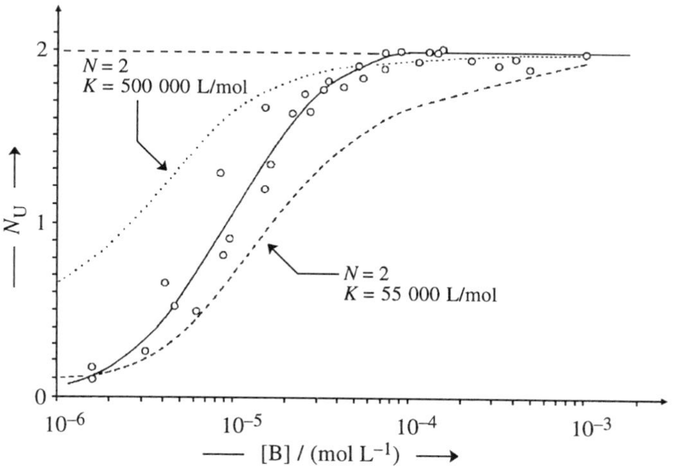

Fig. 10-30 Number N_U of complexed U sites as a function of the logarithm of concentrations [B] of uncomplexed B molecules for the binding of L-leucine to the enzyme α-isopropylmaleate synthetase [34]. Dotted and dashed curves correspond to the Klotz equation.

The general case is a polymer with N independent binding sites with binding capabilities that vary with the extent of binding to the polymer. Such a polymer would have a maximum of $2^{N-1}N$ equilibrium constants (for example, 12 if $N = 3$) that are not simply related to the three stoichiometric equilibrium constants.

Such complexations are difficult to analyze mathematically. The problem can be simplified somewhat if the complexation is assumed to be affected only by the nearest neighbor. The binding of a B molecule to a U site next to an unoccupied U site is then characterized by an equilibrium constant K_o and the binding of a B molecule to a U site next to an already occupied U site by K. The equilibrium constants K and K_o are interconnected by the Gibbs interaction energy E, i.e., $K = K_o \exp(-E/RT)$. With $s \equiv [B]K_o \exp(-E/RT) = [B]K_o\sigma$, the fraction $f_U = N_U/N$ of occupied U sites of infinitely long polymer chains is then obtained by a lengthy calculation as

$$(10\text{-}76) \qquad f_U = \frac{1}{2} + \frac{s-1}{2[(s-1)^2 + 4\sigma s]^{1/2}} \quad ; \qquad \sigma = \exp(-E/RT)$$

Plotting f_U as a function of $\ln([B]K_o) = \ln s - E/RT$ delivers a series of S-shaped curves that vary with E and E/RT, respectively (Fig. 10-31). The binding is cooperative for $E < 0$, uncooperative for $E = 0$, and anti-cooperative (negative-cooperative) for $E > 0$. The isotherm has a double S shape for large values of E. In this case, but not in the case of small values of E, one can infer the presence of two different binding sites from the shape of the function $f_U = f(\ln [B] + \ln K_o)$.

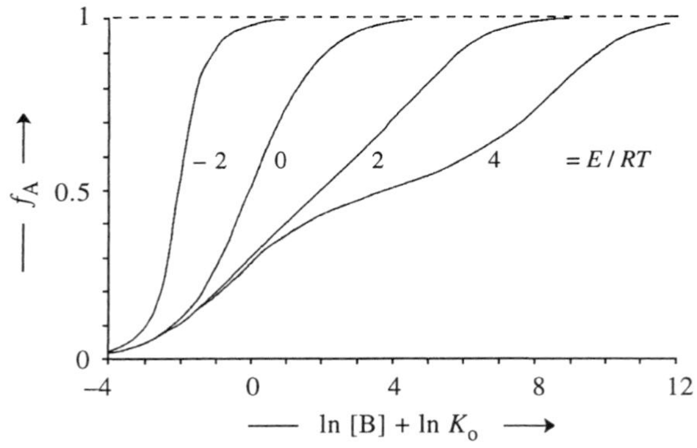

Fig. 10-31 Fraction f_A of bound B groups as a function of $\ln [B] + \ln K_o$.

10.5.5 Polymer-Polymer Complexes

Complexes between constitutionally and/or configurationally different polymers are subdivided according to the type of binding between complexing groups:

- **Symplexes** are complexes between **polycations** and **polyanions** that are held together by electrostatic bonds. 1:1 complexes of these **polyions** are **polysalts**. An example is the complex between sodium carboxymethylcellulose and poly(dimethylallylammonium chloride). Precipitated symplexes are called **coacervates** (L: *co* = together, *acervare* = to heap).

  Polycations have many cationic sites per polymer molecule, polyanions many anionic ones. Water-soluble polycations and polyanions are called **polyelectrolytes**. **Macroions** (**macrocations** and **macroanions**) have only one ionic site per molecule.

- **Hydrogen-bridged complexes** are complexes with hydrogen bonds between two types of polymer molecules. An example is the complex between poly(acrylic acid) and poly(vinyl alcohol).

- **Stereo complexes** are complexes with van der Waals bonds between polymer molecules of different chemical configuration but identical chemical constitution. Example: stereocomplex between isotactic and syndiotactic poly(methyl methacrylate). In solid state, their formation is known as "**racemic crystallization**."
- **Hydrophobe complexes** are caused by the so-called entropy bonds, for example, between bovine serum albumin and poly(4-vinyl pyridine). These "bonds" are not bonds at all but rather ordering phenomena of polar entities such as water molecules near hydrophobic regions.

Polymer-polymer complexes may be completely ordered ("ladder type") or completely disordered ("scrambled-egg type") (Fig. 10-32) or have any type of order between these two borderline types, depending on polymer constitution and concentration, forces between polymers and solvents, and methods of preparation of complexes. Complexation does not need to be 1:1 with respect to complexing groups if some or all of the groups are not ionized, are inaccessible, or are in too short constitutional and/or configurational sequences.

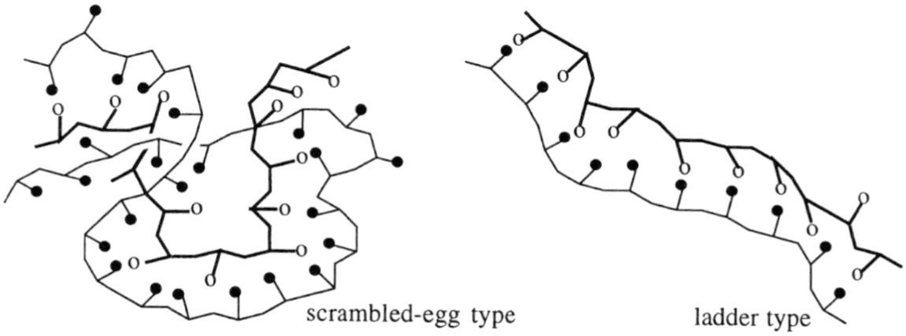

Fig. 10-32 Complexes between two constitutionally and/or configurationally different polymers.

High concentrations of charged polymers usually lead to insoluble *symplexes*; for example, between silicic acids, [$SiO_2 \cdot n\ H_2O$], and poly(2-vinylpyridinium-1-oxide) (I). Such complex formations prevent silicosis which is caused by the formation of fibrous clusters in lungs after inhalation of very fine silicogenic dusts (quartz, cristobalite, etc.).

Mixing of equal-molar solutions of poly(sodium styrene sulfonate) (II) and poly(4-vinyl pyridinium hydrobromide) (III) leads to a stoichiometric complex between side groups, regardless of the ratio II:III and the sequence of mixing (II to III or III to II) (Fig. 10-33) because excess amounts of one compound do not participate in complex formations. Such complexes do not dissolve in water because neutralization of charges increases hydrophobicity.

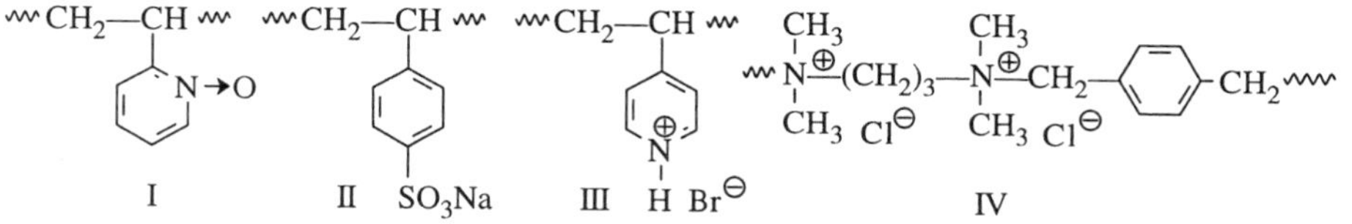

I II III IV

A different picture arises if the charges of polycations are in the main chains and not in sidechains (Fig. 10-33). Addition of II to IV leads here first to a stoichiometric 1:1

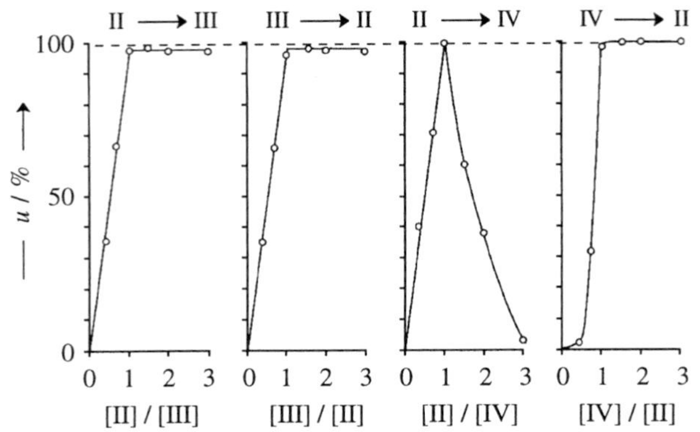

Fig. 10-33 Conversion u of the minority component as a function of the ratio of molar concentrations on coacervation (forming of coacervates) (II → III or III → II) and solubilization (II → IV or IV → II) [34]. Aqueous solutions of poly(sodium styrene sulfonate) (II), poly(4-vinyl pyridinium hydrobromide) (III), or ionenes (IV). Ionenes: see Volume II, p. 450.

complex which then dissolves on further addition of II until everything is dissolved at [II]:[IV] = 3:1. The same phenomenon is observed for the addition of IV to II where a stoichiometric complex is formed only if [IV]:[II] ≥ 1:3. In general, stoichiometries of such complex formations do not refer to molar concentrations of anionic groups $[A^{\ominus}]$ and cationic groups $[C^{\oplus}]$ *per se*, i.e., $[A^{\ominus}] = [C^{\oplus}]$, but need to consider degrees of dissociation α because of the required electroneutrality, i.e., $[A^{\ominus}]\alpha_A = [C^{\oplus}]\alpha_C$.

Dilute solutions may lead to soluble complexes with ladder structures. However, the stoichiometry obtainable with respect to complexing groups depends on the degree of polymerization. In general, logarithms of equilibrium constants increase in an S-shaped manner with logarithms of the degrees of polymerization.

Hydrogen-bridged complexes are formed, for example, between COOH groups of poly(methacrylic acid) (PMAA), $\text{-[}CH_2\text{–}C(CH_3)COOH\text{]}_n\text{-}$, and ether groups of poly(oxyethylene)s (PEOX), $\text{-[}O\text{–}CH_2\text{–}CH_2\text{]}_n\text{-}$. The equilibrium constant is small and of the same magnitude as for hydrogen-bond formations of low-molar mass molecules, i.e., $K \approx 3$ L/mol, if low-molar mass PEOX is present in dilute aqueous solutions of PMAA with the same molar concentration of groups. Equilibrium constants increase strongly with degrees of polymerization of $X > 25$ until they approach a constant value of $K \approx 6000$ L/mol at $X > 100$. The strong increase at $X > 25$ is obviously caused by cooperative effects whereas the limiting value at $X > 100$ points to an inaccessibility of binding groups (maximum degree of binding is $f_A \approx 0.85$).

Stereocomplexes between configurationally different macromolecules of the same constitution have smaller binding energies per group than hydrogen-bridged complexes. Their formation requires the presence of long stereosequences because only then can the entropy loss be compensated by an enthalpy gain from the cooperative effect.

For example, specific enthalpies of stereocomplex formation from isotactic and syndiotactic poly(methyl methacrylate)s, Δh, depend linearly on mass fractions w_{st} of syndiotactic diads; they reach a maximum at $w_{sf} = w_2 = 0.58$ (Fig. 10-34). However, if Δh refers to the mass fraction of syndiotactic heptads (thus assuming that stereocomplex formation from diads to hexads is absent), then the diagram $\Delta h = f(w_{st})$ becomes symmetric and the maximum is now at $w_{st} = w_7 = 0.5$. Hence, stereocomplexes can form

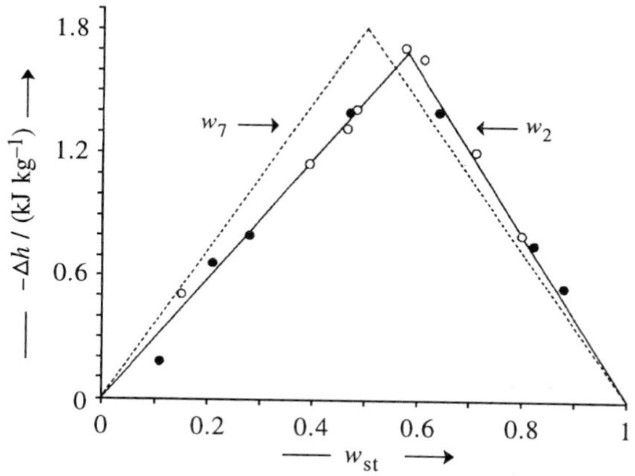

Fig. 10-34 Negative specific enthalpy of stereocomplex formation from 0.01 g/mL solutions of st- and it-poly(methyl methacrylate) in *o*-xylene (O) or *N,N*-dimethylformamide (●) at 25°C as a function of the mass fraction w_{st} of syndiotactic diads (——) or heptads (- - -), respectively [35].

only from syndiotactic heptads, i.e., eight monomeric units. If stereosequences are too short and polymer concentrations sufficiently high, stereocomplexes will not precipitate but the system will rather form a gel instead (Section 10.7).

10.6 Polyelectrolytes

10.6.1 Structure of Polyelectrolyte Solutions

Polyelectrolytes are water-soluble polymers with ionic groups on main-chain atoms or in side-chains (Volume I, p. 47). Synthetic homopolymeric polyelectrolytes and naturally occurring mucopolysaccharides (Volume II, p. 408) have high charge densities. Polyelectrolytes in the wider sense are also polymers with relatively low charge densities such as spherical proteins which can be modeled as electrically charged latex particles.

The structure of aqueous solutions and dispersions of such polyelectrolytes differs with respect to the charge density of their molecules or particles. X-ray measurements of dilute solutions of *weakly charged* particles showed that experimental average distances $d = 2\ d_{exp}$ between two particles practically equal the distances $d = 2\ d_{theor}$ that can be calculated from particle concentrations if particles are homogeneously distributed in solution. An example is a dilute dispersion (ϕ = 0.004) of spherical poly(styrene) latex particles (diameter: 341 nm) with less than 1000 charges per particle. The experimental distance was found to be $2\ d_{exp}$ = 1800 nm whereas the theoretical distance was calculated as $2\ d_{theor}$ = 1940 nm. This observation supports the assumption that *weakly* charged particles repel each other in dilute solution if they carry the same positive or negative electrical charge.

The situation is different for *strongly charged* particles where experimentally measured distances between two particles are far lower than the theoretical ones that were calculated from concentrations. For example, a poly(styrene) latex (d = 419 nm) with ca. 200 000 charges per particle showed a $2\ d_{exp}$ = 1260 nm at a volume fraction of 0.0075 whereas the theoretical value was 2 d_{theor} = 1940 nm. Similarly, a value of 2 d_{exp} = 15.7 nm *versus* 2 d_{theor} = 23.1 nm was found for a poly(sodium styrene sulfonate) latex ($\overline{M}_w$ = 74 000 g/mol) in a 0.01 g/mL aqueous dispersion.

The highly charged particles are therefore much nearer to each other than a purely statistical distribution would predict. Polyelectrolytes must therefore associate. However, charges with equal sign should repel each other, which they would do, of course, if they were all alone and not associated with a polymer chain. In reality, however, there are also equivalent amounts of counterions of opposite sign present which must be near the polyion charges according to the principle of microscopic electroneutrality.

This presence of low-molar mass counterions between two polymer ions gives rise to Coulomb attraction forces as one can see from the potential energies of free ions (⊕ and ⊖), a free cation (⊕) and an ion pair (⊕ ⊖), and a triple ion (⊕⊖⊕) from two cations and one anion. The potential difference is zero for free ions ⊕ and ⊖ but equal to $-e^2/r$ for the ion pair where e = elementary charge and r = distance between the anion and cation. For the triple ion (⊕⊖⊕), there is a repulsion of $+e^2/r$ between the cations and an attraction of $-2e^2/r$ between a cation and the anion, i.e., a potential energy of $-(3/2)e^2/r$. The triple ion is therefore more stable than the ion pair and also more stable than the two single ions. The same reasoning applies to higher ion aggregates: this is why ion crystals such as NaCl are stable.

Highly charged polyelectrolytes in dilute solution must therefore exist as highly ordered ion aggregates, similar to ion lattices (Fig. 10-35). The same phenomenon is observed for high concentrations of weakly charged particles, for example, by small angle X-ray scattering of highly concentrated solutions of bovine serum albumin. Even 20 % solutions of tobacco mosaic virus showed perfect crystalline reflections. For the same reason, dispersions of equal-sized latex particles show iridescent colors which is caused by Bragg diffraction at the highly ordered particle aggregates.

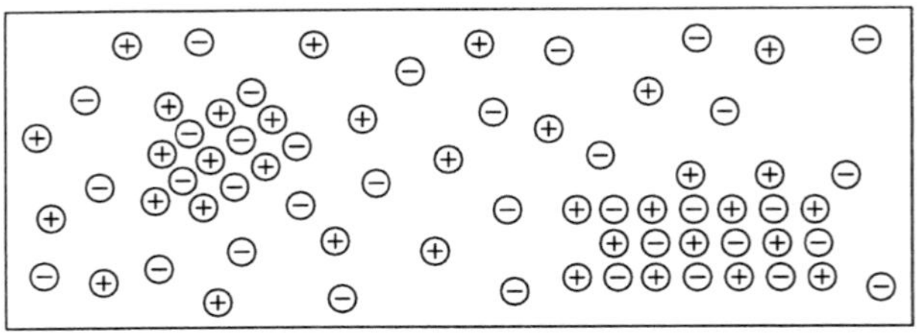

Fig. 10-35 Highly ordered ion aggregates in dilute polyelectrolyte solutions (schematic).

10.6.2 Thermodynamic Activity

Thermodynamic activities of polyelectrolytes can be determined directly by measuring electromotive forces or osmotic pressures and indirectly from thermodynamic activities with the help of the Gibbs-Duhem equation, Eq.(10-5). Activities a_1 of solvents are obtained from isopiestic measurements of vapor pressure (G: *isos* = equal; *pizein* = to compress, to press tight). **Activity coefficients** $\gamma_1 = a_1/x_1$ are calculated from activities and molar concentrations x_1 of solvents; they are also called **osmotic coefficients** because they are the ratios of osmotic pressures in non-ideal and ideal solutions.

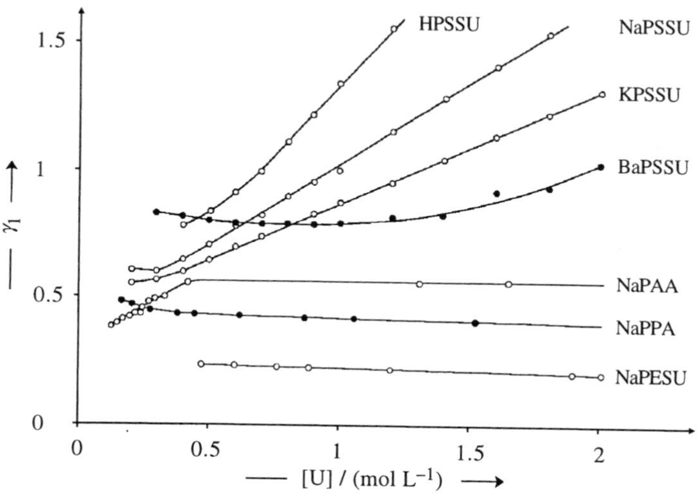

Fig. 10-36 Activity coefficient γ_1 of water as a function of the molar concentration [U] of monomeric units of poly(styrene sulfonic acid) (HPSSU, X = 2500) and its sodium, potassium, and barium salts; and of poly(sodium phosphate) (NaPPA, $\overline{X}_n$ = 600), poly(sodium acrylic acid) (NaPAA, $\overline{X}_w$ = 1640), and poly(sodium ethylene sulfonate) (NaPESU, $\overline{X}_v$ = 770) at 25°C [36, 37].

Osmotic coefficients are strongly affected by the constitution of macroions. Sodium salts of poly(acrylic acid) (NaPAA), poly(ethylene sulfonic acid) (NaPESU), and poly-(styrene sulfonic acid) (NaPSSU) have approximately the same charge density but very different concentration dependences of activity coefficients (Fig. 10-36). For NaPPA, NaPAA, and NaPESU, activity coefficients γ_1 of water are practically constant in the concentration range $0.5 \leq [U]/(\text{mol L}^{-1}) \leq 2$ whereas NaPSSU and KPSSU show a practically linear increase with [U] at [U] ≥ 0.4 mol/L. At higher unit concentrations, activity coefficients of water in solutions of NaPSSU, KPSSU, and HPSSU become even greater than unity. The aromatic compound NaPSSU must therefore affect the water structure quite differently than its aliphatic counterparts NaPAA and NaPESU.

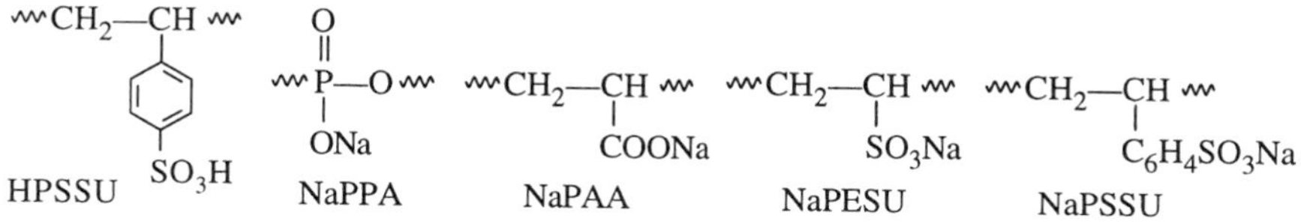

Activity coefficients of polyelectrolytes, γ_2, calculated from activity coefficients of solvents, γ_1, (Eq.(10-5)), agree very well with γ_2 by direct measurements of electromotive forces (Fig. 10-37). For NaPAA and NaPPA, logarithms of γ_2 depend practically linearly on the cubic root of molar concentrations of monomeric units:

(10-77) $\lg \gamma_2 = A - B[\text{U}]^{1/3}$; A, B = polymer-specific constants

The logarithm of γ_2 indicates the deviation from the ideal state. For electrolytes, this deviation is mainly caused by the electrostatic potential which is inversely proportional to the distance r between ions. Hence, $\lg \gamma_2 \sim 1/r \sim [\text{U}]^{-1/3}$ if the ions reside on a three--dimensional lattice.

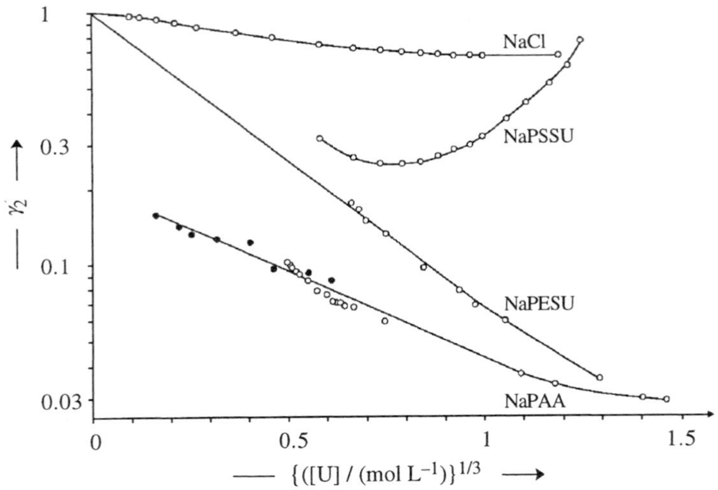

Fig. 10-37 Logarithm of average activity coefficients of sodium chloride (NaCl) and the polyelectrolytes poly(sodium styrene sulfonate) (NaPSSU), poly(sodium ethylene sulfonate) (NaPESU), and poly(sodium phosphate) (NaPPA) as a function of the cubic root of molar concentrations of monomeric units and equivalent molar concentrations of NaCl [36-38].

● Measurement of electromotive forces, O calculated from isopiestic measurments using the Gibbs-Duhem equation. For clarity, γ_2 values of NaPAA have been multiplied by 0.02.

10.6.3 Osmotic Pressure

The dictate of electroneutrality prevents the permeation of low-molecular weight counterions of *salt-free solutions* of polyelectrolytes through osmotic membranes to the pure solvent. Both polyelectrolytes and their counterions thus add to the osmotic pressure but this is practically only the osmotic pressure of the counterions because their molar concentration is far greater than that of the polyelectrolyte. Calculated molar masses are therefore independent of the true molar masses of polyelectrolytes. However, they are far higher than the molar masses of the counterions, which indicates that polyelectrolyte molecules are not totally dissociated: a fraction of the counterions is "bound" by the polyelectrolyte and therefore osmotically inefficient.

In osmotic measurements of polyelectrolytes in *salt-containing solutions*, dissociation of polyelectrolyte molecules decreases because of the presence of low-molar mass gegenions (counterions). Since these counterions may also permeate through the membrane, electroneutrality is established by ion permeation from the solvent chamber to the solute chamber. Since polyelectrolyte molecules do not permeate but add to the chemical potential in the solute chamber, salt concentrations will be different on both sides of the membrane and there will be a **Donnan equilibrium** on both sides of the membrane. This effect will be smaller, the higher the concentration of the added salt.

Donnan equilibria are not specific for charged molecules but are always established in systems with permeating and non-permeating solutes, i.e., also for non-electrolytes (see p. 327).

All kinetically independent entities (molecules, associates, ions) add to the osmotic pressure depending on their activities. The osmotic pressure Π of ideal solutions of neutral polymers is given by the molar concentration [P] of the polymer, $\Pi = RT[\mathrm{P}] = RTc_2/M_2$ (Eq.(10-38)). In solutions of polyelectrolytes with degrees of polymerization, X, gegenions also add to the osmotic pressure according to their degree of dissociation α and their activity coefficient γ. The total concentration of species in very dilute solution is therefore the sum of the molar concentration $[\mathrm{P}] = [\mathrm{M_u}]/X$ of polyelectrolyte molecules and an active concentration $[\mathrm{P}]X\alpha\gamma = [\mathrm{M_u}]\alpha\gamma$ of counterions. Eq.(10-38) becomes

(10-78) $$\Pi / (cRT) = (1/M) + \alpha\gamma X / M \approx \alpha\gamma X / M = \alpha\gamma / M_\mathrm{u}$$

Measurements of osmotic pressures of salt-free solutions of polyelectrolytes do not deliver the inverse molar mass of the polyelectrolyte at infinite dilutions but a quantity that contains the degree of association α of the polyelectrolyte, the activity γ of the gegenions, and the molar mass of the monomeric units.

The total osmotic pressure of polyelectrolytes P in solutions of salt 1 is the sum of osmotic pressures that each component would exert if it were present alone, $\Pi_\mathrm{total} = \Pi_\mathrm{P} + \Pi_\mathrm{salt}$. In an ideal system ($\gamma_1 = 1$) in Donnan equilibrium, the solution side will contain polyions with a concentration [P], dissociated counterions with a concentration $[\mathrm{M_u}]\alpha$, and added salt with a concentration 2 $[\mathrm{S}]_\mathrm{in}$ (assuming monovalent cations and anions). The solvent side of the membrane will contain a salt solution with the concentration $[\mathrm{S}]_\mathrm{ex}$. The **Donnan pressure** (**oncotic pressure**; G: *onkos* = mass, tumor) is therefore

(10-79) $$\Pi_\mathrm{Donnan} = RT\{[\mathrm{M_u}]/X + [\mathrm{M_u}]\alpha + 2\,[\mathrm{S}]_\mathrm{in} - 2\,[\mathrm{S}]_\mathrm{ex}\}$$

In the absence of polyions, ions of added salt would distribute evenly on both sides of the membrane. However, the polyelectrolytes that are present also contribute to counterions, so that a lower concentration of the latter is necessary on the solution side in order to compensate the concentration of salt counterions on the solvent side. This gegenion concentration will be a fraction $(1 - \beta)$ smaller than the concentration $[\mathrm{M_u}]\alpha$ of gegenions in the absence of the polyelectrolyte, hence,

(10-80) $$[\mathrm{S}]_\mathrm{ex} = [\mathrm{S}]_\mathrm{in} + [\mathrm{M_u}]\alpha(1 - \beta)$$

From Eqs.(10-79) and (10-80) and $[\mathrm{M_u}] = cX/M$, one obtains

(10-81) $$(\Pi_\mathrm{Donnan} M)/(cRT) = 1 + X\alpha(2\,\beta - 1)$$

For salt-free polyelectrolyte solutions, the fraction $(1 - \beta)$ will be zero and Eq.(10-81) reduces to Eq.(10-78) with $\gamma = 1$. For high salt concentrations, $\beta \rightarrow 1/2$ and the Donnan pressure becomes identical with the osmotic pressure of neutral polymers. Osmotic measurements on polyelectrolyte solutions thus deliver true number-average molar masses of polyelectrolytes at high concentrations of added salt but not at low ones.

Donnan effects also play a large role in light-scattering experiments. Repulsion between polyions leads to very large second virial coefficients in salt-free solutions and, at higher polyelectrolyte concentrations, also to a decrease in scattering intensity. However, correct mass-average molar masses are obtained from measurements at sufficiently high, constant salt concentrations after extrapolation to zero polyelectrolyte concentration if the effect of Donnan pressure on refractive increments is duly considered.

Light scattering is caused by local fluctuations of refractive indices (Section 5.2.1) and in solutions mainly by concentration fluctuations. In polyelectrolyte solutions, the fluctuating unit is not the polyion itself but an electroneutral region composed of the polyion and the surrounding ion atmosphere. Light scattering is therefore produced by both polyions and their counterions but reduced by repelled salt ions. The latter effect must be considered for refractive index increments which are different in Donnan equilibria and in the original polyelectrolyte solution; it is the former that must be used.

10.7 Gels

10.7.1 Survey

A **gel** is a chemically or physically crosslinked polymer that is highly swollen by a liquid. In gels, molecules of the liquid are so strongly bound to polymer molecules that the liquid does not sweat out. Gels deform relatively little by hydrostatic pressure but very easily by shearing. An example is Jello®, basically a protein (plus sugar and dyestuffs) that is highly swollen by water.

The word "gel" is derived from "gelatin(e)", the degradation product of the protein collagen. Gelatine got its name because very dilute gelatine solutions (ca. 0.6 %) congeal on cooling, i.e., gel (L: *gelare* = to freeze, congeal).

Hydrogels contain water as the swelling agent and **lyogels** contain other liquids. **Microgels** are swollen, crosslinked polymer particles with colloidal dimensions. **Xerogel** is the name of materials that result from gels that have lost their liquids by evaporation or vacuum removal and are below their glass temperatures (G: *xeros* = dry). Xerogels are not gels in the traditional meaning of the word but are rather hard foams; their polymer structure is filled with air. The best known xerogel is silica gel (Volume II, p. 561).

All true gels consist of a polymer and a liquid, both of which are more or less continuously distributed throughout the gel. Polymers are usually present in small concentrations of 0.1-5 %, which makes it easy for other molecules to enter the gel.

For this reason, gel states are preferred states in nature; examples are collagen and cartilage. Food technology exploits gel properties by chemically crosslinking soy and wheat proteins via disulfide bridges. Physically crosslinked gels result from gelation of pectins (marmalades, conserves, jams) (Volume II, p. 416); beating of egg white, the protein albumen of eggs (meringues); and thickening of starch, a mixture of the polysaccharides amylose and amylopectin (puddings) (Volume II, p. 376).

Gels from synthetic polymers include ion exchange resins from crosslinked poly(styrene) (Volume I, p. 580), soft contact lenses from poly(2-hydroxyethyl methacrylate) containing a small percentage of ethylene glycol methacrylate units as crosslinking sites (Volume II, p. 296 ff.), and implants of crosslinked silicones (Volume II, p. 572).

Gels are formed only if three structural demands are fulfilled:

- regular chain sections are long enough for stable associations or complexations;
- some segments are irregular in order to prevent precipitation or crystallization; and
- irregular segments are flexible in order to allow swelling.

10.7.2 Swelling of Chemically Crosslinked Neutral Gels

Swelling of a chemically crosslinked xerogel proceeds similarly to dissolution of an uncrosslinked polymer, i.e., the diffusion is controlled by the mutual diffusion coefficient of swelling agent and network segments. This diffusion coefficient can be obtained from the penetration rate of the swelling agent or from measurements of dynamic light scattering (p. 368).

The swelling of a gel stops at a certain degree of swelling because of elastic retraction forces. In swelling equilibrium, the Gibbs energy of mixing of polymer + swelling agent equals the Gibbs energy of elasticity, $\Delta G_{\text{mix}} = -\Delta G_{\text{el}}$, i.e., a thermodynamically good solvent will swell a weakly crosslinked polymer more than a bad solvent.

The chemical potential $\Delta\mu_{1,\text{gel}}$ of the solvent in the gel is given by the first derivatives of the Gibbs energy of mixing and Gibbs energy of elasticity with respect to the amount $n_1 = N_1/N_A$ of solvent,

(10-82) $$\Delta\mu_{1,\text{gel}} = N_A(\partial\Delta G_{\text{mix}}/\partial N_1)_{N_2} - N_A(\partial\Delta G_{\text{el}}/\partial\lambda)_{N_2}(\partial\lambda/\partial N_1)_{N_2} = 0 \;\; ; \; p, T = const.$$

where N_1 = number of solvent molecules and $\lambda = L/L_0$ = the fractional linear expansion.

The first term of the right side of this equation is the chemical potential of the solvent in the gel, $\Delta\mu_1$ (Eq.(10-25)). At infinitely high degrees of polymerization (which applies to gels), this chemical potential is given by

(10-83) $$\Delta\mu_1 = RT[\chi\phi_2{}^2 + \ln(1 - \phi_2) + \phi_2]$$

The first factor of the second term of Eq.(10-82) contains the Gibbs energy of elasticity, $\Delta G_{\text{el}} = -T\Delta S$. For the isotropic swelling of a network, $\lambda = \lambda_x = \lambda_y = \lambda_z$, with tetrafunctional crosslinks, Eq.(16-46) predicts for $f = 4$

(10-84) $$\Delta G_{\text{el}} = (1/2)\, k_B T N_c \cdot (3\,\lambda^2 - 3 - \ln \lambda^3)$$

and after differentiation,

(10-85) $$N_A(\partial\Delta G_{\text{el}}/\partial\lambda)_{N_2} = (1/2)\, RTN_c(6\,\lambda - 3\,\lambda^{-1})$$

The calculation of the second factor of the second term on the right side of Eq.(10-82) assumes additivity of the volumes V_2 of polymers and V_1 of solvent. The volume fraction ϕ_2 of the polymer is therefore $\phi_2 \equiv V_2/(V_1 + V_2) = V_2/(N_1 V_{1,m} N_A{}^{-1} + V_2)$ where $V_{1,m}$ = molar volume of the solvent. The polymer volume V_2 is the true volume V_o of the non-swollen xerogel without empty volume (see p. 177). Differentiation of λ with respect to N_1 and setting $\lambda^3 = V/V_o = 1/\phi_2$ delivers

(10-86) $$(\partial\lambda/\partial N_1)_{N_2} = V_{1,m}/(3\,\lambda^2 V_0 N_A)$$

Combination of Eqs.(10-82)-(10-86) delivers the effective number N_c of network chains

(10-87) $$N_c = \frac{V_o N_A}{V_{1,m}}\left(\frac{\chi\phi_2^2 + \ln(1-\phi_2) + \phi_2}{\phi_2^{1/3} - (1/2)\,\phi_2}\right)$$

Eq.(10-87) describes reasonably well the behavior of weakly crosslinked polymers in weakly swelling (i.e., bad) solvents. However, many crosslinked polymers still contain uncrosslinked polymer chains, for example, from crosslinking polycondensations with incomplete monomer conversions or from post-crosslinking of linear or branched polymers. In xerogels, these uncrosslinked polymer chains adopt their unperturbed coil dimensions. Since the interaction parameter of uncrosslinked polymers in crosslinked polymers is zero, a quasiternary system crosslinked polymer + uncrosslinked polymer + solvent will swell less than a binary system crosslinked polymer + solvent, i.e., after the uncrosslinked polymer has been extracted.

For strongly crosslinked polymers, contributions by the interaction of polymer and solvent and by the dilution by the solvent are negligible compared to the elasticity term. Hence, the swelling of highly crosslinked polymers is not affected by the thermodynamic goodness of the solvent. Such polymers swell only a little because of short network chains between crosslinking sites.

Addition of a precipitant to a chemically weakly crosslinked polymer will cause network chains to contract and the gel to shrink. The transition swollen gel → shrunken gel is gradual and not abrupt. An example is polymers from acrylamide, $CH_2{=}CH{-}CONH_2$, that have been crosslinked by bisacrylamide, $(CH_2{=}CH{-}CONH)_2CH_2$, as comonomer. These gels swell in water to a degree of swelling of $Q = V_{gel}/V_{gel,0} = 1.25$ (Fig. 10-38, lower curve). Increasing volume fractions of the precipitant acetone lowers the degree of swelling until it reaches only $Q = 0.074$ at $\phi_{acetone} = 0.7$. The gel has therefore shrunk by a factor of 1.25/0.074 = 17.

10.7.3 Swelling of Electrically Charged, Chemically Crosslinked Gels

Electrically charged poly(acrylamide) gels behave differently. Such gels are obtained from the corresponding neutral gels by hydrolysis which converts a part of the acrylamide groups, $-CH_2-CH(CONH_2)-$, to acrylic acid groups, $-CH_2-CH(COOH)-$. The degree of hydrolysis and thus the number of electrical charges per chain increases with time; after 60 days, ca. 25 % of groups are hydrolyzed.

The degree of swelling of partially hydrolyzed gels, $V_{gel}/V_{gel,0}$, increases with time because of increased formation of negatively charged acrylic acid groups that repel each other, leading to a stiffening of the chain segments and a swelling of the gels. In contrast to electrically neutral gels (hydrolysis time: zero days), an incremental addition of acetone suffices here to collapse the gel at a certain critical volume fraction $\phi_{acetone}$ (Fig. 10-38). This critical concentration depends on the concentration of the hydrolyzing agent and can be quite dramatic: the 60-day specimen shrinks at $\phi_{acetone} = 0.58$ to $V_{gel}/V_{gel,0} = 0.083$ from $V_{gel}/V_{gel,0} = 34.6$, i.e., by a factor of 416.

The time to shrink after an infinitesimal addition of precipitant is controlled by the diffusion of the precipitant into the interior of the specimen and thus by the thickness of the specimen. Only very thin specimens thus collapse "immediately."

The sharp transition swollen gel → shrunken gel is reminiscent of a true phase transition. It is not only caused by organic solvents, but also by salts, changes of temperature or pH, or application of electric fields of a few volts per centimeter. Bivalent metal ions are much more efficient than monovalent ones. A collapse of the poly(acrylamide) of

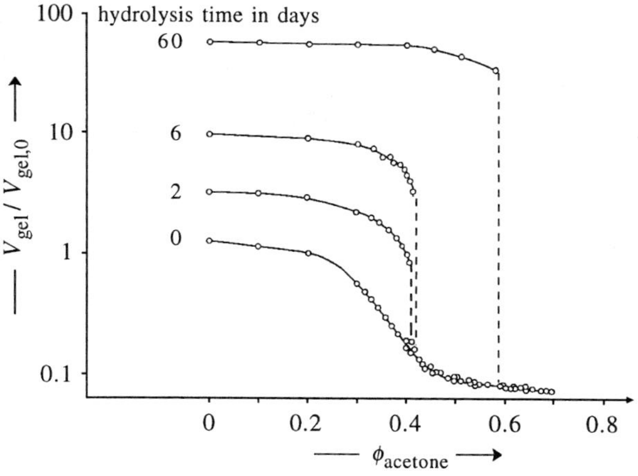

Fig. 10-38 Degree of swelling, $V_{gel}/V_{gel,0}$, of water-swollen poly(acrylamide) networks at 25°C as a function of the volume fraction of acetone in water, $\phi_{acetone}$, before (t = 0) and after hydrolysis of the networks by 0.4 vol% aqueous solutions of TEMED = *N,N,N,N*-tetramethylethylenediamine (pH = 12) [39]. Concentration of TEMED: ϕ_{TEMED} = 0.004 (2 and 6 days) and 0.04 (60 days).TEMED is washed out after the indicated hydrolysis times and before swelling measurements.

Fig. 10-38 by addition of aqueous salt solutions required a critical NaCl concentration of $3 \cdot 10^{-2}$ mol/L but a critical $MgCl_2$ concentration of only $3.6 \cdot 10^{-6}$ mol/L (not shown in Fig. 10-38). Such a sudden collapse of a charged polymer gel is of interest for industrial applications, for example, a collapse by an electric field.

10.7.4 Physically Crosslinked Gels

Formation of complexes between sufficiently long stereosequences or long hydrophobic sequences lead to self-association in dilute solutions (Section 10.5) and to gel formation in concentrated ones. Gels (but not necessarily self-associations) are also formed by crystallizable polymer segments if these segments are long, but not too long. Instead of precipitating or crystallizing, the system remains in the gel state, sometimes for thermodynamic reasons and sometimes for kinetic ones.

Like gels from chemically crosslinked polymers, physically crosslinked gels consist of a strongly swollen three-dimensional network. Depending on the constitution and chemical configuration of polymers, "crosslinks" are bundles of segments that consist either of parallel segments of the ladder type (Fig. 10-32) or of bundles of helical, double-helical, or more complex segments.

Phase diagrams of both non-gel-forming and gel-forming solutions show a binodal that separates one-phase (clear) regions from two-phase (turbid) ones (Fig. 10-39). The counterpart of the critical solution temperature here is the **critical concentration for gel formation,** c_{ccg}. In contrast to non-gel-forming solutions (Fig. 10-15), this quadruple point separates solutions of low concentrations from gels at higher ones and correspondingly one-phase regions at higher temperatures from two-phase regions at lower ones. A temperature increase thus converts physical gels to solutions and vice versa. Such gels are therefore also called **thermoreversible gels**.

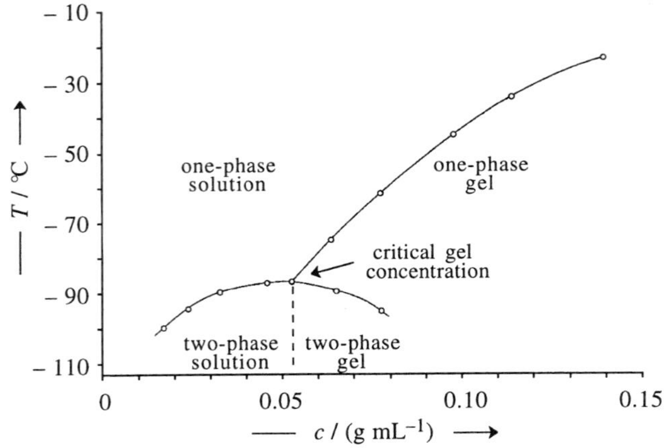

Fig. 10-39 Phase diagram of solutions of an atactic poly(styrene) ($\overline{M}_w$ = 30 000 g/mol) in CS_2 [40].

Critical concentrations for gel formation decrease with the inverse square root of molar mass, $c_{ccg} = KM^{-1/2}$, which indicates that gel formation occurs at a sufficiently great overlap between two polymer coils. A prerequisite for gel formation must be a sufficiently negative enthalpy of gel formation since physical gelation occurs only in certain solvents.

Historical Notes to Chapter 10

Heat of mixing (first use of contact statistics)
A.J.Staverman, Rec.Trav.Chim.Pays-Bas **56** (1937) 885

Statistical thermodynamics of polymer solutions
(These papers did not become well known because they were published in the German-occupied Netherlands during World War II.)
A.J.Staverman, J.H. van Santen, Rec.Trav.Chim.Pays-Bas **60** (1941) 76
A.J.Staverman, Rec.Trav.Chim.Pays-Bas **60** (1941) 640

Lattice theory
P.J.Flory, J.Chem.Phys. **10** (1942) 51
M.L.Huggins, J.Phys.Chem. **46** (1942) 151; Ann.N.Y.Acad.Sci. **41** (1942) 1; J.Am.Chem.Soc. **64** (1942) 1712

Lattice theory of solutions of rods
See Historical Notes to Chapter 8.

Literature to Chapter 10

10.1.a GENERAL SURVEYS

E.A.Guggenheim, Mixtures, Clarendon Press, Oxford 1952

T.L.Hill, Introduction to Statistical Thermodynamics, Addison-Wesley, Reading (MA) 1960

C.Tanford, Hydrophobic Effect, Wiley, New York, 2nd ed. 1973

I.Prigogine, The Molecular Theory of Solutions, Interscience, New York 1975

A.Ben-Naim, Hydrophobic Interactions, Plenum, New York 1980

A.Ya.Kipnis, B.E.Yavelov, J.S.Rowlinson, Van der Waals and Molecular Science, Oxford University Press, New York 1996

T.A.Witten (with P.A.Pincus), Structured Fluids: Polymers, Colloids, Surfactants, Oxford Univ. Press, New York 2004

10.1.b POLYMER SOLUTIONS (general)

P.J.Flory, Principles of Polymer Chemistry, Cornell University Press, Ithaca (NY) 1953

H.Tompa, Polymer Solutions, Butterworth, London 1956

H.Yamakawa, Modern Theory of Polymer Solutions, Harper and Row, New York 1971

M.Kurata, Thermodynamics of Polymer Solutions, Harwood Academic Publ., Chur, Switzerland 1982

R.G.Larson, Constitutive Equations for Polymer Melts and Solutions, Butterworth, Boston 1988

G.Jannink, J. des Cloizeaux, Polymers in Solution, Oxford University Press, Oxford 1990

K.Kamide, Thermodynamics of Polymer Solutions, Elsevier, Amsterdam 1990

H.Fujita, Polymer Solutions, Elsevier, Amsterdam 1990

J. des Cloiseaux, G.Jannink, Polymers in Solution. Their Modelling and Structure, Oxford Univ. Press, New York 1990

M.Van Dijk, A.Wakker, Concepts of Polymer Thermodynamics, Technomic, Lancaster (PA) 1997

W.W.Graessley, Polymeric Liquids and Networks. Structure and Properties, Garland Science, New York (2003)

10.1.c SOLUTION DATA

Ch.Wohlfarth, CRC Handbook of Enthalpy Data of Polymer-Solvent Systems, CRC Press, Boca Raton (FL) 2001

Ch.Wohlfarth, CRC Handbook of Thermodynamic Data of Copolymer Solutions, CRC Press, Boca Raton (FL) 2001

Ch.Wohlfarth, CRC Handbook of Thermodynamic Data of Aqueous Polymer Solutions, CRC Press, Boca Raton (FL) 2004

Ch.Wohlfarth, CRC Handbook of Thermodynamic Data of Polymer Solutions at Elevated Pressures, CRC Press, Boca Raton (FL) 2005

Ch.Wohlfarth, CRC Handbook of Enthalpy Data of Polymer-Solvent Systems, CRC Press, Boca Raton (FL) 2005

10.1.3 SOLUBILITY PARAMETERS

C.M.Hansen, The Three-Dimensional Solubility-Parameter and Solvent Diffusion Coefficient. Their Importance in Surface Coating Formulation, Danish Technical Press, Copenhagen 1967

O.Fuchs, Gnamm-Fuchs: Lösungs- und Weichmachungsmittel, Wissenschaftliche Verlags-gesellschaft,

Stuttgart 1980, Volumes I and II (German language books on solvents and plasticizers for polymers with 7241 references); Volume III (1980) (properties of solvents and plasticizers)

A.F.M.Barton, CRC Handbook of Solubility Parameters and Other Cohesion Parameters, CRC Press, Boca Raton (FL), 2nd ed. 1992

C.M.Hansen, Hansen Solubility Parameters. A User's Handbook, CRC Press, Boca Raton (FL), 2nd ed. 2007

10.1.4 RATE OF DISSOLUTION

B.Narasimhan, N.A.Peppas, The Physics of Polymer Dissolution: Modeling Approaches and Experimental Behavior, Adv.Polym.Sci. **128** (1997) 157

10.2 STATISTICAL THERMODYNAMICS (general theory)

P.J.Flory, Statistical Mechanics of Chain Molecules, Interscience, New York 1969

R.A.Orwell, The Polymer-Solvent Interaction Parameter χ, Rubber Chem.Technol. **50** (1977) 451

R.P.Danner, M.S.High, Handbook of Polymer Solution Thermodynamics, Wiley, New York (1993)
T.Sato, A.Teramoto, Concentrated Solutions of Liquid Crystalline Polymers, Adv.Polym.Sci. **126** (1996) 85

10.2.3 PHASE SEPARATION

L.H.Tung, Ed., Fractionation of Synthetic Polymers, Dekker, New York 1977
O.Olabisi, L.M.Robeson, M.T.Shaw, Polymer-Polymer Miscibility, Academic Press, New York 1979
S.Krause, Polymer-Polymer Compatibility, in D.R.Paul, S.Newman, Eds., Polymer Blends, Academic Press, New York 1978
K.Solc, Ed., Polymer Compatibility and Incompatibility: Principles and Practices, Harwood Academic Publ., Chur, Switzerland 1982
D.J.Walsh, S.Rostami, The Miscibility of High Polymers: The Role of Specific Interactions, Adv.Polym.Sci. **70** (1985) 119
K.Kamide, Thermodynamics of Polymer Solutions. Phase Equilibria and Critical Phenomena, Elsevier, Amsterdam 1990
L.Wild, Temperature Rising Elution Fractionation, Adv.Polym.Sci. **98** (1990) 1
E.Frankuskiewicz, Polymer Fractionation, Springer, Berlin 1994
M.Jiang, M.Li, M.Xiang, H.Zhou, Interpolymer Complexation and Miscibility Enhancement by Hydrogen Bonding, Adv.Polym.Sci. **146** (1999) 121
R.Koningsveld, W.H.Stockmayer, E.Nies, Polymer Phase Diagrams. A Textbook, Oxford Univ. Press, New York 2001

10.3 OSMOTIC PRESSURE

P.-G. de Gennes, Scaling Concepts in Polymer Physics, Cornell University Press, Ithaca (NY) 1979
K.F.Freed, Renormalization Group Theory of Macromolecules, Wiley, New York 1987

10.3.2 MEMBRANE OSMOMETRY

H.Coll, F.H.Stross, Determination of Molecular Weights by Equilibrium Osmotic Pressure Measurements, in Characterization of Macromolecular Structure, Natl.Acad.Sci.U.S. Publ. 1573, Washington (DC) 1968
H.-G.Elias, Dynamic Osmometry, in Characterization of Macromolecular Structure, Natl.Acad.Sci.U.S., Publ. 1573, Washington (DC) 1968
M.P.Tombs, A.R.Peacocke, The Osmotic Pressure of Biological Macromolecules, Clarendon Press, Oxford 1977
H.Coll, Nonequilibrium Osmometry, J.Polym.Sci. D [Macromol.Revs.] **5** (1971) 541

10.3.3 EBULLIOSCOPY AND CRYOSCOPY

R.S.Lehrle, Ebulliometry Applied to Polymer Solutions, Progr.High Polymers **1** (1961) 37
M.Ezrin, Determination of Molecular Weight by Ebulliometry, in Characterization of Macromolecular Structure, Natl.Acad.Sci.U.S., Publ. 1573, Washington (DC) 1968

10.3.4 VAPOR PHASE OSMOMETRY

W.Simon, C.Tomlinson, Thermoelektrische Mikrobestimmung von Molekulargewichten, Chimia **14** (1960) 301 (vapor pressure osmometry of molecular weights)
J. van Dam, Vapor-Phase Osmometry, in Characterization of Macromolecular Structure, Natl.Acad.Sci.U.S., Publ. 1573, Washington, D.C. 1968

10.5 ASSOCIATION AND SELF-ASSOCIATION

H.-G.Elias, Association and Aggregation as Studied via Light Scattering, in M.B.Huglin, Ed., Light Scattering from Polymer Solutions, Academic Press, London 1972
L.Tanford, The Hydrophobic Effect. Formation of Micelles and Biological Membranes, Wiley, New York 1973
I.Klotz, Protein Interactions with Small Molecules, Acc.Chem.Res. **7** (1974) 162
H.-G.Elias, Association of Synthetic Polymers, in K.Solc, Ed., Order in Polymer Solutions, Gordon and Breach, New York 1975
F.Oozawa, S.Asakura, Thermodynamics of the Polymerization of Proteins, Academic Press, London 1975

Z.Tuzar, P.Kratochvil, Block and Graft Copolymer Micelles in Solution, Adv.Colloid Interface Sci. **6** (1976) 201
D.Poland, Cooperative Equilibria in Physical Biochemistry, Oxford Univ. Press, Oxford 1978
A.Ben-Naim, Hydrophobic Interactions, Plenum, New York 1980
V.DeGiorgio, M.Corti, Eds., Physics of Amphiphiles: Micelles, Vesicles and Microemulsions, North Holland, Amsterdam 1985
I.Piirma, Polymeric Surfactants, Dekker, New York 1992
A.Laschewsky, Molecular Concepts, Self-Organization and Properties of Polysoaps, Adv.Polym.Sci. **124** (1995) 1
S.E.Weber, P.Munk, Z.Tuzar, Eds., Solvents and Self-Organization of Polymers, Kluwer Academic, Dordrecht, NL 1996 (NATO ASI Series, Series E)
M.Rosoff, Ed., Vesicles, Dekker, New York 1996

10.5.4-10.5.5 COMPLEXATION
E.A.Bekturov, L.A.Bimendina, Interpolymer Complexes (in Russ.), Nauka, Alma Ata 1977; Adv.Polym.Sci. **41** (1981) 99 (in Engl.)
E.Tsuchida, K.Abe, Interactions Between Macromolecules in Solution and Intermacromolecular Complexes, Adv.Polym.Sci. **45** (1982) 1

10.6 POLYELECTROLYTES
A.Veis, Ed., Biological Polyelectrolytes, Dekker, New York 1970
F.Oosawa, Polyelectrolytes, Dekker, New York 1971
B.E.Conway, Solvation of Synthetic and Natural Polyelectrolytes, J.Macromol.Sci. **C 6** (1972) 113
E.Sélégny, M.Mandel, U.P.Strauss, Eds., Charged and Reactive Polymers, Reidel, Dordrecht 1974 (several volumes)
H.Eisenberg, Biological Macromolecules and Polyelectrolytes in Solution, Clarendon Press, Oxford 1976
M.Mandel, The Physical Chemistry of Polyelectrolyte Solutions, Angew.Makromol.Chem. **123/124** (1984) 63
N.Ise, Ordering of Ionic Solutes in Dilute Solutions Through Attraction of Similarly Charged Solutes - A Change of Paradigm in Colloid and Polymer Chemistry, Angew.Chem. **98** (1986) 323; Angew.Chem.Int.Ed.Engl. **25** (1986) 323
J.R.MacCallum, C.A.Vincent, Eds., Polymer Electrolyte Revs., Elsevier, New York 1987
K.S.Schmitz, Macroions in Solution and Colloidal Suspension, VCH, Weinheim 1993
M.Hara, Ed., Polyelectrolytes. Science and Technology, Dekker, New York 1993
H.Dautzenberg, W.Jaeger, J.Kötz, B.Philipp, C.Seidel, D.Stscherbina, Polyelectrolytes. Formation, Characterization and Application, Hanser, Munich 1994
S.Förster, M.Schmidt, Polyelectrolytes in Solution, Adv.Polym.Sci. **120** (1995) 51
J.L.Barrat, J.-F.Joanny, Polyelectrolyte Solutions, Adv.Chem.Phys. **94** (1996) 1
E.M.Gray, Polymer Electrolytes, Royal Chemical Society, London 1998
S.E.Kudaibergenov, Recent Advances in the Study of Synthetic Polyampholytes in Solution, Adv.Polym.Sci. **144** (1999) 115
M.Jiang, M.Li, M.Xiang, H.Zhou, Interpolymer Complexation and Miscibility Enhancement by Hydrogen Bonding, Adv.Polym.Sci. **146** (1959) 121
T.Radeva, Ed., Physical Chemistry of Polyelectrolytes, Dekker, New York 2001
S.K.Tripathy, J.Kumar, H.S.Nalwa, Eds., Handbook of Polyelectrolytes and Their Applications, Amer.Sci.Publ., Stevenson Ranch (CA) 2004 (3 volumes)
M.Schmidt, Polyelectrolytes with Defined Molecular Architecture, Adv.Polym.Sci. **165** and **165** (2004)

10.7 GELS
A.H.Clark, S.B.Ross-Murphy, Structural and Mechanical Properties of Biopolymer Gels, Adv.Polym.Sci. **83** (1987) 57
W.Burchard, S.B.Ross-Murphy, Eds., Physical Networks, Elsevier Appl.Sci., London 1990
D. De Rossi, K.Kajiwara, Y.Osada, A.Yamauchi, Eds., Polymer Gels: Fundamentals and Biomedical Applications, Plenum, New York 1991
J.-M.Guenet, Thermoreversible Gelation of Polymers and Biopolymers, Academic Press, London 1992
J.P.Cohen-Addad, Ed., Physical Properties of Polymeric Gels, Wiley, New York 1996

K. te Nijenhuis, Thermoreversible Networks, Adv.Polym.Sci. **130** (1997) 1
T.Okano, Biorelated Polymers and Gels, Academic Press, San Diego 1998
Y.Osada, K.Kajiwara, Eds., Gels Handbook, Academic Press, San Diego (CA) 2000 (4 vols.)
Y.Osada, A.R.Khokhlov, Eds., Polymer Gels and Networks, Dekker, New York 2002

References to Chapter 10

[1] B.A.Wolf, Makromol.Chem. **178** (1977) 1869, Fig. 1
[2] G.M.Bristow, W.F.Watson, Trans.Faraday Soc. **54** (1958) 1567 (Table 1), 1731 (Tables 1, 2)
[3] A.Dondos, H.Benoit, Makromol.Chem. **133** (1970) 119, Tables 1, 3, and 8
[4] K.Ueberreiter, F.Asmussen, J.Polym.Sci. **32** (1957) 75, Fig. 2; s.a. Makromol.Chem. **44-46** (1961) 324
[5] J.S.Vrentas, J.L.Duda, S.T.Hsieh, Ind.Eng.Chem., Prod.Res.Dev. **22** (1983) 326, Figs. 1, 3
[6] G.Rehage, D.Möller, J.Polym.Sci. **C 16** (1967) 1787, data taken from Fig. 1
[7] H.-G.Elias, Makromol.Chem. **33** (1959) 140, data of Fig. 2
[8] C.F.Cornet, H. van Ballegooijen, Polymer **7** (1966) 293, Fig. 1
[9] H.-J.Cantow, O.Fuchs, Makromol.Chem. **83** (1965) 244, Table 1
[10] S.Saeki, N.Kuwakara, S.Komno, M.Kaneko, Macromolecules **6** (1973) 246, data of Fig. 1
[11] K.G.Siow, G.Delmas, D.Patterson, Macromolecules **5** (1972) 29, Fig. 1
[12] S.Saeki, N.Kuwahara, M.Nakata, Polymer **17** (1976) 685, data of Fig. 2
[13] Y.C.Bae, S.M.Lambert, D.S.Soane, J.M.Prausnitz, Macromolecules **24** (1991) 4403, Fig. 3
[14] H.-G.Elias, Theta Solvents, in J.Brandrup, E.H.Immergut, E.A.Grulke, Eds., Polymer Handbook, Wiley, New York, 4th ed. (1999), p. VII-291
[15] R.B.Richards, Trans. Faraday Soc. **42** (1946) 10, data of Fig. 10
[16] P.Smith, A.J.Pennings, Polymer **15** (1974) 413, Fig. 6
[17] P.L.Jorkowicz, A.Muñoz, J.Barrera, A.J.Müller, Macromol.Chem.Phys. **196** (1995) 385, Table 1 and Fig. 3
[18] T.A.Ritscher, H.-G.Elias, Makromol.Chem. **30** (1959) 48, Table 10
[19] C.Strazielle, R.Dick, Makromol.Chem. **142** (1971) 146, Table 3
[20] H.-G.Elias, H.P.Schlumpf, Makromol.Chem. **85** (1965) 118, Table 10
[21] G.V.Schulz, H.Doll, Z.Elektrochem. **56** (1952) 248, Table 4 (Fig. 1)
[22] J.C.Meunier, R. van Leemput, Makromol.Chem. **147** (1971) 191, Table 4
[23] J.-G.Zilliox, Makromol.Chem. **156** (1972) 121, Tables 2, 3, and 8
[24] J.E.L.Roovers, S.Bywater, Macromolecules **7** (1974) 443, Table IV
[25] T.Sato, T.Norisuye, H.Fujita, J.Polym.Sci.-Polym.Phys.Ed. **25** (1987) 1, Table II
[26] B.A.Wolf, H.-J.Adams, J.Chem.Phys. **75** (1981) 4121, Tables 4, 7, and 9-12
[27] P.Stepanek, R.Perzynski, M.Delsanti, M.Adam, Macromolecules **17** (1984) 2340, Fig. 2
[28] I.Noda, N.Kato, T.Kitano, M.Nagasawa, Macromolecules **14** (1981) 668, Fig. 7, Table II
[29] H.-G.Elias, Int.J.Polym.Mater. **4** (1976) 209, Fig. 5
[30] K.Šolc, H.-G.Elias, J.Polym.Sci.-Polym.Phys.Ed. **11** (1973) 137
[31] H.-G.Elias, J.Gerber, Makromol.Chem. **112** (1968) 122, Fig. 1
[32] A.Sikora, Z.Tuzar, Makromol.Chem. **184** (1983) 2049; numerical data: private communication
[33] (a) C.Booth, T.D.Naylor, C.Price, N.S.Rajab, R.B.Stubbersfield, J.Chem.Soc.Faraday [1] **74** (1978) 2353, Plate A; (b) C.Price, E.K.M.Chan, A.L.Hudd, R.B.Stubbersfield, Polym. Commun. **27** (1986) 197, Fig. 1b
[34] E.Teng-Leary, G.B.Kohlhaw, Biochim.Biophys. Acta **410** (1975) 210, data taken from Fig. 2
[35] G.Rehage, D.Wagner, in P.Dubin, Ed., Microdomains in Polymer Solutions, Plenum, New York 1985 (= Polym.Sci.Technol. **30** (1985) 87)
[36] N.Ise, T.Okubo, J.Phys.Chem. **71** (1967) 1287; **72** (1968) 1361, 1370, various Tables
[37] N.Ise, K.Asai, J.Phys.Chem. **72** (1968) 1366, various Tables
[38] N.Ise, T.Okubo, Acc.Chem.Res. **13** (1980) 303, various Tables
[39] T.Tanaka, D.Fillmore, S.-T.Sun, I.Nishio, G.Swislow, A.Shah, Phys.Rev.Lett. **20** (1980) 1636, data of Fig. 1
[40] H.-M.Tan, A.Moet, A.Hiltner, E.Baer, Macromolecules **16** (1983) 28, Fig. 2

11 Transport in Solutions

Matter, energy, impulse (= product of force and time), and torsional impulse can all be transported. Matter is transported by diffusion in Earth's gravity field, by sedimentation in a centrifugal field, and, for example, by electrophoresis in an electrical field. Energy is transported by conduction of heat (Section 13.2.3) while the viscosity of gases is caused by a transport of impulse.

This chapter discusses the transport of single macromolecules in *dilute* solutions by translational diffusion, sedimentation, and electrophoresis. The discussion is mainly concerned with phenomenological aspects; molecular theories will be treated in Chapter 12 together with those for dilute solution viscosities. Transport in melts and through solid polymers (permeation) is dealt with in Chapter 14

11.1 Translational Diffusion

11.1.1 Introduction

Diffusion is the equilibration of matter by Brownian movements, either isothermally by a concentration gradient (**translational diffusion**; usually just called **diffusion**), a flow gradient (**rotational diffusion**), or non-isothermally in a temperature gradient (**thermal diffusion**) (L: *dis* = apart, *fundere* = to pour).

Diffusions are manifestations of **molecular dynamics**, which comprises time-dependent fluctuations of structures and properties in equilibrium and on approach to equilibrium. For example, three different regions can be distinguished with respect to the action of heat impulses: (I) shifts of whole molecules, (II) movements of chain segments, and (III) interconversions of microconformations.

Interconversions of microconformations such as gauche → trans require only a low activation energy (Volume I, p. 138; this Volume, p. 42 ff.). Lifetimes of trans and gauche conformers are therefore short, only 10^{-9}–10^{-11} s in liquid hydrocarbons. The fraction of such conformers can be determined by vibrational spectroscopy (infrared, Raman) which responds to time intervals of ca. 10^{-13} s. Nuclear magnetic resonance spectroscopy, on the other hand, works at times of 10^{-6}–10^{-7} s, which allows one to study site-exchange processes of segments within polymer coils (**micro-Brownian movement**). The movement of molecules themselves (**macro-Brownian movement**), such as of polymer molecules in solvents and solvents in polymers, leads to translational diffusion (this Section) whereas site-exchanges between like molecules or segments thereof gives rise to **self-diffusion** (Section 14.2) without net flow between entities.

Site exchanges between unlike molecules or segments lead to concentration or temperature gradients that generate a net flow, for example, between a solution 3 of polymer 2 in a solvent 1 against pure 1 in which the diffusion of molecules 2 from 3 into the pure 1 is counterbalanced by diffusion of 1 into 3.

The flow of a component is proportional to the force that causes that flow according to the thermodynamics of irreversible processes. The proportionality coefficient is the **diffusion coefficient**. Flows and forces may be defined differently, however, so that the diffusion coefficient is not unambiguously defined. For example, flows may be related to masses or molar masses and also relative to the coordinate system or to average velocities of masses m, molar masses M, or volumes V.

For example, the mass flow of component i in binary mixtures of components i and j may be relative to the average velocity of the volume V of the mixture ($j_{i,V}$), relative to the average velocity of the mass $m = m_i + m_j$ of the mixture ($j_{i,m}$), or relative to the fixed coordinates x, y, and z ($j_{i,x,y,z}$). The Nabla operator describes the gradient, for example, $\nabla c_i = \partial c_i/\partial L$:

$$(11\text{-}1) \qquad j_{i,V} = c_i(R_i - \overline{R}_V) = -D_{ij,a}\nabla c_i$$

$$(11\text{-}2) \qquad j_{i,m} = c_i(R_i - \overline{R}_m) = -D_{ij,a}\rho\nabla w_i$$

$$(11\text{-}3) \qquad j_{i,x,y,z} = c_iR_i = -D_{ij,a}\rho\nabla w_i + w_i(j_{i,x,y,z} + j_{j,x,y,z})$$

ρ	= density (mass density) of a mixture of i and j	ρ	$= (m_i + m_j)/V$
c_i	= mass concentration of component i	c_i	$= m_i/V$
w_i	= mass fraction of component i	w_i	$= m_i/(m_i + m_j)$
x_i	= mole fraction of component i	x_i	$= n_i/(n_i + n_j)$
R_i	= velocity of component i	R_i	$= L_i/t$
$\overline{R}_V$	= average velocity of the volume of the mixture	$\overline{R}_V$	$= L_V/t$
$\overline{R}_m$	= average velocity of the mass of the mixture	$\overline{R}_m$	$= L_m/t$

The simplest (classic) version of diffusion processes assumes that the flow is caused only by the concentration gradient and not by additional velocity fields. However, such a field arises, for example, if volumes are not additive for two different concentrated solutions with the same components. In such a case, a volume change of ca. 4 % leads to a difference of ca. 50 % between true and calculated diffusions coefficients if the latter were obtained from $c_iR_i = -D_{ij}\nabla c_i$ instead of Eqs.(11-1) or (11-3).

The right-hand sides of Eqs.(11-1)-(11-3) are always negative because component i diffuses in the direction of the less concentrated solution (or pure solvent). The proportionality constant $D_{ij,a}$ of Eqs.(11-1)-(11-3) is the diffusion coefficient D_{ij} that has been corrected for the change of thermodynamic activity a with the mole fraction x_i of component i, $D_{ij,a} = (\partial \ln a/\partial \ln x_i)D_{ij}$. Diffusion coefficients have the physical unit of area per time. According to the **Einstein-Smoluchowski equation**, $L^2 = 2\,D\Delta t$, they correspond to the average of the square of particle shifts L in the time interval Δt.

Diffusion rates are connected with gradients of chemical potentials via frictional coefficients and concentrations or mass fractions, respectively. Since the chemical potential of component i can be expressed by the chemical potential of component j according to the Gibbs-Duhem equation (p. 298), diffusion coefficients D_{ij} of components i relative to that of component j equals diffusion coefficients D_{ji} of j relative to i. Hence, these diffusion coefficients are **mutual diffusion coefficients** of solute and solvent, $D \equiv D_{ij} = D_{ji}$.

11.1.2 Measurements

Classic Method

In the classic diffusion experiment, the translational diffusion of a solute in a solvent is described by its diffusive flux J_m as the transport of mass m per time t in the x direction through an area A, $J_m = dm/(Adt)$. In general, a solvent with density $\rho_1 < \rho$ is placed on top of a solution with density ρ and concentration c which triggers an immediate diffusion of polymer molecules from the polymer solution to the solvent and a concurrent diffusion of solvent molecules from the solvent to the polymer solution, both caused by Brownian movements of polymer and solvent molecules. The progress of diffusion can be measured by interferometry, absorption spectroscopy, etc. The process continues until the whole volume of the liquid is homogenized.

During the diffusion process, a concentration gradient $\partial c/\partial x$ is established that is proportional to the diffusional flow of polymer concentration from the solution section to the solvent section, $J_c = -D(\partial c/\partial x)$; the proportionality constant is the (mutual) **diffusion coefficient**. Mass flux J_m and concentration flux J_c must be identical, $J_m = J_c$, so that

(11-4) $\partial m/\partial t = -DA(\partial c/\partial x)$ **Fick's first law**

Fick's second law describes the temporal and spatial change of concentration, for example, for a one-dimensional diffusion in x direction:

(11-5) $$\left(\frac{\partial c}{\partial t}\right)_x = \left[\frac{\partial}{\partial x}\left(D\frac{\partial c}{\partial x}\right)\right]_t$$ **Fick's second law**

For concentration-independent diffusion coefficients, it reduces to

(11-6) $\partial c/\partial t = D(\partial^2 c/\partial x^2)$

This equation can be solved for various conditions; for example, for the diffusion of a solution of concentration c' against a solvent with concentration $c'' = 0$. At the initial interface solution/solvent ($x = 0$), the initial solute concentration is $(c' + c'')_0/2 = c_0/2$. The process is allowed to proceed in such a way that the initial concentrations c' and c'' are still present at the two ends of the diffusion chamber. If the diffusion coefficient is independent of time and space, then integration of Eq.(11-6) delivers for the concentration $c_{x,t}$ at distance x after time t:

(11-7) $$c_{x,t} = \frac{c_0}{2}\left[1 - \frac{2}{\pi^{1/2}}\int_0^y \exp\left(-y^2\right)dy\right] \quad ; \quad y \equiv x/\{2\ (Dt)^{1/2}\}$$

The integral in Eq.(11-7) is the error integral. Differentiation of Eq.(11-7) delivers $c = f(x)$ which is a bell curve that is described by a Gaussian function:

(11-8) $$\frac{dc}{dx} = \frac{c_0}{2(\pi Dt)^{1/2}} \exp\left(-\frac{x^2}{4Dt}\right)$$

The diffusion coefficient D is obtained from the position dependence of the concentration at time t or from the time dependence of the concentration at $x = const.$ The bell curve has a maximum $h = c_0/[2\ (\pi\ Dt)^{1/2}]$ at $x = 0$. The calculated diffusion coefficient always refers to one-half of the sum of the two initial concentrations, i.e., to $c_0/2$ if a solution of concentration c_0 diffuses against pure solvent.

Diffusion coefficients D are then extrapolated to zero solute concentration (Section 11.1.5) to give D_0. The value of D_0 depends on the evaluation method. Literature data are usually either mass averages (also called moment averages), $\overline{D}_w = \Sigma_i w_i D_i$, or so-called area averages, $\overline{D}_A = (\Sigma_i w_i D_i^{-1/2})^2$. These two averages are practically identical for random coils and for rods with not-too-broad molar mass distributions.

Dynamic Light Scattering

Brownian movements produce local fluctuations of concentrations which can be followed by **dynamic light scattering**, also known as **quasielastic light scattering** or **photon correlation spectroscopy**. Dynamic light scattering is now the most important method for the determination of diffusion coefficients in solution.

Thermal fluctuations generate swarms (clusters) of molecules and segments with lifetimes of ca. 10^{-6} s that scatter laser light like true molecules. The phase of the scattered light depends on the position of the clusters. The intensity of the scattered light is not constant, though; rather it fluctuates with time because of the variation of the relative phases from the various clusters.

The time t for the formation and decomposition of a cluster approximately equals the time that is needed by a pair of molecules to change the distance between them by about one-half of the wavelength of the laser light, $\lambda = \lambda_0/n_1$, where n_1 = refractive index of the solvent. The larger the time t, the smaller the diffusion rate.

In a time interval $\lambda = \lambda_0/n_1$, A_i photons arrive at the detector. The numbers A_i and A_{i+1} of photons arriving at two consecutive time intervals i and i+1 are multiplied and stored in a computer; this is repeated 10^5-10^7 times. A second channel of the computer stores the products A_iA_{i+2} from time intervals i and i+2, etc. The averaged products A_iA_{i+j} decrease exponentially with the number of channels (Fig. 11-1); this function is called the **autocorrelation function**.

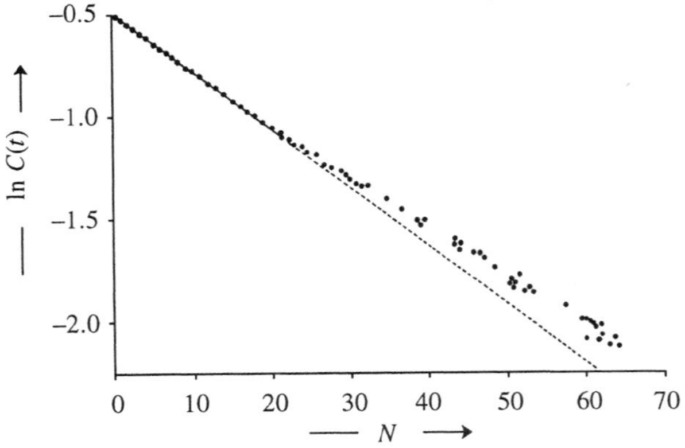

Fig. 11-1 Normalized autocorrelation function $C(t)$ as a function of the number N of channels (which is proportional to the time) for a poly(methyl methacrylate) with $\overline{M}_w / \overline{M}_n$ = 1.14 in acetone [1].
By kind permission of the Neue Schweizerische Chemische Gesellschaft, Basel.

Time intervals are small at short times (small number of channels) so that most molecules had no chance to collide with other molecules. They "remember" therefore the ve-

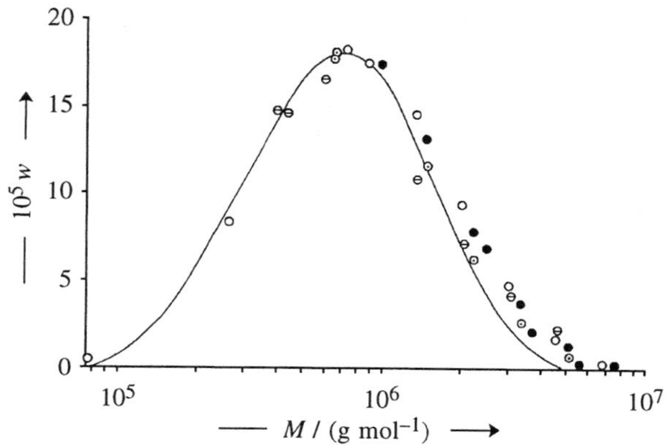

Fig. 11-2 Differential distribution of molar masses of a poly(methyl methacrylate) in tetrahydrofuran by dynamic light scattering (○,⊖,⊕,●) and size-exclusion chromatography (——) [2].

locities and the directions of their movements in the previous time period, i.e., their movements are strongly correlated. At long time intervals, molecules collide often and their movements are therefore uncorrelated.

The normalized autocorrelation function $C(t)$ is a simple exponential function of time for single-sized hard spheres regardless of their size and also for *small* single-sized entities regardless of their shape,

(11-9) $$C(t) = B\exp(-\Gamma t) = B\exp(-Dq^2t) \quad ; \quad q \equiv (4\,\pi n_1/\lambda_o)\sin(\vartheta/2)$$

where $B = const$, ϑ = scattering angle, and Γ = half-width of the distribution of frequencies (**first cumulant**). The time t is proportional to the number of channels N so that ln $C(t)$ is a linear function of N for single-sized hard spheres and *small* single-sized entities with other shapes.

The function ln $C(t)$ = f(N) becomes non-linear at higher channel numbers for non-uniform polymer molecules and/or large entities with internal segmental movements (Fig. 11-1). Eq. (11-9) allows one to calculate diffusion coefficients in the former case (Fig. 11-2) but not in the latter.

The frequency of scattered light shifts to larger values if scattering centers move toward the detector and to smaller ones if they move away from the detector (**Doppler effect**). However, molecules in solution move in all directions and they also have a velocity distribution. Hence, the line width is enlarged but the frequency is not shifted.

The line width is proportional to the diffusion coefficient but it also measures portions of rotations and vibrations. Effects of rotations and vibrations become prominent at higher scattering angles, i.e., at higher values of q (Eq.(11-9), right side) so that the the z-average diffusion coefficient, $\overline{D}_z$, and the mean-square radius of gyration, $\langle s^2\rangle$, are obtainable from the intercept and the slope, respectively, of the function $\Gamma/q^2 = \overline{D}_z(1 + K_q\langle s^2\rangle_z q^2 + \ldots)$.

z-Averages of diffusion constants depend on average molar masses according to $\overline{D}_z = K_D K_{corr} M_{av}{}^{\delta}$ where K_D and δ refer to $D = K_D M^{\delta}$ of *molecularly uniform* polymers and K_{corr} = correction factor for $\overline{M}_w/\overline{M}_n \neq 1$. K_{corr} is very large, for example, $K_{corr} = 0.6647$ for $M_{av} = \overline{M}_n$, $\delta = -0.50$ (see p. 377), and SF distributions with ($\overline{M}_w/\overline{M}_n = 2$).

11.1.3 Friction Coefficients of Translation

Diffusion increases with increasing temperature, i.e., the diffusion coefficient is proportional to the added energy k_BT. According to the **Einstein-Sutherland equation** (less frequently called the **Nernst-Einstein equation**), the proportionality coefficient is the inverse **friction coefficient of diffusion**, $\xi_D \equiv \Psi/v$, which is defined as the frictional resistance Ψ per velocity v. Friction coefficients measure the mass that is transported in unit time; they are thus the hydrodynamic resistance to translation:

(11-10) $D_0 = k_BT/\xi_D$ **Einstein-Sutherland equation**

Spheres

Friction coefficients are relatively easy to calculate for the translational diffusion of spheres. A sphere with the hydrodynamic radius R_{sph} (= **Stokes radius**) moves in the x direction with a velocity v in a continuous viscous medium of viscosity η_1 without formation of eddies. Such a sphere experiences a friction $\Psi_{sph} = \eta_1\int(\partial v/\partial y)da$ where y is the direction perpendicular to the x direction. The area elements da are expressed in fractions of the surface of the sphere, $A = 4\ \pi\ R_{sph}^2$, whereas the velocity gradient (shear rate) of the medium, $\partial v/\partial y$, perpendicular to the direction x of the movement is measured in fractions of v/R_{sph}.

The friction Ψ_{sph} is obtained by integration of $\eta_1\int(\partial v/\partial y)da$ over the total area A of the sphere, delivering $\Psi_{sph} = \eta_1(v/R_{sph})(4\ \pi\ R_{sph}^2)K_{int}$. For spheres, the integration constant has a value of $K_{int} = 3/2$. The **friction coefficient** of spheres is therefore

(11-11) $\xi_{sph} \equiv \Psi_{sph}/v = 6\ \pi\ \eta_1 R_{sph}$ **Stokes equation**

The derivation of the Stokes equation assumes that the liquid immediately near the sphere moves as fast as the sphere itself, i.e., that there is a *sticky boundary*. The numerical factor 6 has to be replaced by a factor of 4 if the boundary slips from the surface.

Because the solvent is treated as a continuum, moving entities must also be large compared to solvent molecules. The Stokes equation has to be modified for entities with dimensions that are similar to those of solvent molecules. It was found experimentally for highly viscous solvents with viscosities $\eta_1 > 0.01$ Pa s that the friction coefficient ξ_{sph} is no longer directly proportional to the macroscopic viscosity η_1 but to $\eta_1^{2/3}$. Alternatively, the η_1 in Eq.(11-11) may be treated as a microviscosity that the entity experiences; it is no longer the macroscopic viscosity.

The Stokes equation applies if the velocity dL/dt of the spheres is smaller than $\eta_1/(2\ R_{sph}\rho)$ as has been found experimentally (ρ = density). For polymers, this condition is always fulfilled.

Ellipsoids

Moving ellipsoids, rods, and other non-spherical entities experience an additional resistance that depends on the orientation of their longest axes. Rods with long axes parallel to flow directions will have a smaller friction than rods with long axes perpendicular to it. The latter rod will thus try to orient itself in the direction of flow. However, forces acting on the rods will be very small at small flow velocities so that Brownian movements may still cause a random distribution of long axes.

Translational friction coefficients of ellipsoids, $\xi_{D,ell}$, are not given as such but always as multiples $P = \xi_{D,ell}/\xi_{D,sph}$ (**Perrin factor**) of the corresponding translational friction

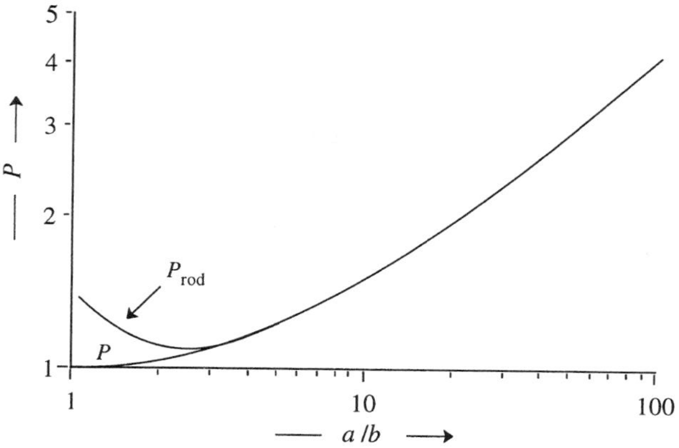

Fig. 11-3 Perrin factors P of translational diffusion as function of the axial ratio a/b of prolate ellipsoids. For large axial ratios a/b, prolate ellipsoids may be approximated as cylindrical rods with $P_{rod} = \Lambda_{rod}^{2/3}/[\ln (2\ \Lambda_{rod})]$ where $\Lambda_{rod} = a/b$. P_{rod} is greater than P by 4.4 % at $a/b = 3$, by 0.4 % at $a/b = 10$, and by 0.1 % at $a/b = 20$.

coefficient of spheres, $\xi_{D,sph}$. Perrin factors of prolate ellipsoids with semi-axes $a > b = c$ and oblate ellipsoids with semi-axes $a < b = c$ were calculated for sticky boundaries as

(11-12) $P = (1 - p^2)^{1/2}/[p^{2/3} (\ln \{1 + (1 - p^2)^{1/2}\}/p)]$ $(p = b/a < 1$; prolate)

(11-13) $P = (q^2 - 1)^{1/2}/[q^{2/3} \tan^{-1}(q^2 - 1)^{1/2}]$ $(q = a/b > 1$; oblate)

Perrin factors are always greater than 1. Ellipsoids therefore have greater translational friction coefficients than spheres of the same volume, $\xi_{D,ell} = P\xi_{D,sph}$. At the same axial ratio, oblate ellipsoids always have greater friction coefficients of rotation than prolate ellipsoids (Fig. 11-3).

Rods

Very long prolate ellipsoids are practically cylindrical rods with $1/p \to \infty$, i.e., $p \to 0$. In this limit, p^2 can be ignored and Eq.(11-12) becomes $P = [p^{2/3} \ln (2/p)]^{-1}$ which, with the axial ratio $\Lambda = L_{rod}/d_{rod}$, can be written as $P_{rod} = \Lambda_{rod}^{2/3}/[\ln (2\ \Lambda_{rod})]$. The friction coefficient $\xi_{D,rod} = P\xi_{D,sph} = 6\ \pi\ \eta_1 R_{sph}$ of cylinders with the same friction-equivalent volume as spheres ($4\pi R_{sph}^3/3 = \pi R_{rod}^2 L_{rod}$) is therefore

(11-14) $$\xi_{D,rod} = 3\pi\,\eta_1 d_{rod} \frac{(3/2)^{1/3} \Lambda_{rod}}{\ln (2\Lambda_{rod})}$$

The Kirkwood-Riseman theory (see p. 373) arrives at a similar equation for rods. For $\Lambda = 100$, this theory delivers a value of $\xi_{D,rod}$ that deviates from $\xi_{D,rod}$ of Eq.(11-14) by only 0.5 %:

(11-15) $$\xi_{D,rod} = 3\ \pi\,\eta_1 d_{rod} \frac{\Lambda_{rod}}{\ln \Lambda_{rod}}$$

Random Coils

The movement of liquids and particles in liquids, respectively, can only rarely be described by simple theories. An example is the Stokes equation, Eq.(11-11), for the friction coefficients of hard spheres. Instead, one has to rely on the postulates and theories of **hydrodynamics (fluid dynamics, fluid mechanics)**. The basic equation here is the **Navier-Stokes equation** which describes the energy balance of moving elements of the fluid. The velocity v of an element with density ρ and dynamic viscosity η is a vector function of space and time that depends on pressure p and external forces $\boldsymbol{F}_{\text{ext}}$:

$$(11\text{-}16) \qquad (\partial \boldsymbol{v}/\partial t) = \boldsymbol{F}_{\text{ext}} - (1/\rho)\nabla p + (\eta/\rho)\nabla^2 \boldsymbol{v}$$

This equation can be solved for various conditions, for example, for the case that the velocity of the liquid is zero immediately next to a *solid* surface. *A priori*, one would not assume that random coils have a surface and especially not a hard one against the surrounding solvent. Monomeric units of random coils are furthermore coupled to each other so that the flow units are not monomeric units but larger segments consisting of several monomeric units. Such segments are not rigid either because Brownian movements and changes of microconformations can cause the ends of segments to move closer to or farther from each other. This stretching and springing back of the end-to-end distances of segments resembles the movements of a spring.

In **spring-and-bead models (bead-rod chains, pearl-necklace models)** segments are replaced by elastic dumbbells (Fig. 11-4). The spring is assumed to have no mass; the mass is concentrated solely in the beads which are assumed to be the friction elements. The dynamics of such a dumbbell is described by the Hooke elasticity of a spring (Chapter 16). The resulting mathematical problem thus resembles that of a Brownian movement of a system of coupled harmonic oscillators.

The **Rouse model** subdivides the whole chain into "Rouse segments." In each segment, the spatial distribution of monomeric units is assumed to obey Gaussian statistics, an assumption that applies to Kuhnian segments. Each Rouse segment is represented by a bead and a spring, so that in the steady state, the friction force F_ξ equals the elastic retraction force, F_{el}. The friction force F_ξ is the product of the velocity $\text{d}\boldsymbol{r}_j/\text{d}t$ and the Rouse friction coefficient ξ_R per bead, whereas the retraction force F_{el} is given by the product of the shift $\Delta \boldsymbol{r}$ of a bead j and the spring constant K_{el}. The effect of inertia is neglected (no overshooting, no flowing back):

$$(11\text{-}17) \qquad \xi_R(\text{d}\boldsymbol{r}_j/\text{d}t) = K_{\text{el}}(\boldsymbol{r}_{j+1} - \boldsymbol{r}_j) + K_{\text{el}}(\boldsymbol{r}_{j-1} - \boldsymbol{r}_j)$$

The same equation applies to all other sets of beads and springs, which results in a system of differential equations. The solution of this system must consider that all spring-bead units of a linear macromolecule are coupled. Spring-bead units must therefore move in a *coordinated* manner, an effect that is very important for the description of processes that depend on time or frequency such as the creep of plastics (Section 17.4.3). Because of these coordinated thermal movements of segments, random coils progress steadily toward their most probable macroconformations.

Diffusion reflects only the resultant of these coordinated movements of segments. In very dilute solutions, the movement of a coil is not affected by that of other coils, i.e., there are no hydrodynamic interactions between segments of different coils. The solvent

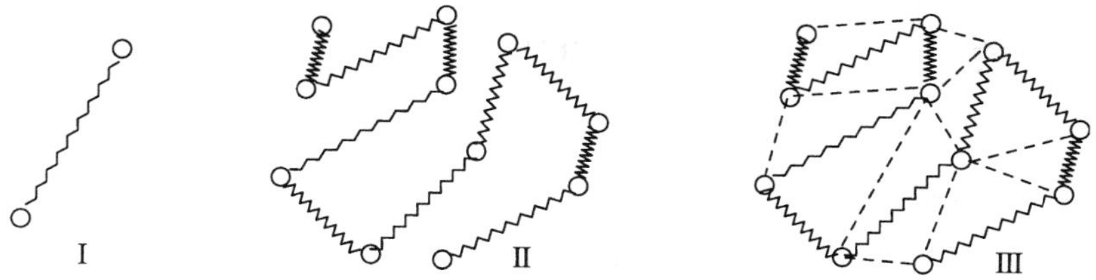

Fig. 11-4 Models for the dynamic behavior of random coils. I: elastic dumbbell with a massless spring between two segments O. II: pearl-necklace model without hydrodynamic interactions between segments (Rouse model). III: pearl-necklace model with (some!) hydrodynamic interactions (Kirkwood-Riseman model).

can therefore flow unimpeded through the coil: the coil is *free-draining*. But solvent molecules do hinder segmental movements, which causes friction. If each segment has a friction coefficient ξ_R and if there are a total of N_K Kuhnian segments, then the friction coefficient of the total molecule is $\xi_D = N_K\xi_R$.

The **Zimm theory** is based on the Kirkwood-Riseman theory. It assumes that the hydrodynamic interactions between segments are so strong that solvent molecules within the coils are dragged along with the coil (but *not* because of thermodynamic bonding to monomeric units). Hence, solvent molecules within the coil cannot flow *through* the coil which is therefore completely *non-draining* and can be treated as a hard sphere.

The **Kirkwood-Riseman theory** allows for different hydrodynamic interactions, for example, for coils with homogeneous segment densities (Debye-Bueche theory) or for coils with Gaussian segmental distributions (Kirkwood-Riseman). The Navier-Stokes equations are then solved using Oseen's method.

The theory makes the following physical assumptions. The chain consists of N_K beads with radius R_{pl} (index pl for "pearl") that are interconnected by massless and frictionless "bonds." A bead (pearl) may be a monomeric unit or a Kuhnian segment of several monomeric units.

The friction coefficient of a bead is assumed to be the Stokes friction coefficient, $\xi_{pl} = 6\,\pi\eta_1 R_{pl}$. Since there are no hydrodynamic interactions between beads, the friction coefficient of the whole chain is simply $\xi_D = 6\,\pi\eta_1 R_{pl}N_K$. However, if there are hydrodynamic interactions, then the friction coefficient of the chain will be greater, the larger the chain volume, i.e., the greater the average distance r_{ij} between any two beads i and j. The expression for ξ_D therefore has to be divided by a function of the *average inverse* distance of the beads.

This function is obtained by summation of averages of inverse spatial distances, $\langle r_{ij}^{-1}\rangle$ of all pairs of beads, i.e., for all values of i and j except $i = j$. Since the resulting double sum appears in the denominator, it has to be multiplied by a length for dimensional reasons. The multiplying length must be the radius R_{pl} of beads because smaller radii of beads will lead to larger interactions per chain for the same number of pairs of beads and the same interactions per chain.

Conversely, the double sum has to be divided by the number N_K of beads because a greater number of beads also means a greater friction coefficient per chain. To the term in the denominator one also has to add "1" since the expression must change to that for a coil without hydrodynamic interactions for a large number of beads, small radii of beads, and large distances between beads.

The friction coefficient of a coil with hydrodynamic interactions thus becomes

$$(11\text{-}18)\qquad \xi_D = \frac{6\,\pi\,\eta_1 N_K R_{pl}}{1 + \dfrac{R_{pl}}{N_K}\sum_{i=1}^{N_K}\sum_{j=1}^{N_K}\langle r_{ij}^{-1}\rangle}$$

The double sum can be solved with the help of the radial distribution function of chain ends of unperturbed chains for the three-dimensional case, Eq.(4-51). The spatial average of the inverse end-to-end distance is

$$(11\text{-}19)\qquad \langle r_{ij}^{-1}\rangle = \frac{\int_0^1 r^{-1}p(r)\mathrm{d}r}{\int_0^\infty p(r)\mathrm{d}r} = \left(\frac{6}{\pi\langle r^2\rangle_o}\right)^{1/2}$$

A Kuhnian chain consisting of N_K segments with Kuhnian length L_K has an unperturbed mean-square end-to-end distance of $\langle r^2\rangle_o = N_K L_K^2$ (Eq.(4-39)), i.e., $\langle r^{-1}\rangle = \{6/(\pi N_K L_K^2)\}^{1/2}$. The same relationship must also apply to the average $\langle r_{ij}^{-1}\rangle$ between the ith and the jth bead, i.e., $\langle r_{ij}^{-1}\rangle = \{6/(\pi N_K L_K^2)\}^{1/2}$. Introducing this expression in Eq.(11-18), replacing the sums by integrals, and integrating over the shell of the sphere delivers the **Kirkwood-Riseman equation** for the friction coefficient:

$$(11\text{-}20)\qquad \xi_D = \frac{6\,\pi\,\eta_1 N_K R_{pl}}{1 + \left(\dfrac{8}{3\,\pi^{1/2}}\right)\dfrac{6^{1/2}\,R_{pl} N_K^{1/2}}{L_K}}$$

This equation reduces to that of a completely draining molecule, $\xi_D = 6\,\pi\eta_1 N_K R_{pl} = \xi_{seg} N_K$, if the second term of the denominator is very small, i.e., $N_K^{1/2} << R_{pl}/L_K$.

Eq.(11-20) reduces to $\xi_D = 6^{1/2}\pi\eta_1 N_K^{1/2} L_K (3\,\pi^{1/2}/8)$ if the second term of the denominator is much greater than unity, i.e., $N_K^{1/2} >> R_{pl}/L_K$. Since $\langle r^2\rangle_o^{1/2} = N_K^{1/2} L_K$ (Eq.(4-39)) and $\langle r^2\rangle_o^{1/2} = 6^{1/2}\langle s^2\rangle_o^{1/2}$ (Eq.(4-43)), friction coefficients of diffusion for unperturbed random coils are given by

$$(11\text{-}21)\qquad \xi_D = 6\,\pi\,\eta_1 \langle s^2\rangle_o^{1/2}(3\,\pi^{1/2}/8) = 6\,\pi\eta_1 R_{h,D}$$

where $R_{h,D} \equiv (3\,\pi^{1/2}/8)\,\langle s^2\rangle_o^{1/2} \approx 0.6647\,\langle s^2\rangle_o^{1/2} \approx (2/3)\,\langle s^2\rangle_o^{1/2}$ is the radius of a sphere that is hydrodynamically equivalent to the unperturbed random coil. Thus, according to the Kirkwood-Riseman theory, unperturbed random coils with large molar masses should behave as equivalent spheres.

Coils with Gaussian distributions of chain segments have smaller segment concentrations at the periphery than near the center (Fig. 4-26). Solvent molecules will therefore flow relatively freely through the outer shell of the sphere. However, the flow is impeded in the interior of the equivalent sphere and the solvent is therefore transported with the internal volume near the center of the sphere.

Since hydrodynamic radii $R_{h,D}$ are only 2/3 of the radii of gyration, s, and radii of gyration are relatively deep in the interior of the coil (Fig. 4-15), non-draining is restricted to a core zone in the interior of the coil. That core zone becomes proportionally greater with increasing degree of polymerization, which can be described by a function $\langle s^2 \rangle_o^{1/2} = f(X) \cdot R_{h,d}$ which varies from $f(X) = 0$ at $X = 1$ to $f(X) \approx (1/0.6647) \approx 1.505$ at $X \to 0$. This variation of $f(X)$ with the degree of polymerization shows that non-draining is solely hydrodynamic and not thermodynamic.

11.1.4 Diffusion Coefficients

Equations for the dependence of diffusion coefficients on the molar masses of polymers are obtained by combining friction coefficients with the Einstein-Sutherland equation, Eq.(11-10), into equations of the type

(11-22) $$D_0 = K_D M^{\delta}$$

Exponents δ are controlled by the shapes of molecules and for coils also by the hydrodynamic interactions whereas proportionality constants are affected by various other quantities (see below). Corrections have to be applied for experimental averages of D and M if polymer molecules are not uniform with respect to molar mass.

Spheres

The molar mass dependence of hard spheres is obtained by combining Eqs.(10-10) and (10-11), setting $\xi_D = \xi_{D,sph}$, and introducing $R_{d,sph} = (3\ V_{sph}/4\pi)^{1/3}$, $V_{sph} = m_{sph}/\rho_{sph}$, and $m_{sph} = M/N_A$ (Eq.(11-23)). Experimental results for some spherical macromolecules are shown in Table 11-1.

(11-23) $$D_0 = \frac{(4/3)^{1/3}}{6\,\pi^{2/3}} k_B N_A^{1/3} \left(\frac{\rho^{1/3} T}{\eta_1} \right) M^{-1/3} = 0.08550\, k_B N_A^{1/3} \left(\frac{\rho^{1/3} T}{\eta_1} \right) M^{-1/3}$$

Table 11-1 Molar mass M, partial specific volume $\tilde{v}$, length $L = 2\ a$ and diameter $d = 2\ b = 2\ c$ from electron microscopy (* if calculated from M and $\tilde{v}$), and translational diffusion coefficients D_0 of spheroidal proteins and the bushy stunt virus in dilute salt solutions at 20°C at $c \to 0$. For comparison: diffusion coefficients of three random coil-forming poly(methyl methacrylate)s in the theta solvent butyl chloride at $\Theta = 35.6$°C (assumption: $L = d = 2\ R_{h,D} = (2/1.505)\ \langle s^2 \rangle_o^{1/2}$).

Polymer	M / g mol^{-1}	$\tilde{v}$ / mL g^{-1}	L / nm	d / nm	$10^7 D_0$ / cm^2 s^{-1}
Ribonuclease	13 863	0.728	3.18*	3.18*	11.9
Lysozyme (egg white)	14 211	0.688	3.12*	3.12*	10.4
Albumen (bovine serum)	66 296	0.734	7.5	3.8	5.94
Hemoglobin (human)	67 209	0.749	6.5	5.50	6.3
Catalase	247 600	0.73	8.0	6.4	4.1
Apoferritin	467 200	0.747	12.2	12.2	3.61
Hemocyanin (*Helix pomatia*)	8 994 000	0.738	32.0	32.0	1.06
Bushy stunt virus (tomato)	10 665 000	0.739	31.0	31.0	1.15
Poly(methyl methacrylate)	197 000	0.82			7.18
	1 220 000	0.82	36.3	36.3	4.09
	6 740 000	0.82	81.1	81.1	1.15

Table 11-1 shows that at about the same molar masses and partial spcific volumes, dimensions of unperturbed random coils of poly(methyl methacrylate) molecules are considerably greater than those of spherical proteins. However, in spite of these size differences, diffusion coefficients of these two types of macromolecules are not very different which indicates a fairly strong immobilization of solvent molecules in random coils.

According to Eq.(11-23), diffusion coefficients of compact spheres (axial ratios $a = b = c$) and spheroids ($a \approx b \approx c$) should be inversely proportional to the cube root of their molar masses (i.e., $\delta = -1/3$), which is indeed approximately found (Fig. 11-5). Predicted and observed values of k_D also agree satisfactory.

Protein data of Table 11-1 deliver $D_0 = 3.07 \cdot 10^{-10} M^{-0.348}$ m² s⁻¹ with a correlation coefficient of 0.994 (dilute salt solutions, 20°C) whereas Eq.(11-23) predicts $K_D = 3.22 \cdot 10^{-10}$ m² (kg mol⁻¹)^{1/3} s⁻¹ and $\delta = -1/3$, assuming 25°C, a viscosity of water of $\eta_1 = 1.0087$ Pa s, and an average density of $\rho \approx 1/\tilde{v} = 1.367$ g/cm³ for all proteins.

Eq.(11-23) cannot be exactly valid for spheroidal proteins because these molecules are spheroids and not ideal spheres; they also have no "hard" surface and are not compact but are hydrated. The deviations between theoretical and experimental diffusion coefficients are especially significant if hydrodynamic radii $R_{D,sph}$ are assumed to be one-half of the diameter of a compact sphere with $\rho = 1/\tilde{v}$; for apoferritin, the discrepancy amounts to 22 %. The deviation is less severe for molar masses because of $R_{D,sph} \sim M^{1/3}$; it is here only $(22)^{1/3}$ % ≈ 2.8 %.

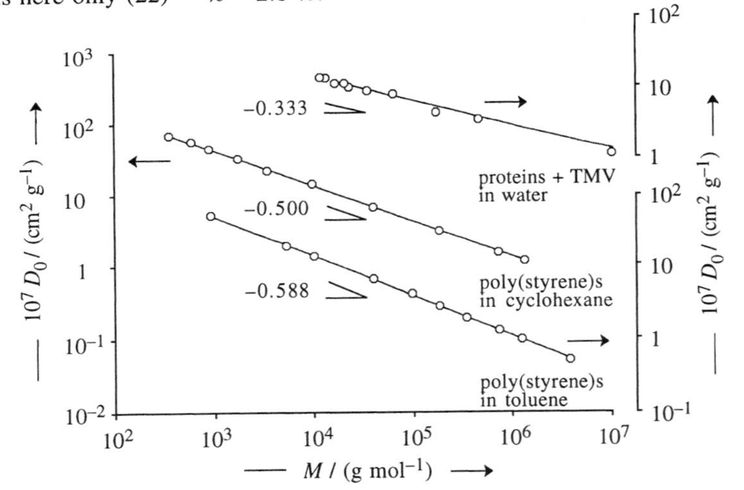

Fig. 11-5 Molar mass dependence of mutual diffusion coefficients D_0 of spheroidal proteins with axial ratios smaller than 1.13 and tobacco mosaic virus (TMV) in dilute salt solutions at 20°C [3]. For comparison (see Section "Coils"): D_0 of random coil-forming poly(styrene)s in cyclohexane at Θ = 34.5°C (unperturbed coils) [4] and in toluene at 15°C (perturbed coils) [5].

Rigid Rods

The diffusion coefficient of rigid rods can be calculated from the Einstein-Sutherland equation, Eq.(11-10), and the Kirkwood-Riseman equation, Eq.(11-15), as

$$(11\text{-}24) \qquad D_0 = \frac{k_B T}{3\pi \eta_1 d_{rod}} \cdot \frac{\ln \Lambda_{rod}}{\Lambda_{rod}}$$

The axial ratio Λ_{rod} of rigid rods with constant diameter is directly proportional to the molar mass M. Diffusion coefficients D_0 must therefore be a function of molar masses because of $D_0 \sim (\Lambda_{rod}/\ln \Lambda_{rod})^{-1} \sim (M/\ln M)^{-1}$. Because of the effect of ln M on the ratio $(M/\ln M)^{-1}$, the diffusion coefficient approaches only slowly the limiting value of $D_0 \sim M^{-1} = M^{\delta}$ at high molar masses. Experimental values of δ vary therefore with molar masses. For example, $\delta = -0.809$ at $10^2 < \Lambda < 10^3$ and $\delta = -0.96$ at $10^8 < \Lambda < 10^9$ instead of $\delta = -1$ at $\Lambda \to \infty$.

Random Coils

The molar mass dependence of diffusion coefficient D_0 of unperturbed random coils is calculated from Eqs.(11-10) and (11-21), using $\langle s^2 \rangle_o^{1/2} = K_{s,o}M^{1/2}$ (Eq.(4-44)),

(11-25) $$D_0 = \frac{(8/3\pi^{1/2})}{6\pi} \cdot \frac{k_B T}{\eta_1} \cdot \frac{1}{\langle s^2 \rangle_o^{1/2}} = 0.0798 \cdot \frac{k_B T}{\eta_1 K_{s,o}} \cdot M^{-1/2} = K_{D,o} M^{-1/2}$$

which is confirmed experimentally for the slope (Fig. 11-5). Experimental and theoretical values of $K_{D,o}$ differ by ca. 7 % which is probably caused by effects of non-uniformity since experimental data for D_0, M, and $\langle s^2 \rangle_o$ in Fig. 11-5 were all mass averages which are *not* the correct corresponding averages for $D_0 = f(1/M_2)$ (see p. 21 ff.). Use of the weight averages of D_0 and M in Eq.(11-24) produces errors in $K_{D,o}$ of ca. 3.6 % if $\overline{M}_w/\overline{M}_n = 1.1$ and 9.8-10.3 % if $\overline{M}_w/\overline{M}_n = 1.3$, depending on the type of molar mass distribution.

Exponents δ of molar mass dependences of diffusion coefficients are therefore −1/3 for hard spheres, −1/2 for unperturbed coils, and −1 for rigid rods, all in the limit of $M \to \infty$. These values lead to the conclusion that δ is the inverse negative value of the fractal dimension, $\delta = -1/\overline{d}_m$ (Section 4.8), i.e., $\overline{d}_m = 3$ for compact spheres, 2 for unperturbed coils, and 1 for rigid rods.

11.1.5 Concentration Dependence, Dilute Solutions

The dependence of diffusion coefficients D on concentration can be described for dilute solutions by a power series, similar to the virial expansion of reduced osmotic pressures, Eq.(10-43):

(11-26) $$D = D_0(1 + k_D c + ...) \quad ; \quad k_D = 2A_2M - k_s - 2\tilde{v}_{2,0} - k_1$$

The proportionality constant k_D contains a thermodynamic contribution $2A_2M$ from the change of chemical potential with concentration, a term $\tilde{v}_{2,0}$ from the concentration dependence of the partial specific volume $\tilde{v}_2$ of the polymer, $\tilde{v}_2 = \tilde{v}_{2,0}(1 + k_2c + ...)$, a term k_1 from the concentration dependence of the partial specific volume $\tilde{v}_1$ of the solvent, $\tilde{v}_1 = \tilde{v}_{1,0}(1 + k_1c + ...)$, and a hydrodynamic contribution k_s from the concentration dependence of friction coefficients ξ, i.e., $\xi = \xi_0(1 + k_sc + ...) \approx \xi_0(1 - k_sc)^{-1}$.

Since k_s, $\tilde{v}_{2,0}$, and probably also k_1 are positive, a negative k_D is expected for diffusion in theta solvents which is indeed observed (Fig. 11-6). A positive k_D arises for

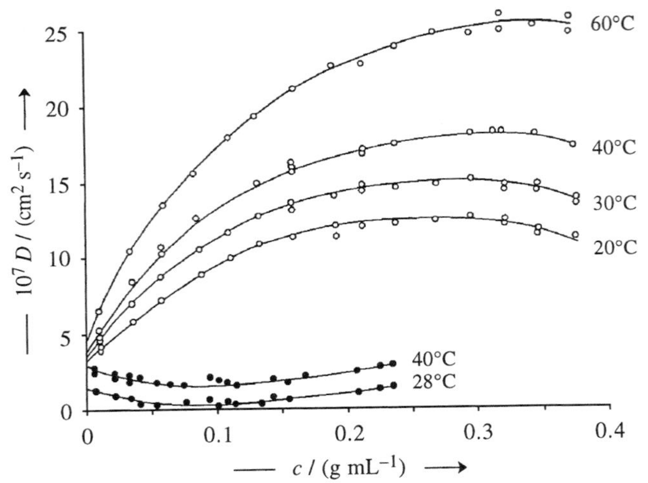

Fig. 11-6 Diffusion coefficients D of a poly(styrene) ($\overline{M}_n$ = 180 000 g/mol) as a function of the concentration c in the good solvent ethyl benzene (O) and the poor solvent cyclohexane (●) [6]. Note that the initial slope is not zero but negative for cyclohexane near the theta temperature of Θ = 34.5°C.

such solvents where the thermodynamic goodness of the solvent, as expressed by 2 A_2M, is great enough to overcome the negative contribution from friction, i.e., k_2, Eq.(11-26). The two density-related quantitites $\tilde{v}_{2,0}$ and k_1 seem to be small correctional factors.

A poly(methyl methacrylate) of M = 200 000 g/mol in the theta solvent butyl chloride at 35.6°C delivered $A_2 = 0$, k_s = 22 mL/g, and $\tilde{v}_{2,0}$ = 0.82 mL/g which leads to $k_D + k_1$ = –24 mL/g which compares favorably with the direct experimental value of k_D = –30 mL/g if one takes into account the probably positive value of k_1. For the same polymer in the good solvent acetone at 20°C, the direct experimental value of k_D = 18 mL/g agrees well with the value of $k_D + k_1$ = 16.4 mL/g that was calculated from the experimental $A_2 = 2.25 \cdot 10^{-4}$ mol mL g^{-2}, k_s = 72 mL/g, and $\tilde{v}_{2,0}$ = 0.798 mL/g.

11.1.6 Concentration Dependence, Semi-concentrated Solutions

Random coils start to overlap at a certain critical overlap concentration (Section 6.3). This overlapping does not affect the local movement of segments which are still present in low concentrations but it does influence the movement of the centers of gravity of the entire coil. At the overlap concentration c_2*, the function $D = f(c_2)$ will therefore change (Fig. 11-7). The transition from one flow regime to the other will not be sharp since the transition from the dilute to the semi-concentrated solution is also not abrupt. It is also difficult to determine since the two linear regions of $D = f(c_2)$ above and below c_2* are relatively narrow (compare Figs. 11-6 and 11-7).

Below the critical concentration c_2*, one measures the diffusion of molecules but above, it the diffusion of segments. The diffusion coefficient $D_{seg,0}$ of segments is obtained by extrapolating diffusion coefficients in the linear region above c_2* to the concentration $c_2 = 0$ (Fig. 11-7). In good solvents, diffusion coefficients $D_{seg,0}$ of segments are greater than translational diffusion coefficients D_0 of whole coils. Friction coefficients $\xi_{seg,0}$ and calculated radii of segments are correspondingly lower but it is questionable whether one can use macroscopic viscosities to calculate radii by Eq.(11-11).

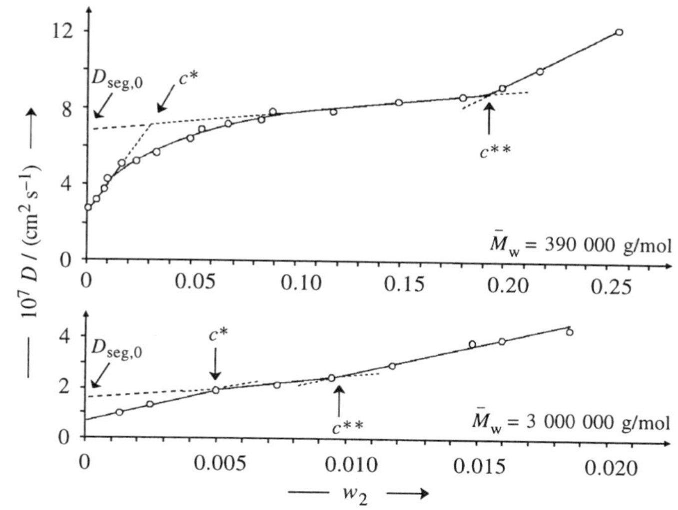

Fig. 11-7 Concentration dependence of diffusion coefficients of two poly(styrene)s in tetrahydrofuran at 30°C [7]. c^* = critical overlap concentration, c^{**} = critical entanglement concentration.

At still higher concentrations c^{**}, high-molar mass chains form entanglements (p. 180) that impede diffusion. Such entanglements act as temporary networks with mesh sizes that are determined by the average distance between two neighboring contact points. This distance is a **screening length** or **correlation length**, L_{sl}, because intermolecularly acting effects of excluded volume are screened by the much more effective contacts. Segments between two contact points behave as blobs with diameters d_{bl} that equal screening lengths (Section 6.3.4).

Unperturbed coils of linear polymer chains start to overlap at higher polymer concentrations than perturbed ones because coil dimensions of the former are smaller than those of the latter. Because chain molecules in the theta state are so strongly coiled, they will not entangle much with other chain molecules so that the entanglement concentration is not much higher then the overlap concentration, i.e., $(c_2^{**})_\Theta \approx (c_2^*)_\Theta$. Perturbed coils, on the other hand, are strongly expanded and begin to overlap at lower concentrations, $c_2^* < (c_2^*)_\Theta$ but start to entangle at higher ones, $c_2^{**} > (c_2^{**})_\Theta$.

Screening lengths L_{sl} of unperturbed and perturbed coils depend therefore differently on polymer concentrations. Scaling theory predicts $L_{sl} = d_{bl} \sim c_2^{-b}$ where b is given by the Flory exponent ν for the molar mass dependence of the radii of gyration and the dimensionality d of the coils:

(11-27) $$L_{sl} \sim c_2^{-b} = c_2^{-\nu/(\nu d-1)}$$

For unperturbed coils (d = 3) in theta solvents (ν = 1/2), screening lengths should be inversely proportional to the polymer concentration, $L_{sl} \sim 1/c_2$. The friction coefficient of a blob, ξ_{bl}, is obtainable from the Stokes equation, $\xi_{bl} = 6\,\pi\,\eta_1 L_{sl}$, since the screening length is also the diameter of the blob ($L_{sl} = d_{bl}$). From the Einstein-Sutherland equation, $D_{bl} = k_B T/\xi_{bl}$, it follows that the diffusion coefficient of a blob in a theta solvent must also be inversely proportional to the solute concentration, $D_{bl} \sim c_2^{-1}$. For perturbed coils, ν = 0.588 (Section 4.5.4), diffusion coefficients in good solvents should therefore be proportional to $c_2^{-0.77}$.

These diffusion coefficients, like those from mass transport, average over the coupled movements of segments and centers of mass. However, theories are usually concerned with diffusion coefficients of centers of resistance against the movements. For this reason, diffusion coefficients from quasi-elastic light scattering are somewhat smaller than the values predicted by the Kirkwood-Riseman theory for the diffusion of unperturbed coils in theta solvents.

Diffusion of polymer molecules in solutions and melts is scientifically, technologically, and biologically extraordinarily important. Diffusion processes control the coupling of polymer chains in polymerizations, the tack of rubbers, and often the physical structure and properties of block polymers.

11.1.7 Structured Flow

The "normal" transport of matter by translational diffusion in fluids is not the only way by which polymers can be transported in gravitational fields. Certain macromolecules (hereafter called "tracers") are rather able to "diffuse" with high speeds through concentrated solutions of other macromolecules, sometimes up to 10^4 times faster than conventional diffusion (Fig. 11-8). For example, such tracers may be macromolecules that contain a small proportion of chromophores or radioactive isotopes.

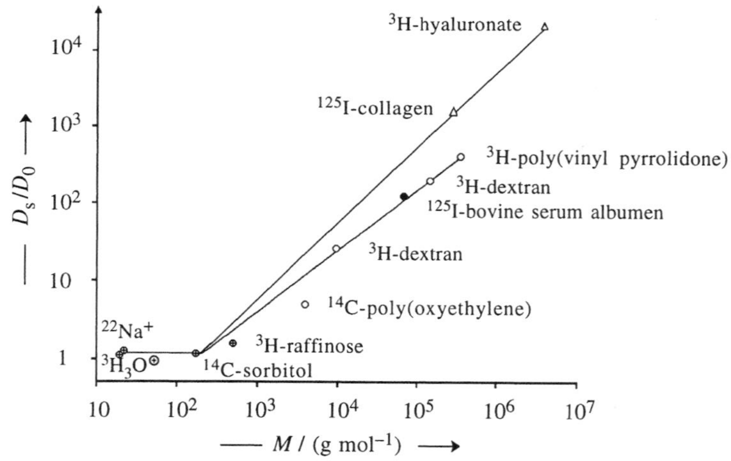

Fig. 11-8 Ratio of diffusion coefficients D_s of tracers in the structured flow system and D_0 in water as a function of the molar mass of the tracer [8]. The structured flow system consisted of a concentration gradient of c = 0.005 g/mL poly(*N*-vinyl pyrrolidone) (M_r = 360 000) in a solution of 0.135 g/mL dextran (M_r = 10 000). Δ Rod-like molecules, O random coils, ● spheroids, ⊕ low molar mass compounds. Lines: empirical with slopes of 1 and 0.8, respectively.

On layering a dilute solution of a tracer on a concentrated solution of another polymer in the same solvent, "fingers" are formed in test tubes and "rings" in Petri dishes. After a while, the fingers and rings disappear and the fluid becomes homogeneous, indicating the absence of phase separation. The transported mass per area is proportional to time and not to the square root of time as in conventional translational diffusion. The phenomenon is also not a **droplet sedimentation** which occurs if a more dense solution S_2 is layered on a less dense solution S_1 because it proceeds with the same speed whether S_2 is placed on S_1, or S_1 is placed on S_2.

Structured flow is the result of the interplay of many factors. Initially, conventional diffusion takes place which then leads to instability at the interface of the two solutions. This instability causes a density inversion which, under the influence of gravity, produces the "fingers." The fingers dissolve by conventional diffusion. Such transportation processes can be very important in biological systems since tissues and cells always contain highly concentrated solutions of polymers (Volume I, Section 14.1.).

A related process is **ionotropic precipitation** which takes place between two compatible components of the solutes and not between two incompatible ones as in structured flows. The diffusion of copper ions into a solution of a polyelectrolyte also produces "fingers". The copper ions then diffuse out of and perpendicular to the long axes of the fingers. The resulting polyelectrolyte salts precipitate and form very regularly arranged tubes if one avoids convection by temperature instabilities, shaking, and the like.

11.2 Sedimentation

Sedimentation and translational diffusion both measure the transport of molecules but the former proceeds under the action of a strong gravitational field and is therefore often considered a "dynamic" method in contrast to the "static" method of translational diffusion. However, the kinetic energy of sedimenting macromolecules is less than 10^{-11} $k_B T$ so that friction coefficients of sedimentation and diffusion are practically identical. At least with respect to friction coefficients, the static/dynamic distinction is unjustified.

11.2.1 Fundamentals

Sedimentation experiments are performed on polymer solutions and colloid dispersions in analytical ultracentrifuges with up to 150 000 revolutions per minute and accelerations of up to 900 000 times the gravitational field of the Earth. The liquid is contained in a sector-shaped compartment in a cylindrical cell that is covered at the top and the bottom with quartz plates that allow one to follow the progress of ultracentrifugation by measuring light absorption, refractive index, or interference as a function of the distance from the center of rotation of a titanium rotor These quantities are then converted to concentrations or concentration gradients, depending on the applied technique.

The ultracentrifuge cell is filled with the liquid (solution, dispersion) consisting of "particles" with density ρ_2 in a solvent with density ρ_1. The centrifugal field causes *sedimentation* of particles in the field direction (towards the bottom of the cell) if $\rho_2 > \rho_1$ and *flotation* in the direction of the center of rotation if $\rho_2 < \rho_1$ (F: *flot* = flood, from L: *fluere* = to flow). After some time, a layer of pure solvent appears in sedimentation rate experiments at the meniscus of the cell and sedimented solute at the bottom (Fig. 11-9).

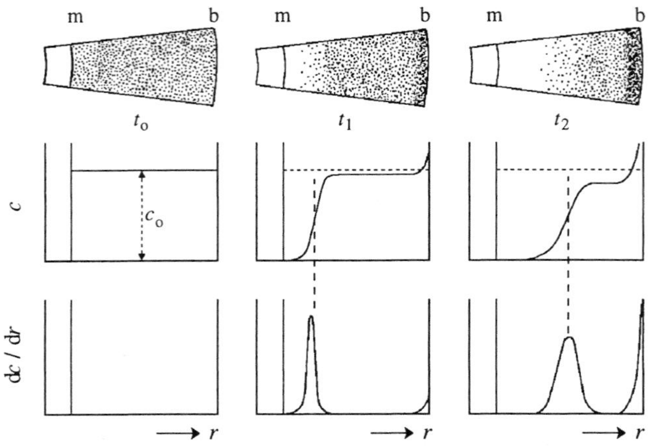

Fig. 11-9 Progress of sedimentation of particles (or molecules) as a function of time in sector-shaped cells (left: initial conditions; center and right: at longer times) and distance r from the center of rotation. Top: visual representation of concentration change; center: change of concentration; bottom: change of concentration gradient. m = meniscus of solution, b = bottom of cell.

The molecular sedimentation rate v_S is proportional to the centrifugal field $\omega^2 r$ where ω = angular velocity and r = distance of sedimentating particles to the center of rotation. The proportionality factor S is called the **sedimentation coefficient** (in literature, usually as "s" which conflicts with the IUPAC designation of s = radius of gyration). The sedimentation coefficient indicates the sedimentation rate per field:

$$(11\text{-}28) \qquad v_S = \mathrm{d}r/\mathrm{d}t = S\omega^2 r \quad ; \quad S = (\mathrm{d}r/\mathrm{d}t)/\omega^2 r$$

The value $S = 1 \cdot 10^{-13}$ s ≡ 1 S is called the **Svedberg unit** S, a non-SI unit.

The change of concentration with time is calculated from the flux $J_S = cv_S$ which is the product of solute concentration c and molecular sedimentation velocity v_S. The mass of the solute that flows in unit time through an area at a distance r_A from the center of rotation to another area at a distance r_B must be equal to the change in the remaining mass with time

$$(11\text{-}29) \qquad (rJ)_A - (rJ)_B = \partial\left(\int_{r_A}^{r_B} rc\mathrm{d}r\right)/\partial t$$

Division of both sides by $\Delta r = r_B - r_A$ delivers for the limiting case $\Delta r \to 0$

$$(11\text{-}30)) \qquad (\partial c/\partial t)_r = -(1/r)[\partial(Jr)/\partial r]_t$$

The total flow $J = J_S + J_D$ is the sum of the forward flow by sedimentation and the backward flow by diffusion caused by the Brownian movement. In the simplest case, the flow by diffusion is described by Fick's first law, i.e., $J_D = -D(\partial c/\partial r)$, Eq.(11-4). The sedimentation term is obtained from $J_S = cv_S$ with $v_S = S\omega^2 r$, Eq.(11-28). Combination of these expressions with Eq.(11-30) leads to the simple **Lamm equation** of ultracentrifugation:

(11-31) $$(\partial c / \partial t)_r = \frac{-\partial[S\omega^2 rc - D(\partial c/\partial r)]}{\partial r}$$

Consideration of Fick's second law, Eq.(11-5), with $r = x$, leads to a Lamm equation that can be solved by numerical integration but not analytically:

(11-32) $$\frac{\partial c}{\partial t} = -\omega^2 S\left(\frac{1}{r}\frac{\partial c}{\partial r} + 2c\right) + D\left(\frac{\partial^2 c}{\partial r^2} + \frac{1}{r}\frac{\partial c}{\partial r}\right)$$

Ultracentrifugation is used for four types of experiments (Table 11-2): determination of sedimentation rates (Section 11.2.2), measurement of sedimentation equilibrium in solution (Section 11.2.3) or in a density gradient of the solvent (Section 11.2.4), and synthetic boundary experiments for the determination of diffusion coefficients. In the latter type of experiments, a sharp initial boundary is produced by a special cell where, at a certain ultracentrifuge speed, a spring contracts or the solvent can overcome capillary forces. Diffusion coefficients are then calculated from peak broadening (Fig. 11-9), for example, by the area method.

Table 11-2 Ultracentrifugation experiments. A_2 = second virial coefficient, D = diffusion coefficient, K = equilibrium constant of association, M = molar mass, S = sedimentation coefficient.

Method	Rotational speed	Rates	Delivered quantity
Sedimentation rate	high	sedimentation >> diffusion	S
Sedimentation equilibrium	low to moderate	sedimentation ≈ diffusion	M, A_2, K
Density gradient	moderate to high	sedimentation ≈ diffusion	density differences
Synthetic boundary	low	diffusion >> sedimentation	D

11.2.2 Sedimentation Rate

Sedimentation Coefficients

A dissolved polymer molecule with the hydrodynamically effective mass m_H and the hydrodynamically effective volume V_H is subjected to a centrifugal force $m_H\omega^2 r$ if it travels in a centrifugal field $\omega^2 r$. The centrifugal force is counteracted by a buoyancy force $V_H\rho_1\omega^2 r$ where ρ_1 = density of pure solvent. The movement with a velocity dr/dt creates a resistance $\xi_S(dr/dt)$ where ξ_S = friction coefficient of the solvated molecule. In the steady state, the resistance is

(11-33) $$\xi_S(dr/dt) = m_H\omega^2 r - V_H\rho_1\omega^2 r$$

The hydrodynamically effective mass, $m_H = m_P + m_S$, is the sum of the mass m_P of the "dry" macromolecule with the molar mass $M = m_P N_A$ and the mass m_S of the solvent molecules that are transported with the macromolecule, i.e., the "solvating" molecules. Defining the degree of solvation as $\Gamma_S \equiv m_S/m_P$, the hydrodynamically effective mass is $m_H = M(1 + \Gamma_S)/N_A$.

The hydrodynamically effective volume, $V_H = V_P + V_S$, is calculated similarly from the specific volume $v_P = V_P/m_P$ of the dry macromolecule and the specific volume $v_S = V_S/m_S$ of the solvating molecules, resulting in $V_H = M(v_P + \Gamma_S v_S)/N_A$.

The specific volume v_S of the solvating molecules differs from the specific volume v_1 of the pure solvent because of the interaction macromolecule-solvating solvent. It furthermore contains a deviation from additivity if polymer and solvent are mixed. v_S is therefore calulated as follows:

The total volume V is the sum of the volumes V_P of dry macromolecules, V_S of solvating molecules, and $V_1 - V_S$ of free solvent molecules. Introduction of specific volumes $v_i \equiv V_i/m_i$ (with i = 1, P, S) leads to

(11-34) $$V = m_P v_P + m_S v_S + (m_1 - m_S)v_1 = m_P v_P + m_1 v_1 + \Gamma_S m_P(v_S - v_1)$$

Very dilute solutions contain far more solvent molecules than are necessary for the solvation of macromolecules. The degree of solvation Γ_S is therefore independent of the polymer concentration c. Differentiation of Eq.(11-34) with respect to the mass m_P of macromolecules delivers the partial specific volume $\tilde{v}_2$ of the polymer:

(11-35) $$\tilde{v}_2 = (\partial V/\partial m_P)_{p,T,m(1)} = v_P + \Gamma_S(v_S - v_1)$$

This partial specific volume of the polymer, $\tilde{v}_2$, is obtained from the concentration dependence of solution densities, $\rho = \rho_1 + (1 - \tilde{v}_2\rho_1)c$. The specific volume of the pure solvent is defined as the inverse of its density, $v_1 \equiv 1/\rho_1$.

Introduction of $\tilde{v}_2$ and v_1 into $V_H = M(v_P + \Gamma_H v_S)/N_A$ leads to

(11-36) $$V_H = (M/N_A)\{\tilde{v}_2 + (\Gamma_S/\rho_1)\}$$

Combination of Eqs.(11-36), (11-28), and (11-33) with $m_H = M(1 + \Gamma_S)/N_A$ delivers an expression for the molar mass M of the "dry" macromolecules:

(11-37) $$M = \xi_S S N_A(1 - \tilde{v}_2\rho_1)^{-1}$$

Eq.(11-37) no longer contains hydrodynamic masses and volumes. However, these quantities are implicit in the friction coefficient ξ_S of sedimenting macromolecules. The density ρ_1 in Eq.(11-37) must be that of the pure solvent and not that of the solution since only solvent molecules flow back if macromolecules sediment.

In Eq.(11-37), sedimentation coefficient S and friction coefficient ξ_S both refer to the same polymer concentration c. In Eq.(11-37), M is the true molar mass for $c \to 0$ and an apparent molar mass, M_{app}, if $c \neq 0$.

Concentration Dependence

Sedimentation coefficients S usually do not depend on rotor speeds U except for very high molar masses. Examples are deoxyribonucleic acids where the ends and the centers of molecules experience different friction forces because the centers are more strongly screened hydrodynamically, so that on centrifugation the ends drag behind the center.

Hydrodynamic screening decreases with increasing rotor speed as do the sedimentation coefficients of very high molar mass deoxyribonucleic acids.

Sedimentation coefficients are measured at finite concentrations and must therefore be extrapolated to zero polymer concentration. According to Eq.(11-37), sedimentation coefficients S are inversely proportional to friction coefficients ξ_S at $M = const.$ Friction coefficients, in turn, are proportional to viscosities η according to the Stokes equation, Eq.(11-11), and viscosities are proportional to concentrations c of dilute solutions. Hence, $1/S \sim \xi_S \sim \eta \sim c$ and one can write

$$(11\text{-}38) \qquad S^{-1} = S_0^{-1}(1 + k_S c + ...)$$

Like the slope coeffficient k_D of diffusion, Eq.(11-26), k_S is not zero for theta conditions. Instead, the slope coefficient rather depends on the molar mass, for example, according to $k_S = 0.052\ M^{1/2}$ for atactic poly(styrene)s in cyclohexane at $\Theta = 35.4°C$.

At low polymer concentrations, sedimentation coefficients S of unperturbed random coil molecules vary with molar masses but only a little with concentration. At higher concentrations, S becomes independent of the molar mass but depends strongly on the concentration (Fig. 11-10).

The transition from one concentration range to the other is very sharp for high molar masses. It is therefore thought to be caused by the onset of the entanglement of coil molecules. Entangled coils no longer sediment as single molecules but rather as a network. Back-flowing solvent molecules then flow through the network and no longer around polymer coils.

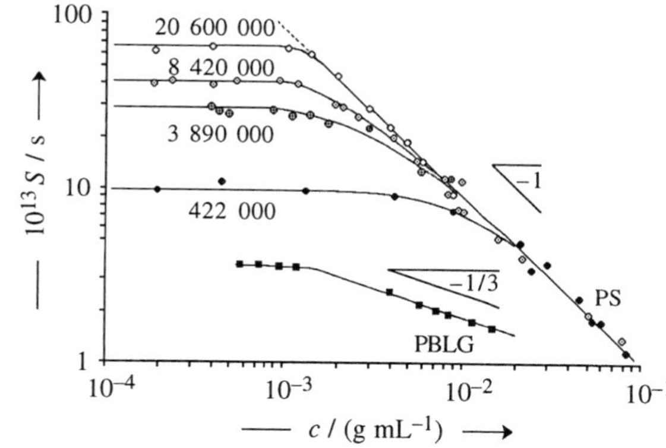

Fig. 11-10 Concentration dependence of sedimentation coefficients S of unperturbed random coils of atactic poly(styrene)s (PS) with different molar masses (in g/mol) in cyclohexane at $\Theta = 35.4°C$ [9] and of a rodlike poly(γ-benzyl-L-glutamate) (PBLG) in N,N-dimethylformamide at $T = 25°C$ [10].

The solvent flow through the network of entangled polymer chains can be described by a kind of **Darcy law** for the flow through porous media (Section 14.4.1). The permeability coefficient λ of such a network,

$$(11\text{-}39) \qquad \lambda = \frac{m_{\text{blob}}}{6\,\pi\, c L_{\text{cl}}} = \frac{L_{\text{cl}}^{\ 2}}{6\,\pi}$$

increases with the mass of blobs, $m_{bl} = cL_{bl}^3$, and is therefore proportional to the square of the correlation length (screening length) L_{cl}. Screening lengths scale with $c^{-\nu/(\nu d-1)}$, (Eq.(11-27)), and the permeability coefficient should thus vary with c^{-2} for theta solutions ($\nu = 1/2$, $d = 3$) and with $c^{-1.54}$ for good solvents ($\nu = 0.588$; $d = 3$). The sedimentation coefficient is furthermore directly proportional to the permeability coefficient:

$$(11\text{-}40) \qquad S \sim c\lambda \sim cL_{cl}^2 \sim c^{[\nu(d-2)-1]/[\nu d-1]}$$

For theta conditions, one expects $S \sim c^{-1}$ (Fig. 11-10). In good solvents, a proportionality $S \sim c^{-0.539}$ is expected but experiments usually find $S \sim c^{-0.7}$.

Two concentration regions are also found for relatively stiff molecules such as poly(γ-benzyl-L-glutamate) (PBLG) in the helicogenic solvent *N,N*-dimethylformamide (Fig. 11-10). These molecules are relatively rigid as can be inferred from the exponent $\alpha_v = 1.82$ of the intrinsic viscosity-molar mass relationship (Section 12.3.5). Such rigid macromolecules cannot entangle, except at very high molar masses. The transition between the two $S = f(c)$ regimes occurs at $c \approx 0.0016$ g/mL which is far below the value of $\phi_2{}^* = 0.056$ that can be expected for a rigid rod with an axial ratio $\Lambda = 141.5$. A possible explanation may be an orientation of these rod molecules by the centrifugal field but such an effect is usually small. There is also no change in the function $\lg S = f(\lg c)$ near the critical concentration for the formation of mesophases.

A similar effect is also found for $\lg S = f(c)$ of spherical molecules. Some researchers therefore doubt the concept of entanglement of molecules in semi-concentrated solutions, at least as far as slow (static) processes are concerned. Instead, it has been proposed to describe the concentration dependence of sedimentation coefficients (and also of diffusion coefficients) by **stretched exponentials**, for example

$$(11\text{-}41) \qquad S = S_0 \exp(-Kc^p) \quad ; \quad K, p = \text{empirical constants}$$

Molar Mass Dependence

Molar masses depend on the product of sedimentation coefficients and friction coefficients of sedimentation, Eq.(11-37). In analogy to the molar mass dependence of diffusion coefficients, Eq.(11-22), one can therefore expect that the sedimentation coefficient at infinite dilution is a power function of the molar mass:

$$(11\text{-}42) \qquad S_0 = K_S M^{\varsigma}$$

The constants K_S and ς are obtained empirically. For molecularly non-uniform polymers, the M in Eq.(11-42) is an exponent average (Eq.(2-19)).

Molar masses can also be obtained from sedimentation coefficients with calibrations if the friction coefficient in Eq.(11-37) is replaced by a directly measurable quantity. Such a quantity is the diffusion coefficient because friction coefficients of sedimentation and diffusion are identical within the limits of experimental error and, according to theoretical calculations, even within 10^{-10} %. With $\xi_S = \xi_D$, one obtains from Eqs.(11-10) and (11-37) the **Svedberg equation**, Eq.(11-43), which contains only the directly measurable quantities S, D, $^*v_2$, ρ_1, and T (Th. Svedberg, Nobel prize 1926):

(11-43) $M = \dfrac{SRT}{D(1-\tilde{v}_2\rho_1)}$; **Svedberg equation**

The M of Eq.(11-43) is an apparent molar mass M_{app} for S and D at finite concentrations and a true molar mass M for S_0 and D_0 at infinite dilution. However, this true molar mass is a mixed average for molecularly non-uniform polymers (Section 2.3.5) that varies with the averages of S_0 and D_0, the molecular shape of the polymer, and the thermodynamic goodness of the solvent. The latter two factors can be expressed by the exponent α of the intrinsic viscosity-molar mass relationship, $[\eta] = K_v M^{\alpha}$ (Chapter 12), using $\alpha = 0$ (spheres), 1/2 (unperturbed coils), 0.588 (perturbed coils), and 2 (rigid rods) (Table 11-3). The exponent $\alpha = -1$ applies to discs with constant diameter and a height that is proportional to the molar mass.

Table 11-3 shows that a wide variety of molar mass averages can be obtained, depending on the averages of S_0 and D_0. Experimental values of these averages agree reasonably well with the ones calculated from experimental mass-average molar masses of parent polymers I and II that were assumed to be molecularly uniform. The latter assumption causes errors but these are in the right direction: calculated values of M are too low for $\overline{M}_z$ (from SE), about right for $\overline{M}_w$ (from LS), and too high for $\overline{M}_n$ (from OP).

For molecularly non-uniform polymers, corrections can be substantial. They always depend on the shape of the macromolecule and its interaction with the solvent (as measured by α_η), the width of the molar mass distribution, and sometimes also on the type of distribution.

Table 11-3 Theoretical moments and averages of some hydrodynamic molar mass averages [11] and some experimental data for mixtures of two narrowly distributed atactic poly(styrene)s I and II in the theta solvent cycohexane at 34.5°C [12]. For calculations, these two polymers were assumed to be molecularly uniform ($\overline{M}_w/\overline{M}_n \equiv 1$). Measurements by sedimentation equilibrium (SE), static light scattering (LS), membrane osmometry (OP), intrinsic vicosity $[\eta]$, and various averages of sedimentation and diffusion coefficients. α = exponent in $[\eta] = K_v M^{\alpha}$ (see Chapter 12).

Theoretical data				Experimental M/(g mol^{-1})	
Method	α	Moment	Average	Calculated	Measured
SE	1/2	$\mu_z^{(1)}$	$\overline{M}_z$	538 000	562 000
LS	1/2	$\mu_w^{(1)}$	$\overline{M}_w$	428 000	426 000
$\overline{S}_n + \overline{D}_n$	2	$\mu_{n-1}^{(1)}$	$\overline{M}_{n-1}$		
$\overline{S}_n + \overline{D}_w$	any	$\mu_n^{(1)}$	$\overline{M}_n$		
$\overline{S}_w + \overline{D}_w$	2	$\mu_n^{(1)}$	$\overline{M}_n$		
	1/2	$\mu_w^{(1/2)}/\mu_n^{(1/2)}$	$\sum_i m_i M_i^{1/2}/\sum_i n_i M_i^{1/2}$	368 000	314 000
	−1	$\mu_w^{(1)}$	$\overline{M}_w$		
$\overline{S}_w + \overline{D}_z$	any	$\mu_w^{(1)}$	$\overline{M}_w$		
$\overline{S}_w + [\eta]$	2	$(\mu_w^{(2)})^{1/2}$	$(\overline{M}_w\overline{M}_z)^{1/2}$		
	1/2	$(\mu_w^{(1/2)})^2$	$(\sum_i w_i M_i^{1/2})^2 \equiv \overline{M}_{v,\Theta}$	398 000	395 000
OP	1/2	$\mu_n^{(1)}$	$\overline{M}_n$	317 000	297 000

An example is the combination of mass-average molar masses of sedimentation coefficients with intrinsic viscosities $[\eta]$ (see Eq.(11-46)), which leads to the following equations for Schulz-Zimm distributions SZ (with coupling factors ς) and logarithmic normal distributions LN:

(11-44) $$\overline{M}_{s_w,\eta} = \frac{\overline{M}_n}{\varsigma}\left(\frac{\Gamma^{3/2}\{[3\lambda+5-\alpha_\eta]/3\}\,\Gamma^{1/2}[\lambda+1+\alpha_\eta]}{\Gamma^3(\lambda+1)}\right) \quad ; \text{ SZ distribution}$$

(11-45) $$\overline{M}_{s_w,\eta} = \overline{M}_n(\overline{M}_w/\overline{M}_n)^{(2\alpha_\eta^2-2\alpha_\eta+5)/6} \quad ; \text{ LN distribution}$$

The friction coefficient ξ_S in Eq.(11-37) can also be replaced by other experimental quantities. In the Svedberg equation, Eq.(11-43), ξ_S was substituted by the expression for the friction coefficient of diffusion, $\xi_D = k_B T/D_0$, Eq.(11-10), which contains the Stokes radius of a hydrodynamically equivalent sphere, $\xi_{eq} = 6\pi\eta_1 R_D$, Eq.(11-11). The radius of a hydrodynamically equivalent sphere can also be obtained as the Einstein radius R_v from intrinsic viscosities $[\eta]$ of dilute solutions (Section 12.3.2), leading to

(11-46) $$M = \left[\frac{6^2\pi N_A R_D^{3/2}}{20^{1/2}R_v^{3/2}}\right]\frac{\eta_1}{1-\tilde{v}_2\rho_1}s^{3/2}[\eta]^{1/2} = \left(\left\{\frac{N_A}{100^{1/3}\beta}\right\}^{3/2}\frac{\eta_1}{1-\tilde{v}_2\rho_1}\right)s^{3/2}[\eta]^{1/2}$$

For historic reasons, the resulting **Mandelkern-Flory-Scheraga equation** (Eq.(11-46), first expression after the equality sign) is often written differently (second expression after the equality sign of Eq.(11-44)), using

$$\beta = \beta'/100^{1/3};\quad \beta' = \Phi^{1/3}/(6\pi\zeta_{D,G});\quad \Phi = 20\pi N_A(\zeta_{v,G})^3/6;\quad \zeta_{D,G}/\zeta_{v,G} = \zeta_{D,v}$$

Stokes radii and Einstein radii are identical for *compact spheres*. In Eq.(11-46), the factor $(6^2\pi N_A/20^{1/2})$ becomes $1.523\cdot 10^{25}$ mol^{-1} and the Flory-Mandelkern invariant is then $\beta = [N_L/(5\cdot 6^4\pi^2)]^{1/3} \approx 2.112\cdot 10^6$ mol$^{-1/3}$.

Stokes and Einstein radii are not necessarily the same for all other kinds of particles. The β value is $2.084\cdot 10^6$ mol$^{-1/3}$ for *porous spheres* and $2.344\cdot 10^6$ mol$^{-1/3}$ for *non-draining coils* (see Table 12-6).

For *rotational ellipsoids*, β depends on the axial ratio Λ: very little variation of β with Λ for oblate ellipsoids ($\beta = 2.115$ at $\Lambda = 2$ and $\beta = 2.15$ at $\Lambda = 300$) but very large ones for prolate ellipsoids ($\beta = 2.13$ at $\Lambda = 2$ and $\beta = 1.81\cdot\Lambda^{0.126}$ at $3 \le \Lambda \le 300$).

Mixtures of a few types of macromolecules can be separated by ultracentrifugation into components if back-diffusion is minor. The areas under the bell-like gradient curves are proportional to the mass fractions of components. This type of experiment is often used in protein chemistry.

Molar mass distributions are obtained from the broadening of gradient curves with time which gives the sedimentation coefficients for the different mass fractions of the solute. The resulting sedimentation coefficients are then extrapolated to infinite time in order to correct for curve broadening by diffusion. The resulting function $w_i = f(S_i)$ is then converted to $w_i = f(M_i)$. Such experiments are advantageously performed in theta solvents in order to reduce corrections for molecular non-homogeneity.

11.2.3 Sedimentation Equilibrium

In sufficiently small gravitational fields, sedimentation rates equal diffusion rates and a sedimentation equilibrium is established. In equilibrium at each position of the ultracentrifuge cell, concentration gradients become constant, $(\partial c/\partial t)_r = 0$, and Eq.(11-31) converts to

(11-47) $$S/D = (\partial c/\partial r)/(\omega^2 rc)$$

Replacing S/D with the corresponding expression of the Svedberg equation, Eq.(11-43) leads to an equation for the sedimentation equilibrium:

(11-48) $$M = \frac{RT}{\omega^2(1-\tilde{v}_2\rho_1)} \cdot \frac{\partial c/\partial r}{rc}$$

Experimentally, it is advantageous to work with small filling heights Δr of ultracentrifuge cells since then the concentration gain between the meniscus of the liquid and the center of the cell equals the concentration loss between the center and the bottom of the cell. For molar mass determinations, it is sufficient to determine the concentration gradient dc/dr at the average fill height r but this procedure loses all information about the molar mass distribution which is contained in the function $dc/dr = f(r)$.

The M in Eq.(11-48) must be a mass average since solving this equation for dc/dr and summation of variables shows that the average concentration gradient of molecules of all molar masses must be proportional to the sum of the products of concentration and molar mass. The same sum is contained in the definition of the mass average of molar mass, Eq.(2-15):

(11-49) $$\langle dc/dr \rangle = \omega^2 r(1 - \tilde{v}_2\rho_1)(RT)^{-1}(\Sigma_i c_i M_i) = \omega^2 r(1 - \tilde{v}_2\rho_1)(RT)^{-1} c\,\overline{M}_w$$

Eqs.(11-48) and (11-49) apply only to infinite dilution. At $c \neq 0$, they deliver apparent molar masses that need to be extrapolated to $c \rightarrow 0$ by plotting $1/M_{w,app} = f(c)$.

11.2.4 Sedimentation Equilibrium in Density Gradients

Sedimentation equilibria in density gradients allow one to determine density differences of macromolecules. They have been used to study the replication of ^{15}N tagged deoxyribonucleic acids but can also distinguish between true copolymers and polymer mixtures, at least in principle. However, experiments of the latter type very often run into problems because of broad molar mass distributions and considerable back-diffusion. Both effects broaden the gradient curves considerably, which leads to difficult-to-resolve strong overlap of curves from substances with different densities. Thus, for synthetic polymers, this type of experiment is only suitable for very high molar masses.

Ultracentrifugation generates density gradients if the solvent is a mixture of substances with very different densities, such as cesium chloride in water or the mixture of benzene and carbon tetrabromide. In such mixtures, both components sediment at dif-

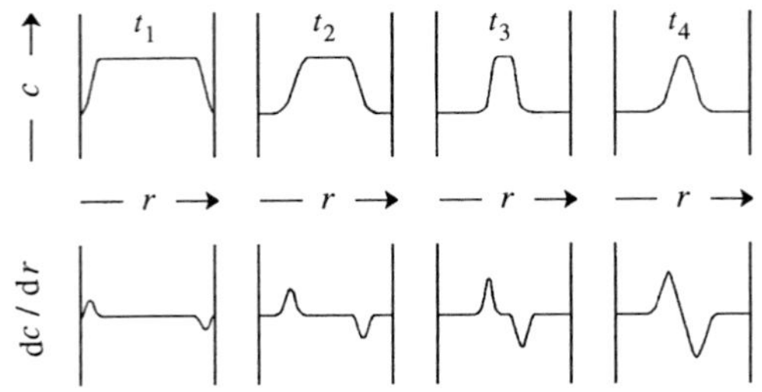

Fig. 11-11 Time dependence of polymer concentration c and concentration gradient dc/dt on the approach to equilibrium in a density gradient (schematic).

ferent rates and form a density gradient. The concentration gradient of the polymers is thus overlapped by the concentration gradient of the solvent mixture.

The density of the solution at the bottom of the ultracentrifuge cell is ρ_b and at the meniscus, ρ_m. If the density ρ of a polymer is in the range $\rho_m < \rho < \rho_b$, the polymer will sediment from the meniscus, float from the bottom, and collect at that position in the gradient where the density of the solvent mixture equals that of the polymer (Fig, 11-11). However, the density ρ of the polymer is not the density of the "dry" polymer but that of a solvated one because one component of the mixed solvent will solvate the polymer better than the other.

The solvent consists of the components 1 + 3. If the polymer is solvated exclusively by component 1, then the solvated polymer molecule will have the mass $m_S = m_P + m_1$ and the extent of preferential solvation will be $\Gamma_1 = m_1/m_P$. Introduction of molar masses from $M_i = m_i M_i$ (i = H, 1, P) delivers $M_S = M_P(1 + \Gamma_1)$. The maximum of the gradient curve (Fig. 11-11, top right) thus indicates that density ρ^* of the density gradient at which the polymer with its solvating shell and its molar mass M_S resides.

The extent of preferential solvation is obtained from

$$\text{(11-50)} \qquad \Gamma_S = (\tilde{v}_2 \rho^* - 1)/(1 - \tilde{v}_1 \rho^*)$$

where $\tilde{v}_2$ and $\tilde{v}_1$ are the partial specific volumes of the polymer 2 and solvent component 1 and ρ^* = density of the solvent mixture at the position of the maximum of the polymer gradient curve, assuming very dilute polymer solutions.

The partial specific volume of the polymer, $\tilde{v}_2$, is obtained with good approximation from the dependence of the density ρ of the polymer solution on the polymer concentration c in the mixed solvent x, $\rho = \rho_x + (1 - \tilde{v}_2 \rho_x)c$, where ρ_x is the density of the mixed solvents consisting of components 1 and 3.

11.3 Field-Flow Fractionation

Field-flow fractionation (**FFF**) is the fractionation of a polymer in a force field and **flow FFF** a field-flow fractionation in the presence of a second solvent.

In FFF, a polymer solution flows through a narrow channel to which a force field is applied perpendicular to the flow direction. The force field can be a temperature difference, a difference in the gravity field caused by centrifugation, a difference in electrical potential, or a crosswise flow of a second solvent. These flow fields generate gradients which separate molecules according to their respective properties in the field.

FFF resembles chromatography since the flow distributes matter between two regions, a relatively stationary one near the wall and a relatively mobile one in the center of the channel. FFF differs from chromatography because the distribution is not between two phases. Despite its many methodical variations, FFF is rarely used.

11.4 Electrophoresis

Electrophoresis is the movement of electrically charged particles of mass m and electrical charge q in a homogeneous electrical field of field strength $\boldsymbol{E}$ (G: *electron* = amber; G: *pherein* = to bear). Such charged particles may be biological cells, association colloids, polyelectrolytes, or other low-molar mass electrolytes and also electrically neutral particles that have been made electrophoretically mobile by complexation.

In free electrophoresis (**Tiselius electrophoresis**; A.W.K.Tiselius, Nobel prize 1948), charged particles move in a salt solution and, in **carrier electrophoresis**, in a swollen inert carrier such as paper, starch gel, agarose, crosslinked poly(acrylamide), etc.

The electrophoretic movement of the particles is forced by a power $q\boldsymbol{E}$ that is opposed by a frictional force $\xi(\mathrm{d}L/\mathrm{d}t)$ where ξ = friction coefficient, $\mathrm{d}L/\mathrm{d}t$ = linear velocity, and $\boldsymbol{E}$ = electric field. According to Newton's law, the resultant of these two forces is

(11-51) $$m(\mathrm{d}^2L/\mathrm{d}t^2) = q\boldsymbol{E} - \xi(\mathrm{d}L/\mathrm{d}t)$$

(11-52) $$\mathrm{d}L/\mathrm{d}t = (q\boldsymbol{E}/\xi)[1 - \exp(-\xi t/m)]$$

The ratio ξ/m of friction coefficient ξ to mass m is ca. $(10^{12}\text{-}10^{14})\ \mathrm{s}^{-1}$ for molecules; hence, for times larger than ca. 10^{-11} s, Eq.(11-52) reduces to

(11-53) $$\mathrm{d}L/\mathrm{d}t \approx q\boldsymbol{E}/\xi$$

The **electrophoretic mobility**, $\mu \equiv (\mathrm{d}L/\mathrm{d}t)/\boldsymbol{E}$, is defined as the moving rate per electrical field of 1V/cm. Introduction of the Einstein-Sutherland equation $D = k_\mathrm{B}T/\xi$ [Eq.(11-10) with $D_0 = D$ and $\xi_\mathrm{D} = \xi$] leads to

(11-54) $$D = k_\mathrm{B}T(\mu/q) \quad ; \qquad \mu = (\mathrm{d}L/\mathrm{d}t)/\boldsymbol{E}$$

In research, electrophoresis is used to analyze and separate particles with respect to their electrophoretic mobilities. The resulting relative proportions of components depend on concentrations c and ionic strength I_c and are therefore extrapolated to the ratio $c/I_\mathrm{c} \to 0$.

In industry, electrophoresis is used in electrodeposition coating where the article serves as the anode. Application of an electric field causes negatively charged latex particles to move to the anode where they deposit as a film. Subsequent electroosmosis removes water from the film, resulting in a solids content of up to 95 %. The remaining water and dissolved ions are eliminated by a follow-up electrolysis.

Electrodeposition coating allows even coating of corners and edges. Since the process works with aqueous dispersions, no recovery of solvent vapor is necessary. Electrodeposition coating is therefore increasingly used for the spray painting of cars.

Historical Notes to Chapter 11

DIFFUSION

Derivation and treatment of Navier (1827) and Stokes (1845) equations, see, for example, L.Prandtl, O.G.Tietjens, Fundamentals of Hydro- and Aeromechanics, McGraw-Hill, New York 1934

F.Perrin, J.Phys.Rad., Ser. VII, **5** (1934) 497
Friction coefficients of ellipsoids

S.H.Koenig, Biopolymers **14** (1975) 2421
Corrections of Perrin equations

Spring-and-bead models

J.G.Kirkwood, J.Riseman, J.Chem.Phys. **16** (1948) 565
Theory of transport of coils with Gaussian segment distributions.

P.E.Rouse, J.Chem.Phys. **21** (1953) 1272
Spring-and-bead model without hydrodynamic interactions (free-draining coils). In the free-draining limit, diffusion coefficients of coil molecules are inversely proportional to the molar mass, which has never been found experimentally.

B.H.Zimm, J.Chem.Phys. **24** (1956) 269
Spring-and-bead model with hydrodynamic interactions (non-draining coils). In the non-draining limit, random coils behave as solvent-filled hard spheres (no flow of solvent molecules through the coil), an assertion that was first made by
P.J.W.Debye, A.M.Bueche, J.Chem.Phys. **16** (1948) 573

ULTRACENTRIFUGATION

T.Svedberg, J.B.Nichols, J.Am.Chem.Soc. **45** (1923) 2910; T.Svedberg, H.Rinde, J.Am.Chem.Soc. **46** (1924) 2677
First analytical ultracentrifuges

O.Lamm, Ark.Math.Astron.Fysik, **21B**/2 (1929) 1
Derivation of Lamm equation

Literature to Chapter 11

11.a GENERAL REVIEWS (polymers in solution)

H.Yamakawa, Modern Theory of Polymer Solutions, Harper and Row, New York 1971

H.Morawetz, Macromolecules in Solution, Interscience, New York, 2nd ed. 1975

W.C.Forsman, Ed., Polymers in Solution. Theoretical Considerations and Newer Methods of Characterization, Plenum, New York 1987

H.Fujita, Polymer Solutions, Elsevier, Amsterdam 1990

J. des Cloiseaux, G.Jannink, Polymers in Solution. Their Modelling and Structure, Oxford Univ. Press, New York 1990

W.W.Graessley, Polymeric Liquids and Networks. Structure and Properties, Garland Science, New York 2003

11.b GENERAL REVIEWS (polymer dynamics)

P.-G.de Gennes, Scaling Concepts in Polymer Physics, Cornell University Press, Ithaca (NY) 1979

R.B.Bird, R.C.Armstrong, O.Hassager, Dynamics of Polymeric Liquids, Vol. 1; R.B.Bird, C.F.Curtiss, R.C.Armstrong, O.Hassager, ditto, Vol. 2, Wiley, New York, 2nd ed. 1987

M.Doi, S.F.Edwards, The Theory of Polymer Dynamics, Oxford University Press, Oxford 1987

K.F.Freed, Renormalization Group Theory of Macromolecules, Wiley, New York 1987

11.1 TRANSLATIONAL DIFFUSION

J.Crank, G.S.Park, Eds., Diffusion in Polymers, Academic Press, London 1968

J.S.Vrenta, J.L.Duda, Molecular Diffusion in Polymer Solution, AIChE J. **25** (1979) 1

B.Nyström, J.Roots, Scaling Concepts in the Interpretation of Diffusion and Sedimentation Phenomena in Semidilute Polymer and Polyelectrolyte Solutions, Progr.Polym.Sci. **8** (1982) 333

B.D.Freeman, Mutual Diffusion in Polymeric Systems, in S.L.Aggarwal, S.Russo, Eds., Comprehensive Polymer Science, First Supplement, Pergamon Press, Oxford 1992, p. 167

P.H.Lloyd, Optical Methods in Ultracentrifugation, Electrophoresis, and Diffusion, Oxford University Press, Oxford 1994

11.1.2 MEASUREMENTS: DYNAMIC LIGHT SCATTERING

R.J.Berne, R.Pecora, Dynamic Light Scattering, Wiley, New York 1976

W.Burchard, New Aspects of Polymer Characterization by Dynamic Light Scattering, Chimia **39** (1985) 10

K.S.Schmitz, An Introduction to Dynamic Light Scattering by Macromolecules, Academic Press, San Diego 1990

W.Brown, Ed., Dynamic Light Scattering, Clarendon, Oxford 1993

W.Brown, Light Scattering, Oxford University Press, New York 1996

11.1.7 STRUCTURED FLOW

W.D.Comper, B.N.Preston, Rapid Polymer Transport in Concentrated Solutions, Adv.Polym.Sci. **55** (1984) 105

11.2 SEDIMENTATION

T.Svedberg, K.O.Pedersen, Die Ultrazentrifuge, Steinkopff, Dresden 1940; -, The Ultracentrifuge, Clarendon Press, Oxford 1940

H.K.Schachman, Ultracentrifugation in Biochemistry, Academic Press, New York 1959

R.L.Baldwin, K.E.van Holde, Sedimentation of High Polymers, Fortschr.Hochpolym.-Forschg. **1** (1960) 451

H.-G.Elias, Ultrazentrifugen-Methoden, Beckman Instruments, Munich, 2nd ed. 1961; -, ditto, with A.Fritsch, Méthodes de L'Ultracentrifugation Analytique, Institut Pasteur, Paris, 3rd ed. (1964)

J.Vinograd, J.E.Hearst, Equilibrium Sedimentation of Macromolecules and Viruses in a Density Gradient, Fortschr.Chem.Org.Naturstoffe **20** (1962) 372

J.W.Williams, Ed., Ultracentrifugal Analysis in Theory and Experiment, Academic Press, New York 1963

C.H.Chervenka, A Manual of Methods for the Analytical Ultracentrifuge, Beckman Instruments, Palo Alto (CA) 1969

T.J.Bowers, A.J.Rowe, An Introduction to Ultracentrifugation, Wiley-Interscience, New York 1970

H.Fujita, Foundations of Ultracentrifugal Analysis, Wiley, New York 1975
R.Hinton, M.Dokata, Density Gradient Centrifugation, North Holland, Amsterdam 1976
C.A.Price, Centrifugation in a Density Gradient, Academic Press, New York 1982
B.Nyström, J.Roots, Scaling Concepts in the Interpretation of Diffusion and Sedimentation Phenomena in Semidilute Polymer and Polyelectrolyte Solutions, Progr.Polym.Sci. **8** (1982) 333
S.E.Harding, A.J.Rowe, J.C.Horton, Analytical Ultracentrifugation in Biochemistry and Polymer Science, The Royal Society of Chemistry, Cambridge 1992
G.Ralston, Introduction to Analytical Ultracentrifugation, Beckman Instruments, Fullerton (CA) 1993
T.M.Schuster, T.M.Laue, Eds., Modern Analytical Ultracentrifugation, Birkhäuser, Basel 1994

11.3 FIELD-FLOW FRACTIONATION
J.C.Giddings, Field Flow Fractionation, Analytical Chem. **53** (1981) 1170 A
J.Janca, Field-Flow Fractionation, Dekker, New York 1987
M.E.Schimpf, Advances in Field-Flow Fractionation for Polymer Analysis, Trends in Polym.Res. **4** (1996) 114
M.E.Schimpf, Thermal Field-Flow Fractionation, Polym. News **24** (1999) 78
H.Cölfen, M.Antonietti, Field Flow Fractionation Techniques for Polymer and Colloid Analysis, Adv.Polym.Sci. **150** (2000) 67

11.4 ELECTROPHORESIS
Ö.Gaal, G.A.Medgyesi, L.Vereczkey, Electrophoresis in the Separation of Biological Macromolecules, Wiley, New York 1980
A.T.Andrews, Electrophoresis, Clarendon Press, Oxford 1986

References to Chapter 11

[1] W.Burchard, Chimia **39** (1985) 10, Fig. 4b
[2] W.Burchard, J.Bauer, P.Lang, Macromol.Symp. **61** (1992) 25, Fig. 2
[3] J.T.Edsall, in H.Neurath, K.Bailey, Eds., The Proteins, Academic Publ., New York 1953, Volume I, Part B, Table VIII
[4] T.Yamada, T.Toshizaki, H.Yamakawa, Macromolecules **25** (1992) 377, Tables I and II
[5] T.Arai, F.Abe, T.Yoshizaki, Y.Einaga, H.Yamakawa, Macromolecules **28** (1995) 3609, Tables 1 and 2
[6] G.Rehage, O.Ernst, Kolloid-Z.Z.Polym. **197** (1964) 64, Figs. 1 and 2
[7] T.L.Yu, H.Reihanian, J.G.Southwick, A.M.Jamieson, J.Polym.Sci.-Polym.Phys.Ed. **18** (1980) 178, Figs. 4 and 5
[8] W.D.Comper, B.N.Preston, Adv. Polym.Sci. **55** (1984) 105
[9] P.Vidakovicz, C.Allain, F.Rondelez, Macromolecules **15** (1982) 1571, Fig. 4
[10] L.-O.Sundelöf, B.Nyström, J.Polym.Sci.-Polym.Lett.Ed. **15** (1977) 377, Table IV
[11] H.-G.Elias, Pure Appl.Chem. **43**/1-2 (1975) 115; see also H.-G.Elias, R.Bareiss, J.G.Watterson, Adv.Polym.Sci. **11** (1973) 111
[12] A.Kotera, T.Saito, K.Takemura, IUPAC Congress Boston 1971, Macromolecular Preprints, p. 1029

12 Viscosity of Dilute Solutions

Viscosities of dilute solutions of polymer homologs increase at constant concentration with the molecular weight of the polymer. Since such viscosities can be determined fast and with simple devices, their measurement became the most popular method for the determination of molecular weight. The theoretical interpretation of such dilute viscosity data is not simple, however.

12.1 Fundamentals

12.1.1 Definitions

Rheology is concerned with the time dependence of deformation of all sorts of matter (Chapter 15). **Viscometry**, a subfield of rheology concerned with fluid matter, characterizes this dependence by the product of stress (or pressure) and time, the **viscosity** (L: *viscum* = mistletoe, bird-lime from mistletoe berries; L = *viscos* = tough, sticky).

There are three different types of viscosity: shear viscosity caused by shearing, extensional viscosity caused by elongation, and volume viscosity caused by hydrostatic pressure. Viscometry of dilute polymer solutions is only concerned with the first of these three viscosities, **shear viscosity**, and here usually only with **dynamic** (shear) **viscosities** η and not with **kinematic** ones, η/ρ, where ρ = density.

Dynamic viscosity, $\eta = \sigma_{21}/\dot{\gamma}$, is the ratio of shear stress σ_{21} perpendicular to the flow direction to shear rate, $\dot{\gamma} = \partial v/\partial y$, the change of flow rate v with distance y perpendicular to the flow direction (see Fig. 15-4). Fluids are called **Newtonian** if the ratio $\eta = \sigma_{21}/\dot{\gamma}$ does not depend on either the shear rate or time.

Viscosities of dilute polymer solutions usually do not depend on time, but those of high-molecular weight polymers may be a function of shear rate. This effect is undesired if one is interested in properties such as molar masses, molecular dimensions, or friction coefficients. Hence, dynamic viscosities of polymer solutions must be measured at low shear rates or be extrapolated to zero shear rate (Section 12.1.2).

In general, one is not interested in the viscosity η of a dilute solution *per se* but in its **viscosity ratio** (IUPAC), $\eta_r = \eta/\eta_1$, the ratio of viscosity η of solution to viscosity η_1 of solvent, both at zero shear rate. The viscosity ratio is traditionally called **relative viscosity**, η_{rel}, which it indeed is.

The **specific viscosity**, $\eta_{sp} = (\eta - \eta_1)/\eta_1 = (\eta/\eta_1) - 1$, now has the IUPAC name **relative viscosity increment**, η_i, since it is not related to mass and therefore not a *specific* quantity (see Chapter 19). However, the proposed symbol η_{i} can be easily mistaken for η_i, the symbol for the viscosity of compound i.

The ratio of specific viscosity and mass concentration c of the solute, $\eta_{red} \equiv \eta_{sp}/c$, was known as the **viscosity number**; IUPAC calls it "**reduced viscosity**." It is indeed a *reduced* quantity but not a reduced *viscosity* and certainly not a "number" since it has the unit of a specific volume (= volume/mass). The former **inherent viscosity**, η_{inh}, is now called **logarithmic viscosity number**, $\eta_{ln} = (\ln \eta_{rel})/c$. It also has the unit of a specific volume and is neither a logarithmic number nor a number and certainly not a viscosity.

Extrapolation of reduced viscosities to zero polymer concentration delivers the **limiting viscosity number** (IUPAC), $\lim_{c\to 0} \eta_{red} = \lim_{c\to 0} (\eta_{sp}/c) = [\eta]$ which is conventionally called **intrinsic viscosity** in English speaking countries and was once called the **Staudinger index** in German speaking ones (in the old literature with symbol Z_η). Far less used is the designation "**hydrodynamic virial coefficient**."

Like reduced viscosity, $[\eta]$ has the physical unit of a specific volume. Its units are cm^3/g = mL/g in modern scientific literature, 100 cm^3/g in older writings, and L/g in the pioneering publications of Hermann Staudinger. Engineers often use m^3/kg.

12.1.2 Experimental Methods

Viscosities of dilute polymer solutions are usually determined with capillary viscometers and, for the study of the influence of shearing, also with rotational viscometers (Fig. 12-1). Falling ball viscometers require too much solution. Capillary viscometers produce shear rates of $\dot{\gamma}/s^{-1}$ = 1-10^5, rotational viscometers, 10^{-2}–10^4, and cone-and-plate viscometers (melts only), 10^{-3}–10^2. Capillary viscometers of the Ostwald, Cannon-Fenske, or Ubbelohde type have shear rates of ca. 10^{-3} s^{-1} (see Volume I, p. 90).

In capillary viscometry, one measures the time t in which a defined volume V of the liquid with density ρ flows a distance L in a capillary with radius R at a pressure $p = \rho g L$. The viscosity is calculated with the **Hagen-Poiseuille Law**, $\eta^* = [\pi R^4/(8\,LV)]\rho g t$, where g = acceleration of free fall. The true viscosity η is obtained from η^* after the so-called **Hagenbach correction** for the partial conversion of potential energy to frictional energy and the production of eddies, $\eta = \eta^* - [kV/(8\,\pi L t)]$.

Capillaries should be selected so that the flow times t exceed 100 seconds since otherwise percent errors in specific viscosities become to large. Relative viscosities should also be larger than 1.2 because very often anomalies are observed for $\eta_{rel} < 1.2$ which are thought to be caused by an adsorption of macromolecules on capillary walls.

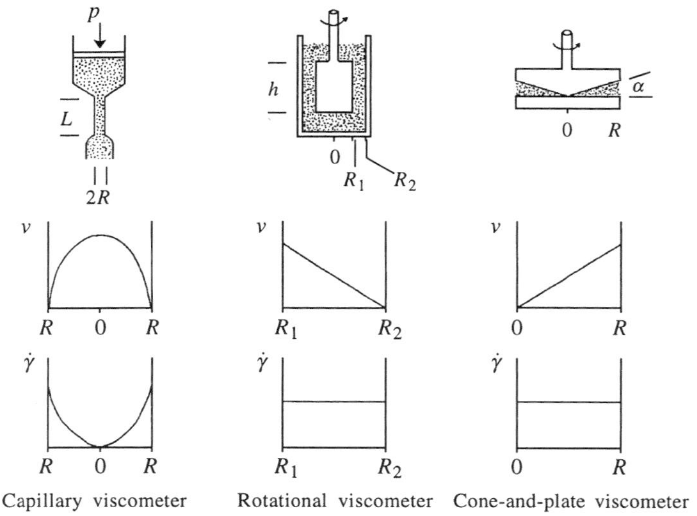

Fig. 12-1 Viscometers (top), flow rates v (center), and shear rates (bottom) as functions of the radii.

Velocity profiles are parabolic in capillary viscometers (Fig. 12-1, center left) but linear in rotational viscometers (Fig. 12-1, center middle) if rotational speeds are sufficiently small and the rotor axis is well centered. In rotational viscometers of the **Couette type** (Fig. 12-1, top middle), centering is achieved by a mechanical axis whereas in the **Zimm-Crothers viscometer** (not shown), surface tension automatically centers a buoyant freely floating rotor (Volume I, Fig. 3-11). The latter viscometer allows one to work at very low shear stresses of $\sigma_{21} \approx 4 \cdot 10^{-4}$ Pa and shear rates of $\dot{\gamma} \approx 0.2\ s^{-1}$. Such low shear rates are required for experiments with long rod-like molecules such as DNA.

At low shear rates, viscosities of polymer solutions are independent of shear rates (Newtonian behavior). The Newtonian range decreases with increasing polymer concentration (Fig. 12-2) and increasing molecular weight (Fig. 12-3).

At higher shear rates, calculated viscosities change from $\eta \neq f(\dot{\gamma})$ to a variation of lg η with lg $\dot{\gamma}$, i.e., the flow becomes non-Newtonian and the calculated viscosity is now an **apparent viscosity**. The linear *de*crease of lg η with increasing lg $\dot{\gamma}$ (Fig. 12-2) is called **shear thinning** or **pseudo plasticity** and the linear *in*crease of lg η with lg $\dot{\gamma}$, **shear thickening** (not shown in Fig. 12-2), provided the function $\eta = f(\dot{\gamma})$ depends only on shear rates (shear gradients) and does not vary with time (see Chapter 15).

With increasing concentration, slopes of $\lg \eta = \lg K_{\omega} + \omega \lg \dot{\gamma}$ become more negative and reach a limiting value of $\omega = -0.8$ at high concentrations. This limiting value seems to be universal for random coils in good solvents. In solvents producing theta states of coil molecules, a slope $\omega = -0.5$ is observed.

The literature sometimes reports so-called second Newtonian ranges at very high shear rates. Such a second Newtonian behavior was not observed for the polymer of Fig. 12-2. The second Newtonian range of the literature seems to be caused by the onset of turbulent flow and/or the degradation of polymers to lower molar masses.

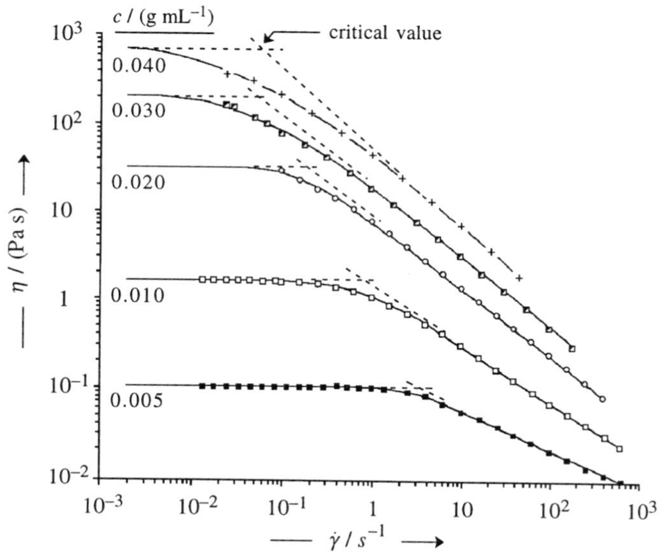

Fig. 12-2 Double-logarthmic plot of viscosity η as a function of shear rate $\dot{\gamma}$ for various concentrations c of a narrow-distribution poly(styrene) of molar mass $\overline{M}_w$ = 23 600 000 g/mol in the good solvent toluene at 25°C [1a]. By kind permission of Steinkopff-Verlag, Darmstadt.

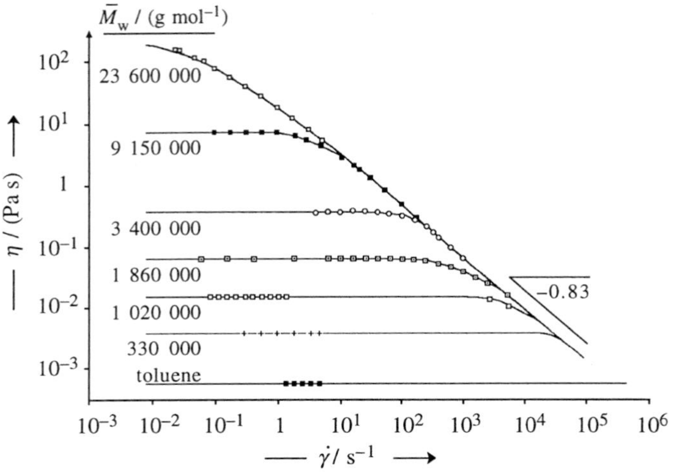

Fig. 12-3 Shear rate dependence of viscosities of 0.030 g/mL toluene solutions of narrow-distribution poly(styrene)s with various mass-average molar masses $\overline{M}_w$ [1b]. By kind permission of Steinkopff-Verlag, Darmstadt.

Similar behavior is observed for the dependence of viscosities on shear rates for constant concentrations of polymers with different molar masses (Fig. 12-3). The higher the molar mass, the lower the shear rate at which the fluid becomes non-Newtonian. The slope of the non-Newtonian function $\eta = K_\omega \dot{\gamma}^\omega$ is ω = –0.83 regardless of molar mass. It is somewhat more negative than the universal limiting value of –0.8, probably because the concentration c = 0.030 g/mL was too low (see Fig. 12-2).

All these data can be combined in a master curve if the logarithm of relative viscosity, η/η_0, is plotted against the logarithm of the reduced shear rate, $\beta = \dot{\gamma}/\dot{\gamma}_{crit}$, where $\dot{\gamma}_{crit}$ = critical concentration for the transition from the Newtonian range (index 0) to the non-Newtonian range (see Fig. 12-2). In the master curve (not shown), this transition is at a value of $\beta = \dot{\gamma}/\dot{\gamma}_{crit} \approx 0.1$ that is independent of molar mass and concentration. The slope of the function $\eta/\eta_0 = K\beta^\omega$ is ω = –0.8. The inverse of $\dot{\gamma}_{crit}$ corresponds to a critical relaxation time t_{crit}.

The reduced shear rate β

$$\beta = \frac{(\eta_0 - \eta_1)M^\alpha \dot{\gamma}}{RTc^{1/\alpha}}\left(\frac{(\eta/\eta_1) - 1}{c}\right) \qquad (12\text{-}1)$$

is a function of the viscosity η_0 in the Newtonian range, the viscosity η_1 of the solvent, the mass M of the polymer, the shear rate $\dot{\gamma}$, the molar gas constant R, the thermodynamic temperature T, the mass concentration c of the polymer, and the exponent α in the intrinsic viscosity–molar mass relationship, $[\eta] = K_v M^\alpha$ (Eq. (12-11).

The rest of this Chapter is concerned only with solution viscosities at sufficiently small shear rates, i.e., in the Newtonian range. Such small shear rates are required because dissolved random-coil-forming macromolecules deform, stretch, and finally break in strong shear fields (Section 15.5.3). The force to break the polymer is proportional to the square of the molecular weight. Fairly stiff macromolecules such as high-molar mass deoxyribonucleic acids even break while flowing out of pipettes.

12.2 Concentration Dependence

12.2.1 Non-Electrolytes

Neither viscosity ratios, $\eta_{rel} = \eta/\eta_1$, nor reduced viscosities, $\eta_{red} = (\eta_{rel} - 1)/c$, increase linearly with concentration for sufficiently large concentration ranges (Volume I, Fig. 3-14). Arrhenius thus tried in 1881 to describe the concentration dependence of viscosities of electrolyte solutions by $\eta_{rel} = \exp(Kc)$. Because this equation did not work for solutions of non-electrolytes, both **Bungenberg de Jong**, **Kruyt**, and **Lens** (BKL) on one hand and **Staudinger** and **Heuer** on the other replaced it by $(\eta_{rel} - 1)/c = [\eta] \exp (k_{BKL}c)$ which can also be written as $\ln \{(\eta_{rel} - 1)/c\} = \ln [\eta] + k_{BKL}c$ where k_{BKL} is an empirical constant that is independent of $[\eta]$. This equation is known as the **Martin equation** if $k_{BKL} = k_M[\eta]$:

(12-2) $\ln \eta_{red} = \ln [\eta] + k_M[\eta]c$ **Martin equation**

The BKL and Martin equations apply, for example, to large ranges of *high* concentrations of PAMAM dendrimers (Fig. 12-4). These dendrimers can be modeled as rough ellipsoids that become more like spheres at higher molar masses (p. 164 ff.).

Deviations from the linear dependence of $\ln \eta_{red}$ on c appear at concentrations greater than ca. 0.65 g/mL, in part because mass densities are used instead of the more appropriate volume fractions, and in part because the maximum packing densities are approached which are, for example, $\phi_{2,max} = 0.601$ for statistically loosely packed unimodal spheres and $\phi_{2,max} = 0.637$ for statistically tightly packed ones (see Volume IV, Section 9.3.3).

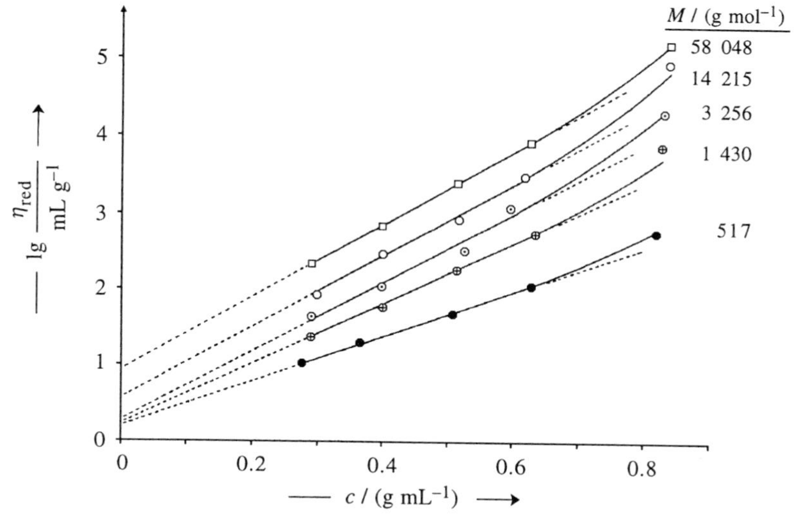

Fig. 12-4 Concentration dependence of reduced viscosities, $\eta_{red} \equiv \eta_{sp}/c$, of highly concentrated solutions of practically molecularly uniform PAMAM dendrimers in ethylene diamine at 20°C, plotted according to the Martin equation, $\lg \eta_{red} = \lg [\eta] + (k_M/2.303)[\eta]c$ [2]. Molar masses correspond to generations 0, 1, 2, 4, and 6. It is not clear whether such extrapolations from high concentrations deliver the same intrinsic viscosities as those from very low concentrations.

The exponential of the Bungenberg de Jong (BKL) equation, exp $(k_M[\eta]c) \equiv e^x$, can be developed in a series $e^x = 1 + x + x^2 + x^3 + ...$ which leads to the **Huggins equation**

$$(12\text{-}3) \qquad \eta_{red} = \eta_{sp}/c = [\eta] + k_H[\eta]^2c + k_{H,2}[\eta]^3c^2 + ...$$

if one identifies the first k_{BKL} as k_H and the second one as $k_{H,2}$. In the past, the somewhat analogous **Schulz-Blaschke equation** was also used:

$$(12\text{-}4) \qquad \eta_{red} = \eta_{sp}/c = [\eta] + k_{SB}[\eta]\eta_{sp} + ...$$

The natural logarithm of $\eta_{rel} = (1 + \eta_{sp})$ with $\eta_{sp} \equiv x$ may be also developed in a Taylor series, ln $(1 + x) = x - (1/2)x^2 + (1/3)x^3 - (1/4)\ x^4 + ...$ which is valid for $-1 < x < +1$, i.e., $\eta_{rel} \leq 2$. Combination of this equation with Eq.(12-3) delivers an equation for the concentration dependence of logarithmic viscosity numbers (= inherent viscosities),

$$(12\text{-}5) \qquad \eta_{inh} \equiv (\ln \eta_{rel})/c = [\eta] + (k_H - (1/2) + \{(1/3) - k_H)[\eta]c\})[\eta]^2c + \ ...$$

which, in the form of Eq.(12-5a), is known as the **Kraemer equation**:

$$(12\text{-}5a) \qquad \eta_{inh} \equiv (\ln \eta_{rel})/c = [\eta] + k_K[\eta]^2c + k_{K,2}[\eta]^3c + ...$$

The parameters k_M, k_H, $k_{H,2}$, k_{SB}, k_K, and $k_{K,2}$ of Eqs.(12-2)-(12-5a) are empirical constants that are controlled by the shapes of the molecules and the goodness of the solvent. Comparison of Eqs.(12-3), (12-5), and (12-5a) shows that k_H and k_K are interconnected via $k_K = k_H - (1/2) - \{(1/3) - k_H\}[\eta]c$. If Eq.(12-3) is linear, then Eq.(12-5a) cannot be linear except for $k_H = 1/3$ which corresponds to perturbed coils.

The practice of linearly extrapolating both $\eta_{sp}/c = f(c)$ and ln $\eta_{rel}/c = f(c)$ to a common intercept may thus lead to severe errors if $k_H \neq 1/3$. Note also that theoretical work indicates that the functions, Eqs. (12-3)-(12-5a), are not simple power series for values of η_{rel} greater than 2.

Combination of Eqs.(12-3) and (12-5a) with $k_H + k_K = 1/2$ delivers the **Solomon-Ciuta equation,** $[\eta] = [2\ (\eta_{sp} - \ln \eta_{rel})]^{1/2}/c$, which, in principle, allows one to calculate intrinsic viscosities from viscosity measurements at a single concentration. However, this equation can only be valid if $k_H = 1/3$ (see above). In industry, it is common practice to measure η_{sp}/c or (ln $\eta_{rel})/c$ at a single concentration and to take these values as measures of intrinsic viscosities or even molecular weights.

The variation of viscosity with concentration can be generalized if one treats the dimensionless relative increase of viscosity, $\eta_{sp} = (\eta - \eta_1)/\eta_1$, as a function of the dimensionless quantity $c[\eta]$. Since the intrinsic viscosity, $[\eta]/(\text{mL g}^{-1})$, measures the volume that is occupied by the mass of the polymer, if follows that $c[\eta]$ is the volume fraction that the polymer would have if the polymer molecules are separated from each other.

However, coil-forming polymer molecules start to overlap with increasing polymer concentration (p. 183). The total volume that is available to polymer molecules is then smaller than the sum of the volumes of isolated polymer molecules, and the product $c[\eta]$ then becomes greater than unity. The inverse product, $1/(c[\eta])$, is therefore a measure of the interpenetration of coil-forming polymer molecules.

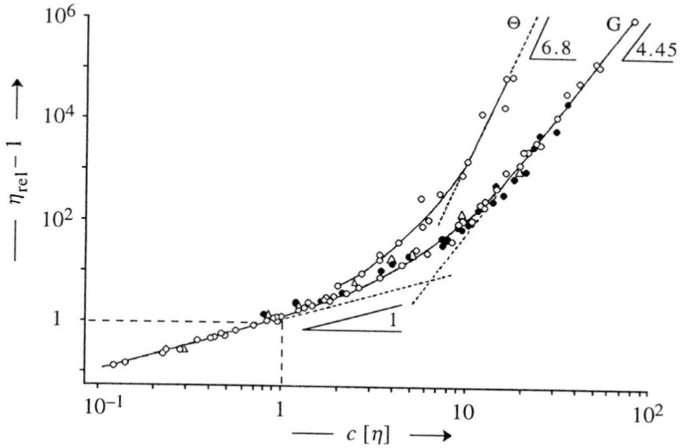

Fig. 12-5 Logarithm of relative viscosity increment (= specific viscosity) as a function of the logarithm of the dimensionless parameters $c[\eta]$ for some polymers in good (G) and poor (Θ) solvents.
O Poly(styrene)s in *trans*-decalene (Θ) or toluene (G) at 25°C [1c]; ● *cis*-1,4-poly(isoprene)s in toluene at 34°C (G) [3]; Δ hyaluronates in water at 25°C (G) [4].

At sufficiently low concentrations, η_{sp} should be proportional to $c[\eta]$ according to Eq.(12-3). A plot of lg $\eta_{sp} = f(\lg c[\eta])$ should thus deliver a straight line with a slope of unity, regardless of the type of polymer, solvent, and temperature, as is indeed found for $c[\eta] < 1$ (Fig. 12-5).

Beyond $c[\eta] = 1$, lg η_{sp} increases strongly with lg $c[\eta]$. The asymptotic range here can be described by $\eta_{sp} \sim (c[\eta])^q$ with q > 1 (Fig. 12-5). Since intrinsic viscosities depend on molecular weights according to $[\eta] = K_v M^{\alpha}$ (Eq.(12-11)), one can write

$$(12\text{-}6) \qquad \eta_{sp} \approx \eta_{rel} = \eta/\eta_1 \sim (c[\eta])^q \sim c^q M^{q\alpha}$$

At high concentrations, reduced viscosity increments practically equal relative viscosities, $\eta_{sp} \equiv (\eta/\eta_1) - 1 \approx \eta/\eta_1$, since $\eta/\eta_1 >> 1$. Such highly concentrated solutions have essentially the consistency of melts. Since melt viscosities at rest (i.e., at very low shear rates) are proportional to the 3.4th power of the molar mass at high molar masses (Section 15.3.4), αq in Eq.(12-6) becomes 3.4.

In melts, linear polymer molecules are present as unperturbed random coils (Chapter 6). Intrinsic viscosities of unperturbed random coils are furthermore proportional to the square root of molar masses, $[\eta] = K_v M^{\alpha}$ with $\alpha = 1/2$ (p. 411 ff.). Since $[\eta]c \approx \eta_{spec} \approx M^{q\alpha}$ and $\alpha q = 3.4$, it follows $q = 3.4/(1/2) = 6.8$. At sufficiently high values of $c[\eta]$, $\eta_{sp} \equiv \eta_{rel} - 1$ should thus be proportional to the 6.8th power of $c[\eta]$ as is indeed found (Fig. 12-5). For perturbed coils in good solvents, $\alpha = 0.764$ and $\eta_{rel} - 1$ should increase with the 4.45th power of concentration (Fig. 12-5).

This dependency of η_{sp} on c and M depends only on the type of coil (unperturbed *versus* perturbed) but not on the type of polymer, solvent, and temperature. The same theoretical reasoning predicts that zero-shear viscosities of high-concentration solutions of rod-like macromolecules should vary with $\phi^{5/3}M^8$. Unfortunately, experimental data to verify this prediction do not seem to exist.

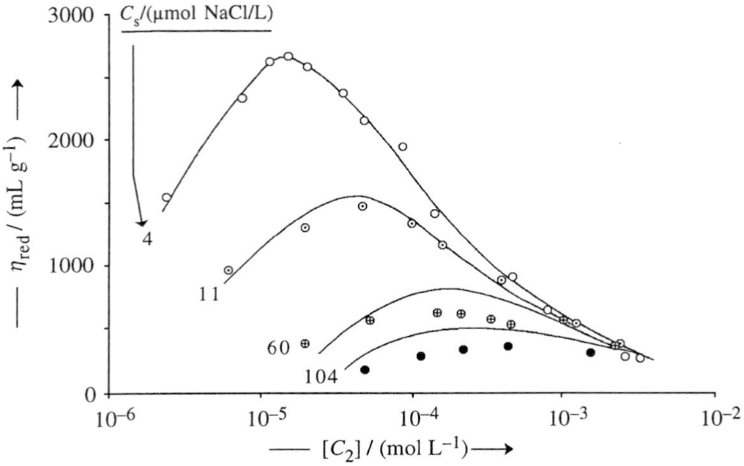

Fig. 12-6 Reduced viscosities at $\dot{\gamma} = 600\ s^{-1}$ as function of the molar concentration C_2 of a sodium poly(styrene sulfonate) ($\overline{M}_w$ = 16 000 g/mol) in aqueous solutions with various constant concentrations C_s/(μmol/L) of added NaCl. Solid lines: calculated with Eq.(12-8) [5]. By kind permission of the American Institute of Physics, Melville (NY).

12.2.2 Polyelectrolytes

Reduced viscosities, $\eta_{red} = \eta_{sp}/c$, of polyelectrolytes increase strongly with increasing polymer concentration, pass through a maximum, and then decline (Fig. 12-6). The maximum of η_{red} is most pronounced in the absence of an added low-molar mass salt. η_{red} decreases with increasing salt concentration C_s and becomes independent of C_s at very high salt concentrations. This type of behavior is observed not only for polyelectrolytes such as sodium poly(styrene sulfonate), $\text{+}CH_2CH(C_6H_4SO_3Na)\text{+}_n$ and poly(acrylic acid), $\text{+}CH_2CH(COOH)\text{+}_n$, but also for polyampholytes such as the copolymer with the monomeric units $-CH_2CHSO_3^{\ominus}-$ and $-CH_2CH(CONH_3)^{\oplus}-$.

This behavior is explained as follows. Polyelectrolytes are little dissociated at both high polyelectrolyte concentrations without added salt and low polyelectrolyte concentrations with high concentrations of added salt. For such systems, the concentration of counterions is greater in the interior of the polymer coils than at the exterior. The resulting osmotic effect causes additional water to enter the coils which therefore expand.

The lower the polyelectrolyte concentration, the more polymer groups will dissociate and the more they will repel each other. As a result, molecule dimensions and therefore also reduced viscosities η_{red} increase strongly with decreasing polymer concentration c, which can be described by the empirical **Fuoss equation** where the intercept A_{FS} is *not* the inverse intrinsic viscosity as was assumed originally (see pp. 404 and 423):

$$(12\text{-}7) \qquad 1/\eta_{red} = A_{FS} + K_{FS}c^{1/2} \qquad \textbf{Fuoss equation}$$

Addition of salt increases the ionic strength outside of the coils relative to the interior; it also reduces the thickness of the ion cloud. Both effects reduce the coil diameter and therefore the reduced viscosity compared to the coil dimensions in salt-free solutions.

These cooperative and counteractive effects are difficult to cast into a quantitative theory. As a result, many theoretical approaches exist. One group of theories starts with rigid chains (because of the repulsion between ionized groups) and then introduces factors that decrease rigidities. Another group of theories takes the opposite approach: flexible chains become more rigid because of electrostatic interactions.

For example, the theoretical curves of Fig. 12-6 are based on the theoretical equation

$$(12\text{-}8) \qquad \eta_{\text{red}} = K_t \xi L_B^2 z_p^2 / \kappa^3$$

where K_t = constant, ξ = friction coefficient of the polyion, $L_B = e^2/(4\,\pi\varepsilon_r\varepsilon_0 k_B T)$ = Bjerrum length, e = elementary charge, ε_r = relative permittivity (= dielectric constant) of the solution, k_B = Boltzmann constant, T = thermodynamic temperature, ε_0 = permittivity of vacuum, z_p = valence of the polyion, $\kappa^2 = 4\,\pi L_B(z_g{}^2 C_g + \Sigma_i z_i C_i)$ = inverse Debye screening length, C_q = number concentrations (q = g: gegenion; q = i: ith type of added ions), and z_q = valence (q = p: polyion; q = g: gegenion; q = i: ith type of added ions).

Scaling theories do without detailed assumptions and rather concentrate on global aspects for three different regions of polyelectrolyte concentrations: (I) very low polyelectrolyte concentrations in the absence of added ions; (II) dilute polyelectrolyte solutions in the presence of added salt; and (III) concentrated polyelectrolyte solutions in the presence of salt (Fig. 12-7). The results of these theories are summarized in Table 12-1.

These theories assume for *semi-concentrated* polyelectrolyte solutions that polyelectrolyte molecules in thermodynamically poor solvents (e.g., water) behave like strings of interconnected electrostatic blobs (Section 6.3.4). The blobs behave like collapsed spheres and the specific viscosity is predicted to increase with the square root of the number concentration of the polymer, $[C] = [P]N_A$, where [P] = molar concentration of the polymer. Such behavior is found for the range $10^{-2} \leq [P]/(\text{mol L}^{-1}) \leq 10^{-1}$.

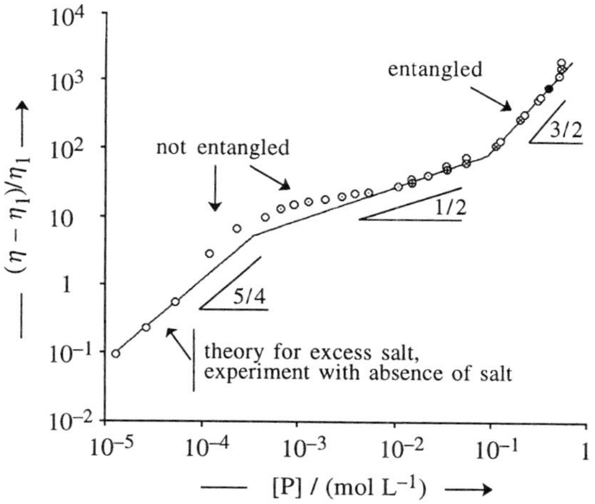

Fig. 12-7 Logarithm of relative viscosity increments (= specific viscosities), $\eta_{sp} = (\eta - \eta_1)/\eta_1$, as a function of the logarithm of molar concentration [P] of a sodium poly(styrene sulfonate) (degree of sulfonation: 92 %; M_r = 1 200 000) in water at 25°C without addition of NaCl (O) or the following NaCl concentrations: 10^{-5} mol/L (⊙), 10^{-4} mol/L (⊕), 10^{-3} mol/L (⊗), and 10^{-2} mol/L (●) [6a]. Solid lines: prediction of scaling theory for Newtonian solutions.

Table 12-1 Predictions of scaling theory for Newtonian viscosities η of semi-concentrated solutions of polyelectrolytes in the undissociated (neutral) state and the dissociated state in the presence of low or high concentrations of added salt [6b]. $C = [P]N_A$ = number concentration of polymer, [P] = molar concentration of polymer, N_A = Avogadro constant, c_s = mass concentration of added salt, X = degree of polymerization of polymer, L = screening length (correlation length).

State	Screening length	Viscosity of coil molecules not entangled	entangled
General relationships	–	$\eta \sim C^{-1}XL^{-3}$	$\eta \sim C^{-3}X^3L^{-9}$
Special functions for polyelectrolytes			
Not dissociated (neutral)	$L \sim C^{-3/4}$	$\eta \sim C^{5/4}X$	$\eta \sim C^{15/4}X^3$
Dissociated, high salt concentrations	$L \sim C^{-3/4}c_s^{1/4}$	$\eta \sim C^{5/4}Xc_s^{-3/4}$	$\eta \sim C^{15/4}X^3c_s^{-9/4}$
Dissociated, low salt concentrations	$L \sim C^{-1/2}$	$\eta \sim C^{1/2}X$	$\eta \sim C^{3/2}X^3$

However, this range with $\eta \sim C^{1/2}$ and $\eta_{sp} \sim [P]^{1/2}$, respectively, should extend down to $[P] \approx 3{\cdot}10^{-4}$ mol L^{-1}, which is not found experimentally (Fig. 12-7). The deviation from the predicted behavior in the range $3{\cdot}10^{-4} \leq [P]/(\text{mol L}^{-1}) \leq 10^{-2}$ is as yet unexplained. However, it was found experimentally that non-Newtonian shear viscosities do fall on the predicted line (not shown in Fig. 12-7).

Since viscosities of many literature reports are indeed non-Newtonian ones (although not recognized as such!), it is understandable why the Fuoss equation has been "confirmed" so often. Because of the effect of shear rate, the Fuoss equation, $1/\eta_{red} = f(c^{1/2})$, cannot deliver the true intrinsic viscosity at $c \to 0$.

Strong repulsion forces reign between ionic groups of the same charge at lengths that are larger than the diameters of blobs. The lengths of these rod-like structures should extend from the diameter of a blob to the correlation length (screening length) L_{cl} which is thus assumed to be proportional to the persistence length L_{ps}. The chain is flexible beyond the correlation length because it is screened by the surrounding chains in semi-concentrated solutions.

For bad solvents, the correlation length L_{cl} is calculated from the ratio r_{cont}/r of conventional contour length to actual end-to-end distance of the chain in dilute, salt-free solutions, the number concentration C of the polymer, the effective bond length (p. 82), the effective number N_{eff} of monomeric units between two charge carriers, and the Bjerrum length L_B as

$$(12\text{-}9) \qquad L_{cl} = (r_{cont}/rb_{eff}C)^{1/2} \quad ; \quad r_{cont}/r = (N_{eff}b_{eff}/L_B)^{2/3}$$

The ratio $r_{cont}/r = 3.23$ is calculated from the effective number $N_{eff} = 4$ of monomeric units between two charge carriers, the effective bond length (b_{eff} = 0.254 nm), and the Bjerrum length in water (L_B = 0.7 nm). At a molar polymer concentration of $[P] = C/N_A$ = 0.1 mol/L, the correlation length is L_{cl} = 0.46 nm and thus only about double the effective bond length of a monomeric unit.

Polymer coils start to entangle at a critical molar polymer concentration (in Fig. 12-7: $[P] \approx 0.1$ mol/L). In this concentration range, the experiment confirms the theoretical prediction of 3/2 for the slope of lg η_{sp} = f(lg [P]0 (Fig. 12-7). At still higher concentrations of ca. 0.7 mol/L, systematic deviations are observed

For very dilute, salt-free solutions of polyelectrolytes, one finds experimentally the same proportionality $\eta \sim C^{5/4}$ that was predicted for semi-concentrated solutions of non-dissociated (neutral) polyelectrolytes (Fig. 12-7). Though the surrounding salt concentration here is larger than the concentration of counterions (which suppresses dissociation) and the effect of solvent is eliminated by using η_{sp} instead of η, it is still not clear why these *dilute* polyelectrolyte solutions should behave like semi-concentrated ones.

12.3 Intrinsic Viscosities

12.3.1 Averages

The extrapolation of values of η_{sp}/c to zero concentration delivers the intrinsic viscosity $[\eta]$ of neutral polymers (Eq.(12-3)) which is a mass average as can be seen from the following. In the limit of $c \to 0$, one has $\eta_{sp} \approx [\eta]c$. Experiments have shown that specific viscosities of homologs of neutral polymer molecules are additive in the Newtonian range: $\eta_{sp} = \Sigma_i \eta_{sp,i} = \Sigma_i [\eta]_i c_i$ (Philippoff). Introduction of this expression into $[\eta] = \lim_{c\to 0} \eta_{sp}/c$ and using $c \equiv \Sigma_i c_i$ and $w_i \equiv c_i/c$ shows that intrinsic viscosities of neutral polymers, *as calculated with the Huggins equation*, are **mass averages**:

(12-10) $[\eta] = \{\Sigma_i [\eta]_i c_i\}/c = \Sigma_i w_i[\eta]_i \equiv \overline{[\eta]}_w$ (Huggins equation only)

Intrinsic viscosities have the physical unit of a specific volume, for example mL/g. They thus indicate which volume is taken up per mass. Since volumes of homologous polymer molecules vary systematically with their molar masses (Sections 4.2 and 4.4.3), intrinsic viscosities can be expressed by a power of molar masses

(12-11) $[\eta] = K_v M^{\alpha}$; $K_v = const.$; $-1 \leq \alpha \leq +2$ (theory, see below)

Eq.(12-11) is known in the literature as the **Kuhn-Mark-Houwink-Sakurada** equation (KMHS equation), **Mark-Houwink-Sakurad** equation, **Mark-Houwink** equation, or **Staudinger** equation (note: Staudinger used only $\alpha = 1$. He abhorred $\alpha \neq 1$).

The molar mass M in Eq.(12-11) is an **exponent average** for molecularly non-uniform macromolecules or particles. Solving Eq.(12-11) for the molar mass, introducing $[\eta] = \Sigma_i w_i[\eta]_i$ and $[\eta]_i = K_v(M_i)^{\alpha}$ for the component i shows that the M in Eq.(12-11) is a (solution) **viscosity-average** of the molar mass, $\overline{M}_v$;

(12-12) $M = \{[\eta]/K_v\}^{1/\alpha} = \{(\Sigma_i w_i[\eta]_i)/K_v\}^{1/\alpha} = \{\Sigma_i w_i M_i^{\alpha}\}^{1/\alpha} \equiv \overline{M}_v$

K_v and α are constants for a polymer homologous series in a solvent at constant temperature. They are obtained from data on polymers with known type and width of molar mass distribution. Conversely, molar masses calculated from Eq.(12-12) are only true viscosity averages if calibrations were done with molecularly uniform polymers or with polymers with known viscosity-average molar masses. Otherwise, serious errors may occur and Eq.(12-11) will deliver molar masses with undefined averages.

Calibration of Eq.(12-11) with averages $\overline{M}_g$ (g = n, w, etc.) other than $\overline{M}_v$ requires a non-uniformity correction factor q_{KMHS} for the effects of molar mass distributions:

(12-13) $[\eta] = K_{\mathrm{v}}\overline{M}_{\mathrm{v}}^{\alpha} = K_{\mathrm{v}}q_{\mathrm{KMHS}}\overline{M}_{\mathrm{g}}^{\alpha}$

Correction factors vary with the type of molar mass distribution (Schulz-Zimm, logarithmic normal, etc.). These factors can be considerable, for example, for $\alpha = 0.764$, SZ distributions, and $\overline{M}_{\mathrm{w}}/\overline{M}_{\mathrm{n}} = 1.1$: 8 % if $\overline{M}_{\mathrm{w}}$ and 15.7 % if $\overline{M}_{\mathrm{n}}$ is used for calibration (Table 12-2). In the equations below, symbols ς and α have the following meanings: $\varsigma = \overline{M}_{\mathrm{n}}/(\overline{M}_{\mathrm{w}} - \overline{M}_{\mathrm{n}})$ and α = exponent in the intrinsic viscosity = f(molar mass), Eq.(12-11):

	Schulz-Zimm distribution	logarithmic normal distribution
with $\overline{M}_{\mathrm{g}} = \overline{M}_{\mathrm{w}}$	$q_{\mathrm{KMSH}} = \dfrac{\Gamma(\varsigma+\alpha+1)}{(\varsigma+1)^{\alpha}\,\Gamma(\varsigma+1)}$	$q_{\mathrm{KMSH}} = (\overline{M}_{\mathrm{w}}/\overline{M}_{\mathrm{n}})^{(\alpha^2-\alpha)/2}$
with $\overline{M}_{\mathrm{g}} = \overline{M}_{\mathrm{n}}$	$q_{\mathrm{KMSH}} = \dfrac{\Gamma(\varsigma+\alpha+1)}{\varsigma^{\alpha}\,\Gamma(\varsigma+1)}$	$q_{\mathrm{KMSH}} = (\overline{M}_{\mathrm{w}}/\overline{M}_{\mathrm{n}})^{(\alpha^2+\alpha)/2}$

Table 12-2 Correction factor q_{KMHS} for molecularly non-uniform polymers with Schulz-Zimm (SZ) or logarithmic normal (LN) mass distributions.

$\overline{M}_{\mathrm{w}}/\overline{M}_{\mathrm{n}}$	Correction factors q_{KMHS} if $\overline{M}_{\mathrm{g}} = \overline{M}_{\mathrm{w}}$ and									
	SZ distributions for					LN distributions for				
$\alpha \rightarrow$	0	0.500	0.764	1.000	2.000	0	0.500	0.764	1.000	2.000
1.1	1.000	0.989	0.992	1.000	1.091	1.000	0.988	0.991	1.000	1.100
1.3	1.000	0.971	0.980	1.000	1.231	1.000	0.968	0.977	1.000	1.300
1.5	1.000	0.959	0.971	1.000	1.333	1.000	0.951	0.964	1.000	1.500
2.0	1.000	0.940	0.958	1.000	1.250	1.000	0.917	0.939	1.000	2.000
3.0	1.000	0.921	0.946	1.000	1.667	1.000	0.872	0.906	1.000	3.000
5.0	1.000	0.907	0.912	1.000	1.800	1.000	0.818	0.865	1.000	5.000

$\overline{M}_{\mathrm{w}}/\overline{M}_{\mathrm{n}}$	Correction factors q_{KMHS} if $\overline{M}_{\mathrm{g}} = \overline{M}_{\mathrm{n}}$ and									
	SZ distributions for					LN distributions for				
$\alpha \rightarrow$	0	0.500	0.764	1.000	2.000	0	0.500	0.764	1.000	2.000
1.1	1.000	1.037	1.157	1.100	1.320	1.000	1.036	1.066	1.100	1.331
1.3	1.000	1.108	1.196	1.300	2.080	1.000	1.103	1.193	1.300	2.197
1.5	1.000	1.175	1.324	1.500	3.000	1.000	1.164	1.314	1.500	3.375
2.0	1.000	1.329	1.627	2.000	6.000	1.000	1.297	1.595	2.000	8.000
3.0	1.000	1.596	2.071	3.000	15.000	1.000	1.510	2.097	3.000	27.000
5.0	1.000	2.028	3.201	5.000	45.000	1.000	1.829	2.958	5.000	125.000

12.3.2 Hydrodynamic Volumes

Theory predicts that the viscosity η of dilute dispersions of small hard spheres 2 with radius R_{sph} in a solvent 1 with molecules of radius R_1 can be expressed as a power series of the volume fraction ϕ_2 of the spheres:

(12-14) $\eta = \eta_1(1 + K_1\phi_2 + K_2\phi_2^2 + ...)$

The proportionality factor K_1 has been calculated by **Einstein** as 5/2 for non-aggregating, unsolvated, uncharged, inflexible, large ($R_{sph} >> R_1$) spheres in a continuum of an incompressible solvent without interaction between spheres (K_2, K_3, ... = 0), without slip between spheres and solvent, and without wall effects. **Batchelor** calculated $K_2 = 6.2$ for binary interactions between spheres. Experiments to determine these numerical values for K_1 and K_2 are by no means trivial. They did confirm $K_1 = 2.5$ and $K_2 = 6.2$ for dispersions of glass and gutta percha spheres with specific volumes v_2 in solvents with density ρ_1 and viscosity η_1 provided that **Reynolds numbers** $Re = \rho_1 v R_{sph}/\eta_1 << 0.1$.

The volume fraction $\phi_2 = V_2/V = N_2 V_v/V$ is defined as the ratio of volume V_2 of solute to total volume V where the volume of the solute is the product of the number N_2 of solute molecules with a viscometrically effective volume V_v per molecule. This volume is $V_v = 4\,\pi\,R^3/3$ for hard spheres with a radius $R = R_{sph}$. For particles with other shapes (ellipsoids, rods, platelets, coils, etc.), R is the so-called **Einstein radius** which is the radius of a hydrodynamically (viscometrically) equivalent sphere.

The number N_2 of dispersed or dissolved particles can be expressed by the number concentration, $C_2 = N_2/V_2 = c_2 N_A/M_2$, where c_2 = mass concentration, N_A = Avogadro constant, and M_2 = molar mass. Introduction of these expressions, the reduced viscosity, $\eta_{red} = (\eta - \eta_1)/(\eta_1 c_2)$, and $K_1 = 5/2$ and $K_2 = 6.2$ in Eq.(12-14) leads to

(12-15) $$\eta_{red} = (5/2)(V_v N_A/M_2) + 6.2\,(V_v N_A/M_2)^2 c_2 + ... = [\eta] + 0.992\,[\eta]^2 c_2 + ...$$

The right side of Eq.(12-15) is identical with the empirical Huggins equation, Eq.(12-3), if one sets $(5/2)(V_v N_A/M) \equiv [\eta]$ and $(2/5)^2\,K_2 \equiv k_H$.

The intrinsic viscosity $[\eta]$ is a *mass average,* Eq.(12-10), and the molar mass from $[\eta]$ is a *viscosity average*, Eq.(12-12), whereas the viscosimetric volume V_v from intrinsic viscosities, Eq.(12-15), is a *number average* as one can see from the following. The intrinsic viscosity of a mixture of polymer molecules with intrinsic viscosities $[\eta]_i$ that are present in concentrations $c_{2,i}$ is $[\eta] \equiv (\Sigma_i\,c_{2,i}[\eta]_i)/\Sigma_i\,c_{2,i}$ (Eq.(12-10)) where $c_{2,i} = m_{2i}/V = n_{2,i}M_{2,i}/V$, $n_2 = \Sigma_i\,n_{2,i}$, and $\Sigma_i\,c_{2,i} \equiv n_2\,\overline{M}_n/V$. Since $[\eta]_i = (5/2)N_A(V_{v,i}/M_{2,i}$ (Eq.(12-15)), it follows that

(12-16) $$[\eta] \equiv \frac{\Sigma_i\,c_{2,i}[\eta]_i}{c_2} = \left(\frac{5\,N_A}{2\,\overline{M}_n}\right)\frac{\Sigma_i n_{2,i}V_{2,i}}{\Sigma_i n_{2,i}} = \frac{K_n}{\overline{M}_n}(\overline{V}_v)_n$$

12.3.3 Spheres

The volume $V_{sph} = (4\,\pi/3)\,R_{sph}{}^3$ of a rigid, compact sphere is identical with its hydrodynamic volume V_v. However, the radius of that sphere, R_{sph}, is greater than the radius of gyration s by a factor of $Q_{sph} = (5/3)^{1/2}$. In the zero concentration limit, Eq.(12-15) thus becomes

(12-17) $$[\eta] = \frac{5\,N_A V_v}{2\,M} = \frac{10\,\pi\,N_A R_{sph}^3}{3\,M} = \Phi_{sph,R}\frac{R_{sph}^3}{M} = \Phi_{sph,R}\frac{(5/3)^{3/2}s^3}{M} = \Phi_{sph,s}\frac{s^3}{M}$$

where $\Phi_{sph,R} \approx 6.306{\cdot}10^{24}$ mol^{-1} and $\Phi_{sph,s} \approx 13.57{\cdot}10^{24}$ mol^{-1}.

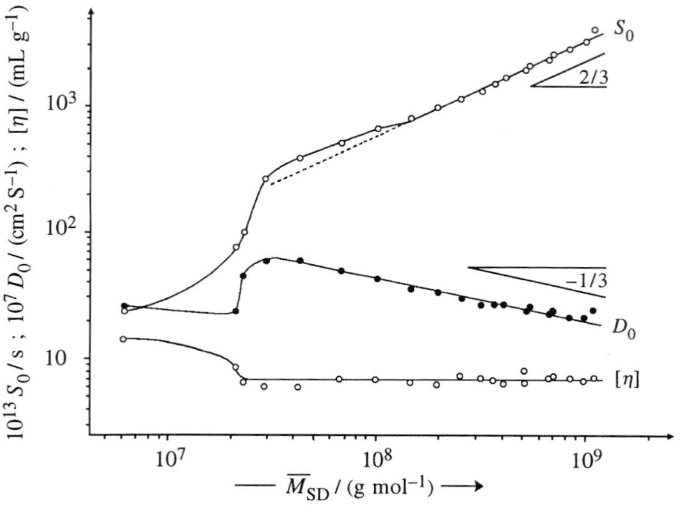

Fig. 12-8 Dependence of sedimentation coefficients S_0, diffusion coefficients D_0, and intrinsic viscosities $[\eta]$ of fractions of glycogen in water at 20°C on the molar mass from sedimentation and diffusion coefficient, Eq.(11-45) [7]. Lines are empirical (best fits).

The discontinuity at $M \approx 2 \cdot 10^7$ g/mol is caused by the transition from single spherical molecules to spherical aggregates at higher molar masses. The slopes $\varsigma = 2/3$ of $S_0 = K_s M^{\varsigma}$ (Eq.(11-42)), $\delta = -1/3$ of $D_0 = K_D M^{\delta}$ (Eq.(11-22)), and $\alpha = 0$ of $[\eta] = K_v M^{\alpha}$ (Eq.(12-11)) follow the prediction of Eq.(2-24), $\alpha = 2 - 3\,\varsigma = -(1 + 3\,\delta)$, with $\alpha = 0$ for spheres (Eq.(12-18)).

The hydrodynamic volumes of homogeneous rigid spheres can be expressed as $V_v = m/\rho = M/(\rho N_A)$ and Eq.(12-17) becomes

$$(12\text{-}18) \qquad [\eta] = 5/(2\,\rho)$$

Intrinsic viscosities of such spheres are independent of their molar masses and just controlled by the density of spheres as shown in Fig. 12-8 for spherical aggregates of glycogen molecules in the range $2 \cdot 10^7 \leq M/(\text{g mol}^{-1}) \leq 10^9$ and in Fig. 12-16 for poly(α,ε-lysine) dendrimers in the range $10^3 \leq M/(\text{g mol}^{-1}) \leq 2 \cdot 10^5$.

12.3.4 Ellipsoids

Intrinsic viscosities are proportional to the ratio V_v/M of viscometric volume V_v and molar mass M (Eq.12-15)). The proportionality factor is $(5/2)N_A$ for viscosimetrically equivalent spheres which can be generalized to ΨN_A for molecules of other shapes so that $[\eta] = \Psi N_A(V_v/M)$. The proportionality factor Ψ is called the **Simha factor**.

For rotational ellipsoids, Simha factors depend on the type of ellipsoid (prolate, oblate) and on the axial ratio $\Lambda = L_{long}/L_{short}$ of the length of the long semi-axis to that of the short semi-axis (Table 12-3). For axial ratios greater than ca. 20, Simha factors can be expressed by the asymptotic relationships Eqs, (12-19) and (12-20), which show that for $\lambda > 20$, logarithms of Simha factors increase practically linearly with the logarithms of the axial ratios.

Table 12-3 Simha factors Ψ of compact rotational ellipsoids as a function of the ratio Λ of long and short semi-axes, $\Lambda = a/c = L/(2\,R)$ for prolate ellipsoids, and $\Lambda = c/a = 2\,R/d$ for oblate ones.

Λ	Ψ_{pr}	Ψ_{ob}	Λ	Ψ_{pr}	Ψ_{ob}
1	2.500	2.500	6	7.098	5.367
2	2.908	2.854	8	10.103	6.700
3	3.685	3.430	10	13.634	8.043
4	4.663	4.059	15	24.65	11.42
5	5.806	4.708	20	38.53	14.80

$$\Psi_{pr} = \frac{\Lambda^2}{5[\ln(2\Lambda) - (1/2)]} + \frac{\Lambda^2}{15[\ln(2\Lambda) - (3/2)]} + \frac{14}{15} \quad (12\text{-}19)$$

$$\Psi_{ob} = \frac{16}{15} \cdot \frac{\Lambda}{\mathrm{tg}^{-1}\,\Lambda} \quad (12\text{-}20)$$

Simha factors depend only on axial ratios and not on any other physical parameter. Since the combination of $[\eta] = \Psi N_A(V_v/M_2)$ with $\rho = m_2/V_v = n_2 M_2/V_v$ leads to $[\eta] = \Psi N_A(n_2/\rho) = \Psi(N_2/\rho)$, it follows that intrinsic viscosities do not allow one to determine molar masses of similar ellipsoids. Similar ellipsoids of equal density thus have the same intrinsic viscosity regardless of their size. This behavior is similar to that of spheres.

The hydrodynamic volume V_v from intrinsic viscosities can be used to calculate axial ratios of ellipsoids if the hydrodynamic volume V_h is known from other experimental data (such as diffusion with $V_h = V_d$) and the two hydrodynamic volumes are the same, for example, $V_v = V_d$. These axial ratios are upper limits for *solvated* spheroids. Indeed, they are always found to be greater than the axial ratios from electron microscopy which delivers ratios of *dry* ellipsoids (Table 12-4).

The degree of solvation, Γ_S, can be obtained from either (I) $[\eta] = \Psi N_A(V_v/M)$ and Eq.(11-36) with $V_S = V_v$ or (II) from Eq.(11-36), $V_S = 4\,\pi\,R_S^3/3$, Eq.(11-11), and Eq.(11-10) which leads to Eq.(12-21) (from I) and Eq.(12-22) (from II).

Table 12-4 Axial ratios Λ, Simha factors Ψ, and degree of solvation (here: hydration), Γ_S, of biological macromolecules from electron microscopy (EM), hydrodynamic measurements such as intrinsic viscosity, $[\eta]$, or diffusion (D), or from calorimetry (C) or nuclear magnetic resonance (NMR).

	Λ		Ψ	Γ_S from			
	EM	$[\eta]$, D	EM	$[\eta]$	D	C	NMR
Spherical molecules							
Hemocyanin	1	3.4	2.5	1.34	0.79	-	-
TBS virus	1	3.4	2.5	0.64	0.27	-	-
Prolate ellipsoids							
Lysozyme	1.5	2.3	2.6	0.34	0.28	0.30	0.34
Albumen	2.0	3.1	2.9	0.52	0.31	0.40	0.40
Hemoglobin	1.2	3.7	2.55	0.65	0.20	0.32	0.42
Catalase	1.3	4.0	2.58	0.76	0.20	-	-

(12-21) $\Gamma_S = \{([\eta]/\Psi) - \tilde{v}_2\}\rho_1$ (from approach I)

(12-22) $$\Gamma_S = \rho_1 \left\{ \frac{4\pi N_A}{3M} \left(\frac{k_B T}{6\pi \eta_1 D_0} \right)^3 - \tilde{v}_2 \right\}$$ (from approach II)

Degrees of solvation from Eqs.(12-21) and (12-22) agree resonably well with those from calorimetry and NMR (Table 12-4). Amounts of bound water are obtained from NMR by measuring the proportion of mobile water molecules that remains after highly concentrated solutions have been frozen. In calorimetry, the proportion of bound water is obtained from the difference between calculated and measured heat on freezing. Results of all four methods ([η], D, C, NMR) agree reasonably well.

12.3.5 Rods

The **Kirkwood-Riseman theory** (p. 373) approximates rods of molar mass M as linear arrangement of spheres that each have a hydrodynamic volume V_h. The spheres are assumed to be the friction elements; their diameter d is therefore the distance between friction elements. Since there are N_{sph} spheres which each have a molar mass $M_{sph} = M/N_{sph}$, the rod has a total length of $L = N_{sph}d = 2\ N_{sph}R_{sph}$ and an axial ratio of $\Lambda = L/d = L/(2\ R_{sph})$. The theory predicts the intrinsic viscosity of rigid rods to be

(12-23) $$[\eta] = \frac{2\pi N_A L^3}{(45\,M + a)[\ln \Lambda + K]} \cdot \frac{3\,V_h}{4\pi R_{sph}^3} = \left(\frac{4\,\Lambda^2}{15 \ln \Lambda} \right) \cdot \frac{N_A V_h}{M_{sph}}$$

where $K = a = 0$ (Kirkwood-Riseman theory). The newer **Doi-Edwards theory** calculates $K = (2 \ln 2) - (7/3)$ and $a = 3$.

For very long rigid rods, $(4\,\Lambda^2)/(15 \ln \Lambda)$ in Eq.(12-23) becomes practically identical with the Simha factor $\Psi_{pr} = 4\,\Lambda^2/(15 \ln (2\,\Lambda))$ for prolate ellipsoids (7 % difference for $\Lambda = 10^4$) since then the terms 1/2, 3/2, and 14/15 can be neglected in Eq.(12-19). In this case, Eq.(12-23) reduces to $[\eta] \approx \Psi_{pr} N_A (V_h/M_{sph})$.

The molar mass M of rods with constant diameter is proportional to the length L and thus to the axial ratio $\Lambda = L/d$. The exponent in the KMHS equation $[\eta] = K_v M^\alpha$ is then given by $\alpha = \mathrm{d}\ln[\eta]/(\mathrm{d}\ln \Lambda)$. Eq.(12-23) thus delivers for the exponent

(12-24) $$\alpha = \mathrm{d} \ln (\Lambda^2/\ln \Lambda) \,/\, (\mathrm{d} \ln \Lambda) = 2 - (\ln \Lambda)^{-1}$$

α becomes 1.78 (for $\Lambda = 10^2$), 1.89 ($\Lambda = 10^4$), 1.93 ($\Lambda = 10^6$), and 2.00 ($\Lambda \to \infty$).

Only for very large axial ratios can intrinsic viscosities become approximately proportional to the square of molar mass. However, macromolecules are rarely so rigid that they can be considered rigid rods for a large range of axial ratios (Fig. 12-9). Only imogolite, a tube-like aluminum silicate with composition $SiO_2 \cdot Al_2O_3 \cdot 2\ H_2O$, behaves approximately like a rod ($\alpha \approx 1.85$). Even deoxyribonucleic acids (persistence length L_{ps} = 63 nm; Table 4-7) and poly(hexylisocyanate)s (L_{ps} = 42 nm) are not rigid enough to behave as rods in certain molar mass ranges. At high molar masses, they all become perturbed coils in good solvents ($\alpha = 0.764$; Section 12.3.7, see also Fig. 4-28).

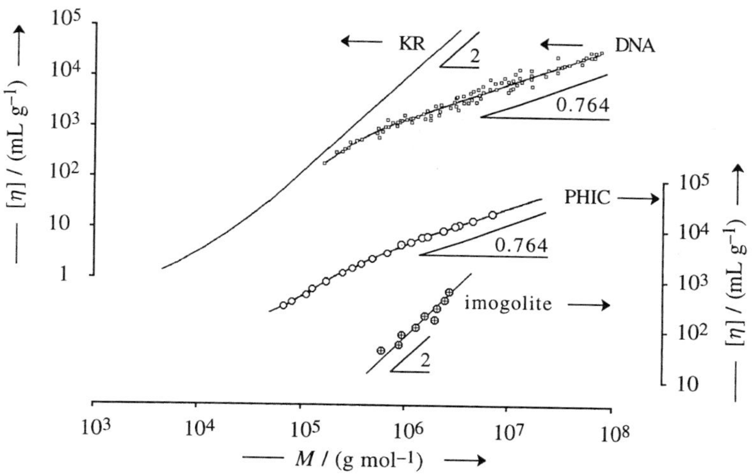

Fig. 12-9 Molar mass dependence of intrinsic viscosities. DNA: deoxyribonucleic acids in aqueous salt solutions at 20°C [8]; PHIC: poly(hexylisocyanate)s in hexane at 25°C [9]; and imogolite in dilute acetic acid (+ 0.02 w% NaN_3; pH = 3) at 30°C [10]. Slopes: theoretical values of α for rigid rods (α = 2) and perturbed coils (α = 0.764). KR: solid line represents prediction of Kirkwood-Riseman theory for rigid rods of DNA molecules (M_u = 3460 g/mol, d = 2 nm).

Deoxyribonucleic acids form double helices with diameters of ca. 2 nm and a large periodicity of 3.4 nm (Volume I, Fig. 14-5). Each strand contains 10 nucleotides per periodicity with average molar masses of ca. 294 g/mol. A "sphere" of d = 2 nm contains 2·2·10/3.40 = 11.76_5 nucleotides. The molar mass of a hydrodynamic unit is therefore $M_{sph} = 11.76_5 \cdot 294$ g/mol ≈ 3459 g/mol.

Logarithms of intrinsic viscosities, as calculated with Eq.(12-23) for the various lengths L and molar masses $M = N_{sph}M_{sph}$, are predicted by the Kirkwood-Riseman theory to increase practically linearly with logarithms of molar masses for axial ratios $\Lambda > 20$ (Fig. 12-9). The theoretical slope is ca. 1.86 for the range $10^2 < \Lambda < 10^4$ which corresponds to molar masses of $1.73 \cdot 10^5$-$1.73 \cdot 10^7$ g/mol. At high molar masses, double helix structures remain but the molecules become perturbed coils.

12.3.6 Unperturbed Coils

The dependence of intrinsic viscosities $[\eta]$ on molar masses M is described empirically by the KMHS equation, Eq.(12-11). For each type of polymer, K_v and α are constants that depend on constitution, configuration, molar mass distribution, and temperature, and, at low molar masses, also on the type of endgroups. Exponents α are related by theory (see below) to exponents ν of the dependence of radii of gyration on molar mass, $\langle s^2 \rangle = K_s M^\nu$ (Eq.(4-69)).

Exponents α

Theories (see below) predict exponents α = 1/2 for non-draining unperturbed coils (ν = 1/2) and α = 0.764 (renormalization theory, ν = 0.588) or 0.8 (mean-field theory, ν = 3/5), respectively, for perturbed coils. An α = 1 is predicted for freely draining coils.

Experimentally, α = 1/2 is indeed found for flexible polymer molecules of sufficiently high molar mass ($X \geq 100$) in theta solvents and $\alpha \approx 0.77$ in thermodynamically good solvents (Fig. 12-10). At small molar masses, intrinsic viscosities in bad and good

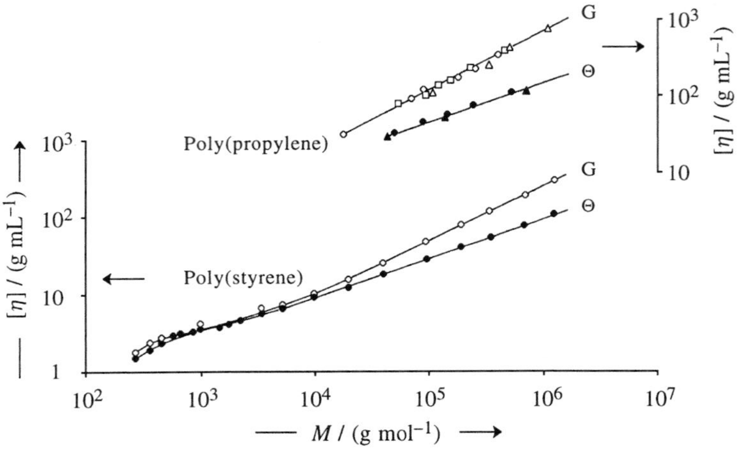

Fig. 12-10 Intrinsic viscosity-molar mass relationships. Top: isotactic (Δ,▲), syndiotactic (□), and atactic (O,●) poly(propylene)s (PP) in the good solvent (G) Decalin® at 135°C (Δ,□,O) or in the theta solvent (Θ) diphenyl at 129°C (●) or 125°C (▲) [11-14]. Bottom: atactic poly(styrene)s ($x_s = 0.59$) in the good solvent toluene at 15°C (O) or in the theta solvent cyclohexane at 34.5°C (●) [15].

solvents are practically identical. At very small molar masses, differences show up again for good and bad solvents; intrinsic viscosities may here even be negative.

Experimental α values deviate from theoretical ones at lower molar masses for three reasons. (1) Endgroups and monomeric units have different constitutions and this influence becomes more prominent at lower molar mass. (2) Chains are no longer self-similar because short chains cannot adopt ideal coil structures. (3) Deviations from self-similarity also occur because the macroconformation of short chains is dominated by sequences of certain microconformations. The latter effect is observed if longer tactic sequences are present in helical conformation and these helices are interconnected by non-helical segments, leading to the macroconformation of a coil.

It seems that flexible chains always lead to exponents $\alpha = 0.500$ *or* $\alpha = 0.764$ if the data cover a sufficiently large range of high molar masses. Such coils of flexible chains are always non-draining. Exponents $\alpha > 0.764$ are observed for semiflexible, worm-like, or rod-like polymer chains in some ranges of molar masses (see Fig. 12-9).

Values of K_v

Intrinsic viscosities of flexible coils become practically identical for good and bad solvents at sufficiently low molar mass (Fig. 12-10). It should thus be possible to obtain values of $K_{v,\Theta}$ in theta solvents by extrapolating K_v values in good solvents to zero molar mass. Corresponding equations are mostly semi-empirical; they can be written as

$$\text{(12-25)} \qquad \frac{[\eta]^a}{M^b} = K_{v,\Theta}^c + K_J \frac{M^d}{[\eta]^e}$$

The various exponents a-e are summarized in Table 12-5. A corresponding plot for one of these extrapolations is shown in Fig 12-11.

Table 12-5 Semi-empirical equations for the determination of $K_{v,\Theta}$ from molar masses M and intrinsic viscosities $[\eta]$ in good solvents. a, b, c, d, e = see exponents in Eq.(12-25).

Authors		a	b	c	d	e
Stockmayer-Fixman-Burchard	SFB	1	1/2	1	1/2	0
Inagaki-Suzuki-Kurata	ISK	4/5	2/5	4/5	1/3	0
Kurata-Stockmayer-Roig	KSR	2/3	1/3	2/3	2/3	1/3
Flory-Fox-Schaefgen	FFS	2/3	1/3	2/3	1	1
Berry	Be	1/2	1/4	1/2	1	1
Bohdanecky (worm-like chains)	Bo	2/3	1/3	1	1/2	0

Non-linear plots can often be avoided by plotting $[\eta]^{1/2}/M^{1/4} = f(M/[\eta])$ according to Berry (not shown). However, other linearizations of $[\eta] = f(M)$ may be better suited depending on the system polymer-solvent-temperature-molar mass range.

Hydrodynamic Radii

Intrinsic viscosities can be used to calculate hydrodynamic radii R_h which are radii of viscometric equivalent spheres, i.e., **Einstein radii** R_v. In general, radii R_v are not identical with hydrodynamic radii from diffusion coefficients, i.e., **Stokes radii** R_D (see also Table 12-7). Similarly to spheres, Eq.(12-17), both Einstein and Stokes radii are related to the radii of gyration s by $R_h = Q_h s$. The proportionality factor Q_h depends on the segment distribution within the molecules (for coils, see Fig. 4-24), i.e., on the molecule shape and the interaction with the solvent.

The *mean-field theory* (p. 111) maintains that long-range interactions are present in thermodynamically good solvents but not in theta solvents. The effects of these interactions are assumed to be the same for molecules at rest (i.e., for radii of gyration) and in flow fields (i.e., diffusion, viscosity, sedimentation). According to this theory, exponents α of Eq.(12-11) can only be 0.50 or 0.80.

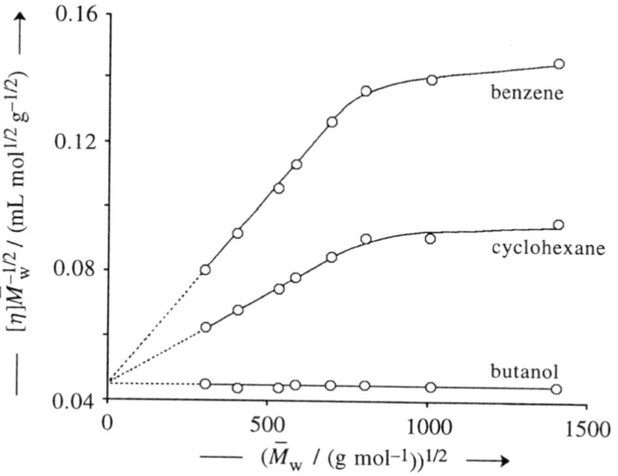

Fig. 12-11 Determination of $K_{v,\Theta}$ of poly(cyclohexyl methacrylate)s in benzene or cyclohexane at 25°C and in butanol at 23°C according to the Stockmayer-Fixmann-Burchard method [16].

Draining theories (p. 372) assume that coils drain differently in good and bad solvents because they are more expanded in the former than in the latter. These theories model chains of coil-forming molecules as collections of beads that are interconnected by massless springs. The *Rouse theory* (Section 15.3.2) assumes that hydrodynamic interactions are absent between beads. The solvent thus flows unimpeded through the coil where it causes friction. The viscosity $\eta = \xi_{seg}F_v$ of the polymer solution is then the product of the friction coefficient ξ_{seg} per segment and a global factor F_v that describes the effect of macroconformation, $F_v = (\rho_{cl}N_A/6)(\langle s^2\rangle_o/M)N_{seg}$ (Section 15.3.2).

In dilute solutions, the density of the coil, $\rho_{cl} = m_{cl}/V_{cl}$, equals the mass concentration c_2 of the polymer, $\rho_{cl} = c_2 = m_2/V$. For dilute solutions, viscosities η can be furthermore written as $\eta = \eta_1(\eta/\eta_1) \approx \eta_1[(\eta - \eta_1)/\eta_1] = \eta_1\eta_{sp}$ and the viscosity/density ratio as $\eta/\rho \approx \eta_1\eta_{sp}/c \approx \eta_1[\eta]$. Introduction of all these equations into $\eta = \xi_{seg}F_v$ delivers

$$(12\text{-}26) \qquad [\eta] = \frac{N_A\xi_{seg}N_{seg}}{6\eta_1}\left(\frac{\langle s^2\rangle_o}{M}\right) = \frac{N_A\xi_{seg}}{6\eta_1 M_{seg}}\left(\frac{\langle s^2\rangle_o}{M}\right)M = K_vM \quad ; \quad \langle s^2\rangle_o/M = const.$$

i.e., a direct proportionality between $[\eta]$ and M for freely draining coils. Such a proportionality, $[\eta] = K_vM$, has never been found experimentally for coil molecules.

The *Kirkwood-Riseman theory* thus assumes hydrodynamic interaction between beads (p. 373) which causes the friction factor to vary with the extent of this interaction. Very strong interactions lead to non-draining coils for which the KR theory predicts

$$(12\text{-}27) \qquad [\eta] = \pi^{3/2}N_A[Q\cdot f(Q)]\frac{\langle s^2\rangle_o^{3/2}}{M} = \Phi\frac{\langle s^2\rangle_o^{3/2}}{M} = \Phi\left(\frac{\langle s^2\rangle_o}{M}\right)^{3/2}M^{1/2}$$

The Kirkwood-Riseman function $Q\cdot f(Q)$ is controlled by the friction coefficient of segments. It varies from very small values for freely draining coils to $Q\cdot f(Q) = 1.259$ for non-draining ones according to the Auer-Gardner revision of the KR function. For coils in theta solvents, the revised KR theory predicts $\Phi = \Phi_\Theta = 4.22\cdot10^{24}$ mol^{-1}.

This value is identical with that from the mean-field theory which writes Eq.(12-27) for the theta state as $[\eta]_\Theta = \Phi_\Theta[\langle s^2\rangle_o^{3/2}/M]$ with $\Phi_\Theta = 10\,\pi N_A(Q_{v,\Theta})^3/3$ where $Q_{v,\Theta} \equiv R_{v,\Theta}/s_o$ relates $s_o \equiv \langle s^2\rangle^{1/2}$ to the viscometric radius $R_{v,Q}$. It is assumed to be universally valid since segment distributions in unperturbed coils are not affected by the chemical structure of segments (Eq.(4-47)). At sufficiently high molar mass, ratios $\langle s^2\rangle_o/M$ are furthermore independent of solvent and molar mass of the polymer (Section 4.4.6). All constants can thus be united to a new constant $K_{v,\Theta}$ and Eq.(12-27) becomes

$$(12\text{-}28) \qquad [\eta]_\Theta = \Phi_\Theta\frac{\langle s^2\rangle_o^{3/2}}{M} = \Phi_\Theta\left(\frac{\langle s^2\rangle_o}{M}\right)^{3/2}M^{1/2} = K_{v,\Theta}M^{1/2}$$

For high-molar mass, unperturbed coils, the **Flory constant** Φ_Θ should therefore be a *universal constant* with a value of $4.22\cdot10^{24}$ mol^{-1} according to the Kirkwood-Riseman theory. This value leads to $Q_{v,\Theta} \equiv R_{v,\Theta}/s_o = 0.874$ for unperturbed coils and to $Q_{sph} = R_{sph}/s = (5/3)^{1/2} = 1.291$ for hard, compact spheres.

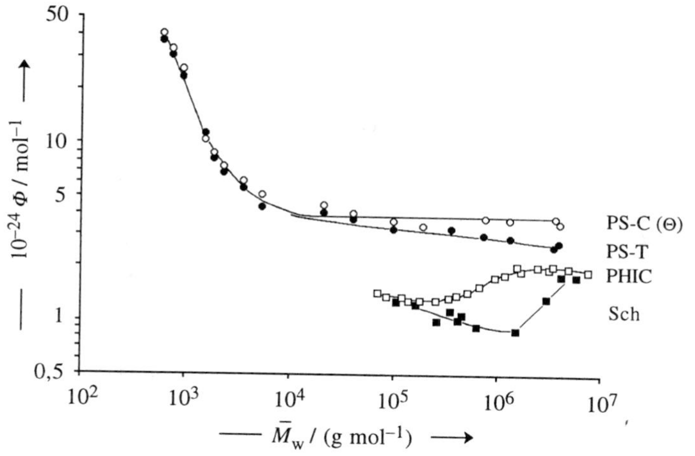

Fig. 12-12 Molar mass dependence of Φ of poly(styrene)s (PS) in the theta solvent cyclohexane (C) at 34.5°C [17] and in the good solvent toluene (T) at 15°C [18], poly(hexylisocyanate) (PHIC) in hexane at 25°C [19], and schizophyllan (Sch) in 0.01 mol/L NaOH at 25°C [20].

Flory constants for high-molar mass coils in theta solvents are indeed independent of molar masses (Fig. 12-12). However, they are lower than the value of $4.22 \cdot 10^{24}$ mol^{-1} that is predicted by the Kirkwood-Riseman theory. A revision of the KR theory by Zimm leads to a lower value of $10^{24}\Phi_\Theta$/mol^{-1} = 3.69 instead of 4.22 (if related to radius of gyration) and to 0.251 instead of 0.287 (if related to end-to-end distance). The value of $3.69 \cdot 10^{24}$ mol^{-1} corresponds to an "average" experimental value as data show for some polymers in theta solvents (Table 12-6; see also Appendix to Chapter 12).

Literature uses different values of Φ_Φ, often without physical units and sometimes with other physical units than are used in this book and in part because they refer to end-to-end distances instead of radii of gyration. The Flory constant with respect to end-to-end distance is $\Phi_{\Theta,r} = 2.87 \cdot 10^{23}$ mol^{-1} since $\Phi_{\Theta,r} = [\eta]M/\langle r^2\rangle_o^{3/2} = [\eta]M/(6^{3/2}\,\langle s^2\rangle_o^{3/2})$.

However, any such agreement seem to be fortuitous because it rests on the assumption that there are no "specific" solvent interactions. Indeed, a universal value of Φ_Θ for unperturbed coils of all linear polymers implies that the ratio $[\langle s^2\rangle_o/M]^{3/2}$ in Eq.(12-28) should be a constant for a polymer regardless of the theta solvent, i.e., that $K_{v,\Theta}$ = constant at constant temperature. This is not found, however (Fig. 12-13), since $\ln K_{v,\Theta} = f(T)$ varies differently for different classes of solvents. Furthermore, $K_{v,\Theta}$ and thus $[\eta]_\Theta$ decrease with decreasing temperature whereas $[\langle s^2\rangle_o/M]$ from direct measurements of unperturbed dimensions is independent of temperature for this polymer (Fig. 6-1).

Table 12-6 Experimental Flory constants Φ_Θ for some polymers in theta solvents.

Polymer	Solvent	Θ/°C	$10^{-24}\Phi_\Theta$/mol^{-1}
Poly(styrene)	cyclohexane	34.5	3.94 ± 0.06
Amylose, synthetic	dimethylsulfoxide	25	3.59 ± 0.50
Poly(isobutylene)	*i*-amyl valerate	25	3.58 ± 0.12
Poly(methyl methacrylate)	acetonitrile	44	3.33 ± 0.11
Poly(α-methyl styrene)	cyclohexane	34.5	2.99 ± 0.13

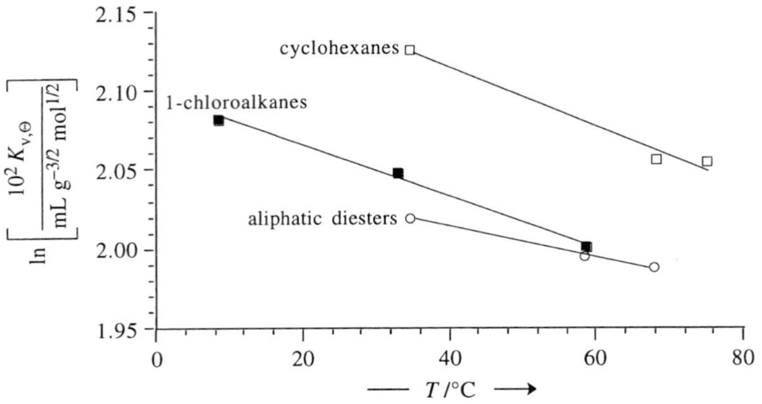

Fig. 12-13 Temperature dependence of ln $K_{v,\Theta}$ of atactic poly(styrene) in various classes of theta solvents [21].

Flexible polymer chains show a monotonous variation of lg Φ_Θ with lg $\overline{M}_w$ but worm-like chains do not (Fig. 12-12). Instead, values of lg Φ_Θ rather decrease with lg $\overline{M}_w$, then pass through a minimum, increase, and then decrease again. This behavior indicates the transition from worm-like to coil-like behavior.

Another indication of the non-equivalency of theta states can be seen from Table 12-6 where the ratios $s_o/R_{D,\Theta}$, $s_o/R_{v,\Theta}$, and $R_{D,\Theta}/R_{v,\Theta}$ decrease from poly(α-methyl styrene) to poly(styrene) and poly(methyl methacrylate) although they were all obtained at approximately the same theta temperature. Fortuitously, experimental averages of these quantities for the three polymers do agree with the predictions of the KR theory.

12.3.7 Perturbed Coils

Local concentrations of monomeric units are much smaller in perturbed coils than in unperturbed ones (Fig. 4-26). One should thus expect perturbed coils to show some draining which, in turn, means that the exponent α of the KMHS equation should adopt values between 0.5 and 1.0 for flexible chains. However, this is not found. Careful measurements over a wide range of molar masses (at least two decades) rather seem to indicate that flexible chains of sufficiently high molar masses adopt either the unperturbed state (α = 1/2) or the perturbed state (α = 0.764) as predicted by the mean-field theory. Writing Eq.(12-27) for the general case with s instead of s_o and introducing $\langle s^2\rangle^{1/2} = K_s M^\nu$ leads to

$$(12\text{-}29) \qquad [\eta] = \Phi\langle s^2\rangle^{3/2}/M = \phi K_s{}^3 M^{3\nu-1} = K_v M^\alpha \quad ; \quad \alpha = 3\nu - 1$$

The upper values of ν = 3/5 (mean-field theory) or 0.588 (renormalization theory) lead to upper values of α = 0.8 (mean-field theory) and 0.764 (renormalization theory), respectively. The theory also predicts for good solvents that Φ does not become constant at high molar masses, which is indeed found (Fig. 12-12).

Radii of gyration of perturbed coils are larger than those of unperturbed ones by an expansion factor of $\alpha_s = (\langle s^2\rangle/\langle s^2\rangle_o)^{1/2}$. A similar equation, $\alpha_v = ([\eta]/[\eta]_\Theta)^{1/3}$, should

Table 12-7 Constants $K_{s,o}$ of molar mass dependency of sedimentation coefficients, $s_o = K_{s,0}M^{1/2}$; $K_{D,\Theta}$ from the molar mass dependency of friction coefficients of diffusion, $\xi_{D,\Theta} = K_{D,\Theta}M^{1/2}$; and $K_{v\Theta}$ from the molar mass dependency of intrinsic viscosities, $[\eta]_\Theta = K_{v,\Theta}M^{1/2}$, for poly($\alpha$-methyl styrene) (PAMS), poly(styrene) (PS), and poly(methyl methacrylate) (PMMA) in the theta state.
Stokes radii were calculated from $R_{D,\Theta} = \xi_\Theta/(6\,\pi)$ and Einstein radii of equivalent spheres from $R_{v,\Theta} = (6\,[\eta]_\Theta M/(20\,\pi\,N_A))^{1/3}$.

Physical property	Physical unit	PAMS Cyclohexane Θ = 34.5°C	PS Cyclohexane Θ = 34.5°C	PMMA Butyl chloride Θ = 35.4°C	Experimental average	Modified Kirkwood-Riseman theory
K_{so}	nm g$^{-1/2}$ mol$^{1/2}$	0.0290	0.0290	0.0219	-	-
$K_{D\Theta}$	nm g$^{-1/2}$ mol$^{1/2}$	0.385	0.432	0.365	-	-
$K_{v\Theta}$	mL g$^{-3/2}$ mol$^{1/2}$	0.073	0.090	0.053	-	-
$s_o/R_{D\Theta}$	1	1.42	1.27	1.13	1.27 ± 0.14	1.28
$s_o/R_{v\Theta}$	1	1.28	1.20	1.08	1.19 ± 0.10	1.20
$R_{D\Theta}/R_{v\Theta}$	1	1.11	1.05	0.95	1.04 ± 0.08	1.07

therefore also apply to viscometric radii, $R_v \sim [\eta]^{1/3}$. Eq.(12-29) becomes

$$(12\text{-}30)\qquad [\eta] = \alpha_v^3[\eta]_\Theta = \alpha_v^3\Phi_\Theta\langle s^2\rangle_o^{3/2}/M = \Phi_\Theta[\langle s^2\rangle/M](\alpha_v/\alpha_s)^3M^{1/2}$$

Radii of gyration do not equal viscometric radii and neither do the respective expansion coefficients (Table 12-7). Theories predict $\alpha_v^3 = \alpha_s^q$ with q = 2.43 for equivalent spheres and q = 2.18 for equivalent ellipsoids. For poly(styrene) in the good solvent toluene at 15°C, these values lead to $10^{-24}\,\Phi_\Theta/\text{mol}^{-1} = 3.76 \pm 0.17$ for equivalent spheres and 4.21 ± 0.26 for equivalent ellipsoids. Since the experimental value for poly(styrene) in the theta solvent cyclohexane is 3.94 ± 0.06, it follows that this procedure does not allow one to distinguish between equivalent spheres and equivalent ellipsoids.

12.3.8 Branched Polymer Molecules

Branched macromolecules are usually characterized by a viscometric branching parameter, $g_v = [\eta]_{br}/[\eta]_{lin} < 1$ (the *greater* the branching, the *smaller* is g_v because of the smaller coil volume). The viscometric branching parameter g_v is related to the branching parameter based on radii of gyration, $g_s = \langle s^2\rangle_{br}/\langle s^2\rangle_{lin}$ (Section 4.7), via $[\eta] = \Phi_\Theta(\langle s^2\rangle_o^{3/2}/M)\alpha_v^3$ which is written for both branched and linear polymers, resulting in

$$(12\text{-}31)\qquad g_v = g_{s,\Theta}^{3/2}\left(\frac{\Phi_{br,\Theta}}{\Phi_{lin,\Theta}}\right)\left(\frac{\alpha_{v,br}}{\alpha_{v,lin}}\right)^3$$

$g_{s,\Theta}^{3/2}$ measures the effect of molecule architecture on the shrinking of molecule sizes whereas $\Phi_{br,\Theta}/\Phi_{lin,\Theta}$ is a measure of the change of intramolecular hydrodynamic interactions and $(\alpha_{v,br}/\alpha_{v,lin})^3$ that of the effect of excluded volumes.

Alternatively, one can also define a viscometric branching parameter h_v via Einstein radii, i.e., via $[\eta] = 10\,\pi\,N_AR_v^3/(3\,M)$ (Eq.(12-17)), for $M = constant$:

$$(12\text{-}32)\qquad g_v \equiv [\eta]_{v,br}/[\eta]_{v,lin} = R_{v,br}^3/R_{v,lin}^3 \equiv h_v^3 \qquad (\text{h}^3\text{ rule})$$

Viscometric branching parameters $g_{v,\Theta}$ are related to branching parameters $g_{s,\Theta}$ since $\alpha_{v,br}/\alpha_{v,lin} \equiv 1$ for the unperturbed state and $\Phi_{br,\Theta}/\Phi_{lin,\Theta} = 1$ if intramolecular hydrodynamic interactions are absent. Eq.(12-31) thus becomes $g_{v,\Theta} = g_{s,\Theta}^{3/2}$ for the theta state (**Thurmond-Zimm theory**).

In perturbed random coils, intramolecular hydrodynamic interactions are present so that $\Phi_{br,\Theta}/\Phi_{lin,\Theta} \neq 1$. In this case, one can therefore write $g_{v,\Theta} = g_{s,\Theta}^{\omega}$ if the coils are non-draining ($\omega = 1$ for the theta state).

The same expression is furnished by the **Zimm-Kilb theory** for star molecules with arms of equal length. This theory is based on the Rouse model; it delivers $\omega = 1/2$ for non-draining molecules and $\omega = 1$ for freely draining ones.

The dependence of the branching parameter $g_{s,\Theta}^{\omega}$ of randomly branched polymer molecules from the number N_{br} of branch points per molecule can be described by the semi-empirical **Bohdanecky equation**, $1/g_{s,\Theta}^{\omega} = 1/g_{g,\Theta} = [\eta]_{lin,\Theta}/[\eta]_{br,\Theta} = A + BN_{br}^{1/2}$ where B can be obtained from the **Kurata-Fukatsu** theory for polymers that are uniform with respect to molar masses and the distribution of branches. The number N_{br} is obtained from $M = M_{bb} + N_{br}M_{br}$ where M = molar mass of the polymer, M_{bb} = molar mass of the backbone, and M_{br} = average molar mass of branches:

$$[\eta]_{lin,\Theta}/[\eta]_{br,\Theta} = A + (B/M_{br})^{1/2}(M - M_{bb})^{1/2} \approx A + (B/M_{br})^{1/2}M_{bb}^{1/2} \quad (12\text{-}33)$$

Fig. 12-14 shows the corresponding plots for two self-branching polymers (LDPE, PDMA) and a copolymer P(S-TCDVB) from styrene S and the multifunctional monomer TCDVB. All three polymers deliver straight lines with a common intercept of A = 0.81. The average molar masses of branches were calculated as $M_{br}/(\text{g mol}^{-1})$ = 7140 (LDPE), 250 000 (S-TCDVB), and 1 030 000 (PDMA), the latter value being in good agreement with the molar mass of branches (1 140 000) that was obtained by light scattering. Eq.(12-33) seems to apply if the molar mass M of the molecule is much larger than the molar mass of the backbone.

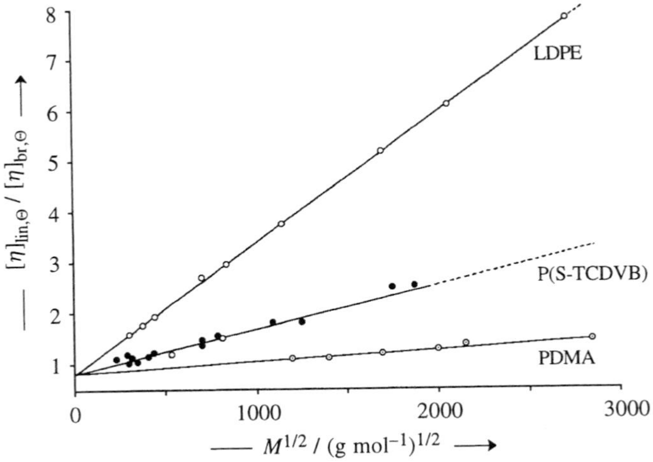

Fig. 12-14 Bohdanecky plot for $[\eta]_{lin,\Theta}/[\eta]_{br,\Theta} = f(M^{1/2})$ for low-density poly(ethylene)s (LDPE), poly(dodecyl methacrylate)s (PDMA), and copolymers of styrene and tetrachlorodivinyl benzene) (S-TCDVB) [22]. Theory predicts an intercept of $[\eta]_{lin,\Theta}/[\eta]_{br,\Theta} = 1$.

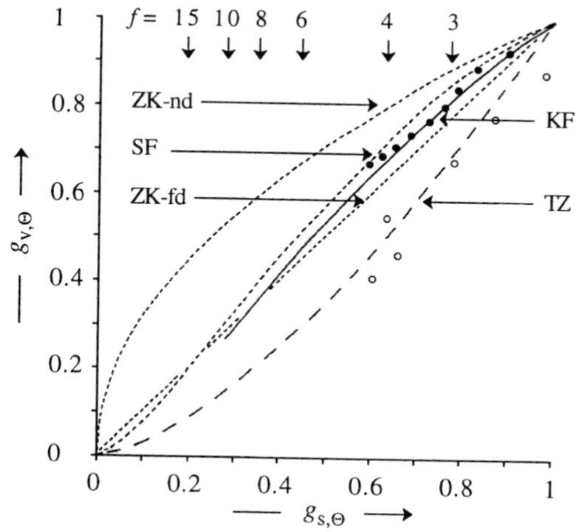

Fig. 12-15 $g_{v,\Theta}$ as a function of $g_{s,\Theta}$ according to the theories of Thurmond-Zimm (TZ; — —) and Kurata-Fukatsu (KF; ——) for randomly branched polymers (see text) and according to the theories of Stockmayer-Fixman (SF) and Zimm-Kilb (ZK) for star molecules with f arms (- - -). fd = free-draining, nd = non-draining. Experimental data [22]: randomly branched copolymers from styrene and tetrachlorodivinyl benzene in various theta solvents (O) and randomly branched polymers from dodecyl methacrylate in the theta solvent pentanol (●) (see also Fig. 12-14).

Fig. 12-15 compares the results of the Thurmond-Zimm theory (TZ) for randomly branched polymers and the Kurata-Fukatsu (KF) theory for polymers with randomly placed branches of equal length with the theories for star molecules according to Stockmayer-Fixman (SF) and Zimm-Kilb (ZK), the latter for both free-draining (fd) and non-draining (nd) molecules. The TZ theory describes very well the dependence of viscometric branching parameters, $g_{v,\Theta}$, on the corresponding branching parameter from radii of gyration, $g_{s,\Theta}$. Viscometric branching parameters for randomly branched polymers with branches of different lengths (TZ) are lower than those for polymers with branches of equal length (KF), which seems to apply to polymers from dodecyl methacrylate.

Randomly Branched Polymers

Randomly branched polymer molecules result from chain polymerizations of bifunctional monomers with chain transfer to polymer molecules and from polycondensations and polyadditions of multifunctional monomers (see Volume I for details). The type and degree of branching depends on the reaction conditions used.

For example, linear poly(ethylene)s (HDPE) in good solvents obey the KMHS equation with $\alpha = 0.74$ (theory: $\alpha = 0.764$) (Fig. 12-16). For LDPEs, lg $[\eta] = f(\lg M)$ is not a straight line ($\alpha = f(\lg M)$) since the fraction of short and long branches increases systematically with increasing molar mass, which leads to successively lower intrinsic viscosities. Such randomly branched polymer molecules are not self-similar.

At very high molar masses, intrinsic viscosities become independent of molar masses, i.e., $\alpha = 0$ in $[\eta] = K_v M^{\alpha}$. Since $\alpha = 0$ and $[\eta]_\infty = 210$ mL/g, coils of these highly branched polymers behave as highly solvated spheres since compact spheres would have $[\eta] = (5/2\ \rho)$ (Eq.(12-18)), i.e., ca. 2.5 mL/g if $\rho = 1$ g/mL.

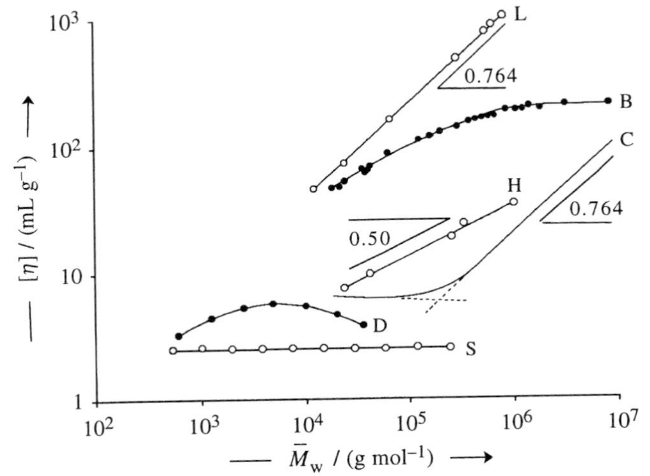

Fig. 12-16 $[\eta] = f(\overline{M}_w)$ for various branched polymers (see below). Slopes are theoretical (see text).
B = randomly branched poly(ethylene)s (LDPE) in tetralin™ at 120°C [23].
C = comb polymers, poly(methyl methacrylate)-*comb*-poly(styrene) in toluene at 25°C [24].
D = dendrimers with 3,5-dioxybenzylidene units in tetrahydrofuran at 30°C [25].
H = hyperbranched polymers from 3,5-diacetoxybenzoic acid in tetrahydrofuran at 25°C [26].
L = linear poly(ethylene)s of high density (HDPE) in tetralin™ at 120°C [23].
S = poly(α,ε-lysine)s dendrimers in *N,N*-dimethylformamide at 25°C; ρ = 1.18 g/mL [27].

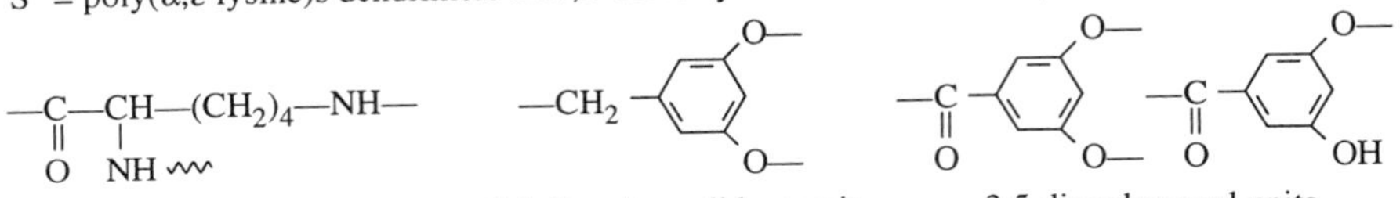

α,ε–lysine unit 3,5-dioxybenzylidene unit 3,5-dioxybenzoyl units

Dendrimers

Dendrimers are polymer molecules with a regular branch-upon-branch structure that usually contain only short segments between branch points (p. 11; see also Vol. I, p. 619; Vol. II, p. 453). Radii of gyration of the three most studied types of dendrimers increase approximately with the cubic root of molar masses (p. 124) which is what one would expect for spherical molecules. Intrinsic viscosities should thus be independent of molar masses (Section 12.3.3), which is indeed found for dendritic poly(α,ε-lysine)s (Fig. 12-16). For other dendrimers, intrinsic viscosities either increase with increasing molar mass (not shown) or even pass through a maximum (Fig. 12-16).

The low intrinsic viscosities of poly(α,ε-lysine)s, $[\eta]$ = 2.5 mL/g, and their independence of molar masses indicate that these dendrimers behave approximately like compact spheres. Indeed, one calculates from $[\eta] = 5/(2\ \rho)$ (Eq.(12-18)) and the experimental density of dry dendrimers ((ρ = 1.18 g/mL) that such compact spheres should have intrinsic viscosities of $[\eta]$ = 2.12 mL/g. The latter value is lower then the experimental value of 2.5 mL/g, which indicates that these spheres must be somewhat solvated.

Poly(α,ε-lysine)s probably behave as fairly compact spheres because of many internal hydrogen bonds between amide groups. If such intramolecular bonds are weak or absent, other molecular shapes evolve that vary with the molar mass as do $[\eta]$ values.

Hyperbranched Polymers

In dendrimers, all monomeric units are connected to other monomeric units (see chemical structures of α,ε-lysine and 3,5-dioxybenzylidene units in Fig. 12-16), except, of course, for the core units and the end units (surface units). In hyperbranched polymers, some monomeric units contain functional groups that are not connected to other monomeric units (see the two units of 3,5-dioxybenzoyl in Fig. 12-16). The random reactions of these AB_2 monomers lead to broad molar mass distributions with types and widths that are usually not known. The dependence of intrinsic viscosities on molar masses contains therefore a systematic (but unknown) error.

The hyperbranched polymers from 3,5-diacetoxybenzoic acid behave as unperturbed random coils with $\alpha = 1/2$ (Fig. 12-16). It is unclear whether this finding can be generalized. Experiments are very time-consuming since the broad molar mass distributions of hyperbranched polymers require large corrections for molecular non-uniformities.

Comb Polymers

Comb polymers consist of a backbone polymer chain to which other polymer chains are connected. An example is a poly(methyl methacrylate) chain of a degree of polymerization X to which styrene chains with N monomer units are bound in regular intervals:

PMMA–*comb*–PS

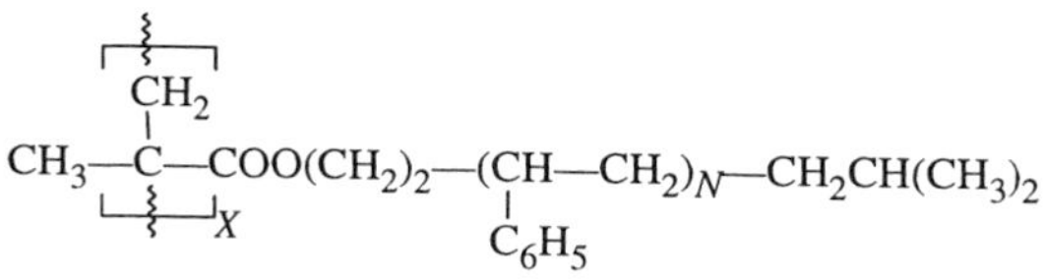

The function $[\eta] = f(M)$ of such a polymer with $N = 28$ and varying X shows a peculiar behavior. At low molar masses, intrinsic viscosities are practically independent of the molar mass. After a fairly sharp break, $[\eta]$ increases with the 0.764th power of the molar mass (Fig. 12-16). This behavior is explained as follows.

At the lowest molar mass of ca. 18 000 g/mol, the degree of polymerization is only $X \approx 5.8$, i.e., the side chains are much longer ($N = 28$) than the main chain. Since the side chains also repel each other, the molecules behave as somewhat solvated spheres with $[\eta] \approx 8$ mL/g.

Intrinsic viscosities then decrease slightly with increasing M until, above $X \approx 58$, $[\eta]$ increases linearly with $M^{0.764}$. The transition from spheroids to coils happens if the degree of polymerization of the main chain becomes about twice as large as that of the side chains ($X_{main} \approx 58$; $X_{side} \approx 28$).

Since the side chains stiffen the chain, the molecule must be worm-like. However, large worm-like molecules behave like perturbed chains with excluded volumes, hence $\alpha = 0.764$ which is indeed the theoretical value for perturbed random coils.

Star Polymers

A star polymer has a central core from which three or more linear arms radiate. Short arms on small cores are very crowded and cannot coil whereas long arms on bigger cores can. The function $[\eta] = f(M)$ is therefore not easy to predict.

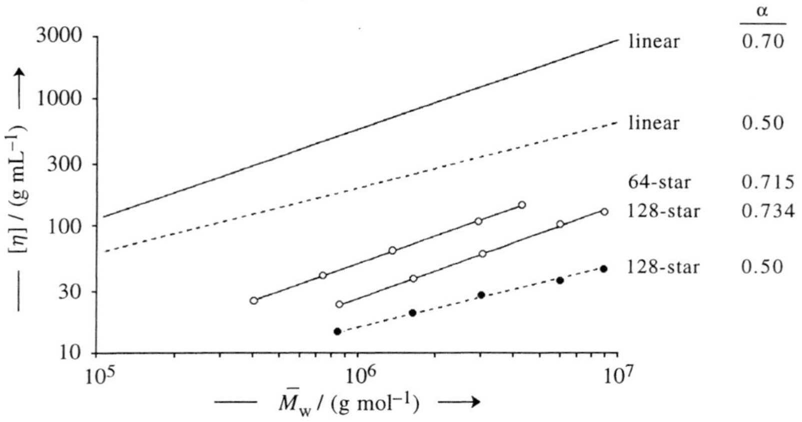

Fig. 12-17 Molar mass dependence of intrinsic viscosities of linear 1,4-poly(butadiene)s and star polymers with 64 or 128 1,4-poly(butadiene) arms in the good solvent cyclohexane at 25°C (——) and in the theta solvent 1,4-dioxane at 26.5°C (- - - -) [28]. Cores of the stars are vinyl carbosilane dendrimers of generation G3 (64-arm) and G4 (128-arm), respectively. Values of α are experimental.

Intrinsic viscosities of star molecules are certainly affected by a purely geometric effect. The Stockmayer-Fixman theory assumes that solvent flow through or at the star molecules differs from that for linear ones. This flow is thought to be the same one that controls translational diffusion. Branching parameters for translational diffusion should therefore also apply to dilute-solution viscosities of these polymers if one considers the h^3 rule (Eq.(12-32)). For theta solvents, the Stockmayer-Fixman theory predicts

(12-34) $$h_d = g_v^{1/3} = f^{1/2}[(2-f) + 2^{1/2}(f-1)]^{-1}$$

where f is the number of arms per core. Calculated theoretical values for star molecules with various numbers of arms are shown in Fig. 12-17. Eq.(12-34) predicts very small g values for large numbers of arms per molecules, for example, $g_v = 0.850$ for 3 arms, $g_v = 0.0257$ for 64 arms, and $g_v \approx 0.0094$ for 128 arms.

For a constant number of arms, g values in theta solvents are expected to decrease with increasing lengths of arms (see p. 118) and a corresponding nonlinearity of lg $[\eta]$ = f(lg M). Instead, one finds for large numbers of arms on large cores that lg $[\eta]$ = f(lg M) is not only linear but delivers the same exponent $\alpha = 1/2$ for linear polymers and 128-arm stars in the theta solvent and practically the same α-values of 0.70 (linear), 0.715 (64-arm), and 0.734 (128-arm) in the good solvent (Fig. 12-17). This behavior resembles that of perturbed coils of flexible linear polymers. It was explained for star molecules by a blob model in which the size of the blobs increases continuously from small diameters near the core of the molecule to large diameters at the periphery.

12.3.9 Polyelectrolytes

Most reduced viscosities of polyelectrolytes described in the literature are obtained with capillary viscometers at shear rates of 500 s^{-1} or more. Since non-Newtonian ranges of polyelectrolyte solutions start at values that are several decades lower than those of neutral polymers, this means that most reported values η_{red} of polyelectrolyte solutions are not Newtonian. Furthermore, these shear-dependent η_{red} values are then linearized by the Fuoss equation, Eq.(12-7), whose intercept does not deliver the true inverse intrinsic viscosity, however (see p. 404).

Truly Newtonian values of η_{rel} and η_{sp}/c are strongly affected by the concentration of added low-molar mass electrolytes and the concentration range of the polyelectrolyte (Figs. 12-6 and 12-7). In turn, this produces very different values of K_v and α in $[\eta] = K_v M^{\alpha}$ (Table 12-8). Neither α nor K_v are therefore universal values for a system polyelectrolyte/solvent/added electrolyte/temperature; instead these values apply only to certain types of added electrolytes and ranges of molar masses of polyelectrolytes. In general, values of α decrease with increasing concentration of added electrolytes.

Table 12-8 Constants K_v and α of the KMHS equation for sodium hyaluronate [29] and poly(*N*-methyl-2-vinylpyridinium chloride) [30] in aqueous sodium chloride solutions at 25°C. Molecular weight range: 100 000 to 3 000 000.

Polymer	[NaCl]/(mol L^{-1})	α	$10^3\ K_v$/(mL g^{-1})
Sodium hyaluronate	0.01	0.916	12.1
	0.06	0.830	20.9
	0.3	0.785	25.3
Poly(*N*-methyl-2-vinylpyridinium chloride)	0.01	0.86	0.0786
	0.1	0.77	0.0840
	0.5	0.63	0.268

A-12 Appendix to Chapter 12: Flory Constants

Particle shape	Λ	$10^{-24}\ \Phi$/mol^{-1}	$10^{-6}\ \beta'$/mol$^{-1/3}$
Sphere, compact, unsolvated	1	13.57	9.802
Ellipsoid, compact, prolate	2	7.336	9.874
Ellipsoid, compact, prolate	3-300	-	$8.867 \cdot \Lambda^{0.1124}$
Ellipsoid, compact, oblate	2	7.199	9.831
Ellipsoid, compact, oblate	3-300	-	$9.828 \cdot \Lambda^{0.002667}$
Coil, unperturbed	-	3.69	9.806
Coil, perturbed	-	2.31	13.04

Historical Notes to Chapter 12

Concentration Dependence of Solution Viscosities

S.F.Arrhenius, Z.Physik.Chem **1** (1887) 285

Logarithmic equation, $\ln \eta_{rel} = Kc$, for the concentration dependence of the viscosity of electrolyte solutions. The same equation was used later by several research groups to describe $\eta = f(c)$ for solutions of non-electrolytes.

H.G.Bungenberg de Jong, H.R.Kruyt, J.Lens, Kolloid-Beih. **37** (1933) 395
H.Staudinger, W.Heuer, Z.Physik.Chem. **A** 171 (1934) 129

Replacement of the Arrhenius equation, $\ln \eta_{rel} = Kc$, by $(\eta_{rel} - 1)/c = [\eta] \exp (k_{BKL}c)$.

A.F.Martin, 103rd Am.Chem.Soc.Meeting (Memphis), Div. of Cellulose Chem., C 1, **23** (1942) 4

Replacement of k_{BKL} by $k_M[\eta]$. k_M is a constant for a series of homologous polymers.

M.L.Huggins, J.Am.Chem.Soc. **64** (1942) 2716

Introduction of the Huggins equation, $\eta_{red} = [\eta] + k_H[\eta]^2c + ...$

Averages

W.Philippoff, Ber.Dtsch.Chem.Ges. **70** (1937) 827

Intrinsic viscosities of molecularly non-uniform polymers are mass-averages of intrinsic viscosities of their components.

E.O.Kraemer, W.D.Lansing, J.Phys.Chem. **39** (1935) 153

The molar mass in $[\eta] = K_vM$ is a mass-average.

P.J.Flory, J.Am.Chem.Soc. **65** (1943) 372

The molar mass in the KMHS equation, $[\eta] = K_vM^\alpha$ is a viscosity average.

Theories for $[\eta] = K_vM^\alpha$ (see also Volume I, p. 104)

The empirical Kuhn-Mark-Houwink-Sakurada equation was independently proposed by

W.Kuhn, Kolloid-Z. **68** (1934) 2 (α was estimated as 0.84 for what is now called perturbed coils).
H.Mark, in E.Saenger, Ed., Der feste Körper (lectures at the 50th anniversary of the Physical Society of Zurich, 1937), Hirzel Verlag, Leipzig 1938 ($0.5 \leq \alpha \leq 1.0$).
I.Sakurada, Proceedings of the Congress of Nippon Kagaku Seni Kenkyu-sho **5** (1940) 33, see T.Saegusa, Makromol.Symp. **98** (1995) 1199 (α not specified)
R.Houwink, J.Prakt.Chem. **157** (1941) 15 (α not specified)

J.G.Kirkwood, J.Riseman, J.Chem.Phys. **16** (1948) 565

$[\eta] = f(M)$ for freely draining and non-draining coils.

P.J.Flory, T.G.Fox, J.Am.Chem.Soc. **73** (1951) 1904

$[\eta] = \Phi(\langle r^2\rangle^{3/2}/M)$

P.E.Rouse, J.Chem.Phys. **21** (1953) 1272

Spring-bead model for coils without hydrodynamic interactions. Freely draining coils lead to $\alpha = 1$.

B.H.Zimm, J.Chem.Phys. **24** (1956) 269

Spring-bead model for coils with hydrodynamic interactions. Non-draining coils lead to $\alpha = 1/2$.

P.J.Flory, Principles of Polymer Chemistry, Cornell Univ. Press, Ithaca (NY) 1953, p. 519

Perturbed coils of flexible chains of linear polymers have $\alpha = 4/5$ because of the presence of ex cluded volumes. In theta solvents, coils adopt their unperturbed dimensions ($\alpha = 1/2$).

J.G.Kirkwood, P.L.Auer, J.Chem.Phys. **19** (1951) 281

Rigid rods are modeled as a row of beads.

Literature to Chapter 12

(see also Literature to Chapter 15)

H.Yamakawa, Modern Theory of Polymer Solutions, Harper and Row, New York 1971
M.Bohdanecky, J.Kovar, Viscosity of Polymer Solutions, Elsevier, Amsterdam 1982
H.Fujita, Polymer Solutions, Elsevier, Amsterdam 1990
J. des Cloiseaux, G.Jannink, Polymers in Solution. Their Modelling and Structure, Oxford Univ. Press, New York 1990
M.Hara, Ed., Polyelectrolytes, Dekker, New York 1993
H.Yamakawa, Helical Wormlike Chains in Polymer Solutions, Springer, Berlin 1997
R.E.Bareiss, Polymolecularity Correction Factors, in J.Brandrup, E.H.Immergut, E.Grulke, Eds., Polymer Handbook, Wiley, New York, 4th ed. 1998
P.R.Dvornic, S,Uppuluri, in J.M.J.Fréchet, D.A.Tomalia, Eds., Dendrimers and Other Dendritic Polymers, Wiley, New York 2002
W.W.Graesssley, Polymeric Liquids and Networks. Vol. I, Structure and Properties, Garland Science, New York (2003)
J.Furukawa, Physical Chemistry of Polymer Rheology, Springer, Berlin 2003
W.-M.Kulicke, C.Clasen, Viscosimetry of Polymers and Polyelectrolytes, Springer, Berlin 2005

References to Chapter 12

[1] W.-M.Kulicke, R.Kniewske, Rheologica Acta **23** (1984) 75, (a) Fig. 1, (b) Fig. 2, (c) Fig. 5
[2] Calculated from data of Table 2 of S.Uppuluri, S.E.Keinath, D.A.Tomalia, P.R.Dvornic, Macromolecules **31** (1998) 4498
[3] D.S.Pearson, A.Mora, W.E.Rochefort, ACS Polymer Preprints **22**/1 (1981) 102, Table 1
[4] E.R.Morris, A.N.Cutler, S.B.Ross-Murphy, D.A.Rees, J.Price, Carbohydrate Polymers **1** (1981) 5
[5] J.Cohen, Z.Priel, Y.Rabin, J.Chem.Phys. **88** (1988) 7111, Fig. 3
[6] D.C.Boris, R.H.Colby, Macromolecules **31** (1998) 5746, (a) data of Fig. 2, (b) Table 1
[7] R.Geddes, J.D.Harvey, P.R.Wills, Biochem.J. **163** (1977) 201
[8] Compilation of literature data: J.Eigner, P.Doty, J.Mol.Biol. **12** (1965) 549
[9] H.Murakami, T.Norisuye, H.Fujita, Macromolecules **13** (1980) 345, Tables 1 and 2
[10] N.Donkai, H.Inagaki, K.Kajiwara, H.Urakawa, M.Schmidt, Makromol.Chem. **186** (1985) 2623, Table 1
[11] F.Danusso, G.Moraglio, Makromol.Chem. **28** (1958) 250 (at-PP)
[12] R.Chiang, J.Polym.Sci. **28** (1958) 235 (it-PP)
[13] H.Inagaki, T.Miyamoto, S.Ohta, J.Phys.Chem. **70** (1966) 3420 (st-PP)
[14] G.Moraglio, G.Gianotti, U.Bonicelli, Eur.Polym.J. **9** (1973) 623 (at-PP and it-PP)
[15] F.Abe, Y.Einaga, H.Yamakawa, Macromolecules **26** (1993) 1891
[16] H.Hadjichristidis, M.Devaleriola, V.Desreux, Eur.Polym.J. **8** (1972) 1193, data of Fig. 1
[17] F.Abe, Y.Einaga, H.Yamakawa, Macromolecules **26** (1993) 1891, Tables I and III
[18] F.Abe, Y.Einaga, T.Yoshizaki, H.Yamakawa, Macromolecules **26** (1993) 1884, Tables I and II
[19] H.Murakami, T.Norisuye, H.Fujita, Macromolecules **13** (1980) 345, Tables I and II
[20] T.Kashiwagi, T.Norisuye, H.Fujita, Macromolecules **14** (1981) 1220, Table I
[21] J.W.Mays, N.Hadjichristidis, L.J.Fetters, Macromolecules **18** (1985) 2231, Table VI (Fig. 2)
[22] M. Bohdanecký, Macromolecules **10** (1977) 971, data of Figs. 3, 4, and 8
[23] R.Kuhn, H.Krömer, G.Rossmanith, Angew.Makromol.Chem. **40/41** (1974) 361, Fig. 3
[24] T.H.Moury, S.R.Turner, M.Rubinstein, J.M.J.Fréchet, C.J.Hawker, K.L.Wooley, Macromolecules **25** (1992) 2401, Table I
[25] R.S.Turner, B.I.Voit, T.H.Mourey, Macromolecules **26** (1993) 4617, Table II
[26] M.Wintermantel, M.Schmidt, Y.Tsukahara, K.Kajiwara, S.Kohjiya, Macromol. Rapid Commun. **15** (1994) 279, Fig. 3
[27] S.M.Aharoni, N.S.Murthy, Polym.Commun. **24** (1983) 132, recalculated from Table 1

[28] J.Roovers, L.-L.Zhou, P.M.Toporowski, M. van der Zwan, H.Iatrou, N.Hadjichristidis, Macromolecules **26** (1993) 4324, Tables I and III
[29] E.Foussac, M.Milas, M.Rinaudo, R.Borsali, Macromolecules **25** (1993) 5613, Table V
[30] M.Yamaguchi, Y.Yamaguchi, Y.Matsushita, I.Noda, Polym.J. **22** (1990) 1077, Table 2

13 Thermal Properties

13.1 Fundamentals

13.1.1 Introduction

With increasing temperature, crystalline low-molecular weight substances convert to liquids at the melting temperature and from liquids to gases at the boiling temperature. These changes of physical states can usually be seen directly since they are accompanied by strong changes of interactions of molecules which in turn lead to widely different viscosities. For example, the melting of such crystals is observable by the naked eye because crystals have very high viscosities (solids) and melts very low ones (liquids).

Polymers very often behave differently since temperature changes not only lead to abrupt viscosity changes such as the transition crystal ⇄ liquid but also to more subtle ones that are caused by changes in mobilites of segments and groups. Polymers may thus exhibit many more thermal "transitions" then low-molecular weight compounds. In addition, viscosities of polymers are relatively high so that thermal equilibria are slow to establish. Furthermore, some polymer "transitions" are not caused by thermodynamics but by kinetics. Effects can be detected not only optically and rheologically but also thermally (Section 13.2) and frequently also electrically (Volume IV).

It is important to distinguish between thermal transitions and thermal relaxations. In **thermal transitions**, substances are in equilibrium both below and above the transition temperature, such as the true melting temperature, T_M. Equilibrium states are not affected by the rate of measurement. A thermal transition of a substance composed of uniform molecules has only one transition temperature, for example, a single melting temperature T_M for the melting of crystals (Section 13.3) and a single clearing temperature for the transition nematic liquid crystal ⇄ melt (Section 13.4). The reversal of the melting process is crystallization (Section 7.3).

Thermal relaxations are kinetic phenomena that vary with the frequency and thus with the time scale. Relaxations are caused by the onset of movements of chemical groups, molecular segments, and whole molecules. In contrast to thermal transitions, the system is *not* in thermodynamic equilibrium below and above the relaxation temperature. Whether or not a relaxation can be observed depends on the measuring method.

Some experimental methods work with such frequencies (time scales) that relaxations appear to be thermal transitions. The best known example is the measurement of the glass ("transition") temperature by differential scanning calorimetry (Section 13.5). At this temperature, glass-like amorphous substances convert to softer and often rubber-like materials. There are also other thermal effects that cannot be easily related to either transitions or relaxations.

13.1.2 Thermodynamic States

A thermodynamic state is described by its Gibbs energy G and its partial first derivatives with respect to temperature T and pressure p, respectively, i.e., by its enthalpy H, entropy S, and volume V:

(13-1) $H = G + TS = G - T(\partial G/\partial T)_p$
(13-2) $S = -(\partial G/\partial T)_p$
(13-3) $V = (\partial G/\partial p)_T$

Partial second derivatives lead correspondingly to the isobaric heat capacity C_p (formerly: specific heat at constant pressure), the cubic expansion coefficient β, and the isothermic (cubic) compressibility κ:

(13-4) $C_p = (\partial H/\partial T)_p = T(\partial S/\partial T)_p = -T(\partial^2 G/\partial T^2)_p$
(13-5) $\beta = V^{-1}(\partial V/\partial T)_p$
(13-6) $\kappa = -V^{-1}(\partial V/\partial p)_T$

IUPAC recommends the symbols α, α_V, or γ for the cubic expansion coefficient and the symbol α_l for the linear one. The polymer literature uses mainly α for the cubic and β for the linear expansion coefficient. Since the literature employs α exclusively for the linear expansion coefficient of coil molecules, α will be used in this book for the linear expansion coefficient of matter and β for the cubic expansion coefficient.

Theoretically important isochoric heat capacities C_V (i.e., at constant volume) can be calculated from experimental isobaric heat capacities with $C_V = C_p - (TV\beta^2/\kappa)$ (G: *baros* = weight; *chora* = room, space). Division of C_V by the amount-of-substance n leads to the molar isochoric heat capacity, $C_{V,m} = C_V/n$, and division of C_V by the mass m the specific isobaric heat capacity $c_V = C_V/m$. The molar isobaric heat capacity, $C_{p,m} = C_p/n$, and the specific isobaric heat capacity, $c_p = C_p/m$, are obtained similarly.

13.1.3 Order of Thermal Transitions

Thermal transitions are characterized by changes in the variation of properties of state (V, H, S, α, C_V) with temperature. Thermodynamic thermal transitions are characterized by thermodynamic equilibria on *both* sides of the transition temperature. Such transitions can be of first, second, ..., Nth order.

An Nth order transition is characterized by a discontinuity of the Nth derivative of the Gibbs energy. **First-order transitions** show the corresponding discontinuities of the *first derivatives* of the Gibbs energy with respect to temperature or pressure, i.e., of the volume V, the entropy S, and the enthalpy H at the first-order transition temperature according to Eqs.(13-1)-(13-3) (Fig. 13-1). Such discontinuity is observed in the melting of infinitely large, perfect crystals. The first derivatives of H and V with respect to temperature (i.e., C_p, C_V, and α) or pressure (i.e., κ) lead correspondingly to infinitely large signals. For imperfect and/or small crystals, discontinuities of H, S, and V degenerate to S-shaped curves and the "infinite" signals of C_V, α, and κ to bell-shaped curves.

Thermodynamic **second-order transitions** are characterized by discontinuities in the *second derivatives* of the Gibbs energy with respect to temperature or pressure, i.e., in α, C_V, or C_p (Fig. 13-1). Examples of true second-order thermodynamic transitions are the so-called lambda transition of liquid helium at 2.2 K, the rotational transitions of crystalline ammonium salts, and the disappearance of ferromagnetism at the Curie point, all of which are one-phase transitions. Other second-order transitions are that of smectic LCs to other smectic LCs and smectic LCs to nematic LCs (Section 13.4).

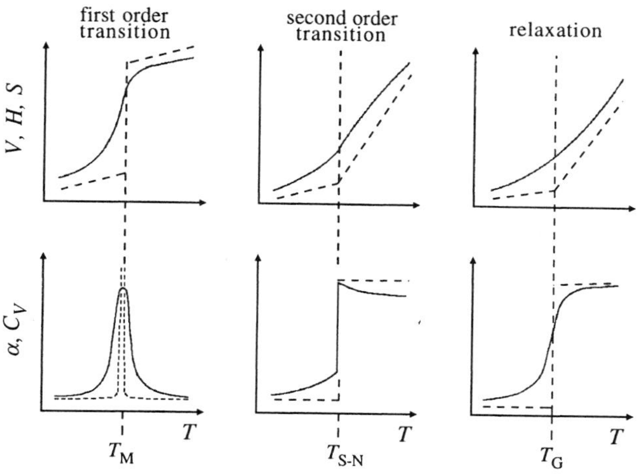

Fig. 13-1 Schematic representation of (top) temperature dependence of the first derivatives of the Gibbs energy with respect to temperature (*H*, *S*) or pressure (*p*) and (bottom) temperature dependence of the partial second derivatives of the Gibbs energy with respect to temperature (β (and also α), C_V) at thermodynamic transitions (left and center) and at relaxations. Examples:

First-order thermodynamic transition: melting process with melting temperature T_M of infinitely large, perfect crystals (- - - -) or of imperfect crystals, including those of polymers with broad molar mass distributions (——).

Second-order transition: conversion $T_{S\text{-}N}$ of smectic LC to a nematic LC with dominating (- - - -) or weak intermolecular cooperative effects (——).

Relaxation: glass transformation with glass temperature T_G from infinitely slow (——) or conventional (- - - -) measurements.

One-phase transitions are characterized by the **Ehrenfest equations**, Eqs.(13-7) and (13-8), that are obtained from Eqs.(13-4)-(13-6) with $\Delta H_{tr} = 0$:

$$(13\text{-}7) \qquad (\mathrm{d}p/\mathrm{d}T)_{tr} = \Delta\beta_{tr}/\Delta\kappa_{tr}$$

$$(13\text{-}8) \qquad (\mathrm{d}p/\mathrm{d}T)_{tr} = \Delta C_{p,tr}/(T_{tr}V_{tr}\Delta\beta_{tr})$$

The right sides of Eqs.(13-7) and (13-8) equal each other for true thermodynamic transitions, i.e., the so-called **Prigogine-Defay ratio** *r* is unity:

$$(13\text{-}9) \qquad r = \Delta\kappa_{tr}\Delta C_{p,tr}/(T_{tr}V_{tr}\Delta\beta_{tr}^2) = 1$$

True second-order thermodynamic transitions thus have only one order parameter, and *r* is therefore unity. The formal thermodynamic classification of thermal transitions corresponds to the phase behavior. All thermodynamic first-order transitions involve two phases, all true thermodynamic second-order transitions, only one.

However, this classification is not clear cut for molecular processes such as the transformation of isomorphs, i.e., polymorphs that are generated by conformational isomers (see Sections 7.1.7 and 7.1.8). If such isomers are neither isoenergetic nor kinetically restricted, then the high-energetic conformer will be converted to the low-energetic conformer at a transition temperature T_{tr} that is lower than the melting temperature T_M; an example is the transformation $(TG)_{3i} \rightleftarrows (TTG)_{2i}$ that proceeds by rotation around

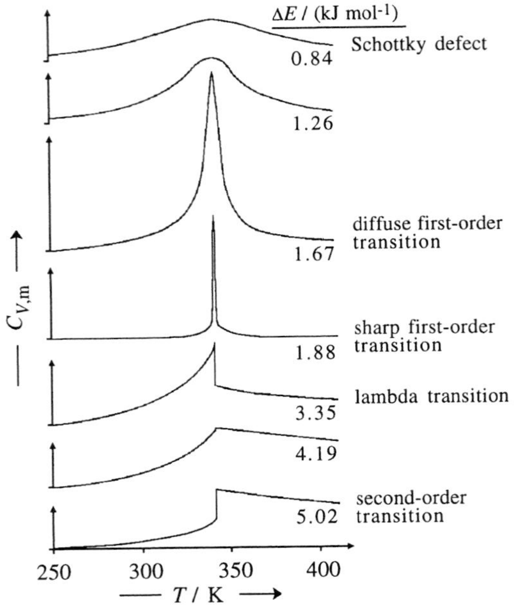

Fig. 13-2 Temperature dependence of isochorous molar heat capacities $C_{V,m}$ calculated for various agglomeration energies ΔE [1]. Ordinates are not to scale.

chain bonds. In loosely packed chains, such rotations are caused by local *intra*molecular-cooperative movements. For tightly packed chains, it also involves *inter*molecular movements of neighboring chains.

The movements proceed more easily, the more crystal defects are present. A chain may have one defect (D) or no defects (O) and pairs of chains therefore two defects (DD), one defect (DO or OD), or no defects (OO). The energy difference per mole of chains is therefore $\Delta E = (E_{DO} - E_{OO}) - (E_{DD} - E_{DO})$. Intramolecular rotational transformations thus exist if ΔE is small and intermolecular ones exist if ΔE is large. The energy difference ΔE thus measures the tendency of defects to agglomerate which favors intermolecular-cooperative movements.

According to model calculations of the temperature dependence of the isochorous molar heat capacity, $C_{V,m}$, various transformations are noted (Fig. 13-2). With increasing agglomeration energy, one first observes a **Schottky defect**, a diffuse and then a sharp first-order transition, a so-called lambda transition, and finally a true second-order transition. According to these calculations, first-order rotational transitions are intramolecular whereas second-order rotational transitions are intermolecular.

13.1.4 Experimental Methods

Thermal transitions and transformations are classically determined from the temperature dependence of thermodynamic quantities such as volume, expansion coefficient, enthalpy, or heat capacity, one data point at a time. Because such experiments are time-consuming and costly, they have mainly been replaced by **thermal analysis** which mea-

Table 13-1 Important thermoanalytical methods. T = temperature, H = enthalpy (indices: S = specimen, R = reference).

Common name	Symbol	Quantity = f(time)
Thermogravimetric analysis	TGA	mass of specimen
Evolved gas analysis	EGA	analysis of liberated gases
Differential thermal analysis	DTA	$T_S - T_R$
Differential scanning calorimetry	DSC	$H_S - H_R$
a. power-compensated DSC		
b. heat-flux DSC		
Thermodilatometry	-	expansion of specimen
Thermomechanical analysis	TMA	deformation of specimen under static load
Dynamic-mechanical analysis	DMA	
a. Modulus	DMA	forced vibration of specimen
b. Damping	DMA	free vibration of specimen
c. Torsional braid analysis	TBA	specimen on vibrating carrier

sures physical properties as a function of temperature by using controlled temperature programs (Table 13-1). These programs also comprise **thermogravimetric analysis (TGA)** and **evolved gas analysis (EGA)** which measure the presence or formation of low-molar mass compounds that were either components of the specimen or resulted from thermal degradation; both TGA and EGA say little or nothing about thermal transitions and transformations.

Heat capacities are usually obtained by **differential scanning calorimetry (DSC)** in which the specimen and a standard are separately heated in such a way that both are always at the same temperature. Caloric effects on heating the specimen are, for example, compensated by an increase or decrease of the electric power (**power-compensated differential scanning calorimetry**); this method thus determines dQ/dt, the change of added heat ΔQ with time. Positive signals indicate exothermic processes (crystallization, exothermic chemical reactions, etc.), negative signals endothermic ones (melting, endothermic chemical reactions, etc.). The added heat, ΔQ, equals the heat difference between the specimen and the standard and, if the heat capacity of the standard is known, also the heat, enthalpy, and heat capacity of the specimen.

In **differential thermal analysis (DTA)** temperatures of both specimen and standard are increased at *constant rate*. At a first-order thermodynamic transition, for example, a melting, heat is taken up until the whole specimen is melted. The temperature of the specimen remains constant whereas that of the standard changes continuously so that the temperature difference between the specimen and the standard is proportional to the heat flow Q_p.

Besides melting temperatures T_M, various other characteristic temperatures are observed by DSC, DTA, etc. (Fig. 13-3). Examples are solid-solid transformations T_{SS} (in Fig. 13-3 shown as first-order transition), glass ("transition") temperatures T_G, liquid-liquid transformations T_{LL} (controversial), onset of crystallization at T_{cryst}, onset of chemical reactions at T_{react} (such as oxidations or crosslinking reactions), and decomposition of the specimen at T_{decomp}.

Base lines of DSC and DTA diagrams are usually not equally high on both sides of the signals since specimens have different heat capacities below and above transforma-

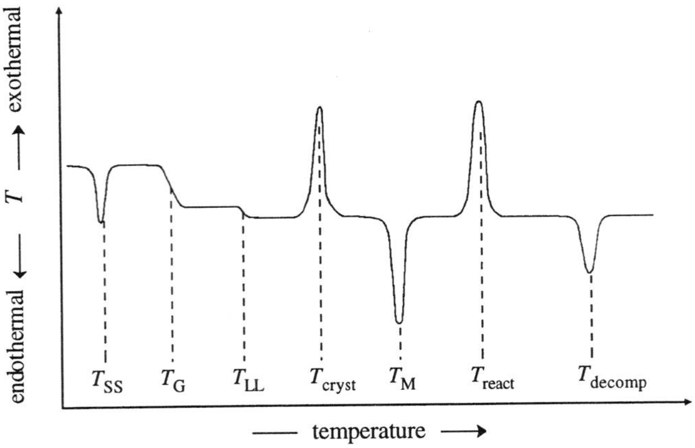

Fig. 13-3 Thermogram of a semicrystalline polymer (schematic) with temperatures of physical transition (M = melting), physical transformations (SS = solid/solid, G = glass-liquid, LL = liquid-liquid), and chemical conversions such as polymer-analog reactions and decompositions.

tion temperatures. The size and shape of signals also depend on the rate of heating (Fig. 13-4). For example, the crystallization of an amorphous specimen may be observed at low heating rates but not at high ones. At high heating rates, steps indicating glass temperatures may be replaced by peaks. Larger sizes of specimens lead to larger temperature gradients and thus to slower temperature equilibrations, etc.

Some scientific and industrial methods determine thermal transitions and transformations of polymers via mechanical properties of specimens. Predominant scientific methods are **thermomechanical analysis (TMA)** and **dynamic mechanical analysis (DMA)**. TMA measures the deformation of the specimen under load whereas DMA studies the free or forced vibration of a specimen as a function of temperature (Chapters 16-18).

Industry mostly uses simple and fast empirical methods that are standardized with respect to the dimensions and shapes of specimens, types of instruments, and experimental protocol (Fig. 13-5). In general, such methods measure the temperature dependence of a combination of several quantities. As a result, "transition temperatures" obtained by these methods differ and cannot be compared with each other or with physically defined transition/transformation temperatures.

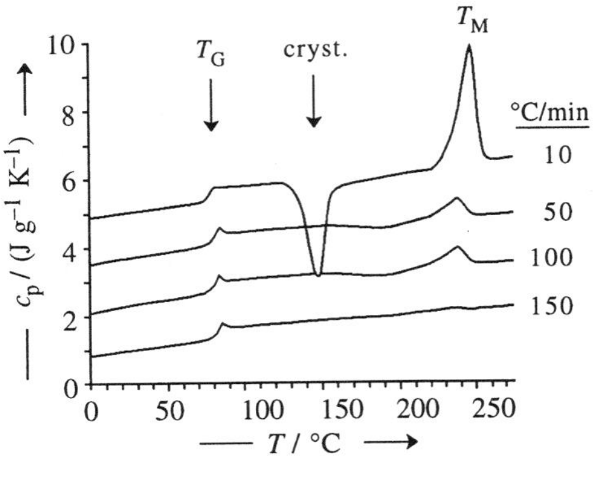

Fig. 13-4
Specific heat capacity of an amorphous poly(ethylene terephthalate) at four heating rates. Data by DSC. Curves have been shifted vertically for clarity [2].

A sharp exothermic crystallization peak is observed at a heating rate of 10°C/min. Crystallization is barely detectable at heating rates of 50 and 100°C/min and absent at a heating rate of 150°C/min.

The glass temperature is shifted slightly to higher temperatures at higher heating rates.

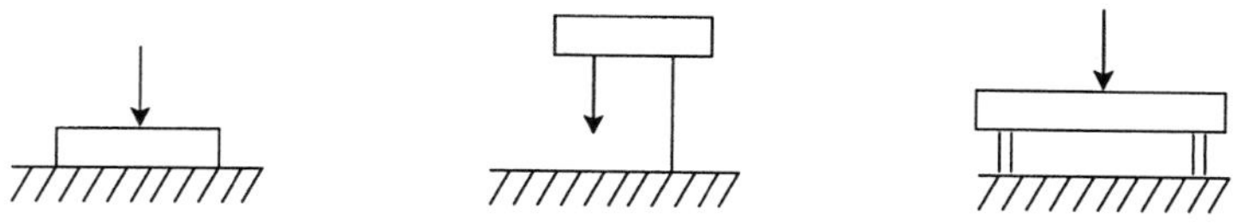

	Vicat A	Vicat B			Martens	HDT-A	HDT-B	HDT-C
F/N =	9.81	49.05	;	σ/MPa =	4.9	1.80	0.45	8
v/(K h^{-1}) =	50	120	;	v/(K h^{-1}) =	50	120	120	120
ΔL/mm =	1	1	;	ΔL/mm =	6	0.21	0.33	

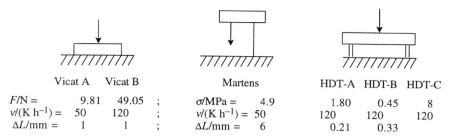

Fig. 13-5 Industrial thermoanalytical methods. F = force, v = velocity, ΔL = deformation.

In **Vicat methods** A and B, a defined force F causes a needle to penetrate the specimen which is heated at a constant rate (velocity) $v = dT/dt$ (Fig. 13-5). The Vicat temperature is the temperature at which a penetration depth of ΔL = 1 mm is attained.

The **Martens method** applies a constant flexural stress of σ = 4.9 MPa to one side of a specimen that is heated at a rate of 50 K/h. It determines the temperature at which the specimen is deformed by ΔL = 6 mm.

The three **heat distortion** (or **heat deflection**) methods also apply various flexural stresses albeit to the center of the specimen and not to the end as in the Martens method. The **heat distortion temperature (HDT)** is then obtained at a certain deformation ΔL.

Transition temperatures can also be determined by **inverse gas chromatography** in which the polymer is precipitated on a carrier where it serves as a stationary phase. One then measures the retention time t_r or retention volume V_r of a volatile solvent as a function of temperature. In the resulting lg V_r = f(1/T) diagram, transition/transformation temperatures are indicated by maxima, minima, or changes of slopes.

13.2 Mobility of Molecules

13.2.1 Thermal Expansion

On heating, isotropic bodies expand regularly in the three spatial directions because of increasing thermal movements of atoms, molecular segments, or molecules. This **thermodilatometry** delivers the **cubic expansion coefficient**, $\beta = V^{-1}(\partial V/\partial T)_p$, which is then converted to the **linear expansion coefficient**, $\alpha = L^{-1}(\partial L/\partial T)_p$, by setting $\beta = 3\alpha$. This conversion is not permitted for anisotropic bodies because they expand differently in the three spatial directions. An example is crystalline poly(ethylene) where the linear expansion coefficient is negative along the chain axis because the increasing amplitude of lateral movements of chain sections causes the chain to contract in the chain direction.

Linear expansion coefficients of polymers depend on the change of forces between atoms with temperature. Such forces are large for covalent bonds but small for van der Waals interactions. A quartz crystal will therefore expand only a little on heating but liquid CS_2 will expand very much. Polymer chains have covalent bonds in the chain direction but exert only relatively weak dispersive, polar, and/or hydrogen-bond forces in the two other spatial directions. The linear expansion of polymers will therefore be between that of solid metals and low-molecular weight liquids (Table 13-2).

Table 13-2 Effect of the bond type in the three spatial directions on linear expansion coefficients α, isobaric specific heat capacities c_p, and heat conductivities λ of various matter. c = covalent bond, m = metallic bond, h = hydrogen bridge, p = dipole/dipole interaction, d = dispersion force. Natural quartz contains up to ca. 0.01 % water.

Material	Unit	Bonds	$\frac{10^6 \alpha}{K^{-1}}$	$\frac{c_p}{J K^{-1} g^{-1}}$	$\frac{\lambda}{W m^{-1} K^{-1}}$
Quartz	SiO_2	c c c	1	0.72	10.5
Diamond (^{13}C)	(C)	c c c	1.34		2300
Iron	(Fe)	m m m	12	0.54	58
Copper	(Cu)	m m m	17	0.38	350
Aluminum	(Al)	m m m	23	0.88	234
Water	H_2O	h h h	70	4.2	
Polyamide 6	$-NH-(CH_2)_5-CO-$	c h d	60	1.6	0.31
Poly(styrene), at	$-CH_2-CH(C_6H_5)-$	c p p	70	1.3	0.16
Poly(vinyl chloride), at	$-CH_2-CHCl-$	c p p	80	0.85	0.18
Poly(tetrafluoroethylene)	$-CF_2-CF_2-$	c p p	99	0.42	0.37
Poly(ethylene), high density	$-CH_2-CH_2-$	c d d	120	2.7	0.44
Poly(isoprene), *cis*-1.4-	$-CH_2C(CH_3{=}CHCH_2)-$	c p d	223	1.9	0.13
Poly(ethylene), amorphous	$-CH_2-CH_2-$	c d d	287	2.1	0.35
Carbon disulfide	CS_2	d d d	380	0.60	

The very different expansion coefficients of polymers on one hand and metals and glass on the other can lead to severe problems if these types of materials are combined and the resulting articles are subjected to thermal demands. For **dimensional stability**, polymers must also not recrystallize because density differences between amorphous and crystalline regions may lead to warping on processing.

13.2.2 Heat Capacity

For polymers, only heat capacities at constant pressure, C_p, (formerly: specific heats) are experimentally accessible whereas heat capacities at constant volume, C_V, are required for theoretical considerations. Both quantities are interconnected via the cubic expansion coefficient β and the isothermal compressibility, κ:

$$C_p = C_V + (TV\beta^2/\kappa) \tag{13-10}$$

For crystalline polymers, molar heat capacities at constant volume, $C_{V,m}$, can be calculated from the frequency spectrum. In the crystalline state, atoms oscillate harmonically around their equilibrium positions. Each oscillation contributes a value of

$$E(\Theta/T) = \Theta^2 [\exp(\Theta/T)] / [1 - \exp(\Theta/T)] \quad ; \quad \textbf{Einstein function} \tag{13-11}$$

to the heat capacity where $\Theta = h\nu/k_B$ = **Einstein temperature**. The molar heat capacity at constant volume is then simply the sum of all these contributions:

$$C_{V,m} = R \Sigma_i E(\Theta/T)_i \tag{13-12}$$

At very low temperatures, heat capacities stem almost completely from lattice vibrations. At somewhat higher temperatures, anharmonicities of vibrations have to be considered. At even higher temperatures, vibrations of whole groups and rotations about chain bonds become important. Another contribution may come from lattice defects.

In static equilibrium, the molar energy is (1/2) RT per degree of freedom according to the **equipartition law**. In crystals, atoms vibrate around their equilibrium positions. Their three degrees of freedom of translation lead to an average molar kinetic energy of $E_{kin} = (3/2)\ RT$. Each elastic vibration is accompanied by potential energy which, on average, equals that of kinetic energy. The total molar energy is thus 3 RT and the molar heat capacity should therefore be $C_{V,m} = 3\ R$ per mole of atoms (**Dulong-Petit law**).

Because some degrees of freedom are frozen, molar heat capacities of polymers are often near 1 R. For example, the polymer of Fig. 13-6 has a specific heat capacity of $c_p = 1.22$ J K^{-1} g^{-1} at 25°C. The heat capacity per mole of monomeric units with M = 120.06 g/mol, $-C_8H_8O-$, is therefore $C_{p,m}$ = 146.47 J K^{-1} mol^{-1}. Since there are 17 atoms per monomeric unit, the heat capacity per mole of atoms is $C_{p,m}$ = 8.62 J K^{-1} mol^{-1} which is practically identical with R = 8.314 J K^{-1} mol^{-1}.

Specific heat capacities of polymers, c_p, range between 0.85 J K^{-1} g^{-1} (poly(vinyl chloride)) and 2.7 J K^{-1} g^{-1} (high-density poly(ethylene) (Table 13-2)) whereas those of mineral fillers are often ca. 0.9 J K^{-1} g^{-1}. Mineral-filled plastics thus need less energy for thermal processing (injection molding, extrusion, etc.) than unfilled ones.

Heat capacities at temperatures sufficiently below glass temperatures are the same for amorphous and semicrystalline polymers (Fig. 13-6). They start to differ as one approaches the glass temperature where new vibrations set in. Since these vibrations of segments are coupled, heat capacities of amorphous polymers "jump" at the glass temperature to higher values and then continue to increase.

Semicrystalline polymers also experience the onset of additional vibrations. Since segments are now less restricted in their movements, they can assume more ordered positions and the chain **recrystallizes** before it melts. The maximum in c_p indicates the melting temperature T_M of the *specimen*. The melting temperature of the *polymer*, $T_{M,o}$, is at the upper end of the melting range where the largest, most perfect crystallites melt.

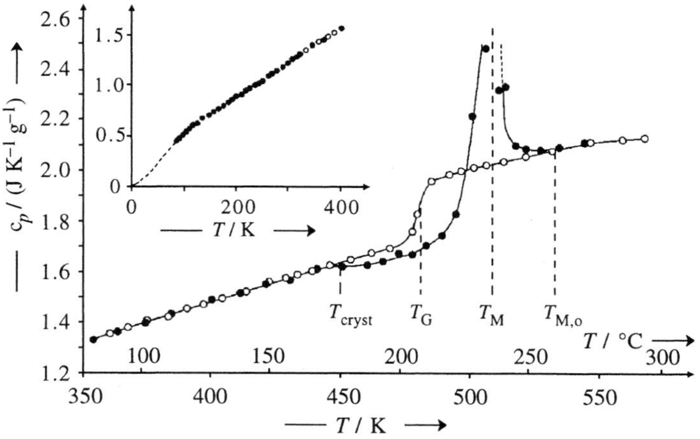

Fig. 13-6 Temperature dependence of the specific heat capacity at constant pressure, c_p, of a semicrystalline (●) and an amorphous (O) poly(oxy-(2,6-dimethyl)-1,4-phenylene) [3]. T_{cryst} = beginning of recrystallization, T_G = glass temperature, T_M = conventional melting temperature, $T_{M,o}$ = ideal melting temperature of the largest crystallites.

13.2.3 Thermal Conductivity

Common polymers do not conduct electricity (Volume IV, Chapter 12). Heat is therefore not transported by electrons as in metals but by phonons (= elastic waves). The distance over which the intensity of elastic waves is reduced to 1/e-th is called the free wavelength of phonons. At not too low temperatures, this distance is practically independent of temperature and is ca. 0.7 nm for glasses, amorphous polymers, and liquids.

The weak decrease of thermal conductivity λ of natural rubber (Fig. 13-7) and other amorphous polymers with decreasing temperature below the glass temperature is therefore practically only caused by the decrease of heat capacity. At very low temperatures of 5-15 K, a plateau is reached (not shown) which is followed by another decrease until the temperature dependence of thermal conductivities becomes $\lambda \sim T^2$ at $T < 0.5$ K.

At temperatures above ca. −123°C, heat is mainly transported by collisions between polymer segments and molecules. Since densities decrease with increasing temperature, thermal conductivities λ should decrease with increasing temperature at $T > T_G$. But since the packing of segments does not vary much below and above the glass temperature, thermal conductivities do not vary much either and have only a weak maximum at T_G.

The situation is very different for semicrystalline polymers where the packing density decreases dramatically as the increase in temperature approaches the melting temperature (Fig. 13-7). The drop of λ is greater, the higher the crystallinity of the polymer. The change of λ begins at far lower temperatures than that of other properties, for example, specific volumes (cf. Fig. 13-9).

Data (not shown) indicate that the heat conductivity of semicrystalline polymers passes through a maximum with decreasing temperature and then decreases to very low values near $T = 0$ K. For high-density poly(ethylene), the maximum is $\lambda \approx 1$ W m^{-1} K^{-1} at a temperature of −170°C.

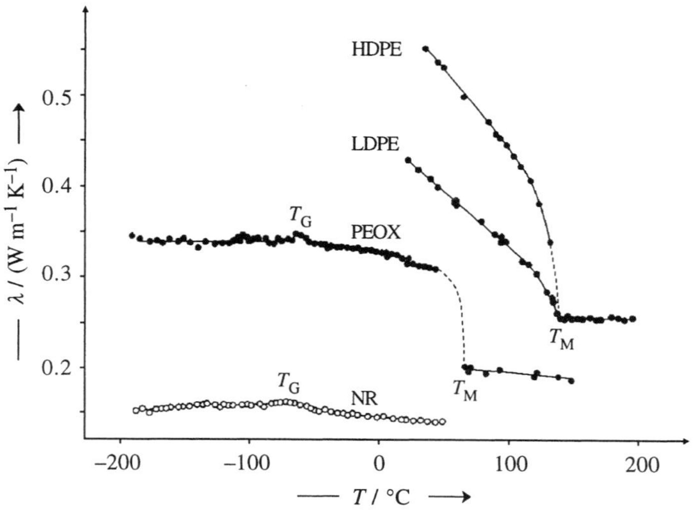

Fig. 13-7 Temperature dependence of thermal conductivities of natural rubber (NR), poly(oxyethylene) (PEOX), high-density poly(ethylene) (HDPE), and low-density poly(ethylene) (LDPE). T_G = glass temperature, T_M = melting temperature. From a compliation of [4].

Table 13-3 Some methods for the study of molecular movements. t_{corr} = correlation time. TSD = thermally stimulated discharge; TSC = thermally stimulated current.

Method		t_{corr} / s
Quasi-elastic neutron scattering		10^{-12} - 10^{-8}
NMR spin-lattice relaxation		10^{-12} - 10^{-5}
Brillouin scattering	(hypersound propagation)	10^{-10} - 10^{-9}
ESR line shape		10^{-10} - 10^{-7}
Dielectric relaxation	(including TSD and TSC)	10^{-10} - 10^{-5}
Kerr effect relaxation	(electrical birefringence)	10^{-8} - 10^{4}
ESR saturation transfer		10^{-7} - 10^{-5}
NMR line shape		10^{-6} - 10^{-3}
Quasi-elastic light scattering	(photon correlation spectroscopy)	10^{-4} - 10^{2}
Relaxation of dipole and quadrupole orders	(NMR)	10^{-3} - 1

13.2.4 Thermal Relaxations

The onset and ending of movements of polymer molecules and segments can be studied by many physical methods, some of which are listed in Table 13-3. These methods work with frequencies $\nu = 1/t_{corr}$ that correspond to correlation times t_{corr} between 10^{-12} s and ca. 10^{4} s. Slow methods are called static or quasi-equilibrium, fast ones, dynamic.

These methods detect characteristic signals at various temperatures, for example, peaks in the temperature dependence of the loss tangent (dissipation factor), tan δ (Fig. 13-8, insert). Since these signals often cannot be immediately correlated with molecular processes, they are labeled with *decreasing* temperature by small Greek letters in the sequence of the Greek alphabet (α, β, γ, δ, ε, ζ, ...), starting with the melting temperature (crystalline polymers, index c) or the glass temperature (amorphous polymers, index a).

For semicrystalline polymers, α_c indicates a signal at or near the melting temperature, β_c the first relaxation below the melting temperature, γ_c the second relaxation below the melting, and δ_c a relaxation at an even lower temperature.

Relaxation temperatures of amorphous polymers are similarly labelled: α_a = relaxation temperature at or near the glass temperature and β_a, γ_a, and δ_a the next relaxations at decreasing temperatures.

Relaxations at higher temperatures than T_M or T_G should be characterized by capital Greek letters in reverse order, starting with Ω_c and Ω_a, respectively, and proceeding to Ψ, X, and Φ.

The temperature at the maximum of the signals is the **relaxation temperature**. This temperature varies with the measuring frequency ν which can be described by the general equation for relaxation processes,

$$\nu = [k_B T/(2\,\pi h)] \exp(-\Delta H_m^{\ddagger}/RT) \exp(\Delta S_m^{\ddagger}/R) \qquad (13\text{-}13)$$

where k_B = Boltzmann constant, h = Planck constant, R = general gas constant, $\Delta H_m^{\ddagger}$ = molar activation enthalpy, and $\Delta S_m^{\ddagger}$ = molar activation entropy. Eq.(13-13) can be rewritten as

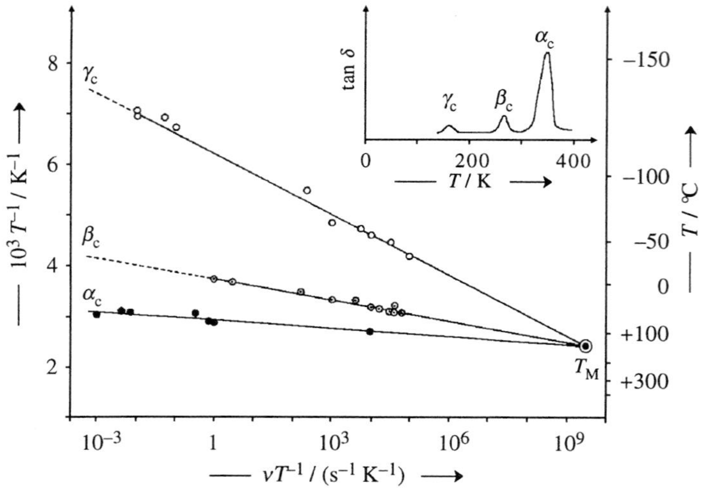

Fig. 13-8 Dependence of inverse relaxation temperatures, $1/T$, of a low-density poly(ethylene) on the logarithm of reduced frequency, ν/T [5]. The lines for the α, β, and γ relaxation processes intersect at $3.32 \cdot 10^9$ s^{-1} K^{-1} and T_M = 131°C. This common point is apparently found only for non-helical polymers. Insert: Mechanical relaxation spectrum of poly(ethylene) at a frequency of ν = 1000 Hz.

$$(13\text{-}14) \qquad \frac{1}{T} = \frac{R}{\Delta H_m^{\ddagger}}\left[\frac{\Delta S_m^{\ddagger}}{R} + \ln\frac{k_B}{2\pi h} - \ln\frac{\nu}{T}\right]$$

A plot of $1/T = f(\lg(\nu/T))$ thus delivers straight lines for relaxation processes (Fig. 13-8).

The various relaxation phenomena are often difficult to trace to molecular processes. Mechanical relaxations usually indicate translations and rotations of polymer molecules and segments whereas dielectric relaxations are caused by rotation of dipoles and the limited transport of charges. Ultrasound is sensitive to relaxations from changes of microconformations and to processes with changes of volumes or enthalpies.

In NMR experiments, a sudden application of an electric field is in time followed by a magnetic polarization. The magnetization is usually an exponential function of time with a constant that is called the **spin-lattice relaxation time**, T_1. These NMR experiments thus correspond macroscopically to dielectric relaxation experiments. However, there are molecular differences which result in values of T_1 that are usually much larger than molecular relaxation times from dielectric relaxations.

13.3 Melting

13.3.1 Fundamentals

Melting is the thermal conversion of crystalline regions to a liquid. At the melting temperature T_M, crystallites and melt are in thermal equilibrium so that the Gibbs activation energy of melting becomes zero, $\Delta G^{\ddagger} = \Delta H^{\ddagger} - T\Delta S^{\ddagger} = 0$, from which it follows $T_M = \Delta H^{\ddagger}/\Delta S^{\ddagger}$ and $\nu/T_M = k_B/(2\pi h) = 3.32 \cdot 10^9$ s^{-1} K^{-1} (cf. Eq.(13-13) and Fig. 13-8). The melting temperature of poly(ethylene) (T_M = 143°C) thus corresponds to a frequency of $\nu = 1.4 \cdot 10^{12}$ Hz which is roughly the natural vibration frequency of chain segments.

This correspondence does not mean that the crystal "explodes" on melting. There are three observations which rather point to a surface melting as the main process. (I) At the melting temperature, the relaxation temperatures of the γ_c and β_c processes correspond to the same frequency as the α_c process (Fig. 13-8); only the latter can comprise the movement of large polymer segments, however. (II) Some polymer single crystals can be heated far above their melting temperatures without "exploding." (III) Melting temperatures depend on the thickness of the lamellae at low degrees of polymerization but become independent of the degree of polymerization at high molar mass (Fig. 7-10).

These observations indicate that melting starts at the surfaces of crystallites, especially at corners and edges, which are less ordered and therefore need less energy for melting than the faces of the crystallite. In contrast to crystallization, nucleating agents are not required for melting processes.

Melting requires that a chain segment becomes disordered. The number N of chain atoms in this section can be estimated from the ratio $\Delta H_m{}^{\ddagger}/\Delta H_{M,m}$ of the activation enthalpy $\Delta H_m{}^{\ddagger}$ of the relaxation process per mole of segments to the melting enthalpy $\Delta H_{M,m}$ per mole of chain atoms. Table 13-4 indicates that the melting process involves ca. 60-160 chain atoms (α_c process), the glass transformation ca. 20-50 chain atoms (β_c process), and the γ_c process ca. 10-20 chain atoms. The numbers N of chain atoms participating in melting processes are in the same range as the numbers of chain atoms in stems of chain-folded crystals.

Table 13-4 Melting enthalpies $\Delta H_{M,m}$ per mole of chain atoms and molar activation enthalpies $\Delta H_m{}^{\ddagger}$ per mole of segments for three relaxation processes in semicrystalline polymers [5]. N = calculated numbers of participating chain atoms. α_c = melting process, β_c = glass transformation.

Polymer	$\dfrac{\Delta H_{M,m}}{\text{kJ mol}^{-1}}$	$\dfrac{\Delta H_m{}^{\ddagger}}{\text{kJ mol}^{-1}}$			N		
		α_c	β_c	γ_c	α_c	β_c	γ_c
Poly(ethylene), low density	4.11	661	142	50	161	35	12
Poly(ethylene), linear	4.11	-	188	44	-	46	11
Poly(chlorotrifluoroethylene)	2.51	222	84	-	88	53	-
Poly(ethylene terephthalate)	3.32	272	75	-	82	23	-
Polyamide 6	3.71	222	84	-	60	23	-
Poly(tetrafluoroethylene)	3.42	218	130	63	64	38	18

13.3.2 Effect of Morphology

Definition of Melting Temperature

Crystals of low-molar mass chemical compounds have very narrow melting ranges. For example, the molecularly uniform alkane $C_{44}H_{90}$ melts at T_M = 86.4°C within a temperature interval of only 0.25°C (Fig. 13-9). The longer chains of $C_{94}H_{190}$ crystallize less perfectly, which produces lattice defects. Because of these defects, some chain

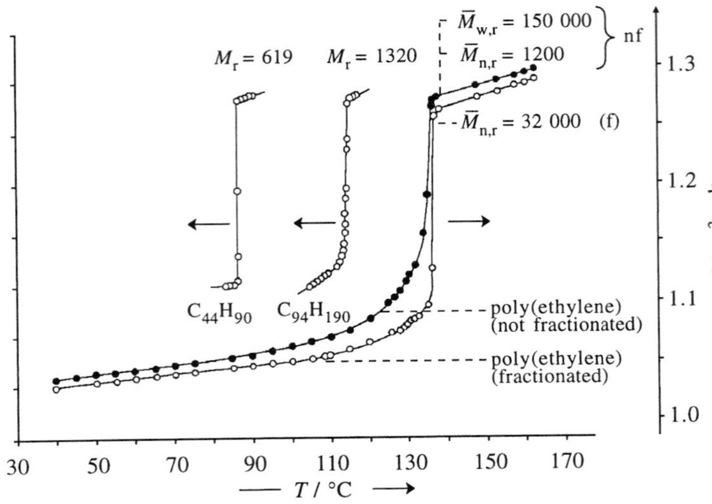

Fig. 13-9 Temperature dependence of specific volumes of two alkanes, $C_{44}H_{90}$ and $C_{94}H_{190}$, (no vertical scales) [6], an unfractionated poly(ethylene) with a broad molar mass distribution, and a fraction of this poly(ethylene) with a number-average molecular weight of 32 000 [7]. Before the measurement, poly(ethylene)s were crystallized just below the melting temperature for 40 days at 131.5°C.

segments and chain ends are more mobile than others. During melting, they shift constantly between crystalline and non-crystalline regions, which causes the melting to start at ca. 110°C and to end at ca. 114.6°C (ΔT = 4.6°C).

An imperfect crystal structure produces a **melting range**. The largest and most perfect crystallites melt at the upper end of this range which, for low-molecular weight chemical compounds, is characterized by a sharp transition to the melt (Fig. 13-9). The temperature at this transition is the **thermodynamic melting temperature** of the *specimen*, $T_{M,o}$.

Endgroups, branches, and chain folds produce additional crystal defects. Polymers thus have broader melting ranges than their oligomeric counterparts, especially polymers with broad molar mass distributions (Fig. 13-10). At the melting temperature T_M of such polymers, the sharp steps of specific volumes v or enthalpies H degenerate to S-shaped curves. The sharp signals of the first derivatives, $(\partial V/\partial T)_p = \beta V$ and $(\partial H/\partial T)_p = C_p$, convert to bell-shaped curves (cf. Figs. 13-1 and 13-4) and the polymer now has a broad melting range. Reported melting temperatures T_M of polymers usually indicate the mid-range of the melting range since the upper end is not easy to determine. Midpoint melting temperatures T_M are therefore usually lower than the **thermodynamic melting temperatures** $T_{M,o}$, of specimens which are lower than those of perfect crystals, $T_{M,\infty}$.

Effect of Heating Rate

Midpoint melting temperatures T_M may also be higher then $T_{M,o}$ if large crystallites are overheated (Fig. 13-10). Such large, **extended-chain crystals** are obtained by crystallization under pressure (Fig. 7-33). This effect is greater, the bigger the extended-chain crystals, the larger the heating rate, and the higher the molar mass of polymers. The effect indicates that no equilibrium is established between still crystallized and already molten chains.

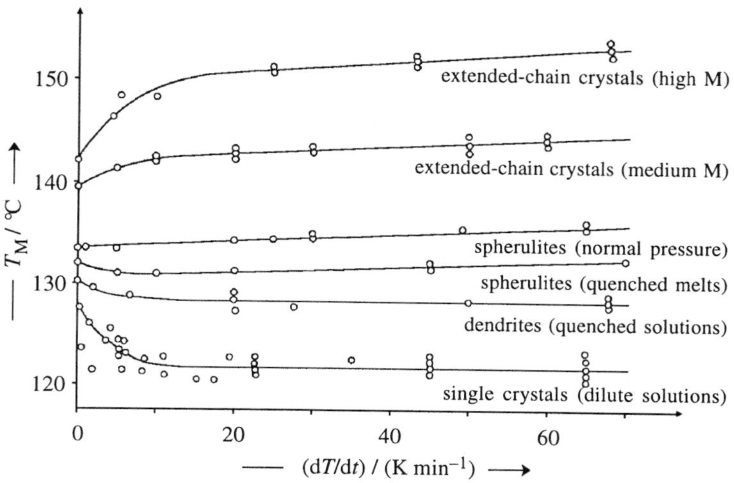

Fig. 13-10 Melting temperatures of differently crystallized poly(ethylene)s as a function of heating rates [8]. Crystallization by high pressure to extended-chain crystals, under normal pressure, in quenched melts or solutions, or from dilute solutions, resulting in extended chain crystals, spherulites, dendrites, or single crystals.

The melting temperature of single crystals from crystallization of dilute solutions decreases with increasing heating rate and finally becomes constant, obviously because chains have sufficient time to reorganize themselves during the melting process.

Effect of Crystallite Size

The **thermodynamic melting temperature** (= **thermodynamic fusion temperature**) $T_{M,o}$ of polymer crystals from infinitely long ($N_u \rightarrow \infty$), extended chains (i.e., infinitely thick lamellae) is given by the ratio of melting enthalpy to melting entropy, $T_{M,o} = N_u\Delta H_{M,u}/N_u\Delta S_{M,u} = \Delta H_{M,u}/\Delta S_{M,u}$.

Folded chains have lower melting temperatures, $T_M = \Delta H_M/\Delta S_M$, since enthalpies of segments are lowered by the surface enthalpy of the folds, $\Delta H_\sigma = (\sigma_e/L_f)$, where σ_e = surface energy and L_f = fold length (see also Fig. 7-15):

$$(13\text{-}15) \qquad \Delta H_M = N_u\Delta H_{m,u} - 2\,\Delta H_\sigma = N_u\Delta H_{m,u} - 2\,(\sigma_e/L_f)$$

Introduction of $\Delta H_M = T_M\Delta S_M$, $\Delta H_{M,u} = T_{M,o}\Delta S_{M,u}$, and $\Delta S_M = N_u\Delta S_{M,u}$ and neglecting entropy differences by setting $\Delta S_M \approx \Delta S_{M,o}$ delivers the **Gibbs-Thomson equation (Thomson's rule)** for the melting temperature of a lamella with the thickness $L_{c,o} = N_uL_u$:

$$(13\text{-}16) \qquad T_M = T_{M,o}\left(1 - \frac{2\,\sigma_e}{L_f\Delta H_{M,u}} \cdot \frac{1}{N_u}\right) = T_{M,o}\left(1 - \frac{2\,\sigma_e L_u}{\Delta H_{M,u}L_f} \cdot \frac{1}{L_{c,o}}\right)$$

Extrapolation of melting temperatures T_M to $1/L_{c,o} \rightarrow 0$ delivers the melting temperature of an infinitely thick lamella ($L_{c,o} \rightarrow \infty$).

Lamella thicknesses $L_{c,o}$ are not easy to determine. Much easier to probe is another equation that is obtained from Eq.(13-16) as follows. The surface energy σ_e is expressed by Eq.(7-7), $\sigma_e = L_{c,o}\Delta H_{M,o}(T_{M,o} - T_{cryst})/(2\ T_{M,o})$. Furthermore, a thickening factor $\gamma \equiv L_f/L_u$ is introduced as the ratio of the height of the fold to the length of a unit in the stem of the lamella. The resulting **Hoffman-Weeks** equation predicts that melting temperatures increase with increasing crystallization temperature,

$$(13\text{-}17) \qquad T_M = (1 - \gamma^{-1})T_{M,o} + \gamma^{-1}T_{cryst}$$

which is indeed found experimentally for values of $\gamma \approx 2\text{-}3$.

Eq.(13-17) ascribes the observed increase of melting temperatures to the thickening of lamellae. However, such a process can occur only if polymer segments are mobile, i.e., if the polymer has a so-called α-relaxation which is observable by NMR experiments or dielectric or dynamic mechanical measurements (Section 13.1.4).

Crystalline α-relaxations are not found for aliphatic polyamides, poly(1,4-phenylene sulfide), poly(ethylene terephthalate), and isotactic poly(styrene) although the melting temperatures of all these polymers increase with inreasing crystallization temperatures. In other polymers, α-relaxations are suppressed by non-crystallizable monomeric units or large entanglement densities in interlamellar regions. A causal relationship between linear function $T_M = f(T_{cryst})$, thickening factor γ, and segmental mobility is therefore not necessarily present. Indeed, the Hoffman-Weeks equation often delivers thickening factors that are too large, which results in too low melting temperatures $T_{M,o}$.

13.3.3 Molar Mass Dependence

Melting temperatures T_M of polymers of a homologous series increase with increasing molar mass and finally approach a limiting value $T_{M,\infty}$ (Volume I, Fig. 1-1). This behavior is caused by endgroups which cannot be fitted into the crystal lattice that is generated by monomeric units. The endgroups thus act as "foreign" substances that cause a depression of the melting temperature.

Classical thermodynamics predicts that the inverse melting temperature depends on the mole fraction x_{end} of endgroups according to $(1/T_M) = (1/T_{M,\infty}) - (R/\Delta H_{M,u,m}) \ln x_{end}$ where $H_{M,u,m}$ = molar melting enthalpy per monomeric unit. For Schulz-Flory distributions of molar masses with a number average degree of polymerization, $\overline{X}_n$, this equation converts to

$$(13\text{-}18) \qquad \frac{1}{T_M} = \frac{1}{T_{M,\infty}} + \frac{R}{\Delta H_{M,u,m}} \cdot \frac{2}{\overline{X}_n} = \frac{1}{T_{M,\infty}} + \frac{R}{\Delta H_{M,u,m}} \cdot Q$$

Such a function $1/T_M = f(1/\overline{X}_n)$ is indeed found for molecularly uniform alkanes, $H(CH_2)_NH$ ($\overline{X}_n \equiv N$) (Fig. 13-11) and also for cycloalkanes with $N > 60$ (Fig. 13-11).

The derivation of Eq.(13-18) assumes implicitly that all molecules adopt the same macroconformation in the crystal, regardless of N, X, or $\overline{X}_n$. This assumption cannot apply to cycloalkanes with $N < 60$ since macroconformations of these rings vary widely with the number of chain atoms per molecule because of ring strain or Pitzer tension.

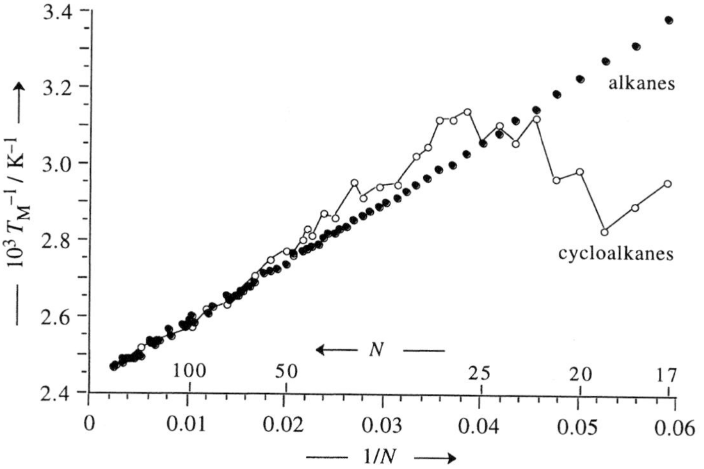

Fig. 13-11 Inverse melting temperatures T_M as a function of inverse numbers N of chain atoms in alkanes, $H(CH_2)_NH$, and cycloalkanes, cyclo$(CH_2)_N$ [9]. The intercept at $1/N \to 0$ corresponds to a melting temperature of $T_{M,0}$ = 140°C.

Eq.(13-18) also does not apply to low-molar mass alkanes with $N < 16$ which have alternating melting temperature for odd/even numbers of methylene groups because of the packing effects of CH_3 endgroups (Volume I, Fig. 1-1). Lower homologs with $17 \leq N \leq 80$ do form extended-chain crystals but chains start to fold at $N > 80$ (Fig. 7-10).

At about the same degree of polymerization, higher cycloalkanes have the same melting temperatures as alkanes (Fig. 13-11). The macroconformation of medium-sized cycloalkanes corresponds to a "natural folding" with parallel stems and a fold with 6 methylene groups in the conformational sequence ...T(GGTGG)T... according to X-ray data for cycloalkanes with N = 24, 26, 34, or 36 CH_2 groups.

Depressions of melting temperatures are caused not only by endgroups but also by (a) added solvents (melting point depression), (b) mixtures with another polymer that mixes in the melt (but not in the solid state), (c) random copolymers with long sequences of crystallizable units, and (d) amorphous regions in semicrystalline homopolymers. Cases (a)-(c) can be described by relatively simple modifications of Eq.(13-18):

(13-19a) $Q = (\tilde{V}_{u,m} / \tilde{V}_{1,m})[-\chi(1-\phi_2)^2 + (1-\phi_2)]$ (addition of solvents)

(13-19b) $Q = (\tilde{V}_{u,m} / \tilde{V}_{3,m})[-\chi(1-\phi_2)^2]$ (addition of polymer)

(13-19c) $Q = \ln x_u - 2x_u(1-x_u)$ (incorporated comonomers)

where $\tilde{V}_{x,m}$ = partial molar volume of the parent polymeric units (x = u), added solvent (x = 1), or comonomeric units (x = 3); ϕ_2 = volume fraction of the parent monomeric units; χ = interaction parameter (Eq.(10-16)); and x_u = mole fraction of monomeric units 2. In the **Baur model**, the second term in Eq.(13-19c) is the inverse degree of polymerization of homosequences of type 2, i.e., $2\,x_u(1-x_u) = 1/\overline{X}_2$.

A newer theoretical approach also considers the different melting enthalpies of monomeric units and endgroups and the mixing entropy of 2 endgroups and N monomeric

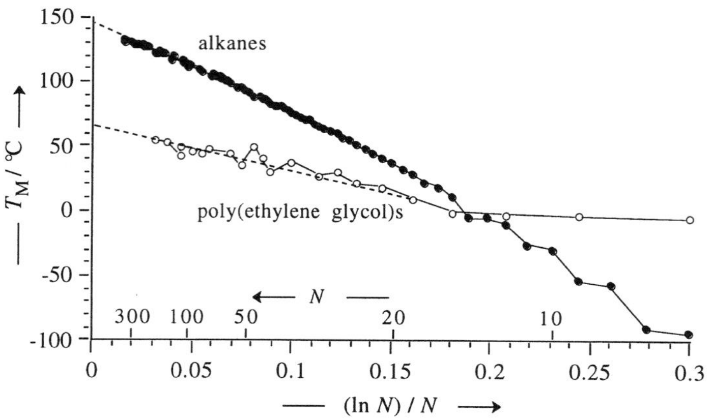

Fig. 13-12 Melting temperature T_M of alkanes, $H[CH_2]_NH$, and poly(ethylene glycol)s, $H[OCH_2CH_2]_XOH$ (as a function of [ln N]/N [9]. N = number of chain atoms; $X = N/3$. - - - Linear extrapolations for the range $0 \leq (\ln N)/N \leq 0.1$.

Linear extrapolations of melting temperatures to $Q = [\ln N]/N = 0$ of chain atoms are somewhat hazardous; for alkanes, they would furnish the following melting temperatures for a poly(ethylene) with infinitely large molar mass: 148.3°C ($Q < 0.15$), 144.4 ($Q < 0.1$), and 141.8 ($Q < 0.05$) with correlation coefficients of 0.994, 0.994, and 0.962, respectively.

units of crystallizing sequences. The resulting **Flory-Vrij equation** is complicated but several small terms may be neglected to give

$$T_M = T_{M,\infty}[1 - (2RT_{M,\infty}/\Delta H_{M,u,m})\{(\ln N)/N\}] \qquad (13\text{-}20)$$

where N = number of chain atoms, $\Delta H_{M,u,m}$ = molar melting enthalpy of monomeric units, and $T_{M,\infty}$ = melting temperature of infinitely long chains. Unfortunately, extrapolation to infinite numbers of chain atoms is not straightforward (see Fig. 13-12).

13.3.4 Effect of Constitution

Melting temperatures of polymers cannot be calculated from primary data for chemical and physical structures although group increment methods (Section 4.3.2) allow some good estimates.

Melting entropies and melting enthalpies deliver the following information. **Melting entropies** measures entropy changes caused by different positions, orientations, microconformations, and volumes of molecules, and monomeric units, respectively. The melting of methane causes "spherical" CH_4 molecules only to change their positions; the molar melting entropy is $\Delta S_{M,m} \approx 10.3$ J K^{-1} mol^{-1}. Melting of ethane, CH_3–CH_3, is accompanied by changes of positions and orientations which leads to a much higher melting entropy of 32.3 J K^{-1} mol^{-1}. Additional conversion of microconformations during melting results in much higher melting entropies per molecule; examples are 118.2 J K^{-1} mol^{-1} (decane) and 138.8 J K^{-1} mol^{-1} (dodecane).

Melting entropies should not be related to molecules *per se* but to the number of monomeric units or chain atoms because it is these structural elements that control the

Table 13-5 Melting temperatures T_M, cubic expansion coefficients β, and compression coefficients κ of melts; changes of specific volumes, Δv_M, during melting; molar melting entropies $\Delta S_{M,u,m}$; and molar melting entropies per chain unit, $\Delta H_{M,u,m}$, of polymer molecules with N chain units per repeating unit.

Repeating unit	N	$\frac{T_M}{°C}$	$\frac{10^4 \beta}{K^{-1}}$	$\frac{10^{10} \kappa}{Pa^{-1}}$	$\frac{\Delta v_M}{cm^3\ g^{-1}}$	$\frac{\Delta S_{M,u,m}}{J\ K^{-1}\ mol^{-1}}$	$\frac{\Delta H_{M,u,m}}{J\ mol^{-1}}$
$-CF_2-$	1	327		4.0	0.065	5.7	3.42
$-CH_2-$	1	144	7.97	1.67	0.173	9.9	4.13
$-OCH_2-$	2	184	2.85		0.085	10.7	4.89
$-OCH_2CH_2-$	2	69		1.75	0.081	8.4	2.87
$-O(CH_2)_5CO-$	7	64			0.041	6.9	2.33
$-NH(CH_2)_5CO-$	7	260	6.4	1.96	0.077	7.0	3.73
$-CH_2CH(CH_3)-$, it	2	187		2.86	0.112	7.6	3.50
$-CH_2CH(C_6H_5)-$, it	2	242			0.061	9.7	5.00
$-CH_2C(CH_3){=}CHCH_2-$, *cis*	3	30		5.0	0.108	4.8	1.46

number of possible conformations per molecule. If one substracts the melting entropy of methane from the melting entropies of higher alkanes, one obtains entropy increments per mole of chain units of $\Delta S_{M,u,m}/(J\ K^{-1}\ mol^{-1})$ = 10.95 (ethane), 10.79 (decane), and 10.71 (dodecane) with an average of 10.82. Changes in segment orientations can be considered by subtracting the value for ethane which leads to 8.6 (decane) and 8.9 (dodecane) with an average of 8.75.

Poly(methylene) has an experimental melting entropy of $\Delta S_{M,u,m}/(J\ K^{-1}\ mol^{-1}) = 9.9$ per chain unit (Table 13-5) which is in the middle of the estimates for conformational changes (10.82) and conformational changes plus segment orientations (8.75). If the melting of poly(methylene) results in 3 possible microconformations with equal probability and if the change of conformational energy is the only contribution to the melt entropy, then a value of $\Delta S_{M,u,m} \approx \Delta S_{conf,u,m} = R \ln 3 = 9.12\ J\ K^{-1}\ mol^{-1}$ would result. For the formation of one trans and two gauche conformations, one would obtain $\Delta S_{conf,u,m} = 7.41\ J\ K^{-1}\ mol^{-1}$.

Hence, experimental molar melting entropies are larger than changes of conformational entropies. However, melting is always accompanied by an increase of volume which is controlled by Δv_M = change of specific volume at the melting temperature,

$$(13\text{-}21) \qquad \Delta H_{M,u,m}/T_{M,o} = \Delta S_{M,u,m} \approx \Delta S_{conf,u,m} + (\beta/\kappa) M_u \Delta v_M$$

where M_u = average molar mass of the chain unit, β = cubic expansion coefficient, and κ = cubic compression coefficient (β and κ just above T_M).

The expansion on melting contributes $(\beta/\kappa)M_u\Delta v_M = 11.56\ J\ K^{-1}\ mol^{-1}$ according to the data of Table 13-5 which, with $\Delta S_{conf,u,m} = 7.41\ J\ K^{-1}\ mol^{-1}$, would lead to a much too high value of $\Delta S_{M,u,m} = 19.0\ J\ K^{-1}\ mol^{-1}$.

The difference between theoretical and experimental values for changes of conformational entropies points to local order in melts which reduces conformational entropies. Changes of conformational entropy that are lower than theoretically expected may also result from high segmental mobilities below the melting temperature, which was indeed found for *cis*-1,4-poly(isopene) by broad-line NMR spectroscopy.

Melting enthalpies are the products of melting entropies and melting temperatures, usually with values of 1-5 kJ per mole chain unit. Low values are caused by high segment mobilities below melting temperatures and high values by both dense chain packing in crystals and strong interactions between chains.

It has therefore been suspected that high melting temperatures are primarily caused by large cohesive energies between chains. However, cohesive energies refer to liquid-to-gas transitions whereas melting is a solid-to-liquid process. For example, infrared spectroscopy has shown that hydrogen bonds between polyamide chains in the crystalline state are mostly preserved in the molten state at temperatures just above the melting temperature. Thus, cohesive energies of polymers in melts and in crystals should not differ very much.

If cohesive energy is the dominant factor for the melting temperature, then one would expect melting temperatures to decrease if the proportion of "diluting" methylene groups per monomeric unit increases. Such a behavior is indeed found for the series of polyamides with the monomeric unit $-NH(CH_2)_NCO-$ (Fig. 13-13) where the cohesive energy of an amide group is 35.6 kJ/mol and that of a methylene group only 2.85 kJ/mol. However, for polyoxides with the monomeric unit $-O(CH_2)_N-$ and a cohesive energy of 4.19 kJ/mol, melting temperatures pass through a minimum with increasing N. A minimum is also observed for polyesters of the types $-O(CH_2)_iO-OC(CH_2)_jCO-$ and $-O(CH_2)_iCO-$ with a cohesive energy of 12.1 kJ/mol for the ester group (not shown). At high proportions of methylene groups, melting temperatures of all of these polymer types approach the melting temperature of poly(ethylene).

Melting temperatures of isotactic poly(1-olefin)s, $\text{+}CH_2-CH(CH_2)_N\text{+}_n$, also decrease with increasing number of methylene side groups, pass through a minimum, and increase again (Fig. 3-13). Polymers with very long side groups never approach the melting temperature of poly(ethylene), however, because long side chains form a somewhat ordered hull around the main chain (so-called side-chain crystallization).

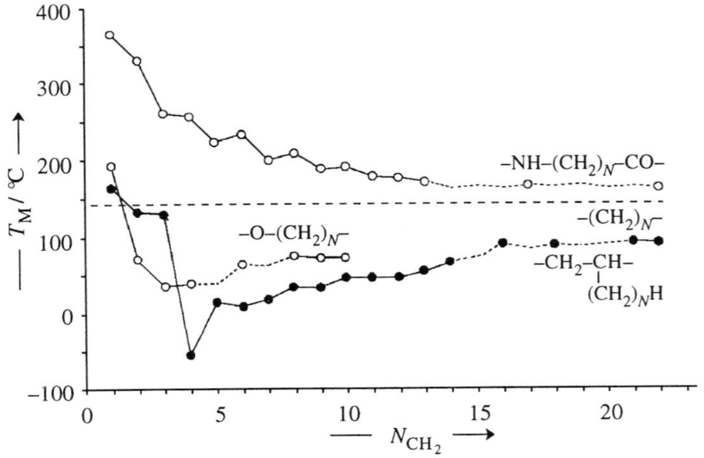

Fig. 13-13 Melting temperature as a function of the number of methylene groups in the main-chain (polyamides, $\text{+}NH(CH_2)_NCO\text{+}_n$, and aliphatic polyoxides, $\text{+}O(CH_2)_N\text{+}_n$) or in the side chain (isotactic poly(1-olefin)s, $\text{+}CH_2CH(CH_2)_NH\text{+}_n$). The horizontal broken line indicates the melting temperature of a high-molar mass linear poly(ethylene).

Important factors for the variation of melting temperatures with polymer constitution are the macroconformation and packing of chains in the crystalline state and the flexibility of chains in the melt. The latter factor is controlled by potential barriers $\Delta E^{\ddagger}_{m}$ which are low for hetero groups as chain atoms but large for methylene groups (Section 3.2.3; Volume I, Section 5.2): values of potential barriers, $\Delta E^{\ddagger}_{m}/(\mathrm{kJ\,mol^{-1}})$, for various chain units are 4.1 (–O–), 5.0 (–COO–), and 8.8 (–S–), but 12.3 for $-CH_2-$.

Melting temperatures are also affected by the packing of chains, for example, in polymers $\text{+}O(CH_2)_N\text{+}_n$ (Fig. 13-13). Poly(oxymethylene) (N = 1) forms a helix in the crystalline state but poly(tetrahydrofuran) (N = 4) does not, which leads to a much higher melting temperature of the former (184°C) compared with the latter (40°C).

Another important factor is the packing of chain units in the helix itself. For example, chain units of poly(oxymethylene) are tightly packed (conformational sequence ~G^+G^+~ and ~G^-G^-~) but those of poly(oxyethylene) much more loosely (conformational sequence ~TTG~) (Fig. 13-14). Molecules of the former are therefore much more rigid and a have a considerably higher melting temperature (184°C versus 69°C).

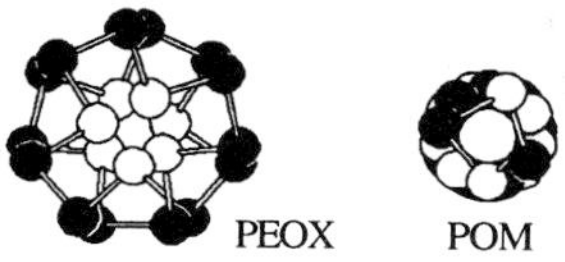

Fig. 13-14 Cross-sections of helices of crystallized poly(oxyethylene) (PEOX) (left) and poly(oxymethylene) (POM) (right). A = cross-sectional area, c = identity period, d = diameter.

Polymer	Helix	c/nm	d/nm	A/nm^2
PEOX	7_2	1.93	0.38	0.216
POM	10_2	1.73	0.20	0.172

Isotactic poly(1-olefin)s, $\text{+}CH_2\text{–}CH[(CH_2)_NH]\text{+}_n$, show the same effect: a decrease in melting temperatures from N =1 to N = 7 (Fig. 13-13) because of less and less tightly packed helices. However, melting temperatures increase with increasing N for $N > 7$ since side chains are increasingly more tightly packed near the main chain (so-called **side-chain crystallization**). For the same reason, it-poly(1-pentene), $\text{+}CH_2\text{–}CH(C_3H_7)\text{+}_n$, with a linear side chain, has a much lower melting temperature than the isomeric it-poly(3-methyl-1-butene), $\text{+}CH_2\text{–}CH(CH(CH_3)_2)\text{+}_n$, (130°C versus 304°C).

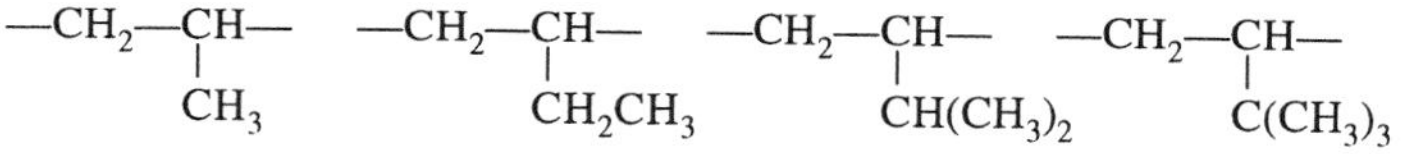

T_M = 186°C   136°C   304°C   > 320°C

Packing effects are also responsible for the melting behavior of constitutionally or configurationally different *copolymers*. In general, melting temperatures of tactic polymers decrease with decreasing tacticity. With few exceptions, completely atactic polymers do not crystallize at all and therefore have no melting temperature.

Melting temperatures of bipolymers increase monotonically with increasing molar proportion of the higher-melting monomeric units if these units are isomorphous and follow each other at random. An example is copolymers with hexamethylene terephthalamide and hexamethylene adipamide units (Fig. 13-15). Melting temperatures of bipolymers with two non-isomorphous comonomeric units pass through minima, however. Examples are bipolymers with hexamethylene terephthalamide and hexamethylene sebacamide units on one hand and terpolymers with 1:1 ethylene glycol and terephthalic acid units and varying proportions of p-hydroxybenzoic acid units on the other.

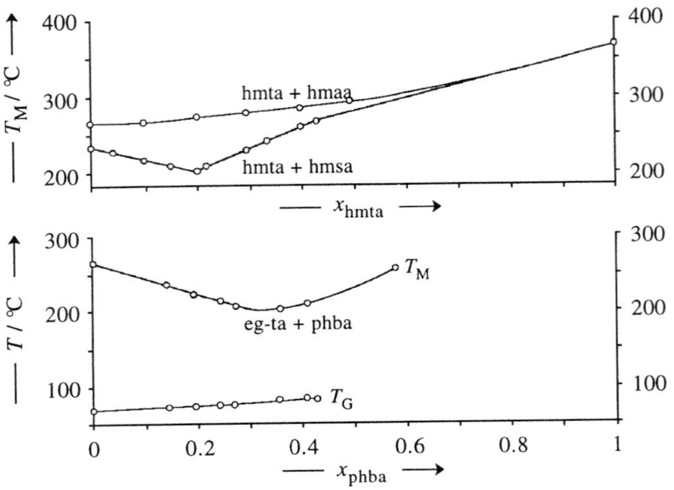

Fig. 13-15 Top: Melting temperatures of bipolymers with hexamethylene terephthalamide units (hmta) and hexamethylene adipamide units (hmaa) or hmta and hexamethylene sebacamide units (hmsa) as a function of the mole fraction x_{hmta} of hmta units [10]. Bottom: Melting and glass temperatures of copolymers with 1:1 ethylene glycol and terephthalic acid units (eg-ta) and varying proportions of *p*-hydroxy benzoic acid units (phba).

13.4 Transitions of Liquid Crystals

13.4.1 Thermal States

On heating, ideal polymer crystals convert to melts at their melting temperatures (Section 13.3) and completely amorphous polymers to melts at their glass temperatures (Section 13.5). In both cases, melts are completely disordered in the ideal case. Correspondingly, semi-crystalline polymers may exhibit both melting and glass temperatures.

Far more thermal states, transitions, and transformations are possible if melts are not isotropic but ordered (Fig. 13-16), i.e., if melts are mesophases = liquid-crystalline phases (LC) since LC in Fig. 13-16 may refer to any of many liquid-crystalline states (Chapter 8). With increasing temperature, the polymer may in general pass through the following states:

glass → crystalline → smectic (||) → smectic (//) → nematic → isotropic

Depending on chemical structure and/or thermal conditions, there may be even more states or also fewer states than the sequence above indicates. For example, the crystalline state may be absent or there may be only one smectic state or none, or only one type of mesophase, etc. More highly ordered smectic structures (||) are often of the S_A type and less ordered smectic structures (//) of the S_C type (Section 8.2.2).

Thermal transitions of thermotropic LC polymers from crystal C to a smectic phase S or from a nematic phase N to an isotropic phase I are thermodynamic first-order transitions. For C ⇄ S and N ⇄ I, enthalpies, entropies, and volumes have a discontinuity jump

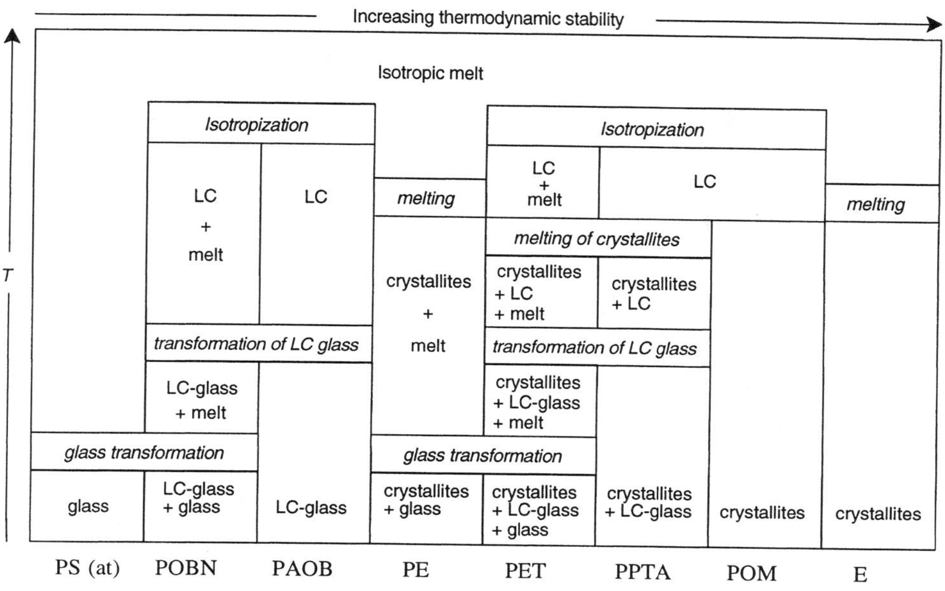

Fig. 13-16 Classification [11] and thermal transitions and transformations of condensed matter [12]. Examples (|| stretched; ⇔ annealed; ⇓ quenched):

E	= simple melting of crystallites	POBN	= poly(oxybenzoate-*co*-naphthoate) ⇓
PAOB	= poly(acryloyl oxybenzoate) ⇓	POM	= poly(oxymethylene) fibers ||
PE	= poly(ethylene)	PPTA	= poly(*p*-phenylene terephthalamide) ⇔
PET	= poly(ethylene terephthalate) ||	PS (at)	= atactic poly(styrene)

and expansion coefficients and molar heat capacities a peak at these transition temperatures (Figs. 13-1 and 13-17). Conversion of one mesophase into another, such as $S_C \rightleftarrows S_A$ or $S_A \rightleftarrows N$, are now believed to be thermodynamic second order transitions.

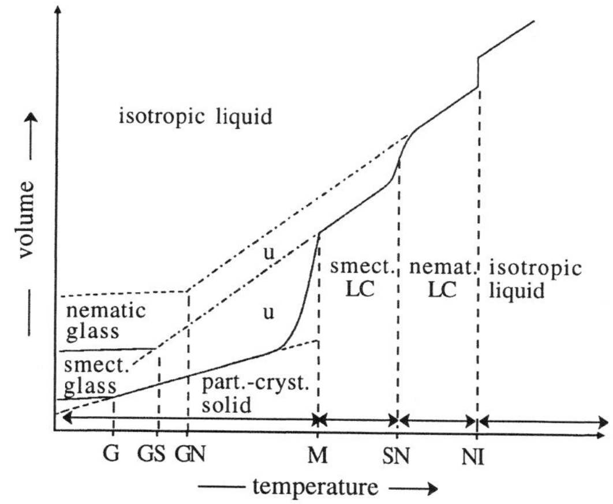

Fig. 13-17 Volume-temperature diagram of a liquid-crystalline polymer (schematic). ——— experimentally observed, - · · - - possible undercooling (u) from liquid-crystalline phases to undercooled smectic or nematic LC phases and further to the corresponding glasses. G = glass transformation, I = isotropic, M = melt transition, N = nematic, S = smectic.

At sufficiently high temperatures, all liquid-crystalline polymers form isotropic liquids (I) provided they do not decompose. Below the transition temperature LC → I, mesophases are milky but above, optically clear. The **isotropization temperature** T_I is therefore also called the **clearing temperature.**

Thermotropic LC polymers may form several types of LC phases or only one of them. In the most general case, cooling of an isotropic liquid first leads to the formation of a nematic phase below the isotropization temperature T_{NI}, then to a smectic phase at a transition temprature $T_{S(C),N}$, followed by another, more ordered smectic phase at the transition temperature $T_{S(A)N}$, and finally to three-dimensionally ordered crystals at the fusion temperature = melting temperature T_M. In general, these temperatures depend on the molar mass of the LCP (Fig. 13-18).

Thermodynamically stable phases are called **enantiotropic**. Such phases exist only between the melting temperature and the higher clearing temperature ($T_I > T_M$) in mesomorphic systems consisting of a single component.

Mesophases will form only as dispersions in undercooled melts if the clearing temperature is lower than the melting temperature ($T_I < T_M$). Such phases are called **monotropic**; they are thermodynamically unstable with respect to the crystalline phase.

If crystallizations can be suppressed, smectic phases will convert below T_M into undercooled smectic liquids, and nematic phases below T_{SN} into undercooled nematic liquids. Formation of undercooled liquids below T_{NI} has never been observed.

At still lower temperatures T_{GN} and T_{GS}, respectively, undercooled liquids freeze and become anisotropic nematic (nG) and smectic (sG) glasses, respectively. Some of these transformation temperatures cannot be observed directly but their existence has been deduced from extrapolations of transformation temperatures to 100 % of the components. Such transformation temperatures have been called **virtual transition temperatures**.

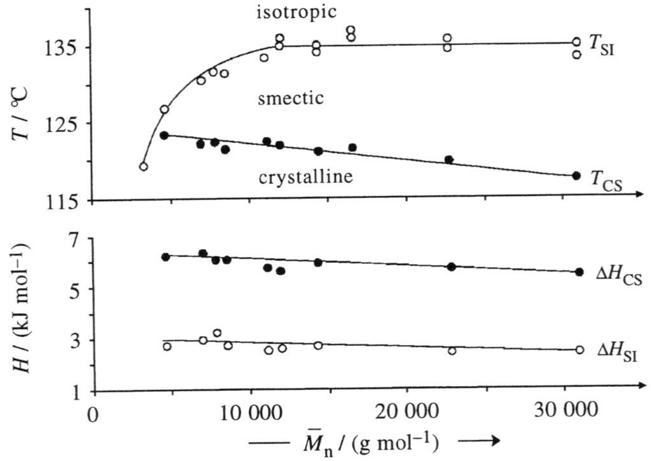

Fig. 13-18 Dependence of the transition temperature crystalline-smectic (T_{CS}), the clearing temperature (T_{SI}), and the transition enthalpies ΔH_{CS} and ΔH_{SI} per mol monomeric unit on the number-average molar mass $\overline{M}_n$ of liquid crystalline methacrylate side-chain polymers that were obtained by polymerization of $CH_2{=}C(CH_3)–COO–(CH_2)_6–O–(p\text{-}C_6H_4)–(p\text{-}C_6H_4)–OCH_3$ [13]. A molar mass of 10 000 g/mol corresponds to a degree of polymerization of ca. 26.

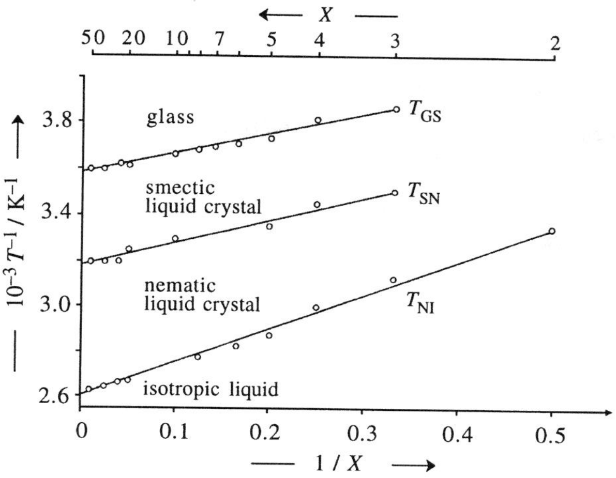

Fig. 13-19 Dependence of inverse transformation temperatures T_{GS} (glass → smectic) and transition temperatures T_{SN} (smectic-nematic) and T_{NI} (nematic-isotropic) on the inverse degree of polymerization, $1/X = 1/N$, of oligomeric and polymeric siloxanes, $(CH_3)_3SiO[SiR(CH_3)\text{-}O]_NSi(CH_3)_3$, with the substituent $R = (CH_2)_6\text{-}O\text{-}(p\text{-}C_6H_4)\text{-}COO\text{-}(p\text{-}C_6H_4)\text{-}O\text{-}CH_3$ [14].

13.4.2 Molar Mass Dependence

Clearing temperatures T_{SI} of smectic phases of side-chain liquid crystalline polymers, $+\!CH_2\text{–}C(CH_3)COO(CH_2)_6O(p\text{-}C_6H_4)_2OCH_3\!+_n$, increase with increasing molar mass until they become constant (Fig. 13-18). However, transition temperatures T_{CS} for the transition crystal-smectic phase decrease linearly with increasing molar mass because longer main chains prevent mesogens from occupying their ideal positions in the LC domain (cf. Fig. 8-16).

However, temperatures of the G ⇄ S transformation and the S ⇄ N and N ⇄ I transitions of oligomeric siloxane side-chain polymers show the same dependence on the degree of polymerization as the melting temperatures (cf. Figs. 13-19 and 13-11).

13.4.3 Thermodynamic Quantities

Slopes of plots of the dependence of inverse transition temperatures on inverse degrees of polymerization (Figs. 13-11 and 13-19), $1/T = f(1/X)$, are controlled by inverse transition enthalpies (Eq.(13-18)). The slopes for the GS transformation and the SN transition are the same but that for the NI transition is larger (Fig.13-19), presumably because fewer contacts need to be broken (smaller transition enthalpy ΔH_{NI}).

The transition enthalpy ΔH_{CS} is larger than the transition enthalpy ΔH_{SI} (Fig. 13-18) because interactions between segments or molecules, respectively, decrease with decreasing state of order in the series crystal → smectic liquid-crystalline domain → isotropic melt.

Spacers allow better orientations of mesogens in both side-chain and main-chain liquid-crystalline polymers. Transition enthalpies and transition temperatures thus increase

Table 13-6 Transition entropies ΔS per monomeric unit [12].

Class	Number of types smectic	nematic	$\Delta S/(J\ K^{-1}\ mol^{-1})$ S → N	N → I
Low-molar mass LCs	4	11	14.3 ± 8.0	2.1 ± 1.5
High-molar mass side-chain LCPs	16	5	9.8 ± 6.6	3.5 ± 2.5
High-molar mass main-chain LCPs	-	64	-	15.6 ± 7.3

strongly with increasing number of CH_2 groups in spacers, both for the series with even numbers of such groups and the series with odd numbers. Because of the higher degrees of order, transition entropies are higher for (a) low-molecular weight LCs than for their high-molar mass counterparts, (b) C →S transitions than S → I transitions, and (c) main-chain LCPs than side-chain LCPs (Table 13-6).

Transition entropies for the transitions S → N and N → I are relatively small (Tables 13-6 and 13-7). For example, they are only ca. 1.4 J K^{-1} mol^{-1} (in moles of chain units), which is caused by the high mobility of groups in the nematic phase. For the same group L in the side chain, practically the same transition entropy is found (Table 13-7) regardless of the type of the main chain II and III in Table 13-7: only the side chain and not the main chain affects the transition.

Entropies of transitions N → I of polymers II and III are only one-half as large as the transition entropy of the constitutionally similar low-molecular weight compound I for the same transition (Table 13-7), which points to smaller mobilities of II and III in the isotropic phase. As expected, the transition entropy for the process S→N is smaller than that for the process N→I since the nematic phase is much more ordered than the isotropic one whereas the degree of order differs only relatively little between the smectic and nematic phases.

Table 13-7 Transformation temperatures T_q and molar transition entropies $(\Delta S_{u,m})_q$ per repeating unit of liquid crystals of the type R-L-X with molecular weights $M_{u,r}$ of repeating units containing the group L = $(CH_2)_6$-O-(*p*-C_6H_4)-COO-(*p*-C_6H_4)-O-CH_3. G = glass transformation, M = melting transition, SN = smectic-nematic transition , NI = nematic-isotropic transition.

Liquid crystal No.	Repeating unit	$(M_u)_r$	T_q / K for q = G	M	SN	NI	$(\Delta S_{u,m})_q/(J\ K^{-1}\ mol^{-1})$ SN	NI
I	C_5H_{11}–L	382.5	-	339	-	363	-	5.61
II	–CH_2–C(CH_3)(COO–L)–	396.5	323	-	-	377	-	2.77
III	–O–Si(CH_3)(L)–	381.5	281	330	320	385	0.94	2.85

13.5 Glass Transformations

Many aspects of the transformation of a *glass-like* amorphous polymer to a high-viscosity, *rubber-like* melt resemble a thermodynamic second-order transition, for example, the discontinuities of C_p, β, and v at the so-called glass (transition) temperature T_G (Figs.

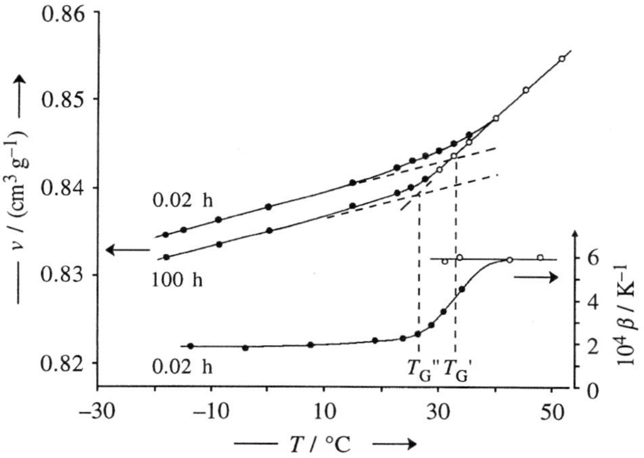

Fig. 13-20 Temperature dependence of specific volumes v and cubic expansion coefficients β of a poly(vinyl acetate) ($\overline{M}_n$ = 60 000 g/mol) that was quenched to –20°C from the melt and then annealed for 0.02 h or 100 h [15]. o Liquid state, ● glass. T_G' and T_G" are both softening temperatures T_S.

13-1 and 13-3). However, such transformations are not true thermodynamic processes since there are no equilibria on *both* sides of the transformation temperature. The magnitude of the transformation temperature T_G is affected by the heating (or cooling) rate and the thermal history of the specimen (Fig. 13-20): the T_G is *not* a *transition.*

For these reasons, industrial DSC and DTA experiments are standardized. Still, very different DSC curves are obtained from different instruments of various manufacturers (Fig. 13-21). Interpolations, as shown for II, gave too low values of T_G for experiments III and IV, correct ones for II and V, and a too high one for I. However, the standard practice is to take the temperature at one-half of the step height as the glass temperature. With this criterion, all values of T_G are too high except that from IV. Endothermic peaks appear at $T > T_G$ if the heating rate is not much smaller than the cooling rate before the experiment; it is caused by a reduction of the enthalpy of the specimen.

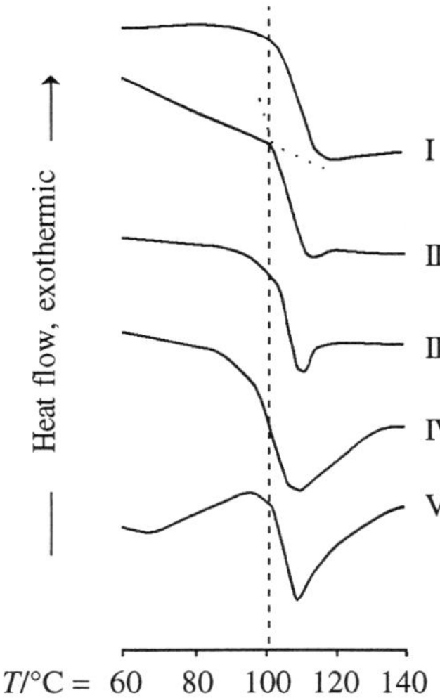

Fig. 13-21 Heat flow curves generated by five of a total of ten different DSC instruments from seven manufacturers that participated in a round-robin trial using standardized conditions (DIN 53765) [15]. In each experiment, a 1.5 mm thick molded sheet of the same high-molecular weight, atactic poly(styrene) specimen ($\overline{M}_w/\overline{M}_n$ = 2.5, $\overline{M}_w$ = 374 000 g/mol) was heated above the glass temperature and then cooled at 20 K/min before the DSC experiments.

The vertical broken line indicates the literature value for the glass temperature of poly(styrene). For the endothermic peaks at II-V, see text.

Note: instrument manufacturers are inconsistent on whether the positive direction of the ordinate should indicate exothermic effects (this figure) or endothermic ones.

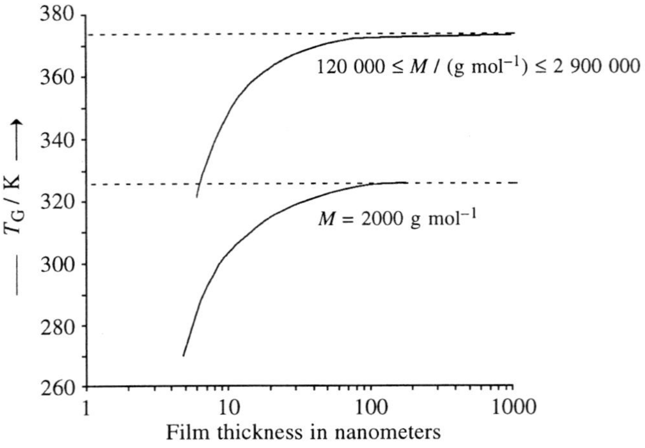

Fig. 13-22 Glass temperature of thin films as a function of the logarithm of film thickness for various high molar mass at-poly(styrene)s [17] and a low-molar mass at-poly(styrene) [18]. - - - - Glass temperatures of bulk poly(styrene)s of "infinite" molar mass and M = 2000 g/mol, respectively [18]. Radii of gyration are ca. 10 nm for M = 130 000 g/mol and ca. 1.3 nm for M = 2000 g/mol.

Experimental glass temperatures T_G are also affected by the rate $\dot{q}$ at which the specimen is cooled from the melt before the DSC experiment. Glass temperatures $T_{G,0}$ at a heating rate of $\dot{q} = 0$ can be obtained from $T_G = T_{G,0} + K \lg [\dot{q}/(K \min^{-1})]$ where K typically adopts values between 2 K and 4 K.

Glass temperatures by DSC may also vary somewhat with the amount, thickness, and appearance (powder, platelet, etc.) of the specimens because these factors affect the heat flow and give rise to a temperature gradient. These effects are usually small and positive, raising glass temperatures by ca. 1-3 K for most specimens and operating conditions.

Much larger lowerings of glass temperatures are observed in ultrathin layers (Fig. 13-22). It is noteworthy that glass temperatures start to decrease from their values in the bulk if the thickness of the films becomes lower than ca. 100 nm, regardless of the radii of gyration of the polymers. The reason for this effect does not seem to be clear yet.

13.5.1 Free Volumes

Phenomenologically, glass temperatures are characterized by a transformation of a somewhat hard, "glassy" body to a rubbery or highly viscous matter. At the temperature of this transformation, viscosities are ca. 10^{12} Pa s, regardless of the substance. For this reason, glass transformations were formerly assumed to be characterized by an "isoviscous" behavior. More modern interpretation identifies glass transformations as the temperature at which all amorphous matter has the same proportion of the so-called **free volume** which is generated by the movement of chain segments (Section 6.1.2).

Volumes of matter change practically linearly with temperature both above and below the temperature of the glass transformation (Figs. 13-1 and 13-3). With the definition of the cubic expansion coefficient, $\beta \equiv V^{-1}(\partial V/\partial T)_p$, one thus obtains equations for the free volumes V_f in the liquid (l) and amorphous (am) states at temperatures of 0 K and T:

(13-22) $(V_{f,l})_T = (V_{f,l})_0 + (V_{f,l})_T\beta_l T$

(13-23) $(V_{f,am})_T = (V_{f,am})_0 + (V_{f,am})_T\beta_{am} T$

To a first approximation, volumes of the liquid and the amorphous matter equal each other at $T = T_G$, i.e., $(V_{f,l})_G = (V_{f,am})_G$. Equating Eqs.(13-22) and (13-23) and again using Eq.(13-23) for $T = T_G$ leads to

(13-24) $\{[(V_{f,am})_0 - (V_{f,l})_0]/(V_{f,am})_0\}\cdot[1 - \beta_{am}T_G] = [\beta_l - \beta_{am}]T_G$

Both volumes must be the same at $T = 0$ K. The term in braces must therefore be the fraction f_{exp} of the free (expansion) volume;

(13-25) $f_{exp} = [\beta_l - \beta_{am}]T_G / [1 - \beta_{am}T_G] \approx [\beta_l - \beta_{am}]T_G$

These fractions f_{exp} from expansion coefficients of liquids and glasses agree reasonably well with those of Table 6-3 that were calculated from specific volumes of liquids and crystals. According to the empirical **Boyer-Simha rule**, $[\beta_l - \beta_{am}]T_G \approx 0.11$ can be assumed for many polymers (Table 13-8).

Table 13-8 Cubic expansion coefficients β in the liquid (l) and amorphous (am) state, fractional free volumes f_{exp} according to Eqs.(13-25) and Table (6-3), and Boyer-Simha factors, $(\beta_l - \beta_{am})T_G$. Deviation from the Boyer-Simha rules are observed for semi-crystalline polymers and polymers with relaxation mechanisms below the glass transformation.

Polymer	T_G/K	$10^4\beta_l$/K^{-1}	$10^4\beta_{am}$/K^{-1}	f_{exp} Eq.(13-25)	f_{exp} Table 6-3	$(\beta_l - \beta_{am})T_G$
Poly(ethylene)	193	7.97	2.87	0.104	-	0.098
Poly(isobutylene)	200	5.79	1.86	0.082	0.125	0.079
Poly(styrene)	373	5.65	2.09	0.144	0.127	0.133
Poly(vinyl acetate)	300	6.53	2.26	0.137	0.14	0.128
Poly(methyl methacrylate)	378	5.28	2.16	0.128	0.13	0.118
Average						0.11 ± 0.02

13.5.2 Molecular Interpretations

Glass temperatures increase with increasing cooling rates from the melt, often ca. 3 K per decade of cooling rate. Because of this kinetic effect, freezing-in temperatures T_F on cooling melts are not identical with softening temperatures T_S on heating glasses. The glass temperature is defined as that temperature at which the two "linear" parts of the experimental curves (e.g., $v = f(T)$, see Fig. 13-20) intersect below T_S and above T_F. In most cases, the difference between T_S and T_F is not important conceptually or numerically, however.

According to small-angle neutron scattering, macroconformations of polymer molecules do not change at the glass temperature (see Section 6.1). The cooling of highly

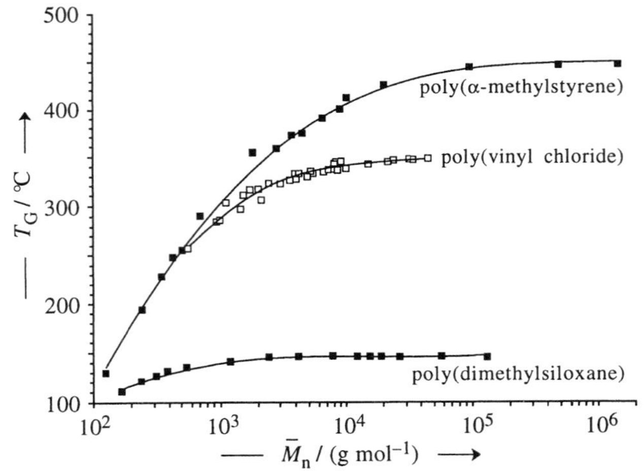

Fig. 13-23 Dependence of glass temperatures of poly(α-methyl styrene)s [19], poly(vinyl chloride)s [20], and poly(dimethylsiloxane)s [21] on number-average molar mass.

viscous polymer melts thus does not freeze-in the movements of whole molecules but only those of chain segments, both intermolecular and intramolecular. This freezing-in probably involves trans-gauche conversions that proceed cooperatively over larger distances since both deuterium NMR and MAS-NMR spectroscopy detect only small changes of rotational angles at T_G (MAS = magic angle sample spinning).

The following observations speak for the freezing-in of segmental movements:

- Both activation enthalpies of β_c relaxations of semicrystalline polymers and of α_a processes of amorphous polymers indicate the participation of segment lengths with 25-50 chain atoms (Table 13-4).
- Glass temperatures of crosslinked polymers are higher, the smaller the number of chain units between crosslinking sites (see below). Very highly crosslinked polymers do not have glass transformations.
- Glass temperatures T_G of linear polymers increase with increasing degree of polymerization and become constant for polymer molecules with $N_{ca} \approx$ 90-600 chain atoms per molecule (molar masses of thousands to hundreds of thousands (Figs. 13-22 and 13-23) which can be described by a function $T_G = f(1/\overline{M}_n)$ (Fig. 13-24)

The function $T_G = f(1/\overline{M}_n)$ can be derived as follows. The free volume of a chain end is assumed to be larger than that of the otherwise chemically identical monomeric units of the linear chain. The excess free volume is ΔV_{exc} per chain end and thus 2 $\Delta V_{exc}N_A$ per amount of linear chains with two endgroups each. The excess free volume per molar mass is therefore 2 $\Delta V_{exc}N_A/M$. Multiplication of 2 $\Delta V_{exc}N_A/M$ by the density ρ of the polymer delivers the proportion of the excess free volume, 2 $\rho\Delta V_{exc}N_A/M = \Delta V_{exc}/V$, which can be expressed by the cubic expansion factor if the latter is written for differences, $\Delta V_{exc}/V = \Delta\beta\Delta T$.

The expansion coefficient can be approximated by the difference $\Delta\beta$ of the cubic expansion coefficients of the glass and melt. The higher the concentration of endgroups, the stronger must be the undercooling if low- and high-molar mass polymers have the same free volume. Hence, the temperature difference ΔT equals the difference between

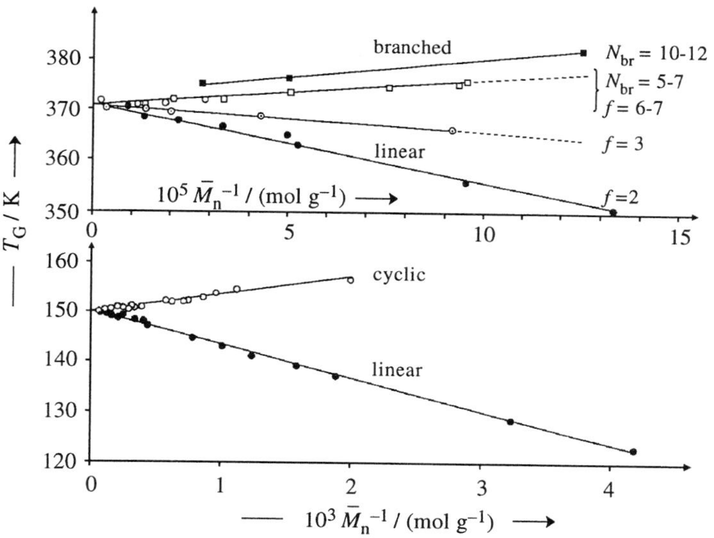

Fig. 13-24 Dependence of glass temperature on the inverse number-average molar mass for poly(styrene)s (top) and poly(dimethylsiloxane)s (bottom). The poly(styrene)s are either linear (f = 2), star-like with f = 3 or f = 6-7 arms, or randomly branched with N_{br} = 5-7 or 10-12 branches per molecule [22]. Poly(dimethylsiloxane)s are either linear or cyclic [23].

the glass temperature $T_{G,\infty}$ of a polymer with no endgroups and $M = \infty$, and the glass temperature T_G of the specimen with $M < \infty$. From $2\,\rho \Delta V_{exc} N_A / M = \Delta V_{exc}/V$ one obtains, with $\Delta V_{exc}/V = \Delta\beta\Delta T$ and $\Delta T = (T_{G,\infty} - T_G)$, the equation

(13-26) $$T_G = T_{G,\infty} - \frac{2\rho N_A \Delta V_{exc}}{\Delta\beta} \cdot \frac{1}{M}$$ **Fox-Flory equation**

Examples of the variation of T_G with $1/\overline{M}_n$ for linear, cyclic, star-like, and randomly branched polymers are shown in Fig. 13-24. Linear (L) and cyclic (C) macromolecules show opposite signs of the function $T_G = f(1/\overline{M}_n)$ because (a) the free volume fraction $\Delta V_{exc}/V$ decreases with increasing $\overline{M}_n$ for L because of the decreasing proportion of endgroups while it is independent of $\overline{M}_n$ for C because of the absence of endgroups, and (b) the average number of microconformations per chain unit is independent of $\overline{M}_n$ for L whereas it increases with increasing $\overline{M}_n$ for C (see also Fig. 3-11). The latter effect increases the flexibility of rings and thus causes T_G to decrease with increasing $\overline{M}_n$. The effect of branching is discussed in the next section.

Flexibilities of segments and molecules, respectively, affect both glass and melting temperatures and one should therefore expect T_G to be a function of T_M for semi-crystalline polymers. Indeed, a common integral curve is obtained for $\Sigma_i N_i = f(T_G/T_M)$ where N_i is the number of polymers that have the same value of T_G/T_M (Fig. 13-25). Deviations from the common curve are only obtained for very low values of T_G/T_M, which belong to unsubstituted polymers such as poly(ethylene), poly(oxyethylene), etc. The median of the curve corresponds to the empirical **Beaman-Boyer rule**:

(13-27) $$T_G \approx (2/3)\,T_M$$

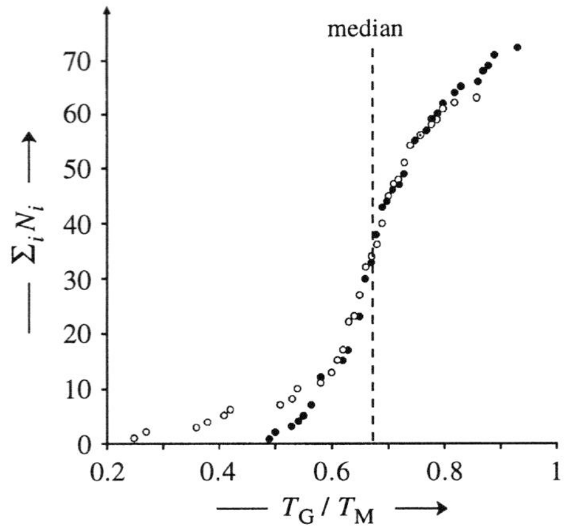

Fig. 13-25 Sum of the number of polymers with the same ratio of T_G/T_M as a function of T_G/T_M for polymer molecules with (O) symmetric monomeric units (for example, with $-CH_2-CR_2-$) or (●) asymmetric monomeric units (for example, $-CH_2-CHR-$) [24].

Deviations from the Beaman-Boyer rule are numerous (see Fig. 13-25). An example is ethylene copolymers where $T_G/T_M < 2/3$ is observed when large neutral substituents reduce the length of crystallizable ethylene segments and thus the melting temperature (Table 13-9). Values of $T_G/T_M > 2/3$ are obtained when substituents increase the barriers for rotation about chain bonds, especially if they create physical crosslinks e.g., by –COOH and -COONa sidegroups.

Crystalline regions may act as physical crosslinks and one would expect them to have the same effect on glass temperatures as chemical crosslinks. Indeed, glass temperatures of semicrystalline st-1,2-poly(butadiene)s, poly(oxyethylene)s, and poly(vinyl chloride)s all increase with increasing degree of crystallinity. However, the glass temperatures of poly(ethylene terephthalate)s pass through a maximum with increasing degree of crystallinity whereas no effect of crystallinity was observed for the glass temperatures of semicrystalline it-poly(propylene)s and poly(chlorotrifluoroethylene)s.

Table 13-9 Melting and glass temperatures of some ethylene copolymers (T_G/T_M in K/K).

Polymer	Monomeric units	T_M/°C	T_G/°C	T_G/T_M
Poly(ethylene)	$-CH_2-CH_2-$	144	−80	0.46
Poly(ethylene-*co*-methyl acrylate)	$-CH_2-CH_2-$ + $-CH_2-CH(COOCH_3)-$	90	−35	0.66
Poly(ethylene-*co*-acrylic acid)	$-CH_2-CH_2-$ + $-CH_2-CH(COOH)-$	103	0	0.73
Poly(ethylene-*co*-sodium acrylate)	$-CH_2-CH_2-$ + $-CH_2-CH(COONa)-$	95	55	0.89

13.5.3 Effect of Constitution

Linear Chains of Homopolymers

Glass temperatures of linear polymer chains are affected by both intermolecular and intramolecular interactions. Potential barriers for the rotation about –C–O– bonds are

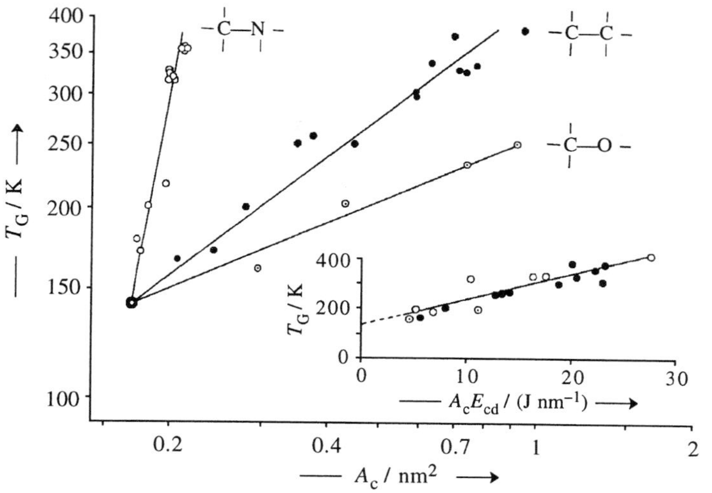

Fig. 13-26 Logarithm of glass temperature T_G as a function of the logarithm of cross-sectional area, A_c, of carbon chains, carbon-oxygen chains, and carbon-nitrogen chains. Insert: glass temperature as a function of the product of cross-sectional area and cohesive energy density [25].

considerably lower than those about –C–C– bonds whereas those about –C–N– bonds are noticeably higher (Volume I, Section 5.2.3). For the same cross-sectional area of chains, glass temperatures thus increase in the order –C–N– > –C–C– > –C–O– (Fig. 13-26). In first approximation, glass temperatures increase linearly with the product of cohesive energy density and cross-sectional area of chains (Fig. 13-26, insert). High cohesive energy densities are caused, for example, by intermolecular hydrogen bonds between polyamide chains and intramolecular hydrogen bonds of cellulose chains.

Straight lines of functions of lg T_G = f(lg A_c) in Fig. 13-26 meet at T_G = 141 K and $A_{c,o}$ = 0.17 nm². The latter value is lower than that of a poly(ethylene) chain in a crystalline lattice (A_c = 0.183 nm², Table 7-4) which includes van der Waals interactions between poly(ethylene) chains. A value of $A_{c,o}$ = 0.17 nm² should thus correspond to the cross-sectional area of an isolated poly(ethylene) chain with a diameter of 0.465 nm, a value that agrees well with the one calculated from the radius of a methylene group (0.20 nm) and one-half of the distance between atoms C¹ and C³ in a carbon zigzag chain –C¹–C²–C³–. A temperature of 141 K should therefore be the lowest possible glass temperature of a carbon atom-containing chain. It agrees well with the lowest reported glass temperature of any polymer, 134 K for poly(diethylsiloxane), and also with that of glassy H_2O (136 K).

Glass temperatures are lowered if chains are packed less tightly in the amorphous state. For example, densities $\rho_{25°C}$/(g cm⁻³) of poly(alkyl methacrylate)s with monomeric units –CH_2–C(CH_3)(COO(CH_2)$_i$H)– decrease from 1.17 (i = 1) to 1.08 (i = 3) and 0.92 (i = 18) and so do their glass temperatures T_G (Fig. 13-27). Brittleness temperatures T_B of the same polymers also first decrease with increasing i but then start to increase again at i > 12. The same behavior is observed for poly(alkyl acrylate)s: first a decrease of T_G and then an increase at $i \geq 9$.

This so-called **side-chain crystallization** is brought about by a parallelization of side chains which prevents them from moving on sudden impact and causes the specimen to

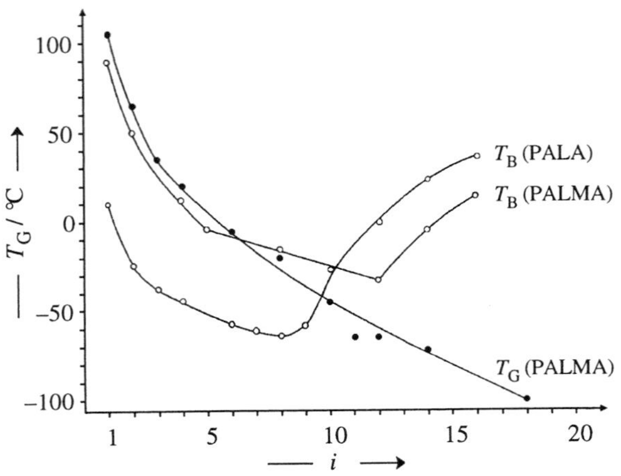

Fig. 13-27 Glass temperatures T_G of poly(alkyl methacrylate)s (PALMA) and brittleness temperatures T_B of PALMA and poly(alkyl acrylate)s (PALA) as a function of the number i of methylene groups in the side chains.

break. Side-chain crystallizations do not affect glass temperatures, which are mainly controlled by the mobility of the main chain. The side chains themselves are too short, even at $i = 18$, to produce an increase of T_G at high i. Such increases require very long and/or rigid side chains, for example, liquid-crystalline groups L of polymer II in Table 13-7 which has a glass temperature of 50°C.

Glass temperatures are also affected by *regioisomerism*. High-molecular weight head-to-head poly(styrene) has practically the same glass temperature as its conventional atactic head-to-tail counterpart (97°C *vs.* 100°C). However, head-to-tail poly(vinyl cyclohexane) has a much lower glass temperature than the head-to-tail polymer (88°C *vs.* 137°C) but this may be caused mainly or in part by different tacticities. Unambiguous is the strong effect of neighboring groups in poly(isobutylene) (PIB): T_G = 87°C for head-to-head PIB but –61°C for head-to-tail PIB.

Glass temperatures of asymmetrically substituted polymer chains can be discussed only as a function of the constitution of repeating units if the various types of chains have the same type and degree of tacticity. For example, glass temperatures of syndiotactic poly(methyl methacrylate)s increase linearly with the mole fraction of syndiotactic triads from 55°C at $x_{ss} = 0$ to 135°C at $x_{ss} = 1$.

Branched Polymers

Star-like and randomly branched polymers have higher glass temperatures than their linear counterparts (Fig. 13-24) because their relatively short segments or arms cannot produce as many macroconformations as linear chains. Glass temperatures for polymers with large numbers of branching points decrease with increasing molar mass.

At high molar masses, glass temperatures are the same for linear polymers, star molecules with any number of arms, and randomly branched polymers with small but constant numbers of branching points. However, very highly branched polymers have only very short segments with very low mobilities. These polymers never reach the glass temperatures of linear polymers, not even at very high molar masses.

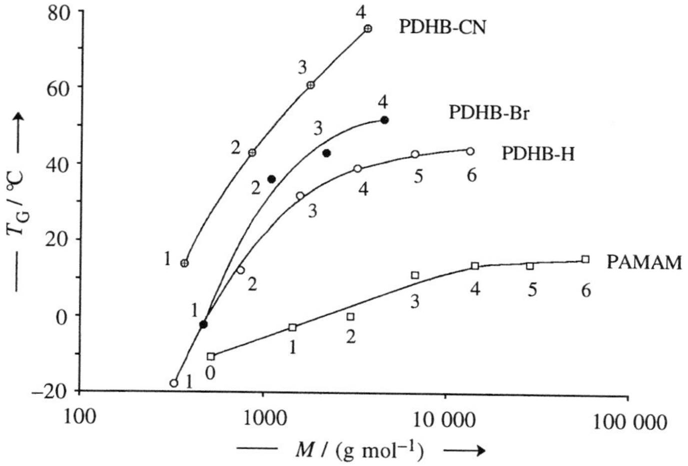

Fig. 13-28 Dependence of glass temperatures of dendrimers on molar masses. PDHB = *dendro*-poly-(3,5-dihydroxy benzylalcohol)s with endgroups –CN, –Br, or –H [26] and NH_2-terminated PAMAM dendrimers [27]. Numbers indicate numbers of generations.

Dendrimers

Glass temperatures of dendrimers increase with increasing molar mass (Fig. 13-28) but T_G is not a linear function of the inverse molar mass as in linear polymers or star molecules with three arms (Fig. 13-24). Glass temperatures rather become constant at higher generations, unlike those of star molecules, because the number of branching points increases with each new generation and does not stay constant.

The final value $T_{G,\infty}$ at higher generation numbers of dendrimer is strongly affected by the type of endgroups (Fig. 13-28). At the same generation number, glass temperatures are the higher the more polar the endgroups.

13.5.4 Plasticizing

Plasticizing is the "softening" of polymers (lowering of the glass temperature) by addition of plasticizers ("external plasticizing") or by incorporating comonomeric units ("internal plasticizing"). The resulting plasticized polymers are technologically characterized by higher flexural strengths, larger extensibilities, and smaller melt viscosities as well as scientifically by lower glass temperatures.

Plasticizing (*plasticization*) refers to a softening by addition of plasticizers or incorporation of comonomeric units. The softening of a polymer by heating is known as *plastifying* and that by contact heat and friction, *plastication* or *plastification*.

External Plasticizing

External plasticizing by adding plasticizers is usually the method of choice for reducing or eliminating the brittleness of certain polymers because it allows one to modify the material by mixing a polymer and a plasticizer. It also has a cost advantage because one can work with a standard polymer and need not synthesize various copolymers.

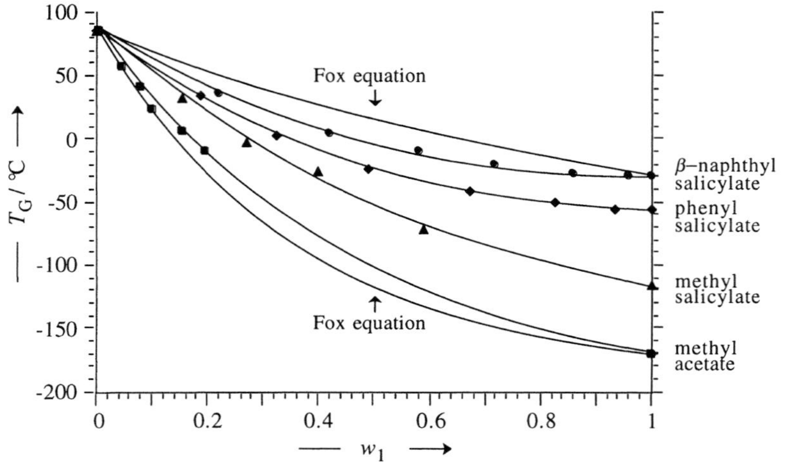

Fig. 13-29 Dependence of the glass temperature on the mass fraction w_1 of added plasticizers for an atactic poly(styrene) [28]. Solid lines through data points: empirical. Top and bottom solid lines without data points: calculations by Fox equation with glass temperatures of polymer (T_G = 85.5°C) and plasticizers β-naphthyl salicylate (T_G = –29.5°C) and methyl acetate (T_G = –170°C).

The smaller the molar mass of plasticizer with similar chemical constitution, the less plasticizer is needed to produce the same lowering of the glass temperature. An example is the series of similar esters in Fig. 13-29:

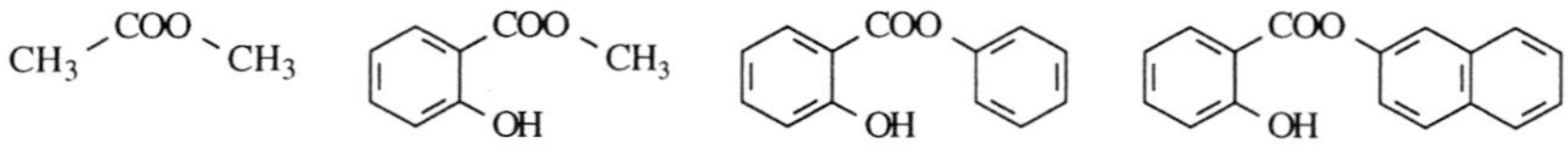

methyl acetate methyl salicylate phenyl salicylate β–naphthyl salicylate

The decrease in glass temperature with increasing proportion of plasticizer is usually described by the **Fox equation**:

(13-28) $1/T_G = (w_1/T_{G,1}) + (w_2/T_{G,2})$

For the systems of Fig. 13-29, the Fox equation delivers too high values for β-naphthyl salicylate (shown) and phenyl salicylate (not shown), approximately correct ones for methyl salicylate (not shown), and too low ones for methyl acetate (shown).

Water is a very efficient plasticizer for hydrophilic polymers such as cellulose and aliphatic polyamides. Dry cellulose has a glass temperature of ca. 225°C but wet cellulose one of less than 0°C. This decrease of the glass temperatures causes sweat-stained cotton fabrics to crease; conversely, the same effect can by used to iron cotton fabrics with steam irons. Even quartz can be plasticized by water. Synthetic quartz takes up ca. 0.1 % H_2O and can be deformed at 400°C without fracture; dry quartz melts at 1713°C.

Even gases can act as plasticizers for glassy polymers. At a pressure of 0.68 MPa (≈ 6.8 atm), CO_2 lowers the glass temperature of polycarbonate A by ca. 9°C. However, such large effects are only produced by gases with high critical pressures. Plasticizing by gases with low critical pressures requires much higher pressures of several hundreds of megapascals (thousands of atmospheres) for marked effects.

Internal Plasticizing

Copolymerization of two monomers leads to **internal plasticizing** if the glass temperature of the parent polymer is decreased. However, glass temperatures of copolymers may also increase relative to that of the parent polymer or even pass through a maximum or minimum with increasing content of comonomeric units.

Many theories have been advanced to describe the variation of glass temperatures with the proportion of comonomeric units. The Gordon-Taylor theory is based on the volume demands of components whereas the Couchman theory assumes entropy changes as the cause of the variation of glass temperatures with the content of comonomeric units.

The **Gordon-Taylor theory** assumes that specific volumes v_G of copolymers in the glass "state" ($T < T_G$) and rubber state ($T > T_G$) both depend on the mass fraction of the two types of comonomeric units, i.e., $v_G = w_1 v_{G,1} + w_2 v_{G,2}$ and $v_R = w_1 v_{R,1} + w_2 \cdot v_{R,2}$, respectively. Additivity of volumes leads with the cubic expansion coefficients of the homopolymers in the rubber, $\beta_{R,1}$ and $\beta_{R,2}$, and the glass, $\beta_{G,1}$ and $\beta_{G,2}$, to

$$(13\text{-}29) \qquad T_G = \frac{w_1 T_{G,1} + k w_2 T_{G,2}}{w_1 + k w_2} \quad ; \quad k = \frac{\beta_{R,2} - \beta_{G,2}}{\beta_{R,1} - \beta_{G,1}} \quad ; \quad w_1 + w_2 \equiv 1$$

For example, the dependence of the glass temperatures of styrene-butadiene copolymers on the mass fraction x_1 of butadiene units (Fig. 13-30, left) can be described by $k = 1.67$ if mole fractions are converted to mass fractions by $w_1 = x_1/[x_1 + x_2(M_2/M_1)]$ where M_2 and M_1 are the molar masses of the styrene and butadiene units, respectively.

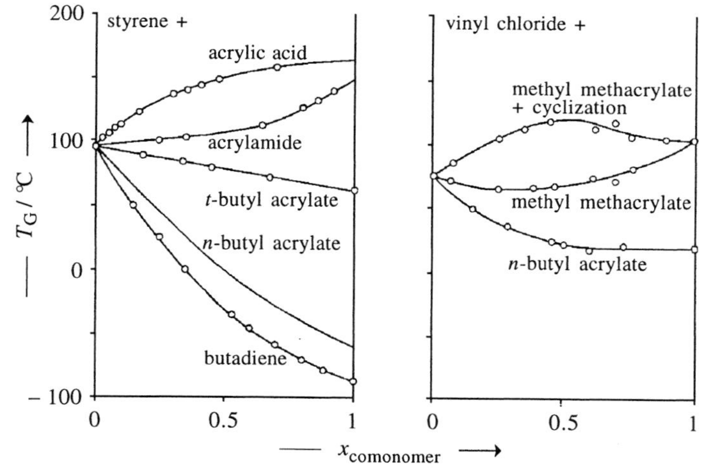

Fig. 13-30 Glass temperatures of copolymers from the free-radical polymerization of styrene (left) [29] and vinyl chloride (right) as a function of the mole fraction of comonomeric units in the copolymer. The copolymer of vinyl choride and methyl methacrylate was subsequently cyclized.

Eq.(13-29) can be linearized to $T_G = T_{G,1} + k[T_{G,2} - T_G](w_2/w_1)$ which is only useful for small values of w_2 because of the propagation of errors for large values of w_2/w_1. For $k = 1$, Eq.(13-29) reduces to Eq.(13-30):

$$(13\text{-}30) \qquad T_G = w_1 T_{G,1} + w_2 T_{G,2}$$

The **Couchman theory** treats glass temperatures of copolymers as thermodynamic second-order transitions. It neglects enthalpic terms and calculates the molar entropy S_m of the system from the contributions of the components and the excess entropy $\Delta S_{m,exc}$:

$$(13\text{-}31) \qquad S_m = x_1 S_{1,m} + x_2 S_{2,m} + \Delta S_{m,exc}$$

At their glass temperatures, the two components 1 and 2 have molar entropies of $S_{G,1,m}$ and $S_{G,2,m}$, respectively. Introduction of the molar heat capacities $C_{p,m}$ of the corresponding homopolymers leads to

$$(13\text{-}32) \qquad S_m = x_1 \left[S_{G,1,m} + \int_{T_{G,1}}^{T} C_{p,1,m} \mathrm{d} \ln T \right] + x_2 \left[S_{G,2,m} + \int_{T_{G,2}}^{T} C_{p,2,m} \mathrm{d} \ln T \right] + \Delta S_{m,exc}$$

Eq.(13-32) is written for the glass "state" at $T < T_G$ and also for the rubber state at $T > T_G$. Both equations are then equated because the resulting expressions for $S_{G,m}$ and $S_{R,m}$ must equal each other at T_G. Some other parameters also equal each other at T_G, i.e., the molar entropies and mole fractions of each of the components, e.g., $\Delta S_{G,1,m} = \Delta S_{R,1,m}$, and $x_{G,1} = x_{R,1}$. For a regular mixture, the two mixing entropies $\Delta S_{mix,G}$ and $\Delta S_{mix,R}$ must equal each other. Also, the two isobaric molar heat capacities must be the same in the two "states", i.e., $\Delta C_{p,i,m} = \Delta C_{p,i,m\text{-},G} = \Delta C_{p,i,m,R}$. Hence, Eq.(13-32) converts to

$$(13\text{-}33) \qquad x_1 \int_{T_{G,1}}^{T} \Delta C_{p,1,m} \mathrm{d} \ln T + x_2 \int_{T_{G,2}}^{T} \Delta C_{p,2,m} \mathrm{d} \ln T = 0$$

Integration of Eq.(13-33) and conversion of mole fractions to mass fractions as well as conversion of molar heat capacities to specific isobaric heat capacities, Δc delivers the **Couchman equation,**

$$(13\text{-}34) \qquad \ln T_G = \frac{w_1 \ln T_{G,1} + k w_2 \ln T_{G,2}}{w_1 + k w_2} \quad ; \quad k = \Delta c_{p,2}/\Delta c_{p,1} \quad ; \quad w_1 + w_2 \equiv 1$$

Eq.(13-34) describes linear, concave, and convex dependencies of glass temperatures on mass fractions (Fig. 13-30). For $k = 1$, it reduces to the **Pochan equation**

$$(13\text{-}35) \qquad \ln T_G = w_1 \ln T_{G,1} + w_2 \ln T_{G,2}$$

The Couchman equation considers only entropic effects. However, enthalpic effects can produce maxima and minima in $T_G = f(w_1)$ curves. For example, bipolymers of vinyl chloride (VC) and methyl methacrylate (MMA) have at $w_1 = 0.33$ a $T_G = 63°C$ that is lower than the glass temperatures of both homopolymers (Fig. 13-30). Conversely, the cyclized VC-MMA copolymer shows a maximum in the $T_G = f(x_{comonomer})$ curve.

Enthalpic interactions are mainly intramolecular between adjacent monomeric units of the same chain; hence, they are affected by sequence statistics. The glass temperature depends here not only on the mass fractions w_1 and w_2 of monomeric units 1 and 2 but, in the simplest case, also on the probability of the presence of diads of monomeric units, p_{11}, p_{12}, p_{21}, and p_{22}, with $p_{11} + p_{12} + p_{21} + p_{22} \equiv 1$. Eq.(13-29) becomes then

Table 13-10 Glass temperatures of homopolymers $(a)_n$ and $(b)_n$ from monomers A and B, their alternating copolymers $(a\text{-}alt\text{-}b)_n$, and their 50/50 statistical copolymers $(a_{50}\text{-}stat\text{-}b_{50})_n$.

Comonomers A	B	T_G/°C $(a)_n$	$(a\text{-}alt\text{-}b)_n$	$(b)_n$	$(a_{50}\text{-}stat\text{-}b_{50})_n$
Ethylene	propylene	–80		–10	–65
	acrylic acid	–80	71	106	
	tetrafluoroethylene	–80	110	127	–13
	acrylamide	–80	120	165	
Methyl methacrylate	butyl methacrylate	105	13	20	
	vinyl chloride	105	50	77	70
	acrylonitrile	105	84	97	82
	α-methyl styrene	105	141	177	
Acrylonitrile	ethylene	97	23	–80	
	propylene	97	52	–10	
	styrene	97	114	100	102
	α-methyl styrene	97	122	177	123
Formaldehyde	ethylene oxide	–83	–63	–67	

$$\frac{1}{T_G} = \frac{w_1 p_{11}}{T_{G,11}} + \frac{w_1 p_{12} + w_2 p_{21}}{T_{G,12}} + \frac{w_2 p_{22}}{T_{G,22}} \qquad (13\text{-}36)$$

The glass temperature $T_{G,12}$ of the alternating bipolymer thus determines the presence of maxima and minima in the composition dependence of glass temperatures of bipolymers. As a result, glass temperatures of 50:50 statistical bipolymers may differ markedly from those of alternating ones (Table 13-10).

Crosslinked Polymers

Glass temperatures are caused by the onset of movements of polymer segments consisting of 20-60 chain atoms (Table 13-4). Crosslinked polymers with larger segments between crosslinks therefore have the same glass temperatures as uncrosslinked polymers. The less flexible shorter segments cause the glass temperature of the crosslinked specimen to increases with the inverse segment length if the segment length exceeds the segment length for unimpeded segmental movements (ca. 60-100 chain units, see Table 13-4). However, the segment length L_{seg} between crosslinking sites is not a directly accessible quantity. It can be calculated theoretically from swelling equilibria, though.

A measurable quantity is the specific volume. Crosslinking replaces the long-range van der Waals interactions between chains by shorter covalent bonds within the copolymer, which reduces the specific volume and probably also the free volume. Changes of specific volumes with temperature can be expressed by cubic expansion coefficients β.

The problem with this approach is that the constitution of crosslinking units usually differs from that of non-crosslinking units; an example is the crosslinking of styrene by unsaturated polyester molecules. The variation of glass temperature with increasing concentration of crosslinking sites is therefore affected by two factors: a *constitutional effect* caused by the difference in cubic expansion coefficients of the two homopolymers (for example, poly(styrene) and poly(divinyl styrene)) and a *crosslinking effect* resulting from the shortening of segments between crosslinking sites.

This approach involves a lengthy derivation. It delivers the **Fox-Loshaek equation** which, after ignoring a minor additive term, can be written to a first approximation as

$$(13\text{-}37) \qquad T_{G,c,x} = T_{G,c,u} + \left(\frac{(\beta_{c,u} - \beta_{c,x}) T_{G,c,u} \overline{M}_{m,c,u}}{(\beta_{c,u} - \beta_{c,x})[1 - \overline{M}_{m,c,u}(n_x/m_c)] - \beta_{G,c,u}} \right) \left[\frac{n_x}{m_c} \right]$$

where T_G = glass temperature and β = cubic thermal expansion coefficient. The indices refer to the glass temperature (G) and the copolymer (c) as an uncrosslinked (u) or crosslinked (x) substance. The expression in parentheses is *not* a proportionality constant since $\overline{M}_{m,c,u}$ is the average molar mass of the monomer mixture (m) which varies with the amount n_x of the crosslinking comonomer (in moles) per total mass m_c (in grams) of the copolymer. Note that $T_{G,c,u}$ is not the glass temperature of the *main chain* of the uncrosslinked (u) copolymer (c) and not the glass temperature of that copolymer *per se* since it refers to the absence of crosslinks.

Hence, Eq.(13-37) simplifies if the two types of comonomeric units have nearly identical chemical constitutions. An example is the copolymer of methyl methacrylate and ethylene glycol dimethacrylate:

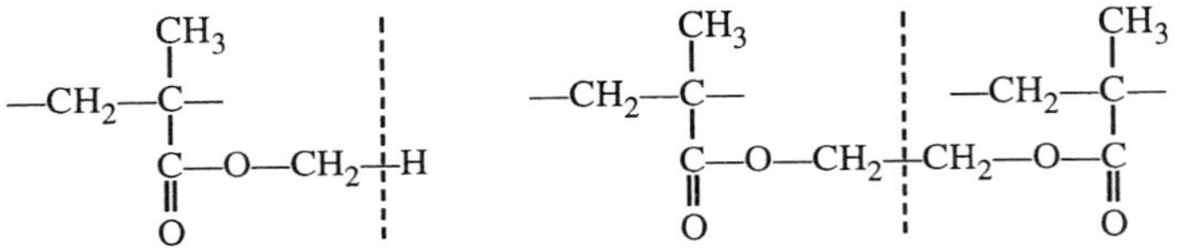

methyl methacrylate unit ethylene glycol dimethacrylate unit

In this case, the glass temperature $T_{G,c,u}$ of the uncrosslinked copolymer is (nearly) identical with the glass temperature $T_{G,0}$ of the homopolymer, e.g., poly(methyl methacrylate), and the average molar mass of the monomer mixture, $\overline{M}_{m,c,u}$, is identical with the molar mass M_0 of the uncrosslinkable comonomer, e.g., methyl methacrylate. The term in parentheses becomes a material constant and Eq.(13-36) for *high*-molecular weights where the proportion of tangling ends can be neglected becomes

$$(13\text{-}38) \qquad T_{G,c,x} = T_{G,0} + \left(\frac{(\beta_{c,u} - \beta_{c,x}) T_{G,0} M_0}{(\beta_{c,u} - \beta_{c,x})[1 - M_0(n_x/m_c)] - \beta_{G,c,u}} \right) \left[\frac{n_x}{m_c} \right] = T_{G,0} + K \left[\frac{n_x}{m_c} \right]$$

Plots of $T_{G,c,x} = f(n_x/m_c)$ of high-molecular weight polymers do indeed deliver straight lines for two pairs of similar comonomeric units such as methyl methacrylate + ethylene glycol dimethacrylate on one hand and styrene + divinyl benzene on the other (Fig. 13-31). At $(n_x/m_c) \rightarrow 0$, they give $T_{G,0}$ as predicted by Eq.(13-38).

13.5.5 Static and Dynamic Glass Temperatures

The previous sections were concerned with so-called **static glass temperatures** which are obtained from slow measurements of the temperature dependence of physical quantities such as heat capacities (including thermal analysis), volume expansion (via densities

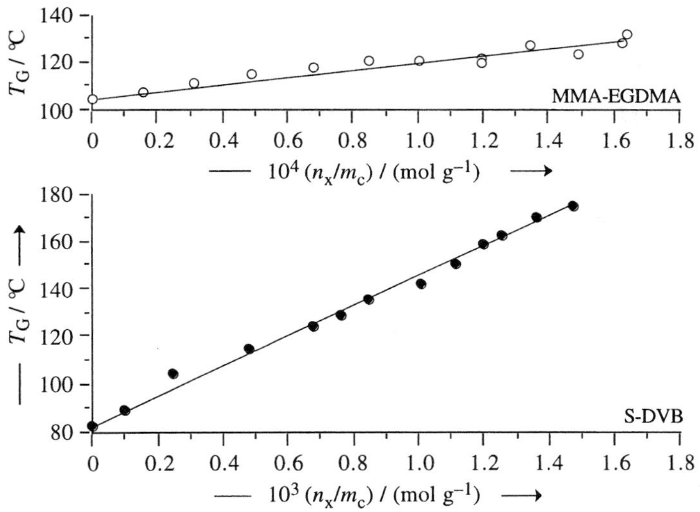

Fig. 13-31 Fox-Loshaek plot [30] of glass temperatures T_G of crosslinked polymers as a function of the amount n_x of crosslinking agent (in moles) per mass m (in grams) of polymer.

Top: copolymers of methyl methacrylate and ethylene glycol dimethacrylate [31].

Bottom: copolymers of styrene and divinyl benzene [32]; the glass temperature of this specimen of poly(styrene) was reported as 82.5°C which is far lower than the presently accepted value of 100°C for *high*-molecular weight polymers.

or X-ray measurements), or refractive indices (because of the Lorentz-Lorenz relationship between volume and refractive index). These measurements are (quasi)static because segments have sufficient time to adjust to static deformations or to low fequencies of dynamic methods.

Conversely, **dynamic glass temperatures** are obtained by high-frequency methods such as broad-line nuclear magnetic resonance and mechanical or dielectric spectroscopy. In these methods, movements of polymer segments cannot follow the rapidly changing field and the segments appear as relatively rigid ones with a correspondingly higher glass temperature.

Static glass temperatures can be converted to dynamic ones and vice versa by the **Williams-Landel-Ferry equation (WLF equation)**. In this approach, glass transformations are assumed to be relaxations that depend on the free volume V_f in the same manner as the melt viscosity η. The dependence of η on V_f is empirically described by the **Doolittle equation**

$$(13\text{-}39) \qquad \ln \eta = \ln A + B(V - V_f)/V_f \quad ; \quad \eta = A \exp B\,[(V - V_f)/V_f\}$$

where the free volume V_f and the total volume V are actually related to masses and A, B = constants. The volume fractions are $\phi_f = V_f/V$ at the temperature T and $\phi_{f,o} = V_{f,o}/V_o$ at a reference temperature T_o. The change of viscosity with temperature is described by a **shift factor**, $a_T = (\eta T_o \rho_o)/(\eta_o T \rho)$, that corrects for thermal expansion by introducing the densities ρ and ρ_o (in mass/volume) for the two temperatures T and T_o. The shift factor corresponds to the ratio t/t_o of relaxation times.

Inserting the Doolittle equation for temperatures T and T_o in the shift factors leads to

$$\text{(13-40)} \qquad \lg a_T = \frac{B}{2.303}\left(\frac{1}{\phi_f} - \frac{1}{\phi_{f,o}}\right) + \lg\left(\frac{T_o \rho_o}{T\rho}\right) \approx \frac{B}{2.303}\left(\frac{1}{\phi_f} - \frac{1}{\phi_{f,o}}\right)$$

It is also assumed that the volume fraction ϕ_f of free volume increases *linearly* with temperature (in K) according to $\phi_f = \phi_{f,o} + \beta_f(T - T_o)$. The expansion coefficient of the free volume, ϕ_f, approximates the true cubic expansion coefficient, $\beta = (1/V)(\partial V/\partial T)$, which describes an *exponential* increase of volume with temperature. Because of this approximation, the WLF equation, Eq.(13-41), is limited to a temperature range of $T_o < T < (T_o + 100\text{ K})$.

Introduction of $\beta_f = (\phi_f - \phi_{f,o})/(T - T_o)$ in Eq.(13-40) delivers the **WLF equation**:

$$\text{(13-41)} \qquad \lg a_T = \frac{-[B/(2{,}303\,\phi_{f,o})][T - T_o]}{[\phi_{f,o}/\beta_f] + [T - T_o]} = \frac{-K[T - T_o]}{K' + [T - T_o]} = \lg t - \lg t_o$$

Eq.(13-41) applies to all relaxation processes. The adaptable parameters K, K', and $\phi_{f,o}$ are often considered to be *universal parameters* for all polymers. For a temperature range of $T_o = T_G + 50$ K, they were originally assumed as $K = 17.44$ K, $K' = 51.6$ K, and $\phi_{f,o} = 0.025$ and more recently as $K = 8.86$ K, $K' = 101.6$ K, and $\phi_{f,o} = 0.025$. In fact, different parameters have to be used for each polymer, for example,

Poly(isobutylene)	$K = 16.6$ K	$K' = 104.4$ K	$T_G = 205$ K
Poly(acetaldehyde)	$K = 14.5$ K	$K' = 24.0$ K	$T_G = 243$ K
Poly(styrene)	$K = 13.5$ K	$K' = 48.7$ K	$T_G = 373$ K
Poly(dimethylsiloxane)	$K = 6.1$ K	$K' = 69.0$ K	$T_G = 150$ K

The WLF equation allows one to calculate the static glass temperature from dynamic glass temperatures that were obtained at various frequencies. The shift factor a_t is obtained from the difference of logarithms of deformation times, t_d.

The higher the frequency of the method, the shorter is the deformation time and the less flexible the polymer becomes as indicated by the higher glass temperatures (Table 13-11). For example, poly(methyl methacrylate) behaves at 140°C as a glass if subjected to a fast process (dropped steel sphere) but as a rubber to a slowly penetrating needle.

Table 13-11 Frequencies ν of various methods and deformation times $t_d = 1/\nu$ and glass temperatures T_G of linear poly(ethylene) (PE), poly(methyl methacrylate) (PMMA), and poly(3,3-bischloromethyl-oxacyclobutane) (PCMOB).

Method	ν/Hz	t_d/s	T_G/°C PE	PMMA	PCMOB
Thermal expansion (dilatometry)	10^{-4}	10000	−125	110	
Penetrometry	10^{-2}	100		120	7
Slow tensile testing	3	0.3	−123		15
Mechanical vibration	90	0.011			25
Electrical spectroscopy	10^3	0.001	−100		32
Rebound elasticity	10^5	0.00001		160	

13.6 Other Transformations and Relaxations

Polymers often show several other characteristic temperatures T_{ch} besides melting and glass temperatures (Section 13.2.4). The temperatures T_{ch} can either be higher or lower then T_M or T_G but in all cases, inverse temperatures are functions of inverse frequencies (Fig. 13-8). Molecular interpretations are difficult and usually controversial.

One of the few clear-cut cases is the behavior of poly(cyclohexyl methacrylate). At a frequency of 10^{-4} Hz, this polymer shows a mechanical loss maximum at the relaxation temperature T_{rlx} = –125°C (Fig. 13-32). With increasing frequency, the maximum shifts to higher temperatures; for $\nu = 8 \cdot 10^5$ Hz, it is at ca. T_{rlx} = 80°C. As usual, $1/T_{rlx}$ is a linear function of the logarithm of frequencey (insert in Fig. 13-32).

Investigations of many different chemical compounds showed that these loss maxima are specific for the cyclohexyl group since, for example, the same function $1/T_{rlx} = f(1/\nu)$ is found for poly(cyclohexyl methacrylate), poly(cyclohexyl acrylate), and cyclohexanol but not for poly(phenyl acrylate). The loss maxima must therefore be caused by the boat-chair conversion of cyclohexane rings.

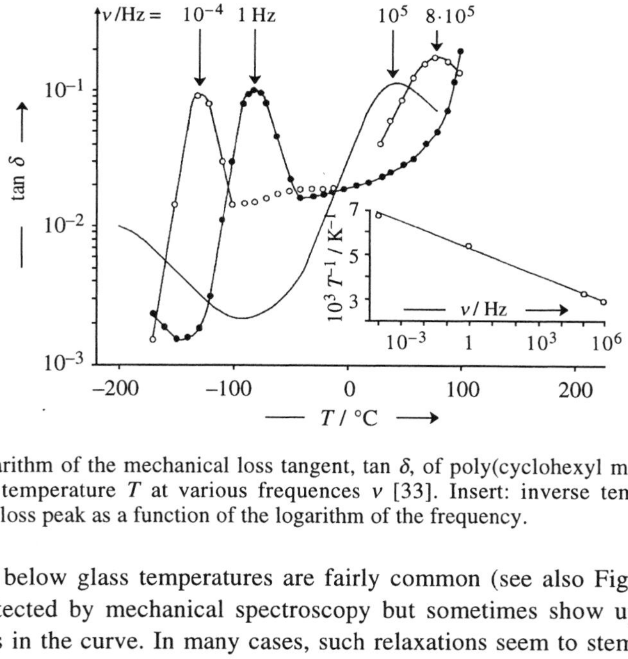

Fig. 13-32 Logarithm of the mechanical loss tangent, tan δ, of poly(cyclohexyl methacrylate) as a function of the temperature T at various frequences ν [33]. Insert: inverse temperature of the maximum of the loss peak as a function of the logarithm of the frequency.

Relaxations below glass temperatures are fairly common (see also Fig. 17-26). They are usually detected by mechanical spectroscopy but sometimes show up in DSC diagrams as bends in the curve. In many cases, such relaxations seem to stem from coupled movements of very short-chain segments. Other relaxations are caused by moisture that was taken up from humid air. The latter effect decreases with increasing crystallinity of polymers because water molecules can only be taken up by amorphous regions.

Some crystalline polymers show relaxations (or transformations?) above their melting temperatures, for example, at ca. 1.2 T_M (in K). This transformation has been interpreted as the desegregation of smectic structures in melts.

Historical Notes to Chapter 13

Crystallization kinetics:
M.Avrami, J.Chem.Phys. **7** (1939) 1103; **8** (1940) 212; **9** (1941) 117

Discovery of chain folding and single crystals:
see Historical Notes to Chapter 7

Literature to Chapter 13

13.1 FUNDAMENTALS (overviews and methods)

J.K.Gillham, Torsional Braid Analysis - A Semimicro Thermomechanical Approach to Polymer Characterization, Crit.Revs.Macromol.Sci. **1** (1972) 83

J.-M.Braun, J.E.Guillet, Study of Polymers by Inverse Gas Chromatography, Adv.Polym.Sci. **21** (1976) 107

W.W.Wendland, Thermal Analysis, Wiley, New York 1986

B.Wunderlich, Thermal Analysis, Academic Press, Boston 1990

Y.Godovsky, Thermophysical Properties of Polymers, Springer, Berlin 1991

–, The Brandon Worldwide Monomer Reference Guide and Sourcebook, Brandon Associates, Merrimack (NJ), 3rd ed. 1993 (data on 3800 monomers, including T_G of their homopolymers)

V.B.F.Mathot, Ed., Calorimetry and Thermal Analysis of Polymers, Hanser, Munich 1994

V.A.Bershstein, V.M.Egorov, Differential Scanning Calorimetry of Polymers, Harwood, New York 1994

G.Höhne, W.Hemminger, H.-J.Flammersheim, Differential Scanning Calorimetry, Springer, Heidelberg 1996

E.A.Turi, Ed., Thermal Characterization of Polymeric Materials, Academic Press, San Diego, 2nd ed. 1997

T.Hatakeyama, L.Zhenhai, Thermal Analysis. Fundamentals and Applications to Polymer Science, Wiley-VCH, Weinheim, 2nd ed. 1999

S.Z.D.Cheng, Ed., Handbook of Thermal Analysis and Calorimetry, Vol. 3, Applications to Polymers and Plastics, Elsevier, Amsterdam 2003

G.W.Ehrenstein, G.Riedel, P.Trawiel, Thermal Analysis of Plastics; Principles and Practice, Hanser, Munich 2004

13.2 MOBILITY OF MOLECULES

R.F.Boyer, S.E.Keinath, Eds., Molecular Motion in Polymers by ESR, Harwood Academic Publ., Chur, Switzerland, 1980

R.T.Bailey, A.M.North, R.A.Pethrick, Molecular Motion in High Polymers, Clarendon Press, Oxford 1981

J.M.O'Reilly, M.Goldstein, Ed., Structure and Mobility in Molecular and Atomic Glasses, Ann.N.Y.Acad.Sci. **371** (1981)

13.2.2 HEAT CAPACITY

B.Wunderlich, H.Baur, Heat Capacities of Linear High Polymers, Adv.Polym.Sci. **7** (1970) 151

13.2.3 THERMAL CONDUCTIVITY

D.R.Anderson, Thermal Conductivity of Polymers, Chem.Revs. **66** (1966) 677

W.Knappe, Wärmeleitung in Polymeren, Adv.Polym.Sci. **7** (1971) 477

D.Hands, The Thermal Transport Properties of Polymers, Rubber Chem.Technol. **50** (1977) 480

D.M.Bigg, Thermal Conductivity of Heterophase Polymer Compositions, Adv.Polym.Sci. **119** (1995) 1

13.3 MELTING

H.G.Zachmann, Das Kristallisations- und Schmelzverhalten hochpolymerer Stoffe, Fortschr.Hochpolym.Forschg.-Adv.Polym.Sci. **3** (1961/64) 581

B.Wunderlich, Macromolecular Physics, Vol. **3**, Crystal Melting, Academic Press, New York 1980

13.4 TRANSITIONS OF LIQUID CRYSTALS (see also Chapter 8)
B.Wunderlich, S.Grebowicz, Thermotropic Mesophases and Mesophase Transitions of Linear, Flexible Macromolecules, Adv.Polym.Sci. **60/61** (1984) 1

13.5 GLASS TRANSFORMATIONS
R.F.Boyer, The Relation of Transition Temperatures to Chemical Structure in High Polymers, Rubber Chem.Technol. **36** (1963) 1303
A.J.Kovacs, Transition vitreuse dans les polymères amorphes. Etude phénoménologique, Fortschr. Hochpolym.Forschg.-Adv.Polym.Sci. **3** (1961/64) 394
N.W.Johnston, Sequence Distribution - Glass Transition Effects, J.Macromol.Sci.-Revs.Macromol. Chem. **C 14** (1976) 215
M.Goldstein, R.Simha, Eds., The Glass Transition and the Nature of the Glassy State, Ann.N.Y.Acad.Sci. **279** (1976) 1
Y.Lipatov, The Iso-Viscous State and Glass Transitions in Amorphous Polymers: New Development of the Theory, Adv.Polym.Sci. **26** (1978) 63
M.Pietrallam, W.Pechhold, Relaxation in Polymers, Springer, New York 1990

13.6 OTHER TRANSFORMATIONS AND RELAXATIONS
A.Hiltner, E.Baer, Relaxation Processes at Cryogenic Temperatures, Crit.Revs.Macromol.Sci. **1** (1972) 215
G.M.Bartenev, Yu.V.Zenlenev, Eds., Relaxation Phenomena in Polymers, Halsted, New York 1974
A.M.North, Relaxations in Polymers, Internat.Rev.Sci.-Phys.Chem. [2] **8** (1975) 1
D.J.Meier, Ed., Molecular Basis of Relaxations and Transitions of Polymers (= Midland Macromolecular Monographs **4**), Gordon and Breach, New York 1978
S.Matsuoka, Relaxation Phenomena in Polymers, Hanser, Munich 1993

References to Chapter 13

[1] B.Wunderlich, M.Möller, J.Grebowicz, H.Baur, Adv.Polym.Sci. **87** (1988) 1, Figs. 2.7-2.9
[2] R.B.Cassel, M.Wiese, American Laboratory **35**/1 (2003) 13, Fig. 1
[3] F.E.Karasz, H.E.Bait, J.M.O'Reilly, General Electric Report 68-C-001 (1968)
[4] W.Knappe, Adv.Polym.Sci. **7** (1971) 477, Figs. 2, 18, and 19
[5] H.W.Starkweather, Jr., J.Macromol.Sci.-Phys. **B 2** (1968) 781; data of Fig. 1
[6] S.H.Kim, L.Mandelkern; reported by L.Mandelkern in G.Allen, J.C.Bevington, Eds., Comprehensive Polymer Science **2** (1989) 363, Fig. 4
[7] R.Chiang, P.J.Flory, J.Am.Chem.Soc. **83** (1961) 2857, Fig. 2
[8] B.Wunderlich, Kunststoffe **55** (1965) 333, Fig. 1
[9] From a compilation of M.Rothe, in J.Brandrup, E.H.Immergut, E.A.Grulke, Eds., Polymer Handbook, J.Wiley & Sons, New York, 4th ed. (1999), IV/2, Tables 1.1.1 and 1.1.2
[10] O.B.Edgar, R.Hill, J.Polym.Sci. **8** (1952) 1, Fig. 10
[11] Y.Fu, ATHAS, Eighth Report, University of Tennessee, Knoxville (TN) 1995; reported in [12]
[12] B.Wunderlich, J.Grebowicz, Adv.Polym.Sci. **60-61** (1984) 1
[13] H.Yamada, T.Iguchi, A.Hirao, S.Nakahama, J.Watanabe, Macromolecules **28** (1995) 50, Table 3
[14] H.Stevens, G.Rehage, H.Finkelmann, Macromolecules **17** (1984) 851, data recalculated from Fig. 6
[15] A.J.Kovacs, J.Polym.Sci. **30** (1958) 131, Fig. 5
[16] J.Rieger, Kunststoffe **85** (1995) 528, data of Fig. 3
[17] J.L.Keddie, R.A.L.Jones, R.A.Cory, Europhys.Letters **27**/11 (1994) 59; J.A.Forrest, K.Dalnoki-Veress, Adv.Coll.Interface Sci. **94** (2001) 167, curve taken from Fig. 1 (smoothed experimental data of many authors)
[18] S.Herminghaus, R.Seemann, D.Podzimek, K.Jacobs, Nachr.Chem. **49**/12 (2001) 138, curve of Fig. 3 (data points omitted)
[19] J.M.G.Cowie, P.M.Toporowski, Eur.Polym.J. **4** (1968) 621, Table 2
[20] G.Pezzin, F.Zilio-Grandi, P.Sanmartin, Eur.Polym.J. **6** (1970) 1053, Table 1
[21] J.M.G.Cowie, I.J.McEwen, Polymer **14** (1973) 423, Table 2

[22] F.Rietsch, D.Daveloose, D.Froelich, Polymer **17** (1976) 859, Tables 1 and 2
[23] S.J.Clarson, K.Dodgson, J.A.Semlyen, Polymer **26** (1985) 930, Tables 1 and 2, Fig. 3
[24] W.A.Lee, G.J.Knight, Brit.Polym.J. **2** (1970) 73, Tables I and II
[25] T.-B.He, J.Appl.Polym.Sci. **30** (1985) 4319, Fig. 1, Table 1, and additional data
[26] K.L.Wooley, C.J.Hawker, J.M.Pochan, J.M.J.Fréchet, Macromolecules **26** (1993) 1514, Table 1
[27] S.Uppuluri, P.R.Dvornic, N.C.Beck Tan, G.Hagnauer, Report ARL-TR-1774, Army Research Laboratory, Aberdeen Proving Ground (MD) 1999
[28] E.Jenckel, R.Heusch, Kolloid-Z. **130** (1953) 89, Table 1
[29] K.H.Illers, Ber.Bunsenges. **70** (1966) 353, data of Fig. 1
[30] T.G.Fox, S.Loshaek, J.Polym.Sci. **15** (1955) 371, Eq.(16) with revised physical units of terms
[31] S.Loshaek, J.Polym.Sci. **15** (1955) 391, Table II
[32] K.Ueberreiter, G.Kanig, J.Chem.Phys. **18** (1950) 399, recalculated data of Fig. 4
[33] J.Heijboer, in D.J.Meier, Ed., Molecular Basis of Transitions and Relaxations, Gordon and Breach, London 1978, p. 75, data of Fig. 5

14 Transport in Polymers

14.1 Introduction

Brownian movements cause small molecules to collide, which leads them to rotate about their axes and to change places with surrounding molecules. Polymer molecules experience the same *global* effects but, in addition, also several local ones which, for flexible macromolecules, range from conversions of microconformations (for example, trans → gauche) to changes of macroconformations of chain segments and deformations of whole chains. These changes are caused not only by thermal effects, i.e., micro-Brownian and macro-Brownian movements, but also by applied mechanical and electrical forces such as shear deformations and extensions.

Deviations from preferential positions of chain units and chain segments by such forces are reversed by relaxation to equilibrium positions. Required relaxation times range from picoseconds for microconformations to seconds for macroconformations. Global effects on whole macromolecules can be investigated by applying static or dynamic mechanical (Chapter 17) or electrical (Volume IV) force fields. Local effects can be studied by special spectroscopic techniques (such as pulsed field gradient spin-echo NMR), quasi-elastic light scattering (Section 11.1.2), neutron spin-echo spectroscopy, and many more.

The various *coordinated* movements of units and segments are treated as distinguishable types with defined relaxation times τ_q. Relaxations of the first kind ($q = 1$) comprise a movement of the whole molecule without any change of its shape (Fig. 14-1). Such a movement has the longest relaxation time since it requires the greatest number of coordinated movements. It is generally assumed that Rouse relaxation times (Section 11.1.3) correspond to such coordinated movements of whole chains.

Relaxations of the second kind ($q = 2$) arise if chains ends move in opposite directions (Fig. 14-1). In relaxations of the third kind ($q = 3$), both chain ends proceed in the same direction but opposite to the movement of the center of the chain, etc. According to classical mechanics, spectra of relaxation times of a molecularly uniform polymer must therefore show a series of discrete signals. However, polymers always give rise to continuous spectra because (I) the various types of relaxations, q, are not coupled linearly and (II) polymers are generally not molecularly uniform.

In *polymer solutions*, additional coupled movements are present between polymer segments and solvent molecules because a force acting on a segment generates a movement of the liquid surrounding the segment which in turn acts on the velocities of the other solvent molecules. The movement of one entity is therefore affected by the forces acting on other entities. This **hydrodynamic interaction** is not considered by the **Rouse theory** (pp. 372, 414). However, this theory is applicable to melts because all entities here are of the same kind.

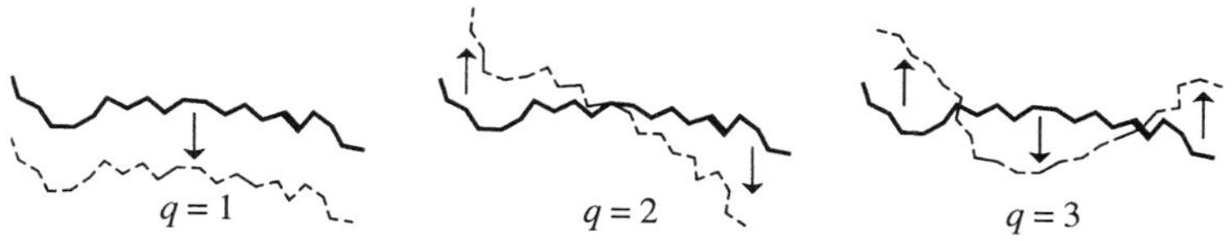

Fig. 14-1 The first three types of movements of a whole chain ($q = 1$ and 2) and its segments ($q \geq 3$).

14.2 Transport in Fluid Polymer Phases

14.2.1 Solvents in Concentrated Polymer Solutions

In polymer solutions, Brownian movements lead to exchanges of positions between different solvent molecules, solvent and polymer molecules, and different polymer molecules even if concentration gradients and/or temperature gradients are absent. This **self-diffusion** is not accompanied by a net transport of mass as it is in the mutual diffusion of, for example, polymer molecules from a polymer solution into a solvent.

The product of solvent viscosity η_1 and self-diffusion coefficient $D_{1,0}$ of *pure solvents* has the unit of a force. For non-self-associating solvents, it is practically independent of the constitution of solvent molecules; examples of $10^7\ \eta_1 D_{1,0}/(\text{g cm s}^{-2})$ at 25°C are 1.35 (CH_2Cl_2), 1.36 ($CH_3COOC_4H_9$), and 1.32 (C_6H_{12}). Higher values are observed for self-associating solvents such as toluene (1.46) and water (2.15).

Self-diffusion coefficients D_1 of solvents in polymer solutions decrease with increasing polymer concentration (Fig. 14-2). In dilute solution, their values have the same order of magnitude as mutual diffusion coefficients in low-molecular weight solutions (Fig. 11-5). In isopropylbenzene solutions of poly(styrene), the self-diffusion coefficient of the solvent is independent of the molecular weight of the polymer (Fig. 14-2). In toluene solutions of the same polymer, however, the high-molecular weight polymer reduces the self-diffusion coefficient of the solvent to the same extent as a crosslinked poly(styrene) does. This difference between isopropylbenzene (IPB) and toluene may be caused by differences in the self-association of the two solvents (probably none in IPB) and similarities and dissimilarities of solvent and polymer structures (isopropylbenzene $(H{-}CH_2)_2CH(C_6H_5)$ is more similar to the monomeric unit $-CH_2CH(C_6H_5)-$ of poly(styrene) than toluene $H{-}CH_2(C_6H_5)$).

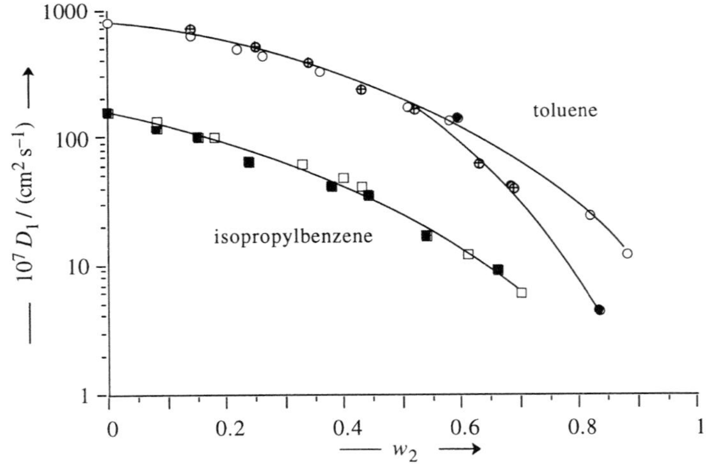

Fig. 14-2 Self-diffusion coefficients D_1 of solvents as a function of the mass fraction w_2 of the polymers. Data for toluene in linear poly(styrene)s of molecular weights 18 000 (○) and 280 000 (⊕) [1] and in a poly(styrene) that was crosslinked by divinylbenzene (●) [2], all at 23°C (all values of D_1 multiplied by 3 for clarity). Data for isopropylbenzene in linear poly(styrene)s of molecular weights of 18 000 (□) and 280 000 (■) at 25°C [3].

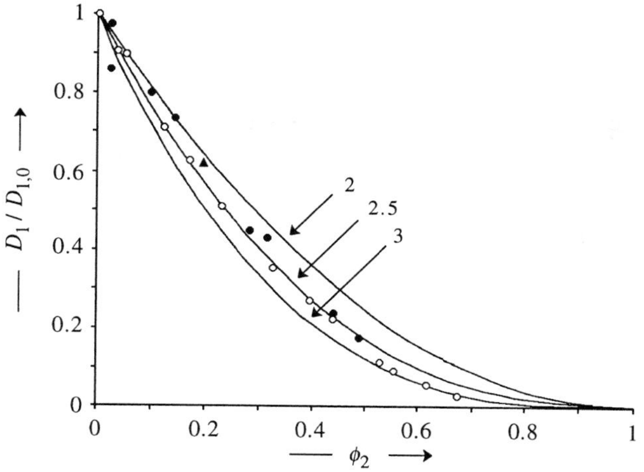

Fig. 14-3 Relative self-diffusion coefficients $D_1/D_{1,0}$ of (O) toluene in poly(styrene) of molecular weight 280 000 (data of Fig. 14-2), (●) water in poly(vinyl pyrrolidone) [4], and (▲) water in an agarose gel [5]. Numbers indicate exponents in Eq.(14-1).

Concentration dependences of self-diffusion coefficients are practically parallel (Fig. 14-2) but shifted proportionally to the solvent viscosity. Since viscosity affects both self-diffusion coefficients of pure solvents, $D_{1,0}$, and self-diffusion coefficients of solvents in polymer solutions, relative self-diffusion coefficients, $D_1/D_{1,0}$, are independent of the solvent (Fig. 14-3). They scale with the volume fraction ϕ_2 of the polymer according to

(14-1) $$D_1/D_{1,0} = (1 - \phi_2)^{5/2} = \phi_1{}^{2.5}$$

14.2.2 Polymers in Melts

In melts, unperturbed coils of chain molecules are filled with segments of other coils (Section 6.2) because of the very low density of their own segments (Fig. 4-24). As a consequence, the diffusion of a segment occurs by changing places with a segment of another molecule. The resulting self-diffusion of the polymer molecule can be determined by NMR spectroscopy similarly to that of solvent molecules in polymer solutions.

According to the Einstein-Sutherland equation, Eq.(11-10), diffusion coefficients are inversely proportional to the friction coefficient ξ_D of the diffusing molecule, $D_0 = k_BT/\xi_D$. The Rouse theory (p. 372) assumes that the molecule consists of N_{seg} segments with friction coefficients ξ_{seg} each. Since N_{seg} is proportional to the degree of polymerization or the molecular weight, self-diffusion coefficients should be inversely proportional to molecular weights, $D_0 = f(M^{-1})$.

However, a dependence of diffusion constants on the inverse *square* of molar mass is found experimentally for the whole range of molecular weights from several hundreds to several hundred thousands as shown by data for linear poly(ethylene)s (Fig. 14-4) and poly(butadiene)s (Fig. 14-6). Since poly(ethylene) chains are entangled at molar masses of $M > 3480$ g/mol (corresponds to ca. 250 methylene units), an individual poly(ethylene) chain must wind its way through a tangle of chain segments (Section 14.2.3).

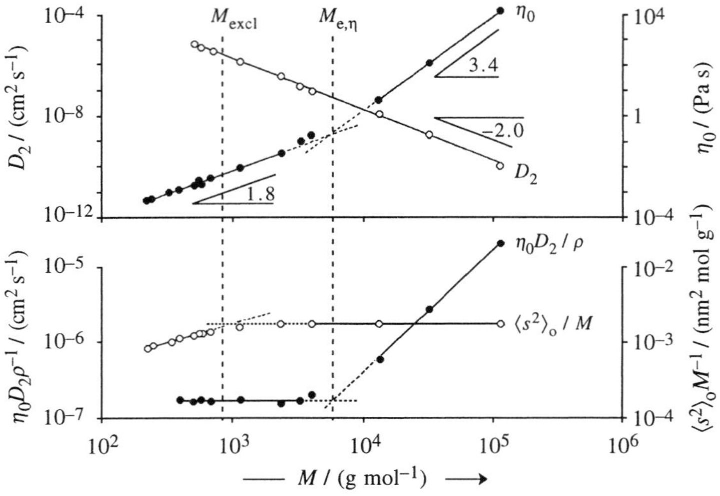

Fig. 14-4 Molar mass dependence of self-diffusion coefficients D_2 [6]; melt viscosities η_0 at zero shear rate [6]; reduced unperturbed mean-square radii of gyration, $\langle s^2\rangle_o/M$, as calculated by the RIS method [7]; and ratios $\eta_0 D_2/\rho$ [8] of melts of alkanes and narrowly distributed linear poly(ethylene)s at 175°C. ρ = density of melt, M_{excl} = molar mass at the onset of excluded volume effects, $M_{e,\eta}$ = viscometric entanglement molar mass (= crossover entanglement molar mass, see also p. 604).

The variation of self-diffusion coefficients with the inverse square of molar masses contrasts with the molar mass dependence of other properties. Reduced unperturbed mean-square radii of gyration, $\langle s_2\rangle_o/M$, of poly(ethylene)s become constant above a molar mass of $M_{ent} \approx 860$ g/mol because the degree of polymerization (ca. 61) is large enough for the formation of unperturbed coils (see Section 4.4.6).

Above this molar mass, coils can be modeled as hydrodynamically equivalent spheres with radii $R_{sph} = \langle s^2\rangle_o^{1/2}$. The self-diffusion of such spheres through the melt as "solvent" with a zero-shear viscosity η_0 can be described by the Einstein-Sutherland equation, $D_2 = k_B T/\xi_{sph}$ (Eq.(11-10)), and the Stokes equation, $\xi_{sph} = 6\,\pi\,\eta_0\langle s^2\rangle_o^{1/2}$ (Eq.(11-11)). Expansion of the right side of $D_2\eta_0 = k_B T/[6\,\pi\,\langle s^2\rangle_o^{1/2}]$ by $\langle s^2\rangle_o/\langle s^2\rangle_o$ and introduction of $k_B = R/N_A$, the molecular volume $V_{mol} = 4\,\pi\,\langle s^2\rangle_o^{3/2}/3$, the density $\rho = m_{mol}/V_{mol}$, and the molar mass $M = m_{mol}N_A$ leads to

$$(14\text{-}2) \qquad D_2\eta_0/\rho = (2/9)\,RT(\langle s^2\rangle_o/M)$$

Since the right side of Eq.(14-2) is independent of the molar mass M, the left side of this equation should also not depend on M and both sides should be numerically identical, which is indeed found for poly(ethylene)s with molar masses in the range between the two vertical broken lines in Fig. 14-4: $D_2\eta_0/\rho = (1.68 \pm 0{,}06)\cdot 10^{-7}$ (cm² s⁻¹)² *versus* (2/9) $RT(\langle s^2\rangle_o/M) = 1.67\cdot 10^{-7}$ (cm² s⁻¹)².

Eq.(14-2) was obtained by a quasi-static derivation. It agrees with the result of hydrodynamic theories with the exception of a numerical factor of 2/9 instead of 1/6 which amounts to a deviation of 12/9 = 4/3.

Above a critical molar mass M_{crit}, $D_2\eta_0/\rho$ is no longer constant (Fig. 14-4) because η_0 increases and D_2 decreases with increasing molar mass. These variations of D_2 and η_0 are explained by a so-called reptation of chain molecules (Section 14.2.3).

Fig. 14-5 Reptation of a test chain (black) through the tangle of other chains (white) of a melt.

14.2.3 Reptation of Polymer Chains

Eyring et al. noted in 1958 that a polymer chain must move in a kind of slalom fashion through a physical network of entangled chains. The movement resembles that of a reptile through underbrush and was thus called **reptation** be de Gennes (1971). It can be described by the **tube model** of Edwards and Doi (1978) (see Historical Notes to Chapter 14).

Entanglements (Section 6.3.3) are considered fairly long-lived. They impose topological constraints on the chain and it is assumed that these constraints control the movement of chains in a melt. Hydrodynamic effects are assumed to be minor and it is also postulated that the fraction of free volume is independent of the molar mass.

According to the **Doi-Edwards theory**, the movement of a test chain (Fig. 14-5) can be described by "reptation" in a kind of "tube" which is formed by the segments of other chains. The tube is ca. 5 nm wide; its center line is called the **primitive chain**. The length of the tube, $L_{tube} = N_{seg}L_{seg}$, is given by the number of segments, N_{seg}, it contains; a segment has a length L_{seg} and a friction coefficient ξ_{seg}.

The test chain needs a certain time to wriggle itself out of the tube, the **reptation time**, t_{rep}, which is given by the diffusion equation, $D_{rep} = \langle L^2\rangle_{tube}/(2\ t_{rep})$, where $\langle L^2\rangle_{tube}$ is the spatial average of the square of the shift length. In the time interval t_{rep}, the test chain with a diffusion coefficient $D_{rep} = k_BT/N_{seg}\xi_{seg}$ moves a distance $L_{tube} = N_{seg}L_{seg}$ which leads to a reptation time of

$$(14\text{-}3) \qquad t_{rep} = \frac{L_{tube}^2}{2\ D_{rep}} = \left(\frac{L_{seg}^2\xi_{seg}}{2\ k_BT}\right)N_{seg}^3 = t_0N_{seg}^3$$

The reptation time is thus proportional to the third power of the number N_{seg} of segments and a proportionality constant $t_0 = L_{seg}{}^2\xi_{seg}/(2\ k_BT)$ which is a microscopic time of the order of 10^{-10} s.

The shift of the tube also leads to a shift of the position of the test chain which, by definition, has the same radius of gyration as the primitive chain. The shift must therefore take place within the reptation time t_{rep} and the diffusion coefficient of reptation, D_{rep}, must equal the macroscopic self-diffusion coefficient D_2:

$$(14\text{-}4) \qquad D_2 = D_{rep} = \frac{\langle s^2\rangle_o}{2\ t_{rep}} = \frac{k_BT\langle s^2\rangle_o}{N_{seg}^3L_{seg}^2\xi_{seg}}$$

The segments of Eq.(14-4) are Kuhnian segments, $N_{seg} = N_K$, the number of which can be expressed by $N_K = \langle r^2\rangle_o/L_K{}^2$ (Eq.(4-39)). For unperturbed random coils, mean-square end-to-end distances of chains, $\langle r^2 \rangle_o$, can be converted to mean-square radii of gyration, $\langle s^2 \rangle_o = \langle r^2 \rangle_o/6$, (Eq.(4-43)), which, in turn, are directly proportional to the molar mass, $\langle s^2 \rangle_o = K_{s,o}M$, (Eq.(4-44)). Eq.(14-4) thus converts to

$$(14\text{-}5) \qquad D_2 = \frac{k_B T \langle s^2 \rangle_o}{N_{seg}^3 L_{seg}^2 \xi_{seg}} = \frac{k_B T L_{seg}^4}{6^3 K_{s,o}^4 \xi_{seg}} \cdot M^{-2} = K_{rep} M^{-2} = D_{rep}$$

The self-diffusion coefficient D_2 should thus be proportional to the inverse square of the molar mass, as is found experimentally (Fig. 14-4). Surprisingly, the same function $D_2 = f(M^{-2})$ is also followed by polymers with molar masses below the molar mass $M_{e,\eta}$ for the onset of entanglements and even for those below M_{excl} (Fig. 4-14). A possible explanation may be an increase in the free volume caused by endgroups.

Self-diffusion coefficients of melts of low-molar mass compounds are approximately 10^{-6} cm²/s, i.e., in the same range as that of mutual diffusion coefficients of low-molar mass compounds in low-molar mass solvents. The dependence on the inverse square of molar masses causes self-diffusion coefficients to decrease rapidly with increasing molar mass: at molar masses of ca. 10^6 g/mol, they are only 10^{-12}-10^{-13} cm²/s (Fig. 14-6).

A poly(ethylene) with a molar mass of $1 \cdot 10^6$ g/mol at 175°C has a self-diffusion coefficient of $D_2 = 1.78 \cdot 10^{-12}$ cm²/s (experimental) and a mean-square radius of gyration of $\langle s^2 \rangle_o = 1.76 \cdot 10^{-11}$ cm² (RIS calculations). Acccording to Eq.(14-4), its reptation time is therefore $t_{rep} \approx 5$ s.

Star-like polymers reptate considerably more slowly than their linear counterparts (Fig. 14-6). For three-arm poly(butadiene)s, deviations from the function $D_2 = f(M^{-2})$ start at a molar mass of ca. 20 000 g/mol which corresponds to arm lengths of $N_{u,arm} \approx$ 125 butadiene units = 500 chain units. This number of chain units agrees with the critical number of chain units for the onset of entanglements.

With increasing lengths of arms, deviations from $D_2 = f(M^{-2})$ become increasingly larger until, at $\overline{M}_w \approx$ 100 000 g/mol, self-diffusion coefficients are approximately proportional to the inverse 10th power of the molar mass (Fig. 14-6). So far, such large exponents have defied any explanation.

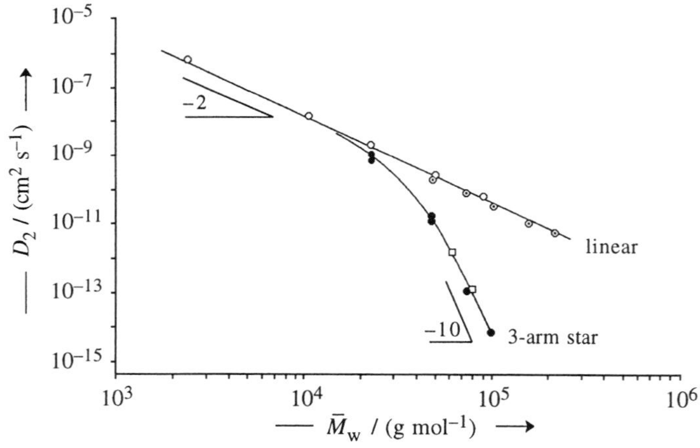

Fig. 14-6 Self-diffusion coefficients of linear (⊙ [9], ○ [10]) and three-arm [11] poly(butadiene)s with narrow (●) or broad (□) molar mass distributions. Measurements at 176°C [9] or 165°C [10, 11].

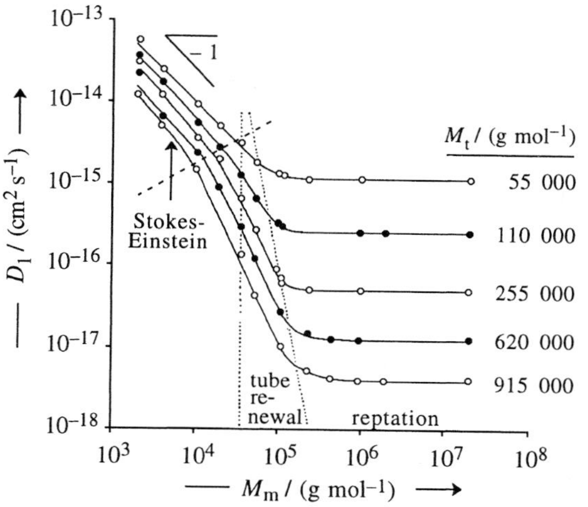

Fig. 14-7 Diffusion coefficients D_t of deuterated poly(styrene)s (tracers) of molar mass M_t in matrices of non-deuterated poly(styrene)s of molar mass M_m at temperatures that correspond to a constant free volume fraction of 0.042. Measurements at 174°C are corrected for a free volume at $T - T_G = 74$ K with the help of the WLF equation. Eq.(13-39). Calculations with Eq.(14-8), $M_e = 18\,000$ g/mol, and $q_z = 3.5$ [12]. By permission of the American Chemical Society, Washington (DC).

14.2.4 Polymer Chains in Polymer Matrices

The self-diffusion of polymer molecules in melts changes if test chains move through a tangle of chains with the same constitution and configuration but different molar masses. This kind of self-diffusion can only be studied with tracers as test chains such as deuterated polymers in their non-deuterated counterparts of the same chemical constitution and configuration. In these experiments, it is assumed that deuteration does not change the intrinsic properties of tracer molecules, an assumption that is not fully justified (see pp. 123, 168, 181, 333). Because of these differences in deuterated and non-deuterated ("protonated" in the language of physics) molecules, data must be compared for constant free volumes (p. 177) and not at constant temperatures.

Diffusion coefficients of tracers with molar masses M_t decrease with increasing molar mass M_m of the matrix and then become constant (Fig. 14-7). Three ranges can be distinguished, depending on the molar mass range:

1. The matrix behaves as a solvent with a viscosity η_m if the molar mass M_t of the diffusing tracer is much larger than the molar mass of the matrix M_m. The matrix is treated as a theta solvent since the matrix and tracer have the "same" chemical structure.

In this case, the diffusion coefficient D_t of the tracer molecule with a hydrodynamic radius R_t of an equivalent sphere is given by the Einstein-Sutherland Eq.(11-10), $D_t = k_B T/\xi_D$, where $\xi_D = N_{seg}\xi_{seg} = 6\,\pi\,\eta_m R_t$ (Stokes Eq.(11-11)) and η_m = viscosity of the matrix. The diffusion proceeds in the Rouse range (Section 15.3.2) so that the viscosity of the matrix is directly proportional to the molar mass, $\eta_m = K_m M_m$. The hydrodynamic radius of the tracer, $R_t = K_t M_t^{1/2}$, varies with the square root of the molar mass, similar to the molar mass dependence of the radius of gyration of unperturbed coils, $\langle s^2\rangle_o^{1/2} = K_{s,o}M^{1/2}$, Eq.(4-44), in diffusion in unentangled matrices (Stokes-Einstein range).

The diffusion coefficient of the tracer is therefore inversely proportional to the square root of the molar mass M_t of the tracer and inversely proportional to the molar mass M_m of the matrix (Fig. 14-7):

$$(14\text{-}6) \qquad D_t = \frac{k_B T}{6\pi \eta_m R_t} = \frac{k_B T}{6\pi K_m K_t} M_t^{-1/2} M_m^{-1}$$

2. In the second limiting case, the molar mass of the matrix is much greater than the molar mass of the tracer ($M_m >> M_t$). According to reptation theory, Eq.(14-5), diffusion coefficients of tracers should not depend on the molar mass of the matrix (Fig. 14-7) but should vary with the inverse square of the molar mass of the tracer.

3. The "walls" of the tube can no longer be assumed to be rigid if tracer and matrix molecules have comparable molar masses. Matrix chains near a tracer chain rather move away from the regions where they act on the tracer. The movement of the tracer is no longer inhibited and the tube is renewed. The theory then assumes that **tube renewal** and reptation are independent of each other:

$$(14\text{-}7) \qquad D_t = D_{rep} + D_{renew}$$

The theory also predicts that the diffusion coefficient of tube renewal, D_{renew}, is given by the molar masses M_m of the matrix and M_e of segments between entanglement sites, a factor q_z that depends on the number z of chain segments which impede the movement of the diffusing tracer chain, and the proportionality constant K_{rep} of Eq.(14-5):

$$(14\text{-}8) \qquad D_{renew} = q_z K_{rep} M_e^2 M_t^{-1} M_m^{-3} \quad ; \quad q_z = (48/25)(12/\pi^2)^{z-1} z$$

Introduction of Eqs.(14-5) (as $D_2 = D_{rep} M_t^{-2}$) and (14-8) in Eq.(14-7) then leads to an expression for the experimentally accessible diffusion coefficient D_t of the tracer:

$$(14\text{-}9) \qquad D_t = D_{rep}[1 + q_z M_e^2 M_t M_m^{-3}] \approx D_{rep} q_z M_e^2 M_t M_m^{-3}$$

D_{rep} and M_e are constants for a given system. Eq.(14-9) can therefore be written as

$$(14\text{-}10) \qquad D_t = K M_m^x M_t^y \quad ; \quad K = D_{rep} q_z$$

The predicted exponents x and y agree well with the observed ones (Table 14-1).

Table 14-1 Exponents x and y of molar masses of matrices (M_m) and tracers (M_t) in Eq.(14-10) for the molar mass dependence of diffusion coefficients D_t of deuterated poly(styrene) as the tracer in non-deuterated poly(styrene) at a constant free volume of 0.042 [12].

Range	Exponent x of M_m		Exponent y of M_t	
	Theory	Experiment	Theory	Experiment
Unentangled matrix	– 1	– 1.0 ± 0.1	– 0.500	– 0.57 ± 0.05
Tube renewal	– 3	– 2.8 ± 0.3	– 1	– 1.00
Reptation	0	0	– 2	– 2.0

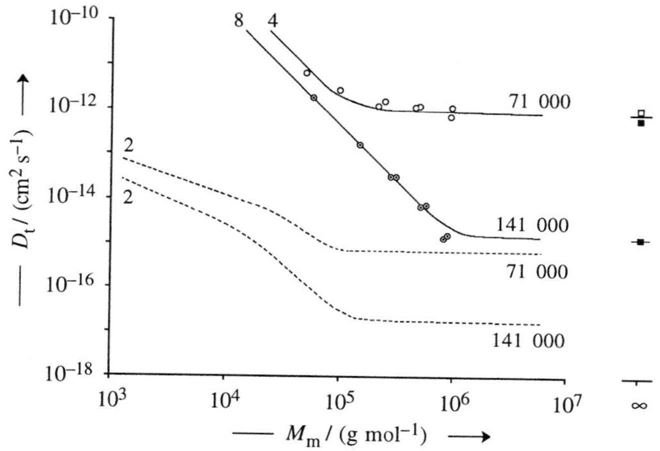

Fig. 14-8 Dependence of diffusion coefficients D_t of star-like poly(styrene)s with $f = 4$ (O) or $f = 8$ (⊙) arms on the molar mass M_m of linear poly(styrene)s as matrix molecules [13]. For comparison: self-diffusion coefficients of linear poly(styrene)s ($f = 2$) with the same molar mass. Microgels (□) and interpenetrating networks (■) as matrices behave like macromolecules of infinitely high molar mass.

In matrices of linear polymers with the same monomeric units, star-like macromolecules diffuse faster than linear macromolecules (Fig. 14-8). It is assumed that in this case self-diffusion proceeds not only by tube renewal but also by a contraction of arms. Both processes are thought to be independent of each other. Arm retraction may be responsible for the dependence of diffusion coefficients on the –2.5th power of the molar mass of the matrix instead of the –3rd power called for by Eq.(14-8).

14.3 Permeation Through Polymers

14.3.1 Overview

In contrast to metals, polymers are permeable to gases and many low-molecular weight liquids. Such a transport of substances into or through polymers is often desired, for example for the dying of fibers and fabrics by dyestuffs or for the controlled application of pharmaceuticals through the skin. In many other cases, permeation is undesirable; examples are the migration of plasticizers in polymers or the loss of carbon dioxide from carbonated beverages through walls of plastic bottles.

Permeation through polymeric materials can occur by two different mechanisms: by diffusion or flow through pores or by transport of substances that are molecularly dissolved in the polymer matrix. Accordingly, polymer membranes are classified as either pore membranes or solubility membranes. Pores are defined as channels with diameters that are much larger than the diameters of the permeating molecules. For permeation through pores, interactions between permeating substances and membrane materials can be neglected. Conversely, the usefulness of solubility membranes depends strongly on the interaction between transported compounds and the polymer matrix.

Table 14-2 Permeability coefficients P of nitrogen in poly(ethylene) and parchment paper (= a cellulose paper that was treated with sulfuric acid).

Polymeric material	$10^{14}\ P/(\text{cm}^2\ \text{s}^{-1}\ \text{Pa}^{-1})$ at T =			
	0°C	30°C	50°C	70°C
Poly(ethylene) film	25	210	740	2200
Parchment paper	1120	940	930	840

The two types of transport through matter can often be distinguished by the temperature or pressure dependence of permeability coefficients P which, in turn, depend on diffusion and solubility and have the physical unit of diffusion coefficient per pressure (Section 14.3.2). The permeation of non-plasticizing gases through solubility membranes or films is not affected by pressure; a pressure effect indicates pores or very fine cracks.

An example is the permeation of nitrogen through poly(ethylene) film or parchment paper (Table 14-2). Nitrogen is soluble in poly(ethylene); it diffuses through poly(ethylene) films by site-hopping processes. The diffusion increases with increasing temperature and so do the permeability coefficients.

Parchment paper contains true pores in the nanometer to micrometer range. Since diffusion coefficients of gases are inversely proportional to their viscosities and viscosities increase with increasing temperature, permeability coefficients of nitrogen decrease with increasing temperature.

The counteraction of these two types of permeation is used to prepare laminates from two different films, for example, from poly(ethylene) and aluminum. Oxygen permeates through poly(ethylene) by a solubility mechanism but through pores in aluminum; such pores may accidentally be produced during the fabrication of aluminum foils. The rate of oxygen permeation through an aluminum foil of 25 μm thickness with pores of 1 μm diameter is ca. $5 \cdot 10^{-5}$ cm³/s at a pressure difference of 10^5 Pa (= 1 bar). After lamination of the aluminum foil with a 25 μm thick poly(ethylene) film, the permeation rate dropped to $5 \cdot 10^{-13}$ cm³/s.

14.3.2 Permeability Coefficients

For simplicity, the permeating substance will be called **permeant** and the permeated material, the **matrix**. Permeants may be gases, liquids (including plasticizers), or dissolved substances (for example, dyestuffs, pharmaceuticals, etc.). The matrix may be a film, foil, sheet, thick-walled container, etc.

A permeant will permeate through a matrix as long as there is a concentration difference between the two sides of the matrix. Examples are a gas with a pressure p against a vacuum or a solution of concentration c against the pure solvent or air. If a constant external concentration difference Δc is maintained, for example, a pressure difference Δp between two sides of a matrix, then a concentration difference Δw will result for the permeant on both sides of the matrix. In the case of permanent gases, the function $\Delta w = f(\Delta p)$ is described by **Henry's law** where S = **solubility coefficient** of the permeant:

(14-11) $\Delta w = S \Delta p$

The solubility coefficient has the physical unit of an inverse pressure if Δp is a pressure and Δw is a mass fraction. For liquids, Δp and Δw are often measured as fractions (volume, mass, etc.). In this case, S is a quantity with the physical unit of unity and its numerical value depends on the physical meaning of Δp and Δw.

Suppose a permeating gas is present on one side of a permeable matrix but not on the other. The pressure difference Δp between the two sides of the matrix will cause the gas to enter the matrix but it will be a time t_1 before the first gas molecules appear on the other side of the matrix (Fig. 14-9). The diffusive flux of the mass of permeant through an area A is $J_m = dm/(A dt)$ (p. 367, first line) so that

(14-12) $m = J_m A(t - t_1)$

A long theoretical calculation has shown that the time lag t_1 is controlled by the thickness L_m of the matrix and the diffusion coefficient D of the permeant according to a modified shift law, $D = (1/3)[L^2/(2\ t)]$, which allows one to calculate the diffusion coefficient of the permeant from the time lag t_1:

(14-13) $t_1 = L_m{}^2/(6\ D)$

The steady state is reached after a time $t \approx 3\ t_1$. For concentration-independent diffusion coefficients, one can replace the concentration change dc in $J_c = -D(dc/dx) = J_m$ (p. 367) by the decrease of mass fractions of the permeant, Δw, and dx by the thickness L_m of the matrix so that $J_m = -D\Delta w/L_m$. Introduction of Eq.(14-11) leads to

(14-14) $J_m = DS(\Delta p/L_m) = P(\Delta p/L_m)$

where the **permeability coefficient** P is the product of the diffusion coefficient D and the solubility coefficient S. Further introduction of Eqs.(14-12) and (14-13) results in

(14-15) $$m = \frac{PA\Delta p}{L_m}\left(t - \frac{L_m^2}{6D}\right)$$

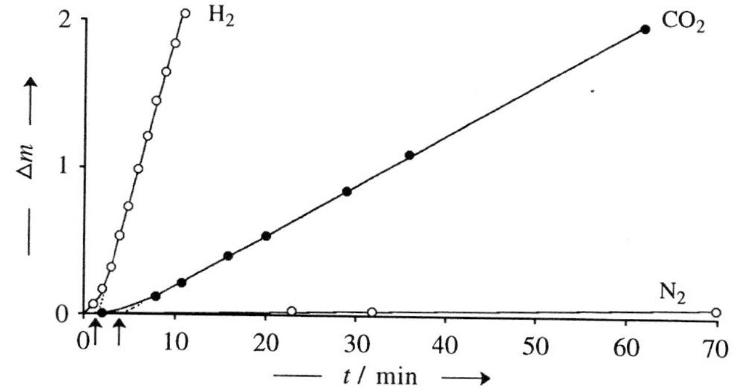

Fig. 14-9 Change of mass of gases, Δm (in arbitray units), with time t on permeation through a film of a styrene copolymer at 25°C [14]. Arrows indicate t_1 for H_2 and CO_2.

The permeability coefficient $P = mL_m/(A\Delta pt)$ (at $t > t_1$) has the physical unit of diffusion coefficient per pressure if m is measured as a volume as is typical for gases. In this case, the solubility coefficient has the unit of an inverse pressure.

Industrial literature uses about 30 different physical units for permeability coefficients, in part with incompatible physical units. In the United States, a "Barrer unit" or "barrier unit" = cc·mil/(100 in^2·atm·day) is often used for the permeation of gases (cc = cm^3, mil = 10^{-3} inch). Also used for gases are m in cm^3, L_m in mm, A in m^2, Δp in atm, and t in 24 h which leads to a physical unit of (cm^3 mm)/(m^2 24 h atm). The rationale for using (cm^3 mm)/m^2 instead of m^2 is the perceived feeling that one can immediately see how the permeation changes if one decreases the thickness L_m of the matrix, increases the pressure difference Δp, and so forth. Unfortunately, units such as (cm^3 mm)/(m^2 24 h atm) are also often reported as cm^3·cm/m^2/24 h/atm which makes the confusion complete. One can reason that the "physical meaning" of P gets lost if P is reported in, e.g., cm^2 s^{-1} Pa^{-1}. However, this is always the case if a physical property is calculated from various other physical properties.

After the initial time lag, the volume (mass, etc.) of permeated material increases linearly with time if the diffusion coefficient D does not depend on concentration (see Fig. 14-9). An increase of D with concentration leads to a function $m = f(t)$ that is convex to the time axis.

14.3.3 Permeation of Gases

Permeability coefficients of gases in polymers vary widely (Table 14-3). For example, the permeability coefficient of oxygen in poly(dimethylsiloxane) is about 20 million times greater than in poly(vinylidene chloride). These differences are very important for applications of polymers such as bottles, containers, and separation membranes.

Table 14-3 Approximate values of permeability coefficients P^* of gases and P of water vapor in polymers at 35°C. For the same polymer, literature data may vary widely because polymers may differ in crystallinity, orientation, and/or proportion of absorbed water. $P^* = 1{\cdot}10^{-14}$ cm^2 s^{-1} Pa^{-1} corresponds at normal pressure ($p = 1{\cdot}10^5$ Pa) to $P = 1{\cdot}10^{-9}$ cm^2 s^{-1}.

Polymer	10^{14} P^*/(cm^2 s^{-1} Pa^{-1}) O_2	CO_2	10^9 P/(cm^2 s^{-1}) H_2O
Poly(dimethyl siloxane)	7000	35000	40
Natural rubber	180	1100	0.3
Butyl rubber (copolymer of isobutene + 4 % isoprene)	100	500	0.1
Poly(tetrafluoroethylene)	32	75	0.0006
Poly(styrene), normal	19		0.14
biaxial orientation	0.02	0.8	0.08
Poly(ethylene), high density	3	3	0.02
Poly(vinyl chloride), unplasticized	0.3	1.2	0.02
Poly(ε-caprolactam) (PA 6)	0.3	0.7	0.0001
Poly(vinyl alcohol) (0 % relative humidity)	0.07	0.09	
(100 % relative humidity)	0.0005	650	53
Poly(ethylene terephthalate), normal	0.04	0.02	0.01
biaxial orientation	0.02	0.01	0.2
Cellulose hydrate	0.02	0.04	1.9
Poly(vinylidene chloride) (actually a copolymer)	0.0004	0.002	0.0004
Upper limit for bottles for carbonated drinks, cola	1	0.5	0.14
beer	0.05	0.5	0.14

For example, carbonated liquids require plastic bottles with very low permeabilities for CO_2 (permeation from the inside to the outside) and O_2 (permeation from the outside to the inside), the former in order to maintain the fizz and the latter in order to preserve the aroma. Membranes for artificial lungs and other air-generation systems must be highly permeable for oxygen and packaging films for fresh fruits and vegetables permeable for oxygen, carbon dioxide, and ethylene (produced by the plants).

Because of increased segment mobility, permeability coefficients are greater at application temperatures $T > T_G$ than at $T < T_G$. Elastomers with low glass temperatures T_G are therefore more permeable to gases than are plastics (see the first three polymers of Table 14-3). Butyl rubber has the lowest permeability for oxygen and nitrogen; it is therefore used for inner tubes of tires.

Permeabilities of thermoplastics are affected by bulky substituents, matrix/permeant interactions, and humidity. Crystalline regions and filler particles increase the length of the path a permeating molecule must travel. Therefore, partially crystalline polymers such as poly(vinylidene chloride) (for packaging purposes a copolymer with 15-20 % vinyl chloride) or poly(vinyl alcohol) and highly oriented plastics such as poly(ethylene terephthalate) are used as **barrier plastics** for packaging films; a grade of the latter with higher molecular weight is also used or bottles (see Volume II, p. 352).

Permeability coefficients are controlled by both diffusion coefficients and solubility coefficients. Diffusion coefficients are usually the smaller, the greater the diameter of the permeating gas molecules (Table 14-4). Crystallinity reduces diffusion coefficients of gases usually according to $D = D_{am}(1 - x_{cr})$ where D_{am} = diffusion coefficient of the amorphous polymer and x_{cr} = degree of crystallinity of the polymer. Solubility coefficients depend not only on crystallinities and free volumes but also on interactions between polymers and gases.

The temperature dependence of diffusion coefficients is controlled by the activation energy $E_D^{\ddagger}$ of diffusion and that of solubility coefficients by solution enthalpies ΔH where D_∞ and S_∞ are the coefficients at infinitely high temperature:

(14-16) $$D = D_\infty \exp(-E_D^{\ddagger}/RT)$$

(14-17) $$S = S_\infty \exp(-\Delta H/RT)$$

Solubility cofficients usually decrease with increasing temperature whereas diffusion coefficients increase. Permeability coefficients can therefore become greater or smaller at higher temperatures. These trends change at the glass temperature.

Table 14-4 Diffusion coefficients, solubility coefficients, and permeability coefficients of gases with molar masses M and molecule diameters d in crosslinked *cis*-1,4-poly(isoprene) at 25°C.

Gas	$\frac{M}{\text{g mol}^{-1}}$	$\frac{d}{\text{nm}}$	$\frac{10^7 D}{\text{cm}^2\ \text{s}^{-1}}$	$\frac{10^5 S}{(\text{cm}^3/\text{cm}^3)\ \text{Pa}^{-1}}$	$\frac{10^{14} P}{\text{cm}^2\ \text{s}^{-1}\ \text{Pa}^{-1}}$
H_2	2	0.234	85	0.040	340
O_2	32	0.292	21	0.070	150
N_2	28	0.315	15	0.035	51
CO_2	44	0.323	11	0.90	1000

14.3.4 Permeation of Liquids

Permeation of liquids in polymers is both a severe problem for certain polymer applications and a desired feature for others. Permeating water from the humidity of air affects the weatherability of polymers with hydrolyzable chemical bonds as well as the dielectric properties of electrical insulators. Additives can migrate out of plastic containers and conversely food components into them. Dyeing of fibers is accelerated by "carriers" which are liquids. In transdermal applications of pharmaceuticals, active components should move through the skin in a controlled manner (see also Volume IV).

Permeation of liquids out of polymers is also an analytical problem in the drying of polymers. Evaporation of solvents by heating polymer solutions very often does not remove all liquids, even at temperatures above the boiling temperature of the solvent (Volume I, p. 188). Such **solvent inclusions** can be severe; an example is the removal of carbon tetrachloride from poly(styrene) where up to 20 % of CCl_4 may remain in the polymer after it is dried in an oven. More successful for solvent removal is azeotropic distillation or freeze-drying.

The time dependence of permeation of non-dissolving liquids into polymers can be described empirically by

(14-18) $$m_t/m_\infty = KAt^n$$

where m_t, m_∞ = mass of liquid in the polymer at times t and infinity, A = surface area of specimen, t = time, and K and n system-dependent constants.

Exponents n are controlled by the dimensionless **Deborah number** $De = t_{rlx}/t_D$, where t_{rlx} is the relaxation time of polymer chains and t_D is a characteristic diffusion time of the permeating liquid, $t_D = L^2/(2D_1)$, where D_1 = diffusion coefficient of the liquid in the polymer. This number got its name from the prophetess Deborah (Judges 5:5: "The mountains flowed before the Lord ...").

Permeation of liquids into a polymer disturbs the original microconformations and macroconformations of polymer molecules. Polymer chains try to adopt new equilibrium conformations but this takes time. If the relaxation time of polymer segments is much smaller than the characteristic time of the liquid ($De < 0.1$), i.e., if the relaxation rate of the segments is much larger than that of the permeant, then the movement of the permeant will cause "immediate" conformational changes of the chains. Both permeant and polymer behave as viscous liquids (Case I). The system can be described by Fick's equations: the exponent adopts a value of n = 1/2 and the product KA becomes $4\ D/\pi$.

In the so-called Case II, Deborah numbers are greater than 10, i.e., mobilities of permeants are much larger than the relaxation rates of polymer segments. The macroconformation of the polymer chains does not change during the permeation, i.e., the polymer behaves as an elastic body for the permeant. The permeation is characterized by a sharp boundary between a swollen zone and the glassy body in the interior. The swollen zone moves with constant speed and the permeated amount of the liquid is directly proportional to the time t, i.e., n = 1.

The range $0.1 < De < 10$ is the range of the so-called *anomalous diffusion* or *viscoelastic diffusion*. Relative movements of permeant molecules and conformational changes of polymer molecules are practically simultaneous and the exponent now adopts values of $1/2 < n < 1$.

14.4 Transport of Polymers Through Porous Membranes

14.4.1 Membranes

Porous membranes are usually subdivided according to the pore size (microporous, macroporous, fibrillar) or to the filtration process (ultrafiltration, nanofiltration, reverse osmosis, etc.). However, transports through such membranes are controlled by many more parameters than just average pore size: chemical structure of membranes, shape and size distribution of pores, electrical charge of membranes, and many others.

Microporous membranes, also called *dense membranes*, have "pores" of ca. 3 nm which are really not pores in the conventional sense of the word but spaces between polymer chains. The transport through such membranes is therefore similar to that through solubility membranes.

Microporous membranes are manufactured by direct polymerization, casting, drawing, or blow molding. They are used for dialyses, electrodialyses, piezodialyses, and pervaporations.

Macroporous membranes are true pore membranes with diameters of pores ranging from ca. 5 nm to ca. 1 μm. The macropores of these membranes are interconnected; for efficiency, the pore volume should exceed 40 vol%.

Macroporous membranes are prepared from polymers in solvent-non-solvent mixtures that phase-separate on evaporation of liquids or similarly from block polymers that form microdomains. Macroporous membranes are also obtained from polyelectrolyte complexes and ionotropic gels or by high-energy irradiation of polymers.

Fibrous membranes have pore diameters of more than ca. 2 μm. They consist of a "felt" of randomly oriented short fibers, for example, in papers or fleeces.

Polymers for porous membranes include celluloses, cellulose acetates, poly(vinyl alcohol), polycarbonates, polysulfones, poly(ethylene)s, poly(acrylonitrile), etc.; non-polymeric membranes may be porous glass or sintered metal powders.

For compact permeant particles, porous membranes act as sieves; examples are the removal of bacteria by fibrous membranes and filtration of spheroidal proteins by macroporous ones. Microporous membranes act as sieves for low-molecular weight ions. Water passes through such membranes as single molecules. Cations are hydrated and cannot pass through the interstitial volumes between polymer chains.

All porous membranes have in common that the flow through the membrane is not only controlled by the viscosity η (as it is in solubility membranes) but also by the pressure gradient dp/dL. The rate of flow, v, is given by the **Darcy law:**

(14-19) $$v = -P_{\text{eff}}\eta^{-1}(dp/dL) \quad ; \quad P_{\text{eff}} = \text{effective permeability}$$

14.4.2 Diffusion Through Pores

Diffusion in pores is unimpeded if the radii of the pores are much greater than the hydrodynamic radii of the polymer molecules, $R_p >> R_h$. In this case, diffusion coefficients in pores, D_p, practically equal those in free solution, D_o. The ratio D_p/D_o becomes unity (Fig. 14-10). For two reasons, this ratio decreases if R_h approaches R_p.

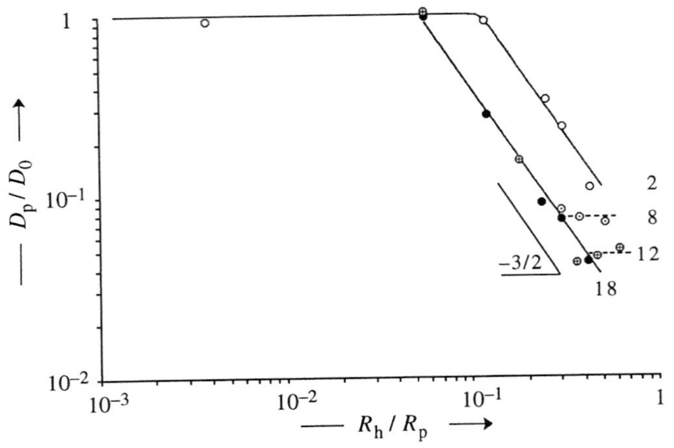

Fig. 14-10 Ratio of diffusion coefficients, D_p/D_0, of amyl acetate solutions of linear (O) and 8- (⊖), 12- (⊕), and 18-arm (●) poly(isoprene)s in cylindrical pores ($43 \leq R_p$/nm ≤ 4850) as a function of the ratio of hydrodynamic radii of polymer molecules and radii of pores, R_h/R_p [15]. D_0 = diffusion coefficient of polymers at zero polymer concentration.

First, coil molecules cannot attain their equilibrium macroconformations in narrow pores, which leads to a decrease of segment concentrations near pore "walls" if polymer adsorption is absent. This polymer-poor wall layer extends over a constant fraction of the pore diameter if pore sizes are comparable to polymer dimensions. Since the equilibrium concentration c_p of polymers in pores is smaller than the polymer concentration c_0 outside the pores, driving forces for diffusion are reduced by a factor $K = c_p/c_0$.

Second, pore walls impede the hydrodynamic flow of polymer molecules, which causes the friction coefficient ξ_p of polymer molecules in pores to be smaller than the friction coefficient ξ_0 of polymer molecules in solution. Since friction coefficients are proportional to radii, Eq.(11-11), and inversely proportional to diffusion coefficients, Eq.(11-10), one obtains

$$(14\text{-}20) \qquad D_p/D_0 = K(\xi_0/\xi_p)^q = (c_p/c_0)(R_h/R_p)^q$$

D_p/D_0 begins to decrease at $R_h/R_p \approx 0.1$ for linear poly(isoprene)s but at $R_h/R_p \approx 0.05$ for star molecules (Fig 14-10). It seems that chains of linear poly(isoprene)s can meander through the pore system like a snake but that tentacles of star-molecules cannot. In both cases, the exponent q in Eq.(14-20) is −3/2 which is as yet unexplained. However, the ratio D_p/D_0 of star molecules is independent of the number of arms since the star molecules are very compact and have practically the same packaging density.

Ratios D_p/D_0 of star-like poly(isoprene)s become independent of R_h/R_p if R_h/R_p approaches unity: 8-arm molecules at $R_h/R_p \approx 0.3$ and 12-arm molecules at $R_h/R_p \approx 0.45$. This observation also remains unexplained. However, theory predicts for $R_h/R_p > 1$ that the exponent should become q = −2/3 for good solvents and q = −1 for theta solvents.

In the experiments described above, membranes act as barriers for permeants. Such barrier membranes may be "solid" (i.e., below the glass temperature) or "liquid" (i.e., above the transition temperature liquid crystal-glass). Membranes composed of lipids are the best known example of the "liquid" type.

The transport of permeants through liquid membranes can be accelerated by so-called carriers. In biological membranes, most carriers are simple proteins and the permeants are ions, oxygen, amino acids, saccharides, etc. For example, potassium ions are transported by valinomycin, cyclo[(D-val)(L-lac)(L-val)(D-hyiv)]$_3$ (val = valine, lac = lactic acid, hyiv = α-hydroxyisovaleric acid). Dissolved valinomycin has a central cave of 0.6-0.7 nm diameter in which the approximately equal-sized potassium ion is complexed by the six carbonyl oxygens of the valine units, $-NH-CH[CH_2CH(CH_3)_2]-CO-$. The complex travels through the lipid membrane, is discharged on the other side of the membrane, and returns to the cell.

Transport through biological membranes can also occur in channels instead of by carriers. In lipid membranes, such channels are formed by embedded gramicidin A, a linear peptide with the constitution

OCH(L-val)gly(L-ala)(D-leu)(L-ala)(D-val)(L-trp)(D-leu)(L-trp)(D-leu)(L-trp)NHCH_2CH_2OH

Historical Notes to Chapter 14

Reptation

H.Eyring, T.Ree, N.Hirai, Proc.Natl.Acad.Sci. (US) **44** (1958) 1213 (slalom through obstacles)

P.G. de Gennes, J.Chem.Phys. **52** (1971) 572 (reptation)

M.Doi, S.F.Edwards, J.Chem.Soc., Faraday Trans. [2] **74** (1978) 1789, 1802, 1818 (tube model)

Literature to Chapter 14

14.1 GENERAL REVIEWS

P.G.de Gennes, Scaling Concepts in Polymer Physics, Cornell University Press, Ithaca (NY) 1979

R.B.Bird, R.C.Armstrong, O.Hassager, Dynamics of Polymeric Liquids, Vol. 1; R.B.Bird, C.F.Curtiss, R.C.Armstrong, O.Hassager, ditto, Vol. 2, Wiley, New York, 2nd ed. 1987

M.Doi, S.F.Edwards, The Theory of Polymer Dynamics, Oxford University Press, Oxford 1987

K.F.Freed, Renormalization Group Theory of Macromolecules, Wiley, New York 1987

14.2 TRANSPORT IN FLUID POLYMER PHASES

J.Klein, The Self-Diffusion of Polymers, Contemp.Phys. **20** (1979) 611

R.T.Bailey, A.M.North, R.A.Pethrick, Molecular Motion in High Polymers, Clarendon Press, New York 1981

M.Tirrell, Polymer Self-Diffusion in Entangled Systems, Rubber Chem.Technol. **57** (1984) 523

14.3 PERMEATION THROUGH POLYMERS

B.J.Hennessy, J.A.Mead, T.C.Stening, The Permeability of Plastics Films, Plastics Institute, London 1966

J.Crank, G.S.Park, Eds., Diffusion in Polymers, Academic Press, New York 1968

H.B.Hopfenberg, Ed., Permeability of Plastic Films and Coatings, Plenum Press, New York 1974

H.Yasuda, Units of Permeability Constants, J.Appl.Polym.Sci. **19** (1975) 2529

T.R.Crompton, Additive Migration from Plastics into Food, Pergamon Press, Oxford 1979

M.B.Huglin, M.B.Zakaria, Comments on Expressing the Permeability of Polymers to Gases, Angew.Makromol.Chem. **117** (1983) 1

H.L.Frisch, S.A.Stern, Diffusion of Small Molecules in Polymers, Crit.Revs.Solid State Mater.Sci. **11** (1983) 123
J.Comyn, Ed., Polymer Permeability, Elsevier Appl.Sci.Publ., London 1985
G.E.Zaikov, A.P.Jordanskii, V.S.Markin, Diffusion of Electrolytes in Polymers, VNU Science Press, Utrecht, Netherlands 1987
L.A.Errede, Molecular Interpenetration of Sorption in Polymers, Pt. I, Adv.Polym.Sci. **99** (1991) 1
T.Matsuura, Synthetic Membranes and Membrane Separation Processes, CRC Press, Boca Raton (FL) 1993
P.Neogi, Ed., Diffusion in Polymers, Dekker, New York 1996
Yu.Yampolskii, I.Pinnau, B.D.Freeman, Eds., Materials Science of Membranes for Gas and Vapor Separation, Wiley, Hoboken (NJ) 2006

14.4 TRANSPORT OF POLYMERS THROUGH POROUS MEMBRANES

K.Lakshminarayanaiah, Transport Phenomena in Membranes, Academic Press, New York 1969
P.Meares, Membrane Separation Processes, Elsevier Scientific, Amsterdam 1976
R.J.Kostelnik, Polymeric Delivery Systems, Gordon and Breach, New York 1978
S.G.Shultz, Basic Principles of Membrane Transport, Cambridge University Press, London 1980
M.Starzak, The Physical Chemistry of Membranes, Academic Press, New York 1984
R.E.Kesting, Synthetic Polymeric Membranes - A Structural Perspective, Wiley, New York, 2nd ed. 1985
P.Tyle, Ed., Drug Delivery Devices, Dekker, New York 1988

References to Chapter 14

[1] F.D.Blum, S.Pickup, K.R.Foster, J. Colloid Interface Sci. **113** (1986) 336, (a) Fig. 1a, (b) Fig. 1b; numerical data: see [3]
[2] S.Pickup, F.D.Blum, W.T.Ford, M.Periyasamy, J.Am.Chem.Soc. **108** (1986) 987, Table II
[3] F.D.Blum, private communication, February 7, 1987
[4] F.D.Blum, S.Pickup, reported by [3]
[5] W.Derbyshire, I.D.Duff, Faraday Discuss.Chem.Soc. **57** (1974) 243
[6] D.S.Pearson, G. Ver Strate, E. von Meerwall, F.C.Schilling, Macromolecules **20** (1987) 1133
[7] A.Tonelli, quoted in [6]
[8] H.-G.Elias, An Introduction to Plastics, VCH, Weinheim 1993, Fig. 6-2
[9] J.Klein, D.Fletcher, L.J.Fetters, Nature **304** (1983) 526, Fig. 1
[10] C.R.Bartels, B.Crist, W.W.Graessley, Macromolecules **17** (1984) 2702, Table III
[11] C.R.Bartels, B.Crist, Jr., L.J.Fetters, W.W.Graessley, Macromolecules **19** (1986) 785, Table III
[12] P.F.Green, E.J.Kramer, Macromolecules **19** (1986) 1108, Fig. 5
[13] K.R.Shull, E.J.Kramer, G.Hadziioannou, M.Antonietti, H.Sillescu, Macromolecules **21** (1988) 2578, data of Fig. 1
[14] P.Goeldi, H.-G.Elias, unpublished data
[15] M.P.Bohrer, L.J.Fetters, N.Grizzuti, D.S.Pearson, M.V.Tirrell, Macromolecules **20** (1987) 1827, Tables 1-III

15 Melt Viscosity

15.1 Deformation

Two limiting cases exist for the deformation of matter: viscous and elastic. Typical fluids like water deform irreversibly under their own weight and start to flow. They show a typical **viscous** behavior (p. 395). The shear stress of ideal viscous liquids (= **Newtonian liquids**) is directly proportional to the rate of deformation but not to the deformation itself (Chapter 12).

Typical solids such as steel resist deformation. Small forces cause only small deformations from which the deformed body recovers "instantaneously" after removal of the load. Such bodies behave **elastically**; in the ideal case of **Hookean bodies**, the shear stress is proportional to the deformation but not to the deformation rate (Chapter 16).

Ideal-viscous and ideal-elastic behavior are two borderline cases since each body has both viscous and elastic properties according to a rheological axiom. In macromolecular compounds, both types of behavior often become comparable at relatively small deformations and small deformation rates. An especially peculiar behavior is shown by nutty putty (Silly putty®) which is a silicone elastomer that contains an inorganic filler such as iron oxide. A lump of this material flows slowly on a flat surfce under its own weight, i.e., it shows a typical viscous behavior. However, it rebounds like a rubber ball if a sphere of silly putty is dropped on a surface, i.e., it behaves elastically. Conversely, it is also brittle because a sheet of this material breaks if it is bent fast.

Many polymers show similar behavior although not as pronounced as nutty putty. Molten polymers do flow under their own weight but the melts are also somewhat elastic. Amorphous and semi-crystalline polymeric solids behave elastically under small loads but deform irreversibly under larger ones. Polymers thus combine viscous and elastic behavior: they are **viscoelastic**. Whether the viscous or the elastic behavior is more pronounced depends on the magnitude, the duration, and the speed of deformation.

The interrelationship between force, deformation, and time is studied by **rheology**, the science of flow (G: *rheos* = flow; *logos* = speech, reason). Rheology describes the flow behavior by **constitutive equations** that describe the interplay between force, deformation, and time. In many cases, the rheological behavior of polymers is much too complex to be described by rigorous constitutive equations. Therefore, one usually uses ideal equations and then introduces empirical correction factors.

Forces produce simple deformations by shearing, linear drawing, pressurizing or compressing, and more complicated deformations by bending or twisting (Fig. 15-1). For comparison, forces are usually related to the areas on which they act.

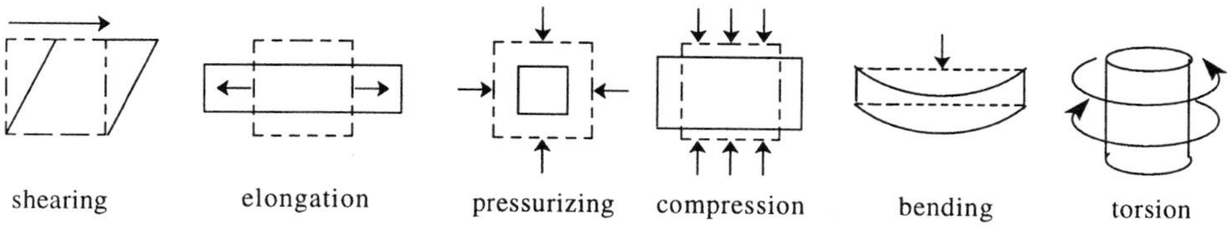

Fig. 15-1 Simple deformations of bodies. A body with contours – – – is deformed by forces (⟶) to a new shape with contours ——.

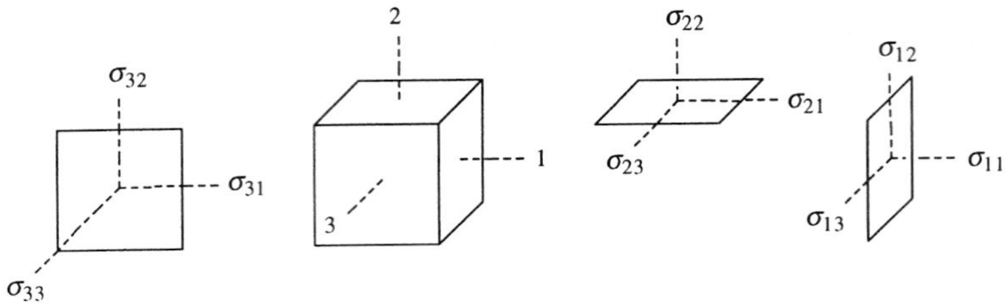

Fig. 15-2 Definition of stresses on a body (spatial directions indicated by either 1,2,3 or x,y,z).

Forces $\boldsymbol{F}$ acting upon areas A are called **stresses,** $\sigma = F/A$. One usually considers only two types of stresses: those from elongations and those from shearing. Stresses from elongating (drawing) are $\sigma = \boldsymbol{F}/(H_0W)$ and those from shearing, $\sigma = \boldsymbol{F}/(L_0W)$ (Fig. 15-3).

A body may experience a total of nine different simple stresses in the three spatial directions 1, 2, and 3 (Fig. 15-2). Stresses perpendicular (normal) to areas are called **normal stresses,** σ_{11}, σ_{22}, and σ_{33}. They are usually designated as positive for elongation and negative for pressurizing or compression. The sum of the three normal stresses equals zero, $\sigma_{11} + \sigma_{22} + \sigma_{33} = 0$. The six other stresses, σ_{12}, σ_{13}, σ_{23}, σ_{32}, σ_{31}, and σ_{21}, refer to stresses that are perpendicular to one of the planes 1, 2, or 3.

By definition, a body with three areas i = 1, 2, and 3 is elongated by a **tensile stress (nominal stress, engineering stress)** σ_{11} in 1→1 direction and sheared by a **shear stress** $\sigma_{21} \equiv \tau$ in 2→1 direction. The difference, $\sigma_{11} - \sigma_{22}$, is called the **first normal stress difference** and the difference, $\sigma_{22} - \sigma_{33}$, the **second normal stress difference. Shear strain** γ_e is the ratio of the first normal stress difference to the shear stress, $\gamma_e = (\sigma_{11} - \sigma_{22})/\sigma_{21}$, shown in Fig. 15-3 as $\gamma = \Delta L/H$, also called **(elastic) shear deformation.**

The ratio of shear stress σ_{21} and elastic shear strain γ_e is the **shear modulus**:

$$G = \tau/\gamma_e = \sigma_{21}/\gamma_e = \sigma_{21}{}^2/(\sigma_{11} - \sigma_{22}) \qquad (15\text{-}1)$$

The change of shear strain γ with time t is the **shear rate** or **velocity gradient,** $\dot{\gamma} = \mathrm{d}\gamma_e/\mathrm{d}t$ (for many different designations of rheological quantities, see Table 15-1).

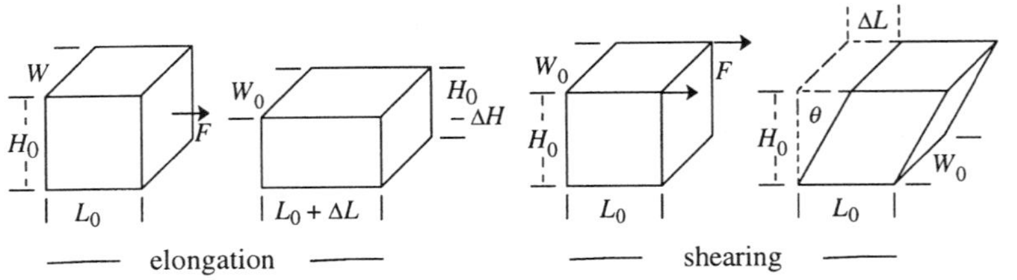

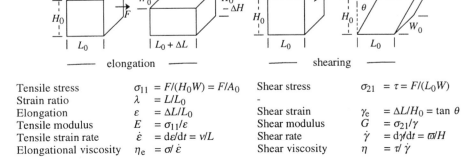

Tensile stress	σ_{11}	$= F/(H_0W) = F/A_0$	Shear stress	σ_{21}	$= \tau = F/(L_0W)$
Strain ratio	λ	$= L/L_0$	-		
Elongation	ε	$= \Delta L/L_0$	Shear strain	γ_e	$= \Delta L/H_0 = \tan\theta$
Tensile modulus	E	$= \sigma_{11}/\varepsilon$	Shear modulus	G	$= \sigma_{21}/\gamma$
Tensile strain rate	$\dot{\varepsilon}$	$= \mathrm{d}\varepsilon/\mathrm{d}t = v/L$	Shear rate	$\dot{\gamma}$	$= \mathrm{d}\gamma/\mathrm{d}t = \varpi/H$
Elongational viscosity	η_e	$= \sigma/\dot{\varepsilon}$	Shear viscosity	η	$= \tau/\dot{\gamma}$

Fig. 15-3 Deformations of a body with initial length L_0, width W_0, and height H_0 by a force $\boldsymbol{F}$ and the attacked areas H_0W_0 (elongation) and L_0W_0 (shearing). θ = shearing angle.

Table 15-1 Names of stresses (S), length ratios (A), deformations (D), moduli (M), deformation rates (R), and viscosities (V) of elongations and shearings. * Sometimes erroneously called "torsional modulus" (for the meaning of "torsion", see Fig. 15-1).

	Elongation Usual name	other name(s)	Shearing Usual name	other name(s)
S	Tensile stress nominal stress	engineering stress	shear stress	shearing stress
A	Strain ratio	draw ratio	-	-
D	Elongation	tensile strain, linear strain, engineering strain, Cauchy elongation	shear strain	elastic shear deformation
M	Tensile modulus	Young's modulus, modulus of elasticity	shear modulus *	modulus of rigidity
R	Tensile strain rate	-	shear rate	velocity gradient
V	Extensional viscosity	Trouton viscosity, tensile viscosity	shear viscosity	dynamic viscosity, viscosity

The **dynamic (shear) viscosity** η (absolute viscosity, mostly only "viscosity") is the ratio of **shear stress** $\tau = \sigma_{21}$ to shear rate $\dot{\gamma}$. It measures the energy E per volume V that is dispersed by the liquid in the time interval t, $(E/V)t$ in, e.g., (J/m^3)s = Pa s.

(15-2) $$\eta = \tau / \dot{\gamma} = \sigma_{21} / \dot{\gamma} = G\gamma_e / \dot{\gamma}$$ [energy/(area × velocity)]

The inverse of viscosity is called **fluidity**, $v = 1/\eta$. The ratio of dynamic viscosity η to density ρ (= mass of solute per volume of liquid) is the **kinematic viscosity**, $\nu = \eta/\rho$.

The ratio of the first normal stress difference to the square of the shear rate is known as the **first normal stress coefficient**, ψ_1:

(15-3) $$\psi_1 = \frac{\sigma_{11} - \sigma_{12}}{\dot{\gamma}^2} = \frac{\sigma_{21}\gamma_e}{\dot{\gamma}^2} = \eta\frac{\gamma_e}{\dot{\gamma}} = G\left(\frac{\gamma_e}{\dot{\gamma}}\right)^2$$

The **elastic shear time**, t_s, is defined as the ratio of elastic shear deformation γ_e (= shear strain) to the shear rate $\dot{\gamma}$, as ratio of dynamic viscosity η to the shear modulus G, or as ratio of the first normal stress coefficient ψ_1 to the dynamic viscosity η :

(15-4) $$t_s = \frac{\gamma_e}{\dot{\gamma}} = \frac{\sigma_{11} - \sigma_{22}}{\sigma_{21}\dot{\gamma}} = \frac{\sigma_{21}}{G\dot{\gamma}} = \frac{\eta}{G} = \frac{\psi_1\dot{\gamma}}{\sigma_{21}} = \frac{\psi_1}{\eta} \quad ; \quad t_{rlx} = \eta_0/G_0$$

At a shear rate of zero, it becomes the **structure relaxation time**, t_{rlx}.

Stresses (S), strains (deformations D), moduli (M), deformation rates (R), and viscosities (V) of elongations can be defined similarly to those of shearing (Table 15-1). In terms of naming, "shear" is then mostly replaced by "tensile" (see Fig. 15-3). The whole field hosts a plethora of non-systematic terms which are in part historical, in part stem from industrial testing procedures, and in part come from various national and international standardization attempts (ISO, DIN, ASTM, IUPAP, IUPAC, etc.).

15.2 Viscometry

15.2.1 Types of Viscosities

"Rheology" was originally the science of the flow of liquids and gases. However, the field now covers the time dependence of deformations of all types of matter, including solids (Chapters 16-18). Depending on the type of stress, one distinguishes three types of viscosities:

Shear viscosity	= shear stress	/ shear rate	;	$\eta = \sigma_{21}/\dot{\gamma}$
Extensional viscosity	= tensile stress	/ tensile strain rate	;	$\eta_e = \sigma_{11}/\dot{\varepsilon}$
Volume viscosity	= hydrostatic pressure	/ deformation rate	;	$\eta_v = p/\dot{v}$

Shear processes (Sections 15.3 and 15.4) and extensional processes (Sections 15.5) differ in their action on coil molecules (Fig. 15-4). Shear viscosities are important for the processing of polymers by extrusion and injection molding and extensional viscosities for fiber spinning and film blowing (Volume IV). Little is known about volume viscosities = bulk viscosities.

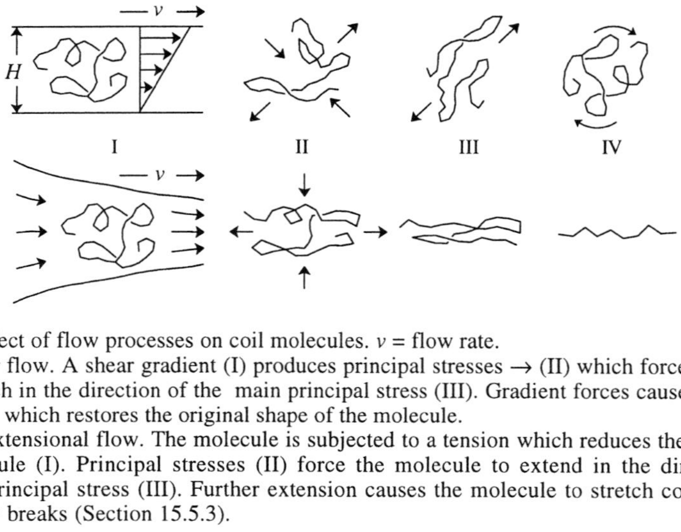

Fig. 15-4 Effect of flow processes on coil molecules. v = flow rate.
Top: shear flow. A shear gradient (I) produces principal stresses → (II) which force the coil molecule to stretch in the direction of the main principal stress (III). Gradient forces cause the molecule to rotate (IV), which restores the original shape of the molecule.
Bottom: extensional flow. The molecule is subjected to a tension which reduces the cross-section of the molecule (I). Principal stresses (II) force the molecule to extend in the direction of the dominating principal stress (III). Further extension causes the molecule to stretch completely (IV) until it finally breaks (Section 15.5.3).

15.2.2 Viscometers

Viscosities η are calculated acording to **Newton's law**, Eq.(15-2), from the ratio of shear stress, σ_{21}, and shear rate (velocity gradient), $\dot{\gamma}$. At small shear rates, η is independent of both the shear rate $\dot{\gamma}$ and time t. Such viscosities η_0 are called **Newtonian viscosities, stationary viscosities, zero-shear viscosities**, or **viscosities at rest**.

Some other rheological quantities are also independent of shear rate, time, and deformation: shear modulus G_0 (Eq.(15-1)), first normal stress difference $N_{1,0}$, elastic shear time $t_{s,0}$ (Eq.(15-4)), and first normal stress coefficient $\psi_{1,0}$ (Eq.(15-3)). All these quantities are material constants.

Kinematic Newtonian **viscosities** are now reported in pascal×second = kg m^{-1} s^{-1} and no longer in poise (1 poise = 0.1 Pa s). Viscosities measure the energy E in joule (1 J = 1 kg m^2 s^{-2}) that is dissipated by a volume flow $\dot{V} = dV/dt$ in m^3 s^{-1}. Newtonian viscosities, η_0/(Pa s), vary over tens of decades, for example, at room temperature, from 10^{-5} (air) and 10^{-3} (water) to 1 (glycerol), 10^2-10^6 (polymer melts), 10^9 (pitch), and 10^{21} (silicate glasses). Depending on the viscosity, very different viscometers (viscosimeters) have to be used for its determination.

In **band viscometers**, an "infinitely long" band runs with a velocity v_{band} through the specimen that is confined between two "infinitely long" parallel plates that are at a distance 2 R. The specimen is at rest near the surface of the plates but moves near the band with the same velocity, v_{band}, as the band itself. The velocity v of the specimen thus changes linearly with the distance to the band from $v = v_{band}$ at $R = 0$ to $v = 0$ at $R = R$. The velocity gradient (= shear rate) is therefore constant, $\dot{\gamma} = dv/dR = const$, which is a distinct advantage over other viscometers that have, for example, parabolic velocity profiles. Unfortunately, band viscometers can only be used for specimens with very high viscosities because of sealing problems.

Rotational viscometers approximate the conditions of a band viscometer by letting a rotor rotate *in* a stator (**Couette type**, Fig. 12-1; see also Fig. 3-13 in Volume I). Rotational viscometers with rotors moving *around* stators are used less often. If the rotor is centered exactly and if the gap between the walls of the rotor and the stator is very narrow, then the velocity gradient is constant within the gap. The resulting raw data need to be corrected for the finite length of the central rotor, the variation of the velocity gradient with the radius, etc. Rotational viscometers typically work with velocity gradients of (10^{-2}-10^4) s^{-1}.

In **oscillation viscometers**, a measuring probe is moved about its center position in the fluid to be measured. An example is a stator that hangs from a torsional wire in a liquid within a rotor. Here, the torsional angle of the wire is a measure of the torsional moment that is imposed on the liquid by the rotor. The viscosity is then calculated from the torsional moment. These viscometers can work with very small torsions at which the applied energy does not disturb the physical structure of the fluid and thus also neither the physical structure of the polymer molecules nor their associates.

Cone-and-plate viscometers consist of a cone that rotates in the melt on a plate (Fig. 12-1). Uniform shear rates are maintained at small angles between the cone and the plate, usually between 0.5° and 3°. Shear rates of these viscometers vary commonly between 10^{-3} s^{-1} and 10^2 s^{-1}. Cone-and-plate viscometers are used for very viscous substances and specimens. For pigmented systems, **plate-plate viscometers** are suitable.

Capillary viscometers are used for low-viscosity solutions (Section 12.1.2) and also for high-viscosity solutions and melts (this Section). The movement of melts and solutions of high concentrations of high-molecular weight polymers requires substantial driving pressures. For this reason, such capillary viscometers for melt viscosities often consist of metal tubes of length L and radius R in a pressure container.

Shear rates are usually between ca. 1 s^{-1} and 10^5 s^{-1}. In laminar flow of liquids in capillaries, they are not constant. The velocity profile is rather parabolic (Fig. 12-1); it ranges from zero at the center of the capillary to high values at the capillary wall. The average shear rate across the diameter of the capillary is $\langle \dot{v} \rangle = pR/(3\ \eta L)$ and the maximum shear rate at the wall is $\dot{v}_{max} = pR/(2\ \eta L)$ (Appendix to Chapter 15).

Industrial viscometers. In industry, flow behavior of melts and concentrated solutions is often characterized by viscosity related numbers that are obtained using standardized equipment and conditions. Since test conditions are invariant, neither shear rates nor shear strains are known.

Falling-ball viscometers measure the time that a sphere needs to roll through a *solution* between two marks in a slanted tube. In **Cochius tubes**, the ascendance time of an air bubble is a measure of viscosity and in **Ford cups**, the time needed by a volume of the liquid to flow under its own pressure through a hole in the bottom of a cup.

Melts of *thermoplastics* are usually characterized by the **melt flow index** (**MFI**) which indicates the grams of polymer that are extruded win 10 min by a standard plastometer under standard loads at a defined temperature. It is proportional to an inverse melt viscosity at an unknown shear rate. The **melt volume index** (**MVI**) measures the distance that a plunger moves in 10 min under a defined load (see Volume IV, p. 85).

Rubbers are often characterized by their **Mooney viscosities**. Here, the polymer is deformed in a cone-and-plate viscometer at constant temperature (e.g., 100°C) and constant rotational velocity. The resilience after a defined time (e.g., 4 min) is reported as the Mooney viscosity, for the example as ML-4/100.

15.3 Newtonian Shear Viscosities

15.3.1 Melt Viscosities of Linear Polymers

Newtonian melt viscosities η_0 of mixtures of polymer molecules with the same constitution and configuration but different molar masses are simply mass averages of the melt viscosities η_i of its i components, $\eta_0 \equiv \overline{\eta}_{w,0} = \sum_i w_i \eta_{0,i} / \sum_i w_i$.

Empirically, it has been found that Newtonian melt viscosities are power functions of molar masses:

$$(15\text{-}5) \qquad \eta_0 = K_\eta M^\varepsilon$$

The exponent ε adopts a low value ε_l at low molar masses and a large value ε_h at high ones (Fig. 14-4). It changes fairly abruptly from ε_l to ε_h at a certain critical molar mass. The lower exponent ε_l depends on the type of polymer; its experimental values vary between ca. 0.85 and 1.8, for example, 1.8 in Fig. 14-4. The higher value seems to be a universal value of $\varepsilon_h \approx 3.4$ for linear polymers (see Figs. 14-4 and 15-5).

A derivation similar to that for (solution) viscosity averages, $\overline{M}_v$, shows that the molar mass M in Eq.(15-5) is a (melt) viscosity average $\overline{M}_\eta$:

$$(15\text{-}6) \qquad M = \{\eta_0/K_\eta\}^{1/\varepsilon} = \{(\Sigma_i w_i \eta_{0,i})/K_v\}^{1/\varepsilon} = \{\Sigma_i w_i M_i^\varepsilon\}^{1/\varepsilon} \equiv \overline{M}_\eta$$

The molar mass M in Eq.(15-6) thus becomes a mass average for $\varepsilon = 1$. If $\overline{M}_w$ is used in Eq.(15-5) instead of the correct $\overline{M}_\eta$, a correction factor q_η has to be introduced that corrects for the width of the molar mass distribution of the polymer:

$$(15\text{-}7) \qquad \eta_0 = K_\eta \overline{M}_\eta^\varepsilon = K_\eta q_\eta \overline{M}_w^\varepsilon$$

The correction factor can be calculated from the exponent ε and the type and width of the distribution. For example, the width of the distribution is characterized by $\overline{M}_w/\overline{M}_n$ for logarithmic normal distributions of molar masses and $\varsigma = \overline{M}_n/(\overline{M}_w - \overline{M}_n)$ for Schulz-Zimm distributions of molar masses where ς = degree of coupling of chains during termination (see Eq.(2-65)). If zero-shear viscosities η_0 are correlated with mass-average molar masses $\overline{M}_w$ instead of viscosity-average molar masses, $\overline{M}_\eta$, the following correction factors have to be used (Γ = gamma function):

(15-8) Schulz-Zimm distribution (SZ) $$q_\eta = \frac{\Gamma(\varsigma+\varepsilon+1)}{(\varsigma+1)^\varepsilon\,\Gamma(\varsigma+1)}$$

(15-9) logarithmic normal distribution (LN) $$q_\eta = (\overline{M}_w/\overline{M}_n)^{(\varepsilon^2-\varepsilon)/2}$$

These corrections are not trivial (Table 15-2). For ε = 3.4 and $\overline{M}_w/\overline{M}_n$ = 1.1, correction factors are already $q_\eta \approx 1.41$ (SZ) and 1.48 (LN). For $\overline{M}_w/\overline{M}_n$ = 2, they increase to $q_\eta = 4.23$ (SZ) and $q_\eta = 16.91$ (LN), respectively.

A survey of literature data shows that melt viscosity = f(molar mass) correlations always refer to dynamic viscosities and not to kinematic ones, that reported dynamic viscosities are not always *measured* zero-shear viscosities, that viscosities are mostly correlated with mass-average molar masses and not with (melt) viscosity averages, and that molar masses have almost never been corrected for molecular non-uniformity.

Table 15-2 Correction factors q_η for the use of mass-average molar masses instead of melt-viscosity average molar masses in Eq.(15-5) (see Eqs.(15-8) and (15-9)).

ε		Schulz-Zimm distribution			Logarithmic normal distribution		
	$\overline{M}_w/\overline{M}_n \rightarrow$	1.1	1.25	2	1.1	1.25	2
↓	$\varsigma \rightarrow$	10	4	1	–	–	–
0.8		0.993	0.984	0.963	0.992	0.982	0.946
3.0		1.289	1.680	3.000	1.330	1.993	8.000
3.4		1.408	1.997	4.225	1.475	2.485	16.910

15.3.2 Rouse Theory

The curious molar mass dependence of Newtonian melt viscosities has been the subject of several theories. **Eyring** applied his theory of rate processes (originally developed for small molecules) to chain molecules in melts by considering that a polymer chain must run a kind of slalom around the segments of other chains. This theory led to exponents of $\varepsilon_l = 4/3 \approx 1.33$ for the low-molar mass region and $\varepsilon_h = 10/3 \approx 3.33$ for the high-molar mass one; these exponents were predicted to be the same for linear and star molecules. Experimental data for 11 linear and star molecules led to $\varepsilon_l = 1.45 \pm 0.20$ and $\varepsilon_h = 3.36 \pm 0.21$. However, the theory was never really accepted.

The **Rouse theory** was more successful. This theory assumes that each hydrodynamic property P_h of a molecule can be described by the product of a local property f_h and a

Table 15-3 Results of Rouse theory and reptation theory (Section 15.3.4) for self-diffusion coefficients D, Newtonian viscosities η_0, and ratios $D\eta_0/\rho_2$. $\nu_0 = \eta_0/\rho_2$ = kinematic viscosity; ρ_2 = density of melt, N_{seg} = number of segments per molecules, ξ_{seg} = friction coefficient of a segment.

	Self-diffusion	Newtonian viscosity	$D\eta_0/\rho$
General	$D = \dfrac{k_B T}{\xi_{seg}} F_D$	$\eta_0 = \xi_{seg} F_\eta$	$D\nu_0$
Rouse	$F_D = \dfrac{1}{N_{seg}}$	$F_\eta = \dfrac{1}{6} N_A \rho_2 \dfrac{\langle s^2 \rangle_o}{M} N_{seg}$	$D\nu_0 = \dfrac{1}{6} RT \dfrac{\langle s^2 \rangle_o}{M}$
Reptation	$F_D = \dfrac{1}{6\, N_{seg}^2}$	$F_\eta = 6 N_A \rho_2 \dfrac{M_{seg}}{M_c} \left(\dfrac{\langle s^2 \rangle_o}{M} \right) N_{seg}{}^3$	$D\nu_0 = RT \dfrac{M_{seg}}{M_c} \left(\dfrac{\langle s^2 \rangle_o}{M} \right) N_{seg}$

global property F_h. The local property is controlled by the friction coefficient ξ_{seg} of a segment whereas the global property is given by factors F_D (self-diffusion) and F_η (viscosity) which describe the macroconformation of the molecule (Table 15-3).

The Rouse theory uses tensors in order to arrive at the results depicted in Table 15-3. Practically the same expression can be obtained by a "back-of-the-envelope" approach:

Chains of macromolecules are assumed to consist of N_{seg} segments with a molar mass M_{seg} and a friction coefficient ξ_{seg} each. The total friction coefficient of the molecule is $\xi_{mol} = N_{seg}\xi_{eg}$ which is related to the effective viscometric radius R_v of the molecule by the **Stokes equation**, $\xi_{mol} = 6\,\pi\eta_0 R_v$, Eq.(11-11). Multiplication of both sides of the Stokes equation by R_v^2 and introduction of the volume $V_{mol} = 4\,\pi\, R_v^3/3$ of an equivalent sphere, the density of the melt $\rho = m_{mol}/V_{mol}$, the molar mass $M = m_{mol}N_A$, and the molecular friction factor $\xi_{mol} = N_{seg}\xi_{seg}$ delivers

$$(15\text{-}10) \qquad \eta_0 = (4/3)(1/6)\, \rho \cdot N_A \xi_{seg} [\, R_v^2/M] N_{seg}$$

The viscometric radius $R_{v,\Phi}$ of the molecule is related to the unperturbed radius of gyration by $\langle s^2 \rangle_0{}^{1/2} = kR_{v,\Phi}$ which converts Eq.(15-10) to

$$(15\text{-}11) \qquad \eta_0 = \zeta_{seg} \left(\frac{4}{3\,k^2} \right) \left(\frac{1}{6} \right) \rho N_A \left[\frac{\langle s^2 \rangle_0}{M} \right] N_{seg}$$

Eq.(15-11) is identical with the exact Rouse equation in Table 15-3 if $k^2 = 4/3$. Introduction of the molar mass $M = N_{seg}M_{seg}$ of the polymer shows that Newtonian viscosities should be directly proportional to the molar mass, which is sometimes found (Fig. 15-5):

$$(15\text{-}12) \qquad \eta_0 = \left(\frac{N_A}{6} \right) \left\{ \frac{\zeta_{seg}}{M_{seg}} \right\} \left[\rho \frac{\langle s^2 \rangle_0}{M} \right] M = K_R M$$

In Eq.(15-12), round parentheses contain general constants, the term in braces indicates local parameters, and the term in square brackets comprises global parameters. K_R is therefore a system-specific constant.

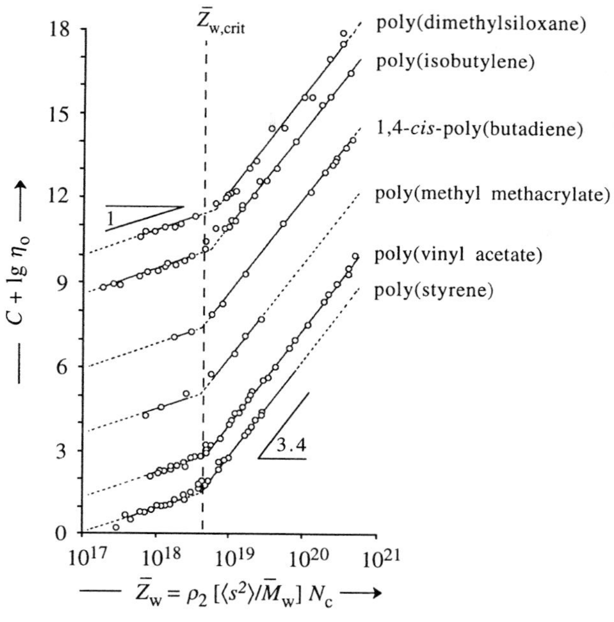

Fig. 15-5 Dependence of the Newtonian viscosity on a parameter $\bar{Z}_w$ that is proportional to the number of chain units per molecule, $N_c = N_a N_{seg}$, the density ρ_2 of the polymer, and the reduced mean-square radius of gyration, $\langle s^2 \rangle / \bar{M}_w$, which is a polymer-specific constant for high molar masses. For clarity, viscosities are shifted vertically by a polymer-dependent constant C. The author [1] did not report the magnitudes of C and also not the physical units of η_0 and $\bar{Z}_w$ which is proportional to $\bar{M}_{e,\eta}$, the viscometric entanglement molar mass (= crossover entanglement molar mass) (see p. 604).

Since $\langle s^2 \rangle_0^{1/2} = kR_{v,\Theta}$ and $k = (4/3)^{1/2}$, it follows that $\langle s^2 \rangle_0^{1/2}/R_{v,Q} \approx 1.155$. The radius of gyration should thus be ca. 16 % greater than the viscometric radius. Experimentally, a value of $\langle s^2 \rangle_0^{1/2}/R_{v,Q} \approx 1.19 \pm 0.10$ was found (Table 12-7).

The temperature dependence of Newtonian viscosities frequently follows an Arrhenius type equation, $\eta_0 = A \exp(-E^{\ddagger}/RT)$, where $E^{\ddagger}$ = activation energy and A = system-specific constant for the constant thermodynamic state (isotropic, nematic, smectic A, etc.). The temperature dependence can also be expressed by the semi-empirical WLF equation, Eq.(13-40), where η and η_I are the Newtonian viscosities at temperature T and a reference temperature T_I and B^* and C polymer-specific constants. The lowest possible reference temperature T_I is the lower transformation temperature of the particular thermodynamic state, for example, the glass temperature:

$$\text{(15-13)} \qquad \lg \eta_{0,\mathrm{rel}} = [-B^*(T - T_I)]/[C + (T - T_I)]$$

15.3.3 Corrections for the Rouse Range

Rouse theory thus predicts that Newtonian viscosities η_0 should be directly proportional to the number N_{seg} of chain segments per molecule (Eq.(15-11)) and thus also to the molar mass, $M = N_{seg}M_{seg}$ (Eq.(15-12)), for the whole range of molar masses. However, this is not found experimentally since there are two regions with different power expressions, $\eta_0 = K_\eta M^\varepsilon$ (Eq.(15-5), Fig. 14-4 and 15-5). In older experiments, exponents $\varepsilon_l \approx 1$ were found for low-molar mass regions (Fig. 15-5) but more recent experiments (for example, Fig. 14-4) show that the exponent ε_l for the lower molar mass region is generally not unity. This is true even if the density ρ in Eq.(15-11) is treated as a variable and one plots $\eta_0/\rho = f(M)$ instead of $\eta_0 = f(M)$.

A trivial reason for $\varepsilon_l \neq 1$ may be the common use of mass-average instead of melt viscosity-average molar masses (Table 15-2), which may cause grave errors in data correlations. Not so trivial is the assumption that the friction coefficient ξ_{seg} in Eq.(15-12) is indeed that of a segment and that a molecule consists of many segments, with the *same* friction coefficient regardless of the size of the molecule.

This assumption is true for entangled molecules where relaxation times are that of a segment between two entanglement sites and not that of a whole molecule. However, small molecules are not entangled and their "segments" move in a more or less coordinated way (see Fig. 14-1). The friction coefficient of a segment of an unentangled molecule will thus increase with the molar mass (because the "segment" is getting larger) until entanglement sets in and the segment size becomes constant since it is now the segment between two entanglement sites. However, Rouse theory assumes that friction coefficients ξ_{seg} are independent of the size of molecules and thus independent of N_{seg}, Z, or M. Values of M must thus be corrected for the effect of variable segment length by replacing $\xi_{seg}M$ in Eq.(15-12) by $(\xi_{seg}/\xi_{seg,crit})^{\varepsilon_i}M$. An example is the correction for poly(dimethylsiloxane) at 25°C: ε_i = 1.00 (uncorrected: ε_i = 1.267) for low M and ε_i = 3.34 (uncorrected: 3.342) for high M.

It was also argued alternatively that one should compare constant segment mobilities, i.e., that melt viscosities of polymers should not be compared at constant temperature T but instead at a characteristic temperature T^*. That characteristic temperature is the glass temperature if one assumes that all polymers have the same free volume fraction at T_G. A plot of lg η (at $T^* = T_G$) as a function of lg $\overline{M}_w$ shows indeed that linear and cyclic poly(dimethylsiloxane)s have the same critical molar masses, $\overline{M}_{w,crit}$, for the transition from the Rouse region to higher molar masses (Fig. 15-6). The function lg η = f(lg $\overline{M}_w$) has practically the same exponent ε_h for linear and cyclic polymers: 3.21 (linear) versus 3.47 (cyclic). However, only linear polymers showed a Rouse exponent of $\varepsilon_l \approx 1$ whereas cyclic ones had a much lower value of $\varepsilon_l = 0.57$.

15.3.4 Reptation

Reptation theory explains the presence of a bend in the function lg η = f(lg M) of linear polymers (Figs. 14-4, 15-5, 15-6) by the onset of chain entanglements (Fig. 11-4) at a critical molar mass. This critical mass varies with the constitution and configuration of polymer chains as well as the density ρ of the melt. Polymer chains that are very expanded per molar mass (those with a high ratio $\langle s^2\rangle_o/M$) can entangle very easily with other chains, i.e., the critical number $N_{c,crit}$ of chain atoms between entanglement points is small. Less expanded coils are more compact (small ratio $\langle s^2\rangle_o/M$); they do not entangle easily and the critical number N_c of chain atoms is high. As a result, changes in the slopes of η = f(M) functions occur at approximately the same value of the product $\rho_2[\langle s^2\rangle_o/\overline{M}_w]N_c$ (Fig. 15-5) where ρ_2 corrects for the varying mass densities.

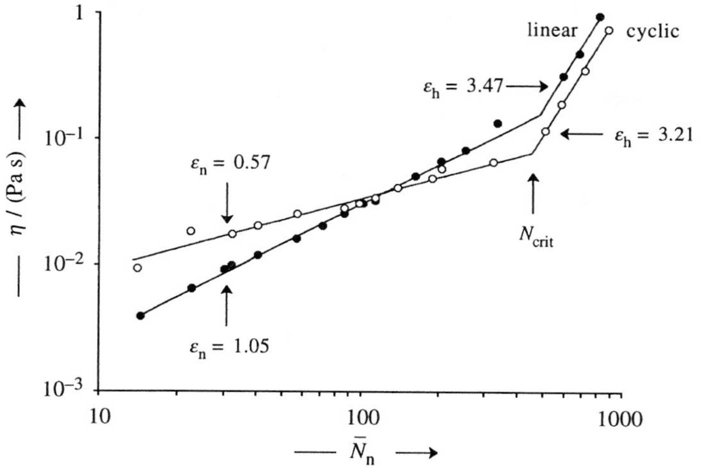

Fig. 15-6 Melt viscosities η as a function of the number-average number $\overline{N}_n$ of chain units of linear and cyclic poly(dimethyl siloxane)s that were measured at a shear rate of $\dot{\gamma}$ = 8.11 s^{-1} and temperatures of $T^* = T_G$ + 148.3 K [2]. By kind permission of Elsevier Science, Oxford.

Entanglements behave like physical crosslinks. At low shear rates (Newtonian range) their number is constant and the melt behaves like a network. The elasticity of such a network is described by the shear modulus G which has the physical unit of pressure. Since the Newtonian shear viscosity (in Pa s) is the product of a pressure (Pa) and a time (s), one can write $\eta_0 = Gt_s$ (Eq.(15-4), where t_s is a characteristic shear time. The shear modulus of the network is assumed to be independent of time.

The reptation theory identifies the characteristic shear time t_s as the reptation time t_{rep}, i.e., the time a chain needs to wriggle itself out of the tube (Fig. 14-5). The reptation time is given by $t_{seg} = [(L_{seg}^2 \xi_{seg})/(2k_BT)]\, N_{seg}^3$ (Eq.(14-3)) where the number N_{seg} of segments per molecule is expressed by $N_{seg} = M/M_{seg}$ so that

$$(15\text{-}14) \qquad \eta_0 = Gt_{rep} = G\left[\frac{L_{seg}^2 \xi_{seg}}{2\, k_B T M_{seg}^3}\right] M^3 = K_\eta M^3$$

According to reptation theory, Newtonian melt viscosities should increase with the third power of molar mass above the critical entanglement molar mass. However, experiments always deliver higher exponents than 3.0, usually exponents of ca. 3.4 (Figs. 14-4, 15-5, 15-6).

A possible reason for the discrepancy between reptation theory and experiment could be a "breathing" of the tube. Such a breathing would push unentangled chain loops back into the matrix.

Chain ends may also be subject to additional relaxations which would shorten the tube length. The resulting fluctuation of the tube should then cause the exponent to adopt a value of 3.4 instead of 3.0.

For infinitely high molar masses, "breathing" of tubes and chain end relaxations can be neglected and this expanded theory therefore expects an exponent of 3.0. However, such high molar masses have never been obtained.

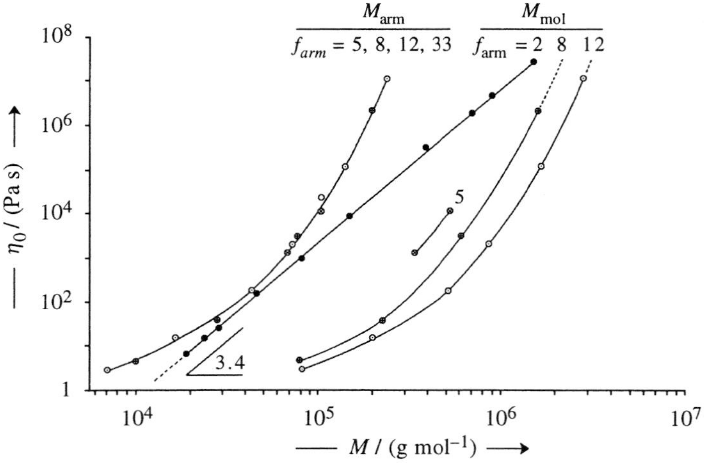

Fig. 15-7 Newtonian melt viscosities of linear (●) and 5- (⊗), 8- (⊕), 12- (⊙), and 33-arm (O) poly(isoprene)s (73 % cis-1,4, 20 % trans-1,4, 7 % 3,4) at 60°C as a function of the molar mass M_{mol} of molecules (right side of graph) and M_{arm} of arms (left side of graph) [3]. Lines obey Eq.(15-15).

15.3.5 Non-linear Macromolecules

The molar mass dependence of Newtonian melt viscosities of star polymers differs considerably from that of linear ones where lg η_0 = f(lg M) is linear for the range $2 \cdot 10^4 \leq M/(\text{g mol}^{-1}) < 1.5 \cdot 10^6$ (Fig. 15-7). In contrast, the function lg η_0 = f(lg M) of star polymers is rather curved and it is also shifted to higher molar masses with increasing numbers of arms per molecule.

According to reptation theory, branching points hinder reptation. The reptating molecule must retract its arms, which is more difficult the longer the arm. The movement of an arm is therefore practically independent of that of other arms. The larger the molar mass M_e between entanglement sites, the more easily the molecule can move through the entangled mass of molecules. Reptation theory predicts that zero-shear viscosities should depend on the ratio of the molar masses of arms, M_{arm}, to entanglement segments, M_e, but not on the molar mass of molecules, M, nor on the number f_{arm} per star molecule. Newtonian viscosities are furthermore greater the higher the polymer concentration c and the larger the Rouse relaxation time of the arm:

$$(15\text{-}15) \qquad \eta_0 = \frac{2t_{\text{arm}}}{5K}\frac{cRT}{M}\left(\frac{M_{\text{arm}}}{M_e}\right)^2 \exp\left(\frac{KM_{\text{arm}}}{M_e}\right) \quad ; \quad K = \text{constant near unity}$$

This equation describes very well the molar mass dependence of Newtonian viscosities of star-like polymer molecules (Fig. 15-7).

Instead of plotting the logarithm of Newtonian melt viscosities as a function of the logarithm of molar mass, one can also plot them as a function of the logarithm of the viscometric radius of molecules from intrinsic viscosities. Again, two straight lines that intersect at a radius of $R_v \approx 4$ nm (Fig. 15-8) are observed for styrene polymers .

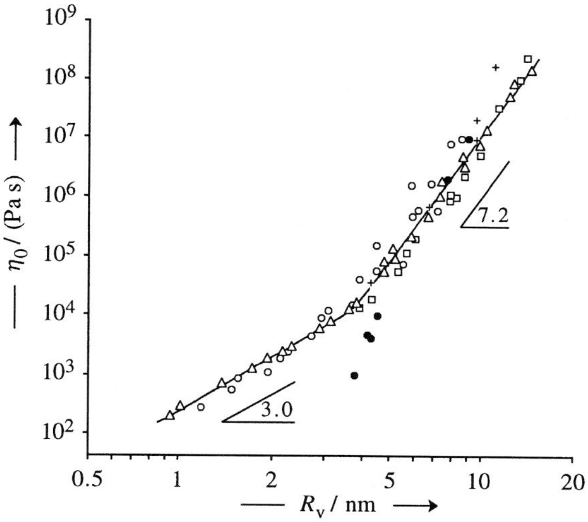

Fig. 15-8 Zero-shear viscosities η_0 of poly(styrene)s at 443 K as a function of viscometric radii R_v [4]. Symbols: Δ Linear (R_v from $[\eta]_\theta$); O cyclic (R_v at $M/2$); □ 3-, 4-, and 6-arm stars (R_v at 2 M_{arm}); + H-shaped (R_v at $M/(5/3)^{1/2}$); ● spherical microgels.
By permission of the American Chemical Society, Washington (DC).

The exponents of the function $\eta_0 = K_R R_v^q$ have values of $q \approx 3$ (low molar masses) and $q \approx 7.2$ (high molar masses), respectively. The functions are independent of molar mass and architecture (linear, cyclic, star, and H-shaped); they even cover larger spherical microgels but not smaller ones with $R_v < 4$ nm. It was not tested whether the observed functions apply also to other polymer structures. It must also be noted that η_0 (melt) and R_v (solution) were obtained at very different temperatures.

15.4 Non-Newtonian Shear Viscosity

15.4.1 Overview

Newtonian viscosities η_0 are given by the ratio of shear stress, σ_{21}, to shear rate, $\dot{\gamma}$, i.e., $\eta_0 = \sigma_{21}/\dot{\gamma}$ (Eq.(15-2)). The slope of a plot of $\sigma_{21} = f(\dot{\gamma})$ thus delivers a straight line with a slope of η_0 (Fig. 15-9 left); η is independent of $\dot{\gamma}$ (Fig. 15-9, right).

All non-Newtonian liquids are characterized by a non-linear dependence of σ_{21} on $\dot{\gamma}$, at least at higher shear rates. The ratio $\sigma_{21}/\dot{\gamma}$ is therefore not constant and the viscosity η calculated from this ratio is an **apparent viscosity.**

There are three major groups of non-Newtonian liquids. *Group I* comprises liquids where the shear stress is linearly dependent on shear rate at *low* shear rates but not at higher ones (Fig. 15-9, left), $\sigma_{21} = 0$ at $\dot{\gamma} = 0$, and apparent viscosities do not depend on the duration of the experiment.

Most liquids of this group are **shear thinning** (Fig. 15-9, right): a constant η at low shear rates $\dot{\gamma}$ is followed by a decrease of apparent viscosities η with increasing shear rate. Shear thinning is observed when asymmetric rigid particles orient themselves in a

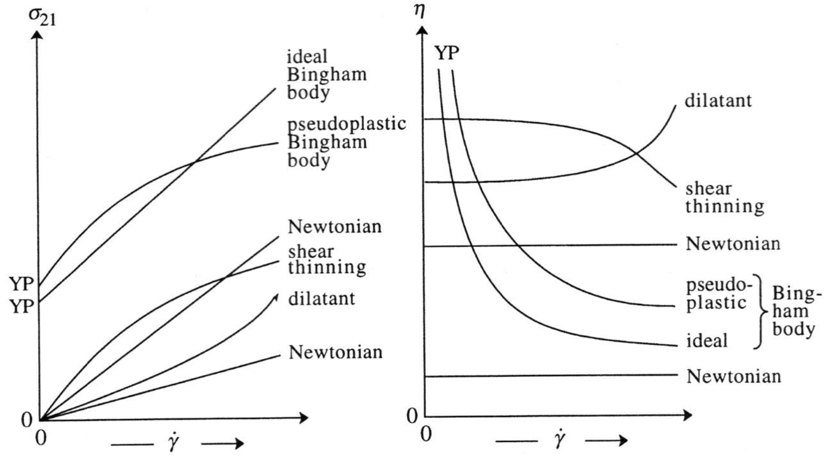

Fig. 15-9 Schematic representation of shear stress σ_{21} (left) and (apparent) viscosity η (right) as a function of shear rate $\dot{\gamma}$ for Newtonian, dilatant, and shear-thinning liquids and ideal and pseudoplastic Bingham bodies. The initial slope of $\sigma_{21} = f(\dot{\gamma})$ of shear thinning and dilatant liquids is the Newtonian viscosity which is independent of the shear rate (see $\eta = f(\dot{\gamma})$). YP = yield point.

flow field and/or flexible molecules are deformed by the velocity gradient. The entities thus adopt a "structure" and shear thinning is therefore also called "structural viscosity." Shear thinning is especially prominent at high molar masses. It is more pronounced the broader the molar mass distribution.

Shear thinning is very desirable for paints. At the high shear rates during painting, paints should flow easily (low apparent viscosity) so that the paint film forms a smooth paint layer. However, paints should not drop from the brush under their own weight which corresponds to low shear rates.

Shear-thinning liquids are sometimes said to have a second Newtonian region at very high shear rates. However, careful experiments have never substantiated this claim; the apparent second Newtonian region is probably caused by turbulent flow. However, because of the presence of this apparent second Newtonian region, shear-thinning liquids are sometimes called **pseudoplastic** since the resulting function $\eta = f(\dot{\gamma})$ resembles that of a pseudoplastic body at high shear rates where η becomes independent of $\dot{\gamma}$.

Far less common are liquids with **dilatancy** (L: *dilatare* = to enlarge, extend), i.e., liquids where the shear stress first increases strongly with increasing shear rate and then less strongly (Fig. 15-9, left) and the viscosity behaves in a Newtonian manner at low shear rates but then increases strongly with the shear rate (Fig. 15-9, right).

Examples of dilatant liquids are certain ionomers in non-polar organic solvents where dilatancy is said to be caused by increasing intermolecular interactions and/or chain extensions. Dilatancy is also observed for some dispersions. It is very important for extensional viscosities (Section 15.5).

Group II materials show **plasticity**, i.e., they have a yield point which is characterized by a finite value of shear stress if the shear rate approaches zero (Fig. 15-9, left). Correspondingly, viscosities are "infinitely large" below a certain shear rate (Fig. 15-9, right). Plasticity is thought to arise from association phenomena. An example is tomato ketchup which flows from the bottle only after it is vigorously shaken.

Materials with plasticity are called **Bingham bodies**. *Ideal Bingham bodies* behave like Newtonian liquids once the shear rate surpasses the yield point. *Pseudoplastic Bingham bodies*, on the other hand, exhibit shear thinning beyond the yield point.

Group I and Group II materials have in common that the application of a shear stress leads immediately to the corresponding shear rate and that the resulting viscosities do not change with time as Group III materials do. The latter group comprises two types:

Thixotropic materials have apparent viscosities that decrease with increasing time at constant shear rate (G: *thixis* (from *thinganein* = to touch); *tropos* = a turn, change). Examples are supensions of silicates with platelet-like shapes such as bentonite or montmorillonite where thixotropy is interpreted as the collapse of a house-of-cards like structure. Thixotropy is a desired property of drilling fluids in oil recovery.

Antithixotropic or **rheopectic materials** show the opposite effect: an increase of viscosity with time at constant shear rate (G: *pexis* = solidified, curdled). An example is the solidification of suspensions of clay or gypsum on rhythmic beating or vibration.

Flow properties can also be affected by **wall effects**. Application of shear stress to certain dispersions and gels causes liquid to be squeezed out which then acts as a lubricant. The resulting plug flow is sometimes very desirable, for example, for toothpaste.

Another complication is caused by turbulent flow which sets in for shear-thinning liquids at far lower Reynolds numbers than for Newtonian ones.

15.4.2 Rheometry

Flow properties of non-Newtonian and Newtonian liquids are usually measured with the same intruments. For capillary viscometers, shear stress σ_{21} is independent of non-Newtonian properties since it depends only on the radius R and length of the capillary L and the driving pressure p, $\sigma_{21} = pR/(2\,L)$ (Eq.(A 15-1)).

However, the shear rate is a much more complicated function of the shear stress for non-Newtonian liquids than for Newtonian ones. The maximum shear rate at the capillary wall is no longer given simply by $\dot{v}_{max} = pR/(2\,\eta_0 L)$ for Newtonian viscosities (Eq.(A 15-3) but has to be calculated from apparent viscosities $\eta_{app} = \sigma_{21}/\dot{\gamma}$ and the change of shear rate $\dot{\gamma}$ with shear stress σ_{21}, $\mathrm{d}\dot{\gamma}/\mathrm{d}\sigma_{21}$, from

(15-16) $$1/\eta = (3/4) + (1/4)\sigma_{21}(\mathrm{d}\dot{\gamma}/\mathrm{d}\sigma_{21})$$ **Rabinowitsch-Weissenberg equation**

The required data are taken from a plot of lg $\dot{\gamma}$= f(lg σ_{21}), the so-called **flow curve** (Fig. 15-10). In this plot, shear-thinning liquids show a ("first") Newtonian range at low values of σ_{21}, followed by shear-thinning, and finally a "second" Newtonian range at high values of σ_{21}.

The existence of a second range with $\sigma_{21} = \eta\,\dot{\gamma}$ is not disputed but it is questionable whether this is a true second *Newtonian* range. High shear rates often cause turbulence so that the flow is no longer a shear flow. High shear rates may also lead to aggregation or to shear degradation and sometimes to a combination of all of these effects so that the polymer in this range is no longer the initial polymer.

The flow behavior in the non-Newtonian range is often described by empirical **flow laws**, for example:

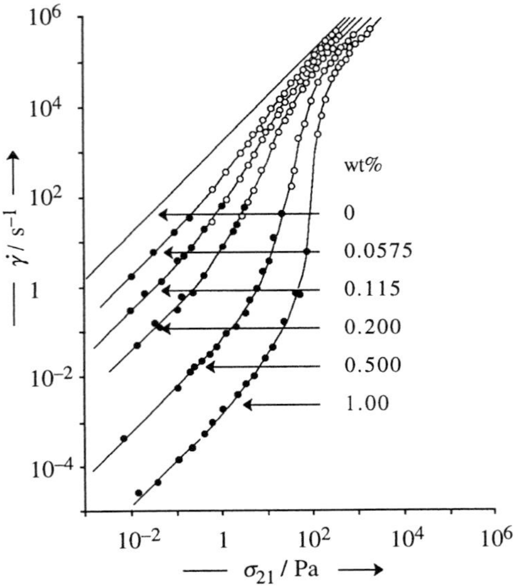

Fig. 15-10 Flow curves of butyl acetate solutions of a cellulose nitrate at 25°C as measured with capillary viscometers (O) or rotary viscometers (●) [5].
By kind permission of the American Physical Society, Melville (NY).

(15-17) $\dot{\gamma} = K_1\ \sigma_{21}{}^{\mathrm{m}}$ **Ostwald-de Waele equation**

(15-18) $\dot{\gamma} = K_2 \sinh (\sigma_{21}/K_3)$ **Prandtl-Eyring equation**

(15-19) $\dot{\gamma} = K_4\sigma_{21} + K_5\sigma_{21}{}^3$ **Rabinowitsch-Weissenberg equation**

where K_1-K_5 are empirical constants. The exponent m is called the **flow exponent** or **pseudo-plasticity index**; it is unity for Newtonian liquids and smaller than unity for shear-thinning fluids. Eq.(15-17) is sometimes written as $\sigma_{21} = K_6\,\dot{\gamma}^{\mathrm{n}}$ where K_6 is the **consistency index** and n the **power law index**. When plotted as lg σ_{21} = f(lg $\dot{\gamma}$), it shows two regions for shear-thinning melts: at low shear rates, the Newtonian range with lg σ_{21} = K_6' + lg $\dot{\gamma}$, and at high shear rates another linear function, lg σ_{21} = K_6'' + n lg $\dot{\gamma}$ with n ≠ 1. In the empirical Vinogradov-Malkin presentation, one plots lg (η/η_0) as a function of lg $\eta_0\,\dot{\gamma}$ which also shows two regions (Fig. 15-11): the Newtonian at low values of $\eta_0\,\dot{\gamma}$ where the logarithm of η/η_0 is independent of $\eta_o\dot{\gamma}$ and the shear-thinning region where lg η/η_0 depends on a negative power of lg $\eta_o\dot{\gamma}$.

Newer flow laws try to cover the whole flow range, often including the experimentally found but physically probably nonsensical "second Newtonian range." These modern flow laws can all be written as

(15-20) $\eta = \eta_\infty + (\eta_0 - \eta_\infty)[1 + K(\dot{\gamma}t_{\mathrm{rlx}})^{\mathrm{a}}]^{-1/\mathrm{b}}$

where $t_{\mathrm{rlx}} = 1/\dot{\gamma}_{\mathrm{crit}}$ is the characteristic relaxation time and $\dot{\gamma}_{\mathrm{crit}}$ = critical shear rate. The constants K, a, and b have different meanings, depending on the model or empirical finding (Table 15-4). The Carreau equation is used industrially by the CAMPUS system to characterize the flow properties of plastics (see pp. 5, 518; see especially Volume IV, Section 5.1.2).

Table 15-4 Constants in Eq.(15-20). q, n = constants.

Author	η_∞	K	t_{rlx}	a	1/b
Cross	η_∞	K	1	2/3	1
Carreau	η_∞	1	t_{rlx}	2	(n – 1)/2
Gaidos-Darby	η_∞	4 q(1 – q)	t_{rlx}	2	1

Interesting results are obtained if the Carreau equation is modified and the resulting expression plotted according to Vinogradov-Malkin. Because a true second Newtonian region is highly improbable (p. 397), one can set $\eta_\infty = 0$. Introducing the standard equation $\dot{\gamma}_{crit} = \sigma_{21,crit}/\eta_0$ (Eq.(15-2), replacing the relaxation time by $t_{rlx} = 1/\dot{\gamma}$, and setting a = by and (n – 1)/2 ≡ 1/b, one obtains

$$\text{(5-21)} \qquad \lg \frac{\eta}{\eta_0} = -\frac{1}{b} \lg \left(1 + \left[\frac{\eta_0 \dot{\gamma}}{\sigma_{21,crit}} \right]^{by} \right)$$

This equation allows one to interpret the curve in the plot of lg η/η_0 as a function of $\eta_0\dot{\gamma}$ (Fig. 15-11). At very small shear rates $\dot{\gamma}$, the term in square brackets approaches zero and so does lg (η/η_0); i.e., one observes Newtonian behavior ($\eta/\eta_0 = 1$). At very large shear rates, Eq.(15-21) reduces to lg $(\eta/\eta_0) = -$ y lg $[\eta_0\,\dot{\gamma}/\sigma_{21,crit}]$.

According to experimental data (Fig. 15-11), poly(ethylene) melts behave in a Newtonian manner up to $\eta_0\,\dot{\gamma} \approx 3{\cdot}10^2$ Pa and then, after a fairly sharp transition at $\sigma_{21,crit} \approx 10^4$ Pa, turn non-Newtonian with an exponent of y = –2/3. These values of $\sigma_{21,crit}$ and y are independent of the polymer constitution (poly(ethylene), poly(styrene), poly(isobutylene)), the molar mass (various poly(ethylene)s), and the temperature (poly(isobutylene)). It seems that these values of $\sigma_{21,crit}$ and y are universal for linear polymers.

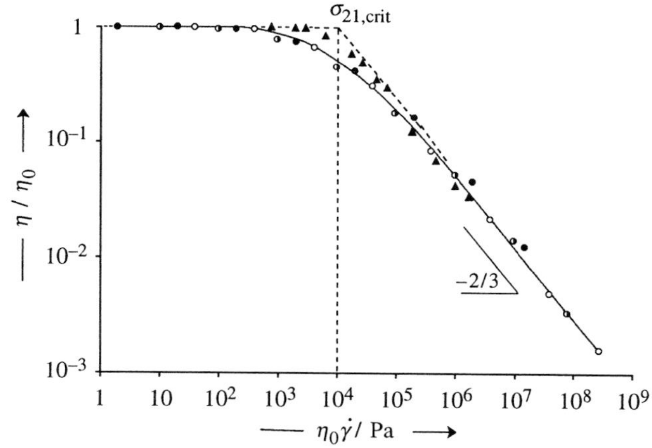

Fig. 15-11 Normalized melt viscosities as a function of the shear stress $\eta_0\,\dot{\gamma}$ for poly(ethylene)s with melt flow indices of 1.7 (O), 7 (◑), and 20 (●) [6] and a glass fiber reinforced atactic poly(styrene) (▲). A poly(isobutylene) delivers the same curve as poly(ethylene) regardless of temperature (–20°C to 80°C) (not shown) [7]. The higher the melt flow index, the smaller is the molar mass.

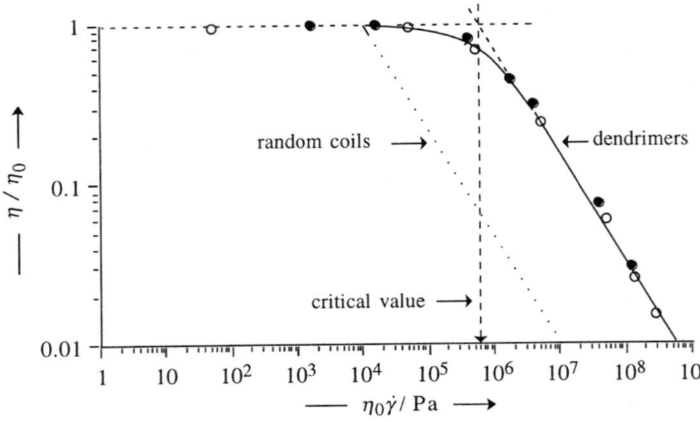

Fig. 15-12 Normalized melt viscosity as a function of the shear stress for the higher generations of polyamidoamine dendrimers. Data for generation G5 (○, M = 28 826 g/mol) and G6 (●, M = 58 048 g/mol) at 40°C from steady shear viscosities a low shear rates and complex viscosities at higher frequencies [8]. The dotted slanted line indicates the ideal non-Newtonian line of random coil-forming polymers in Fig. 15-11. The broken vertical line shows the position of the critical value of $\eta_0\,\dot{\gamma}$ for dendrimers which is ca. 60 times greater than that of random coil polymers (Fig. 15-11).

High-molar mass polyamidoamine dendrimers show similar behavior (Fig.15-12). Dendrimer molecules have a symmetrical branch-upon-branch structure with short segments between branch points. Their deformation thus requires far higher critical stresses which are in this case ca. 60 times higher than those of random coil-forming polymers. However, they have the same exponent of $y \approx -2/3$ in Eq.(15-21).

Low-generation dendrimers G0-G4 behave similarly (not shown). Their critical stress of $\sigma_{21,\text{crit}} = 10^5$ Pa is lower than that of high-generation dendrimers ($6{\cdot}10^5$ Pa) but higher than that of linear polymers (10^4 Pa). However, the exponent y is approximately minus 1 and not ca. – 2/3 as for G5-G6 dendrimers and random coils.

15.4.3 Melt Elasticity

Entanglements act like temporary physical crosslinks and, as a result, polymer melts are somewhat elastic. The flow of entangled polymers generates elastic vibrations which are dampened at small shear rates but less and less at higher ones. At a certain critical shear rate, the polymer melt separates in part from the wall and the flow leads to elastic turbulence at Reynolds numbers that are far lower than those of low-molar mass substances.

Turbulence roughens the surface of the melt. On extrusion of such melts to sheets or pipes or on melt-spinning to fibers, these turbulences are frozen on cooling and the surface of the article or fiber appears "fractured." The term "**melt fracture**" refers to this phenomenon and not to the fracture (breaking) of the polymer strand.

Melt fracture usually does not affect the dependence of melt viscosity on shear rate. However, it may create a jump in the flow curve at very high molar masses.

Extrusion processes deform entangled polymer coils. The entanglements prevent molecule segments from slipping past each other and a normal stress is built up. Upon exiting the nozzle or spinneret, the stress is relieved and the molecules return to their thermodynamically more favorable shape of an unperturbed coil. The melt therefore expands normal to the flow direction and the cross-section of the extruded strand becomes larger than the diameter of the nozzle. This phenomenon is known by many names such as **Barus** or **memory effect** (melts), **parison swell** (extrusion), or **swelling** (formation of hollow articles) (see Volume IV).

The entropy-elastic (viscoelastic) properties are also responsible for the **Weissenberg effect**. Stirring a Newtonian liquid causes the liquid level near the stirrer to be lower than the liquid level at the container wall. However, shear-thinning liquids will climb the drive shaft. This **rod climbing** can be quite dramatic: a 1 % aqueous solution of a high-molecular weight poly(acrylamide) has been reported to climb 30 cm!

Memory effects may be positive (swelling) or negative (contraction). The latter effect occurs if rodlike molecules crystallize after extrusion.

15.5 Elongational Viscosity

15.5.1 Fundamentals

In contrast to low-molecular weight liquids, polymer fluids can be considerably extended without breaking. This property allows fiber spinning from polymer melts and concentrated solutions, blowing of hollow bodies from melts, and vacuum forming of sheets. The extensibility of polymer melts is also important for injection molding.

Extensibility is not a unique property of polymers. It is also found for honey and concentrated soap solutions because the molecules of these substances are physically crosslinked by intermolecular association. However, such extensibilities are short-lived because of the strong fluctuation of the physical crosslinking sites.

Extensional flow always occurs if viscoelastic liquids are elongated. In the simplest case of extensional flow, flow gradients are in the flow direction, whereas they are perpendicular to the flow direction in simple shear flow.

The **extensional viscosity (elongational viscosity)** is defined as the ratio of the tensile stress σ_{11} in the elongational direction to tensile strain rate $\dot{\varepsilon}$:

(15-22) $$\eta_e \equiv \sigma_{11}/\dot{\varepsilon}$$

The elongational viscosity is also called the **Trouton viscosity** after its discoverer and **tensile viscosity** because it depends on tensile properties.

The tensile strain rate, $\dot{\varepsilon} \equiv \mathrm{d}\varepsilon_H/\mathrm{d}t = \mathrm{d}\ln(L/L_0)/\mathrm{d}t$, is defined as the change of **Hencky strain**, $\varepsilon_H \equiv \ln(L/L_0)$, with time where L_0 is the original length. It is not given by the change of the **nominal strain**, $\varepsilon \equiv (L - L_0)/L_0$, with time.

In contrast to shear viscosity, elongational viscosities must always be identified by the type of deformation. The three main deformation rates in 11, 22, and 33 direction (xx, yy, and zz) are defined in such a way that $\dot{\varepsilon}_{11} \geq \dot{\varepsilon}_{22} \geq \dot{\varepsilon}_{33}$. The constant ratio $K = \dot{\varepsilon}_{22}/\dot{\varepsilon}_{11}$ characterizes a special type of elongational flow. These ratios are $K = -1/2$ for uniaxial extensions, $K = 1$ for equal biaxial ones, and $K = 0$ for planar ones (pure

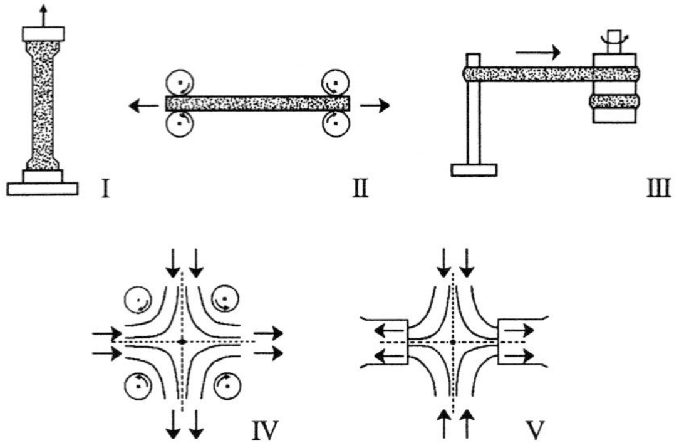

Fig. 15-13 Schematic representation of some extensional viscometers.

Top row: measurement of rubberlike states (including melts of high entangled polymers) with one-sided uniaxial extension (I), uniaxial extensions with clamps or rotating rolls (II), or winding of a band fastened at one side on a rotating drum (III).

Bottom row: measurement of liquids by rotating rolls (IV) or suction nozzles (V). —— Flow lines; - - - symmetry lines; ● stagnation point.

shear). Uniaxial extensions are exploited in the characterization of liquids; they are important for fiber spinning (Volume IV). Biaxial elongations are significant for blow molding and vacuum forming.

15.5.2 Elongational Viscosity of Melts

Elongational viscosities are very difficult to measure. In the simplest case, a clamped "solid" specimen is vertically stretched (Fig. 15-13, I). In order to avoid gravitational effects, the specimen must reside in a liquid that has the same density as the specimen. The maximum extension is controlled by the length of the rheometer. Data must be corrected for neck formation.

Horizontal arrangements of specimens avoid the requirement of equal densities of the specimen and surrounding liquid (Fig. 15-13, II). Neck formation is prevented by rotating rolls. There are also no length restrictions on rheometer dimensions. Such rheometers with rotating clamps allow strain ratios of up to L/L_0 = 1100 for high-molar mass poly(ethylene) melts. The elongational viscosity of elastomers can be measured by attaching one end of an elastomer band to a pole and then winding the other end on a rotating drum (Fig. 15-13, III).

Similar principles guide the design of rheometers for the measurement of elongational viscosities of liquids. In one apparatus, two liquids flow past each other at angles of 0° and 90°; the flow is then redirected by rollers (Fig. 15-13 IV). In another instrument, flows in the 0° and 180° directions are redirected by suction nozzles (Fig. 15-13, V). In both cases, the flow rate is zero at the stagnation point at the center of the instruments but increases in the axial regions. The flow corresponds to a uniaxial extension if cylindrical nozzles are used but to pure shear if the nozzles have sets of parallel slits.

Extensional and shear rates are affected differently by deformation rates (Fig. 15-14). Shear viscosities behave in a Newtonian manner at low shear rates ($\eta_s \neq f(\dot{\gamma})$) and then usually decrease at higher ones (shear thinning). At small deformation rates, extensional viscosities are also independent of uniaxial deformation rates (**Trouton range** with $\eta_e \neq f(\dot{\varepsilon})$) but then increase with increasing deformation rate, pass through a maximum, and then decrease. Deviations from the ideal behavior occur at the same deformation rate for both shear and elongational viscosities (Fig. 15-14). Troutonian uniaxial elongational viscosities are three times as large as Newtonian shear viscosities, $\eta_{e,0} = 3\ \eta_{s,0}$, whereas Troutonian biaxial elongational viscosities are six times as large.

Above the critical deformation rate, the same polymer shows shear thinning but initially dilatant elongational flow. The subsequent maximum of elongational viscosities is higher, the broader the molar mass distribution and the longer the branches that are present per molecule in low-density poly(ethylene)s. The height and width of the maximum is practically independent of the presence of short chain branches.

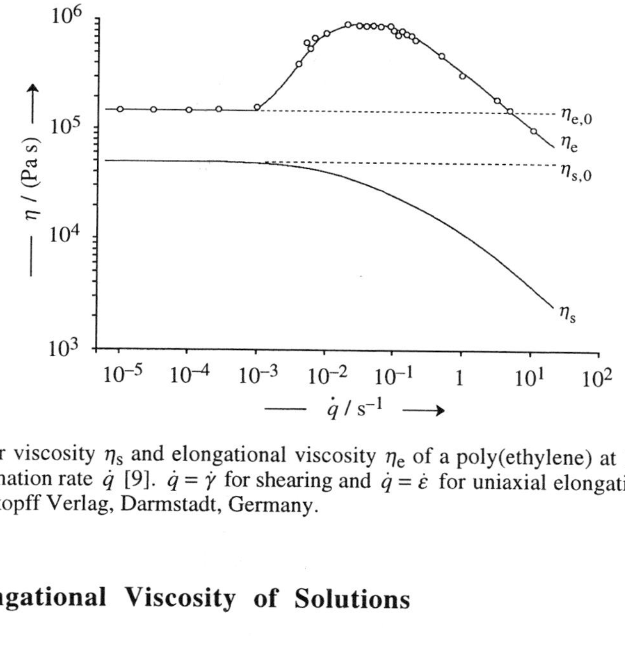

Fig. 15-14 Shear viscosity η_s and elongational viscosity η_e of a poly(ethylene) at 150°C as a function of the deformation rate $\dot{q}$ [9]. $\dot{q} = \dot{\gamma}$ for shearing and $\dot{q} = \dot{\varepsilon}$ for uniaxial elongation. By kind permission of Steinkopff Verlag, Darmstadt, Germany.

15.5.3 Elongational Viscosity of Solutions

Rodlike polymers such as tobacco mosaic virus and very rigid macromolecules with large Kuhnian lengths such as poly(*p*-phenylene-2,6-benzobisthiazole) orient themselves in dilute solution more in the flow direction the larger the extensional flow rate. Molecule axes are then no longer randomly distributed in space and the solution becomes anisotropic. The resulting optical birefringence, Δn, increases first strongly with increasing deformation rate (Fig. 15-15, left) but then becomes independent of the deformation rate (not shown). The asymptotic value of Δn indicates complete orientation of the major molecule axes of all molecules in the flow direction.

Flexible chains like those of atactic poly(styrene) form random coils that deform and orient only a little or not at all at large deformation rates (Fig. 15-15, right) because elastic (entropic) forces cause oriented segments to return fast to the thermodynamically more favorable random coils. However, these elastic retraction forces can be overcome above a (high!) critical elongation rate $\dot{\varepsilon}_{crit}$. Since the major axis is then more or less

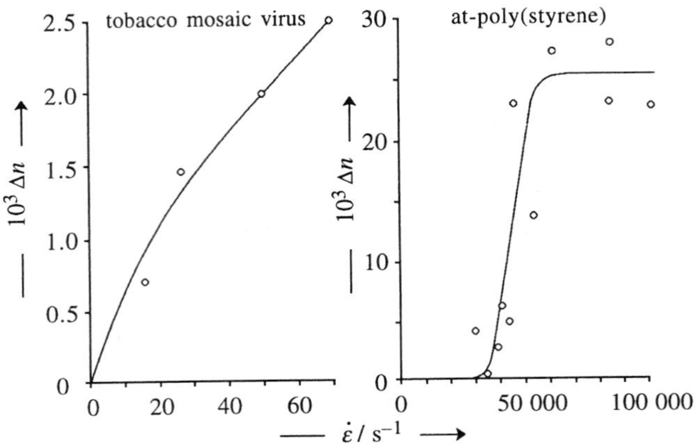

Fig. 15-15 Optical birefringence Δn as a function of the elongational rate $\dot{\varepsilon}$ for (left) a $2{\cdot}10^{-4}$ g/mL solution of tobacco mosaic virus in a 50/50 = V/V water/glycerol solution [10] and (right) a $50{\cdot}10^{-4}$ g/mL solution of an atactic poly(styrene) ($M = 2{\cdot}10^6$ g/mol) in (probably) Decalin® [11].

oriented in the flow direction, only relatively small additional increases of deformation rates suffice to orient the major molecule axes completely. The optical birefringence therefore increases in an S-like fashion with increasing deformation rate.

The critical deformation rate $\dot{\varepsilon}_{crit}$ is an inverse characteristic time, i.e., the time t_{rlx} for the conformational relaxation of coils (*not* the relaxation time of already elongated coils). According to the Zimm theory, relaxation times t_{rlx} of non-draining coils of chains with bond lengths b and bond angles τ between chain atoms in solvents with viscosity η_1 should increase with the 3/2th power of the number N_u of chain units (Eq.(15-23)) or the molar mass, respectively. This is indeed found (Fig. 15-16).

(15-23) $$t_{rlx} = (2\,\pi)\,(\eta_1 k_B T)\,(b^3 \cot^{5/2}\tau/2)\,N_u^{3/2} = 1/\sigma_{21}$$

The Rouse theory, on the other hand, predicts a dependence of relaxation times on the square of the degree of polymerization N_u (Eq.(15-24)):

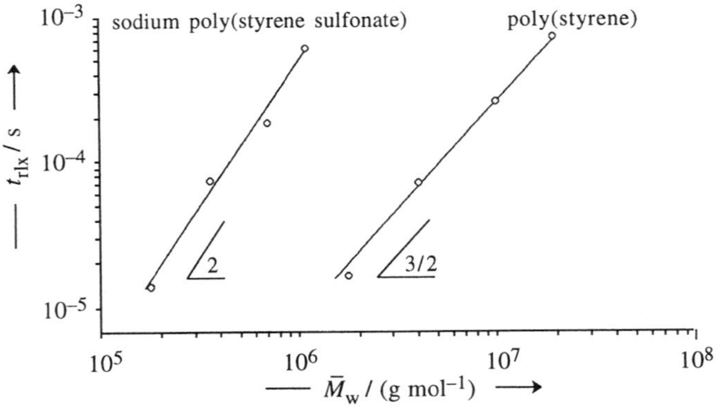

Fig. 15-16 Relaxation time $t_{rlx} = 1/\dot{\varepsilon}$ as a function of the mass-average molar mass of a 0.1 % solution of sodium poly(styrene sulfonate)s in 0.004 mol/L aqueous NaCl [12] and a 0.1 % solution of poly(styrene) in o-xylene [13], both at 25°C. $\overline{M}_w = N_u M_u$, M_u = molar mass of unit.

(15-24) $t_{\text{rlx}} = (\pi/2)\,(\eta_1/k_{\text{B}}T)\,(b^3 \cot^2 \tau/2)\,N_{\text{u}}^2$

Such free-draining coils are formed by polyelectrolyte molecules in dilute salt solutions, for example, by sodium poly(styrene sulfonate)s (Fig. 15-16, left). Relaxation times are in the microsecond to millisecond range.

Above critical elongation rates, molecules are more and more stretched until chain bonds break. The fracture occurs in the center of the molecules so that the resulting molecules consist of 1/2, 1/4, 1/8, etc. of the original molar mass.

The critical elongation rate for a fracture must be inversely proportional to the square of the molar mass as can be derived from the Stokes equation. The reason for this behavior is the fact that the rate of the flow of liquid within the molecule must vary relative to the various monomeric units although the whole molecule moves with the same speed as the liquid itself;

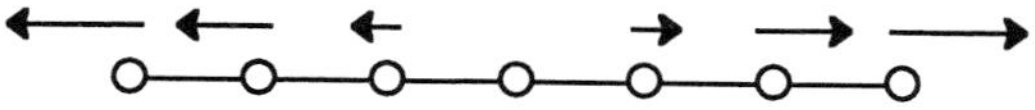

According to Stokes, the force F_i acting on the ith monomeric unit is given by

(15-25) $F_i = 6\,\pi\,\eta_i R_{\text{u}} v_{1,i} Q \quad ; \quad v_{1,i} = i L_{\text{u}}\,\dot{\varepsilon}$

where η_1 = viscosity of solvent, R_{u} = radius of monomeric unit (or subunit in general), $v_{1,i}$ = velocity of solvent relative to the ith unit, Q = screening factor, L_{u} = length of repeating unit, and $\dot{\varepsilon}$ = elongation rate. Summation over all i from $i = 0$ to $i = N/2$ delivers the fracture force for one-half of the molecule. The fracture force for the complete molecule is therefore

(15-26) $F_{\max} = (3\,\pi\,\eta_1 R_{\text{u}} L_{\text{u}} Q\,\dot{\varepsilon} N_{\text{u}}^2)/4$

The critical elongation rate, $\dot{\varepsilon}_{\text{crit}} = \text{f}(N_{\text{u}}^{-1/2})$, for the fracture of the molecule is therefore proportional to the inverse square of the degree of polymerization, N_{u}, or the molar mass, respectively, (Eq.(15-26)), although the critical elongation rate for the elongation itself varies with $N_{\text{u}}^{-3/2}$ (Eq.(15-23)). There must be therefore a molar mass for which $\dot{\varepsilon}_{\text{crit}} = \dot{\varepsilon}_{\text{B}}$. Above this molar mass, molecules can no longer be elongated without fracture. For poly(styrene)s, this molar mass is $30{\cdot}10^6$ g/mol.

This critical molar mass varies with the constitution of macromolecules. It is higher for "smooth" unsubstituted chains such as poly(oxyethylene)s than for chains with bulky substituents. This effect is responsible for the fact that pipetting of dilute solutions of high-molar mass deoxyribonucleic acids leads to polymer degradation although the shear rates are very small.

Chain degradation to lower molar masses by elongational flow is not caused by turbulence since it happens at Reynolds numbers that are ca. 1/3 of that of pure solvents. However, it occurs by turbulence on shear flow of very dilute polymer solutions at concentrations as low as 10^{-4} g/mL. This degradation reduces the friction of solutions by up to 75 %. The phenomenon is called the **Toms effect**. It eases the flow of crude oil through pipe lines, increases the pumping height and width of the water stream for fire fighting, and may even increase the speed of ships.

A-15 Appendix to Chaptor 15: Capillary Viscometry

In capillary flow, the force $\boldsymbol{F}_{\rm f}$ caused by the driving pressure p on a capillary with length L and internal radius R, $\boldsymbol{F}_{\rm f} = \pi R^2 p$, is opposed by a friction force, $\boldsymbol{F}_{\rm r} = 2\,\pi\,RL\sigma_{21}$ where σ_{21} = shear stress. In the steady state, $\boldsymbol{F}_{\rm f}$ equals $\boldsymbol{F}_{\rm r}$ and the the shear stress at the capillary wall is

(A 15-1) $\sigma_{21} = pR/(2\,L)$

Shear stresses are the same for Newtonian and non-Newtonian liquids because they depend only on the measuring system (p, R, L) and not on the properties of the fluid.

The flow v in the capillary changes with the distance from the wall. The change is given by $\mathrm{d}v = \dot{\gamma}\mathrm{d}R$ where $\dot{\gamma}$ is obtained from Eqs.(15-2) and (A 15-1) as $\dot{\gamma} = pR/(2L\eta)$. Integration with the condition $v_{\rm R} = 0$ delivers the velocity v at a distance $y \leq R$ from the wall as

(A 15-2) $$v = \frac{(R^2 - y^2)p}{4\pi L}$$

The shear rate $\dot{\gamma}$ is calculated by modeling the liquid volume in the capillary as a system of concentric hollow cylinders. The volume flow per time through a hollow cylinder with the radii y + (y + dy) is $V_i/t = 2\,\pi\,vy\mathrm{d}y$. Integration over all volume elements, i.e., from $y = 0$ to $y = R$ delivers the total flow volume (**Hagen-Poiseuille law**):

(A 15-3) $$\frac{V}{t} = \int_{y=0}^{y=R} 2\pi v y\mathrm{d}y = \frac{\pi p}{2\eta L}\int_{y=0}^{y=R}(R^2 - y^2)y\mathrm{d}y = \frac{\pi p}{2\eta L}\left[\frac{R^2y^2}{2} - \frac{y^4}{4}\right]_0^R = \frac{\pi p R^4}{8\eta L}$$

The maximum shear rate at $y = R$ is calculated from Eqs.(A 15-1), (A 15-3), and (15-2) as

(A 15-4) $$\dot{\gamma}_{\max} = \left(\frac{\mathrm{d}v}{\mathrm{d}y}\right)_{\max} = \frac{4V}{\pi R^3 t} = \frac{pR}{2\eta L}$$

whereas the average one for the total capillary diameter is

(A 15-5) $$\langle\dot{\gamma}\rangle = \frac{pR}{3\eta L} = \frac{2}{3}\dot{\gamma}_{\max}$$

Eqs.(A 15-4) and (A 15-5) apply only to Newtonin liquids (see Section 15.4.2 for non-Newtonian ones).

Historical Notes to Chapter 15

See Chapters 11, 12, and 14.

Literature to Chapter 15

15.1 GENERAL REVIEWS

P.-G. de Gennes, Scaling Concepts in Polymer Physics, Cornell University Press, Ithaca (NY) 1979

R.T.Bailey, A.M.North, R.A.Pethrick, Molecular Motion in High Polymers, Oxford University Press, Oxford 1981

M.Doi, S.F.Edwards, The Theory of Polymer Dynamics, Oxford University Press, Oxford 1986

R.I.Tanner, K.Walters, Rheology: An Historical Perspective, Elsevier Science, Amsterdam 1994

R.K.Gupta, Polymer and Composite Rheology, Dekker, New York 2000

15.1a NOMENCLATURE

J.M.Dealy, Official Nomenclature for Material Functions Describing the Response of a Viscoelastic Fluid to Various Shearing and Extensional Deformations, J.Rheol. **28**/3 (1984) 181

15.2-15.4 VISCOMETRY AND SHEAR VISCOSITY

M.Reiner, Deformation and Flow, Lewis and Co., London 1949

F.R.Eirich, Ed., Rheology, Theory and Applications, Academic Press, New York 1956-1969 (5 volumes)

A.S.Lodge, Elastic Liquids, Academic Press, London 1964

V.Semjonov, Schmelzviskositäten hochpolymerer Stoffe, Adv.Polym.Sci. **5** (1968) 387 (melt viscosities of polymers)

S.Middleman, The Flow of High Polymers, Interscience, New York 1968

J.A.Brydson, Flow Properties of Polymer Melts, Iliffe Books, London 1970

W.W.Graessley, The Entanglement Concept in Polymer Rheology, Adv.Polym.Sci. **16** (1974) 1

G.Astarita, G.Marruci, Principles of Non-Newtonian Fluid Mechanics, McGraw-Hill, London 1974

J.Schurz, Struktur-Rheologie, Berliner Union, Stuttgart 1974

R.Darby, Viscoelastic Fluids, Dekker, New York 1976

C.D.Han, Rheology in Polymer Processing, Academic Press, New York 1976

L.E.Nielsen, Polymer Rheology, Dekker, New York 1977

R.B.Bird, R.C.Armstrong, O.Hassager, Dynamics of Polymeric Liquids, Vol. 1, Fluid Mechanics; R.B.Bird, O.Hassager, R.C.Armstrong, C.F.Curtiss, ditto, Vol. 2, Kinetic Theory, Wiley, New York 1977

R.S.Lenk, Polymer Rheology, Appl.Sci.Publ., Barking, Essex 1978

W.R.Schowalter, Mechanics of Non-Newtonian Fluids, Pergamon, Oxford 1978

R.S.Porter, A.Casale, Polymer Stress Reactions, Academic Press, New York 1978/79 (2 Vols.)

K.Murakami, K.Ono, Chemorheology of Polymers, Elsevier, Amsterdam 1979

J.D.Ferry, Viscoelastic Properties of Polymers, Wiley, New York, 3rd ed. 1980

K.Walters, Rheometry: Industrial Applications, Res. Studies Press (Wiley), Chichester 1980

G.Astarita, Ed., Rheology, Plenum, New York, 3 Vols. 1980

G.V.Vinogradov, A.Ya.Malkin, Rheology of Polymers, Mir Publ., Moscow; Springer, Berlin 1980

R.W.Whorlow, Rheological Techniques, Harwood, Chichester 1980

F.N.Cogswell, Polymer Melt Rheology, Wiley, New York 1981

W.W.Graessley, Entangled Linear, Branched and Network Polymer Systems - Molecular Theories, Adv.Polym.Sci. **47** (1982) 67

H.Janeschitz-Kriegl, Polymer Melt Rheology and Flow Birefringence, Springer, Berlin 1983

M.Ballauf, B.A.Wolff, Thermodynamically Induced Shear Degradation, Adv.Polym.Sci. **85** (1988) 1

R.Larsen, Constitutive Equations for Polymer Melts and Solutions, Butterworths, London 1988

S.W.Churchill, Viscous Flows: The Practical Use of Theory, Butterworths, Stoneham (MA) 1988

N.W.Tschoegl, The Phenomenological Theory of Linear Viscoelastic Behavior, Springer 1989

J.Klafter, J.M.Drake, Eds., Molecular Dynamics in Restricted Geometries, Wiley, New York 1989

H.A.Barnes, J.F.Hutton, K.Walters, An Introduction to Rheology, Elsevier Science, Amsterdam 1989 and 1993

P.-G. de Gennes, Introduction to Polymer Dynamics, Cambridge Univ. Press, New York 1990

N.P.Cheremisinoff, An Introduction to Polymer Rheology & Processing, CRC Press, Boca Raton (FL) 1993

C.W.Macosko, Rheology. Principles, Measurements and Applications, VCH, Weinheim 1994

C.L.Rohn, Analytical Polymer Rheology. Structure-Processing-Property Relationships, Hanser, Munich 1995

A.V.Shenoy, D.R.Saini, Thermoplastic Melt Rheology and Processing, Dekker, New York 1996; Hall, London 1996
D.C.Rapaport, The Art of Molecular Dynamics Simulation, Cambridge Univ. Press, New York 1997
P.J.Carreau, D.C.R. De Kee, R.P.Chhabra, Eds., Rheology of Polymeric Systems: Principles and Applications, Hanser/Gardner, Cincinnati 1997
S.Middleman, Introduction to Fluid Dynamics, Wiley, New York 1999
R.G.Larson, The Structure and Rheology of Complex Fluids, Oxford Univ.Press, New York 1999
F.A.Morrison, Understanding Rheology, Oxford Univ. Press, New York 2001
J.Furukawa, Physical Chemistry of Polymer Rheology, Springer, Berlin 2003
J.M.Dealy, R.G.Larson, Structure and Rheology of Molten Polymers, Hanser Gardner, Cincinnati (OH) 2006

15.5 ELONGATIONAL VISCOSITY
J.W.Hill, J.A.Cuculo, Elongational Flow Behavior of Polymeric Fluids, J.Macromol.Sci.Revs. **C 14** (1976) 107
J.C.S.Petrie, Elongational Flows (Research Notes in Mathematics, Bd. 29), Piman, London 1979
J.Meissner, Alte und neue Wege in der Rheometrie der Polymer-Schmelzen, Chimia **38** (1984) 35, 65 (old and new ways in rheometry of polymer melts)
J.Ferguson, N.E.Hudson, Extensional Flow of Polymers, in R.A.Pethrick, Ed., Polymer Yearbook **2**, Harwood Academic Publ., Chur, Switzerland 1985, p. 155
A.Keller, J.A.Odell, The Extensibility of Macromolecules in Solution: A New Focus for Macromolecular Science, Colloid Polym.Sci. **263** (1985) 181
J.Meissner, Makromol.Chem.-Makromol.Symp. **56** (1992) 25
T.Q.Nguyen, H.H.Kausch, Flexible Polymer Chain Dynamics in Elongational Flow, Springer, Berlin 1999

References to Chapter 15

[1] T.G.Fox, J.Polym.Sci. **C 9** (1965) 35, Fig. 2
[2] D.J.Orrah, J.A.Semlyen, S.B.Ross-Murphy, Polymer **29** (1988) 1452, Table 2 and Fig. 1
[3] D.S.Pearson, S.J.Mueller, L.J.Fetters, ACS Polymer Preprints **23**/2 (1982) 21; L.J.Fetters, private communication, 21 May 1982
[4] M.Antonietti, T.Pakula, W.Bremser, Macromolecules **28** (1995) 4227, Fig. 7
[5] W.Philippoff, F.H.Gaskins, J.G.Brodnyan, J.Appl.Phys. **28** (1957) 1118, Fig. 2
[6] Numerical data from a graph in BASF company literature
[7] Fig. 2.54 in G.V.Vinogradov, A.Ya.Malkin, Rheology of Polymers, Mir Publ., Moscow, and Springer-Verlag, Berlin 1980 (translation of Reologia Polimerov, Khimia, Moscow 1977). The figure was first published by E.Mustafaev, A.Ya.Malkin, E.P.Plotnikova, G.V.Vinogradov, Vysokomol.Soedin. **6** (1964) 1515
[8] P.R.Dvornic, S.Uppuluri, Rheology and Solution Properties of Dendrimers, in J.M.J.Fréchet, D.A.Tomalia, Eds., Dendrimers and Other Dendritic Polymers, Wiley, New York 2001, Fig. 14.10; data points randomly selected from published curves.
[9] H.M.Laun, H.Münstedt, Rheol.Acta **17** (1978) 415, Fig. 5
[10] D.P.Pope, A.Keller, Colloid Polym.Sci. **255** (1977) 633, Fig. 8b
[11] D.P.Pope, A.Keller, Colloid Polym. Sci. **256** (1978) 751, Table 1
[12] A.Keller, J.A.Odell, Colloid Polym.Sci. **263** (1985) 181, Fig. 22
[13] C.J.Farrell, A.Keller, M.J.Miles, D.P.Pope, Polymer **21** (1980) 1292, Fig. 3

16 Elasticity

16.1 Introduction

Experience shows that mechanical forces cause bodies to change their shape and/or their size; for example, by compression, shearing, drawing, etc. (Fig. 15-1). The forces $\boldsymbol{F}$ causing these deformations are related to the affected areas A and reported as mechanical stresses, $\sigma = \boldsymbol{F}/A$. Conversely, deformations lead to mechanical stresses.

For mechanical properties of thermoplastics, thermosets, elastomers, fibers, etc., the relevant physical quantities and their physical units are

energy, work	E	in	J
force	$\boldsymbol{F}$	in	J m^{-1} = N
interfacial tension	γ	in	J m^{-2} = N m^{-1}
mechanical stress, modulus	$\sigma = E/V = F/A = p$	in	J m^{-3} = N m^{-2} = Pa

These quantities are also used for industrial fibers. Textile fibers are usually characterized by

specific modulus (textile modulus)	E/m	in	J kg^{-1}
linear mass (titer)	$m/L = \rho A$	in	g km^{-1} = tex

The modulus σ (in GPa) is the product of textile modulus (in N/tex) and density (in g/cm^3).

An **ideal-elastic body** is deformed by a load but returns "immediately" to its original shape if the load is removed. Examples are the dropping of a steel sphere on a steel plate or the moderate stretching of a rubber band. Steel and rubber differ in the accompanying heat effects, however. Steel becomes cooler but rubber gets warmer.

This distinct behavioral difference is produced by different molecular mechanisms by which these materials dissipate the stress. The deformation of steel causes iron atoms to leave their positions at rest. The deviation from these positions is affine: the entropy does not change but the enthalpy does because the energy for the shifting of iron atoms from their positions at rest is taken from the system, i.e., the steel becomes cooler. Steel is therefore an **energy-elastic body** as are also polymer crystals (Section 16.3).

Strong deformation of elastomers, however, causes chain segments to slip past each other; this internal friction produces heat and less random microconformations in chain segments. Thus, elastomers are **entropy-elastic bodies** (Section 16.4).

Bodies are enthalpy-elastic (commonly called "energy elastic") or entropy-elastic only at small deformations. Larger deformations are not completely reversible and the body remains somewhat deformed after the load is removed. Such behavior is characteristic of **viscoelastic bodies**, i.e., materials that behave both elastically (reversible deformation) and in a viscous manner (irreversible deformation) (Chapter 17). Under most deformation conditions, practically all plastics are viscoelastic.

For all mechanical measurements it is decisive how and to what extent the applied force is converted to a deformation or *vice versa*. For example, classical theory of elasticity assumes that the applied stress σ is directly proportional to the resulting deformation ε, i.e., that stress and deformation are linearly correlated. The proportionality constant for stress = f(deformation) is the decisive **descriptor**.

The interdependence of stress and strain can be studied by three groups of methods:

Macroscopic methods treat matter as a homogeneous continuum and describe the interrelationship between stress and deformation by the equations of classical mechanics. These equations are formulated in such a way that the resulting descriptors can also be used for other types of deformation. For example, data from torsional experiments should also be able to describe the stress-strain behavior of the same body. These descriptors are specific and need to be determined separately for each body since they are descriptors of the **specimen** and not of the **material** *per se.*

Microscopic methods treat the material as a heterogenous mixture of various components that have different properties. Each component is homogeneous (but not necessarily isotropic). The material therefore consists of different phases. The descriptors are now both the properties of each phase and the interrelationships between phases.

Molecular methods do away with the assumption of a continuum. Instead, properties are described by intramolecular and intermolecular force fields.

The application of the various theoretical methods frequently suffers from the fact that polymeric bodies are often in non-equilibrium states. In these cases, mechanical properties are determined by both the thermal history of the specimen and the type and speed of the measurement. In addition, measurements should often simulate the behavior of polymers under *practical* conditions, which means that the specimen is subjected simultaneously to various forces and deformations, for example, a combination of tensile and torsional strains.

In order to obtain comparable descriptors, shapes, and dimensions of specimens as well as the type, load, duration, etc., of the test, experiments need to be standardized. Such standards have been developed by the *International Standardization Organization* (ISO) and many national organizations such as the *American Society for Testing and Materials* (ASTM), the *British Standards Institution* (BSI), and the *Deutsches Institut für Normung* (DIN; German Standardization Institute). The European Community has developed *European Norms* (EN) that are supposed to supersede the various national standards but unfortunately often deviate from the standards of ISO.

There are approximately 5000 standards for plastics which vary from country to country. In addition, many standards, including ISO, allow the use of various types of test specimens or test methods. Since this again leads to many data that are not comparable (and to high costs of testing!), more than 40 important producers of plastics have agreed to adhere to the CAMPUS® system of testing (see Volume IV).

CAMPUS® is the acronym for *C*omputer *A*ided *M*aterials *P*reselection by *U*niform *S*tandards, a system with ca. 50 different characteristic descriptors for mechanical, electrical, optical, thermal, and rheological properties. It must be remembered, however, that this system, like all other testing standards, delivers descriptors for *specimens* and not material constants *per se* because the resulting descriptors depend on the shape and dimensions of the test specimen as well as the type of testing. For example, heat distortion temperatures (obtained with a standard load) are not identical with the glass temperature (zero load) and melt flow indices are neither inverse Newtonian viscosities nor well-defined inverse non-Newtonian viscosities. In addition, it must be remembered that one measures the properties of *substances* and not that of *molecules* (but see Section 16.4.2).

It is therefore convenient to treat the behavior of polymers on deformation in three chapters. The present Chapter 16 discusses the response of polymers to very small deformations of short durations where polymers show practically an ideal elastic behavior. Chapter 17 discusses effects at larger deformations where polymers behave viscoelastically and Chapter 18 treats the fracture of polymers. Chapters 16 and 17 are mainly concerned with principles; practical aspects of plastics, elastomers, and fibers are discussed in greater detail in Volume IV together with electrical and optical properties.

16.2 Tensile Properties

16.2.1 Basic Terms

Nominal Tensile Properties

Tensile tests are the most important mechanical mesurements. The **test specimen** is standardized; it is either a rectangular (flat) bar or a dumbbell ("dog bone") (Fig. 16-1).

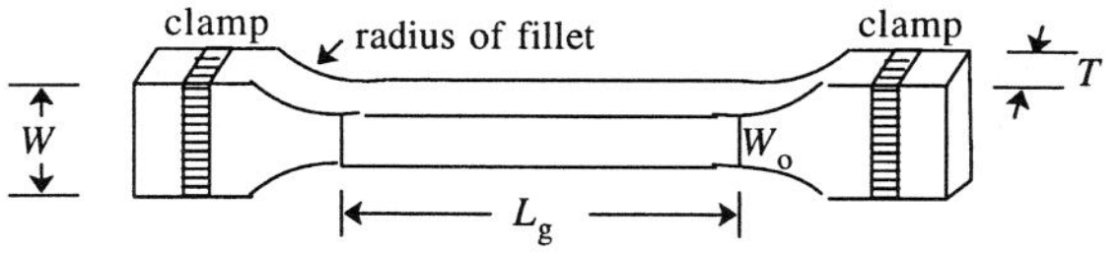

Fig.16-1 Dumbbell. L_g = gauge length, W = overall width, W_o = gauge width, T = thickness.

Test specimens with initial length L_0 are clamped on both sides and drawn with constant speed in the longitudinal direction to various lengths L. For engineering purposes, the applied force $\boldsymbol{F}$ is divided by the smallest *original* cross-sectional area $A_0 = W_0T$ of the test specimen and reported as **nominal tensile stress = engineering stress**, $\sigma_{11} = F/A_0$, usually just called **tensile stress**. Nominal tensile stresses are followed as a function of time t, **draw ratio (strain ratio)** $\lambda = L/L_0$, or **nominal elongation (elongation, Cauchy elongation, tensile strain, engineering strain)**, $\varepsilon = \lambda - 1 = (L - L_0)/L_0 = \Delta L/L_0$ (Fig. 16-2). A lengthening of the test specimen to three times its original length ($L = 3\ L_0$) thus produces an elongation of $\varepsilon = 2$ which is usually reported as 200 %.

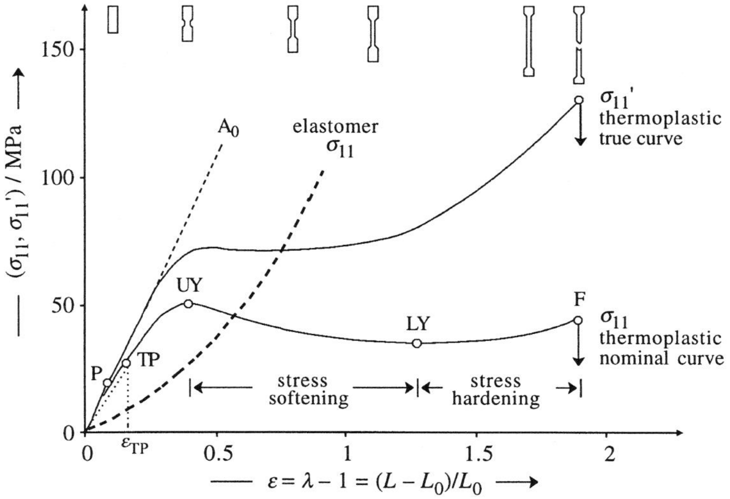

Fig. 16-2 Nominal and true stress-strain curves produced by a rectangular test specimen of a thermoplastic (——) with E = 200 MPa, σ_Y = 50 MPa at ε_{UY} = 0.38 (38 %), and σ_F = 45 MPa at ε_F = 1.9 (190 %) and an elastomer (- - -, schematic). Nominal tensile stress σ_{11} and true stress σ_{11}' are both shown as functions of nominal elongation ε. The true stress-strain curve was calculated from the nominal one taking into account the telescope effect (= tapering of test bar during drawing (see top).

Letter designations at the function $\sigma_{11} = f(\varepsilon)$ indicate A_o = initial asymptote, P = proportionality limit, TP = technical proportionality limit, UY = upper yield point, LY = lower yield point, F = fracture limit. For the meaning of these terms, see next page.

Science and industry mostly use the same terminology, albeit often with different meanings. The **proportionality limit** P (or **proportionality point**) of science refers to the point where the stress-strain curve deviates from Hooke's law (see Section 16.2.2), $\sigma_{11} = E\varepsilon$, where σ_{11} = tensile stress, E = modulus of elasticity, and ε = elongation. For thermoplastics, this point is difficult to determine since it is at elongations of only several hundredths of a percent. The **technical proportionality limit** TP (elastic limit) of thermoplastics is therefore obtained by using a defined nominal elongation of tens of a percent, ε_0, and then drawing a vertical line to the actual stress-strain curve (see Fig. 16-2). The (**technical**) **tensile modulus** E_t (usually just called E) is the slope given by the horizontal axis and the straight line between the origin and the point TP. It is somewhat smaller than the true modulus of elasticity which is calculated from the line between point P and the origin. The reported moduli of elastomers are never true moduli of elasticity since they are either secant moduli or tangent moduli at 100 % or 300 % elongation (Fig. 17-1).

With increasing elongation, elastomers are more and more stretched which requires more and more force and leads to smaller and smaller cross-sections. Stress-strain curves of elastomers are therefore convex to the ε-axis (Fig. 16-2).

Stress-strain curves of thermoplastics, on the other hand, are initially concave to the ε-axis. Thermoplastics and some thermoplastic fibers then show **necking**, i.e., formation of a neck in the specimen (top of Fig. 16-2) which leads to a maximum in the nominal stress-strain curve, the **upper yield point**, UY. This neck formation is caused by small differences in the cross-section of the test specimen. These differences act as "weak points": the reduced cross-sections lead to slightly increased tensile stresses at these points which causes the viscosity at these points to decrease, resulting in a neck, and so forth.

Necks are first formed near clamps since clamps exert a pressure on the specimen which leads to an increased stress concentration near the clamps. Since the elongation of the specimen causes an internal friction and thus a production of heat, the temperature of the test specimen will rise especially at the points with higher stress concentrations; temperature increases up to 50 K have been observed near the clamps. However, neck formation is not *caused* by the developed heat since necking is also observed for isothermal conditions.

Once formed, the neck travels along the specimen and gets longer and longer (**telescope effect**) but the cross-section of the necks remains constant. The nominal tensile stress becomes smaller (**stress softening**) or stays constant (very rare) and the specimen gets longer without being externally heated. Because this was contrary to the behavior of other materials, the effect was dubbed **cold flow**.

Necking and the subsequent onset of telescoping produce a maximum in the nominal stress-strain curve (Fig. 16-2), the coordinates of which are ε_Y (**yield point**) and $\sigma_{11,Y}$ (**yield stress**). Amorphous polymers show yield points at temperatures below the glass temperature whereas semicrystalline polymers have yield points between the glass temperatures and melting temperatures.

The specimen fractures at an **elongation at break** (or **fracture elongation**) ε_B with a **tensile strength at break** (or **fracture strength**) σ_B. The "tensile strength" of literature, if unspecified, refers to the stress at that eleongation where the material fails. For materials that yield, it is the yield stress σ_Y, whereas for non-yielding polymers it is the fracture strength σ_B. The former materials are usually thermoplastics whereas the latter ones are mainly thermosets.

True Stress-Strain Curves

Neither tensile stresses σ_{11} nor elongations ε are true stress-strain data because the drawing reduces the cross-sectional area of the specimen from A_0 to A even if telescope effects are absent. However, the total volume V may remain constant, $V_0 = A_0L_0 = AL$, on drawing the specimen to a length L from L_0 by a constant force $\boldsymbol{F}$. The true tensile stress σ_{11}' is therefore

$$\text{(16-1)} \qquad \sigma_{11}' = F/A = (F/A_0)(L/L_0) = \sigma_{11}(L/L_0) = \sigma_{11}\lambda$$

The volume of the specimen becomes larger if the drawing produces voids and/or crazes (see below). It becomes smaller if the drawing causes crystallization and segmental orientations.

The draw ratio $\lambda = L/L_0$ and the Cauchy elongation $\varepsilon = (L - L_0)/L_0$ are only nominal. If a body with an original length L_0 is extended continuously by ΔL and if this extension can be thought of as the sum of two extensions, $\Delta L = \Delta L_1 + \Delta L_2$, then the total extension is given by $\varepsilon_{12} = (\Delta L_1 + \Delta L_2)/L_0$. However, this extension differs from that at two consecutive extensions, $\varepsilon_1 + \varepsilon_2 = (\Delta L_1/L_0) + (\Delta L_2/(L_0 + \Delta L_1))$. Because of $\varepsilon_{12} \neq \varepsilon_1 + \varepsilon_2$, the elongation $\varepsilon = (L - L_0)/L_0$ cannot be the true elongation (**Hencky elongation**) ε'. Instead, the Hencky elongation is obtained by the summation of infinitesimally small length changes per momentary length, i.e., by integration of all relative length changes:

$$\text{(16-2)} \qquad \varepsilon' = \int_{L_0}^{L'} (\mathrm{d}L / L) = \ln(L'/L_0) = \ln(1+\varepsilon)$$

Stress Softening

Stress softening is the decrease of nominal stress with nominal extension (Fig. 16-2). It is caused by shear flow, formation of crazes, and/or shear bands and shear zones.

Shear flow is caused by the movement of chain segments relative to each other. The volume of the specimen does not change. Shear flow may be global (homogenous flow comprising the whole specimen) or local.

Local shear flow is caused by inhomogeneities in the specimen. It leads to the formation of **shear bands** at angles of 38-45° to the direction of stress (Fig. 16-3). Chain segments are arranged at angles to the shear bands and the direction of stress. Correspondingly, shear flow in thin films lead to **shear zones**.

Crazes are up to 100 µm long and up to 10 µm wide (Fig. 16-3). Their long axes are perpendicular to the stress direction. They are not cracks since their interior is not empty but filled with amorphous microfibrils of 0.6-30 nm diameter that are anchored to the continuous matrix. These microfibrils are oriented in the stress direction and are therefore perpendicular to the long axes of the crazes.

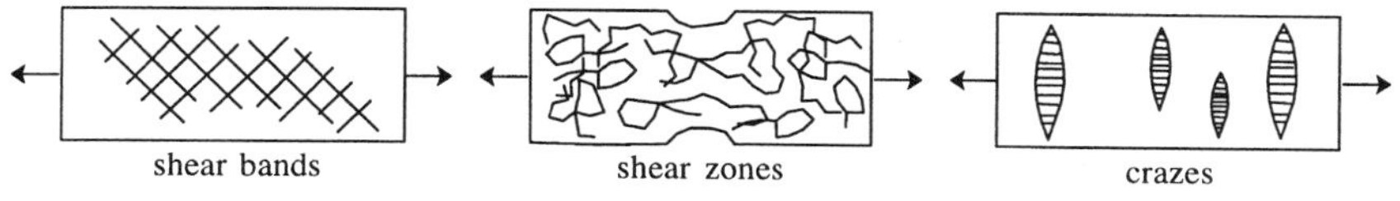

Fig. 16-3 Microdeformations in polymers. Arrows indicate stress directions.

The dimensions of crazes are larger than one-half of the wavelengths of incident light (0.4 μm to 1.1 μm). Since also the refractive indices of the matrix, the microfibrils, and the air all differ, an intense light scattering is produced that leads to a whitening of the specimen. This **stress whitening** is not accompanied by a fracture of the specimen.

On further elongation, more and more crazes are formed. The concentration of microfibrils increases, which leads to an increased resistance to deformation which counteracts the stress softening. The nominal stress thus passes through a minimum and increases again (**stress hardening, strain hardening**) (Fig. 16-2) until the specimen finally fractures.

16.2.2 Hooke's Law

The longitudinal elongation of an isotropic rod of length L_0 and cross-sectional area A_0 requires a force $\boldsymbol{F}$. At *very small* loads, the change of length, $\Delta L = L - L_0$, is proportional to both the initial length L_0 and the applied force F. The change of length is inversely proportional to the initial cross-sectional area A_0:

(16-3) $$\Delta L = const\ (L_0/A_0)\boldsymbol{F} = D(L_0/A_0)\boldsymbol{F} = (1/E)(L_0/A_0)\boldsymbol{F}$$

Introduction of the Cauchy elongation (= engineering strain), $\varepsilon = \Delta L/L_0$, and the engineering stress, $\sigma_{11} = \boldsymbol{F}/A_0$, leads to **Hooke's law**, which is a limiting law for infinitesimally small elongations:

(16-4) $$\sigma_{11} = E\varepsilon = (1/D)\varepsilon \quad ; \quad E = \lim_{\varepsilon \to 0} (\mathrm{d}\sigma_{11} / \mathrm{d}\varepsilon)$$

The proportionality constant D in Eq.(16-3) is called the **tensile compliance**, the inverse of which is the **tensile modulus** E, usually called **Young's modulus** or **modulus of elasticity** (but see below; L: *modulus* = small measure, diminutive of *modus* = measure).

According to Hooke's law, nominal stresses σ_{11} are directly proportional to nominal elongations ε in the limiting case of diminishingly small elongations. This proportionality no longer applies to larger elongations, but the upper range of proportionality, the **proportionality limit**, is not easy to determine. For this reason, tensile moduli are often determined as the secant from the origin of the coordinate system to a defined **elasticity limit** and not as a tangent of the curve $\sigma_{11} = \mathrm{f}(\varepsilon)$. For plastics, the elasticity limit is defined as the elongation of 0.01 % that remains if the stress is removed. In the CAMPUS® system, tensile moduli of plastics are obtained from secants to elongations between 0.05 and 0.25 %. For elastomers, secants to 100 % elongation (100 % modulus) or 300 % elongation (300 % modulus) are often used (for details see Section 17.1.2).

Elongation is a deformation in the [11]-direction; it has counterparts in shearing (deformation in the [21]-direction) and pressurizing (= all-sided compression; Section 15.1). In static tangentional shearing of isotropic bodies, **shear stresses** $\sigma_{21} \equiv \tau$ and **elastic shear deformations** γ_e are correspondingly interconnected by the **shear modulus** $G = \sigma_{21}/\gamma_e$, the inverse of which is **shear compliance** J. In analogy, an all-sided static pressure p causes an **all-sided compression (volume strain, bulk strain)**, $\theta = \Delta V/V_0$, of *isotropic* bodies; the proportionality constants here are the **bulk modulus (compression modulus)** K and the **bulk compliance** B.

Moduli		*Compliances*	
Tensile modulus	$E = \sigma_{11}/\varepsilon$	Tensile compliance	$D = 1/E$
Shear modulus	$G = \sigma_{21}/\gamma_e$	Shear compliance	$J = 1/G$
Bulk modulus	$K = p/(-\Delta V/V_o)$	Bulk compliance	$B = 1/K$

The moduli E, G, and K are all measures of elasticity albeit for different types of deformation: elongation (E), shearing (G), and pressurizing (K). For this reason, it is preferable not to call E "modulus of elasticity" but rather tensile modulus (describes type of deformation) or Young's modulus (historical but not descriptive). In special cases, E can be converted to G or K (see Section 16.2.3).

The shear modulus G is often called "torsional modulus" which is etymologically incorrect for simple shear since the Latin word *torquēre* does not mean "to shear" but rather "to twist" (Fig. 15-1). However, the twisting of a transversally isotropic body around its longitudinal axis is characterized by a modulus perpendicular to the long axis which is indeed a shear *torsional* modulus (Section 16.2.3).

Either bulk compliances *or* bulk moduli are often called *compressibilities*. Symbols K and B are also often interconverted in the literature. Note that K is the IUPAC symbol for the bulk modulus.

16.2.3 Poisson's Ratio

The volume of a body may or may not change upon deformation. If the volume does change, it will decrease on pressurizing or compression and increase on elongation. In the latter case, a square cylinder will increase its original length L_0 by ΔL but decrease each perpendicular diameter d_0 by Δd. The original volume V_0 of the cylinder thus increases by ΔV:

$$\begin{aligned}(16\text{-}5)\qquad \Delta V &= (d_0 - \Delta d)^2(L_0 + \Delta L) - d_0{}^2L_0 \\ &= d_0{}^2\Delta L - 2\,\Delta d d_0 L_0 + [(\Delta d)^2L_0 + d_0{}^2\Delta L - 2\,d_0\Delta d\Delta L + (\Delta d)^2\Delta L]^2\end{aligned}$$

The terms in square brackets are small and can be neglected, which reduces Eq.(16-5) to $\Delta V \approx d_0{}^2\Delta L - 2\,\Delta d d_0 L_0$. Introduction of the Cauchy elongation, $\varepsilon = (L - L_0)/L_0$, and Hooke's law, $\varepsilon = \sigma_{11}/E$, delivers an expression for the relative change of volume:

$$(16\text{-}6)\qquad \frac{\Delta V}{V_0} \approx \frac{d_0^2\Delta L - 2\,\Delta d \cdot d_0 L_0}{d_0^2 L_0} = \varepsilon - \frac{2(\Delta d/d_0)\varepsilon}{\Delta L/L_0} = \varepsilon(1 - 2\,\mu) = (\sigma_{11}/E)(1 - 2\,\mu)$$

The term $\mu = (\Delta d/d_0)/(\Delta L/L_0)$ is called the **relative strain contraction** or **Poisson's ratio**. As indicated by its derivation, it is constant only for very small deformations and not for larger ones.

Pressurizing leads to a negative tensile stress σ_{11} that is three times as large as the applied pressure p. Introduction of $\sigma_{11} = -3\,p$ and the bulk modulus, $K = p/(-\Delta V/V_0)$, in $\Delta V/V_0 = \sigma_{11}/E)(1 - 2\,\mu)$, Eq.(16-6), results in the following expression for E,

$$(16\text{-}7)\qquad E = 3\,K(1 - 2\,\mu) = 2\,G(1 + \mu) \quad ; \quad \mu = \frac{3K - E}{6K} = \frac{3K - 2G}{2(3K + G)}$$

if all parameters are constant and independent of deformation, volume change, and stress direction. The interrelationship between tensile modulus E and shear modulus G (right expression for E) is more complicated to derive (not shown).

Table 16-1 Density ρ, Poisson's ratio μ, tensile modulus E, shear modulus G, and bulk modulus K of materials at room temperature. Moduli are experimental unless noted. ‖ In fiber direction; a) tricalcium silicate hydrate; UHMW = ultrahigh molar mass.

Materials	ρ/(g cm^{-3})	μ	G/GPa	K/GPa	E/GPa
Theoretical upper limit (exact)		0.50	0	∞	0
Water (4°C)	1.000	≈ 0.50	0	2.04	≈ 0
Natural rubber	0.92	0.4999	0.00035	2	0.001
Gelatin gel (80 % water)	1.01	0.50			
Mercury	13.59	0.50	0	25	≈ 0
Poly(ethylene), low density	0.92	0.49	0.070	3.3	0.20
Polyamide 6.6	1.14	0.44	0.70	5.1	1.9
Poly(styrene)	1.05	0.38	1.2	5.0	3.4
Aluminum	2.702	0.34	27	75	72
Ice (–4°C)	0.917	0.33	3.7	10.0	9.9
Granite		0.30	12	25	30
Steel (V2A)	7.86	0.28	80	170	195
Glass (E glass)	2.54	0.23	25	37	72
Concrete	2.61[a)]	≈ 0.10			34
Quartz	2.65	0.07	47	39	101
Poly(ethylene), lattice modulus (‖)	1.00				354
Poly(styrene), expanded		0.03			
Cork	< 0.25	0.00			
Graphite (in the layer direction)	2.25	0	500	333	1000
Diamond, [110]-direction	3.515	0			1160
Al_2O_3 fibers (‖)	3.97	0	1000	667	2000
Bodies without contraction at right angles		0	E/3	E/2	∞
Pyrite, single crystal		–0.14			
Isotropic materials (lower limiting value)		–1			
Poly(ethylene), UHMW, microporous		> –1.2			
Poly(tetrafluoroethylene), microporous		> –12			

For *isotropic bodies*, Poisson's ratio can vary only between 1/2 (if $E = 0$ or $G = 0$) and –1 (if $K = 0$) (Eq.(16-7)). The upper limit of $\mu = 1/2$ is obtained for deformations with constant volume ($\Delta V = 0$) and contractions crosswise to the deformation ($\Delta d \neq 0$; Eq.(16-6)). Examples are liquids such as water and mercury or highly swollen gels (Table 16-1).

Poisson's ratio becomes zero if crosswise contractions are absent (Eq.(16-6)). Theory predicts this special case for all ideal energy-elastic bodies; examples are diamond (in the [110]-direction), graphite (in the layer direction), and Al_2O_3 fibers (in the fiber direction). Steel is not a true energy-elastic body (Table 16-1). Polymers in the isotropic state behave more like viscous liquids ($\mu \rightarrow 1/2$) than elastic bodies ($\mu \rightarrow 0$) with respect to Poisson's ratio and the three moduli E, G, and K (Table 16-1).

Eq.(16-7) does not apply to measurements on polymers with different deformation times since viscoelastic properties become noticeable (Chapter 17). This equation is also not applicable to *anisotropic polymers* such as semicrystalline polymers, oriented specimens, fibers, or fiber-reinforced polymers.

Poisson's ratios of anisotropic bodies differ in the three spatial directions. An example is (semi)crystalline poly(p-phenylene terephthalate) with $\mu = 0.31$ in the [100]-direction, $\mu = 0.20$ in the [010]-direction, and $\mu = 0.24$ in the [110] direction. Anisotropic bodies may also have Poisson's ratios greater than 1/2 or smaller than –1. Poisson's ratios of less than −1 have been observed for some microporous polymers (Table 16-1). These **auxetic** materials expand laterally when stretched (G: *auxanein* = to increase). An example of a material with $\mu > 1/2$ is an orthogonally woven cloth with a 45° direction of warp and woof (see Volume IV, Section 6.4.2) where $\mu = 1$.

The bulk modulus K of polymers is generally greater than the shear modulus G (Table 16-1). A deformation by compression is thus more difficult than that by shearing; as a result, polymers deform mainly by shearing. Hence, the shear modulus G is the most important modulus for elastomers, thermoplastics, and thermosets, followed by the bulk modulus.

Tensile moduli E are less important theoretically than shear moduli G and bulk moduli K since E can be expressed by G and K. Insertion of $\mu = (3\ K - E)/(6\ K)$ into $E = 2\ G(1 + \mu)$, Eqs.(16-7), delivers

$$(16\text{-}8) \qquad 3/E = 1/G + 1/(3\ K)$$

However, bulk moduli are difficult to obtain experimentally so that only E and G are usually measured. Tensile moduli are furthermore the most important moduli for highly oriented fibers which are deformed little by shearing, pressurizing, or compression.

For theoretical considerations, the relative effects of bulk and shear moduli on deformation are often characterized by the **Lamé constant** λ:

$$(16\text{-}9) \qquad \lambda = K - \frac{2}{3}G = \frac{EG - 2G^2}{3G - E} = \frac{9K^2 - 3KE}{9K - E} = \frac{3\mu K}{1+\mu} = \frac{2\mu G}{1-2\mu} = \frac{\mu E}{(1+\mu)(1-2\mu)}$$

Isotropic bodies are characterized by two of the three moduli E, G, and K since the extension in one spatial direction leads to equivalent tensiles stresses in the other two other directions, which are dissolved by either shearing or compression. The situation is much more complicated for the deformation of anisotropic bodies (see Section 16.3.1).

16.2.4 Testing

Tensile moduli can be obtained by macroscopic methods (stress-strain measurements, flexural tests) and microscopic ones (for example, X-ray diffraction, Raman scattering, or coherent inelastic neutron scattering). Tensile moduli can also be calculated by various methods (Section 16.3.4).

Tensile Moduli

In tensile testing, a standardized test bar (p. 519) is drawn with constant speed and the resulting stress σ is recorded as a function of the change in length, $\Delta L = L - L_0$, draw ratio $\lambda = L/L_0$, or Cauchy elongation, $\varepsilon = (L - L_0)/L_0$ (Fig. 16-2). For isotropic bodies, the initial slope of the curve $\sigma_{11} = f(\varepsilon)$ delivers the tensile modulus E which usually does not deviate much from the true modulus of elasticity of the specimen (see p. 522). For anisotropic bodies, the initial slope provides the longitudinal tensile modulus $E_{\parallel}$ if the applied stress is parallel to the preferential axes of the molecules of the specimen; for example, the chain axes of linear polymer molecules.

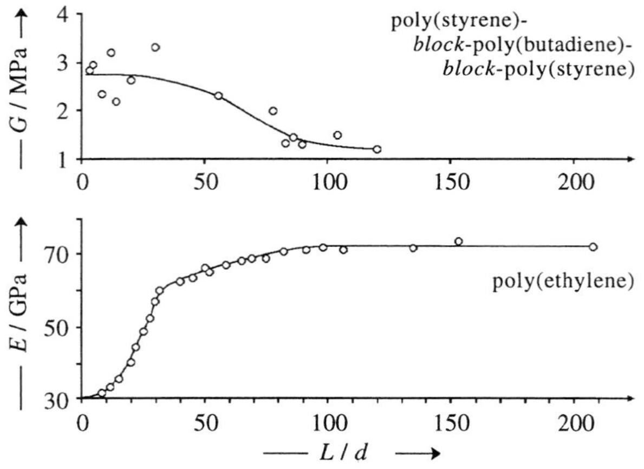

Fig. 16-4 Moduli of anisotropic polymers as a function of the aspect ratio L/d of rectangular test specimens with length L and diameter d [1].

Top: shear modulus G of test bars of a poly(styrene)-*block*-poly(butadiene)-*block*-poly(styrene) with hexagonally arranged cylindrical poly(styrene) domains in a continuous poly(butadiene) matrix.

Bottom: tensile modulus E of films of a semicrystalline poly(ethylene). The films were drawn to a draw ratio of $l/l_0 = 28$ before the test where l_0 = original length. Note that the tensile modulus becomes constant at $L/d > 100$; test bars for tensile testing usually have ratios of only $L/d = 8$.

With kind permission by Elsevier Science, Oxford.

The resulting modulus is usually that of the test bar and not that of the *material* of the specimen, especially if the material is anisotropic. The reason is that clamping of test bars produces additional stresses near the clamps which are balanced out at far greater distances in anisotropic specimens than in isotropic ones. Since rectangular test bars will therefore not receive the full load, it is advantageous to use dumbbells as test specimens.

Fig. 16-4 shows the variation of modulus with increasing aspect ratio of rectangular test specimens. The tensile modulus E of highly drawn semicrystalline poly(ethylene) films increased first strongly and then less strongly with increasing aspect ratio L/d of test specimens (Fig. 16-4, bottom). It became independent of specimen dimensions at values of $L/d > 100$ which are far larger than the usual aspect ratios of $L/d = 8$ of test bars. Conversely, shear moduli of test bars of a thermoplastic elastomer (Fig. 16-4, top) first decreased with increasing aspect ratio and then became constant. True moduli of materials can therefore be obtained only be measurements of very long test bars or extrapolations to very large aspect ratios.

Fibers have very large aspect ratios which consequently do not influence the values of moduli. However, test results of fibers are sensitive to variations in fiber diameters. Both natural and synthetic fibers always have distributions of fiber diameters within a single fiber and from fiber to fiber. Thicker fibers have larger surfaces, which leads to greater surface defects, such as porosities, and for some spinning processes also to a gradient of chemical composition from the interior to the exterior of the fiber. For inorganic high-modulus fibers, both tensile moduli and tensile strengths increase with decreasing fiber diameter (Fig. 16-5). The effect is less pronounced for organic fibers.

Tensile tests are short-time tests so that test results depend in part on testing speed, time, or frequency. According to ISO 527 and CAMPUS®, the following testing speeds should by used for plastics: 1 mm/min for tensile moduli, 5 mm/min for fracture

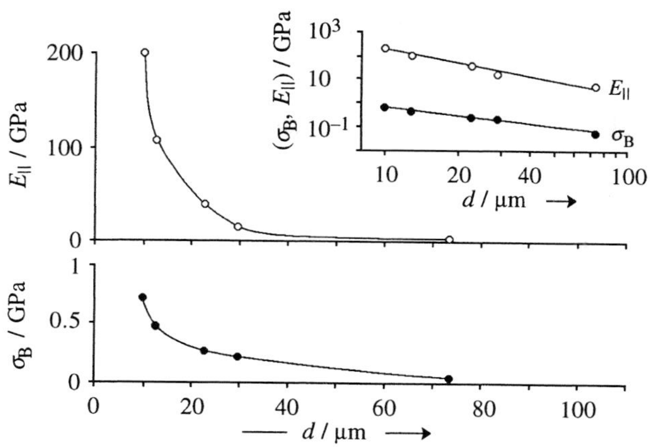

Fig. 16-5 Dependence of longitudinal tensile moduli $E_{\parallel}$ and tensile strengths σ_B on the diameter d of $Si_xN_yC_z$ fibers. Data of [2].

strengths and fracture elongations, and 50 mm/min for yield strengths and elongations at yield (point Y in Fig. 16-2). Tensile moduli are practically not affected by testing speeds since they are true material constants that are calculated from initial slopes of the force $\boldsymbol{F}$ or tensile stress σ_{11}, respectively, as a function of elongation, $\sigma_{11} = \boldsymbol{F}/A_0 = f(\varepsilon)$, where A_0 = initial cross-sectional area of the test specimen.

Above the proportionality limit, stress and strain are no longer directly proportional (Fig. 16-2). As a consequence, functions $\boldsymbol{F} = f(\varepsilon)$ and $\sigma_{11} = f(\varepsilon)$ vary with the rate of extension, $\dot{\varepsilon} = d\varepsilon/dt$ (Fig. 16-6): the larger $\dot{\varepsilon}$, the more difficult it is to elongate the polymer. Pre-stretching by a magnitude σ_e before testing reduces the extensibility further. The rate of extension therefore has a relatively small influence on the tensile modulus, a moderate one on elongation, and a very strong one on fracture strength (Fig. 16-6).

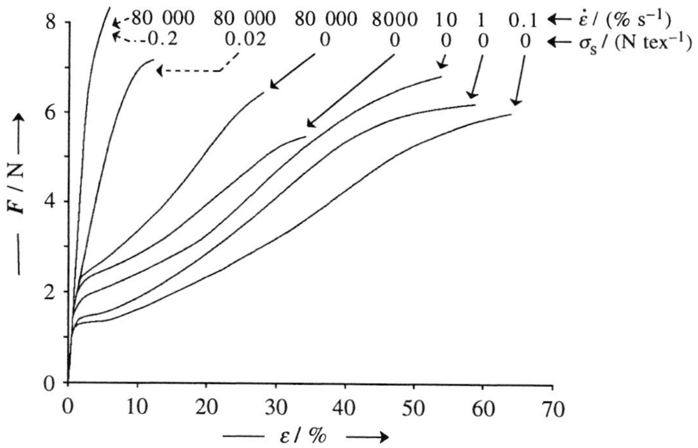

Fig. 16-6 Effect of the rate of elongation, $\dot{\varepsilon}$, and applied pre-stress, σ_e, on the dependence of tensile force $\boldsymbol{F}$ on elongation ε at room temperature for multifilament yarns from poly(ethylene terephthalate) [3]. Yarns were tested without thermal pre-stressing ($\sigma_e = 0$) or with heat setting (= thermal pre-stressing) of 0.02 N/tex or 0.2 N/tex at 220°C. 1 tex = 1 g/(1000 m).

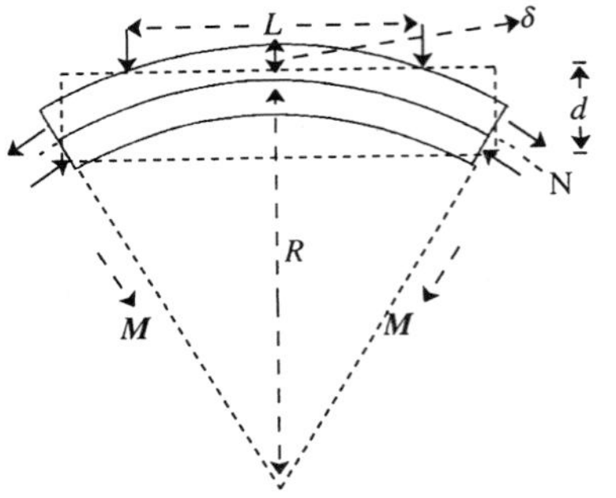

Fig. 16-7 Bending of a rectangular specimen of width b and thickness d by forces → (3-point method). - - - Original shape, —— shape after bending. The flexural moment $\boldsymbol{M}$ bends the specimen about the neutral axis N to a sector of a ring with radius R if the span L is much greater than the thickness d (ca. $L/d >> 16$). δ = deflection.

Flexural Moduli

Tensile moduli can also be determined by measuring flexibilities: the flexural response of specimens. In such measurements, a rectangular specimen is either clamped on one side and the free end of the specimen subjected to a load (1-point method) or clamped on both ends and loaded in the center (3-point method).

3-point measurements are more complicated then simple elongation or pressurizing because the convex side of the specimen is elongated above the so-called neutral axis whereas the concave side below this axis is compressed (Fig. 16-7). Along the span L on the upper side of the bent specimen with width b and thickness d, the maximum tension is $\sigma_{max} = (3\ \boldsymbol{F}L)/(2\ bd^2)$ and the maximum extension, $\varepsilon_{max} = 6\ \delta d/L^2$, where $\boldsymbol{F}$ = force at the center of the specimen and δ = deflection. The **flexural modulus** is therefore $E_f = \sigma_{max}/\varepsilon_{max} = (\boldsymbol{F}L^3)/(4\ bd^3\delta)$.

For isotropic materials, flexural moduli should be the same as tensile moduli, which is found to be true within ± 20 % for amorphous polymers (Fig. 16-8). Semicrystalline polymers are anisotropic; their flexural moduli are often much larger then their tensile moduli ($E/E_f \leq 1$). Conditioned polyamides and some liquid-crystalline polymers (LCPs) have $E/E_f > 1$, obviously from plasticizing effects by water (polyamides) and molecules that are not in liquid-crystalline domains (LCPs).

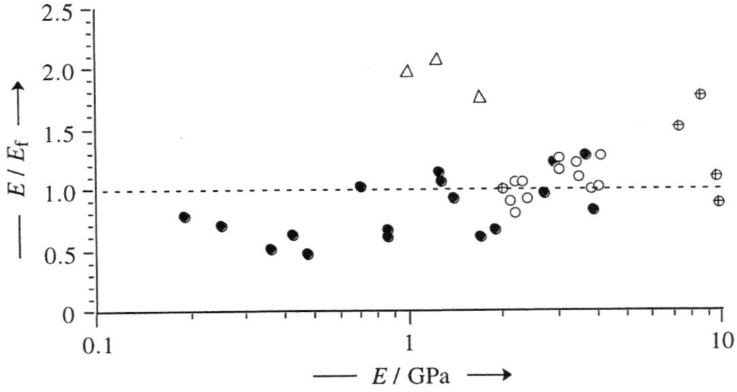

Fig. 16-8 Ratio E/E_F of tensile moduli E and flexural moduli E_F as a function of the tensile moduli of amorphous polymers (O), semicrystalline polymers (●), conditioned semicrystalline polymers (aliphatic polyamides at 50 % relative humidity) (Δ), and liquid-crystalline polymers (⊕). Data of [4].

Lattice Moduli

Tensile moduli and flexural moduli are material constants but cannot be correlated directly with molecular parameters because chain atoms (amorphous polymers) and segmental axes (semicrystalline polymers) are more or less distributed at random. The applied force is therefore not distributed evenly to all chain atoms but only to a few of them. As a result, tensile moduli of conventionally processed polymers are much lower than theoretically calculated ones (Section 16.3.5) and are related directly neither to elastic stiffness tensors ("stiffnes constants") nor to elastic compliance tensors ("compliance constants") (Section 16.3.3). Oriented polymers do have higher tensile moduli than conventionally processed ones but these are generally still much lower than theoretically calculated ones (Section 16.3.5).

However, the moduli of elasticity of semicrystalline polymers can also be obtained "microscopically"; for example, by *X-ray diffraction*. Application of a force $\boldsymbol{F}$ per cross-sectional area A_0 of the specimen changes Bragg angles $\Delta\theta$, i.e., the distances between lattice points, and therefore also the distances between atoms (Section 7.1.4). The macroscopic tension, $\sigma_{11} = \boldsymbol{F}/A_0$, is assumed to be identical with the microscopic tension at the lattice points. The tensile modulus is calculated from Hooke's law, $E = \sigma_{11}/\varepsilon$, where the elongation, $\varepsilon = \Delta d/d = -\cot\theta/\Delta\theta$, is calculated from Bragg's law, Eq.(7-1). This method allows one to determine elastic moduli parallel ($E_{\parallel}$) and perpendicular ($E_{\perp}$) to the chain axis (Section 16.3.5); these moduli are also called **lattice moduli**.

The method depends critically on the microscopic distribution of the macrocopically applied stress. This stress can be assumed to be homogeneously distributed among the crystalline and amorphous regions of semicrystalline polymers if the observed lattice moduli are independent of the thermal and mechanical history of the specimen.

However, some polymers show a strong variation of observed lattice moduli $E_{\parallel}$ with both temperature T and draw ratio $\lambda = L/L_0$ in certain temperature intervals (Fig. 16-9). At both very low and very high temperatures, lattice moduli are independent of both draw ratios and temperature, which indicates that the stress distribution is indeed homogeneous at these conditions.

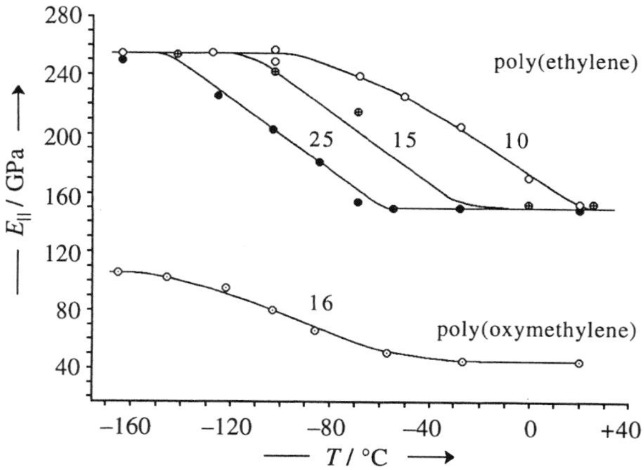

Fig. 16-9 Dependence of lattice moduli $E_{\parallel}$ from X-ray measurements on temperature T at which the specimens were drawn [5]. Numbers indicate draw ratios $\lambda = L/L_0$ as a measure of the degree of crystallinity.

At very low temperatures (< –140°C), lattice moduli of poly(ethylene) become independent of temperature, no longer affected by draw ratios, and identical with theoretical moduli (Tables 16-4 and 16-5). The lattice modulus of poly(oxymethylene) also seems to become constant at low temperature but is far lower than the theoretical one (85 GPa versus 220 GPa).

Alternatively, lattice moduli can be obtained by Raman spectroscopy or coherent inelastic neutron scattering. Both methods measure the velocity of photons in crystalline regions. For example, the low-frequency *Raman spectrum* of paraffins, $H(CH_2)_nH$, shows a series of signals whose vibration frequencies v decrease systematically with increasing length L of molecules.

The frequency of very long, elastic rods is given by $v = (1/2L)(E_{\parallel}/\rho)^{1/2}$. In semicrystalline polymers, these "rods" are the stems of folded micelles (Fig. 7-11). Since $L < \infty$ and $v = f(1/L)$, lattice moduli by Raman spectroscopy are found to be too high; for example, $E_{\parallel}$ = 368 GPa for poly(ethylene). If L is identified with the length of stems of folded micelles, $E_{\parallel}$ is reduced to ca. 285 GPa which is very close to the value of 258 GPa that is calculated by the valence force-field method (VFF method) (Section 16.3.3).

Coherent inelastic neutron scattering is a very expensive method. It determines the interrelationships between the frequency and the dispersion of the phase angle both along and perpendicular to the chain direction. Lattice moduli E (= $E_{\parallel}$ or $E_{\perp}$) are calculated from $E = \rho v^2$ where v = velocity of photons.

16.2.5 Types of Deformation

Elastomers, thermoplastics, and thermosets behave very differently on deformation. On drawing, tensile stresses of *elastomers* increase with elongations, at first slowly and then more rapidly, until the specimen finally is torn apart (Fig. 16-2). The molecular reason for this behavior is as follows. Segments between crosslinking points are fairly mobile because elastomeric behavior is observed only at temperatures above the glass temperature. The drawing of the elastomer first transforms microconformations of lower energy to those of higher energy (for example, gauche → trans). Further drawing requires more and more energy which causes tensile stresses to increase strongly with only small increases in elongation. Finally, energies of chemical bonds have to be overcome to sever the chemical bonds of network chains.

In *thermoplastic elastomers*, covalent crosslinking points are replaced by "hard" domains, i.e., domains that are below the glass temperature T_G (p. 262). Similarly to conventional elastomers, elongation of the specimen causes transformations of microconformations in the elastomeric matrix where $T_G > T$. However, since "hard" domains ($T_G < T$) are connected by many more network chains (N >> 3-4) than the crosslinking points of elastomers (N = 3-4), tensile stresses increase, at first slowly and then only a little, with increasing elongation (SBS in Fig. 16-10).

Many *common thermoplastics* yield, i.e., show a maximum in the stress-strain curve (Fig. 16-2) before they can be elongated further (Fig. 16-10; fracture elongation increases in the order PTFE, POM, PC, PE, PET-u). Other thermoplastics (PS) and practically all *thermosets* (for example, PF) fracture without showing a yield point. The stress-strain behavior changes if thermoplastics are biaxially drawn before tensile testing (PET-str): an initial steep increase of σ_{11} is followed by a plateau and then another increase.

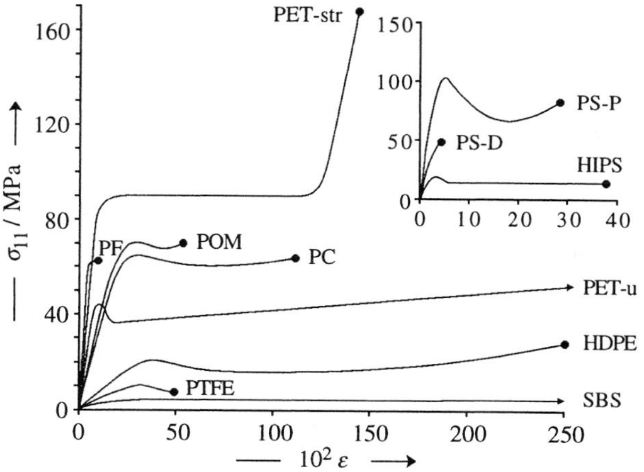

Fig. 16-10 Stress-strain diagrams of some polymers at room temperature:
PC = bisphenol A polycarbonate; HDPE = high-density poly(ethylene)
PET = poly(ethylene terephthalate); PF = phenol-formaldehyde resin
POM = poly(oxymethylene); PTFE = poly(tetrafluoroethylene) (Teflon)
SBS = poly(styrene)-*block*-poly(butadiene)-*block*-poly(styrene) (thermoplastic elastomer)
u = unstretched film, str = biaxially stretched film, ● fracture, → additional elongation.
Insert: behavior of poly(styrene) (PS) on drawing (PS-D) and on pressurizing (PS-P) as compared to a high-impact strength, rubber-reinforced poly(styrene) (HIPS).

The stress-strain behavior is also affected by the type of deformation. On drawing, conventional (atactic) poly(styrene) appears as a rigid, brittle material since it responds by forming crazes (Fig. 16-10, insert; PS-D). Pressurizing prevents the formation of microcavities and the same poly(styrene) appears now as rigid-ductile (PS-P).

According to their stress-strain behavior, polymers are usually subdivided into six classes (Table 16-2). The first term indicates the magnitude of the tensile modulus (rigid *versus* soft) and the second term the magnitude of the fracture elongation (brittle–strong–ductile). In the literature, " hard" is often used instead of "rigid" but "hard" should be avoided since "hardness" denotes the resistance against surface deformation. Literature also often uses "tough" instead of "ductile" although "toughness" indicates the behavior on impact (Section 18.3.5). An "elastic" polymer can be drawn to large lengths; it is certainly not "weak" which indicates early rupture.

Table 16-2 Classification of polymers according to their stress-strain behavior (cf. Fig. 16-10). E = tensile modulus, σ_Y = yield strength, ε_B = fracture elongation. PMMA = at-poly(methyl methacrylate), LDPE = low-density poly(ethylene); other abbreviations, see Fig. 16-10.

Correct classification	Conventional classification	E	σ_Y	ε_B	Example
Rigid-brittle	hard-brittle	large	–	small	PS, PF
Rigid-strong	hard-strong	large	large	small	PMMA
Rigid-ductile	hard-tough	large	large	large	POM, PC
Soft-strong	soft-strong	small	small	small	PTFE
Soft-ductile	soft-tough	small	small	large	LDPE
Soft-elastic	soft-weak	small	–	large	SBS

Rigid polymers have high moduli of elasticity, E. According to ASTM convention, plastics are called "rigid" if $E > 700$ MPa, "semi-rigid" if $700 \geq E$/MPa ≥ 70, and "soft" (or "non-rigid") if $E < 70$ MPa.

The stress-strain behavior between the elongation at yield ("yield point") ε_Y, and the fracture elongation, ε_B, indicates whether a polymer is brittle, strong, ductile, or elastic. Non-yielding polymers do not flow on elongation. They do not absorb energy on drawing and are therefore "brittle" (for the behavior on impact, see Chapter 18). A polymer is conventionally called "brittle" if the fracture elongation is less than 20 % (United States: < 10 %). Polymers with large yield strengths σ_Y are either strong (small fracture elongation) or ductile (large fracture elongation).

Rigid polymers with high moduli of elasticity have their counterparts in *soft polymers* (Table 16-2). Soft polymers can obviously not be brittle but they may be strong, ductile, or elastic, depending on the range between yield point (if any) and fracture elongation.

All these designations apply only to standard testing conditions because a polymer can be rigid or soft and brittle, strong, ductile, or elastic depending on temperature and deformation rate. These properties are not controlled *per se* by the chemical and physical structure of the polymer but rather by the testing temperature, the type of deformation, and the deformation rate.

Poly(styrene) appears as a rigid-brittle polymer in conventional tensile testing but as a rigid-ductile one on pressurizing (Fig. 16-10, insert). All polymers are rigid at fast deformation, low temperature, and large tensile stress and all polymers are soft at very slow deformation, high temperature, and small tensile stress (Fig. 16-6). "High" and "low" with respect to temperature do not relate here to the temperature scale *per se* but to the testing temperature with respect to the glass temperature (Fig. 16-11). In all cases, useful tensile properties are obtained only if polymer molecules are "interconnected": either by chemical crosslinking (elastomers, thermosets) or by physical crosslinking through entanglements in bulk (thermoplastics) or in domains (thermoplastic elastomers).

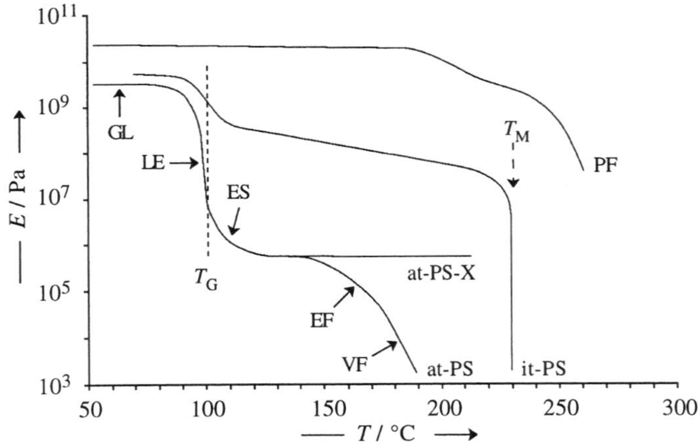

Fig. 16-11 Schematic representation of the temperature dependence of tensile moduli of an amorphous at-poly(styrene) (at-PS), its lightly crosslinked counterpart (at-PS-X), a semicrystalline isotactic poly(styrene) (it-PS), and a strongly crosslinked phenol-formaldehyde polymer (PF). GL = glass state, LE = leather-like behavior, ES = elastomeric behavior, EF = elastomeric flow, VF = viscous flow, T_G = glass temperature, T_M = melting temperature.

Tensile moduli of solid *amorphous polymers* are usually in the range $10^9 < E/\text{Pa} < 10^{10}$; they are practically independent of temperature. The moduli decrease drastically to values of ca. 10^5-10^6 Pa at the glass temperature (Fig. 16-11) where the polymers show leather-like behaviors. At still higher temperatures, the behavior depends on the extent of entanglement, i.e., on molar mass and molar mass distribution. At high entanglement densities (high molar masses), a plateau with temperature-independent moduli is observed ($E \approx 10^5$-10^6 Pa) where the polymers are rubber-like. The plateau is broad for high molar masses but non-existent for low molar masses that cannot entangle. At still higher temperatures, moduli decrease and the polymers behave as rubber-like liquids. Finally, moduli of $E \approx 10^3$ Pa are reached when the polymers melt to viscous liquids.

Chemically lightly crosslinked polymers have similar moduli to those of thermoplastics if $T < T_G$ but are elastomeric at $T > T_G$. Network chains between crosslinks are fairly long; they can be deformed but cannot flow away, a typical elastomeric behavior. Hence, tensile moduli of elastomers have broad plateaus at $T > T_G$. Above degradation temperatures, elastomers decompose chemically and the modulus drops.

Chemically strongly crosslinked polymers (thermosets) have very high tensile moduli (PF in Fig. 16-11) which, at $T < T_G$, are usually one decade higher than those of thermoplastics. Changes of moduli at the glass temperature are small since network chains are short and thus not very mobile. At $T > T_G$, thermosets are somewhat leather-like but the rubber-like plateau is not very distinct because thermosets are highly crosslinked and decompose chemically at even higher temperatures.

Semicrystalline polymers usually have relatively small amorphous regions, so their moduli drop less dramatically at T_G than those of amorphous polymers. In crystalline regions, mobilities of chain segments are drastically reduced so that crystalline regions act as large physical crosslinking sites. For this reason, moduli of semicrystalline polymers are much larger than those of amorphous polymers with the same chemical constitution and configuration. Between T_G and the melting temperature T_M, smaller and less ordered crystalline regions begin to melt and moduli slowly decrease (Fig. 16-11: it-PS). All remaining crystalline regions melt at T_M, which causes the moduli to drop catastrophically.

16.3 Energy Elasticity

Deformations are only purely energy-elastic if they are small and reversible; an example is the dropping of a steel ball on a steel plate. For this reason, polymers can only be meaningfully characterized by tensile moduli E if deformations are small. Examples are thermosets, highly crystalline thermoplastics, and strongly oriented polymeric fibers.

For strongly deformable polymers, shear moduli G are much more appropriate. Examples are elastomers and also most thermoplastics because they exhibit viscoelasticity.

The simple relationships between tensile modulus E, shear modulus G, and bulk modulus K (Section 16.2.2) apply only to isotropic bodies. The situation is much more complicated for anisotropic bodies (as many polymers are) where one needs to apply linear elasticity theory. The following pages try to outline the basics of this theory so that one becomes acquainted with the terminology and some of the results.

16.3.1 Generalized Hooke Equation

The interdependence of stress and strain in anisotropic bodies is much more complicated than suggested by Hooke's equation for isotropic bodies, Eq.(16-4). Each force acting on one of the three spatial planes of a body (Fig. 15-2) produces a response (i.e., a stress) in each of the three spatial directions at that plane so that a total of 9 stresses σ_{ij} exist (i,j = x,y,z or 1,2,3). The body then attempts to dissolve the stresses by deformations, i.e., changes of sizes and/or shapes.

Deformations cause the energy per volume of a system to change from U_0 to U which can be written as a Taylor series for the deformations $\boldsymbol{e}_{ij}$ and $\boldsymbol{e}_{kl}$:

$$(16\text{-}10) \qquad U = U_0 + \sum_{i=1}^{3}\sum_{j=1}^{3} \boldsymbol{B}_{ij}\boldsymbol{e}_{ij} + \sum_{i=1}^{3}\sum_{j=1}^{3}\sum_{k=1}^{3}\sum_{l=1}^{3} C_{ijkl}\boldsymbol{e}_{ij}\boldsymbol{e}_{kl} + O(\boldsymbol{e}^3)$$

The deformations $\boldsymbol{e}_{ij}$ and $\boldsymbol{e}_{kl}$ are tensors that are not identical with elongations ε (Section 16.2) that are so-called **engineering properties**. Note that books on theoretical mechanics often use e as the symbol for engineering deformation and ε for tensors!

The coefficients $\boldsymbol{B}_{ij}$ and $\boldsymbol{C}_{ijkl}$ describe the first and the second derivative, respectively, of the energy per volume after the deformation:

$$(16\text{-}11) \qquad \boldsymbol{B}_{ij} = (\partial U/\partial \boldsymbol{e}_{ij})|_{\boldsymbol{e}_{ij}=0} \quad ; \quad \boldsymbol{C}_{ijkl} = (\partial^2 U/\partial \boldsymbol{e}_{ij}\partial \boldsymbol{e}_{kl})|_{\boldsymbol{e}_{ij}=\boldsymbol{e}_{kl}=0}$$

In equilibrium, $U_0 = 0$ and $(\partial U/\partial \boldsymbol{e}_{ij})|_{\boldsymbol{e}_{ij}} = 0$. For *small* deformations, terms with higher orders of $\boldsymbol{e}$ can be neglected $[O(\boldsymbol{e}^3) = 0]$ and Eq.(16-10) thus reduces to

$$(16\text{-}12) \qquad U = \sum_{i=1}^{3}\sum_{j=1}^{3}\sum_{k=1}^{3}\sum_{l=1}^{3} C_{ijkl}\boldsymbol{e}_{ij}\boldsymbol{e}_{kl}$$

The first derivative of U with respect to deformation $\boldsymbol{e}_{ij}$ delivers the stress σ_{ij}:

$$(16\text{-}13) \qquad \sigma_{ij} = \partial U/\partial \boldsymbol{e}_{ij} = \sum_{k=1}^{3}\sum_{l=1}^{3} C_{ijkl}\boldsymbol{e}_{kl} \qquad ; \qquad \boldsymbol{C}_{ijkl} = \partial^2 U/(\partial \boldsymbol{e}_{ij}\partial \boldsymbol{e}_{kl})$$

Eq.(16-13) is the **generalized Hooke equation** which relates stresses σ_{ij} and deformations $\boldsymbol{e}_{ij}$ and $\boldsymbol{e}_{kl}$, respectively, to the energy U of the material via the so-called **stiffness constants $\boldsymbol{C}_{ijkl}$**.

Eq.(16-13) connects continuum theories with molecular theories because the energy U depends on intramolecular and intermolecular interactions between atoms and atomic groups of polymer molecules, respectively. If these interactions and their interplays are known, descriptors $\boldsymbol{C}_{ijkl}$ can be calculated and therefore also deformations and tensile stresses. For an example, see Table 16-4 in Section 16.3.4.

The three-dimensional case is described by 81 stiffness constants which are reduced to 21 because of the symmetry rules, $\boldsymbol{C}_{ijkl} = \boldsymbol{C}_{jikl}$, $\boldsymbol{C}_{ijkl} = \boldsymbol{C}_{ijlk}$, and $\boldsymbol{C}_{ijkl} = \boldsymbol{C}_{klij}$. For higher symmetries such as in orthotropic, oriented, and isotropic bodies, this number is reduced even further (Section 16.3.3).

16.3.2 Linear Elasticity Theory

In isotropic bodies, a force $\boldsymbol{F}_x$ acting in the x direction will cause an elongation in the x direction. In anisotropic bodies, the same force will lead to additional deformations in the y and z directions. Conversely, an elongation in one direction will lead to stresses in all three spatial directions (Fig. 16-12).

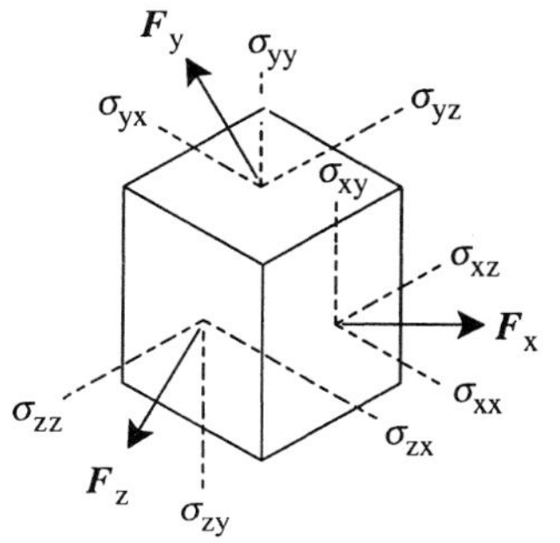

Fig. 16-12 Components σ_{ij}, of stresses if the forces $\boldsymbol{F}_x$, $\boldsymbol{F}_y$, and $\boldsymbol{F}_z$ are not perpendicular to the areas formed by x and y, x and z, and y and z.

Stresses

For small deformations, the total deformation of an anisotropic body can by thought of as a linear combination of deformations in various directions. **Linear elasticity theory** thus assumes that each stress component is linearly connected with each component of elongation, $\boldsymbol{e}_{ij}$ (i,j = 1,2,3). The stress in the [xx] direction (= [11]-direction) is therefore written as

$$\sigma_{11} = \boldsymbol{a}_{11}\boldsymbol{e}_{11} + \boldsymbol{b}_{11}\boldsymbol{e}_{22} + \boldsymbol{c}_{11}\boldsymbol{e}_{33} + \boldsymbol{d}_{11}\boldsymbol{e}_{12} + \boldsymbol{f}_{11}\boldsymbol{e}_{13} + \boldsymbol{g}_{11}\boldsymbol{e}_{21} + \boldsymbol{h}_{11}\boldsymbol{e}_{23} + \boldsymbol{k}_{11}\boldsymbol{e}_{31} + \boldsymbol{l}_{11}\boldsymbol{e}_{32}$$

Similar equations apply to stresses (σ_{12} .... σ_{33}) in the other eight directions; each with nine descriptors per direction (from $\boldsymbol{a}_{12}$ .... $\boldsymbol{l}_{12}$ to $\boldsymbol{a}_{33}$ .... $\boldsymbol{l}_{33}$).

The nine components of stress are commonly written as elements of a matrix, the **stress tensor**, σ_{ij}:

$$(16\text{-}14) \qquad \sigma_{ij} = \begin{pmatrix} \sigma_{xx} & \sigma_{xy} & \sigma_{xz} \\ \sigma_{yx} & \sigma_{yy} & \sigma_{yz} \\ \sigma_{zx} & \sigma_{zy} & \sigma_{zz} \end{pmatrix} \quad \text{or} \quad \sigma_{ij} = \begin{pmatrix} \sigma_{11} & \sigma_{12} & \sigma_{13} \\ \sigma_{21} & \sigma_{22} & \sigma_{23} \\ \sigma_{31} & \sigma_{32} & \sigma_{33} \end{pmatrix}$$

Mass, temperature, density, etc., are independent of the direction and thus only chacterized by a number and a physical unit. Such physical quantities are called **scalars** (L: *scālae* = stairs). They are zero-rank tensors (see below).

Forces, speeds, etc., are additionally characterized by a direction in a plane. As **vectors** (L: *vehere* = carrier, from *vehere* = to carry), they are **first-rank tensors** (L: *tensus* = tense, from *tendere* = to stretch) with physical symbols that are **bold** and in *italics*. In d-dimensional space, vectors have *d* components.

A mechanical stress is defined as a force per area; it is the ratio of two vectors. Quantities such as stresses or elongations that relate one vector to another vector are **second-rank tensors**. In three--dimensional space, they are composed of $N^2 = 3^2 = 9$ components. Stiffness constants (discussed below) are fourth-rank tensors with $3^4 = 81$ components in three-dimensional space.

The first index i of the components σ_{ij} of the stress defines the direction of the *normal* (perpendicular) to the plane that is acted on by the force. The second index j indicates the direction of the stress. The three tensors σ_{ii} (σ_{11}, σ_{22}, σ_{33}) are therefore ***normal* stresses** that are independent of each other.

Shearing in the 11 (xx) direction shifts the two areas A_{11} (or A_{xz}) parallel to each other. The stress σ_{12} (or τ_{12} or simply τ) is therefore called the ***shear*(ing) stress**. For this stress, $\sigma_{13} = \tau_{13}$ is the **tangential stress**. Shear and tangential stresses are **deviatoric stresses** (L: *deviare*, *de* = away from, *via* = street) since they deviate from the **dilatational stress** (L: *dilatare* = to extend) that is caused by extension of the specimen.

In general, forces are not perpendicular to the corresponding planes but act under certain angles. Each slanted plane is subjected to three stresses which are related to the corresponding normal stresses via the cosine of the angles that the axes form with the normals to the planes. Rotation of the coordinate system shows that tensors σ_{ij} ($i \neq j$) must be identical. There are therefore maximally 6 independent, normalized projections of forces (σ_{11}, σ_{22}, σ_{33}, $\tau_{12} = \tau_{21}$, $\tau_{13} = \tau_{31}$, $\tau_{23} = \tau_{32}$) and 6 independent relative deformations ($\boldsymbol{e}_{11}$, $\boldsymbol{e}_{22}$, $\boldsymbol{e}_{33}$, $\boldsymbol{e}_{12} = \boldsymbol{e}_{21}$, $\boldsymbol{e}_{13} = \boldsymbol{e}_{31}$, $\boldsymbol{e}_{23} = \boldsymbol{e}_{32}$).

Deformations

For isotropic bodies, only two types of strain are considered in the simplest case: the **extensional strain**, $\varepsilon = (L - L_0)/L_0$, and the **shear strain**, $\gamma = \tan \theta$ (Fig. 15-3), which are independent of each other. In anisotropic bodies, extension and shearing are present simultaneously in all directions.

Suppose a point P_1 with the initial Cartesian coordinates x, y, and z is near another point P_2 with the initial coordinates $x + \mathrm{d}x$, $y + \mathrm{d}y$, and $z + \mathrm{d}z$. These points P_1 and P_2 are separated by a vector $\mathrm{d}\boldsymbol{r} = (\mathrm{d}\boldsymbol{x}, \mathrm{d}\boldsymbol{y}, \mathrm{d}\boldsymbol{z})$.

If shifting or turning moves the point P_1 to a new position P_1' with the coordinates $x + u$, $y + v$, and $z + w$, then the point P_2 will also move to a new position P_2' with the coordinates $x + \mathrm{d}x + u + \mathrm{d}u$, $y + \mathrm{d}y + v + \mathrm{d}v$, and $z + \mathrm{d}z + w + \mathrm{d}w$. The change of position is therefore given by $u + \mathrm{d}u$, $v + \mathrm{d}v$, and $z + \mathrm{d}z$.

If the vector $\mathrm{d}\boldsymbol{r}$ is linearly related to the relative shifts $\mathrm{d}u$, $\mathrm{d}v$, and $\mathrm{d}w$, then the shift of coordinates of u is given by

(16-15) $$u + \mathrm{d}u = u + (\partial u/\partial x)\mathrm{d}x + (\partial u/\partial y)\mathrm{d}y + (\partial u/\partial z)\mathrm{d}z$$

and similarly for the shifts $v + \mathrm{d}v$ and $w + \mathrm{d}w$. Nine differential quotients $\partial u/\partial x$, $\partial u/\partial y$ .... $\partial w/\partial z$ are thus required for the description of a spatial deformation. $\partial u/\partial x$ must indicate the elongation (or contraction) $\boldsymbol{e}_{xx}$ since x and u belong to the same coordinate direction. The same is true for $y + v$ and $z + w$, respectively, so that

(16-16) $$\boldsymbol{e}_{xx} = \frac{\partial u}{\partial x} \quad ; \quad \boldsymbol{e}_{yy} = \frac{\partial v}{\partial y} \quad ; \quad \boldsymbol{e}_{zz} = \frac{\partial w}{\partial z}$$

For three-dimensional deformations, elongations $\boldsymbol{e}_{xy} = \boldsymbol{e}_{yx}$, $\boldsymbol{e}_{yz} = \boldsymbol{e}_{zy}$, and $\boldsymbol{e}_{zx} = \boldsymbol{e}_{xz}$ must also be known:

(16-17) $$\boldsymbol{e}_{xy} = (\partial v/\partial x) + (\partial u/\partial y) \; ; \; \boldsymbol{e}_{yz} = (\partial w/\partial y) + (\partial v/\partial z) \; ; \; \boldsymbol{e}_{zx} = (\partial u/\partial z) + (\partial w/\partial x)$$

Deformation ε_{ij} $(i \neq j)$ of Eq.(16-17) are engineering properties. For tensors, deformations are defined as one-half of the sum of both contributions, for example, $e_{xy} \equiv (1/2)[(\partial v/\partial x) + (\partial u/\partial y)]$. Deformations are therefore identical for tensors and engineering quantities. However, they are not equal since engineering quantities ε_{ij} are twice as large as tensors, e_{ij}.

The **deformation tensor** e_{ij} and the corresponding engineering deformation are thus

$$(16\text{-}18) \qquad e_{ij} = \begin{pmatrix} e_{xx} & e_{xy} & e_{xz} \\ e_{xy} & e_{yy} & e_{yz} \\ e_{xz} & e_{yz} & e_{zz} \end{pmatrix} = \begin{pmatrix} \varepsilon_{xx} & (1/2)\varepsilon_{xy} & (1/2)\varepsilon_{xz} \\ (1/2)\varepsilon_{xy} & \varepsilon_{yy} & (1/2)\varepsilon_{yz} \\ (1/2)\varepsilon_{xz} & (1/2)\varepsilon_{yz} & \varepsilon_{zz} \end{pmatrix}$$

$$(16\text{-}19) \qquad \varepsilon_{ij} = \begin{pmatrix} \varepsilon_{xx} & \varepsilon_{xy} & \varepsilon_{xz} \\ \varepsilon_{xy} & \varepsilon_{yy} & \varepsilon_{yz} \\ \varepsilon_{xz} & \varepsilon_{yz} & \varepsilon_{zz} \end{pmatrix} \quad \text{with} \quad \varepsilon_{ij} = 2\, e_{ij} \ (\text{if } i \neq j)$$

16.3.3 Stiffness Constants and Compliance Constants

Types of stress tensors and deformation tensors as well as stiffness constants are characterized by four subscripts (for example, see C_{ijkl} in Eq. (16-10)). In order to decrease the number of subscripts, both stress tensors and deformation tensors are reduced to vectors with 6 components. For stress components, the symbol σ is maintained. All components σ_{ij} with i = j are then written as σ_p (with p = 1, 2, 3) and all components σ_{ij} with i ≠ j = 1, 2, 3 are converted to σ_p with p = 4, 5, 6:

Tensors	xx	yy	zz	yz	zy	xz	zx	xy	yx
	11	22	33	23	32	13	31	12	21
Matrix indices	1	2	3	4		5		6	

Deformation components e_{kl} (kl = 11, 22, 33) convert similarly to e_q. However, shear components e_{kl} (with kl = 23, 32, 13, 31, 12, 21) are redefined as $2\, e_{kl} = \varepsilon_q$, i.e., they are replaced by the corresponding engineering quantities. Descriptors C_{ijkl} are now written C_{pq} ; for example, C_{2233} becomes C_{23} and C_{1113} is now C_{15}.

In matrix notation, the generalized Hooke equation is therefore written as

$$(16\text{-}20) \qquad \sigma_p = C_{pq} e_q \quad ; \quad e_p = C_{pq}^{-1}\sigma_q = S_{pq}\sigma_q$$

where σ_p = elements of stress tensors, ε_q = elements of deformation tensors, C_{pq} = **(elastic) stiffness tensor (elastic stiffness constant, elastic modulus, Voigt elasticity constant)**, and S_{pq} = **(elastic) compliance tensor (compliance constant, Reuss elasticity constant)**. C_{pq} and S_{pq} are not components of tensors, since they no longer follow the transformation laws for tensors because of the shift of deformations to engineering quantities. As components of a matrix, C_{pq} equals S_{pq}^{-1} only in special cases (see also Table 16-3). Unfortunately, symbols do not conform to words: *C* indicates *s*tiffness whereas *S* is the symbol for *c*ompliance!

Stresses and deformations remain invariant if p and q are exchanged. In the general case, there are therefore 21 stiffness constants C_{pq} and 21 compliance constants S_{pq}. The number of these constants is reduced if the material has certain symmetries. The general case with 21 stiffness constants and 21 compliance constants corresponds to a triclinic system (Table 7-1). The number of stiffness constants (and correspondingly also the number of compliance constants) is reduced further for other bodies: to 13 (monoclinic), 9 (orthorhombic), 7 or 6 (tetragonal), 5 (hexagonal), 3 (cubic), and 2 (isotropic).

Orthotropic Bodies

Bodies with orthotropic symmetry have the same properties in the three planes that are perpendicular to each other; for crystals, these are the cubic, tetragonal, and orthorhombic systems (Table 7-1). However, such symmetries are not confined to crystalline bodies. Fiber-reinforced plastics also show orthotropy if the fibers are more densely packed in the [12] plane than in the [23] plane. Another example is oriented film.

According to the generalized Hooke equation, Eq.(16-13), tensile properties of orthotropic bodies are described by a 6×6 symmetric matrix, $\sigma = C_{pq}\boldsymbol{e}$, where σ and $\boldsymbol{e}$ are tensile properties and τ and γ are shear properties:

$$\begin{pmatrix} \sigma_1 \\ \sigma_2 \\ \sigma_3 \\ \tau_{23} \\ \tau_{13} \\ \tau_{12} \end{pmatrix} = \begin{pmatrix} C_{11} & C_{12} & C_{13} & 0 & 0 & 0 \\ C_{21} & C_{22} & C_{23} & 0 & 0 & 0 \\ C_{31} & C_{32} & C_{33} & 0 & 0 & 0 \\ 0 & 0 & 0 & C_{44} & 0 & 0 \\ 0 & 0 & 0 & 0 & C_{55} & 0 \\ 0 & 0 & 0 & 0 & 0 & C_{66} \end{pmatrix} \begin{pmatrix} \boldsymbol{e}_1 \\ \boldsymbol{e}_2 \\ \boldsymbol{e}_3 \\ \gamma_{23} \\ \gamma_{13} \\ \gamma_{12} \end{pmatrix} \tag{16-21}$$

Because of symmetries in orthotropic bodies, only nine independent stiffness constants C_{ij}, and correspondingly nine independent compliance constants S_{ij}, are required for a description of such a body.

However, mechanical properties are conventionally not described by stiffness and compliance constants but by engineering properties E, μ, and G. These properties are connected with the descriptors S_{pq} and C_{pq} as follows.

Application of pure stress σ_1 in the [1]-direction does not involve tensile stresses in the two lateral directions [2] and [3] ($\sigma_2 = \sigma_3 = 0$). The orthotropic body responds not only with an elongation ε_1 in the [1]-direction but also with narrowings ε_2 and ε_3 in the two lateral directions. Since no shear stresses have been applied ($\tau_{23} = \tau_{13} = \tau_{12} = 0$), shear deformations will be absent ($\gamma_{23} = \gamma_{13} = \gamma_{12} = 0$).

Compliance constants S are ratios of elongations and applied stress. Since here the deformation in the [1]-direction is caused by the tensile stress $\sigma_1 = E_1\varepsilon_1$, the compliance constant is given by $\boldsymbol{S}_{11} = E_1^{-1}$.

Deformations in the [2]-direction and [3]-direction are also caused only by the tensile stress σ_1 in the [1]-direction. However, the elongation of the body in the [1]-direction causes it to shrink in the [2]-direction and [3]-direction. The two other compliance constants are therefore $S_{22} = -\varepsilon_2/\sigma_1$ and $S_{33} = -\varepsilon_3/\sigma_1$. Because of the cross-deformations, Poisson's ratios will be $\mu_{12} = \varepsilon_2/\varepsilon_1$ and $\mu_{13} = \varepsilon_3/\varepsilon_1$, which in turn leads to $S_{12} = -\varepsilon_2/\sigma_1 = -\mu_{12}\varepsilon_1/\sigma_1 = -\mu_{12}/E_1$ and therefore also to $S_{13} = -\mu_{13}/E_1$.

Table 16-3 Relationships between engineering quantities E, μ, and G and compliance constants, S_{pq}, and stiffness constants, C_{pq}, respectively, of orthotropic bodies.

Index	Tensile properties S_{pq}	C_{pq}	Index	Shear properties S_{pq}	C_{pq}
11	$1/E_1$	$ZE_1[1 - \mu_{23}^2(E_3/E_2)]$	44	$1/G_{23}$	G_{23}
12	$-\mu_{12}/E_1$	$Z[\mu_{12}E_2 + \mu_{13}\mu_{23}E_3]$	55	$1/G_{13}$	G_{13}
13	$-\mu_{13}/E_1$	$ZE_3[\mu\text{-}_{12}\mu_{23} + \mu_{13}]$	66	$1/G_{12}$	G_{12}
22	$1/E_2$	$ZE_2[1 - \mu_{13}^2(E_3/E_1)]$			
23	$-\mu_{23}/E_2$	$Z(E_3/E_1)[\mu_{23}E_1 + \mu_{12}\mu_{13}E_2]$			
33	$1/E_3$	$ZE_3[1 - \mu_{12}^2(E_2/E_1)]$			

$\mu_{ij}E_i = \mu_{ji}E_j$; $Z^{-1} = 1 - \mu_{13}(E_3/E_1)[2\,\mu_{12}\mu_{23} - \mu_{13}] - \mu_{23}^2(E_3/E_2) - \mu_{12}^2(E_2/E_1)$

Similarly, pure stress may be applied in the [2]-direction with resulting elongations in the three spatial directions. The same calculations are then repeated for pure stress in the [3]-direction. The result shows that pure stress reduces the 9 compliance constants to 6 independent descriptors because of $S_{12} = S_{21}$, $S_{13} = S_{31}$, and $S_{23} = S_{32}$ and therefore also $\mu_{12}E_2 = \mu_{21}E_1$, $\mu_{13}E_3 = \mu_{31}E_3$, and $\mu_{23}E_3 = \mu_{32}E_{32}$ (Table 16-3). However, orthotropic bodies require 9 descriptors and not 6 so that the remaining three engineering properties, the shear moduli G_{12}, G_{13}, and G_{23}, must be obtained by three shear experiments, $\tau_p = G_{pq}\gamma_q$.

Interrelationships between engineering quantities and compliance constants S_{pq} are simpler than those between engineering quantities and stiffness constants C_{pq}. Mechanics thus prefers compliance constants. Molecular theories, on the other hand, usually calculate stiffness constants since they are much easier to calculate by potential functions, etc. than compliance constants.

Oriented Bodies

Some polymers and polymer systems have greater symmetries than orthotropic bodies. In biaxially oriented films and sheets, chain segments and/or crystallites are oriented in the [12]-plane. In stretched fibers or bundles of fibers, the long axes of molecules, and/or long molecule segments are oriented in the [3]-direction (Fig. 16-13). Such bodies are **transversely isotropic**. True isotropic bodies do not have preferred axes.

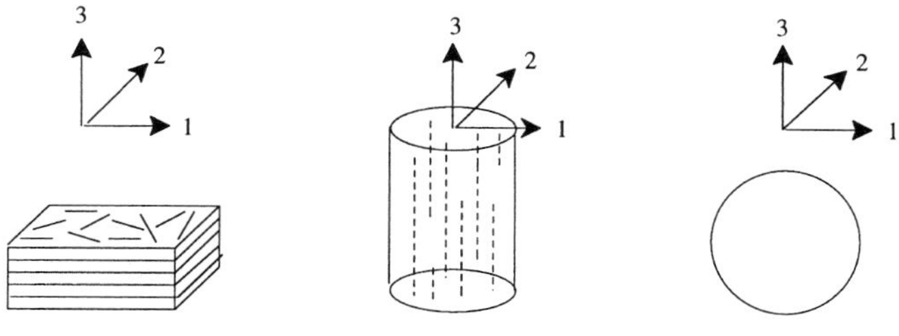

Fig. 16-13 Biaxially oriented (left), uniaxially oriented (center), and isotropic bodies (right). The symmetry axis of biaxially oriented bodies is parallel to the *plane* of orientation whereas in uniaxial oriented bodies, it is normal (perpendicular) to the *direction* of orientation.

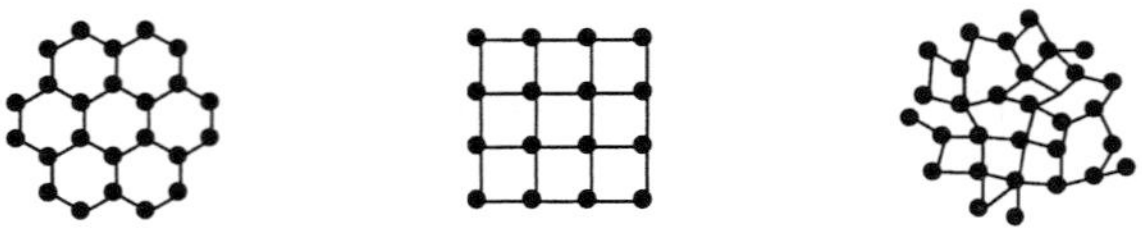

Fig. 16-14 Cut through the [12]-plane of bundles of fibers or molecules that are perpendicular to the paper plane. Left: hexagonal array; center: square arrangement; right: disordered. Interaction between fibers or molecules are indicated by solid lines.

Six descriptors are required if cross-sections of fibers in fiber bundles or chain segments in semicrystalline polymers are regularly arranged (Fig. 16-14, left and center). Only five descriptors are needed for biaxially oriented bodies with random arrangements of fibers with respect to the [12]-plane (Fig. 16-14, right).

Since random orientations of cross-sections of fibers in the [12]-plane require only 5 descriptors, 4 descriptors must be idential with other ones. Because of symmetries (see Fig. 16-13), one has the following identities: $S_{11} = S_{22}$, $S_{44} = S_{55}$, and $S_{13} = S_{23} = S_{32}$. Solving the matrix for the corresponding inverse generalized Hooke's law leads to

(16-22)
$$\begin{aligned} \varepsilon_1 &= S_{11}\sigma_1 + S_{12}\sigma_2 + S_{13}\sigma_3 \quad ; \quad \gamma_{23} = \gamma_{13} = S_{44}\tau_{13} \\ \varepsilon_2 &= S_{12}\sigma_1 + S_{11}\sigma_2 + S_{13}\sigma_3 \quad ; \quad \gamma_{12} = S_{66}\tau_{12} = 2(S_{11} - S_{12})\tau_{12} \\ \varepsilon_3 &= S_{13}\sigma_1 + S_{13}\sigma_2 + S_{33}\sigma_3 \end{aligned}$$

Such bodies thus have five moduli of elasticity:

E_L = longitudinal (||) tensile modulus in the [3]-direction;
E_T = transverse (⊥) modulus E_T in the [1]- and [2]-directions, respectively;
G_{TT} = transverse shear modulus in the [12]-direction;
G_{LT} = longitudinal shear modulus in the [23]-direction; and
K = bulk modulus.

Moduli, stiffness constants, and compliance constants, respectively, can be expressed by two Poisson's ratios, μ_{LT} and μ_{TL}. Poisson's ratio $\mu_{LT} = -\varepsilon_T/\varepsilon_L$ measures the transverse elongation that is caused by the imposed longitudinal elongation ε_L. Poisson's ratio μ_{TL}, on the other hand, indicates the longitudinal elongation ε_L that is produced by the elongation ε_T in the transverse direction. Poisson's ratio $\mu_{TT} = (1/2)(E_T/G_{TT}) - 1$ is not an independent quantity since it can only be obtained from tensile moduli E_T and shear moduli G_{TT} that in turn depend on the two Poisson's ratios, μ_{LT} and μ_{TL}:

(16-23)
$$\begin{aligned} E_L &= E_3 = 1/S_{33} = C_{33}(1 - \mu_{LT}\mu_{TL}) = C_{13}(1 - \mu_{LT}\mu_{TL})/\mu_{LT} \\ E_T &= E_1 = 1/S_{11} = C_{11}(1 - \mu_{LT}\mu_{TL}) \\ \\ G_{LT} &= 1/S_{44} = C_{44} = C_{55} \\ G_{TT} &= 1/S_{66} = C_{66} = E_T/[2(1 + \mu_{TT})] \quad ; \quad S_{66} = 2(S_{11} - S_{12}) \\ \\ \mu_{LT} &= \mu_{23} = -S_{13}/S_{33} = \mu_{TL}E_L/E_T \\ \mu_{TL} &= \mu_{13} = -S_{13}/S_{11} = \mu_{LT}E_T/E_L \\ \mu_{TT} &= \mu_{12} = -S_{12}/S_{11} = (1/2)(E_T/G_{TT}) - 1 \end{aligned}$$

Isotropic Bodies

Linear-elastic isotropic bodies have no preferential axes. The inverse generalized Hooke law thus reduces to

$$\text{(16-24)}\quad \begin{pmatrix} \varepsilon_1 \\ \varepsilon_2 \\ \varepsilon_3 \\ \gamma_{23} \\ \gamma_{13} \\ \gamma_{12} \end{pmatrix} = \begin{pmatrix} S_{11} & S_{12} & S_{12} & 0 & 0 & 0 \\ S_{12} & S_{11} & S_{12} & 0 & 0 & 0 \\ S_{12} & S_{12} & S_{11} & 0 & 0 & 0 \\ 0 & 0 & 0 & 2(S_{11}-S_{12}) & 0 & 0 \\ 0 & 0 & 0 & 0 & 2(S_{11}-S_{12}) & 0 \\ 0 & 0 & 0 & 0 & 0 & 2(S_{11}-S_{12}) \end{pmatrix} \begin{pmatrix} \sigma_1 \\ \sigma_2 \\ \sigma_3 \\ \tau_{23} \\ \tau_{13} \\ \tau_{12} \end{pmatrix}$$

An elongation in the [1]-direction thus leads to $\varepsilon_1 = S_{11}\sigma_1$ and therefore to $E_1 = 1/S_{11}$. Since the body is isotropic, one also has only one tensile modulus: $E_1 = E_2 = E_3 = E$. The body has also only one shear modulus, i.e., $G_{12} = 1/[2\ (S_{11} - S_{12})]$ with $G_{12} = G_{13} = G_{23} = G$.

The six possible Poisson's ratios are also identical: $\mu = \mu_{12} = \mu_{21} = \mu_{13} = \mu_{31} = \mu_{23} = \mu_{32}$. However, because of $\mu_{12} = -\varepsilon_1/\varepsilon_2 = -S_{12}/S_{22}$, etc., one has also $G = E/[2\ (1 + \mu)]$. In contrast to anisotropic bodies, shear moduli of isotropic bodies can be calculated from the corresponding tensile moduli if Poisson's ratio μ is known.

Amorphous polymers are usually isotropic bodies. In such bodies, tensile stresses σ are not coupled with shear stresses τ and neither are shear stresses τ coupled with tensile strains ε. A stress σ_{11} produces in the [1]-direction an elongation $\varepsilon_{11} = \sigma_{11}/E$ but contractions in directions [2] and [3]. These contractions ε_{22} and ε_{33} are negative elongations. They are measured by Poisson's ratio of isotropic bodies, $\mu = \varepsilon_{22}/\varepsilon_{11} = \varepsilon_{33}/\varepsilon_{11}$, which indicates the relative lateral strain contraction (Section 16.2.3).

A tensile stress σ_{11} thus produces lateral contractions $-\varepsilon_{22} = \mu(\sigma_{11}/E)$ and $-\varepsilon_{33} = \mu(\sigma_{11})$. The deformations are thus

$$\text{(16-25)}\quad \begin{aligned} \varepsilon_{11} &= E^{-1}[\sigma_{11} - \mu(\sigma_{22} + \sigma_{33})] \\ \varepsilon_{22} &= E^{-1}[\sigma_{22} - \mu(\sigma_{11} + \sigma_{33})] \\ \varepsilon_{33} &= E^{-1}[\sigma_{33} - \mu(\sigma_{11} + \sigma_{22})] \end{aligned}$$

The shear strain γ_{ij} is connected to the shear stress τ_{ij} by the shear modulus G. One thus obtains with $\varepsilon_{12} \equiv \gamma_{12}$, $\sigma_{12} \equiv \tau_{12}$, etc.

$$\text{(16-26)}\quad \gamma_{12} = G^{-1}\tau_{12} \;\;;\;\; \gamma_{13} = G^{-1}\tau_{13} \;\;;\;\; \gamma_{23} = G^{-1}\tau_{23}$$

Pressurizing (= all-sided compression) produces a relative volume change of $-\Delta V/V_o = 3\ \varepsilon = \varepsilon_{11} + \varepsilon_{22} + \varepsilon_{33}$. The elongation ε is calculated from Eq.(16-25) by using the tensile modulus $E = 1/S_{11}$, Poisson's ratio $\mu = -S_{12}/S_{11}$, and the pressure $p \equiv \sigma_{11} = \sigma_{22} = \sigma_{33}$, resulting in $\varepsilon = (S_{11} + 2\ S_{12})p$. The bulk modulus K of isotropic bodies is therefore

$$\text{(16-27)}\quad K = \frac{p}{-\Delta V / V_0} = \frac{1}{3(S_{11} + 2S_{12})} = \frac{E}{3(1 - 2\mu)}$$

16.3.4 Theoretical Moduli

Theoretical moduli can be calculated by various methods. The **valence force-field method (VFF)** considers in the simplest case a single polymer chain containing N chain bonds of length b. The conventional contour length of this chain is $r_{\text{cont}} = Nb \sin(\tau/2) = Nb \cos \beta = L_0$ where τ = bond angle of chain atoms and $\beta = \alpha/2 = (180° - \tau)/2$ = one-half of the complementary angle α to τ (see also Fig. 4-14). The chain is in the all-trans conformation with the chain axis in the draw direction.

The chain is drawn but not bent or twisted. The force $\boldsymbol{F}$ acting on the cross-sectional area A_c of the chain causes the bond angle β to be extended by $\Delta\beta$ and the bond length b by Δb. The chain is therefore elongated by a length ΔL,

$$(16\text{-}28) \qquad \Delta L = \Delta[Nb \cos \beta] = N[\Delta b \cos \beta - b\Delta\beta \sin \beta]$$

The bond elongation is calculated from $\Delta b = (\boldsymbol{F} \cos \beta)/K_b$ where $\boldsymbol{F} \cos \beta$ = component of force $\boldsymbol{F}$ and K_b = force constant, a quantity that is available from infrared or Raman spectroscopy.

The enlargement of the angle $\beta = 90° - (\tau/2)$ by $\Delta\beta = -\Delta\tau/2$ requires knowlege of $\Delta\tau$ which is given by the force constant K_τ and the angular momentum $\boldsymbol{M}$ that acts on every bond angle, $\Delta\tau = \boldsymbol{M}/K_\tau$. The angular momentum equals the momentum of the force acting normal to the angle, i.e., $\boldsymbol{M} = (1/2)\,\boldsymbol{F}b \sin \beta$. The enlargement $\Delta\beta$ of the angle β is therefore $\Delta\beta = -\Delta\tau/2 = -(\boldsymbol{F}b \sin \beta)/(4\,K_\tau)$.

Introduction of the expressions $\Delta b = (\boldsymbol{F} \cos \beta)/K_b$, $\Delta\beta = -(\boldsymbol{F}b \sin \beta)/(4\,K_\tau)$, the simple Hooke equation, $E_\parallel = (\boldsymbol{F}/A_c)(L_o/\Delta L)$, and $L_o = Nb \cos \beta$ leads to

$$(16\text{-}29) \qquad E_\parallel = \frac{b\cos\beta}{A_c}\left[\frac{\cos^2\beta}{K_b} + \frac{b^2\sin^2\beta}{4\,K_\tau}\right]^{-1} = \frac{b\sin(\tau/2)}{A_c}\left[\frac{\sin^2(\tau/2)}{K_b} + \frac{b^2\cos^2(\tau/2)}{4\,K_\tau}\right]^{-1}$$

The greater the cross-sectional area A_c of the chain, the smaller the longitudinal tensile modulus $E_\parallel$. A systematic variation of $E_\parallel$ with A_c can therefore be expected for a series of polymers with the same helix conformation (Fig. 16-15).

Eq.(16-29) was derived by assuming constant torsional angles. However, the longitudinal elongation of helical chains is accompanied by changes of torsional angles because of changes in microconformations; for example, from gauche to trans. For this case, VFF delivers equations that are more complex than Eq.(16-29). However, these equations still predict a decrease of tensile moduli with increasing cross-sectional area of chains (see Fig. 16-15).

Tensile moduli can also be calculated by the **dynamic lattice theory (Born method)** which traces physical properties of crystals to the thermal movements of their chain atoms. This theory requires the exact knowledge of potential functions (for example, Lennard-Jones, Urey-Bradley, Buckingham, etc.), force constants, lattice constants, lattice angles, etc. The theory has been little used for polymers because these quantities are rarely known with the required precision and calculations are complicated or even impossible for certain crystal structures like the one of $\text{+NH(CH}_2)_5\text{CO+}_n$ (PA 6) with intermolecular hydrogen bonds in pleated sheet structures (Figs. 7-7 and 7-8).

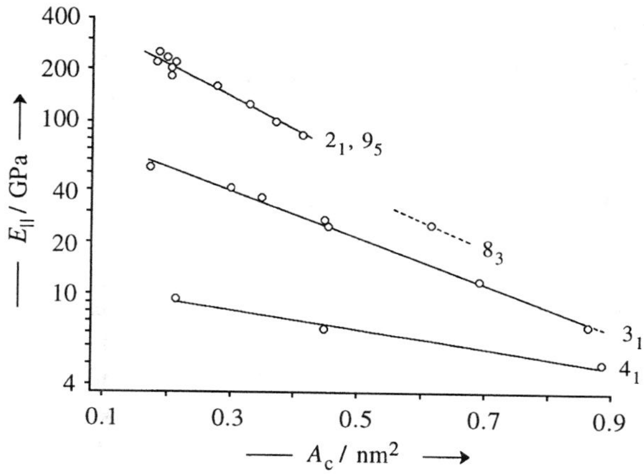

Fig. 16-15 Longitudinal tensile moduli $E_{\|}$ as a function of the cross-sectional area A_c of C–C, C–O, and C–N polymer chains in all-trans conformation (2_1) or various helical conformations (9_5, 8_3, 3_1, 4_1). Tensile moduli are either experimental lattice moduli (see below) or theoretically calculated tensile moduli in the chain direction, both at low temperatures (< –180°C).

An example is the application of Born theory to crystalline poly(ethylene). The unit cell of this polymer contains 2 monomeric units, $-CH_2-CH_2-$. All 12 atoms of the unit cell are involved in intramolecular or intermolecular interactions so that one has to solve 3×12 equations for the three spatial directions. However, the number of equations reduces to the 6 matrix equations of Eq.(16-21) because of the symmetries for orthorhombic poly(ethylene) crystals (space group D_{2h}).

The orthotropy of poly(ethylene) crystals leads to 9 stiffness constants and 9 compliance constants (Table 16-4). Stiffness constants $C_{33} = 1/S_{33}$ for the chain axis are not much affected by the choice of lattice constants (columns 2 and 3) or temperature (columns 3 and 4) but very much by the various potential functions (columns 4-6).

Table 16-4 Stiffness constants C_{ij} of poly(ethylene) as calculated by valence force field (VFF) or molecular mechanics (MM) for various temperatures, lattice constants a and b, respectively, and/or temperatures [7-10]. Lattice constants were assumed to be a = 0.695 nm and b = 0.475 nm (I) and a = 0.72 nm and b = 0.495 nm (III); experimental values for $C_{36}H_{74}$ are a = 0.742 nm and b = 0.496 nm [6]. Inverse numerical values of compliance constants, S_{ij}^{-1}, are included for comparison.

Index ij	C_{ij}/GPa – 196°C I [7] VFF	C_{ij}/GPa – 196°C III [7] VFF	C_{ij}/GPa 20°C III [7] VFF	C_{ij}/GPa 27°C [8] MM	C_{ij}/GPa [9] VFF	S_{ij}^{-1}/GPa^{-1} [10] VFF
11	7.33	9.27	6.28	8.5	13.75	9.44
12 = 21	2.26	3.68	2.18	5.0	7.34	–16.08
13 = 31	3.14	3.63	2.90	4.5	2.46	–22730
22	10.0	10.93	9.35	9.0	12.50	8.56
23 = 32	6.34	6.67	6.07	6.4	3.96	–1053
33	257.3	257.4	257.2	250	325.4	324.1
44	3.31	2.46	2.93	2.8	3.19	3.19
55	1.13	1.27	0.88	1.7	1.98	1.98
66	3.54	4.99	2.97	3.4	6.24	6.24

Stiffness constants C_{33} are practically independent of temperature because bond lengths C–C and bond angles C–C–C do not vary much with temperature (p. 198). Calculated values agree reasonably well with experimental ones of 250-320 GPa (see also Table 16-5). All other descriptors are very sensitive to the choice of potential functions, force constants, lattice parameters, etc.

Shear stiffness constants C_{44}, C_{55}, and C_{66} also agree reasonably well with the corresponding inverse shear compliance constants S_{ij}^{-1} (Columns 6 and 7). In contrast to C_{33} and S_{33}, shear stiffness constants as well as values of C_{11}, C_{12}, C_{13}, C_{22}, and C_{23} (and the corresponding compliance constants) are very sensitive to the choice of potential functions, lattice constants, etc.

Other theoretical methods are based on energy minimizations, for example, **molecular mechanics (MM method)**. This method calculates the potential energy of the whole molecule by adding all interatomic interactions. Energy minimization then leads to the most stable physical structure of the entire molecule. The resulting chain is then somewhat extended and the structure with the smallest energy is evaluated again. After some hundred thousand of such energy minimizations for a sufficiently large lattice consisting of several thousand atoms, one obtains a limiting value for the potential energy as a function of elongation and finally stress-strain curves (Fig. 16-16). Results by the MM method are compared to the ones by the VFF method in Table 16-4, column 5.

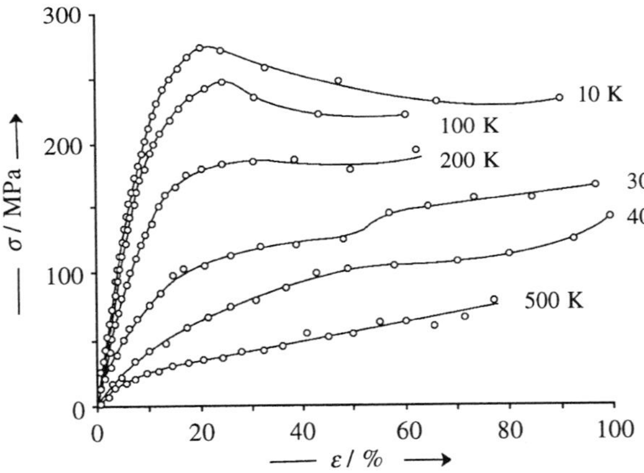

Fig. 16-16 Stress-strain curves of a poly(ethylene) chain consisting of 1000 carbon atoms as calculated by molecular mechanics for a change of stress of 0.5 MPa ps^{-1} = 0.5 EPa s^{-1} = $0.5 \cdot 10^{18}$ Pa/s with time [11]. Circles indicate averages of 5 independent calculations; lines are empirical.

Tensile moduli $E_{\parallel}$ calculated by the VFF method agree usually within ±10 % with those from experimental methods such as X-ray diffraction, Raman scattering, or inelastic neutron scattering (Table 16-5). The largest moduli $E_{\parallel}$ are therefore found for polymer molecules with small cross-sections of chains. At the same cross-sectional area, helical polymer chains have smaller moduli than polymers in all-trans conformation (Fig. 16-15) which can be rationalized as stretching of a rod *versus* extension of a spiral spring. The largest tensile moduli are found for somewhat twisted chains such as poly(*p*-phenylenebenzbisoxazole) or those with strong intramolecular interactions such as hydrogen bridges as in it-poly(vinyl alcohol) or cellulose I.

Table 16-5 Calculated and experimental lattice moduli in the chain direction ($E_{\parallel}$) and perpendicular to the chain direction ($E_{\perp}$) compared to tensile moduli E of conventionally processed polymers.
Methods: INS = inelastic neutron scattering, Raman scattering, SS = stress-strain experiments, VFF = valence force-field method, X-ray diffraction. VFF values apply to 0 K, experimental values to either low temperatures (usually 77 K) or 25°C (*). Poly(*p*-phenylenebenzbisoxazole) is the *cis* polymer and poly(*p*-phenylenebenzbisthiazole) the *trans* compound. [a] Ramie at 65 % relative humidity.

Polymer		A_c/nm²	$E_{\parallel}$/GPa VFF	$E_{\parallel}$/GPa X-ray	$E_{\parallel}$/GPa Raman	$E_{\parallel}$/GPa INS	$E_{\perp}$/GPa X-ray	$E_{\perp}$/GPa INS	E/GPa SS
Poly(ethylene)		0.183	316	260	285	329	≤3.8	6	< 1.6
Poly(vinyl alcohol),	st-	0.216	287	250*			≤8.8*		< 7
	it-		323						
Polyamide 6,	γ-Mod. (7_2 helix)	0.192	54	27				1.9	
	α-Mod. (zigzag)	0.186	312	270					
			175*	165*					< 3.8
Poly(propylene), it	(3_1 helix)	0.343	41	40	37		<3.1		< 1.8
Poly(oxymethylene),	orthorhombic	0.183	220	220	189	149	7.8	6	< 3.1
Poly(oxyethylene)		0.216	9	10			4.3		
Poly(*p*-phenylenebenzbisoxazole)		0.201	460	477					
Poly(*p*-phenylenebenzbisthiazole)		0.206	405	399					
Cellulose I		0.328	168	140					< 28.5[a]
Poly(tetrafluoroethylene)				156	203	222			

Moduli $E_{\perp}$ perpendicular to the chain direction are much smaller than moduli $E_{\parallel}$ in the chain direction because intermolecular forces between chains are much smaller than intramolecular ones along the chain (Table 16-5). Transverse moduli are strongly increased if dispersion forces between molecules are replaced by intermolecular hydrogen bonds (cf. poly(ethylene) *versus* poly(vinyl alcohol)).

16.3.5 Real Moduli of Elasticity

In general, tensile moduli of plastics and fibers are not only much lower than longitudinal lattice moduli of the same polymers but also often much lower than transverse lattice moduli (Table 16-5). This observation points to an insufficient orientation of chain segments (amorphous polymers) and crystalline regions (semicrystalline polymers), respectively. However, it does not indicate complete absence of orientations since test specimens of plastics are usually prepared by injection molding which promotes orientation of chain segments in the lateral direction. Fiber spinning also leads to orientation of chain segments and crystallites.

Testing conditions affect moduli since stress-strain measurements are often performed (a) with too short test bars, (b) at too high elongations, and (c) with too long times. These effects reduce the observed tensile moduli. Very high strain rates, on the other hand, will increase tensile moduli (Fig. 16-6). Moduli are practically independent of degrees of polymerization, and thus on molar mass distributions, if the degree of polymerization is greater than ca. 100. They usually decrease with increasing temperature (Figs. 16-9 and 16-11).

Table 16-6 Names of mixing rules. * After some mathemtical manipulations.

Field	Exponent in Eq.(16-30) n = 1	(n → 0)*	n = −1
Mathematics	arithmetic average	geometric average	harmonic average
Chemical engineering	mixing rule	logarithmic average	inverse mixing rule
Mechanics	Voigt model	–	Reuss model
Electricity	parallel	–	in series
Materials science	upper limit	–	lower limit

Mixing Rules

Conventionally processed *amorphous* polymers are often macroscopically and microscopically isotropic bodies. *Semicrystalline* polymers may appear isotropic but are certainly not microscopically isotropic since a chain runs through many lamellae and lamellae are oriented in different directions. Semicrystalline polymers can therefore be treated as two-component bodies. Tensile moduli of such bodies are controlled by the proportions and moduli of the amorphous and crystalline regions.

In the simplest case, properties P such as tensile moduli E vary with the volume fraction ϕ_i of the components according to the **mixing rule**

$$E^n = E_A{}^n\phi_A + E_B{}^n\phi_B \quad ; \quad \phi_A + \phi_B \equiv 1 \tag{16-30}$$

where n = exponent with values of +1 or −1 and A, B = components, for example, A = amorphous and B = crystalline or M = matrix and F = filler (or fiber). Mixing rules carry different names, depending on the field in which they are used (Table 16-6).

The arithmetic average (n ≡ 1) corresponds to the **Voigt model** of mechanics which assumes that all regions are equally extended but experience different stresses. A macroscopic analog is a composite consisting of a thermoplastic matrix M with embedded fibers F whose long axes are all in the strain direction. Matrix and fibers are both extended to the same extent but the less pliable fibers experience much greater stresses than the more pliable matrix. In electrical engineering, this corresponds to a parallel connection with the simple mixing rule, $E_{\parallel} = \phi_M E_M + \phi_F E_F$.

The **Reuss model** assumes the opposite: uniform stresses but non-uniform elongations. A macroscopic analog is a dispersion of fibers in a continuous matrix where the long axes of all fibers are perpendicular to the direction of elongation. In electrical engineering, this model corresponds to a series connection and in chemical engineering to the inverse mixing rule: $1/E_{\perp} = (\phi_M/E_M) + (\phi_F/E_F)$. Table 16-7 shows an example.

Table 16-7 Models for a poly(ethylene) [12]. Moduli are extrapolated to 100 % crystallinity (25°C).

Model	Stress	Strain	E/GPa	G/GPa	K/GPa
Reuss model (⊥)	uniform	non-uniform	5.05	2.0	1.95
Experiment	-	-	5.82	2.23	1.99
Voigt model (‖)	non-uniform	uniform	15.8	18.8	0.33

According to Table 16-7, calculated Reuss moduli (*E*, *G*, *K*) agree much better with experimental moduli than Voigt moduli. The reason is the unrealistic assumption of the Voigt model that stiff crystallites can be elongated to the same extent as the soft matrix. The Reuss model is much more realistic in its assumption that crystallites and the matrix are elongated to approximately the same extent but experience different stresses because of their vastly different rigidities. Indeed, the Reuss bulk modulus *K* (from all-sided compression) shows the least difference between experimental and model moduli (2 %) and the tensile modulus *E* (on unidirectional extension) the largest (15 %). In shearing (modulus *G*), some stresses are relieved perpendicular to the shear direction, and the deviation of the Reuss modulus from the experimental modulus is therefore less than that for *E* but higher than that for *K* (11 %).

Takayanagi Models

In many cases, experimental data can be much better decribed by the two Takayanagi models than the simple Reuss (two types of phases in series) and Voigt models (two types of phases in parallel) (Fig. 16-17). The models assume that the material consists of two types of matter, A (amorphous) and C (crystalline) (as in semicrystalline polymers), or, similarly, F (fiber) in M (polymer matrix) (as in fiber-reinforced thermoplastics), or T (thermoplastics) in R (rubber) (as in thermoplastic elastomers). Of course, there may be A in a continuous matrix of C or C in a continuous matrix of A, etc.

If one type of matter such as C exists in both a continuous and a discontinuous phase, then the two phases C may be in series (model I) or parallel (model II). C and A may then be in parallel (model I) or in series (model II).

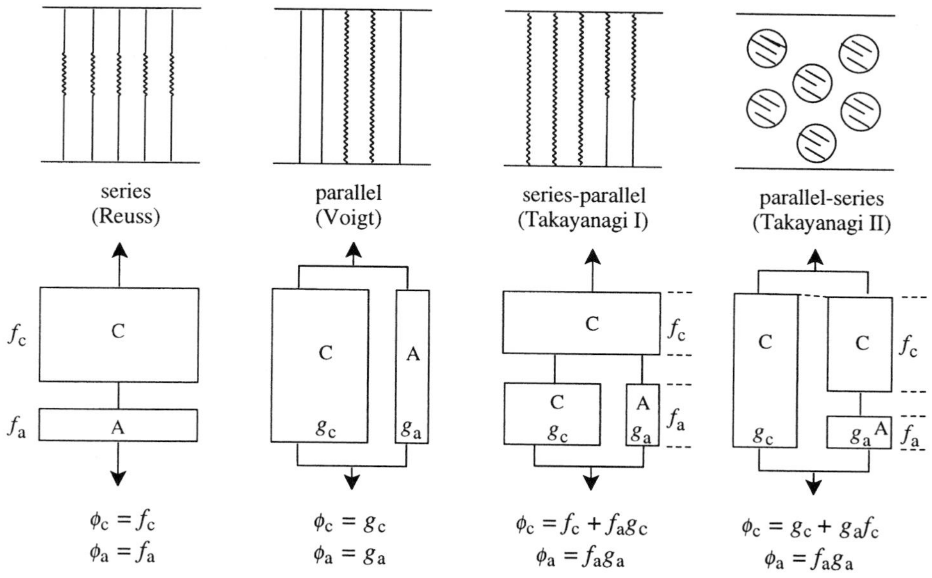

Fig. 16-17 Reuss, Voigt, and Takayanagi models for heterogeneous systems, for example, with crystalline regions C and amorphous regions A (or fibers or fillers F in a matrix M). ϕ, f, and g are volume fractions.

The volume fraction of *all* crystalline phases C is ϕ_c and that of *all* amorphous phases A, is ϕ_a, so that $\phi_c + \phi_a \equiv 1$. These fractions may be either in series (symbol *f*) or in parallel (symbol *g*) so that the totals of these volume fractions are $f_c + f_a \equiv 1$ and $g_c + g_a \equiv 1$ (Fig. 16-17). If the crystalline fractions are in series (and the subsequent amorphous fraction in parallel to the crystalline one) (model I), then the total volume fraction of crystalline regions will be $\phi_c = f_c(g_c + g_a) + f_a g_c = f_c + (1 - f_c)g_c = f_c + f_a g_c$. For model II (parallel-series), the volume fraction of the crystalline regions is $\phi_c = g_c(f_c + f_a) + g_a f_c = g_c + (1 - g_c)f_c$. Both models are restricted to a range $f_c < \phi_c < 1$ if $f_c = \text{const} > 0$.

The series-parallel model I predicts for the tensile modulus the function

$$\frac{1}{E_{SP}} = \frac{f_c}{E_c} + \frac{f_a}{g_c E_c + g_a E_a} \tag{16-31}$$

It converts to the Reuss model for $g_c = 0$ (i.e., $g_a = 1$) where $f_a = \phi_a$ and $f_c = \phi_c$. For $f_c = 0$, it becomes the Voigt model with $g_c = \phi_c = 1 - \phi_a$.

The parallel-series model II delivers for the modulus

$$E_{PS} = g_c E_c + \frac{g_a}{\dfrac{f_c}{E_c} + \dfrac{f_a}{E_a}} = g_c E_c + \frac{g_a E_a E_c}{f_c E_a + f_a E_c} \tag{16-32}$$

It becomes the Voigt model for $f_c = 0$ and the Reuss model for $g_c = 0$.

Takayanagi models describe well the dependence of tensile moduli on the volume fraction of crystalline regions (Fig. 16-18). However, curve fits are not only sensitive to experimental data and the choice of $f_c = const$ or $g_c = const$, but also to the type of plot (cf. PS data for $E = f(\phi_c)$ and $\lg E = f(\phi_c)$).

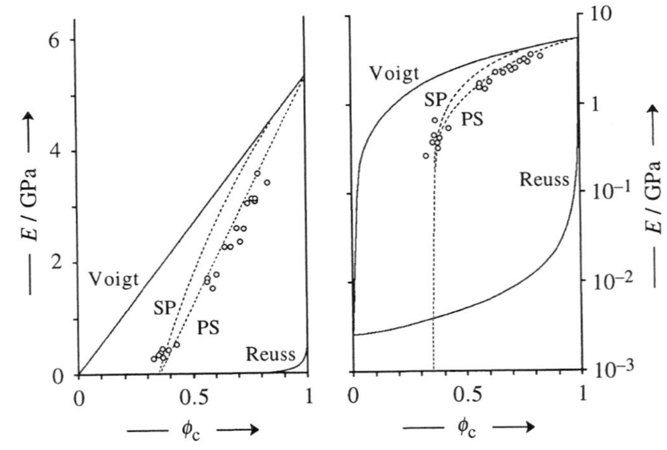

Fig. 16-18 Tensile moduli E (left) and their logarithms (right) of semicrystalline poly(ethylene)s at 25°C as a function of the degree of crystallinity ϕ_c [13]. Tensile moduli from sound velocity, degrees of crystallinity from density. O Experimental data; —— predictions of Reuss and Voigt models; - - - calculations for Takayanagi models SP (||) and PS (⊥) using E_c = 5400 MPa at $\phi_c = 1$ (cf. Table 16-7), E_a = 2.6 MPa at $\phi_c = 0$ (corresponds approximately to the plateau modulus $G_N{}^o$ (Table 17-6)), and $\phi_c = 0.35$. Lines - - - are only for demonstration; they are not optimized for best curve fitting.

Effect of Processing

Longitudinal tensile moduli $E_{\parallel}$ of *flexible* polymers can be increased strongly by draw processes such as the stretching of polymer sheets, the extrusion of polymer melts, and the gel spinning of polymer solutions. For example, the longitudinal tensile modulus of poly(oxymethylene) (theoretical $E_{\parallel}$ = 220 GPa) is ca. 2 GPa from conventional extrusion but 24 GPa from hydrostatic extrusion and ca. 60 GPa from tensile drawing with simultaneous heating by microwaves. The longitudinal tensile modulus of high molecular weight, high-density poly(ethylene) is only ca. 1 GPa by injection molding but can be increased to 40 GPa by fiber extrusion with subsequent drawing, to 130 GPa by extrusion of a 5 % gel followed by removal of the solvent, and to 220 GPa (which is almost the theoretical value of 260 GPa) by multiple extrusion through capillaries. However, such processing may lead not only to high tensile moduli and fracture strengths in the longitudinal direction but also to brittleness perpendicular to the strain direction.

The orientation of chains and chain segments of flexible polymers can only be obtained by external fields. *Semiflexible*, *liquid-crystalline* polymers, on the other hand orient their anisotropic chain segments spontaneously. Fast quenching of nematic LC phases leads to nematic LC glasses with high longitudinal tensile moduli, both from thermotropic and lyotropic polymers (Table 16-8).

Tensile moduli also depend on the environment during processing. Humidity acts as a plasticizer for polar polymers; it reduces the moduli because chain segments become more mobile. The diffusion of water molecules into the polymer produces time-dependent tensile moduli. For example, a polyamide 6 had a tensile modulus of 2.75 GPa (dry), 1.7 GPa (after 24 h at 25°C), and 0.86 GPa (after 4 months in air).

Table 16-8 Tensile moduli of thermotropic (TT) and lyotropic (LT) polymeric LC glasses longitudinal ($E_{\parallel}$) and transverse ($E_{\perp}$) to the drawing direction and tensile moduli E of isotropic glasses.

	Polymer	$E_{\parallel}$/GPa	$E_{\perp}$/GPa	E/GPa
TT	Poly(*p*-hydroxybenzoate-*co*-ethylene terephthalate) [= X7G™]	54.1	1.38	2.21
TT	Poly(*p*-hydroxybenzoate-*co*-2-hydroxy-6-naphthalate) [= Vectra™]	10.6	2.6	5.0
LT	30 % Poly(*p*-phenylenebenzbisthiazole) in poly(2,5-benzimidazole)	120	16.8	62
LT	Poly(*p*-phenylene terephthalamide) [= Kevlar 49™], bulk	138	7	
	ditto, fiber spun from concentrated sulfuric acid	83		
	ditto, stretched film from concentrated sulfuric acid	8.9	0.6	250

16.4 Entropy Elasticity

16.4.1 Phenomena

Energy-elastic and entropy-elastic bodies differ considerably in deformation and the resulting properties. Metals are energy-elastic; their atoms are arranged in a three-dimensional crystalline lattice with relatively small interatomic distances. The increase of these distances by drawing requires high energies. Tensile moduli of metals are therefore very large. Furthermore, relatively small deformations of ca. 0.1 % cause lattice planes to slip past each other, become spirally relocated, etc., causing the deformation to increase faster than the stress. The deformation becomes irreversible and the elastic body becomes "plastic."

The deformation of thermoplastic polymers ($T < T_G$) changes torsional and bond angles and therefore intermolecular distances and, at very high elongations, may also change bond lengths. Because of large tensile moduli, strong forces are required for deformations. At elongations of more than ca. 0.1-0.2 %, segments start to slip past each other and the deformation becomes irreversible (Chapter 17).

Elastomers ($T > T_G$) can be reversibly extended, often to several hundred percent, before the deformation becomes irreversible. Such large deformations require (a) weakly crosslinked polymers that (b) consist of flexible polymer chains. The resulting elastomers (rubbers) have simultaneously the properties of solids, liquids, and gases. Like solids, they have dimensional stability: after not too large deformations, they return to their initial state. Like liquids, they have large expansion coefficients (Table 13-2). Similarly to compressed gases, stresses increase with increasing temperature if $T > T_G$ (Fig. 16-19). However, stresses also increase with decreasing temperature if $T < T_G$ since such crosslinked polymers now behave like thermosets.

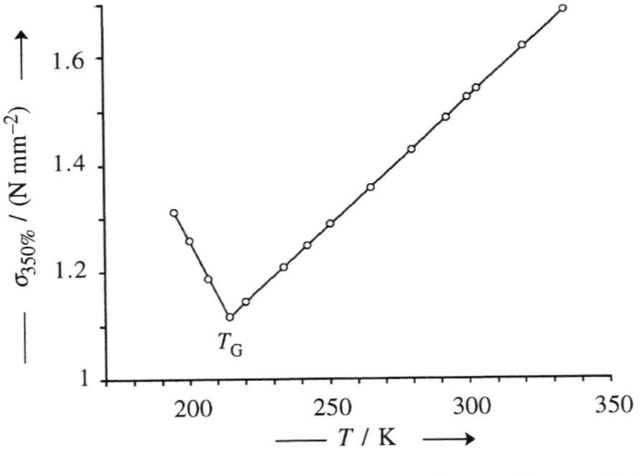

Fig. 16-19 Temperature dependence of tensile stress, $\sigma_{350\%}$, of a weakly crosslinked natural rubber at 350 % elongation [14]. Tensile stress is proportional to the force $\boldsymbol{F}$ since $\sigma = \boldsymbol{F}/A$.

The gas-like behavior is characteristic for entropy-elastic bodies. A single coil *molecule*, when attached at one end of an immobile body, will stretch on elongation but will return to its original macroconformation if the load is removed (Section 16.4.2). In a *substance* consisting of many molecules with flexible segments, segments and molecules may slip past each other on sufficiently large deformations and the body is deformed irreversibly. The slippage is prevented if chains are intermolecularly interconnected ("crosslinked"); on release of stress, segments return to their initial positions.

This **rubber elasticity** can be described in various ways. Molecularly, it is a deformation of macroconformations. Energetically, it is a change of entropy that can be described by both phenomenological and statistical thermodynamics. The **entropy elasticity** of rubbers differs from the elasticity of metals which is an **energy-elasticity**. There are chemically lightly crosslinked polymers with *rigid* segments that are predominantly energy-elastic. However, they are presently only of academic interest.

Entropy-elastic and energy-elastic bodies differ characteristically in their macroscopic properties:

	energy-elastic bodies	*entropy-elastic bodies*
Reversible deformation	small (ca. 0.1 %)	large (several 100 %)
Elastic moduli	large	small
Temperature change on deformation	cooling	warming
Heating of bodies under constant load	expansion	contraction
Heating of undeformed bodies	expansion	expansion

Undeformed entropy-elastic bodies expand on heating but once deformed, such bodies contract on warming. A certain concentration must exist therefore at which these two effects compensate each other, i.e., where the expansion coefficient is zero. This point is usually at a concentration of 5 % to 10 %.

16.4.2 Entropy-elasticity of Single Molecules

The elongation of single chain molecules can be investigated experimentally by atomic force microscopy (AFM) (Fig. 16-20) where the force $\boldsymbol{F}$ is recorded as a function of the length L. The "roughness" of the resulting curve for $\boldsymbol{F} = \mathrm{f}(L)$ is created by the subsequent response of single bonds to the applied force (Fig. 16-21). The sudden increases and drops as well as the various plateaus of the force are caused by various molecular processes (Fig. 16-21) that can be simulated by molecular dynamics.

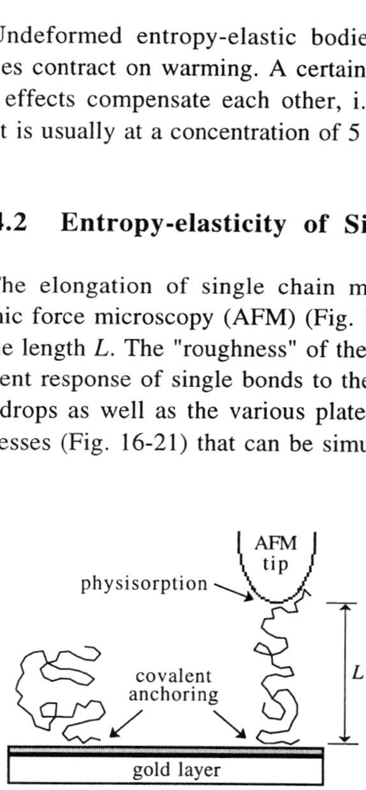

Fig. 16-20 Determination of the elasticity of coil molecules by atomic force microscopy. The segment at one end of the molecule is bound covalently to alkane thiols whose highly ordered crystal-like layers are covalently connected to a flat, polycrystalline layer of gold.

The other end of the polymer molecule is physically adsorbed by the 10-50 nm wide tip of the lever arm of the atomic force microscope. The tip is coupled to a piezoelectric measuring device via a soft lamellar spring which allows one to measure length changes of 0.1 nm or more.

Experiment and simulation agree well for the drawing of single dextran molecules (a polymer with glucose rings that are interconnected by glycosidic bonds $>^5CH{-}O{-}^6CH_2{-}$ (see Volume II, p. 386) except for the change of conformation at $-O{-}^6CH_2{-}^5CH<$ which was found experimentally to take place at a force of ca. 300 pN (Fig. 16-21) and at ca. 600 pN by simulation. This difference was caused by the vastly different speeds of drawing used: it took place in only 1 ps in simulation but required 10^{12} ps = 1 s in the experiment. Hence, the thermal fluctuations in the "slow" experiment were not "seen" by the "fast" simulation. As a result, structures appeared much more stable in simulation and conversions of structures required more force.

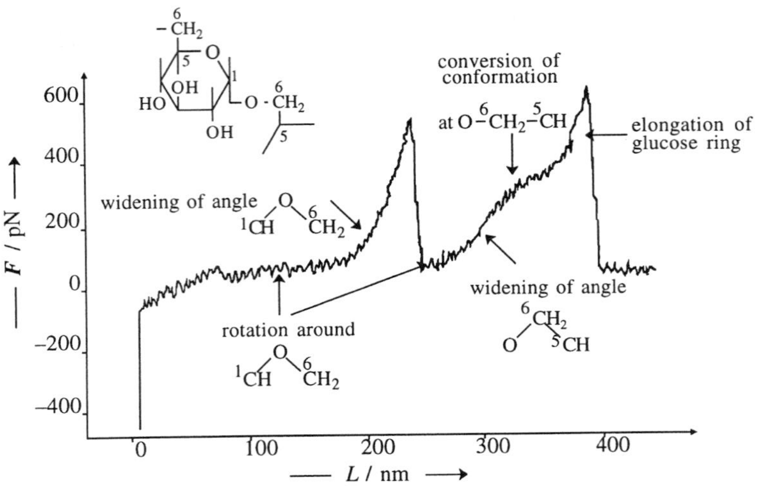

Fig. 16-21 Force $\boldsymbol{F}$ as a function of the extension of single dextran molecules to a length L [15]. Experimental data by AFM. Text refers to the results of simulations by molecular dynamics: increasing extension of the chain causes first a rotational conversion of the microconformation at the glycosidic bond $>^{1}CH{-}O{-}^{6}CH_2{-}$, then a widening of this bond angle, followed by a widening of the bond angle of $-O{-}^{6}CH_2{-}^{5}CH<$, a conversion of the microconformation of this structure, and finally an elongation of the glucose rings. The insert shown a section of the dextran chain.

The force-extension behavior of coil molecules can be described with conventional chain models. The retraction force $\boldsymbol{F}_{\text{chain}}$ for the re-establishment of the equilibrium structure of freely jointed chains (p. 88) is given by

$$(16\text{-}33) \qquad \boldsymbol{F}_{\text{chain}} = k_B T / [L_{\text{seg}} £^{*}(r_{oo}/r_{\text{cont}})]$$

where r_{oo} = end-to-end distance of the coiled chain, $r_{\text{cont}} = N_{\text{seg}} L_{\text{seg}}$ = conventional contour length, N_{seg} = number of segments of length L_{seg}, and $£^{*}$ = inverse Langevin function (p. 102). The retraction force of wormlike chains is

$$(16\text{-}34) \qquad \boldsymbol{F}_{\text{chain}} = (k_B T / L_{\text{ps}})[\{1/4\}\{1 - (r_o/r_{\text{cont}})\}^{-2} - \{1/4\} + \{r_o/r_{\text{cont}}\}]$$

where r_o = end-to-end distance of the chain and L_{ps} = persistence length. Eqs.(16-33) and (16-34) also apply if valence angles of chain atoms are stretched to 180°. In this case, the historic contour length L_{chain} replaces the conventional contour length r_{cont}.

A plot of $\boldsymbol{F}_{\text{chain}}$ as a function of the normalized length L/r_{cont} delivers universal curves if the adoptable quantities N_{seg} and L_{seg} are suitably selected (Fig. 16-22). For the graphs of Fig. 16-22, the same segment length of L_{seg} = 0.30 nm was chosen, i.e., approximately the crystallographic bond length of b_{cr} = 0.254 nm.

16.4.3 Chemical Thermodynamics

Changes of the thermodynamic state of entropy-elastic bodies can be described quantitatively by the basic equations of phenomenological thermodynamics. In analogy to

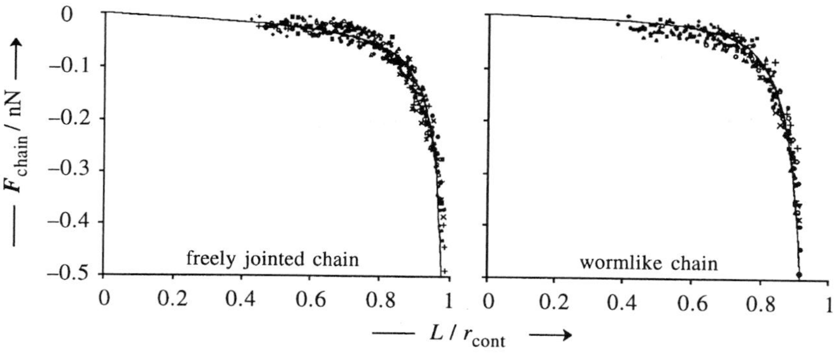

Fig. 16-22 Retraction force of poly(methacrylic acid) ($\overline{M}_w$ = 66 200 g/mol; 1 thiol endgroup) in water as a function of the reduced length, L/r_{cont}. The chain was modeled as a freely jointed chain with L_{seg} = 0.33 nm and many measurements for each of eight investigated chains with different bridging segments ($61 \leq N_{seg} \leq 295$) or as a wormlike chain with L_{seg} = 0.28 nm ($77 \leq N_{seg} \leq 372$) [16].
With kind permission of the American Chemical Society, Washington, DC.

the expression of chemical thermodynamics for the pressure p (a force per area) as a function of the change of internal energy U with volume V and a term for the change of pressure with temperature T, one can express the force $\boldsymbol{F}$ in tensile tests as the sum of the change of internal energy U with length L and a term for the change of force with temperature (Eq.(16-35)). Note the sign change.

	Chemical thermodynamics		Tensile testing
(16-35)	$p = -(\partial U/\partial V)_T + T(\partial p/\partial T)_V$	;	$\boldsymbol{F} = (\partial U/\partial L)_T + T(\partial \boldsymbol{F}/\partial T)_L$

Similarly, the change of the Helmholtz energy A with volume V has a counterpart in the change of the Helmholtz energy with length L (Eq.(16-36)):

	Chemical thermodynamics		Tensile testing
(16-36)	$(\partial A/\partial V)_T = (\partial U/\partial V)_T - T(\partial S/\partial V)_T$	;	$(\partial A/\partial L)_T = (\partial U/\partial L)_T - T(\partial S/\partial L)_T$

Introduction of Eq.(16-36), right, into Eq.(16-35), right, leads to

(16-37) $$(\partial A/\partial L)_T + T(\partial S/\partial L)_T = \boldsymbol{F} - T(\partial \boldsymbol{F}/\partial T)_L$$

Thermodynamics delivers $(\partial S/\partial V)_T = (\partial p/\partial T)_V$ and, by analogy, $(\partial S/\partial L)_T = (\partial F/\partial T)_L$. Introduction of the latter equality into Eq.(16-37) leads to the **thermodynamic equation of state** of entropy-elastic bodies:

(16-38) $$\boldsymbol{F} = (\partial A/\partial T)_L$$

For moderately stretched elastomers, forces $\boldsymbol{F} = A_0\sigma$ are proportional to the temperature, i.e., $\boldsymbol{F} = const\ T$ and $(\partial \boldsymbol{F}/\partial T)_L = const$, and therefore also $\boldsymbol{F}/T = (\partial \boldsymbol{F}/\partial T)_L$. Introduction of the latter equation into Eq.(16-35) results in $(\partial U/\partial L)_T = 0$: internal energies U do not change on isothermal elongation. Entropy-elastic bodies thus differ dramatically from energy-elastic ones where internal energies do change.

The change of internal energy, $\partial U = \boldsymbol{F}\mathrm{d}L + C_p\mathrm{d}T$, is calculated from the force $\boldsymbol{F}$ and the heat capacity C_p at constant pressure for the changes in length, dL, and temperature, dT. For entropy-elastic bodies, it does not change on heating so that $\partial U = 0$. The heat capacity at constant pressure does not change on heating entropy-elastic bodies, i.e., C_p = const. However, on heating such a body from temperature T_I to T_II, the length of an entropy-elastic body is increased from L_I to L_II. Integration for the change from state I to state II delivers $\boldsymbol{F}(L_\mathrm{II} - L_\mathrm{I}) = -\, C_p(T_\mathrm{II} - T_\mathrm{I})$. Since $T_\mathrm{II} > T_\mathrm{I}$, it follows that $L_\mathrm{II} < L_\mathrm{I}$: an entropy-elastic body contracts on heating under constant load.

Because of the constant load, stress must increase on heating: a rubber band becomes more taut on heating if it is clamped at both ends. This behavior follows directly from the total differential of the length change:

$$\text{(16-39)} \qquad \mathrm{d}L = (\partial L/\partial \boldsymbol{F})_T\mathrm{d}\boldsymbol{F} + (\partial L/\partial T)_{\boldsymbol{F}}\mathrm{d}T$$

Heating while maintaining the length (dL = 0) results in

$$\text{(16-40)} \qquad (\mathrm{d}\boldsymbol{F}/\mathrm{d}T)_L = -\,(\partial L/\partial T)_{\boldsymbol{F}}/(\partial L/\partial \boldsymbol{F})_T$$

The term $(\partial L/\partial \boldsymbol{F})_T$ is positive since the length L increases with increasing force $\boldsymbol{F}$. However, the term $(\partial L/\partial T)_{\boldsymbol{F}}$ is negative since the linear thermal expansion coefficient, $(1/L)(\mathrm{d}L/\mathrm{d}T)$, is negative because of $\boldsymbol{F}(L_\mathrm{II} - L_\mathrm{I}) = -\, C_p(T_\mathrm{II} - T_\mathrm{I})$ for $T_\mathrm{II} > T_\mathrm{I}$. The change of force with temperature at constant length, $(\partial \boldsymbol{F}/\partial T)_L$, must therefore be positive.

The statements above apply to ideal entropy-elastic bodies. Real entropy-elastic bodies always contain an energy-elastic component (see Fig. 16-23). For a uniaxial deformation, the force $\boldsymbol{F}_\mathrm{en}$ from this component is given by

$$\text{(16-41)} \qquad \boldsymbol{F}_\mathrm{en} = (\partial U/\partial L)_{T,V} = \boldsymbol{F} - T(\partial \boldsymbol{F}/\partial T)_{V,L}$$

The fraction of the energy-elastic component, $\boldsymbol{F}_\mathrm{en}/\boldsymbol{F}$, can therefore be obtained from force-temperature measurements at constant volume.

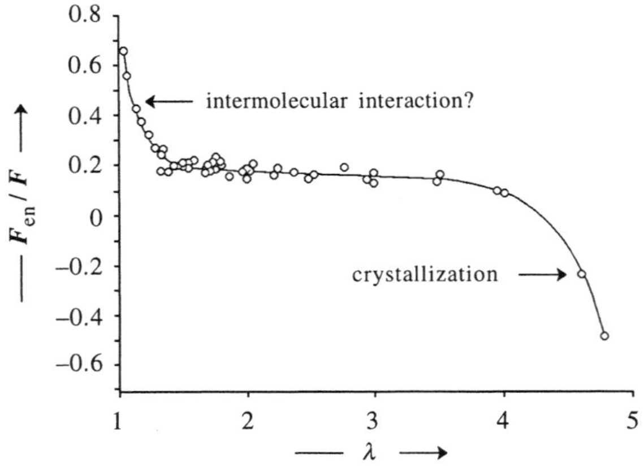

Fig. 16-23 Energy-elastic ratio $\boldsymbol{F}_\mathrm{en}/\boldsymbol{F}$ of natural rubber as a function of the uniaxial strain ratio, $\lambda = L/L_0$ [17]. Natural rubber consists of ca. 95 % *cis*-1,4-isoprene units (Volume II, p. 246).
By kind permission of the American Institute of Physics, Melville (NY).

Such measurements are experimentally difficult and one usually measures how the force changes with temperature at constant length. The data are then converted to those for constant volume.

As demanded by theory, energy-elastic ratios, F_{en}/F, are indeed practically constant for a considerable range of strain ratios, $\lambda = L/L_0$ (Fig. 16-23). In this range, energy-elastic ratios are also independent of the measuring method, the crosslinking conditions, the degree of crosslinking, the type of deformation (elongation, twisting), the nature of swelling-causing liquids, and the degree of swelling.

Energy-elastic ratios decrease at greater strain ratios (Fig. 16-23) if elastomers such as natural rubber crystallize under stress. They increase at small strain ratios, probably because contributions by intermolecular interactions become significant.

Energy-elastic ratios F_{en}/F can be positive (as in Fig. 16-23) or negative (Table 16-9). The energy of a network is lowered–and the energy-elastic ratio becomes negative–if microconformations change from gauche to trans on stretching *and* trans conformations are those with the lowest energy (examples on the right side of Table 16-9). The molecular reasons for positive energy-elastic ratios F_{en}/F (left side of Table 16-9) are not clear.

Table 16-9 Energy-elastic proportions F_{en}/F of crosslinked polymers. * See also Fig. 16-23.

Polymer	F_{en}/F	Polymer	F_{en}/F
Poly(vinyl alcohol)	0.42	Poly(isobutylene)	–0.06
Poly(dimethylsiloxane)	0.19	Poly(isoprene), *trans*-1,4	–0.09
Poly(isoprene), *cis*-1,4	0.17 *	Poly(butadiene), *trans*-1,4	–0.25
Poly(styrene), at	0.16	Poly(ethylene)	–0.42
Poly(butadiene), *cis*-1,4	0.12	Poly(ethylene-*co*-propylene)	–0.43
Poly(oxyethylene)	0.08	Poly(tetrahydrofuran)	–0.47

16.4.4 Statistical Thermodynamics

Chemical thermodynamics of elastomers allows predictions about the change of thermodynamic states but not about the effect of chemical and physical structures on the properties of elastomers. Structural effects can be evaluated by statistical thermodynamics.

A linear macromolecule may adopt many different macroconformations (Chapter 4). A polymer consisting of macromolecules in various macroconformations (shapes) is called **flexible** if the shapes respond fast to applied external forces. For example, a tensile stress will cause an elongation of a polymeric material whereas a shear stress will lead to a flow. Large tensile stresses will cause irreversible deformations of polymers (Section 16.2). This irreversible deformation can be prevented if all polymer chains are interconnected by chemical bonds, resulting in a crosslinked **network** (Fig. 16-24).

The connection points are called **junctions** or **crosslinks**. They are usually trifunctional as in Fig. 16-24, I but may also be tetrafunctional; junctions with higher functionalities are rare. The chain sections between two junctions are called **network chains**. In elastomers, network chains usually consist of many monomeric units.

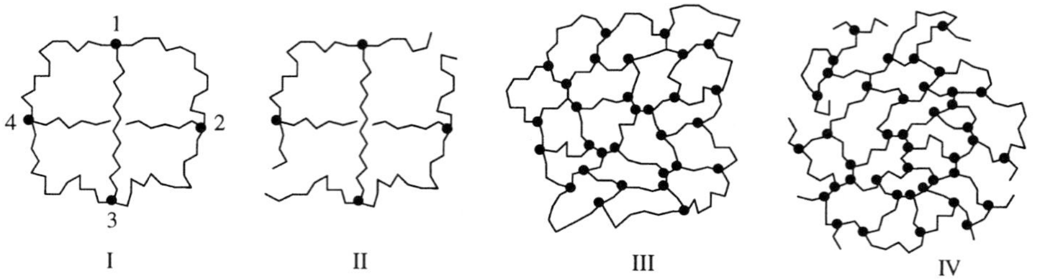

Fig. 16-24 Some networks with trifunctional ($f = 3$) crosslinks (•) and various numbers of network chains (N_c) and junctions (N_x). A perfect network consists of $N_c = (f/2)\, N_x$ network chains.

I: Perfect network with $N_x = 4$ junctions (1, 2, 3, 4) and $N_c = 6$ network chains (1-2, 1-3, 1-4, 2-3, 2-4, 3-4) consisting each of 13 monomeric units (–) and thus a mesh size (loop size) of 39 units (for example, the ring formed by junctions 1-2-4. Dangling ends (loose ends) are absent.

II. The same "network" as in I with loose ends.

III. Perfect network with loops consisting of 14 monomeric units each and no loose ends. Network chains are of different lengths.

IV. Irregular network with different sized network chains and loops, containing loose ends.

In a **perfect network**, all junctions have the same functionality, all network chains the same number of monomeric units, and all meshes (loops) the same number of monomeric units. Loose ends (dangling ends) are absent as are twists, knots, entanglements, and excluded volumes.

The Gibbs energy $G(r)$ of a network consisting of many network chains changes if the network is elongated. Statistical thermodynamics calculates first the Gibbs energy of a single network chain, $G'(r) = H - TS$. The entropy of this network chain is given by the distribution function $W(r)$ of the vector $\boldsymbol{r}$ from one end of the network chain to the other, $S = k_B \ln W(\boldsymbol{r})$. This function can be approximated by a Gaussian distribution of the chain ends r (Section 4.4.8), i.e., $W(\boldsymbol{r}) = [3/(2\,\pi\,\langle r^2\rangle_o)]^{3/2} \exp\,[-\,3\,r^2/(2\,\langle r^2\rangle_o)]$ (cf. Eq.(4-47) for the radius of gyration if the network chain is assumed to be a single segment with the degree of polymerization, $X = 1$). Introduction of the expression for $W(\boldsymbol{r})$ into $G'(r) = H - k_BT \ln W(\boldsymbol{r})$ leads to

$$(16\text{-}42)\qquad \begin{aligned} G'(r) &= H - k_BT \ln\,[3/(2\,\pi\,\langle r^2\rangle_o)]^{3/2} &&+ k_BT[3\,r^2/(2\,\langle r^2\rangle_o)] \\ &= C'(T) &&+ k_BT[3\,r^2/(2\,\langle r^2\rangle_o)] \end{aligned}$$

where the first and the second term of the right side are combined into a temperature-independent constant $C'(T)$. The Gibbs energy of a network consisting of N_c network chains with an average end-to-end distance $r^2 = \langle r^2\rangle$ of *all* network chains is then

$$(16\text{-}43)\qquad G(r) = C(T) + N_ck_BT[3\,\langle r^2\rangle/(2\,\langle r^2\rangle_o)]$$

On drawing an elastomer, macroscopic deformations of junctions are transferred to network chains. The change of the elastic Gibbs energy of the network is therefore the sum of two contributions: (a) conformational changes within each network chain from $G(r_o)$ to $G(r)$ (conformational term) and (b) spatial redistributions of junctions (dispersion term). The conformational term is obtained from Eq.(16-43) minus the corresponding term for the initial (isotropic) state with $\langle r^2\rangle = \langle r^2\rangle_o$. The dispersion term indicates the increase of volume from V_o to V similar to that of ideal gases, i.e., $\Delta G_{disp} = -N_x k_B T \ln (V/V_o)$. In a *perfect* network composed of N_c network chains and junctions of functionality f, the number of junctions is $N_x = 2\,N_c/f$. The change of elastic Gibbs energy is therefore given by

$$\begin{aligned}(16\text{-}44)\qquad \Delta G_{el} &= \Delta G_{conf} + \Delta G_{disp} = [G(r) - G(r_o)] + \Delta G_{disp}\\ &= N_c k_B T[3\,\langle r^2\rangle/(2\,\langle r^2\rangle_o)] - N_c k_B T[3/2] - N_x k_B T \ln (V/V_o)\\ &= (3\,N_c/2)\,k_B T\,[(\langle r^2\rangle/\langle r^2\rangle_o) - 1] - (2\,N_c/f)\,k_B T \ln (V/V_o)\end{aligned}$$

16.4.5 Models

The deformation of an elastic network causes a shift of the mean-square average of end-to-end distances of network chains that cannot be measured directly. However, the shift can be expressed by the macroscopic strain ratio, $\lambda = L/L_o$, if one assumes certain models. At present, the following three models are usually discussed: simple-affine model, affine model, and phantom network model (Table 16-10).

The common names of these models are confusing since all three of them assume phantom networks, i.e., infinitely thin network chains (no excluded volume of chain segments). They differ in the type of affinity and the presence/absence of fluctuations about junctions.

The **simple-affine model** (symbol S) assumes a phantom chain whose *chain segments* are shifted affine (i.e., linear) to the corresponding macroscopic deformation. It is thus a "segment-affine phantom chain model."

The **affine model** (symbol N) is used most often. It assumes an affine shift of *network junctions* of phantom chains (i.e., also chain vectors $\boldsymbol{x}, \boldsymbol{y}$ and $\boldsymbol{z}$ in the three spatial directions, i = x, y, z) with respect to the macroscopic draw ratio, $\lambda_i = L_i/L_{i,o}$. The correct name of this model is thus "junction-affine phantom chain model."

Since $\lambda_x^2 = \langle x^2\rangle/\langle x^2\rangle_o = \langle x^2\rangle/[\langle r^2\rangle_o/3]$ for the x direction, one has for all three spatial directions

$$(16\text{-}45)\qquad \langle r^2\rangle = \langle x^2\rangle + \langle y^2\rangle + \langle z^2\rangle = (\lambda_x^2 + \lambda_y^2 + \lambda_z^2)\langle r^2\rangle_o/3$$

The volume V after drawing is calculated from the draw ratio, $\lambda_i = L_i/L_{i,0}$ (i = x, y, z), the initial lengths, $L_{x,0} = L_{y,0} = L_{z,0} = L_0$, and the initial volume, $L_0^3 = V_0$, resulting in $V = L_x L_y L_z = \lambda_x L_{x,0} \lambda_y L_{y,0} \lambda_z L_{z,0} = L_0^3(\lambda_x\lambda_y\lambda_z) = V_0(\lambda_x\lambda_y\lambda_z)$. Introduction of Eq.(16-45) into $V = V_0(\lambda_x\lambda_y\lambda_z)$ delivers a substitute equation for Eq.(16-44):

$$(16\text{-}46)\qquad \Delta G_{el} = N_c(k_B T/2)\{[\lambda_x^2 + \lambda_y^2 + \lambda_z^2 - 3] - [(4/f) \ln (\lambda_x\lambda_y\lambda_z)]\}$$

Table 16-10 Models for rubber elasticity.

Symbol	Conventional name	Excluded volume of segments	Affinity	Fluctuations
S	simple-affine	no	segments	no
N	affine	no	junctions	no
F	phantom network	no	junctions	junctions

According to this model, the change of elastic Gibbs energy on drawing of an elastomer is controlled only by the number N_c of network chains, the temperature T, the three draw ratios λ_i, and the functionality f of the network junctions but not by their structure. The junctions may be chemical and physical; it is only required that their number does not change on drawing.

The affine model N also assumes that fluctuations about junctions are suppressed by adjacent network chains. This assumption is well obeyed by non-swollen, undeformed networks but not by highly swollen networks in thermodynamically good solvents.

The so-called **phantom network model** (symbol F) allows a "free" *fluctuation of junctions* about their average positions. These fluctuations are assumed to follow Gaussian statistics. Since elastomers have high polymer concentrations, a fluctuation of a junction must be coupled with the fluctuation of other junctions. The model F is therefore a "junction-affine phantom chain model with coupled fluctuations of junctions."

For the elastic Gaussian energy, model F delivers a similar expression to Eq.(16-46). However, there is no dispersion of the junctions because their average positions are given by the macroscopic deformation. The dispersion term $[(4/f) \ln (\lambda_x\lambda_y\lambda_z)]$ of Eq.(16-46) is therefore zero. Furthermore, the front factor $N_c(k_BT/2)$ of Eq.(16-46) has to be replaced by $N_c(k_BT/4)$ because the deformation of the network changes only one-half of the mean-square distance while the other half stems from fluctuations.

All three models predict the same dependence of tensile strength σ_{11} on the deformation parameter $(\lambda - \lambda^{-2})$ (see below), i.e., $\sigma_{11} = K(\lambda - \lambda^{-2})$. Stress-strain experiments thus do not allow one to distinguish between them.

However, the three models do vary in their predictions of the dependence of the expansion factor of radii of gyration, α_s, as a function of the macroscopic draw ratio λ. Expansion factors parallel (||) and perpendicular ($\perp$) to the draw direction can be obtained from inelastic neutron scattering of partially deuterated networks by measuring the mean-square radii of gyration before (index o) and after drawing (index || or $\perp$): $(\alpha_s)_{||} = (\langle s^2\rangle_{||}/\langle s^2\rangle_o)^{1/2}$ and $(\alpha_s)_{\perp} = (\langle s^2\rangle_{\perp}/\langle s^2\rangle_o)^{1/2}$, respectively.

Longitudinal expansion factors $(\alpha_s)_{||}$ depend directly (model S) or in a more complicated way (models N and F) on the draw ratio λ and in model F also on the functionality f of the junctions (Table 16-11). Experiments on poly(dimethylsiloxane) networks showed that the functions $(\alpha_s)_{||} = f(\lambda)$ and $(\alpha_s)_{\perp} = f(\lambda)$, respectively, are independent of crosslinking conditions (bulk or swollen) (Fig. 16-25) and that the simple-affine model S did not apply at all. Factors $\alpha_{s,\perp}$ followed model $F_{\perp}$ for short network chains but seemed to approach model $N_{\perp}$ for longer ones. Expansion factors $\sigma_{s,||}$ of networks with short network chains adhere to model $N_{||}$, those with medium-sized network chains to model $F_{||}$, and those with very long network chains to none of the models.

Table 16-11 Longitudinal and transverse expansion factors of radii of gyration as a function of the macroscopic draw ratio λ as predicted by the three models S, N, and F for phantom networks.

Model		$(\alpha_s)_{\|\|} =$	$(\alpha_s)_{\perp} =$
S	Segment-affine	λ	$\lambda^{-1/2}$
N	Junction-affine	$[(\lambda^2 + 1)/2]^{1/2}$	$[(\lambda + 1)/(2\,\lambda)]^{1/2}$
F	Junction-affine with fluctuation	$\{[f + 2 + (f-2)\lambda^2]/[2f]\}^{1/2}$	$\{[f + 2 + (f-2)\lambda^{-1}]/[2f]\}^{1/2}$

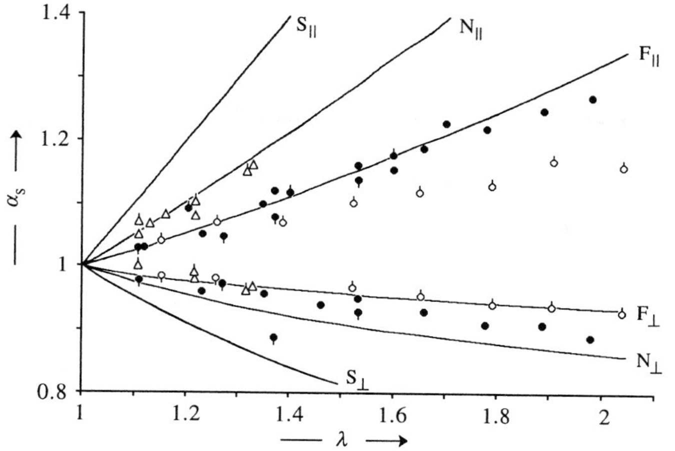

Fig. 16-25 Dependence of expansion factor α_s of radii of gyration parallel (||) or perpendicular (⊥) to the draw direction. Lines: predictions of models (model F: junctions with functionality $f = 4$) (see Table 16-11). Symbols: experimental data for unswollen crosslinked poly(dimethylsiloxane)s with molar masses of network chains of 6000 (Δ), 10 000 (●), or 25 000 g/mol (O) [18]. Same symbols with small vertical lines: swollen specimen with volume fractions $\phi_2 = 0.71$ (upper lines), or $\phi_2 = 0.60$ (lower lines).

16.4.6 Uniaxial Extension

The standard model for uniaxial extension is usually model N which, however, seems to apply only to relatively strongly crosslinked polymers (Fig. 16-25). Uniaxial drawing of such a polymer in the x direction leads to $\lambda_x = \lambda = L/L_0$, and, because of a possible volume change to V from V_0 also to $\lambda_y = \lambda_z = [(V/V_0)(1/\lambda_x)]^{1/2}$. Eq.(16-46) converts to $\Delta G_{el} = (N_c/2)k_BT\{[\lambda^2 + 2(V/V_0)\lambda^{-1} - 3] - (4/f) \ln (V/V_0)\}$. Differentiation of this equation with respect to length $L = L_0\lambda$ delivers the force $\boldsymbol{F}$,

$$(16\text{-}47) \qquad \boldsymbol{F} = (\partial \Delta G_{el}/\partial L)_{T,V} = (\partial \Delta G_{el}/\partial \lambda)_{T,V}/L_0 = N_c k_B T(\lambda - (V/V_0)\lambda^{-2})/L_0$$
$$= N_c k_B T (V/V_0)^{2/3}(\alpha - \alpha^{-2})/L_{i,V}$$

where $\alpha = L/L_{i,V} = L/[L_0(V/V_0)^{1/3}] = \lambda(V/V_0)^{-1/3}$ is the linear expansion of the expanded volume V relative to the initial volume V_0 of the unstretched isotropic body.

Eq.(16-47) can be used to calculate the tensile stress $\sigma_{11} = F/A_0$ as a function of the concentration of network chains, $[M_c] = N_c/(N_A V_0)$, the linear expansion, $\alpha = L/L_{i,V}$, and the volume ratio, V/V_0, where $V_0 = A_0 L_0$:

$$(16\text{-}48) \qquad \sigma_{11} = \boldsymbol{F}/A_0 = RT[M_c](\lambda - (V/V_0)\lambda^{-2}) = RT[M_c](\alpha - \alpha^{-2})(V/V_0)^{1/3}$$

V/V_0 becomes unity if the volume does not change on extension and Eq.(16-48) converts to

$$(16\text{-}49) \qquad \sigma_{11} = RT[M_c](\lambda - \lambda^{-2}) = E_{app}(\lambda - \lambda^{-2}) \qquad \text{(model N)}$$

indicating that the tensile stress σ_{11} is proportional to a dimensionless extension parameter, $(\lambda - \lambda^{-2})$. Comparison with Hooke's law, $E = \sigma_{11}(\lambda - 1)^{-1}$, shows that the front factor $RT[\mathrm{M_e}]$ corresponds to an apparent modulus, $E_{app} = \sigma_{11}(\lambda - \lambda^{-2})^{-1}$.

The effect of swelling of a network can be evaluated by remembering that the ratio $V_0/V = V_0/(V_0 + \Delta V)$ of unswollen networks (Eq.(16-48)) corresponds to the volume fraction $\phi_2 = V_2/V$ of swollen networks. For swollen networks, Eq.(16-49) thus converts to $\sigma_{11} = RT[\mathrm{M_c}](\lambda - \lambda^{-2})\phi_2^{1/3}$.

In Eqs.(16-48) and (16-49), the molar concentration $[\mathrm{M_c}]$ of network chains indicates the *initially present* concentration of *chemical* junctions. However, experiments deliver the *actual* concentration of all *effective* junctions, both chemical and physical. Chemical junctions are not necessarily all equally effective since elongations are controlled by the *shortest* network chains. A part of the total functionality of chemical junctions is furthermore wasted in the formation of loose ends and loops. Physical junctions increase the number of effective junctions, especially if networks crystallize on stretching.

Eq.(16-48) describes well the stress-strain behavior of crossliked natural rubber on pressurizing ($\lambda < 1$) and on uniaxial drawing with strain ratios of $1 \leq \lambda < 5$ (Fig. 16-26). At larger strain ratios, crosslinked natural rubber begins to crystallize. The crystalline regions act as physical junctions and the tensile stress increases strongly.

Model F considers additional coupled fluctuations of network chains, which leads to a decrease of the molar concentration of network chains, $[\mathrm{M_c}]$, by the molar concentration of junctions, $[\mathrm{M_x}]$. Eq.(16-49) for model N is thus replaced by Eq.(16-50) if the deformation is invariant with respect to volume. Hence, model N and model F differ only in the front factor:

$$\sigma_{11} = RT([\mathrm{M_c}] - [\mathrm{M_x}])(\lambda - \lambda^{-2}) \qquad \text{(model F)} \tag{16-50}$$

16.4.7 Biaxial Extension

An equal elongation of a volume-constant elastomer ($\lambda_x\lambda_y\lambda_z \equiv 1$) in the two directions x and y leads to $\lambda_x = \lambda_y \equiv \lambda$ and $\lambda_z = 1/\lambda^2$. Introduction of these values into Eq.(16-46) results in $\Delta G_{el} = N_c(k_BT/2)[2\lambda^2 + \lambda^{-4} - 3]$ which, upon differentiation with respect to $L = L_0\lambda$, delivers $\boldsymbol{F} = (\partial\Delta G_{el}/\partial L)_{T,V} = (2N_c/L_0)(k_BT)[\lambda - \lambda^{-5}]$. Introduction of $\sigma = F/A_0$, $V_0 = L_0A_0$, $k_B = R/N_A$, and $[\mathrm{M_c}] = N_c/(N_AV_0)$ results in

$$\sigma = 2RT[\mathrm{M_c}](\lambda - \lambda^{-5}) \tag{16-51}$$

The stress in each of the two directions x and y is thus greater than the uniaxial one (cf. Eq.(16-48)).

16.4.8 Elongation of Real Networks

None of the two classic junction-affine models N and F describes *all* stress-strain curves of real networks, especially not those that are swollen by liquids (Fig. 16-25). One of the reasons for this failure seems to be that, at small deformations, fluctuations around

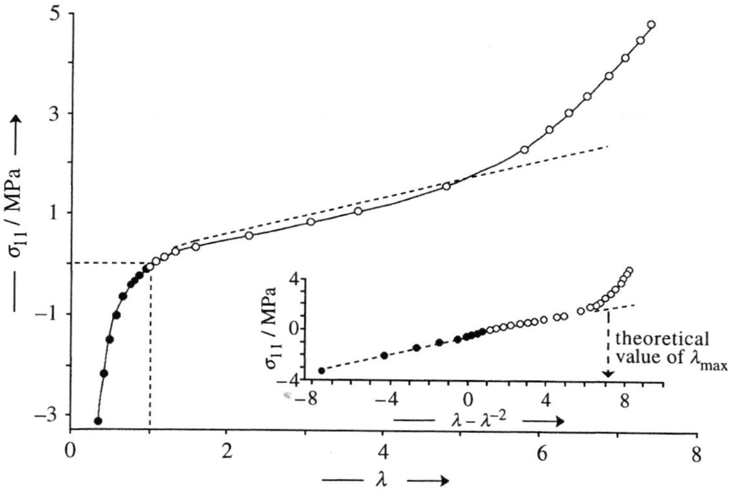

Fig. 16-26 Tensile stress σ_{11} of a crosslinked natural rubber at 25°C as a function of the draw ratio $\lambda = L/L_0$ [19]. o Extension ($\lambda > 1$), ● pressurizing ($\lambda < 1$). Insert: determination of $RT[M_c]$ from the initial slope of the function $\sigma_{11} = f(\lambda - \lambda^{-2})$. - - - Calculated from Eq.(16-48) with $RT[M_c] = 0.364$.

The molar concentration of network chains is $[M_c] = 1.47 \cdot 10^{-4}$ mol/cm³ since $T = 298$ K and $R \approx 8.314$ J K^{-1} mol^{-1}. The density of natural rubber, $\rho = 0.91$ g/cm³, is affected only a little by cross-linking. The average molar mass of network chains is $M_c = \rho/[M_c] = 6190$ g/mol which corresponds to a degree of polymerization of $X_c = M_c/M_u = 6190/68.12 \approx 90.9$ and an average number of chain atoms between junctions of $N_c = 4 \times 90.9 = 364$ (= number N_{bd} of bonds).

Natural rubber has a reduced end-to-end distance of $\langle r^2 \rangle_o/M = 6.79 \cdot 10^{-3}$ nm² mol g^{-1}. The average end-to-end distance of network chains is therefore $\langle r^2 \rangle_o^{1/2} = [6.79 \cdot 10^{-3} \times 6190 \text{ nm}^2]^{1/2} = 6.48$ nm if the network chains have Gaussian distributions.

Network chains can be extended to their (average) contour length $r_{cont} = N_c b \sin(\tau/2)$ (Eq.(4-16)). The average length of carbon-carbon chain bonds of a *cis*-1,4-poly(isoprene) chain with the repeating unit $-CH_2-C(CH_3)=CH-CH_2-$ is $b \approx 0.152$ nm. The conventional contour length of a network chain is $r_{cont} = 364 \times 0.152 \sin(111.5/2)$ nm $= 45.7$ nm if an average bond angle of 111.5° is assumed. Unperturbed coils can be extended maximally to $\lambda_{max} = r_{cont}/\langle r^2 \rangle_o^{1/2}$. In this example, this value is $\lambda_{max} = 45.7/6.48 = 7.05$ which agrees well with the value in the insert.

the mean positions of junctions are suppressed by entanglements of network chains. In this case, the model N applies. At large deformations and small deformation rates, entanglements can disentangle, which allows fluctuations of junctions. The stress-strain behavior then approaches that of model F for coupled junction-affine fluctuations.

The transition from model N to model F involves a decrease of the apparent modulus $E_{app} = \sigma_{11}/(\lambda - \lambda^{-2})$ since the front factor of model N contains the molar concentration $[M_c]$ of network chains whereas the front factor of model F includes the difference between the molar concentrations of network chains and junctions, $[M_c] - [M_x]$. This decrease is often described by the semi-empirical Mooney-Rivlin equation (next page).

The elastic Gibbs energy modeled by a symmetry condition with the empirical constants C_1' and C_2' (cf. the first summand of Eq.(16-46)) is given by:

(16-52) $$\Delta G_{el} = C_1'[\lambda_x^2 + \lambda_y^2 + \lambda_z^2 - 3] + C_2'[\lambda_x^{-2} + \lambda_y^{-2} + \lambda_z^{-2} - 3]$$

The elastic Gibbs energy can be written as a function of the elongation invariants, i.e., functions of λ_1, λ_2, and λ_3 that are independent of the choice of axes,

(16-53) $\Delta G_{el} = C_1'[I_1 - 3] + C_2'[I_2 - 3] + O(I_3 - 3) \ldots$

where $I_1 = \lambda_x^2 + \lambda_y^2 + \lambda_z^2$ = first elongation invariant, $I_2 = \lambda_x^2\lambda_y^2 + \lambda_y^2\lambda_z^2 + \lambda_z^2\lambda_x^2$ = second elongation invariant, and C_1', C_2' = constants. To a first approximation, higher orders can be neglected ($O(I_3 - 3) \approx 0$). One then introduces repeatedly the condition $\lambda_x\lambda_y\lambda_z = 1$ for an incompressible material, for example, $\lambda_x = 1/(\lambda_y\lambda_z)$.

The elastomer should be furthermore volume-invariant ($\lambda_x\lambda_y\lambda_z \equiv 1$) and extended in the x direction ($\lambda_x = \lambda$; $\lambda_y = 1/\lambda_x^{1/2}$; $\lambda_z = 1/\lambda_x^{1/2}$). The change of elastic Gibbs energy is therefore $\Delta G_{el} = C_1'[\lambda^2 + (2/\lambda) - 3] + C_2'[\lambda^{-2} + 2\lambda - 3]$. Differentiation of this equation with respect to $L = L_0\lambda$ delivers the force

(16-54) $\boldsymbol{F} = (\partial \Delta G_{el}/\partial L)_{T,V} = C_1'[2\lambda - 2\lambda^{-2}]/L_0 + C_2'[-2\lambda^{-3} + 2]/L_0$

Insertion of $\sigma_{11} = \boldsymbol{F}/A_0 = L_0\boldsymbol{F}/V_0$, $C_1 = C_1'/V_0$, and $C_2 = C_2'/V_0$, as well as rearrangement leads to the **Mooney-Rivlin equation**:

(16-55) $\sigma_{11}/(\lambda - \lambda^{-2}) = 2\,C_1 + 2\,C_2\lambda^{-1}$

A plot of $\sigma_{11}/(\lambda - \lambda^{-2}) = f(1/\lambda)$ delivers straight lines for both unswollen and swollen rubber networks (Fig. 16-27). The slopes decrease with increasing solvent concentration. At $\lambda^{-1} \to 0$, all lines have a common intercept of $2\,C_1$ which is usually assumed to be the front factor of the equations of statistical thermodynamics, Eq.(16-48) or Eq.(16-50).

Slopes $2\,C_2$ are independent of the chemical structure of the networks (insert in Fig. 16-27) but depend on crosslinking conditions, i.e., chain orientations, degrees of crosslinking, etc. The slopes seem to be affected by chain entanglements since the ratio $C_2/C_1 = K_1A_c^{-2}$ decreases with increasing cross-sectional area A_c of polymer chains: the larger A_c, the smaller is the tendency to entangle at the same chain flexibility.

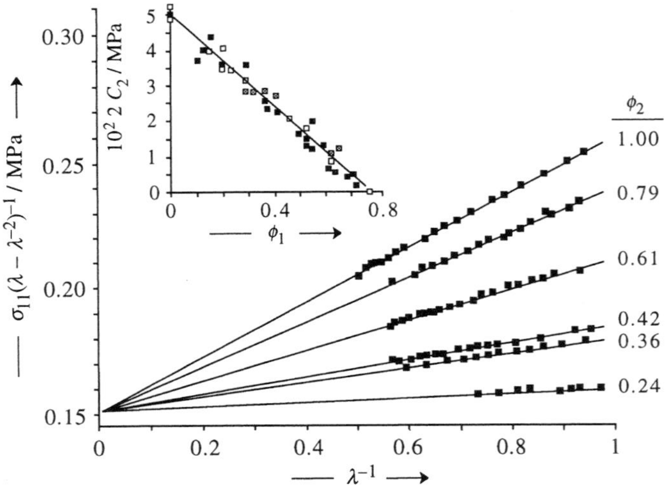

Fig. 16-27 Reduced stress, $\sigma_{11}/(\lambda - \lambda^{-2})$, of a crosslinked natural rubber as a function of the inverse draw ratio, $1/\lambda$, at 45°C in the unswollen ($\phi_2 = 1$) and swollen state in decane ($\phi_2 < 1$) (data of [20]).
Insert: slopes $2\,C_2$ as a function of the volume fraction ϕ_1 of the solvent 1 in swollen, crosslinked natural rubbers(■), butadiene-styrene-rubbers (□), and butadiene-acrylonitrile-rubbers (⊠) [21].

16.4.9 Shearing of Networks

On shearing, bodies are deformed in the [21]-direction (xy direction) but not in the [3]-direction (z direction) ($\lambda_z = const$) (Figs. 15-2 and 15-3). Shearing extends a volume-constant body ($\lambda_x\lambda_y\lambda_z = 1$) in the x direction and contracts it in the y direction ($\lambda_y = 1/\lambda_x = 1/\lambda$).

In Eq.(16-46), the term $(4/f)\ \ln(\lambda_x\lambda_y\lambda_z)$ becomes zero and the term for elongation, $[\lambda_x^2 + \lambda_y^2 + \lambda_z^2 - 3]$, becomes $[\lambda^2 + \lambda^{-2} - 2] = [\lambda - \lambda^{-1}]^2$ for shearing. Eq.(16-46) converts to

(16-56) $$\Delta G_{el} = (k_BT/2)N_c[\lambda^2 - \lambda^{-2} - 2] = (k_BT/2)N_c\gamma^2$$

Differentiation of Eq.(16-56) with respect to the *dimensionless* shear deformation, $\gamma \equiv [\lambda - \lambda^{-1}]$, delivers the shear energy E_s:

(16-57) $$E_s = (\partial\Delta G_{elast}/\partial\gamma)_{T,V} = k_BTN_c[\lambda - \lambda^{-1}] = k_BTN_c\gamma$$

Shearing involves the whole volume. The shear stress is therefore $\sigma_{21} = E_s/V_0$ which corresponds to the expression for tensile stress, $\sigma_{11} = \boldsymbol{F}/A_0$, where a tensile force $\boldsymbol{F}$ acts on an area A_0.

Introduction of the shear energy, $E_s = k_BTN_c\gamma$, and the molar concentration of network chains, $[M_c] = N_c/(N_AV_0)$, into the expression $\sigma_{21} = E_s/V_0$ for the shear stress shows that $RT[M_c]$ is the shear modulus G (compare with Eq.(15-1)):

(16-58) $$\sigma_{21} = RT[M_c]\gamma = G\gamma$$

For perfect networks, the molar concentration of network chains is calcuulated as a ratio of density ρ and molar mass M_c of network chains, $[M_c] = \rho/M_c$, regardless of whether the network is swollen or not. However, networks are not perfect and one has to correct for loose ends. Crosslinking of primary linear chains (2 chain ends) with molar mass M_0 introduces a correction factor $(M_0 - 2\,M_c)/M_0$ and the shear modulus becomes

(16-59) $$G = RT[M_c] = \frac{RT\rho}{M_c}\left(\frac{M_0 - 2\,M_c}{M_0}\right) = \frac{RT\rho}{M_c}\left(1 - \frac{2\,M_c}{M_0}\right)$$

In polycondensations and polyadditions of f-functional monomer molecules, crosslinking is simultaneous with the polymerization to larger chains, which leads to

(16-60) $$G = \frac{RT\rho}{M_c}\left(1 - \frac{2}{f}\right)$$

According to Eqs.(16-59) and (16-60), shear moduli G increase linearly with temperature (Fig. 16-28). The shear modulus is also inversely proportional to the molar mass M_c of network chains.

The equation derived above applies to junction-affine deformations of phantom networks (model N). For such networks with coupled fluctuations of junctions (model F),

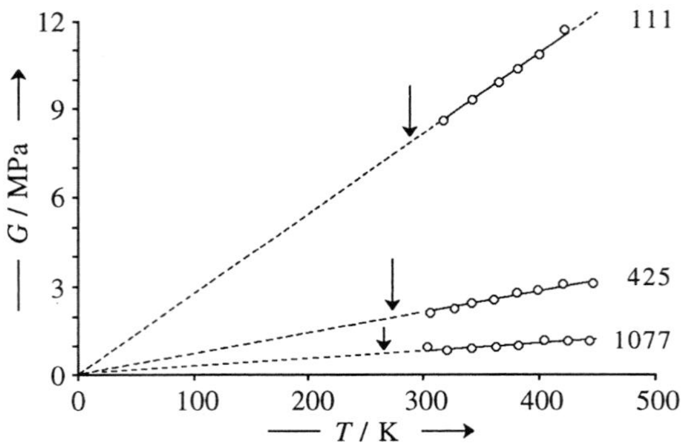

Fig. 16-28 Shear modulus G at a frequency of $\nu = 0.01$ Hz as a function of temperature T for cross-linked polymers from 1,4-butanediol diglycidyl ether, N,N'-dimethylethylenediamine, and 1,4-diaminobutane [22]. Numbers show molar masses M_c of network chains that were calculated from $M_c = \rho RT\{1-(2/f)\}/G$ (Eq.(16-60)) using a functionality $f = 3$ and a density $\rho = 1.1$ g/cm^3. Arrows indicate glass temperatures.

molar concentrations of network chains, $[M_c]$, have to be replaced by $[M_c] - [M_x]$ (see Eq.(16-50)). For tetrafunctional junctions ($f = 4$) in perfect networks ($N_e = 0$), the equation $N_c = (f/2)N_x - (N_e/2)$ (see Fig. 16-24) reduces to $N_c = 2\ N_x$ which leads to $[M_c]/2 = [M_x]$: in this case, the shear modulus of a perfect network in model F is just one-half of that for model N.

For *small* shear deformations, shear moduli can be converted to tensile moduli E. Model N predicts $RT[M_c] = \sigma_{11}/(\lambda - \lambda^{-2})$ (Eq.(16-49)), which is confirmed experimentally for small elongations, i.e., for $\lambda \to 1$ and $\varepsilon \to 0$, respectively (cf. Fig. 16-26). Because $\lambda \equiv \varepsilon + 1$, the extension factor therefore becomes

$$(16\text{-}61) \qquad \lambda - \lambda^{-2} = 1 + \varepsilon - (1+\varepsilon)^{-2} \approx 1 + \varepsilon - (1 - 2\,\varepsilon) = 3\,\varepsilon$$

Introduction of Hooke's law, $\sigma_{11} = E\varepsilon$, shows that the front factor of model N is just 1/3 of the tensile modulus E:

$$(16\text{-}62) \qquad RT[M_c] = \lim_{\lambda\to 0}\,[\sigma_{11}/(\lambda - \lambda^{-2})] = \sigma_{11}/(3\,\varepsilon) = E/3$$

Further comparison of $RT[M_c] = E/3$ (Eq.(16-62)) with $RT[M_c] = G$ (Eq.(16-58)) delivers $E = 3\ G$ as is also predicted by Eq.(16-7) for volume-invariant deformations if Poisson's ratio is $\mu = 1/2$.

Hence, on shearing to small shear deformations, elastomers behave as Hookean bodies. However, they are not Hookean on elongation because the tensile stress is not proportional to the elongation, $\varepsilon = \lambda - 1$, but to $\lambda - \lambda^{-2}$ (cf. Eq.(16-48)).

Tensile moduli E are directly proportional to the molar concentration $[M_e]$ of network chains (Eq.(16-62)). For highly swollen networks, $[M_e]$ should be inversely proportional to the third power of the mesh width L_e of the network, $[M_c] \sim 1/L_c^3$. Scaling theory postulates that mesh widths should depend on segment concentrations c according to $L_c \sim c^{-\nu/(\nu d-1)}$ where ν = Flory exponent and d = dimensionality (Eq.(11-27)). Networks swollen in good solvents should have $d = 3$ and $\nu \approx 3/5$ so that $E \sim [M_c] \sim 1/L_c^3 \sim c^{9/4}$.

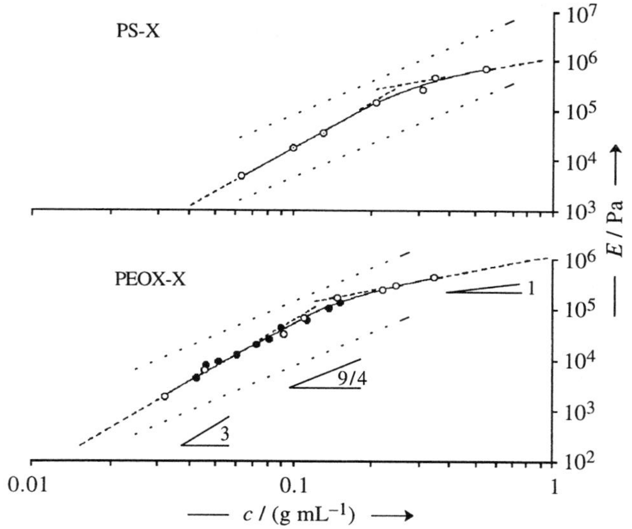

Fig. 16-29 Moduli E by uniaxial compression of swollen polymer networks as functions of polymer concentration c [23]. Dotted lines indicate slopes of 9/4 = 2.25 that are predicted by scaling theory for low polymer concentrations.

Top: benzene-swollen poly(styrene)s crosslinked with divinyl benzene (⊙) or ethylene dimethacrylate (O).

Bottom: water-swollen poly(oxyethylene)s of molar masses 1800 g/mol (O) and 5600 g/mol (●), respectively, crosslinked with an aliphatic "poly"isocyanate.

Experimentally, the exponent of C was indeed found to have an "average" value of 9/4 (Fig. 16-29). However, the exponent approaches a value of 3 at very low polymer concentrations and a value of 1 at very high ones, which remains an unexplained finding.

Historical Notes to Chapter 16

Mechanical Models

W.Voigt, Lehrbuch der Kristallphysik, G.Teubner, Leipzig 1910
A.Reuss, Z.Angew.Math.Mech. **9** (1929) 49
M.Takayanagi, Mem.Fac.Engng. Kyushu Univ. **23** (1963) 1, 41; M.Takayanagi, S.Uemura, S.Minami, J.Polym.Sci. **C 5** (1964) 113

Rubber Elasticity

First Molecular Interpretations

K.H.Meyer, G. von Susich, E.Valko, Kolloid-Z. **59** (1932) 208
E.Guth, H.Mark, Mh.Chem. **63** (1934) 93; Z.Elektrochem. **43** (1937) 683
W.Kuhn, Kolloid-Z. **76** (1936) 258 (contrary to many literature citations, an earlier paper in Kolloid-Z. **68** (1934) 2 is not concerned with rubber elasticity but with single-chain molecules).

Affine Model N
F.T.Wall, J.Chem.Phys. **11** (1943) 527; P.J.Flory, F.T.Wall, J.Chem.Phys. **19** (1951) 1435

Phantom Network F
H.M.James, E.Guth, J.Chem.Phys. **15** (1947) 651, 669; **21** (1953) 1039; E.Guth, J.Polym.Sci. **C 12** (1966) 89

Mooney-Rivlin Equation
M.Mooney, J.Appl.Phys. **11** (1940) 582
Empirical use of Eq.(16-52) without –3 in the second summand.
R.S.Rivlin, Phil.Trans.Royal Soc. **A 240** (1948) 459, **A 241** (1948) 379; R.S.Rivlin, D.W.Saunders, Phil.Trans.Royal Soc. **A 243** (1948) 251
Semi-empirical derivation of Eq.(16-52).

Literature to Chapter 16

16.1.a GENERAL LITERATURE ABOUT MECHANICAL PROPERTIES
R.G.C.Arridge, Mechanics of Polymers, Clarendon Press, Oxford 1975
J.F.Vincent, J.D.Currey, The Mechanical Properties of Biological Materials, Cambridge University Press, Cambridge 1980
I.I.Perepechko, Low-Temperature Properties of Polymers, Pergamon, London 1980
J.S.Hearle, Polymers and Their Properties. Vol. 1: Fundamentals of Structure and Mechanics, Wiley, New York 1982
R.A.Pethrick, R.W.Richards, Eds., Static and Dynamic Properties of the Polymeric Solid State, Reidel, New York 1982
N.J.Mills, Plastics - Microstructure & Engineering Applications, Halsted Press, New York, 2nd ed. 1993
L.E.Nielsen, R.F.Landel, Mechanical Properties of Polymers and Composites, Dekker, New York, 2nd ed. 1993
I.M.Ward, J.Sweeney, An Introduction to the Mechanical Properties of Solid Polymers, Wiley, Chichester 2004

16.1.b STRUCTURE-PROPERTY RELATIONSHIPS
D.W. van Krevelen, Properties of Polymers - Correlation with Chemical Structures, Elsevier, Amsterdam, 3rd ed. 1990
J.Bicerano, Ed.., Computational Modeling of Polymers, Dekker, New York 1992
J.Bicerano, Prediction of Polymer Properties, Dekker, New York 1993

16.1.c TESTS
G.C.Ives, J.A.Mead, M.M.Riley, Handbook of Plastics Test Methods, Iliffe, London 1971
S.Turner, Mechanical Testing of Plastics, Butterworths, London 1973
A.Y.Malkin, A.A.Askadsky, V.V.Kovriga, A.E.Chalykah, Experimental Methods in Polymer Physics, Mir Publ., Moscow, and Prentice Hall, Englewood Cliffs (NJ), 2 vols. 1978 and 1979
J.F.Rabek, Experimental Methods in Polymer Chemistry. Physical Properties, Wiley-Interscience, New York 1980
R.P.Brown, Ed., Handbook of Plastics Test Methods, Godwin Ltd., London 1981
V.Shah, Handbook of Plastics Testing and Failure Analysis, Wiley, Hoboken (NJ), 3rd ed. 2007

16.3 ENERGY ELASTICITY
J.H.Weiner, Statistical Mechanics of Elasticity, Wiley, New York 1983
A.E.Zachariades, R.S.Porter, Eds., High Modulus Polymers–Approaches to Design and Development, Dekker, New York 1987
A.E.Zachariades, R.S.Porter, Eds., The Strength and Stiffness of Polymers, Dekker, New York 1993
G.Bartenev, Mechanical Strength and Failure of Polymers, Prentice-Hall, Englewood Cliffs (NJ) 1993
M.G.Northolt, P. den Decker, S.J.Picken, J.J.M.Balthussen, The Tensile Strength of Polymer Fibers, Adv.Polym.Sci. **178** (2005) 1

16.4 ENTROPY ELASTICITY
L.Bateman, Ed., The Chemistry and Physics of Rubber-Like Substances, McLaren, London 1963
L.R.G.Treloar, The Physics of Rubber Elasticity, Clarendon Press, Oxford, 3rd ed. 1975
G.Heinrich, E.Straube, G.Helmis, Rubber Elasticity of Polymer Networks: Theories, Adv.Polym.Sci. **85** (1988) 33
O.Kramer, Ed., Biological and Synthetic Polymer Networks, Elsevier, New York 1988
J.E.Mark, B.Erman, Rubberlike Elasticity: A Molecular Primer, Wiley, New York 1988
B.Erman, J.E.Mark, Structures and Properties of Rubberlike Networks, Oxford Univ. Press, Oxford 1997

References to Chapter 16

[1] R.G.C.Arridge, M.J.Folkes, Polymer **17** (1976) 496, Figs. 1 and 2
[2] B.G.Penn, F.E.Ledbetter, III, J.E.Clemons, J.G.Daniels, J.Appl.Polym.Sci. **27** (1982) 3751, Table VI
[3] M.Beier, E.Schollmeyer, Angew.Makromol.Chem. **60/61** (1989) 53, selected data of Figs. 3a and 3b
[4] H.-G. Elias, Makromoleküle, Vol. 2, Technologie, Hüthig & Wepf, Basel, 5th ed. (1992), data of Tables 15-8, 15-11, !5-13, 15-14, 15-15, 15-16, and 15-18
[5] J.Clements, R.Jakeways, I.M.Ward, Polymer **19** (1978) 639, Fig. 2 (PE); B.Brew, J.Clements, G.R.Davies, R.Jakeways, I.M.Ward, J.Polym.Sci.-Polym.Phys.Ed. **17** (1979) 351, Fig. 2 (POM)
[6] P.E.Teare, Acta Cryst. **12** (1959) 294
[7] A.Odajima, T.Maeda, J.Polym.Sci. **C 15** (1966) 55, Table VI
[8] A.A.Gusev, M.M.Zehnder, U.W.Suter, Macromol.Symp. **90** (1995) 85
[9] G.Wobser, S.Blasenbrey, Kolloid-Z.Z.Polym. **247** (1970) 985
[10] R.W.Gray, N.G.McCrum, J.Polym.Sci. [A-2] **7** (1969) 1329
[11] D.Brown, J.H.R.Clarke, Macromolecules **24** (1991) 2075, Fig. 2a; J.H.R.Clarke, in K.Binder, Ed., Monte Carlo and Molecular Dynamics Simulations in Polymer Science, Oxford University Press, New York 1995, Fig. 5.8 (modified)
[12] A.Odajima, T.Maeda, J.Polym.Sci. **15** (1966) 55, Table VIII, Set I
[13] P.D.Davidse, H.I.Westerman, J.B.Westerdijk, J.Polym.Sci. **59** (1962) 389, every third value of the data in Table III
[14] K.H.Meyer, C.Ferri, Helv.Chim.Acta **19** (1935) 570, data taken from Fig. 6
[15] M.Radmacher, Nachr.Chem.Techn.Lab. **47** (1999) 393, Fig. 3 (in this paper, the correct literature reference 19) is M.Rief, F.Oesterhelt, B.Heymann, H.E.Gaub, Science **275** (1997) 1295).
[16] C.Ortiz, G.Hadziioannou. Macromolecules **32** (1999) 780, Figs. 9 and 10
[17] M.C.Shen, D.A.McQuarrie, J.L.Jackson, J.Appl.Phys. **38** (1967) 791, Table 1, Fig. 4
[18] M.Beltzung, C.Picot, J.Herz, Macromolecules **13** (1984) 663, data taken from various figures and tables
[19] L.R.G.Treloar, Trans.Faraday Soc. **40** (1944) 59, data taken from Fig. 5
[20] G.Allen, M.J.Kirkham, J.Padget, C.Price, Trans.Faraday Soc. **67** (1971) 1278, Fig. 3a
[21] S.M.Gumbrell, L.Mullins, R.S.Rivlin, Trans.Faraday Soc. **49** (1953) 1495, Figs. 7 and 8
[22] M.Fischer, D.Martin, M.Pasquier, Macromol.Symp. **93** (1995) 325, Fig. 2
[23] G.Hild, R.Okasha, M.Macret, Y.Gnanou, Makromol.Chem. **187** (1986) 2271, Tables 1-6

17 Viscoelasticity

17.1 Introduction

17.1.1 Overview

Depending on their chemical and physical structure, polymers may exhibit one or more of the four ranges of tensile stress–strain functions (Figs. 16-2 and 16-10):

- the elastic range from the origin to the technical proportionality limit TP with the coordinates σ_P and ε_P;
- the ductile range from the technical proportionality limit TP to the upper yield point UY with the coordinates σ_{UY} and ε_{UY};
- the stress softening range from the upper yield point UY to the lower yield point LY with the coordinates σ_{LY} and ε_{LY}; and
- the stress hardening range from the lower yield point LY to the fracture limit F with the coordinates σ_B and ε_B.

Plastics have small elastic ranges that are controlled by energy elasticity (Section 16.3) whereas the large elastic ranges of elastomers are caused by entropy elasticity (Section 16.4). The subsequent ductile range indicates the deformation of polymers. Brittle polymers break between the technical proportionality limit TP and the upper yield point UY whereas ductile polymers show a maximum of stress at UY (Section 17.2). This maximum appears because of the onset of flow processes that are dominated by viscoelastic effects (Section 17.3) under both static (Section 17.4) and dynamic conditions (Section 17.5). These effects also influence the hardness, friction, and abrasion of polymers (Volume IV). A separate Chapter 18 treats the fracture of polymers.

17.1.2 Definitions

According to Hooke's law, Eq.(16-4), tensile stresses $\sigma_{11} = E\varepsilon$ of uniaxial extensions of energy-elastic bodies are directly proportional to Cauchy elongations $\varepsilon = (L - L_0)/L_0 = \lambda - 1$ where $\lambda \equiv L/L_0$. This **proportionality range** is extremely small, however, for plastics, only several hundredths of a percent. Entropy-elastic bodies, on the other hand, can be elongated often to several hundred percent. However, they do not follow Hooke's law, $\sigma_{11} = E(\lambda - 1)$, but other functions, for example, $\sigma_{11} = RT[\mathrm{M_e}](\lambda - \lambda^{-2})$.

Tensile moduli are therefore extracted in a different manner from experimental data on plastics and elastomers. An additional problem is the correct determination of tensile moduli from stress-strain curves. For example, the tensile modulus E of energy-elastic bodies is defined as the *tangent* of the function $\sigma_{11} = \mathrm{f}(\varepsilon)$ at elongations of $0 \leq \varepsilon \leq \varepsilon_P$ where ε_P = proportionality limit (Fig. 17-1, I). Since this limit is very small and difficult to determine, reported tensile moduli of plastics are usually not tangents but *secants*. Of course, this requires a convention about the elongation for which the secant is calculated. This elongation is often called the **elasticity limit**. For plastics, it is defined as ε_{EL} = 0.01 % but it is often taken as a point in the range 0.05 % $\leq \varepsilon_{EL} \leq$ 0.25 % (CAMPUS® system) (Fig. 17-1, II). This secant modulus is often called the **technical modulus of elasticity** in the plastics industry and **initial modulus** in the textile industry.

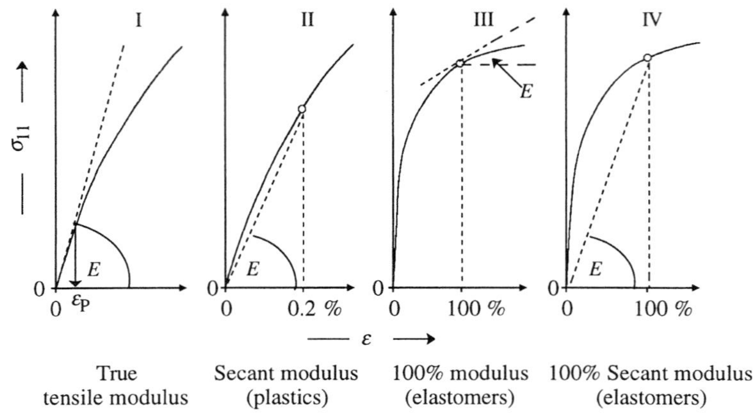

Fig. 17-1 Four definitions of tensile moduli (not to scale).

Elastomers, on the other hand, are usually characterized by *tangent* moduli, either from elongations to ε = 100 % (**100 % modulus**) (Fig. 17-I, III) or to 300 % (**300 % modulus**) (not shown). Sometimes, **secant moduli** are used (Fig. 17-1, IV for an elongation of 100 %).

The proportionality limit is *not* an elasticity limit since the body behaves elastically even at elongations greater than the proportionality limit. The true **elasticity limit** is reached if the first indication of an irreversible deformation (**plasticity**) becomes noticeable. This elasticity limit is, however, difficult to determine. For plastics, a **conventional elasticity limit** is defined as the stress ε_{el} at which the specimen retains an irreversible increase in length of 0.01 % after removal of the load. For load-bearing articles, this elasticity limit should not be surpassed.

"Plasticity" has various meanings. In mechanics, it denotes a *deviation* from Hooke's law that is caused by the onset of flow, i.e., in the function tensile stress = f(strain). In rheology, "plasticity" refers to the abrupt *onset* of flow processes above a certain yield point in the function shear stress = f(shear rate) (p. 505).

Tensile stress-strain curves of *metals* allow one to distinguish between an elastic (reversible) deformation and a plastic (irreversible) one fairly well. In this case, the true elasticity limit is a well-defined yield point since it indicates the minimum stress above which a defined permanent deformation is retained, for example, 0.01 %.

The **yield strength** of metals is sometimes defined as the intersection of the two tangents at the initial and the final slopes of the stress-strain curve (Fig. 17-2, center). The final slope here indicates the plasticity range where the tensile stress rises only a little with increasing elongation until the metal finally breaks.

Polymers behave in more complicated ways. The plasticity range of rigid-brittle polymers is not well defined (see PS-D in Fig. 16-10) and it is problematic to draw a tangent at the end range.Therefore, a line is drawn from a strain of ε = 0.2 % at σ_{11} = 0 (the so-called **off-set strain**) parallel to the initial slope of the $\sigma_{11} = f(\varepsilon)$ curve (Fig. 17-2, I). The intersection of this off-set line with the stress-strain curve is the **off-set yield stress,** $\sigma_{0.2}$, also called **tensile stress at off-set yield point** or **proof stress**. $\sigma_{0.2}$ is therefore *not* the tensile stress at an elongation of 0.2 %!

Similarly to metals, rigid-brittle and rigid-strong polymers (for definitions, see Table 16-2) have only small plasticity ranges. Their stress-strain curves resemble those shown

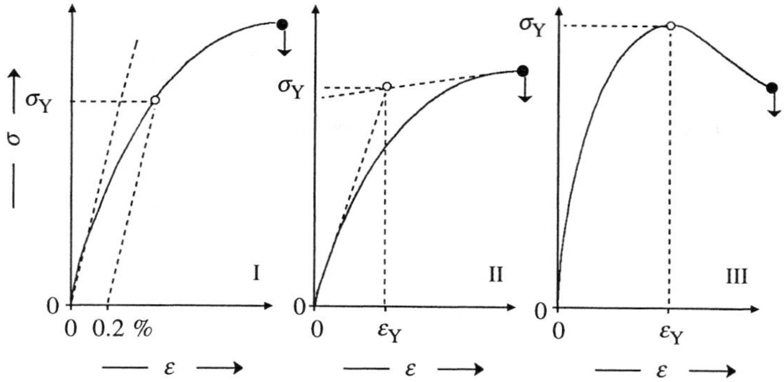

Fig. 17-2 Three definitions of tensile stresses at yield. I: off-set yield stress (here at 0.2 % elongation); II: Yield stress at the intersection of two tangents; III: yield stress as a maximum of the stress-strain curve. ● Fracture.

in Fig. 17-2, left and center. The end-point B of the curves is given by the **elongation at break (fracture elongation)**, ε_B, with the **tensile strength** (at break) σ_B.

In older technical literature, "tensile strength" refers to the *largest* load that a polymer can carry. This load may be either the tensile strength at yield or the fracture strength (see Fig. 16-10).

Stress softening (strain softening) between upper and lower yield points (Fig. 16-2) is caused by shear flow and the formation of shear bands, shear zones, and/or crazes (p. 521). Beyond the lower yield point, some polymers show **stress hardening (strain hardening)** which is caused by increased orientation of chain segments (such as microfibrils in crazes) and/or stress-induced crystallization.

As mentioned before (p. 519 ff.), engineering data deliver engineering stresses (nominal tensile stresses), $\sigma_{11} = F/A_o$, and Cauchy strains (nominal strains), $\varepsilon = (L - L_o)/L_o$. For volume-invariant specimens, true stresses are given by $\sigma_{11}' = \sigma_{11}(L/L_o) = \sigma_{11}(\varepsilon + 1)$ and true strains (Hencky strains) by $\varepsilon' = \ln (1 + \varepsilon)$ (p. 521). Sometimes, other measures of strains are used, such as the **Kirchhoff strain**, $\varepsilon_K = [(L/L_o)^2 - 1]/2$, and the **Murnaghan strain**, $\varepsilon_M = [1 - (L_o/L)^2]/2$. Fig. 17-3 compares nominal and true stress/strains.

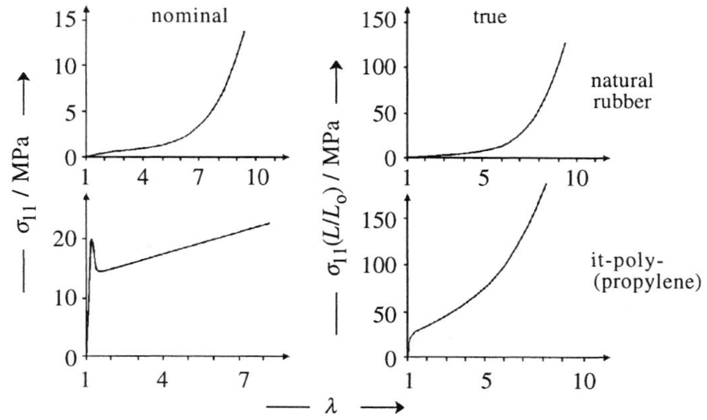

Fig. 17-3 Stress-strain diagrams of an elastomer (natural rubber; top) and a plastic (it-poly(propylene); bottom) at room temperature [1]. Left: nominal (engineering) curves; right: true curves (assuming volume invariance).

17.2 Yield Point

17.2.1 Considère Construction

The elongation of a *ductile* material is initially homogeneous: the same nominal stress $\sigma = F/A_0$ is present at each elongation ε. However, the deformation becomes inhomogeneous after the onset of a telescope effect. Since the nominal stress σ of *volume-invariant* materials is related to the true stress σ' by $\sigma = \sigma'/(L/L_0) = \sigma'/\lambda$, nominal stresses change with the strain ratio according to

$$(17\text{-}1) \qquad \frac{d\sigma}{d\lambda} = \frac{d(\sigma'/\lambda)}{d\lambda} = \frac{d\sigma'}{\lambda\, d\lambda} - \frac{\sigma'}{\lambda^2} = \frac{1}{\lambda^2}\left[\lambda \frac{d\sigma'}{d\lambda} - \sigma'\right]$$

The telescope effect starts at $d\sigma/d\lambda = 0$. At this point, one has $d\sigma'/d\lambda = \sigma'/\lambda = \sigma'/(\varepsilon + 1) = \sigma_{max}$ according to Eq.(17-1). The maximum stress σ_{max} can be obtained from the **Considère construction** in which the nominal stress σ and the true stress σ' are plotted as functions of the elongation and a tangent to the curve is drawn from the point $\varepsilon = -1$ (i.e., $\lambda = L/L_0 = 0$). The contact point of the curve and the tangent is the maximum stress σ_{max}. Three cases can be distinguished (Fig. 17-4):

I. $d\sigma'/d\lambda > \sigma'/\lambda$: No tangent to the curve $\sigma' = f(\lambda)$ can be drawn from the point $\lambda = 0$. The specimen is drawn uniformly; no neck is formed.

II. $d\sigma'/d\lambda = \sigma'/\lambda$ (one point): The specimen elongates homogeneously, starts necking at the upper yield point, and shows stress softening. This behavior is common to many metals and polymers.

III. $d\sigma'/d\lambda = \sigma'/\lambda$ (two points): Uniform elongation to the upper yield point, cold flow, lower yield point, and stress hardening. Metals may also show a second *upper* yield point.

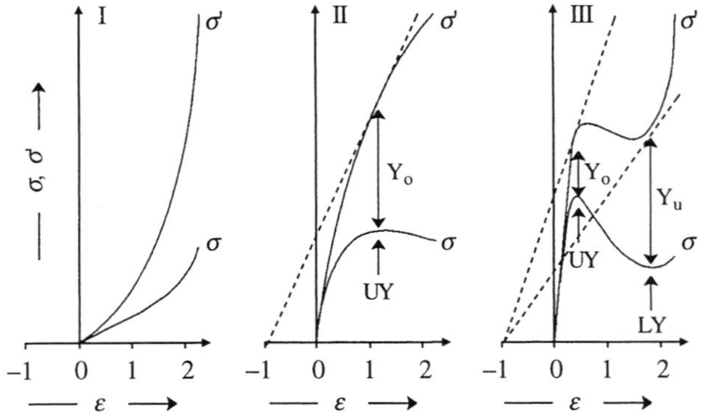

Fig. 17-4 Three types of Considère constructions for volume-invariant bodies.
I: Neither necking nor cold flow.
II: Necking starts at the upper yield point Y_o, followed by stress softening.
III: Neck formation, upper yield point Y_o, stress softening, lower yield point Y_u, followed by strain hardening. Metals (but not polymers) may show two maxima in $\sigma' = f(\varepsilon)$ (not shown).

The maximum stresses correspond to upper yield stresses. For ductile, volume-invariant specimens, they are given completely by the strain dependence of true stresses. Since upper yield stresses indicate the failure of the specimen under load, it is often concluded that the failure is solely due to flow and not to strength (provided the specimen does not fracture before it yields). This assumption is probably wrong (see Section 17.2.2).

17.2.2 Molecular Reasons for Yielding

All polymers, except highly crystalline and highly oriented ones, show a range where the stress is no longer proportional to the strain. These deviations from Hooke's law start at the proportionality limit and end at the fracture strength (rigid polymers) or the upper yield strength (ductile polymers) (Fig. 16-2). Fracture and upper yield strengths thus indicate when certain critical distances between segments (polymers) or atoms (metals) are surpassed.

Critical distances depend on the structure of matter. For atomic lattices, they indicate a certain critical distance between atoms. For extended chain molecules that are drawn in the chain direction, critical distances indicate critical bond lengths between two adjacent chain atoms. For amorphous polymers, critical distances indicate that a certain intermolecular distance is surpassed.

In general, forces per area (= stresses σ) between atoms or molecules, respectively, vary with atomic distances L as shown in Fig. 17-5. In equilibrium, stresses are zero and the distance between atoms is L_o. According to **Frenkel**, stress = f(distance) can be approximated by a sine function with the wavelength λ:

(17-2) $$\sigma = \sigma^o \sin [2 \pi (L - L_o)/\lambda] = \sigma^o \sin [2 \pi (L_o\varepsilon)/\lambda]$$

where the amplitude equals the theoretical strength σ^o and $\varepsilon \equiv (L - L_o)/L_o$. Differentiation of Eq.(17-2) yields

(17-3) $$(\mathrm{d}\sigma/\mathrm{d}L)_{L_o\varepsilon=0} = \sigma^o (2\pi/\lambda)[\cos (2\pi L_o\varepsilon/\lambda)]_{L_o\varepsilon=0} = 2\pi\sigma^o/\lambda$$

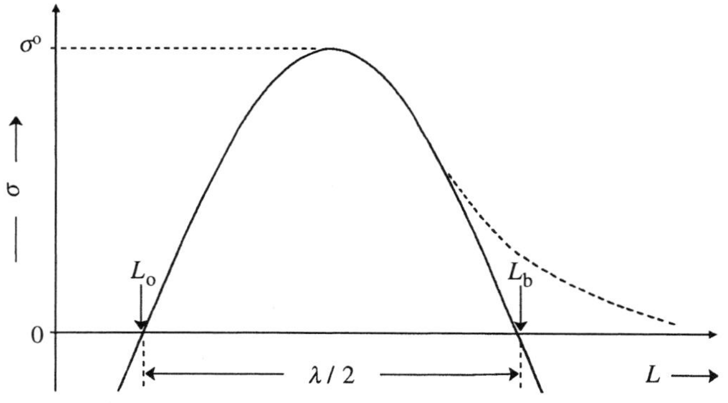

Fig. 17-5 The dependence of stress σ on the distance L between atoms (dotted line) can be approximated by a sine function (solid line) [2]. The true equilibrium stress is zero at atomic distances L_o. The theoretical strength is given by the maximum of the curve. Bonds between atoms lose their strength if the distance $L_b - L_o = \lambda/2$ is surpassed.

Hooke's law applies for small elongations, i.e., $\sigma = E\varepsilon = E[(L/L_o) - 1]$. The change of stress with distance is therefore $d\sigma/dL = E/L_o$. Equating this expression with $(d\sigma/dL)_{L_o\varepsilon=0}$ and setting $\lambda/2 = L_b - L$ leads to

$$(17\text{-}4) \qquad \sigma^o = \frac{E\lambda}{2\pi L_o} = \frac{E(L_b - L_o)}{\pi L_o} = KE \qquad \textbf{Frenkel equation}$$

Frenkel derived Eq.(17-4) for the theoretical shear stress $\tau^o = (bG)/(2\pi H)$ of atomic crystals, where G = shear modulus, b = repeating distance in the shear direction, and H = distance between shear planes. The ratio τ^o/G for crystals varies between 0.24 (diamond) and 0.034 (gold, zinc).

According to Eq.(17-4), theoretical strengths σ^o should be directly proportional to the tensile modulus E. The proportionality constant, $K = (L_b - L_o)/(\pi L_o)$, is predicted to depend only on the critical strain ratio, L_b/L. *Intermolecular bonds* should separate at $L_b/L_o < 1.30$ since intermolecular forces are smaller than intramolecular ones.

Theoretical strengths of zigzag and helical C–C chains have indeed been found to be directly proportional to the tensile modulus (Fig. 18-13) with a proportionality constant of $\sigma^o/E_{||} = K_{||} \approx 0.095$. The theoretical strength in the *chain direction* should thus be ca. 1/10 of the modulus of elasticity. Because $1 + \pi K_{||} = L_b/L_o \approx 1.30$, *intra*molecular bonds should separate if the bond is stretched to more than 30 % of its length at rest. *Inter*molecular bonds should separate at $L_b/L_o < 1.30$ since intermolecular forces are smaller than intramolecular ones.

Chain molecules are separated from each other at the upper yield strain ε_Y with the upper yield stress σ_Y. Since σ_Y should be directly proportional to E according to Eq.(17-4), one would expect $\sigma_Y = K_I E^b$ with b = 1. Such a dependence of σ_Y on E is indeed found for polymers with glass temperatures that are higher than the testing temperatures (Fig. 17-6). This group, (Group I), comprises all amorphous thermoplastics (ABS polymers, bisphenol A polycarbonates, etc.) as well as some dry (= unplasticized by water) semicrystalline polymers (polyamides 6, polyamides 610, etc.) (Table 17-1).

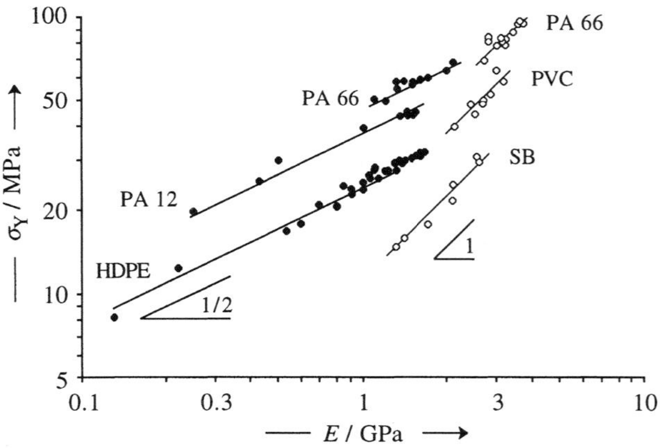

Fig.17-6 Tensile stresses at yield, σ_Y, as a function of tensile moduli E at T = 23°C.
o Group I commercial polymers with $T_G > T$: PA 66 = dry polyamides 6.6 (T_G = 50°C); PVC = poly(vinyl chloride)s (T_G = 81°C); SB = thermoplastic styrene-butadiene copolymers ($T_G \approx$ 86°C).
● Group II commercial polymers with $T_G < T$: HDPE = high-density poly(ethylene)s (T_G = –80°C) including Surlyn® ionomers; PA 12 = conditioned poly(laurolactam)s ($T_G \ll$ 42°C); PA 66 = conditioned poly(hexamethylene adipamide)s ($T_G \ll$ 50°C).

Table 17-1 Tensile moduli E, theoretical tensile stresses $\sigma^o = E/30$, tensile stresses at yield σ_Y, and ratios σ_Y/E and σ_Y/σ^o of some thermoplastics [3] and metallic elements [4]. Values σ_Y/σ^o of *all* metals are between those of antimony and vanadium.

a = amorphous polymers, a' = slightly crystalline polymers (degree of crystallinity < 5 %), c = semicrystalline polymers with unknown degree of crystallinity.

Material	Trade name	E/MPa	σ^o/MPa	σ_Y/MPa	σ_Y/E	σ_Y/σ^o
Antimony		80 000	2 670	11	0.000 138	0.0041
Iron		204 000	6 800	250	0.001 23	0.037
Titanium		113 000	3 770	140	0.001 24	0.037
Aluminum		63 000	2 100	110	0.001 75	0.052
Vanadium		134 000	4 470	840	0.006 27	0.19
Poly(styrene), at-	Vestyron 114	3 300	110	60	0.018	0.55 a
Poly(vinyl chloride), at-	Hostalit Z 2060 C	2 700	90	50	0.019	0.56 a'
Poly(ethylene)	Lupolen 6031 M	1 650	55	32	0.019	0.58 c
Poly(styrene-*co*-acrylonitrile)	Tyril 602	3 900	130	82	0.021	0.63 a
Poly(oxymethylene)	Delrin 500 NC-10	3 100	103	72	0.023	0.70 c
Polycarbonate A	Macrolon 2400	2 400	80	63	0.033	0.79 a
Poly(propylene), it-	Hostalen PP, 41050	1 190	40	32	0.027	0.80 c
Poly(ε-caprolactam)	Ultramid B 35	3 200	107	90	0.028	0.84 c

Group II of polymers does not follow $\sigma_Y = K_I E$ but $\sigma_Y = K_{II} E^{1/2}$ (Fig.17-6). This group comprises polymers with glass temperatures that are lower than the temperature of measurement. By necessity, all polymers of this group are semicrystalline since polymers with $T > T_G$ can only be solids if $T_M > T$. Examples of polymers of this group are poly(oxymethylene) and it-poly(propylene) as well as the conditioned polyamides 6, 6.6, and 12, i.e., specimens that are plasticized by water. The same polymer can therefore belong to either Group I or II, depending on its glass temperature (see also Chapter 18).

The tensile strength at yield, σ_Y, of Group I polymers is directly proportional to the tensile modulus, $\sigma_Y = K_{I,Y} E$ (Fig. 17-6), which allows one to calculate L_b/L_o if one assumes that the separation of chains by yielding can be expressed by $K_{I,Y} = (L_b - L_o)/(\pi L_o)$, i.e., by the value for K in Eq.(17-4).

For the polymers of Fig. 17-6, L_b/L_o was found to increase with the strength of intermolecular forces: by 1.09 for polyamides 6.10 (hydrogen bonds between amide groups), by 1.06 for at-poly(vinyl chloride)s (strong dipole-dipole interactions between Cl groups), and by 1.05 for at-poly(styrene)s (π-π interactions between phenyl groups). σ_Y should thus have values between ca. $E/30$ and ca. $E/60$, which corresponds to the theoretical estimate of $\sigma^o \approx E/30$ that was derived from the Grüneisen relationship between bulk modulus and internal energy.

Polymers thus differ fundamentally from metals with respect to tensile stress at yield or at the off-set yield point. Assuming a theoretical strength of $\sigma^o = E/30$, yield strengths of metals are only between 0.4 % and 19 % of their theoretical strengths (Table 17-1). Without exception, yield strengths of both amorphous and semicrystalline polymers deviate far less from their theoretical strengths. For the polymers in Table 17-1, ratios of yield strengths to theoretical strengths, σ_Y/σ^o, vary between 0.55-0.79 for amorphous polymers and 0.58-0.84 for the semicrystalline ones. For polymers with high cohesive energy densities as a measure of intermolecular forces, this ratio may approach unity.

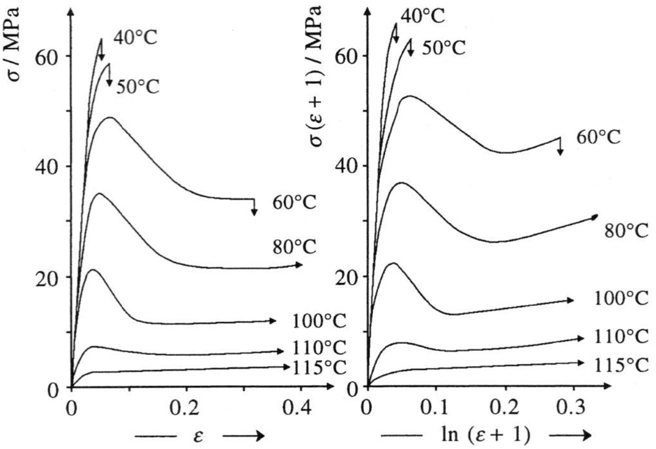

Fig. 17-7 Effect of temperature on the dependence of tensile stress σ on elongation ε. Experimental data for an atactic poly(methyl methacrylate) that was drawn with a velocity of $d\varepsilon/dt = 1.67 \cdot 10^{-3}$ s^{-1}. At this rate, the glass temperature T_G is ca. 115°C.
Left: nominal values [5]; right: true values calculated for invariant volumes.

17.2.3 Yield Points and Necking

Practically all polymers can be extended to elongations where Hooke's law no longer applies. Whether or not a polymer yields depends on the temperature or, more precisely, on the difference between the glass temperature and the measuring temperature of amorphous polymers (Fig. 17-7) and the difference between the melting temperature and the measuring temperature in the case of semicrystalline ones.

Rigid polymers can only be elongated to a small extent before they break; an example is atactic poly(methyl methacrylate) at 40°C (Fig. 17-7). The same polymer behaves elastomerically at temperatures near the glass temperature of 115°C but is ductile between 60°C and 110°C where it yields and shows a telescope effect.

Necking is also observed for semicrystalline polymers (Fig. 16-10) at temperatures between the glass temperature and the melting temperature. Hence, differences in polymer morphologies *per se* cannot be the primary reason for necking.

Possible reasons for necking may be local temperature increases or geometric effects. Such temperature increases can be caused by internal friction. For adiabatic elongations, they should amount to $\Delta T = \Delta p/(c_p \rho)$ where $\Delta p = \eta \dot{\gamma} =$ pressure difference or stress difference $\Delta\sigma$, η = dynamic viscosity, $\dot{\gamma}$ = elongation rate, ρ = density, and c_p = specific heat capacity. At stresses of ca. 100 MPa, densities of $\rho \approx 1$ g/cm^3, and specific heat capacities of $c_p = 2$ J g^{-1} K^{-1}, temperature increases of $\Delta T = 50$ K can be expected for the necking zone.

Such temperature increases would decrease the viscosity in the necking zone and thus increase the flow. However, temperature increases of less than 10 K have been observed for the usual elongation rates of $\dot{\gamma} \approx 100$ s^{-1} which is too small to lower the very high viscosities at temperatures of $T < T_G$. The same is true for local melting as a potential cause of yield points of semicrystalline polymers.

Geometric effects may be another reason for necking, i.e., microscopically small, local differences in the cross-section of the test specimens. At such smaller cross-sections, the same force will produce greater local tensile stresses which would lead to locally larger extensional viscosities. A local build-up of heat liberated from the deformation would then stabilize the flow zone which subsequently travels along the test specimen. However, geometric effects cannot be the primary reason for necking either since yielding is also found for compressions (Fig. 16-10).

17.2.4 Flow Criteria

The very low yield stress of metals led to theories of dislocations in atomic lattices and to flow criteria for the appearance of yield points, i.e., to elongational stresses for rigid bodies and shear stresses for ductile ones. Such flow criteria should apply not only to simple tensile and shear stresses but also to multiaxial stress fields.

A multiaxial stress field can be described by a suitable combination of the three principal stresses σ_1, σ_2, and σ_3 of the system. These principal stresses are the normal stresses σ_x, σ_y, and σ_z in a coordinate system with the axes x, y, and z.

According to these criteria, yield points should appear if a certain critical value is surpassed. Depending on theory, this critical value is either that of the (I) maximum shear stress, (II) maximum principal stress, (III) maximum shear deformation, (IV) maximum principal deformation, or (V) maximum deformation energy.

The maximum shear stress as critical value (I) is the **Tresca criterion**. According to the Tresca theory, a yield point with an elongational stress σ_y should appear if the shear stress at the yield point becomes $\sigma_{Y,s} = (1/2)(\sigma_1 - \sigma_3)$. For a simple tensile experiment with drawing in the [1] direction (x direction), one has $\sigma_3 = 0$. At the yield point, the applied stress σ_1 becomes $\sigma_{Y,t}$ so that the maximum shear stress should be one-half of the tensile stress, $\sigma_{Y,s} = (1/2)\ \sigma_{Y,t}$.

The **von Mises theory**, on the other hand, assumes that yield points appear if a critical deformation energy is surpassed (V). Matter is modeled as a regular octahedron with 12 side lengths a and a surface area of $A = 2\ a^2 \cdot 3^{1/2}$. The normals on the areas relative to the three main axes have the values $1/3^{1/2}$ each so that the normal stress to an octahedral area is $\sigma = (1/3)(\sigma_1 + \sigma_2 + \sigma_3)$.

The shear stress at an octahedral area is called **octahedral stress**. It is calculated from the differences of principal stresses as

$$\tau_{oct} = (1/3)[(\sigma_1 - \sigma_2)^2 + (\sigma_2 - \sigma_3)^2 + (\sigma_3 - \sigma_1)^2]^{1/2} \qquad (17\text{-}5)$$

The von Mises criterion assumes that a flow point is observed if the term in square brackets in Eq.(17-5) becomes constant, i.e., if $\tau_{oct,y} = (1/3)\ const^{1/2}$. An expression for the *const* is obtained as follows. Elongation of the specimen by a tensile stress $\sigma_1 = \sigma$ involves stresses of $\sigma_2 = \sigma_3 = 0$ for the other two principal stresses so that the flow stress, $\sigma = \sigma_{Y,t}$, adopts a value of $\sigma_{Y,t} = (0.5\ const)^{1/2}$.

On the other hand, extension by a shear stress, $\sigma_2 = -\sigma_3$, leads to $\sigma_1 = 0$. At the yield point, shear stress assumes the value $\sigma_2 = \sigma_{Y,s}$ so that $const = 2\ \sigma_{Y,t}^2$. Eq.(17-5) thus converts to $\sigma_{Y,s} = (1/3)^{1/2}\sigma_{Y,t}$.

Table 17-2 Observed flow stresses σ_Y for compression ($\sigma_{Y,c}$), tensile elongation ($\sigma_{Y,t}$), and shearing ($\sigma_{Y,s}$) [6] and ratios of flow stresses calculated therefrom.

Polymer	$\sigma_{Y,c}$	$\sigma_{Y,t}$	$\sigma_{Y,s}$	$\frac{\sigma_{Y,c}}{\sigma_{Y,t}}$	$\frac{\sigma_{Y,c}}{\sigma_{Y,s}}$	$\frac{\sigma_{Y,t}}{\sigma_{Y,s}}$
ABS polymer	6.2	6.5	3.5	0.95	1.77	1.86
Polyamide	8.9	9.7	5.9	0.92	1.51	1.64
Poly(vinyl chloride)	9.8	8.3	6.0	1.18	1.63	1.38
Poly(propylene), it-	6.3	4.7	4.0	1.34	1.58	1.18
Poly(ethylene)	2.1	1.6	1.4	1.31	1.50	1.14
Poly(tetrafluoroethylene)	2.1	1.7	1.6	1.24	1.31	1.06
Average				1.16 ± 0.18	1.55 ± 0.15	1.38 ± 0.32
von Mises criterion				1.00	1.73	1.73
Tresca criterion				1.00	2.00	2.00

Both Tresca and von Mises assume that flow points are the same for compression (column 2 of Table 17-2) and tension (column 3), i.e., ($\sigma_{Y,c} = \sigma_{Y,t}$), which seems to be true within (the rather large) limits of error (column 5). The ratios of flow stresses for compression and shear, $\sigma_{Y,c}/\sigma_{Y,s}$ (column 6), and tension and shear, $\sigma_{Y,t}/\sigma_{Y,s}$ (column 7), are much lower than predicted by the Tresca criterion. The high end of their range is almost identical with the von Mises criterion of $3^{1/2} \approx 1.73$.

Deviations from the von Mises criterion are more severe the higher the crystallinity (and thus the anisotropy) of the polymers; poly(tetrafluoroethylene) is an example. Amorphous polymers, on the other hand, almost obey the von Mises criterion; the ABS polymer is an example and also the value of $\sigma_{Y,c}/\sigma_{Y,s}$ of the only slightly crystalline poly(vinyl chloride). It may be surprising that the polyamide of Table 17-2 has relatively high values of $\sigma_{Y,c}/\sigma_{Y,s}$ and $\sigma_{Y,t}/\sigma_{Y,s}$ although it is fairly crystalline. However, this specimen may have been plasticized by humidity as is fairly common for aliphatic polyamides.

Both criteria assume that hydrostatic effects are absent (otherwise, one has to consider $\sigma_1 + p$, etc). This condition is fairly well fulfilled for metals for which the Tresca and von Mises criteria were originally developed. However, stress-strain curves of thermoplastics are strongly influenced by pressure (Table 17-3) whereas the effect is relatively small for unfilled, void-free elastomers. The tensile modulus of rigid thermoplastics as well as the yield strength increases strongly with pressure (Table 7-3) but the elongation at yield is often little affected. With increasing pressure, tensile strengths at break and fracture elongations increase for rigid polymers but decrease for ductile ones.

Table 17-3 Influence of pressure p on the tensile modulus E, the maximum nominal stress (= nominal strength at yield) σ_Y, and the nominal fracture elongation ε_B of an isotactic poly(propylene) [7].

p/MPa	E/GPa	σ_Y/MPa	ε_B/%
0.1 (atmospheric pressure)	1.4	30	> 200
276	5.9	87	50
690	7.7	172	24

17.3 Flow Range

17.3.1 Crazes and Shear Bands

Beyond the proportionality and elasticity limit, respectively, polymers start to flow and deform (Fig. 16-2). *Elastomers* (where $T_G < T$) deform homogeneously if the deformation is not very large: all volume elements are deformed to the same extent. Such homogeneous deformations are characterized by a narrowing of the cross-section of the specimen between the clamps.

The tapering is accompanied by a homogeneous elongation, which points to an isotropic distribution of stress in the specimen. The elongation is caused by a conversion of some microconformations from gauche to trans, which lengthens polymer segments. The conversions come to an end if the shortest segments between crosslinking sites are completely stretched.

The scenario is somewhat different for *amorphous thermoplastics* (where $T > T_G$) where movements of larger chain segments are frozen, which prevents conversion of microconformations of large segments. On elongation of the specimen, these large segments cannot simply stretch and perform large-volume cooperative movements that give way to a homogeneous elongation. Instead, local impediments, for example, entanglements, lead to local deformations and not to global ones and the specimen responds by formation of shear bands and/or crazes (p. 521).

The type of local deformation is controlled by the distance between entanglements. Low entanglement densities lead to long segments between entanglements that can stretch to some extent and form amorphous microfibrils in the stretched direction. These microfibrils have diameters of up to 30 nm; they are anchored at both ends in the remaining continuous matrix of the specimen. Since the density of the microfibrils is greater than the density of the matrix (molecules are more densely packed in fibrils), empty space is formed around the matrix (Fig. 16-3, center). Because of the direction and size of the interal stress produced, this empty space has its largest extension perpendicular to the length of the fibrils (Fig. 16-3, center). The resulting **crazes** can be up to 100 μm long and up to 10 μm wide. Differences in the refractive indices of the fibrils and the surrounding air produce reflexions which lead to turbidities of the specimens, which is especially noticeable for originally glass-clear polymers.

Larger entanglement densities lead to shorter distances between entanglements. Segments can move only a little from their positions at rest; they can no longer form fibrils. Instead, internal stresses are dissolved by formation of **shear zones** in thin films (2D bodies) (Fig. 16-3, left) and of **shear bands** at 38-45° to the stress direction in 3D bodies (Fig. 16-3, right). In shear bands, the long axes of chain segments are at angles between those of shear bands and the direction of stress.

17.3.2 Effect of Entanglement Density

Entanglements are local obstacles for the homogeneous dissolution of stresses. Deformations are therefore controlled by the number concentration of entanglements per volume, the **entanglement density**, $\nu_e = \rho N_A / M_e$, where ρ = density of the specimen and M_e = molar mass of segments between two entanglements.

Chemical crosslinks are also local obstacles for deformation. The total number concentration of junctions in a chemically crosslinked polymer is therefore $\nu = \nu_e + \nu_x$ where ν_x = number concentration of chemical crosslinks.

Molar masses M_e of chain segments between junctions are relatively small, usually between ca. 1000 g/mol and 15 000 g/mol. In most cases, such segments are too small to obey Gaussian coil statistics (Section 4.4.7). Instead, segments behave as worm-like chains with end-to-end distances $\langle r^2\rangle_o^{1/2}$ and true contour lengths $L_{chain} = Nb = N_{ps}L_{ps}$ where N_{ps} = number of rigid segments with persistence lengths L_{ps}. Such chains should elongate to a draw ratio of $\lambda_{max} = L_{chain}/\langle r^2\rangle_o^{1/2}$. Experimentally, one obtains draw ratios λ that are either the fibril lengths in crazes or the size of necks from shear zones.

According to measurements of *thin films*, crazes and shear deformations differ in the size and dependence of the ratio $(\ln \lambda)/(\ln \lambda_{max})$ on the number concentration of junctions ν, i.e., the sum of number concentrations of entanglements, ν_e, and chemical crosslinks, ν_x. These dependences do not differ in the type of junctions (physical entanglements versus chemical crosslinks).

Crazes are observed at small values of ν and shear zones at large ones (Fig. 17-8) and both crazes and shear zones in a small range of medium values of ν. At very small concentrations of junctions, values of $(\ln \lambda)/(\ln \lambda_{max})$ may even be larger than unity which means that crazed specimens can obviously be elongated to greater lengths than predicted by the contour lengths of worm-like chain sections between junctions.

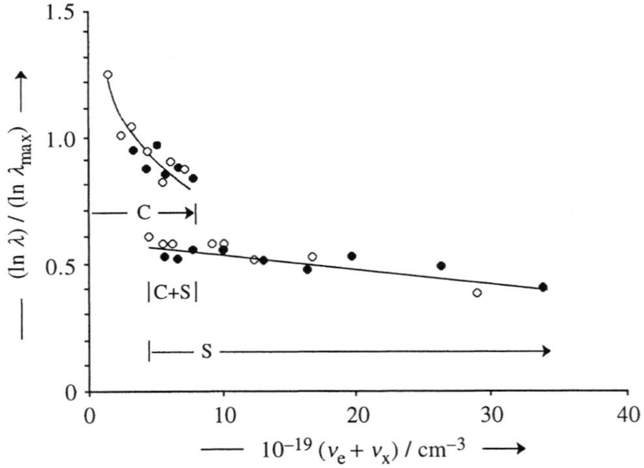

Fig. 17-8 Ratio of natural logarithms of true elongations, λ, to natural logarithms of maximally possible elongations, λ_{max}, as a function of the total number concentration of junctions, $\nu_e + \nu_x$, for uncrosslinked polymers (poly(styrene)s and others) (○; $\nu_x = 0$) and chemically crosslinked poly(styrene)s with different degrees of crosslinking (●) [8]. Formation of crazes (C), or shear zones (S).

17.3.3 Proportion of Crazes and Shear Bands

While the various types of deformation can be detected by the naked eye (necking, turbidity) or microscopically (shear bands, crazes), their relative proportions can be evaluated by creep experiments.

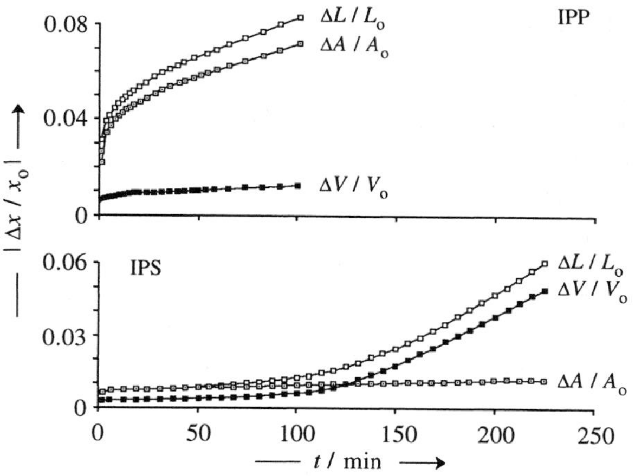

Fig. 17-9 Change of dimensions, $|\Delta x/x_0|$, with time at 30°C for two impact-modified polymers, poly(propylene) (IPP) and poly(styrene) (IPS) [9a]. x may be a length L, an area A, or a volume V.

In such experiments, a constant axial stress is applied and the elongations $\Delta L/L_0$, area contractions $\Delta A/A_0$, and volume changes $\Delta V/V_0$ are measured as functions of time (Fig. 17-9). The volume does not change on pure shear deformation with formation of shear bands since the increase of length to L from L_0 is compensated by a decrease of cross-sectional area to A from A_0 so that $V_0 = L_0A_0 = LA = V$. Ideally, $\Delta V/V_0 = const$, which is almost followed by impact-modified isotactic poly(propylene) (IPP) (Fig. 17-9, top) (it-poly(propylene) is a semicrystalline polymer).

Crazing, on the other hand, is accompanied by an increase in volume to $V = LA$ from $V_0 = L_0A_0$ whereas the cross-sectional area stays constant, $A_0 = A$. The relative change of volume here equals the relative change of length, $(V - V_0)/V_0 = (L - L_0)/L_0$, so that $\Delta A/A_0 = const$. Such a behavior is almost shown by impact-modified (rubber-modified) atactic poly(styrene) (IPS) (Fig. 17-9, bottom) (at-poly(styrene) is an amorphous polymer).

A plot of relative volume change, $\Delta V/V_0$, as a function of relative length change, $\Delta L/L_0$, should therefore be quite illuminating. In such a plot, deformation by crazing should deliver a straight line with a slope of unity whereas deformation by formation of shear bands and/or homogeneous tapering should result in a straight line with a slope of zero.

However, ideal behavior is not observed since experimental slopes vary between 0.12 and 0.96 (Fig. 17-10). Rubber-modified poly(styrene) (IPS) deforms about 96 % by crazing and about 4 % by other mechanisms whereas rubber-modified isotactic poly-(propylene) deforms only 12 % by crazing. Independent experiments showed that these proportions are not influenced by the applied stress.

The situation is different for rubber-modified (impact-modified) poly(oxymethylene) (IPOM) where an 8.4 % increase in applied stress caused a ca. 70 % increase in the deformation by crazing (Fig. 17-10). This behavior seems to indicate for this case that deformations by crazing and shear deformation are not independent of each other.

The different behavior of IPP and IPS is caused by the position of the glass temperature T_G relative to the temperature of the measurement, $T = 30°C$. Impact-modified isotactic poly(propylene) consists of amorphous rubber domains ($T_G < T$) that are

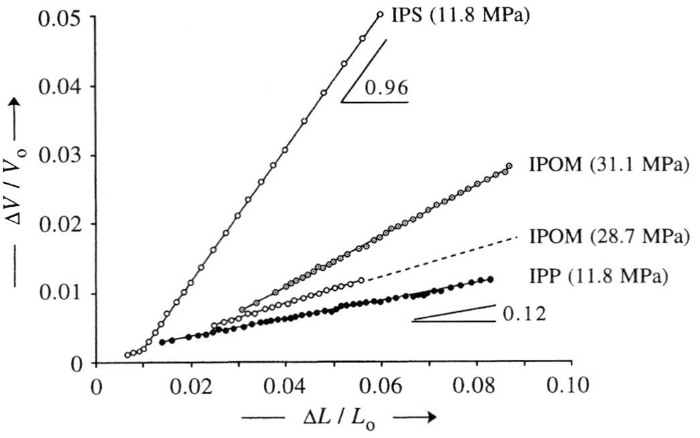

Fig. 17-10 Relative volume change, $\Delta V/V_o$, as function of the relative length change, $\Delta L/L_o$, for impact-modified (rubber-modified) polymers at 30°C under the indicated constant axial stresses [9b]. IPS = poly(styrene), IPP = poly(propylene), IPOM = poly(oxymethylene).

dispersed in a poly(propylene) matrix composed of crystalline ($T_M \approx 160$°C) and amorphous domains ($T_G \approx -10$°C). In the two amorphous domains, chain segments are mobile since $T > T_G$. The application of stress in these domains leads to cooperative movements of segments which results in shear flow of the specimen.

Shear flow is impossible in rubber-modified poly(styrene) because the glass temperature of the poly(styrene) matrix ($T_G \approx 90$°C) is greater than the measuring temperature ($T = 30$°C). Local internal stresses can be resolved here only by formation of crazes.

17.3.4 Large Elongations

In tensile experiments, a specimen is elongated from a length L_o to a length $L = \lambda L_o$ by an applied force $\boldsymbol{F}$ per actual area A. The specimen is thus subjected to work W which results in a change of the original stress σ_o' to a new true stress $\sigma_1' = \boldsymbol{F}/A$ at the new state 1 and similarly to a state 2 (see next page):

$$(17\text{-}6) \qquad \sigma_1' = \sigma_o' + \lambda_1(\mathrm{d}W/\mathrm{d}\lambda_1)$$

To a first approximation, the deformation of the entangled physical network can be treated like that of a chemically crosslinked one. The work W then equals the elastic Gibbs energy, ΔG_{el}, (see Eq.(16-46)).

Eq.(16-46) can be simplified by assuming an incompressible system (i.e., $V = V_o$ and therefore $\lambda_x \lambda_y \lambda_z = 1$) and a uniaxial elongation ($\lambda_x \equiv \lambda$ and therefore $\lambda_y = \lambda_z = 1/\lambda^{1/2}$). The elastic Gibbs energy in state 1 is therefore (N_c = concentration of network chains)

$$(17\text{-}7) \qquad \Delta G_{el,1} = N_c(k_B T/2)[\lambda_1^2 + 2\,\lambda_1^{-1} - 3]$$

and similarly for state 2. Differentiation of the elastic Gibbs energy with respect to the deformation λ_1 leads to $\mathrm{d}\Delta G_{el,1}/\mathrm{d}\lambda_1 = N_c k_B T[\lambda_1 - \lambda_1^{-2}]$ for state 1 and similarly for state 2. Incompressible systems ($V_o = const$) allows one only to determine stress differences. For $\lambda \equiv \lambda_1 - \lambda_2$ and systems without change of internal energy, one obtains

(17-8) $$\sigma' = \sigma_1' - \sigma_2' = \lambda(\mathrm{d}\Delta G_{el}/\mathrm{d}\lambda)/V_o = (N_c/V_o)k_BT[\lambda^2 - \lambda^{-1}] = E_h[\lambda^2 - \lambda^{-1}]$$

The front factor $(N_c/V_o)k_BT$ has the same physical unit energy per volume as the front factor $RT[M_c]$ in Eq.(16-61), i.e., modulus, and can therefore be identified as the **strain-hardening modulus**, E_h.

Eq.(17-8) was derived for entropic (irreversible) deformations which set in after the yield point is surpassed. At small deformations before the yield point, enthalpic deformations prevail which can be considered by adding the true yield stress σ_Y' to the right side of Eq.(17-8) which then becomes the **Argon-Haward equation**

(17-9) $$\sigma' = \sigma_Y' + E_h[\lambda^2 - \lambda^{-1}]$$

which can be written in engineering terms as

(17-9a) $$\sigma = \sigma'/\lambda = \sigma_Y'/\lambda + E_h[\lambda^2 - \lambda^{-1}]/\lambda = \sigma_Y + E_h[\lambda - \lambda^{-2}]$$

At large deformations, a linear dependence of the true stress σ' on $\lambda^2 - \lambda^{-1}$ is indeed observed (Fig. 17-11, right). However, true intercepts σ_Y' are larger than predicted by the Considère rule (Fig. 17-11, left).

Strain-hardening moduli decrease with increasing temperature (Fig. 17-11). However, they are considerably lower then shear storage moduli G_N^o (see Section 17.5). For example, the strain-hardening modulus of poly(ethylene) at 90°C is only E_h = 0.28 MPa whereas the shear storage modulus at 100°C was found to be $G_N^o \approx$ 2.8 MPa (see data in Table 17-7 (p. 605)).

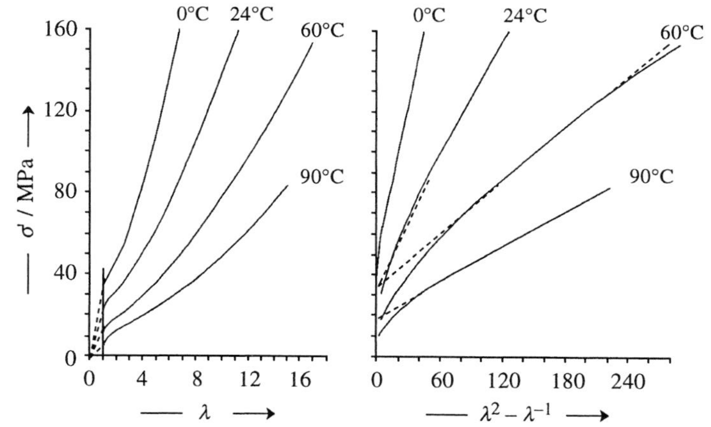

Fig. 17-11 Left: True tensile strength σ' of a high-density poly(ethylene) as a function of the draw ratio λ at four temperatures [10]. Broken lines correspond to the Considère rule.

Right: True tensile strength σ' of the same polymer plotted as a function of $\sigma' = f(\lambda^2 - \lambda^{-1})$ according to the Argon-Haward equation (17-9) [11]. The slopes correspond to values of E_h/MPa of 2.50 (0°C), 0.95 (24°C), 0.42 (60°C), and 0.28 (90°C).

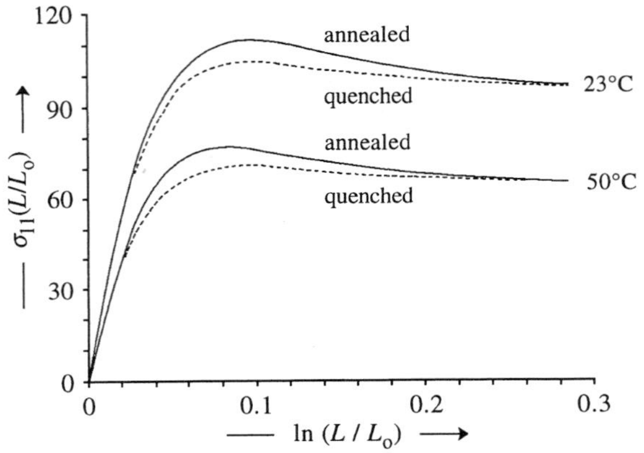

Fig. 17-12 True tensile strength σ_{11} of a poly(methyl methacrylate) as a function of the Hencky strain, ln (L/L_o) [12]. The polymer was either annealed or quenched before it was drawn at 25°C or 50°C with a tensile strain rate of $d\varepsilon/dt = 10^{-3}$ s^{-1}.

17.3.5 Yielding as a Flow Process

Flow points are found for elastomers and soft-ductile thermoplastics although there is neither necking nor cold flow (Fig. 17-4, I). Rigid-brittle thermoplastics, on the other hand, break before they reach a flow point. Other types of thermoplastics have yield points, i.e., maxima in the dependence of engineering stress on elongation (Fig. 17-4).

Yield points are affected by the history of the specimens (Fig. 17-12). For example, annealed polymers have higher yield stresses than quenched ones. To a first approximation, the relative increase of yield stresses seems to be independent of temperature. On subequent elongation, however, the dependences of true tensile stresses on elongation become independent of the history of the specimen.

The onset of yielding beyond the proportionality limit is strongly affected by the speed of drawing (Fig. 16-5). Flow points and yield points must therefore be controlled mainly by kinetic effects, i.e., by flow processes. In contrast to low-molecular weight molecules, the flow units of polymers are not whole molecules but rather "segments" which may be monomeric units or sequences of such units.

At rest, flow units reside in a lattice that is formed by the other flow units, for example, a cubic lattice with distance L between the nearest corner points (see Fig. 7-1) where the flow units reside. For the movement of a segment from one corner point to another, work is needed to overcome the resistance of other segments and to form a sufficiently large "hole" into which the segment can be placed. The required energy is provided by the thermal activation energy, $\Delta E^{\ddagger}$. According to the **Eyring rate theory**, the rate of "jumps" is given by $\dot{N} = dN/dt = k \exp[-\Delta E^{\ddagger}/(k_B T)]$ where $\Delta E^{\ddagger}$ = activation energy and k = rate constant with the physical unit of a frequency (= inverse time).

Jumps are fostered by the applied tensile stress, $\sigma = \boldsymbol{F}/A$, which acts on the segment with a force of $\boldsymbol{F} = A\sigma = L^2\sigma$. Since the maximum of the activation energy is at $L/2$, work (energy) of $E_s = F(L/2) = (1/2)\sigma L^3$ is achieved that lowers the activation energy from $-\Delta E^{\ddagger}$ to $-(\Delta E^{\ddagger} - \Delta E_s)$.

The rate of jumps in the forward direction, $\dot{N}_{\rightarrow} \equiv dN/dt$, increases correspondingly to $\dot{N}_{\rightarrow} = k \exp[-(\Delta E^{\ddagger} - E_s)/(k_B T)]$. In the backward direction, the rate of jumps is decreased to the same extent so that $\dot{N}_{\leftarrow} = k \exp[-(\Delta E^{\ddagger} + E_s)/(k_B T)]$. The net rate of jumps (in travelled length per time) is thus

$$(17\text{-}10) \qquad L\dot{N} = L(\dot{N}_{\rightarrow} - \dot{N}_{\leftarrow}) = Lk \exp[-\Delta E^{\ddagger}/(k_B T)]Q$$
$$\text{where } Q = \{\exp[\sigma L^3/(2\,k_B T)] - \exp[-\sigma L^3/(2\,k_B T)]\}$$

The rate $\dot{N}$ equals the rate of elongation, $\dot{\varepsilon} = \sigma/\eta_e$, where η_e = extensional viscosity. It follows that $L\dot{N} = \sigma L/\eta_e$. Setting $x \equiv \sigma L^3/(2\,k_B T)$, one can write the expression in braces in Eq.(17-10) as $\{\exp(+x) - \exp(-x)\} = 2 \sinh x$. Eq.(17-10) then becomes

$$(17\text{-}11) \qquad \sigma = 2\,k\eta_e \exp[-\Delta E^{\ddagger}/(k_B T)] \sinh[\sigma L^3/(2\,k_B T)]$$

Inspection of Eq.(17-11) shows that $x = \sigma L^3/(2\,k_B T)$ is much larger than unity. An example, x = 4.94 for a segment length of only L = 1 nm and T = 293 K, σ = 40 MPa = $40 \cdot 10^6$ J m^{-3}, and $k_B \approx 1.381 \cdot 10^{-23}$ J K^{-1}. Since one can write $\sinh x = (1/2) \exp(x)$ for $x > 2.3$, Eq.(17-11) converts to

$$(17\text{-}11a) \qquad \sigma \approx k\eta_e \exp[-\Delta E^{\ddagger}/(k_B T)] \exp[\sigma L^3/(2\,k_B T)]$$

The tensile stress σ can be replaced by the product $\eta_e \dot{\varepsilon}$ of extensional viscosity η_e and tensile strain rate $\dot{\varepsilon}$. After taking the logarithm and rearranging, Eq.(17-11) becomes

$$(17\text{-}12) \qquad \sigma/T = (2/L^3)[(\Delta E^{\ddagger}/T) - k_B \ln k] + (2\,k_B/L^3) \ln \dot{\varepsilon} = K + B \ln \dot{\varepsilon}$$

where σ may be the tensile stress at yield, σ_Y. According to Eq.(17-12), σ_Y/T should increase linearly with the natural logarithm of the tensile strain rate (Fig. 17-13). The slopes are independent of temperature so that $2\,k_B/L^3 = const$: the jump length is independent of temperature.

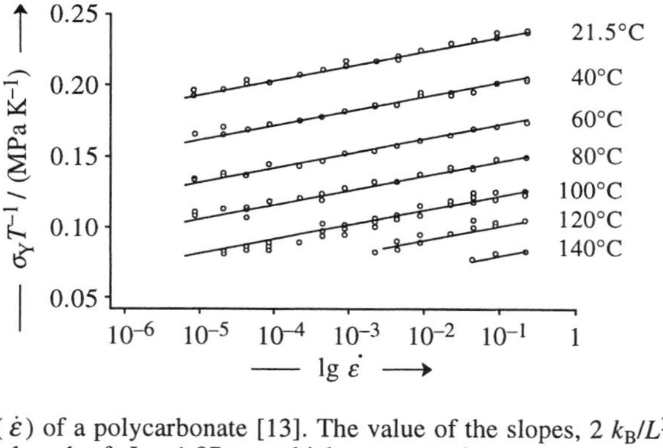

Fig. 17-13 $\sigma_Y/T = f(\dot{\varepsilon})$ of a polycarbonate [13]. The value of the slopes, $2\,k_B/L^3$ = 4205 Pa K^{-1}, at 80°C leads to a jump length of L = 1.87 nm which corresponds approximately to the length of two monomeric units. Hence, the flow process must consist of a cooperative movement of larger segments.

17.4 Creep and Relaxation

17.4.1 Two-parameter Models

The discussion of energy-elastic and entropy-elastic behavior in Chapter 16 assumed that bodies return immediately and completely to their initial states after loads are removed. For polymers, this return to initial positions requires some time, however. Polymers may also be irreversibly deformed by flow processes.

Newtonian viscosities and Hookean elasticities are ideal cases of mechanical behavior that can be represented by simple mechanical models. These mechanical analogs can then describe many experimental results such as stress as a function of deformation, deformation rate, time, or frequency. In many cases, analogs allow one to convert test results into various macroscopic properties that are usually obtained by different testing procedures. However, these mechanical models say nothing about molecular processes.

The mechanical model for *elasticity* is an elastic spring S (Fig. 17-14) which extends on elongation according to Hooke's law, $\sigma_{11} \equiv \sigma_S = E\varepsilon_S$, Eq.(16-2), and on shearing according to $\sigma_{21} \equiv \tau_S = G\gamma_S$, Eq.(15-2). On removal of the load, the spring returns immediately to its initial position.

The mechanical model for *viscosity* is a viscous liquid in a dash pot D with a (perforated) plunger (Fig. 17-14). The liquid should be Newtonian: shearing should be described by $\sigma_{21} \equiv \tau_D = \eta\,\dot{\gamma}_D$ and $\gamma_D = (\tau_D/\eta)t$, respectively, and elongation by $\sigma_D = \eta_e\,\dot{\varepsilon}$ and $\varepsilon_D = (\sigma_D/\eta_e)$, respectively (Fig. 15-3). On removal of the load, Newtonian liquids do not return immediately to their initial state.

Bodies are called **viscoelastic** if time-independent elastic properties and time-dependent viscous properties act simultaneously. In the simplest cases, viscoelastic properties can be modelled by simple combinations of Hookean and Newtonian behavior. The combination of a spring and a dashpot in a series leads to the so-called **Maxwell element** whereas the parallel combination of a dashpot and a spring results in the **Voigt-Kelvin element**. The Maxwell element is a model for relaxations and the Voigt-Kelvin element is a model for retardations. Both models describe **linear viscoelasticities** since they predict linear interrelationships between stresses, deformations, and deformation rates.

Combinations of more than two of these models lead to more complicated elements such as the four possible **3-parameter models** or the **Burgers element** which is a combination of a Maxwell body and a Voigt-Kelvin element in a series (Section 17.4.4).

In the Maxwell element, the deformations of the spring and dashpot are additive, $\varepsilon = \varepsilon_S + \varepsilon_D$, whereas the spring and dashpot experience the same stress, $\sigma = \sigma_S = \sigma_D$. However, deformations are time-dependent because the viscous dashpot does not respond immediately to an applied stress. The deformation changes with time according to $\dot{\varepsilon} = \mathrm{d}\varepsilon/\mathrm{d}t = \mathrm{d}\varepsilon_S/\mathrm{d}t + \mathrm{d}\varepsilon_D/\mathrm{d}t = \dot{\varepsilon}_S + \dot{\varepsilon}_D$. Inserting into this equation the expressions for the time dependence of the elongation of a Hookean spring, $\sigma_S = E\varepsilon_S \rightarrow \mathrm{d}\varepsilon_S/\mathrm{d}t = \mathrm{d}(\sigma_S/E)/\mathrm{d}t \rightarrow \dot{\varepsilon}_S = \dot{\sigma}_S/E$, and the elongation of a Newtonian liquid, $(\sigma_D/\eta_e) = \dot{\varepsilon}_D$, leads to $\dot{\varepsilon} = (\dot{\sigma}/E) + (\sigma/\eta_e)$ and therefore also to $\sigma = \eta_e\dot{\varepsilon} - (\eta_e/E)\dot{\sigma}$. The equation for the shearing of a Maxwell element, $\tau = \eta\dot{\gamma} - (\eta/G)\dot{\tau}$, is derived similarly (see Fig. 17-14).

The resulting equations are not indexed for elongation and shearing. In tensile experiments, the elongational viscosity η_e is simply seen as "viscosity." Since the deformation is mainly by shearing (Section 16.2.3), tensile moduli E are replaced by shear moduli G. The rate of elongation is therefore expressed by $\dot{\varepsilon} = \mathrm{d}\varepsilon/\mathrm{d}t = G^{-1}(\mathrm{d}\sigma/\mathrm{d}t) + (\sigma/\eta)$.

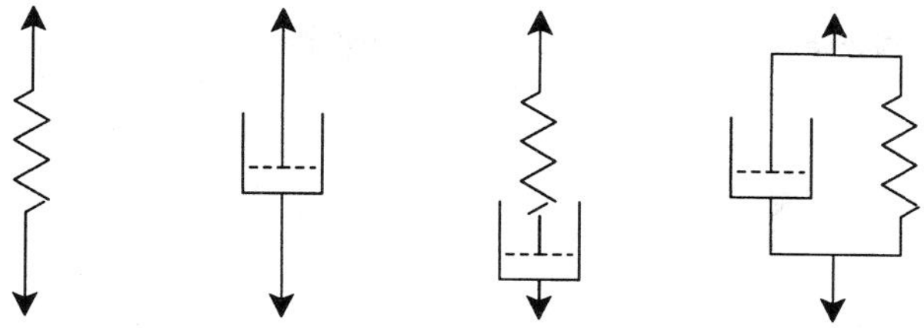

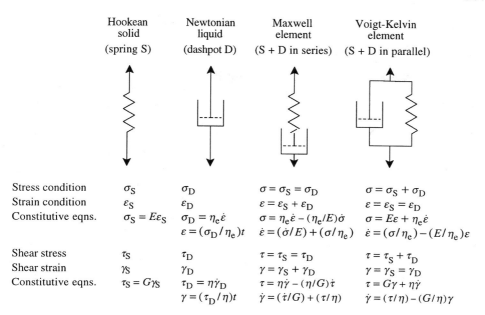

	Hookean solid (spring S)	Newtonian liquid (dashpot D)	Maxwell element (S + D in series)	Voigt-Kelvin element (S + D in parallel)
Stress condition	σ_S	σ_D	$\sigma = \sigma_S = \sigma_D$	$\sigma = \sigma_S + \sigma_D$
Strain condition	ε_S	ε_D	$\varepsilon = \varepsilon_S + \varepsilon_D$	$\varepsilon = \varepsilon_S = \varepsilon_D$
Constitutive eqns.	$\sigma_S = E\varepsilon_S$	$\sigma_D = \eta_e \dot{\varepsilon}$	$\sigma = \eta_e \dot{\varepsilon} - (\eta_e/E)\dot{\sigma}$	$\sigma = E\varepsilon + \eta_e \dot{\varepsilon}$
		$\varepsilon = (\sigma_D/\eta_e)t$	$\dot{\varepsilon} = (\dot{\sigma}/E) + (\sigma/\eta_e)$	$\dot{\varepsilon} = (\sigma/\eta_e) - (E/\eta_e)\varepsilon$
Shear stress	τ_S	τ_D	$\tau = \tau_S = \tau_D$	$\tau = \tau_S + \tau_D$
Shear strain	γ_S	γ_D	$\gamma = \gamma_S + \gamma_D$	$\gamma = \gamma_S = \gamma_D$
Constitutive eqns.	$\tau_S = G\gamma_S$	$\tau_D = \eta\dot{\gamma}_D$	$\tau = \eta\dot{\gamma} - (\eta/G)\dot{\tau}$	$\tau = G\gamma + \eta\dot{\gamma}$
		$\gamma = (\tau_D/\eta)t$	$\dot{\gamma} = (\dot{\tau}/G) + (\tau/\eta)$	$\dot{\gamma} = (\tau/\eta) - (G/\eta)\gamma$

Fig. 17-14 Simple mechanical models for stresses and deformation rates of viscoelastic bodies. In the final equations for the Maxwell element and the Voigt-Kelvin model, indices S (spring) and D (dashpot) have been omitted since one cannot trace the various contributions. For simplification, the symbol for tensile stress $\sigma_S = \sigma_{11}$ was simply written as σ. In order to avoid a mix-up of tensile and shear stresses, the symbol $\sigma_D = \sigma_{21}$ for shear stress was replaced by the traditional symbol τ.

Tensile data: σ = shear stress, $\dot{\sigma}$ = dσ/dt = shear rate, E = tensile modulus, ε = elongation, $\dot{\varepsilon}$ = dε/dt = tensile strain rate, η_e = extensional viscosity (Trouton viscosity).

Shear data: η = (shear) viscosity, τ = shear stress, $\dot{\tau}$ = dτ/dt = shear deformation rate, G = shear modulus, γ = shear deformation, $\dot{\gamma}$ = dγ/dt = shear rate.

The spring and dashpot are parallel in the **Voigt-Kelvin element**. A macroscopic analog is the MacPherson strut, a car suspension where a helical spring surrounds a shock absorber. Elongations here are the same for both the spring and dashpot ($\varepsilon = \varepsilon_S = \varepsilon_D$) whereas tensile stresses are additive ($\sigma = \sigma_S + \sigma_D$). With the expressions for σ_S and σ_D (Fig. 17-14), one obtains $\sigma = E\varepsilon + \eta_e\dot{\varepsilon}$ [alternatively: $\sigma = G\varepsilon + \eta\,\dot{\varepsilon}$] for the stress and $\dot{\varepsilon} = (\sigma/\eta_e) - (E/\eta_e)\varepsilon$ [alternatively: $\dot{\varepsilon} = (\sigma/\eta) - (G/\eta)\varepsilon$] for the strain rate. The equations for the shear stress, τ, and the shear rate, $\dot{\gamma}$, are derived in a similar manner (Fig. 17-14).

Differential equations for both Maxwell and Voigt-Kelvin elements contain three variables each (σ, e, and t, or τ, γ, and t) and two adoptable quantities each (E and η_e or G and η). Consequently, these equations cannot be solved in general but only for certain conditions. Table 17-4 contains the relevant equations for tensile and shear experiments.

For a Maxwell element with constant elongation as an example, the equations of Table 17-4 are derived as follows. The general equation (first line of Table 17-4) is rearranged and written as dε/dt – G^{-1}(dσ/dt) = (σ/η). Since $\varepsilon \equiv \varepsilon_o = const$ and thus dε/dt = 0, one obtains for the ratio of stress and viscosity $\sigma/\eta = -G^{-1}(\mathrm{d}\sigma/\mathrm{d}t)$ and, after integration, for the stress $\sigma = C\exp[-(G/\eta)t]$ where the integration constant C must have the physical unit of a stress. Hence, at constant elongation $\varepsilon = \varepsilon_o$, the body is under a stress $C = G\varepsilon_o$. The ratio η/G has the physical unit of time: it is the relaxation time t_{rlx}. It also follows from $\sigma = G\varepsilon_o \exp[-(t/t_{rlx})$ that the stress decreases exponentially with time.

Table 17-4 Time dependence of tensile stress σ and elongation ε of Maxwell and Voigt-Kelvin elements at constant elongation ($\varepsilon = \varepsilon_0$), constant stress ($\sigma = \sigma_0$), constant tensile strain rate ($d\varepsilon/dt = const$), or constant variation of stress with time ($d\sigma/dt = const$). C, C' = various constants.
Conditions for $t = 0$: 1) $C = G\varepsilon_0$; 2) $C = \sigma_0/G$; 3) $\varepsilon = \varepsilon_0$; 4) $C = -\sigma_0/G$ if $\varepsilon = 0$.

	Maxwell element		Voigt-Kelvin element	
–	$\sigma = \eta[(d\varepsilon/dt) - (\eta/G)(d\sigma/dt)]$		$\sigma = G\varepsilon + \eta(d\varepsilon/dt)$	
$\varepsilon = \varepsilon_0$	$\sigma = G\varepsilon_0 \exp[-(G/\eta)t]$	1)	$\sigma = G\varepsilon_0$	
$\sigma = \sigma_0$	$\varepsilon = \sigma_0[G^{-1} + \eta^{-1}t]$	2)	$\varepsilon = (\sigma_0/G)\{1 - \exp[-(G/\eta)t]\}$	4)
$\varepsilon = \varepsilon_1 t$	$\sigma = \varepsilon_1\eta + C\exp[-(G/\eta)t]$		$\sigma = \varepsilon_1\eta + \varepsilon_1 Gt$	
$\sigma = \sigma_1 t$	$\varepsilon = \varepsilon_0 + (\sigma_1/G)t + [\sigma_1/(2\eta)]t^2$	3)	$\varepsilon = -(\sigma_1\eta/G^2) + (\sigma_1/G^2)t + C'\exp[-(G/\eta)t]$	

These equations allow one to evaluate two types of experiments: the stress relaxation experiment where the body is under constant stress ($\sigma = \sigma_0$) (Section 17.4.2) and the creep experiment with constant elongation of the body ($\varepsilon = \varepsilon_0$) (Section 17.4.3). Each of these two types of experiments, and also the dynamic tests in Section 17.5, allow one to predict the behavior of a body under certain other conditions.

17.4.2 Stress Relaxation

Relaxation (L: *re* = backward, again; *laxare* = to solve, from *laxus* = loose) is defined by mechanics as the decrease of stress with time at constant deformation. For example, the deformation of a liquid by shearing generates a stress that opposes the deformation. In a liquid, such a stress decreases rapidly because molecules can find new positions fast.

Such behavior is well described by the Maxwell model. On elongation, a spring lengthens fast to a final value. At that point, one can either hold the elongation constant (I) or remove the stress (II). If the elongation is held constant (Case I), then the plunger will move slowly through the viscous liquid until the whole stress is removed. If, on the other hand, the stress is removed once the final position is reached (II), then the spring will contract immediately to its initial condition.

Because shear moduli and tensile moduli are interrelated (cf., $G = E/3$ for small deformations (Eqs.(16-61) and (16-58)), Maxwell elements can be used for both elongation and shearing whereas Voigt-Kelvin models are only useful for creep experiments.

Instead of elongating, one can also compress. In stress relaxation experiments, a specimen is suddenly compressed by a value of ε_0 at a time t_0 (Fig. 17-15, I). An example is a plastic gasket between a bottle neck and a metal cap. This deformation is then held constant ($\varepsilon = \varepsilon_0$) during the time interval between $t = 0$ and $t = t_E$.

The body responds to this compression by formation of a stress σ_0 at t_0. Since the deformation rate is for the compression, one obtains $\dot{\varepsilon} = (\dot{\sigma}/G) + (\sigma/\eta) = 0$ (Fig. 17-14 and small print on p. 586), The integration of this equation results in (Table 17-4)

(17-13) $$\sigma = G\varepsilon_0 \exp[-(G/\eta)t] = \sigma_0 \exp[-(t/t_{rlx})]$$

where $t_{rlx} \equiv \eta/G$ is the relaxation time. The body thus reacts to a deformation ε_0 at $t = 0$ by an "immediate" increase of the stress to σ_0. The stress then decreases exponentially

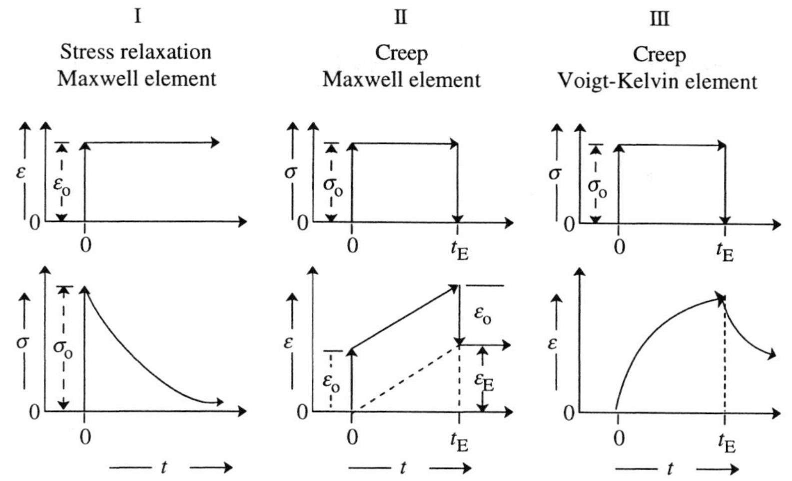

Fig. 17-15 Stress relaxation (left) and creep (center) of a Maxwell element compared to the creep of a Voigt-Kelvin element (right). In the *relaxation experiment*, a stress σ_o is applied at the time t_o (bottom left) to which the body responds by an "immediate" increase of elongation from $\varepsilon = 0$ to $\varepsilon = \varepsilon_o$ (top left). If the deformation is kept constant (top left), the body will relax and the stress decreases exponentially with time (bottom left).

Conversely, an initial stress is applied in the *creep experiment* (top center) which leads to an immediate deformation of the Maxwell element (bottom center) but to a slow one with time in the Voigt-Kelvin element. After a time t_E, the stress σ_o is suddenly removed upon which the elongation of the Maxwell element relaxes immediately from ε_o to ε_E (bottom center) whereas the Voigt-Kelvin body relaxes more slowly.

until it becomes again zero at infinite time. This relaxation is the reason why thermoplastic seals have to be retightened from time to time.

The relaxation time t_{rlx} is the time at which the stress has dropped to (1/e)th ≙ 36.8 % of the initial value. The viscosity η in Eq.(17-13) is an elongational viscosity and not a shear viscosity (see Fig. 17-14). Because of the different models, the modulus G is a **relaxation modulus** and not the shear modulus.

The ratio of the relaxation time t_{rlx} and a characteristic time scale of the experiment is called the **Deborah number** *De*. This relaxation Deborah number is not identical with the diffusion Deborah number (p. 486).

By definition, the relaxation Deborah number is zero for true Newtonian liquids, unity for glass temperatures, and infinity for true Hookean bodies. Its values are ca. 10^9-10^{11} for glasses, ca. 0.1-10 for polymer melts, ca. 10^{-6} for mineral oil, and ca. 10^{-12} for water (which is therefore a slightly elastic liquid (for proof, just jump from a 10 m diving board and fall on your belly)).

Polymers do not have *one* relaxation time but a spectrum of relaxation times because of distributions of distances between entanglement junctions (amorphous polymers), crosslinking sites (crosslinked polymers), and/or crystalline regions (semicrystalline polymers). Only idealized, perfect elastomers have equal distances between crosslinks. Short deformation times allow stresses to be dissolved by rotation of chain units about chain bonds within ca. 10^{-5} s. Large deformation times will cause shifts of junctions relative to each other which requires large relaxation times which also impede a viscous flow at small deformation times. Between these two very different relaxation times, there exists a range in which the relaxation modulus is practically constant.

Real elastomers, on the other hand, have a distribution of distances between junctions and therefore also distributions of the two types of relaxation processes which give rise to a spectrum of relaxation times. This spectrum can be modeled by a series of parallel Maxwell elements.

Relaxations cannot be modeled by the Voigt-Kelvin model, however, since a sudden elongation can be followed by the spring but not by the dashpot that needs to overcome an "infinitely" large (viscous) resistance. Such a resistance requires an infinitely large stress which is not realistic.

17.4.3 Creep Experiment

Retardation is defined as the increase of deformation with time caused by constant stress. Such retardation processes reveal themselves by a creep or an afterflow of matter. Since the phenomenon was first observed at room temperature without heating the apparently solid materials, it was also called **cold flow**.

In creep experiments, bodies are subjected to a tensile stress σ_0 at time zero. Because spring and dashpot are arranged in series in the Maxwell element and the spring reacts instantaneously, the element immediately adopts an elongation of $\varepsilon_0 = \sigma_0/E$ (Fig. 17-15). Under uniform loading (= constant stress σ_0), the change of stress with time is zero, $\dot{\sigma} = d\sigma/dt = 0$, and the element elongates with constant speed, σ_0/η, because of the dashpot in series, i.e., it creeps and has an **afterflow**, respectively (Fig. 17-15, center).

The total deformation ε at time t is obtained by integration of the differential equation, $\dot{\varepsilon} = (\dot{\sigma}/G) + (\sigma/\eta)$ (Fig.17-14, for $E \rightarrow G$, see p. 586, bottom):

$$(17\text{-}14) \qquad \int \dot{\varepsilon}\,dt = \int (d\varepsilon/dt)dt = \varepsilon = (\sigma_0/G) + (\sigma_0/\eta)t = \varepsilon_0 + (\sigma_0/\eta)t \quad ; \quad \sigma = \sigma_0 = const$$

After removal of the load at time $t = t_E$, the body will contract by $\varepsilon_0 = \sigma_0/G$ but remains constantly deformed by a permanent set of $\varepsilon_E = (\sigma_0/\eta)t_E$. The Maxwell element thus behaves as an elastic solid at the beginning of the experiment but as a viscous liquid during the experiment: it is a **viscoelastic liquid** (Table 17-5).

Table 17-5 Comparison of simple mechanical models for deformations. σ = stress, G = modulus, η = viscosity, ε = strain, S = solid, L = liquid.

Model	Stress function	Behavior at Start	Behavior at End
Hookean body	$\sigma = G\varepsilon$	S	S
Voigt-Kelvin element	$\sigma = G\varepsilon + \eta\,(d\varepsilon/dt)$	L	S
Newtonian liquid	$\sigma = \eta\,(d\varepsilon/dt)$	L	L
Maxwell element	$\sigma + (\eta/G)(d\sigma/dt) = \eta\,(d\varepsilon/dt)$	S	L
Jeffreys model	$\sigma + (\eta/G)(d\sigma/dt) = G\varepsilon + \eta\,(d\varepsilon/dt)$	S	S

Contrary to the predictions of the Maxwell element, polymers at constant stress do not elongate linearly with time. Their increase of elongation with time can be described better by the *Voigt-Kelvin model* (Fig. 17-15, III). For this model, the rheological equation of state for tensile deformation is $\dot{\varepsilon} = (\sigma/\eta) - (G/\eta)\varepsilon$ (Fig. 17-14, Table 17-5; for $E \rightarrow G$, see p. 586, bottom). The solution of this differential equation is

(17-15) $\varepsilon_{VK} = (\sigma_o/G_{VK})[1 - \exp(-G_{VK}t/\eta)] = (\sigma_o/G_{VK}) - (\sigma_o/G_{VK})[\exp(-t/t_{rtd})]$

According to this model, the deformation ε is zero immediately after application of the stress at $t = 0$ (Fig. 17-15), which is not found for polymers. The deformation then increases rapidly with time, a typical behavior of liquids, and then more slowly.

The **retardation time** t_{rtd} indicates the time at which the retarding part of deformation at constant load reaches a value of (1 – 1/e) = 0.632 of the final deformation, ε_E, at the time t_E. The Voigt-Kelvin body thus behaves as a **viscoelastic solid**: initially as a liquid and later as a solid (Table 17-5).

The retardation time t_{rtd} is of the same order of magnitude as the relaxation time. However, retardation times and relaxation times are not identical because they depend on different models.

After infinite time at constant stress, the deformation approaches a value of $\varepsilon_o = \sigma_o/G$. Removal of the stress after a time t_E leads to a fast and then a slow decrease of elongation with time according to $\varepsilon = (\sigma_o/G)[1 - \exp(-t_E/t_{rtd})][\exp\{-(t - t_E)/t_{rtd}\}]$ (right part of the curve in Fig. 17-15, bottom). At infinite time, the initial state is reached with $\varepsilon = 0$.

Shear deformations are treated similarly. From the differential equation for shear processes, $\dot{\gamma} = (\tau/\eta) - (G/\eta)\gamma$, one obtains $\gamma = (\eta/G)[1 - \exp(-t/t_{rtd})]$. The **retardation modulus** G of this equation is neither identical with the retardation modulus of Eq.(17-15) nor with the shear modulus of Eq.(16-57) since the true viscoelastic behavior is semiquantitavely described by the Voigt-Kelvin and Maxwell models.

17.4.4 Three- and Four-Parameter Models

Models with more than two mechanical devices are obviously more realistic than the Maxwell and Voigt-Kelvin models with just two. There are four possible 3-parameter models (Fig. 17-16), consisting either of two springs and one dashpot (Group A) or of one spring and two dashpots (Group B). Model 3A' with parallel spring and Maxwell element is called a **standard linear solid (SLS)**. **Jeffrey's model** (Table 17-5) is Model 3B: a dashpot and a Kelvin element in series. An often used 4-parameter model (4) is the **Burgers model** with a Maxwell and a Voigt-Kelvin element in series (Fig. 17-16).

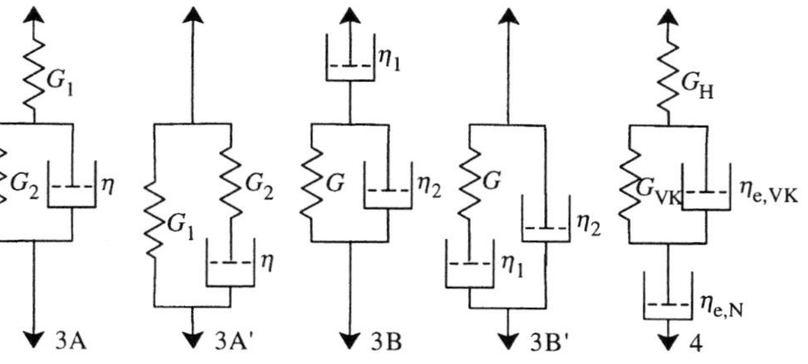

Fig. 17-16 The four possible 3-parameter models and the Burgers 4-parameter model.

The two A-group models have the same type of dependence on variables σ, η, and t and the two B-group models another type:

$$(17\text{-}16,\ 3A) \qquad (G_1 + G_2)\sigma + \eta \frac{d\sigma}{dt} = G_1 G_2 \varepsilon + G_1 \eta \frac{d\varepsilon}{dt}$$

$$(17\text{-}16,\ 3A') \qquad G_2\sigma + \eta \frac{d\sigma}{dt} = G_1 G_2 \varepsilon + (G_1 + G_2)\eta \frac{d\varepsilon}{dt}$$

$$(17\text{-}16,\ 3B) \qquad G\sigma + (\eta_1 + \eta_2)\frac{d\sigma}{dt} = G\eta_1 \frac{d\varepsilon}{dt} + (\eta_1 \eta_2)\frac{d^2\varepsilon}{dt^2}$$

$$(17\text{-}16,\ 3B') \qquad G\sigma + \eta_2 \frac{d\sigma}{dt} = G(\eta_1 + \eta_2)\frac{d\varepsilon}{dt} + (\eta_1 \eta_2)\frac{d^2\varepsilon}{dt^2}$$

Since the **Burgers model** combines a Maxwell element and a Voigt-Kelvin element in series (Fig. 17-16), the two rheological equations of state

$G_M(d\varepsilon_M/dt) = (d\sigma/dt) + (G_M/\eta_M)\sigma$ (Maxwell element)

$(\sigma/\eta_{VK}) = (d\varepsilon_{VK})/dt + (G_{VK}/\eta_{VK})\varepsilon_{VK}$ (Voigt-Kelvin element)

are combined in such a way that ε_M und ε_{VK} can be eliminated. The rheological equation of state for this model becomes a second-order differential equation:

$$(17\text{-}17) \qquad \frac{d^2\sigma}{dt^2} + \left\{\frac{G_M}{\eta_M} + \frac{G_M}{\eta_{VK}} + \frac{G_{VK}}{\eta_{VK}}\right\}\frac{d\sigma}{dt} + \left(\frac{G_M G_{VK}}{\eta_M \eta_{VK}}\right)\sigma = G_M \frac{d^2\varepsilon}{dt^2} + \left(\frac{G_M G_{VK}}{\eta_{VK}}\right)\frac{d\varepsilon}{dt}$$

This equation as well as the simpler equations of state for the four 3-parameter models can only be solved for special cases.

For tensile tests with an applied initial stress σ_o, the total deformation ε of a Burgers body is the sum of deformations caused by the Hookean body (ε_H), Newtonian liquid (ε_N), and Voigt-Kelvin element (ε_{VK}), i.e., $\varepsilon = \varepsilon_H + \varepsilon_N + \varepsilon_{VK}$. Introduction of $\varepsilon_H = \sigma_o/G$ and $\varepsilon_N = (\sigma_o/\eta)t$ (Fig. 17-14) and the expression for ε_{VK} from Eq.(17-15) delivers for the creep experiment

$$(17\text{-}18) \qquad \varepsilon = \frac{\sigma_o}{G_H} + \frac{\sigma_o}{\eta_N}t + \frac{\sigma_o}{G_{VK}}\left[1 - \exp(-G_{VK}t/\eta_{VK})\right] = \frac{\sigma_o}{G(t)}$$

The time-dependent ratio σ_o/ε is a creep modulus, $\sigma_o/\varepsilon = G(t)$. The recovery after removal of the load is therefore

$$(17\text{-}19) \qquad \varepsilon = \frac{\sigma_o}{\eta_N}t_1 + \frac{\sigma_o}{G_{VK}}\left[1 - \exp(-G_{VK}t_1/\eta_{VK})\right]\left[\exp(-G_{VK}(t - t_1)/\eta_{VK})\right]$$

In Fig. 17-17, the time dependence of deformation by creep according to the Burgers model is compared to that of the Maxwell, Voigt-Kelvin, and Hooke elements, using reasonable model parameters (see caption). Also included in the graph is the prediction of the Nutting model (see Eq.(17-20)).

Creep curves are usually not analyzed in detail because of the unknown contributions to deformation by elasticity (ε_H), viscosity (ε_N), and viscelasticity (ε_{VK}). The time-depen-

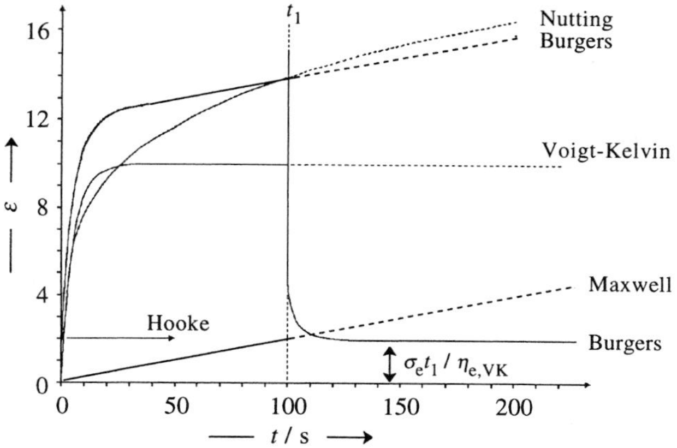

Fig. 17-17 Time-dependence of deformation ε by creeping, i.e., at constant stress of σ_o = 100 MPa for bodies with G_H = 500 MPa, G_{VK} = 100 MPa, $\eta_{VK} = 5 \cdot 10^8$ Pa s, and $\eta_N = 5 \cdot 10^{10}$ Pa s. The Maxwell body, Voigt-Kelvin body, and Burgers body would expand according to the broken lines if the experiment was not stopped at t_1 = 100 s. After removal of the applied stress after 100 s, the Burgers body recovers fast but remains constantly deformed by a slightly time-dependent amount of $\sigma_o t_1/\eta_{VK}$ (solid line). The Hookean contribution to the elongation of the Burgers body is independent of time. The relaxation time is $t_{rlx} = \eta_{VK}/G_{VK}$ = 5 s for this set of data. The Nutting function was fitted to ε_1 at t_1 using β = 1 and n = 1/4 in Eq.(17-20).

dent viscous and viscoelastic contributions are often combined to a new parameter ε_k whose time dependence is described by the **Findlay law**, $\varepsilon_k = \varepsilon_{k,o} t^m$. The creep curve, ε = f(t) can then be described by the **Nutting equation**:

(17-20) $$\varepsilon = \varepsilon_H + \varepsilon_N + \varepsilon_{VK} = \varepsilon_H + \varepsilon_k = \varepsilon_H + \varepsilon_{k,o} t^m \approx K \sigma_o{}^{\beta} t^n$$

According to Eq.(17-20), logarithms of elongation should vary linearly with the logarithm of time, which is indeed found (Fig. 17-18). Also found is the predicted depen-

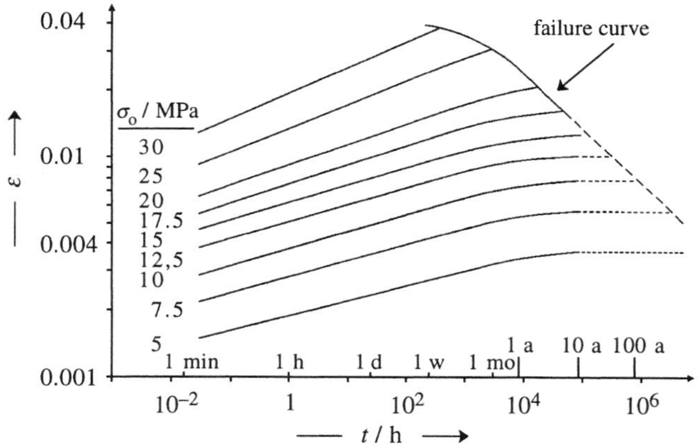

Fig. 17-18 Plot of lg $\varepsilon = f$(lg t) at various applied stresses σ_o for creep curves of the acetal copolymer Hostaform C 9021 at 20°C [14a] according to the Nutting equation, Eq.(17-20). The exponent n (slope) increases slightly with increasing σ_o. By kind permission of Hanser-Verlag, Munich.

dence of the front factor $K\sigma_o^{\beta}$ on the applied stress σ_o. The Nutting equation frequently allows one to extrapolate results of short-term measurments to long-term behavior, especially for large values of σ_o. Another empirical equation for creep processes is the hyperbolic equation $\varepsilon = K_t \sinh(\sigma/\sigma_{crit})$.

17.4.5 Boltzmann Superposition Principle

In tensile experiments, tensile stress is zero at the beginning of the experiment and the specimen is then elongated with constant elongation rate, $\dot{\varepsilon} = \dot{\varepsilon}_o$. Integration of $\dot{\sigma} = d\sigma/dt = E\dot{\varepsilon}_o - (\sigma E/\eta_e)$ for the stress rate of a Maxwell body (see Fig. 17-14) delivers

$$(17\text{-}21) \qquad \sigma = \eta_e \dot{\varepsilon}_o[1 - \exp\{-E\varepsilon/(\eta_e \dot{\varepsilon}_o)\}]$$

where η_e and E have replaced the conventional η and G for shear. For $\eta_o\dot{\varepsilon}_o = const < \infty$, the tensile stress thus increases with increasing elongation first strongly and then less strongly until it asymptotically reaches a value of $\sigma_\infty = \eta_o\dot{\varepsilon}_o$ at $\varepsilon \to \infty$ (Fig. 17-19, top). The two limiting functions $\sigma = f(\varepsilon)$ for large and small elongation rates are obtained by setting $E\varepsilon/(\eta_o\dot{\varepsilon}_o) \equiv x$ and developing the exponential term, exp (–x), in an infinite series, $1 - x + x^2/2! - ...$ At infinitely large elongational rates (i.e., for $\eta_o\dot{\varepsilon}_o \to \infty$), one obtains Hooke's law, $\sigma = E\varepsilon$.

The initial slope of the function $\sigma = f(\varepsilon)$ thus delivers the modulus whereas the stress σ becomes zero at diminishingly small extensional rates, i.e., $\eta_o\dot{\varepsilon}_o \to 0$. If the elongational viscosity is independent of the elongational rate (Troutonian liquid; $\eta_e = const$), then the modulus will decrease with decreasing elongational rate $\dot{\varepsilon}_o$ (and $\eta_o\dot{\varepsilon}_o$, respectively) (Fig. 17-19, top). The specimen thus behaves as a rigid material at large elongational rates (large E) but as an elastomer at small ones (low E).

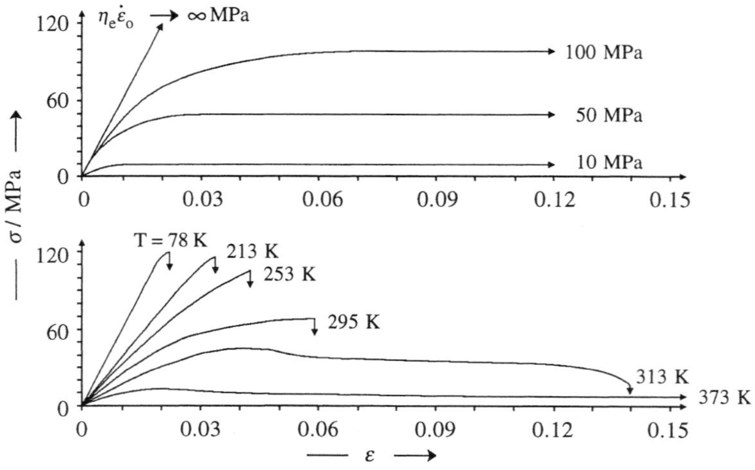

Fig. 17-19 Stress-strain behavior of a poly(methyl methacrylate) at (bottom) various temperatures and (top) products, $\eta_e\dot{\varepsilon}_o$, of extensional viscosities, η_e, and initial elongational rates, $\dot{\varepsilon}$. ↓ Fracture.
Bottom: Experiments. At 373 K, the specimen can be drawn to $\varepsilon = 2.84$ without breaking.
Top: Calculated with the Maxwell model for $E = 6000$ MPa and various values of $\eta_e\dot{\varepsilon}_o$.

At variable extensional rates and constant temperature (Fig. 17-19, top), calculated functions $\sigma = f(\varepsilon)$, and thus also the modulus E, behave similarly to these functions at constant extensional rate and variable temperatures if the yield point is neglected (yield points appear only in a certain temperature interval; see Fig. 17-7). The two functions, $E = f(t)$ and $E = f(T)$, can therefore be interconverted if the relaxation mechanism does not vary with temperature. The same is true for shear moduli G, tensile compliances D, and shear compliances J.

The interconversion makes use of the Boltzmann superposition principle which says that a deformation by an additional load (or a recovery after removal of a load) is independent of the preceding loadings. The Boltzmann superposition principle is derived as follows. Application of a constant shear stress $\tau_{o,a}$ at time $t_o = 0$ (Fig. 17-20, top) leads to an increase of the shear deformation γ (Fig. 17-20, bottom), first fast, and then more slowly. After a time t_1, the shear deformation has reached a value of $\gamma_1 = \tau_{o,a}/G_1 = \tau_{o,a}J_1$. Removal of $\tau_{o,a}$ at t_1 leads to a decay of γ which reaches a value of $\gamma = \gamma_1 - \Delta\gamma_1$ at a time t_2.

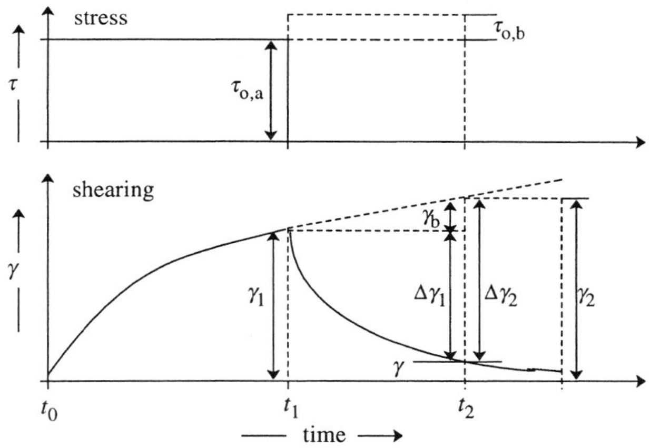

Fig. 17-20 Demonstration of the Boltzmann superposition principle for the creep experiment (see text). τ = shear stress, γ = shear deformation.

Without removal of $\tau_{o,a}$ at t_1, the stress would have increased at t_2 to a value of γ_2. However, the actual value of γ at t_2 is $\gamma = \gamma_2 - \Delta\gamma_2$. The recovered deformation is therefore $\Delta\gamma_2 = \gamma_2 - \gamma_1 + \Delta\gamma_1 = \gamma_b + \Delta\gamma_1$ which is identical with the one that would be present if an additional shear stress τ_b had been applied at time t_1. The deformation by an additional load is therefore independent of the previous loadings.

An example of the application of the Boltzmann superposition principle is the conversion of the time dependence of shear moduli $G(t)$ at various temperatures (Fig. 17-21, left) to a reference temperature T_o with the help of the shift factor a_T of the WLF equation (p. 467). This conversion presumes the spectrum of relaxation times to be independent of temperature. The resulting standard curve for shear moduli comprises 16 decades of hours from 10^{-11} h to 10^5 h, i.e., 40 ns to 11.4 years (Fig. 17-21, right). Such a large time span cannot be covered by any single experimental method but requires overlapping techniques, for example, various dynamic methods.

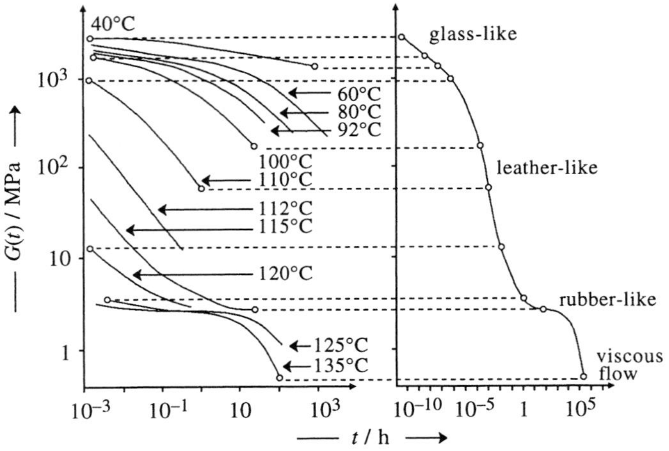

Fig. 17-21 Left: Time dependence of shear moduli $G(t)$ from measurements of stress relaxation of a poly(methyl methacrylate) ($\overline{M}_\eta$ = 3.6·10^6 g/mol) at temperatures of 40 ≤ T/°C ≤ 135 [15].

Right: The same experimental values for a reference temperature of T_o = 388 K, obtained by using the shift factor of the WLF equation (p. 467) and a glass temperature of T_G = 378 K. Some shifts are marked by small circles o. The polymer flows during long deformation times but appears as a glass during short ones. The true glass state (horizontal line) is not obtained, not even at deformation times of 10^{-11} seconds.

17.5 Dynamic Deformations

All methods discussed so far are called "static" in the literature although they are neither "static" (for example, creep methods) nor equilibrium procedures. Methods are called "dynamic" it the specimen is subjected to *periodic* deformations. Such dynamic methods can be subdivided into two groups: those with forced oscillations (for example, with the Rheovibron®) and those with free ones (torsion pendulum). These mechanical methods work at fairly low frequency ranges. They are complemented by methods for higher frequency ranges such as nuclear magnetic resonance, ultrasound, or dielectric methods so that a large range of frequencies and temperatures can be investigated.

17.5.1 Forced Oscillations

In the simplest case, *forced oscillations* of specimens are sinusoidal. Since the Rheovibron® instrument works with continuous forced tensile stresses, the resulting flexural moduli are tensile moduli (for flexural moduli from static measurements, see p. 528). In the torsion pendulum (Section 17.5.2), specimens are subjected to a torque and then allowed to oscillate freely with decreasing amplitudes. The resulting deformations are shear deformations and the obtained moduli are therefore shear moduli.

On uniaxial shearing with constant circular frequency, $\omega = 2\pi\nu$, shear stresses τ change with time according to $\tau(t) = \tau_0 \sin \omega t$ where τ_0 = amplitude (maximum value of stress) and ν = frequency (Fig. 17-22). The resulting shear strain (shear deformation) γ

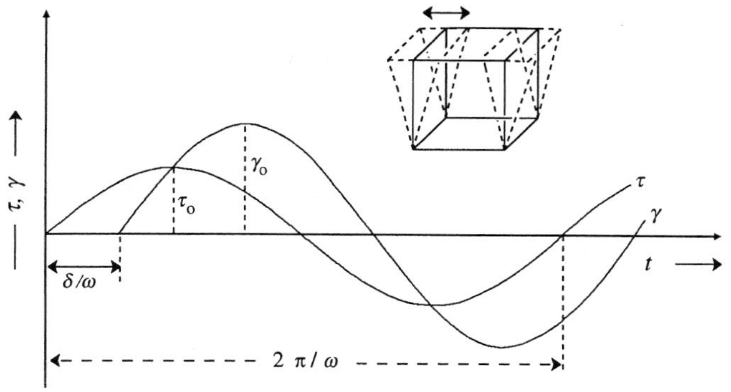

Fig. 17-22 Applied shear stress τ and resulting out-of-phase deformation γ for a sinusoidal uniaxial shearing as function of time t.
τ_o, γ_o = amplitudes, δ = phase angle, ω = circular frequency (= angular velocity).

is also a sinusoidal function of time. For Hookean bodies, γ follows the applied stress instantaneously so that $\gamma(t) = \gamma_o \sin \omega t$. For viscoelastic bodies, the shear strain lags behind the applied shear stress by a phase angle δ. The viscoelastic amplitude γ_o differs from the Hookean amplitude τ_o (Fig. 17-22) because of differences in the stored energy.

Ideally, the phase angle δ is constant and one obtains from the vector diagram

(17-22) $\gamma(t) = \gamma_o \sin(\omega t - \delta)$

The phase angle ϕ between applied shear stress and resulting shear strain, the phase shift, is also known as (mechanical) **loss angle,** $\delta \equiv \phi$. It differs from the dielectric loss angle which is defined as $\delta \equiv 90° - \phi$ (Volume IV, Section 12.2.5).

The stress vector can be thought of as a sum of two components: a component that is in phase with the deformation, $\tau' = \tau_o \cos \delta$, and a component that is out of phase with the deformation, $\tau'' = \tau_o \sin \delta$. Each component is characterized by a modulus.

The **real modulus** G' **(in-phase modulus, elastic modulus)** measures the stiffness and shape stability of the specimen. It is also called **shear storage modulus** because it measures the energy that is stored during deformation. The real modulus is calculated from the in-phase component τ' and the amplitude γ_o:

(17-23) $G' = \tau'/\gamma_o = (\tau_o/\gamma_o) \cos \delta = G^* \cos \delta$

where $G^* = \tau_o/\gamma_o$ is the **complex modulus** (Section 17.5.3).

The **imaginary modulus** G'' **(shear loss modulus, 90° modulus, out-of-phase modulus, viscous modulus)** is defined in the same way:

(17-24) $G'' = \tau''/\gamma_o = (\tau_o/\gamma_o) \sin \delta = G^* \sin \delta$

The shear loss modulus describes the mechanical energy that has been lost because it has been dissipitated as heat. It is directly proportional to the heat, $Q = \pi G''\gamma_o^2$, that has been liberated per cycle; γ_o = maximum value of shear deformation.

The terms "complex modulus," "real modulus," and "imaginary modulus" are used because one can write these moduli as complex quantities, $G^* = G' + iG''$, where $i^2 = -1$. Despite their name, imaginary moduli G'' are true physical quantities, i.e., products of a number and a mechanical stress.

Similar equations can be derived for the **tensile storage modulus**, $E' = (\sigma_o/\varepsilon_o) \cos \delta$, and the **tensile loss modulus**, $E'' = (\sigma_o/\varepsilon_o) \sin \delta$.

The ratio of shear loss modulus to storage modulus is the (mechanical) loss **tangent** $\Delta = (\sin \delta)/(\cos \delta) = \tan \delta$ which is approximately (but not exactly) the same for tensile and shear moduli:

$$(17\text{-}25) \qquad \Delta = \tan \delta = G''/G' \approx E''/E' \; ; \qquad \text{exact: } \frac{E''}{E'} = \frac{G''}{G'}\left[\frac{1}{1+[G'/(3K)][1+(G''/G')^2]}\right]$$

Pressurizing (= all-sided compression) has no shear component. Hence, the ratio of imaginary and real bulk moduli does not equal the ratio of the corresponding shear moduli. One rather has $K''/K' < G''/G'$.

Instead of applying a shear stress $\tau(t)$ and then measuring the resulting deformation $\gamma(t)$, one can also deform the specimen and then measure the stress. In this case, stress does not lag brhind the deformation but precedes it. This is immediately clear from the analogous electrical experiment: an electric current can flow only after an electric potential has been established.

The physical quantities outlined above allow one to define a number of rheological parameters: the **dynamic stationary shear viscosity** η_o', the **out-of-phase dynamic shear viscosity** η'', and the **complex dynamic stationary shear viscosity** η_o^*:

$$(17\text{-}26) \qquad \eta_o' = \lim_{\omega \to 0}(G''/\omega) \quad ; \quad \eta'' = G'/\omega \quad ; \quad \eta_o^* = \lim_{\omega \to 0}(G^*/\omega)$$

The **elasticity coefficient** A_G is the ratio of the shear storage modulus G' and the square of the circular frequence (angular frequency), ω, in the limit of very low frequencies. The ratio of the elasticity coefficient and the square of the dynamic shear viscosity is the **stationary shear compliance** J_e^o (Section 17.5.3):

$$(17\text{-}27) \qquad A_G = \lim_{\omega \to 0}(G'/\omega^2) \quad ; \quad J_e^o = A_G/(\eta_o')^2$$

17.5.2 Free Oscillations

A **torsion pendulum** works with *free* oscillations. The specimen is fastened on top of a stationary plate and its top twisted by an angle θ relative to the position at rest (Fig. 17-23). On release, the upper part of the specimen will then twist back to a position beyond its resting position and oscillate freely about the resting position with decreasing amplitude. The oscillation frequency is independent of the amplitudes but it cannot be varied at will because it depends on the damping properties of the specimen.

The torsion pendulum delivers shear moduli G that are calculated from the frequency ($\nu = 1/t$), the moment of inertia I (in mass times area), and a geometric factor K_{geom} of the specimen. The geometric factor is calculated from the geometric dimensions of the specimen (length L, radius R, width W, thickness d, form factor μ):

circular rods: $K_{geom} = 8\,\pi L/R^4$
rectangular rod: $K_{geom} = 64\,\pi^2 L/(\mu W d)$

where $\mu = 2.249$ ($W/d = 1$), 3.659 ($W/d = 2$), 4.493 ($W/d = 4$), and 5.232 ($W/d = 40$).

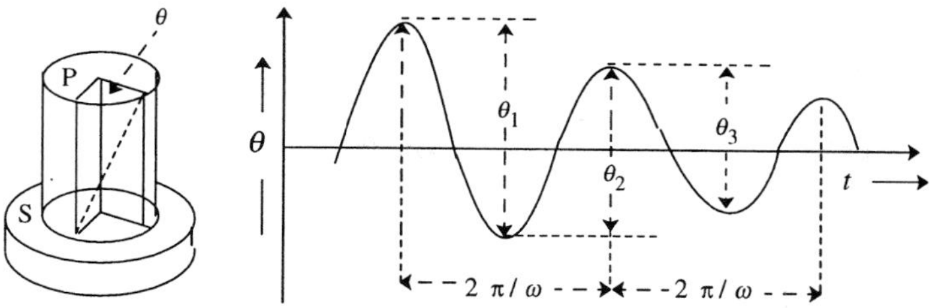

Fig. 17-23 Torsion pendulum. Left: a polymer specimen P is clamped on a stationary support and the top of the specimen is twisted by an angle θ. Right: The specimen is allowed to oscillate freely with a frequency ν about the position at rest. The time span between two maxima is constant, $t = 1/\nu = 2\pi/\omega$, but the amplitudes θ_1, θ_2 ... decrease with time. ω = circular frequency.

The logarithm of the ratio of two successive amplitudes is called the **logarithmic decrement,** $\Lambda = \lg(\theta_n/\theta_{n+1})$, a quantity that indicates the stored energy which is lost per cycle. The logarithmic decrement is not identical with the mechanical loss factor since it is also controlled by the geometry of the specimen. For example, it adopts a value of $\Lambda \approx \pi \tan \delta$ if cylindrical specimens are subjected to small dampings.

17.5.3 Complex Moduli

Real moduli G' and imaginary moduli G'' can be derived from complex variables such as the **complex deformation,** $\varepsilon^* = \varepsilon_o \exp[i(\omega t)]$, instead of the deformation ε, and the **complex stress,** $\sigma^* = \sigma_o \exp[i(\omega t + \delta)]$, instead of the stress σ. The **complex shear modulus** is then

(17-28) $$G^* = \sigma^*/\varepsilon^* = (\sigma_o/\varepsilon_o)\exp(i\delta) = (\sigma_o/\varepsilon_o)(\cos\delta + i\sin\delta) = G' + iG''$$

Quantities from sinusoidal experiments can also be represented by the **complex shear compliance,** J^*, which is the ratio of **complex shearing,** $\gamma^* = \gamma_o \exp[i(\omega t - \delta)]$, instead of shearing γ, and **complex shear stress,** $\tau^* = \tau_o \exp(i\omega t)$, instead of shear stress τ:

(17-29) $$1/\boldsymbol{G}^* = \gamma^*/\tau^* = \boldsymbol{J}^* = J' - iJ'' \quad ; \quad J''/J' = G''/G' = \tan\delta$$

The complex shear modulus $\boldsymbol{G}^*$ is the inverse of the complex shear compliance $\boldsymbol{J}^*$ but this inverse relationship does not apply to the components, G' and G'' and J' and J'':

(17-30) $$G' = \frac{J'}{(J')^2 + (J'')^2} = \frac{1/J'}{1 + \tan^2\delta} \quad ; \quad J' = \frac{G'}{(G')^2 + (G'')^2} = \frac{1/G'}{1 + \tan^2\delta}$$

$$G'' = \frac{J''}{(J')^2 + (J'')^2} = \frac{1/J''}{(1 + \tan^2\delta)^{-1}} \quad ; \quad J'' = \frac{G''}{(G')^2 + (G'')^2} = \frac{1/G''}{1 + (\tan^2\delta)^{-1}}$$

In contrast to static measurements, the storage modulus G' from dynamic measurements does *not* equal the inverse storage compliance, J', and the loss modulus G'' is *not* the inverse of the loss compliance, J''.

These equations can be applied to mechanical models, for example, the *Maxwell element* with $d\varepsilon/dt = (\sigma/\eta) + G^{-1}(d\sigma/dt)$ (Table 17-4). Introduction of the relaxation time, $t_{rlx} \equiv \eta/G$, leads to $t_{rlx}G(d\varepsilon/dt) = \sigma + t_{rlx}(d\sigma/dt)$. Replacing ε by $\varepsilon^* = \varepsilon_o \exp[i(\omega t)]$ and σ by $\sigma^* = \sigma_o \exp[i(\omega t + \delta)]$, considering $de^u/dx = e^u(du/dx)$, rearranging, and introducing Eq.(17-28) leads to

$$\text{(17-31)} \qquad \frac{i\omega t_{rlx}G}{1 + i\omega t_{rlx}} = \frac{\sigma_o \exp[i(\omega t + \delta)]}{\varepsilon_o \exp(i\omega t)} = \frac{\sigma^*}{\varepsilon^*} = G^* = G' + iG''$$

Multiplication of both the nominator and the denominator of Eq.(17-31), $\sigma^*/\varepsilon^* = (i\omega t_{rlx}G)/(1 + i\omega t_{rlx})$ by $(1 - i\omega t_{rlx})$, and introduction of $i^2 = -1$ leads to

$$\text{(17-32)} \qquad \frac{\sigma^*}{\varepsilon^*} = \frac{G\omega^2 t_{rlx}^2}{1 + \omega^2 t_{rlx}^2} + i\frac{G\omega t_{rlx}}{1 + \omega^2 t_{rlx}^2}$$

According to Eq.(17-31), the first summand of Eq.(17-32) can be identified as the shear storage modulus G' and the second summand as iG'' so that

$$\text{(17-33)} \qquad G' = \frac{G\omega^2 t_{rlx}^2}{1 + \omega^2 t_{rlx}^2} \; ; \; G'' = \frac{G\omega t_{rlx}}{1 + \omega^2 t_{rlx}^2} \; ; \; G^* = \frac{G i\omega t_{rlx}}{1 + i\omega t_{rlx}}$$

The loss factor becomes $\Delta = \tan \delta = 1/(\omega t_{tlx})$ since $\tan \delta = G''/G'$. Equations for the Voigt-Kelvin element can be derived in a similar manner.

The Maxwell model shows correctly that shear storage moduli G' increase with increasing logarithm of ωt in an S-shaped manner (Fig.17-24): the shorter the time between periods, the less the body can relax and the more rigid it is. The model also predicts correctly that the shear loss modulus G'' passes through a maximum with increasing values of lg ωt. However, the Maxwellian loss tangent does not pass through a maximum, as is found experimentally, but instead decreases exponentially from a finite value to zero. The Voigt-Kelvin model is worse: it predicts $G' = const$ and increases of both G'' and $\tan \delta$ with ωt from zero to infinity (Fig.17-24).

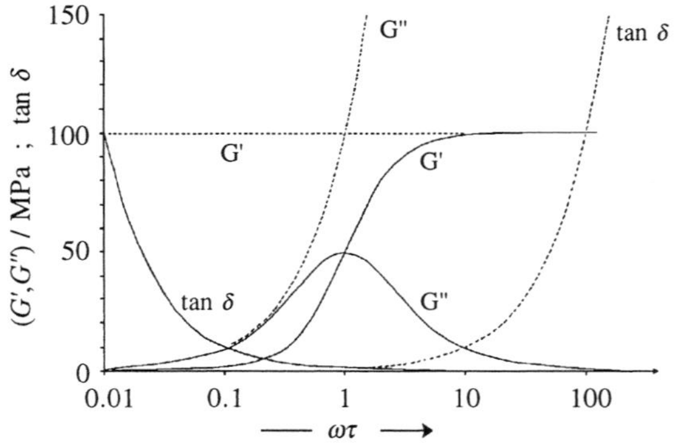

Fig. 17-24 Shear storage moduli G', shear loss moduli G'', and loss tangents $\tan \delta$ as functions of the logarithm of the product $\omega\tau$ of circular frequency ω and relaxation time τ for a body with a shear modulus $G = 100$ MPa. Solid lines: Maxwell model; dotted lines: Voigt-Kelvin model.

Table 17-6 Shear storage moduli G', shear loss moduli G'', and the corresponding shear compliances J' and J'' of the Maxwell element, the Voigt-Kelvin element, and the standard linear solid.

	Maxwell	Voigt-Kelvin	Standard linear solid
G'	$\dfrac{G\omega^2 t_{rlx}^2}{1+\omega^2 t_{rlx}^2}$	G	$\dfrac{G_1 G_2}{G_1 + G_2 - \eta\omega \tan\delta}$
G''	$\dfrac{G\omega t_{rlx}}{1+\omega^2 t_{rlx}^2}$	$G\omega t_{rlx} = \omega\eta$	$\dfrac{G_1 G_2 \tan\delta}{G_1 + G_2 - \eta\omega \tan\delta}$
J'	J	$\dfrac{J}{1+\omega^2 t_{rlx}^2}$	$\dfrac{1}{G_1} + \dfrac{1}{G_2}\left(\dfrac{1}{1+\omega^2 t_{rlx}^2}\right) = J_u + \dfrac{(J_r - J_u)}{1+\omega^2 t_{rlx}^2}$
J''	$J/\omega t_{rlx} = 1/(\omega\eta)$	$\dfrac{J\omega t_{rlx}}{1+\omega^2 t_{rlx}^2}$	$\dfrac{1}{G_2}\left(\dfrac{\omega t_{rlx}}{1+\omega^2 t_{rlx}^2}\right) = \dfrac{(J_r - J_u)\,\omega t_{rlx}}{1+\omega^2 t_{rlx}^2}$
$\tan\delta = \dfrac{J''}{J'}$	$\dfrac{G''}{G'} = 1/\omega t_{rlx}$	$\dfrac{G''}{G'} = \omega t_{rlx}$	$\dfrac{G''}{G'} = \dfrac{(J_r - J_u)\,\omega t_{rlx}}{J_r + J_u \omega^2 t_{rlx}^2}$
$J(t)$	$J + t/\eta$	$J[1 - \exp(-t/t_{rlx})]$	$J_u + (J_r - J_u)[1 - \exp(-t/t_{rlx})]$

The standard linear solid model (SLS) delivers better results (Table 17-6). At very long times (very low frequencies), an SLS relaxes completely. The shear storage compliance becomes $J' \approx (1/G_1) + (1/G_2) = J_r$ and the shear loss compliance, $J'' = 0$. At very large frequencies (very small times), the body remains in the unrelaxed state, i.e., $J' \approx 1/G_1 = J_u$ and $J'' \approx 0$. In the general case, one therefore obtains the **Debye equations**:

$$J' = J_u + \frac{J_r - J_u}{1+\omega^2 t_{rlx}^2} \quad ; \quad J'' = \frac{(J_r - J_u)\,\omega t_{rlx}}{1+\omega^2 t_{rlx}^2} \tag{17-34}$$

These equations apply to bodies with high shear storage compliances J' at low frequencies (rubber-like behavior) and low shear storage compliances J' at high frequencies (glass state) (Fig. 17-25).

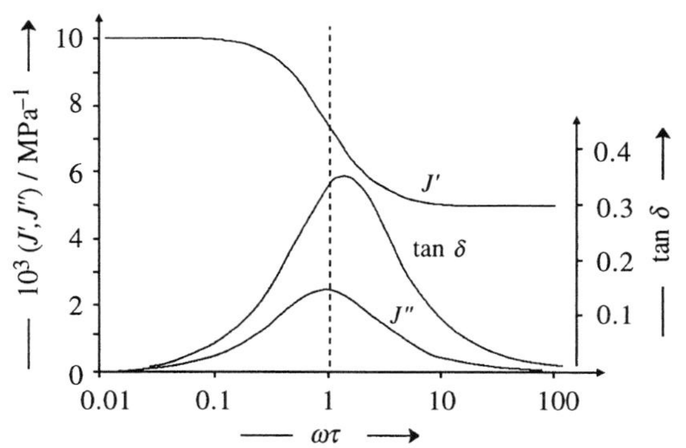

Fig. 17-25 Shear storage compliance J', shear loss compliance J'', and loss tangent $\tan\delta$ of an anelastic SLS with $J_r = 0.01$ MPa^{-1} and $J_u = 0.005$ MPa^{-1}.

In such SLS, shear loss compliances J'' pass through a maximum at $\omega t_{rlx} = 1$: the loss is thus the greatest if the circular frequency equals the inverse relaxation time. However, the loss tangent does not have its maximum at $\omega\tau = 1$ but at somewhat greater values of $\omega\tau$ as one can also see by inserting the expressions for J' and J'' into $\tan \delta = J''/J'$.

The behavior of bodies such as the standard linear solid is called "anelastic" (G: *an* = not, *elastos* = beaten, from *elaunein* = to drive). Although such bodies behave Hookean in one respect (doubling the stress doubles the strain at any time), they do not respond instantaneously to a change in stress as Hookean bodies do. The stress-strain relationship of anelastic bodies is unique because after sufficient time, each stress leads to a unique strain: $J_r\sigma + t_{rlx}J_u\dot{\sigma} = \varepsilon + t_{rlx}\dot{\varepsilon}$.

Experiments with low-molecular weight compounds confirm molecular models for this type of behavior. For chain molecules, however, the curve $J' = f(\omega)$ is more flat than predicted by theory and the maxima at $J'' = f(\omega)$ and $\tan \delta = f(\omega)$ are broader. The reason for this deviating behavior is that polymers do not have *one* relaxation time but a whole spectrum of relaxation times (see p. 589).

17.5.4 Dynamic Moduli of Solid Polymers

Dynamic-mechanical mesurments of polymer properties provide a host of information since such experiments determine moduli at various frequencies, rates, or times and not just at time etc. as static experiments do. They also deliver more realistic test results since polymers are subjected to various stresses, strain rates, and frequencies in daily use.

Dynamic-mechanical experiments deliver relatively small deformations. According to Eq.(16-62) ff., real shear moduli G' should be (approximately) three times as large as real tensile moduli, $E' \approx 3\ G'$ which is indeed found experimentally for low temperatures (Fig. 17-26). With increasing temperature, however, both E' and G' decrease and the gap between them becomes greater ($E' > 3\ G'$).

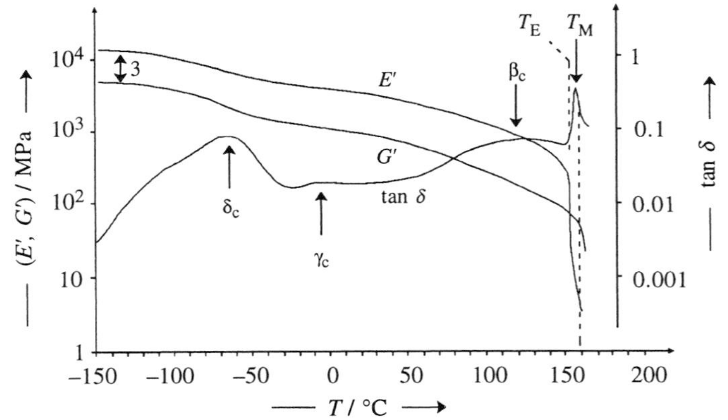

Fig. 17-26 Temperature dependence of tensile storage moduli E', shear storage moduli G', and mechanical loss factors, $\tan \delta$, of an acetal copolymer (= copolymer of trioxane with a small proportion of a cyclic ether) [14b]. Data were obtained with frequencies between 0.3 Hz and 15 Hz. T_M = melting temperature (α_c); β_c, γ_c, δ_c = solid state relaxations (see text; for symbols, see p. 437).

The tensile storage modulus E' drops catastrophically at a temperaure $T_E < T_M$ whereas the shear storage modulus G' does this only at the melting temperature T_M. At this temperature T_E, G' remains relatively high because the many entanglements cause the melt to behave like a physical network (Section 17.4.4). As a result, the mechanical loss factor, tan δ, increases dramatically at T_E and then passes through a maximum at the melting temperature T_M.

Strong effects are not only observed for thermal transitions such as the melting temperature (α_c relaxation) but also for some other relaxations such as δ_c (Fig.17-26). Below the melting temperature of $T_M \approx 159°C$, the loss tangent has a broad maximum at $T_\beta \approx 120°C$ that is caused by the onset of segmental movements in the crystalline regions.

The δ_c relaxation at $T_\delta \approx -65°C$ is caused by the relaxation of chain segments in amorphous regions; it is the dynamic glass temperature at this frequency. The very weak maximum at ca. –5°C is present in off-the-shelf acetal copolymers but disappears on drying. It is probably caused by regions that are plasticized by water.

Relaxations depend very strongly on the applied frequencies as indicated by the loss tangents of an amorphous poly(cyclohexyl methacrylate) (Fig. 13-33) and the relaxation behavior of a semicrystalline poly(ethylene) (Fig. 13-8).

17.5.5 Shear Storage Moduli of Polymer Melts

Shear storage moduli G' of polymer melts vary with circular frequencies ω in an interesting manner (Fig. 17-27). In the so-called end zone at small values of ω, lg G' is directly proportional to lg ω for high molar mass polymers. With increasing frequency, the modulus then becomes independent of the frequency (plateau zone) until finally lg G' is again directly proportional to lg ω (transition zone). The size of the plateau decreases with decreasing molar mass so that low molar mass polymers only show transition zones but not the plateau and end zones.

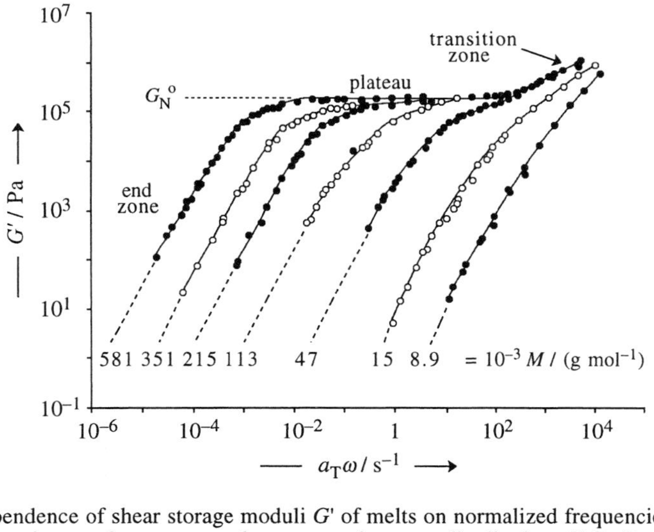

Fig. 17-27 Dependence of shear storage moduli G' of melts on normalized frequencies $a_T\omega$ of poly(styrene)s with various molar masses and narrow molar mass distributions [16]. Data from various temperatures and circular frequencies were recalculated for a temperature of 160°C with the help of the WLF shift factor a_T (see Section 13.5.5). G_N^o = plateau modulus.

In the **transition zone**, values of G' asymptotically join a function $G' = f(a_T\omega)$ that is independent of the molar mass M. Since frequencies $\omega = 1/t$ are high, times t between periods are short so that only short segments can relax completely, which explains $G \neq f(M)$. The transition zone characterizes the viscous behavior of the polymer.

In the **end zone**, the function $G' = f(a_T\omega)$ varies with the molar mass, which can be represented by a function $G' = K_M(a_T\omega)^\gamma$ where both K_M and γ are dependent on the molar mass. For poly(styrene) at 160°C, one finds $\gamma = 3.28 - 1.35 \lg M$, which may or may not be caused by a molar mass dependence of the shift factor a_T. The front factor K_M depends very strongly on the molar mass, $K_M = 5{\cdot}10^{-20} M^5$, which is typical for gels. The low frequencies correspond to large times, i.e., to the relaxation of complete chains. In the end zone, the relaxation spectrum must reflect changes of macroconformations; it indicates viscoelastic properties.

Shear storage values in the end zone are very much affected by the width of molar mass distributions. Reptation theory predicts that shear storage compliances should depend on molar mass averages according to $J_e = C[(\overline{M}_{z+2}\overline{M}_{z+3}\overline{M}_{z+4})/(\overline{M}_w\overline{M}_z\overline{M}_{z+1})]$. Because of the interrelationships between the various molar mass averages (p. 18 ff.), this equation reduces to $J_e = C(\overline{M}_z/\overline{M}_w)^9$ for logarithmic normal distributions. However, mixing experiments found $J_e = C(\overline{M}_z/\overline{M}_w)^3$, which may have been caused by incorrect polymolecularity corrections.

Since the transition zone characterizes the viscous behavior and the end zone the viscoelastic one, the **plateau zone** must characterize the elastomeric properties (cf. Fig. 16-11). In this range, shear storage moduli $G_N{}^o$ are independent of frequency and molar mass if molar mass distributions are narrow. For broad distributions, however, plateaus may not be very well developed; they even may be absent.

The elastomeric behavior of polymer melts, as evidenced by the existence of the plateau modulus $G_N{}^o$, must be caused by *temporary* (physical) crosslinks. Such physical crosslinks seem to be mainly true entanglements but may also be cohesive contacts (see p. 180). The greater the molar mass M_e between two such physical crosslinks, the smaller the stored stress will be and therefore also $G_N{}^o$. The plateau modulus will increase with increasing temperature (the greater the molar energy RT, the greater is the stored stress) and also with increasing density ρ (= mass per volume) of the melt, and, in the case of concentrated solutions, also with increasing volume fraction ϕ_P of the polymer ($\phi_2 = 1$ for melts). The entanglement molar mass is therefore given by

(17-35) $$M_e = QRT\rho\phi_P/G_N{}^o$$

where Q is a numerical factor which is usually assumed to equal unity. It is also often assumed that the entanglement molar mass from $G_N{}^o$ equals that from melt viscosity but experimental data disagree (Table 17-7). Reptation theory maintains that Q reflects fluctuations of tube lengths (Section 14.2.3), which leads to a better relaxation of deformations in dynamic measurements ($Q = 1$ for melt viscosity but $Q = 4/5$ for shear moduli).

There are three different critical molar masses for the "abrupt" change of physical properties of uncrosslinked polymers with molar mass. The critical molar mass M_{excl} from the change of slopes of the function $\lg \langle s^2\rangle^{1/2} = f(M)$ (Fig. 4-23) or $\lg (\langle s^2\rangle/M) = f(M)$ (Fig. 14-4) indicates the onset of excluded volume effects. It differs from the so-called crossover molar mass $M_{e,\eta}$ which indicates the change of exponents in plots of $\lg \eta_o = f(\lg M)$ (see Fig. 14-4) or similar plots (see Fig. 15-5); this critical molar mass is also called a (viscometric) entanglement molar mass. Finally, there is another entanglement molar mass $M_{e,G}$ that is calculated from the plateau modulus with Eq.(17-35).

Table 17-7 Crossover (entanglement) molar masses $M_{e,\eta}$ from Newtonian melt viscosities η_o, entanglement molar masses $M_{e,G}$ from shear storage plateau moduli $G_N{}^o$ (using $Q = 1$), and calculated entanglement densities C_e (experimental data [17-20]), and packing lengths L_p (p. 180). The "atactic poly(propylene)" is a hydrogenated 1,4-poly(2-methyl-1,3-pentadiene) and the poly(vinyl cyclohexane) is a hydrogenated atactic poly(styrene). The poly(isoprene) contained 20% trans units, 75 % cis units, and 5 % 3,4 units. In the literature, $M_{e,\eta}$ is often called M_c and $M_{e,G}$ is called M_e.

Polymer	$\frac{T}{°C}$	$\frac{\rho}{g\ cm^{-3}}$	$\frac{G_N{}^o}{MPa}$	$\frac{M_{e,\eta}}{g\ mol^{-1}}$	$\frac{M_{e,G}}{g\ mol^{-1}}$	$\frac{M_{e,\eta}}{M_{e,G}}$	$\frac{10^{-19}\,C_{e,G}}{cm^3}$	$\frac{L_p}{nm}$
Poly(ethylene)	170	0.768	2.46	3 480	1 150	3.03		0.179
	140	0.785	2.6		1 040		50.4	0.169
	25	0.851	3.5		602			0.137
Poly(propylene), st	190	0.762	1.35		2 170			0.210
it	190	0.766	0.43		6 860			0.312
at	190	0.765	0.42		7 010			0.313
at	140	0.791	0.47		4 620		10.3	
at	25	0.852	0.48		3 520		14.6	
Poly(isobutylene)	140	0.849	0.32		7 390		7.02	
	25	0.918	0.32	15 200	5 700	2.67	9.72	
Poly(butadiene), *cis*-1,4	25	0.895	1.11	6 380	2 000	3.19		0.212
Poly(isoprene)	140	0.849	0.32		7 290		7.02	
	25	0.918	0.40	15 200	5 700	2.67	9.72	0.310
	–30	0.950	0.30	10 000	6 400	1.56		0.300
Poly(styrene), at	270	0.897	0.20		14 020		4.17	0.399
	190			35 000	13 580	2.58		
	140	0.969	0.20		16 640		4.38	0.392
Poly(vinyl cyclohexane)	160	0.920	0.068	68 000	49 000	1.39		0.559
Poly(oxyethylene)	140	0.996	1.20		2850		26.3	0.200
	80	1.081	1.59	5 870	2 000	2.94		0.191
Poly(dimethylsiloxane)	140	0.895	0.20		12 300		4.38	
	25	0.970	0.214	24 500	8 980	2.55	6.30	0.406

The critical molar mass $M_{e,G}$ allows one to calculate the number of entanglements per volume, the **critical entanglement density** $C_{e,G}$:

$$C_{e,G} = N_A\rho/M_{e,G} \qquad (17\text{-}36)$$

The critical entanglement densities seem to be in the range $(1\text{-}50)\cdot10^{19}$ cm^{-3}. According to Eqs.(17-35) and (17-36), the ratio of critical entanglement densities must be $C_{e,G,1}/C_{e,G,2} = [(G_N{}^o)_1T_2]/[(G_N{}^o)_2T_1]$. Experimentally, the ratio $G_N{}^o(1)/G_N{}^o(2)$ at two temperatures 1 and 2 seems to be practically independent of the temperature since it was found for 11 polymers that $G_N{}^o(413\ \text{K})/G_N{}^o(298\ \text{K}) = 0.99 \pm 0.11$. Critical entanglement densities are therefore inversely proportional to the temperature, $C_{e,1}/C_{e,2} = T_2/T_1$.

There have been several attempts to correlate the plateau modulus $G_N{}^o$ (or the entanglement molar mass $M_{e,G}$, see Eq.(17-35)) with molecular data such as end-to-end-distances of unperturbed coil molecules, $\langle r^2\rangle_o$, packing densities p, or combinations of packing lengths L_p and Kuhnian lengths L_K. These correlations have been either empirical, semi-empirical, or guided by theory.

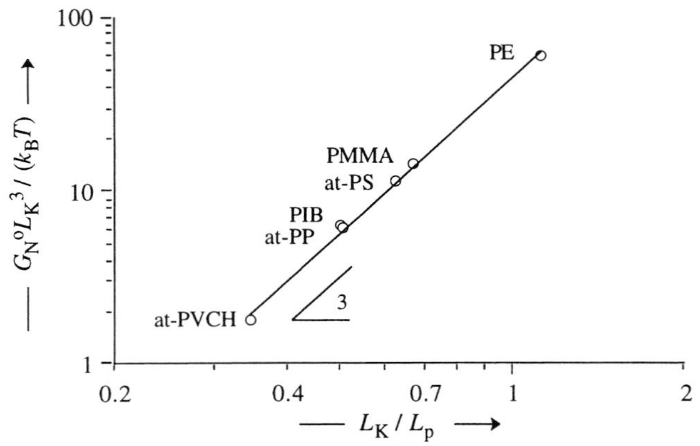

Fig. 17-28 Graessley-Edwards diagram: logarithm of the reduced plateau modulus, $G_N{}^oL_K{}^3/(k_BT)$, as a function of the logarithm of the ratio of Kuhnian and packing lengths, L_K/L_p, for atactic poly(vinyl cyclohexane) (at-PVCH), atactic poly(propylene) (at-PP), poly(isobutylene) (PIB), atactic poly-(styrene) (at-PS), atactic poly(methyl methacrylate) (at-PMMA), and poly(ethylene) (PE) (data from a compilation of [21]).

An example is a plot of the reduced plateau modulus, $G_N{}^oL_K{}^3/(k_BT)$ as a function of the ratio of Kuhnian and packing lengths, L_K/L_p (Fig. 17-28), which delivers a straight line with a slope of 3. This function seems to be independent of temperature.

17.5.6 Shear Storage Moduli of Solutions

The logarithm of the (shear storage) plateau modulus of polymer solutions, $G_N{}^o$, decreases with decreasing logarithm of the volume fraction ϕ_p of the polymer (Fig. 17-28):

(17-37) $$G_N{}^o = (G_N{}^o)_o \phi_p{}^b$$

For each type of polymer, the function lg $G_N{}^o$ = f(lg ϕ_p) seems to be independent of the molar mass. The slope is b = 2.23 for both linear poly(butadiene)s and their hydrogenation products (Fig. 17-29) and b = 2.09 for atactic poly(styrene)s.

The shear storage compliance $J_e{}^o$ in the steady state follows a similar function albeit with a negative slope; for example, b = –2.26 for linear poly(butadiene)s. At very low volume fractions of the polymer, values of $G_N{}^o$ become independent of ϕ_p.

Since shear storage moduli $G_N{}^o$, shear storage compliances $J_e{}^o$, and Newtonian shear viscosities describe different moments of the end zone of relaxation spectra, $H_1(t_{rlx})$,

(17-38) $$G_N^o = \int_{-\infty}^{+\infty} H_1(t_{rlx})\, d \ln t_{rlx}$$

(17-39) $$\eta_o = \int_{-\infty}^{+\infty} t_{rlx} H_1(t_{rlx})\, d \ln t_{rlx}$$

(17-40) $$J_e^o = \eta_o^{-2} \int_{-\infty}^{+\infty} t_{rlx}^2 H_1(t_{rlx})\, d \ln t_{rlx}$$

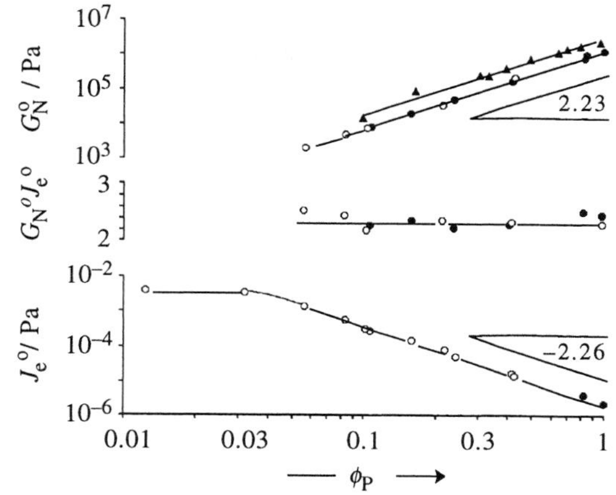

Fig. 17-29 Shear storage plateau modulus $G_N{}^o$, shear storage compliance $J_e{}^o$ in the steady state, and the product $G_N{}^oJ_e{}^o$ of linear poly(butadiene)s with molar masses of 350 000 g/mol (○) or 200 000 g/mol (●) and various hydrogenated poly(butadiene)s (▲) in industrial hydrocarbon oils at 25°C as a function of the volume fraction ϕ_p of the polymers [22a].

the product $G_N^oJ_e^o$ will be a measure of the width of the spectra in the end zone:

$$(17\text{-}41) \qquad G_N^oJ_e^o = \frac{\int_{-\infty}^{+\infty} \tau^2 H_1(\tau)\,\mathrm{d}\ln t_{rlx} \Big/ \int_{-\infty}^{+\infty} \tau H_1(\tau)\,\mathrm{d}\ln t_{rlx}}{\int_{-\infty}^{+\infty} \tau H_1(\tau)\,\mathrm{d}\ln t_{rlx} \Big/ \int_{-\infty}^{+\infty} H_1(\tau)\,\mathrm{d}\ln t_{rlx}} = \frac{\langle t_{rlx,w}\rangle}{\langle t_{rlx,n}\rangle}$$

The product $G_N{}^oJ_e{}^o$ should thus be a constant, which is indeed found experimentally (Fig. 17-29). Numerical values of $G_N{}^oJ_e{}^o$ vary between 2.0 and 2.3; for example, the value of $G_N{}^oJ_e{}^o$ is 2.15 for linear poly(butadiene)s.

Shear storage compliances of linear and branched polymers differ significantly. They can be compared if one uses a reduced form of the shear storage compliance, for example,

$$(17\text{-}42) \qquad J_{e,red} = \frac{\rho_2\phi_2 RTJ_e{}^o}{gM}\left(\frac{\eta_o}{\eta_o - \eta_1}\right)^2$$

where η_o = zero-shear viscosity of the polymer (in melt or solution), η_1 = viscosity of solvent, ρ_2 = density of polymer, ϕ_2 = volume fraction of polymer (ϕ_2 = 1 in the melt), and g = branching factor (Section 4.7). The branching factor $g = (15f - 14)/(3f - 2)^2$ applies to a special distribution of the lengths of branches of star molecules that was not discussed in Section 4.7.1.

In both melts and solutions, the reduced shear compliance, $J_{e,red}$, of linear polymer molecules is proportional to $[M_{e,G}(\phi)/M]^{-1}$. For star molecules, it is independent of this ratio (Fig. 17-30). It is also independent of $[M_{e,G}(\phi)/M]^{-1}$ for very dilute solutions where $J_{e,red}$ = 0.4 which corresponds to the Rouse prediction of $J_e{}^o = (2\,M)/(5\,\rho RT)$ for unbranched polymers in the melt.

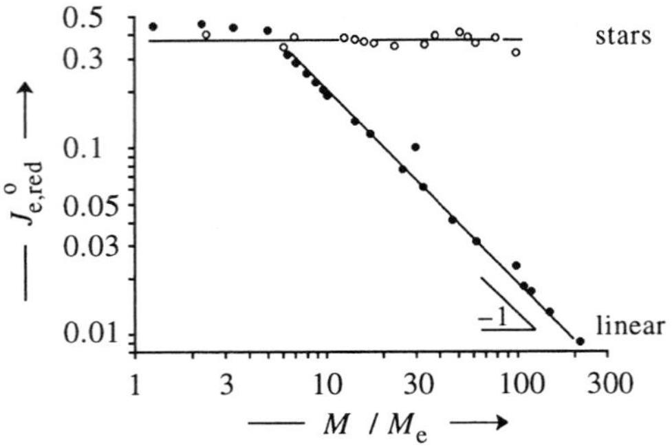

Fig. 17-30 Reduced shear compliance $J_{e,red}$ as a function of $M/M_e\phi_2$ of linear polymers and stars with 3 or 4 arms [22b].

A-17 Appendix: Alternating Deformations

The equation $\varepsilon = \varepsilon_o \sin \omega t$ describes a deformation ε that varies sinusoidally with the product of circular frequency ω (e.g., in rad/s) and time t. The amplitude of such a deformation is ε_o and the deformation rate is $\dot{\varepsilon} = d\varepsilon/dt = \varepsilon_o\omega \cos \omega t$. This equation allows one to calculate the variation of stress with time for the *Maxwell model* as follows.

Equating this expression for $\dot{\varepsilon}$ with the Maxwell equation for the deformation rate, $\dot{\varepsilon} = E^{-1}(d\sigma/dt) + (\sigma/\eta_e)$ (Fig. 17-14, line 4), delivers the differential equation $\varepsilon_o\omega E\cos \omega t = (d\sigma/dt) + \sigma(E/\eta_e)$ which is an equation of the type A = (dy/dx) + By. The general solution of this equation is

$$\text{(A 17-1)} \qquad y \exp\left(\int B\, dx\right) = \int \exp\left(\int B dx\right) A dx + C$$

if A and B depend only on t. The integral $\int$ Bdx will become Et/η_e since $B = E/\eta_e$ and $x = t$. The general equation is therefore

$$\text{(A 17-2)} \qquad \sigma \exp(t/t_{rlx}) = \varepsilon_o\omega E \int \exp(t/t_{rlx}) \cos \omega t\, dt + C$$

$$\text{(A 17-3)} \qquad \sigma \exp\left(\frac{t}{t_{rlx}}\right) = \frac{\varepsilon_o\omega E t_{rlx}}{1+\omega^2 t_{rlx}^2}(\cos \omega t + \omega t_{rlx} \sin \omega t) \exp\left(\frac{t}{t_{rlx}}\right) + C$$

$$\text{(A 17-4)} \qquad \sigma = \frac{\omega t_{rlx}}{1+\omega^2 t_{rlx}^2} \varepsilon_o E (\cos \omega t + \omega t_{rlx} \sin \omega t) \exp\left(\frac{t}{t_{rlx}}\right) + C \exp\left(\frac{-t}{t_{rlx}}\right)$$

where t_{rlx} = relaxation time and C = integration constant. The steady state is reached if $t/t_{rlx} >> 1$. The term $\exp(t/t_{rlx})$ will then become unity and the term $C \exp(-t/t_{rlx})$ will be zero.

The tangent of the phase angle is given by $\tan \vartheta = \sin \vartheta/\cos \vartheta = 1/(\omega t_{rlx})$ and the sine by $\vartheta = (1 + \omega^2 t_{rlx}^2)^{-1/2}$. Substituting these expressions into the trigonometric relationship

(A 17-5) $$\cos \omega t + \omega t_{rlx} \sin \omega t = \frac{\cos \omega t(\sin \vartheta) + \sin \omega t(\cos \vartheta)}{\sin \vartheta} = \frac{\sin(\omega t + \vartheta)}{\sin \vartheta}$$

delivers

(A 17-6) $$\cos \omega t + \omega t_{rlx} \sin \omega t = (1 + \omega^2 t_{rlx}{}^2)^{1/2} \sin(\omega t + \vartheta)$$

Introduction of Eq.(A 17-6) into Eq.(A 17-4) for the steady state leads to an equation for the dependence of stress σ of a *Maxwell body* on circular frequency ω, time t, and phase angle ϑ:

(A 17-7) $$\sigma = \frac{\omega t}{(1 + \omega^2 t_{rlx}^2)^{1/2}} \varepsilon_0 E \sin(\omega t + \vartheta)$$

A similar derivation delivers the corresponding expression for the *Voigt-Kelvin* body:

(A 17-8) $$\sigma = (1 + \omega^2 t_{rlx}^2)^{1/2} \varepsilon_0 E \cos(\omega t - \vartheta)$$

Historical Notes to Chapter 17

Mechanical Models

J.C.Maxwell, Phil.Mag. (IV) **35** (1868) 124

Lord Kelvin, Elasticity, Encyclopedia Brittanica, 9th Ed., London 1875

W.Voigt, Abh.Königl.Ges.Wiss.Göttingen, Math. **36**/1 (1890)

H.Jeffreys, The Earth, Cambridge University Press, Cambridge 1929, p. 265
quoted by R.B.Byrd, C.F.Curtiss, R.C.Armstrong, O.Hassager, in C.Booth, C.Price, eds., Polymer properties (= Volume 2 of Comprehensive Polymer Science (G.Allen, J.C.Bevington, eds.), Pergamon Press, Oxford 1989

J.M.Burgers, Verh.K.Akad.Wet. **16** (1938) 8

Superposition Principle

L.Boltzmann, Poggendorf's Ann.Phys.Chem. **7** (1876) 624

Telescope Effect

Effect of temperature: F.H.Müller, Kolloid-Z. **114** (1949) 59; **115** (1949) 48; **126** (1952) 65

Geometric reasons: P.I.Vincent, Polymer **1** (1960) 7

Flow Criteria

C.A.Coulomb, Mem.Math.Phys. **7** (1773) 343

H.Tresca, C.R.Acad.Sci. [Paris] **59** (1864) 754

R. von Mises, Göttinger Nachr.Math.-Phys. Klasse (1913) 582

Theory of Rate Processes

H.Eyring, J.Chem.Phys. **4** (1936) 283

Literature to Chapter 17

17.1-17.3 GENERAL LITERATURE (see also Chapters 16 and 18)

R.A.Pethrick, R.W.Richards, Ed., Static and Dynamic Properties of the Polymeric Solid State, Reidel, New York 1982

D.W. van Krevelen, Properties of Polymers - Correlation with Chemical Structures, Elsevier, Amsterdam, 3rd ed. 1990

J.Bicerano, Ed., Computational Modeling of Polymers, Dekker, New York 1992

J.Bicerano, Prediction of Polymer Properties, Dekker, New York 1993

N.J.Mills, Plastics - Microstructure & Engineering Applications, Halsted Press, New York, 2nd ed. 1993

L.E.Nielsen, R.F.Landel, Mechanical Properties of Polymers and Composites, Dekker, New York, 2nd ed. 1993

E.Riande, R.Dïaz-Calleja, M.Prolongo, R.Masegosa, C.Salom, Polymer Viscoelasticity, Dekker, New York 2000

Yn-H.Lin, Polymer Viscoelasticity: Basics, Molecular Theories and Experiments, World Scientific, River Edge (NJ) 2003

W.W.Graessley, Polymeric Liquids and Networks, Garland Science, New York 2003; Vol. I: Structure and Properties (2003), Vol. II: Dynamics and Rheology (2006)

I.M.Ward, J.Sweeney, An Introduction to the Mechanical Properties of Solid Polymers, Wiley, New York, 2nd ed. 2004

M.T.Shaw, W.J.MacKnight, Introduction to Polymer Viscoelasticity, Wiley-Interscience, Hoboken (NJ), 3rd ed. 2005

17.4 CREEP AND RELAXATION and 17.5 DYNAMIC DEFORMATIONS

N.G.McCrum, B.R.Read, G.Williams, Anelastic and Dielectric Effects in Polymeric Solids, Wiley, London 1967

R.M.Christensen, Theory of Viscoelasticity: An Introduction, Academic Press, New York 1970

T.Murayama, Dynamic Mechanical Analysis of Polymeric Materials, Elsevier, Amsterdam 1978

J.D.Ferry, Viscoelastic Properties of Polymers, Wiley, New York, 3rd ed. 1980

R.T.Bailey, A.M.North, R.A.Pethrick, Molecular Motion in High Polymers, Clarendon Press, Oxford 1981

R.A.Pethrick, R.W.Richards, Eds., Static and Dynamic Properties of the Polymeric Solid State, Reidel, New York 1982

J.J.Aklonis, W.J.MacKnight, Introduction to Polymer Viscoelasticity, Wiley-Interscience, New York, 2nd ed. 1983

M.Doi, S.F.Edwards, The Theory of Polymer Dynamics, Oxford University Press, Oxford 1986

M.Nagasawa, Ed., Molecular Conformation and Dynamics of Macromolecules in Condensed Systems (First Toyota Conference), Elsevier, Amsterdam 1988

N.W.Tschoegl, The Phenomenological Theory of Linear Viscoelastic Behavior, Springer, Berlin 1989

S.Matsuoka, Ed., Relaxation Phenomena in Polymers, Hanser Gardner, Cincinnati (OH) 1992

S.V.Bronnikov, V.I.Vettegren, S.Y.Frenkel, Kinetics of Deformation and Relaxation in Highly Oriented Polymers, Adv.Polym.Sci **125** (1996) 103

R.S.Lakes, Viscoelastic Solids, CRC Press, Boca Raton (FL) 1998

E.Riande, R.Dïaz-Calleja, M.G.Prolongo, R.M.Masegosa, C.Salom, Dekker, New York 1999

K.P.Menard, Dynamic Mechanical Analysis. A Practical Introduction, CRC Press, Boca Raton (FL) 2000

References to Chapter 17

[1] P.I.Vincent, Encycl.Polym.Sci.Technol. **7** (1967) 292, Figs. 2, 3, and 5. The original paper does not contain numerical data for the lower left graph of Fig. 17-3.
[2] J.Frenkel, Z.Phys. **37** (1926) 572
[3] CAMPUS® data of various industrial companies
[4] E.Rabinowicz, Friction and Wear, Wiley, New York, 2nd ed. (1995)
[5] Y.Nanzai, Progr.Polym.Sci. **18** (1993) 437, Fig. 1c
[6] R.G.C.Arridge, Mechanics of Polymers, Clarendon Press, Oxford 1975, from data of Table 7.1
[7] K.D.Pae, D.R.Mears, J.A.Sauer, J.Polym.Sci. **B** (Polymer Letters) **6** (1968) 773, Table 1
[8] C.S.Henkee, E.J.Kramer, J.Polym.Sci.-Polym.Phys.Ed. **22** (1984) 721, Fig. 8
[9] F.Kloos, Angew.Makromol.Chem. **133** (1985) 1, (a) Figs. 1 and 2, (b) Figs. 3 and 9
[10] G.Meinel, A.Peterlin, J.Polym.Sci. [A-2] **9** (1971) 67, Fig. 7
[11] A.Argon, J.Macromol.Sci.-Phys. **B 8** (1973) 373; see R.N.Haward, Macromolecules **26** (1993) 5860
[12] O.A.Hasan, M.C.Boyce, Z.S.Li, S.Berko, J.Polym.Sci.-Polym.Phys. **B 31** (1993) 185, Fig. 5
[13] C.Bauwens-Crowet, J.C.Bauwens, G.Homès, J.Polym.Sci. [A-2] **7** (1969) 735, Fig. 1
[14] H.Domininghaus, Plastics for Engineers, Hanser, Munich 1993, (a) modified Fig. 212, (b) Fig. 211 plus other literature data
[15] J.R.McLoughlin, A.V.Tobolsky, J.Colloid Sci. **7** (1952) 555, data of Fig. 5
[16] S.Onogi, T.Masuda, K.Kitagawa, Macromolecules **3** (1970) 109, selective data of Fig. 2
[17] L.J.Fetters, D.J.Lohse, D.Richter, T.A.Witten, A.Zirkel, Macromolecules **27** (1994) 4369, Tables 1 and 2
[18] J.Roovers, P.M.Toporowski, Rubber Chem.Technol. **63** (1990) 734
[19] L.J.Fetters, D.J.Lohse, W.W.Graessley, J.Polym.Sci.-Polym.Phys. **B 37** (1999) 1023 Tables I-IV
[20] L.J.Fetters, D.J.Lohse, S.T.Milner, W.W.Graessley, Macromolecules **32** (1999) 6847, Tables I and II
[21] Data from compilations of W.W.Graessley, Polymeric Liquids and Networks, Structure and Properties, Garland Science, New York 2004, Tables 4.2 and 9.3. The plot was suggested by W.W.Graessley, S.F.Edwards, Polymer **22** (1981) 1329
[22] V.R.Raju, E.V.Menezes, G.Marin, W.W.Graessley, L.J.Fetters, Macromolecules **14** (1981) 1668, (a) Table III, (b) Fig. 15

18 Fracture

18.1 Introduction

18.1.1 Definitions

Failure is defined as the loss of useful properties for an intended application. It can take many forms that range from yielding (Chapter 17) to actual fracture. The fracture behavior of polymers depends on their chemical and physical structure; the temperature, environment, and load; and the type, duration, and frequency of deformation. At certain conditions, some polymers break instantly whereas others survive for months or years. The fracture surface can be smooth or rough and the elongation at fracture less than 1 % or more than 1000 %.

There are two borderline types of fracture. In **brittle fracture**, the specimen breaks at a right angle to the draw direction whereas in **tough fracture** it fractures in the direction of shear stress (Fig. 18-1). By convention, a polymer is brittle if its elongation at break is less than 20 % (Europe) or less than 10 % (US).

A true brittle fracture of a perfect solid such as an atomic crystal is very difficult because it comprises the simultaneous breaking of many bonds. However, most of the crystals are not perfect since they contain lattice voids, interstices, and other imperfections that act as nucleating entities for the formation of microcracks. The growth of these microcracks leads to fissures and finally to a catastrophic failure of the crystal.

Brittle polymers practically always contain "natural" microvoids. Additional microvoids result from the formation of crazes or shear bands (Fig. 16-3) and/or the separation of lamellae or spherulites of semicrystalline polymers.

The fracture of ductile polymers involves slip processes that are caused by viscous flow and include the gliding of chain segments (in unoriented polymers) and in crystalline regions (in semicrystalline polymers). Long deformation times may also lead to a dissolution of chain entanglements, even at small stresses. Although yielding of a thermoplastic by formation of shear bands or crazes is a failure of the specimen as far as working materials are concerned, it is also a very useful property for packaging films.

The same processes and several other ones are involved in the failure of fiber-reinforced plastics (Volume IV) where thermoplastic polymeric matrices can fracture in a brittle or ductile manner or form shear bands. Fibers of these composites may bend or form kinks. Fiber-reinforced thermosets may also fail because of the formation of kink bands but such composites may also delaminate or form steps on compression.

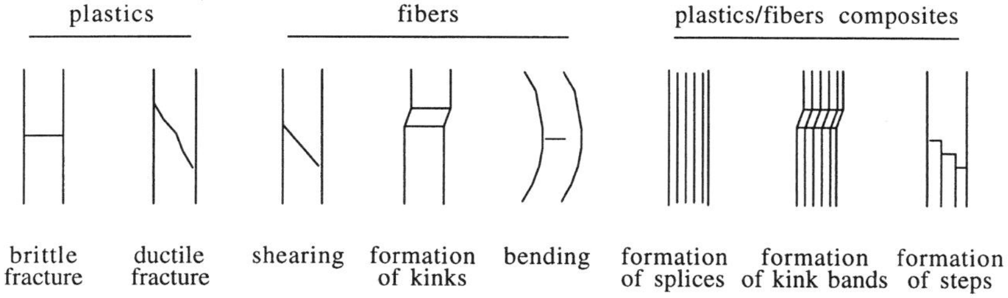

Fig. 18-1 Types of failure of plastics, fibers, and composites of plastics and fibers.

18.1.2 Test Methods

The various types of failure are responses to different types of experiments. Industry uses mainly relatively simple test methods that try to simulate one or more real (complex) types of stresses by using simplified test conditions. By necessity, these test methods have to be standardized (Section 16.1). Most common are short-term tests, for example, tensile tests with constant drawing rates.

These short-term tests are not as revealing as long-term tests of **creep (rupture) strength** in which very many test specimens of the same polymer are subjected to various loads which correspond to tensile strengths σ. The times t_B are recorded at which these specimens break. The logarithm of tensile strength of the specimen (which is the original load), σ_B, is then plotted against the time t_B at which the specimen fractured (Fig. 18-2).

Amorphous polymers such as UP-GF, SAN, and PS usually show a linear decrease of lg σ_B as a function of time to break, t_B. The slopes of the lines depend very much on the type of polymer. Semicrystalline polymers such as poly(ethylene) often show a break in the dependence of σ_B on t_B which is caused by the transition from viscous fracture at small t_B to brittle fracture at large t_B (see Fig. 18-2).

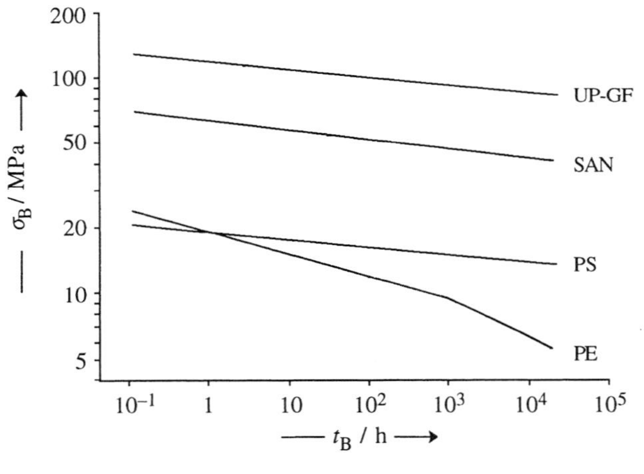

Fig. 18-2 Time dependence of tensile creep rupture strengths of a glass-fiber reinforcd unsaturated polyester (UP-GF), a styrene-acrylonitrile copolymer with high impact strength (SAN), an atactic poly(styrene) (PS), and a semicrystalline poly(ethylene) [1a]. Experimental conditions were not specified. Note that the longest time of $t_B \approx$ 20 000 h corresponds to 833 days, i.e., 2-1/4 years.
By kind permission of BASF, Ludwigshafen.

Alternatively, one can also subject specimens to different loads and then test their tensile behavior at various times. These tests require a great number of initial specimens and cover months to years. For example, if three specimens each are tested each month with five different loads and the test series is continued for two years, a total of 3·5·24 = 360 specimens are initially required!

An example is the creep rupture test of carbon black-filled poly(ethylene) tubes at various temperatures (Fig. 18-3). Transitions from ductile to brittle failure happen earlier, the higher the temperature but well below the melting temperature of this semicrystalline polymer.

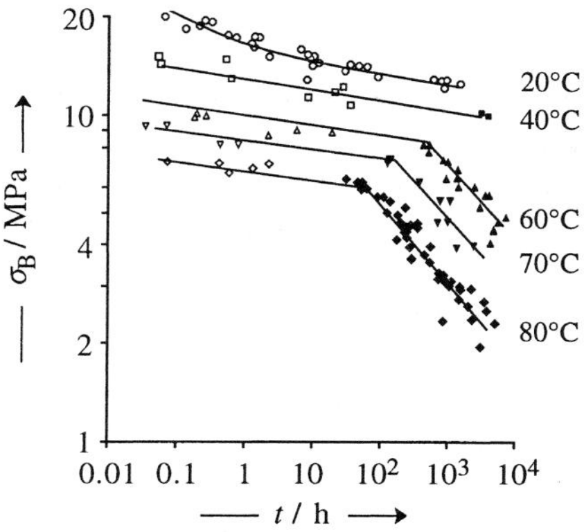

Fig. 18-3 Results of creep rupture tests: tensile strength as a function of time to rupture tubes composed of a carbon black-filled poly(ethylene [1b]. Dark symbols indicate brittle fracture.
By kind permission of BASF.

Another type of testing subjects the specimen to periodic loading where the flexural strength at break, σ_B, is measured as a function of the number N of loadings (Fig. 18-4). The decline of flexural strengths at break with the logarithm of loadings, N, is known as the **Wöhler curve** since it was observed first more than 120 years ago by August Wöhler for the failure of railroad tracks. For example, such tests revealed for flexural stresses of 40 MPa that common atactic poly(styrene)s break after 300 loadings whereas so-called impact poly(styrene)s (= copolymers of styrene and acrylonitrile (SAN)) survive one million loadings.

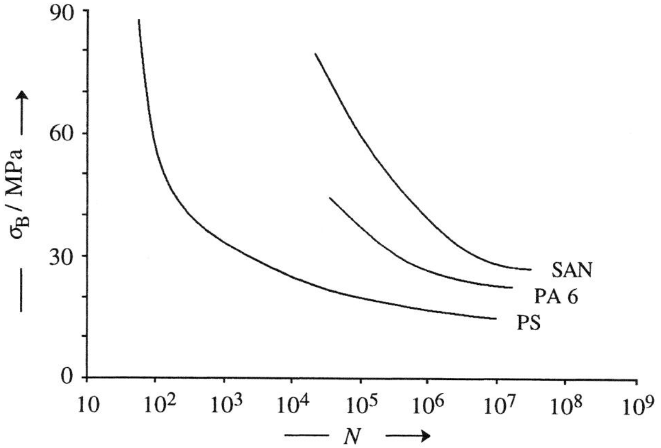

Fig. 18-4 Flexural strength at break as a function of the number of loadings for a styrene-acrylonitrile copolymer (SAN), a conditioned polyamide 6 (PA 6), and an atactic poly(styrene) [1a]. By kind permission of BASF, Ludwigshafen.

18.1.3 Effect of Molar Mass

Theories of tensile strength and impact strength usually apply only to infinitely large macromolecules in perfect polymer crystals or to completely disordered random coils in amorphous polymer glasses. Such ideal physical structures have ideal stress distributions. Real polymers, however, are always disordered, in part for chemical reasons (low-molar masses, endgroups, tacticity) and in part for non-ideal physical structures (chain orientations, lattice disorders, different morphologies, "natural" microcracks, etc.).

Polymers with low-molar masses have diminishingly small tensile strengths at break and very small fracture elongations (Fig. 18-5). Above a certain critical molar mass, fracture strengths increase linearly with increasing number-average molar mass. These critical molar masses are highest for compression molded specimens, lower for injection-molded ones with narrow molar mass distributions (MMD), and lowest for injection-molded ones with broad MMD. The lower critical molar masses of injecteion-molded specimens are not surprising since injection molding leads to orientation of chain segments (as indicated by interference colors) and therefore greater strength. The critical molar mass for the onset of strength in these injection-molded specimens with broad molar mass distributions ($M_{e,\eta} \approx 40\ 000$ g/mol) corresponds to the critical molar mass for entanglements from Newtonian viscosities (Table 17-7).

Above this lower critical molar mass, fracture strengths increase until they fairly suddenly become independent of molar mass. This second, higher critical molar mass is (almost?) independent of the type of specimen preparation (injection and compression molded). This behavior is found for both amorphous thermoplastics (Fig. 18-5) and crosslinked rubbers (Fig. 18-6).

At the same number-average molar mass, fracture strengths are larger, the broader the molar mass distribution (Fig. 18-5): polymers with broader distributions have proportionally more longer chains, hence a greater likelihood to entangle. If fracture strengths are plotted as a function of $(\overline{M}_n + \overline{M}_w)/2$ instead of $\overline{M}_n$, a single curve is obtained for injection molded specimens and another one for compression molded ones (not shown). These two functions, $\sigma_B = f[(\overline{M}_n + \overline{M}_w)/2]$, have a common intersection with the abcissa at $M_o = [(\overline{M}_n + \overline{M}_w)/2]_{\sigma=0} \approx 65\ 000$ g/mol.

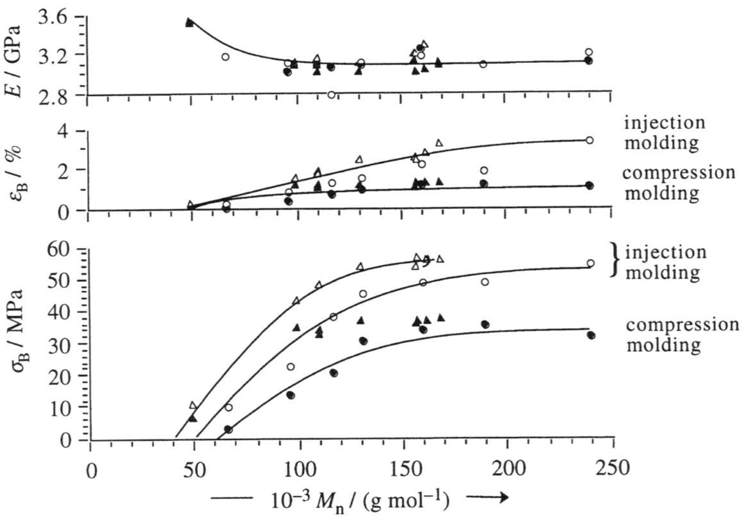

Fig. 18-5 Tensile moduli E, fracture elongations ε_B, and tensile strengths at break, σ_B, of compression-molded (●,▲) and injection-molded (O,Δ) specimens at 23°C and 50 % relative humidity as functions of the number-average molar masses of at-poly(styrene)s with narrow (O,●; $\overline{M}_w / \overline{M}_n$ = 1.1 ± 0.1) or broad (Δ;▲; $\overline{M}_w / \overline{M}_n$ = 2.2 ± 0.4) molar mass distributions [2].

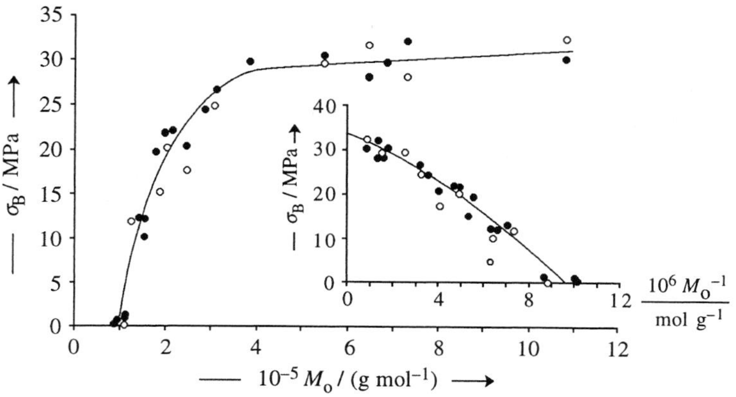

Fig. 18-6 Dependence of tensile strength at break, σ_B, of crosslinked butyl rubbers (= copolymers of isobutene and ca. 4 % isoprene) on the molar mass M_o of primary molecules before crosslinking [3a]. The primary molecules were fractions of the uncrosslinked copolymer that were crosslinked (vulcanized) in such a way that network chains had the same average molar mass of M_c = 37 000 g/mol. The function $\sigma_B = f(M_o)$ is not affected by the vulcanization time of 30 min (O) or 60 min (●). Insert: Dependence of σ_B on $1/M_o$ [3b].

Similarly to thermoplastics (Fig. 18-5), elastomers (Fig. 18-6) have noticeable fracture strengths σ_B only above a critical molar mass of their primary chains ($M_o \approx$ 100 000 g/mol), which in this case is about 2.7 times greater than the average molar mass of network chains between crosslinks (M_c = 37 000 g/mol). The subsequent increase of σ_N with increasing M_o is usually described by a function $\sigma_B = \sigma_{B,o} - B\overline{M}_n^{-1}$ but is represented better by a function $\sigma_B = (\sigma_B)_o - B\overline{M}_n^{-1} - C\overline{M}_n^{-2}$ (insert of Fig. 18-6). Both the linear and the quadratic function have the same correlation coefficient of 0.961.

Because of the alleged dependence of σ_B on the inverse molar mass, it is often assumed that the decrease in fracture strength with decreasing molar mass is caused by endgroups. Because of the relatively high critical molar masses for the onset of $\sigma_B = f(M) > 0$, one would rather suspect that polymers have fracture strengths only if there are more than 2-3 crosslinks per chain, either chemical ones as in elastomers ($M_o/M_c \approx 2.7$ for crosslinked butyl rubbers) or physical ones (entanglements) for thermoplastics ($M_o/M_{e,\eta} > 2$ for thermoplastic poly(styrene) at 25°C).

18.1.4 Stress Cracking

The simultaneous action of mechanical forces and chemicals on metals or plastics may lead to damage that is often called **stress-induced corrosion**. For plastics, this "corrosion" usually consists of the formation of small cracks on the surface of the specimen. Since chemical reactions usually play only a small role in this process or none at all, the phenomenon is better called **stress cracking**. Damage by stress cracking is especially important for stressed bottles, tubes, cables, etc. that are in contact with chemical agents, especially surface-active ones.

The onset and extent of stress cracking depends on the polymer/agent interaction and on the magnitude of stress or deformation, respectively. For bisphenol A polycarbonates (PC) at stresses of less than 10 MPa or elongations of less than 5 %, no damage is observed for long times. PCs are also not damaged if unstressed specimens are submerged in mixtures of toluene and 1-octane. However, if one dips the polymer into a toluene/octanol mixture and then applies a stress of 10 MPa, crazes appear within minutes. The crazes becomes cracks and the specimen disintegrates.

The effects are sometimes baffling. Containers made from poly(methyl methacrylate) (PMMA) can be filled with alcohol or cleaned in dishwashers without damage. However, stress cracking appears if alcohol is poured into PMMA containers freshly out of dishwashers. The reason is that PMMA absorbs up to 2 % water on prolonged washing at elevated temperatures. The subsequent treatment with alcohol extracts this water, which leads to a concentration gradient which in turn generates stresses that are dissolved by the formation of crazes.

The extent of stress cracking depends on the environment. Effects are small for non-wetting agents, where stress cracking proceeds in three phases. First, defects grow to visible crazes that then deepen to a limiting value. Finally, the material strengthens again.

Such limiting depth of crazes also exists for wetting agents, where crazes grow to cracks and the material swells and finally disintegrates. Depending on the solubility parameter (a measure of the thermodynamic goodness of the solvent (Section 10.1.3)), the disintegration requires a certain critical elongation which corresponds to a critical tensile stress (Fig. 18-7).

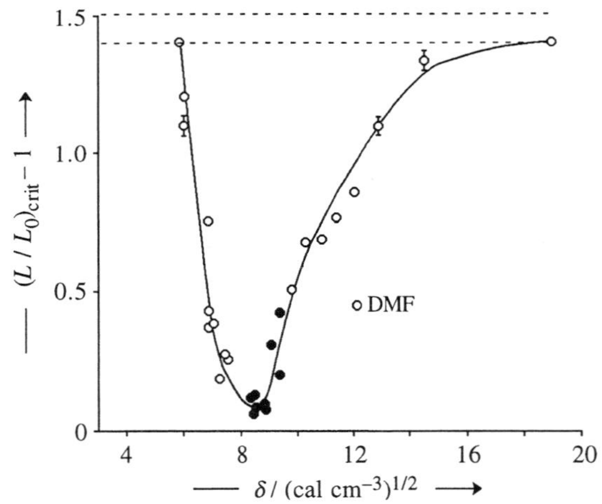

Fig. 18-7 Critical elongations for the formation of crazes (○) or cracks (●) in a poly(2,6-dimethyl-1,4-phenylene oxide) as a function of the solubility parameter δ [4]. - - - - Upper and lower limits for fracture elongations in the absence of solvents. 1 (cal/cm^3)$^{1/2}$ = 2.05 (J/cm^3)$^{1/2}$.
With kind permission by American Chemical Society, Washington (DC).

The lowest critical elongations are observed for those solvents that have the same solubility parameter as the polymer. Such solvents always swell the polymer but do not necessarily dissolve it. On elongation, the resulting deformations are subjected to a concentration gradient which leads to stresses that are then resolved by stress cracking.

The susceptibility to stress cracking decreases with increasing molar mass of the polymer. At higher molar masses, polymer molecules have more entanglements whose elastic response to stresses leads to relaxations. For this reason, chemically crosslinked elastomers show far less stress cracking than thermoplastics, and thermosets have practically none at all. Addition of plasticizers increases the mobility of chain segments, which leads to equilibration of stresses and reduced stress cracking. The mobility of chain segments is also the reason why stress cracking is absent at temperatures above T_G.

18.2 Fracture Strengths

18.2.1 Introduction

Most common thermoplastics have tensile strengths at break between 20 MPa and 100 MPa despite very different chemical and physical structures. Fracture strengths of fibers are generally higher and those of elastomers usually lower.

For example, at temperatures of 23°C = $T << T_G \approx$ 100°C, atactic poly(styrene) (PS) is a rigid-brittle amorphous polymer that fractures perpendicularly to the drawing direction without significant elongation at break (ε_B < 1.4 %). Crosslinked butyl rubber (PIB), on the other hand, is a soft-elastic polymer that can be drawn to several hundred percent at temperatures $T >> T_G \approx -65$°C before it ruptures. Yet, despite their very different fracture elongations, both polymers have practically the same fracture strength in the limit of very high molar masses: σ_B = 35.7 MPa for compression molded PS (Fig. 18-5) and σ_B = 35.8 MPa for crosslinked PIB (Fig. 18-6). Semicrystalline poly(ethylene) (PE) at room temperature has practically the same tensile strength at break, $\sigma_B \approx$ 33 MPa, although it can also be drawn to several hundred percent. All these polymers consist of carbon chains of the type $\text{-[-CH}_2\text{-CR'R''-]}_n\text{-}$ with R', R" = H (PE); R' = R" = CH_3 (crosslinked PIB); or R' = H and R" = C_6H_5 (atactic PS).

Neither crystallinity nor chemical or physical crosslinking can therefore be the main factor for similar magnitudes of tensile strengths at break of unoriented polymers. One rather may suspect that similar fracture strengths are caused by similar chemical structures, i.e., similar energies for the breaking of carbon-carbon bonds of main chains. This energy can be estimated as follows.

The breaking of a C–C bond of a carbon chain with the cross-sectional area A_c requires a force $\boldsymbol{F}_c$. The bond is severed when the distance L_b between carbon atoms becomes ca. 30 % greater than the equilibrium distance, the crystallographic length c (Section 17.2.2), i.e., $L_b = 0.3\ c$. The required force for bond separation is $\boldsymbol{F}_e = E_c/L_b = E_m/(N_A L_b)$ where E_m = molar bond energy (E_m = 348 kJ/mol for carbon chains).

The cross-sectional area of a chain is $A_c = ab/N_u$ for a rectangular unit cell with a, b = lattice constants and N_u = number of chains per unit cell. In the chain direction, the maximum tensile strength of a carbon chain is therefore $\sigma_{||} = (N_u E_m)/(0.3\ c N_A ab)$. A poly(ethylene) chain with a = 0.742 nm, b = 0.495 nm, c = 0.254 nm, and N_u = 2 (Table 7-4) should therefore have a maximum fracture strength of $\sigma_{||}$ = 41.2 GPa.

This theoretical fracture strength in the draw direction is about 1000 times greater than the tensile strength at break of amorphous or semicrystalline polymers with carbon chains. Hence, it was concluded that such polymers *a priori* contain many defects at which stresses become concentrated: the higher the stress concentration, the lower the tensile strength. This scenario led to the adoption of Griffith's fracture theory that was first derived for silicate glasses and then applied to other brittle materials (next section).

18.2.2 Fracture of Brittle Polymers

Griffith observed in 1920 that small glass beads have higher strengths than bigger ones. He concluded that this phenomenon must be caused by natural defects in the glass. Such defects should be randomly distributed in the beads and the defects should have a size distribution. A large glass bead will therefore contain more of the larger defects than a smaller bead. Since fracture starts first at bigger defects, bigger beads will have less strength than smaller beads. The same reasoning applies to the effect of fiber diameter on the strength of fibers (Fig. 16-5).

However, randomly distributed small defects can lead only to macroscopic fractures if applied stress forces them to become larger and finally to congregate to one large hole. Alternatively (or additionally), applied stress may cause the development of new defects that finally become so large in number that the specimen breaks.

Two conditions must be fulfilled if defects are to grow to macroscopic holes. (A) The energy of the system must decrease if defects grow and (B) the stress at fracture must equal or be greater than the theoretical strength.

Condition (B) had already been investigated by Inglis in 1913 for different geometric shapes of defects (holes, cracks, etc.). Stresses around a defect are not equal; they are rather much greater at the tip of a crack or at the point of highest curvature of a non-round hole than stresses at other positions at the surface of the defect. The defect thus acts as a **stress concentrator**. The ratio of the maximum stress and the average stress at the defect is called the **stress concentration factor** $\boldsymbol{K_t}$.

If a plate with an elliptical hole is stressed perpendicular to the main axis of the ellipse, the stress concentration factor at the tip of the ellipse will be

$$(18\text{-}1) \qquad K_t = \sigma_{max}/\sigma_o = 1 + 2\,(a/R)^{1/2} \qquad \textbf{Inglis equation}$$

where σ_{max} = maximum stress, σ_o = applied stress, 2 a = length of the main axis, and R = curvature radius at each end of the main axis (Fig. 18-8, I). The same stress concentration factor is present at a semielliptical nick with a depth a (Fig. 18-8, II). For a wedge-like nick, the half-length a will be much greater than the curvature radius R and the stress concentration factor becomes $K_t \approx 2\,(a/R)^{1/2}$ (Fig. 18-8, III).

A body under unidirectional tension is elastically stressed; it contains elastic deformation energy. This energy is changed if already existing defects grow. The growth of defects also leads to larger surfaces of defects. The work W required to enlarge these surfaces is provided by the applied stress, which leads to a change of the elastically stored energy by an amount U. The difference of energies, $W - U$, is available for the formation of new surfaces. Hence, a defect can grow only if $W - U \geq 0$.

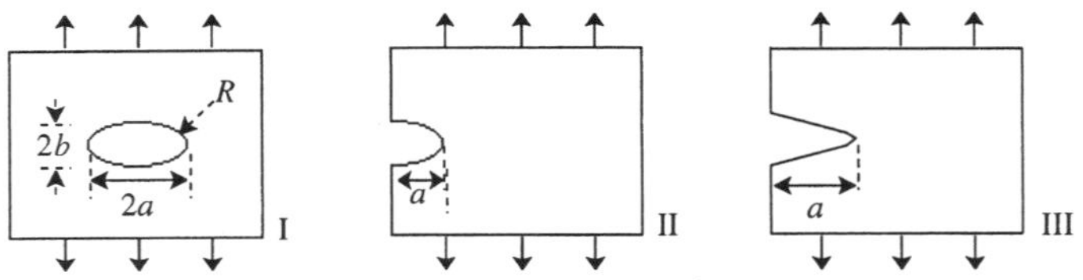

Fig. 18-8 Plates with an elliptical hole (left), a half-elliptical nick (center), and a sharp wedge-like nick (right). $2a$ = length of major axis, $2b$ = length of minor axis, R = radius of curvature. Plates have a thickness d. ↑,↓ direction of force.

The work $W = O\gamma$ is calculated from the surface area O of the elliptical hole and the surface energy E_S per area A, i.e., the surface tension $\gamma = E_S/A$. The circumference of an ellipse is $C = 4\,a\mathrm{E}$ where a = large half-axis and E = elliptic integral which becomes approximately unity if a is much greater than the small half-axis b. The work for a plate with thickness d is therefore $W = 4\,ad\gamma$.

The lowering of the elastic energy U of a very large plate with an elliptical hole relative to that of the intact plate is given by $U = -\,\pi a^2 d\sigma^2/E^\#$ according to Inglis where σ = stress and $E^\#$ = reduced modulus of elasticity (see below). If the hole grows in the direction of its long axis, the change of total energy will be

$$\text{(18-2)} \qquad \frac{\mathrm{d}(W-U)}{\mathrm{d}a} = \frac{\mathrm{d}}{\mathrm{d}a}\left(4ad\gamma - \frac{\pi a^2 d\sigma^2}{E^\#}\right) \leq 0 \quad ; \quad 4\,d\gamma - \frac{2\pi a d\sigma^2}{E^\#} \leq 0$$

A very long elliptical hole of length 2 a or a surface crack of depth a can therefore only grow in a very large plate if the applied constant stress exceeds a certain value:

$$\text{(18-3)} \qquad \sigma > \left(\frac{2E^\#\gamma}{\pi a}\right)^{1/2} \qquad \textbf{Griffith equation}$$

The same condition also applies to constant deformation.

The stress σ in Eq.(18-3) is a critical tensile strength σ_B that depends on the square roots of the so-called **reduced modulus** $E^\#$, surface tension γ, and large half-axis a. The reduced modulus equals the tensile modulus for thin plates with plane stress, $E^\# = E$, but becomes $E^\# = E/(1-\mu)^2$ for plane strain (two-dimensional elongation of thick plates) where μ = Poisson's ratio.

The Griffith theory describes well the behavior of silicate glasses because they are practically 100 % energy-elastic. The predicted dependence of tensile strength at break, σ_B, on the square root of the crack length a is also found for poly(styrene)s that have been provided with artificial surface cracks (Fig. 18-9).

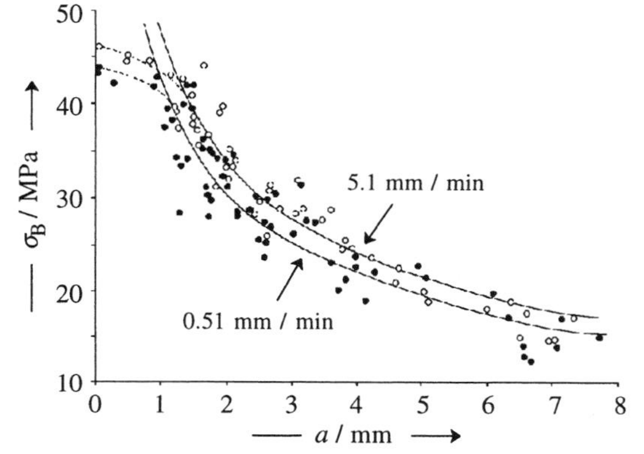

Fig. 18-9 Dependence of tensile strengths at break, σ_B, on the length a of artificial surface cracks in poly(styrene) rods with square cross-sections between 0.3 cm × 0.5 cm and 2.8 cm × 0.5 cm, respectively, that were drawn with velocities of 0.51 mm/min (●) and 5.1 mm/min (O), respectively [5]. Solid lines correspond to the prediction of the Griffith theory, $\sigma_B = K(E^{1/2})$, using empirical proportionality constants K. Deviations from Griffith's theory occur at $a < 1$ mm.

Table 18-1 Experimental tensile moduli E, critical surface tensions γ_{crit}, tensile strengths σ_B, and fracture elongations ε_B of commercial polymers at 20°C after molding (M) or processing to high modulus fibers (F). "Natural crack lengths" a were calculated from these data using Eq.(18-3).

Polymer		$\frac{E}{\text{GPa}}$	$\frac{\gamma_{crit}}{\text{mJ m}^{-2}}$	$\frac{\sigma_B}{\text{MPa}}$	ε_B	$\frac{a}{\text{nm}}$
Poly(styrene), at	M	3.00	34	40	0.025	40.6
Poly(methyl methacrylate), at	M	3.3	41	70	0.06	17.6
Poly(vinyl chloride), at	M	2.70	39	49	0.15	27.9
Poly(oxymethylene)	M	3.7	36	70	0.45	33.6
Polyamide 6.6	M	2.50	46	74	< 1.00	13.4
Polycarbonate A	M	2.3	43	60	1.10	17.5
Poly(tetrafluoroethylene)	M	0.48	18.5	24	3.0	9.8
Poly(ethylene terephthalate)	M	0.13	43	54	3.0	1.2
Poly(ethylene), low density	M	0.15	33	23	9.0	6.0
Wool (protein fiber)	F	4	45	200	0.41	2.9
Poly(acrylonitrile)	F	5	50	280	0.35	4.0
Poly(ethylene terephthalate)	F	18.5	43	1400	0.20	0.26
Polyamid 6.6	F	12.5	46	1000	0.17	0.36
Cotton (cellulose fiber)	F	15	42	500	0.14	1.60
Hemp (cellulose fiber)	F	29	42	850	0.02	1.07
Poly(oxymethylene)	F	38.4	36	2000	0.05	0.69

However, higher drawing speeds produce higher fracture strengths, which indicates that the deformation of poly(styrene) iss not completely energy-elastic. Furthermore, fracture strengths become almost independent of crack lengths for crack lengths smaller than $a \approx 1$ mm, which led some authors to conclude that such lengths are the lengths of natural cracks (or, more precisely: flaws). However, ocular and microscopic inspection of specimens show that natural cracks must be much smaller than 1 mm.

The length of natural cracks can be calculated with the Griffith equation, Eq.(18-3), using $E^{\#} = E$. Tensile moduli E and tensile strengths σ_B are obtained from stress-strain diagrams. The surface tension γ of Eq.(18-3) is replaced by the critical surface tension for the following reason.

The surface tension γ of perfect brittle materials equals the interfacial tension $\gamma_{sv}{}^{o}$ between the solid surface (s) and the surrounding vapor (v). In equilibrium, the interfacial tension, $\gamma_{sv}{}^{o} = \gamma_{sv} - \Pi_{eq}$, is given by the difference between the measured interfacial tension, γ_{sv}, and the spreading pressure Π_{eq} of the saturated solvent vapor on the surface of the solid (Section 9.2). Π_{eq} becomes zero for diminishingly small contact angles; hence, interfacial tensions γ_{sv} about equal the critical surface tensions γ_{crit}.

Equation $a = (2\,E\gamma_{crit})/(\pi\sigma_B{}^2)$ thus allows one to calculate "natural crack lengths" a from experimental data of σ_B, E, and γ_{crit} (Table 18-1). All these crack lengths are in the nanometer range. For molding materials, they decrease with increasing fracture elongation, i.e., for greater contributions of flow processes. For highly stretched fibers, crack lengths approach the bond lengths of covalent bonds (0.154 nm for C–C bonds). Hence, it does not make much sense to talk about *surface* tensions because one can talk only of macroscopic surfaces if the longest length of the phase is a multiple of atomic distances, i.e., if that length is ca. 5 nm or more.

18.2.3 Initiation of Fracture

The presence of natural flaws (cracks) as the sole reason for the low fracture strengths of conventionally processed polymers is not very convincing because yield strengths of polymers almost reach theoretical values (Table 17-1; $\sigma_Y/\sigma^\phi \geq 0.55$). Much more important seems to be the influence of morphology.

Theoretical fracture strengths apply to the breaking of covalent bonds of chains with chain axes completely in the draw direction whereas only a fraction of segments are oriented that way in compression molded thermoplastics. Indeed, fracture strengths increase with the degree of orientation; drawn poly(ethylene) fibers have ca. 9 times greater fracture strengths than their compression-molded counterparts (ca. 270 MPa *vs.* 32 MPa). The fracture strength of ultra-stretched fibers from poly(ethylene)s with ultrahigh molar masses approaches ca. 4000 MPa.

In conventionally processed polymers, chain axes of segments are more or less randomly oriented, which causes local stresses to vary with the degree of orientation of chain segments. For example, high stresses in ductile polymers can be reduced by flow processes, which is not possible for rigid polymers with little mobility of chain segments where stresses can be resolved only by breaking chain bonds.

The fracture process then proceeds as follows. Isotactic poly(propylene), for example, has an infrared absorption peak at 976 cm^{-1}. On application of a stress of 500 MPa, this peak becomes asymmetric and shifts to 974 cm^{-1}. Hence, the stress distribution becomes asymmetric since ca. 90 % of all chain bonds are now deformed. Simultaneously, a new band appears at ca. 955 cm^{-1}. Ca. 10 % of all deformed bonds must therefore be under considerably higher stress than the others. The maximum stress is therefore ca. 10 times greater than the average one.

Thermal fluctuations about rest positions of chains increase stresses. Covalent bonds break if the stress at these bonds surpasses a certain critical value. The bonds fracture homolytically so that the two new chain ends carry one primary radical each. According to ESR measurements, the number concentration of such radicals is ca. $C_{rad} \approx 10^{14}$ radicals per cm^3 for unstretched polymers and ca. 10^{16} radicals/cm^3 for stretched ones. Since the number concentration of chains is $C_{chain} = N_A\rho/M \approx 10^{-18}$ cm^{-3} for a polymer with $M = 10^6$ g/mol and density $\rho = 1$ g/cm^3, about 0.01-1 % of all chains are broken.

The primary radicals react with adjacent chains in a series of chain transfer reactions

(18-4)

$$\sim CH_2{-}CH_2{-}CH_2{-}CH_2\sim \;\longrightarrow\; \sim CH_2{-}\dot{C}H_2 + \dot{C}H_2{-}CH_2\sim$$

$$\sim CH_2{-}\dot{C}H_2 + \sim CH_2{-}CH_2{-}CH_2\sim \;\longrightarrow\; \sim CH_2{-}CH_3 + \sim CH_2{-}\dot{C}H{-}CH_2\sim$$

$$\sim CH_2{-}\dot{C}H{-}CH_2{-}CH_2{-}CH_2\sim \;\longrightarrow\; \sim CH_2{-}CH{=}CH_2 + \dot{C}H_2{-}CH_2\sim$$

which cause more and more chains to break whereas the concentration of primary radicals remains constant. The newly formed chain ends are stabilized as vinyl or methyl groups. According to IR data, one primary radical leads to 10^3-10^4 new endgroups.

Small-angle X-ray measurements have shown that these processes lead to disc-like nanocracks of ca. 20 nm width and ca. 10 nm length in the draw direction in concentrations of ca. $C_B \approx 10^{-16}$ nanocracks per cm^3. The number concentration C_B of nanocracks is therefore about as large as the number concentration C_{rad} of primary radicals.

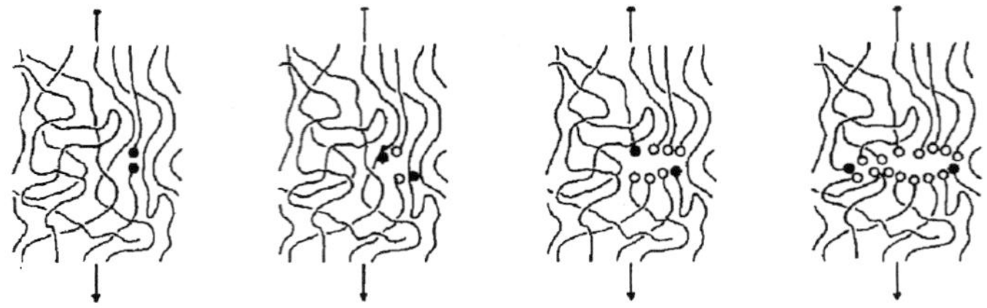

Fig. 18-10 Formation of microcracks through the generation of primary radicals ● at a stressed chain and subsequent chain transfer of radicals with formation of new endgroups O [6].

According to these experiments, microcracks are formed as shown in Fig. 18-10. It does not seem to be known which morphology favors the formation of the first radicals. In semicrystalline polymers, severely stressed chains may break first, for example, the stretched chains in crystal bridges that interconnect lamellae. However, the concentration of such chains is much too low for the observed strengths.

18.2.4 Fracture Propagation

The formation of microcracks must be followed by other processes that lead to the macroscopic fracture of the specimen. These processes are not very well understood.

Further drawing may certainly lead to more nanocracks. It is not clear, however, how these nanocracks then enlarge to microcracks (crazes). For example, small-angle X-ray measurements showed for the elongation of fibers that dimensions of microcracks stayed constant but that their number increased. The fibers broke after the ratio $\Delta L_{mc}/d$ of the average distance ΔL_{mc} between two microcracks to the diameter d of microcracks reached a critical value. The average distance ΔL_{mc} is proportional to the cube of the inverse number concentration C_{mc} of microcracks, $\Delta L_{mc} \sim (C_{mc})^{-1/3}$. For many fibers, it was found that $(C_{mc})^{-1/3}/d = 3 \pm 0.5$. Hence, fibers break if the average distance between two microcracks is about triple the diameter of a microcrack.

Fibers are relatively rigid entities that often undergo brittle fracture because they cannot release stresses by shearing. Ductile polymers are sheared on elongation, which shifts the primary nanocracks so that macroscopic fracture occurs at much higher elongations. However, crazing leads to microfibrils with increased tensile strengths and moduli but these are embedded in microvoids with zero strengths and moduli.

It is therefore convenient to unite the competing effects of tensile moduli (E), tensile strength at break (σ_B), and elongation at break (ε_B) in one dimensionless number, the **Hooke number**, $He = \sigma_B/(E\varepsilon_B)$, which can be viewed as the reduced fracture strength.

According to Hooke's law, Hooke numbers become unity at small elongations (Fig. 18-11). At greater fracture elongations, Hooke numbers of amorphous and semicrystalline thermoplastics decrease sharply, which can be described by

$$(18\text{-}5) \qquad He \equiv \frac{\sigma_B}{E\varepsilon_B} = \frac{1}{[1+(\varepsilon_B/\varepsilon_{crit})^{ab}]^{1/b}}$$

with empirical exponents of a = 0.92 and b = 3. The transition from $He = 1$ at small values of ε_B to values of $He < 1$ is fairly sharp and occurs at $\varepsilon_{B,crit} = 0.0168$.

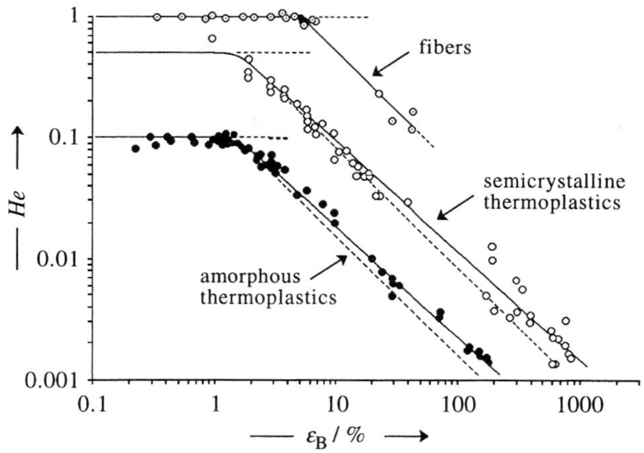

Fig. 18-11 The dependence of logarithms of dimensionless Hooke numbers, *He*, on the logarithms of fracture elongations ε_B of thermoplastics and fibers [7]. For clarity, values of *He* of semicrystalline thermoplastics were shifted downward by a factor of 2 and those of amorphous thermoplastics by a factor of 10. Solid lines were calculated with Eq.(18-5), b = 3, and a = 1 (fibers) or a = 0.92 (thermoplastics), respectively. Broken lines for thermoplastics correspond to a = 1. All polymers were commercial materials (data from CAMPUS® files).

● Amorphous thermoplastics: poly(styrene)s, poly(styrene-*co*-acrylonitrile)s, poly(acrylonitrile-*co*-styrene-*co*-acrylic esters), poly(methyl methacrylate)s, bisphenol A polycarbonates.

○ Semicrystalline thermoplastics: poly(ε-caprolactam)s, poly(hexamethylene adipamide)s, poly(butylene terephthalate)s, linear poly(ethylene)s with low densities.

⊙ Natural fibers: hemp, cotton, jute, flax, ramie, silk, wool; inorganic fibers from SiC, quartz, ceramics, glass, or carbon.

Fibers show similar behavior except that the critical fracture elongation is now increased to $\varepsilon_{B,crit} = 0.0566$ from $\varepsilon_{B,crit} = 0.0168$ for thermoplastics and the exponent becomes a = 1 instead of a = 0.92 for thermoplastics. For both fibers and thermoplastics, the other exponent is b = 3. Since unstressed fibers generally contain oriented chain segments but thermoplastics do not, the decrease of the exponent a from unity (fibers) to 0.92 (thermoplastics) must be due to an onset of additional segment orientations at fracture elongations of $\varepsilon_B > \varepsilon_{B,crit}$.

For a = 1, Eq.(18-5) can be written as $\sigma_B/E = \varepsilon_B[1 + (\varepsilon_B/\varepsilon_{crit})^b]^{1/b}$. With an error of less than 1 %, the right-hand side of this equation becomes ε_{crit} if b = 3 and $\varepsilon_B/\varepsilon_{crit} \geq 3$. Such thermoplastics show $\sigma_B/E = \varepsilon_{crit}$ if $\varepsilon_B \geq 0.05$ at $\varepsilon_{crit} = 0.0168$ and therefore also $\sigma_B = 0.0168\ E \approx E/60$, which agrees well with the prediction for the strength of plastics based on yield strengths (Table 17-1). σ_B/E increases with increasing ε_B if specimens become oriented during drawing (a < 1) since now $\sigma_B/E = (\varepsilon_{crit})^a \varepsilon_B{}^{1-a}$ and $He = \sigma_B/(E\varepsilon_B) = (\varepsilon_{crit}/\varepsilon_B)^a$ for $\varepsilon_B/\varepsilon_{crit} > 5$.

In all of these cases, Hooke numbers are only a function of fracture elongations which depend on a plethora of processes: shear processes, orientations of chain segments, formation of fibrils and voids in crazes, generation of nanocracks and microcracks, etc. Most of these microprocesses are local, i.e., they occur within the meshes of the physical network that is formed by the entanglement of chains. The larger the mesh, the greater will be the volume in which these microprocesses can take place. Hooke numbers must therefore depend on the mesh size of the physical network.

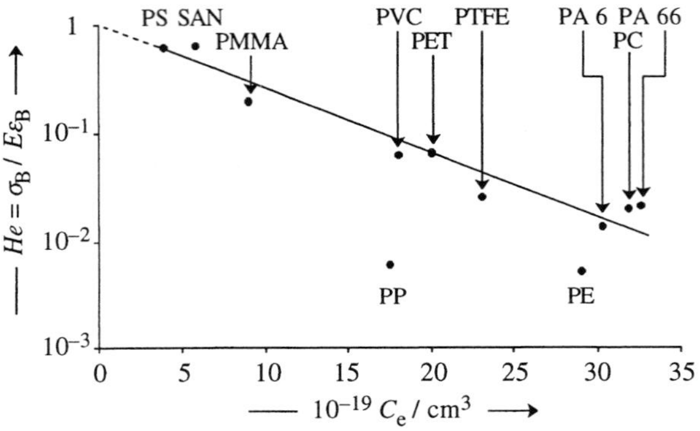

Fig. 18-12 Logarithm of Hooke numbers He as a function of the number concentration C_e of entanglements in amorphous polymers (PS, SAN, PMMA) and semicrystalline polymers with different degrees of crystallinity [8].

The mesh size is characterized by the number concentration of entanglements, $C_e = \rho N_A/M_e$, where ρ = density and M_e = molar mass of segments between entanglement points. Hooke numbers depend on C_e according to $He = \exp(-K\nu_e)$ (Fig. 18-12) where K = empirical constant. The smaller the molar mass M_e, the smaller is also the Hooke number, the greater is ε_B (at $E = const$ and $\sigma_B = const'$), and the smaller is ε_B (at $E = const''$ and $\varepsilon_B = const'''$).

18.2.5 Theoretical Fracture Strengths

Theoretical fracture strengths of polymer molecules in the chain direction can be estimated by using Eyring's theory of rate processes which maintains that the decisive step for fracture is the scission of a covalent bond by thermal fluctuations. The number N of such scissions changes with time t according to $dN/dt = k \exp[-\Delta E_0^{\ddagger}/(k_BT)]$ where k = rate constant with the physical unit of frequency (= inverse time). Since $k \sim 1/t_0$ and thus $dN/dt \sim 1/t$, one can also write $t = t_0 \exp[\Delta E_0^{\ddagger}/(k_BT)]$ where t = lifetime of chain, $t_0 = 1/\nu_0$ = period of thermal oscillations of bound atoms, and ν_0 = frequency of molecular oscillations (10^{12} s^{-1} $< \nu_0 < 10^{14}$ s^{-1}).

The thermal energy for the scission of a chain decreases by $E_B^{\ddagger} = \sigma\Delta V^{\ddagger}$ if the chain is subjected to a stress σ. Since stresses in solids are inhomogeneous, activation volumes $\Delta V^{\ddagger}$ for homogeneous stresses have to be replaced by $\Delta V^{\ddagger} = \Delta V_0^{\ddagger}q$ where q = stress concentration factor. From $t = t_0 \exp[(\Delta E_0^{\ddagger} - \sigma\Delta V_0^{\ddagger}q)/(k_BT)]$ one therefore obtains at temperature T for an inhomogeneously stressed polymer

(18-6) $$\sigma = \frac{1}{\Delta V_0^{\ddagger} q}\left[\Delta E_0^{\ddagger} - k_BT \lg\left(\frac{t}{t_0}\right)\right] = \sigma_{max}\left[1 - \frac{k_BT}{\Delta E_0^{\ddagger}} \lg\left(\frac{t}{t_0}\right)\right]$$

At T = 0 K, the maximum possible tensile strength is $\sigma_{max} = \Delta E_0^{\ddagger}/(\Delta V_0^{\ddagger}q)$ and the theoretical tensile strength at homogeneous stress is $\sigma^o = \sigma_{max}q = \Delta E_0^{\ddagger}/\Delta V_0^{\ddagger}$.

The activation energy $\Delta E_0^{\ddagger}$ for the scission of covalent bonds approximately equals the activation energy for the thermal degradation of polymers. The activation volume at

Table 18-2 Cross-sectional areas A_c, lattice moduli $E_{\|}$, thermal activation energies $\Delta E_o^{\ddagger}$ of degradation, activation lengths $L_o^{\ddagger} = V_o^{\ddagger}/3$, theoretical longitudinal tensile strengths σ^o at $T = 0$ K, stress concentration factors q, and calculated tensile strengths σ of some polymers at 295 K [9]. σ_B = experimental tensile strength of commercial polymers at 298 K.

Chains in all-trans conformations: PE = poly(ethylene), PA 6.6 = poly(hexamethylene adipamide), PVC = st-poly(vinyl chloride); in slightly helical conformation: PTFE = poly(tetrafluoroethylene); in helix conformation: POM = poly(oxymethylene), PP-it = it-poly(propylene), PS-it = it-poly(styrene).

Polymer	$\frac{A_c}{nm^2}$	$\frac{E_{\|}}{GPa}$	$\frac{10^{20}\Delta E_o^{\ddagger}}{J}$	$\frac{L_o^{\ddagger}}{nm}$	q	$\frac{k_B T}{\Delta E_o^{\ddagger}} \lg \frac{t}{t_o}$	$\frac{\sigma^o}{MPa}$	$\frac{\sigma}{MPa}$	$\frac{\sigma_B}{MPa}$
PE	0.182	340	50.0	0.249	470	0.24	32 500	52	23
PA 6.6	0.203	200	30.0	0.257	5	0.43	17 500	1960	74
PTFE	0.277	156	56.3	0.332	-	0.22	15 300	-	54
PVC	0.286	200	65.9	0.331	80	0.19	18 200	185	49
POM	0.172	150	18.9	1.104	-	0.65	14 000	-	70
PP-it	0.343	42	20.7	1.745	7	0.30	3 900	573	30
PS-it	0.698	12	38.2	3.018	32	0.32	1 400	30	40

0 K has been calculated as $\Delta V_o^{\ddagger} = A_c[(3\ N\Delta E_o^{\ddagger})/(4\ bE_{\|})]^{1/2}$ where A_c = cross-sectional area of the polymer chain, N = number of atoms participating in the expansion, b = bond length, and $E_{\|}$ = longitudinal lattice modulus. At 0 K, the theoretical longitudinal tensile strength is therefore $\sigma^o = (2/A_c)[(bE_{\|}\Delta E_o^{\ddagger})/3]^{1/2}$.

Experimental data for A_c, $E_{\|}$, and $\Delta E_o^{\ddagger}$ and theoretical longitudinal tensile strengths σ^o calculated therefrom are collected in Table 18-2 using the length of a covalent bond as the bond length b and setting the number of atoms participating in the expansion as N = 1. Calculations of the extent of thermal fluctuation, $[k_B T/\Delta E_o^{\ddagger}]\ \lg (t/t_o)$, at T = 295 K apply to t = 1 s and $1/\nu = t_o = 1 \cdot 10^{-13}$ s. Stress concentration factors have been obtained from the time until fracture at the applied stress.

According to these calculations, the theoretical tensile strength σ^o of polymers is a linear function of the lattice modulus $E_{\|}$, regardless of whether the chains are in the all-trans (zigzag) or helical conformation (Fig. 18-13). The observed dependence of theoretical tensile strength on longitudinal modulus, $\sigma_o = 0.095\ E_{\|}$, indicates that at 0 K, theoretical strengths should be 9.5 % of the moduli in the chain direction.

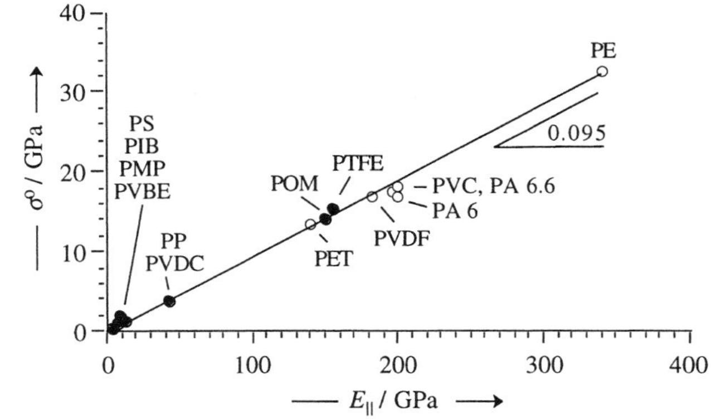

Fig. 18-13 Theoretical tensile strengths σ^o as a function of experimental longitudinal lattice moduli $E_{\|}$ for polymers in the all-trans conformation (O) and in the helix- conformation (●) [9].

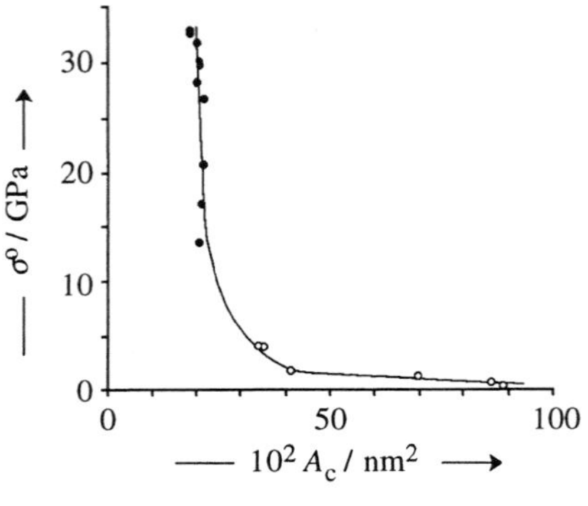

Fig. 18-14 Theoretical fracture strength σ^0 in the longitudinal direction as a function of the cross-sectional area A_c of polymer chains in (●) all-trans conformation or (O) helix conformation [9].

Calculated tensile strengths for 295 K are much lower because of non-negligible thermal fluctuations and large stress concentration factors. For zigzag chains, stress concentration factors increase with decreasing intermolecular forces from PA 6.6 (hydrogen bonds) to PVC (dipole-dipole interactions) and PE (dispersion forces). They are lower for helical macroconformations (it-PS, it-PP *versus* PE).

Since tensile moduli and tensile strengths in the lateral chain direction are interrelated and the tensile moduli are affected by cross-sectional areas and macroconformations of chains, one can expect that tensile strengths are also affected by cross-sections and macroconformations. The theoretical tensile strength of chains in an all-trans conformation decreases rapidly with increasing cross-sectional area (Fig. 18-14). Tensile strengths of helical chains are much lower and vary far less with increasing cross-sectional area. The differences are easy to understand: the elongation of helical chains changes conformational angles whereas the elongation of zigzag chains widens valence angles, which requires much more energy.

18.3 Real Fracture Strengths

18.3.1 Introduction

It seems that conventionally prepared testing specimens and processed thermoplastics either do not contain "natural cracks" (previous sections) or that dimensions of natural cracks, if present, are so small that Griffith's theory does not apply (see Fig. 18-9). However, "cracks" (microcracks, etc., see Section 16.2.1) may form and grow during tensile testing. The tensile strength is then the result of a series of overlapping effects.

Since cracks act as **stress concentrators**, resistance against fracturing can be measured more reliably if the specimen contains a well defined artificial crack of known shape, for example, a notch at the side of the specimen (see Figure 18-8, center and right). Such notched specimens are used for tensile and flexural tests (Sections 18.3.2-18.3.4) and tests of impact resistance (Section 18.3.5), using different testing methods for rigid and ductile polymers.

The polymer reaches its upper yield point if the stress concentration at the tip of the notch becomes sufficiently high. The stress is then relieved by stress softening which is caused by cooperative movements of chain segments. These movements lead to extensive changes of macroconformations. In semicrystalline polymers, such changes can only happen in amorphous regions, which leads, for example, to fracture between spherulites, or radial to spherulites in spherulitic polymers.

The cooperative movement of segments leads to either shear compliance or normal stress compliance. Stress energies are dissipated mainly by the formation of crazes (Fig. 16-3), which is much more effective than the dissipation of energy by shear flow. This effect is utilized for the impact strengthening of thermoplastics (Volume IV, Section 5.3) which is achieved by blending thermoplastics with elastomers.

18.3.2 Critical Stress Intensity Factors

The mechanisms for the absorption of energy described above are the reason why the Griffith theory consistently delivers too low tensile strengths. For example, the Griffith equation, Eq.(18-3), predicts for a poly(styrene) with $E = 3$ GPa, $\gamma_{crit} = 34$ mJ m^{-2}, and a crack length of 2 mm a tensile strength at break of 0.18 MPa whereas the experimental fracture strength is 40 MPa (Table 18-1).

Since other energies have to be considered besides the energy to create new surfaces, the terms in the numerator of the Griffith equation, Eq.(18-3), can be united to a new factor, the **stress intensity factor,** $K_I = (2\ E^{\#}\gamma)^{1/2}$. The factor K_I considers stresses that are perpendicular to the width a of the crack. Similarly, there is also a stress concentration factor K_{II} for shear processes in a plane.

The stress concentration factor K_I becomes the **critical stress intensity factor** K_{IC} (= **fracture toughness**) if the stress becomes the fracture strength. The Griffith equation converts then to the **Irwin-Orowan equation**:

(18-7) $\sigma_B = K_{IC}/(\pi a)^{1/2}$

Eq.(18-7) was originally derived for stress fields around an idealized notch in energy-elastic materials such as steel. The theory delivers explicit equations for stresses in the three spatial directions.

Eq.(18-7) applies to an infinitely large plate with a crack that is much longer than it is wide. In real specimens, cracks are present in many other arrangements, for example, as a series of collinear cracks with an average distance of $2d$ (Fig. 18-15, left). In tensile and flexural experiments, one therefore has to consider the geometries of both cracks and the specimen itself.

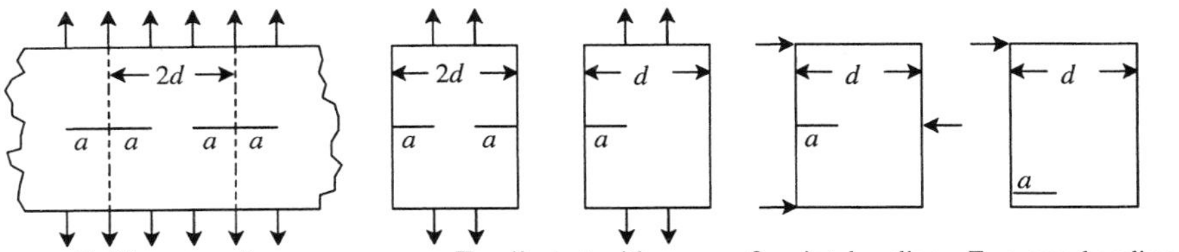

Fig. 18-15 Collinear cracks with length a in and geometries of some testing experiments.

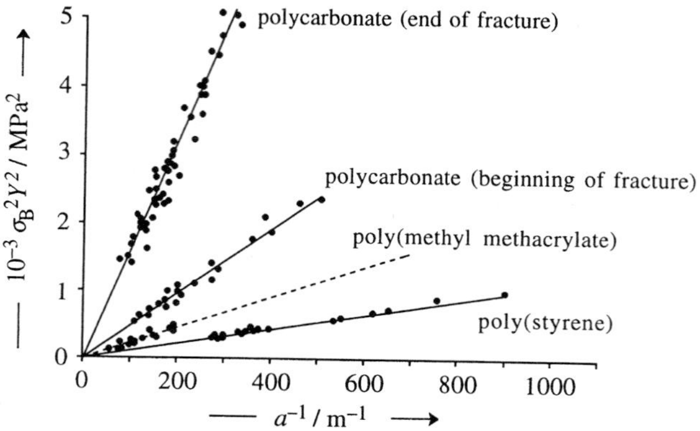

Fig. 18-16 The square of the reduced fracture strength $\sigma_B{}^2Y^2$ as a function of the inverse length a of the notch [10a]. Data at 20°C for elongation rates $[dL/dt]/[\text{cm min}^{-1}]$ of 1 (PMMA), 0.05 (PS), 0.05 and 0.5 (PC, beginning of fracture), and 0.05, 0.5, 5, 20, and 50 (PC, end of fracture), respectively [8]. Y = geometric factor.

The squared Eq.(18-7) can be written as

(18-8) $$K_{IC}{}^2 = \sigma_B{}^2\pi a = \sigma_B{}^2Y^2(a/d)a = \sigma_B{}^2Y'a \quad ; \quad Y' \equiv Y^2(a/d)$$

where $Y^2 = \pi d/a$ = geometric correction factor. Such geometric correction factors have been calculated for many geometries of cracks and specimens. For example, plates with double notches have geometric correction factors of $Y^2 \approx [2\,d/a]\tan[\pi a/(2d)]$, which leads to $Y^2 \approx 3.40$ for $a/d = 0.3$.

Rearrangement of Eq.(18-8) delivers $\sigma_B{}^2Y^2 = [K_{IC}{}^2(d/a)]a^{-1}$. A plot of $\sigma_B{}^2Y^2$ as a function of a^{-1} for data at constant d/a should thus lead to straight lines that pass through the origin (Fig. 18-16). The slopes deliver the critical stress intensity factor, K_{IC}. As found experimentally, elongation rates do not affect the values of K_{IC} of thermoplastics (not shown) but do so for rubber-modified poly(styrene) (Fig.18-17).

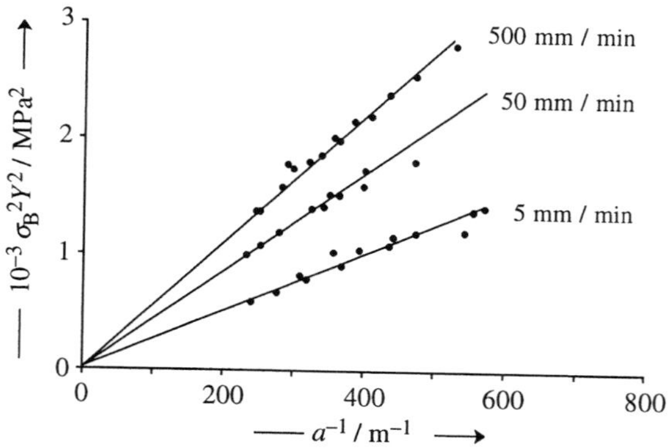

Fig. 18-17 The square of the reduced fracture strength, $\sigma_B{}^2Y^2$, as a function of the inverse diameter a of the notch for various elongation rates. Data for a rubber-modified poly(styrene) (rubber particles of ca. 1 mm diameter) [10b]. The slopes deliver critical stress concentration factors $K_{IC}/(\text{MPa m}^{1/2})$ of 2.4 (at 500 mm/min), 2.13 (at 50 mm/min), and 1.6 (at 5 mm/min).

18.3.3 Fracture Toughness

Critical stress intensity factors can be determined in the laboratory and used without field testing for the determination of the maximum loading of energy-elastic materials. Instead of the critical stress intensity factor, a so-called **fracture surface energy** (= **critical strain release rate**) is often used. The fracture surface energy is obtained by equating $\sigma_B = (2\ \gamma E^{\#}/\pi)^{1/2}a^{-1/2}$ (Eq.(18-3)) and $\sigma_B = (K_{IC}^2/Y')^{1/2}a^{-1/2}$ (Eq.(18-8)) to yield

(18-9) $$G_{IC} \equiv K_{IC}^2/E^{\#} = 2\ \gamma Y'/\pi$$

with the symbol G in honor of Griffith. The critical strain release rate (not a rate!) indicates the energy that is required to form a new surface. It is directly connected with the molecular and microscopic processes during fracture. Table 18-3 shows examples.

Table 18-3 Static glass temperature T_G, tensile modulus E, critical stress inensity factor K_{IC} (plane stress), and critical strain release rates $G_{IC,1}$ (plane stress) and $G_{IC,2}$ (plane elongation) of different materials at 23°C [10c]. 1 MN m$^{-3/2}$ = 1 MPa m$^{1/2}$. For comparison: the critical surface tension of poly(styrene) is γ_{crit} = 0.034 J m^{-2}. ABS = acrylonitrile-butadiene-styrene polymer.

Material	T_G / °C	E / GPa	K_{IC} / MN m$^{-3/2}$	$G_{IC,1}$ / kJ m^{-2}	$G_{IC,2}$ / kJ m^{-2}
Steel alloy		210	150	107	
Poly(ethylene), high density			0.7	6.0	89
medium density			1.3	11.9	17
low density	−80	0.89	5.0	35	
Natural rubber, vulcanized	−73	0.003	0.2	13	
Poly(propylene), it-	−15	1.4	3.0-4.5	0.5	
Polyamide 6	50	3.0	2.8	2.6	
Polyamide 6.6	50		2.5-3.0	0.25	4.15
Poly(vinyl chloride), at-	82	3.6	2.0-4.0	1.23	1.44
Polycarbonate, bisphenol A-	150	3.2	2.2	3.5	5.0
Poly(styrene), at-	100	3.3	0.7-1.1	0.35	0.90
Poly(styrene), at-, rubber-modified		2.0	2.0	1.0	15.0
ABS (acrylonitrile-butadiene-styrene)			2.0	1.6	27
Poly(methyl methacrylate), at-	115		0.7-1.6	1.06	1.28
Epoxy resin, hardened		2.8	0.6	0.089	
rubber-modified		2.4	2.2	2.0	
Wood		2.1	0.5	0.12	
Glass	70	4	0.7	0.007	

18.3.4 Fracture of Ductile Polymers

Application of a stress normal to the large axis (symmetry axis) of an elliptical crack will induce the largest stresses at both ends of this axis. The polymer will try to reduce the stresses by stress relaxation which leads to plastic zones at the ends of the crack (Fig. 18-18). These plastic zones have lengths L_{pz} that are much larger than their average widths $\overline{\delta}$; ratios of $L_{pz}/\overline{\delta} \approx 40$ are not uncommon.

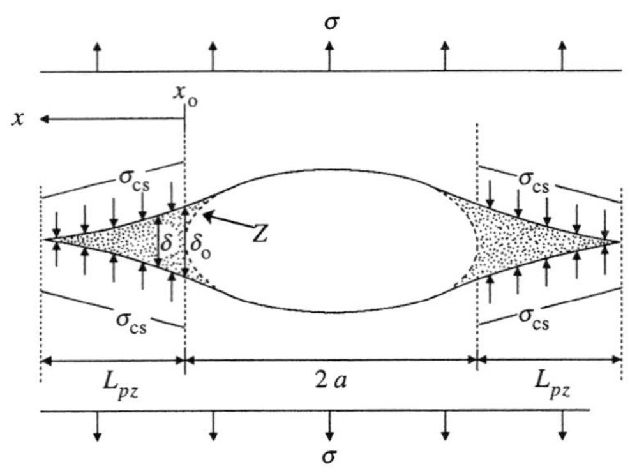

Fig. 18-18 Dugdale model of plastic zones (dotted) at both ends of the symmetry axis of an elliptical crack with length L_{pz} and width δ. Note that the picture is not to scale: since experimental ratios of $L_{pz}/\delta \approx 40$ are not uncommon, real plastic zones practically do not extend into the non-plastic range beyond the δ_o line. σ = tensile stress, σ_{cs} = cohesive stress, L_{pz} = length of plastic zone, 2 a = length of symmetry axis of elliptical crack.

In Fig. 18-18, areas Z can be neglected because of the large ratio $L_{pz}/\bar{\delta}$. The border between the plastic zone and the ellipsoid can thus be described by a straight line with the length δ_o, the so-called **crack tip opening displacement.**

The growth of the plastic zone is opposed by a constant cohesive energy, σ_{cs}, which tries to keep the material together. The stress intensity factors K_I increase with increasing distance δ from the crack tip opening δ_o. At each distance δ from δ_o, there is a *negative* stress σ and therefore also a stress intensity factor $K_I' = -\sigma_{cs}[2/(\pi\delta)^{1/2}]$ according to the Irwin-Orowan equation, Eq.(18-7). This expression contains a factor 2 because one has to consider a *pair* of forces. The total stress intensity factor is obtained by integrating all these factors from the distance δ_o to the distance $\delta = L_{pz}$ at the tip of the plastic zone:

$$(18\text{-}10) \qquad K_I = -\int_{\delta=0}^{\delta=L_{pz}} K_{I,i}\,d\delta = -\int_{\delta=0}^{\delta=L_{pz}} \sigma_{cs}\left(\frac{2}{\pi\delta}\right)^{1/2} d\delta$$

$$= -\sigma_{cs}\left(\frac{2}{\pi}\right)^{1/2}[2\delta^{1/2}]_0^{L_{pz}} = -\sigma_{cs}\left(\frac{8L_{pz}}{\pi}\right)^{1/2}$$

The stress at the ends of the crack cannot become infinitely large, i.e., the two stress intensity factors $K_{I,o}$ before yielding and K_I must compensate each other, i.e., $K_{I,o} + K_I = 0$. Substitution of K_I by $-K_{I,o}$ in Eq.(18-10) delivers an expression for the length L_{pz} of the plastic zone which applies for the condition $L_{pz} << 2a$:

$$(18\text{-}11) \qquad L_{pz} = \frac{\pi}{8}\left(\frac{K_{I,o}}{\sigma_{cs}}\right)^2$$

Eq.(18-10) contains the distance δ which can be calculated from

$$(18\text{-}12) \qquad \delta = \frac{8}{\pi E^{\#}}\sigma_{cs}L_{pz}\left[\zeta - \frac{x}{2L_{pz}}\lg\left(\frac{1+\zeta}{1-\zeta}\right)\right] \quad ; \quad \zeta = [1-(x/L_{pz})]^{1/2}$$

where x = length of the tip (see Fig. 18-18). At δ_o, $x = 0$ and also $K_{I,o} = K_{IC}$ so that

(18-13) $\delta_o = 8\ \sigma_{cs}L_{pz}/(\pi E^{\#}) = K_{IC}^2/(\sigma_{cs}E^{\#})$

The combination of Eq.(18-13), Eq.(18-9), and $\sigma_{cs} \approx \sigma_Y$ (as found experimentally) shows that the fracture toughness is controlled by two parameters, the yield strength σ_Y and the crack tip opening displacement δ_o:

(18-14) $G_{IC} = K_{IC}^2/E^{\#} = \delta_o\sigma_Y$

According to experiments, δ_o depends on the molar mass of the polymer (via entanglement density?) but not on temperature T and elongation rate $\dot{\varepsilon}$. Since σ_Y is a function of T and $\dot{\varepsilon}$, it follows that the critical strain release rate G_{IC} is also dependent on temperature and elongation rate. Furthermore, K_{IC} is not an independent quantity since it is a function of $E^{\#}$ which is also controlled by temperature and elongation rate.

18.3.5 Impact Strength

Conventional tensile tests subject polymers to speeds between 0.06 m/s and 3 m/s (between 1 mm/min and 50 mm/min). In daily life, strain rates are much higher. Doors are slammed shut with speeds of ca. 3 m/s and cars may hit stationary objects with speeds of 33 m/s (= 75 miles per hour). Such high speeds can only be achieved in testing with very expensive high-speed tensile testing machines that are able to generate speeds of up to 250 m/s.

Testing for impact strength is far less expensive. Impact strength is defined as the resistance against fracture on an impact. All tests of impact strengths are standardized. Elongation rates and impact speeds of various impact tests are compared in Table 18-4.

Most methods determine the energy that is required to fracture the specimen. In the **Izod** (and the related **Dynstat**) **method**, a pendulum hits the unnotched or notched specimen which is clamped at one side; the specimen is thus predominantly subjected to flexural stresses and less dominantly to shear stresses.

In the **Charpy method**, the unnotched or notched specimen is clamped at both sides which leads to compression at the top zone of impact, flexural stresses in the center, and strong tensile stresses at the bottom zone.

Table 18-4 Approximate working range of impact tests of unnotched specimens. $v_\varepsilon = \varepsilon/t$ = speed of elongation, $v_s = L/t$ = velocity of impact.

Method		v_e/s^{-1}	$v_s/(m\ s^{-1})$
High-speed tensile tests	pneumatic	100 - 10 000	20 - 240
	hydraulic	1 - 100	0.08 - 4
Flexural tests	Izod	60	2
	Charpy	10	3
Falling weight tests		0.1 - 10	1 - 4
Conventional tensile tests		0.001 - 0.1	0.000 01 - 0.1

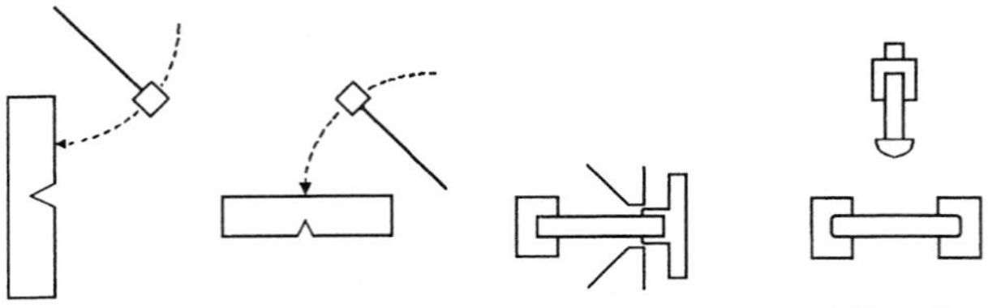

Fig. 18-19 Tests of notched impact strength (Izod and Charpy) and impact strength (tensile impact and falling dart methods). Izod and Charpy methods are also used with unnotched specimens.

Impact strengths can also be determined by dropping a standardized sphere or dart from a standardized height (**falling ball test, falling dart test** (Fig. 18-19, right)). For **tensile impact tests**, specimens are clamped at both ends and hit on both sides of the clamp (Fig. 18-19, second from right).

Similar tests can be performed with notched specimens (Fig. 18-19, first and second from left). These tests of **notched impact strength** produce elongation speeds of up to 5000 s^{-1}. The smaller the radius of the notch, the higher is the stress concentration at the tip of the notch and the smaller is the impact strength (Fig. 18-20).

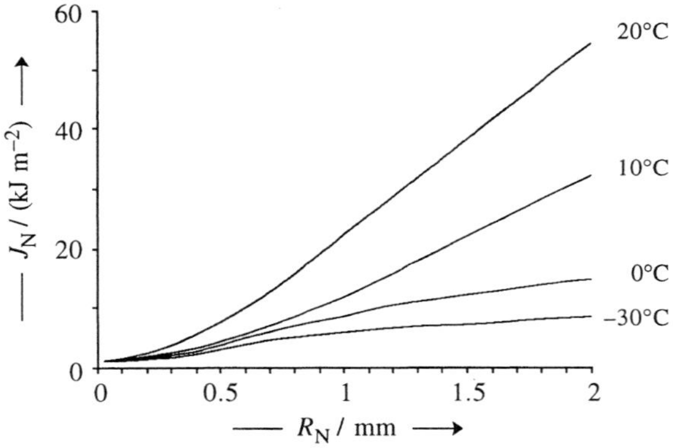

Fig. 18-20 Notched impact strengths J_N of a poly(vinyl chloride) as a function of the radius R_N of the notch (Charpy tests) [11]. By kind permission of Chapman and Hall, London.

All polymers are brittle at low temperatures. At higher temperatures, stresses can be reduced by formation of shear bands or crazes because of the increased mobility of chain segments. Impact strengths increase therefore with increasing temperature and especially near the glass temperature. For the same reason, polymers with additional transformation temperatures below the glass temperature are practically always more impact resistant than polymers without such transformations.

Semicrystalline polymers also have high impact strengths if their glass temperatures are far below the testing temperatures. For the same reason, amorphous polymers can be made much more impact resistant by modification with rubbers, called toughening (Volume IV, Section 10.4.7). Polymers with molecular weights below the entanglement molecular weight have very low impact strengths since they can form few, if any, crazes.

The notched impact strength of *infinitely thin* specimens is practically given only by the energy for the initiation of a fracture. The relevant property is here the energy per width of the notch. For *infinitely thick* specimens, the energy for fracture initiation can be neglected and the required energy is that of the fracture propagation. The relevant property is then the energy per width of notch *and* thickness of specimen.

In Europe, notched impact strengths refer to infinitely thick specimens (usually given in kJ m^{-2}). In the United States, notched impact strengths apply to infinitely thin specimens and data are mostly reported in ft lbf/in ≈ 0.0535 kJ m^{-1} (usually erroneously as ft lb/in). European and US values cannot be compared (Table 18-5).

Table 18-5 Impact strengths γ_i, notched impact strengths $\gamma_{i,N}$, and notched tensile impact strengths $\gamma_{i,N,t}$ of some European industrial thermoplastics. nf = no fracture.

Thermoplastic			γ_i / kJ m^{-2}		$\gamma_{i,N}$ / kJ m^{-2}		$\gamma_{i,N,t}$ / kJ m^{-2}
Symbol	Trade name		23°C	−30°C	23°C	−30°C	23°C
PE-HD	Hostalen GA 7260		-	-	2.1	3.0	45
PE-HD	Lupolen 6031 H		nf	nf	15	9	120
PS	Polystyrol 143 E		9	9	2	2	-
ABS	Magnum 3153		nf	89	11	9	47
PMMA	Degalen 6		14	14	2.1	2.0	-
PBT	Vestodur 1000 nf		135	130	4.5	4.3	55
POM	Delrin 100 NC-10		250	200	12	7	-
PA 66	Ultramid A3, low mol. wt.,	dry	nf	280	6	6	-
		air dry	nf	300	12	5.5	-
PA 66	Ultramid A5, high mol. wt.,	dry	38	30	3.5	2.5	-
		air dry	64	30	3.5	2.5	-

Historical Notes to Chapter 18

Griffith Equation
C.E.Inglis, Trans.Inst.Naval Archit. **55** (1913) 219
A.A.Griffith, Phil.Trans.Royal Soc. **A 221** (1920) 163

Irwin-Orowan Equation
G.R.Irwin, in Fracturing of Metals, Amer.Soc.Metals, Cleveland 1948, p. 147; J.Appl.Mech. **24** (1957) 361; Fracture, Encycl. Physics **6**, 551, Springer, Berlin 1958
E.Orowan, Trans.Inst.Eng.Shipbuild.Scotl. **89** (1946) 165; -, Rept.Progr.Phys. **12** (1949) 185; E.Orowan, in W.M.Murray, Ed., Fatigue and Fracture of Metals (MIT Symp., June 1950), Wiley, New York 1952, p. 139; E.Orowan, Weld.J., Res.Suppl. **34** (1955) 157-S

Dugdale Model
D.S.Dugdale, J.Mech.Phys.Solids **8** (1960) 100
G.I.Barenblatt, Adv.Appl.Mech. **7** (1962) 55

Literature to Chapter 18

(see also Literature to Chapters 16 and 17)

REVIEWS and MONOGRAPHS

L.Nielsen, Mechanical Properties of Polymers and Composites, Dekker, New York, 2 Vols. 1974 and 1976

R.G.C.Arridge, Mechanics of Polymers, Clarendon Press, Oxford 1975

D.W. van Krevelen, Properties of Polymers–Correlation with Chemical Structures, Elsevier, Amsterdam, 3rd ed. 1990

A.Y.Malkin, A.A.Askadsky, V.V.Kovriga, A.E.Chalykah, Experimental Methods in Polymer Physics, Mir Publ., Moscow, and Prentice Hall, Englewood Cliffs (NJ), 2 vols. 1978 and 1979

J.G.Williams, Stress Analysis of Polymers, Wiley, Chichester, 2nd ed. 1980

J.F.Rabek, Experimental Methods in Polymer Chemistry. Physical Properties, Wiley-Interscience, New York 1980

I.I.Perepechko, Low-Temperature Properties of Polymers, Pergamon, London 1980

J.S.Hearle, Polymers and Their Properties. Vol. 1: Fundamentals of Structure and Mechanics, Wiley, New York 1982

R.A.Pethrick, R.W.Richards, Eds., Static and Dynamic Properties of the Polymeric Solid State, Reidel, New York 1982

R.B.Seymour, C.E.Carraher, Structure-Property Relationships in Polymers, Plenum, New York 1984

J.I.Kroschwitz, Ed., Polymers, An Encyclopedic Sourcebook of Engineering Properties (= reprint of sections in H.F.Mark et al., Eds., Encyclopedia of Polymer Science and Engineering, Wiley, New York, 2nd ed. 1987

I.M.Ward, D.W.Hadley, An Introduction to the Mechanical Properties of Solid Polymers, Wiley, New York 1993

W.Brostow, Ed., Performance of Plastics, Hanser Publishers, Munich 2001

TESTING

G.C.Ives, J.A.Mead, M.M.Riley, Handbook of Plastics Test Methods, Iliffe, London 1971

S.Turner, Mechanical Testing of Plastics, Butterworths, London 1973

R.P.Brown, Ed., Handbook of Plastics Test Methods, Wiley, New York, 3rd ed. 1989

V.Shah, Handbook of Plastics Testing and Failure Analysis, Wiley, Hoboken (NJ), 3rd ed. 2007

FRACTURE

E.H.Andrews, Fracture in Polymers, Oliver and Boyd, Edinburgh 1968

S.Rabinowitz, P.Beardmore, Craze Formation and Fracture in Glassy Polymers, Crit.Revs.Macromol.Sci. **1** (1972) 1

R.P.Kambour, A Review of Crazing and Fracture in Thermoplastics, J.Polym.Sci. [Revs.] **D 7** (1973) 1

J.A.Manson, R.W.Hertzberg, Fatigue Failure in Polymers, Crit.Revs.Macromol.Sci. **1** (1973) 433

E.H.Andrews, P.E.Reed, Molecular Fracture in Polymers, Adv.Polym.Sci. **27** (1978) 1

L.G.E.Struik, Physical Aging in Amorphous Polymers and Other Materials, Elsevier, Amsterdam 1978

R.W.Hertzberg, J.A.Manson, Fatigue of Engineering Plastics, Academic Press, New York 1980

H.H.Kausch, Ed., Crazing in Polymers, Adv.Polym.Sci. **52/53** (1983)

A.E.Zachariades, R.S.Porter, Eds., The Strength and Stiffness of Polymers, Dekker, New York 1983

A.J.Kinlock, R.J.Young, Fracture Behaviour of Polymers, Appl.Sci.Publ., London 1983

A.S.Argon, R.E.Cohen, O.S.Gebizlioglu, C.E.Schwier, Crazing of Block Copolymers and Blends, Adv.Polym.Sci. **52/53** (1983) 275

J.G.Williams, Fracture Mechanics of Polymers, Wiley, New York 1984

W.Brostow, R.D.Corneliussen, Eds., Failure of Plastics, Hanser, Munich 1986

A.G.Atkins, Y.W.Mai, Elastic and Plastic Fracture, Halsted Press, New York 1986

H.H.Kausch, Polymer Fracture, Springer, Heidelberg, 2nd ed. 1987

H.H.Kausch, Ed., Crazing in Polymers II, Adv.Polym.Sci. **91/92** (1990)

G.Bartenev, Mechanical Strength and Failure of Polymers, Prentice-Hall, Englewood Cliffs (NJ) 1993

B.Crist, The Ultimate Strength and Stiffness of Polymers, Annu.Rev.Mater.Sci. **25** (1995) 295

D.Miannay, Fracture Mechanics, Springer, Berlin 1998
J.Scheirs, Ed., Compositional and Failure Analysis of Polymers, Wiley, New York, 2000

References to Chapter 18

[1] –, Kunststoff-Physik im Gespräch, BASF, Ludwigshafen, 7th edition 1998, (a) p. 38, (b) p. 45
[2] H.W.McCormick, F.M.Brower, L.Kin, J.Polym.Sci. **39** (1959) 87, Tables I-III
[3] P.J.Flory, Ind.Eng.Chem. **38** (1946) 417, data of (a) Fig. 105 and (b) Fig. 106, respectively
[4] G.A.Bernier, R.P.Kambour, Macromolecules **1** (1968) 393, Fig. 4
[5] J.P.Berry, J.Polym.Sci. **50** (1961) 313, data of Fig. 3
[6] S.N.Zhurkov, V.A.Zakrevskyi, V.E.Korsukov, V.S.Kuksenko, J.Polym.Sci. [A-2] **10** (1972) 1509; see also S.N.Zhurkov, Vestnik Akad. Nauk SSSR –/11 (1957) 78; S.N.Zhurkov, Int. J.Fracture Mechanics **1** (1965) 311
[7] H.-G.Elias, J.Polym.Sci. **B** (Polym.Phys.) **33** (1995) 955, Fig. 5
[8] H.-G.Elias, Macromol.Chem.Phys. **195** (1994) 3117, Table 4
[9] T.He, Polymer **27** (1986) 253, Tables 1 and 2
[10] J.G.Williams, Fracture Mechanics of Polymers, Ellis Horwood, Chichester 1984, (a) Figs. 6.2, 6.9, 6.12, and 6.13; (b) Fig. 6-23 (modified) (the reference to this figure (K.Nikpur, J.G.Williams, J.Mater.Sci. **14** (1979) 467) does not contain the figure); (c) Tables 6.3 and 8.1, expanded by scattered literature data
[11] W.V.Titow, PVC Plastics, Elsevier, Amsterdam 1990, Fig. 8.2

19 Appendix

19.1 SI Physical Quantities and Units

According to **Maxwell**, many physical properties can be described in quantitative terms by *quantity calculus*. The value of a physical quantity (symbol in *italics*) equals the product of a numerical value (Roman (upright) type) and a physical unit (symbol in upright letter):

$$\text{physical quantity} = \text{numerical value} \times \text{physical unit}$$

This equation can be manipulated by the ordinary rules of algebra. If, for example, a certain item has a length L of 0.002 meters (m) = 2 millimeters (mm), then this may be written as

$$L = 0.002 \text{ m} \quad \text{or} \quad L = 2 \cdot 10^{-3} \text{ m} \quad \text{or} \quad 10^3 \, L/\text{m} = 2 \quad \text{or} \quad L = 2 \text{ mm} \quad \text{or} \quad L/\text{mm} = 2$$

but *not* as $10^{-3}\,L/\text{m} = 2$. For example, a column head $10^2\,F/(\text{N m}^{-2})$ for a column entry of 7.35 thus indicates a force $\boldsymbol{F} = 7.35 \cdot 10^{-2}$ N/m^2. Literature data often do not follow these SI (Systéme International) rules by the International Standardization Organization (ISO), which are adopted by IUPAP (International Union of Pure and Applied Physics), IUPAC (International Union of Pure and Applied Chemistry), etc. Instead one finds various nonrational notations such as $\boldsymbol{F}$, N m^{-2} or $\boldsymbol{F}$ [N m^{-2}] for a column entry of $7.35 \cdot 10^{-2}$, often with wrong algebraic statements such as $10^{-2}\,\boldsymbol{F}$, N m^{-2}.

The International System of Units (système international; **SI system**) uses seven **base physical quantities** and seven **SI base units** (Table 19-1). It replaces the formerly used CGS and MKS systems; it is *not* "the metric system". The SI system also defines several **derived units** (Table 19-2) which now include the radian and steradian that were formerly considered "supplementary units." Several older units may still be used for the time being (Table 19-3).

In order to avoid cumbersome numbers, symbols of units may be prefixed with SI prefixes (Table 19-4). The decimal sign between digits in a number is either a period (US) or a comma (e.g., Europe). No commas or periods should be used to separate groups of three digits for better reading. Instead, spaces should be applied (exception: groups of four digits may be written without space). Example:

1 234.567 8 or 1234.5678 or (Europe) 1234,5678 or 1 234,567 8 but *not* 1,234.5678.

Numbers and units are separated by a gap and groups of quantities or units by spaces or multiplication signs (· or ×).

All countries except the United States of America, Liberia, and Myanmar have adopted the SI system. In many countries, the SI system is the only system of weights and measures that can be lawfully used in trade and commerce. American technical literature and sometimes also American scientific literature still uses American and old Imperial British units although the metric system was introduced in 1896 by an act of Congress and by law all U.S. government units were supposed to convert to the SI system by the end of 1992. The following tables list names and symbols of base and derived SI units, temporarily allowed non-SI units, and SI prefixes. This book follows IUPAC/IUPAP recommendations; exceptions are noted in the list of symbols of physical quantities.

Table 19-1 Names and symbols of physical quantities and their SI base units.

Physical quantity		SI base unit		SI symbol
SI symbol	SI name	English name	American name	
l	length	metre	meter	m
m	mass	kilogramme	kilogram	kg
t	time	second	second	s
I	electric current	ampere	ampere	A
T	thermodynamic temperature	kelvin	kelvin	K
n	amount of substance	mole	mole	mol
I_v	luminous intensity	candela	candela	cd

Table 19-2 Derived SI units and their symbols (recommendations by IUPAC and IUPAP).

Physical quantity Symbol	Name	SI unit SI name	SI symbol and unit(s)			
α, β, γ	plane angle	radian	rad	= m/m	= 1	
ω, Ω	solid angle	steradian	sr	$= m^2/m^2$	= 1	
ω	angular velocity	-		rad/s		
	angular acceleration	-		rad/s^2		
ν	frequency	hertz [1]	Hz	$= s^{-1}$		
v, u, w	speed [2], velocity [3]	-		$m\ s^{-1}$		
$a, (g)$	acceleration	-		$m\ s^{-2}$		
P	power, radiant flux	watt	W	= V A	$= J\ s^{-1}$	$= m^2\ kg\ s^{-3}$
E	energy, work, heat	joule	J	= N m		$= m^2\ kg\ s^{-2}$
$\boldsymbol{F}$ [3]	force	newton	N	$= J\ m^{-1}$		$= m\ kg\ s^{-2}$
-	impact strength (US)	newton		$J\ m^{-1}$		$= m\ kg\ s^{-2}$
G	weight	newton	N	$= J\ m^{-1}$		$= m\ kg\ s^{-2}$
-	impact strength (Europe)	-		$J\ m^{-2}$		$= kg\ s^{-2}$
γ	interfacial tension	-		$J\ m^{-2}$	$= N\ m^{-1}$	$= kg\ s^{-2}$
p, σ	pressure, stress	pascal	Pa	$= J\ m^{-3}$	$= N\ m^{-2}$	$= m^{-1}\ kg\ s^{-2}$
	impulse, momentum	-		N s		$= m\ kg\ s^{-1}$
Q	electric charge	coulomb	C	= A s		
U	electric potential, electromotive force	volt	V	$= W\ A^{-1}$	$= J\ C^{-1}$	$= m^2\ kg\ s^{-3}\ A^{-1}$
R	electric resistance	ohm	Ω	$= V\ A^{-1}$		$= m^2\ kg\ s^{-3}\ A^{-2}$
G	electric conductance	siemens	S	$= \Omega^{-1}$		$= m^{-2}\ kg^{-1}\ s^3\ A^2$
C	electric capacitance	farad	F	$= C\ V^{-1}$		$= m^{-2}\ kg^{-1}\ s^4\ A^2$
ε	relative permittivity [4]	-	1			
Φ	magnetic flux	weber	Wb		= V s	$= m^2\ kg\ s^{-2}\ A^{-1}$
L	magnetic inductance	henry	H	$= V\ s\ A^{-1}$		$= m^2\ kg\ s^{-2}\ A^{-2}$
B	magnetic flux density	tesla	T	$= Wb\ m^{-2}$		$= kg\ s^{-2}\ A^{-1}$
	magnetic field strength	-		$A\ m^{-1}$		
Φ_v	luminous flux	lumen	lm	= cd sr		
E_v	illuminance	lux	lx	$= cd\ sr\ m^{-2}$		
A	radioactivity	becquerel	Bq	$= s^{-1}$		
D	absorbed dose (radiation)	gray	Gy	$= J\ kg^{-1}$		$= m^2\ s^{-2}$
$\dot{D}$	(absorbed dose rate)	-		$Gy\ s^{-1}$	$= W\ kg^{-1}$	$= m^2\ s^{-3}$
x	(exposure)	-		$C\ kg^{-1}$		
$\dot{x}$	(exposure rate)	-		$A\ kg^{-1}$		
-	dose equivalent	sievert	Sv		$= J\ kg^{-1}$	$= m^2\ s^{-2}$
t, θ [5]	Celsius temperature	degree Celsius [6]	°C			

1) The physical unit "hertz" should *only* be used for "frequency" in the sense of "cycles per second". The unit of radial (circular) or angular velocity is rad/s, which can be written as s^{-1} but *not* as Hz.

2) Nonvectorial; the velocity of light usually has the symbol *c*.

3) Vectorial; the symbols are then in bold letters (***u***, ***v***, ***w***).

4) Formerly: dielectric constant.

5) This book uses *T* as a symbol for both the kelvin and the Celsius temperature (see footnote 7 on next page).

6) For symbols and the writing of names of units that are derived from the names of famous scientists, see footnote 7 on next page.

Table 19-3 Older units. * Units that may be used with SI prefixes or SI units.

Physical quantity	Physical unit SI name	Physical unit SI symbol	Physical unit Value in SI units	Notes
Time	minute	min	= 60 s	1)
Time	hour	h	= 3600 s	1)
Time	day	d	= 86 400 s	1)
Length	ångstrøm	Å	= 10^{-10} m = 0.1 nm	2)
Area	barn	b	= 10^{-28} m^2	
Volume	liter *	l, L	= 10^{-3} m^3 ≡ 1 L	
Mass	ton(ne) *	t	= 10^3 kg	3)
Mass	unified atomic mass unit 4)	u = $m_a(^{12}C)/12$	≈ 1.660 54·10^{-27} kg	4,5)
Energy	electronvolt *	eV	≈ 1.602 18·10^{-19} J	6)
Pressure	bar *	bar	= 10^5 Pa	2)
Plane angle	degree	°	= (π/180) rad	
Plane angle	minute	'	= (π/10 800) rad	
Plane angle	second	"	= (π/648 000) rad	
Temperature	Celsius temperature	°C	= θ/°C = (T/K) – 273.15	7)

1) IUPAC allows the use of the non-SI units "minute", "hour", and "day "in appropriate contexts although these three physical units are not part of the SI system. These units should not be used with SI prefixes.

"Month" and "year" are not scientific units. In commercial data, the symbol for "month" is often "mo" and the symbol for "year" either "yr" or, preferably, "a" (L: *annus* = year).

2) This unit is approved for "temporary use with SI units" in fields where it is presently used.

3) IUPAC allows the use of the physical unit "ton(ne)" = 1000 kg (especially for technical and commercial data) which is, however, not an SI unit. "Ton(ne)" (symbol: ton) is not to be confused with "long ton" (≈ 1016.047 kg) and "short ton" (≈ 907.185 kg); both "ton(ne)s" are often used in commerce without the adjectives "long" and "short".

4) The value of this unit depends on the experimentally determined value of the Avogadro constant N_A; the value of the corresponding SI unit is therefore not exact.

5) The unified atomic mass (physical unit: kg) is sometimes called the dalton (symbol Da). In the biosciences, and recently also in polymer science, "dalton" has erroneously come to mean the relative molecular mass (physical unit: 1) or the molar mass (physical unit: g/mol)!

6) The value of this unit depends on the experimentally determined value of the elementary charge e; the value of the corresponding SI unit is therefore not exact.

7) The SI unit of the Celsius temperature *interval* is the degree Celsius (symbol of the unit: °C), which is equal to the kelvin (*not* "degree kelvin" or "degree Kelvin"). The *symbol* of the unit kelvin (small k!) is K, *not* °K. Celsius is always written with a capital C.

When quoting temperatures in kelvin, a space should be written between the numerical value and the symbol K as it is customary for all physical properties. However, if temperatures are given in degrees Celsius, no space should exist between the numerical value and the symbol. Thus: T = 298 K but also T = 25°C (this book).

IUPAC now recommends the symbol θ for the physical property "Celsius temperature". In polymer science, a capital theta (Θ) is the traditional symbol for the property "theta temperature". The experience of this author has shown that θ and Θ are easily mixed up, even by experienced polymer scientists. This book thus uses the symbol T for both thermodynamic temperatures and Celsius temperatures; the possibility of a mix-up is remote since physical units are always given.

8) Names of units derived from persons' names are not capitalized. Exceptions are "degree Celsius" and two-letter symbols for dimensionless quantities (Reynolds number, Deborah number, etc.).

Table 19-4 SI prefixes for SI units.
Origin: D = Danish, G = Greek, I = Italian, L = Latin, N = Norwegian.
a) ISO added the letter "y" because the prefix "o" would be misleading.
b) ISO replaced "s" by "z" in order to avoid the double use of "s" ("s" is also the symbol for "second").

Factor	Prefix	Symbol a)	Common name American	Common name European 1)	Origin of prefix	
10^{24}	yotta a)	Y	septillion	quadrillion 2)	L:	*octo* = eight $[10^{24} = (10^3)^8]$
10^{21}	zetta b)	Z	sextillion	1000 trillion 3)	L:	*septem* = seven $[10^{21} = (10^3)^7]$
10^{18}	exa	E	quintillion	trillion 4)	G:	*hexa* = six $[10^{18} = (10^3)^6]$
10^{15}	peta	P	quadrillion	1000 billion 5)	G:	*penta* = five $[10^{15} = (10^3)^5]$
10^{12}	tera	T	trillion	billion 6)	G:	*teras* = monster
10^9	giga	G	billion	1000 million 7)	G:	*gigas* = giant
10^6	mega	M	million	million	G:	*megas* = big
10^3	kilo	k	thousand	thousand 8)	G:	*khilioi* = thousand
10^2	hekto 9)	h	hundred	hundred	G:	*hekaton* = hundred
10^1	deka 10)	da	ten	ten	G:	*deka* = ten
10^{-1}	deci	d	one tenth	one tenth	L:	*decima pars* = one tenth
10^{-2}	centi	c	one hundredth	one hundredth	L:	*pars centesima* = one hundredth
10^{-3}	milli	m	one thousandth	one thousandth	L:	*pars millesima* = one thousandth
10^{-6}	micro 11)	μ	one millionth	one millionth	G:	*mikros* = small
10^{-9}	nano	n	one billionth	one milliardth	G:	*nan(n)os* = dwarf
10^{-12}	pico	p	one trillionth	one billionth	I:	*piccolo* = small
10^{-15}	femto	f	one quadrillionth	one billiardth	D, N:	*femten* = fifteen
10^{-18}	atto	a	one quintillionth	one trillionth	D, N:	*atten* = eighteen
10^{-21}	zepto b)	z	one sextillionth	one trilliardth	L:	*septem* = seven $[10^{-21} = (10^{-3})^7]$
10^{-24}	yocto a)	y	one septillionth	one quadrillionth	L:	*octo* = eight $[10^{-24} = [10^{-3})^8]$

1) Most European countries (Germany: see 3), 5), 7)). England reverted to the US system in 1974.
2) France: called "septillion" before 1948.
3) France: called "sextillion" before 1948. Germany: Trilliarde.
4) France: called "quintillion" before 1948.
5) France: called "quadrillion" or "quatrillion" before 1948. Germany: Billiarde.
6) France: called "trillion" before 1948.
7) France: "milliard"; was called "billion" besides "milliard" before 1948. Germany: Milliarde.
8) France: "mille".
9) NIST recommends "hekto" (etymologically correct, see right column) but ISO uses "hecto".
10) NIST recommends "deka" (etymologically correct, see right column) but ISO uses "deca".
11) USA: μ as the symbol for "micro" is neither known to the general public nor to newspapers and magazines; it is also not on typewriter keyboards. The prefix "μ" is therefore sometimes replaced by the non-SI prefix "mc" (from "micro"; 1 mcg = 1 microgram = 1 μg). The non-SI prefix "ml" is then substituted for the SI prefix "m" = "milli" (1 mlg ≡ 1 milligram = 1 mg).

Table 19-5 Fundamental constants used in this book (CODATA: 2006 adjustment).

Physical quantity	Symbol	= number × physical unit	IUPAC symbol
Boltzmann constant	k_B	$= 1.380\ 6504 \cdot 10^{-23}$ J K^{-1}	k
Avogadro constant	N_A	$= 6.022\ 141\ 79 \cdot 10^{23}$ mol^{-1}	N_A or L
Loschmidt constant	N_A/V_m	$= 2.686\ 7774 \cdot 10^{25}$ m^{-3}	n_o
Molar gas constant	R	$= 8.314\ 472$ J mol^{-1} K^{-1}	R
Atomic mass constant	m_u	$= 1.660\ 538\ 782 \cdot 10^{-27}$ kg (= $m(^{12}C)/12$)	m_u

19.2 Common Physical Quantities and Units

All countries except the United States, Liberia, and Myanmar have adopted the **International System of Units** (systéme international: **SI system**). United States governmental, commercial, and technical literature sometimes still uses American and old Imperial (British) units although the forerunner of the SI system, the metric system, was introduced by an act of Congress in 1896.

F.Cardarelli, Encyclopedia of Scientific Units, Weights and Measures: Their SI Equivalences and Origins, Springs, Berlin 2003, 846 pp.

Table 19-6 Conversion of non-SI units to SI units.
Symbols ≡ identical by definition, = equal, ≈ approximately equal, ∴ equivalent to.

Name of non-SI unit	Non-SI unit	= SI unit
Length		
Yard	1 yd	= 0.914 4 m (exact)
Foot	1 ft = 1' = 12"	= 0.304 8 m (exact)
Inch	1 in = 1"	= 2.54 cm (exact)
Mil	1 mil	= 25.4 μm (exact)
Micron	1 μ	= 10^{-6} m = 1 μm
Millimicron	1 mμ	= 10^{-9} m = 1 nm
Ångstrøm	1 Å	= 10^{-10} m = 0.1 nm
Area		
Square yard	1 sq. yd.	= 0.836 127 36 m^2
Square foot	1 sq. ft.	= 9.203 04·10^{-2} m^2
Square inch	1 sq. in.	= 6.451 6·10^{-4} m^2
Volume		
Cubic yard	1 cu. yd.	= 0.764 554 857 m^3
US barrel petroleum	1 bbl = 42 US gal	= 0.158 987 m^3
US barrel	1 barrel	= 0.119 m^3 = 119 L
Cubic foot	1 cu. ft.	= 2.381 685·10^{-2} m^3
Gallon (British or Imperial)	1 gal	= 4.545 96·10^{-3} m^3 = 4.545 96 L
Gallon (US dry)	1 gal	= 4.405·10^{-3} m^3
Gallon (US liquid)	1 gal = 4 US qt.	= 3.785 412·10^{-3} m^3 = 3.785 412 L
Liter (cgs)	1 L	= 1.000 028·10^{-3} m^3
Liter (SI)	1 L	≡ 1.000 000·10^{-3} m^3
Quart (US dry)	1 qt.	= 1.101 L
Quart (US liquid)	1 qt. = 2 US pints	= 0.946 335 L
Pint (US liquid)	1 pt. = 2 US cups	= 0.473 168 L
Pint (US dry)	1 pt.	= 0.550 6 L
Cup (US)	1 cup = 8 fluid oz.	= 0.236 534 L
Ounce (British liquid)	1 oz.	= 0.028 413 L
Ounce (US fluid ounce)	1 oz. = 2 table sp.	= 0.029 574 L
Cubic inch	1 cu. in.	= 0.016 387 064 L
Mass		
Long ton (UK)	1 l.t. = 2240 lb	= 1016.046 909 kg
Short ton (US)	1 sh.t. = 2000 lb	= 907.184 74 kg
Pound (international)	1 lb	= 453.592 37 g
Pound (avoirdupois) (US)	1 lb = 16 oz.	= 453.592 427 7 g
Pound (apothecaries' or troy, US)	1 lb = 8 drams	= 373.242 g
Ounce (avoirdupois) (US)	1 oz.	= 28.349 52 g
Ounce (troy)	1 oz.	= 31.103 5 g

Table 19-6 (continued)

Name of non-SI unit	Non-SI unit	= SI unit
Time		
Year	1 a (US: 1 yr)	≡ 365 days (statistics only)
Month	1 mo (US: 1 mon)	≡ 30 days (statistics only)
Day	1 d	≡ 24 h = 86 400 s
Hour	1 h (US = 1 hr)	≡ 60 min = 3600 s
Minute	1 min	≡ 60 s (US: 60 sec)
Temperature		
Degree Celsius	y°C – 273.16°C	= x K
Degree Fahrenheit	(z°F – 32°F)(5/9)	= y°C
Density (1 kg m^{-3} = $1 \cdot 10^{-3}$ g cm^{-3})		
Specific gravity	1 lb/cu.in.	= 27.679 904 71 g cm^{-3}
	1 oz/cu.in.	= 1.729 993 853 g cm^{-3}
	1 lb/cu.ft.	= $1.601\ 846\ 337 \cdot 10^{-2}$ g cm^{-3}
Energy, work (1 J ≡ 1 N m ≡ 1 W s) (for commercial units, see Volume II)		
Kilowatt hour	1 kWh	= 3.6 MJ
Cubic foot-atmosphere	1 cu.ft.atm.	= 2.869 205 kJ
British thermal unit	1 Btu_{mean}	= 1.055 79 kJ
British thermal unit, international	1 Btu_{IT}	= 1.055 056 kJ
-	1 cu.ft.lb(wt)/sq.in.	= 195.237 8 J
Liter atmosphere (cgs)	1 L atm	= 101.325 0 J
Meter kilogram-force	1 m kgf	= 9.806 65 J
Calorie, international	1 cal_{IT}	= 4.186 8 J
Calorie, thermochemical	1 cal_{th}	= 4.184 J
Erg	1 erg	= 1.10^{-7} J = 0.1 μJ
Force (1 N ≡ 1 J m^{-1} ≡ 1 kg m s^{-2})		
-	1 ft-lbf/in. notch	= 53.378 64 N
Kilogram force	1 kgf	= 9.806 65 N
Pound-force	1 lbf	= 4.448 22 N
Gram-force	1 gf	= $9.806\ 65 \cdot 10^{-3}$ N
Pond	1 p	= $9.806\ 65 \cdot 10^{-3}$ N
Dyne	1 dyn	= $1 \cdot 10^{-5}$ N
Length-related force		
-	1 kp/cm	= 980.665 N m^{-1}
-	1 lbf/ft	= 14.593 898 N m^{-1}
-	1 dyn/cm	= $1 \cdot 10^{-3}$ N m^{-1}
Pressure, mechanical stress (1 MPa ≡ 1 MN m^{-2} ≡ 1 N mm^{-2})		
Physical atmosphere	1 atm ≡ 760 torr	= 0.101 325 MPa
Bar	1 bar	= 0.1 MPa
Technical atmosphere	1 at = 1 kg/cm^2	= 0.098 065 MPa
Pound-force per square inch	1 psi = 1 lbf/sq. in.	= $6.894\ 76 \cdot 10^{-3}$ MPa
Inch mercury (32°F)	1 in. Hg	= $3.386\ 388 \cdot 10^{-3}$ MPa
Inch water (39.2°F)	1 in. H_2O	= 249.1 Pa
Torr	1 torr	= (101 325/760) Pa ≈ 133.322 Pa
Millimeter mercury	1 mm Hg	= 13.5951 · 9.806 65 Pa ≈ 133.322 Pa
-	1 dyn/cm^2	= $1 \cdot 10^{-5}$ MPa
Millimeter water	1 mm H_2O	= $9.806\ 65 \cdot 10^{-6}$ MPa

Table 19-6 (continued)

Name of non-SI unit	Non-SI unit	= SI unit
Power (1 W = 1 J s^{-1})		
Horsepower (boiler)	1 hp	= 9810 W
Horsepower (electric)	1 hp	= 746 W
Horsepower (UK)	1 hp	= 745.700 W
Horsepower (metric)	1 PS	= 735.499 W
British thermal unit per hour	1 Btu/h	= 0.293 275 W
-	1 cal/h	= $1.162\ 222 \cdot 10^{-3}$ W
Thermal conductivity		
-	1 cal/(cm s °C)	= 418.6 W m^{-1} K^{-1}
-	1 Btu/(ft h °F)	= 1.731 956 W m^{-1} K^{-1}
-	1 kcal/(m h °C)	= 1.162 78 W m^{-1} K^{-1}
Heat transfer coefficient		
-	1 cal/(cm^2 s °C)	= $4.186\ 8 \cdot 10^{4}$ W m^{-2} K^{-1}
-	1 BTU/(ft^2 h °F)	= 5.682 215 W m^{-2} K^{-1}
-	1 kcal/(m^2 h °C)	= 1.163 W m^{-2} K^{-1}
Length-related mass (= linear density)		
Fineness (metric)	1 tex	= $1 \cdot 10^{-6}$ kg m^{-1}
Fineness (denier)	1 den	= $0.111 \cdot 10^{-6}$ kg m^{-1}
Fracture length		
-	1 g/den	= $9 \cdot 10^{3}$ m
Textile strength		
Tenacity	1 gf/den = 1 gpd	= 0.082 599 N tex^{-1} = 0.082 599 m^2 s^{-2} = 98.06 MPa · (density in g cm^{-3})
Dynamic viscosity		
Poise	1 P	= 0.1 Pa s
Centipoise	1 cP	= 1 mPa s
Kinematic viscosity		
Stokes	1 St	= $1 \cdot 10^{-4}$ m^2 s^{-1}
Heat capacity		
Clausius	1 Cl	= 1 cal_{th}/K = 4.184 J K^{-1}
Molar heat capacity		
Entropy unit	1 e.u.	= 1 cal_{th} K^{-1} mol^{-1} = 4.184 J K^{-1} mol^{-1}
Relative permittivity		
"Dielectric constant"	1	1
(Electrical) conductance		
Inverse ohm	1 mho	= 1 S
Electrical field strength		
-	1 V/mil	= $3.937\ 008 \cdot 10^{4}$ V m^{-1}

19.3 Concentrations

Concentrations measure the abundance of substance 1 in all substances i present; $i = 1, 2, 3, ...$

Mass fraction $= w_1 = m_1/\Sigma_i m_i = m_1/m = c_1/c$. Mass m_1 of substance 1 divided by the sum of masses m_i of all substances i. Since all masses reside in the same gravity field, a mass fraction can also be called a **weight fraction** ("weight" is not an accepted ISO term; it was formerly the name of a mass in a gravity field).

The value of 100 w_1 is called weight percent (wt-%) and the value of 1000 w_i is called weight promille (wt-‰). The English language literature also uses part per hundred (1 pph = 1 %), part per million (1 ppm = 10^{-4} %), part per (American) billion (1 ppb = 10^{-7} %), and part per (American) trillion (1 ppt = 10^{-10} %).

Volume fraction $= \phi_1 = V_1/(\Sigma_i V_i = V_i/V$. Volume V_1 of substance 1 divided by the total volume of all substances i *before* the mixing process.

Mole fraction, amount fraction, number fraction $= x_1 = n_1/(\Sigma_i n_i) = n_1/n$. Amount n_1 of substance 1 divided by the sum of amounts n_i of all substances i. "Amount-of-substance" (short: "amount") is measured in moles, never in kilograms; it is not a mass! The amount-of-substance was (and still is) erroneously called "mole *number*" in the literature.

Mass concentration = mass density. In polymer science, generally as $c_1 = m_1/V$, i.e., as mass m_1 of substance 1 per volume V of mixture *after* mixing. IUPAC recommends the symbols γ_1 or ρ_1 instead of c_1 but ρ_1 may be confused with the same symbol for the mass density of a *neat* substance. Mass concentrations are usually called "concentrations" in the literature.

Number concentration = number density of entities. $C_1 = N_1/V$. Number N_1 of entities of type 1 (molecules, atoms, ions, etc.) per volume V of mixture *after* mixing. IUPAC recommends the name "concentration" for this physical quantity but this may be confused with the much more common "concentration" = mass concentration.

Amount-of-substance concentration = amount concentration. In polymer science, this is usually defined as $[1] = n_1/V$, i.e., amount-of-substance 1 per volume V of mixture *after* mixing. IUPAC recommends $c_1 = n_1/V$, which may be confused with the symbol c_1 for the mass concentration. The amount concentration is often called "**mole concentration**" or **molarity** and given the symbol M; the latter symbol is not recommended by IUPAC and should not be used with SI prefixes (i.e., not mM for an amount concentration of "millimole" per Liter).

Molality of a solute. $a_1 = n_1/m_2$, i.e., amount n_1 of substance 1 per mass m_2 of solvent 2. Molalities are often denoted by m, which should not be used as the symbol for the unit mol kg^{-1}.

19.4 Ratios of Physical Quantities

The terms "normalized", "relative", "specific", and "reduced" are sometimes used with different meanings although they are clearly defined.

Normalized requires that the quantities in the numerator and denominator are of the same kind. A normalized quantity is always a fraction (quantity of the subgroup divided by the quantity of the group); the sum of all normalized quantities equals unity.

Relative also refers to quantities that are of the same kind in the numerator and denominator but the quantity in the denominator may be in any defined state. Example 1: relative viscosity $\eta_r = \eta/\eta_2$ as the ratio of the viscosity η of a solution to the viscosity η_2 of the solvent 2. Example 2: relative humidity = ratio of moisture content of air to moisture content of air saturated with water (both at the same temperature and pressure).

Specific refers to a physical quantity divided by the mass. The symbol of a specific quantity is the lower case form of the symbol of the quantity itself. Example: specific heat capacity $c_p = C_P/m =$ = heat capacity (in heat per temperature) divided by mass.

The so-called specific viscosity $\eta_{sp} = (\eta - \eta_1)/\eta_1$ is *not* a *specific* quantity. Indices are italicized only if they refer to a constant physical quantity.

Reduced refers to a quantity that is divided by a specified other quantity. Example: reduced osmotic pressure Π/c = osmotic pressure Π divided by the mass concentration c.

Dimensionless quantity: a product or a ratio of two or more different physical quantities that are combined in such a way that the resulting physical quantity has the physical unit of unity (i.e., is "dimensionless").

19.5 Names and Constitutions of Polymer Molecules

The following list contains constitutional line formulae of polymer molecules, their common chemical names, and predominantly used symbols (abbreviations and acronyms) of polymers that are mentioned in this volume. Configurations and/or tacticities, if any, of common polymers are included in names but not in line formulae. For more details about chemical structures, especially complex ones that are mentioned in this book but not listed below, see Volumes I and II, and for an extensive list of abbreviations and acronyms, see also Volume IV.

Carbon Chains

Repeating unit with substituents R	R'	Common name	Symbol
$-CH_2-CRR'-$		**poly(1-olefin)s**	
H	H	poly(ethylene)	PE
H	CH_3	poly(propylene), it-	PP
H	C_2H_5	poly(1-butene), it-	PB
H	C_3H_7	poly(1-pentene), it-	-
H	$CH_2CH(CH_3)_2$	poly(4-methyl-1-pentene), it-	PMP
CH_3	CH_3	poly(isobutylene)	PIB
CH_3	C_6H_5	poly(α-methylstyrene), at-	PAMS
$-CH_2-CHR'-$		**vinyl polymers**	
H	F	poly(vinyl fluoride), at-	PVF
H	Cl	poly(vinyl chloride), at-	PVC
H	C_6H_5	poly(styrene), at-	PS
H	C_6H_{11}	poly(vinyl cyclohexane), at	-
H	OH	poly(vinyl alcohol), at-	PVAL
H	$OCOCH_3$	poly(vinyl acetate), at-	PVAC
H	OCH_3	poly(vinyl methyl ether), at-	PVME
H	$CH=CH_2$	1,2-poly(butadiene)	-
H	$COCH_3$	poly(vinyl methyl ketone), at-	-
$-CH_2-CRR'-$		**vinylidene polymers**	
F	F	poly(vinylidene fluoride)	PVDF
Cl	Cl	poly(vinylidene chloride)	PVDC
$-CH_2-CRR'-$		**acrylics**	
H	COOH	poly(acrylic acid), at-	PAA
H	$COOCH_3$	poly(methyl acrylate), at-	PMA
H	$CONH_2$	poly(acrylamide), at-	-
H	$CONHCH(CH_3)_2$	poly(isopropyl acrylamide), at-	-
H	CN	poly(acrylonitrile), at-	PAN [1)]
$-CH_2-CRR'-$		**methacrylics**	
CH_3	$COOCH_3$	poly(methyl methacrylate), at-	PMMA
CH_3	$COOC_4H_9$	poly(butyl methacrylate), at-	PBMA
CH_3	$COOC_8H_{17}$	poly(octyl methacrylate), at-	POMA
CH_3	$COOC_{10}H_{21}$	poly(decyl methacrylate), at-	-
CH_3	$COOC_{12}H_{25}$	poly(dodecyl methacrylate), at-	PDMA
CH_3	$COOC_{16}H_{33}$	poly(hexadecyl methacrylate), at-	-
CH_3	$COOC_{18}H_{37}$	poly(octadecyl methacrylate), at-	-
CH_3	$COOC_{22}H_{25}$	poly(docosyl methacrylate), at-	-
CH_3	$COOCH_2CH_2OH$	poly(2-hydroxyethyl methacrylate), at-	PHEMA
CH_3	$COOC_6H_{11}$	poly(cyclohexyl methacrylate), at-	-

[1)] PAN is a registered trademark in Europe.

Carbon Chains (continued)

Repeating unit with substituents R	R'	Common name	Symbol
$-CH_2-CHR=CH-CH_2-$		**polydienes**	
H	-	1,4-poly(butadiene), *cis*	BR
CH_3	-	1,4-poly(isoprene), *cis*	IR
Cl	-	poly(chloroprene)	CR
$-CRR'-CR_2-$		**various**	
F	F	poly(tetrafluoroethylene)	PTFE
F	Cl	poly(chlorotrifluoroethylene)	PCTFE

Various other carbon chain polymers

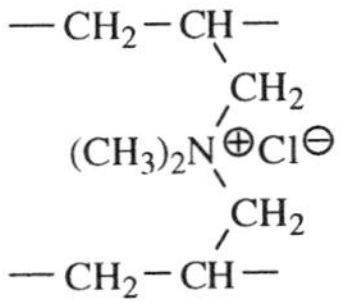

poly(*N*,*N*-diallyl dimethyl ammonium chloride)

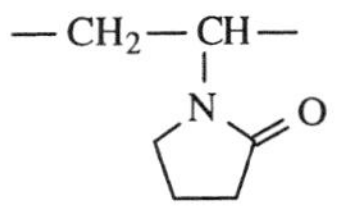

poly(*N*-vinyl pyrrolidone)

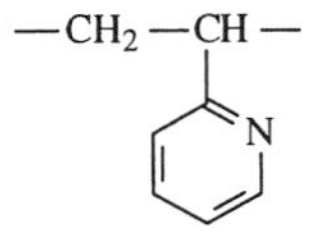

poly(2-vinyl pyridine)

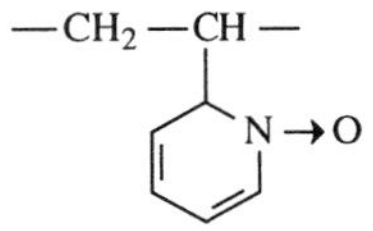

poly(2-vinyl pyridine-1-oxide)

Heterochains

Repeating unit with chain units -Z-	-Z'-	Common name	Symbol
$-O-Z-$		**polyethers**	
$-CH_2-$		poly(oxymethylene)	POM
$-(CH_2)_2-$		poly(oxyethylene), poly(ethylene oxide),	PEOX
		poly(ethylene glycol) (if both endgroups are OH)	PEG
$-(CH_2)_4-$		poly(tetrahydrofuran)	PTHF
$-CH(CH_3)-$		poly(acetaldehyde)	-
$-CH_2-CH(CH_3)-$		poly(propylene oxide)	PPOX
$-C_6H_2[2,6\text{-}(CH_3)_2]-$		poly(oxy-2,6-dimethyl-1,4-phenylene) (commonly called "polyphenylene oxide." The commercial product is a polymer blend!)	PPO
$-O-Z-CO-$		**AB polyesters**	
$-CH_2-$		poly(glyolide)	PGA
$-CH(CH_3)-$		poly(L-lactide)	PLA
$-CH(CH_3)-CH_2-$		poly(β-D-3-hydroxybutyrate)	PHB
$-CH(CH_3)-CH_2-$		poly(D,L-β-methyl-β-propiolactone)	-
$-(CH_2)_5-$		poly(ε-caprolactone)	PCL
$-p\text{-}C_6H_4-$		poly(p-hydroxybenzoic acid)	POB

Heterochains (continued)

Repeating unit with chain units –Z–	–Z'–	Common name	Symbol
–O–Z–O–CO–		**AB polyesters**	
$–C_6H_4–C(CH_3)_2–C_6H_4–$		polycarbonate A	PC-A
–O–Z–O–CO–Z'–CO–		**AABB polyesters**	
$–(CH_2)_2–$	$–p\text{-}C_6H_4–$	poly(ethylene terephthalate)	PET
$–(CH_2)_3–$	$–p\text{-}C_6H_4–$	poly(trimethylene terephthalate) (also called poly(propylene terephthalate))	PTT
$–(CH_2)_4–$	$–p\text{-}C_6H_4–$	poly(butylene terephthalate)	PBT
$–p\text{-}C_6H_4–$	$–p\text{-}C_6H_4–$	poly(*p*-phenylene terephthalate)	PPT
–NH–Z–CO–		**AB polyamides**	
$–CH_2–$		poly(glycine)	PA 2
$–(CH_2)_5–$		poly(ε-caprolactam)	PA 6
$–(CH_2)_{11}–$		poly(laurolactam)	PA 12
$–CH(CH_3)–$		poly(L-alanine)	-
$–CH[CH_2–CH(CH_3)_2]–$		poly(L-leucine)	-
$–CH[CH_2CH_2COOCH_2C_6H_5]–$		poly(γ-benzyl-L-glutamate)	PBLG
$–CH[(CH_2)_4NH]–$		poly(α,ε-L-lysine)	-
$–p\text{-}C_6H_4–$		poly(*p*-benzamide)	PPBA
–NH–Z–NH–CO–Z'–CO–		**AABB polyamides**	
$–(CH_2)_6–$	$–(CH_2)_5–$	poly(hexamethylene adipamide)	PA 6.6
$–p\text{-}C_6H_4–$	$–p\text{-}C_6H_4–$	poly(*p*-phenylene terephthalamide)	PPTA
various repeating units		**Various polymers**	
$–(CH_2)_3N<$ *)		poly(propylene imine) dendrimers	PPI
$–(CH_2)_2COONH(CH_2)_2N<$ [2])		polyamidoamine dendrimers	PAMAM
$–N(C_6H_{13})–CO–$		poly(hexylisocyanate)	-
$–O–Si(CH_3)_2–$		poly(dimethylsiloxane)	PDMS

[2]) Branching unit; core unit and endgroups of this dendrimer can vary.

Various Other Aromatic Heterochain Polymers

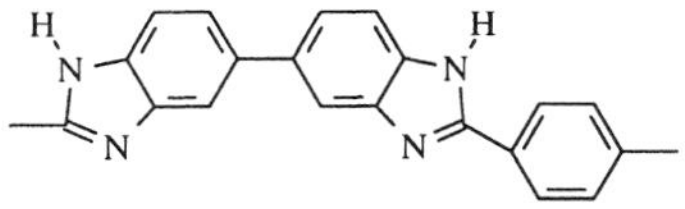

polybenzimidazole

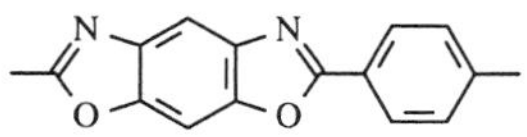

poly(*p*-phenylene-2,6-benzoxazoldiyl

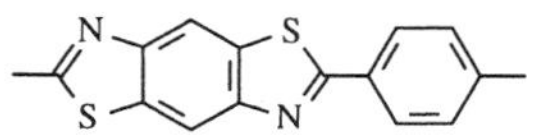

poly((*p*-phenylene-2,6-benzobisthiazoldiyl

Glucoses and Polysaccharides

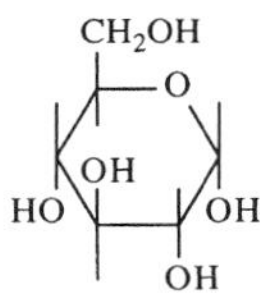

α-D-glucose

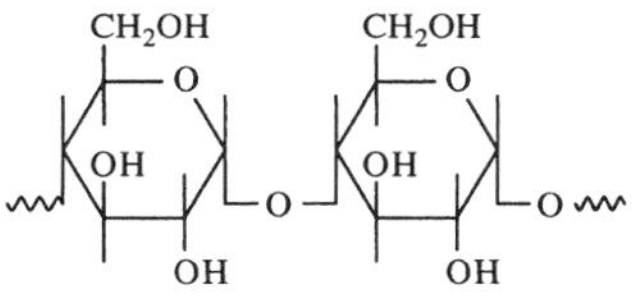

amylose

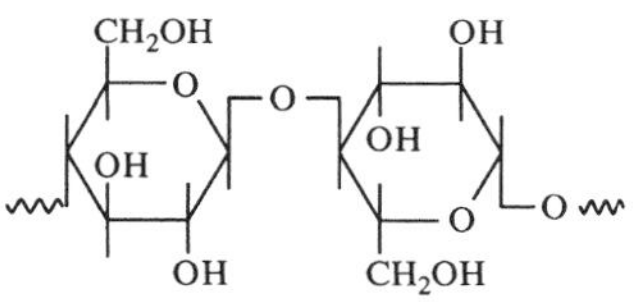

cellulose

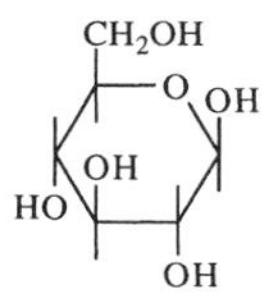

β-D-glucose

20 Subject Index

Entries are listed in strict alphabetical order; they may consist of a single word, abbreviations, acronyms, or combinations thereof. For alphabetization, technical terms consisting of two nouns were considered to be one word, whether written as two words (example: acetal polymer), with a hyphen (for example, tension-thinning), or in parentheses, brackets, or braces (example: "Catalyst, def.", comes before "Catalyst efficiency"). Qualifying numbers and letters as well as hyphens, parentheses, brackets, and braces in names of chemical compounds such as 1-, 1,4-, α-, β-, *o*-, *m*-, *p*-, L-, D-, etc., also have been disregarded for alphabetization. Terms consisting of an adverb and a noun are arranged according to the noun (example: Molar mass → Mass, molar).

The following abbreviations are used: abbr. = abbreviation, def. = definition; eqn. = equation; ff. = and following; PM = polymerization, ZN = Ziegler-Natta.

-, molecular reasons for 573
Young's equation 276
Young's modulus, def. 493, 522
YSS theory 110

z-average molar mass, def. 20